Handbook of Construction Equipment Maintenance

OTHER McGRAW-HILL HANDBOOKS OF INTEREST

Azad • Industrial Wastewater Management Handbook
Baumeister • *Marks'* Standard Handbook for Mechanical Engineers
Brater and King • Handbook of Hydraulics
Callender • Time-Saver Standards for Architectural Design Data
Chow • Handbook of Applied Hydrology
Conover • Grounds Maintenance Handbook
Crocker and King • Piping Handbook
Croft, Carr, and Watt • American Electricians' Handbook
Davis and Sorensen • Handbook of Applied Hydraulics
DeChiara and Callender • Time-Saver Standards for Building Types
Fink and Carroll • Standard Handbook for Electrical Engineers
Harris • Handbook of Noise Control
Harris and Crede • Shock and Vibration Handbook
Havers and Stubbs • Handbook of Heavy Construction
Hicks • Standard Handbook of Engineering Calculations
Higgins and Morrow • Maintenance Engineering Handbook
Hopf • Handbook of Building Security Planning and Design
Juran • Quality Control Handbook
Karassik, Krutzsch, Fraser, and Messina • Pump Handbook
LaLonde and Janes • Concrete Engineering Handbook
LeGrand • The New American Machinist's Handbook
Lewis • Management Handbook for Plant Engineers
Lewis and Marron • Facilities and Plant Engineering Handbook
Machol • System Engineering Handbook
Mantell • Engineering Materials Handbook
Maynard • Handbook of Business Administration
Merritt • Building Construction Handbook
Merritt • Standard Handbook for Civil Engineers
Merritt • Structural Steel Designers' Handbook
O'Brien • Scheduling Handbook
Perry • Engineering Manual
Rothbart • Mechanical Design and Systems Handbook
Smeaton • Switchgear and Control Handbook
Streeter • Handbook of Fluid Dynamics
Urquhart • Civil Engineering Handbook
Woods • Highway Engineering Handbook

Handbook of Construction Equipment Maintenance

LINDLEY R. HIGGINS, P.E.　*Consulting Engineer*

Senior Editor, *Construction Contracting*
(Formerly *Construction Methods and Equipment*)
McGraw-Hill, Inc., New York, N.Y.

McGRAW-HILL BOOK COMPANY
New York   St. Louis   San Francisco   Auckland   Bogotá
Düsseldorf   Johannesburg   London   Madrid
Mexico   Montreal   New Delhi   Panama
Paris   São Paulo   Singapore
Sydney   Tokyo   Toronto

Library of Congress Cataloging in Publication Data

Main entry under title:

Handbook of construction equipment maintenance.
   Includes bibliographies and index.
   1. Construction equipment—Maintenance and repair.
I. Higgins, Lindley R.
TH900.H28   621.8   79-12393
ISBN 0-07-028764-3

Copyright © 1979 by McGraw-Hill, Inc. All rights reserved.
Printed in the United States of America. No part of this
publication may be reproduced, stored in a retrieval system,
or transmitted, in any form or by any means, electronic,
mechanical, photocopying, recording, or otherwise, without
the prior written permission of the publisher.

1234567890 HDHD 7865432109

*The editors for this book were Harold B. Crawford and Margaret Lamb,
the designer was Naomi Auerbach, and the production supervisor
was Teresa F. Leaden. It was set in Caledonia
by University Graphics, Inc.*

Printed and bound by Halliday Lithograph.

The pronouns "he" and "his" and terms such as "foreman" have been used in a purely generic sense in this book to accommodate the text to the limitations of the English language and avoid awkward grammatical constructions.

Contents

Contributors vii
Preface xi
Reference Tables xiii

Section 1. Establishing and Managing the Maintenance Function
    1. Introduction to the Theory and Practice of Construction Equipment Maintenance .. 1-1
    2. Establishing the Maintenance Function in Construction 1-3
    3. A System for Organizing Maintenance Information 1-11
    4. Developing and Training Maintenance Personnel 1-15
    5. Human Factors in Construction Equipment Maintenance 1-35

Section 2. Establishing the Maintenance Facility
    1. The Construction Maintenance Shop 2-1
    2. Heavy Tools for Construction Equipment 2-9
    3. Portable Power Tools for Construction Equipment Maintenance 2-17
    4. Basic Hand Tools of Maintenance 2-31

Section 3. The Economics of Construction Equipment Maintenance
    1. The Maintenance Budget 3-1
    2. Standard Accounting Practice for Maintenance Cost Recording 3-3
    3. The Role of the Computer in Maintenance Cost 3-9
    4. The Role of Preventive Maintenance 3-15

Section 4. Basic Maintenance Technology
    1. Corrosion Control 4-1
    2. Equipment Parts Cleaning 4-5
    3. Lubrication 4-11
    4. Centralized Lubrication Systems 4-21
    5. Arc Welding in Maintenance 4-27
    6. Gas Welding in Maintenance 4-79
    7. Metal Resurfacing 4-107
    8. Maintenance of Plain Bearings 4-125
    9. Maintenance of Rolling Bearings 4-135
    10. Maintenance of Mechanical Power Transmission Equipment 4-155
    11. Maintenance of Scaffolds and Ladders 4-171
    12. Chain Hoists and Chain Slings 4-187
    13. Maintenance of Belt Conveyors and Conveying Equipment 4-203
    14. Maintenance of Hydraulic Hose and Fittings 4-223
    15. Steam and Hot-Water Cleaning 4-245

Section 5. Maintenance of Power Systems
    1. Maintenance of Electrical Power Systems 5-1
    2. Maintenance of Diesel Power Systems 5-9

Contents

    3. Maintenance of Gasoline Power Systems 5-33
    4. Principles and Maintenance of Hydraulic Systems 5-39
    5. Maintenance of Electric Motors 5-53

Section 6. Maintenance of Ground Contact Elements
    1. Upkeep and Maintenance of Tires for Construction Equipment 6-1
    2. Wear and Maintenance of the Undercarriage 6-15
    3. Dozer Moldboard and Ripper Tooth Maintenance 6-23
    4. Maintenance of Earthmoving Buckets and Bucket Teeth 6-31
    5. Maintenance of Vibratory Compactor Drums 6-43

Section 7. Maintenance Programs for Equipment Entities
    1. Diesel Tractor Maintenance 7-1
    2. Scraper and Scraper Blade Maintenance 7-29
    3. Motor Grader Maintenance 7-53
    4. Static Weight Compactor Maintenance 7-91
    5. Hydraulic Excavator and Backhoe Maintenance 7-107
    6. Power Shovel Maintenance 7-121
    7. Asphalt Paving Equipment Maintenance 7-161
    8. Hydraulic Crane Maintenance 7-177
    9. Crushing Equipment Maintenance 7-207
    10. Asphalt Plant Maintenance 7-269
    11. Trenching and Ditching Equipment Maintenance 7-303
    12. Drills and Drilling Equipment Maintenance 7-333

Section 8. Pump and Compressor Maintenance in Construction
    1. General Pump Maintenance 8-1
    2. Centrifugal Pump Maintenance 8-7
    3. Concrete Pump Maintenance 8-19
    4. Rotary Helical-Screw Air Compressor Maintenance 8-33
    5. Reciprocating Air Compressor Maintenance 8-55

Index follows Section 8

Contributors

ADE, ROGER *Marketing Services, Hyster Company, Construction Equipment Division, Kewanee, Ill.* (SECTION 7, CHAPTER 4)

ALLEN, HARRY W. *Manager, Rotary Compressor Product Maintenance, Gardner-Denver Company, Quincy, Ill.* (SECTION 8, CHAPTER 4)

ANDERSON, K. D. *Terex Division, General Motors Corporation, Hudson, Ohio* (SECTION 6, CHAPTER 2)

ARBORE, LOUIS A. *Manager, Michelin Tire Corporation, Technical Group, Earthmoving Dept., Lake Success, N.Y.* (SECTION 6, CHAPTER 1)

ARNOLD, R. E. *Manager, Product Engineering, General Electric Company, Industrial Motor Division, Schenectady, N.Y. Mr. Arnold is now retired.* (SECTION 5, CHAPTER 5)

AXELSON, WILLIAM *Vice-President, The Hotsey Corporation, Englewood, Colo.* (SECTION 4, CHAPTER 15)

BAECKER, M. F. *Engineering Department, Gardner-Denver Company, Quincy, Ill. Mr. Baecker is now retired.* (SECTION 8, CHAPTER 5)

BENNETT, J. L. *Safety Assurance Manager, The Black & Decker Manufacturing Company, Towson, Md.* (SECTION 2, CHAPTER 3)

BISHOP, A. *Product Manager, Mounted Bearings, Dodge Division, Reliance Electric Company, Mishawaka, Ind.* (SECTION 4, CHAPTER 10)

BITTEL, LESTER R. *Director, Academy Hall, Inc., Strasburg, Va.* (SECTION 1, CHAPTER 5)

BOEYER, G. *Product Manager, Shaft Mounted Speed Reducers, Flexible Couplings, Dodge Division, Reliance Electric Company, Mishawaka, Ind.* (SECTION 4, CHAPTER 10)

BREWER, ALLEN F. *Consultant in Lubrication, St. Lucie County, Fla.* (SECTION 4, CHAPTER 3)

BURROWS, LINCOLN *Sperry Vickers, Division of Sperry Rand Corporation, Troy, Mich.* (SECTION 5, CHAPTER 4)

BURTON, H. *Product Manager, Dry Fluid Drums and Couplings, Dodge Division, Reliance Electric Company, Mishawaka, Ind.* (SECTION 4, CHAPTER 10)

CALLAHAN, JAMES J. *Vice-President, Research and Engineering, Trabor Lubricating Systems, Lubriquip Division, Houdaille Industries, Inc., Cleveland, Ohio.* (SECTION 4, CHAPTER 4)

CATERPILLAR TRACTOR COMPANY *Peoria, Ill.* (SECTION 7, CHAPTER 1)

CLEVELAND TRENCHER COMPANY *Staff, A Division of the American Hoist and Derrick Company, St. Paul, Minn.* (SECTION 7, CHAPTER 11)

DEICHEL, ARNOLD *Manager, Product Services, Ingersoll-Rand Company, Compactor Division, Shippensburg, Pa.* (SECTION 6, CHAPTER 5)

THE COMPRESSED AIR AND GAS INSTITUTE *Staff, Cleveland, Ohio.* (SECTION 7, CHAPTER 12)

D'HOORE, M. *Product Specialist, Babbitted and Bronze Sleeve Type Bearings, Dodge Division, Reliance Electric Company, Mishawaka, Ind.* (SECTION 4, CHAPTER 10)

ELSON, R. *Product Specialist, Mounted Bearings, Dodge Division, Reliance Electric Company, Mishawaka, Ind.* (SECTION 4, CHAPTER 10)

ESCO CORPORATION *Engineering Staff, Portland, Ore.* (SECTION 6, CHAPTER 4)

FORSYTHE, GEORGE O. *General Service Manager, Northwest Engineering Company, Green Bay, Wisc.* (SECTION 7, CHAPTER 6)

FRICKE, C. L. *Engine Service Manager, Briggs & Stratton Corporation, Milwaukee, Wisc.* (SECTION 5, CHAPTER 3)

GLEEKMAN, LEWIS W., DR. *Materials and Corrosion Engineering Services, Smithfield, Mich.* (SECTION 4, CHAPTER 1)

GOLDENBOGEN, W. N. *Supervisor, Material Quality, Gould, Inc., Clevite Engine Parts Division, Cleveland, Ohio.* (SECTION 4, CHAPTER 8)

GOOD, W. RALPH *SKF Industries, King of Prussia, Pa.* (SECTION 4, CHAPTER 9)

HAGELIN, VERNON D. *Technical Services, Deere & Company, Moline, Ill. Mr. Good is now retired.* (SECTION 5, CHAPTER 2)

HIGGINS, LINDLEY R., P.E. *Consulting Engineer, Senior Editor, "Construction Contracting" (Formerly "Construction Methods & Equipment"), McGraw-Hill, Inc., New York, N.Y.* (SECTION 1, CHAPTERS 1, 2, and 3)

HINKEL, J. E. *Manager, Application Engineering, The Lincoln Electric Company, Cleveland, Ohio.* (SECTION 4, CHAPTER 5)

HOLLIDAY, H. E. *Service Manager, CM Hoist Division, Columbus McKinnon Corporation, Tonawanda, N.Y. Mr. Holliday is now deceased.* (SECTION 4, CHAPTER 12)

HOPPENRATH, R. A. *Product Application Manager, Conveyors, Barber-Greene Company, Aurora, Ill.* (SECTION 4, CHAPTER 13)

HUEMMER, C. *Product Manager, V-Belt Drives, Tapered Bushings, Dodge Division, Reliance Electric Company, Mishawaka, Ind.* (SECTION 4, CHAPTER 10)

HULLMANN, HERMAN A. *Technical Services Representative, Fleet Services, Fiat-Allis Construction Machinery Inc., Springfield, Ill.* (SECTION 6, CHAPTER 3)

ISENBERGER, CHARLES *Grove Manufacturing Company, Shady Grove, Pa.* (SECTION 7, CHAPTER 8)

JOHNSON, JESS *Kohler Company, Kohler, Wisc.* (SECTION 5, CHAPTER 1)

JOHNSON, ROBERT M. *Manager of Trades Training, Dow Chemical USA, Midland, Mich.* (SECTION 1, CHAPTER 4, PART I)

KELLEY, K. N. *Manager, Marketing Communications, Stoody Company, Industry, Calif. Mr. Kelley is now retired.* (SECTION 4, CHAPTER 7)

LINNEMAN, H. W. *Gardner-Denver Company, Quincy, Ill.* (SECTION 8, CHAPTER 2)

LOUCKS, CHARLES M. *Consulting Chemist, Arlington Heights, Ill.* (SECTION 4, CHAPTER 2)

MACY, BARGER K. *Technical Publications Supervisor, Blaw-Knox Construction Equipment Inc., Matoon, Ill.* (SECTION 7, CHAPTER 7)

MARTIN, RICHARD A. *Service Training Supervisor, Austin-Western Division, Clark Equipment Company, Aurora, Ill.* (SECTION 7, CHAPTER 3)

MORGAN, JAMES P., JR. *Technical Services Manager, The Black & Decker Manufacturing Company, Towson, Md.* (SECTION 2, CHAPTER 3)

OLSON, ROBERT L., P.E. *Product Application Department, The Gates Rubber Co., Englewood, Colo.* (SECTION 4, CHAPTER 14)

PORTER, ROBERT J. *General Service Manager, The Gorman-Rupp Company, Mansfield, Ohio.* (SECTION 8, CHAPTER 1)

RITTER, G. F. *Service Manager, Asphalt and Stabilization Products, Barber-Greene Company, Aurora, Ill.* (SECTION 7, CHAPTER 10)

SAFIER, LEONARD *President, Patent Scaffolding Company, A Division of Harsco Corporation, Ft. Lee, N.J.* (SECTION 4, CHAPTER 11)

SAMEK, DAVID J. *Iowa Manufacturing Company, Cedar Rapids, Iowa.* (SECTION 7, CHAPTER 9)

SAREEN, A. *Product Specialist, Dry Fluid Drives and Couplings, Dodge Division, Reliance Electric Company, Mishawaka, Ind.* (SECTION 4, CHAPTER 10)

SENRICK, JAMES F. *Division Manager, Owatonna Tool Company, Owatonna, Minn.* (SECTION 2, CHAPTER 2)

SHERMAN, MARVIN *Sperry Vickers, Division of Sperry Rand Corporation, Troy, Mich.* (SECTION 5, CHAPTER 4)

SMITH, JIM D. *General Service Manager, Austin-Western Division, Clark Equipment Company, Aurora, Ill.* (SECTION 7, CHAPTER 3)

STORR, ALLEN *Manager of Training Resources, Dow Chemical USA, Midland, Mich.* (SECTION 1, CHAPTER 4, PART II)

STREJC, W. *Product Specialist, Chain Drives, Dodge Division, Reliance Electric Company, Mishawaka, Ind.* (SECTION 4, CHAPTER 10)

TEREX DIVISION *General Motors Corporation, Hudson, Ohio.* (SECTION 7, CHAPTER 2)

THOMAS, GARY J. *Vice-President and Equipment Manager, S. J. Groves & Company, Minneapolis, Minn.* (SECTION 2, CHAPTER 1)

TRW/J. H. WILLIAMS DIVISION *Staff, Buffalo, N.Y.* (SECTION 2, CHAPTER 4)

UNION CARBIDE CORPORATION *Engineering Staff, Linde Division, New York, N.Y.* (SECTION 4, CHAPTER 6)

VIERS, ANTHONY *Controller, Advance Construction Company Inc., Hinsdale, Ill.* (SECTION 3, CHAPTERS 1, 2, and 3)

WOLTERSDORF, ROBERT W. *Technical Services Manager, Crane and Excavation Group, Koehring Corporation, Milwaukee, Wisc.* (SECTION 7, CHAPTER 5)

WEATHERTON, ROBERT P. *General Manager, Concrete Pump Division, Challenge-Cook Bros., Inc., Industry, Calif.* (SECTION 8, CHAPTER 3)

WYDER, CARL G. *Senior Maintenance Editor, McGraw-Hill Inc., New York, N.Y. Mr. Wyder is now retired.* (SECTION 3, CHAPTER 4)

ZAJAC, J. D. *Product Standards and Service, CM Hoist Division, Columbus McKinnon Corporation, Tonawanda, N.Y.* (SECTION 4, CHAPTER 12)

ZAMBELAS, GEORGE *Engineer, Michelin Tire Corporation, Technical Group, Earthmoving Department, Lake Success, N.Y.* (SECTION 6, CHAPTER 1)

Preface

Three decades ago maintenance, as a distinct identifiable activity, won recognition in the field of industrial manufacturing. At that time construction was at the beginning of its phenomenal growth. Repair work on earthmoving and related machinery was performed strictly on an as needed, when needed, and where needed basis. A few of the largest contractors were aware that equipment upkeep cost money, although only a handful had even a glimmer of an idea of how much money might be involved. Nobody worried very much though. Maintenance costs were simply tacked onto the bid, openly or buried in a mass of other figures.

Meanwhile, over at the manufacturing plant spread, maintenance was being measured and codified, standards were being developed, and records of machine history and repair costs were being tabulated and—even more important—being compared and evaluated against past events in a search for potential savings. And why not? It soon became quite evident that American industry was spending well in excess of $2 billion on the somewhat unrewarding activity of restoring machinery to its original operating condition.

Preventive maintenance, corrective maintenance, engineered maintenance, and maintenance cost control became techniques and slogans of the hour. Conferences and associations began to sprout, all dedicated to the idea of exchanging information and improving processes through sound maintenance. In short, maintenance became respectable. Skilled engineers took over key jobs within maintenance. The unpopular but essential industrial engineer—known to our fathers and grandfathers as the ubiquitous efficiency expert—moved in as overseer. The result? In nearly every company where maintenance became "managed," costs were cut in half—and usually much more than that!

Today, the construction industry finds itself where industry stood in the late 1940s, with regard to the cost of maintenance. But there's a difference. In our present tight profit, high inflation world, the potential for gain is greater and the number of dollars is considerably more.

So, as the future of construction equipment maintenance is mirrored in industrial manufacturing's maintenance past, so too is this handbook—to some degree—a mirror of the *Maintenance Engineering Handbook,* (Higgins & Morrow, McGraw-Hill Book Company, 1977).

And because of this mirroring, I have abstracted key chapters from the earlier industry-based work and revised them to suit the day-to-day problems of the construction industry.

But there's a great deal more here than that. For the first time ever, specific upkeep problems on specific units of construction equipment are given full-dress coverage. Moreover, contributing authors have been carefully culled from prominent technologists at leading construction equipment manufacturing concerns. In short what you have here is a compendium of expertise by a collection of experts.

I want to pay particular thanks to *Construction Contracting*, the watershed of information about construction equipment and its upkeep, and to its editors and publisher, without whose constant efforts little would have been done to advance the cause of this vital element of the construction industry. I want too to thank the great McGraw-Hill organization for its role as repository for the vast bulk of technical information extant in America today.

Let me also express my sincerest thanks to Rosalie Hermanns and Kathleen Hermanns for their long hours and constant care over the pages of this work.

Lindley R. Higgins, P.E.

Reference Tables

Table of Conversion Factors

Multiply	By	To obtain	Multiply	By	To obtain
Btu	778	Foot-pounds	Horsepower	0.7457	Kilowatts
Btu per hour	0.000293	Kilowatts	Inches	2.54	Centimeters
Centimeters	0.3937	Inches	Kilograms	2.20462	Pounds
Cubic centimeters	0.06102	Cubic inches	Kilograms per square centimeter	14.22	Pounds per square inch
Cubic centimeters	0.0002642	Gallons	Kilometers	0.62137	Miles
Cubic feet	0.028317	Cubic meters	Kilowatts	1.341	Horsepower
Cubic feet	1,728.0	Cubic inches	Kilowatts	56.92	Btu per minute
Cubic feet	7.48052	Gallons	Kilowatt-hours	3415.0	Btu
Cubic feet	28.32	Liters	Liters	0.03531	Cubic feet
Cubic inches	16.39	Cubic centimeters	Liters	61.02	Cubic inches
			Liters	0.2642	Gallons
Cubic meters	35.3145	Cubic feet	Meters	3.28083	Feet
Cubic meters	61,023.0	Cubic inches	Meters	39.37	Inches
Cubic meters	264.2	Gallons	Meters	1.094	Yards
Cubic meters	1.308	Cubic yards	Meters per minute	3.281	Feet per minute
Cubic yards	0.76453	Cubic meters			
Feet	30.48	Centimeters	Meters per second	3.281	Feet per second
Feet	0.3048	Meters			
Feet per minute	0.508	Centimeters per second	Meters per second	196.8	Feet per minute
Feet per minute	0.01667	Feet per second	Miles	1.60935	Kilometers
Gallons (US)	3,785	Cubic centimeters	Pounds	0.45359	Kilograms
			Pounds of water (60°F)	0.010602	Cubic feet
Gallons (US)	0.1337	Cubic feet	Pounds of water (60°F)	27.68	Cubic inches
Gallons (US)	231	Cubic inches			
Gallons (US)	0.003785	Cubic meters			
Gallons (US)	3.7853	Liters	Pounds of water (60°F)	0.1198	Gallons
Gallons of water	8.3453	Pounds of water (at 60°F)			
			Pounds per sq. ft.	4.883	Kilograms per square meter
Gallons per minute	0.060308	Liters per second	Pounds per sq. in.	0.0703	Kilograms per square centimeter
Grams	15.432	Grains			
Horsepower	33,000	Foot-pounds per minute	Quarts	0.946	Liters
			Square centimeters	0.1550	Square inches
Horsepower	2,546	Btu per hour			
Horsepower	42.42	Btu per minute			
Horsepower	1.014	Horsepower (metric)	Square feet	929	Square centimeters

Table of Conversion Factors *(Continued)*

Multiply	By	To obtain	Multiply	By	To obtain
Square feet	0.0929	Square meters	Watts	0.01434	Kilogram-calories per minute
Square inches	6.452	Square centimeters			
Square meters	10.765	Square feet	Watt-hours	3.415	Btu
Watts	3.415	Btu per hour	Watt-hours	0.8605	Kilogram-calories

Table of Temperature Equivalents—Fahrenheit and Celsius Scales

(Expressed in increments of 9°F and 5°C)

Fahrenheit	Celsius	Fahrenheit	Celsius
−35	−35	95	35
−24	−30	104	40
−13	−25	113	45
−4	−20	122	50
5	−15	131	55
14	−10	140	60
23	−5	149	65
32	0	158	70
41	5	167	75
50	10	176	80
59	15	185	85
68	20	194	90
77	25	203	95
86	30	212	100

Beaufort Scale of Wind Force

(Compiled by U.S. Weather Bureau, 1955)

Beaufort number	Miles per hour	Knots	Wind effects observed on land	Terms used in USWB forecasts
0	Less than 1	Less than 1	Calm; smoke rises vertically	Light
1	1–3	1–3	Direction of wind shown by smoke drift; but not by wind vanes	Light
2	4–7	4–6	Wind felt on face; leaves rustle; ordinary vane moved by wind	Light
3	8–12	7–10	Leaves and small twigs in constant motion; wind extends light flag	Gentle
4	13–18	11–16	Raises dust, loose paper; small branches are moved	Moderate
5	19–24	17–21	Small trees in leaf begin to sway; created wavelets form on inland waters	Fresh

Beaufort number	Miles per hour	Knots	Wind effects observed on land	Terms used in USWB forecasts
6	25–31	22–27	Large branches in motion; whistling heard in telegraph wires; umbrellas used with difficulty	Strong
7	32–38	28–33	Whole trees in motion; inconvenience felt walking against wind	Strong
8	39–46	34–40	Breaks twigs off trees; generally impedes progress	Gale
9	47–54	41–47	Slight structural damage occurs; (chimney pots, slates removed)	Gale
10	55–63	48–55	Seldom experienced inland; trees uprooted; considerable structural damage occurs	Whole gale
11	64–72	56–63	Very rarely experienced; accompanied by widespread damage	Whole gale
12 or more	73 or more	64 or more	Very rarely experienced; accompanied by widespread damage	Hurricane

Handbook of Construction Equipment Maintenance

Section **1**

Establishing and Managing the Maintenance Function

Chapter **1**

Introduction to the Theory and Practice of Construction Equipment Maintenance

LINDLEY R. HIGGINS, P.E.
Consulting Engineer,
Senior Editor, *Construction Contracting*
(Formerly *Construction Methods & Equipment***),**
McGraw-Hill, Inc., New York, N.Y.

Equipment maintenance is not simply preventive maintenance, although preventive maintenance is an important ingredient. It is not lubrication, although lubrication is a function of maintenance. It is not a hasty rush to repair a broken machine part or replace a faltering bearing, although these are maintenance activities. Equipment maintenance is a science, an exercise in economics, an art, and a philosophy.

Because maintenance is all these things and because too many practitioners treat it wholly as one of its components, this handbook came into being. It's 55 chapters, organized into 8 broad divisions of construction equipment upkeep, are designed to show contractors, maintenance managers, master mechanics, and even front-office accountants the scope of the whole construction equipment maintenance problem and the keys to its solution in every area.

Evolution of maintenance A week or so after the invention of the wheel, its performance dropped. The round edge flattened; the axle became enlarged; or, perhaps, the wheel simply cracked. And maintenance came into being. Almost as long as machinery has existed, it has required repair or restoration of parts.

When labor and materials were relatively cheap compared with profits, maintenance received little attention. Machinery was simply pulled off the line or taken out of service, and new units were brought in. That maintenance of equipment might be an important ingredient of conducting business never occurred to managers—in construction or industry. The recognition that it might be more economic to repair a machine than to discard it evolved slowly—often spurred by disaster or economic need.

But, by the close of World War II, technology had become so complex and the cost of new equipment so enormous that manufacturing management began taking a look at the spiraling costs. Meanwhile construction management merely tacked another item onto job bids and let it go at that.

Today, nearly three and a half decades later, maintenance management in industry has become a major, respectable, and vital sector. Management of construction firms is also, finally, beginning to take a closer look at the problem. Of course, the construction giants have always done so. Perhaps this is, in part, why they are giants.

But, in general, construction has failed to keep pace with industry in the maintenance sector. This handbook hopes to restore the balance and to bring construction firms' maintenance programs into the twentieth century.

Interaction—key to maintenance philosophy Equipment maintenance in the construction industry is a little like a child's teeter-totter—two elements are required, each interacting with the other. The contractor, at one end, strives to provide repair service in just the right amount to restore the unit to full operation without excessive cost. The machine, at the other end, has a level of maintainability to set the degree of contractor effort and an index of durability to determine the frequency of this effort. Directly between them exists the fulcrum of interaction that dovetails contractor and machine into a cost-curbing, downtime-diminishing entity.

Both major ingredients are presently in a state of flux. Maintenance management techniques are on a third-generation level. Contractors, borrowing heavily from their counterparts in manufacturing, can learn to measure the effectiveness of work performance, to improve both the quality and quantity of finished work, and to deploy preventive maintenance selectively to minimize cost and maximize results. With access to computer services—their own or a time-shared arrangement—contractors harness electronic surveillance to scan repair dollar outlay or record tasks as done. They can even program a computer to print out orders for action to be taken at specific times and in specific places.

Heavy equipment manufacturers are learning more about maintenance too. Although none have attained the ideal level, reports indicate that high-failure areas of most machines are becoming easier to reach, sub-assembly modules that can slide out and be exchanged for new ones are becoming more available, parts service is somewhat better, and local service centers are growing in number and convenience.

Machine maintainability can get a big boost from the contractor who wisely borrows a technique called corrective maintenance (CM) that has achieved wide industrial acceptance. The chief tenet of CM is that each repair job should leave the unit a little more durable, at least at the prior failure point, than it was before failure.

Maintenance tools and materials, too, are moving upward, in a spiral of improvement. Easier-to-operate shop hoists take up less room, do their jobs faster, and respond better to controls. New-design welding rods make possible the joining of all kinds of bronze, aluminum, and cast iron. Lubricating oils have hit new plateaus of sophistication and flexibility; they work well and without breaking down in extremes of heat and cold or other adverse conditions.

Ideas and improvements are everywhere at hand and are available for the taking. Yet before a contractor applies them, self-examination is essential. What are the goals for profit maximization? What maximum or minimum level of maintenance will the contractor accept?

The handbook as a tool This handbook is a maintenance tool—a guide to most of the problems you encounter, and a reference for every user. Remember to use the index as well as the table of contents. You will find repetition here—planned repetition; it is designed to give you more than one view of the problem at hand. Engines or tracks may be covered in individual chapters but also covered as components of a machine entity. Hydraulic systems are discussed in a separate chapter but are touched on elsewhere as vital elements of many construction machines.

Material is here for top management of a construction firm, for financial officers, for those officers responsible for personnel selection and development, for the materials managers or storekeepers, and for central shop managers.

Basic maintenance technologies such as corrosion control, lubrication, welding, metal resurfacing, and mechanical power transmission are covered, as are highly technical phenomena that often lie at the root of machine failure.

But the limitations of this or of any handbook are important to remember. Each construction spread is different from others. Nowhere is this difference more evident than in maintenance. Your approach, your shop, your solutions, all must mesh with your needs. Learn to shape this handbook's general technology to the specifics of your operation. Remember that maintenance is a highly individualized activity. This handbook can point you toward the answer, but how you reach it is your decision.

Chapter **2**

Establishing the Maintenance Function in Construction

LINDLEY R. HIGGINS, P.E.
Consulting Engineer
Senior Editor, Construction Contracting
(Formerly *Construction Methods & Equipment***)**
McGraw-Hill, Inc., New York, N.Y.

Maintenance management is a science of alternatives, not a study of technology. Taken to the limit it offers a choice between total equipment overhaul and total repair inactivity—with an infinite number of variations along the way.

This battery of options exists for every level of repair sophistication from the time for replacing a V-belt to the frequency of checking an electronic control circuit.

But to reach this idealistic plateau, maintenance managers have to look at their operation. Which machines are breaking down? How often? Why? What does an hour of downtime really cost? How much preventive maintenance is enough? Is the true cost of a spare part what was paid for it? When does an operator's hourly wage become a maintenance cost? What part does depreciation play in the value placed on a repair tool? On a maintenance shop?

In short, maintenance must be managed through a plan of action, a system of records, and a process of analysis. These—interconnected logically—constitute a sound maintenance program.

Do you need such a program? Do you already have one? Is it working for you? Or, perhaps, against you?

Plan a Course of Action

Maintenance programs, like suits of clothes, can rarely be picked from the rack and put into service without a little tailoring. There are universal truths in maintenance that are as basic as a pair of sleeves to a jacket. Yet the most successful programs are those built around a specific organization and its unique set of problems.

The key idea, then, in program installation or improvement is to measure. You can't begin to manage maintenance until you learn the attitudes of your company and yourself toward it, determine the dimensions of the repair problem in terms of equipment and its current condition, and evaluate the capabilities of your tools of solution—mechanics, shop, support facilities. Hence you must begin, like a management consultant, by measuring everything that has an effect on maintenance.

Assess goals However redundant it may seem, you must always remember that any maintenance program will fail if you and key members of your organization don't believe in the essential nature of maintenance. This does not mean adopting a philosophy that maintenance is better than anything. Rather, you must assess and adopt a positive attitude toward equipment upkeep, and build your program around it.

Breakdown maintenance (no repairs until failure occurs), continuous preventive maintenance, or a combination of each are all valid solutions to equipment upkeep. But you must choose your route, and stick to it. In most cases, the last alternative is the preferred one, since equipment elements and operations differ. So, for maximum effectiveness, your approach should be flexible.

Your organization, front-office people, field supervisors, equipment operators, maintenance managers, and repair mechanics must think the same way—your way—about this activity. To achieve a sense of unity, keep every member of the team aware of what is being done, why it is being done, and how it is being done; apply this dictum across the board from office clerk to executive vice president.

Measure the target Maintenance programs begin with equipment. Take an inventory of what you own. Major equipment is easy to pinpoint. Establish a numbering system and mark the equipment with either a metal identifying tag or a paint-on stencil. But what else should get numbers? Among the logical items to include are electric motors (even when they are an integral part of something else), conveyor sections, pumps, compressors, diesel and gas engines, optional attachments like special buckets or scarifiers, drill rigs, and storage bins or tanks. Items better left out of your inventory include equipment tires and tracks, hose, valves, sprockets, and earthmoving equipment sub-assemblies other than engines and attachments. Don't tag roll-over protection systems (ROPS), engine covers, side rails, or hydraulic cylinders. These items are difficult to isolate, unique in the way they're maintained, or simply too frequently replaced to record in this manner.

Inventory your maintenance personnel Look at skills and abilities. The ideal mechanic is one who works in many craft areas (machinist, carpenter, electrician, welder). But these are scarce and not always employable within the strictures of union contracts.

Next to versatility, a mechanic's greatest asset lies in familiarity with your equipment. A mechanical or an electrical genius who can't tell a dozer from a grader may prove worse than useless on a construction spread. General skills in maintenance are about three times as important as specific ones, except perhaps in internal combustion engine analysis or electronic-circuitry overhaul.

It is also vital to determine just how hard your maintenance people are working. Some fancy ways of getting this information require time, management experts, and expenditures that the typical construction company cannot justify. Simpler, less expensive indexes are (1) your own observations and (2) a basic ratio-delay study.

Making fair and thorough observations demands time and effort on your part. But it gets you out where the action is, always a source of sound managerial input.

For a basic ratio-delay study, assemble a small team of management people (three or four) who can recognize your mechanics on sight. Send them into the shop and out on jobs at random times and in random patterns. Have them jot down the number, but never the identities, of those working and those idle. After about five work days, sum up the results of the readings and calculate the percent of those classified as idle compared with the total maintenance crew. A value of less than 35 percent means that your mechanics are not as productive as they should be. The reasons vary, but it may be that supervision is lax, material is slow in arriving, equipment is rarely available, or some personnel simply are not very good employees.

Check out your maintenance tools The largest and perhaps most important tools of maintenance are the shops—central or in the field.

The main or central shop must house maintenance management, the records and administration of this activity, spare parts, the heavy tools of repair (hoists, jacks, etc), and, at times, some quite large pieces of equipment. The shop must be big and strong enough to handle its tasks. For maximum effectiveness, design the shop from the inside out. The simplest way is to make cardboard templates of heavy tools, benches, cabinets, and representative equipment to be serviced (¼ in. to equal 1 ft is ideal). With templates set in place, with adequate work space and usable aisles, wrap four walls around what you have.

Of course it's not quite that simple; there are several other important considerations. Chiefly, be sure that your largest piece of equipment can get through the door. Be equally certain that the heaviest machine will not destroy the shop floor (and that it can be raised with a hoist without causing building failure). Be sure that there is sufficient electric power with many convenient outlets. You may need compressed air or steam (usually 100 psi, in both cases). Steam means condensate. Isolate this and provide the necessary curbs, drains, and pitched floors to carry the water away. Be sure there's enough light (in maintenance use a 100-footcandle minimum).

Plan for safety by providing at least two exits, suitable fire control (including extinguishers for chemical and electric fires), and rapid access to first aid (kits, stretchers). Ventilate liberally. Mechanics who are accustomed to outdoor construction need a more-than-ordinary level of air when working inside.

Field shops are much simpler, often set up in open areas or in trailers. They need to be tailored only to work on the site. But they must have an electric supply (frequently from a portable generator set), a complement of tools and spare parts sufficient for problems that arise at the site, and some form of communication with the central shop if fast extra service should be necessary. Of course, safety provisions are as important here as at home base, increasing with distance from the center of operations and the difficulty of access.

Shop tools and test equipment also must match your spread. Any good repair shop requires at least one hoist, sized to match the heaviest anticipated load. An overhead beam for the hoist should extend through the door opening so that a sub-assembly may be brought in easily. There must be a means, inside or out, of working safely beneath machines. A steel ramp and platform arrangement or a drive-over pit is the safest and most practical. While there are many tools you could keep available, some of the basically essential ones are drill press, grinder, small lathe, powered impact wrenches, power saw, and a complete bench welding facility.

Two other elements requiring consideration are communications and mobility. Pocket pagers and public address systems help locate mechanics at a single large spread. Two-way radios, telephones (publicly or privately owned), or other longer-range devices are essential as activities disperse. Vehicles are an important adjunct to any maintenance operation. Be sure yours are capable of carrying heavy spare parts and are suited to the comfortable transport of maintenance crews.

The final in-house help you need to complete your maintenance program is administrative capability. This includes one or more competent clerks, an area and a system for files, and possibly a method for data collection. This last, whether mechanical or electric, must be approached with caution. The ultimate solution, the so-called computer, is ideal for maintenance record keeping only so long as it remains just a tool and does not become an end in itself. Never modify work patterns, schedules, or priorities to suit data-processing practice. Never sanction decision making by those tending data-processing equipment unless they are independently qualified as maintenance managers. In many localities you can hire time on an available computer (time sharing).

Harness available help The best maintenance help can usually be supplied by the company that made the machine. Major manufacturers have available not only detailed and specialized repair manuals and recommended spare parts lists, but lubrication and overall preventive maintenance schedules, checklists, and tips on equipment upkeep. Dealers, too, usually can provide a more knowledgeable and experienced mechanic's touch, a more complete spare parts inventory, and more sophisticated diagnostic equipment.

Contracted maintenance service provides another source of auxiliary assistance. Its advantages include the assistance of skilled, experienced mechanics, elimination from your books of some expensive tools and parts, and relief from the necessity of maintaining a staff of in-house mechanics who are frequently idle.

The disadvantages of hired repair are the extra hourly cost of such service, lack of involvement with your special problems, and, if you have an exclusive contract with one firm, a certain disregard for prompt action on the part of the maintenance contractor. In hiring this kind of service, be guided by three P's—proximity, parts, and promptness. A fourth P—price—should be obvious.

Schedule work for action The final measure of your maintenance operation is the repair work schedule. Essentially, this merely formalizes your mental and verbal decisions. Begin by scheduling on a weekly basis over a 3-month period (13 weeks). First determine your straight-time capability—the number of mechanics in your workforce times 40 hr each week, or 520 hr for the entire cycle per workers. Overtime should be avoided on all work not involving emergencies or loss of job production time (downtime).

From this 520-hr slug, deduct all routine or special maintenance time (toolbox safety talks, union meeting time, floating coverage at specific sites, planned preventive maintenance and lubrication time, and holidays and vacations). What remains is the work hours available for equipment repair. Make an educated guess on each job as it comes in and log it into the schedule. This will also show you whether your workforce is balanced. If these schedules reveal either an underused or an overused figure in excess of 20 percent, run the scheduling check for a second 13-week period. If the same result occurs, it probably means that you should decrease or increase your crew. An alternate solution for overuse is to consider temporary overtime. But this should be measured against the cost of adding an extra mechanic or two.

Record for Posterity

The construction industry has a reputation for antipathy toward paperwork. Unfortunately, sound maintenance management rests firmly upon this unpopular base. Equipment spreads are getting too large and too expensive to depend on old-time methods of memory, friendly advice, or inflexible beliefs. Contractors must learn that good records, well kept, return the cost and effort invested in very material terms.

Such an effort is not overwhelming. There are only three basic rules for maintenance record keeping: (1) repair work must be asked for, (2) it must be paid for, and (3) its history must be written. With these rules, record keeping will do the job it's supposed to do without growing into a separate, expensive, homegrown empire.

Work must be asked for A formal work request (sometimes called a work order or ticket) is not, as its name might imply, merely a beginning; it is also a source of the more permanent data needed to manage maintenance correctly. There probably are as many different designs for work orders as there are firms maintaining equipment. Each is tailored to the special needs of the company using it—as your forms should be. But whatever the size, color, or layout, there are some basic ingredients.

First, a description of the work to be done is needed. It should be as accurate as possible before the repair is undertaken. At least one copy of this request should go to the mechanic assigned to the job.

Second, there should be a similar amount of space on the form, adjacent to the work request, for what was actually done. This is particularly important if the work varied from the original request. The form should also contain the date requested, date started, date completed, equipment tag number of the machine involved, the hours spent by each mechanic, and the spare parts and material used.

Third, for ease of handling and subsequent data compilation and filing, the request should be numbered and of the snap-out variety. The number of copies should be tailored to the needs of the firm. But copies in excess of four are difficult to read and often simply lead to extra file drawers and useless records. Use a ballpoint pen for these forms, both for legibility and permanence.

Fourth, at some point you may want to tie such forms into a data collection system. The standard electronic data-processing setup (a computer) usually limits job titles to 20 to 30 characters. In this case, a row of boxes of the appropriate number will accommodate these letters and help to keep things straight (see Fig. 2-1).

Work must be paid for For computing the direct cost of maintenance (the job itself only), there are essentially two ingredients—objects and people.

Objects include materials, spare parts, and sub-assemblies. Materials include accountable items (metal or plastic sheet, steel shapes, and fittings) and expendable items (welding rods, fasteners, pipe nipples, and electric wire). It is seldom necessary to charge the expendable material to specific jobs or machines. Costs for these are usually allocated to a general overhead account. The only difference between spare parts and sub-assemblies is complexity—a dozer engine is a sub-assembly: a dozer engine crankshaft is a spare part. Each of the three—materials, spare parts, sub-assemblies (except only the expendables)—are part of the cost of a specific repair job and must be recorded.

People costs are the dollar values of the times spent by mechanics paid hourly. Like object costs, they must be recorded by job and by machine. They require a more intricate system of data gathering since mechanics may be working on more than one job during any 8-hr day—or even during a single working hour. They may have dead time (traveling to the next job, unassigned waiting, or contractual free time). For each such instance, you must have an overhead account (with account number) to which a mechanic's time can be charged. Straight-time overtime (the actual working time duration of a job in an overtime period) must be separated from premium-time overtime. This premium should never be

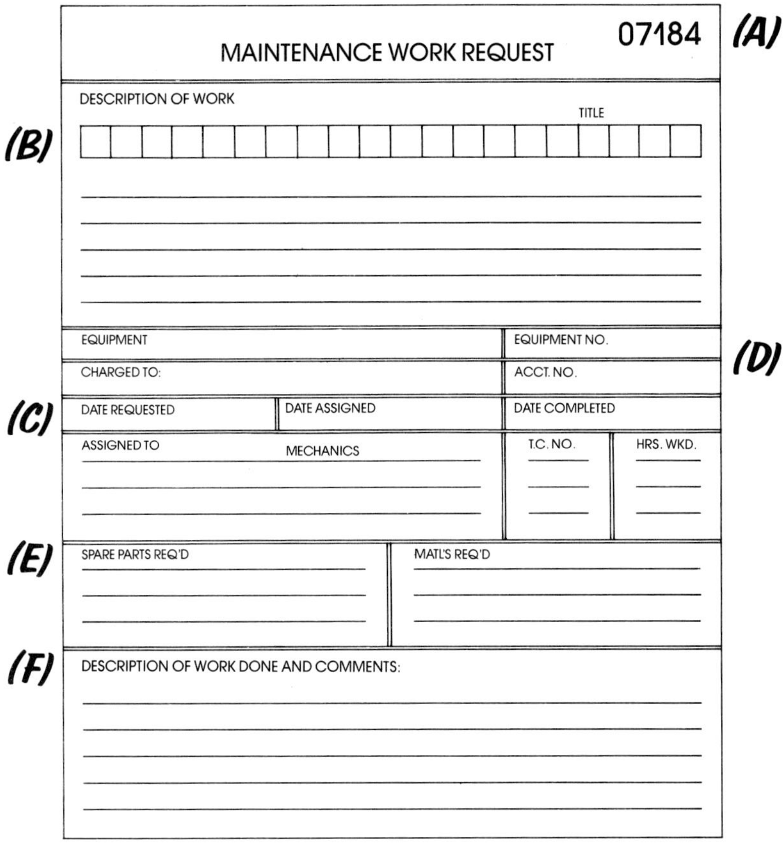
MAINTENANCE WORK REQUEST
07184
(A)
DESCRIPTION OF WORK
TITLE
(B)
EQUIPMENT
EQUIPMENT NO.
(D)
CHARGED TO:
ACCT. NO.
(C)
DATE REQUESTED
DATE ASSIGNED
DATE COMPLETED
ASSIGNED TO
MECHANICS
T.C. NO.
HRS. WKD.
(E)
SPARE PARTS REQ'D
MAT'L'S REQ'D
(F)
DESCRIPTION OF WORK DONE AND COMMENTS:

Form follows function

(A) Prenumbered forms give you a check on the quantity of repair requests being made and act as a preventive measure against these being overlooked.

(B) Clarity in the job title fed into the electronic data processing system assures you subsequent clarity in your permanent records.

(C) Wide disparity between requested and assigned dates may mean you're shorthanded and should consider increasing the number of maintenance mechanics.

(D) Making sure that repairs are charged to both the equipment being fixed and the job where repair was needed are vital ingredients to meaningful records and equitable charging.

(E) Careful itemization of required parts and raw materials *before* the job is assigned can save hours of delay.

(F) Work performed often differs considerably from work asked for. This space helps you learn what's happening to equipment and how to prevent recurrence.

Fig. 2-1 Steps in designing the typical work request form.

charged against a machine (although it should be charged against a job) when you are summing up a unit's maintenance cost history.

People costs should always be calculated in hours first and converted to dollars only at the end. This simplifies record keeping and lets mechanics become part of the system by recording on their time cards what they have done. An arbitrary cutoff in time increment (½ hr is ideal) should limit the degree of time fragmenting permitted for mechanics working on several jobs during an 8-hr period.

A special overhead account should be maintained for mechanics furnishing area coverage or troubleshooting types of maintenance where a number of machines or jobs are involved. The intangibles of managing maintenance make it poor practice to add supervisory time to the specific machine repair cost.

To collect all these people charges, it is best to begin with the time card or an equivalent employee time report. The time charged by mechanics needs approval (initialing) by central shop or field supervision. Policy here must clearly indicate the way this is to be done.

There are two ways that object and people maintenance costs must be assigned. The first is by job. Whether or not you back charge repair expense to your customer, it is important to isolate and assign these charges to specific jobs as an aid to further bidding. The second way to assign costs is by machine. A unit's repair costs are a vital index of its present worth, its need for modification, or its imminent arrival at the end of the road on your spread.

History must be written Repair cost related to specific machines becomes a key ingredient in determining the fate of current units and in making decisions about acquiring new ones. One of those universal truths in maintenance is that identical machines, purchased at the same time, with identical modifications made on each, and given identical assignments, are quite capable of developing widely differing repair histories. Before you can correct for this phenomenon, you have to know it exists. This is the rationale behind all record keeping.

Begin the establishment of your paperwork system by setting up an equipment record file that is keyed to the numbers assigned to machines and sub-assemblies. Each file should contain any data you have on the item—repair manuals, modification drawings, and the equipment record sheet.

This equipment record sheet—preferably an 8½ × 11 in. heavy stock card—performs two functions: It profiles the machine (make, model number, purchase price and date, modifications, and auxiliary attachments), and it itemizes repair history. Itemizing should include the date of each repair, cost, parts or sub-assemblies replaced, and comments of mechanics or their bosses at the time of repair. Like everything else in a maintenance program, equipment records must match your goals, your needs, and your organization. For example, if you prefer a manual entry system to typing collected data, design the record card for this. If your system is computerized, select a form that meshes with that setup.

Analyze for Profit

Up to this point, everything has been aimed at establishing a sound maintenance program. But once it has been established, how do you use it?

Since construction is a business geared to profit, the answer to this question is that you must use the program to reduce your maintenance cost and increase your profit. To accomplish this, your new or reconstructed program must tell you what a maintenance cost is or, perhaps more important, what it is not.

In simple terms, such cost is not only the sum of a mechanic's time, translated into dollars, and the cost of materials used—although these are both components of the cost. A maintenance cost also includes the costs of the machine's time lost to the job, operator idled by the downtime, possible shifting of a second unit to the job, penalties incurred, job schedule disruption, and administration of all these events. In short, a maintenance cost is a repair cost plus a downtime cost.

PM: key to downtime control Preventive maintenance (PM) is the single most effective tool for curbing downtime. Regrettably, it is misunderstood and misapplied more often than it is harnessed correctly. PM is not a virtue, or an ideal, or a universal necessity; it is also not simply a continuous lubrication program. It is a tool to control downtime and hence to reduce downtime cost. It is what you do before breakdown to prevent breakdown. But, like any activity on your spread, preventive maintenance costs money.

For a graphic look at this, consider the two curves in Fig. 2-2. The curve at the left shows how downtime costs decrease as PM hours increase. The right-hand curve shows how PM costs increase as PM hours spent increase. As the curves move to the right, you see how a great deal of PM effort and expense can still fail to produce any significant drop in downtime costs. For each machine, you must establish the optimum, and try to keep that unit's PM coverage near that point. And you must remember that the optimum will probably move to the right as time goes by and the equipment naturally ages.

Frequency, the time between specific PM checks, should initiate with the equipment manufacturer. As feedback and age develop, this frequency must be adjusted. The important thing is that PM must match each machine, not be applied as a security blanket equally draped over all units, identical or not.

Because lubrication is so frequently equated by so many to preventive maintenance, it is important that you recognize that this activity is just a part of the overall endeavor. Get your lubricant supplier to survey all your equipment (usually a free service) and make up a program of frequencies and materials use for you to follow. Renew this service about every third year (more often if you add a lot of new equipment). Keep in mind that downtime reduction is your primary goal. Standardized lubricants can save you money, but machine breakdown will dispel that gain quickly.

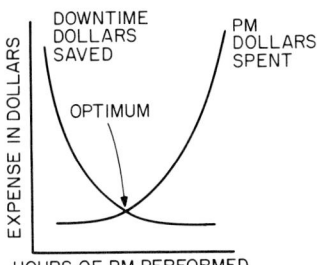

Fig. 2-2 PM as a mixed blessing.

Maintainability: path to profit Getting equipment out of the downtime doldrums quickly depends heavily on a machine's ability to be maintained easily. The passenger car radio is a classic example of a relatively simple unit whose maintainability is out of all proportion to its importance as a machine component or even to the often minor repair that it may require.

Construction equipment must be viewed, prior to purchase, in terms of its maintainability and reviewed during each repair for ways to improve this factor for the future. The two key elements of maintainability are access and modularization.

Access, that is, getting at machine components, also includes toolability, or ease of opening or removing something with basic maintenance tools. As much as half a day can be spent locating a special wrench to remove one minor element. Moreover, the more frequently a component must be checked or fixed, the easier it should be to reach and repair.

Modularization is the use of sub-assemblies that are easy to remove and replace. Of course, it is only as effective as the availability of the replacement sub-assembly—at your dealer's or in your shop. Most modularization today must originate with the manufacturer, so it becomes a major factor during selection. But it can also occasionally be built in during major overhaul.

Corrective maintenance: a big help This activity, performed at the time a repair is made, attempts to leave a piece of machinery a little better than it was before breakdown. The concept might be stated: As long as I have gotten this far, why not take the opportunity to see it doesn't happen again? But it may be a lot simpler than that.

Every failure has a reason. Repeatedly rupturing hose lines may be collecting corrosive oil splatters. Broken bucket teeth may mean they need additional hardening. Cast-iron parts that crack on schedule may be experiencing excessive shock. Belts may be too tight or too loose. Worn gears may not be getting enough lubricant. Some reasons are traceable to application, some to design. In the latter case, especially, corrective maintenance can prevent recurrence. The oil can be stopped from hitting the hose, or an oil-resistant hose can be substituted. Teeth can be hardened; cast-iron parts can be replaced by costlier but sturdier steel ones. The best advice possible for this aspect of maintenance is "harden up," "beef up," and "step up" as a route to improvement.

Spare parts: area for strategy Replacement parts and sub-assemblies regularly mark the difference between high and low downtime expense since time itself is such an important ingredient here. Failure to keep the right part on hand can mean anything from an hour to a week on inoperative equipment. Yet every dollar you have invested in spare parts is nonworking, non-interest-gathering capital. Frequently, the total runs to tens of thousands of dollars even at the small- to medium-size spread.

At some point in your program installation or revamping effort, you must begin a study of this necessary evil. First, inventory everything in your spare parts group. Sum up both what you payed for it and what it's worth now. This will give you an indication of the way inflation is affecting your private resources. The inventory should also reveal the amount of duplication that exists among these parts.

Next, set up a continuous program of parts accounting that tells you how long something lies on the shelf and how long it takes to replace at your dealer's or from the manufacturer. Just where you cut off reordering a specific part is, of course, a decision that applies only to you and to your operation. But a $100 item that can be ordered and received in 1 hr means that, to justify its existence in your shop as a spare part, it must produce more than $100/hr in downtime cost. As a rule, you must question parts carried by your shop in direct ratio to their cost.

Another area of effective cost strategy is to take advantage of the occurrence of standard parts on the equipment maker's list of recommended spares. Such items as bearings, shaft collars and seals, V-belts, and a number of engine parts have hardware store or mill supply house counterparts. These items are frequently cheaper when they are bought over the counter this way than when they are ordered from manufacturers, using their part numbers.

Again, effective strategy has its roots in facts—facts gleaned from sound record keeping. For the very reason that spare parts are nonworking capital and at the same time are a hedge against downtime disaster, it is important to have all possible data at your fingertips.

Chapter **3**

A System for Organizing Maintenance Information

LINDLEY R. HIGGINS, P.E.
Consulting Engineer
Senior Editor, Construction Contracting
(Formerly *Construction Methods & Equipment***)**
McGraw-Hill, Inc., New York, N.Y.

What the 1950s meant to industrial maintenance—in recognition, management, and technological gains—the current decade seems destined to mean to construction equipment maintenance. In both cases, the underlying cause can be traced to mounting costs—both as a net dollar outlay and as a vital percent of overall business expense.

More important, however, is the current trend among constructors to think in terms of true downtime cost. This means putting meaningful price tags on machine unavailability, idle operator time, cost of borrowing equipment from other parts of a job or from other jobs, and the rental of stand-in replacements. It is important to remember that, unlike industry, where it is less expensive to add second and third shifts or to build standby machinery, construction must work within a rigid time-and-bid framework that is expensive to adjust.

For these reasons, managing maintenance is a serious challenge for everyone in the construction industry, and one that cannot long wait for answers.

To help find, classify, and utilize these answers, I have developed a numbered coding system that will define the limits, the content, and the organization of data for construction equipment maintenance.

Reasons for developing the code Can simply assigning numbers to the key problems inherent in maintenance make them go away? How can such a system ease the cost crunch in equipment repair and upkeep? This coding system won't make your troubles vanish, but here are some of the positive things it can do right now.

- It can organize the contractor approach, as well as that of dealers and manufacturers, to the scope, the divisions, and the techniques of sound maintenance practice.
- It can break down and classify repair problems and solutions and prevent overlooking vital machine areas or components.
- It can provide the basis for building up data files and for cross-filing useful solutions by arranging activities and equipment into groups and subgroups.
- It can establish a common language for construction equipment maintenance so that all those involved in this area can communicate needs and remedies—the cornerstone of all problem-solving endeavors.

How the code works The Construction Equipment Maintenance Code divides all contractor maintenance activity into ten groups—100 through 000—and then into subgroups, with even finer categories available to match systems refinement to specific need and technological development. Because it can be easily expanded beyond the three-digit initial characteristic, its capacity for data storage is limitless. And currently unassigned (open) numbers in prime groups assure this extra capacity characteristic.

The ten principal categories of construction equipment maintenance plus a general description of their content follow:

100 Maintenance Management. This group sets the control stage, the logical starting point, covering such subtopics as proven management techniques, personnel training, record keeping, cost control, preventive maintenance, spare parts management, and other aspects of maintenance management, all in organized detail.

200 Maintenance Tools and Facilities. The 200 group defines the physical plant and equipment that any contractor needs to accomplish repair work easily and economically. Specific categories include shop planning (fixed, mobile, temporary job site), measuring and testing equipment, hand and power tools, and materials for maintenance.

300 Basic Maintenance Technology. This group covers the techniques and practices common to all repair and upkeep activities, including welding, cleaning, corrosion control, coating and hardening, and vibration control.

400 Power and Propulsion System Maintenance. This group codifies the analysis and repair of diesel and gasoline engines plus electric motor and power drive system maintenance.

500 Ground Contact Element Maintenance. Such maintenance goes to the heart of the wear point problem and organizes repair of machine elements receiving the greatest punishment. These elements include equipment tires and tracks and compactor working ends, blades, teeth, and buckets.

600 Maintenance of Equipment Entities. Maintenance analysis is organized into specific programs for specific construction machines. In distinct groupings, it covers dozers and loaders, graders, scrapers, cranes, and paving equipment.

700 Maintenance of Support Elements. The special auxiliary units of construction are separated into repair and upkeep classifications. Headings include pump and compressor maintenance, drill and tool repair, and conveyor maintenance.

800 Vehicular Maintenance. The unique needs for keeping on- and off-the-road equipment operative are recognized in this group. It organizes the approach to motor preventive maintenance, chassis and power system repair, and general body work.

900 Lubricants and Lubrication. This code isolates this important maintenance operation into a group of its own, taking into account the need for lubricant analysis and its proper application and stressing the need to employ both the correct equipment and the proper lubrication procedure.

000 Safety and Environment. This maintenance group underlines the growing importance of these two areas of managerial concern in the repair shop and on the job site. The category itemizes legal controls, safety practices, protective equipment, first aid, and the various subdivisions of operating in and with an awareness of surroundings.

Detailed breakdown of categories At present this system is partially complete, with gaps that will, in most cases, be filled at a later date. But it is important for the reader to realize that any such system must have initial flexibility built into it. These gaps represent that flexibility. Remember, too, that this code is of use to you only so long as it helps. When it no longer does so or when you find yourself or your subordinates serving the code instead of enjoying its services, it is time to scrap it.

100 MAINTENANCE MANAGEMENT
110 Maintenance Management Techniques
120 Selecting, Training, and Supervising Maintenance Personnel
130 Maintenance Record Keeping
140 Managing Maintenance Costs
150 Preventive Maintenance Practices and Systems
160 Equipment Spare Parts and Stores Keeping
170 Contracted Maintenance
180 Use of Electronic Data-Processing in Maintenance Management
190 (Open)
200 MAINTENANCE TOOLS AND FACILITIES
210 Fixed Maintenance Facilities

220 Mobile and At-Site Maintenance Facilities
230 Maintenance Measuring and Testing Equipment
240 Maintenance Hand Tools
250 Maintenance Heavy and Power Tools
260 Maintenance Kits and Portable Carts
270 Maintenance Materials
280 Lifts, Hoists, Platforms, and Scaffolds
290 (Open)
300 BASIC MAINTENANCE TECHNOLOGY
310 Welding
320 Cleaning and Corrosion Control
330 Coating and Hardening
340 Mechanical Power Transmission
350 Maintenance of Piping, Tubing, and Hose
360 Maintenance of Hydraulic and Pneumatic Systems
370 Maintenance of Electrical and Electronic Systems
380 Vibration
390 (Open)
400 POWER AND PROPULSION SYSTEM MAINTENANCE
410 Diesel Engine Maintenance
420 Gasoline Engine Maintenance
430 Power Drive and Transmission Maintenance
440 Electric Motor and Generator Maintenance
450 (Open)
460 (Open)
470 (Open)
480 (Open)
490 (Open)
500 GROUND CONTACT ELEMENT MAINTENANCE
510 Tire Care and Maintenance
520 Track Maintenance
530 Compactor Contact Elements Maintenance
540 Blade and Tooth Care and Maintenance
550 Bucket and Shovel Care and Maintenance
560 (Open)
570 (Open)
580 (Open)
590 (Open)
600 MAINTENANCE OF EQUIPMENT ENTITIES
610 Dozer and Loader Maintenance
620 Grader Maintenance
630 Scraper Maintenance
640 Crane Maintenance
650 Paver and Paving Equipment Maintenance
660 Compactor Maintenance
670 Excavator and Backhoe Maintenance
680 Tunneling Equipment Maintenance
690 (Open)
700 MAINTENANCE OF SUPPORT ELEMENTS
710 Pump and Compressor Maintenance

720 Drill and Drill Tool Maintenance
730 Conveyor Maintenance
740 Crushing Equipment Maintenance
750 Concrete- and Asphalt-making Equipment Maintenance
760 Pile Driving Equipment Maintenance
770 Maintenance of Offshore and Underwater Equipment
780 (Open)
790 (Open)
800 VEHICULAR MAINTENANCE
810 (Open)
820 (Open)
830 (Open)
840 (Open)
850 (Open)
860 (Open)
870 (Open)
880 (Open)
890 (Open)
900 LUBRICANTS AND LUBRICATION
910 Lubricants and Coolants
920 Lubrication Application Equipment
930 Lubrication Practices
940 (Open)
950 (Open)
960 (Open)
970 (Open)
980 (Open)
990 (Open)
000 SAFETY AND ENVIRONMENT
010 Safety Laws and Regulations
020 Safety Practices
030 Protective Wear and Equipment
040 First Aid
050 (Open)
060 (Open)
070 (Open)
080 Environmental Control Practices
090 Maintenance of Environmental Control Equipment

Chapter **4**

Developing and Training Maintenance Personnel

Part 1 Maintenance-Trades Training

ROBERT M. JOHNSON
Manager of Trades Training, Dow Chemical U.S.A., Midland, Mich.

Part 2 Supervisory Training

ALLEN STORR
Manager of Training Resources, Dow Chemical U.S.A., Midland, Mich.

PART 1 MAINTENANCE-TRADES TRAINING

Introduction Today, the training function must be consistent with all other functions within any enterprise such as construction as far as its cost effectiveness is concerned. All training, whether at the apprentice, journeyman, or supervisory level, must be carefully examined and directed at those situations where it will have the greatest impact on profit improvement. It is with this emphasis in mind that the material for Part 1 is presented.

Research
Initially, training personnel should be concerned with what has been going on in craft training and what people are saying about it. This knowledge and understanding are important so that mistakes in the design and development of training programs can be eliminated. In this way, more meaningful craft training programs will become available for the future. All too often a review of the past is overlooked.

Three studies of state and national scope, by Drew, Crabtree, and Strauss, which deal with the attitudes of apprentices and journeymen toward training, are highly recommended reading. Reviewed are some significant findings within these studies which emphasize the importance of what apprentices and journeymen are saying about the training function.

What is apprentice training? Where is it given? The Strauss study entitled "Apprenticeship: Problems and Policies" was conducted primarily to provide background so that the reader could intelligently evaluate the role of apprenticeship in America today.

Strauss explains that in many instances apprenticeship training had been traditionally thought of as training provided on the job. Today, many consider there is a strong need for classroom-related instruction in combination with on-the-job training. This combination provides the effective training necessary for becoming a journeyman craftsman. Strauss also points out that most related instruction has been given at local high schools. This location is now rapidly changing to include many junior colleges or other post-high-school institutions. Also, many trade unions are building their own facilities and providing their own instructors.

What are attitudes of apprentices toward training? Strauss's findings included the attitudes of apprentices toward classroom-related instruction. In general, they were unhappy with this phase compared with other phases of their training. Apprentices evaluated such training as "impractical, irrelevant, boring and a waste of time." A substantial number complained about instructional methods.

Instructional techniques used in one state primarily consisted of a workbook, which in turn meant that the student progressed at his own learning rate. Apprentices participating in this kind of learning considered it similar to a supervised study hall and believed that they could accomplish just as much if they studied the book at home. Also, the workbooks were out of date.

There is still an argument between those who advocate and those who oppose classroom-related instruction. According to Strauss, those who believe classroom-related instruction should assume a larger role base their belief on the following:

    1. Apprentices do not realize a well-rounded education, since firms that hire apprentices are becoming more specialized.

    2. Training on the job is done in a haphazard way. Though many trades require apprentices to keep books listing their work experience and require foremen to countersign these books, it is not uncommon for both parties to distort the truth.

    3. Classroom-related-instruction learning may be more effective than learning on the job because it is full-time, whereas learning on the job is largely secondary when compared with getting the job done.

    4. An apprentice learns sloppy ways of doing things on the job. (However, opponents of classroom-related-instruction training argue that the instructor may not know the right way himself.)

    5. Many of the old job skills required a high degree of manual dexterity and knack. These are being made obsolete by newer technologies which require more cognitive skills, for example, the field of electronics. Obviously, these skills are better learned in the school setting.[1]

Tradesmen's view of training Drew's apprenticeship study of the machinist, tool and diemaker, printer, culinary, and pipe trades was made in order that recommendations could be made for improving the training of skilled craftsmen.

Some suggestions were made by the pipe-trade respondents for improvement of apprenticeship training. Three of the most important in a list of twelve were the real need for qualified instructors who can teach, training information that is relevant, and an on-the-job rotation system to cover all aspects of the job.

Journeymen attitudes toward training indicated they were concerned about the problem of obsolescence. They felt a need for keeping up with technological change. Thirty-seven percent indicated that at least once a month they received work assignments for which they did not feel qualified.

Journeymen respondents who indicated that training was mandatory stated that it was required by the union, the employers, or both. However, 80 percent of the respondents were not required to participate in training or schooling as a condition of employment.

The most common methods for keeping up to date with new technology were
    1. Trade journals
    2. Other workers in the trade
    3. Manufacturing service bulletins
    4. Foremen and supervisors
    5. Special union reports

Although three of the five methods were directly related to reading abilities, it was found

---

[1]George Strauss, Apprentice-related Instruction: Some Basic Issues, *Journal of Human Resources*, vol. 3, no. 2, pp. 213–236, spring 1968.

that 55 percent of the respondents had not participated in any organized education and training for the 2-year period prior to the survey.

Record keeping on training after the employee became a journeyman was lacking. Forty-two percent of the journeymen respondents said that no records were kept; another 29 percent did not know of any record keeping. Only 2 percent indicated that employers kept records.[2]

Changes in apprentice training Crabtree's study, "The Structure of Related Instruction in Wisconsin Apprenticeship Programs," investigated related instruction offered to apprentices in Wisconsin vocational schools. It had the following conclusions:

1. The lack of statewide standards in most aspects of related instruction has encouraged the development and implementation of a whole array of fragmented policies and procedures at the community level.

2. It is highly unlikely that the present structure of related instruction will see any drastic changes for the immediate future.

3. Efforts to legislate changes in the apprenticeship law to provide for statewide standards probably would not receive much support, particularly from employers and labor.

4. Much of what we can do to improve related instruction depends on how we presently perceive it.

5. What is urgently needed is a change in our conceptualization of apprentice training.

Such changes as those previously suggested cannot come about by merely supplying curriculum outlines and work processes to counselors or by publication of attractive bulletins. Comprehensive manpower planning for apprenticeship is a prerequisite to all that follows. In view of this basic requirement and in light of the fragmented leadership and administrative policies and procedures, some method must be found that will tend to strengthen these relationships.

A number of considerations for improving and strengthening related instruction and apprenticeships are presented below. Some of these are aimed at the administrative and coordinating function. Others focus attention on the related-instruction function in the schools.

1. In an effort to strengthen the leadership and administrative relationships of related instruction, consideration should be given to expanding the participating membership of the local and/or advisory committees to include representatives from the vocational schools.

2. Enrollment policy represents another basic consideration for improving related instruction.

3. A change in enrollment is fundamental to providing increased flexibility in instructional techniques.

4. There is definite need for a standardized core curriculum for all trade programs.

5. The present structure of related-instruction credit is based on an hour-for-hour equivalency. Crediting means little or nothing outside of apprenticeship training.

6. The practice of allowing credit for previous education and/or work experience, home assignments, and correspondence courses is in need of review.

7. In the course of the present study, it became clear that a few schools are not providing related instruction for certain trades, primarily because of small enrollments.

8. Finally, several aspects of related instruction need special attention in the form of research at the state level.[3]

References

Crabtree, James S.: The Structure of Related Instruction in Wisconsin Apprenticeship Programs, Ph.D. dissertation, University of Wisconsin, Madison, Wis., 1967.

Drew, Alfred S.: *Educational and Training Adjustments in Selected Apprenticeable Trades*, vols. 1, 2, Purdue Research Foundation, Lafayette, Ind., 1969.

[2] Alfred S. Drew, "Educational and Training Adjustments in Selected Apprenticeable Trades," vol. 1, pp. 27–39, Purdue Research Foundation, Lafayette, Ind., 1969.

[3] James S. Crabtree, "The Structure of Related Instruction in Wisconsin Apprenticeship Programs," pp. 191–201, Ph.D. dissertation, University of Wisconsin, Madison, Wis., 1967.

Research in Apprenticeship Training, *Proceedings of a Conference,* University of Wisconsin, Center for Studies in Vocational and Technical Education, Madison, 1967.

Strauss, George: "Apprenticeship Problems and Policies," Institute of Industrial Relations, University of California, Berkeley, 1969.

Strauss, George: Apprentice-related Instruction: Some Basic Issues, *Journal of Human Resources,* vol. 3, no. 2, pp. 213–236, spring 1968.

U.S. Department of Labor: "Apprenticeship Schedules," Trade and Industrial Publication No. 2, Manpower Administration, Bureau of Apprenticeship and Training, Washington, D.C., 1962.

U.S. Department of Labor: "The National Apprenticeship Program," Manpower Administration, Bureau of Apprenticeship and Training, Washington, D.C., 1968.

Needs and Basic Provisions for Training

Need for training Your company may be one of the many that are considering (or who will be considering) training programs for your tradesmen. Perhaps you are unable to hire qualified men to carry out daily maintenance requirements—or your tradesmen are not properly trained in the fundamentals involving a particular trade since maintenance and construction work is becoming increasingly complex owing to technological advancements in equipment. Maybe they have never received any formal training. Whatever the reason, it is important that you identify the need for establishing a training program. Your next move is to convince your management. And be absolutely sure they are sold on this venture.

Basic provisions After you have identified the need for training, develop a list of general provisions for your program. Give consideration to the following:

1. Selecting qualified men to be trained
2. Determining standards of performance for the trainees
3. Measuring the performance of trainees against standards of performance
4. A job-progression schedule showing job classification and wages related to a training schedule
5. A job description for each classification
6. Settling individual trainee problems such as what to do when a trainee is doing failing work and what happens to the trainee who misses work because of illness
7. Determining the number of men to be trained
8. Determining how, when, and where training is to be given
9. Delegating responsibility for carrying out the training program
10. Correlating classroom training with shop and/or field training
11. Rotating trainees to provide them with varied shop and/or field experiences
12. Evaluating training to ensure that it is always kept current
13. Determining what happens to the trainee who fails the program
14. Determining who is going to buy supplies used in the program—the trainee or the company
15. A concise definition of the skills which the trainee is required to be able to perform upon completion of training

Training Agreements

Writing training agreements The training agreement covers the detailed wording of the basic provisions which have been stated. After you have established the need for training and have carefully analyzed the basic provisions to be included in your program, the training department is confronted with the task of writing the training agreement. Whoever has this task should prepare a rough draft. Have the draft completed and reviewed by management. At this point, the cooperation of the union is essential. In the interest of keeping them informed it is well to have union representation review the rough draft. They may have suggestions for improvement. After you and the union representatives have agreed to the contents of the agreement, submit it to the company and union bargaining committees for approval.

Contract A practice that works effectively is to include in your company-union contract only general provisions, for example, the privilege to conduct training as required. Included in these general provisions is a statement to the effect that training agreements for each trade may be drawn up between the company and union provided that the training agreements contain nothing contrary to the terms of the company-union contract. Using this approach, both you (the company) and stewards are free to prepare a training agreement for each trade and tailor it to specific needs. Revisions can be made with a minimum of negotiating. In the event you have a joint apprenticeship committee, the committee would be the first and, in some situations, the final approving group.

Bureau of apprenticeship and training As part of the U.S. Department of Labor, the Bureau of Apprenticeship and Training stimulates and assists businesses in the development, expansion, and improvement of apprenticeship and training programs designed to provide the skills required by an economy that is becoming increasingly complex.

The Bureau of Apprenticeship and Training's principal functions are to encourage the establishment of sound apprenticeship and training programs and to provide technical assistance to industry in setting up such programs. In the performance of these functions, the bureau is guided by the Federal Committee on Apprenticeship, made up of leaders of management, labor, and vocational education. The bureau works closely with state apprenticeship agencies, trade and industrial education institutions, and management and labor.

Through its field staff, with offices in every state, the bureau works with local employers and employees in developing apprenticeship and industrial training programs to meet specific needs.

Joint apprenticeship committees These committees, made up of representatives of management and labor, work together for the development and administration of local apprenticeship training programs. In addition to local groups, there are national trade committees representing national organizations. The national committees formulate, with assistance from the Bureau of Apprenticeship and Training, national policies on apprenticeship in the various trades and issue basic standards to be used by affiliated organizations. Typical duties of these committees are to

1. Provide that each apprentice be interviewed, placed under the agreement, and impressed with the responsibilities he is about to accept and the benefits he will be entitled to receive. The committee may designate interviewers other than committee members.

2. Determine whether the apprentice's scheduled wage increases shall be approved in the event that he is delinquent in fulfilling all apprenticeship obligations.

3. Offer constructive suggestions for the improvement of the apprenticeship program.

4. Certify the names of graduate apprentices to the registration agencies and recommend that a certificate of completion of apprenticeship be awarded upon satisfactory completion of the requirements of apprenticeship. No certificates are issued by the committee.

5. Assist in setting up the requirements that each man trained under a nonregistered plan will have to meet in order to become eligible for certification under the standards of apprenticeship.

6. Assist in answering questions concerning the training of apprentices and senior apprentices.

7. Review an apprentice's previous related training and/or work experience to adjust his starting rate upward equal to his related training and/or work experience.

8. In general, be responsible for the successful completion of the apprenticeship by the apprentices under the standards.

Basic standards for apprenticeship Programs registered by the Bureau of Apprenticeship and Training must provide that

1. The starting age of an apprentice is not less than sixteen.
2. There is full and fair opportunity to apply for apprenticeship.
3. Selection of apprentices is based on qualifications alone.
4. There is a schedule of work processes in which an apprentice is to receive training and experience on the job.
5. The program includes organized instruction designed to provide the apprentice with knowledge in technical subjects related to his trade (a minimum of 144 hr per year is normally considered necessary).
6. There is a progressively increasing schedule of wages.
7. Proper supervision of on-the-job training with adequate facilities to train apprentices is ensured.
8. The apprentice's progress, in both job performance and related instruction, is evaluated periodically and appropriate records are maintained.

9. There is employee-employer cooperation.
10. Successful completions are recognized.
11. There is no discrimination in any phase of apprenticeship employment and training.

Selecting Qualified Men

Need for selection   You cannot afford to overlook a sound selection program for men entering your trades training program. Always keep in mind that each man entering any training program is a potential foreman. By the time a man completes a training program for a particular trade you are going to have a sizable monetary investment. A good selection program will aid in hiring a high-caliber tradesman, and this man will be able to grow as maintenance and construction work becomes more complex. A sound selection program will contribute in lowering overall costs. It will lower the turnover of manpower; lower the cost of training; decrease the chance of lowering the standards of a particular trade; and improve employee morale, productivity on the job, and quality of work performed.

Many companies are using a series of tests which aid them in their selection of tradesmen. Frequently, a series of tests has been used in the selection of men at the point of hire. In addition, a group of tests can be selected which will indicate the man's ability to achieve in a particular trade. In general, it should be remembered that the reason for using a test battery in selecting tradesmen is that it serves as an aid in attempting to see the man as he actually is. As a result, you will have a better understanding of this man's potential relative to the job he is seeking.

Types of tests   Several types of tests have been found useful in the selection of maintenance trainees. Examples are:

General Ability Tests.   Measure ability to learn.

Interest Tests.   Indicate areas of activities in which a man shows an interest. Individuals are more likely to be motivated to achieve levels of satisfactory performance in areas which are of interest to them.

Personality Tests.   Measure the pressures that produce behavior and may possibly help in predicting behavior. Certain things about behavior either support the general activity or retard its success. Professional interpretation of personality tests will provide this kind of information.

Achievement Tests (Equivalency Tests).   Indicate the degree of mastery or retention of a subject field as compared with what other people have retained in the same subject field.

Aptitude Tests.   Attempt to measure special individual abilities which will indicate if the man is likely to perform a given job satisfactorily.

When existing tests cannot be validated, outside consulting firms are now developing and validating selection tests for the private sector. This is to eliminate any concern in employment practices as applied to selection tests.

Tests and test scores   You may believe that the union would object to the use of a test battery for the selection of tradesmen. However, many companies have installed test batteries and in many cases have accomplished this through the help of the union. While you are considering a training agreement for a particular trade, work with your union in developing the selection program. If the union is unfamiliar with tests and testing procedures, have a representative take the test you are thinking of using and have a qualified person explain the results. This is one of the better ways to achieve union acceptance of a testing program.

It may also be helpful to have the men who are already tradesmen take your proposed test or test battery. This approach will enable you to evaluate the test and will be helpful in establishing minimum test scores. Your job of selling the union on a test-selection program will be much easier if members of the union have a part in its development.

Retesting procedure   A basic question you may have to resolve when setting up a training program is "What happens to the man if he fails the series of tests you are using for selection?" Resolving this question should be a joint effort of the company-union bargaining committees. This will result in establishing a uniform retesting procedure. You might want to develop a retesting procedure similar to the following:

1. A second test will be given to those who fail a test for the first time, provided the

score on the first test is at least 75 percent of the minimum required score. The second test will be given not less than 6 months after the first test.

2. If the score on the first test is less than 75 percent of the minimum required score, the man will be considered for a second test after 6 months only if he successfully completes a recognized educational course related to the subject failed.

3. The second test will be an alternate form of the first test whenever possible.

Conducting Training

Instructors The practice among companies varies greatly. Most trades training programs have one or a combination of the following people do the actual classroom instructing:
1. Shop supervisors
2. Engineers
3. Outside educational institutions
4. Manufacturer or dealer representatives
5. Training department personnel

Using the foremen or superintendents to do the actual instructing has the advantage of making it possible to carry out the important principle that the responsibility for training must rest with supervision. It affords opportunities for supervisors and men to get to know each other as individual personalities, which is important for morale. It gives added insurance that the men will be taught the practical things they need to know on the job. However, it has two major disadvantages (and they are mighty big disadvantages): (1) it is difficult to train most foremen or superintendents to be good instructors; (2) they usually cannot take enough time from their other duties to prepare for their classes.

Engineers usually have the technical know-how that is required for maintenance instruction, but it is difficult to make a good instructor out of an engineer, especially if he does it only part time.

If you do use foremen, shop superintendents, engineers, or other technical men as instructors, give them an intensive instructor-training course with plenty of opportunity to practice teaching techniques before you turn them loose on trainees. It will pay big dividends.

Many manufacturers or dealers offer excellent instruction service whereby a well-qualified representative will come to your plant or you can send your men to their plant. They furnish the required equipment and put on a short course covering the maintenance and operation of the equipment you purchase from them. If manufacturers or dealers bring the course to your plant, they will usually tailor it to fit your specific needs, if you so request.

I believe that instruction is best given by a full-time man with a degree in vocational education and some teaching experience in shopwork in public schools, night schools, or trade schools. This person could be a member of the maintenance or construction organization or training department. In addition, a shop or field supervisor should be available to supplement his work. Thus you have an instruction team of two men who complement each other.

Some of the major advantages of this team approach are
1. The training man knows how to teach—how to get the materials across to the man.
2. The supervisor furnishes "expert" knowledge on specialized subjects.
3. The supervisor knows what is going on in the training program and can thus help keep it geared to needs.
4. The training man can assist the supervisor in planning and organizing the sessions which the supervisor will teach. He can help secure material, visual aids, and demonstrations, as well as sit in on the session with the supervisor to help get the points across to the men.
5. The supervisor is not overburdened with classroom instructing duties.

Classrooms and laboratories Give a great deal of thought to where you are going to hold your classes. Consideration should be given to the classroom regarding air conditioning, lighting, space, size of classes, noise, location, and equipment. Too often classes are conducted where the proper environmental conditions do not exist. It is obvious that poor classroom facilities are not conducive to learning.

When laboratories can be planned so that laboratory work is correlated with the classroom training, you have the ideal training situation. Many times practical work

1-22   Establishing and Managing the Maintenance Function

experiences are assumed to be provided in the shop or field when in fact they have not actually happened. Sometimes proper instruction is not provided, or classroom instruction and work experience do not have the proper timing. One sure way that you can solve this problem is to provide a laboratory with the classroom.

The following are examples of selected apprenticeship schedules:

APPRENTICESHIP SCHEDULE FOR MILLWRIGHT APPRENTICESHIP PROGRAM – 4 YEARS

RELATED INSTRUCTION

	Hours
1. Safety	200
2. Mathematics	60
3. Reading Engineering Drawings	30
4. Sketching	20
5. Hydraulics	50
6. Equipment Training	66
7. Shop Theory	50
8. Lubrication	20
9. Welding	80
Total hours	576
Total months	4

WORK EXPERIENCE

	Months
1. TRADE ORIENTATION	1
Learn shop and field procedures, housekeeping, stock items, and tool identification.	
2. JOB PLANNING	3
Learn to follow instructions, select tools, and materials. Read, draw sketches, and plan job operations.	
3. USE OF TOOLS AND MACHINES	15
(Bench and field) Power and radial drill, tap, ream, counterbore, countersink, hand grind, file, scrape, chip, gage blocks, indicators, lubricants, and equipment inspection.	
4. EQUIPMENT MAINTENANCE AND REPAIR	15
Inspect, troubleshoot, adjust, remove and replace parts of plant machinery.	
5. WELDING	1
Cut, tack, bevel, and form through heating.	
6. MISCELLANEOUS	9
Materials, cleanup, tool-crib experience, and general shop duties.	
Total months	44

APPRENTICESHIP SCHEDULE FOR INDUSTRIAL BOILERMAKER (CHEMICAL) APPRENTICESHIP PROGRAM – 4 YEARS

RELATED INSTRUCTION

	Hours
1. Safety	200
2. Mathematics	60
3. Reading Engineering Drawings	30
4. Welding	80
5. Standards	60
6. Shop Orientation	126
7. Sketching	20
Total hours	576
Total months	4

WORK EXPERIENCE

	Months
1. TRADE ORIENTATION	5
Assist in drilling, punching, grinding, shearing, fitting, etc. Learn safety and good housekeeping practices, names and applications of tools pertinent to boilermaker trade, identification of metals.	
2. JOB PLANNING	9
Learn to follow instructions, receive instructions from foremen, coordinate with building personnel, measure, layout, sketch, order stock. Visualize and plan job from sketch or blueprint, make simple layouts.	
3. USE OF TOOLS AND MACHINES	18
Operate all tools and machines used by boilermakers to fabricate and repair chemical process equipment. Water and air test tanks and condensers. Bend, roll, and position plates. Fit patches.	
4. WELDING	6
Cut, tack, bevel, and form through heating.	
5. RELATED TECHNOLOGY	2
Must have work experience necessary to become familiar with ceramics, glass, lead, porcelain, and carbon as it relates to the boilermaker trade.	

APPRENTICESHIP SCHEDULE FOR INDUSTRIAL BOILERMAKER (CHEMICAL) APPRENTICESHIP PROGRAM – 4 YEARS (Continued)

WORK EXPERIENCE

	Months
6. MISCELLANEOUS	4
Cleanup, tool procurement, safety inspections and precautions, check standards.	
Total months	44

APPRENTICESHIP SCHEDULE FOR CARPENTER APPRENTICESHIP PROGRAM – 4 YEARS

RELATED INSTRUCTION

	Hours
1. Safety	200
2. Mathematics	60
3. Reading Engineering Drawings	60
4. Welding	80
5. Standards	60
6. Hand and Power Tools	116
Total hours	576
Total months	4

WORK EXPERIENCE

	Months
1. TRADE ORIENTATION	2
Good housekeeping, how and where to order stock and tools, name and applications of special tools.	
2. BUILDING CONSTRUCTION	14
Learn standard way to construct buildings including framing, siding, walls, ceiling, roofing, interior and exterior trim and flashing.	
3. CONCRETE	7
Learn to mix, pour, repair bust out, lay forms, add reinforcement, and finish.	
4. USE OF TOOLS AND MACHINES	1
Learn correct use and care for tools and machines for the trade. Includes repair and sharpening of personal and company tools.	
5. STEEL CONSTRUCTION	6
Name of construction steel, how to erect fastening, use in framing, door and window jambs.	
6. FINISH CARPENTRY	2
Rudiments of cabinet and furniture repair.	
7. WELDING	6
Cut, tack, bevel, and form through heating.	
8. MISCELLANEOUS	6
Construction of scaffolding and barricades, cleanup, procedures, boxing and crating.	
Total months	44

APPRENTICESHIP SCHEDULE FOR PIPEFITTER-FABRICATOR APPRENTICESHIP PROGRAM – 4 YEARS

RELATED INSTRUCTION

	Hours
1. Safety	200
2. Mathematics	60
3. Reading Engineering Drawings	30
4. Welding	80
5. Standards	60
6. Sketching	70
7. Field orientation	40
8. Hand and Power Tools Instruction	36
Total hours	576
Total months	4

WORK EXPERIENCE

	Months
1. TRADE ORIENTATION	6
Learn fittings, sizes, materials available, application, standards, how to get out stock, tools of the trade, safety, good housekeeping practices.	
2. JOB PLANNING	10
Receive instructions from foremen, measure, layout, sketch, get out stock, visualize and plan job from sketch or blueprint.	
3. USE OF TOOLS AND MACHINES	18
Abrasive cutoff machines, do-all saws, benders, punches, van stone grinders, welders, squares, scribes, hammers, levels, layouts, combination squares, etc.	
4. WELDING	6
Cut, tack, bevel, and form through heating.	

APPRENTICESHIP SCHEDULE FOR PIPEFITTER-FABRICATOR
APPRENTICESHIP PROGRAM—4 YEARS (Continued)

Work Experience

	Months
5. MISCELLANEOUS	4
Cleanup, tool procurement, review work with foreman, safety, check references and standards, inspect completed work.	
Total months	44

APPRENTICESHIP SCHEDULE FOR MACHINIST
APPRENTICESHIP PROGRAM—4 YEARS

Related Instruction

	Hours
1. Safety	200
2. Mathematics	60
3. Reading Engineering Drawings	30
4. Sketching	20
5. Theory	80
6. Vendor Training	30
7. Machine Instruction	156
Total hours	576
Total months	4

Work Experience

	Months
1. TRADE ORIENTATION	2
Assist journeymen in the shop. Handle minor jobs on the machine. Learn the shop rules and procedures.	
2. JOB PLANNING	3
Learn to follow instructions, cooperate with other tradesmen, select stock, measure, cut off stock, prepare machines and tools for proper machining procedures.	
3. USE OF TOOLS	4
Measuring, filing, drilling, tapping, micrometer reading, gaging, layout, job assembling, and tool sharpening.	
4. USE OF MACHINES	23
Turchan duplicator	1
Keyseater	0.25
Drill machine	0.75
Mills	5
Balancer	2
Bullard turret lathe	1.5
Lathe	12
Shaper	0.5
5. OUTSIDE REPAIR AND BENCH REPAIR	4
(Same as 3 above)	
6. MISCELLANEOUS	8
Cleanup, crane operation crib, assist in general shop operation.	
Total months	44

APPRENTICESHIP SCHEDULE FOR RIGGER
APPRENTICESHIP PROGRAM—3 YEARS

Related Instruction

	Hours
1. Safety	200
2. Mathematics	40
3. Reading Engineering Drawings	30
4. Welding	80
5. Equipment and Fittings	20
6. Use of Tools	48
7. Fork Truck Training	6
8. Defensive Driving	8
Total hours	432
Total months	3

Work Experience

	Months
1. TRADE ORIENTATION	1
Learn safety and good housekeeping habits, and field procedures.	
2. JOB PLANNING	2
Learn to follow instructions, receive instructions from foremen, and coordinate jobs with building personnel. Visualize and plan job from onsite inspection.	
3. MOVING HEAVY MACHINERY	12
Know weight limits of equipment. Know weights of machinery to be moved.	
4. SETTING MACHINERY AND PLANT EQUIPMENT	12
Learn correct procedures of rigging up and placing equipment and machinery.	

APPRENTICESHIP SCHEDULE FOR
RIGGER
APPRENTICESHIP PROGRAM – 3 YEARS (Continued)

WORK EXPERIENCE

	Months
5. USE OF TOOLS AND MACHINES.......................... Correct use and care of tools and machines of the trade.	2
6. WELDING... Cut, tack, bevel, and form through heating.	3
7. MISCELLANEOUS... Cleanup, tool procurement, review work with other crews and trades, safety inspections and precautions, check standards.	1
Total months..	33

Evaluating classroom training It is difficult to put teeth in your training or to hold to the standards you expect of your men unless you have a strong testing program in the following areas:

1. Periodic quizzes throughout the progress of each course
2. Final examinations covering each course or covering a prescribed period of time

A periodic quiz (10 min) can be given at the beginning of each class session. A more common practice is to devote every fourth or fifth class session to a quiz. The latter method is preferable, since the time element is not so critical. Of course, the frequency may vary considerably because of changing conditions, but in general it is wise to have at least four or five periodic quizzes in each course. There are three main advantages to this procedure:

1. You have a measuring stick which tells you how much each trainee is learning during the progress of the course.

2. The trainee knows where he stands and has an opportunity to improve before it is too late. If he fails even one quiz, call him in—together with the steward and instructor—and discuss his progress. You may not find out what his real trouble is, but it is almost a sure bet that he will show marked improvement after such a discussion. Should it happen that the trainee continues to fail his quizzes and you have held several meetings with steward, trainee, and instructor, you have a good case built up to take him out of the trade if you feel that should be done.

3. A quiz is an excellent learning device. Reviewing a quiz, pointing out errors, emphasizing key points—all help to fix the knowledge in the mind of the trainee.

Final exams for the course should be comprehensive and cover only material that the trainee has been taught in the class, field, or shop.

Shop and/or field training The trainee usually spends no more than 4 hr per week in class. In the majority of companies the trainee works in the shop and field at least 90 percent of the time. Therefore, it is important that considerable planning be devoted to seeing that the trainee gets well-rounded work experience in the shop and field.

The training schedule in the shop is fairly easy to set up. You can determine how many hours should be spent on each machine or piece of equipment and on repair work coming into the shop. Then set up a schedule of rotation, and designate the foreman responsible for carrying out the schedule as planned. Perhaps the greatest difficulty is in trying to carry out the rotation schedule in the face of varying work loads and kinds of repair work being done in the shop. It is virtually impossible to follow the schedule exactly.

A rotation schedule in the field is much more difficult to carry out. However, it is essential that this be done so that the trainee has an opportunity to perform all phases of the field work required of his trade. Rotating the trainee among all the foremen and journeymen is the most common method of accomplishing this. Some companies specify the number of hours to be spent on each phase of field work. The important thing is to see that the schedule is practical and is followed as closely as possible. A great deal of cooperation is required to assure that the trainee gets well-rounded field experience.

Job skills Three significant problems may exist in apprentice field training programs. They are

1. Training on the job is accomplished in a haphazard way—the apprentices are asked to keep a record of work experience and foremen are required to countersign the

1-26   Establishing and Managing the Maintenance Function

results. However, this tends to be an accounting of time in a place, rather than a substantiation of a satisfactory skill performance.

2. Training on the job is too often lower in priority than getting the job done.

3. Training on the job sometimes finds the apprentice learning sloppy ways of doing things.

It is most important to eliminate these problems in any training program. As a suggested solution to this problem, consider developing a list of key skills which the apprentice will be required to perform satisfactorily. Advantages of this approach are

1. Establishes job-skill requirements measured by satisfactory performance.

2. Provides a list of job skills so that each apprentice will have a clear-cut understanding of what is expected of him during his shop or field work experience.

3. Provides a "check-off" system whereby each skill is recorded as having been satisfactorily performed; therefore, each apprentice will know precisely his own progress.

4. Provides a list of job skills tied to performance which will assist supervisors of craftsmen in having a better understanding of the job requirements and performance progress of apprentices.

5. Allows for individual differences in apprentice progress.

6. Provides a means for continual improvement of the program, and for keeping it relevant to current needs. Since the skill segments are identified, new skill requirements are easily recognized and added, or outdated ones are changed.

7. Provides a list of job skills which will give greater direction to related instruction requirements.

8. Assists in eliminating the possibility of apprentices staying on the same kind of work for too long a period.

9. Assists the foremen in rating apprentices, since he will know the specific requirements of each trade.

Examples of the job-skill concept for the pipe fitter and millwright trades are shown in Figs. 4-1 and 4-2.

Evaluating shop and/or field training   Examinations are, in themselves, a form of rating your trainees, but certainly only a small part of the total evaluation of your potential tradesmen. Additional forms of written ratings are necessary to enable you to guide and develop the trainee to the stage where he is highly skilled in his trade. But, before plunging into this rating business with both feet, consider a few basic fundamentals and major difficulties that may be encountered.

PIPE FITTER JOB SKILLS

Each apprentice will be provided opportunities to work on all the listed job skills within the 48-month apprenticeship	Has observed or helped on the job.		Has performed the job satisfactorily under close direction or supervision.		Has performed the job satisfactorily entirely on his own.		Exceptionally fine performance.	
MORAF - (FABRICATION OF PLASTIC-LINED STEEL PIPE)	Date	Initials	Date	Initials	Date	Initials	Date	Initials
Identify the various types of plastic liner.								
Demonstrate correct way to adjust and use all MORAF tools.								
Cut pipe to proper length using established cutoff values.								
Apply correct cutback values and cut steel pipe to liner surface.								
Remove steel ring using gear puller.								
Thread pipe using standard pipe dies and procedures.								
Remove grease and oil from pipe with a solvent cleaner.								
Install proper pipe flange on pipe to established dimensions.								
Heat plastic liner (flame or hot air) to recommended temperature.								
Attach MORAF tool, proper die. Mold the liner.								

Fig. 4-1   Job-skill evaluation: pipe fitter.

Developing and Training Maintenance Personnel 1-27

MILLWRIGHT JOB SKILLS

Each apprentice will be provided opportunities to work on all the listed job skills within the 48-month apprenticeship period.	Has observed or helped on the job.		Has performed the job job satisfactorily under close direction on supervision.		Has performed the the job satisfactorily entirely on his own.		Exceptionally fine performance.
PRECISION COUPLING ALIGNMENT	Date	Initials	Date	Initials	Date	Initials	Initials
Properly install shaft couplings.							
Correctly align equipment with straight edge to within 1/64"							
Correctly set up dial indicator and indicator holder.							
Accurately measure shims with micrometers to within 0.001" thick.							
Correctly install shims to bring alignment within 0.002" or 0.003" tolerance.							
Accurately make preset alignment allowance for temperature changes							
Accurately make hot checks and dynamic tests to verify results.							

Fig. 4-2 Job-skill evaluation: millwright.

1. The rater must be willing and able to justify every mark or statement he puts on a rating sheet. The problems encountered in trying to stick to this principle have often caused the rating system either to be thrown out entirely or to degenerate into a meaningless routine of periodically checking rating sheets to make the record look good. Unless supervisors conscientiously and honestly try to carry out this principle of going to bat for every item they rate, it is better not to have a formal rating system.

2. Each trainee should have two or more independent ratings from different foremen. Rotating the trainee among several foremen solves this problem. A common practice is to rate the trainee monthly until he completes classroom training, then semiannually.

3. The rater should discuss the rating with the trainee. The primary purpose of the rating is to bring to light areas wherein the trainee needs improvement and then take proper action to bring about the necessary improvement. Frequent counseling between the rater and trainee is essential.

4. Ratings should be treated as a confidential matter by the supervisors. Embarrassing situations can arise unless considerable caution is exercised.

5. The makeup or content of the rating form itself is not as important as the use of administration of the rating procedure. A smoothly operating rating procedure is not easily attained; in fact, at times the whole procedure may look downright discouraging. But even if you only occasionally spot a difficulty and are able to correct it, the effort is rewarding and worthwhile.

The subject of rating is controversial. There is no one best rating form, no one best rating procedure. Experience may be the best teacher.

Training-material sources Experience has proved that the best practice is to prepare and develop training material within your own organization. Sometimes a textbook is available which will cover general theory courses for a great many maintenance trades. However, you may find that you have to use several textbooks. Using several textbooks is costly, but developing training material is even more costly. In many instances the development of training materials is the only way out. Be prepared for this to happen.

Investigate various textbook companies for new materials. Each year many of these companies revise their training materials, and in many cases new ones are added. It is essential for the instructor to evaluate new training materials periodically. This effort can result in a tremendous amount of time saved in the development of training programs.

Today, many firms have organized and developed maintenance training programs. These firms in many instances will provide you with valuable training materials just for the asking. If you are contemplating training programs, send your instructors to others to see what kinds of training materials are available and discuss training problems with experienced people. This is sound practice.

Manufacturers or dealers of equipment for which you will be conducting training will be extremely helpful. They often will send a representative to your plant and provide you with training material and even help conduct classes. Manufacturers have training films on their equipment which can be loaned to you. It is important that you always include the manufacturer in your investigation of training materials.

Audiovisual aids Include audiovisual aids in your training programs. A tremendous amount of audiovisual-aid equipment and material is available. Evaluate the application of chalkboards, display materials, models, objects, film strips, slides, transparencies, films, charts, posters, records, tape recorders, and possibly closed-circuit television to your training situation. Good audiovisual aids will help the trainees learn more efficiently, with greater retention of the material presented.

It is not necessary to spend a great deal of money to have good visual aids in your training. Excellent movies, film strips, and slides can be rented at nominal fees or secured on a free-loan basis. Many times you can make your own visual aids. If this is done, keep in mind it can be time consuming, especially if you are a novice. Often, the best visual aid of all is overlooked—the piece of equipment itself. Bring it into the classroom for a demonstration and discussion. Trips to equipment sites should always be considered, and you will find they are extremely beneficial to the trainee.

Many public school systems, colleges, universities, and companies have organized audiovisual-aid departments. In addition to this, many companies are producing audiovisual aids in conjunction with their products. Some companies are in the audiovisual-aid business. You should consult all these sources. In doing this, you should be able to obtain help from your purchasing agent.

Trainee recognition Firms vary in their method of giving recognition to trainees who have completed their respective training programs. The method is not so important as the fact that some form of recognition is given. When a trainee satisfactorily completes the requirements of his training program (including both classroom and field instruction), he becomes eligible to enter the journeyman classification. It is extremely worthwhile to present to each trainee a certificate proclaiming this achievement. Certificates may include signatures of the firm president, maintenance shop supervisor, and perhaps the instructor. This will indicate there are others who are aware of his efforts in becoming a tradesman.

You might be interested in holding a graduation ceremony for your tradesmen. At this ceremony speakers from top management and the union can recognize the importance of training and also present certificates to the tradesmen. This is a fine way of highlighting the completion of the training program.

Certificates of completion of apprenticeship are awarded to apprentices when they have finished their training. They are issued by the state apprenticeship agencies in those states having such an agency. Otherwise the Bureau of Apprenticeship and Training awards certificates in accordance with the standards recommended by the Federal Committee on Apprenticeship.

To sum up:
  Be sure you have a well-documented need to train your own tradesmen.
  Be sure your management is committed to the need.
  Read all the research available.
  Learn from the experience of all the other companies you can.
  Gather the best men you can find to do the training.
  Develop the capability of administering the program so that the on-the-job portion and the related training can be mutually reinforcing.

PART 2. SUPERVISORY TRAINING

Today managers in the maintenance function have an increasingly important responsibility. That responsibility is to be aware of the different aspects of training as it applies to supervisors working for them.

There are several ways to approach this training responsibility. An overall look at training will be the beginning. This should help the manager make intelligent training decisions for his supervisory personnel.

The necessity for training is often questioned—as it should be. We may ask, "Why train?" Training changes or increases an individual's knowledge or skills or corrects a fault or a deficiency. Training may not be needed immediately but may be given now to correct a perceived need in the future. The payoff for training can be immediate, in the future, or never (training given but not needed and/or reinforced).

Whether we train for technology, managerial skills, or interpersonal skills, we are training to make maximum use of the human resources in our organization. The importance of this is becoming more apparent as we see the rapid and increasing number of technological and sociological changes in our environment.

There are three major areas of supervisory training, in our opinion, that are important.

First, let us look at technology. We know that technology is rapidly expanding. Great strides have been made in technology as it applies to maintenance functions. Examples are increasing use of automated equipment, electronic control systems, computers tied into maintenance and maintenance information systems, preventive-maintenance vibration analysis, and new techniques for making decisions. To keep up with this new knowledge, an individual could be attending training programs the year round.

The second major area is managerial skills. Again, we are getting increasing numbers and kinds of programs dealing with the basic elements of management—planning, leading, organizing, and controlling. We have seen management by objectives or goal-setting systems instituted in a great many companies. We're seeing increasing use of job-performance reviews or performance-evaluation techniques. The supervisor may need this knowledge and skill to participate effectively in setting and meeting objectives and goals and in evaluating people. These are only two examples of many specific areas of managerial skills. Some sources of these kinds of programs are listed below:

- The Dartnell Corporation, Chicago, Ill.
- University of Utah, Salt Lake City, Utah
- University of Texas, Austin, Tex.
- Michigan State University, East Lansing, Mich.
- Pennsylvania State University, University Park, Pa.
- University of California, Berkeley, Calif.
- Harvard University, Cambridge, Mass.
- Ohio State University, Columbus, Ohio
- Massachusetts Institute of Technology, Cambridge, Mass.
- American Management Association, New York, N.Y.
- American Society for Training and Development, Madison, Wis.
- University of Michigan, Ann Arbor, Mich.

The third major area of training is the area of interpersonal skills. This can be included in the previously mentioned area—managerial skills. However, it should stand alone because of increasing knowledge in the field of human behavior and its applications to the management function.

Today many companies are devoting more time, money, and effort to the "people" aspect of their business than ever before. Seminars and workshops in such areas as motivation, job enrichment, communication networks, sensitivity laboratories, and transactional analysis are some typical programs designed to make a supervisor more effective on the interpersonal level.

Getting Started

All training ultimately has as its goal the production of changes of behavior. Training should either add new behavior to the trainee or get rid of behavior not required or wanted and substitute new behavior. Psychologists say if we want to produce a behavior change, the behavior that we want must be rewarded. In any job situation the most important individual to reinforce the behavior desired is that person's boss. All too often we ignore this fact and then wonder why training seemingly has no effect. The seed of training must fall on fertile ground. To illustrate this—suppose the training department were asked to design and conduct a training program to teach journeymen how to plan their jobs. The training department, being aware that that particular behavior (planning) might not be reinforced back on the job, should check with the journeymen's supervision.

Upon probing, it might be that the journeymen's foremen perceive the planning function as theirs—they would plan the job for the journeymen and do not want the journeymen to plan their own jobs. So rather than spend the time and money to train the journeymen to plan (which, incidentally, they might already know how to do) the training department should recommend that job-planning training for journeymen be dropped.

Determining Training Needs of Supervisors

Many managers see training as being good (like motherhood) and as something that is needed. Let's take a harder look at training and see why it is needed, or if it is needed at all. As mentioned earlier, most training is intended to solve a performance problem or performance deficiency.

Before we train, we need to look at whether or not that particular problem can be solved by training. In other words, what are the rewards for the individual for doing what we want him to do versus what are the rewards for not doing it? Is it a "can he do it" or a "won't he do it" situation? If we were to train an individual or retrain him in a particular skill that he already knows but doesn't use, it's money wasted. We need to look at the on-the-job situation or environment in which he's operating to find out why he is not doing it. As managers we need to be aware that there are problems that are not going to be solved by training, and it's an exercise in futility and a waste of money to think that they will!

Problem Analysis

What makes a manager suspect he has a problem—perhaps solvable by training? There are many sources of data that we can look at. Certainly in the maintenance area, we can look at output, costs, safety records, absentee records, turnover, and grievances. We need to collect all these data to see if they indicate a potential problem solvable by training. At this point, however, a manager should not charge ahead and say, "Yes, we've got a problem; now let's train them." We need to talk to the men to be trained, their bosses, and perhaps their subordinates in order to assess accurately the on-the-job situation in which they are operating. From the information gathered we may be able to determine the causes of the problem. We also can determine at this time the present levels of knowledge and skills, and the on-the-job climate. Now we ought to be able to determine whether or not the proposed training will be reinforced or rewarded by the proposed trainees' bosses.

Objectives and Goals

The chief concerns frequently encountered in training are how to determine if training has had any effect, how to measure that effect, and, sometimes, how to justify or identify the cost. These are the reasons for having training objectives and goals. If you don't know where you are going or what you really want to accomplish, almost any procedure is going to get you there, and then you really won't know if you got there or accomplished anything or not.

A training program should have the objectives of the program spelled out in behavioral terms. By that we mean the performance we expect as a result of the training expressed as to what the trainee will be able to do at the completion of the training.

For example, suppose we send a file clerk to a training program to learn how to type. The objective of that program is to teach the file clerk how to type. That sounds simple—but the objective is not stated in behavioral terms. A behavioral objective has these characteristics:

- Specific
- Quantitative
- Qualitative
- Time-based

All describe the performance we expect

Let's restate the objectives of training the file clerk in behavioral terms:

Upon completion of the course the participant will be able to type a standard company letter format at a minimum of 40 words per minute with a maximum of three errors per page using an electric typewriter in a standard plant office.

With the performance objectives stated this way, we will have a much easier task to measure that performance so that we can evaluate the training more effectively.

Training programs should have the objectives stated behaviorally if at all possible.

Training Methodologies

After you have identified valid training needs, determined the subject areas, and set training goals, then is the time to determine the proper method(s) to reach these goals. Below are described some broad training methodologies.

Programmed learning Programmed learning courses are in the area of technical skills, managerial skills, and interpersonal skills. However, it is quite important in the interpersonal-skill area that there be some structured opportunity for the participant to interact with other participants in a group discussion. This will let the participants try out new interpersonal skills in a controlled situation.

Group discussion This is probably still the most common and popular method of conducting training for supervisors. It certainly makes use of the fact that there is a tremendous amount of learning to be had from the group participants as they discuss the topics and issues in the program. The discussion leader usually will present some theory, raise an issue or problem, and then lead the group into a discussion of the points that he deems important. Negative feelings, restrictions, and counterarguments are also explored.

Role playing Role playing is another technique quite useful in certain kinds of training situations. Normally, the trainer will discuss the technique, assign the roles to be taken, and then have the participants interact with each other using the roles assigned as a guide. Frequently we find participants, in the role-playing situation, interject into their roles their own feelings, beliefs, and attitudes toward the situation being played. This personal involvement gives a good opportunity for the participants to examine how they react in situations like this on the job. Role playing is frequently used as a skill-practice technique in giving effective job-performance reviews or performance evaluations and in teaching interviewing skills.

Case studies In a case study the participants are given information concerning a problem situation. They are not given the conclusion or the ending of the problem but are challenged to analyze the case, to look for underlying causes, and to propose corrective action. Case studies which are derived from actual experiences the participants can identify with are usually the most effective.

Games and simulation techniques A new methodology is the area of business games or simulation techniques. These games can be used in many training areas. In playing business games it is imperative that after the game has finished, proper time be devoted to analyzing the game play. If you want more information about business games, a useful procedure is to get this information from knowledgeable people who have used the technique and follow their recommendations.

Action maze The action maze (developed by Allen A. Zoll III) is an interesting and useful training methodology. It can be used to present a problem situation and alternative solutions. In some regards it is like programmed instruction in that the individual makes choices, turns to a designated page, reads that page, makes another choice, etc. An action maze could be used to explore a discipline problem, maintenance problem, or unsatisfactory job performance. After the initial situation is explained, there are several paths (decision choices) through the maze. The various paths can explore different approaches to problem solving. Participants find this a useful and interesting technique. Within any group usually all the alternative paths are explored. The group discusses the reasons for the variation, and much learning can occur.

Job rotation Often this is not thought of as a training technique, but in reality it is. Many companies have found it to their advantage to provide supervisors and other employees with opportunities to move from department to department to perform jobs which are quite different from their existing ones. This rotation can give supervisors broader understanding and skills in the various areas of operations. Rotation helps to prepare them for areas of greater responsibility at higher supervisory levels.

Laboratory training This is a form of training which is very different from most conventional approaches to training. The major objective includes establishing a condition in the training environment where the participants may look at their own behavior and the behavior of other participants in the processes that are occurring in the group. The participant may learn a great deal about himself, how his behavior affects others, how others affect him, and what makes groups effective.

Programs using this methodology include conflict resolution, interpersonal communications, personal-growth groups, team-building training, T groups, and encounter groups.

Establishing and Managing the Maintenance Function

If you are interested in this type of training, it is advisable that you get some consultation from qualified individuals who are familiar with it or contact the National Training Laboratories in Washington, D.C.

Evaluation of Training

Evaluating training is unquestionably a very difficult task in many situations. It depends primarily on the goals that were set (preferably in behavioral terms), the specific skills and knowledge that the trainee is to have acquired, and the application of that knowledge and skill on the job. There is no one way that could be considered a perfect way to evaluate training. However, training evaluation should be built into the overall program from the planning stage on. Below are several evaluation techniques.

At the completion of a training program the participants fill out a reaction form—it's an immediate-feedback device. An example is shown in Fig. 4-3.

It may be appropriate to follow up a training program at a 3- or 6-month interval. This could be done by interviewing the trainee, the trainee's boss, the trainee's subordinates.

Fig. 4-3 Typical reaction form for rating instructors.

Questionnaires should also be devised to get the information required. Other techniques for measuring training effectiveness deal with performance variables such as costs, grievances, and process improvements.

Finally, some training programs have content that lends itself to tests which can be given at the end of the program to cover knowledge and skills learned. In some instances, tests can be given before the training. The use of the pretest and post-test technique allows measurement of knowledge acquired during the training program. However, not all programs lend themselves to this type of measurement. Many results of training have to be classified as intangible and in all probability cannot be measured accurately.

Resources

The manager who is contemplating training for his supervisors has the following options to think about. Assuming that the manager has minimal knowledge in the training field, it's important then that he utilize some external resources to be sure that the training, as he contemplates it, is the most effective or the most helpful for the situation as he sees it. Certainly, if he is in a company fortunate enough to have its own training department or a training expert, he has a ready-made available source of information, help, and guidance. He can be guided by their suggestions and recommendations.

Lacking an in-house source, the manager must turn to external sources.

The external sources that are available are management consultants specializing in the training function or personnel function. One source of these consultants is the American Society of Training and Development Consultant Directory.

Some corporations today have subsidiaries or divisions whose function is to sell their training skills and knowledge to other corporations.

Books, magazines, periodicals, professional journals, and professional societies all constitute good information sources for the manager. A brief list includes:

Books

American Society for Training and Development: "Training and Development Handbook," 2d ed., McGraw-Hill Book Company, New York, 1976.

McGehee, William, and Paul W. Thayer: "Training in Business and Industry," John Wiley & Sons, Inc., New York, 1961.

Journals

Harvard Business Review, Graduate School of Business Administration, Harvard University, Uxbridge, Mass.

Human Resource Management, Bureau of Industrial Relations, Graduate School of Business Administration, The University of Michigan, Ann Arbor, Mich.

Personnel Journal, Baltimore, Md.

Technical Education News, McGraw-Hill Book Company, New York.

Training and Development Journal, American Society for Training and Development, Madison, Wis.

Training in Business and Industry, Gellert Publishing Corporation, New York,

SUMMARY

If as a manager you are contemplating training, whether to solve a problem, to increase knowledge or skills, or to add information, secure appropriate advice and guidance before committing yourself to the training program. If the training is to solve a problem, make sure that it is the type of problem that training can solve. Engage competent professionals to assist you in determining needs, to help design your programs, to conduct or train those who are going to conduct the program, and to evaluate the program.

Have your training goals well thought out and, if at all possible, stated in behavioral terms—compatible and consistent with the demands of the supervisor's job and department and corporate goals. After all, if you don't know where you are going, any path will get you there.

Finally, recognize your own role, as the manager, in the training environment. As the manager you are the major reinforcer for the behavior you want from your participants as they come from the program.

Chapter **5**

Human Factors in Construction Equipment Maintenance

LESTER R. BITTEL
Director, Academy Hall, Inc., Strasburg, Va.

Human factors that affect construction equipment maintenance and the people who work at it are not unlike human factors in other fields of endeavor. Such factors consist largely of the needs and motivation of individuals and the interplay of the behavior of these individuals as they try singly and in groups to achieve satisfaction of their needs. The important words are *individuals, groups, motivation,* and *behavior.* Because individuals who make up the maintenance force are different from people who work elsewhere—*and because their work is different*—their behavior will also be different. Because maintenance is a job with unique working conditions and unique working goals, behavior of the maintenance force—and the behavior of those who supervise or work with the maintenance force—is especially sensitive. In addition, such behavior is hard to predict and to control.

A UNIQUE KIND OF WORK?

Is construction equipment maintenance work unique? Yes, when compared with routine construction work. Much of it occurs at random and in unexpected places and amounts rather than in a planned, predetermined, and carefully scheduled fashion.

A bricklayer knows exactly where he will work each day, exactly what he will be doing, and about how much he will be expected to produce. Contrasted with this, a maintenance worker may work on one job today and on another site tomorrow. Even when his skills assignment has been narrowed down to a specialty, like repairing loaders, he is not certain which make and model will need attention next.

It is relatively simple to measure the amount of work expected from a bricklayer. But even with various techniques for predetermining and estimating the time needed for a maintenance craftsman to perform a job, maintenance work measurement is cumbersome and complex.

Actual construction work is product-oriented. Workers laying brick can see immediate evidence of the fruits of their labor. Their work is in the mainstream of the enterprise. The work of maintenance employees, on the other hand, tends to be obscured. Their work is performed, not on the building going up, but in support of, or service to, equipment to be used by those performing the actual construction. Thus, maintenance employees are engaged in a form of service work. As such, neither they nor their superiors call the tune for their labors. Regardless of how carefully they anticipate and plan their work, theirs is usually a *reactive* effort.

Because of its repetitive nature, there is a rhythm and a flow to most construction work. For just the opposite reason, maintenance work is often performed in herky-jerk fashion—short bursts of very hard effort followed by periods of standing around, for example, waiting to see if the repaired equipment will function properly.

Much repair work is performed under time pressure. For example, a loader is out of service; it must be fixed as quickly as possible. Consequently, there is little time for the maintenance craftsman to savor his work. For him, the pressure of time is real and harassing.

When speed of performance becomes the dominant factor in the work, quality begins to suffer. A maintenance employee may want to do good work and may know and respect precision and craftsmanship, but time may not permit it.

Similarly, budgeting of many maintenance jobs is done in fits and starts according to availability of funds. Accordingly, a maintenance employee may see a tuneup, for instance, deferred for weeks—only to observe an expensive contract mechanic hired in lush months to catch up.

Instances like these typify maintenance work. They illustrate the significant fact that maintenance work is unique in many ways. And it is the nature of the work itself—as much as any other factor—that shapes the behavior of those who do it.

A UNIQUE WORK FORCE?

Are the *people* employed in construction equipment maintenance work different in many significant ways from those employed in construction skills or in office work? The reply appears to be yes.

Generally speaking, there are three—or possibly four—kinds of maintenance employees: (1) those in higher management and/or staff positions who are engineers or technicians either by education or by experience; (2) skilled craftsmen, either true journeymen or those who have acquired the know-how by on-site training and rigorous experience; (3) semiskilled mechanics who have bid into the maintenance department for various reasons and who may or may not be dedicated to the profession; and (4) bottom-of-the-scale employees who will work out their careers in labor gangs.

At the top of the scale, the trained engineer is inclined to be object-oriented rather than person-oriented. Having chosen a career that stresses inflexibile laws of cause and effect, facts and figures with clear-cut solutions are his strength. But a shift to a responsible position is difficult for some people. And rarely have they been properly prepared for it. So an engineer may be willing and eager to improve his interpersonal relationships, but he will find it difficult.

At the craft level, the picture is not much different. These employees (especially the older ones) have gravitated to jobs that place great value upon hand skills. Where judgment and knowledge are required, their jobs, like those of engineers, are of a fixed rather than of a flexible nature. Then, too, many craftsmen have learned their skills the hard way, acquiring knowledge as they went along. This tends to make them secretive and protective in their dealings with trainees and with employees in other skill areas.

At the semiskilled level, construction equipment maintenance employees may not be much different from general construction employees. They may have chosen maintenance work simply because the pay scale was attractive or because there was less chance of being laid off. They may even dislike the work, especially its heavier and dirtier aspects. And, when they have been drawn from younger ranks, they are more likely to be high-school graduates than dropouts—with higher expectations from both their jobs and their superiors.

At the bottom of the scale, there is a mixed group of men—often from the disadvantaged sectors of the economy—who (1) either have no hope of moving upward or no ambition to do so or (2) who are only temporarily employed at the entry level and are eager to move up into semiskilled work.

In summary, maintenance employees tend to be object-oriented rather than people-oriented, slightly older, possessing greater skills; prouder, perhaps, and more self-disciplined; protective about their hard-earned skills and knowledge—all with the many exceptions noted.

Having acknowledged that both construction equipment maintenance work and workers are different, the biggest mistake a manager could make would be to assume that all maintenance workers are alike. A truism is the fact that each person is different from another, different in many possible ways. People are shorter or taller; blonder or darker; thinner or fatter. They are less talkative or more garrulous; more stupid or smarter; slower or quicker. One person is completely wrapped up in his job; the next with his family. One worker is cooperative; the next resists every try for teamwork. Some are dedicated; others don't care very much. Some are drinkers, some abstainers; some are given to profanity, while some quote the Bible. Quality is the hallmark of one worker; another leaves a slipshod imprint on every job. Surely, nothing is more undeniable than that each person who chooses (or is chosen by) maintenance work for employment thinks and acts in a unique way.

MOTIVATIONAL PATTERNS

Behavior among people varies so greatly that there might sometimes appear to be no reason behind most of it. Fortunately, however, there is. There is a set of personal needs that moves each individual forward. If you use this scale of values and priorities, you can anticipate and understand each person's behavior.

Employees, like most of us, seek to satisfy what a famous psychologist, Dr. A. H. Maslow, calls the "five basic needs." And we seek a good part of these satisfactions at work.

Dr. Maslow outlines the basic needs this way:

We Want to Be Alive and to Stay Alive. We *need* to breathe, eat, sleep, reproduce, see, hear, and feel. But, in America, these needs rarely dominate us. Real hunger, for example, is rare. But, all in all, our number one needs are satisfied. Only an occasional experience— a couple of days without sleep, a day on a diet, a frantic 30 sec under water—reminds us that these basic needs are still with us.

If you occasionally employ workers who have lived in a country where basic needs are not so easily met—World War II refugees from Hungary, for example—you will find they are more apt to produce a high volume of work just in return for the prospect of their salary and the basic needs it can buy. Even though they may no longer fear being hungry, the responses of hungry days remain and motivate them to work harder.

We Want to Feel Safe. We like to feel that we are safe from accident or pain, from competitors or criminals, from an uncertain future or a changing today. None of us ever feels completely safe. Yet most of us feel reasonably safe. To assure this, we have laws, police, insurance, social security, and union contracts.

We Want to Be Social. Group ties have always been strong. We marry, join lodges, and even do our praying in groups. Social need varies widely from person to person—just as other needs do. Few of us want to be hermits. Not too many people have frank and deep relationships within these groups—or even with their wives, husbands, or close friends— but, to a greater or lesser degree, a need to be social operates in all of us.

We Need to Feel Worthy and Respected. When we talk about our self-respect or our dignity, this is the need we are expressing. When an employee isn't completely adjusted to life, this need may show itself as undue pride in achievements, self-importance, boastfulness.

Because so many of our other needs are so easily satisfied in America, this need often becomes the most pressing. Look what we go through to think well of ourselves–and to have others do likewise. When we try to dress well for a party, we're expressing this need. When we buy a new car even though the old one is in good shape, we're giving way to our desire to show ourselves off. We even change our personalities to get the respect of others. No doubt you've put on your company manners when visiting with other people. It's natural, we feel, to act more refined in public than at home—or to cover up our less acceptable traits.

We Need to Do the Work We Like. This is why many people who don't like their jobs turn to hobbies for self-expression. It is also why so many other people get wrapped up in their work. We all know the men who enjoy the hard burden of laboring work—or the carpenter who hurries home from work to his own shop. This need rarely is the be-all and end-all of our lives. But there are very few of us who aren't influenced by it.

INDIVIDUAL MOTIVATION

In assessing each individual's motivation, care must be taken to avoid jumping too quickly to conclusions. For example, it might be presumed that, to the man in the labor gang, pay and security are highest on his priority list. Or that, to the graduate engineer, meaningful, fulfilling work is his main goal. Generally speaking, these conclusions may be true. And as such they provide a starting place in knowing how to respond to others.

But the man in the labor gang may hunger for respect from his associates as much as anyone does. Or the engineer may actually think his work beneath him and seek only higher and higher pay. The point is that each person has his own system of priorities. The extent to which he satisfies them at work determines the quality of his behavior at work.

The more he finds his work well paid and meaningful, the harder he will work. The more he feels that his pay is only adequate or that the work is demeaning, the more difficult it will be to motivate him.

GROUP BEHAVIOR

Few people in construction equipment maintenance work alone. At the least, they work in pairs or as small teams. When they do, a new element in behavior is introduced. It is the influence—or pressure—of the group. Sociologists call this "group dynamics." No matter what it's called, this group effect may take precedence over an individual's own inclinations. An apprentice carpenter, for example, may want to show the boss how quickly he can work, but, if the others in his work gang advise him to "slow down," he may do just that in order to satisfy the third and fourth needs in Maslow's hierarchy, the need to be part of a social group and especially the need to be a respected member of that group.

In working with an individual, it follows that a supervisor must always keep in mind the relationship of that individual to his work group. The supervisor cannot afford to embarrass an individual before the group or expect him to perform in a way that is contrary to the group's norms.

In working with groups, a supervisor must, first of all, respect their collective power. To challenge their positions head-on is to invite disaster. It is better to work with a group to solve a problem than to try to force the group to change from a direction that seems to satisfy their collective needs.

The supervisor must also acknowledge the presence of influential leaders in every group. Many may be quiet and low-key rather than articulate and forceful. Influential leaders of this kind maintain their influence by virtue of their age, experience, know-how, or position in the local community. And, while they should not be manipulated, their assistance should be sought—either directly or indirectly.

THE POWER OF PARTICIPATION

People work hardest at jobs on which they have helped to plan the way the work is to be done. They strive most vigorously toward goals they have helped to set for themselves. That is the practical lesson spelled out by Maslow's fourth and fifth needs. When you ask a man's advice about how his job should be performed, that is a measure of your respect for that individual. When his advice is followed or when the individual is permitted to follow the course he has suggested, that is evidence of respect of an even higher order.

If, when planning the work, an individual helps to set some of the rules (to get it done by 5:00 P.M., to use the least amount of new materials, to fix it this time so there is no callback), he will strive to attain these goals because they represent for him what he believes will make the work worthwhile.

It cannot be stressed too forcefully: It is what is important to the individual, not to his boss or to the company, that really counts. And, in maintenance work in particular, there is more of an opportunity for the supervisor to offer to share with his work group some of his responsibilities for planning their work. This sharing of both goals and methods is called participation.

In genuine participation the employee and the work group join with their supervisors in deciding what kind of work will be done and how it will be done. It doesn't mean the supervisor's giving up his authority. And it need not be done on every job or in every situation. A little participation goes a long way in supplying the motivation needed to spark the efforts of both individuals and groups of individuals.

NONMOTIVATORS

During the years when scientific management methods were being developed, there was an understandable tendency to overemphasize management's appeal to the pay and security needs of its employees. But, as most industrial societies have become more affluent, more and more of the factors which were once truly motivational have become regarded by employees as amenities to be expected from any respectable organization. Accordingly, social scientists describe many formerly sought-after conditions of work now as *maintenance* or *hygiene* factors rather than as *motivators*. Included in this group are wages, salaries, holidays, vacations, paid insurance, sick pay, pension plans; seniority rights and grievance procedures; air conditioning and good ventilation when working indoors, sound equipment and tools, freedom from excessive noise and dirt; bright lighting; services such as cafeterias, parking lots, clean rest rooms; work clothing, dignified job titles and privileges; and many kinds of social activities including office parties, outings, picnics, tours, bowling clubs, and the opportunity to gossip and socialize with other employees.

Employees expect these maintenance factors; few good or potentially good employees will work where they can't be found. But these factors will not inspire from these employees anything but a routine, unimaginative effort in return. Only true *motivational* factors, most of which can be supplied only by an employee's boss, will provide an incentive for well-disciplined, cooperative, dedicated efforts. These motivational factors, as you can see, all derive from Maslow's fourth and fifth needs. They include the nature of the work itself; recognition for achievement; access to information; utilization of one's abilities; challenging assignments; extended responsibilities; freedom to act rather than requiring to be told what to do; and involvement in planning, goal setting, and problem solving.

MOTIVATION IN THE WORK ITSELF

Research into what makes people want to work (as opposed to having to be pushed or manipulated into working) points to a simple conclusion. All those seemingly wonderful conditions provided by company policy do not motivate at all. It is the quality of the job or the work itself that exerts the greatest motivational force. And it is the supervisor who contributes most to the quality of that work. A company may provide all the benefits and comfort it can, but these serve little more than to induce good employees to come to work. Once they are employed, the task of shaping the work so that it motivates employees to greater, more conscientious effort is the primary responsibility of the supervisor.

In the construction equipment maintenance field, the supervisor's options are broad indeed. He cannot—or need not—introduce every motivating factor into each employee's work. But he can draw from the reserve of motivational factors in order to introduce one or two into each work situation.

Recognition for Achievement. Whenever a particularly fine job is done, the worker deserves praise in the presence of his associates. A mechanic or other employee who has completed a particularly dirty or complicated job can be given an extra half-hour to wash up or rest.

Access to Information. A mechanic who has been asked to shorten the time he spends on a job, for instance, can be shown the cost history on similar jobs.

Utilization of One's Ability. A carpenter can be invited to contribute his judgment when planning a major building alteration. A laborer who learned to type in military service can be assigned part time to maintaining equipment records.

Challenging Assignments. Instead of shipping a complex electronic control back to the vendor for repair and thus losing its service for a couple of weeks, an electrician can be asked if he can do it on-site so that it can be placed in service the next day.

Extended Responsibilities. This technique of *job enlargement* or *job enrichment* consists of enlarging the scope of a person's job. It is the opposite of the typical division of labor wherein each person performs a very small segment of the job. For example, an overhaul job that used to be worked on progressively by engine mechanics, body men, and electricians is assigned in its totality to the mechanics. A carpenter who used to build the forms and then wait for the millwrights to pour the concrete takes complete charge of foundation work. Obviously, there are craft and union restrictions and traditions to cope with, but job enlargement works in many shops simply because most employees enjoy such extensions of their routines.

Freedom to Act. A mechanic who is on location two buildings away from the centralized maintenance shop is encouraged to use its assistance when he runs into an unexpected situation.

Involvement in Planning. Workmen are asked to list the deferred jobs that ought to be given priority next month. Mechanics are invited to suggest in what order equipment will be given checkups during the coming month.

Involvement in Problem Solving. Mechanics who have worked in the area or on problem equipment are made an integral part of the group assigned to solve especially difficult or repetitive breakdown problems. A mechanic who is to be sent out to rebuild a motor is asked to meet with an engineer ahead of time to decide how the job should be done rather than to simply be given instructions for the job from the engineering department.

Involvement in Goal Setting. Rather than assigning work and jobs one by one or piece by piece, the supervisor meets with craftsmen or craft teams to decide how much (or how many) can be accomplished during the next work period. Employees who participate in such goal setting accept these goals as their own and strive hard to meet them.

INTERDEPARTMENTAL RELATIONS

As observed earlier, maintenance is a service-type function. For this reason, its activities cross many lines. Because so many areas of work depend upon it for service, the demands upon the maintenance function typically exceed its resources. Since most jobs are done under the pressure of time, results are often unsatisfactory from the user's point of view. Inevitably, friction arises. It may take the form of snide remarks or searing criticism by carpenters and foremen directed at equipment maintenance workers. It may take the form of an all-out shouting match between construction foremen and maintenance foremen. It may take more subtle shape as when the construction superintendent, while speaking to the company president, claims his cost excesses are the result of faulty maintenance.

It is best to bring such aggravations and disharmonies into the open. They grow more abrasive and unmanageable when held in. A shouting match, primitive though it may be, is therapeutic for both sides. In many situations, both construction and equipment maintenance people feel abused and put upon. And the viewpoint of each may be justified. Most organizations *are* pressed for more than they can accomplish. Something has to give, but power plays can lead to continued ineffectiveness, as human energy is consumed in politicking and in-fighting. Open confrontation, however, uses human energies constructively. Direct communications clear the way for compromise. Human relations are essentially a matter of give and take. And each party must both give *and* take if there is to be harmony and effectiveness. The same observation applies to interdepartmental relationships. There must be both give and take—based upon understanding and the reasonableness that knowledge induces.

CONFIDENCE IN OTHERS

More than almost any other factor, management involves human factors. Regardless of the economic and technical resources at his command, most of a manager's leverage is derived through his ability to select, organize, develop, and motivate other people. Quite literally, even a technically skilled equipment maintenance manager accomplishes his work through the work of other people. The secret of this particular power lies in the *concept of confidence.* Unless a manager is genuinely able to develop his own confidence in the capabilities of other people—and express and act upon this confidence—almost every attempt to direct or control their activities will ultimately be unsuccessful.

Confidence, while mainly an emotional characteristic, can, happily, be increased intellectually (see Fig. 5-1). Confidence grows slowly, of course. Confidence begins with the acknowledgments that mistrust of others stems from fear of them and that most of our fears are ungrounded. Typically, our fears are rooted in a mistrust of the unknown. Simply finding out more about people and getting to know them help to remove these misgivings. With mistrust removed, it is natural to develop a respect for an individual, if only a respect for his right as a human being to search for his place in the world.

Once we respect people and recognize the value of their contributions to life, it is easier to communicate with them. Management does this by keeping employees informed of

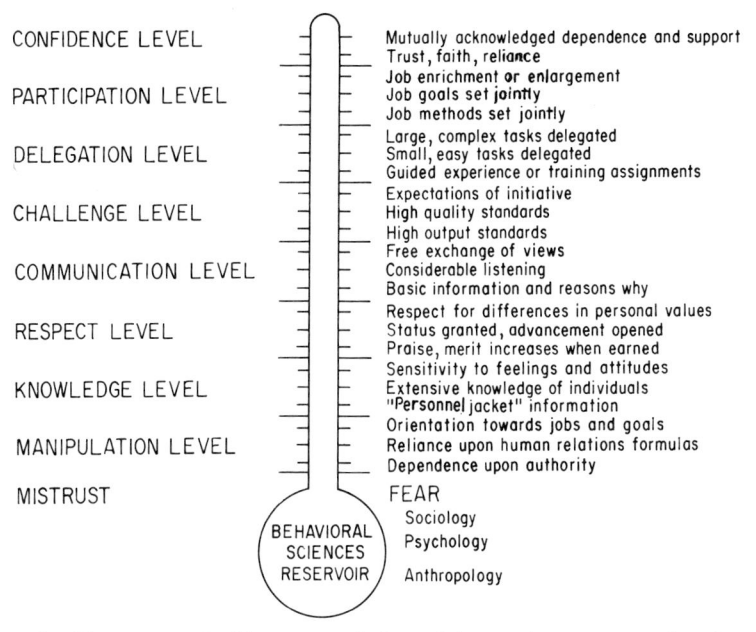

Fig. 5-1 Confidence meter. Confidence in another's capabilities and integrity rises through several levels, beginning with mistrusts and reaching its highest level through participation and job enrichment. ("The Nine Master Keys of Management," Lester R. Bittel, McGraw-Hill Book Company, New York, 1972. By permission of the publisher.)

conditions and developments that may affect their daily work or their careers. It is a bigger step forward, however, when a manager shows the mettle of his respect for an employee by giving him work to do that challenges his skills, his intellect, or his integrity. Work that demands that a man call upon his unfulfilled potential raises his self-esteem. When he accomplishes what is difficult, his regard for himself and for the manager who challenged him also rises. Challenges to a man's inner reserves are best made systematically through the process of delegation. Assignment to an employee of a task that might be considered only the prerogative of his superior is a fine expression of confidence.

The ultimate challenge in the work place is that of participation. In this relationship a manager or foreman offers to share his responsibilities with an employee. An employee develops, together with his boss, the way in which his job should be carried out and, at the highest level, the goals of the job itself.

SUMMARY

A construction equipment foreman or engineer may have the keenest analytic mind and take action in the most systematic, vigorous way but, without mastery of the art of human relationships, his chances of success will be sharply limited. To strengthen these, he must develop an openness and objectivity toward others. Sound relations also require an acknowledgment of mutual dependence, together with abandonment of a pushbutton philosophy that stresses manipulation rather than motivation. Such enlightenment calls for devotion to the improvement of employee's knowledge and skills. And it recognizes that success or failure in relations with other human beings depends in large measure upon the manager's own attitudes and motivation.

References

Basset, Glen A.: "The New Face of Communication," American Management, New York, 1968.
Bittel, Lester, R.: "What Every Supervisor Should Know," 3d ed., McGraw-Hill Book, New York, 1973.
Boyd, Bradford: "Management Minded Supervision," 2d ed., McGraw-Hill Book, New York, 1976.
Davis, Keith: "Human Behavior at Work," 5th ed., McGraw-Hill Book, New York, 1976.
Dowling, William F., and Leonard P. Sayles: "How Managers Motivate," 2d ed., McGraw-Hill Book, New York, 1978.
Likert, Rensis: "The Human Organization," McGraw-Hill Book, New York, 1967.
McGregor, Douglas: "The Human Side of Enterprise," McGraw-Hill Book, New York, 1960.
Myers, M. Scott: "Every Employee a Manager: More Meaningful Work through Job Enrichment," McGraw-Hill, New York, 1970.
Roethlisberger, F. J.: "Management and Morals," Harvard University Press, Cambridge, Mass., 1965.

Section **2**

Establishing the Maintenance Facility

Chapter **1**

The Construction Maintenance Shop

GARY J. THOMAS
Vice President and Equipment Manager, S. J. Groves & Co.,
Minneapolis, Minn.

Two entirely different type shops are required by the contractor: the field shop and the rebuild shop.

The Field Shop

The field shop may range from a large spreading oak tree with one or two highway van trailers for parts storage, telephone, and some type of office facilities for the maintenance people to the other extreme of a very large multibay shop with high overhead clearances, heavy crane service, large parts storage, and office facilities. Each project has certain shop requirements based not so much on dollar volume but on location of project in relation to parts source, major rebuild facilities (either your own or a dealer) and the type, size, age, and number of machines required to complete the project on schedule. Weather conditions in the area where you are working and length of time the project is to run also have great bearing on your decision. Many other factors must be looked at when you plan your field shop, but those mentioned are most important.

One other thing that all too often has a great bearing on what type of field shop is to be built is monies in the bid. Be sure that the estimate includes the right bid figure to give your maintenance people a facility that will enable them to assure your production people of the required machine availability they need to do the job on time and at the unit costs that were in the bid. It is not uncommon to economize on shop facilities to the point where the maintenance crew is severely limited, and the amount of work hours of repairs they can produce for you can become so poor that the profit-and-loss figure on the job might be affected. One cannot expect people to produce to their maximum if their environment is not reasonably protected from rain, extreme cold, wind, and dust conditions.

Highway van trailers are efficient and economical units to use for your field shop parts storage and office facilities. They are readily available in most parts of the country and can be purchased out of a used fleet in multiples. If you use two or more trailers in a single shop facility, it is easier to work with units of the same dimensions. This allows roof structures or floor structures built between the units to fit better (see Fig. 1-1).

If you are going to require only one trailer on the job, a suggested floor plan is shown in Fig. 1-2.

The average contractor, as a rule, will remodel the units with his own workers but care

2-2 Establishing the Maintenance Facility

should be taken; many times the finished product is very expensive and not to the standard that might be available from an outside source. In many large cities firms that specialize in mobile home and camper repair will offer attractive prices for building the office portion of your van. Such firms have available doors and windows of metal frame construction to give a professional look to your trailer. They also have lighting and heating supplies, air-conditioning units, and wall and ceiling materials that are built for the rugged use received by a mobile facility. Be sure that adequate lighting and wall receptacles are installed throughout the trailer and that electric service installed meets your demands, for example, electric heating units, air conditioners, hose press machines, and grinders. The safest way to build the unit is to follow commercial electrical codes. There are no size standards but a readily available unit about 8 ft wide and 36 ft long works very well.

Dimensions for the office are determined by the number of people who might be in the facility at one time. As a general rule, the office should be at least 15 ft in length, but it

Fig. 1-1 Lean-to shop using trailer and existing structure as support for roof. This application is adequate for warm areas.

would be far better if the partition wall is 20 ft back from the front end. Side-door trailers are available but this type of door is not desirable for a personnel door that gets heavy use. The best solution is to cut a new opening at the desired location and install a good-quality metal door prehung in a heavy metal frame. Check with the manufacturer to be sure the structural strength of the trailer is not affected by this door opening.

Cutting a window in each side of the trailer office area is recommended to allow enough work light during the day without generator operation in remote areas without line power. A combination air conditioner and heater is ideal in areas requiring minimum heat but is not adequate if winter temperatures dip to extremes. Thermostatically controlled heaters using fuel oil or gas are better for cold climates.

When selecting a trailer, choose a square-nosed rather than a round-nosed unit if possible, as too much floor space is lost in the round-nosed type. Conventional office furniture gives more room and efficiency than built-in desks, for example. It is also desirable to buy a good-quality extra-heavy metal bin for parts storage. Wood bins work but take up more space and are not readily adaptable to various-sized compartments that seem to change constantly in a parts department. The final costs for wood and metal bins are surprisingly close.

If the parts storage area needs more space, simply use trailers in multiples. With two or more trailers on a long job, building a connecting floor to back the trailer up gives a far easier work area for the crew: They gain extra above-ground storage for heavy parts and also can go from trailer to trailer without climbing stairs.

Smaller jobs or short-term projects can set up trailers as shown in Fig. 1-3.

Leave 20 to 25 ft between the two trailers to be roofed and used as the shop. Prefabricated gabled roof trussing of desired width covered with 8 × 12 ft panelized roof sections can be used for more than one shop facility if care is used in the disassembly of the unit and proper storage techniques are used at the completion of the project.

The shop is important but far more important are the tools your crew uses. Generally, mechanics are very well equipped with required hand tools up to 1¼ in. in combination wrenches and ¾ in. drive to 2¼ in. maximum in socket wrenches. Many are better equipped, but do not rely on that as a standard. The logistics of the job determines how to tool up the shop.

The two most important questions for a mechanic when a machine is not performing properly are: What is the problem? Where is the problem located in the machine? With advanced hydraulics and automatic transmissions, a mechanic must have proper test boxes and gauges to locate the component that has failed. Many thousands of dollars are wasted each year by mechanics who use the buy-and-try method of repairing because the boss won't supply the proper tools. Electric systems are also an important and apparently mysterious part of a machine. More wrong components get changed in an electric system than in any other, usually because the mechanic lacked the proper meter to determine

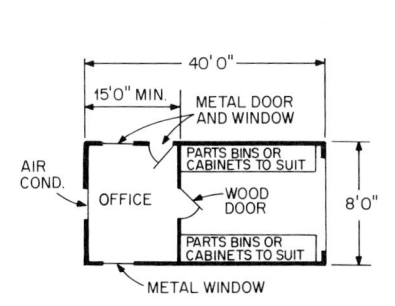

Fig. 1-2 Typical shop floor plan for a trailer shop.

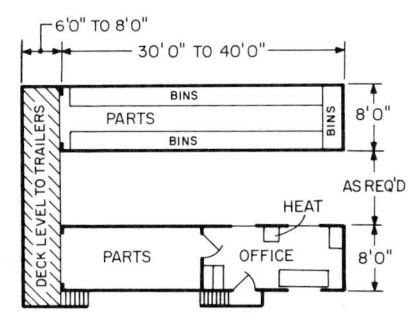

Fig. 1-3 Shop arrangement for two-trailer facility.

where the problem was. You don't need to buy every little test gadget on the market, but the following tools are recommended on every job.

Hydraulic test box or sufficient individual gauges to check out all hydraulic systems, both flow and pressure.

Electric test meters should be available to troubleshoot a faulty machine properly.

End wrenches above 1¼ in. should be available to the mechanic. Maximum size will be determined by your fleet.

1-in. drive socket sets should be also made available; size will be determined by the fleet.

Torque wrenches of ½ and ¾-in. drive are generally all that is required but the fleet you are going to run may require a ⅜-in. drive reading in inch pounds for delicate equipment and a torque multiplier for very high torque requirements should be available.

Every field operation should have a complete set of taps and dies. These tools may not be required every day, but they can save hours of downtime if readily available when needed.

The mechanic should have available a small hand-held electric drill with ¼- to ⅜-in. chuck capacity in addition to a drill motor with ½- to ⅝-in. capacity chuck. Be sure that drill bits sized from ¹⁄₁₆ to 1 in. are included in the tool list.

A tool that is not really a must but can again be a great timesaver if a mechanic must do any amount of drilling with reasonable accuracy is a magnetic drill stand. Be sure the large drill motor and the magnetic press are compatible in their mounting lugs.

Air or electric impact wrenches in small sizes of ½ or ⅝ in. or large sizes of ¾ to 1 in. or splined are tremendous laborsavers if a great deal of nut running is required.

Other electric tools that will be needed are a 7-in. portable angle grinder, a ¾-hp bench grinder, and a battery charger of suitable size to handle your fleet. A few other tools that are not an absolute must but are handy to have and save time are a solder gun, a 7¼-in. portable circle saw, a reciprocal saw, and an electric hand-held engraver.

2-4 Establishing the Maintenance Facility

In addition to those tools mentioned, a shop needs many miscellaneous tools and supplies as indicated below. The amount or number of each will be determined by project size.

18- and 36-in. bolt cutters	Various C-clamps
3-in. through 6-ft crow bars	#2 cable cutter
Hacksaw	Various-sized shovels
3-in. through 20-ft sledgehammers	Welder chipping hammers
Grease guns	18- , 24- , 36-in. pipe wrenches
100-ft steel tape	18-in. street brooms
6-cu ft wheel barrow	Drum hand pumps
1200 Lennox hole saw	Heavy-duty jumper cables
or equivalent	2- and 4-qt measure cans
Shop floor creeper	Hand wire brushes
Oil drain pans	Parts wash brushes
Battery filler jugs	Mechanic wire
Assortment of files	Funnels
Putty knives	Oil cans
5-gal gas cans	Droplights
Radiator filler cans	Extension cords
Garden hoses	Punca-Loc hose clamp kit
Grinding stone grinder	Oil filter wrenches, large and small sizes

Some type of shop air compressor should be included. Don't use too small a unit if you plan to have any air-operated tools such as grinders, wrenches, and chippers. An air compressor built by a reputable firm has almost an infinite life if maintained properly.

Sufficient hydraulic or mechanical jacks should be provided as well as adequate blocking or stands. These are essential for both efficient maintenance and proper safety measures.

The normal field shop is generally not built with overhead crane service because of high installation cost. The floor-mounted jib crane is one way to get limited crane service at a nominal cost. With care in installation, removal, and storage, this unit can give many years of service. If the shop you are building is not to be equipped with crane facilities, be sure that the door and roof are high enough to allow the mechanic to use a mobile hydraulic crane to do what lifting is required in the building.

Today's fleet of construction tools almost dictates that you have a hydraulic hose press on your job along with adequate hose and fittings of various sizes and configurations. Many times a down machine can be put back to work relatively quickly if you have this hose machine on the job. There are many brands on the market, but be sure that you get a machine that is compatible with the type of hose and fittings that you have on your

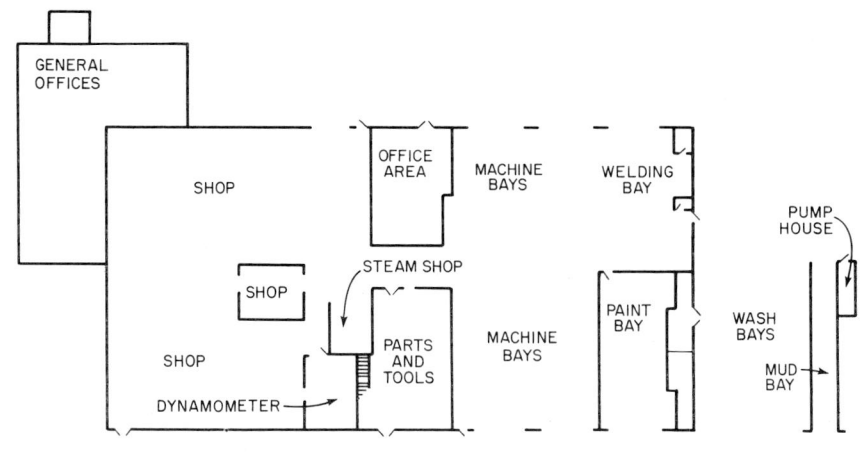

Fig. 1-4 Typical rebuild shop.

Fig. 1-5 Exterior of a large combination repair and component rebuild shop with 12,000 sq ft of component area. The shop is air-conditioned and dirt free. Its repair shop has drive-through bays 120 ft long, 30 ft wide. The two elements of the shop are separated by a parts room and offices.

construction fleet. This will probably be your single largest investment in parts and tools on the job but it is really essential in today's hydraulic machine age.

Electric-driven welders or transformer welders are inadequate in most temporary field locations. Usually, there are not enough large in-shop welders; therefore, use either a skid-mounted or a trailer-mounted welder sitting outside the building for shop welding. This also enables you to hook onto the shop welder quickly and to go to the field for extensive welding repairs if that should be necessary.

You will need a hydraulic-operated service press sometime during the course of a job to remove sprockets or to perform other pulling or pushing jobs such as bearings and gears on and off shafts. There is no substitute for a sprocket puller, if the requirement comes up; it is a very costly investment, but it is certainly a necessity if you are remotely located with a relatively large fleet of machinery and a high machine availability requirement.

We have covered most of the basic tools that a mechanic needs to maintain a relatively large fleet of machinery on the job. From time to time other tools may be required such as micrometers, dial indicators, and more sophisticated tools. Generally, they are not needed in a field shop but are required in area shops that do major overhauling and individual component rebuilding.

The Rebuild Shop

A rebuild shop (Fig. 1-4) may be either of two different kinds of facilities. One shop can rebuild every component on a machine in addition to rebuilding the structural parts of the machine. The other shop sends components, engines, and transmissions to others for rebuilding and does only structural repairing.

The first shop takes much consideration before the decision to have one can be made. Quite a large facility is needed to rebuild components and to have space for repair bays. The environment of the rebuild area has to be clean and very well ventilated, sometimes requiring air conditioning or filtered-makeup air in the building.

Such a shop also requires a large cash outlay for expensive engine, hydraulic, and electric test devices. In addition to a large shop area, you will be faced with a large parts need, both in monies and in space.

If you are considering this type of facility, consulting other contractors on their shops' efficient and inefficient areas is highly recommended. Most equipment manufacturers and their dealers can be of great help.

The very efficient design in Fig. 1-4 can handle almost any type overhaul that might

2-6 Establishing the Maintenance Facility

Fig. 1-6 Typical shop interior.

Fig. 1-7 Typical shop interior showing extra light advantage of fiber glass doors. Note reel-mounted hose for station air and lubrication products.

Fig. 1-8 Project-built structure using form-type lumber and trailers for storage and office. This shop proves very adequate for warm climate. Note canvas rollup door.

come up. The component overhaul area can easily turn two components per day. A double shift could increase this to almost four units per day.

Ideal repair facilities are those with drive-through bays 30 ft wide and 120 ft long with at least 20 ft to the hook of the crane. A pair of bays this size with a third like-size bay split between clean/paint and welding area will provide an ideal shop that can turn out a great deal of work (see Fig. 1-5).

If you elect to build a bay that is used for painting, be prepared for problems and unbelievable costs. In view of the various laws dealing with paint facilities from the standpoint of safety and comfort, it is almost impossible to justify your own paint shop. The problem is aggravated in a cold climate requiring large makeup air-heating units with high capacities. Some refinishing companies in the United States specialize in this type of work; frequently, their quality and cost are very attractive.

A contractor's repair facility probably has as much character and uniqueness as the contractor himself; Figs. 1-6 through 1-10 are typical, but it is almost impossible to

Fig. 1-9 A 40 × 60 ft rigid frame metal building that can be used with van trailers for office and parts storage. This facility will handle a fleet for a good-sized project.

2-8 Establishing the Maintenance Facility

Fig. 1-10 Tandem 40 × 80 ft foldup buildings used in extremely cold, isolated areas. Buildings are insulated and lined. Parts storage and maintenance offices are in one-half of one unit with remaining area used as shop. Building is easy to erect and tear down.

recommend a specific shop to fit a specific job. Only the person putting the job together has this ability.

The point is: Be sure to give the maintenance crew what it needs in tools and structure to keep operating. But don't go overboard! A very large structure is not always the answer; good planning, good parts availability, and the tools required to do the job will produce a good many repairs from the small shop.

Chapter **2**

Heavy Tools for Construction Equipment

JAMES F. SENRICK
Division Manager, Owatonna Tool Company, Owatonna, Minn.

As construction equipment increases in size and complexity of design, the requirement for specialized heavy-duty maintenance tools and equipment becomes apparent. Not only must preventive maintenance programs be followed but when a machine breaks down, it must be repaired quickly to reduce costly downtime and prevent the possibility of shutdown of an entire equipment spread.

Tools and equipment are most often classified as general-purpose and special tools. They are further identified as essential or convenience tools. An essential tool is defined as one that performs a function for which there is no satisfactory alternative that is readily available, effectively reduces service time, or performs to substantially increase the owner's satisfaction or to justify the owner's investment. A convenience tool performs a function for which there is an alternative but this tool handles the job in a more satisfactory manner, reduces service time, performs to substantially increase the owner's satisfaction or to justify the owner's investment.

General tools General tools used in the maintenance and repair of construction equipment consist of basic items found in the mechanic's tool box such as

Combination wrench sets through 1 5/8 in.

Socket wrench sets with 3/8-, 1/2-, and 3/4- in. drive

Miscellaneous tools: punches, chisels, screwdrivers, files, pliers, ball peen and plastic hammers, scrapers, feeler gauge sets, pry bars, retaining-ring pliers, magnetic pickups, and hacksaws

Other tools and equipment with general usage characteristics are found in the repair facilities of contractors and equipment distributors. The size and quantity of the tools and equipment are dependent not only on the size of the repair facility but also on the type of machines being serviced. A further classification of general tools and equipment can be made with relation to need in the main shop and specialization areas such as engine rebuild area, fuel system repair and test area, electric component repair and test area, undercarriage rebuild area, machine shop, fabrication area, and cleaning and painting area.

Typical of the tools and equipment found in the main shop are

Bridge cranes with power hoists
Yard crane or boom truck
Fork truck (2 ton)
Air compressor (up to 15 hp)
Centralized lubricating oil delivery system
Powered floor sweeper

2-10 Establishing the Maintenance Facility

Agitating cleaning tank
Vertical shop press (150 ton)
Arbor press with pedestal
Pedestal grinder (1 hp)
Oxygen-acetylene welding set
Gear and bearing puller set (17½- through 50-ton capacity)
Bearing heater
Battery charger
Battery tester starter and volt-amp meter
Work benches with heavy-duty vises
Parts storage baskets with wheeled bases
Wheeled dolly
Pneumatic grinders
Electric hand drill (½ in.)
High-pressure washer or steam cleaner
Impact wrenches (½- through 1-in. drive with sockets)

The engine rebuild, fuel injection, and electric areas include
Bridge crane with power hoist
Jib crane with power hoist
Agitating cleaning tanks (large and small)
Glass bead cleaning machine
High-pressure washer or steam cleaner
Engine positioning stands
Engine carts
Cylinder head stands
Bearing heater
Arbor press with pedestal
Valve facing machine
Valve seat insert boring tool
Valve seat grinder
Connecting rod boring machine
Dynamometer with attachments
Hydraulic test bench with attachments
Fuel injection pump test stand
Flare nut wrenches
Nozzle and injector test pump
Electric test bench
Armature growler
Armature lathe
Bench drill
Pneumatic wrenches (⅜- through ¾-in. capacity)
Inside and outside micrometer set
Work benches with heavy-duty vises
Parts storage racks and baskets with wheeled bases

An undercarriage area would include
Bridge crane with power hoist
Jib crane with power hoist
Agitating cleaning tank
Glass bead cleaner
Roller welder
Track link welder
Roller and idler press
Track press with accessories (200 ton)
"Lazy Susan" for storage of pins, bushings, and links
Track windup machine
Drill press with tapping attachment
Oxyacetylene cutting outfit
Impact wrench for track bolt removal
Pneumatic wire brush
Air chipping hammer
Inside and outside micrometer set

Parts storage racks and baskets
The machine shop area would include
  Lathe
  Radial drill
  Bench drill
  Electric hand drill (½ in.)
  Tool grinder
  Work bench with vise
  Inside and outside micrometer set
The fabrication area should have
  Bridge crane with power hoist (5 ton)
  Semiautomatic open arc welder
  Air/carbon arc cutting torch
  AC-DC welder (500 amp)
  Portable welder (350 amp)
  Oxyacetylene cutting and welding set
  Welding rod storage oven
  Pneumatic sander
  Pneumatic hand grinder
  Pedestal grinder (1 hp)
  Clamp set
  Air chipping hammer
  Shop press (150 ton)
  Work bench with heavy-duty vise
The cleaning and painting area will have
  Automatic or hand-held paint stripper
  High-pressure wash unit
  Small-capacity heated high-pressure wash or steam gun
  Pneumatic sander
  Electrostatic paint spray
  Airless paint spray
  Paint storage cabinet
  Work bench

Specialized tools In addition to general tools listed above and depending on the type of equipment being serviced, hundreds of additional items may be required to adequately perform the service function. These tools are designed either by experienced tool manufacturing specialists or by the tool development groups of the equipment manufacturer. Cooperative design efforts between the tool manufacturer and the equipment manufacturer are very common and have resulted in availability of specialized tools and equipment for specific-purpose applications. Prominent in the realm of heavy tools designed for specific applications are some designed for crawler tractors.

Crawler Tractor Master Pin Removal and Installation Tool Sets. For those crawler tracks which do not employ a split-design master link, a master pin is used to hold the track together. The press fit is such that up to 150 tons of force may be required to remove the master pin.

A hydraulic pump, either hand or motor powered, and a hydraulic cylinder with appropriate valving and accessories are employed to generate the force necessary. Hydraulic pressure up to 10,000 psi is used.

When coupled to a specially designed C frame or legs with a bolster and properly sized tooling, it is possible not only to remove the master pin and bushing but to install them as well. In most cases the crawler truck can be split without removal of one or more track shoes.

Illustrated in Figs. 2-1 through 2-4 are the four operations performed when servicing crawler track master pins and bushings. The tooling is set up in the press in basically the same manner for all tractor makes and models.

Compared to old-fashioned torch-and-sledgehammer methods of master pin and bushing service, the use of hydraulic power makes the job easy and safe. In fact, it is a one-person operation.

Crawler Tractor Sprocket Removal and Installation Tool Sets. As with the master pin removal process, the force requirements for sprocket removal from crawler tractors have now reached 150 tons on the larger models of machines. A system has been developed

2-12 Establishing the Maintenance Facility

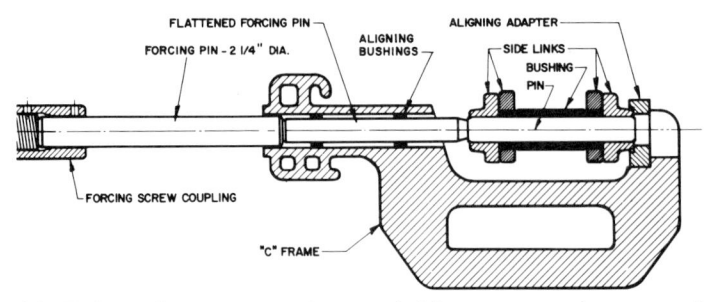

Fig. 2-1 Tooling is shown in position for removal of the master pin and separation of track.

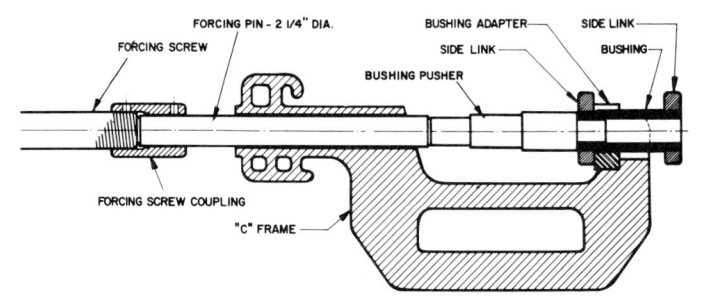

Fig. 2-2 The master bushing is being pressed out of one of the master side links.

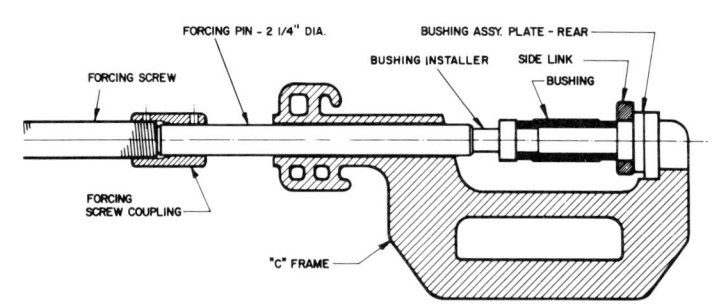

Fig. 2-3 The master bushing is pressed into each of the master side links.

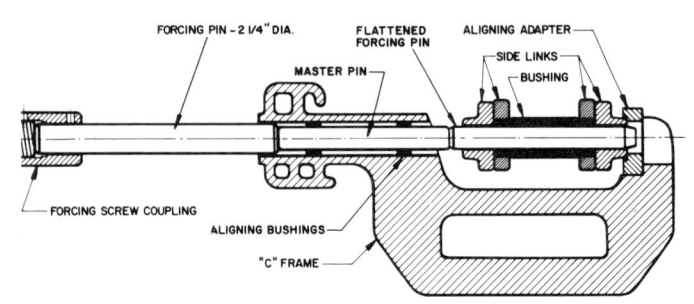

Fig. 2-4 The tooling setup used for installation of the master pin.

that allows the sprocket puller tool set to be built up piece by piece, therefore making a one-person operation possible. The use of various-sized sprocket shaft adapters, removing adapters, sprocket removal spacer tubes, sprocket installation spacer tubes, and puller leg end adapters, with hydraulic cylinder and pump groups, makes general usage over a wide range of crawler tractors possible.

Hydraulic pressures up to 10,000 psi are used. Center-hole hydraulic rams with ram screw, screw crank, and speed nut are preferred in this type of application because of the versatility offered in the buildup of the tool set.

Figures 2-5 and 2-6 are typical of a removal and installation setup. Size and number of adapters will vary depending upon the characteristics of a specific machine.

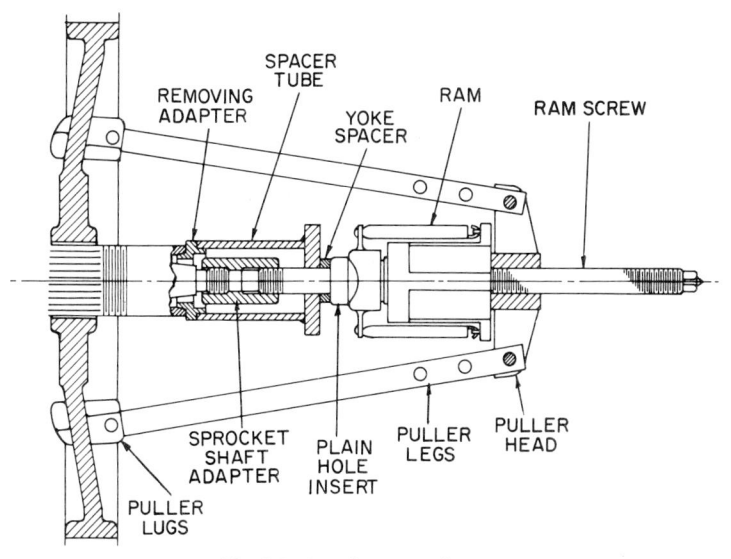

Fig. 2-5 Sprocket removal setup.

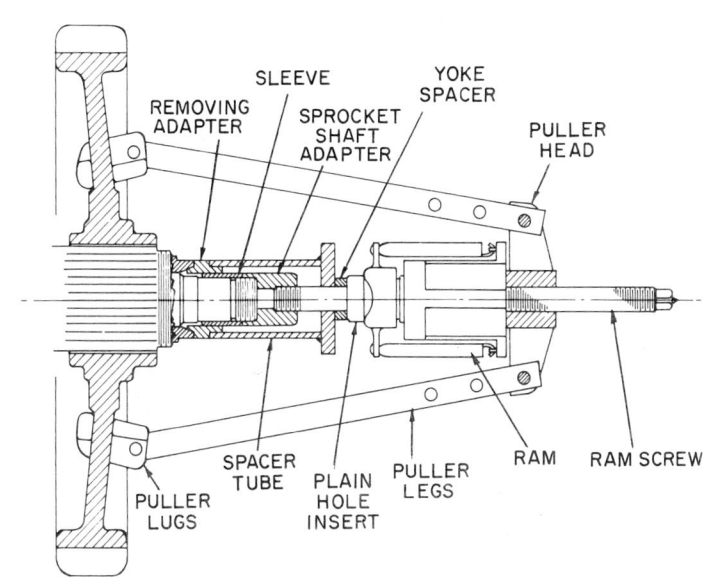

Fig. 2-6 Sprocket installation setup.

2-14 Establishing the Maintenance Facility

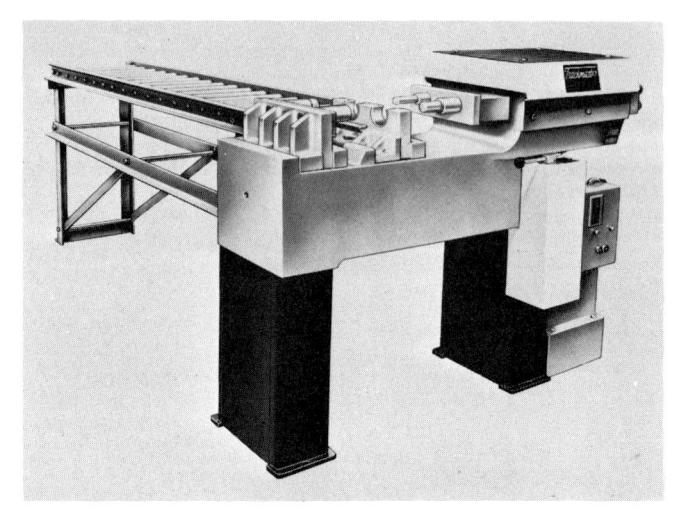

Fig. 2-7 Typical track pin press.

Track Pin Press. Designed specifically to make possible the turning of pins and bushings and/or the replacement of parts in crawler track is the track pin press. Available in capacities ranging from 60 to 200 tons and in single- or double-end configurations, it is a tool of utmost importance in an undercarriage rebuild facility (see Figs. 2-7 and 2-8).

Both internal and external wear occur on track pins and bushings. In addition, wear occurs on the sprocket, idler, and roller assemblies as well as the links and grousers. Consultation with equipment dealer undercarriage specialists will assist in determining the proper time to turn pins and bushings to expose new wear surfaces and allow total pin and bushing life to match link life—the ultimate objective. It must be remembered that the turning of pins and bushings will greatly reduce track stretch and increase life expectancy on the sprocket, roller, and idler assemblies, all resulting in lower operating cost of the crawler tractor.

The 200-ton track pin press design includes a rigid frame to prevent flex under full load; a hydraulic system which offers dual displacement capability for fast ram approach and high-force output at reduced speed; an adjustable pressure control valve to regulate

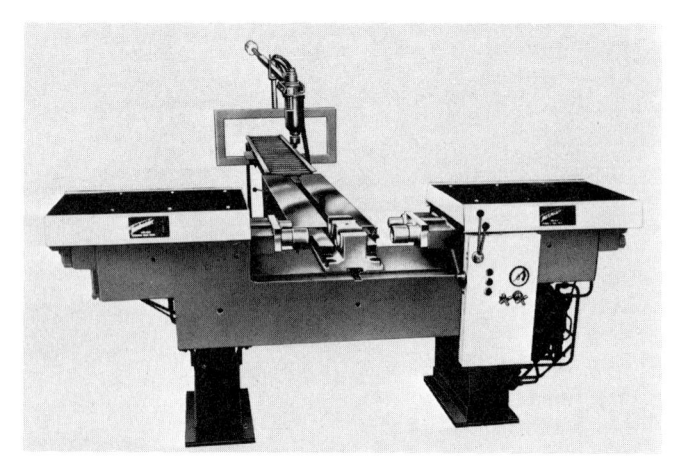

Fig. 2-8 Typical track pin press.

output force when specified by the tractor manufacturer for track assembly; control level movement which logically corresponds with action of the press rams, elevating table, and track indexer to eliminate operator confusion and error; fully supported and guided ram workheads to prevent deflection and maintain tool alignment despite off-center loading; a rigid center-saddle design to eliminate adjustment and prevent broaching of side links; versatility to allow accommodation of all track made today and in the foreseeable future.

To attempt to discuss all tools available for improved maintenance and repair of construction equipment is practically impossible. Rather, it is suggested that reference be made to technical manuals and service bulletins as published by the equipment manufacturer.

Chapter **3**

Portable Power Tools for Construction Equipment Maintenance

J. L. BENNETT
Safety Assurance Manager, The Black & Decker Mfg. Co.,
Towson, Md.

JAMES P. MORGAN, JR.
Technical Service Manager, The Black & Decker Mfg. Co.,
Towson, Md.

Proper maintenance of portable electric and air construction tools is necessary for efficient operation. If not properly taken care of, they can be dangerous to operate. Most manufacturers furnish an instruction book with each tool, indicating its proper use and maintenance. These instructions should be read and followed.

PORTABLE ELECTRIC TOOLS

Establishing preventive maintenance checks on each tool according to the amount of abuse it receives and noting all repairs in a card system provide the best means for guarding against breakdown.

Parts Maintenance

Careful attention to proper maintenance of each component is required for safe operation of electric power tools.

Bearings Motor and gear shafts rotate in bearings; spindles and mechanism components rotate or slide in bearings. Portable electric tool bearings may be of the ball, roller, needle, or sleeve type. Proper lubrication is essential to bearing life, but excess lubrication can cause as many problems as too little. Frequency of lubrication depends on the type of bearing and length of time the tool is in use.

Ball bearings are generally prelubricated by the manufacturer and require no further lubrication. Roller, needle, and sleeve bearings require additional lubricant unless it is automatically supplied in the gear chamber.

When bearings show signs of wear, they should be replaced, as they can damage the tool.

Gearing Common types of gears in portable construction equipment tools include spur, helical, spiral, bevel, and worm. Gearing is generally housed in a gearbox which contains grease and oil in the correct amount. Gear lubricants should be changed periodically in accordance with the tool's instruction or maintenance book.

2-18 Establishing the Maintenance Facility

Worn gears will be noisy and should be replaced when definite signs of undercutting appear on the gear teeth.

Electric cord The electric cord on a portable tool is its lifeline. Cords should be kept clean and free of oil and grease. Tools should be stored so that cords do not have sharp bends. Worn cords should be replaced as they can fail and present shock hazards.

Tools to be grounded are equipped with three-wire cords and three-prong plugs. The green wire is used for grounding. The preventive maintenance schedule should include a check of the continuity of the ground wire in the cable from the grounding prong on the plug to the motor housing.

Double insulated tools have only two-wire cords. In this type of tool, grounding is not necessary since a system of protective insulation provides safety from shock hazards.

When replacing an electric cord, use the manufacturer's cord set of the same size and type of cable originally provided with the tool. Cord protectors where the cable enters the tool housing prolong the life of the cable, and it is important to replace them when installing new cable.

When using an extension cord, it is important that the proper size be used in order to prevent low voltage at the tool. Check the ampere rating of the tool which appears on the

TABLE 3-1 Ampere Ratings for Extension Cords

Extension-cable length, ft.	Ampere rating (on nameplate)					
	0–2.0	2.10–3.4	3.5–5.0	5.10–7.0	7.10–12.0	12.1–16.0
	Wire size (AWG)					
25	18	18	18	18	16	14
50	18	18	18	16	14	12
75	18	18	16	14	12	10
100	18	16	14	12	10	
150	16	14	12	12		
200	16	14	12	10		

nameplate and refer to Table 3-1 for determining the proper size of extension cord. If the tool has a three-wire cord, use a three-wire extension cord to retain continuity of the ground circuit.

Brushes Motor carbon brushes should be checked on a preventive maintenance schedule and should be replaced with the manufacturer's part before they are completely worn out in order to prevent excessive commutator wear. Condition of the commutator, on which the brushes ride, should also be checked. If commutators are deeply grooved, the life of new brushes will be low. Grooved commutators should be turned and reundercut.

Switches Two basic types of switches are used in electric tools: ac-dc, which has quick make-and-break contacts; and ac only, which has slow make-and-break contacts.

Double-pole switches are used on heavy-duty portable electric tools, while single-pole switches are often used on lighter tools. The double-pole switch breaks both sides of the line, whereas the single-pole switch breaks only one side. The black line should be hooked to the switch when connecting a single-pole switch.

Three basic types of terminal connections are used in portable electric tools. The most popular uses a screw connection and must have a proper terminal on the leads for fitting under the screws. Push-in switches are also used. Here, the leads are tinned and forced into slots provided. To release, a wire is inserted into the release slot and the leads are pulled out. A third type of switch has wire leads permanently connected internally. Exterior connections are then made to the motor and cable with splicing connectors.

Repairing small switches is not practical.

Motor Three basic types of motors are used in portable electric construction equipment tools: shunt, series, and induction.

The series motor, which is the most common, consists of a stationary field and a rotating armature equipped with a commutator on which the carbon brushes ride. The stationary field can fail owing to open circuit, short circuit, or ground. Failures may be caused by severe overload, overheating owing to lack of cooling air, or excessive exposure to abrasive or conductive dirt. The armature, or rotating member, may also fail owing to open circuit, short circuit, or ground. Mechanical failure may be due to pinion or bearing journal wear. Excessive sparking at the brushes may be caused by poor commutator surface. Clean this with fine sandpaper or a glass brush. If the commutator is deeply grooved, it should be turned on a lathe.

DC motors in portable electric tools generally use a permanent magnet field. These have no windings and are not subject to the same failures as a wound field. They can, however, lose their magnetism from shock or overheating and will have to be remagnetized by the manufacturer. The armature of a dc permanent magnet shunt motor is identical to that of a series motor and can be maintained in the same manner.

DC permanent magnet shunt motors used in tools intended for operation on ac lines include a rectifier constructed with diodes. Repair of these rectifiers is not practical, and they should be replaced when faulty.

Induction motors have cast or fabricated rotors, which are virtually indestructible, and need no maintenance. The field or stator of an induction motor can fail owing to overloading, overheating, or exposure to conductive dirts and abrasives in the same manner as the field of a series motor. Single-phase induction motors generally have starting windings that are removed from the circuit after the motor reaches full speed. This switching is accomplished by a relay or centrifugal switch, which should be checked if a starting winding fails.

Most portable electric construction equipment tool motors have cooling fans which require ventilation in order to prevent overheating. Ventilating openings and passages should be cleaned periodically.

Cordless Cordless electric tools (see Fig. 3-1) are powered by self-contained rechargeable batteries. It is advisable to have batteries on charge whenever the tools are not in use.

Chargers for batteries of cordless electric tools generally contain a transformer, rectifier, and necessary control devices. Chargers should be returned to the manufacturer for repair.

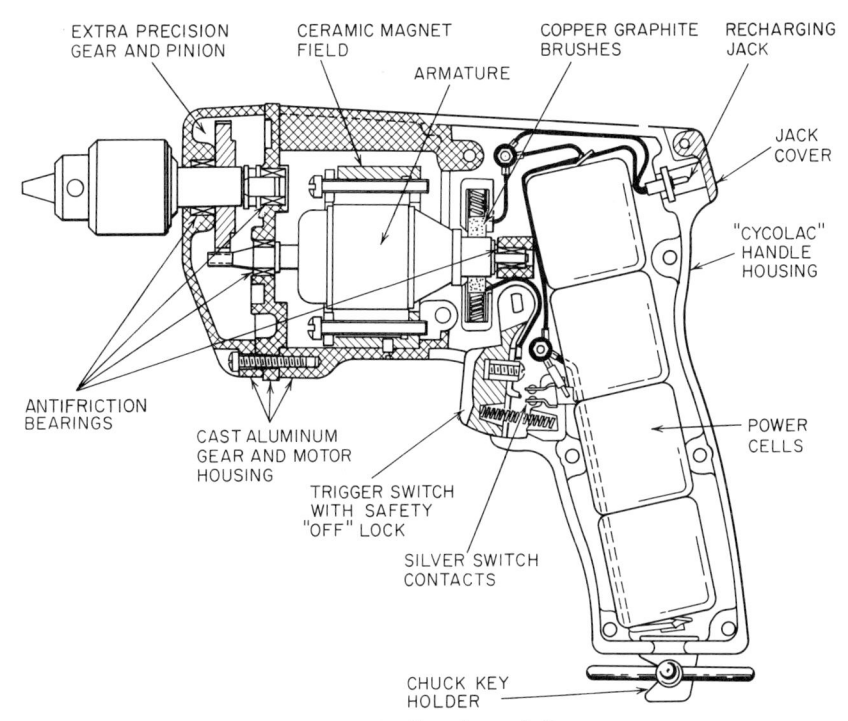

Fig. 3-1 Cordless electric drill.

Motors in cordless electric tools are the dc permanent magnet shunt type, designed for low voltages produced by batteries. Maintenance procedures are the same as for ordinary dc permanent magnet shunt motors. Motors in cordless tools should be well maintained to avoid undue drain on batteries.

Double insulated tools These tools have insulated spindles and insulated housings or protective insulation between life parts and metal housings. The spindle insulation is attained by an extra insulating member on the armature shaft, in the gear train, or at the spindle itself.

Since safe operation depends on the insulation, it must be checked frequently by applying high voltage from each plug terminal to the housing and to the spindle. Damaged housings should be replaced with original parts.

Variable speed tools Speed changes on portable tools are accomplished either mechanically or electrically. Mechanical speed changes use syncromesh drives in a two-speed gearbox. Service procedures are the same as on a normal gearbox.

Speed changing motors employ various methods. The simplest arrangements use series/parallel field coil connections or coil tapping. More sophisticated types use rectification to dc with control of the rectified current. Semiconductor devices are frequently used in the control circuits and, if replaced, should use parts with ratings identical to original equipment.

Inspection of Portable Electric Tools

Periodic inspection of portable electric construction equipment tools should include the following:

Ground test at 500 volts This should be performed with the switch on, testing from each line prong of the plug to the housing.

Continuity test of the ground circuit Perform with a continuity tester from the grounding plug to the housing.

Cable Cable should be checked for worn or frayed spots and replaced if in poor condition.

Brushes and commutator Check and replace worn carbon brushes. Clean commutators, and turn on a lathe if excessively grooved.

Ventilating openings Clean regularly to keep free of dirt.

Conductive dust Wipe or blow out carbon or other conductive material inside the motor housing.

Causes of failure When a portable electric construction equipment tool will not run, the following may be responsible: plug, carbon brushes, electric connections, cord, armature, or field.

If the motor overheats, the problem will generally occur in brushes, commutator, armature, field, or ventilating openings.

If the unit is noisy, gears and bearings should be inspected.

If the unit is grounded, the problem will generally be in the armature, field, brush holder, or cable. Mislocated wires can be pinched or abraded and consequently grounded.

If the motor lacks power, a check should be made of brushes, armature, commutator, and field.

Armatures can be checked for shorts on a growler and for open circuit with a continuity tester. Field coils can be checked with a resistance meter or bridge.

Maintenance of Specific Portable Electric Construction Tools

Drills The portable electric drill is distinguished primarily by its chuck. Chucks are assembled to the spindle with a tapered fit, male thread, or female thread. They are generally the gear type (see Fig. 3-2), although there are other types known as keyless, which can be hand tightened or impact tightened.

Very large drills are often equipped with a Morse taper socket instead of a chuck, and these sockets should be kept clean and free from burrs to maintain proper grip on the drill shank.

Screwdrivers Screwdrivers are similar to drills, but they have ratcheting clutches which provide impact for screw tightening. The positive type of clutch depends on operator pressure for resulting torque. The adjustable type of clutch is provided with a

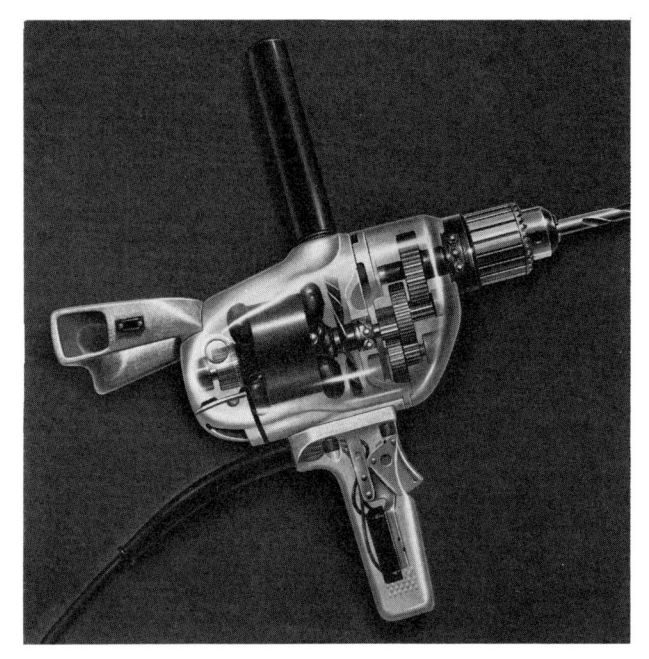

Fig. 3-2 Drill (equipped with gear-type chuck).

spring which can be adjusted for controlled torque. The clutches are made of hardened steel for long life, but they do require lubrication in accordance with manufacturer's recommendations. Worn clutch teeth will result in loss of torque and must be replaced (see Fig. 3-3).

Impact wrenches Impact wrenches have mechanisms to provide high torque on large nuts and bolts. Impact energy is accumulated in a hammer member between blows.

Lubrication is necessary for long life of these mechanisms. Spindle ends on impact wrenches are either square or splined for acceptance of commercial sockets. It is important to use impact classified sockets. Worn sockets are a frequent reason for loss of torque

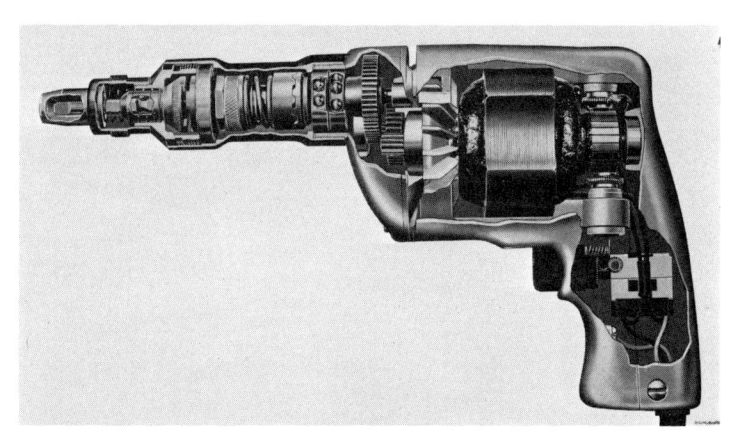

Fig. 3-3 Screwdriver (can be equipped with either positive or adjustable clutch).

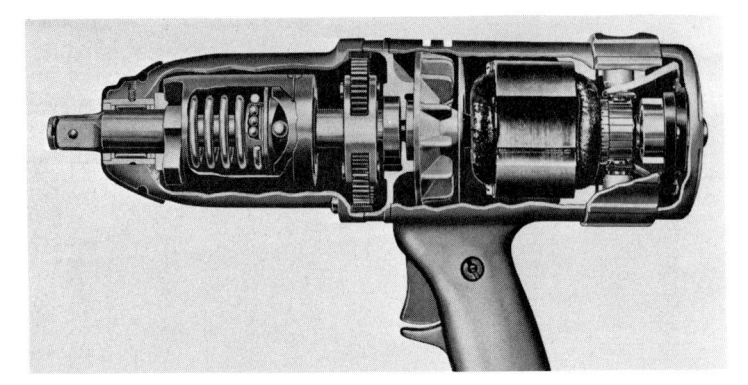

Fig. 3-4 Impact wrench (designed to provide high torque).

Fig. 3-5 Sander (should be cleaned frequently with air hose).

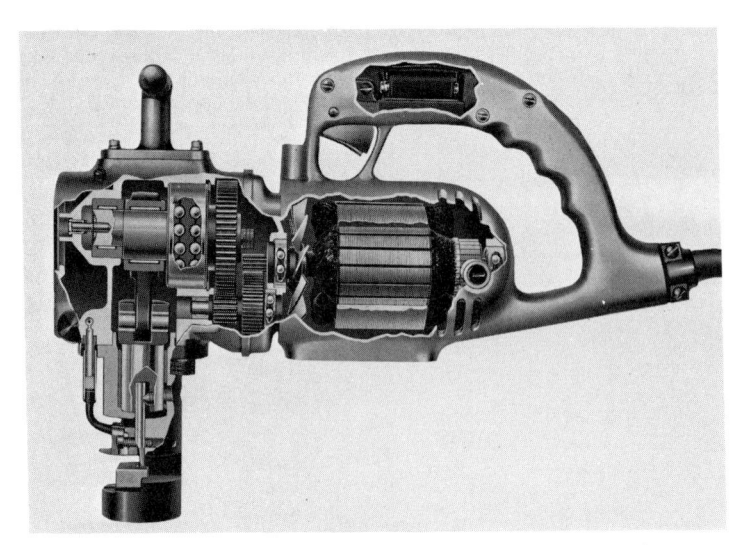

Fig. 3-6 Shears and nibbler (sharp cutting blades most important).

Fig. 3-7 Professional circular saw.

and should be replaced if they are loose on the spindle or if they fit the bolt head or nut sloppily (see Fig. 3-4).

Sanders and portable grinders All tools used with abrasives are subject to damage from the dust. Clean the ventilating system and moving parts with an air hose.

Occasionally sanders and portable grinders should be disassembled and all parts cleaned. Gears should be relubricated at the same time. If units are used continuously, this procedure should be followed every 60 to 90 days (Fig. 3-5).

Shears and nibblers Blades should be adjusted and sharpened to the manufacturer's specifications regularly. Sharp punches and dies are important to the proper performance of nibblers, and sharpening should follow manufacturer's instructions. Worn spindle bearings can cause improper clearance between punch and die, resulting in poor cutting, excessive loads, or damage. Check this clearance frequently (Fig. 3-6).

Circular saws Check lower retracting guards on circular saws so that they will operate freely in order to cover the blade when the saw is removed from the work. Lower shoes on circular saws are frequently bent and should be straightened or replaced (Fig. 3-7).

Keeping saw blades sharp will prevent motor overload and burn-out.

Reciprocating saws Blade holders and clamps should be checked to prevent blade breakage. If spindle bearings are worn, excess play between the reciprocating shaft and the spindle will cause diffi-

Fig. 3-8 Reciprocating saw.

2-24 Establishing the Maintenance Facility

cult cutting, and the bearings should be replaced (Fig. 3-8).

Routers Routers have high-speed spindles driven by the motor. The collet which grips the cutter must be very true to prevent excessive vibrations, and worn or wobbly collets should be replaced. If spindle bearings become rough or sloppy, they should be replaced. Cutters must be kept sharp to avoid motor overloads and burning of the work (Fig. 3-9).

Hammers Demolition hammers, rotary hammers, and percussion drills all produce heavy blows on the accessory being used. Mechanism parts are designed to withstand heavy impact, but they require lubrication to manufacturer's specifications for long life. Periodically this type of tool should be dismantled and thoroughly cleaned. Worn parts should be replaced (Fig. 3-10).

Fig. 3-9 Router (high-speed spindle is driven directly by the motor).

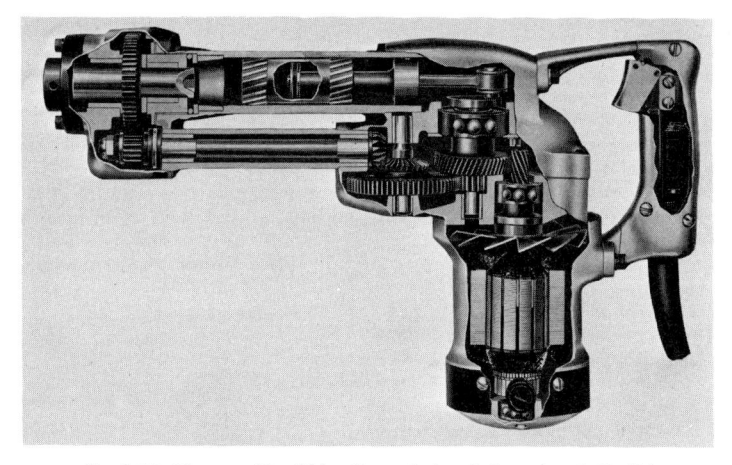

Fig. 3-10 Hammer (should be dismantled and cleaned periodically).

PORTABLE AIR TOOLS

Before planning maintenance, it is necessary to examine the components of portable air tools.

Air Motor Components

These parts include the following (see Fig. 3-11):

End plates Both rear and front end plates act as a seal for the ends of the motor (Fig. 3-12). Units operating with continued starts and stops during usage such as impact wrenches and screwdrivers have holes or starting slots to channel air underneath the blades to move them outward and obtain best efficiency in the acceleration of the motor. Continually rotating units such as grinders and sanders do not need starting slots to obtain normal operating efficiency.

Cylinder liner This liner is a hollow cylinder with holes that allow air to enter the motor and slots or holes that allow exhaust (Fig. 3-13). The most important part of this liner is the seal point which prevents air from leaking into the exhaust side of the motor.

Some liners have a seal radius, which is a machined area in the liner to maintain a very close seal point over a longer length of rotation.

IDENTIFICATION OF AIR MOTOR PARTS

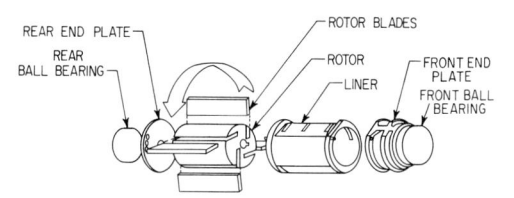

Fig. 3-11 Air motor parts.

Rotors The rotor is a cylindrical metal piece slightly shorter in length than the cylinder liner and with a smaller diameter. It has radial or tangential slots running its entire length to accommodate the blades. The rotor, with the blades, transforms air pressure into mechanical torque.

There are two types of rotors:

1. *Fixed* (locked). Clearance between the end of the rotor and the end plates is critical and maintained with different-sized spacers (see Fig. 3-14). The fixed rotor is used

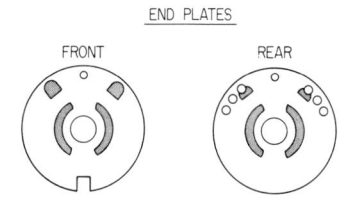

Fig. 3-12 Rear and front end plates.

Fig. 3-13 Motor seal of cylinder liner.

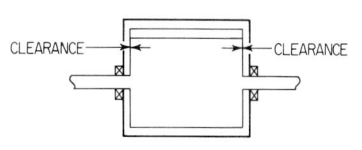

Fig. 3-14 Locked rotor motor.

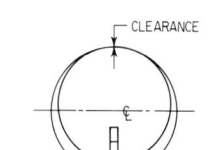

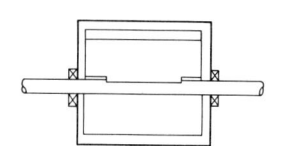

Fig. 3-15 Floating rotor motor.

in extremely high-speed motors such as die grinders. There is no rubbing between the rotor and end plates, so that there is less power loss from friction.

2. *Floating.* The liner is slightly longer than the rotor. The rotor floats endwise (see Fig. 3-15) since it is a slipfit on the rotor shaft, or its shaft has bearing slip and, therefore, floats to a balanced position between the end plates when the motor is turning.

There are two types of blade placement (see Fig. 3-16):

1. *Radial.* The blades radiate from the center, moving uniformly around a central axis.

2. *Tangential.* The blades are at a tangent to the center. The advantage is that more blade area is exposed for the same size motor using radial blades, hence, more power in the same size package. Also, the blades are longer in length, which will require fewer replacements. The disadvantage is that it cannot be reversed.

Blades Blades are made of linen impregnated with resin or aligned fiber glass. These blades are as long as the rotor, not quite as thick as the rotor slots, and about as wide as the depth of the rotor slots.

The function of the blades is to trap air between

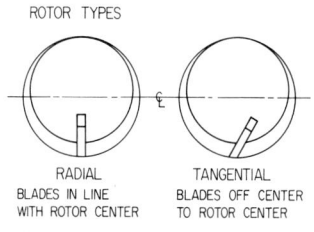

Fig. 3-16 Two types of blade placement.

them and rotate the rotor. When high-pressure air is trapped, it moves each blade toward the lower pressure exhaust port, rotating the motor.

The point where each blade picks up air is called the air seal. When the blade crosses the exhaust port, the air is dumped into the atmosphere with the blade traveling to the seal point again.

Governors Air motors reach their maximum power at approximately 50 percent of their free running speed. In other words, a tool at 0 rpm, or at its full running speed, is doing no work and develops no horsepower. To have a tool run at a free speed (within proper safety limits) and at maximum power, governors are used.

Failure Diagnosis and Troubleshooting Techniques

To test air tools properly, the following equipment is needed:

1. *An air line system.* Usually ½- or ¾-in. minimum line is used from the compressor. Be sure no restrictive connections of a smaller inside diameter are used.

2. *An air filter, regulator, and lubricator.* The filter cleans moisture and dirt from the air. The regulator determines the maximum amount of air which is allowed to pass through the system. The lubricator provides a mist of oil to enter the tool along with air.

3. *An air flow meter to check cfm.* The air flow meter must be exactly perpendicular to the horizontal to get a proper reading, and it should also be positioned at eye level.

4. *A tachometer.* A tachometer is used to test all rotary tools in order to determine the proper rpm.

Tool Maintenance Characteristics

Percussion tools (chipping hammers, scalers) The chipping hammer is a typical percussion tool. In such tools, the piston usually is the most important part; if badly worn, the blow will not be effective. Pistons usually wear out at the working end. Sometimes replacing a worn piston will restore the tool to good operating condition. However, in many cases, the bore of the barrel may be worn. When design permits, the barrel can be rebored and fitted with an oversize piston. The usual procedure is grinding or honing. For hammers with replaceable bushings, bushings should be replaced when wear becomes appreciable. The fit in the chisel shank in the bushing makes an air seal that creates a cushion at the end of the piston stroke.

A worn valve will reduce the effectiveness of the piston blow, and inspections should check for this.

The throttle-valve mechanism in the handle should be inspected frequently, and worn parts should be replaced so that the throttle will start promptly, open fully, and shut off completely.

It is advisable to carry in stock an estimated number of parts, such as chisel bushings, pistons, and valves for the valve block; also, throttle parts such as compression springs, trigger pins, valve stems, and valves. Initial advice can be secured from the manufacturer, but actual experience with the work environment, plus repair records, will dictate the extent of spare-part inventory.

Rotating tools (grinders, drills) The grinder is shown in Fig. 3-17 as a typical rotating tool. Nearly all such tools have ball bearings, and it is advisable to replace these before they are badly worn to prevent misalignment.

The key which holds the rotor to the rotor shaft should be replaced if it shows the slightest wear. There should be a close sliding fit; too much play will cause the rotor shaft, in contact with the rotor bore, to wear too rapidly.

Most air grinders do not have gears, and they operate at relatively high speeds. It is essential that the spindle carrying the grinding wheel run true, and bearings be in good condition; otherwise, the spindle will vibrate considerably, causing rapid wear and increasing the possibility of wheel breakage. Be sure spindle attachments such as chucks, collets, and arbors run true with the spindle to avoid vibration.

Inspect gears and the ball bearings in which they are mounted when changing or replenishing grease (Fig. 3-18). If wear is allowed to go on too long, it will increase at an accelerating pace, and maintenance costs will become unduly high.

All governed tools must have frequent speed and maintenance checks.

Portable Power Tools for Construction Equipment Maintenance 2-27

Fig. 3-17 Vertical air grinder.

Operational Setup

Air pressure All tool maintenance begins with the air supply system. Working air pressure (the pressure at the tool) directly affects tool life. A ½-in. impact wrench takes twice as much time to tighten a bolt at 60 psi as at 90 psi. Longer time means more impact with an accompanying increase in tool wear. It also means more mechanic time. Higher-than-recommended air pressure results in harder-than-normal impact, along with increased wear of internal parts.

Piping Figure 3-19 shows an ideal piping system. Compressed air always contains some moisture, depending on the condition of the air taken into the compressor and on the system of compression. In the process of compressing, heat is created, so that air moisture changes to vapor which is then carried through the reservoir or tank into a

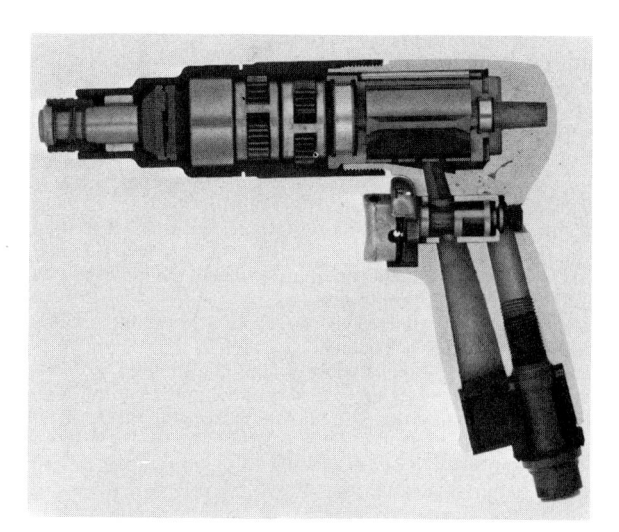

Fig. 3-18 Positive clutch pistol-grip air screwdriver.

2-28 Establishing the Maintenance Facility

pipeline. When the pipeline is long, considerable heat loss takes place and the moisture condenses.

Moisture can be eliminated to a large extent by use of an aftercooler at the compressor, and by water traps along the lines. The main transmission line should be sloped or angled back to the compressor to direct water into the receiver to be drained away. Down pipes should be taken off the top of the transmission line so that a minimum amount of water will be picked up. If the compressor is not equipped with an aftercooler and if the main transmission lines are not installed so that the water will flow by gravity to the receiver, the installation of water traps at low points along the transmission line will be helpful. These may be legs made of a length of large pipe, with a cap on the bottom and a petcock

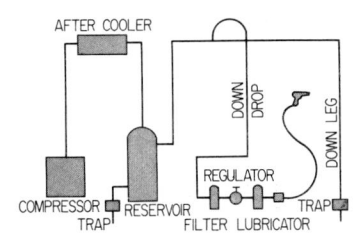

Fig. 3-19 Efficient piping system.

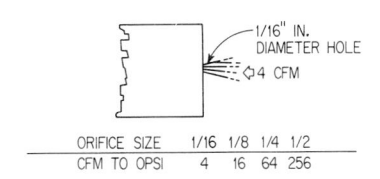

Fig. 3-20 Flow through the orifice.

ORIFICE SIZE	1/16	1/8	1/4	1/2
CFM TO 0PSI	4	16	64	256

for draining; the best are automatic traps, which dump accumulated water and sludge with no attention. All piping water traps should be drained weekly (preferably on Monday) to purge moisture from the lines.

Flow through the orifice The size of the orifice in the hose or hose fittings can severely restrict the air pressure delivered to the tool (see Fig. 3-20).

The hose should pass twice as much as the tool requires:

Tool, cfm	Min hose size, in
0–20	1/4
20–40	3/8
40–80	1/2

A good rule is to use as large a hose as practical and as short a length as possible.

Lubrication Air tools will operate without oiling but not for long. Inadequate lubrication is the most common cause of air tool failures.

Lubricants differ for different kinds of air tools. It is not safe to use just any lubricant, whether the tool is of the rotating or the percussion type. Use the oil and grease recommended by the manufacturer. For most air tools, the best means of lubrication is the line oiler. It is installed in the line just ahead of the hose connection of the air tool and gives a constant flow of lubricant while the air is moving. It can be regulated to give the desired amount. When it has a transparent bowl, there is no excuse for it to run dry. In some types, the oil can be seen to drop into the transmission line. To gauge the flow of oil, hold a piece of white paper near the exhaust of the air tool while it is running, and observe the amount of oil that collects.

Many tools of the rotating type have an oil reservoir in the handle; it will lubricate for a short period of time and can be easily refilled.

For percussion tools, a light oil, in most cases SAE 10 machine oil, is suitable. Heavier oil might make the valves stick or hesitate.

The worst enemy of lubrication is water in the compressed air. Traveling along the pipelines at high velocity, it tends to wash the lubricant out of the tool. This can cause corrosion, particularly if the tool lies idle for some time without being conditioned by an internal bath of lubricant.

Clean, dry incoming air is just as important for long tool life as proper lubrication. Air line filters installed between the compressor and the oiler will keep dirt, rust, and scale from making short work of air motor internal parts. Filters and drop line drains should receive regular attention.

Troubleshooting Clinic

When the air system has been checked out, the reader can concentrate on preventive maintenance and repair. Under no-load conditions, there is little difference in air pressure between inlet and exhaust, and the tool may seem to run satisfactorily. However, when the tool is loaded, the pressure difference increases and power leaks will be noticed. Air can leak through worn "O" rings and throttles, reversing valves and air inlets. Worn or broken rotor blades can scar liners and end plates and cause leaks. Worn blades should be replaced when the blade loses 20 percent of its width. To check a blade, compare it with a new one. After each repair, adjust air line pressure to the level for which the tool was designed. Next, lubricate the tool according to the manufacturer's specifications. Check free rpm against the manufacturer's specifications. Next, load the tool down; then check free rpm again to make sure it hasn't changed. Eventually, even the best-cared-for tool will wear. Drill chucks, power chisel spring retainers, all lose their holding power with years of use and need replacement.

Often the maintenance department needs guidance as to whether a tool should be repaired or replaced. A fairly generally accepted policy states that when the cost of repair is half the cost of a new tool, it should be replaced.

Safety

It is a responsibility of maintenance to see that tools are in safe operating condition. All rotating tools should be checked with a tachometer to assure the proper operating speed before accessories, such as grinding wheels, are adapted to the tool.

With the exception of cone-shaped wheels and small mounted points, all wheels should be operated under or inside of guards. The diameter of the wheel arbor should be of the correct size for the kind and size of grinding wheel it will carry. The wheel washer and collar should be of the correct diameter and thickness and made with the proper recess to grip the wheel firmly. The two should never be of different diameters. The nut which holds the wheel on the arbor and the washer against the wheel must be of ample size and strength to do the job.

A wheel may be chipped or cracked in transportation or in storage. Hold the wheel loosely on a finger slipped through the arbor hole and tap lightly with the wooden handle of a hammer to see if the ring is clear. If it is, mount it on the arbor, drawing the nut up firmly but not so tight as to cause fracture. Hold the tool under a bench and run it; if it is fractured, and failure results, injury will be avoided. If the result is satisfactory, retest the speed of the tool and wheel with a tachometer to be sure it conforms to the safety code provided by the Grinding Wheel Manufacturers Association.

Test Specifications for Air Tools

Specifications should be available from the vendor. The following steps should be observed when checking out a repaired rotary air tool; for example, in an air grinder rated at 4500 rpm:

First, squirt a few drops of moisture air guard into the inlet bushing to assure that all internal parts are properly cleaned and lubricated. Next, turn the tool on and check the free rpm which will be read from a tachometer applied to the spindle of the tool. Minimum reading should be 4200 and maximum should be 4400 rpm. While taking this reading, observe the reading of the air flow meter. This flow rate is the maximum free air reading.

Since this is a governor-operated tool, check to see if the governor is operating by loading the tool down (applying pressure to the output spindle with a piece of wood).

Next, check the stall leak reading by stalling the tool and slowly rotating the spindle. The highest reading on the air flow meter during a 360° rotation is the stall leakage. Stall leak is usually checked on all rotating tools, including those without governors.

Chapter **4**

Basic Hand Tools of Maintenance

THE STAFF
TRW/J. H. Williams Division, Buffalo, N.Y.

Regardless of the size of an operation, lack of an adequate assortment of hand tools for any maintenance job can lead to unnecessary and unwarranted expense.

Lack of proper tools for a job leads to improvising and misapplying the tools at hand. Costly abuse, excessive wear, and premature tool failure inevitably follow. Not only will the cost of worn and broken tools be exceedingly high, but the additional time spent accomplishing the task and the accompanying increased and costly accident rate will far exceed the cost of providing and maintaining an adequate hand tool inventory.

COST REDUCTION FACTORS

Four major factors enter into cost reduction in hand tool investment: (1) careful selection and top quality, (2) adequate range of sizes and styles, (3) correct use, and (4) proper care.

Careful selection While price is important in the selection of any product, value—true value—is the real determining factor. Generally, the best-quality hand tool is the least costly in the long run. It is rare indeed for cheap tools ever to deliver real value. Top-quality tools forestall breakage, excessive wear, and premature failure, all of which result in lost time and high accident rates.

Another important factor in careful selection is buying the right tool for the job.

It is important that tool buyers and users study the latest tool catalogs so that they will know the work-saving and safety features of all the specialized types of tools available today.

Adequate range In selecting wrenches, a study should be made of the range of nut and bolt sizes encountered on the various jobs. Thought should also be given to any clearance or accessibility problems that may exist. As an example, in choosing combination open-end box wrenches, it might be well to provide both the long and short series in the required range of openings. It may be that in many cases the shorter length will overcome accessibility problems. In others, the additional leverage offered in the long series may be of prime importance.

Detachable sockets and their various drive components make it possible for the operator or mechanic to tailor the wrenches to fit the job.

In addition, the proper organization of tools in sets saves a mechanic's time by eliminating needless lost time involved in getting forgotten tools or substituting for lack of proper tool assortment. There is no reason for a mechanic to be a hip-pocket operator, when an adequate range of tools is readily available.

Correct use The assumption that everybody knows how to use the more common wrenches and hand tools is not borne out by accident records. Wrenches that are pushed instead of pulled, screwdrivers that are used as levers or prybars, and pliers used as substitutes for wrenches are just a fraction of the long list of improper uses that appear on accident reports.

A continuing educational program in the correct and safe use of wrenches and tools is essential to hand tool economy and safe working habits. Such a program should begin with instructions that give workers a thorough understanding of the purpose and limitations of each tool. They must be made to realize that a tool, like fingers, is an extension of the hand. The use and care of tools are the basis of their skill and trade.

Proper care Keeping tools clean, properly lubricated, and in a good state of repair is essential to hand tool economy. To promote proper tool care, good housekeeping should be encouraged. One of the best ways to assure good housekeeping is to see that adequate tool boxes, chests, and cabinets are provided for all types of work.

For small jobs to be done away from the work bench out on location, use tool boxes that can be easily carried yet provide space and compartments to keep tools in an orderly fashion. For larger jobs, mobile cabinets should be provided, cabinets that can be easily wheeled to the job. This encourages the use of correct tools for the job by making it easy to have an adequate selection of tools right at hand.

ORGANIZING TOOL SETS

Having the right tool at the right time can best be assured by organizing tools in sets. This should be done by studying the tool requirements for various types of maintenance work. Once this has been done, carefully select the hand tools required in an adequate range of sizes and styles, and provide proper storage and transportation through the right choice of tool boxes and mobile cabinets.

Provide the necessary educational material to encourage correct use and proper care and you will be well on the way to hand tool economy and safety.

Chain wrenches A chain wrench (Fig. 4-1) is a versatile wrench for turning pipe as well as any square, hexagonal, or odd-shaped piece. Where close quarters exist, such as a pipe close to a wall (Fig. 4-2), only enough clearance to pass the chain through is necessary for a chain wrench. Normally, thin-walled copper tubing can be turned as the tubing won't be crushed by a chain wrench.

Chain wrenches come in two handle lengths. Both have the same capacity from

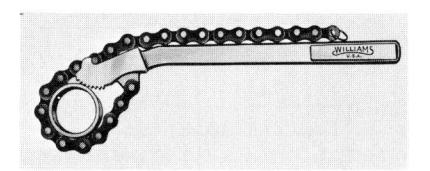

Fig. 4-1 Chain wrench applied to a section of pipe.

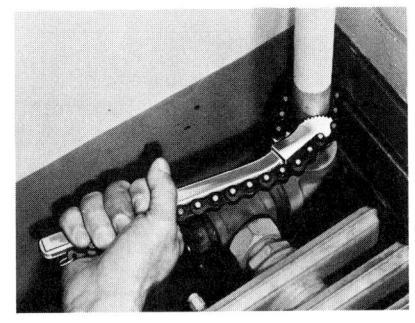

Fig. 4-2 Turning a pipe close to a wall with a chain wrench.

$3/8$ in. (0.95 cm) to 4 in. (10.16 cm) OD. The longer handle provides extra leverage and keeps hands away from hot work.

Put the chain around the part to be turned, and hook it where it will tighten as it is turned. Chains are user replaceable and should be replaced when visual inspection indicates any wear.

Chain pipe tongs Chain pipe tongs vary in length from roughly 14 in. (35 cm) to over 7 ft (2.13 m) for pipe up to 18 in. (46 cm) in diameter. Each style comes in a range of handle and chain lengths to accommodate various diameters.

Basic Hand Tools of Maintenance 2-33

Chain pipe tongs operate in the same way as a pipe wrench. The chain is looped around the piece to be turned and then secured on either side of the jaws. Once the chain is secured, pulling on the handle causes the teeth of the jaw to bite into the piece and further pulling turns the piece.

Four styles of chain pipe tongs differ primarily in length and jaw design. The original style (Fig. 4-3) is available in lengths up to 87 in. (2.21 m). The style shown in Fig. 4-4 has double-ended jaws which are reversible to give twice the jaw life. This style handles pipe to 12 in. (30.5 cm) in diameter. Figure 4-5 shows a pipe tong with a V-shaped recess in the

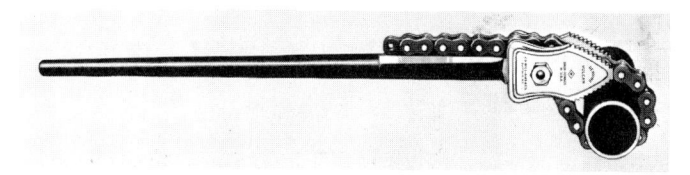

Fig. 4-3 Original-style chain tongs.

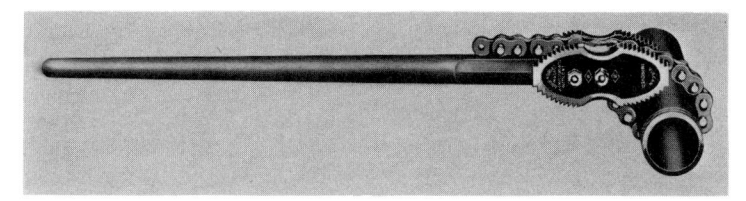

Fig. 4-4 Improved-style chain tongs.

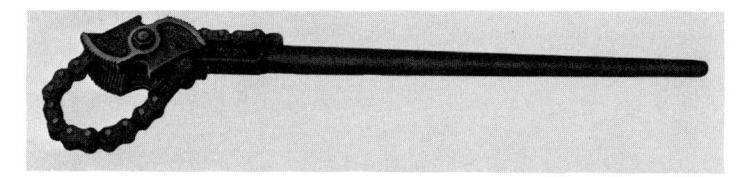

Fig. 4-5 Chain tongs with V-shaped jaw recess for added grip.

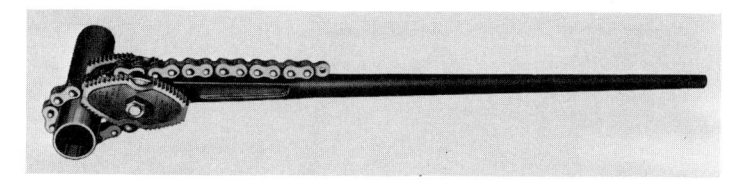

Fig. 4-6 Reversible chain tongs.

jaws. Jaws are reversible as well. The differing feature of the tongs shown in Fig. 4-6 is its reversible feature. Without removing the tongs or loosening the chain, this style will turn pipe instantly in either direction. Also, outer jaws may be removed to make properly fitting tongs for narrow heads and flanges. Jaw removal is accomplished by taking out the single through bolt and self-locking nut used in assembly.

Chain pipe vise Chain pipe vises are used for holding pipe while cutting, reaming, or threading, or while making up pipe assemblies and fittings. The vise may be mounted on a bench (Fig. 4-7) or on a vise stand (Fig. 4-8).

Establishing the Maintenance Facility

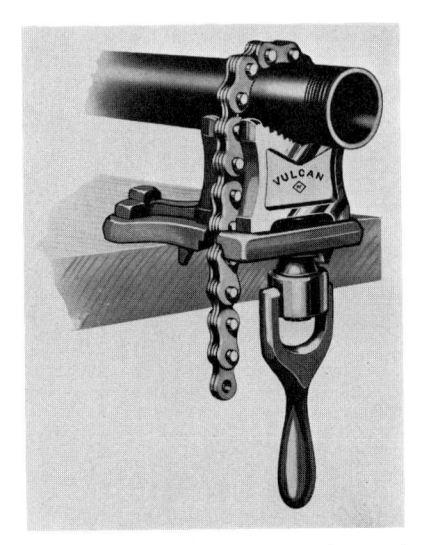

Fig. 4-7 Chain pipe vise mounted on tool bench.

The chain pipe vise is operated in the following manner: Begin with the chain loosened. The pipe is placed on the vise jaws. Next the chain is wrapped around the pipe and secured in the vise base. The handle below the vise is turned until the desired chain tension is obtained. As the chain tension increases, the pipe is pressed against the jaws.

Maintenance is a matter of common sense—keep working parts free of rust, chips, and dirt. The screw should be kept lubricated with a light grease or oil. Replaceable parts include jaws, chain screw, and handle.

Tubing cutter Tubing cutters make cleaner cuts than hacksaw blades, which often leave burred edges or cuttings that can clog lines and fittings. Tubing cutters are available to handle tubing up to 1 in. (2.54 cm) in diameter. A tubing cutter, complete with parts, is shown in Fig. 4-9.

To make a cut, place the cutter on the tube with the cutter wheel on the mark where the cut is to be made. Turn the cutter knob so as to move the wheel into light contact with the tube. Revolve the cutter around the tube, gradually tightening the cutter knob until the tube is cut through. Use a feed pressure that keeps the wheel cutting but doesn't flatten the tubing.

Flaring tool Many fittings used on tubing lines require that the tubing be flared at its ends. Flaring tools form the correct shape easily and accurately to ensure proper sealing.

The tubing end being flared should be cut off squarely and be free of burrs. Before clamping the tube in the flaring bars, the coupling nut should be removed or slid down the tube out of the way. The tubing should be inserted into the hole of the proper size of the flaring bars. The tubing is clamped into place by turning the wing nuts. Be careful to maintain the clearance until the tubing is secured. Next, place the yoke over the flaring

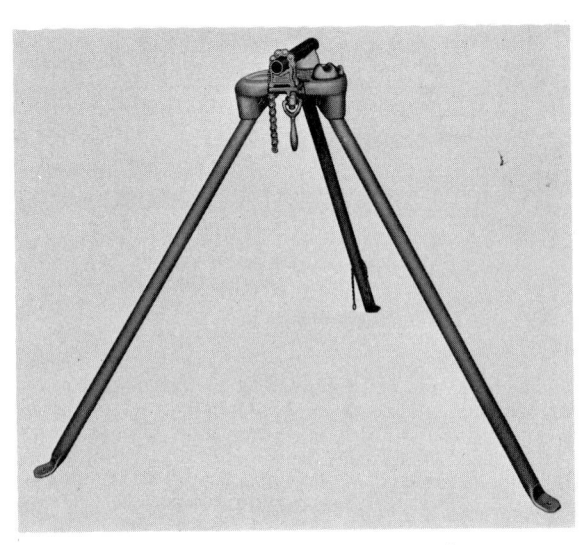

Fig. 4-8 Chain pipe vise on vise stand.

bars. The proper-sized flaring die is then placed over the end of the tube. The yoke is tightened by turning the feed screw. The yoke cone should fit into the recess of the flaring die. Turning the yoke feed screw produces a bell-shaped area with the end of the tubing turned in. The feed screw is then loosened and the flaring die removed. The feed screw is tightened on the end of the tube. The turned-in end is forced down, completing the flare.

Flange-Jacks Flange-Jacks are designed to separate flanged pipe joints at the flanges for purposes of sealing a line or replacing a gasket. Used in pairs, Flange-Jacks keep the pipe in proper alignment without the need to use wedges or hammers.

To use Flange-Jacks, first remove two bolts 180° apart. The jaws are fitted to the empty

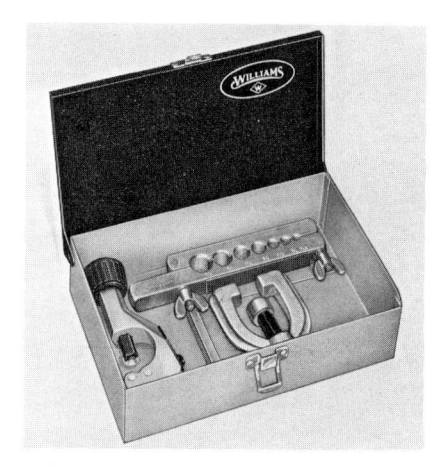

Fig. 4-9 Tubing cutter and flaring tool set.

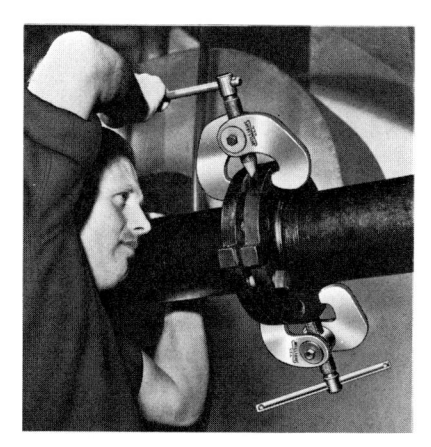

Fig. 4-10 Using Flange-Jacks to replace a damaged gasket.

bolt holes and tightened slightly. The rest of the bolts are loosened or removed to allow the flanges to spread. When the jack screws are tightened, cone-shaped shafts part the flanges in a wedging action (Fig. 4-10).

Feeler gauge A feeler, or thickness gauge, is used to measure the space between two surfaces. They are available in both metric and U.S. Customary calibrations. To use a feeler gauge, be certain surfaces are free from dirt. Insert the right blade until a slight drag is felt; this indicates proper measure. If one blade doesn't provide the proper thickness a combination of two may be necessary. The sum of the figures on the blades equals the thickness. Metric measure is in thousands of millimeters (microns) while U.S. Customary measure is in thousandths of an inch.

Fig. 4-11 Long-bladed feeler gauge.

Feeler gauges are available in blades as long as 12 in. (30.5 cm) as shown in Fig. 4-11 and several types of sets as shown in Fig. 4-12.

Screwdrivers The most used screwdrivers in maintenance are the common straight-tipped and Phillips tip. Common screwdrivers are available with square blades, with or without holders, and round blades, and electricians' types in stubby, long, and pocket sizes as shown in Fig. 4-13. Mechanics' types in either round-blade or square-blade varieties have wider blades than do electricians' screwdrivers. In some applications spring-type screw holders, which hold the screw head to the screwdriver blade, help to start a screw easier.

2-36 Establishing the Maintenance Facility

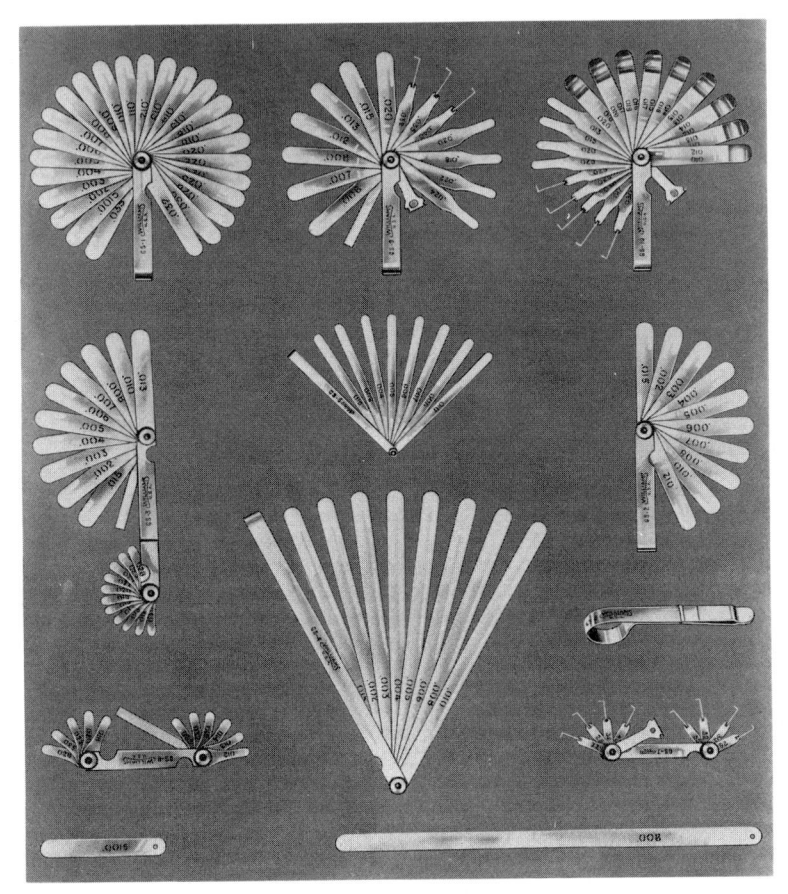

Fig. 4-12 Feeler gauge sets.

Sizing a screwdriver to a screw is just a matter of snug fit. Where a screw head fits into a restricted area or a tight recess, the narrower electricians' style must be used. Sizes of both types are measured by the length of the blade rather than the overall length of the screwdriver. As the length of the blade increases, the diameter also increases. Tips become wider and thicker with larger-diameter blades.

The cross-tip Phillips screwdriver fits Phillips head screws, which are used in production fastening. Phillips screwdrivers have round blades, in a variety of lengths. Sizes are designated as No. 1, No. 2, etc. Matching the screwdriver to the screw reduces the possibility of damage to both members.

Handle designs may affect comfort. Larger-diameter handles provide extra turning power. Cushion grips may also be preferred over plastic handles.

Screwdrivers are most effective when they are kept perpendicular to the screw

Fig. 4-13 Varieties of common screwdrivers.

slot. Additionally, the screwdriver should be pushed as it is being turned to keep the blade in the screw slot.

It is said that a screwdriver is the most often misused tool. Even light prying, like opening a paint can, may damage a screwdriver tip. Hammering a screwdriver for anything heavier than cleaning a screw slot generally deforms the blade.

When extra turnpower is needed, a better solution to a screw turning problem is a screwdriver bit, discussed in the section on socket wrench drivers and accessories.

Pliers Pliers are holding tools that provide the extra gripping power which hands alone aren't capable of. Pliers also isolate hot and cold objects or provide protection from skin-irritating liquids. Pliers can squeeze, twist, pull, cut, and hold. Corresponding to the variety of jobs that can be done with pliers is an assortment of pliers to meet any need.

Slip-joint pliers pivot on one of two opening points, providing a choice of maximum jaw openings to meet holding needs. Good for general work, slip-joint pliers usually contain a shear-type wire cutter (Fig. 4-14). Avoid using such tools as hammers or exposing them to excessive heat. Repairs are not recommended, so cracked, broken, or sprung pliers should be discarded.

Utility pliers have interlocking tongues and grooves to hold jaws parallel over a wide range of openings. Sizes range from midget, for delicate work, to openings of 2⅞ in. for

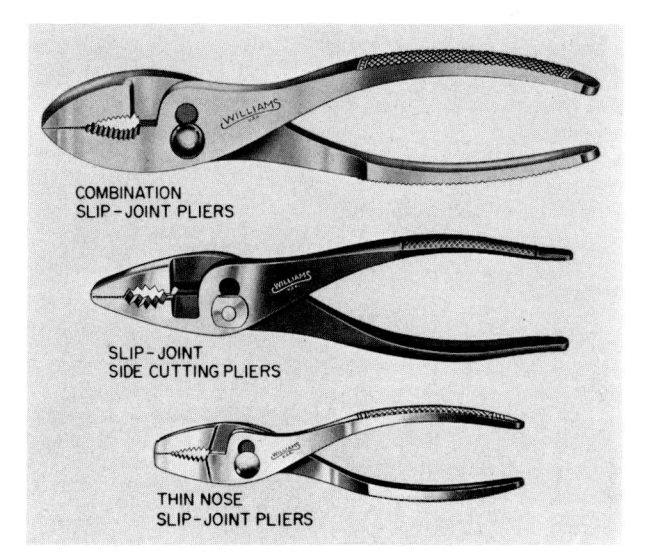

Fig. 4-14 Slip-joint pliers.

heavy-duty style (Fig. 4-15). These pliers have a wider capacity than slip-joint pliers and may be used to grip round, square, flat, and hexagonal objects with limited torque. Repair attempts should be avoided, as with slip-joint pliers.

Diagonal cutting pliers have sloping cutting edges but no actual jaws for gripping. They are primarily intended for cutting wire used in electronic, electrical, communication, and engine work. Other uses include working with cotter pins and cutting soft metal and nails. Hardened nails and hard metals will cause nicks and indentations in cutting surfaces. Sizes range from 4 to 7 in. in length. Several patterns are available from high leverage, heavy-duty styles to midget types for compact solid state electronic repairs (Fig. 4-16). End cutting nippers (Fig. 4-17) are designed for cutting wire and nails close to work.

Long nose pliers may be subdivided into round nose, chain nose and needle nose types. Round nose pliers are useful for forming wire loops necessary for electrical repairs. Chain nose pliers are available in extra long nose styles which are handy for gripping nuts or washers which fall into confined areas. Needle nose pliers are less bulky than chain nose pliers, and useful for very precise small-part retrieval. Some varieties are available in the

2-38 Establishing the Maintenance Facility

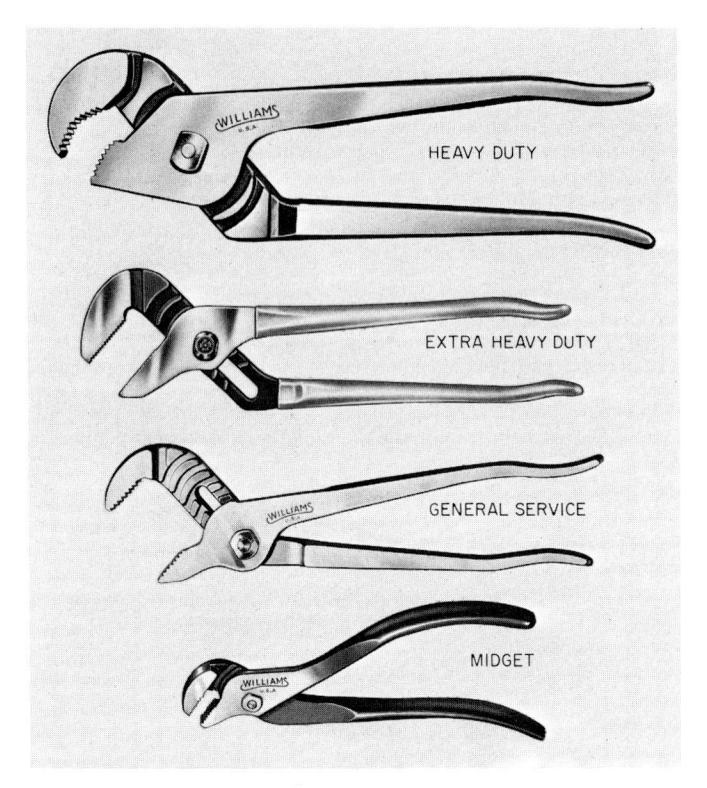

Fig. 4-15 Four styles of utility pliers.

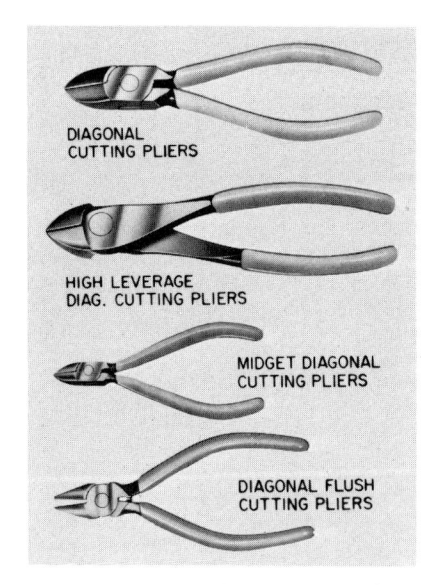

Fig. 4-16 Varieties of diagonal cutting pliers.

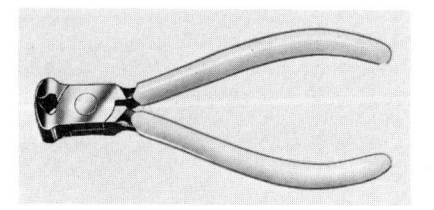

Fig. 4-17 End cutting nippers.

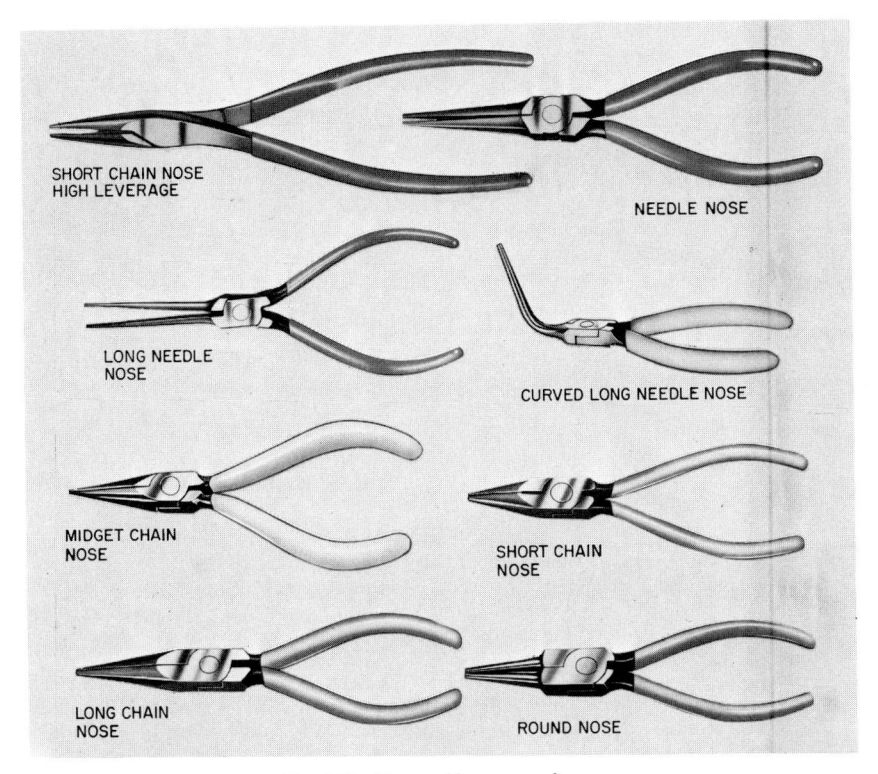

Fig. 4-18 Types of long nose pliers.

smaller midget sizes for electronic applications. Other variations include the curved nose type which bypasses work obstructions as shown in Fig. 4-18.

Duckbill pliers got their name from jaws that resemble the bills of ducks. Their wider, squared-off jaws have more surface contact than long nose pliers for larger objects. Variations include high leverage, short, long, and regular bills. Duckbill pliers with long, thin bills (Fig. 4-19) measure about 2⅝ in.

Pliers are made in a wide range of sizes to handle a variety of applications without strain. Hammering pliers in an effort to cut something is dangerous and damaging. Pliers should not be used as wrenches since rounded nuts often result. Once a nut is well rounded even a wrench is useless.

In selecting pliers consider the cushion handle grips for comfort. They may or may not be good insulators. Some pliers have spring opening jaws which may be useful. Maintenance is generally a commonsense situation. Keeping pliers free of dirt and grit plus lubrication of moving parts keep the tool from becoming stiff. Wiping with oil normally controls rust.

Wrenches The main function of wrenches is holding and turning nuts, bolts, cap screws, plugs, and various regular-shaped threaded parts. Wrenches fit these fasteners better than pliers, and when used correctly do not round corners. Excessive torque can damage threads, however.

Open-End Wrenches. These wrenches are most commonly found in the double head 15° angle type. The double head pattern nearly always has openings of two differing sizes for added flexibility (Fig. 4-20).

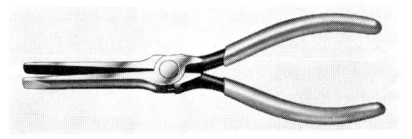

Fig. 4-19 Duckbill pliers with extra long bills.

2-40 Establishing the Maintenance Facility

Heads are offset by 15° to allow hexagonal nut rotation in a swing arc of 30° (Fig. 4-21).

As is common with other wrenches, the length of the open-end wrench is porportional to its opening. Since a larger nut requires greater torque for loosening or tightening than does a smaller nut, wrenches with larger openings are made longer than smaller wrenches. Open-end wrenches may be used as shown in Fig. 4-22.

It's simple to use an open-end wrench correctly. Select the size opening that fits the fastener closely. Apply the wrench so it's seated fully on the fastener with the handle directed to the right. Pull rather than push the wrench. Pulling the wrench toward the body keeps the tool more in control and

Fig. 4-20 Typical double head open-end wrench.

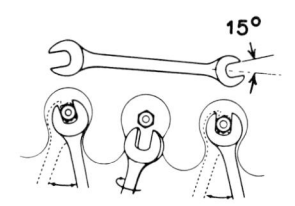

Fig. 4-21 15° angle open-end wrench allows rotation of hex nuts in a 30° swing by flopping the wrench.

thus is safer than pushing. When pushing outward there's a tendency for the arm to straighten, causing the wrench to slip off the fastener.

A sense of how much force is needed to tighten a fastener is acquired from working with a wrench. The farther out on the handle the wrench is gripped, the more torque can be applied.

Variations in open-end wrenches include single head patterns, extreme angled heads, and different shapes. Often, single head patterns are found in larger sizes. There are single head wrenches which have openings up to 7⅝ in. and weigh up to 150 lb. A typical single head pattern open-end wrench is shown in Fig. 4-23.

Open-end wrenches with different angle openings are available for working with small fasteners in cramped areas. Often there is the usual 15° angle at one end with a 60° or 75° angled head at the other (Fig. 4-24). These series of wrenches are very handy for electrical work on engines. The angled head allows access to a hard-to-get-at fastener (Fig. 4-25).

An example of a specially adapted open-end wrench for engine work is the tappet wrench. There are two variations, depending on the reach required. Both have the extra thin heads and 15° angle head. The double head pattern has the advantage of two openings in one wrench while the single head pattern has an extra long handle for difficult access situations and hot engines. Check nut wrenches are made thin for adjusting thin lock nuts used in machinery. These wrenches are made amply strong for their intended service and should not be used for severe service work.

Box Wrenches. These are so named because they totally enclose the nut, boxing it in. Box wrenches come in various configurations, but all are used for the final tightening or loosening. By totally enclosing the fastener, a surer grip is afforded than with an open-end wrench. Because a box wrench fits so securely, it takes a bit more maneuvering to secure the nut and can be slower to work with than an open-end wrench.

There are several variations of box wrenches, each with its own attractive characteristics. Double offsets provide maximum clearance between hand and fas-

Fig. 4-22 Using open-end wrenches on a flanged pipe-pump connection.

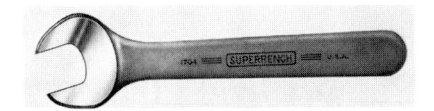

Fig. 4-23 Typical single head pattern open-end wrench.

Fig. 4-24 Open-end wrench with one head angled at 60°.

tener and come in long and short patterns. The long pattern with its long torque arm allows comfortable loosening and tightening torque. The short pattern, while requiring more effort, may be necessitated by a clearance problem. Where hand clearance is not a problem the 15° angle offset pattern box wrench is the best choice. This wrench style is effective in transmitting nearly all the user's effort into the nut because of the absence of the double offset. Again long and short patterns can be selected; the long pattern should be the choice where space permits as shown in Fig. 4-26

Openings of box wrenches may be of the 12-point and 6-point types. Points are the corners in the opening. The 6-point type is more popularly known as a hex box wrench. Either opening type is immediately recognizable by the points in the opening. In general maintenance, the 12-point opening is popular because the opening requires much less wrench maneuvering to get it on the nut compared with hex box wrenches. Hex box wrenches have some favor in the automotive maintenance areas.

Because of the secure fit that a box wrench provides, several variations cover a broad range of maintenance needs. The ratcheting box wrench (Fig. 4-27) is an attempt to provide the convenience of a ratchet wrench in the form of a box wrench. The one-way ratchet mechanism is used in reverse by flopping the wrench.

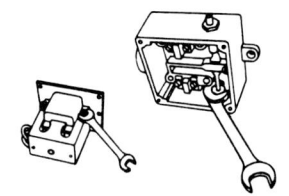

Fig. 4-25 Using 75° angle opening wrench in a close clearance application.

Single head box wrenches are available for larger fasteners. Where the job demands an extra-tough-duty tool, the heavy pattern with its extra strength is the better choice. The heavy pattern does have a weight penalty. For other applications the easier-to-use regular pattern may perform dependably (Fig. 4-28).

Still another variation of box wrench is especially useful when it comes to wrench safety. Fasteners which are frozen on or must be set tightly often tempt mechanics to hammer wrenches or add pipes to increase torque. Both practices are dangerous because in either case the wrench can be overpowered and may fail. The striking face box wrench comes to the rescue to solve the common problem of loosening rusted-on nuts. A striking face wrench has a heavy anvil designed for striking with a hammer or sledge. This style is available in two patterns with straight and offset (Fig. 4-29).

The straight pattern is the most effective but is only useful in applications of suitable clearances. Often the offset pattern is necessary because of obstructions. An offset striking

Fig. 4-26 Typical long 15° angle offset box wrench.

Fig. 4-27 Ratcheting box wrenches are available in a range of openings.

2-42 Establishing the Maintenance Facility

Fig. 4-28 Tightening a valve nut with a regular pattern box wrench.

face box wrench (Fig. 4-30) is needlessly used where the regular pattern would have been more effective.

A special box wrench suited for work on air and hydraulic lines and tubing is the double head split box wrench. Just as the name implies, there is a split or section removed from the complete box which allows application to a hose fitting. While it is possible to make such a wrench from a box wrench, the resulting wrench is seriously weakened and may break on a tough application.

Combination Wrenches. These combine one end of an open-end wrench with a box wrench. In one wrench, the user has the quick tightening capability of the open-end wrench plus the secure final tightening capability of the box wrench. As with other wrenches, long and short patterns are available. A long pattern series combination wrench is shown in Fig. 4-31 and an electronic combination wrench is shown in Fig. 4-32. Flare nut combination wrenches with an open-end wrench at one end and a 6-point split box wrench opening at the other end are available in several sizes.

A wrench style becoming more popular in general maintenance is the flex/open-end wrench. One end has open jaws, while the other has a socket member which swivels 180°. The flex/open-end wrench (Fig. 4-33) has the same size opening at either end.

Adjustable Wrenches. The adjustable wrench (Fig. 4-34) is very useful. Adjustable wrenches are single-end open-end wrenches with a movable jaw, sized from 4 to 24 in. in length with a maximum capacity of nearly 2½ in. Because of its adjustable feature, this one tool can handle a metric or conventional fastener without guesswork. Adjustable wrenches are not built for excessive hard work and are inclined to slip off fasteners easier than standard wrenches. Take care that the pulling force is always applied

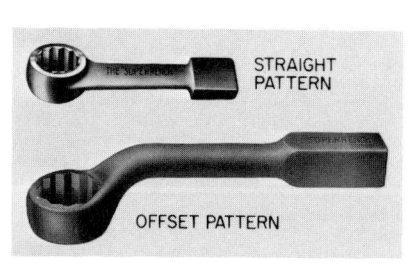

Fig. 4-29 Straight and offset striking face box wrenches.

Fig. 4-30 Using an offset striking face box wrench isn't as effective as the straight pattern.

Fig. 4-31 Typical long pattern combination box–open-end wrench.

through the upper or fixed jaw. The adjustable jaw must always be turned up close to the work so that a firm contact is established between the fastener and the jaws of the wrench.

When a nut is turned with a conventional adjustable wrench, there is a tendency for the jaw setting to loosen. Adjustable wrenches with a locking feature that retains the tight jaw setting reduce the tendency for rounded corners on nuts (Fig. 4-35). By pushing the pin the opening is locked into place.

Adjustable wrenches are available with cushion grips and insulated handles, but not all cushion grips are good dielectric material. When an adjustable wrench with an insulated handle is required, be sure the insulating material is adequate. The insulated handle shown in Fig. 4-36 is rated for up to 5000 volts rms.

Since there are moving parts in adjustable wrenches, repairs to these wrenches should be expected. Most are easily repair-

Fig. 4-32 A combination box–open-end wrench popular in electronic maintenance.

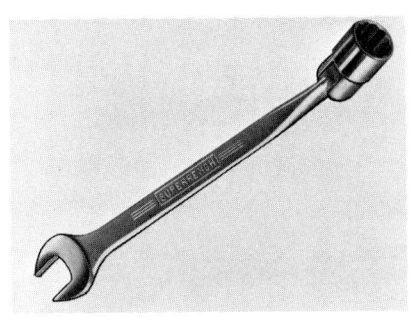

Fig. 4-33 Flex/open-end wrench is gaining favor in maintenance.

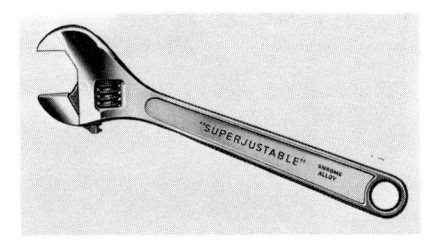

Fig. 4-34 Typical adjustable wrench.

able in the field. When selecting adjustable wrenches note the maximum jaw opening. Certain series have a larger capacity for a given length over others, enabling one wrench to meet the needs of two.

While adjustable wrenches may be the boon to the home mechanic, maintenance professionals recognize that an adjustable wrench cannot substitute for open-end wrenches or box wrenches.

Socket Wrenches, Drive Tools, and Accessories. These satisfy a wide variety of maintenance needs quickly and efficiently. Sockets and related tools are available in five drive sizes: $1/4$, $3/8$, $1/2$, $3/4$, and 1 in. Whether the sockets are of metric or fractional openings the five drive sizes apply to

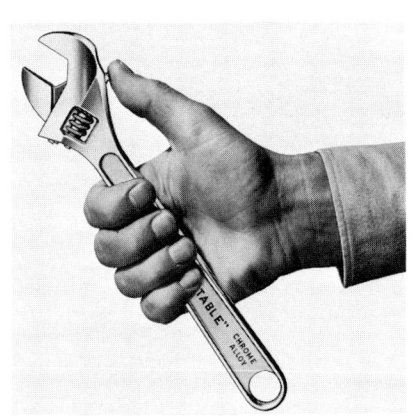

Fig. 4-35 Adjustable wrench with locking feature. Just a push of the pin locks the setting in place. Opening position is unlocked with a push upward from the bottom.

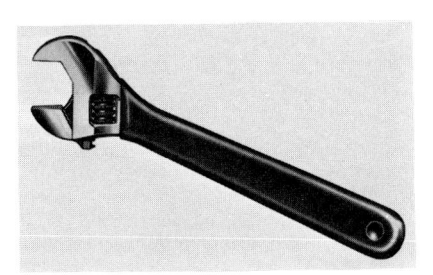

Fig. 4-36 Adjustable wrench with insulated handle rated for 5000 volts rms. Not all handle coverings are good insulators.

2-44 Establishing the Maintenance Facility

Fig. 4-37 Typical socket wrenches.

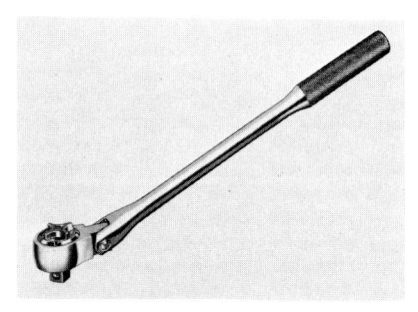

Fig. 4-38 This 3/8-in. drive flex-head ratchet flexes to help bypass obstructions.

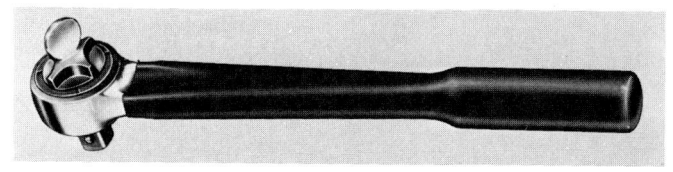

Fig. 4-39 Insulated-handle ratchet with shifter that is easy to use even when wearing gloves.

Fig. 4-40 A 1-in. drive flex-handle driver.

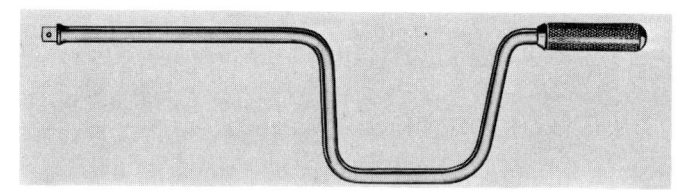

Fig. 4-41 Speeders are sometimes overlooked as the solution for quick nut turning.

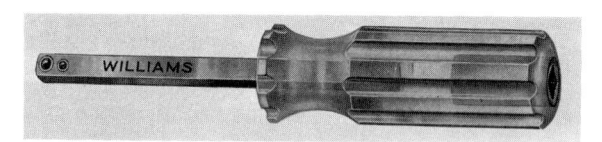

Fig. 4-42 Speeder handles move 1/4-in. drive sockets quickly.

either series. Sockets themselves are thin-walled metal cups with teeth on their inner surfaces made to fit over hexagonal nuts in the same way as the enclosed jaws of box wrenches.

The widest variety of sockets are in the ¼- to ½-in. drive sizes consisting of regular 6-, 8-, and 12-point varieties, and 6- and 12-point extra-deep versions. On the ¾- and 1-in. drive sizes 12-point regular sockets are the most common. Typical sockets are shown in Fig. 4-37. For nuts with long, protruding bolts, an extra-deep socket may be required as also shown in Fig. 4-37. Extra-deep sockets sometimes eliminate the need for a short extension as well. Regular 6- and 12-point sockets seat rapidly and the 12-point style requires only half the swing of hex sockets when used with nonratcheting drivers. Sockets in 1-in. drive have a safety attachment device for positive locking with drivers. Sockets of 6-point size should be used on undersize or badly worn nuts. Regular 4- and 8-point sockets are designed expressly for square nuts and bolt heads, such as found on set, machine, and lag screws, stove bolts, and nuts. Made in ¼-, ⅜-, and ½-in. drives with openings up to 1 in., sockets usually provide a safer, surer grip than open-end wrenches.

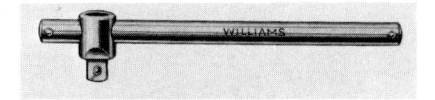

Fig. 4-43 T handles provide more torque than speeder handles in 1/4-in. drive.

The selection of drivers is quite extensive for ¼-, ⅜-, and ½-in. drive sockets. Ratchets have been the most popular type because of their quick resetting action. A ratchet is much faster than other drivers because it does not have to be lifted from the work after each swing. The ratchet is simply moved back and forth while the driving mechanism rotates the socket in the direction selected by a shifter or reversing lever.

When working in confined areas with a ratchet, the amount of swing that a ratchet requires determines whether it can be used. For example, a ratchet that requires an arc of 27° in order to operate cannot be used where clearance would only allow a 20° swing. Ratchets with swing arcs as small as 16° are available and hasten work in confined areas. When access is difficult often a ratchet with a flex head helps clear the way to allow handle swing. An example of a ⅜-in. flex-head ratchet is shown in Fig. 4-38. This style has been a favorite for certain types of engine work. Ratchets are available with special shifters and insulated handles. Special shifters make it easy to reverse directions when wearing gloves. A ½-in. drive ratchet with an insulated handle and special shifter is shown in Fig. 4-39.

While a ratchet is a very popular driver, other drivers save time when nuts must quickly be moved on or off a long threaded area. Flex handles are available in all drive sizes. A flex handle is a handle attached to a pivoted tang which drives the socket. The handle may be pivoted through a wide arc. A 1-in. drive flex handle is shown in Fig. 4-40. Speeders move nuts quickly and are available in the smaller drive sizes. Just as their name implies speeder handles speed up nut turning. A ½-in. drive speeder is shown in Fig. 4-41. In the small ¼-in. drive size, speeder handles shown in Fig. 4-42 and sliding T handles shown in Fig. 4-43 are good drivers for low torque needs.

One reason that sockets are so useful in maintenance work is the availability of drive accessories. Drive accessories bypass obstructions that keep flat wrenches from the work. These accessories include extension bars, flexible extensions, universal joints, and adapters. With a selection of these drive accessories in the maintenance tool section, nuts that are buried in obscure areas are generally accessible. Often just a simple extension allows a driver to clear a congested area. In drive sizes up to ½ in., extensions are available from about 2 in. (5.1 cm) to 2 ft (61 cm). Using a single extension of the proper length is safer than using two shorter extensions. Often it takes some experimenting to find the combination that works. Once a workable combination is found it's worth the extra time to examine the drive combination with an eye to simplifying the number of joints, if possible.

A universal joint such as shown in Fig. 4-44 is a good way to avoid an obstruction. It should only be used up to angles of 65°; the less the angle, the more efficient the operation. Universal joints should not be used at right angles, and this type of work should be done with flex handles. Universal sockets, shown in Fig. 4-45, flex in the same way as universal joints and can work the nut in tightly confined areas as shown in Fig. 4-46.

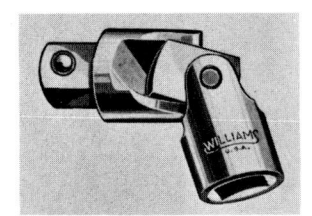

Fig. 4-44 Universal joints like this one flex to avoid obstructions.

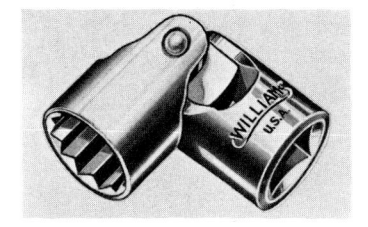

Fig. 4-45 Universal sockets have a built-in universal joint.

2-46　Establishing the Maintenance Facility

TABLE 4-1 Safe Use of Adapters

Originating drive size, in.	Can be adapted down to drive sockets in this drive size, in.	Can be adapted up to drive sockets in this drive size, in.
1/4	—	3/8
3/8	1/4	1/2
1/2	3/8	3/4
3/4	1/2	1
1	3/4	—

Flexible extensions, normally available in smaller sizes, conveniently bend in just about any direction. Flexible extensions should be bent to the minimum for effective turning.

Adapters allow the intermixing of drive sizes. Adapters are for use on sockets with attachments of larger or smaller drive sizes as shown in Table 4-1. The full leverage of drivers should not be used when adapting to sockets of a smaller drive size. Intermixing of drive elements is generally done as an alternative when enough drivers in the same size are not available to meet a need. Once again having the proper selection in the needed drive size saves time on the job.

Nuts in extremely tight spots may become accessible in still another way—by coupling crowfoot attachments with extensions and drivers, such as flex handles. Made in 3/8-in. drive only, with openings from 3/8 in. (0.95 cm) to 1 1/16 in. (2.7 cm), the crowfoot attachment may be the only safe solution as shown in Fig. 4-47.

Ratchet action can be added to a nonratcheting driver with a ratchet adapter. A typical arrangement using a ratchet adapter is shown in Fig. 4-48. In an assembly using a ratchet adapter, it should be placed as close to the operator as possible for safe, convenient access to the shift lever.

Stud extraction may be accomplished by using a stud remover. A typical setup is shown in Fig. 4-49. Liberal use of penetrating oil should be a prerequisite of stud removal.

With the sophistication of drivers and accessories available it just makes sense that there should be bits available to turn other fasteners besides nuts. Hex and screwdriver bits (Fig. 4-50) allow speedy turning of hollow-head, slotted, and Phillips screws. Adapter sockets are made in 1/4-, 3/8-, and 1/2-in. drives.

Chrome sockets are meant to be driven by hand drivers. They should not be used with impact guns. Black-finish power drive sockets are designed to meet the needs of a power driver. They are made of extra-tough alloy steel, specially heat treated to withstand the constant shock and pounding involved in electric and pneumatic impact nut setting and power nut running.

Surface drive sockets are designed to contact the flats rather than the corners of hex

Fig. 4-46 Universal sockets get at tightly confined nuts.

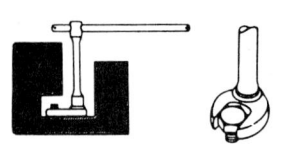

Fig. 4-47 Using a crowfoot attachment in a tightly confined area.

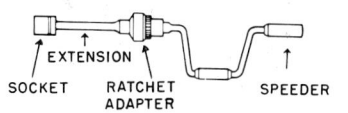

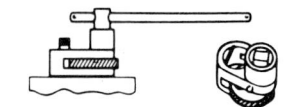

Fig. 4-48 The correct way to use a ratchet adapter with a speeder.

Fig. 4-49 Using a stud remover.

nuts; they have the advantage of self-seating as required on multiple nut runners. Magnetic styles are used for self-tapping screws. Power drive sockets and accessories have roughly the same appearance, as hand-drive sockets.

The moving parts of socket wrench drivers will perform better when occasionally lubricated. Ratchets may be immersed in light machine oil overnight. Rotate or spin ratchet to flush out dirt or hardened lubricant and to remove excess oil. Occasional lubrication of the flex joint of flex handles will reduce wear.

Fig. 4-50 Hex and screwdriver bits and sockets.

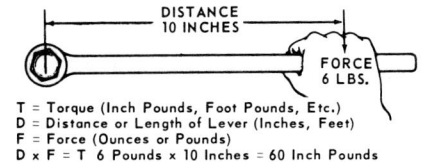

T = Torque (Inch Pounds, Foot Pounds, Etc.)
D = Distance or Length of Lever (Inches, Feet)
F = Force (Ounces or Pounds)
D x F = T 6 Pounds x 10 Inches = 60 Inch Pounds

Fig. 4-51 A visual definition of torque.

Torque wrenches One of the most costly hand tools of maintenance is the torque wrench. Because of its precision construction it is often considered an instrument rather than a tool. Rough handling can ruin the calibration, so calibration checks should be made on a scheduled basis.

When using any style of torque wrench, a firm, steady pull should be applied. Jerking motions could cause inaccurate reading. Most wrenches have a grip for grasping.

An understanding of torque helps to eliminate the confusion over just what a torque wrench does. Torque is technically defined as a force working through a distance.

Units of torque may be newton-meters or foot-pounds. Torque is distinctly different from a force which is measured in pounds per square inch or kilograms per square centimeter. As shown in Fig. 4-51, torque measurement is figured in terms of distance and force. Distance is the length of the lever or wrench. Force is the amount of pull or push applied to the end of the lever or wrench.

All wrenches apply torque but the simple difference between an ordinary wrench and a torque wrench is that a torque wrench indicates torque by sight or sound.

A popular style of torque-indicating wrench is the dial torque wrench (Fig. 4-52). There

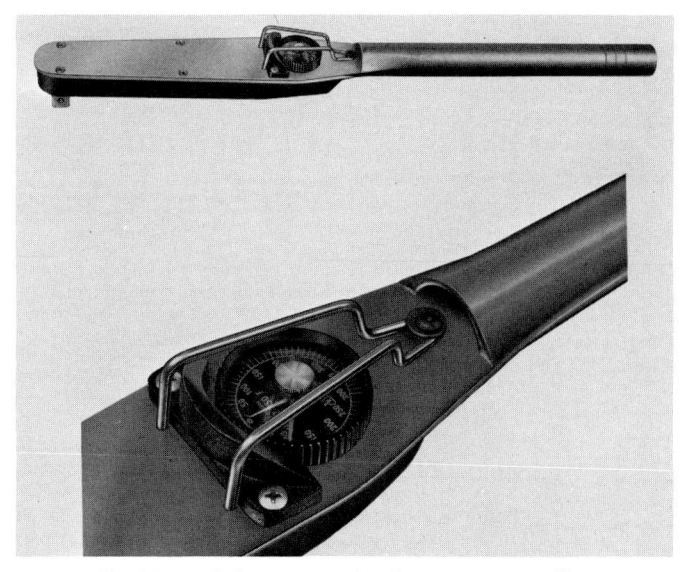

Fig. 4-52 A dial torque wrench indicates torque visually.

2-48 Establishing the Maintenance Facility

are several variations, each usually available in one or more drive sizes. Drive sizes for either metric or U.S. Customary calibrations are ¼, ⅜, ½, ¾, and 1 in.

One variation of the dial torque wrench is the single-pointer dial style. In use, the pointer indicates the actual torque applied and returns to zero as the grip is relaxed on the wrench. The follower-needle style has an additional pointer which remains at the highest torque obtained. This feature is almost a necessity at times when viewing the dial is difficult. Another very useful feature when low light levels or obstructions hinder direct viewing is the electric light dial. With this feature, an adjustment may be made to allow a built-in electric light to flash when the desired torque is reached. Add-on ratchet mechanisms are also available. These adapters speed torquing in cramped quarters.

Certain maintenance personnel prefer the deflecting-beam style of torque measuring wrenches. These wrenches are known for their ability to withstand hard use. When torque is applied to the wrench, a beam deflects relative to a floating bar. The wrench is calibrated such that the amount of deflection can be read directly as torque through use of converging scales. The torque is read where a particular scale gradation and reading edge

Fig. 4-53 An adjustable torque limiting screwdriver.

intersect, so the wrench is said to be of the deflecting-beam converging-scale type. Some models are available with an adjustable sound setting which emits an audible click when the desired torque is reached. Deflecting-beam wrenches are available in ⅜- and ½-in. drive sizes with plain and ratcheting heads.

Torque limiting wrenches are useful in a number of applications where visual indications are not possible or desirable. This style of wrench has a provision for setting a torque value which, when reached, causes an audible snap; it is available in a wide variety of drive sizes. Ratcheting-head or plain-head models are available in preset or adjustable micrometer-calibrated style.

Within a given drive size there is often a choice of calibrations. The calibration that most suitably reflects the need of the job at hand makes reading the torque easier.

Adjustable torque limiting screwdrivers (Fig. 4-53) solve the problem of torquing screws to requirements.

Torque multipliers Torque multipliers provide the operator with nut-turning power that uses only a fraction of the force required when using conventional tools. They offer

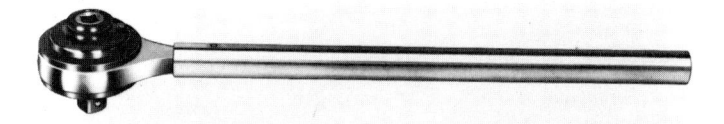

Fig. 4-54 Torque multiplier.

safe, convenient extra turn power when the operator is confronted with the need for unusually high torque requirements within a limited amount of working or leverage space.

A torque multiplier (Fig. 4-54) is a device or tool often referred to as a gear-head wrench which, through the use of a planetary gear train contained in a metal housing, increases or multiplies the output of a drive tool such as a ratchet or ratcheting torque wrench.

Uses. In practice, torque multipliers are used to turn fasteners where working space is limited and unusually high torque must be applied.

Incorporated in the torque multiplying unit is a female square drive opening for engaging a torque wrench or ratchet, a square drive tang to which a socket is attached, and a fitting to accept a reaction bar (a metal bar or rod used to brace the multiplier against a stationary supporting object).

A torque multiplier is generally needed when
1. A predetermined torque measurement must be applied to bolts or nuts having a diameter of 1 in. or more.
2. Bolts or nuts must be tightened to a predetermined torque in a working space that prevents the use of a torque wrench with a long handle.

In use, a standard socket of appropriate size is placed on the male square drive tang. A suitable ratchet or ratcheting torque wrench is inserted in the female square drive input of the torque multiplier as shown in Fig. 4-55. With this assembly placed on the fastener to be turned, it must be remembered that the input and output rotation will be in the same direction. Rotation of the reaction bar will be in the opposite direction. The reaction bar, therefore, must rest securely against an object sturdy enough to withstand the force that will be generated.

Selection. The procedure for selecting a torque wrench/torque multiplier combination is as follows:
1. Determine the torque required.
2. Measure the swing space.
3. Select a torque wrench of sufficient capacity for it to have a lever or handle length and/or a ratchet mechanism to allow a handle swing operable within the available swing space.
4. Choose a torque multiplier that, in combination with the torque wrench, will deliver the desired amount of torque when a force of approximately 100 lb is applied by the operator.
5. Locate a stationary object of sufficient strength and rigidity against which the reaction bar may be conveniently braced.

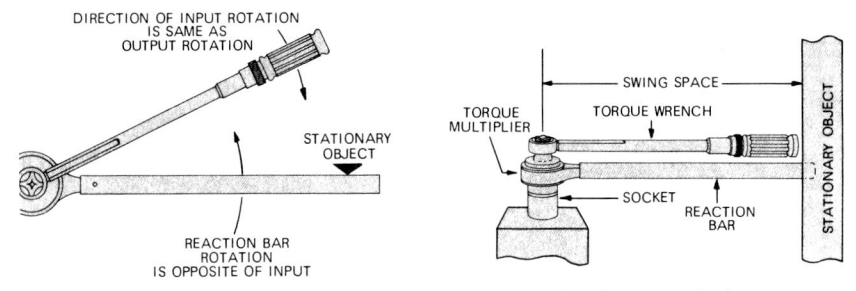

Fig. 4-55 A torquing application using a torque wrench and torque multiplier.

When extreme torque values are required and space or manpower is limited, two or more torque multipliers may be used in tandem. In this case, the net advantages of all or both multipliers are multiplied to arrive at the total net advantage of the stack. Care should be taken not to exceed maximum input torque values.

In a tandem torque multiplier arrangement, an antibacklash tool often solves the windup problem usually experienced. As torque builds, an ever-increasing swing arc becomes necessary to continue the turning process. An antibacklash tool acts as a storage dog to absorb this effect and allow instant turning, without unwinding the setup each time. Since constant pressure is being held, the setup maintains rigidity so that only a short swing arc is required to complete the turning process swiftly and securely. A setup for hopper car maintenance is shown in Fig. 4-56. The antibacklash tool is the L-shaped member.

Torque wrench testers Manufacturers of torque tools generally provide a calibration service through which torque wrenches may be returned to the factory for recalibration. Often maintenance needs do not permit the necessary time for the factory procedure. Torque wrench testers are available which quickly test the accuracy of any manual torque wrench. Typical of the torsion-style torque testers, the unit shown in Fig. 4-57 is available in U.S. Customary or metric calibrations in several ranges. In use, the torque wrench to be checked is attached to the tester using the proper adapter. Torque is applied in a clockwise direction, slowly increasing to the desired value. Readings of the tester and wrench are compared. The pointer is returned to zero and the test is repeated. Findings

2-50 Establishing the Maintenance Facility

are compared to determine if adjustment to the wrench is necessary. Testers are capable of close accuracy (about +1 percent, or one scale gradation, whichever is greater).

When maintenance manuals indicate a torque value there is usually good reason for doing so. Head bolts on engines and compressors have a certain degree of tightness required for correct operation. Certain alloy metals have a limitation above which threads may be changed. Tanks and other pressure vessels require a specific nut tightness for proper sealing. Fasteners themselves have distinct limitations, depending on size and style of nuts and bolts. A correctly torqued fastener can make or break the operation of countless devices no matter how the other aspects of maintenance are performed. When the question is: How tight is right?, the torque wrench is usually the solution.

Other wrenches Other wrenches used in maintenance are traditional tools which are popular with certain industries like railroads and shipyards with specialized toolrooms; they may not be encountered as often as the previous tools.

Set screw wrenches with single and double heads are intended for set screws and for

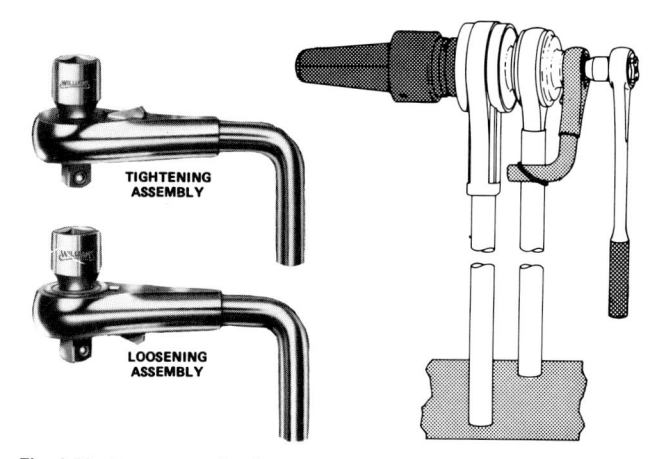

Fig. 4-56 Using an antibacklash tool in a tandem torque multiplier setup.

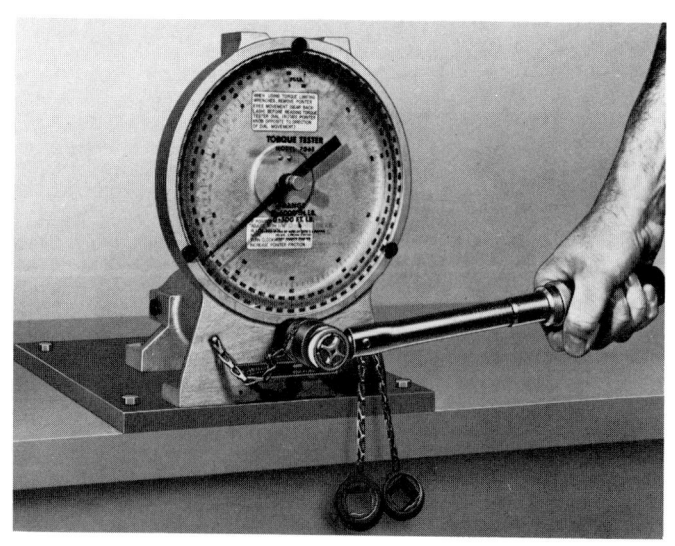

Fig. 4-57 Checking the accuracy of a torque wrench with torque wrench tester.

square-head cap screws and nuts. The head angles (Fig. 4-58) are angled at 22½°. Head heights of set screws are greater than ordinary bolt heads and require wrenches of more than average head thickness (Fig. 4-59). Jigs, fixtures, and machine tools involving frequent adjustment most often use hardened set screws. Using a set screw wrench on set screws usually means greater safety and longer life than with regular open-end wrenches.

Another wrench used where square nuts are found is the S or car wrench (Fig. 4-60). Square nuts require a 90° rotation with most open-end wrenches. The S design permits flopping the wrench to reduce the swing to 45°.

Tool post and square box wrenches are intended for use with square-head nuts. Machine tools and industrial equipment are often supplied with wrenches of this type where numerous set screw adjustments are continuously required. A typical tool post wrench (Fig. 4-61) has its head thickness proportioned to grip the full height of screw heads regardless of size. Substitution with regular open-end wrenches should be avoided.

T handle socket wrenches and offset socket wrenches (Fig. 4-62) are becoming somewhat less popular in maintenance, since detachable sockets with fast-acting drivers may

Fig. 4-58 Set screw wrench with double head.

Fig. 4-59 A set screw wrench is about as thick as a set screw.

Fig. 4-60 By flopping an S wrench, square nuts can be rotated in 45°.

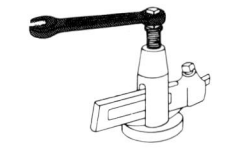

Fig. 4-61 A tool post wrench.

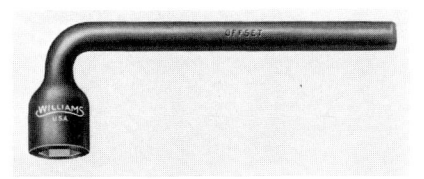

Fig. 4-62 Offset socket wrench.

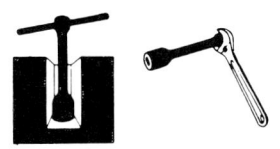

Fig. 4-63 A T handle socket wrench reaches recessed nuts. Shank may be turned with a wrench for extra torque.

be easier and quicker to use in many applications. Nevertheless, the madeup configuration is useful on flanges, caps, and assemblies of many kinds which have unavoidable obstructions (Fig. 4-63). Shanks do have hexagonal ends which permit wrench application for extra leverage. Avoid using pipe for extra leverage. Made with hex or square openings, the offset socket wrench has the correct shape for obscure nuts (Fig. 4-64). Handle lengths are proportional to opening sizes to provide adequate leverage. Extending with pipe is not recommended. A socket with a long flex handle may be a good alternate.

The distinctive appearance of the bulldog wrench allows it to grip pipe, square, hex, or round shapes or any others that will fit between their strong, tough jaws. They are often used for holding while nuts, sleeves, collars, and connectors of various types are being turned. Still popular with railroads and shipyards, this wrench should not be used after teeth have become rounded. The diameter range of bulldog wrenches is from ⅛ in (0.32 cm) to 3½ in. (8.87 cm). Both single and double head patterns are available (Fig. 4-65).

2-52 Establishing the Maintenance Facility

Spanner wrenches vary widely in appearance and are often needed in servicing areas around rotating shafts. Popular versions (Fig. 4-66) include adjustable-face, hook (fixed and adjustable), and pin types (fixed and adjustable). Pin, face, and hook spanners are made to many standard dimensions for turning packing glands and adjustable collars (Fig. 4-67). Adjustable spanner wrenches service several sizes of adjusting collars, locknuts, rings, and bearings (Fig. 4-68).

Hammers There's more to using a hammer safely than just to beat away at something. The often-used ball peen hammer (Fig. 4-69) is easily mistreated. Striking a hardened surface such as another hammer's striking face invites shattering of one of the faces. A ball peen hammer is meant for surfaces such as chisels and punches and for riveting, shaping, and straightening unhardened metal. When striking another tool such as a chisel, the hammer face should be about 3/8 in. (0.95 cm) larger in diameter than the face of the

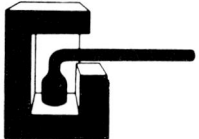

Fig. 4-64 An offset socket wrench is the right choice to reach a nearly hidden nut.

Fig. 4-65 Bulldog wrenches can grip rounds.

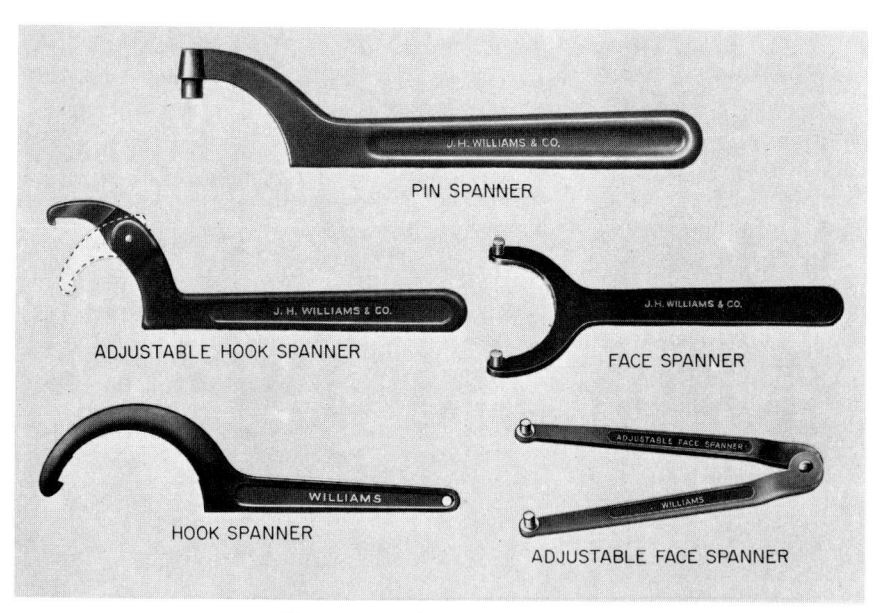

Fig. 4-66 Popular spanner wrenches.

Fig. 4-67 Typical applications for face and pin spanner wrenches.

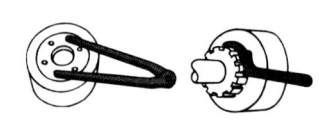

Fig. 4-68 Adjustable spanner wrenches handle special needs.

Basic Hand Tools of Maintenance 2-53

struck tool. Using a hammer of the correct weight class to match the heaviness of the blow conserves effort, adds control of the hammer, and improves the integrity of the struck members. Ball peen hammers are available in eight weight classes to provide the heaviness of blow needed for the job. A hammer blow should be struck squarely so that the full face of the hammer meets the struck surface. Glancing blows should be avoided.

Safety goggles protect eyes against flying metal that may result from using a ball peen hammer. Any hammer head which shows signs of mushrooming (spreading in a mushroom-shaped pattern) or which develops cracks should be discarded. Hammers with loose heads or damaged handles can be safely repaired with a new handle.

Soft-face hammers (Fig. 4-70) are intended for striking blows where surfaces may be damaged by the hardness of a metal-faced hammer. Plastic soft-face hammers have replaceable heads in six grades of hardness to meet needs of rubber, wood, lead, fiber, and rawhide, as shown in Table 4-2. One hammer face can be fitted with a soft tip, the other with a medium tip, just by screwing in the color-coded tip as shown in Fig. 4-71. Soft-face hammers are available in about a dozen size and weight classes, including shot-loaded types which provide more driving power with less effort. Soft-face hammers are often the only answer for seating parts, straightening shafts, and metal forming.

Chisels Chisels for metal working include the cold chisel, cape chisel, round-nose chisel, and diamond-point chisel (Fig. 4-72). The cold chisel, with its flat point, is the most common tool in the chisel family. Cold chisels are measured by the width of the point, which ranges from $\frac{1}{4}$ in. (0.63 cm) to $1\frac{1}{8}$ in. (2.85 cm). A cold chisel is normally struck with

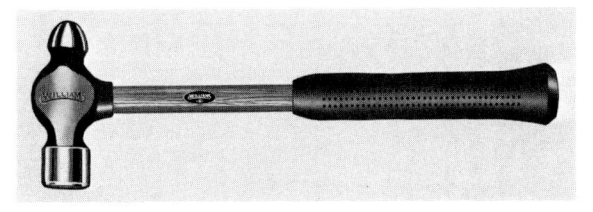

Fig. 4-69 A ball peen hammer with fiber glass handle.

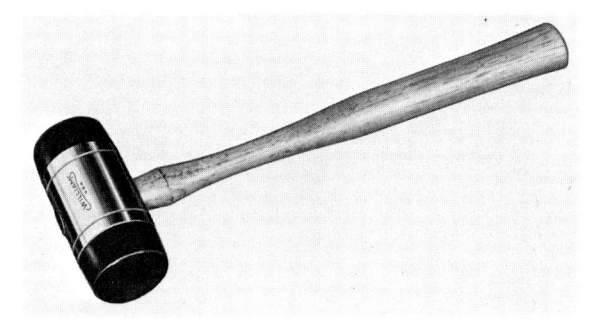

Fig. 4-70 A soft-face hammer with replaceable plastic tips.

a ball peen hammer. Small cold chisels are held between the thumb and fingers. Larger chisels may be held by the fist. Generally, a hammer in the weight class of 12 oz (0.34 kg) is a good choice. Striking should be done by a series of quick blows while wearing eye protection and keeping attention at the blade rather than at the striking face. It does take some practice to gain enough confidence to be sure that the hand holding the chisel is clear of the hammer blow. Cold chisels are effective for chipping, cutting metal parts like screw heads, and splitting rusted-on nuts. A chisel is effective when the work is securely held and firmly blocked so the work receives the impact of the hammer blow. After each blow, the result should be noted and the angle of the chisel adjusted to get the depth of cut required. When starting, it's often good to get a feel of the hardness of the material by

TABLE 4-2 Six Replaceable Tips Meet Soft-Face Hammer Needs

TYPE OF HAMMER	SOFT	MEDIUM	TOUGH	MED. HARD	HARD	X-HARD
SOFT RUBBER	BROWN					
WOOD	BROWN	RED		CREAM		
RUBBER		RED				
HARD WOOD			GREEN			
LEAD			GREEN	CREAM		
PLASTIC			GREEN		BLACK	
RAWHIDE		RED	GREEN	CREAM	BLACK	YELLOW
MICARTA					BLACK	YELLOW
FIBER					BLACK	YELLOW
COPPER						YELLOW

using the chisel at a long angle, nearly parallel with the work. A few test blows set the pattern of what to do next. Sometimes an application of oil makes the cutting surface easier to drive when working on steel. When chipping, the operator should be positioned so that flying objects are directed safely away from people. Eye protection is a must. Cold chisels may be resharpened with a file or whetstone by an experienced workman.

The diamond-point chisel has a diamond-shaped face for use in corners, grooves, and similar shapes where material must be removed. The V-shaped blade of the cape chisel allows access to keyways and slots. Curved recesses and filleted corners can be handled with a round-nose chisel. With any chisel, eye protection is a must.

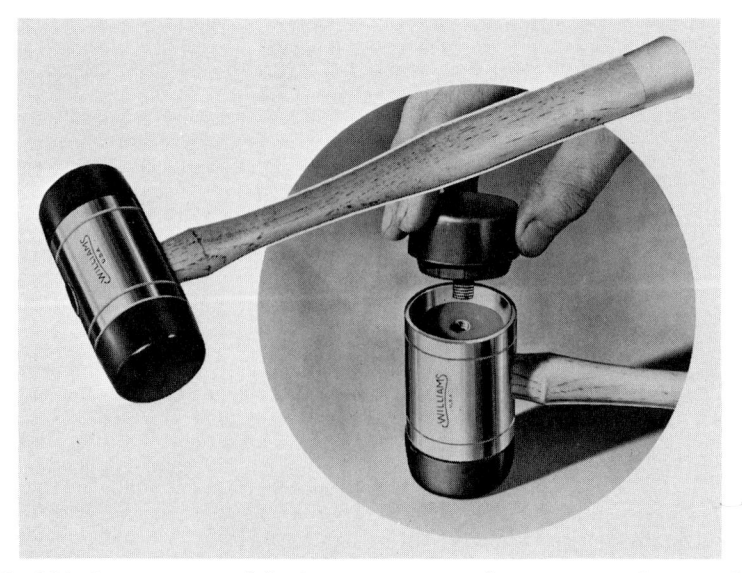

Fig. 4-71 Screw-in tips on soft-face hammer customize softness to meet nearly any need.

Punches Punches are normally available in five styles. They are used much like chisels and are usually struck with a ball peen hammer. A punch should be held perpendicular to the work and struck with a sharp blow; inspect progress after each blow. A solid punch (Fig. 4-73) is able to take the heavy blows required to knock out deheaded rivets and pins which position machinery parts. Once the work has been moved, the taper of the solid punch prohibits further use and the pin punch (Fig. 4-74) is used to complete driving out the object. The slender profile of the pin punch can't take the heavy blows a solid punch can. Select the largest diameter of either solid or pin punch that will fit. Both are available in at least five sizes, starting with a point diameter of 3/32 in. (2.38 mm).

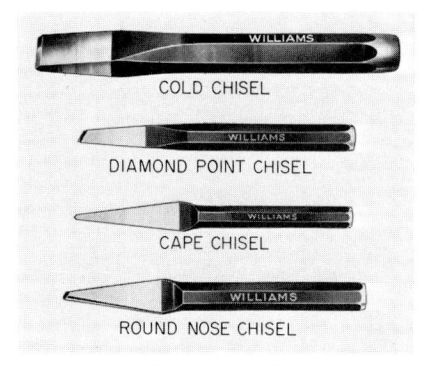

Fig. 4-72 Often-used chisels.

A long taper punch (Fig. 4-75) is used for lining up or shifting parts to locate a common hole where a bolt or pin is to be placed. The long taper limits the abuse it will take without failure. Often a screw or pin alone cannot be fitted into place and perform needed alignment at the same time.

A center punch (Fig. 4-76) is used primarily to make an impression for drilling in metal. The indentation keeps a drill from wandering over the surface and assures the hole will be drilled where intended. The center punch produces a wide shallow dent and is available in four sizes.

Primarily a layout tool, the prick punch (Fig. 4-77) is used for marking holes and scribing lines on sheet metal surfaces. It is smaller and has a much sharper point than a center punch.

Pullers Pullers are special tools for removing rotating elements from shafts, often pulleys and gears. Two- and three-jaw pullers (Fig. 4-78) come in six sizes to meet the capacity and spread necessary to remove the part desired. Specialized pullers include timing gear pullers, steering wheel pullers, and harmonic balancer pullers. For refrigeration maintenance service there is still another version. Most pullers have provisions for pulling gears with tapped holes as well as force fits. Jaws are fitted to the part to be removed and the jacking screw is tightened with a wrench. Three-jaw pullers often have an advantage on tough pulling jobs with the improved force distribution ability of the third jaw.

Tool holders and blades The ingenuity of maintenance personnel is called on when a part must be reshaped or completely remade to get a machine operational. When a lathe is needed, a selection of tool holders and tool bits provides a range for cutting and shaping.

A tool holder is a forged piece of metal which holds the tool bit in contact with the

Fig. 4-73 A solid punch can take heavy blows used in starting pin removal.

Fig. 4-74 A pin punch has a slender body to fit into holes vacated by pins.

Fig. 4-75 The long taper punch is suited to alignment.

Fig. 4-76 A center punch produces a shallow indentation in metal.

Fig. 4-77 A prick punch is used as a layout tool in sheet metalwork.

2-56 Establishing the Maintenance Facility

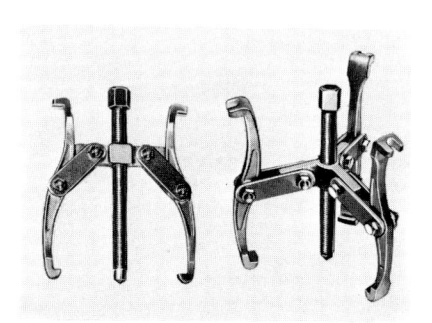

Fig. 4-78 Two- and three-jaw pullers are available in a wide range of sizes.

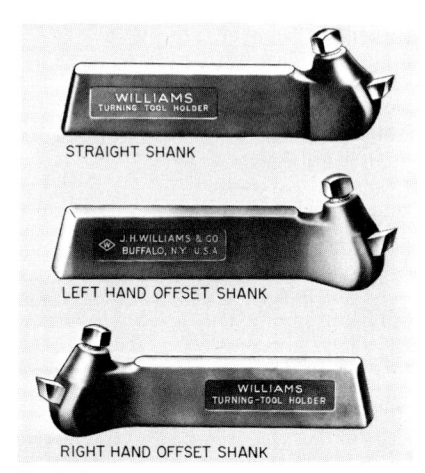

Fig. 4-79 Turning tool holders with tool bits in different shapes.

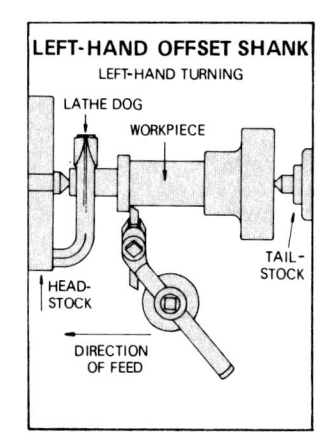

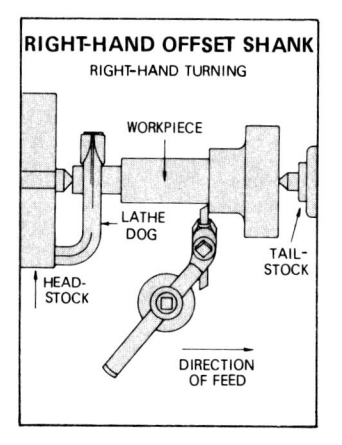

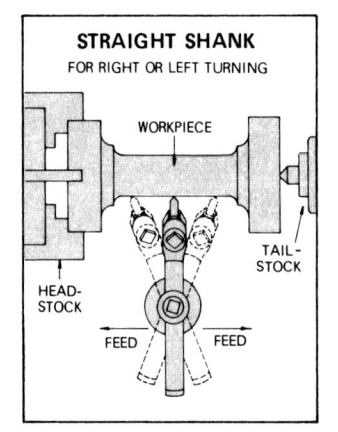

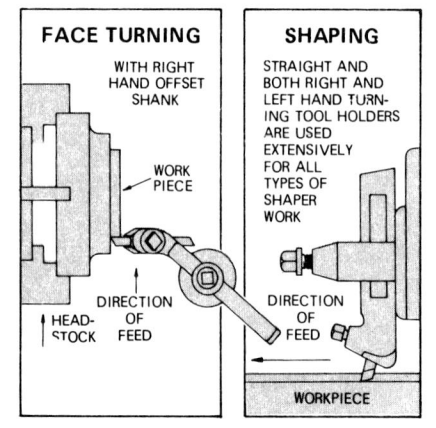

Fig. 4-80 Applications of turning tool holders.

workpiece. For metal turning and shaping, a turning tool holder (Fig. 4-79) is used. Turning tool holders are available in straight shank or in right and left offsets. Typical applications of turning tool holders are shown in Fig. 4-80. Cutting-off and side tool holders are available in the usual rigid style and in the spring version (Fig. 4-81). The spring-head version provides relief from any sudden or excess pressure of the cutting-off blade which it holds. Even an inexperienced operator can cut off work considerably out of round without chatter, climbing, or damage to tool or work. The shock-absorbing qualities of the spring tend to protect lathe bearings and journals and to improve the quality of work done on old and loose lathes. Typical applications of straight and offset versions are shown in Fig. 4-82.

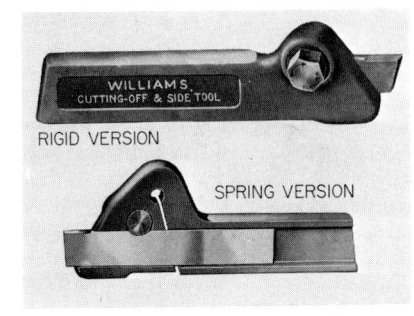

Fig. 4-81 Rigid style and spring style cutting-off and side tool holders in straight shanks.

Restoration of a knurled pattern requires knurls and a knurling tool holder. Popular patterns include the diamond and straight line grooves. Knurling is often used to increase gripping ability on metal handles.

Knurling tool holders are available in revolving-head and self-centering–head versions (Fig. 4-83). Knurling is done at right angles to the work in straight pattern holders only.

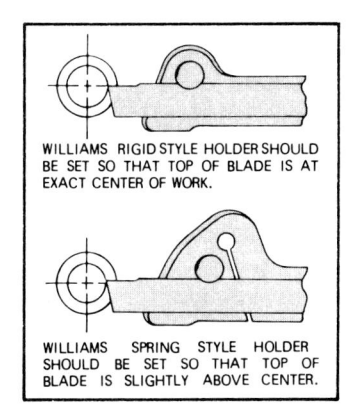

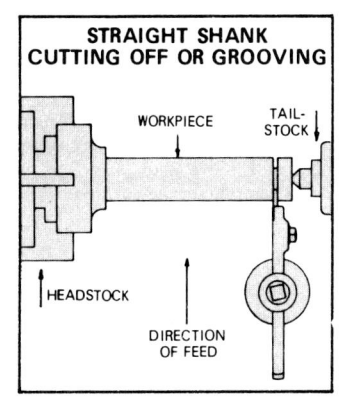

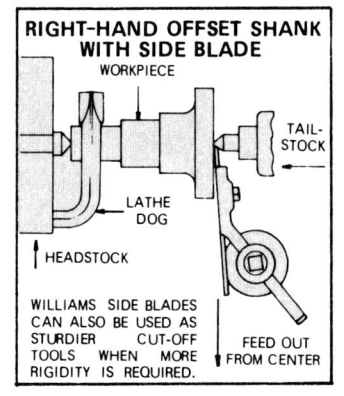

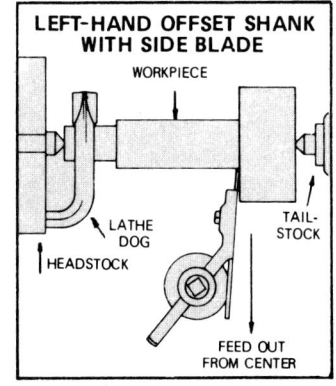

Fig. 4-82 Typical uses of cutting-off and side holders.

Fig. 4-83 Two styles of knurling tool holders.

Formed cutter–style threading tool holders are normally used for cutting coarse and fine 60° sharp V threads from 4 to 32 pitch, depending on size of cutter used. A typical threading tool holder is shown in Fig. 4-84.

The tool holders and boring and shaper items normally cut with high-speed steel tool bits and cutters. For carbide tool bits, carbide turning tool holders may be used (Fig. 4-85). Right-hand and left-hand offsets are available.

Cutting shapes on inside surfaces of objects may be done with a shaper tool fitted with bar and toolbit shown in Fig. 4-86. Boring tool holders allow hollowing out inside surfaces as shown in Fig. 4-87. Three types of boring bar holders are shown in Fig. 4-88.

Tool bits Selection of the correct carbide single-point tool bit depends upon the specific operation at hand. There are different points for chipping, shouldering, turning, chamfering, undercutting, and threading. Each type of point is available in about six shank sizes. Each point shape and shank dimension is available in at least three grades. A utility grade is suitable for matching cast iron and nonferrous metals, as in general-purpose roughing and finishing.

Grades for general finishing of steel and heavy-duty steel roughing are considered premium grades.

General Hints. These tool bit operating hints can promote long tool life.
1. Use flat rigid base on tool holder.
2. Use tool holders designed for carbide tool bit use.
3. Never use a hammer on cutting end of tool bit. If necessary, set short of designed length and adjust from rear.
4. Don't have tool bit against work when tightening clamping screws.
5. Use dog-point or flat clamping screws.
6. Cut overhang to absolute minimum.
7. Don't dip tool bit in any liquid while tool bit is hot.
8. Allow tool bit to cool naturally or in stream of air.
9. Use generous flow of coolant at a high velocity when cutting steel or other materials where coolant is required.
10. Always disengage feed before stopping spindle.
11. Always use silicon carbide or diamond wheels for grinding the carbide tip.
12. Sharpen carbide tool bits at regular intervals to get longest life.

If Tool Bits Chatter:
1. Tool bit overhang may be excessive.
2. End-cutting-edge angle may be insufficient.
3. Tool bit nose radius may be too large.
4. Feed may be insufficient.
5. Tool bit relief angles (clearance) may be too large.
6. Rake angles may be too large (decrease as a last resort).
7. With hand feed on plunge-cut tool bits (grooving or cut-off), increase rate of feed in proportion to increase in rate of speed. Example: if you double the speed, double the rate of feed.

Fig. 4-84 A threading tool holder with formed cutter.

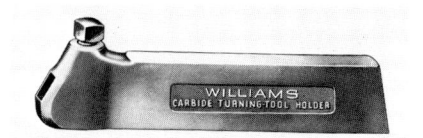

Fig. 4-85 A carbide turning tool holder.

If Tool Bits Wear Rapidly:
1. Feed may be insufficient, causing rubbing action.
2. Tool bit relief angle (clearance) may be insufficient.
3. Chatter and vibration may be excessive.

C-clamps C-clamps are available in specific series of sizes for different jobs. Within a given series, at least seven sizes cover maximum openings from 2 in. (5.1 cm) to at least 12 in. (30.5 cm).

Selection. Choose a size in keeping with the job. One too large presents an obstruction hazard that can slow down work and create unsafe working conditions. The right size will offer adequate clamping action with minimum screw adjustment and leave the work area relatively free of obstructions in performing the task at hand (Fig. 4-89). An adequate range of sizes encourages selection of the proper size and style for all jobs. Where the advantage of a deep throat is not required and the work is fairly thick, try a general service clamp. In Fig. 4-89 a shorter screw offers less projection and minimum screw adjustment. Under some conditions, the clamp frame can be used to align the work. Whatever the job, it pays to know C-clamps.

Three series of C-clamps are shown in Fig. 4-90. General-service C-clamps are used for numerous maintenance needs. Where a wider frame is required, the deep throat series is generally one-third wider than general-service C-clamps.

Heavy-service series C-clamps designed for extra-tough needs do not use the swivels found in other C-clamps. The screw terminates in a specially hardened flat end. Often two lengths of screws are available. Short screws handle most needs but where maximum openings are needed, long screws which reach all the way to the frame pad should be selected.

C-clamps for welding needs are specially coated in zinc to resist welding spatter. Zinc-plated screw shields (Fig. 4-90) protect screw shields from spatter.

C-clamps are used in so many ways it would stagger one's imagination. Bracing, holding, welding, and positioning, in their many ramifications, often subject C-clamps to unusually severe and abusive conditions. Frames of most drop-forged makes are generally sufficiently strong and their design is such that they are not commonly susceptible to serious damage.

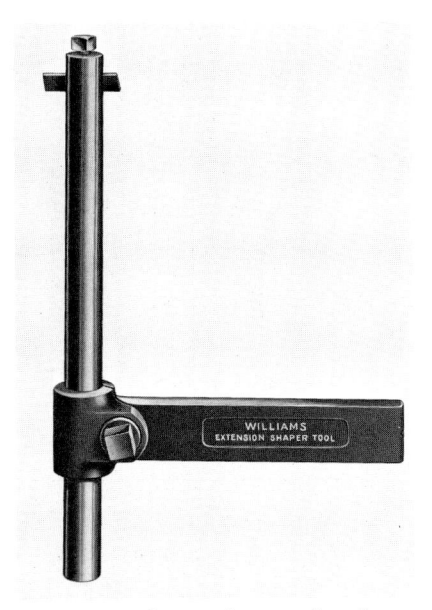

Fig. 4-86 A shaper tool cuts inside surfaces.

Damage to C-clamps involving bent or mutilated screws, banged-up swivels, or a combination of both renders them unsafe for further use, however.

If screws and swivels are not easily or readily replaceable on the job, unsafe practices become a temptation. Rather than discard such a clamp or send it back to the manufacturer for repair, many operators will continue to use clamps minus swivels or clamps with bent screws where alignment of swivel and frame pad is very poor.

To avoid unsafe conditions the answer is obvious—easy field replacement of screws and swivels is often necessary.

A field-repairable swivel assures C-clamp dependability and therefore is a necessary feature in the selection of C-clamps for most applications. Swivels may be replaced easily (Fig. 4-91), yet this design exceeds retentivity of some nonrepairable designs.

Care. C-clamps deserve the same simple care that other good tools should receive. Screw threads should be cleaned regularly and a light film of oil applied. Clamps with bent screws or without swivels should be taken out of service until they have been properly repaired. Protect your C-clamp investment by carrying a stock of replacement

2-60 Establishing the Maintenance Facility

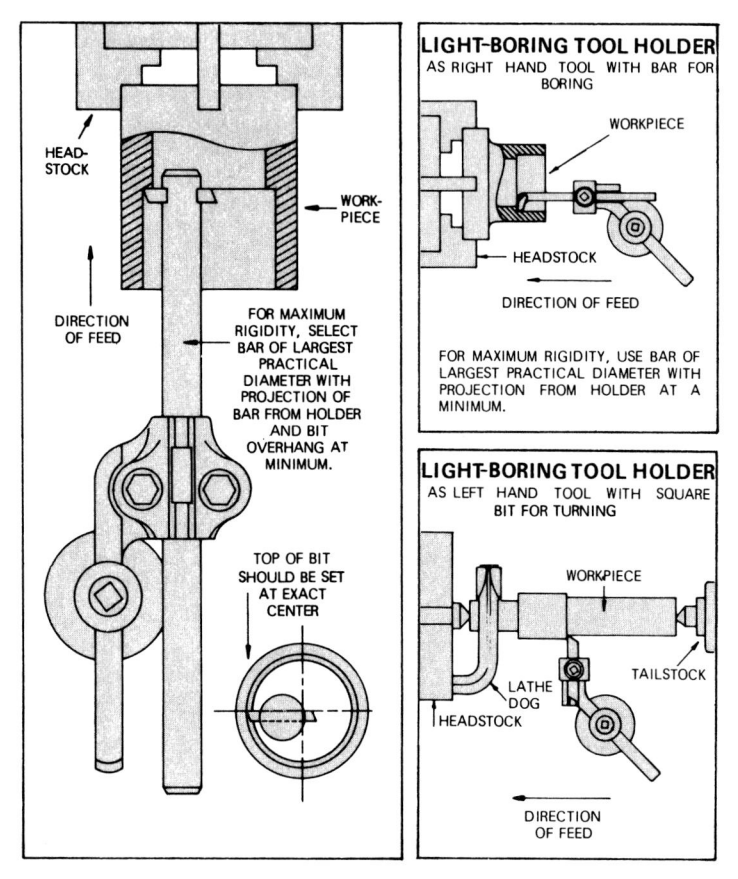

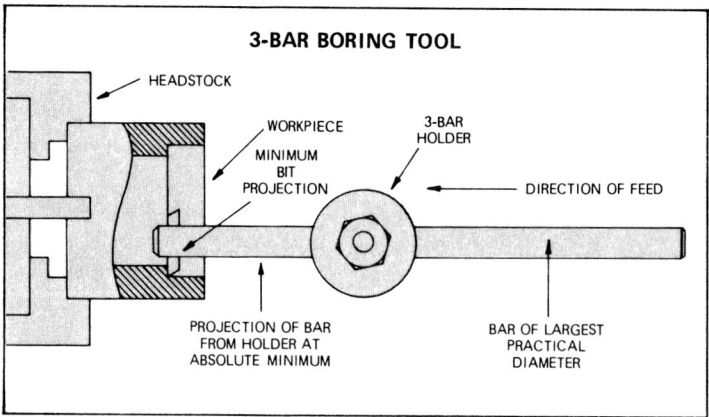

LIGHT-BORING TOOL HOLDER
AS RIGHT HAND TOOL WITH BAR FOR BORING

- WORKPIECE
- HEADSTOCK
- DIRECTION OF FEED

FOR MAXIMUM RIGIDITY, USE BAR OF LARGEST PRACTICAL DIAMETER WITH PROJECTION FROM HOLDER AT A MINIMUM.

- HEADSTOCK
- WORKPIECE
- DIRECTION OF FEED

FOR MAXIMUM RIGIDITY, SELECT BAR OF LARGEST PRACTICAL DIAMETER WITH PROJECTION OF BAR FROM HOLDER AND BIT OVERHANG AT MINIMUM.

TOP OF BIT SHOULD BE SET AT EXACT CENTER

LIGHT-BORING TOOL HOLDER
AS LEFT HAND TOOL WITH SQUARE BIT FOR TURNING

- WORKPIECE
- TAILSTOCK
- LATHE DOG
- HEADSTOCK
- DIRECTION OF FEED

3-BAR BORING TOOL

- HEADSTOCK
- WORKPIECE
- MINIMUM BIT PROJECTION
- 3-BAR HOLDER
- DIRECTION OF FEED

PROJECTION OF BAR FROM HOLDER AT ABSOLUTE MINIMUM

BAR OF LARGEST PRACTICAL DIAMETER

Fig. 4-87 Typical boring setups.

Basic Hand Tools of Maintenance 2-61

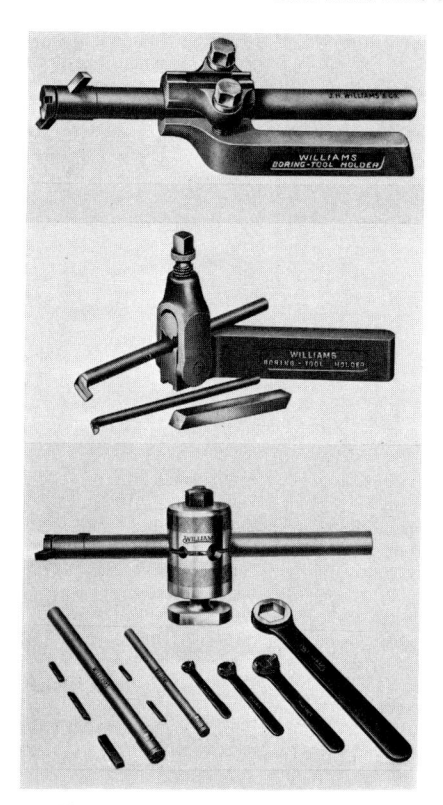

Fig. 4-88 Boring equipment variations.

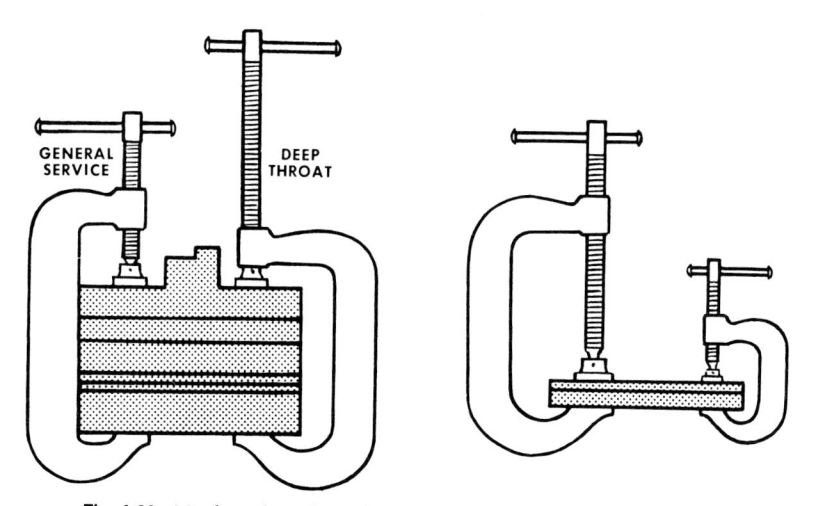

Fig. 4-89 Matching the right C-clamp to the right need gives good clearance.

2-62 Establishing the Maintenance Facility

screws and swivels to make immediate repairs and keep C-clamps in safe-as-new condition at all times.

Other clamps sometimes encountered in maintenance include toolmakers' C-clamps and machinists' clamps (Fig. 4-92). Toolmakers' C-clamps are available with or without swivels in four sizes. Both machinists' clamps and toolmakers' C-clamps cover needs from 1 in. (2.5 cm) to about 4 in. (10.1 cm).

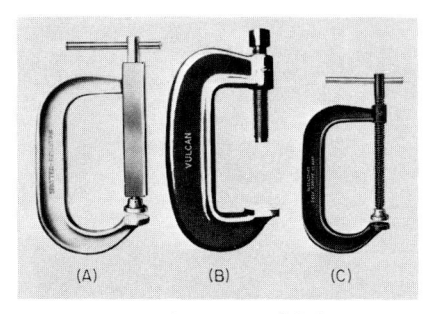

Fig. 4-90 Three series of C-clamps.

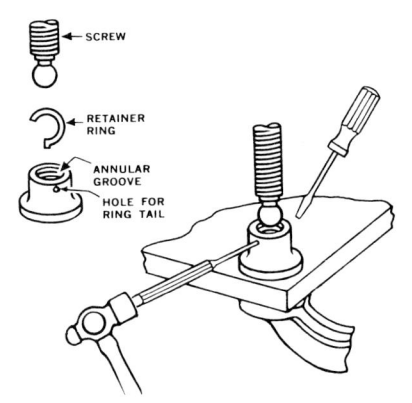

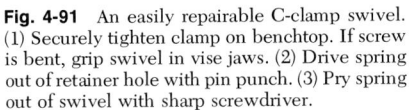

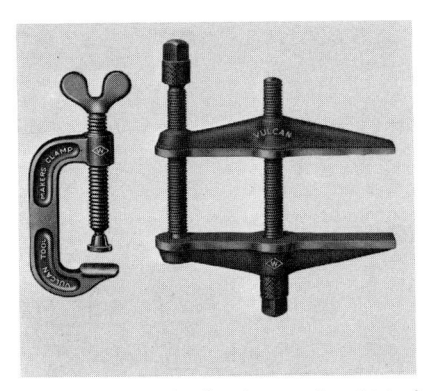

Fig. 4-91 An easily repairable C-clamp swivel. (1) Securely tighten clamp on benchtop. If screw is bent, grip swivel in vise jaws. (2) Drive spring out of retainer hole with pin punch. (3) Pry spring out of swivel with sharp screwdriver.

Fig. 4-92 Toolmakers' C-clamp and machinists' clamps.

Section **3**

The Economics of Construction Equipment Maintenance

Chapter **1**

The Maintenance Budget

ANTHONY VIERS
Controller, Advance Construction Co., Inc., Hinsdale, Ill.

Because the major portion of a construction company's capital structure is invested in equipment, it would be logical to assume that a corresponding amount of management attention would be dedicated to protect that investment. But equipment normally is managed by instinct, which generally does not produce satisfactory results. Sound management practices are needed to ensure the highest return on investment.

When a piece of equipment is purchased the cost does not end. A particular piece of equipment will require proper repair and maintenance to keep it operating efficiently during its useful life, and the cost to accomplish this might be equal to the original purchase price.

It is important that company management establish a policy and budget for repairs and maintenance. The repair and maintenance policy can then be the established guide for the equipment superintendent and the budget becomes the guideline for expenditures.

Management has to answer many questions when establishing a maintenance policy, including the following:

1. What acquisition policy will we establish? Will it be based on equipment age, use, condition, and obsolescence, or will we evaluate all of these areas when making a decision to purchase equipment?

2. Do we have the proper maintenance staff to handle all repairs in our own shop?

3. Will we subcontract major repairs such as main drives and engines?

4. Do we want to minimize maintenance expense by trading in equipment sooner?

5. Will our maintenance policy be such that we will aggressively maintain our equipment? This may extend the useful life and thereby minimize the frequency of capital purchases.

6. What are the normal working conditions of our equipment? Will more or less maintenance effort and expenditure be required to keep equipment operating efficiently and costly downtime at a minimum?

Some of these questions might seem basic or even unnecessary, but there are many approaches to successful and profitable operation of a construction company. For example, consider the case of two contractors working on the same project; each had a different equipment maintenance policy. One contractor employed a vigorous maintenance policy and the other contractor maintained equipment on a demand basis only. Yet, at the completion of the project, both contractors made what they felt was a reasonable profit.

After establishing a maintenance policy it will be necessary to provide a system for equipment cost recording, which will allow the establishment of a maintenance budget. For a maintenance budget to be effective, an equipment costing system must be established that is based on sound cost-recording principles. It would be very difficult to meet budget goals if there were no way to monitor and control expenditures against the budget.

The seasonality of many contracting businesses by itself tends to make it rather difficult to monitor and control maintenance expenses. We can readily understand that maintenance costs are variable and are influenced by construction volumes and equipment use. Such influences make it rather difficult to monitor a maintenance budget on any basis other than a full construction season.

One method used to control and budget maintenance costs effectively is to develop a special report that records maintenance costs in an on-going 12-month basis. To do this, subtract costs from the thirteenth month and add costs of the current month. This system covers a 12-month period and is representative of a full construction season and current construction volume. Manually record the costs in summary form from the company's monthly financial statements. This method allows accounting personnel to establish an average monthly budget for the repair and maintenance personnel. The basic general ledger schedule of accounts necessary to accomplish this is outlined in Chap. 2 of this section.

Equipment maintenance costs must be included in the bid and recovered from project revenues, so control over maintenance expenditures is essential. If we needlessly spend too much on equipment maintenance, our bids are high; competitors, bidding lower, will not allow us to recover this cost through the bidding process, and we lose jobs. If we spend too little, we may needlessly exhaust the useful life of the equipment. Or we may be less efficient in completing a project on time because of equipment failures and we reduce job profit. Balanced maintenance policies and budgets are vital to survival in the highly competitive construction industry.

Chapter **2**

Standard Accounting Practice for Maintenance Cost Recording

ANTHONY VIERS
Controller, Advance Construction Co., Inc., Hinsdale, Ill.

Heavy construction equipment, such as hauling units, dozers, scrapers, cranes, and backhoes, is expensive. It also has a limited useful life, a limited period in which it can be used economically. When the useful life of a piece of equipment is at an end, the equipment may be replaced and it should have produced enough earnings to cover the cost of replacement. It is essential that a contractor be able to estimate what it costs to own and to maintain equipment during its useful life. These aspects of the cost of equipment are identified as the cost of ownership (fixed costs) and the cost of repairs and maintenance (operating or variable costs).

Ownership costs are fixed; they represent costs related to the investment in the machine and are never charged entirely to any single job. Instead they are reduced to an hourly figure and prorated over the assumed useful life of the machine. The useful life of a particular piece of equipment is determined by its wear and obsolescence based on the owner's experience or schedules supplied by the manufacturer of the equipment. At this point, it might be necessary to point out that the established useful life of a particular piece of equipment may be different from that established by Internal Revenue Service guidelines or used for tax reporting purposes on the contractor's accounting records.

Ownership or fixed costs The hourly charge for machine depreciation is found by dividing the machine's delivered price by the anticipated hours of useful life. The delivered price of a machine is FOB (factory price plus freight charges). On rubber tire equipment, tire prices could be deducted from delivered price, since tires are normally treated as operating cost items and therefore have no bearing on machine depreciation. A straight line write-off method is used that in no way dictates what method is used for tax purposes.

Next, a method of depreciating the machine over its useful life must be considered to arrive at an hourly charge rate. Actual useful life of the equipment is difficult to determine since proper preventive maintenance will extend the service life and improper preventive maintenance will shorten the service life. Probably no other cost item is subject to so many influencing factors, such as use and wear, preventive maintenance, quality and timeliness of major repairs, weather, obsolescence, and quality and reliability of the machine. Economic life has a great influence on the final cost per yard figure in the bidding process. The best method for determining the useful life of a given piece of equipment is a history of use based on the actual experience of each contractor. This fact alone emphasizes the need for every contractor to keep historical records of hours used and repair costs.

3-4 The Economics of Construction Equipment Maintenance

If you do not have the historical records of equipment hours of use, it will be necessary to estimate the useful life hours. For a guide refer to the manufacturer's estimate as to the expected useful life hours of a given model of equipment.

The salvage value of a piece of equipment will vary widely, depending on its condition, its obsolescence, and the general market. This may be as little as scrap value or might be 30 percent or more of its original cost. Some owners prefer to assume no salvage value in calculating depreciation by applying the trade-in value it has toward the initial cost of new equipment. Another method used is to consider no resale or salvage value. This is a conservative practice since used-equipment value cannot be considered stable.

The basic formula for calculating hourly depreciation rate for equipment is as follows:

$$\text{Hourly depreciation rate} = \frac{\text{Purchase price} + \text{freight} - (\text{original tire value and salvage})}{\text{Total useful life in hours}}$$

Including salvage in the formula is optional. Interest, taxes, and insurance are the other items of ownership or fixed costs associated with the hourly cost of depreciation. Interest on investment in equipment should be a percentage based on the borrowing experience of the individual contractor. Caution should be exercised here because, when money is borrowed, the actual interest is not the only cost. Return on the capital investment must be considered. Taxes, such as property tax, licensing, and use taxes for highway construction vehicles, must be considered. The individual must refer to his own area to evaluate these taxes for consideration as part of the average hourly cost. Insurance coverage is for fire, theft, and collision damage to the equipment and, again, has to be considered as part of the hourly ownership cost of equipment.

The easiest method for allowing for the costs of interest, taxes, and insurance would be to appraise these items based on actual cost. Then, develop a percentage of purchase cost on an annual basis, using the following formula:

$$\frac{\text{Average annual investment} \times \text{percentage}}{\text{Average operating hours per year}}$$

Operating or variable costs These costs are made up of the following categories: (1) repair parts and labor, (2) fuel, (3) grease and oil, (4) filters and supplies, (5) operators' wages.

Repair Parts and Labor. This category should cover the normal labor plus fringes and parts and supplies for ordinary operation of the machine, as well as provision for periodic overhaul. Repair labor cost should include premium pay, social security, unemployment compensation, taxes (state and federal), workmen's compensation insurance, and employee fringe benefits. Cost, will vary widely, depending on preventive maintenance, severity of service, and skills of the operator. General repair and operating cost experience is, of course, the most reliable guide, but it is not always known to the estimator. A common method for applying the cost of repair parts and labor is estimated as a percentage of the hourly depreciation cost. In the case of cost estimating by this method, the item of repair parts and labor cost is amortized over the estimated useful life of the machine. Conversely, using the actual cost of repair parts and labor would create peaks and valleys and would cause erratic changes in hourly cost of repair parts and labor from year to year. To justify using the average method over the useful life of equipment, Fig. 2-1 illustrates how depreciation and repair costs compensate each other over the useful life of equipment.

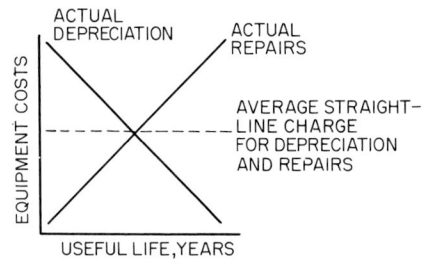

Fig. 2-1 The interaction of depreciation and repair costs.

Repair parts costs include repair and replacement parts for normal operation as well as for periodic overhaul. Labor costs include those related to maintenance and repair but not the costs of the operator or oiler assigned to the machine.

Fuel. Fuel is certainly the most accurate and easiest cost to determine by simply observing it, keeping a record of the fuel used, and applying the cost per gallon times gallons used per hour for an hourly cost of fuel. Another method uses a percentage of the hourly depreciation cost to equal fuel cost per hour. To be able to do this, it is important that the contractor know his yearly fuel cost in relation to his yearly depreciation cost based on usage. The justification for the average method of allocating fuel cost is that the higher the depreciation cost the more fuel the equipment will consume. Although this may not always be true, it does provide an acceptable method for allocating hourly fuel cost.

Grease and Oil. Grease and oil can also be allocated as a percentage of depreciation cost or use actual cost by machine.

Filters and Supplies. Filters and miscellaneous supplies can be included with repair parts and labor cost or actual cost by machine.

Some of the best reference material available regarding fuel, grease, oil, and filter costs can be obtained from suppliers and equipment manufacturers and can be looked to as an excellent source of information regarding these items.

Operators' Wages. The operating manpower for each shift includes an operator and sometimes an oiler for one or more machines. The wage rates vary widely in different parts of the country and local conditions should be followed in establishing the hourly wage rates. To these should be added additional amounts for social security, unemployment compensation, taxes (state and federal), workmen's compensation insurance, and fringe benefits reduced to an hourly cost. Rather than having two sets of rates, it might be more convenient to add a factor for overtime pay when required. If the contractor's accounting records make a distinction between wages and the overtime portion of wages, he could add a factor to the base hourly wage to cover overtime pay.

The foregoing discussion of standard accounting practice for maintenance cost recording is not unique to any individual contractor regardless of size. There are many applications which differ in format, but it is of great importance that each contractor develop his own system for recording equipment costs. Unless an attempt is made to maintain an accurate equipment costing system, an important area of the estimating process, loss of a contract or reduction of profits may result.

An equipment costing data sheet (Fig. 2-2) was designed for simplicity and can be used by most contractors for equipment costing; there are many possible variations of this format to suit the individual circumstances of each contractor.

The next phase of standard accounting practice concerns recording equipment maintenance expenses on a contractor's general ledger schedule of accounts. Generally, it is advisable to collect and record these expenses in the same manner as previously established in the equipment costing method. The actual costs of the record are normally collected and recorded through the payroll and accounts payable function of the company.

An example of an acceptable general ledger schedule of accounts for equipment cost recording is
    Repair and maintenance labor, straight time
    Repair and maintenance labor, premium time
    Shop and yard salaries
    Repair and maintenance payroll taxes
    Repair and maintenance parts
    Repair and maintenance supplies
    Repair and maintenance filters
    Repair and maintenance subcontracted
    Perishable tools
    Diesel fuel expense
    Gasoline fuel expense
    Grease and oil expense
    Shop and yard utilities
    Shop and yard rent
    Shop and yard miscellaneous expenses
    Equipment depreciation expense
    Equipment insurance expense
    Equipment license and permits expense
    Equipment tax expense

EQUIPMENT COSTING DATA SHEET

No. 1   Equipment model _____
No. 2   (a) Useful life _____ years ×
        (b) Estimated hours per year _____ =
        (c) Useful total hours _____
No. 3   Equipment purchase price
        (including sales tax + options) per hour $ _____
No. 4   Depreciation: No. 3 divided by No. 2 (c) =
No. 5   Interest, taxes, insurance: _____ % of
        No. 3 divided by No. 2 (b) =
No. 6   Ownership costs: No. 4 + No. 5 $ _____
No. 7   Repair parts and labor
        (including payroll additives): _____ % of No. 4 _____
No. 8   Fuel _____ gallons per hour @ _____ ¢ per gallon
No. 9   Grease and oil @ _____ % of No. 4
No. 10   Total operating cost:
         No. 6 + No. 7 + No. 8 + No. 9 $ _____
No. 11   Operator wages per hour
         (including payroll additives) _____
No. 12   Total equipment cost per hour
         with operator, No. 10 + No. 11 $ _____

Fig. 2-2   Typical equipment costing data sheet.

It must be emphasized at this time that every equipment costing system has to be uniquely designed for each construction organization, depending on size, type of equipment, and availability of personnel. Another consideration in the schedule of accounts is the accounting method employed; some of the expense accounts could be considered direct job cost expenses and others might be considered overhead cost expenses. To determine how best to implement an equipment cost recording schedule of accounts requires a thorough review of an individual company's accounting records.

Chapter **3**

The Role of the Computer in Maintenance Cost

ANTHONY VIERS
Controller, Advance Construction Co., Inc., Hinsdale, Ill.

Cost keeping for construction equipment maintenance, particularly in heavy construction, constitutes an important application area which is successfully using data-processing systems (the computer). These computer systems are available in many forms from manufacturers and service bureaus.

The manner in which equipment costs are easily tabulated is a computer function rather than a clerical function. Allowing people to focus on decisions, a computer can accumulate and distribute costs which would take many clerical hours, with results that might not be as accurate. Management needs the most accurate and timely reports possible, allowing sound decisions in the interest of the company. At this point it would be wise to evaluate the number of major pieces of equipment a company has. For example, it might be foolish for a contractor who has fifteen major pieces of equipment and who subcontracts all repairs and maintenance to install a computer or enlist a service bureau to record equipment costs.

Equipment costs can be easily reported by use of forms that closely resemble time cards used for labor costing. Such forms allow a contractor to accumulate equipment hours used, repair hours, and repair costs on a daily basis as reported by field superintendents, foremen, and mechanics. The information can then be accumulated by equipment number and a report can be produced weekly, monthly, or as required by the contractor. Current-period, year-to-date, and life-to-date costs can be recorded so that management can see current equipment cost trends and historical costs. Then, these costs can be compared with estimates to be sure that costs are being covered in the bidding process.

The danger of guessing at equipment costs when bidding for work is that if the guess is lower than the actual cost, after completion, profits will be too low or nonexistent. If the guess is too high, the company may not get the contract.

A computer equipment cost system does not have to be complicated. To be a sound costing system, it must reflect company philosophy and most ideally be compatible with the accounting records, bidding process, and job costing systems within the company. If they are designed by employees of the contractor or in conjunction with other systems design people, it is important that they reflect the thinking of the individual contractor so that he is comfortable working with and using the equipment cost system.

Many failures of computer equipment costing systems have been caused by people who lack sufficient knowledge of relationships of equipment costing in the construction industry and by others who fail to communicate the knowledge necessary to obtain a simple but successful computer equipment costing system.

3-10 The Economics of Construction Equipment Maintenance

The systems themselves can be as simple as accumulating the hours of usage times an hourly rate and then applying a percentage against the accumulative dollars of cost by equipment. The percentage rate need be only representative of the repair labor and repair parts cost as shown on the contractor's accounting records.

The information required from a project to establish equipment charges depends upon the particular system used. In general, however, the following information is needed and recorded by the superintendent or foreman: (Fig. 3-1) (1) project or job number, (2) equipment number, (3) cost code, and (4) hours' distribution.

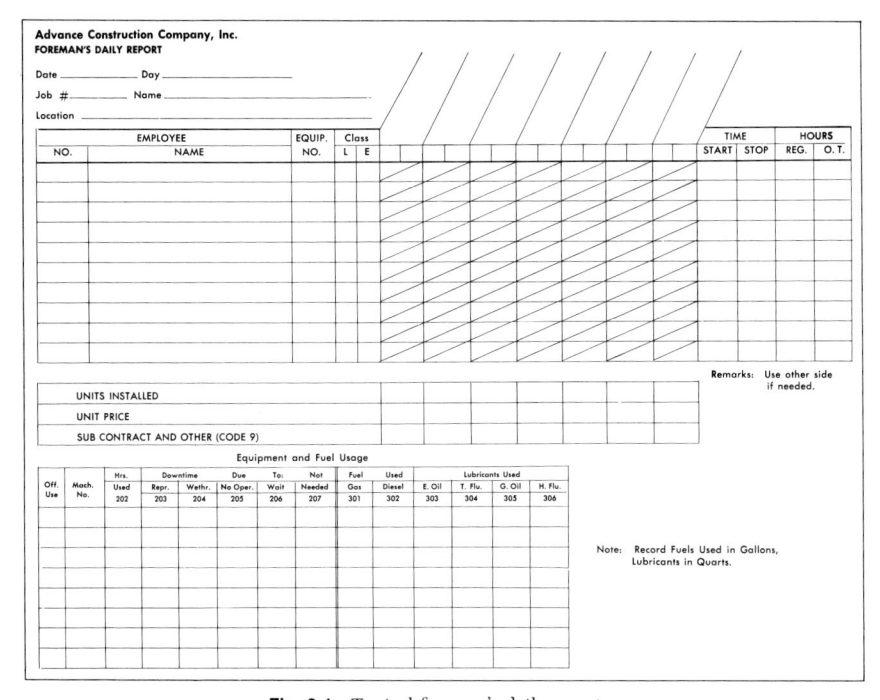

Fig. 3-1 Typical foreman's daily report.

The above information is used basically to record equipment hours of use for the day or period and to record the equipment cost by job cost code (work activity or pay item) for the job cost report.

The repair parts cost for the equipment cost report usually originates from the accounts payable system. When the accounts payable system is set up, thought should be given to recording the parts invoices in a way which will allow them to be charged to the specific equipment number. At this time, if a more detailed parts cost distribution is required, then a coding system can also be set up to distribute these charges to special code categories as follows:

 101 Engine
 102 Power train
 103 Hydraulic control
 104 Attachments
 105 Tire and tracks
 106 Preventive maintenance
 107 Body chassis
 108 Shop miscellaneous
 109 Shop labor and supervision

These categories allow a broader breakdown of equipment parts cost on an equipment cost report. The same procedure can be followed for repair labor. A mechanic's time sheet (Fig. 3-2) can be designed so that, when working on a piece of equipment, he can also distribute his payroll hours according to one of the codes that is compatible with the parts cost distribution. Such a system allows for a uniform breakdown of parts and labor costs on the equipment cost report.

ADVANCE CONSTRUCTION CO.

NAME _____
DATE _____

	Equip. no	Hour meter	Code	Time
MON.				
TUES.				
WED.				
THURS.				
FRI.				
SAT.				

101 - ENGINE 105 - TIRE - TRACKS
102 - POWER TRAIN 106 - PREVENT. MAINT.
103 - HYD. CONTROL 107 - BODY CHASSIS
104 - ATTACH. 108 - SHOP MISC.

Fig. 3-2 Typical mechanic's time sheet.

The next step is to expand the system to include such additional costs as fuel, depreciation, insurance, taxes, interest, and shop overhead; divide these total costs by hours of use to come up with the average hourly cost of a given piece of equipment.

Other reports that can be developed include comparative analysis of productive time and repair time, evaluation of operating rates, and analysis of repair cost. From this type of information management can determine when a piece of equipment should be replaced because of excessive repair costs or downtime.

Many contractors probably have much more sophisticated data-processing systems for the costing and control of equipment. It is important to remember that the individual contractor should have a system which is easy for him and for his employees to understand so that the information, regardless of how detailed it may be, is accurate and represents true costs as reflected on accounting records.

Data-processing systems, as such, should not necessarily be designed as individual

3-12 The Economics of Construction Equipment Maintenance

Fig. 3-3 Typical voucher authorization.

Fig. 3-4 Typical equipment repair cost sheet.

The Role of the Computer in Maintenance Cost

SCHEDULE OF MACHINERY AND EQUIPMENT
12/31/75

ASSET NO.	DESCRIPTION	SERIAL NUMBER	PURCHASE DATE	PURCHASE PRICE	ACCUM. DEPR. 12/31/74	1975 DEPR.	ACCUM. DEPR. 12/31/75	BOOK VALUE 12/31/75
76	HUBER 3-WHEEL 12 TON ROLLER LIFE 4YRS METH SL	S/N 10-1166	4/20/64	3,100.00	2,790.00	.00	2,790.00	310.00
85	LETOURNEAU SHEEPSFOOT SELF PROP ROLLR LIFE 6YRS METH DDB	S/N GP61598UBQ	5/08/65	7,200.00	7,200.00	.00	7,200.00	.00
92	INGERSOLL RAND AIR COMPRESSOR 125 CFM LIFE 6YRS METH DDB	S/N T032085	5/31/65	1,200.00	1,200.00	.00	1,200.00	.00
104	INGERSOLL RAND AIR COMPRESSOR 250CFM LIFE 6YRS METH DDB	S/N 21753	3/17/66	8,320.00	8,320.00	.00	8,320.00	.00
108	TWO MOTOR FORK LIFT LIFE 6YRS METH DDB	S/N 461601044	6/17/66	2,720.00	2,720.00	.00	2,720.00	.00
114	DOUBLE DRUM SHEEPSFOOT ROLLER LIFE 6YRS METH DDB	S/N 73812	9/23/66	1,000.00	1,000.00	.00	1,000.00	.00
115	HAWKEYE PIPEMASTER LIFE 6YRS METH DDB	S/N 711067	11/08/66	2,509.88	2,509.88	.00	2,509.88	.00
115	RACK C GATE ASSEMBLY LIFE 6YRS METH DDB	S/N	2/16/68	900.00	900.00	.00	900.00	.00
116	ARROW G-400 HYDRAULIC HAMMER LIFE 6YRS METH DDB	S/N 4694	12/28/66	6,000.00	6,000.00	.00	6,000.00	.00
119	ROME DOUBLE CUT DISC LIFE 6YRS METH DDB	S/N	2/14/67	3,199.00	3,199.00	.00	3,199.00	.00
125	JAEGER 4PT. TRASH PUMP LIFE 6YRS METH DDB	S/N 193069	4/30/67	1,328.00	1,328.00	.00	1,328.00	.00
126	INGERSOLL RAND 125 COMPRESSOR LIFE 6YRS METH DDB	S/N 125CR6944	5/17/67	1,410.75	1,410.75	.00	1,410.75	.00
134	ELECTRO MAGIC STEAM CLEANER LIFE 6YRS METH DDB	S/N 27401-KY	9/26/67	2,291.00	2,291.00	.00	2,291.00	.00
135	TENNANT #42 SWEEPER LIFE 6YRS METH DDB	S/N 402	9/30/67	750.00	750.00	.00	750.00	.00
135	STREET SWEEPER LIFE 5YRS METH DDB	S/N	8/05/71	825.00	704.40	60.30	764.70	60.30

Fig. 3-5 Typical schedule of machinery and equipment.

units but more efficiently developed as a complete system. This approach allows a uniform comparison among the general ledger, job cost, estimating, and equipment costing systems, thereby recognizing the need for control and the ability to tie in subsidiary cost reports with the financial statements of the contractor.

Figures 3-1, 3-2, and 3-3 show some forms that can be used for recording equipment cost information for later entry into system or equipment records. There can be many variations of these forms to suit the individual contractor. Figures 3-4 and 3-5 are examples of data-processing reports to control equipment costs.

Chapter **4**

The Role of Preventive Maintenance

CARL G. WYDER[1]
Senior Maintenance Editor, McGraw-Hill Inc., New York, N.Y.

WHAT PREVENTIVE MAINTENANCE IS

Ask any 10 construction equipment engineers to define preventive maintenance, and you'll probably get 10 different meanings. Preventive maintenance (PM) varies widely in scope and intensity of application.

Many engineers think of PM only in terms of periodic inspections of equipment to prevent breakdowns before they occur. To this limited view some add repetitive servicing, upkeep, and overhaul. In a more advanced stage are those who include other repetitive maintenance functions such as lubrication, painting, and cleaning. Further along the way are those who also study materials and finishes of the equipment before it is purchased.

Basic definition No matter to what degree of refinement a PM program is developed, all of them contain the following basic activities:
  1. Periodic inspection of equipment to uncover conditions leading to production breakdowns or harmful depreciation
  2. Upkeep of equipment to sterilize such conditions or to adjust or repair such conditions while they are still in a minor stage

This basic concept will be the definition in most of the discussion in this chapter. At the end of the chapter, finer techniques which might be added to any basic program if the extra costs can be justified, will be reviewed.

WHY CONSTRUCTION EQUIPMENT NEEDS PM

Any well-designed PM program will yield benefits far in excess of its cost. I still have to find someone practicing PM who says it doesn't pay. Many have had some doubts before adopting it, but none thereafter.

Not every contractor can expect to derive equal benefits. The type of construction projects and the sophistication of equipment involved are factors in amount and scope of results. The more highly specialized equipment becomes, the more it needs the advantages of PM. Costs of maintenance of modern equipment are higher and costs of work time lost, too. On any site where lost time is important, PM will reduce it. There is no question that lost time will be less with PM than without it but to what extent depends on the individual contractor.

[1]The author is now retired.

3-16 The Economics of Construction Equipment Maintenance

Preventive maintenance is not a cure-all for excessive lost work time or high maintenance costs. There are other maintenance functions with which PM must be integrated to achieve an efficient equipment-maintenance program—a good paperwork system, work planning and scheduling, training, work measurement, control reports, and good shop and tools. Major returns with which PM has rewarded its users include the following:
 1. Less lost work time
 2. Less overtime pay for maintenance men on ordinary adjustments and repairs than for breakdown repairs
 3. Fewer large-scale repairs, and fewer repetitive repairs
 4. Lower repair costs for simple repairs made before breakdowns because less manpower, fewer skills, and fewer parts are needed for planned maintenance
 5. Fewer on-site problems because of properly adjusted equipment
 6. Postponement or elimination of cash outlays for premature replacement of equipment because of better conservation of assets and increased life expectancy
 7. Less standby equipment needed, thus reducing capital investment
 8. Decline of maintenance costs—labor and material—on asset items in the program
 9. Identification of items with high maintenance costs, leading to investigation and correction of causes such as (1) misapplication, (2) operator abuse, and (3) obsolescence
 10. Better spare-parts control, leading to minimum inventory
 11. Better employee relations because construction workers don't suffer involuntary layoffs or loss of incentive bonus from breakdowns
 12. Greater safety for workers, and improved protection for equipment, leading to lower compensation and insurance costs
 13. Lower cost of actual construction

These are all realistic benefits that apply in any economy—peace or war, expanding, stable, or contracting.

BEFORE YOU START A PM PROGRAM

No matter who starts the idea of a PM program, the construction maintenance engineer or manager usually has to run it. But unless he is careful to erect firm foundations of company understanding and policy before applying the program, he will find it tough going afterward. And the program may fail, not because it was unworthy but because it wasn't given a fair chance.

How to sell PM The success of a PM program hinges largely on how well everybody in the company—higher management, foremen, and laborers—is sold on PM.

Most logically, start selling with higher management. This immediately raises the question: What will a PM program cost? So maintenance men clamor for a proven formula for dollars-and-cents savings. For some time I have suggested the following approach: Check over records for the past year or more for all equipment breakdowns. List the total cost of breakdown repairs—labor, material, overtime, any other charges. List what each breakdown has cost in idle time of laborers, spoilage, and rework. To these you might add the cost of operating overhead and other possible losses such as cost of injuries. Next, estimate what the repairs would have cost if they had been made before the breakdowns—if there had been time for planning, getting materials, and making productive use of equipment operators on other sites. The difference is what might be spent on a preventive-maintenance program.

In drawing up any cost versus savings comparison, there is a possibility (at least at the start) that direct maintenance costs will rise. Management must realize PM is an investment that needs extra cash as does new equipment. In the case of PM, the return is highly promising. A good way to document the gross return is to check the list of benefits given earlier under "Why Construction Equipment Needs PM," and evaluate them as best you can. The summation is always impressive. The secret of selling management on PM is to show PM in its overall results of lower cost of equipment repair.

The next step is to win site foremen to the idea. Unless the foreman sees definite gains for his own interests, he will stall or balk whenever it's time to shut down a piece of equipment for a scheduled inspection or overhaul. But if he knows that time lost for PM will be less in the long run than total outages from breakdowns, he will cooperate. The biggest objection to PM might come from a foreman who has a whole crew instead of a single man involved in a shutdown. In this case your strong argument is that you are servicing all equipment simultaneously and at one predetermined time rather than single units in separate breakdowns for a greater total of lost time.

"But you can't always prove these benefits in advance," is the wail of many maintenance engineers. "Without decisive proof we can't get to first base with either management or on-site people." Then you show maintenance engineers how other contractors have profited from PM. Send them a report on defects you have discovered by inspection and corrected, and show what might have happened if the defects had not been discovered. If they won't buy that, you can do as many others have already done—quietly install PM on one or two sites where it should prove most effective. Or find one sympathetic foreman willing to give PM a trial, and use this successful case as a wedge to break down stubborn resistance. In that case be sure to keep a record of breakdowns, lost time, overtime, and similar penalties. The results have always spoken for themselves.

Should the union be advised? By all means, say those who have learned the hard way. By acquainting the union with the program, you gain its confidence; you can also forestall possible complaints and grievances by first clearing up misunderstandings. Show the union that men will not suffer from PM.

Finally, all maintenance supervisors and craftsmen must be briefed. Because these are the men who are most intimately involved, they need more administrative details than others. Craftsmen long schooled in the old order of standby or breakdown maintenance also need help in making a mental about-face under a new regime of PM.

Program takes time Anybody who expects full benefits from PM quickly will be disappointed. All experts agree it takes several years to get rolling. But you will see some progress after several months, and it will keep snowballing.

There are other factors in the time picture—size of company, type of operations, educational qualifications of the maintenance engineer and his assistants, proper clerical help, and the present condition of construction equipment. As a rule, larger contractors require more study and need more highly refined PM programs and methods of administration.

Conditions inventory If a contractor has been plugging along with 80 percent of its maintenance man-hours spent on breakdown repair, the time of conversion to a satisfactory PM basis (it may range from 30 to 80 percent man-hours on PM) will be delayed. Before the maintenance engineer can apply PM to any equipment, he must get it into good shape. It may take 6, 12, or more months to do it. And he must have an equipment record. The maintenance head should not fail to point this out when he's got the green light from management for a PM program. Otherwise, the time and cost of reconditioning gets charged against PM and gives it a black eye right at the start. It is therefore a good plan to take a conditions inventory to size up the cost of reconditioning equipment and the time needed. Such a survey often leads to disposal of high-repair-cost items that should have been junked long ago.

As a result of this inventory some contractors have to add craftsmen to their maintenance forces to catch up with the bigger repair load. Outside mechanics or local repair shops can lighten the burden and are sometimes better equipped to do a cheaper and better job. Often this backlog of repairs stems from an inadequate maintenance force. If so, work force enlargement should be made permanent.

HOW TO START A PM PROGRAM

To many of the uninitiated, PM is a system. They think all they have to do is set up forms, inspection schedules, and a corps of inspectors and let the calendar do the rest. They ask for hard-and-fast rules of conduct to use like dimensional blueprints for building and running the program.

That is not the way to start a PM program. This concept loses sight of the true goal of PM and of all equipment maintenance functions—maintenance for low-cost continuous operation of a piece of construction equipment. This same element of cost dominates every phase of a good PM program and determines what to do. The right and economic PM plan for one contractor might be wrong and uneconomic for another. Take the case of a company that bought equipment solely for one job. It maintained this equipment on that basis. Shortly after the conclusion of the job, the equipment practically fell apart. But it had served the purpose. It was a case of poor maintenance, but economically justified, and therefore good PM.

If a PM program is to succeed, the administrator must learn to let economic considerations guide and even overrule engineering dictates. Any good engineer can set up an airtight PM program to conserve construction equipment, and possibly at minimum maintenance cost. But he's got to learn, right from the start, to examine the effect of all facets of the PM program on costs. It might seem engineering folly to let a $500 motor go to ruin to keep a job going. But when balanced against a loss of $2000 in work in process because of lost time, it makes sense.

Master the principles There is no ready-made, on-the-shelf PM program for any contractor. It must be tailor-made—measured and cut to fit individual requirements. I have studied hundreds of PM programs and never found two exactly the same.

The reason should be clear. There are no two contractors identical in size, age, location, equipment, or types of jobs tackled. They differ in organization, operating policies, and personnel. Problems of maintenance differ.

There are, however, resemblances in objectives and basic principles, if not in engineering or paperwork. To anyone seeking a ready-made program—and the majority of beginners seem to pursue this will-o'-the-wisp—I can only give this advice: Learn PM principles first, and let the paperwork follow. Paperwork is important. But it can cost more than it should if it's the wrong kind.

Where to start PM Consensus is that it's too big a bite to apply PM to the entire company at once. It's best to build up the program in pieces. How fast you do it isn't significant. When one piece is finished, start the next.

Is it best to tackle one job site at a time or one type of equipment over the entire company? Opinion seems divided. So why not decide on the easier? Local conditions probably rule which approach is best. Another factor is how well PM has been sold. If you have to show rapid proof of the value of PM, start where you think it's needed most and therefore will pay the biggest dividend more quickly.

Basic problem For the sake of simplicity, view PM primarily as the function of minimizing breakdowns or harmful depreciation of equipment, through periodic inspections, to discover and correct unfavorable conditions. The entire program hangs on inspections and their related duties of adjustment and repair.

Inspections are costly in labor and sometimes in equipment lost time. They are the key point in the cost of a PM program. The fewer inspections needed, the lower the cost. The problem, therefore, is to strike a favorable balance between this cost and the cost of not utilizing PM.

WHAT TO INSPECT IN PM

By far the most wanted information from those starting a construction equipment PM program has been: "Can you give me a list of what items to include and how often to inspect?" It would be nice if we could, but it isn't that easy. Generally, every piece of equipment in regular use—from dozers to electrical control systems—should be inspected; and backup equipment (outdated but still serviceable graders, for example) should also be included.

In addition, all safety equipment should be inspected frequently both for employee safety and to meet Occupational Safety and Health Administration (OSHA) requirements.

A good PM program will also include special measures for preservation of equipment during off-season layups, scheduled vacations, unexpected labor troubles, or weather catastrophes.

What Not to Inspect For

Up to this point the approach to what to inspect has been made purely for good maintenance or physical or operating conditions. On that basis alone an engineer would be inclined to include everything that wears out or is likely to cause lost time. This is where the economics of PM must step in to sift out unprofitable activities. There is no need to inspect everything.

How do you decide what not to inspect? An overall analysis such as the following will help:

    1. Is this a critical item? If failure will cause a major work stoppage, or costly damage, or harm to an employee, need for PM is almost certain.

2. Is standby equipment available in case of failure? If the duty can be easily shifted to other equipment, need for PM is contingent on other factors, such as cost of breakdown maintenance.

3. Does cost of PM exceed expense of lost time and cost of repair or replacement? If it costs no less to tear down a machine to repair a repetitive wear point than the overall cost of the repair itself, the value of PM is highly questionable.

4. Does the normal life of the equipment without PM exceed needs? If obsolescence is expected sooner than decay, PM may be a waste of money.

CHECKLIST FOR MACHINE INSPECTION INSPECT AND CHECK OFF FOLLOWING ITEMS SHOW ALL DEFECTIVE ITEMS WITH X MARK										
TO BE CHECKED	DEF.	OK	DEF.	OK	DEF.	OK	DEF.	OK	DEF.	OK
Motor		✓		✓		✓		✓		✓
Bearings	X			✓		✓		✓	X	
Gears	X		X		X		X		X	
Clutch		✓		✓		✓	X			✓
Brake		✓	✓	X		✓		✓		✓
Piping		✓		✓		✓		✓		✓
Guards		✓		✓	X			✓		✓
Lubrication	X		X		X			✓		✓
Rolls		✓		✓		✓		✓		✓
Cutter		✓		✓		✓		✓		✓
INSPECTED BY DATE	H.V. 6-1		H.V. 7-1		H.V. 8-1		H.V. 9-1		H.V. 10-2	

Fig. 4-1 PM checklist. Simple checklist can be used for different types of machinery. Items in to-be-checked column are set up permanently for each type and revised when necessary.

What to Inspect For

When drawing up a list of construction equipment items for PM you must have given thought as to why you need to inspect them. Now comes the job of determining what physical parts of each piece of equipment need attention.

Service manuals Experience isn't enough to draw on. One of the best sources is the service manual issued by the equipment manufacturer. It's an invaluable guide to what and when to inspect, as well as how to install, service, and maintain equipment. In appreciation of their value, many contractors accumulate and file service manuals. Purchase orders specify two or more copies. One goes to the central file, another to the maintenance foreman, and often a third to an area file for mechanics. This growing demand for service manuals—incidentally the government insists on them—has put so much pressure on manufacturers that rarely is one not available for new construction equipment.

Checklists After going to all the trouble of developing a list of equipment and their inspection points, how do you make sure they are not overlooked? This is done by a checklist. In principle, a checklist itemizes for the inspector all points to be checked on any one piece or type of equipment. It provides spaces for dates and initials to show when inspected and by whom.

Checklists have other advantages. They assume uniform and complete inspections regardless of who does the job. They are invaluable when new inspectors or substitutes are needed or when rotation of inspectors is practiced.

A simple form is shown in Fig. 4-1. The layout is not important. The problem is more what to put on it. The goal should always be simplicity and a minimum of paperwork. Make sure no machine part or item is omitted that needs attention. But also see that inspection costs are not inflated by needless checks and tests.

There is some danger that a checklist will make an inspector feel it lists everything he needs to inspect, no more. To avoid discouraging his ability or imagination, you can provide a space on the form for extra comments. Challenge the inspector's pride with a question such as "Do we need to include anything else in the next inspection?" Like PM inspection lists, checklists need continual refining and updating. Experience usually cues the need for checking overlooked items.

HOW OFTEN TO INSPECT—FREQUENCY

Other than what to inspect, most people ask for a ready-made list of how often to inspect. This decision probably has the most bearing on costs and savings of a PM program. Overinspection is a needless expense and may involve more lost time than an emergency breakdown. Underinspection results in more breakdowns, earlier replacements. Good balance is needed to bring optimum savings.

No ready-made list is available. You must work out your own values. Age and kind of equipment, environment, types of operation, and similar factors must be considered. No two companies are alike.

But you can get helpful timetables from many equipment manufacturers. They are careful to qualify their recommendations by saying the tables apply to normal conditions. They may describe what these normal conditions are. And they may suggest you trim the normal figures by various percentages for special exposures or types of service.

Start with engineering analysis There is no dearth of data on inspection frequencies for any type of construction equipment. The big job is to assemble data, sort out what you need, and then adjust the data to your company conditions.

First step in gauging the best frequency cycle is an engineering analysis of equipment from the following viewpoints:

1. *Age, condition, and value.* Older and poorer equipment needs more frequent services. But, if ready for the junkpile or soon to be obsolete, equipment may be cheaper to inspect on a skeleton basis or not at all.

2. *Severity of service.* More severe applications of identical equipment require shorter cycles.

3. *Safety requirements.* Allow a wide margin for safety.

4. *Hours of operation.*

5. *Susceptibility to wear.* What is the exposure to dirt, friction, fatigue, stress, or corrosion? What is life expectancy?

6. *Susceptibility to damage.* Is the piece of equipment subject to vibration, overloading, or abuse?

7. *Susceptibility to losing adjustment.* How will madadjustment or misalignment affect it?

It is best to follow manufacturers' recommendations until you have good reason to alter them. If in doubt, err on the safe side. Seek the data involved in carrying out the following procedures:

1. *Service records.* Dig out whatever data on costs and performance you have—equipment records, lost time reports, routine maintenance schedules. They are excellent clues not only as to what to inspect for but also how often.

2. *Maintenance work orders.* Sort out completed orders by individual equipment items or functions, if you don't already have an equipment record. Then analyze the nature of the repairs.

3. *Craftsmen.* Get the benefit of their close experience, including lubrication men.

4. *Foremen.* Ask them how often they think service is needed.

Gradual refinement needed Once you've decided on frequency cycles, you've only begun. You have to check the results continually and be willing to modify cycles to meet operating requirements. This is true of all facets of the PM program—beginning with what to inspect and what to inspect for. You have to add or subtract, even right-about-face. When you find yourself replacing parts that are still good, you may be playing too safe. Cost of discarding good parts adds to the cost of PM.

Periodic appraisal of inspection programs is necessary to keep on an even keel. Many contractors, especially larger ones, have been refining and tightening up their programs for several years or more. Frequency cycles are the chief targets.

To ensure a continuing attack on the validity of a cycle, some contractors use these methods:

1. Cut and try. Whenever a unit is inspected or repaired, decide when next to inspect.
2. Check new equipment more frequently until broken in.
3. Require inspectors to indicate on checklists or inspection reports whether frequency cycle needs a boost or can take a cut.

Statistical checks How can one know whether he is overmaintaining, undermaintaining, or inspecting just the right amount? It's a matter of individual analyses of actual results. On one side of the ledger are number and costs of inspections and services; on the other side, number and overall costs of repairs and breakdowns. If there are no repairs, chances are you're overmaintaining. If there are too many repairs, the inspections aren't getting at the root of the trouble. The dollar-wise appraisal always helps you arrive at a good balance.

Another way to evaluate the success of PM is by a comparison of scheduled maintenance (PM or routine repairs) with unscheduled maintenance (emergency repairs). Good maintenance engineers insist on a monthly report of these types of work in maintenance man-hours. Too much unscheduled work points to lack of PM. This may mean too few units inspected or too low an inspection frequency. No unscheduled work is the other extreme. What is a good ratio? Some contractors boast of 80 to 90 percent man-hours on scheduled work, including repairs turned up by PM inspections.

WHEN TO INSPECT—SCHEDULES

Up to this point we've decided what to inspect and how often. The next step is making up a work schedule to include each PM item. Theoretically, a schedule must be perfect in coverage. If it overlooks a single item, some kind of trouble will pop up later. Failure to check brakes, for example, might invoke costly penalties if they don't function when they should.

Ordinarily, scheduling involves a determination of calendar inspection dates that will fulfill the frequency requirements in the most efficient way. This is not always possible, particularly in the case of on-site construction equipment. In setting up schedules, the maintenance engineer must be continually conscious of his responsibility to keep production going at the lowest overall cost. He must arrange schedules to be adaptable to on-site needs.

Practically every company can divide its PM inspection and service functions into three groups:

1. *Routine upkeep.* This type of work is done at regular short intervals—adjusting lubricating, cleaning—while equipment is operating or productively idle.
2. *Periodic inspections.* Covers work at prescribed intervals on equipment that is running or shut down—visual inspections, teardown inspections, overhauls, and scheduled replacement of parts.
3. *Contingent work.* Includes work at indefinite intervals when equipment is down for other reasons.

Obviously, the more PM work you can squeeze into the contingent category, the less costly it will be. To ensure that the work will be done, you can schedule such items on a which-comes-first basis (1) by listing these PM items on the work-planning sheet kept for repetitive maintenance jobs and (2) by protecting yourself by a so-called tickler item timed for the maximum allowable period between inspections. Along the same lines, you can pile up a list of repair jobs that have been uncovered by running inspections and decide which can be deferred.

In scheduling PM for routine upkeep or periodic inspections, the following are acceptable goals:
1. Handle them on the regular shift, preferably, to minimize overtime.
2. Remember to schedule PM work for slack seasons whenever possible.
3. Shoot for the least amount of productive downtime.

Types of schedules There are many designs and layouts, but schedules generally are of two main types:
1. Overall charts, which list on one large sheet every piece of construction equipment and type of work done
2. Individual cards, which usually means a separate card for each piece of construction equipment or type of work done

3-22 The Economics of Construction Equipment Maintenance

The overall chart is the simpler approach and gives a quick picture of the PM workload. A typical schedule lists days or months across the top and itemizes equipment down the left side. Dates for inspections are shown by a check mark or cross in the appropriate columns all across the chart. Different symbols can be used to show cleaning, adjustment, or overhaul. The chart can be used merely as a master list to originate PM work orders or also as a blanket order for the whole year.

Many small and large contractors are using card schedules. They have the advantage of holding more details on PM requirements and are often combined with equipment records. In this case the card bears a series of dates along one edge for a movable tag which acts as a visual signal when the next inspection is due.

Job scheduling In the true sense the date scheduling by chart or card is actually only preliminary programing. The real scheduling takes place when a definite day has been set and the job has been planned as to method, tools, and equipment. Whoever does the final scheduling—the dispatcher, area-maintenance engineer, or foreman—must analyze the job for the skills needed and the time required. Some contractors have established task limits by an easy historical analysis of past performances; others have used time studies.

Application of good scheduling practices to PM raises some popular questions:

Should every item be scheduled, no matter how small? Consensus is yes. If an inspection is worth making, it's worth scheduling in some way or another.

What should be done when a schedule is not completed on time? If the change is permanent, revise the schedule up or down as needed. If it is temporary, ignore the schedule when activity drops, or slot in a special inspection when it jumps (if length of usage is an important factor).

How do you handle equipment that needs PM servicing more than once a day? Arrange for an area craftsman to service on a standing order. Otherwise, reroute your PM inspector or serviceman.

WHO INSPECTS—ORGANIZATION

Nobody has to upend the maintenance organization chart to install a PM program. It can be made to fit any type of setup with minor changes. For PM is more a philosophy of operation than a method. Other than the possible addition of one or more clerks and engineers, the same personnel as a rule will be able to execute the program.

Some companies like to handle PM by a separate division of inspectors, crafts, and supervisors. They say it protects PM against domination by other maintenance functions. Others prefer all maintenance work—routine and PM—to be done by the same force. But they avoid neglect of PM by giving it priority. One contractor sets up a separate budget for PM manpower. In either case, both views agree on the following principles:

   1. Don't allow PM work to be interrupted by other maintenance work. It's a great temptation to relax at the start to keep repairs moving. But eventual decrease in breakdowns will release enough men from repairs to keep on top of everything.

   2. The work routine generated by PM should follow such administrative principles as regular maintenance for authorization, accumulation of labor and material costs and reports. To eliminate the need for special forms, one company simply stamps PM on regular forms to differentiate.

   3. The PM function should be supervised by the same engineer who directs all other maintenance. He may need assistance, but PM needs his direction.

If you have a maintenance force of 100 or more, you'll probably need a full-time PM administrator. He plans and schedules all necessary inspections, overhauls, services, and repairs. He issues PM work orders for supervisory approval. He follows all PM jobs to completion. In smaller companies, the maintenance engineer usually can supervise the program himself, provided he has the clerical help to process the paperwork details.

What type of inspector? In the switch from breakdown maintenance to PM the new role of inspector raises some questions. What makes a good inspector? Should he be a specialist? Should he inspect full time and do no repairs? Should the job be rotated among qualified craftsmen?

Consensus is that a good PM inspector is generally a craftsman with top skills, one who has the ability to test, adjust, and repair the equipment he inspects. It helps if he's a trained troubleshooter. Lubrication men, say many, cannot always diagnose trouble.

Where do You Get Inspectors? Right from your own force. Train a competent craftsman in the philosophy of PM, and show him the few paperwork requirements—he'll catch on.

As to developing specialists, opinion seems about evenly divided. And both sides claim advantages. Those in favor of full-time men say they get better inspections. They disapprove of men inspecting their own work. Besides, they feel that only trained specialists should be used on hydraulic or electronic equipment.

Those in favor of rotating the inspection duties among qualified craftsmen present these views: Inspectors do a better job because they know someone else will follow. Such men have a fresh viewpoint, and if they don't see the previous report they can be more impartial in checking each other. There is no need to set up a new classification or argue about special pay. They steer clear of jurisdictional disputes by assigning the proper crafts. Also, they have substitutes for illness and vacation absences.

In all cases, pro and con, the practice is fairly unanimous to make simple adjustments during inspections but to avoid lengthy repairs. Some contractors set arbitrary time limits for repairs of 15 min to 2 hr to enable the inspector to complete his inspection schedule.

What About Pay? There's only a general pattern to guide you. If craftsmen are rotated through inspection assignments, usually they earn their regular rate. Full-time inspectors may be on hourly wages or on salary, depending on the level at which they operate. If the inspector is at the foreman's level in job classification, he should get foreman's pay. In short, qualifications and rank should decide.

Inspection reports Reports are indispensable. Your aim is to cut field paperwork to a minimum but not to the point where it doesn't tell the full story. So don't abbreviate inspection reports just to save writing time; you may lose out on full benefits of the cost of inspection.

The simplest report form is one which can be universally applied. A typical form has headlines for spaces for filling in general data. Below this there are ruled lines for filling in items that need attention. It can be used for all types of construction equipment. When using such a form a separate reusable checklist for each type of equipment is needed.

In many cases the checklist and report form are combined. The data are all preprinted. The inspector merely checks or OK's each item on the list and explains in a Remarks column for comments whatever items need attention and how soon.

There are many variations of either type, again an evidence of the tailored-to-meet-the-need approach. Three important questions can go on any form: (1) What is the cause of the failure or defect? (2) What is the remedy? (3) Have you any ideas for improvement in machine, equipment, inspection methods, or tools?

Routing inspection reports Completed inspection reports become the basis for maintenance work orders in the same way that requests originate from regular work production or maintenance people. Many ask who should request a work order for repairs found necessary by a PM inspector. The answer is whoever normally has to authorize that type of work in the organizational setup.

In some companies the reports are screened through maintenance foremen (1) to be kept updated on conditions, (2) to check the need when severe or urgent, (3) to coordinate with on-site foremen for repair, and (4) to decide which one should generate a request for the work order. Other contractors route all inspection reports to a maintenance clerk or dispatcher who automatically issues work orders on minor jobs and refers major jobs to the maintenance engineer for decision. In one large company the inspectors themselves prepare work-order requests at the close of each day.

A good rule is that unsafe conditions or major defects deserve special routing immediately when discovered.

Checking inspections What evidence has a maintenance engineer that his inspectors are doing a thorough job and that PM repairs are not being neglected? I know of one engineer who makes a practice of spot checks to see that firmly established inspection procedures are being followed. Any contractor that has too many breakdowns on equipment covered by the inspection schedule has good reason to suspect the quality of inspections is poor. But first see whether the item is on the checklist and how many inspectors reported it. Was an urgent recommendation postponed?

As to neglect of repairs, usually a good work order will ensure timely completion of work orders arising out of PM. That is one advantage of integrating a PM program with normal paperwork procedures. Anyone who doesn't have this centralized control of all maintenance work certainly needs some kind of separate watchdog on PM jobs. Foremen who schedule their own work without master control are often prone to let PM work slide. One way to keep tabs on PM repairs is to hold inspection reports in a tickler file until repair orders are completed.

Inspection methods Because PM inspections are highly repetitive, good methods and procedures will pay big dividends. Planning (as contrasted with scheduling) is the development of step-by-step procedures needed to do the work. Time studies help effectiveness. An engineer can contribute much in these studies.

The following are some avenues particularly worth exploring:

1. Study of methods of inspection and servicing for a better or faster job or to cut costs. Perhaps inspections can be combined with other maintenance work. How about centralized lubrication?

2. Planning of methods of inspection and of routes to cut walking time. Give the men job writeups for each piece of equipment inspected in any quantity or at great frequency. Sometimes it is possible to arrange checklists in sequence of operations on a specific construction site. Also include safety precautions and safety equipment needed.

3. Planning of overhauls, similar to inspections, to cut lost time. This requires best utilization of crafts and timesaving coordination. Often replacement parts can be prefabricated and made ready to shove in fast.

4. Review of major repetitive overhauls after completion to see where planning for next one can be improved.

5. Provision of better tools and test instruments. How about a work cart, scooter, or mobile unit to carry equipment and spare parts? Will torque wrenches or power tools speed up inspections? Ask the men themselves for suggestions on better tools—they will undoubtedly have some good ideas.

6. Redesign of equipment to speed up inspections. Perhaps you can substitute quick-detachable fasteners or screws or bolts on guards, panels, or access plates.

Inspection manuals Judging by the reports from plants using written PM instructions of one kind or another, manuals are almost a must. A manual goes beyond the checklist stage in that it gives detailed practices and procedures on maintenance of all important construction equipment.

A typical manual tells how to operate and service physical properties, what materials and tools to use, and what safety measures to take. It devotes a single page to each kind or type of equipment and is well indexed for easy and quick reference. Everybody in construction equipment maintenance can use it—foremen, planners, inspectors, and craftsmen.

Manuals can often take several years' work to assemble. Sources of information are manufacturers' catalogs and instructions, company records, and maintenance experience and know-how. A good way to start is gradually to build up a file of procedure sheets on items where simple instructions won't do. Make them as brief as possible and write them in the user's language, or they won't be used. Use lots of close-up pictures or diagrams for arrangements of parts where words will not be as clear.

Standard procedures are also helpful for troubleshooting. An inspector may use know-how to check, adjust, service, or repair equipment and still not be able to troubleshoot. In complex equipment, such as highly specialized hydraulic or electronic equipment, he must put his finger on the trouble. More and more contractors are already issuing procedures for equipment, such as motors, on a symptoms-cause-cure basis. But these won't always work in hydraulics or electronics. In such cases you can ask the maker to devise a troubleshooting sequence of testing where the next step is dictated by the results of the previous one.

PAPERWORK FOR PM

The biggest obstacle in the adoption of a PM program seems to be paperwork. The inexperienced maintenance engineer envisions a mountain of details and wants none of it.

There is no need for paperwork to be top-heavy or burdensome. But both the maintenance engineer and higher management must realize that the cost of good paperwork is justified by the results.

The Role of Preventive Maintenance 3-25

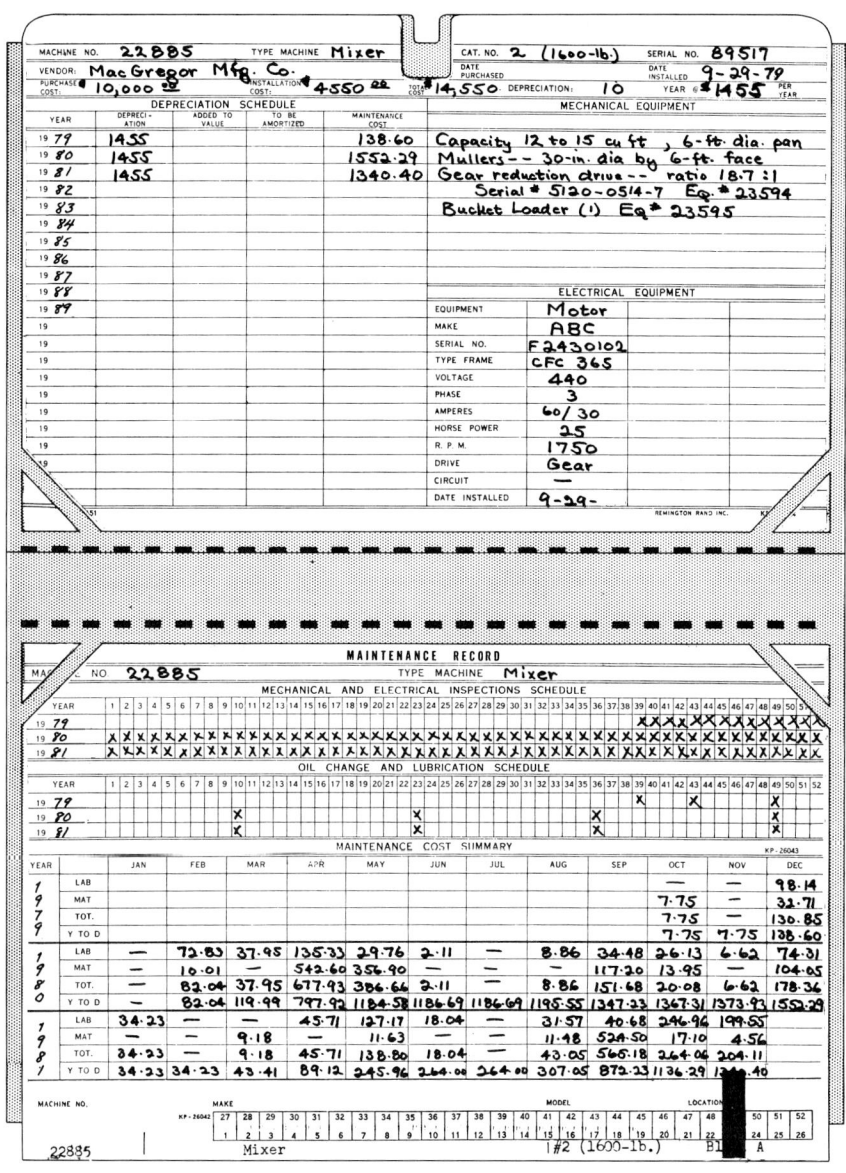

Fig. 4-2 Maintenance-cost record. Many contractors use a manually posted record of maintenance costs in combination with an equipment record and inspection schedule. The setup consists of two forms for each item, kept in one of about 75 hinged cardholders in a flat metal tray. As leaves are flipped over, the top card shows equipment data, the bottom card shows schedule and costs. Tab at bottom is moved to next inspection date. When bottom card is completely filled, it may be retained as a permanent record in the leaf and superimposed by a fresh card. In such case the lower part showing machine number and schedule positions can be trimmed off.

No matter what paperwork system you select—and there are many good ones—there are some basic guides to follow:

1. Minimize the number of forms and entries. Don't try to record information just because it's nice to have. But don't eliminate data to the point where records cannot be interpreted and thus lose their usefulness. Most contractors include total labor and material costs on an equipment record for every job, and note the nature of the job and its number. If more details are wanted, they can refer to the completed-order file.

2. Integrate the PM system with other maintenance paperwork procedures. A good PM program won't stand alone. It has to be meshed with regular maintenance. This is particularly true in paperwork flow. Methods and routines for PM work orders, time reports, material requisitions, and cost accumulations should coincide with regular maintenance procedures.

3. Make sure costs of all primary PM inspection activities are accounted for. Only in this way can you prove to yourself and higher management what the exact costs are and how well you are doing.

4. Arrange for a periodic control report, say, once a week or month, to check on PM performance. Such a report might summarize the number of inspections scheduled, completed, uncompleted (and why), number of work orders originated by PM, and number completed. As the program gets going the number of work orders will slowly drop and smooth out to a fairly even flow. Also, consider using this same control report as the basis for keeping management posted on PM. Include it in a regular report that is good practice to send to higher management on all maintenance activities.

Five forms Any PM program adds a maximum of five basic forms to conventional maintenance paperwork: equipment record, checklist (Fig. 4-1), inspection schedule, inspection report, and equipment- and maintenance-cost record (Fig. 4-2). In most cases some of these are conveniently combined—such as equipment inventory, inspection schedule, and equipment cost record on one form, or checklist and inspection order on another—leaving only two or three forms to cope with at most. By now the function of each form should be clear enough to enable anyone to interpret these forms and design a set for himself.

The bigger problem in designing paperwork, as already hinted, is in the choice of type of paperwork and flow rather than content of forms. Several systems may satisfy requirements, but what about operating cost? The simplest card system will work in a large company as well as in a small one but at tremendous if not prohibitive cost. It may also be too complicated for small contractors but quite manageable for large ones. Size of the company and its operations as well as the number of details wanted largely govern the best methodology.

Section **4**

Basic Maintenance Technology

Chapter **1**

Corrosion Control

DR. LEWIS W. GLEEKMAN
Materials and Corrosion Engineering Services, Southfield, Mich.

To the construction equipment maintenance engineer, corrosion is best defined as the deterioration of any material in contact with its surroundings. This applies not only to metal parts of equipment but also to plastics and protective coatings. While the cost of corrosion damage to equipment and accessories (estimated at $6 billion per year) is huge, more critical costs are involved in lost man-hours caused by equipment failure.

Learning to identify various types of corrosion and knowing what to do about them are important parts of the job for today's construction equipment maintenance engineer or foreman.

Corrosion may be uniform or localized. *Uniform corrosion* means that a piece of equipment is being attacked by wet or dry chemical or electrochemical forces over most or all of its surface. *Localized corrosion* can be divided into two forms, one too small to see and the other visible to the human eye; both usually affect only part of a piece of construction equipment.

TYPES OF CORROSION

The following are basic types of corrosion: galvanic, erosion, crevice corrosion, pitting, selective leaching, exfoliation, intergranular corrosion, and stress cracking.

Galvanic Galvanic corrosion occurs when two different metals are in contact in the presence of a conductive solution. In galvanic corrosion, the more active metal deteriorates while the less active or noble metal is protected. To minimize galvanic corrosion it is best to avoid having dissimilar metals in contact. A second method places insulation between dissimilar metals.

Erosion A selective type of corrosion, erosion, is caused by the motion of a corroding solution over a metal surface. Such motion removes the protective film, usually by mechanical wear. Erosion appears as smooth-bottomed shallow pits with a very directional appearance. It is frequently seen on threaded areas and elbows.

Cavitation, a special form of erosion, is caused by the formation and collapse of vapors at the metal surfaces. The change from high to low pressure disturbs the base metal by removing its normal protective film. *Fretting,* another form of erosion, occurs when metal slides over metal and causes mechanical damage to either one or both of the metals. Fretting is most commonly caused by vibration.

Crevice corrosion Crevices normally exist at lap joints, gaskets, or around bolts and rivets. Crevices are also created by deposits of corrosion products on a surface or scratches in the paint film. Acidity changes in the crevice, lack of oxygen, detrimental metallic-ion buildup, and depletion of a corrosion inhibitor are other causes.

Materials such as stainless steel and titanium, which depend on an oxide film to achieve corrosion resistance, are more susceptible to crevice corrosion than others.

Pitting A highly localized type of attack, pitting can be prevented in many cases by the use of an inhibitor such as sodium or potassium dichromate. Surface cleanliness and selection of materials known to be resistant to pitting are the best ways of avoiding the problem.

Selective leaching Sometimes called *parting corrosion,* selective leaching is the removal of one element from an alloy. It occurs in many alloys, though it is most frequently found in copper-based alloys. The addition of small amounts of arsenic, antimony, or tellurium inhibits this form of corrosion in brass, but the best method of prevention is the use of nonsusceptible alloys.

Exfoliation A variation of selective leaching is exfoliation, which is surface corrosion spreading below the surface. It differs from pitting in that the attack has a laminated appearance with whole layers of material eaten away in the form of a flaky or blistered surface.

Intergranular corrosion This is an attack concentrated at the grain boundaries without appreciable corrosion evident on the grains themselves. Austenitic stainless steels, high-nickel alloys, and aluminum alloys are most frequently involved.

Stress cracking This cracking results from residual or applied stresses plus corrosion and usually occurs without notable loss of metal in the form of uniform corrosion. Most alloys are susceptible to this problem, but the number of combinations of alloy and corrosion that cause it are relatively few. Stresses that cause cracking arise from residual cold work and thermal stresses on contraction after welding or after other thermal treatments; they may also be applied externally during use of the equipment.

METHODS OF STOPPING CORROSION

The following methods are the key to corrosion prevention: change material, change the environment, protect the material.

Change material The practice of using a more corrosion-resistant material can mean a change in alloying or switching to nonmetallic parts—plastics, with or without reinforcement, for example. The matter of changing material is not merely a function of selecting a material that has improved corrosion resistance; other factors such as thermal and electrical properties, ease of fabrication, strength, availability, and cost must be considered.

Change the environment To reduce corrosion, the easiest and most obvious method is to lower the temperature. Corrosion processes are chemical reactions, and every 18°F decrease in temperature reduces the reaction rate by half. Other environmental changes involve agitation, aeration, and velocity.

Protect the material Isolate the metallic surface from the corrosive environment with either organic or metallic coatings on the surface. Organic coatings can be thick or thin, paint film or solid linings, or plastic in the form of tape, sheet, or powder fused to the surface. Metallic coatings can be applied as electroplated materials or deposited by chemical means in an electrodeless deposition similar to silvering glass to make a mirror. A metal part may be coated at moderate temperatures by diffusion as from applying zinc in galvanizing or aluminum in aluminizing or metallizing of the surface by spray application of partially melted materials.

Design factors In addition to these methods, choosing equipment designed to minimize corrosion is recommended. The following design factors should be considered when buying equipment:

1. Butt joints should be used where possible.
2. Equipment that must be washed and drained should have an effective system that allows prompt attention to this detail.
3. Means of access for inspection and maintenance should be provided at all necessary points.

4. Dissimilar metals should not be used in contact with each other where such use can be avoided.

5. Localized turbulence and areas of high velocity at feed and drain connections of fittings and lines should be minimized where possible.

6. Equipment should not be allowed to rest in liquid or damp material (including damp earth) and should be promptly cleaned of such material. Porous material should be waterproofed.

NONMETALLIC MATERIALS

Use of nonmetallic materials for some equipment parts has proven very successful in lowering corrosion.

Plastics Although limited in use in the past, plastics are becoming more useful for construction equipment parts with the development of reinforced and thermoset materials. Some plastics are now capable of withstanding thousands of pounds of pressure and high temperatures. Polyethylene, polypropylene, polyvinylchloride, the styrene–synthetic rubber blends, the acrylics, and the fluorocarbons are thermoplastics which now have relatively high thermal distortion temperatures. Thermoset (materials not softened by heat) reinforced plastics include polyesters, epoxy and furane resins, and phenolic and epoxy resins. Polyvinyl-chloride, polypropylene, and reinforced polyesters are beginning to replace stainless steel, lead, and galvanized steel in some equipment parts—particularly pumps, blowers, and fan wheels. Nylon is becoming more common in bearings and other parts. Plastics are generally not subject to pitting, stress corrosion cracking, or other forms of corrosion common to metal. They also do not require protective painting. However, some plastics do present stress problems when used in combination with metal parts.

A chemically resistant plastic, tetrafluoroethylene (Teflon or Halon), which withstands temperatures to 500°F, is being used for gaskets, "O" rings, seals, and other relatively small molded items. Loose linings, including nozzle linings, also use this material. Polyethylene, a lower-cost plastic, is being used for piping and tubing.

Fire-retardant grades of most plastics are now available and should be used in construction equipment parts where required. This is particularly true of plastics such as urethane and epoxy foam used as insulation.

Rubber and elastomers For many years, these materials have been used where chemical corrosion is common. Natural rubber compounds resist a wide variety of chemical solutions, yet are readily attacked by strong oxidizing acids. The temperature limit for soft rubber compounds is about 140°F and about 180°F for hard rubber compounds. Elastomer compounds extend these ranges, yet still provide natural rubber's corrosion-resistant qualities.

Wood, concrete, and glass are other materials that offer good resistance to corrosion in some cases, but their application in construction equipment is limited.

PROTECTIVE COATINGS

Protective coatings are probably the most widely used and, at the same time, the most controversial material used to minimize corrosion of metal and certain other materials. While protective coating and painting provide protection, such protection is limited by conditions under which the equipment is used. Any coating will have some pinholes or holidays, and these points require continuous maintenance. Faults in a coating can also be minimized by adding extra coats of the protection or by increasing the degree of the coating's cure. Baked phenolics, baked epoxies, air-dried epoxy, vinyl, and neoprene are some of the coatings commonly used.

INHIBITORS

The corrosion of iron and other metals can frequently be minimized or inhibited by the addition of soluble chromates, phosphates, molybdates, silicates, amines, and other chemicals singly or in combination. These are most useful for equipment used in a closed system requiring constant immersion of construction equipment.

REFERENCES

Ailor, W. H.: "Handbook of Corrosion Testing and Evaluation," sponsored by Electrochemical Society, John Wiley & Sons, New York, 1971.
"Corrosion Data Survey, IV Edition," NACE, Houston, Texas, 1968.
Fontana, M. G., and Greene, N. D.: "Corrosion Engineering," McGraw-Hill Book Company, New York, 1967.
LaQue, F. D., and Copson, H. R.: "Corrosion Resistance of Metals and Alloys," 2d ed., Reinhold Publishing Co., New York, 1963.
Seymour, R. B., and Steiner, R. H.: "Corrosion Resistant Plastics," Reinhold Book Corporation, New York, 1955.
Uhlig, H. H. (ed.): "The Corrosion Handbook," sponsored by the Electrochemical Society, John Wiley & Sons, Inc., New York, 1958.

Chapter **2**

Equipment Parts Cleaning

CHARLES M. LOUCKS
Consulting Chemist, Arlington Heights, Ill.

When fouling of construction equipment has occurred because of corrosion and water scaling, chemical solvents may be used for the necessary cleaning. The methods are discussed in this chapter.

A large construction firm may have facilities for cleaning by means of tanks, jets and steam jennies, vapor degreasers, electrolytic cleaners, or ultrasonic transducers. Such facilities are used mainly for removing soil or rust from external surfaces of metal parts.

Quite another problem is the cleaning of inaccessible interior surfaces when equipment items are too big to be handled or disassembled. Such to-be-cleaned construction equipment becomes the containing vessel when solvents are pumped in by means of special truck-mounted tanks, pumps, mixers, and heaters.

Cleaning materials The removal of corrosion products and/or scale from large construction equipment requires large volumes of cleaning solvents, perhaps 1000 to 100,000 gal.

Acids and alkalis Much of such cleaning can be done with dilute solutions of relatively inexpensive acids and alkalis. Solutions of soda ash, caustic soda, phosphates, or silicates, plus synthetic detergents for better wetting and emulsifying, will remove oil, grease, and general soil when applied with heat and turbulent movement. Alkalis are also used after acids have been applied for scale removal. This assures that acid residues have been neutralized.

The most common acid solvent is inhibited muriatic acid. This acid is inhibited, as all acid solvents must be, to reduce chemical attack on metal surfaces to an acceptable level. Muriatic acid is cheap and effective. It forms reaction products that are generally water-soluble so they are removed in the used solvent. Sulfuric acid is seldom used for the opposite reason, insoluble reaction products. Nitric acid cannot be prevented from attacking carbon steels or copper alloys. It has certain special applications where the substrate metal is stainless steel or aluminum. Of special interest for small-scale construction equipment maintenance cleaning is sulfamic acid, mainly because it is a dry, solid product that is safely handled. It has acid properties only after it is dissolved in water. Solid inhibitors and wetting additives can be premixed in the packaged product.

Sequestrants Recently a class of alkaline salts called *sequestrants* have come into general use for prevention of scale formation and for periodic removal of both water scale and corrosion products.[1]* The most useful examples are derived from an organic acid called ethylene-diamine-tetra-acetic acid (EDTA). The sodium salt dissolves water-hardness scale. The ammonium salt is used to remove iron oxides and copper.

Another use for sequestrants such as EDTA and sodium gluconate is in the alkaline

*Superior numbers refer to references at the end of the chapter.

rinse used after a conventional acid stage.[2] The sequestrants prevent the precipitation of dissolved metal ions by the alkali. Less rinsing saves time and rinse water.

Thiourea and its derivatives can form acid-soluble complexes that are used to prevent dissolved copper from plating out of an acid solution onto steel surfaces.

Synthetic detergents and acid inhibitors Although they serve different purposes, both synthetic detergents and acid inhibitors are large organic molecules that are attracted to surfaces. Synthetic detergents are attracted to oil-water interfaces where they promote wetting, emulsion formation, detergency, and foam. Acid inhibitors are attracted to metal surfaces where they interfere with the chemical reaction of acid on metal.

Organic solvents Certain relatively small-volume cleaning jobs require nonwater solvents for removing oil and grease. Shop cleaning of engine parts may be done with Stoddard solvent, kerosene, or diesel fuel. On-site degreasing of larger parts is more likely to employ the chlorinated solvents trichlor or perchlor, which are nonflammable but have toxic vapors. Freons, with both chlorine and fluorine atoms in the molecules, are nonflammable and relatively nontoxic, but the cleaning uses are limited because of cost. They are especially recommended for cleaning electric motors.

Specialty cleaning products There are many packaged proprietary products used for construction equipment maintenance cleaning and housekeeping. The quantities required each day are small. A purchaser buys the special formulations and convenient packaging provided by the vendor.

For cleaning large equipment the vendor is likely to be an outside contractor who handles bulk chemicals by the truckload and does his own formulating.

METHODS OF APPLICATION

Fill and empty In cleaning the intricate internal surfaces of tank trucks or other carrier and storage vessels where exposure to water has led to formation of corrosion products and scale, the equipment can generally be filled with a liquid solvent and, at the proper time, emptied by opening a drain valve.

A vessel is pumped full of hot inhibited acid, for example. The acid reacts with corrosion products and scale; then the drain valve is opened and the solvent is removed. Rinsing and neutralizing solutions are handled in the same way. The mechanical requirements are simple. There are tank trucks to haul a liquid acid, such as inhibited concentrated muriatic acid, to the site. A water line to the truck position supplies water. A steam line furnishes steam for heating as the acid is diluted and pumped into the vessel through a temporary pipeline from the temporary acid pump. Simple connections will be needed at the fill line and, if necessary, on the vessel. A vent line allows air and gases, including hydrogen, to escape from the high point on the equipment tank. A gauge glass will indicate the final solvent level unless the vessel is to be filled to overflowing at the vent.

During the reaction time the service engineer in charge takes samples, runs analyses, records temperatures, and decides when the reaction period is finished. The solvent is drained to waste through another temporary line provided for that purpose. It may have been decided to blanket the internal surfaces with nitrogen gas during the draining and rinsing operations. In that case, nitrogen is admitted at the vent connection as draining proceeds.

Other decisions will have been made regarding the rinsing and neutralizing. It may be agreed to keep the rinse water slightly acid to prevent precipitation of metal salts. This can be done with very small amounts of any acid. Citric acid is often used. Or it may have been decided to reduce the number of water rinses by using immediately an alkaline solution containing the neutralizing agent such as soda ash plus sequestrants of the EDTA, gluconate types. Use of an alkaline solution will also prevent precipitation of metal ions remaining in the acid residuals. The vessel will eventually be opened for inspection and for the removal of any loose, undissolved debris that normally remains after a chemical cleaning operation.

Flow-through vessels Many vessels such as pipelines were designed for mass flow-through circulation rather than for filling and draining. Such a situation presents a problem to the chemical cleaning engineer. The ideal solution to the problem would be to clean the system while operating it in a manner for which it was designed. The present state of the technology, however, may not allow this ideal solution, so cleaning engineers and equipment owners have improvised other methods that work rather well.

One solution is to provide large temporary circulating pumps that, according to the design engineer, will provide flow through all parallel paths. Instead of draining, each fluid is replaced by the next until solvents, rinses, and neutralizers have been put in and then completely removed.

The foamed-acid technique[3] has also been useful because the flow characteristics of foam allow it to enter one end of a pipe from a water box, fill the pipe completely, and emerge into an empty water box on the opposite end.

Some awkward pipeline problems have been approached by using a flow of steam[4] adequate to carry cleaning reagents and loose debris over the ups and downs. Otherwise, each high and low would need to be provided with vents and drain lines. Or solvents, rinses, and neutralizers have been put in lines and held in position by rubber plugs as the train moves along under the pressure of fluid pumped in behind.

When circulating such a system the natural corrosiveness of the fluid used becomes important. Inhibited muriatic acid may be replaced by inhibited organic acids. Higher temperatures are generally used with the organic acids but control of the temperature becomes easier and less critical than with muriatic acid.

Large hollow vessels Vessels of large volume and limited surface area to be cleaned are not adapted to fill-and-empty or flow-through methods. Cleaning reagents have been applied in the form of a gel. Or interior surfaces may be cleaned by using automated spray devices that do not require personnel to remain inside the vessel. In some instances, reactants (both alkalis and acids) have been put into the hollow space by means of steam and allowed to condense on the interior walls. Cleaning tank exteriors chemically has not met with success.

New ideas The chemical cleaning business has traditionally been conservative, partly because cleaning engineers are not at liberty to take chances with contractors' equipment and partly because a common inertia exists which favors using tomorrow the same methods that were used yesterday. Perhaps the most significant trend has been toward the use of reactants such as EDTA-type chelants rather than cheap, aggressive muriatic acid. Use of chelants, plus reducing agents, for both scale and corrosion and ammoniated EDTA for removing iron oxide and copper represent real progress. The cost of such expensive materials can be offset by savings elsewhere. Time is the largest potential saving; indeed, the ultimate goal should be to return the serviced equipment to the construction site as quickly as possible.

The use of noncondensed phases as carriers, the use of chelants to save rinse time and water, better analytical control during cleaning, and the search for entirely new ideas are only a few of the interesting possibilities for future development.

Pigs, plugs, balls, and jets Mechanical devices often are used alone or in connection with chemical solvents. In pipe cleaning, tools can be made to travel through the line by the force of a fluid behind them. The fluid may be water, oil, gas, or a chemical cleaning solution. The tool may have rubber disks that fit the inside of the pipe, the disks being attached to a central shaft to which scrapers and brushes may also be attached. Even a radioactive capsule may be attached to assist in showing the location of the tool if the radiation can be detected outside the line. Sometimes rubber stoppers are used to separate a slug of one fluid from the next. In this manner, cleaning, rinsing, and neutralizing solutions have been put through pipes in proper sequence to avoid filling entirely with first one fluid, then the next. There are rubber balls of any diameter, with chain mesh to fit them when inflated. The balls move with the fluid flow as do the pigs and plugs.

Water jets powered by pumps up to 10,000 psi have become very useful cleaning tools where chemical removal of deposits is not the best method.

Disposal problems In recent years, the problem of disposing of large volumes of cleaning solvents has come to require serious attention.[5] The problem may indeed dictate what reactants will be permitted. For years laboratory tests have been run to determine a degree of dilution which might avoid fish kill. Acid has been dumped onto sludge from the water softener or into pits filled with crushed limestone or neutralized with caustic soda, soda ash, or powdered lime or limestone. Solvents containing ammonia, oxidizing agents, and dissolved copper have been disposed of by dilution. Eventually, more sophisticated and expensive methods will be required to control not only pH but all dissolved solids as well. Methods involving ion exchange, reverse osmosis, electrodialysis, or evaporation suggest themselves.

SUMMARY OF DATA AND DECISIONS

It is necessary to know the nature of deposit phases. Chemical cleaning involves chemical reactions between substances that foul the equipment (corrosion products and scale) and chemicals that are chosen to correct the fouling condition. As a reaction goes on, new substances will be formed. Knowledge of deposit phases reveals what has been going on in a system to cause fouling. Such knowledge helps in selecting the solvent to be used for cleaning and in anticipating the identity of the reaction products, which must be soluble in the cleaning solution to be removed from the system. Reaction products that are flammable, toxic, or corrosive must be anticipated. In addition to identifying deposit ingredients, it is common practice to use deposit samples for solvent trials and corrosion tests.

It is necessary to know engineering materials and design. What engineering materials will be exposed to the cleaning solvent? What are the design details? Both materials and design are described in prints and descriptive bulletins furnished by equipment vendors. If there is any doubt about the chemical properties of a material versus solvents to be proposed, laboratory trials should be made. Specimens of materials can be exposed to solvents and deposits under conditions that simulate the cleaning conditions that are being planned.

After securing the information indicated, there are decisions to be made. If an outside service agency is employed, the decisions may be left to them or they may be reached by consultation between the agency and the contractor. If the contractor's own men do the job, someone must make the following decisions:

1. What reactants, inhibitors, surfactants, or neutralizers should be used?

2. What reactant concentrations and temperatures are necessary? What mixing and heating facilities will provide the chosen conditions?

3. What precautions are necessary to protect people and equipment? This involves safety instructions, clothing and special protective devices for workmen. It involves isolating the equipment to prevent the solvent escaping through forgotten connections or valves that fail to hold. It may mean roping off the work area and posting safety signs. Perhaps most important, it means providing for reaction products that may be hazardous. Hydrogen gas always is anticipated when cleaning involves acids and steel; providing a suitable vent for hydrogen should be routine. The possibility of toxic gases such as hydrogen sulfide or chlorine must be anticipated, either to prevent the reaction or to dispose of the product. Solvents and rinses must be disposed of in a way that avoids pollution problems.

4. What supervision is needed during the operation? What people are needed and what engineering and chemical data so that people can follow the progress of the cleaning process? Someone must decide when each step has been completed and it is time to go to the next until everything has been done and equipment is once again ready for use.

5. What records should be kept and by whom? There are data and records to be kept during planning, performing, and evaluating of the results.

CHEMICAL VERSUS ALTERNATIVE METHODS

To compare costs of chemical methods versus alternatives, consider cleaning time, lost equipment usage time, man-hours, tools, equipment, materials needed, degree of restored efficiency to be expected, and safety to equipment and personnel.

In any case, the objective is to get the most cleaning per dollar of cost. If construction equipment is too large or complex to allow mechanical methods to be used, the answer is to use chemical methods. Still the question is: What is the best way to get the most cleaning for the money? Can equipment maintenance personnel do the work, or is it better to seek outside sources? Are outside services bought by bids? On what are the bids based? Who specifies procedures, materials, results? What assurance is there of competent planning and performance? Some answers depend on whether the purchaser has men who know chemical cleaning technology well enough to furnish specifications describing what is being purchased and to judge the competence of the service that is offered for hire. Without such knowledge, the purchaser is in the position of a layman seeking advice from medical experts. Bids and lowest prices are not necessarily the greatest bargain. The experience and qualifications of whoever will be responsible for the service are most important.

REFERENCES

1. Blake, D. M., J. P. Engle, and C. A. Lesinski: The Use of Chelating Agents in Chemical Cleaning, Proceedings of the 23d International Water Conference, Engineers Society of Western Pennsylvania, pp. 135–142, 1962.
2. U.S. Patent No. 3,067,070.
3. (a) Carroll, D. B., C. L. Eddington, and J. P. Engle: Chemical Cleaning with Foamed Solvents, Proceedings of the 22d Annual Water Conference, Engineers Society of Western Pennsylvania, pp. 35–40, 1961. (b) U.S. Patent No. 3,037,887.
4. (a) Loucks, C. M.: Something New in Chemical Cleaning, *Power Engineering*, vol. 65, pp. 58, 59, June 1961. (b) U.S. Patent No. 3,084,076.
5. Bell, W. E., and E. D. Escher: Disposal of Chemical Cleaning Waste Solvents, *Materials Protection and Performance*, vol. 9, pp. 15–18, December 1970.

Chapter **3**

Lubrication

ALLEN F. BREWER
Consultant in Lubrication, St. Lucie County, Fla.

Lubrication is of interest to the construction equipment maintenance engineer because it has an influence on the costs he must charge to maintenance. Any machine will operate most dependably when it is properly lubricated. Under such conditions the maintenance engineer has only to note that lubrication is being properly maintained and that lubricants most suited to the machinery are being used. This leads to minimum cost of maintenance, fewer headaches for the maintenance engineer, and lower-cost construction.

Conventional tests For lubrication oils such tests include viscosity, flash and fire points, pour points, carbon-residue content, emulsification and demulsibility, acidity or neutralization number, and saponification number.

For greases, the base, penetration, and dropping- or melting-point tests are the most significant.

Viscosity As an indication of the relative fluidity of any lubricating oil viscosity is discussed regardless of the service. Machinery builders develop their lubrication recommendations around viscosity; it is the number one test when purchasing. Viscosity, however, does not denote quality; it indicates simply how the oil will flow at the temperatures under which lubrication must be maintained.

Flash and Fire Points These points are customarily quoted in listing the characteristics of a lubricating oil; but, unless the oil is to be used under very high temperature conditions where vaporization could be a factor, the flash and fire points are of little interest to the construction equipment maintenance engineer.

Pour-Point Test Here is another temperature test that should be considered with respect to the viscosity of an oil. It indicates how fluid an oil will be at very low temperatures. With straight mineral oils (without use of pour-depressant additives or special dewaxing) naphthenic-base oils will show lower pour tests (for the same viscosity) than paraffin-base oils. The pour test is most useful when oils are to be used in circulating systems which may be exposed to low atmospheric temperatures. In small-diameter piping, an oil of inadequate pour test could become so congealed during a cold overnight shutdown as to result in starved lubrication and the need for bearing replacement.

Carbon-Residue Content The carbon-residue content of an oil is a factor in internal combustion engine service. Being an indication of residual matter, it becomes allied with the lubricating ability of the oil. For clean engine performance, it should be as low as possible. With modern detergent and dispersant types of heavy-duty engine oils, engine cleanliness is more positively assured by the quality and additive make-up of the oil than by the carbon-residue content of the base oil. In internal combustion engine maintenance it is very important to watch the cleanliness of the air intake and the water temperature. Dirty air or condensation of moisture in the crankcase can contribute far more to cause a dirty engine and need for frequent overhaul than the fractional percent of carbon residue which may be indicated by the laboratory test.

Emulsification and Demulsibility These are tests which definitely relate to lubricating ability when an oil is to be used in hydraulic operations. Emulsification indicates the tendency of an oil to mix intimately with water to form a more or less stable emulsion. Demulsibility indicates the readiness with which subsequent separation will occur. Best performance of turbine-grade oils occurs when the equipment operates at temperatures of 150 to 160°F or less, and when an oil-reconditioning system is operating to remove water and contaminants.

Neutralization Number This number is related to acidity. It measures the number of milligrams of potassium hydroxide required to neutralize 1 g of oil, normally less than 0.10. An abnormal rise in oil used in turbine or hydraulic equipment can indicate oil oxidation.

Saponification Number Saponification is a chemical reaction involving the action of an alkali on a fat or fatty acid; the resultant combination is called a soap. Saponification will rarely occur in a well-refined mineral oil and then only where fatty acids may have resulted because of oxidation or chemical breakdown. In effect, the principle of saponification is the basis for the manufacture of certain greases.

The tendency which a petroleum lubricating oil may have to saponify is determined by neutralization and is measured by the equivalent amount of caustic potash required to react with or neutralize 1 g of oil under test. In terms of milligrams of caustic the resultant figure is called the saponification number of the oil. Obviously this should be as low as possible. Increase in the tendency toward saponfication may have a like effect upon the tendency toward oxidation and gum formation in the oil.

Lubricating-grease characteristics—tests

Base. Lubricating greases are classified broadly according to the type of soap used in their manufacture. The more conventional products include the calcium (lime) base or general-purpose greases usable to around 160°F, sodium base products usable up to at least 250°F, and soda-lime (mixed base) greases which are so widely used for service where the combined features of their respective elements are enhanced by inclusion of antioxidation additives.

In addition there are the more recently developed multipurpose greases of primarily lithium, barium, or strontium base, as well as the aluminum and lead soap products required along with high-temperature durability up to 275°F (plus).

Among the nonpetroleum materials which are applicable to grease service are the bentones, which can be compounded with certain petroleum products, and molybdenum disulfide, which has excellent high-temperature stability, good film tenacity, and a low coefficient of friction. Table 3-1 shows in detail the conventional types of lubricating greases, their characteristics and, as will be noted, the base of each.

Penetration This word is used to describe the consistency of a grease and to some extent its texture. The construction equipment maintenance engineer is concerned with grease consistency because the grease must continually reach the parts to be lubricated. In a pressure system serving a considerable number of bearings, a grease too heavy for the diameter or pumping ability of the pump could cause clogging of the lines and, again, inadequate lubrication, which could mean only parts renewal later.

Dropping or Melting Point The melting point is a temperature measurement that indicates the tendency of a grease to soften with increase in temperature. The percentage and type of soap used, the viscosity of the mineral oil, and the type of alkali affect the dropping point of the finished grease. As a rule sodium- and lithium-soap greases show higher dropping points than those of calcium base. To some extent this indicates that the former are better suited for higher-temperature equipment, although the dropping point should not be assumed to indicate the maximum usable temperature.

TYPES OF LUBRICANTS

Petroleum lubricants are broadly classified according to the service for which they are most widely used. Some are virtually specialties; others can be successfully applied to such a wide variety of equipment as to become multipurpose in nature.

The maintenance engineer is interested in the following classifications:
Circulating oils
Gear oils
Machine or engine oils
Steam-cylinder oils
Wire-rope lubricants
Greases of calcium, sodium, aluminum, lithium, or barium base
Synthetic and solid lubricants

TABLE 3-1 Classification Chart for Greases

Worked penetration range at 77°F by ASTM Method D 217	NLGI° consist- ency No.	Nature of grease according to consistency	Typical means of application	Typical types of grease within adopted ranges
400–430	00	Semifluid	Packed in special felt bag of good porosity	Nonmelting inorganic base containing MoS_2
355–385	0	Semifluid	Brush	Sodium or calcium
310–340	1	Very soft	Pin-type cup	Sodium, calcium, or aluminum
265–295	2	Soft	Pressure gun or centralized pressure system	Sodium, calcium, lithium, or mixed (sodium-calcium); or inorganic thickened
220–250	3	Light cup		
175–205	4	Medium cup		
130–160	5	Heavy cup	Pressure gun or by hand	Sodium or calcium
85–115	6	Block type	Hand applied, cut to fit lubricator or bearing pocket	Sodium

° National Lubricating Grease Institute.

NOTE: In addition to these conventional products, there are those multipurpose greases prepared with inorganic thickeners to impart distinctive temperature-consistency (nonmelting) relationship, dependable mechanical stability, marked load-carrying ability, excellent water resistance; usable up to 350° plus. A. Gordon Brewer and P. C. Jarvis, "Inorganic Thickened Greases in the Steel Industry," Shell Oil Company, discussed at the AISE Annual Meeting in 1963, published in *Iron and Steel Engineer*, March 1964.

Circulating oils These are probably the highest-quality lubricants available today. They are obtained over a comparatively wide range of viscosities, that is, from about 21 to 550 centistokes or from 100 to 2500 Saybolt Seconds Universal (SSU) viscosity at 100°F. In this category are included steam-turbine-grade oils, hydraulic oils, heavy-duty internal combustion engine oils.

Circulating oils may have a paraffin or naphthenic base according to equipment. For turbine, hydraulic, and similar equipment, the former predominates. Either naphthenic- or paraffin-base oils are used for heavy-duty engine service. The viscosity range at 100°F is given in Table 3-2.

Circulating oils contain additives. Turbine and hydraulic oils are fortified to enable them to resist oxidation and to retard rusting in the system; they also usually contain a foam dispersant.

The modern heavy-duty internal combustion engine oil is specifically refined to function under high engine temperatures and bearing loads. These oils are highly resistant to oxidation and are fortified with detergent and dispersant additives.

Gear oils These may be straight mineral oils of widely varying viscosity or compounded oils containing extreme-pressure additives to improve the film strength and load-carrying ability.

Straight mineral gear oils range normally from SAE 80 to 250. The lower viscosity grades are used for low-temperature equipment; the heavier grades, SAE 140 or 250, are selected for equipment which will normally range above 100°F.

Where gears of the above type are exposed and bath or hand lubricated, the viscosity of the lubricant must be increased to enable the film to resist throwoff. Exposed gears generally do not run too fast, but, because of their location, they may be exposed to wide temperature ranges such as in the swing gear on a power shovel. For such gears a straight mineral residual petroleum product is used which may be compounded with a small percentage of pine tar to improve the adhesiveness. Some such lubricants are cut back with solvents to facilitate application. These thinners later evaporate from the film, but, as some solvents may be flammable, it is well not to use such lubricants in enclosed areas. The question of toxicity also is important.

TABLE 3-2 Viscosity Range of Circulating Oils

Equipment	Viscosity at 100°F	
	Centistokes	SSU
Steam turbine:		
Direct-connected	32–40	150–185
Geared	65–110	300–500
Hydraulic:		
Light-service	21–54	100–250
Machine tools	30–121	140–550
Heavy-duty	Up to 154	Up to 700
Internal combustion engines:		
Heavy-duty SAE 30 to SAE 60	110–370	500–1700

Machine or engine oils The straight mineral red oils come under this classification. They came into usage for general lubrication of external operating parts of engines, pumps, compressors, and general equipment when unit lubrication by oil can or oil cup was practiced. Later they were adapted to ring oilers, but on modern equipment the higher-quality turbine-grade oils are used.

The average so-called machine or engine oil is a good lubricant for once-through lubrication. However, since the resistance to oxidation is lower than that of modern premium-grade oils, ordinary machine oils are not recommended for equipment where formation of sludge or gummy residues could add to the troubles of the maintenance engineer.

Steam-cylinder oils The necessity for lubricating steam cylinders with something more dependable than the time-honored tallow pot became evident when steam engines were operated on high-pressure steam and when multistaging or expansion was adopted. By that time the petroleum industry had perfected methods of refinement and residual lubricating stocks were available so that compounding with a small percentage of fatty (animal) oil such as lard oil, tallow, or wool fat to improve the wetting ability of the finished oil became standard practice. The principle remains the same today, although the petroleum chemist has isolated certain base stocks such as bright stock and the fire stocks dewaxed more or less. These, together with steam-refined stock, make available a variety of products for compounding according to the nature of the steam (i.e., pressure, temperature, moisture content), the utilization of the exhaust, and whether or not very rapid atomization is desired.

Being residual in nature, steam-cylinder oils are necessarily of higher viscosity than distilled oils such as turbine oils. They can be grouped into three broad classifications according to viscosity:
   Light—100 to 120 SSU at 210°F
   Medium—120 to 150 SSU at 210°F
   Heavy—150 and above SSU at 210°F

Steam-cylinder oils are used by injection into the steam line by a hydrostatic or mechanical forced lubricator. When a suitable injection quill is installed in the line, the steam atomizes the drops of oil as they pass onto the quill and thereby carry a so-called fog or mist of oil to all parts of the cylinder walls, pistons, valves, and valve seats.

Wire-rope lubricants Lubrication of wire rope has undergone quite a transition. At one time it was felt that the best protection resulted from using a comparatively heavy residual-type lubricant similar to a heavy gear lubricant. Today the idea of using lighter-bodied oils is popular, applying them by spray to assure better penetration between the rope strands. For this purpose a specially prepared fluid lubricant of about 600 SSU viscosity is adaptable. Inclusion of a small percentage of pine tar gives it added stickiness and also penetrative ability. This type of lubricant is especially suited for wire rope which must be exposed to the weather and to low temperatures as on aerial tramways.

For less severe service a somewhat heavier straight mineral product is satisfactory; this type of lubricant can be applied by drip, brush, or split box.

The construction equipment maintenance engineer is particularly concerned with good wire-rope lubrication because strand breakage can require removing the rope from service and installing a new rope. Safety precautions require rigid inspection.

Greases The American Society for Testing and Materials (ASTM) defines a lubricating grease as "a combination of a petroleum product, and a soap or mixture of soaps, suitable for certain lubrication applications." The metal used in making the metallic soap constituent of a grease denotes its base, for example, calcium, sodium, aluminum, lithium, or barium. In addition a mixture of calcium plus sodium produces what is called a mixed-base grease. Table 3-1 indicates the features and serviceability of modern greases.

Greases are further identified by the type and viscosity of the petroleum oil used in their make-up, by their degree of plasticity, and by their dropping or melting point. A combination of these factors scientifically worked out will produce a grease of remarkable stability and endurance over a wide range of temperature conditions. A multipurpose grease is the ideal, as it reduces the possibility of misapplication and is a factor in storage.

There used to be a popular conception that greases were chosen for a job when a fluid oil could not be retained because of the housing or inadequacy of seals. Modern design has relegated this idea to the past. Today, precision manufacture is so fine and seals are so perfect that lifetime lubrication by just a few grams of grease in a bearing is practicable. Furthermore, greases are available which will function over very wide temperature ranges. But when any such bearing is to be disassembled from the other parts of a piece of equipment during overhaul procedure, the sealed ball or roller bearings must be carefully handled. An effective seal is effective only as long as it is not abused. Careless handling or soaking in solvents may lead subsequently to entry of abrasive dust.

In maintenance work the protection of a grease lubrication system is just as important as protection of the bearings or other parts being lubricated. Careless use of tools around fittings, control outlets, lengths of pipe, or inadvertent striking while moving a beam or scaffold might render one or more outlets inoperative, because of stricture or grease leakage.

Synthetic and solid lubricants Names and characteristics of nonpetroleum lubricants are given in Tables 3-3 and 3-4.

ADDITIVES

Additives serve a variety of useful purposes; accordingly, there are a variety of additives.

Pour-point depressants These depressants are added to lubricating oils to enable them to flow at and be usable at lower temperatures than would be possible with the base oil. Such additives retard or change the action of formation of wax crystals at low temperatures so that they do not interfere with the fluidity of the oil.

Viscosity-index improvers Viscosity-index improvers are added to an oil to improve the viscosity-temperature relationship or, in other words, to obtain as little change in viscosity as possible over the expected service-temperature range.

Antitoxidants Antioxidants, often called *inhibitors*, are widely used to fortify steam-turbine, hydraulic, and circulating oils against oxidation when subject to oxidizing conditions. Oxidation of lubricants vitally concerns the engineer because the resulting gums and sludges generally call for considerable expensive equipment overhaul and cleaning of the working parts.

TABLE 3-3 Characteristics of Synthetic Lubricants*

	Low-temperature flow properties	Inhibitor susceptibility	Resistance to oxidation	Water solubility	Thermal stability	Lubricating ability	Viscosity index
Hydrocarbons	Fair to 0°F, although some are OK to −40°F	Poor to good according to type	Fair to good according to type	Immiscible	Poor to good according to type	Comparable with equivalent petroleum oils	Slightly above petroleum paraffin-base oils
Polyalkylene oxides and polyethers (glycols)	Generally good to −40°F	Generally fair	Low	Mostly water-soluble to insoluble	Suitable for high-temperature work	Good where high-temperature usage requires vaporization with minimum residue	Up to 150
Esters	Generally good to −70°F	Good	Fair	Mostly immiscible	Have high flash point		Up to 150
Silicones	Good to −70°F	Poor	Good up to 390°F; above this they oxidize rapidly	Immiscible	Good	Most effective only when one surface is nonferrous. Additives improve	High
Fluorocarbons	Fair	....	Excellent	Immiscible	Excellent but volatility may be high	Not too well known	Below 100

*From Allen F. Brewer, "Effective Lubrication," Robert E. Krieger Publishing Co., Inc., Huntington, N.Y., 1973.

TABLE 3-4 Solid Lubricants*

Product	Name and characteristics	Lubrication service adaptability
Bentones	Produced by reacting hydrous magnesium aluminum silicate (montmorillonite) or bentone clay with an ammonium salt	Effective in compound with petroleum greases. Prepared by a gelling process. No soap is involved
	Features are stability at high temperatures, water resistance; do not liquefy	Well suited for high-temperature service and extreme water conditions
Boron nitride	Sometimes referred to as white graphite owing to unctuous nature. Produced as a light fluffy white powder in an arc furnace at very high temperature. Has excellent stability at high temperatures. Insoluble in water but decomposes in most acids. Has disadvantage of inability (by itself) to adhere to metal surfaces	Usable as a component with silicone-type lubricants in range of 5 to 25% concentration. Carrier serves chiefly to carry the boron nitride to the surfaces to be lubricated. Film of this material is very durable at high temperatures even after carrier has been dissipated
Fuller's earth	Silica base—finely divided	Can be used dry or mixed with water, light oil, or grease
		Effective in retarding fretting corrosion. High-temperature resistant up to around 700°F
Graphite	Produced from coke or anthracite coal. Milled to obtain colloidal graphite usable for lubrication	Can be used dry or mixed with oil or grease. Its chemical inertness enables its use where high thermal stability is required. Maximum usable temperature is around 1500°F. Not too effective in preventing corrosion when used dry
	The flake nature in form of sheets piled on top of each other imparts the lubricating effect as these sheets slide over one another in motion	
Molybdenum disulfide (MoS_2)	Stable at high temperatures. Good film tenacity. Low coefficient of friction	Effective in reducing friction at high sliding velocities. May be mixed with a solvent for application to parts to be lubricated. To obtain best results from a chemically active lubricant of this type, the metal surfaces should be *clean*
Mica	A natural mineral which is ground very finely	Can be used like talc as a lapping material to obtain high surface finish of machine parts. Sometimes added as a filler or thickener in certain lubricants
Talc	Powdered soapstone	Suitable as a lapping material for finishing or working in machine parts
Zinc oxide (ZnO_2)	White in color. Particle size very small—requires no milling. Has low coefficient of friction	Usable as a component with mineral oil for lubrication of parts where perishable products are being produced as in food handling and meat processing

*From Allen F. Brewer, "Effective Lubrication," Robert E. Krieger Publishing Company, Inc., Huntington, N.Y., 1973.

Foam depressants These additives are useful in turbine and circulating oils to prevent foaming when they are agitated with air; foam depressants also accelerate foam dispersion when it has once formed.

Anticorrosion additives and rust preventives These serve a very useful purpose especially when added to circulating oils and to some types of greases. They retard metal corrosion and rusting when the surfaces are exposed to moist air or to water.

Extreme-pressure additives Extreme-pressure (EP) additives are most widely known probably because they are so closely associated with the automotive hypoid gear. An extreme-pressure additive is a chemical compound which increases the load-carrying ability of the lubricating film when subjected to high rubbing speeds. Extreme-pressure additives reduce friction and wear. When metal-to-metal contact occurs between meshing gear teeth, so-called spot welding develops between the surfaces under the extremely high spot temperatures which prevail. This friction can lead to serious tearing away of surface metal and to serious malfunctioning of the gear set. Then a maintenance problem results, usually calling for gear replacement. An extreme-pressure additive in the gear oil imposes a sufficiently protective and easily sheared lubricating film between the teeth.

Engine-cleanliness additives Detergents and dispersants are included in this category. They are used in the make-up of modern heavy-duty motor and engine oils.

Detergents are cleaners. They ensure most satisfactory performance of a circulating oil by preventing residual nonlubricating matter, such as sludges which result from oil decomposition and fuel combustion, from accumulating around piston rings, in bearing clearances, and elsewhere on the engine parts. Regardless of the original degree of purity of the oil, ultimately it will get dirty because of entry of road dust via the crankcase breather or air cleaner, condensation water, and the natural results of service under high temperatures. The oil filter removes some of these contaminants, but some will remain to develop the sticky, gummy substances which ultimately bake onto the hot metal surfaces. A suitable detergent in the oil prevents this buildup by virtue of its continual dissolving and cleansing action. *Dispersants* are companion additives to assist in this function. A dispersant is included to keep these finely divided insoluble materials dispersed in the oil until drain-out. No harm to the engine results from this action because after filtration the dispersed material is virtually nonabrasive.

CONSTRUCTION EQUIPMENT LUBRICATION MAINTENANCE

While lubrication can retard wear, it cannot entirely prevent it. Wear can result from dust contamination or failure of the lubricating system to maintain a proper film on moving parts. Furthermore, the load on a lubricant under severe operating conditions is equally heavy on the equipment. Adequate maintenance and proper design, however, can control the effects of these loads.

A general inspection should be made once a month. At every third such inspection, the foreman should invite an engineer from the lubricant supplier to accompany the maintenance people on their tour. By working together they can determine where the use of an improved bearing or lubricating device would make the equipment run more smoothly, longer, and more economically; where a more suitable grade of lubricant would reduce the frequency of lubrication; or where housing of overexposed gears would improve safety, reduce fire hazards, and cut the cost of removing dripped lubricant.

Equipment operators have a responsibility in equipment lubrication, too. They should not squirt oil carelessly or turn down grease cups with a wrench. Critical parts may be lubricated, but lubricant is wasted and bearing seals may be broken.

Lubricant Protection

Protection of lubricating oils and greases in service is just as important as selecting products with the right characteristics. Protection is a requirement which is often neglected. Premium-grade products are purchased and then stored in a dirty storeroom or even out-of-doors, drums on end, to accumulate water and dirt. Subsequently, it is almost impossible to draw oil or scoop out grease from such a container without some contamination. Obviously, the answer is to provide a special location *indoors* for storing lubricants and to plan a definite schedule for taking stock, refilling containers or lubricating systems, and cleaning, with assigned personnel responsible for the schedule.

Contractors are taking lubricant protection more and more seriously. Many companies have experienced lubrication engineers on their operating staffs. The lubricating-equipment people, in turn, have developed devices and procedures for handling and distributing lubricants which are equally progressive.

Location and personnel A clean, well-lighted room or building is advisable with provisions for heating in cold weather. It should be specifically kept for lubricant storage and reserve lubricating equipment. In this way the responsibility for cleanliness and proper location of lubricant containers can be assigned to one or two individuals who, in reality, become assistant lubrication engineers. They can be trained by the maintenance engineer to appreciate the problem should a bearing or gear set fail as a result of contaminated lubricant. Likewise, they can be schooled in appreciating the value of quality lubricants and the reasons why such products are virtually specialties for the service to be performed.

Fire protection The possibility of fire in a well-planned lubricant-storage area is remote, assuming that *no smoking* rules are observed, that casual visits from other people are prohibited, that oil drip is prevented or cleaned up promptly, that waste or wiping rags are stored in metal containers and in minimum quantity, that sparking or arcing tools are used only under conditions of good ventilation. Even so, insurance regulations will require installation of suitable fire-extinguishing equipment. The accepted foam-type device for smothering oil fires is best. In a small storeroom one or two hand units may suffice. In a larger area a multiple-gallon foam cart with adequate hose may be required.

Endurance Value of Petroleum Lubricants

The petroleum chemist goes beyond the laboratory procedures in considering tests to predict lubricating or endurance value. He considers those factors which will denote the wetting ability of the product, its film-forming ability, its behavior when exposed to water, and its tendency to form sludge or saponifiable by-products. Service considerations will dictate the relative importance of each of these factors.

Wetting ability The wetting ability of the lubricating film is regarded as its most important function if dependable and protective lubrication is to be maintained. In other words, the extent to which effective lubrication can be expected depends upon the extent to which the lubricating film actually wets the surfaces of the metal parts between which motion is taking place.

Wetting ability as a function of adhesion can be illustrated by wetting steel-strip surfaces with oils which are to be compared. The surface which is wet or coated with a satisfactory oil will retain this film when dipped in water. When a steel strip is coated with an oil of poor wetting ability, the water will displace the oil film. Increase in temperature (use of warm water) hastens this displacement effect.

Surface tension Surface tension in a liquid involves the cohesive action of the component particles. It is related directly to viscosity, temperature, and emulsion-forming tendency. Inasmuch as it is an indication of the relative strength of the lubricating film, higher-viscosity oils can be expected to produce films of greater strength at the same temperature. As the temperature is increased, the surface tension will be reduced.

Interfacial tension Interfacial tension in petroleum oils is affected by oxidation. The compounds formed during this reaction tend to reduce interfacial tension. As these compounds usually have an emulsifying attraction for water, a low interfacial tension may well indicate its presence.

Adhesion Adhesion already has been mentioned as being associated with wetting ability in a lubricating film. If good wetting ability prevails, one may assume that the adhesion property is good. To some extent, however, this will depend on the surface finish of the contact metals and the way in which the lubricant has been refined. Too much polish or surface finish is not conducive to good wetting ability or adhesion. The same holds true for overrefinement of an oil.

Saponification and emulsification These conditions, as they relate to the lubricating value and utilization of petroleum lubricants, have already been discussed as to test procedure. While saponification is a characteristic which is relatively negligible in petroleum lubricating oils, organic acids may exist or may develop to react with an alkali. An oil with a comparatively high saponification number may be susceptible to emulsification, an effect that would be undesirable in a hydraulic system.

Lubrication Procedures

Lubrication procedure involves the means provided for lubrication, timing according to the nature of the lubricants and the requirements of the construction equipment, training of personnel as to resultant benefits, arrangement of records, and analysis of failures which may be traced to faulty lubrication.

Hand-pressure grease guns Modern hand-pressure grease guns provide application pressures as high as 10,000 to 12,000 psi. Such guns can be used either with or without hose connections, according to the fitting and the location of the part to be lubricated. Pressure can be applied either before or after attaching the gun to the fitting. The usual method of developing pressure in a hand gun is to force a plunger against the grease in the barrel, the stem of the plunger being threaded to enable a screw-down action when the handle is turned.

When pressure is to be developed before attachment of the gun to the fitting, a check valve is installed in the tip. In such a gun the act of attachment opens this valve and permits grease to be forced automatically to the bearing.

The purpose of designing rigid-connection guns with check valves is to eliminate the necessity of relieving the pressure before detaching the gun from the fitting and to enable pressure to be raised before attachment; this eliminates the possibility of twisting off the fitting. Direct connection also reduces the possibility of leaks which might develop in flexible hose.

Power guns Where a considerable amount of equipment is involved with numerous grease fittings, the portable power gun is suggested. It hold up to 100 lb of grease. Some of the latest designs work directly from the grease container.

Smaller power guns develop 2000 to 3000 psi and are powered by a pump handle or lever. Later models are electric- and air-powered.

Timing Timing begins with the establishment of a suitable schedule. The schedule would require study of the equipment, the extent to which its parts are housed or protected to conserve lubricant, the speed of various parts, the possibility of lubrication contamination, and the ability of the lubricant to act as a flushing agent.

Lubrication personnel There is a decided trend to pay more attention to the status of lubrication personnel. Organization and training are key factors in developing a lubrication-minded staff. The idea of a so-called grease monkey must be completely discarded. A lubrication mechanic's job is as important as that of any crane operator, for, without the mechanic, the crane could not continue to operate effectively.

Lubrication failures Lubrication failures are most often caused by one of the following:
1. Unsuitable grade or type of lubricant.
2. The lubrication system is not suited to the design of the equipment.
3. The lubricant is contaminated by dust, dirt, water, or dilution by fuel.
4. A suitable lubrication schedule is not followed.

Chapter **4**

Centralized Lubrication Systems

JAMES J. CALLAHAN
Vice President, Research and Engineering, Trabon Lubricating Systems, Lubriquip Division, Houdaille Industries, Inc., Cleveland, Ohio

As labor and equipment replacement costs continue to rise, more and more heavy-duty vehicle owners are looking for ways to improve equipment performance and reduce maintenance costs. One way to accomplish this is through the installation of centralized lubrication systems.

A centralized lubrication system, properly designed and installed, will meter the proper amount of lubricant to all points, with the exception of such rotating points as universal joints. This measurement will minimize wear, even with heavier loads and poor operating environments, resulting in longer equipment life and less downtime.

A centralized system also minimizes problems of lubricant contamination and allows for more economical use of grease, since each lube point receives only its precise requirement. Lubrication also can be centrally monitored, so that the operator will be alerted if a lubrication point becomes damaged or blocked by dirt. Another frequently overlooked advantage is that personnel safety is increased because it is no longer necessary for an operator to climb over a machine to reach lubrication points (Fig. 4-1).

The basic system A basic centralized lubrication system for mobile equipment can be broken down into three functions: (1) lubricant reservoir and pump actuation, (2) proportioning network, and (3) lubrication points.

Of primary concern in the proper design of a system is the quantity and size of the various lubrication points. First, the equivalent area of each point must be calculated. This value is then multiplied by a film thickness to determine the volumetric requirement of each point. The equivalent area for bearings is determined by the following equations:
Plain bearings

$$A = LD$$

where L = length, in.
D = diameter, in.
Antifriction bearings

$$A = D^2 R$$

where D = shaft diameter, in.
R = number of rows

For heavy equipment, the film thickness requirement is 0.001 to 0.002 in. of grease per hour. This value multiplied by the equivalent area will yield the volume of lubricant required by the bearing in 1 hr. By knowing the relative sizes of all bearings, the designer can then determine the proportioning network. This consists of feeders which operate progressively to divide the lubricant. Once the master feeder and secondary feeders are selected, the pump can be selected and the reservoir size determined.

Each time the pump is actuated, either by hand or automatically, lubricant is delivered to the system. The master feeder, or first feeder, divides the lubricant into amounts required by the secondary feeder assemblies. These, in turn, serve any number of lubrication points. The pump will develop only the pressure that is needed.

Since a centralized lubrication system works on a progressive piston-displacement basis, every piston must complete its stroke before the next piston can cycle. If any one

Fig. 4-1 Centralized lubrication lines have access to all key points on this track system.

piston in the system fails to complete its cycle, pressure will build up and warn the operator of a malfunction. When high pressure occurs, the blockage can be traced by checking the indicators on the feeder assemblies.

On automatic systems, the power supply of the lubrication pump can be air, hydraulic, or electric. Selection of power take-off is generally worked out with the maintenance supervisor before designing the system in order to be certain that other service areas are not affected.

In order to better illustrate how a typical centralized system works, let's take a look at two specific examples.

Front end loader A centralized lubrication system on a wheel-type articulated front end loader consists of a grease reservoir, an air-operated pump, and a feeder assembly (Fig. 4-2). Special lubrication points, other than those common to most heavy equipment, include bucket hinge pins, cylinder pivots, and lift arm pins. This particular system is very flexible and both small and large bearings can be automatically lubricated by correct selection of feeder piston size.

Both bearing size and manufacturer's suggested lubrication intervals are considered when designing a proportioning network of this type. Special protection also is given to fittings and components that are subjected to severe use.

Heavily loaded points such as the bucket pivot and tilt bearings are given greater lubrication quantities to help keep out dirt and other foreign matter. High-pressure hose or nylon tubing is used where flexing is required. Feeders are mounted on the vehicle in such a way that flex lines are minimized and only short lines to bearings are required. Use of short lines reduces the possibility of line breakage and also helps keep system pressure low.

The lubrication pump is actuated each time the operator depresses the vehicle's brake pedal. Other take-off points may include a compressor governor unloader valve, or in the case of a double-acting hydraulic pump, opposite ends of a hydraulic cylinder. The pump is adjustable, so that lubricant delivery can be changed to suit all applications. Where necessary, a timer can be used in conjunction with a solenoid valve for added control of lube quantity.

Controlling the frequency of pump actuations under varying conditions often is a problem in installations of this type. The physical environment in which a vehicle

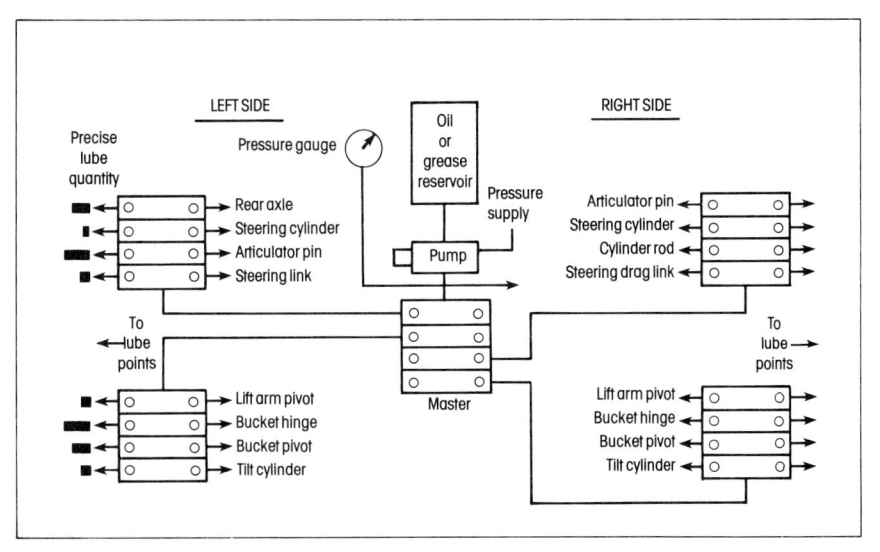

Fig. 4-2 Flow schematic for centralized lubrication system applied to a front end loader.

operates and the operator's habits, such as brake usage and loading procedures, can affect the amount of lubricant being delivered to lubrication points.

A rugged, economical solid state dc program timer has been developed that can be used in conjunction with a solenoid valve to eliminate this problem when it occurs. The time can be adjusted to any one of 12 frequencies per hr—from a minimum of 1 to a maximum of 60 cycles.

By setting the pump at a predetermined output and using the program timer, the designer can precisely control the amount of lubricant used in a given period. This permits an accurate estimate of reservoir-filling frequency based on machine operating hours. The timer was specially designed and tested for heavy-duty mobile equipment.

Feeders on this front end loader are mounted out of the way. However, the master feeder is generally located within sight of the operator so that he can observe the system's performance from the cab. A cycle indicator—an extension of one of the feeder pistons—is included to indicate that the system is working properly. Also, a pressure gauge mounted in the cab alerts the operator when unusually high pressure occurs as a result of a blocked bearing so that he can take corrective action.

The lubricant used in the system should be selected on the basis of prevailing mean ambient temperatures in the geographical area where the vehicle operates.

Large-drill or shovel system Because of the capital outlay required for very large equipment such as a drill or shovel, an automatic, centralized lubrication system is really a necessity. This ensures that all points are lubricated regularly, thereby prolonging the life and improving the performance of the vehicle. The system specified, however, is basically the same as that designed for smaller vehicles. The only difference is the degree of sophistication applied in monitoring and operating the various components.

Since electric power is generally available on larger equipment and because a large quantity of lubricant is needed, barrel pumps are generally employed. This type of system can service hundreds of lubrication points on even the largest off-the-road equipment.

The particular system illustrated in Fig. 4-3 is a large Schramm rotary blast-hole drill. The automatic grease system for the main frame, crawlers, proper bearings on the upper works, main machinery, revolving frame, and boom are served by a common grease reservoir. The system provides continuous lubrication to the lubrication points on the crawlers and propel bearings when the machine is being moved. A diverter valve is used to send the lubricant to different master feeders. Once the machine is stationary and the operator begins drilling, a solenoid valve is energized and lubrication is delivered to the

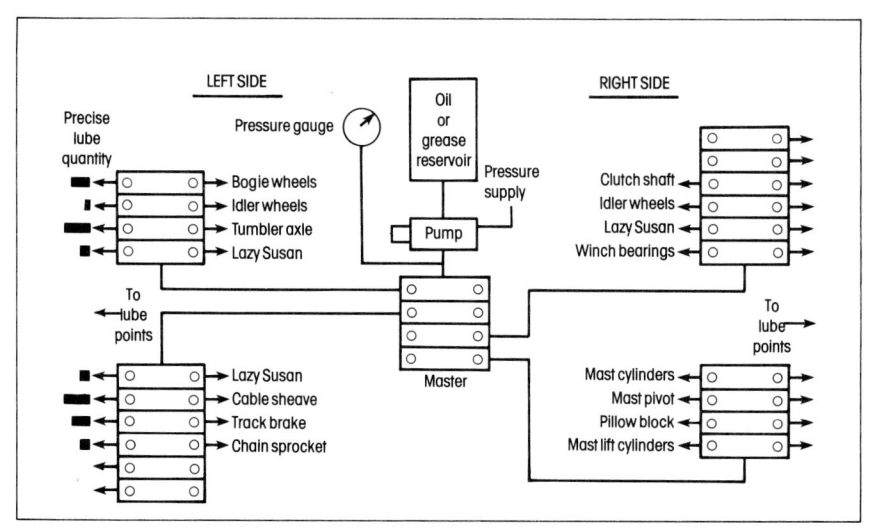

Fig. 4-3 Flow schematic for centralized lubrication system applied to a drill unit.

main-group master feeder. One of the outlets on the master supplies part of the lubricant to the propel-group master, thus keeping the crawlers and other drive points constantly lubricated.

Another microswitch on the master, actuated by piston movement, permits centralized monitoring of system operation. The quantity of grease is controlled by an adjustable timer-counter located in the operator's cab. If, for any reason, the microswitch is not actuated in a predetermined time interval, a red light will go on and warn the operator of a malfunction. The exact location of the difficulty can be traced by using various types of performance indicators. This particular system, therefore, is truly centralized and automatic from the standpoint of both operating and monitoring.

On large equipment of this type, grease is used at the rate of approximately 15 cu in./hr during drilling or digging and at about three times this amount when the machine is moving and lubrication is continuous.

On vehicles with open gears, a separate system that sprays lubricant onto the gear teeth may be installed.

Dollar savings accrue The experience of the G Company with centralized lubrication systems is typical of many heavy-equipment operations. The firm began using automatic lubricators on their mobile equipment in 1963; the results have been impressive.

For example, a 16-yd Auto-Car, used primarily for stocking stone, was equipped with a Trabon system that services 68 lubrication points. All the system's components are in the cab, and it is actuated every time the brakes are applied. The truck uses approximately 2 qt of #90 gear oil per week.

The direct savings in man-hours and materials on this truck, after installation of automatic lubricators, amount to $250 per year. Since the total cost of labor and materials for installing the system was $730, the company reached the breakeven point in only 3 years.

In another instance, a 5-yd 88-B Bucyrus-Erie shovel was equipped with a semiautomatic Trabon lubrication system that services 130 lubrication points. The system consists of three zones: one for the propel mechanism, one for the machinery on the revolving frame, and one for the boom and hook rollers.

Each zone is manually serviced with an air-operated pressure gun, actuated by the shovel air compressor. The gun is held on the master valve until the entire block of feeder valves has cycled the proper number of times. The exact number of cycles is determined by experience. This procedure is then repeated for the other two zones.

This system allows the shovel to be serviced in 15 min as opposed to the usual 70 to 90 min. Direct savings in man-hours and materials amounts to $1375 per year, so that the breakeven point was reached in about 2 years.

As other heavy-equipment companies are discovering, the industry demand for larger, faster, and more heavily loaded pieces of equipment is increasing. As it does, lubrication at the right time and in the correct quantity is becoming more important in order to protect larger investments and to ensure that the equipment will operate up to its rated capacity.

A centralized system offers maintenance personnel a powerful tool in keeping equipment on the job and out of the shop.

Chapter **5**

Arc Welding in Maintenance

J. E. HINKEL
Manager, Application Engineering, The Lincoln Electric Company, Cleveland, Ohio

Among the more important uses of welding in maintenance are repairing and making machinery and equipment. In this respect, welding is an indispensable tool without which operations would soon shut down. Fortunately, welding machines and electrodes have been developed to the point where reliable welding can be accomplished under the most adverse circumstances. Frequently, welding must be done under something less than ideal conditions, and therefore equipment and men for maintenance welding should be the best.

Besides the quick on-the-spot repairs of broken machinery parts, welding offers maintenance a means of making many items needed to meet a particular demand in a required minimum of time. Broken castings, when new ones are no longer available, can be replaced with steel weldments fashioned out of standard shapes and plates. Special machine tools for specific operations can be designed and made for a fraction of the cost that might be needed to buy a standard machine that would have to be adapted to do the job.

The almost infinite variety of this type of welding makes it impossible to do more than suggest what can be done. Figures 5-1 to 5-5 show what the imaginations of some maintenance men have accomplished in this field. As for the welding involved, it should present no particular problems if the welders have the necessary training and background to provide them with a knowledge of the many welding techniques that can be used.

With welding, a maintenance crew can fabricate and erect many of the structures required, even to the extent of making structural steel. Welding can be done either in the maintenance department or at the site. Structures must, of course, be adequately designed to be able to withstand the loads to which they will be subjected. Such loads will vary from those of wind and snow in a simple shed to dynamic loads of several tons where a crane is involved. Materials and joint design must be selected with a knowledge of what each can do. Then the design must be executed with the use of properly trained welders only. Structural welding involves out-of-position work, frequently under awkward conditions, so that a welder must be able to put in good welds under all kinds of conditions.

Standard structural shapes can be used. Frequently pipe makes an excellent structural shape. Scrap materials often can be put to good use. In using scrap, however, it is best to weld with a low-hydrogen E7018-type of electrode, since the analysis of the steel may be unknown and some high-carbon steel may be encountered. The low-hydrogen electrodes will minimize the tendency to crack. This structural scrap frequently comes from old structures, such as elevated railroads being dismantled, which used riveted-quality steel that takes little or no account of the carbon content. Where the quality of the steel is known, an E6010 electrode is used for erection welding. An E7024 or E7014 electrode can be used for fabricating in the shop, if the welding can be done in the flat position. Typical joints that are used in welded structures are shown in Figs. 5-6 to 5-9.

WELDING PROCESSES

Electric-arc welding Electric-arc welding employs the heat of an electric arc to bring metals to be welded to a molten state. In electric-arc welding, the work to be welded is made part of an electric circuit, known as the welding circuit, which has its power source

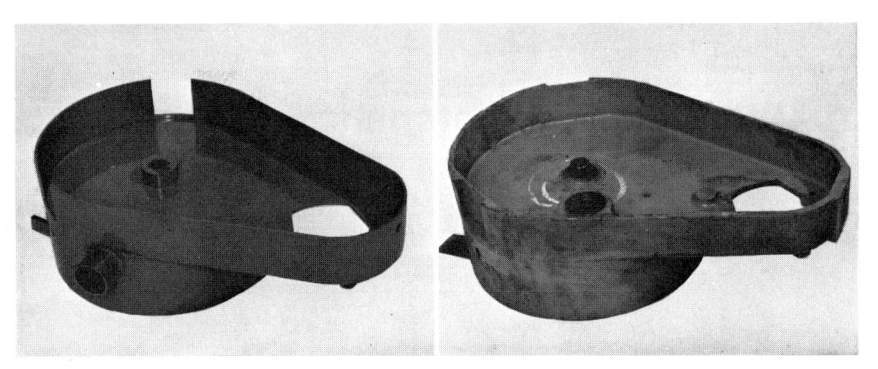

Fig. 5-1 Steel replacement and the cast-iron cover it replaced.

in a welding generator or transformer. One cable carrying current from the power source is attached to the work, and another cable is attached to an electrode holder. An arc is established between the electrode and the work. The arc is moved along the work, melting and fusing the metal as it progresses. Since the arc is one of the hottest commercial sources of heat, this melting takes place almost instantaneously as the arc is applied to the metal.

A variety of welding processes are in common use, employing the electric arc to obtain the welding heat. Each has its particular advantage. All, however, have one problem in

Fig. 5-2 Long delivery prompted welding of this cast-iron punch-press frame.

common—that of shielding the arc. Molten steel has a strong affinity for oxygen and nitrogen. If the arc and molten-metal pool are exposed to the atmosphere during welding, the metal will pick up oxygen and nitrogen, forming oxides and nitrides in the weld as it solidifies. These are impurities which will embrittle the weld and thus weaken it.

All the arc-welding processes familiar to the maintenance welder use some method of

Fig. 5-3 Plant-made racks for holding steel.

Fig. 5-4 Typical welding jig and positioner that can be readily fabricated.

shielding the arc and molten pool from the atmosphere, obtaining welds, when correctly made, that are as strong as, or stronger than, the metal being welded. These processes are variations of shielded-metal-arc welding.

Manual and automatic welding Manual welding, also called hand welding, is welding in which the entire welding operation is performed and controlled by hand. Automatic welding differs from hand welding in that welding equipment mechanically performs the welding operation. The terms *semi* and *fully* are also used to identify automatic welding further in respect to the degree of automation. With semiautomatic welding, the welding

Fig. 5-5 Maintenance-department fabricating trusses. Trusses made from channels and angles. A jig was laid out on plate in the yard.

equipment is traveled manually along the joint. With fully automatic welding, the welding equipment is traveled mechanically along the joint.

SHIELDED-METAL-ARC WELDING[1]

Shielded-metal-arc welding is by far the most widely used method of arc welding. With this welding method, an electric arc is formed between a consumable metal electrode and the work. The intense heat of the arc, which has been measured at temperatures as high as 13,000°F, melts the electrode and the surface of the work adjacent to the arc. Tiny

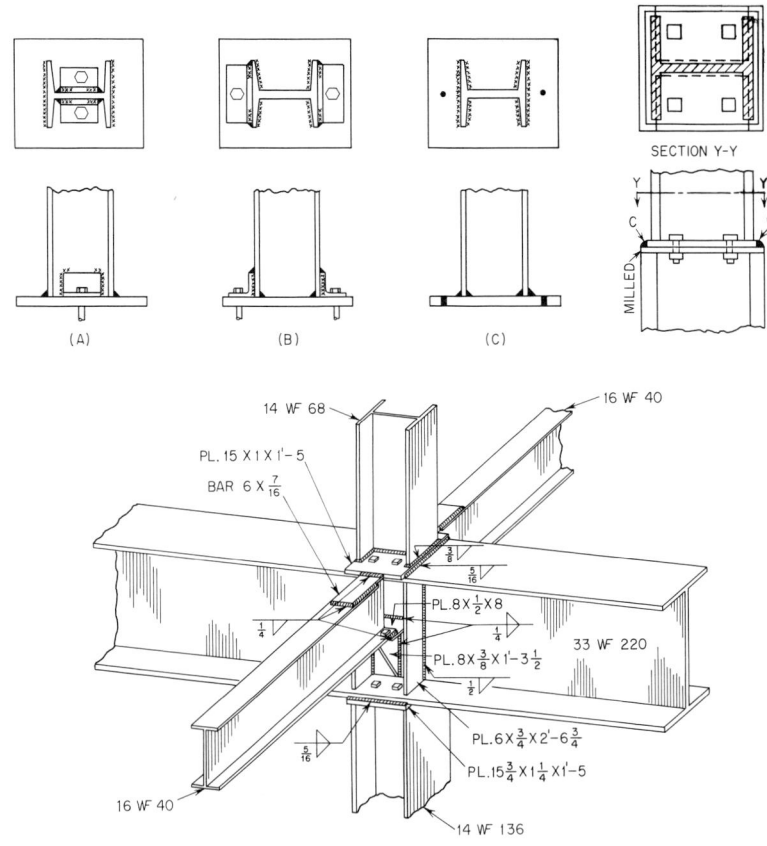

Fig. 5-6 Typical column bases, column splices, and beam-to-column connections that can be used in structural welding.

globules of molten metal rapidly form on the tip of the electrode and transfer through the arc in the *arc stream* into the molten *weld pool* or *weld puddle*, on the work's surface. The actual transfer is induced by the force of gravity, molecular attraction, and surface tension, if the welds are flat or horizontal. Molecular attraction and surface tension are the forces that induce metal transfer from the electrode to the work when the weld is being made in the vertical or overhead position.

[1] Recent welding-process developments and modifications in existing processes are tending to confuse the process classifications established by the American Welding Society. For the most part, the "family grouping" and process name are the same as present American Welding Society designations. In a few instances, however, minor modifications in family-group identification and process name have been made in an attempt to improve clarity and continuity.

In addition to supplying filler metal for the weld deposit, other materials are usually introduced into and/or around the arc; these perform one or all three of the following functions, depending upon the material being welded and the process being used: (1) shielding the arc and preventing atmospheric contamination of the molten metal in the arc stream and the weld puddle; (2) providing scavengers and deoxidizers to protect the molten crater; (3) producing a slag blanket over the very hot but solidified weld. All these functions are necessary to assure the strength and quality of the weld being made.

Self-shielded metal arc welding Electrodes for the shielded metal-arc-welding-process are manufactured by extruding, dipping, or fabricating. The extruded and dipped electrodes, more often referred to as coated, or covered, electrodes, contain the shielding, scavenging, and deoxidizing materials in the covering that surrounds a solid metal core. The fabricated, or cored, electrodes contain the shielding, scavenging, and deoxidizing materials compacted in the electrode core surrounded by a metal sheath. Since both the covered and the fabricated electrodes contain all the materials to accomplish complete arc shielding, they are called self-shieldng electrodes.

The arc-shielding action is essentially the same for both the covered electrodes, as illustrated in Fig. 5-10, and the fabricated electrodes. But the actual method of arc shielding and volume of slag produced will vary with different electrode types.

The bulk of the core or covering materials in some electrodes is converted to a gas by the heat of the arc, and only a small amount of slag is produced. This type of self-shielding electrode, depending largely on a gaseous shield to prevent atmospheric contamination, can be identified by the incomplete or light slag covering of the completed weld.

The other extreme in self-shielding electrode design is the type where the bulk of the covering material is converted into slag in the arc heat with only a small volume of gas being produced. With this type, the tiny globules of metal being transferred in the arc stream are entirely coated with a thin film of molten slag. This slag floats to the surface of the molten weld puddle before solidifying. The electrodes are identified by the heavy slag deposit that completely covers the surface of the finished weld.

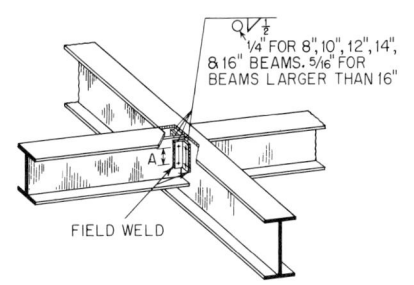

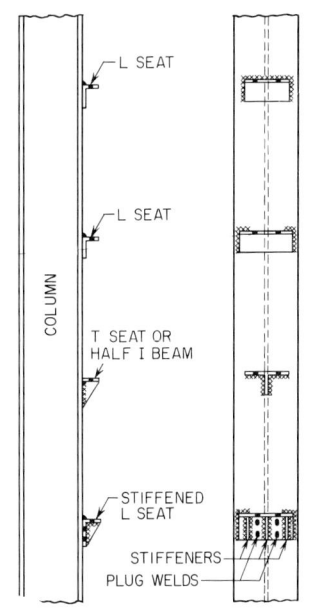

Fig. 5-7 Beam-to-beam framing and methods for seating beams on columns.

Between these extremes there is a wide variety of electrode types with the ability to produce various combinations of gas and slag shielding. These variations in slag action and arc shielding also influence the performance characteristics of the many different types of self-shielding electrodes available for use in maintenance and manufacturing. For example, an electrode that has a heavy slag action is also one which has a high deposition rate and is suited for making large welds in flat positions. An electrode that develops a gaseous arc shield is one which also has a low deposition rate and smaller molten weld puddle and therefore is suited for making welds in the vertical and overhead positions.

These and many other performance characteristics are the reasons why one type of self-shielded electrode is preferred over all others for a specific weld in a specific position.

Manual shielded-metal-arc welding Extruded, dipped, and fabricated electrodes are used for manual shielded-metal-arc welding. These electrodes range in length from 9 to 18 in. The consumable welding electrode is placed in a hand-held clamping device called the electrode holder. Welding begins by touching the tip of the electrode to the work to complete the electric welding circuit, then withdrawing the tip, establishing the arc. As the heat of the arc melts the tip of the electrode, the welding operator, called the welder, manually lowers the tip of the electrode, maintaining a uniform distance between it and the work, thereby maintaining a steady arc. Simultaneously, the welder manually moves

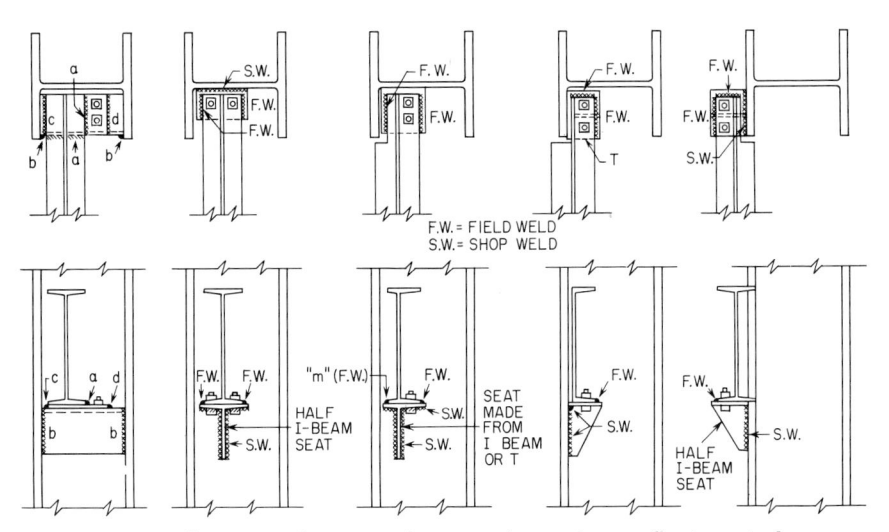

Fig. 5-8 Different ways of connecting beams to columns when an offset is required.

the electrode along the work at a rate of speed that deposits sufficient filler metal to create the needed weld size.

Semiautomatic and fully automatic flux-cored arc welding The electrode used for semiautomatic and fully automatic flux-cored arc welding is mechanically fed through a welding gun or welding jaws into the arc from a continuously wound coil that weighs approximately 50 lb. Only the fabricated flux-cored electrodes are suited for this method of welding, since coiling the extruded flux-cored electrodes would damage the coating. In addition, metal-to-metal contact at the electrode's surface is necessary to transfer the welding current from the welding gun or welding jaws into the welding electrode. This is impossible if the electrode is covered.

Typical applications of semiautomatic and fully automatic equipment for flux-cored arc welding are shown in Figs. 5-11 and 5-12. For a given cross-sectional area of electrode wire, much higher welding amperage can be applied with semiautomatic and fully automatic welding. This is because the current travels only a very short distance along the bare metal electrode, since contact between the current-carrying jaws and the bare metal electrode occurs close to the arc. In hand welding, the welding current must travel the entire length of the electrode, and the amount of current is limited to the current-carrying capacity of the wire. The higher currents used with automatic welding result in a high weld-metal deposition rate. This increases welding speed, reduces welding time, and lowers welding costs.

Submerged-arc welding With submerged-arc welding the arc is completely hidden under a small mound of granular inorganic flux which is automatically deposited around the electrode wire as it is fed to the work (Fig. 5-13). The arc and molten pool are completely blanketed with flux at all times, and there are no visible arc rays or weld spatter.

Under usual welding conditions, the quantity of flux melted weighs approximately the same as the electrode consumed. The unfused flux may be collected and reused. Precautions should be taken to keep the flux and the work clean in order to prevent weld contamination and to maintain weld quality.

The high currents used with submerged-arc welding also develop a deep-penetrating-arc characteristic. Consequently, no groove or a small groove may be used, depending on the thickness of the base metal, with correspondingly smaller additions of filler metal. For example, no chamfering is necessary for two-pass butt joints in steel up to ⅝ in. thick. Complete penetration can also be obtained in fillet welds for material up to ¾ in. thick without chamfering. For joints in thicker material, a double V-groove weld is used. The graph of Fig. 5-14 shows typical relations between penetration and applied current.

With submerged-arc welding, distortion is minimized because of high welding speeds, minimum number of passes, and efficient application of heat. This means that less heat is applied to the weld area and, furthermore, that the heat is applied more uniformly than with hand welding. Distortion caused by an unbalanced heat condition, as in single-groove multiple-pass welded joints, can be corrected by presetting the base-metal parts to offset angular movement. The other methods of controlling distortion, discussed later in this chapter, can also be applied.

Although the submerged-arc-welding process is used primarily for production welding, it also has potential maintenance use which even to this day has been only partially exploited. The process is particularly suited to rebuilding worn surfaces and developing abrasive-resistant surfaces for manufacturing operations encountering severe metal-erosion problems.

Gas-shielded arc welding In gas-shielded arc welding, the arc and weld region is shielded from the air by a protective gas. This gas may or may not be inert. The gases experiencing greatest industrial use are argon, helium, and CO_2. Two variations of the gas-shielded arc-welding process are gas tungsten-arc welding and gas metal-arc welding.

Gas-shielded tungsten-arc welding Gas-shielded tungsten-arc welding with an inert gas was originally developed to weld the corrosion-resistant and other difficult-to-weld metals such as aluminum and copper. Over a period of years, however, its application has expanded to include welding and surfacing operations on practically all commerical metals.

The gas tungsten-arc-welding process obtains the necessary heat for welding by a very intense electric arc which is struck between a virtually nonconsumable tungsten electrode and the metal workpiece (Fig. 5-15). On joints where filler metal is required, a welding rod is fed into the weld zone and melted with the base metal in the

Fig. 5-9 A beam-and-girder connection and a column detail showing craneway.

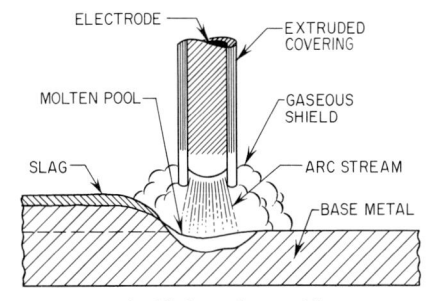

Fig. 5-10 Shielded-metal-arc-welding process.

manner used with oxyacetylene welding. The weld zone is shielded from the atmosphere by an inert gas fed through the welding torch. Either argon or helium may be used.

Inert-gas-shielded tungsten-arc welds, because of 100 percent protection from the atmosphere, are stronger, more ductile, and more corrosion-resistant than welds made with ordinary arc-welding processes. Corrosion due to flux entrapment does not occur,

Fig. 5-11 Fully automatic flux-cored arc welding.

and postwelding cleaning operations are reduced to a minimum. The entire welding action takes place practically without spatter, sparks, or fumes. Fusion welds can be made in nearly all metals used industrially. These include aluminum alloys, stainless steel, magnesium alloys, nickel and nickel-base alloys, copper, silicon copper, copper nickel, brasses, silver, phosphor bronze, plain-carbon and low-alloy steels, cast iron, and others.

Fig. 5-12 Semiautomatic flux-cored arc welding.

Arc Welding in Maintenance 4-35

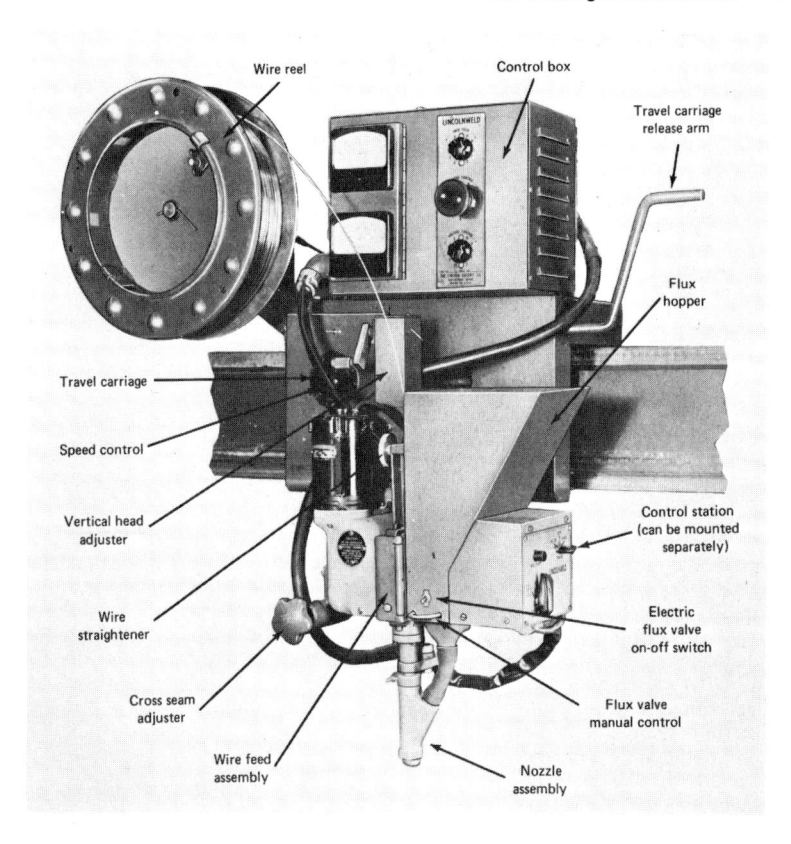

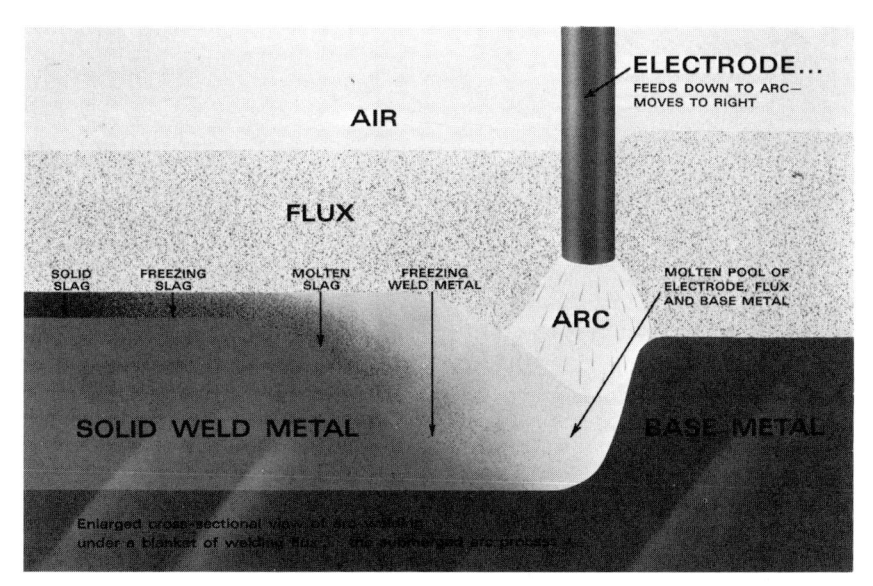

Fig. 5-13 Elements of the submerged-arc-welding process.

4-36 Basic Maintenance Technology

The process is also widely used for welding various combinations of dissimilar metals and for applying hard-facing and surfacing materials to steel.

The power supply for inert-gas-shielded tungsten-arc welding may be either alternating or direct current. However, certain distinctive weld characteristics obtained with each type often make one or the other better suited for a specific application.

In dc welding, the welding-current circuit may be hooked up as either straight polarity or reverse polarity. The connection for dc *straight-polarity* (DCSP) welding is electrode negative and work positive. In other words, the electrons flow from the electrode to the plate or workpiece, as shown in Fig. 5-16. For dc *reverse-polarity* welding (DCRP), the connections are just the opposite; electrons flow from the plate to the electrode, as shown in Fig. 5-17.

In straight-polarity welding, the electrons hitting the plate at high velocity exert a considerable heating effect on the plate. In reverse-polarity welding, just the opposite occurs; the electrode acquires this extra heat, which then tends to melt the end of the electrode. Thus, for any given welding current, DCRP requires a larger-diameter electrode than does DCSP. For example, a $\frac{1}{16}$-in.-diameter pure-tungsten electrode can handle 125 amp of welding current under straight-polarity conditions. If the polarity were reversed, this amount of current would melt off the electrode and contaminate the weld metal. Hence a $\frac{1}{4}$-in.-diameter pure-tungsten electrode is required to handle 125 amp DCRP satisfactorily and safely.

These opposite heating effects influence not only the welding action but also the shape of the weld obtained. DCSP welding will produce a narrow, deep weld; DCRP welding, because of the larger electrode diameter and lower currents generally employed, gives a wide relatively shallow weld.

One other effect of DCRP, the so-called plate-cleaning effect which seems to occur, is

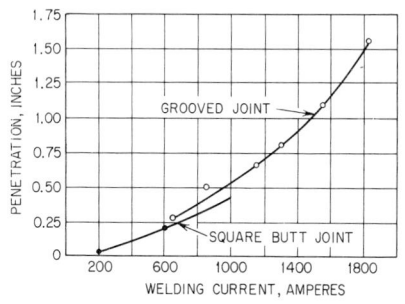

Fig. 5-14 Penetration versus applied current for submerged-arc-welding.

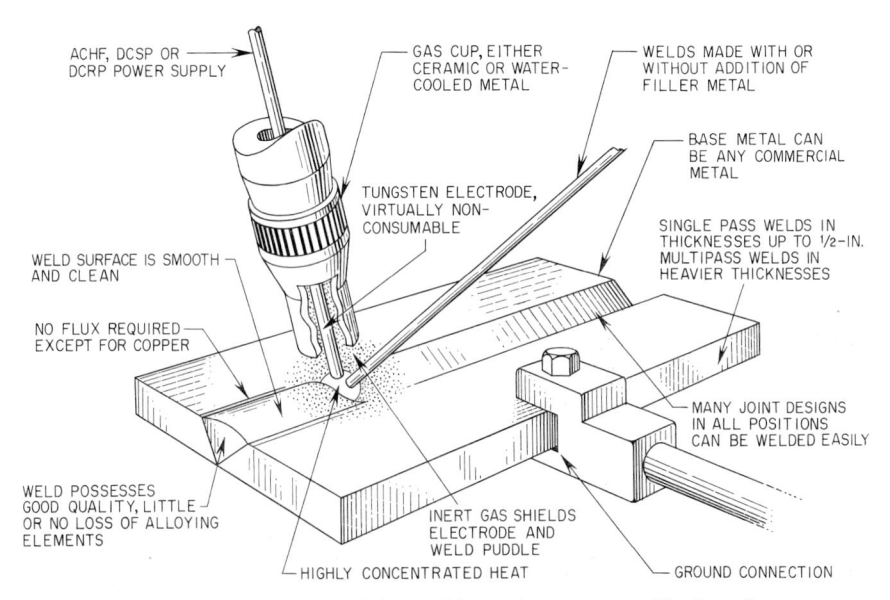

Fig. 5-15 Inert-gas-shielded arc welding with a nonconsumable electrode.

worth mentioning. Although the exact reason for the surface-cleaning action is not known, it seems probable that either the electrons leaving the plate or the gas ions striking the plate tend to break up the surface oxides, scale, and dirt usually present.

Welding with an alternating current is theoretically a combination of DCSP and DCRP welding, since the current flows in one direction and then in the other, or reverse,

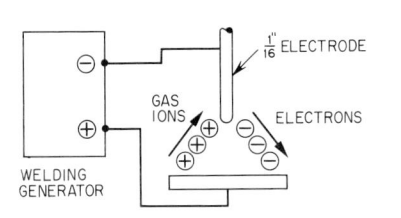

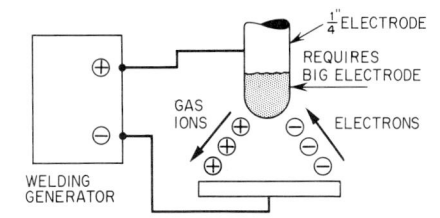

Fig. 5-16 Direct-current straight polarity.

Fig. 5-17 Direct-current reverse polarity.

direction. However, moisture, oxides, and scale on the surface of the plate tend to prevent (partially or completely) the flow of current in the reverse-polarity direction. To ensure proper current flow in the reverse direction when welding with alternating current, it is common practice to introduce into the welding current a high-voltage high-frequency low-power current. This high-frequency current jumps the gap between the electrode and the workpiece and pierces the oxide film, thereby forming a path for the welding current to follow. Superimposing this high-voltage high-frequency current on the welding current gives the following advantages:

1. The arc may be started without touching the electrode to the workpiece.
2. Better arc stability is obtained.
3. A longer arc is possible; this is particularly useful in surfacing and hard-facing operations.
4. Welding electrodes have longer life.
5. The use of wider current ranges for a specific-diameter electrode is possible.

A typical weld contour produced with high-frequency stabilized alternating current is shown in Fig. 5-18, together with DCSP and DCRP welds for comparison.

Tungsten-arc-welding equipment The basic equipment requirement for manual inert-gas tungsten-arc welding is a welding torch plus additional apparatus to supply (1) electric power, (2) argon, and (3) water. Also, certain protective equipment should be employed to protect the operator from the arc rays during welding operations.

The welding current is supplied either by a variable-voltage welding generator or rectifier for dc welding or by a variable-voltage welding transformer for ac welding. When selecting a generator or rectifier, it is important to obtain one which has good current control at the lower end of its current range. This ensures the arc stability required for efficient operation. If you plan to use an older dc welding machine which operates inefficiently in the lower current range, a resistor should be used in the ground line between the generator and workpiece. These resistors are marketed

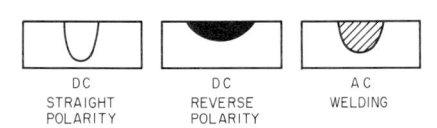

Fig. 5-18 Comparison of weld penetration for the three types of welding current used with inert-gas tungsten-arc welding.

by most manufacturers of dc generating equipment. Several firms manufacture transformers which are special for tungsten-arc welding, some with built-in high-frequency stabilization. Bear in mind that some transformers are designed to produce a balanced wave and can be used at the full rated capacity. Others are not, and should not be used at over 70 percent maximum capacity to avoid overloading the primary. Be certain you know which type you are using. A high-frequency generator used with ac welding can also be obtained from any reputable dealer.

High-purity argon is supplied in steel cylinders, each containing approximately 240 cu ft of argon at a pressure of 2000 psi. A regulator is needed to reduce this pressure to that required for welding, generally about 20 psi. In addition, a flowmeter is required at every

4-38 Basic Maintenance Technology

welding station, since different materials need different flows or amounts of argon for adequate protection. Where a large amount of welding is being done continually, it is advisable to connect a manifold to a bank of cylinders and pipe the argon to each individual work station. Again, a flowmeter is required for each station.

When currents above 130 amp are used, water cooling of the torch and power cable is required. The cooling water for water-cooled torches must be clean; otherwise, restricted or blocked passages may cause excessive overheating and damage to the equipment. Most shops have an adequate supply of cooling water available. However, where welding is done in large shops or outdoor locations, completely self-contained units are available. A typical portable installation is shown in Fig. 5-19.

An inert-gas tungsten-arc-welding torch feeds both the welding current and the inert gas to the weld zone, as shown in Fig. 5-20. The current is fed to the weld zone through the tungsten electrode, which is held firmly in place by the electrode holder. The argon (or helium) is fed to the weld zone through a gas cup at the head of the torch.

The electrode should extend about $1/8$ to $3/16$ in. beyond the end of the gas cup for butt welding and about $1/4$ to $3/8$ in. for fillet welding.

Recommended gas-cup sizes for the various torches and electrode diameters are specified by the manufacturer. Ceramic cups are generally acceptable when the welding current is less than 250 amp. With higher currents or where welding conditions are unusually severe, water-cooled metal gas cups must be used to prevent overheating. Water-cooled cups should never come into contact with the workpiece when the welding current is ON. Conductivity of the hot gases may cause the arc to jump from the electrode to the cup rather than to the workpiece, thus damaging the cup.

As with all industrial equipment, certain common-sense precautions should be

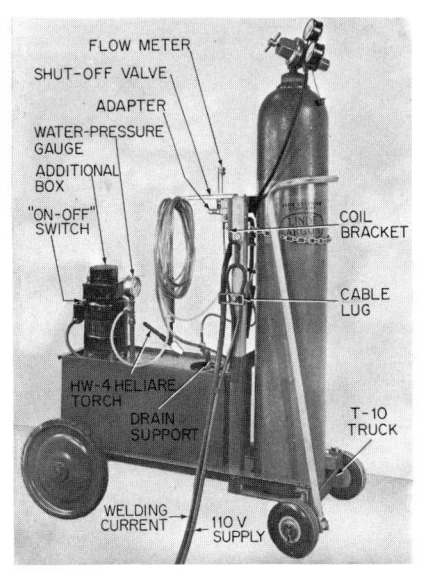

Fig. 5-19 Typical portable installation for inert-gas tungsten-arc welding.

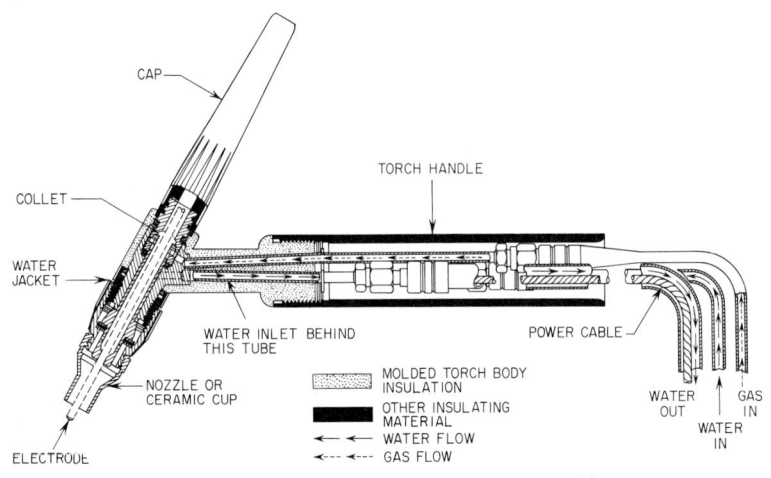

Fig. 5-20 A water-cooled torch for inert-gas tungsten-arc welding.

observed. In the case of tungsten-arc welding, the operator should be properly protected from arc rays. This requires suitable clothing to cover all exposed skin surfaces and a welder's helmet with the proper shade of glass to protect the eyes and face. The shade of the glass lens will depend upon the intensity of the arc. The recommended shades for various current ranges are listed in Table 5-1.

Gas-shielded metal-arc welding Gas-shielded metal-arc welding is commercially called MIG welding when an inert gas is used. An arc between the consumable wire electrode and the workpiece (Fig. 5-21) is maintained in an atmosphere of inert gases, principally argon. The gases shield the weld zone from possible contamination by the atmosphere and eliminate the need for flux. Quality welds can be produced by either manual or machine welding. Welds made by this process are relatively clean and require little or no postweld finishing.

TABLE 5-1 Lens Shades for Current Ranges

Glass No.	Welding current, amp
6	Up to 30
8	30 to 75
10	75 to 200
12	200 to 400
14	Above 400

With inert-gas-shielded arc welding, you can weld such metals as aluminum, magnesium, copper, nickel, silicon bronze, aluminum bronze, stainless steel, low-alloy steel, and carbon steel. A consumable electrode similar to the metal being welded is used.

The average current density used is about twenty times that recommended for carbon-arc welding and about six times that recommended for covered-metal-arc welding. This high current density results in concentrated heat that produces narrow welds with deep penetration, a small heat-affected zone, and reduced distortion. Conventional dc welding or constant-potential power supplies may be used.

A constant-potential power source is preferred for inert-gas-shielded metal-arc welding with a continuously fed bare-wire electrode. As shown in the accompanying graph (Fig. 5-22) constant potential has a flat volt-ampere characteristic rather than the drooping characteristic of conventional dc power. Since the welding voltage remains essentially constant, the speed of wire feed controls the welding current.

A manual welder for inert-gas-shielded arc welding with a consumable electrode is shown in Fig. 5-23. This particular unit, a portable welder, uses welding currents as high as 500 amp. The electrical control box contains the various control circuits for wire feed, gas flow, and application of welding current. The wire drive unit feeds the consumable wire electrode at the required speed. Once the welding conditions have been set up, the trigger switch on the water-cooled torch stops and starts welding. The remote-control box permits the operator to adjust arc length and to inch out wire electrode for arc striking

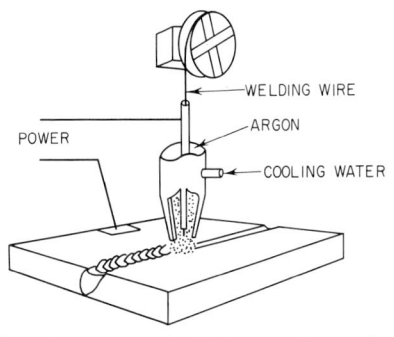

Fig. 5-21 Inert-gas-shielded arc welding with a consumable electrode.

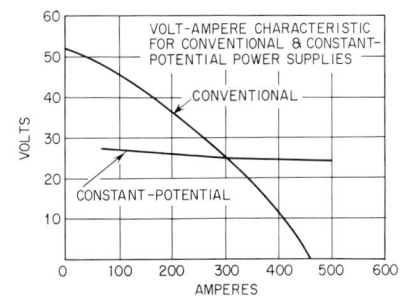

Fig. 5-22 Ampere-volt characteristic of constant-potential and conventional power supplies.

without leaving his welding position. Source of welding current, supply of inert shielding gas (argon, helium, or a mixture of approximately 95 percent argon and 5 percent oxygen), and a supply of cooling water also are required.

The gas-shielded arc-welding processes can successfully weld plain carbon steel, but, in most instances, when compared with other arc-welding processes, cost has prohibited their use.

Gas-shielded metal-arc welding-CO_2 Another version of gas-shielded metal-arc welding uses CO_2 (carbon dioxide) rather than an inert gas to blanket the arc and the surrounding weld area. A typical production welding installation is illustrated by Fig. 5-24. There are two variations of this process. The first uses a solid electrode, the second a fabricated flux-cored electrode. In addition to providing filler metal, these electrodes or the flux contain elements which perform a scavenging and deoxidizing action in the crater to improve weld quality.

The flux-cored-electrode process has the flux within an outer steel sheath.

The gas-shielded metal-arc-welding process with CO_2 is used for production welding of carbon steels and for fabricating industrial piping and sheet metal.

Fig. 5-23 A manual welder for inert-gas-shielded arc welding with a consumable electrode. *(Linde Division, Union Carbide Corporation, New York.)*

Gas-shielded spot welding This method of welding combines either gas-shielded tungsten-arc-welding or gas-shielded metal-arc-welding equipment with an electrical timing-control system that automatically starts and maintains the arc for a controlled time period. Two lapped pieces of metal are spot-welded together by applying heat from an electric arc to the top surface of the joint. Welding action is controlled by the current input to the arc and the time the arc dwells on the material being welded.

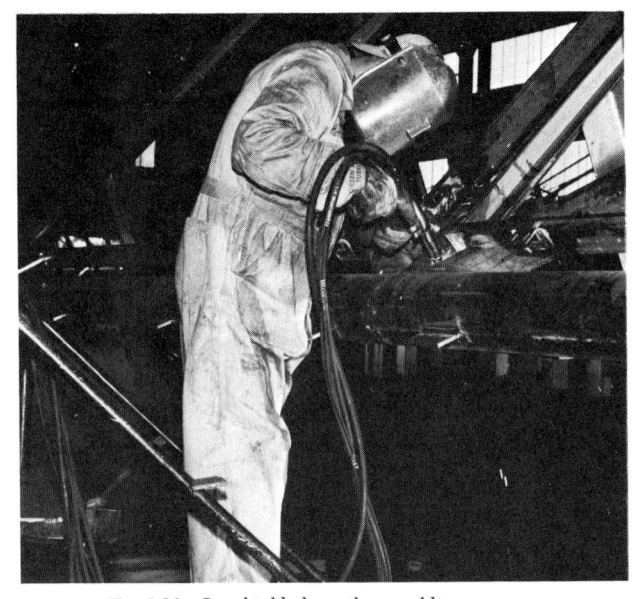

Fig. 5-24 Gas-shielded metal-arc welding.

Shielding of the arc, electrode (consumable metal or nonconsumable tungsten), and fluid weld puddle are similar to those of conventional gas-shielded arc welding.

The resulting spot weld parallels that produced by resistance-welding techniques; however, no electrode pressure is required, and the welding is done from one side of the plate without requiring any weld backup. Both inert-gas and CO_2 spot welding are experiencing expanding industrial use.

RESISTANCE WELDING

Resistance-welding processes are primarily designed for production-welding usage. Nevertheless, a few of the processes can be used effectively by the maintenance department. With this method of welding, the joining of the parts being welded is accomplished by the heat obtained from resistance of the work to the flow of electric current in a circuit of which the work is a part and by the application of pressure.

Fig. 5-25 This resistance spot welder is being used to fabricate sheet metal.

Spot welding Spot welding is the most common resistance-welding process. It is usually employed in the welding of thin metal sheets and is accomplished by placing the sheets between movable copper-alloy electrodes. The electrodes carry the welding current and can be actuated to apply the proper pressure during the welding cycle. A typical production installation in a sheet-metal shop is illustrated by Fig. 5-25. Aluminum presents a special problem because of its high electrical conductivity. So does copper, which has practically the same conductivity as electrode material.

Although maintenance welding departments occasionally have the larger floor-mounted equipment, more often the small portable hand-held spot-welding guns are used for fabricating sheet metal.

OTHER WELDING PROCESSES

The welding processes and equipment described to this point have potential use in the typical maintenance welding department. Many other welding processes are being used which, admittedly, have limited maintenance use. These will be summarized briefly. Additional information about specific processes can be obtained from the American Welding Society (see the references at the end of this chapter).

Atomic-hydrogen welding Atomic-hydrogen welding differs from the other arc-welding methods in that the arc is formed between two tungsten electrodes and the work is not part of the welding circuit. A stream of hydrogen gas is passed through the arc and, in the heat of the arc, changes from molecular to atomic form, giving off an intense heat. The hydrogen acts as an effective heat-transfer medium and results in high heat being applied close to the work. A filler rod is used to supply additional metal to the joint. The process has some advantages in welding thin sheet where a high finish is needed.

Electroslag welding Electroslag welding is the metal-arc-welding process employing the principles of submerged-arc welding. This process involves fusion of the base metal and continuously fed filler metal under a substantial layer of high-temperature, electrically conductive molten flux. By feeding one or a combination of two or three electrodes simultaneously into the arc, plates ranging from 1 to 14 in. thick can be joined in a single pass. Application is generally limited to very heavy weldments. Welds are usually made with the joint vertical and with welding progressing from bottom to top.

Plasma-arc welding Plasma-arc welding exists in several forms. The basic principle is that of an arc or jet created by heating electrically a plasma-forming gas (such as argon with additions of helium or hydrogen) to such a high temperature that its molecules become ionized atoms possessing extremely high energy. When properly controlled, this process results in very high melting temperatures. Plasma-arc welding holds a potential solution to the easier joining of many hard-to-weld materials. When modified for metal cutting, this process achieves unusually high cutting speeds. Another application is the depositing of materials having high melting temperatures to produce surfaces of high resistance to extreme wear, corrosion, or temperature.

Stud welding Stud welding is the end welding of a stud, ordinarily a machine screw, at a particular spot on the work by fusion. An electric arc, struck between the stud serving as the electrode and the baseplate, brings the tip of the stud and the surface of the work adjacent to the stud to a molten state. A light pressure is applied, forcing the stud into the molten weld puddle. Current flow is discontinued, and the stud fuses to the work surface as it cools. A compact unit, called a stud welder, supplies the welding current. The arc may be shielded or unshielded.

Carbon-arc welding Carbon-arc welding employs a carbon rod as an electrode. The arc is formed between the carbon and the work, creating a molten pool on the work surface. This pool is kept molten by playing the arc across it. If extra filler metal is needed to make the weld, it is supplied by introducing a filler rod into the arc, where it is melted into the molten pool. This is a puddling process, and is not applicable to vertical or overhead welding. Shielding may be obtained if desired by introducing a paste, powder, or fibrous flux into the arc.

Carbon-arc welding is used only for specialized applications. The carbon arc is also used for cutting where a precision cut is not necessary or on alloys that cannot be cut by the gas process.

Flash welding Flash welding is a resistance-welding process in which fusion is produced by a high localized heat obtained from the electrical resistance existing between two touching surfaces. This type of resistance is evidenced by a flashing, or shower, of sparks produced by the arcing of current at the joining surfaces. When the temperature of the metal has increased to where the joining surfaces have plasticized, the parts are forced together under pressure to make the weld. A portion of the metal squeezes out (upsets) to form the flash. This must be trimmed off, and the joint then ground or otherwise finished to the section desired.

Percussion welding Percussion welding is a process in which fusion temperature results from an arc created across a gap between two surfaces to be joined, the arc being caused by rapid discharge of electrical energy. A percussive (impact) force is applied during or immediately following the electrical discharge.

Projection welding Projection welding is another method of resistance welding. It differs from those previously described, since it uses projections, or embossments, to localize the current flow and welding heat at predetermined points. These projections, which serve as points of contact, are a part of one or both of the parts to be joined. The parts are supported and pressed together by special dies during welding.

Seam welding Seam welding is fundamentally a spot-welding process. One or two electrode wheels running along a straight line at a fast rate of travel make a series of closely spaced spot welds. When the welded spots are so close that they actually overlap, they form a gastight or watertight seam, as required for a vessel. In other cases, the series of spots may be so spaced that the process becomes a mere tack-welding operation in the assembly of a unit. This is called roll-spot welding and is used to speed up standard spot welding.

Upset welding Upset welding is process in which fusion is produced by the heat obtained from electric resistance through the area of contact of two surfaces held together under pressure. In this case, the force is applied prior to introduction of the electric current and is continued until heating is complete. The continued force produces an upsetting as in flash welding, but since the surfaces are in solid contact with one another, there is no arcing or flashing effect.

Electron-beam welding Electron-beam welding directs a bombardment of electrons at the workpiece placed in a vacuum. Electrons are admitted from a filament, acting as a type of nonconsumable electrode, and are highly accelerated by high-voltage potential between the electrode and the work. The high-velocity energy of the electrons converts to heat when they strike the work. The electron flow is electrically concentrated into a beam by means of an electron gun. Since the operation is carried on in a vacuum, the process can be used to weld highly reactive metals without contamination.

Explosive welding Explosive welding is a process wherein a surface-to-surface bond is achieved by the compressive force of a controlled explosion.

Flow welding This is a process where fusion is produced by heating with molten filler metal poured over the joint until the welding temperature is attained and the required filler metal has fully penetrated the joint.

Hammer welding Hammer welding was commonly employed by the blacksmith of yesteryear; it sees very little industrial usage today. It is also called forge welding.

Friction welding Friction welding is based on the fact that a rapidly moving part in pressure contact with a stationary part generates heat in contacting surfaces. When the fusion temperature is reached, movement is stopped and pressure maintained or increased until the weld is completed.

Induction welding Induction welding depends upon the resistance of the workpiece to the flow of an induced electric current to create heat for fusion. The pieces to be joined are placed within a radio-frequency field, usually developed to the inside of a radiating coil that has been designed to approximate the shape of the intended assembly. Filler metal having a low melting temperature is prepositioned at the joint and distributes through the heated joint by capillary action.

Pressure welding Pressure welding is a process in which two pieces of ductile metal are butt-welded or lap-welded by the application of pressure only, without any of the metal reaching the melting point. Heat, if applied, is sufficient only to facilitate plastic flow of the metal under pressure. Bonding depends upon the ability to bring a large number of atoms on the two surfaces being joined into immediate contact. This requires perfect cleanliness of the surfaces, good alignment, and application of high pressures. The pressure is a squeezing action rather than impact.

Thermit welding Thermit welding is based on the chemical reaction between aluminum and iron oxide. The members to be welded are aligned in proper relation, and a mold is built around the ends to be joined. A pouring gate in the top of the mold receives the molten metal. The Thermit charge is placed in a crucible which has a pouring hole in its bottom. The charge is a mixture of iron oxide and granulated aluminum together with small quantities of alloying elements in the iron oxide. Ignition of the mixture produces a reaction between the iron oxide and aluminum, liberating a large amount of heat. The aluminum combines with the oxygen in the iron oxide and releases free molten steel, which flows into the mold, thus producing the weld.

WELDABILITY OF METALS

The term *weldability* is a relative one. Practically all metals are weldable. Some, however, require special welding procedures in order to preserve the properties and characteristics of the metal for which it was originally alloyed.

Special welding procedures are variants within a limited range of possibilities. If a metal cannot be welded with the regular mild-steel electrodes, E6010 and E6012, for example, some degree of preheat with these electrodes is the next step. Following this, the next alternative is to use a low-hydrogen electrode and finally a stainless-steel type of electrode.

The first aspect of any maintenance welding job is to consider the metal being welded. The behavior of the metal under the heat cycle of welding may or may not be critical. The economy and qualtiy of welding on various metals may be affected by any one or more of the following factors:

1. Oxidation. (*a*) Oxidation producing a gaseous oxide of some one of the elements causing gas holes in the weld metal; (*b*) oxidation producing solid oxides which have a melting temperature higher than the metal, thus causing slag inclusions; (*c*) oxidation producing oxides which are soluble or which are heavier and sink in the molten metal and which render the weld metal brittle or of low strength.

2. Vaporization. Vaporization of some element in the metal which vaporizes at a temperature lower than the melting point of the metal.

3. Nonmetallic inclusions. Some metals may contain finely divided nonmetallic inclusions which have a melting point higher than that of the metal and therefore did not coalesce when the metal was refined but do melt and coalesce under the high temperature of the arc and then form visible slag inclusions.

4. Change of structure. Change of structure or arrangement of elements within the metal may take place during arc welding, causing change of physical properties or change of resistance to corrosion.

5. Gas solubility of metal. (a) Different elements may affect the solubility of various gases at different temperatures, and a decrease in solubility of a gas with a decrease in temperature at the freezing point may cause porosity in weld metal; (b) the fluxing out or eliminating of an element during welding may cause the capacity of the metal for a given gas to decrease and thus cause the gas to be given up, producing porosity in the weld metal; (c) gases are absorbed during welding to form stable compounds with elements in the metal and thus alter the composition and physical properties of the weld metal.

6. High coefficient of thermal expansion, or high contraction of weld metal upon cooling.

7. Hot shortness, or low strength of the metal at high temperatures.

8. Thermal conductivity, or rate of transfer of heat from fusion zone.

9. Hardenability. Tendency of metal to become hard and brittle in the weld or fusion zone during heat cycle of welding.

The foregoing list indicates why some metals are more satisfactory than others. A careful study of the factors listed indicates that most of the possible undesirable characteristics can be corrected by one or more of the following methods:

1. Selection of metal within the permissible class most suitable for arc welding
2. Use of proper shielded arc
3. Use of proper fluxing material
4. Use of proper electrode or filler material
5. Proper welding procedure
6. In some cases, preheat and postheat treatment

In considering the weldability of any metal, it should be borne in mind that the weld largely depends upon the characteristics of the weld metal which may come from two sources, namely, base metal and electrode or filler metal.

If little or no electrode or filler metal is used, the proper selection of the base metal becomes of prime importance. If the weld metal comes mostly from the electrode or filler metal, then selection of the proper filler metal or electrode becomes of prime importance. However, both electrode and base metal are subjected to similar requirements during arc welding, and both should be of best arc-welding quality, although in many cases the electrode or filler metal serves as a corrective for the base metal.

THE CARBON STEELS

The carbon steels are widely used in all types of manufacturing. The weldability of the different types (low, medium, and high) varies considerably. The preferred analysis range of the common elements found in the carbon steels is shown in Table 5-2. Welding metals whose elements vary above or below the range usually call for special welding procedures.

Low-carbon steels (0.10 to 0.30 percent carbon) Steels of low-carbon content represent the bulk of the carbon-steel tonnage used by industry. These steels usually are

TABLE 5-2 Preferred Analysis Range of Carbon Steels

	Low, %	Preferred, %	High, %
Carbon	0.06	0.10 to 0.25	0.35
Manganese	0.30	0.35 to 0.80	1.40
Silicon		0.10 or under	0.30 max
Sulfur		0.035 or under	0.05 max
Phosphorus		0.03 or under	0.04 max

more ductile and easier to form than higher-carbon steels, and for this reason are used in most applications requiring considerable cold forming, such as stampings and rolled or bent shapes in bar stock, structural shapes, or sheet. Steels with less than 0.13 percent carbon and 0.30 percent manganese have a slightly greater tendency for internal porosity than steels of higher carbon and manganese content.

Medium-carbon steels (0.30 to 0.45 percent) The increased carbon content in medium-carbon steel usually raises tensile strength of the material and also hardness and wear resistance. These steels experience selective use by manufacturers of railroad equipment, farm machinery, construction machinery, material-handling equipment, and similar products. The medium-carbon steels can be successfully welded with the E60XX electrode if certain simple precautions are taken, and the cooling rate is controlled to prevent excessive hardness.

High-carbon steels (0.45 percent and higher) The high-carbon steels are generally used in a hardened condition. In this group are most of the steels used in tools for forming, shaping, and cutting. Tools used in metalworking, woodworking, mining, and farming, such as lathe tools, drills, dies, knives, scraper blades, and plowshares, are typical examples. The high-carbon steels are often described as being difficult to weld, and are not suited to mild-steel welding procedures. Usually, low-hydrogen or other special electrodes are required, and controlled welding procedures, including preheating and postheating, are needed to provide welds that are crackfree.

The higher the carbon content steel, the harder it becomes when it is quenched from above the critical temperature. Welding raises steel above the critical temperature, and the cold mass of metal surrounding the weld area creates a quench effect. Hardness and absence of ductility result in cracking as the weld cools and contracts. Preheating from 300 to 600°F and slow cooling will usually prevent cracking. Figure 5-26A and B shows a calculator for determining preheat and interpass temperatures.

For steels in the higher-carbon ranges (over 0.30 percent) special electrodes are recommended. The lime-ferritic low-hyrogen electrodes (E7016 or E7018) can be used to good advantage in overcoming the cracking difficulties in high-carbon steels. A 308 stainless-steel electrode can also be used to give good physical properties to a weld in high-carbon steel.

Cast iron Cast iron is a complex alloy in which the most important element in welding is the very high carbon content. Quickly cooled cast iron is harder and more brittle than slowly cooled cast iron. The metal also naturally has a low ductility, which results in considerable strain on parts of a casting when one local area is heated. The brittleness and the uneven contraction and expansion of cast iron are the principal concerns in welding it.

Each job must be analyzed to predetermine the effect of welding heat, and procedures correspondingly adopted. Welds can be deposited in short lengths, allowing each to cool. Peening of the weld metal while red hot may be used to stretch the weld deposit.

Either steel or cast-iron electrodes may be used as well as carbon electrodes and nonferrous rods. All oil, dirt, and foreign matter must be removed from the joint before welding. With steel electrodes, intermittent welds no longer than 3 in. and light peening should be used. To reduce contraction, the work should never be allowed to get too hot in one spot. Preheating will help to soften the deposit to make it more machinable.

For welds of such machinability, a nonferrous-alloy rod should be used. A two-layer deposit will have a softer fusion zone than a single-layer deposit. When it is practical, heating of the entire casting to a dull-red heat is recommended, further to soften the fusion zone and burn out dirt and foreign matter. A lower heat can be used if necessary. When the weld to be made is in a deep groove, it is general practice to use a steel electrode for welding cast iron to fill the joint to within approximately $\frac{1}{8}$ in. of the surface and then finish the weld with the more machinable nonferrous deposit.

THE ALLOY STEELS

High-tensile low-alloy steels This group of steels is being used increasingly in metal fabricating because their high physical properties permit the use of thinner sections, thus saving metal and reducing weight. They are made with a number of different alloys and can be readily welded with the proper type of electrode designed especially for these metals. Excellent joints of the same high physical properties as the base metal are

4-46 Basic Maintenance Technology

obtained by the use of these electrodes. It is not necessary, as might be suspected, to have a core wire of the same composition as each of the alloys. In some cases, this may even be undesirable, since the electrode metal, in going through the arc, frequently has its analysis and characteristics changed.

Stainless steels Electrodes are made to match various types of stainless steels so that corrosion-resistance properties are not destroyed in welding. The most commonly used types of stainless steels for welded structures are the 304, 308, 309, and 310 groups. Group 304, with 0.08 percent carbon maximum, is a commonly specified type of stainless steel used for weldments.

The general mild-steel welding procedures are used, taking into account the fact of higher electrical resistance, lower thermal conductivity, and higher thermal expansion of the stainless steels. It is important to fit work carefully and clean all edges of foreign material. Light-gauge work must be clamped firmly to prevent distortion and buckling. Small-diameter and short electrodes should be used to prevent loss of chromium and

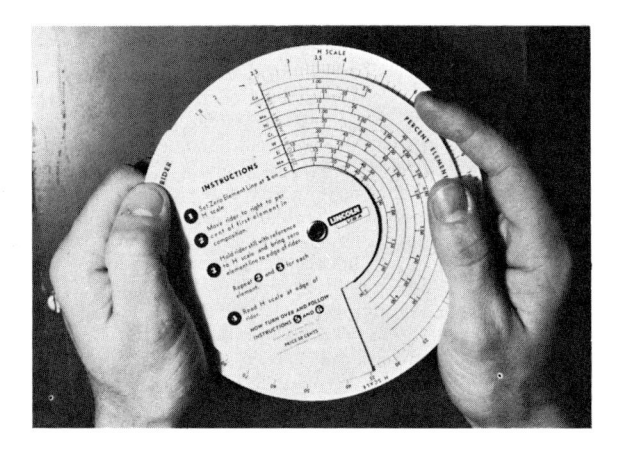

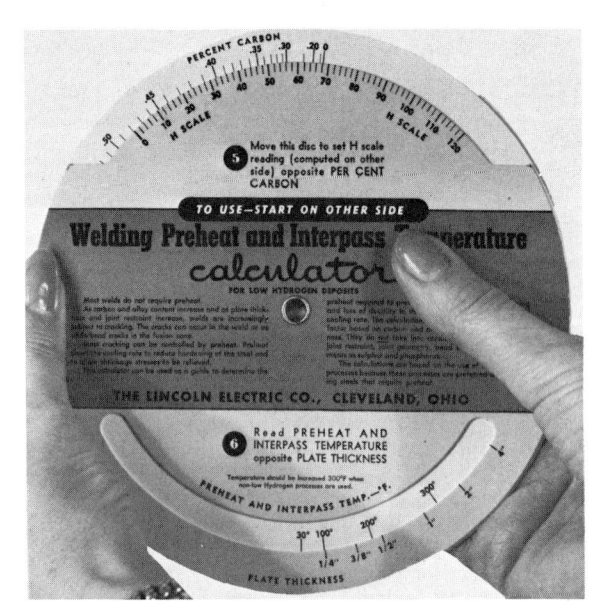

Fig. 5-26 Calculator for determining preheat and interpass temperatures.

undue overheating of the electrode. The weld deposit should be approximately the same analysis as the plate.

Stainless clad steel The significant precautions in welding this material are in joint design, including edge preparation, procedure, and choice of electrode. An electrode should be used of the correct analysis for the cladding being welded. The joint must be prepared and welded to prevent dilution of the clad surface by the steel backing material. The backing material is welded with a mild-steel electrode but in multiple passes to prevent penetration into the cladding. The clad side is also welded in small passes to prevent penetration into the backing material and resulting dilution of the stainless joint. Where in thin-gauge material it is necessary to make the weld in one pass, a 309 stainless electrode should be used for the steel side as well as the stainless side. The design and preparation of the joint can do much to prevent iron pickup as well as reduce labor costs in making the joint.

Chromium steels The intense air-hardening property of these steels, which is proportional to the carbon and chromium content, is the chief consideration in establishing welding procedures. Considerable care must be taken to keep work warm during welding and annealed afterward; otherwise the welds and area adjacent to the welds will be brittle. It is well to consult steel suppliers for specific heat treatment, temperatures, and treatment.

High-manganese steel The tough work-hardening characteristic of this material recommends it for surfaces which must resist abrasion or wear as well as shock. For building up parts of high-manganese steel, an electrode should be used of such type that the physical characteristics of the deposited metal will be approximately the same as those of the base metal.

THE NONFERROUS METALS

Aluminum Most fusion welding of aluminum alloys is done with the inert-gas metal-arc (MIG) process. Weld properties generally are at least equal to those of the base metal at zero temper. Welding speeds are higher than those obtainable with any other arc or gas process. Heat-affected zones are narrower than those with oxyacetylene or covered-electrode arc welding. A dc (reverse-polarity) electric arc, established in an envelope of inert gas between a consumable electrode and the workpiece, is used for welding aluminum by the MIG process.

MIG and another inert-gas shielded-arc process, gas tungsten-arc (TIG), are the principal methods for welding aluminum. The two processes are similar in that an inert gas is used to shield the arc and the weld pool, making flux unnecessary. The chief differences are in the electrodes and the characteristics of the power used.

In MIG welding, the electrode is aluminum filler fed continuously from a reel into the weld pool.

The DCRP action propels the filler metal across the arc to the workpiece in line with the axis of the electrode, regardless of the orientation of the electrode. Because of this and aluminum's density, surface tension, and cooling rate, horizontal, vertical, and overhead welds are made with relative ease. High deposition rates are practical, producing less distortion, greater weld strength, and lower welding costs for a given job than other fusion-welding processes.

TIG welding uses a nonconsumable tungsten electrode, with aluminum-alloy filler material added separately, either from a hand-held rod or from a reel.

Alternating current is preferred by many users of both manual and automatic TIG welding of aluminum. This is because ac TIG achieves an efficient balance between penetration and cleaning.

Copper and copper alloys Copper and its alloys can be welded with shielded-metal-arc, gas-shielded, or carbon-arc welding. Of the three, gas-shielded arc welding with an inert gas is preferred.

Decrease in tensile strength as temperature rises and high coefficient of contraction may make welding of copper complicated. Preheat usually is necessary on thicker sections because of the high heat conductivity of the metal. Keeping the work hot and pointing the electrode at an angle so the flame is directed back over the work will aid in permitting the gas to escape. It is also advisable to put as much metal down per bead as is practical.

CONTROL OF DISTORTION

Distortion in the metal being welded, caused by the heat of welding, may be a problem in welding sheet metal or unrestrained large sections. The following suggestions will help overcome problems of distortion, based on three simple rules applied singly or together:
1. Reduce the effective shrinkage force.
    a. Avoid overwelding. Use as little weld metal as possible by taking full advantage of penetrating effect of arc force.
    b. Use correct edge preparation and fit-up to obtain required fusion at root of weld.
    c. Use few passes.
    d. Place welds near neutral axis.
    e. Use intermittent welds.
    f. Use backstep welding method.
2. Make shrinkage forces work to minimize distortion.
    a. Locate parts out of position so that when weld shrinks they will be in correct position.
    b. Space parts to allow for shrinkage.
    c. Prebend parts so that contraction will pull parts into alignment.
3. Balance shrinkage forces with other forces (where natural rigidity of parts is insufficient to resist contraction).
    a. Balance one force with another by correct welding sequence so that contraction caused by weld counteracts forces of welds previously made.
    b. Peen beads to stretch weld metal. Care must be used not to damage weld metal.
    c. Use jigs and fixtures to hold work in a rigid position with sufficient strength to prevent parts from distorting. Fixtures actually cause weld metal to stretch, thus preventing distortion.

Shielded-metal-arc welding There are two aspects to the problem of selecting the correct electrode for making a good weld under given conditions. The selection must be made according to (1) electrode type as to coating and core-wire analysis and (2) electrode diameter size. In selecting the type of electrode it is necessary to know
 1. The position in which the work is to be welded
 2. The type and thickness of the metal being used
 3. The preparation of the work with regard to fit-up
 4. The type of available welding current
 5. The class of work (that is, whether deep penetration, surface quality, required physical properties, or code requirements) is the chief essential

The American Welding Society has established specifications for the manufacture of welding electrodes to fulfill the above job requirements. The following specifications have been issued, classifying electrodes as follows:
  Mild Steel Covered Arc Welding Electrodes, A5.1-69
  Low Alloy Steel Covered Arc Welding Electrodes, A5.5-69
  Corrosion-resisting Chromium and Chromium-Nickel Steel Covered Welding Electrodes, A5.4-69
  Copper and Copper-alloy Arc Welding Electrodes, A5.6-69
  Nickel and Nickel-alloy Covered Welding Electrodes, A5.11-69

In addition to these classifications, electrodes are also manufactured for hard surfacing, welding cast iron, and other miscellaneous applications.

The mid- and low-alloy-steel electrodes are classified with a numbering system for simple identification. E6010 is a typical four-digit classification number. The prefix E designates a metal-arc-welding electrode; the first two digits stand for the minimum allowable tensile strength of deposits in thousands of pounds per square inch. The third digit stands for the welding position or positions in which the electrodes will make a satisfactory deposit, and the last digit indicates various arc characteristics, among them polarity.

Since at least 90 percent of all arc welding is done in mild steel, the following brief descriptions of mild-steel electrode types are included. The significance of the various classification digits as explained for these electrodes is consistent throughout the E70, E80, E90, E100, and E110XX series of steel electrodes. Table 5-3 gives classification characteristics and uses for steel electrodes.

TABLE 5-3 Steel-Electrode Classification, Characteristics, and Uses

Class no.	Work position	Current supply	Basic application
EXX10	All	dc+	Designed to produce good mechanical properties consistent with good radiographic inspection quality. Application is usually structural where multipass welding is employed, such as shipbuilding, bridges, buildings, and piping and pressure vessels
EXX11	All	ac (dc+)	Designed to do the work of XX10, but to employ an ac source. Slightly higher tensile and yield strength
EXX12	All	dc − ac	Especially recommended for single-pass, high-speed, high-current horizontal fillet welds. It is characteristically easy to handle and useful in cases of poor fit-up, both groove and fillet, where a wide range of currents is used. Class 12 has reduced penetration but can meet radiographic standards with single-pass welds
EXX13	All	ac (dc−)	Designed for light-sheet-metal work, but now used widely as an electrode having light penetration. Frequently used in vertical down-welding, even though it produces a flat bead. Particularly well designed for use with low-voltage ac transformers
EXX14	All	ac (dc−)	An iron-powder electrode designed to do the work of 13 with increased deposit rate, although 14 has lower deposition rates than 24 and 27. In the flat position, 13 and 14 have similar welding speeds. Has improved weld appearance and ease of welding in drag technique
EXX15	All	dc+	Offers good physical properties and x-ray quality. A low-hydrogen electrode for difficult-to-weld material such as high-carbon or low-alloy steel. Also, free machining, high-sulfur-bearing steel. Frequently pre- and postheating may be eliminated or reduced by using low-hydrogen rod. Electrode covering cannot perform properly with included moisture. Electrode should be heated before use as recommended by the manufacturer or stored in a moisture-free area
EXX16	All	ac dc+	An electrode similiar to 15 designed to be used with ac and dc + supply
EXX18	All	ac dc−	A 30% iron-powder titania-type electrode. An electrode similar to 15 with a higher deposition rate and an improved weld appearance. Offers better slag removal and higher usable current than the E7016 type
EXX24	HF-F	ac (dc−)	An iron-powder-type electrode ideal for fillet welds. The iron powder in the electrode coating assists in increasing the deposit rate over the 12 class. Electrode can be used in drag technique with ease of handling and good weld appearance. Requires better fit-up than 12, but is of similar application, although limited as to position
EXX27	HF-F	dc − ac	When this high-iron-powder electrode is used in the drag technique, it is faster than the 18 electrode. It is primarily a downhand deep-groove rod, well suited for heavy sections. Second only to 24 in welding speed, but with properties superior to it. Both are equally easy to handle
EXX28	HF-F	dc − (ac)	A 50% iron-powder lime-type electrode. This one yields the highest deposition rates of the low-hydrogen group. The coating also produces an easy-to-maintain arc with a smooth, wide bead. Can be used only in the flat position

HF—horizontal fillet position; F—flat position.

Cellulose-coated electrodes EXX10 and EXX11 The relatively thin coverings of these electrodes contain a high percentage of cellulose. This type of covering produces a small volume of molten slag in the weld crater and light slag coverage of the solidified weld bead. The EXX10 and EXX11 electrodes can be used in all welding positions, as illustrated in Fig. 5-27.

Types E6010 and E6011 These types may be classified as general-purpose electrodes, since they are used for a wide variety of work and possess high average mechanical characteristics. E6010 is best suited for direct current, electrode positive. In sizes of $\frac{3}{16}$ in. and smaller, in any type of weld, it is suitable in all positions—flat, horizontal, vertical, and overhead (Fig. 5-28). It has deep-penetration qualities and is used very satisfactorily on square-groove butt joints where the electrodes actually scarf or melt the plates. It produces a rather flat bead shape.

Fig. 5-27 EXX10 and EXX11 electrodes are used for all-position welding.

The E6010 electrode has a high cellulose content in the covering. The arc is very penetrating, with a relatively quick solidifying slag and weld-crater action. Protection of the molten metal is obtained principally by gases since only a small amount of slag is produced. The weld metal has excellent physical qualities. Some of the applications are welding pipe, ships, machinery, structures (especially field or erection), and jigs and fixtures.

The E6011 electrode is similar to the E6010 but is designed for ac or either-polarity dc operation. The dc polarity (electrode negative) or reverse (electrode positive) depends upon the type of work being performed. The characterisitics of E6011 electrode design are also high-cellulose covering, penetrating arc, quickly solidifying slag action, similar to E6010, and protection of molten metal obtained principally by gases. As in the case of E6010, this electrode is well suited for making vertical and overhead fillet and butt welds. The applications are the same as for E6010.

These electrodes are generally recommended for use where the weld metal cannot be deposited in the flat, or downhand, position. The deposited metal has good strength and high elongation.

Of the same general characteristics are several electrodes for welding the low-alloy high-tensile steels (E7010, E9010, etc).

Titania-coated electrodes (EXX12, EXX13) The medium-thick coverings of these electrodes contain a relatively high titania content. This type of covering produces a medium volume of molten slag in the weld crater which simply covers the weld bead when it solidifies. These electrodes can be used in all positions, but are more difficult to control out of position than the cellulose types.

Types E6012 and E6013 The E6012 electrode has a medium-thick covering and is used with direct current with the electrode negative or may be used with alternating current. Sizes of 3/16 in. and smaller are suitable for all positions, and larger sizes for welding in flat positions. The electrode may be used for fillet welding, single- or multiple-pass, and can be used for butt welds of the V-groove or U-groove type. Because of its deposition characteristics and ability to build up, it is used to fill gaps in cases of poor fit-up. E6012 coverings are high in titania and low in cellulose. The arc is less penetrating

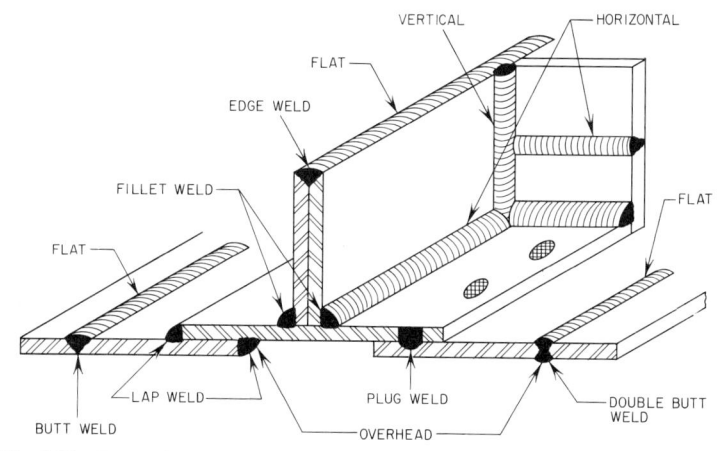

Fig. 5-28 Types of welds encountered in structural and general maintenance welding.

than that of E6010 and E6011, but adequate when correct welding procedures are used. The larger amount of slag gives a better coverage, producing a finer ripple with a more pleasing bead surface.

The E6012 electrode has higher melting rate with lower spatter than E6010 or E6011. It is ideally suited for horizontal and flat fillet welds, for applications where fit-up may be poor, and on steel having characteristics which give poor welding action with electrodes producing greater penetration. It can be used for butt welds of the V-groove or U-groove types. Because it does not penetrate deeply, it is used in cases where dilution of weld and base metal is not desirable. It produces a somewhat convex bead. The weld metal has higher tensile strength and slightly lower elongation than have E6010 and E6011. Some typical applications are welding sheet-metal ducts, tanks, machine guards, and structural work.

E6013 has better ac operation than E6012 and develops a smoother bead appearance. Penetration is similar to that of E6012, so that it works well for poor fit-up. E6013 is more suitable for light-gauge metals than E6012. The bead has a tendency to be convex in making horizontal fillets. The applications are similar to those for E6012.

Iron-powder electrodes The iron-powder-covered electrodes have an exceptionally heavy covering containing a large quantity of iron powder (Fig. 5-29). This type of covering makes welding with these electrodes faster and easier. Welding

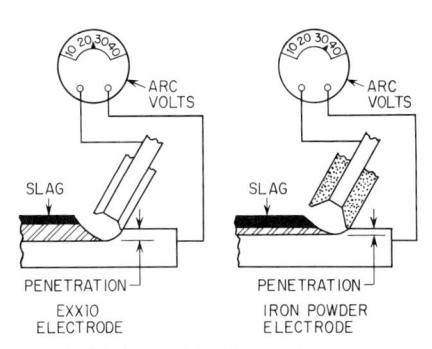

Fig. 5-29 Essential difference between iron-powder and EXX10 electrodes.

speeds are increased as much as 50 percent. Weld appearance is smoother; slag is practically self-cleaning; spatter is eliminated almost completely.

All these advantages result from the nature of the iron-powder coating. The covering more efficiently utilizes the heat of the arc in melting. Welding currents can be increased for a given-diameter electrode, providing greater deposition rates without the difficulties of excessive penetration, gouging, undercutting, and spatter normally encountered when welding with higher currents.

The electrodes operate on either alternating or direct current, but alternating current is preferred. Slightly higher currents than those used with conventional electrodes are required. Also, an electrode one size smaller in diameter is generally used.

They are ideally suited for contact or drag welding techniques, although an arc may be held if desired.

Type E7024 This type of electrode has been designed especially for welding flat and horizontal fillets with either alternating or direct current. It is widely used for production welding in making machinery and structures.

Type E6027 This type of iron-powder electrode has been designed especially for welding flat, deep-groove butt welds with either alternating or direct current. It is also used for flat and horizontal fillets. It is used for code work. The bead has excellent wash-in properties and makes a smooth cover pass. The slag is extremely friable and therefore easily removable under all conditions.

Type E7014 Iron powder has been added to the covering to produce this modified version of an E6013 type of electrode. The result is an excellent electrode having iron-powder characteristics plus the feature of being suited to out-of-position use up to 45° downhill. Although classified as all-position, it is rarely used for vertical and overhead welding.

Buildup and manganese-steel electrodes Several manufacturers are making iron-powder maintenance electrodes for buildup work on worn machinery parts and welding manganese steel. These electrodes carry an official AWS classification, but are usually descriptively named. The high deposition rate of the electrodes results in depositing 35 to 45 percent more metal per minute than is possible with conventional electrodes. This means considerable saving in time when areas being restored to size require the deposition of a large quantity of metal. Properties and characteristics of the electrodes are varied to meet particular service requirements.

Lime-covered low-hydrogen electrode types The low-hydrogen electrode consists essentially of a rimmed-steel core wire upon which a covering of the carbonate of soda and lime type is applied, using other compounds low in hydrogen. This covering is slightly thicker than normal for each diameter, and the electrode is slightly more difficult to use because of the shortness of the arc which must be maintained. A typical analysis of the deposit from this electrode is 0.08 percent carbon, 0.56 percent manganese, and 0.25 percent silicon. The arc is moderately penetrating; the slag heavy, friable, and easily removed; and the deposited metal lies in a flat bead or may be even slightly convex.

The as-welded mechanical and impact properties of deposits made using the low-hydrogen-type electrodes have been found to be superior to those of E6010 and E6011 electrodes depositing weld metal of the same composition. Numerous tests have indicated that the as-welded mechanical and impact properties of deposits from these electrodes approach the properties of stress-relieved deposits of conventional electrodes. Whereas the properties of deposits of conventional electrodes are materially improved when they are stress-relieved, the deposits of low-hydrogen electrodes are changed only slightly. The reduced tendency for underbead cracking and the high quality of as-welded deposits of these electrodes materially reduce the preheat and postheat of weldments, resulting in better welding conditions and reductions in the cost of thermal treatments.

Low-hydrogen electrodes operate best on dc reverse polarity, but most types can also be used with alternating current. They were developed for welding higher-strength high-carbon alloy steels in which the ordinary electrodes are subject to developing *underbead cracking*. These underbead cracks occur along the line of fusion between the parent metal and the weld metal and are caused by the hydrogen present in the conventional electrode covering. Naturally, eliminating the hydrogen tends to help control underbead cracking and permits the welding of the weld steels with little or no preheat, thus making for better welding conditions. Although these cracks do not occur in ordinary steels, they may occur whenever an ordinary electrode is used on high-tensile steels.

Another use for the low-hydrogen-type electrode is the welding of high-sulfur steels. The ordinary electrode deposit on these steels (which contain 0.10 to 0.25 percent sulfur) is badly honeycombed. Low-hydrogen-type electrodes can be used to weld these steels.

Many of the newer high-tensile steels being produced today call for low-hydrogen electrodes ranging up to 110,000 psi tensile strength (E110XX).

The low-hydrogen electrode was developed during World War II for the welding of armor plate, and in addition to its use on alloy steels, high-carbon steels, and high-sulfur steels, it has been found useful on malleable iron, on spring steels, and for welding the mild-steel side of clad plates. Another extensive use has been in the welding of steels which will subsequently be enameled and in all those steels which contain selenium. It is an excellent maintenance electrode, since it can be used with assurance of good welding on steels whose analysis is unknown or may be questioned.

Type E7015 The E7015 electrode was the first of the low-hydrogen types for welding the carbon steels. It was designed exclusively for dc electrode-positive operation. The E7015 electrode can be used in all positions up to and including 5/32-in. diameter. The larger diameters are useful for fillet and butt welds in the horizontal and flat positions.

Type E7016 The E7016 classification of electrode has all the characteristics of the E7015 classification. The core wire and coating are similar, except for the use of a certain amount of potassium silicate or other potassium salts on the E7016 classification to facilitate its use on alternating current. All that has been said of the E7015 electrode applies equally well to the E7016.

Type E7018 Iron powder has been added in the E7018 type of electrode, thus producing the iron-powdered, low-hydrogen electrode manufactured under this classification. The electrodes have the advantage of low-hydrogen properties plus the excellent operating characteristics associated with iron powder. All that has been said of the E7015 and E7016 applies equally to the E7018. This is an excellent maintenance electrode.

Type E7028 The E7028 classification of electrode combines the advantages of the low-hydrogen types and the heavy-covered powdered-iron types. The electrode manufactured under this classification has a high deposition rate, and is limited in application to horizontal- and flat-position welding.

SUBMERGED-ARC WELDING—EQUIPMENT, ELECTRODES, AND FLUX

Welding equipment The welding heads normally used for fully automatic submerged-arc welding perform the triple function of progressively depositing flux along the joint, feeding the electrode, and transmitting welding current to the electrode. The flux is usually supplied from a hopper either mounted directly on the head or connected to the head by tubing. The bare electrode or wire is fed into the welding head from a coil mounted on a reel. The distance between the end of the electrode and the base metal is maintained constant by special controls which automatically regulate the electrode-feed motor speed or welding current.

Equipment manufactured for semiautomatic submerged-arc welding performs the same functions as that for fully automatic welding. The welding head, however, now consists of a welding gun and wire feeder unit. The flux for semiautomatic welding is supplied by a canister mounted on the welding gun or a continuous-flow flux feed from a pressurized flux tank. With semiautomatic welding equipment, the electrode wire is mechanically fed to the work but the welding gun is manually moved along the joint being welded. This procedure gives added flexibility to this method of welding by permitting its use on irregular shapes and contours, thereby promoting expanded use.

Direct current is used with both semiautomatic and fully automatic submerged-arc welding, whereas alternating current is usually limited to fully automatic submerged-arc welding. The welding voltage for submerged-arc welding will range from 28 to as high as 55 volts. Currents generally used for submerged-arc welding are higher than those used for the other arc-welding processes. They range from a low of 200 amp up to as high as 4000 amp.

Alternating current may be supplied from one or more heavy-duty welding transformers. Direct current may be supplied by one or more motor-generator or rectifier welding machines having capacity suitable for the application. The dc power supplies can be constant-potential or variable-voltage types, depending upon the application and manufacturer's recommendations. Installations of semiautomatic and fully automatic welding equipment are illustrated by Figs. 5-30 and 5-31.

Electrodes and fluxes The ferrous and nonferrous electrodes commonly used for submerged-arc welding are bare rods or wires with clean, bright surfaces to facilitate the introduction of relatively high currents. Electrodes are normally used in the form of coils ranging in weight from a minimum of 25 to 200 lb. On very high-production welding

Fig. 5-30 Semiautomatic submerged-arc welding.

installations, the electrode is frequently fed from a coil in a drum. These drums range up to as high as 1000 lb in weight. Ferrous wire of composition that might readily rust is coppercoated to retard rusting and improve the contact surfaces.

The fluxes used with submerged-arc welding are granulated fusible mineral materials which are essentially free from substances that would create large amounts of gases during welding. These fluxes are made to a variety of chemical specifications which develop particular performance characteristics. The flux has a number of functions to perform, including prevention of atmospheric contamination and performing a scavenging-deoxidizing action on the molten metal in the weld crater. Some special fluxes perform the additional function of contributing alloying elements to the weld deposit, thereby developing specific weld-metal characteristics of higher strength or even abrasion resistance. The choice of flux depends on the welding procedure to be employed, the type of joint, and the composition of the material to be welded.

SPECIAL APPLICATIONS

Sheet-metal welding The welding of sheet metal, as illustrated by Fig. 5-32, has frequent application in maintenance. The principles of good welding practice apply in welding sheet metal as elsewhere, but the nature of the work places special emphasis on several aspects. The problem of distortion requires special consideration in welding thin-gauge metals as well as the problems of burning through the metal. Special attention should therefore be given to all the factors involved in controlling distortion: the speed of welding, the choice of proper joints, good fit-up, position, selection of proper current, use of clamping devices and fixtures, number of phases, and sequence of beads.

Within the limits of good welding appearance, the highest arc speeds and the highest currents should be used. In sheet-metal work, however, there is always the limitation imposed by the threat of burn-through. As the gap in the work increases in size, the current must be decreased to prevent burn-through, which, of course, will reduce welding speeds. A clamping fixture will improve the fit-up of joints and thus make possible the

Fig. 5-31 Fully automatic submerged-arc welding.

Fig. 5-32 Typical sheet-metal welding using the shielded-metal-arc-welding process.

higher speeds. If equipped with a copper backing strip, the clamping fixture will make for easier welding by decreasing the tendency to burn through and will also remove some of the heat which causes warpage. Where possible, sheet-metal joints should be welded downhill at about a 45° angle with the same currents as are used in the flat position, or slightly higher. Tables 5-4 and 5-5 offer a guide to the selection of the proper current, voltage, and electrodes for the various types of joints used with sheet metal ranging from 20 to 8 gauge.

Hard surfacing The building up of a layer of metal or a metal surface by electric welding, commonly known as arc-weld surfacing, has an important and useful application in equipment maintenance. Applications of the process are varied and many, such as restoring worn cutting edges and teeth on excavators, building up worn shafts with low- or medium-carbon deposit, lining a carbon-steel bin or chute with stainless-steel corrosion-resistant alloy deposit, putting a tool-steel cutting edge on a medium-carbon-steel base, and applying wear-resistant surfaces to metal machine parts of all kinds. The dragline of Fig. 5-33 is being returned to new condition by rebuilding and hard surfacing.

Arc-weld surfacing includes, but is not limited to, hard surfacing. There are many building-up applications where hard surfacing is not required.

Wear is the gradual impairment of machinery parts through use. Excluding corrosion, wear results from various combinations of abrasion and impact. Abrasive wear results from one material scratching another and impact wear from one material hitting another.

How to resist abrasive wear Abrasive wear is resisted by materials with a high scratch hardness. Sand wears metals with a low scratch hardness at a high rate, but under the same conditions it will wear a metal of high scratch hardness very slowly. Scratch hardness, however, is not necessarily measured by standard hardness tests. Brinell and Rockwell hardness are not reliable measures for determining the abrasive-wear resistance of a metal. A hard-surfacing material of the chromium-carbide type may have a hardness of 50 Rockwell C. Sand will wear this material at a slower rate than it will a steel hardened to 60 Rockwell C. The sand will scratch all the way across the surface of the steel. On the surfacing alloy the scratch will progress through the matrix material and then stop when the sand grain comes up against one of the microscopic crystals of chromium carbide,

TABLE 5-4 Welding Currents for Sheet Metal

Type of welded joint	20 ga			18 ga			16 ga		
	F*	V*	O*	F	V	O	F	V	O
Plain butt	30†	30†	30†	40†	40†	40†	70†	70†	70†
Lap	40†	40†	40†	60†	60†	60†	100	100	100
Fillet				40†	40†	40†	70†	70†	70†
Corner	40†	40†	40†	60†	60†	60†	90†	90†	90†
Edge	40†	40†	40†	60†	60†	60†	80†	80†	80†

*F—flat position; V—vertical; O—overhead.
†Electrode negative, work positive.

TABLE 5-5 Sizes of Electrodes for Sheet Metal

Type of welded joint	20 ga			18 ga			16 ga		
	F*	V*	O*	F	V	O	F	V	O
Plain butt	$3/32$	$3/32$	$3/32$	$3/32$	$3/32$	$3/32$	$1/8$	$1/8$	$1/8$
Lap	$3/32$	$3/32$	$3/32$	$3/32$	$3/32$	$3/32$	$1/8$	$1/8$	$1/8$
Fillet				$3/32$	$3/32$	$3/32$	$1/8$	$1/8$	$1/8$
Corner	$3/32$	$3/32$	$3/32$	$3/32$	$3/32$	$3/32$	$1/8$	$1/8$	$1/8$
Edge	$3/32$	$3/32$	$3/32$	$3/32$	$3/32$	$3/32$	$1/8$	$1/8$	$1/8$

*F—flat position; V—vertical; O—overhead.

which has a higher scratch hardness than sand. If two metals of the same type have the same kind of microscopic constituents, however, the metal having the high Rockwell hardness will be more resistant to abrasive wear.

How to resist impact wear Whereas abrasive wear is resisted by the surface properties of a metal, impact wear is resisted by the properties of the metal beneath the surface. To resist impact, a tough material is used, one which does not readily bend, break, chip, or crack. It yields so as to distribute or absorb the load created by impact, and the ultimate strength of the metal is not exceeded. Included in impact wear is that caused by bending or compression at low velocity without impact, resulting in loss of metal by cracking, chipping, upsetting, flowing, or crushing.

Types of surfacing electrodes Many different kinds of surfacing electrodes are available. The problem is to find the best one to do a given job. Yet because service conditions vary so widely, no universal standard can be established for determining the ability of surfacing to resist impact or to resist abrasion. Furthermore, there is no ideal surfacing material that resists both impact and abrasion equally well. In manufacturing surfacing electrodes, it is necessary to sacrifice somewhat one quality to gain the other. A material that has a high resistance to abrasion will have a low resistance to impact. High impact resistance is gained by sacrificing abrasion resistance.

Price is no index to quality of electrodes. Simply because an electrode contains an expensive ingredient does not necessarily make it superior for wear resistance. Thus the user of surfacing materials must rely upon the manufacturer's recommendations and his own tests to determine the best surfacing material for his purpose.

How to choose hard-facing material The chart of Fig. 5-34 lists the relative characteristics of manual hard-surfacing materials. It shows in the various columns the ability of each of the materials to resist abrasion, metallic friction, impact, and corrosion. It also gives the relative hardness, ductility, and cost of depositing the material, as well as the physical limitations of weld size in applying each one. This chart is a guide to selecting

    1. The hard-surfacing electrode best suited for a job not hard-surfaced before
    2. A more suitable hard-surfacing electrode for a job where present material has not produced desired results

TABLE 5-4 Welding Currents for Sheet Metal *(Continued)*

Type of welded joint	14 ga			12 ga			10 ga			8 ga		
	F	V	O	F	V	O	F	V	O	F	V	O
Plain butt	85†	80	85†	115	110	110	135	120	115	190	130	120
Lap	130	130	130	135	120	120	155	130	120	165	140	120
Fillet	100	90	85	150	140	120	160	150	130	160	160	130
Corner	90	80	75	125	110	110	140	130	125	175	130	125
Edge	110	80	80	145	110	110	150	120	120	160	120	120

TABLE 5-5 Sizes of Electrodes for Sheet Metal *(Continued)*

Type of welded joint	14 ga			12 ga			10 ga			8 ga		
	F	V	O	F	V	0	F	V	O	F	V	O
Plain butt	$1/8$	$1/8$	$1/8$	$5/32$	$5/32$	$5/32$	$5/32$	$5/32$	$5/32$	$3/16$	$5/32$	$5/32$
Lap	$5/32$	$5/32$	$5/32$	$5/32$	$5/32$	$5/32$	$3/16$	$3/16$	$5/32$	$3/16$	$3/16$	$5/32$
Fillet	$1/8$	$1/8$	$1/8$	$5/32$	$5/32$	$5/32$	$3/16$	$5/32$	$5/32$	$3/16$	$5/32$	$5/32$
Corner	$1/8$	$1/8$	$1/8$	$3/16$	$5/32$	$5/32$	$3/16$	$5/32$	$5/32$	$3/16$	$5/32$	$5/32$
Edge	$1/8$	$1/8$	$1/8$	$3/16$	$5/32$	$5/32$	$3/16$	$5/32$	$5/32$	$3/16$	$5/32$	$5/32$

4-58 Basic Maintenance Technology

Example 1. APPLICATION: Dragline bucket tooth, as illustrated by Fig. 5-35. SERVICE: Sandy gravel with some good-size rocks.

Maximum wear that can be economically obtained is the goal of most hard-surfacing applications. Try to use a material that rates as high as possible in the resistance-to-abrasion column unless some other characteristics shown in the other columns make it unsuited for this particular application.

Fig. 5-33 Shielded-metal-arc welding is used to rebuild and to hard-surface worn areas of a dragline bucket.

First, consider the tungsten-carbide types. Notice that they are composed of very hard particles in a softer and less abrasion-resistant matrix. Although such material is the best for resisting sliding abrasion on hard material, in sand the matrix is apt to scour out slightly, and then the brittle particles are exposed. These particles are rated poor in impact resistance, and they may break and spall off when they encounter the rocks.

Next best in abrasion, as listed in the chart, is the high-chromium carbide type shown in the electrode-size column to be a powder. It can be applied only in a thin layer, and also is not rated high in impact resistance. This makes it doubtful for use in this rocky soil.

The rod-type high-chromium carbides also rate very high in abrasion resistance, but do not rate high in impact resistance. However, the second does show sufficient impact rating to be considered if two or three different materials are to be tested in a field test. Since there is a chance that it has enough impact resistance to do this job, we should not like to pass up its very good wearing properties.

Nevertheless, the semiaustenitic type is balanced in both abrasion and impact resistance. It is much better in resistance to impact than the materials that rate higher in abrasion. Thus semiaustenitic is the first choice on this job, considering that the added impact resistance of the austenitic type is not necessary, since the impact in this application is not extreme.

Example 2. APPLICATION: Same dragline tooth used in Example 1. SERVICE: Soil changed to clay and shale.

The semiaustenitic type selected in the first example stands up well, but the teeth wear only half as long as the bucket lip. With double the wear on the teeth, only half the downtime periods for resurfacing would be needed, and both teeth and bucket could be done together.

Since the impact is now negligible with the new soil conditions, go to a material higher in the abrasion column. Choose a material such as the first high-chromium carbide rod, which could give twice the wear by controlling the size bead applied and still be within reasonable cost.

Example 3. APPLICATION: Same dragline tooth as in Examples 1 and 2. SERVICE: Soil changed to contain large rocks.

If the earth has been changed so that it contains many hard and large rocks and the teeth are failing because of spalling under impact, move down in the abrasion-resisting column to a better impact-resistant material, such as the semiaustenitic type.

From the above, it can be seen that where a dragline operates in all kinds of soils, a material that is good in both abrasion and impact, such as a semiaustenitic type, is the best choice when in doubt as to the conditions that will be met.

When this same type of reasoning is followed in checking the important characteristics, a material can be chosen for any application. And if, for any reason, the first choice does not prove satisfactory, it is usually a simple matter to improve the next application by choosing a material that is rated higher in the characteristic that has caused difficulty.

Where failures occur because of cracking or spalling, it usually indicates that a material higher in impact or ductility rating should be used. Where normal wear alone seems too rapid, a material higher in abrasion rating is indicated.

Check welding procedure Often hard-surfacing failures due to cracking or spalling may be caused by improper welding procedures rather than by improper choice of hard-surfacing material. Before changing to a different hard-surfacing material, serious consideration should be given to the question of whether or not the material has been properly applied.

For almost any hard-surfacing application, very good results can be obtained if the following precautions are observed:

1. Do not apply hard-surfacing material over cracked or porous areas. Remove any defective areas down to sound base material.

2. Preheat. Preheating to 400 to 500°F improves the resistance to cracking and spalling. This minimum temperature should be maintained until welding is completed. The exception to this rule is 11 to 14 percent manganese steel, which should be kept cool.

3. Cool slowly. If possible, allow the finished weldment to cool under an insulating material such as lime, asbestos, or sand.

4. Do not apply more than the recommended number of layers.

When more than normal buildup is required, apply intermediate layers of either medium carbon or stainless steel. This will provide a good bond to the base metal and will eliminate excessively thick layers of hard-surfacing material which might otherwise spall off.

Stainless steel is also an excellent choice for intermediate layers on manganese steels or for hard-to-weld steels where preheating is not practical.

Check before total wear Whenever possible, examine a surfaced part when it is only partly worn. Examination of a part after it is completely worn is unsatisfactory. Did the surface crumble off, or was it scratched off? Is a tougher surface needed, or is additional abrasion resistance required? Should a heavier layer of surfacing be used? Should the surfacing be reduced? All these questions can be answered by examination of a partly worn part and with a knowledge of the surfacing costs and the service requirements.

In case it is impossible to analyze carefully the service conditions, it is always on the safe side to choose a material tougher than is thought to be required. A tough material will not knock or chip off and will offer some resistance to abrasion. A hard abrasion-resistant material is more susceptible to chipping, and surfacing material does not do any good when it is knocked off in large pieces.

After some experience is gained in the use of surfacing materials, various combinations of materials can be tried out to improve product performance. For example, on a part which is normally surfaced with a tough, semiaustenitic electrode, it may be possible to get additional abrasion resistance without sacrificing resistance to cracking. Fuse a little of the powdered chromium-carbide material on critical areas where additional protection is needed.

4-60 Basic Maintenance Technology

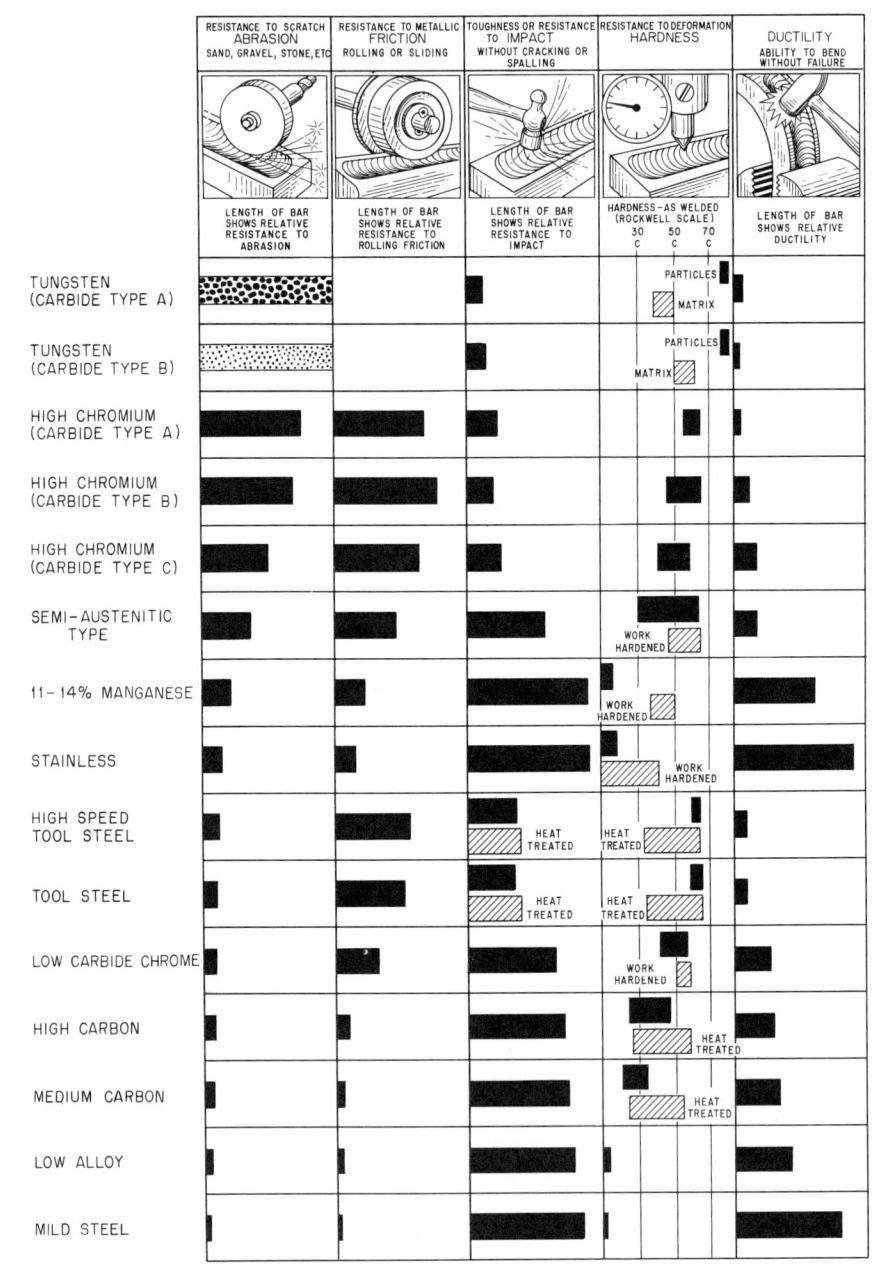

Fig. 5-34 Hard-surfacing guide.

Arc Welding in Maintenance 4-61

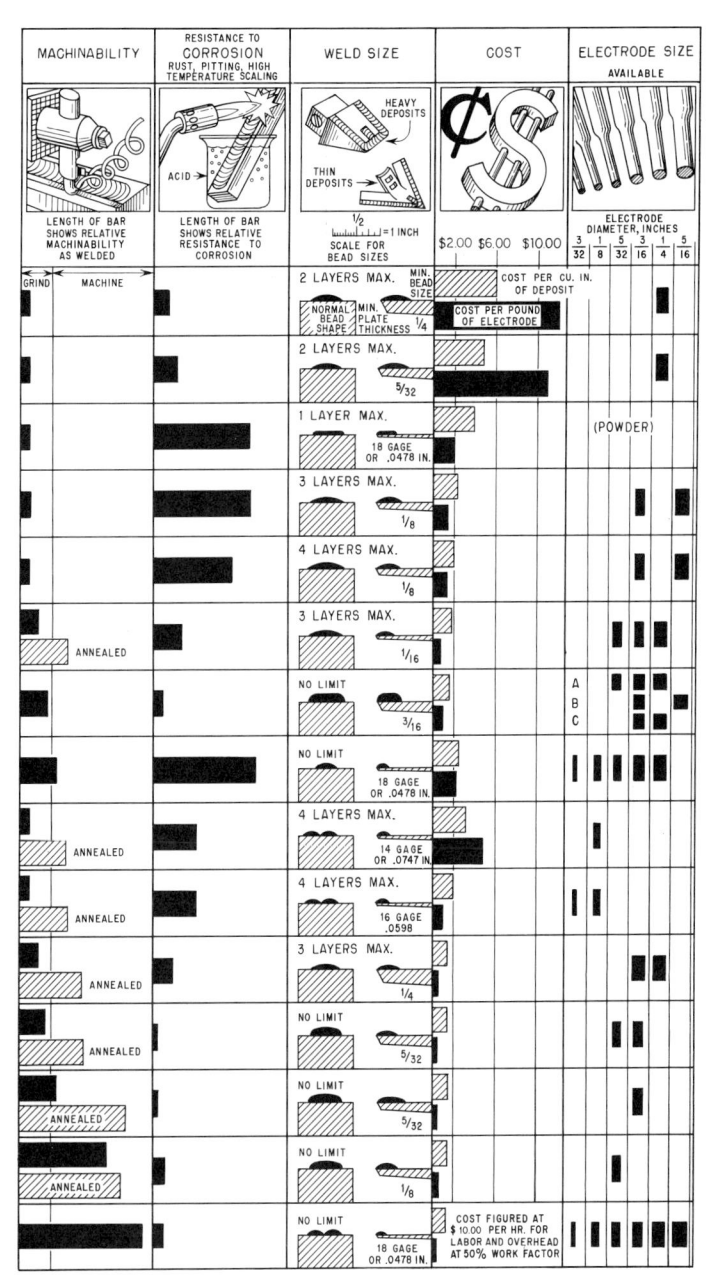

Fig. 5-34 (Continued).

4-62   Basic Maintenance Technology

Many jobs that are badly worn are first built up to almost finished size with a high-carbon electrode. They are then surfaced with an austenitic rod, and finally a few beads of chromium-carbide deposit are placed in spots requiring maximum protection against abrasion.

Regardless of the circumstances, careful analysis of the surfacing problem is well worthwhile. For examples of jobs see Figs. 5-36 to 5-38.

Fig. 5-35 Bucket teeth have been rebuilt and hard-surfaced.

Hard surfacing with submerged-arc process   The submerged-arc process offers several advantages for hard surfacing. The greater uniformity of the surface makes for better wearing qualities. The speed of submerged-arc welding creates major economies in hard-surfacing areas which require the deposition of large amounts of metal. These areas may be either flat or curved surfaces. Mixer bottom plates, scraper blades, fan blades, chutes, and refinery vessels are examples of flat plate to be surfaced. Shafts, blooming-mill spindles, skelp rolls, crane wheels, tractor idlers and rollers, and rams are examples of cylindrical surfaces (Figs. 5-39 to 5-42).

Fig. 5-36 Mild-steel die, on the edge of which tool steel has been deposited by means of tool-steel electrode.

Fig. 5-37 Cone used for uncoiling steel. Hard-facing material has been deposited on mild-steel base. Surface is ready for grinding.

The process can be used with either fully automatic equipment or with semiautomatic equipment, the choice depending upon the economics of the application. It is a relatively simple calculation to determine the savings that will result from using the submerged-arc process and thus arrive at a decision as to which type of equipment is warranted. Fully automatic equipment can be quickly fitted with auxiliary accessories which result in more economical metal deposition. An oscillating device can be added to an automatic head to create a bead up to 3 in. wide in a single pass. Another attachment permits the feeding of two electrode wires through a single head and single contact jaw. Both these attachments are useful in hard surfacing.

Hard surfacing with the submerged arc can be done with several different types of materials. The hard-surfacing deposit can be created by using solid alloy wires and a neutral granular flux. It can also be created by using a solid mild-steel wire and an agglomerated alloy flux, the alloys being added to the deposit through the flux rather than through the wire. Also available are tubular wires which contain alloying material in the hollow portion of the mild-steel tube. All the methods have particular advantages. In considering the submerged-arc process, it is well to consult a qualified field engineer who can recommend methods and procedures. With submerged-arc welding, considerable variation in the hard-surfacing deposit can be made by changing the welding procedure to control admixture and the heat-treatment effect of the welding cycle. Procedures should be established with the help of qualified engineers.

Fig. 5-38 Using mild-steel electrode to build up inside diameter and all teeth of 25-year-old cast-steel gear that could not be replaced.

Carbon arc Manual carbon-arc welding can be used to good advantage for welding of copper and its alloys, cast iron, and galvanized sheets and hard surfacing with alloy powder. The hard carbon can also be used for the cutting of steel, cast iron, and the stainless steels, the last two of which cannot be readily cut with the oxyacetylene torch.

The procedures to be used in welding any particular material will vary with the application. In making an edge weld where no filler metal is to be added, the average speeds given in Table 5-6 should be obtained.

Pointing of the Carbons. The diameter of the point should be approximately half the

Fig. 5-39 Steel-mill coke pushers being hard-faced by submerged-arc process using mild-steel wire and alloy flux. Fully automatic equipment in foreground, semiautomatic in background.

diameter of the carbon used. The taper should be gradual back to the point where it is gripped in the holder.

Position of the Carbon in the Holder. The carbon should be gripped as close to the arc as practical because, if a long length of carbon is exposed, the heating causes the carbon to vaporize and burn very rapidly, resulting in excessive wastage.

TABLE 5-6 Average Conditions for Welding

Metal thickness	Arc volts	Arc amp	Carbon size, in.	Welding speed, fph
16 ga. (0.0598 in.)	25	90–100	3/16	135
14 ga. (0.0747 in.)	25	125–135	1/4	125
12 ga. (0.1046 in.)	25	200–250	1/4–5/16	110
10 ga. (0.1345 in.)	25	250–275	1/4–5/16	100

Polarity. Carbon negative should be used in all cases.

Currents. The proper current to be used depends upon the work to be done. Table 5-7 will serve as a guide. The currents given are about the maximum which should ever be used. Smaller currents may be used, depending upon the weight or thickness of the base metal.

TABLE 5-7 Maximum Currents for Hand Carbon Arc

Size of carbon electrode, in.	Maximum current
5/32	50
3/16	100
1/4	200
5/16	350
3/8	450
1/2	700

Arc torch The development of the carbon-arc torch (Fig. 5-43) has further extended the use of the carbon-arc-welding technique to jobs where the application of heat is desired without melting the base metal. With the arc torch, a high-temperature flame is held between the two carbon electrodes clamped in adjustable jaws. The flame is played over the surface of the work, similiar to a gas flame, and as the carbons are consumed, they can be adjusted to maintain a constant distance between them.

The torch is useful for all brazing, soldering of light or heavy copper and galvanized or tinned parts, preheating localized areas prior to welding, and general heating for bending or straightening.

The carbon-arc torch operates at 35 to 40 volts, and since most welder controls are calibrated in amperes at the average metallic arc of 25 to 30 volts, the machine controls should be set 20 percent above recommended current settings. Copper-coated and copper-cored carbons are generally used in 1/4 to 3/8-in. diameter. The current should never be set so high that the copper coating is burned away more than 1/2 in. ahead of the arc. Only enough current should be used to cause the filler material to flow freely on the work. This will avoid consuming carbons too rapidly. The recommended current is between 50 and 75 amps on 5/16-in. carbons. Best results in brazing are obtained when carbons are 1/4 to 3/8 in. away from the work. When possible the joint should be lying horizontally to secure the best flow of molten filler rod.

Cutting Steel can be readily cut with great accuracy by means of the oxyacetylene torch. Not all metals cut as easily as steel. Cast iron, stainless steel, manganese steels, and nonferrous materials cannot be satisfactorily cut and shaped with the oxyacetylene cutting process because of their reluctance to oxidize. In these cases, arc cutting is often used to good advantage.

The cutting of steel is a chemical action. The oxygen combines readily with the iron to form iron oxide. In cast iron, this action is hindered by the presence of carbon in graphite form. Thus cast iron cannot be cut as readily as steel. Higher temperatures are necessary, and cutting is slower. In steel, the action starts at bright-red heat, whereas in cast iron the temperature must be nearer the melting point in order to obtain a sufficient reaction.

Fig. 5-40 Hard-surfacing wire-mill roll by submerged-arc process. Mild-steel wire and alloy flux. Gas torch keeps roll up to temperature.

Because of the very high temperature, the rate of cutting is usually fairly high. However, as the process is essentially one of melting without any great action, tending to force the molten metal out of cut, some provision must be made for permitting the metal to flow readily away from the cut. This is usually done by starting at a point from which the molten metal can flow readily. This method is followed until the desired amount of metal has been melted away.

Fig. 5-41 Automatic head adapted for oscillating and for two electrodes being used to deposit 3-in. beads on a flat mixer bottom plate.

As an example, the general method is to apply the electric arc on the underside of the work, starting at a lower corner, working toward the center on the lower surface and then up the side, and repeating this action as many times as necessary. This will allow the molten metal to flow out of the cut.

A carbon electrode is generally used. Graphite electrodes are used to some extent

Fig. 5-42 Submerged-arc welding being used to hard-surface a cylindrical surface.

because they permit use of higher currents. Shielded-metal-arc-type electrodes are also effective. In starting a cut, the arc is held at the point selected for the initial cut, as, for example, a lower corner. When the metal begins to flow and run off, the arc is moved along at a rate to permit metal to flow continuously out of the cut.

The width of the cut is dependent upon the ability of the operator to follow a straight line, the size of electrode used, and the thickness of material. The width of the cut is greater on thick sections than on thin.

A process which has come into use for the cutting of materials not readily cut with the oxyacetylene flame is the "oxyarc" cutting process. It cuts by directing a stream of oxygen into a pool of molten metal. The pool is made and kept molten by an arc established between the base metal and the covered tubular cutting electrode, which is consumed during the cutting operation. In addition to providing the arc, the electrode also provides an oxidizing flux and a means of conveying oxygen to the surface being cut.

The tubular cutting electrode is made of mild steel. This is not a detriment when the electrode is used for cutting materials other than mild steel because no contamination of the base metal adjacent to the cut occurs. The possibility of contamination is eliminated by the combination of extremely high heat and oxygen under

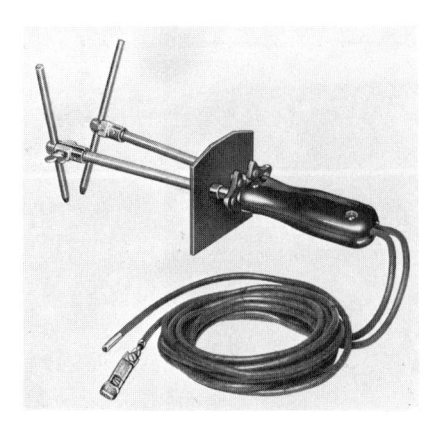

Fig. 5-43 Typical carbon-arc torch.

pressure, which act together to oxidize the electrode and coating at the point of the arc before the electrode metal can fuse with the base metal. The electrode covering helps to maintain arc stability by confining and directing the arc. When in use, it acts as an insulator to prevent the arcing of the rod at undesirable areas.

Quite simple equipment is required for metal cutting by this arc-oxygen process. There is, first of all, a special electrode which is not unlike an arc-welding electrode holder in appearance. This holder serves the double function of conducting current and feeding oxygen for the cutting operation. The tubular cutting electrode can be inserted or removed from the holder with ease. The only other equipment required is an ac or dc arc welder and an oxygen source with usual regulators.

The arc-oxygen process has been used successfully to cut high-chrome and chrome-nickel stainless steels, nickel, cupronickel, Monel, nickel-clad or stainless-clad steel, bronze, copper, brass, aluminum, cast iron, and mild and low-alloy steels. It has not supplanted, however, the oxyacetylene flame for cutting mild steel and other readily oxidized materials, because it is somewhat more expensive.

SELECTION AND MAINTENANCE OF EQUIPMENT

Machines Satisfactory welding can be accomplished with either alternating or direct welding current. Each type of current, however, has a particular advantage which makes it best suited for certain types of welding and welding conditions. The chief advantage of alternating current is its elimination of arc blow, which may be encountered when welding in heavy plate or into a corner. The magnetic fields set up in the plate deflect the path of the arc. Alternating current tends to minimize this deflection and will also sometimes increase the speed of welding with larger electrodes, over $3/16$ in., and with the iron-powder-type electrodes.

The chief advantages of direct current are the stability of the arc and the fact that the current output of the motor-generator-type welder will remain constant in spite of variations of input voltage which affect a transformer-type welder. Direct current, therefore, is a more versatile welding current. Certain electrodes, such as stainless, require a very stable arc, and as yet these electrodes operate much better with direct current. Direct current, because of its stability, is also better for sheet-metal welding where danger of burn-through is present. The dc arc can also be more readily varied to meet different welding conditions. A wider range of control over both voltage and current permits closer adjustment of the arc for difficult welding conditions, such as might be encountered in vertical or overhead welding. Because of its versatility, direct current should be available for maintenance welding.

Direct-current welders (Figs. 5-44 and 5-45) are made either as motor-generator sets or as transformer-rectifier sets. Motor-generator sets are powered by ac or dc motors.

Generators are also powered by small air-cooled gasoline engines (Fig. 5-46). The advantage of this type of set is that, for on-the-spot maintenance welding, it is not necessary to string electric power lines to the welding set, which may have to be used in a location some distance from a power line.

Engine-driven welders powered by gasoline engines are also available and come in larger sizes than the air-cooled engine sets. These are suitable where the size of the plant maintenance operation warrants a larger welder.

For most general maintenance welding, a 250-amp output capacity is ample. Several manufacturers make machines especially for this type of welding, which are compact and readily portable. Higher amperages may be required in particular applications, and, for these, heavy-duty machines should be used. Such a machine is shown in Fig. 5-47.

Another type of welding machine is one which produces both alternating and direct welding current, either of which is available at the flip of a switch (Fig. 5-48). This type of equipment is ideal for maintenance welding, since it makes any kind of welding arc available, giving complete flexibility to the maintenance welding.

Figure 5-49 shows a self-propelled truck with welder driven from power take-off.

Automatic submerged-arc welding is increasingly used as a maintenance process. Both fully automatic and semiautomatic equipment can be used with the process. The chief use of the process in maintenance is for hard surfacing. It permits the rapid deposition of large amounts of uniformly excellent weld metal. Where the maintenance work includes hard

4-68   Basic Maintenance Technology

surfacing, it is well to consider the use of this process. Semiautomatic equipment is relatively inexpensive, and can be adapted to existing welding equipment of larger amperage outputs. Fully automatic equipment is more expensive, and only a large volume of work will justify its installation.

Accessory equipment   The varied and severe service demands made on equipment for maintenance welding require that the best in accessories be used for maximum efficiency. Most maintenance welders make racks for themselves, or other storage conve-

Fig. 5-44   Compact portable welder designed especially for maintenance work. Shows temporary work table for welding angle-iron frame.

niences, which they attach directly to the welding machine for storing and transporting electrodes and accessories. These arrangements will vary to suit individual tastes and needs. The end result of all of them, however, is to have everything immediately available for use.

A part of such accessory equipment mounted on welders should be a fire extinguisher. Maintenance welding may be required in an area containing a fire hazard. At all times, the possibility of fire should be the welder's concern, and in addition to having the proper fire-extinguishing equipment at hand, he should police the area for flammable materials.

Many electrodes holders are available, but only a few combine all the desired features. The operator holds the electrode clamped in the holder, and the current from the welding set passes through the holder to the electrode. The clamping device should be so designed as to hold the electrode securely in position and yet permit quick, easy change of electrodes. It should be light in weight, properly balanced, and easy to handle, yet sturdy enough to withstand rough usage. It should be designed so that it will remain cool enough to be handled comfortably (Fig. 5-50).

Care should be taken in the selection of face or head shields to ensure maximum protection to the operator. These shields are generally constructed of some kind of pressed-fiber insulating material, usually black to reduce reflection. The shield should be light in weight and designed to ensure greatest comfort to the welder. The glass windows in the protective shields should be of such composition as to absorb the infrared rays, the ultraviolet rays, and most visible rays emanating from the arc. The welding lens in the head or face shield should be protected from molten-metal spatter and breakage by a chemically treated clear nonspatter glass covering the exposed side of the lens.

A good protective lens and shield should be used by the operator, and the arc should never be looked at with the naked eye at close quarters. When a new lens is put into the shield, care should be taken that no light leaks in around the glass. If practical, the welding room should be painted a dead black or some other color to prevent reflection. Other workers around an arc can be readily protected by movable or portable screens.

Special goggles are used by welders' helpers, foremen, supervisors, inspectors, and others working close to a welding arc, to protect their eyes from occasional flashes. A good goggle has adjustable elastic headbands and is light, cool, well ventilated, and comfortable. Clear cover glasses and tinted lenses in various shades are available for this type of goggle.

During the arc-welding process, some sparks and globules of molten metal are thrown out from the arc. For protection from possible burns, it is advisable that the operator wear an apron of leather or other protective material. Some operators also wear spats or leggings and sleevelets of leather or other fire-resistant material. Some sort of protection should be provided for the operator's ankles and feet, since a globule of molten metal can cause a painful burn before it can be extracted from the shoe.

A gauntlet type of glove, preferably of leather, is generally used by operators to protect the hands from the arc rays, spatter of molten metal, or sparks. Gloves also provide protection when handling the work.

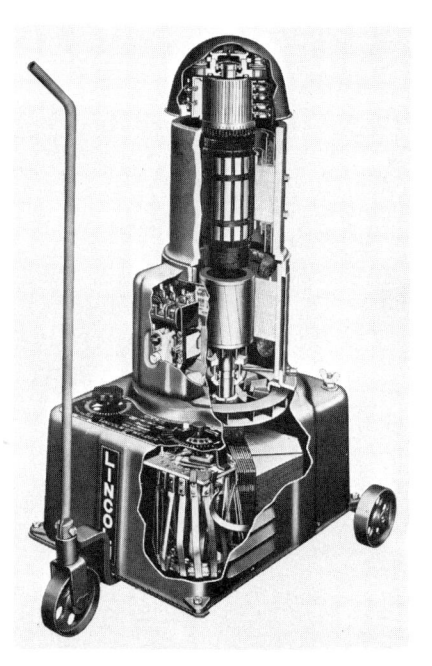

Fig. 5-45 Motor-generator dc maintenance welder.

As a means of protection to other workers from the arc rays, spatter of molten metals, and sparks, the scene of each welding operation should be enclosed by a portable or permanent structure, booth, or screen. Where the welding machine must be taken to the work, it is advisable to surround the scene of the welding operation with portable screens painted dead black to prevent reflection of the arc rays.

Other tools which will prove of value in any shop where welding is done include wire brushes for cleaning the welds, cold chisels for chipping, clamps for holding work in position for welding, wedges, and, where work is large or heavy, a crane or chain block. A drill, air hammer, and grinder are also of value.

INSTALLATION

Good welding begins with proper installation of equipment. Installations should be made in locations that are as clean as possible, and there should be provisions for a continuous supply of clean air for ventilation. It is important to provide separate enclosures if the atmosphere is excessively moist or contains corrosive vapors. If welding must be done where the ambient temperature is high, place the equipment in a different location. Sets operated outdoors should be provided with protection against inclement weather.

4-70   Basic Maintenance Technology

When making an installation, keep the following points in mind:

1. Consult the local power company to ensure adequate supply of the right type of power.
2. Provide adequate and even support for the set.
3. See that there is adequate protection against mechanical abuse and atmospheric conditions.
4. Make proper provisions for large quantities of fresh air for ventilation and cooling.
5. Ground the frame of the welding set solidly.

Fig. 5-46   Small engine-driven combination welder and power supply promotes speedy on-the-spot repair.

Fig. 5-47   Mobile 300-amp dc motor-generator set for work in power plant. Welding cables are fed through wall to weld pipe several hundred feet away. Cable reel is mounted on rear of welder platform. Dual-voltage switch permits use of 220 or 440.

Arc Welding in Maintenance 4-71

6. Check electrical connections to make sure they are tight.
7. The fuses for a motor-generator welder should be of the high-lag type and be rated two or three times the input-current rating of the welder.
8. Make sure that the line and welding leads are of sufficient capacity to handle the required current and are well insulated (Tables 5-8 to 5-11).
9. Check over the set before operating to make sure that no parts are visibly loose or not in good condition.

TABLE 5-8 Recommended Wire Sizes for Input Power Cable for Typical Motor-Generator-Type Welder
(Based on National Electrical Code)

Welder size	60-Hz input voltage	Ampere rating	3 wires in conduit or 3-conductor cable. Type R	Grounding conductor
200	230	44	8	8
	460	22	12	10
	575	18	12	14
300	230	62	6	8
	460	31	10	10
	575	25	10	12
400	230	78	6	6
	460	39	8	8
	575	31	10	10
600	230	124	2	6
	460	62	6	8
	575	50	8	8
900	230	158	1	3
	460	79	6	6
	575	63	4	8

TABLE 5-9 Input Cable Sizes for ac/dc Welder

Welder	Volts input	Amp input		Wire size (3 in conduit)			Wire size (3 in free air)		
		With condsr.	Without condsr.	With condsr.	Without condsr.	Ground conduct.	With condsr.	Without condsr.	Ground conduct.
300	200	84	104	2	1	1	4	4	4
	440	42	52	6	6	6	8	8	8
	550	38	42	8	6	6	10	8	8
400	220	115	143	0	00	00	3	1	1
	440	57.5	71.5	4	3	3	6	6	6
	550	46	57.2	6	4	4	8	6	6
500	220	148	180	000	0000	0000	1	0	0
	440	74	90	3	2	2	6	4	4
	550	61	72	4	3	3	6	6	6

OPERATION AND MAINTENANCE

Careful observance of the following precautions and principles will do much to ensure the maximum of satisfactory service from arc welders.

Keep machine clean and cool Because of the large volume of air pulled through welders by fans in order to keep the machines cool, the greatest enemies of continuous, efficient performance are airborne dust and abrasive materials. Where machines are

4-72 Basic Maintenance Technology

TABLE 5-10 Welding Cable Sizes, Motor-Generator Welder

Machine size, amp	Cable sizes for lengths (electrode plus ground)		
	Up to 50 ft	50–100 ft	100–250 ft
200	2	2	1/0
300	1/0	1/0	3/0
400	2/0	2/0	4/0*
600	3/0	3/0	4/0*
900		Automatic application only	

*Recommended longest length of 4/0 cable for 400-amp welder, 150 ft; for 600-amp welder, 100 ft. For greater distances, cable size should be increased; however, this may be a question of cost—consider ease of handling vs. moving of welder closer to work.

TABLE 5-11 Welding Cable Sizes 11 ac/dc Welder—for Combined Lengths of Electrode and Ground Cable

Machine size, amp	Lengths up to 70 ft	70–150 ft	150–250 ft
300	0	1/0	3/0
400	2/0	2/0	4/0*
500	2/0	3/0	4/0*

*Recommended longest length of 4/0 cable for 400-amp welder is 150 ft and for 500-amp welder, 120 ft. For longer lengths, cable size should be increased; however, it may be a question of cost and flexibility, so that the welder should be moved closer to the work.

subjected to ordinary dust, they should be blown out at least once a week with dry, clean compressed air at a pressure not over 30 psi. Higher pressures may damage windings.

In foundries or machine shops, where cast-iron or steel dust is present, substitute vacuum cleaning for compressed air. Compressed air under high pressure tends to drive the abrasive dust into the windings. If vacuum-cleaning equipment is not available, compressed air may be used at low pressure.

Abrasive material in the atmosphere grooves and pits the commutator and wears out brushes.

Greasy dirt or lint-laden dust quickly clogs air passages between coils and causes them to overheat. Since resistance of the coils is raised and the conductivity lowered by heat, it reduces the efficiency and can result in burned-out coils if the machine is not protected against overload. Overheating makes the insulation between coils dry and brittle.

Do not block the air intake or exhaust vents, because doing so will interrupt the proper flow of air through the machine.

Keep the covers on the welder. Removing them destroys the proper path of ventilation.

Do not abuse it *Never leave the electrode grounded to the work.* This condition creates a so-called dead short circuit. The machine is forced to generate much higher currents than it was designed for, which can result in a burned-out machine.

Fig. 5-48 Unit which produces both alternating and direct welding current.

Do not work the machine over its rated capacity. A 200-amp machine will not do the work of a 400-amp machine. Operating above capacity causes overheating, so that the insulation may be destroyed or the solder in the commutator connections melted.

Use extreme care in operating a machine on a steady load other than arc welding, such as thawing water pipes, supplying current for lighting, running motors, charging batteries, or operating heating equipment. For example, a dc machine, NEMA-rated 300 amp at 40

Fig. 5-49 Self-propelled truck with welding generator driven from power take-off. Has cable reel and cutting torches.

volts or 12 kW should not be used for any continuous load greater than 9.6 kW, and not more than 240 amp. This precaution applies to machines with a duty cycle of at least 60 percent. Machines with lower load-factor ratings must be operated at still lower percentages of the rated load.

Do not handle roughly. A welder is a precisely aligned and balanced machine. Mechancial abuse, rough handling, or severe shock may disturb the alignment and balance of the machine, resulting in serious trouble. Misalignment can cause bearing failure, bracket failure, unbalanced air gap, or unbalance in the armature.

Never pry on the ventilating fan or commutator to try to move the armature. To do so will damage the fan or commutator. If the armature is jammed, inspect the unit for the cause of the trouble. Check for dirt or foreign particles between the armature and frames. Inspect the banding wire on the armature. Look for a frozen bearing.

Do not neglect the engine if the welder is an engine-driven unit. It deteriorates rapidly if not properly cared for. Follow the engine manufacturer's recommendations. Change oil regularly. Keep air filters and oil strainers clean.

Do not allow grease and oil from the engine to leak back into the generator. Grease quickly accumulates dirt and dust, clogging the air passages between coils.

Maintain it regularly

Bearings. The ball bearings in modern welders have sufficient grease to last the life of the machine under normal conditions. Under severe conditions—heavy usage or in a dirty location—the bearings should be greased about once a year. An ounce of grease each year is sufficient for each bearing. A pad of grease, approximately 1 cu in. in volume, weighs close to 1 oz.

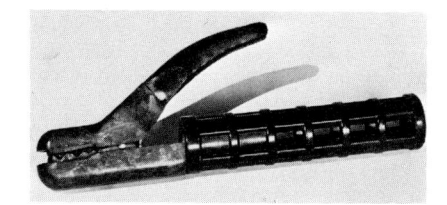

Fig. 5-50 Fully insulated electrode holder designed especially for cool operation.

Dirt is responsible for more bearing failures than any other cause. This dirt may get into the grease cup when it is removed to refill, or it may get into the grease in its original container. Before the grease cup or pipe plug is removed, it is important to wipe it absolutely clean. A piece of dirt no larger than the period at the end of this sentence may cause a bearing to fail in a short time. Even small particles of grit that float around in factory atmospheres are dangerous.

If too little grease is applied, bearings fail.

Too light grease will run out. Grease containing solid materials may ruin antifriction bearings; rancid grease will not lubricate.

Dirty grease or dirty fittings or pipes cause bearing failures.

Bearings do not need inspection. They are sealed against dirt, and it is inadvisable to open them unless necessary.

If it is necessary to pull bearings, it should be done with a special puller designed to act against the inner race. These pullers can be bought.

Never clean new bearings before installing. Handle them with care. Put them in place by driving against the inner race. Make sure that they fit squarely against the shoulders.

Brackets or End Bolts. If it becomes necessary to remove a bracket, to replace a bearing, or to disassemble the machine, do so by removing the bolts and tapping lightly and evenly with a babbitt hammer all around the outside diameter of the bracket ring. Do not drive off with a heavy steel hammer.

The bearing housing may become worn and oversized, because of the pounding of the bearing when the armature is out of balance. Bracket bearing housings may be checked for size by trying a new bearing for fit. The bearing should slide into the housing with a light drive fit. Replace the bracket if the housing is oversize.

Brushes and Brush Holders. Set brush holders approximately $\frac{1}{32}$ to $\frac{3}{32}$ in. above the surface of the commutator. If brush holders have been removed, be certain that they are set squarely in the rocker slot when replaced. Do not force the brush holder into the slot by driving on the insulation. Check to ensure that the brush-holder insulation is squarely set.

Tighten brush holders firmly. When properly set, they are parallel to the mica segments between commutator bars.

Use the grade of brushes recommended by the manufacturer of the welding set. Brushes may be too hard or too soft and cause damage to the commutator. Brushes will be damaged by excessive clearance in the brush holder or uneven brush spring pressure. High commutator bars, high mica segments, excessive brush spring pressure, and abrasive dust also will wear out brushes rapidly.

Inspect brushes and holders regularly. A brush may wear down and lose spring tension. It will then start to arc, with damage to the commutator and other brushes.

Keep the brush contact surface of the holder clean and free from pit marks. Brushes must be able to move freely in the holder. Replace them when the pigtails are within $\frac{1}{8}$ in. of the commutator or when the limit of spring travel is reached.

New brushes must be sanded in to conform to the shape of the commutator. This may be done by stoning the commutator with a stone or by using fine sandpaper (not emery cloth or paper). Place the sandpaper under the brush, and move it back and forth while holding the brush down in the normal position under slight pressure with the fingers.

See that brush holders and springs seat squarely and firmly against the brushes and that the pigtails are fastened securely.

Commutators. Commutators normally need little care. They will build up a surface film of brown copper oxide, which is highly conductive, hard, and smooth. This surface helps to protect the commutator. Do not try to keep a commutator bright and shiny by constant stoning. The brown copper oxide film prevents the buildup of a black abrasive oxide film that has high resistance and causes excessive brush and commutator wear.

Wipe clean occasionally with a rag or canvas to remove grease discoloration from fumes or other unnatural film.

If brushes are chattering because of high bars, high mica, or grooves, stone by hand or remove and turn in a lathe if necessary.

Most commutator trouble starts because the wrong grade of brushes is used. Brushes that contain too much abrasive material or have too high a copper content usually scratch the commutator and prevent the desired surface film from building up. A brush that is too soft may smudge the surface with the same results as far as surface film is concerned. In general, brushes that have a low voltage drop will give poor commutation. Conversely, a brush with high voltage drop commutates better but may cause overheating of the commutator surface.

If the commutator becomes burned, it may be dressed down by pressing a commutator stone against the surface with the brushes raised. If the surface is badly pitted or out of round, the armature must be removed from the machine and the commutator turned in a lathe. It is good practice for the commutator to run within a radial tolerance of 0.003 in.

The mica separating the bars of the commutator is undercut to a depth of $\frac{1}{32}$ to $\frac{1}{16}$ in. Mica exposed at the commutator surface causes brush and commutator wear and poor commutation. If the mica is even with the surface, undercut it.

When the commutator is operating properly, there is very little visible sparking. The brush surface is shiny and smooth with no evidence of scratches.

Generator Frame. The generator frame and coils need no attention other than inspection to ensure tight connections and cleanliness. Blow out dust and dirt with compressed air. Grease may be cleaned off with naphtha. Keep air gaps between armature and pole pieces clean and even.

Armature. The armature must be kept clean to ensure proper balance. Unbalance in the set will pound out the bearings and wear the bearing housing oversize. Blow out the armature regularly with clean, dry compressed air. Clean out the inside of the armature thoroughly by attaching a long pipe to the compressed-air line and reaching into the armature coils.

Motor Stator. Keep the stator clean and free from grease. When reconnecting it for use on another voltage, solder all connections. If the set is to be used frequently on different voltages, time may be saved by placing lugs on the ends of all stator leads. This eliminates the necessity for loosening and resoldering to make connections, since the lugs may be safely joined with a screw, nut, and lock washer.

Exciter Generator. If the machine has a separate exciter generator, its armature, coils, brushes, and brush holders will need the same general care recommended for the welder set.

Keep the covers over the exciter armature, since the commutator can be damaged easily.

Controls. Inspect every time the welder is used to ensure that the ground and electrode cables are connected tightly to the output terminals. Loose connections cause arcing that destroys the insulation around the terminals and burns them.

Do not bump or hit the control handles. It damages the controls, resulting in poor electrical contacts. If the handles are jammed, inspect for the cause.

Check the contact fingers of the magnetic starting switch regularly. Keep the fingers free from deep pits or other defects that will interfere with a smooth, sliding contact. Copper fingers may be filed lightly. All fingers should make contact simultaneously.

Keep the switch clean and free from dust. Blow out the entire control box with compressed air.

Connections of the leads from the motor stator to the switch must be tight. Keep the lugs in a vertical position. The line voltage is high enough to jump between the lugs on the stator leads if they are allowed to become loose and cocked to one side or the other.

Keep the cover on the control box at all times.

CONDENSERS

Condensers may be placed in an ac welder to raise the power factor if required. When condensers fail, it is not often apparent from the appearance of the condenser. Consequently, if it is desired to check to see if they are operating correctly, the following should be done: At rated input voltage and with the welder drawing the rated output load current, the input current reading should correspond to the nameplate amperes. If the reading is 10 to 20 percent more, at least one condenser has failed.

Caution: Never touch the condenser terminals without first disconnecting the welder from the input power source; then discharge the condenser by touching the two terminals with an *insulated* screwdriver.

DELAY RELAYS

The delay relay contacts may be cleaned by passing a cloth soaked in naphtha between them. Do not force the contact arms or use any abrasives to clean the points. Do not file the silver contacts. The pilot relay is enclosed in a dustproof box and should need no attention. Relays are usually adjusted at the factor and should not be tampered with unless faulty operation is obvious.

Table 5-12, a troubleshooting chart, may prove to be a great timesaver.

Basic Maintenance Technology

TABLE 5-12 Arc-Welding Troubleshooting Chart

Trouble	Cause	Remedy
Welder will not start (Starter not operating)	Power circuit dead	Check voltage
	Broken power head	Repair
	Wrong supply voltage	Check name plate against supply
	Open power switches	Close
	Blown fuses	Replace
	Overload relay tripped	Let set cool. Remove cause of overloading
	Open circuit to starter button	Repair
	Defective operating coil	Replace
	Mechanical obstruction in contactor	Remove
Welder will not start (Starter operating)	Wrong motor connections	Check connection diagram
	Wrong supply voltage	Check name plate against supply
	Rotor stuck	Try turning by hand
	Power circuit single-phased	Replace fuse; repair open line
	Starter single-phased	Check contact of starter tips
	Poor motor connection	Tighten
	Open circuit in windings	Repair
Starter operates and blows fuse	Fuse too small	Should be two to three times rated motor current
	Short circuit in motor connections	Check starter and motor leads for insulation from ground and from each other
Welder starts but will not deliver welding current	Wrong direction of rotation	Check connection diagram
	Brushes worn or missing	Check that all brushes bear on commutator with sufficient tension
	Brush connections loose	Tighten
	Open field circuit	Check connection to rheostat, resistor, and auxiliary brush studs
	Series field and armature circuit open	Check with test lamp or bell ringer
	Wrong driving speed	Check name plate against speed of motor or belt drive
	Dirt, grounding field coils	Clean and reinsulate
	Welding terminal shorted	Electrode holder or cable grounded

Arc Welding in Maintenance 4-77

TABLE 5-12 Arc-Welding Troubleshooting Chart (*Continued*)

Trouble	Cause	Remedy
Welder generating but current falls off when welding	Electrode or ground connection loose	Clean and tighten all connections
	Poor ground	Check ground-return circuit
	Brushes worn off	Replace with recommended grade. Sand to fit. Blow out carbon dust
	Weak brush spring pressure	Replace or readjust brush springs
	Brush not properly fitted	Sand brushes to fit
	Brushes in backward	Reverse
	Wrong brushes used	Renewal part recommendations
	Brush pigtails damaged	Replace brushes
	Rough or dirty commutator	Turn down or clean commuator
	Motor connection single-phased	Check all connections
Welder runs but soon stops	Wrong relay heaters	Renewal part recommendations
	Welder overloaded	Considerable overload can be carried only for a short time
	Duty cycle too high	Do not operate continually at overload currents
	Leads too long or too narrow in cross section	Sould be large enough to carry welding current without excessive voltage drop
	Power circuit single-phased	Check for one dead fuse or line
	Ambient temperature too high	Operate at reduced loads where temperatuer exceeds 100°F
	Ventilation blocked	Check air inlet and exhaust openings
Welding arc is loud and spatters excessively	Current setting too high	Check setting and output with ammeter
	Polarity wrong	Check polarity; try reversing or an electrode of opposite polarity
Welding arc sluggish	Current too low	Check output and current recommended for electrode being used
	Poor connections	Check all electrode-holder, cable, and ground-cable connections. Scrap iron is poor ground return
	Cable too long or too small	Check cable voltage drop and change cable

TABLE 5-12 Arc-Welding Troubleshooting Chart (*Continued*)

Trouble	Cause	Remedy
Touching set gives shock	Frame not grounded	Ground solidly
Generator control fails to vary current	Any part of field circuit may be short-circuited or open-circuited	Find faulty contact and repair

REFERENCES

"Procedure Handbook of Arc Welding," 12th ed., The Lincoln Electric Company, Cleveland, 1973.
"New Lessons in Arc Welding," The Lincoln Electric Company, Cleveland, 1973.
Jefferson, T. B., and Gorham Woods: "Metals and How to Weld Them," The James F. Lincoln Arc Welding Foundation, Cleveland, 1962.
Rossi, Boniface E.: "Welding Engineering," McGraw-Hill Book Company, New York, 1954.
Austin, John Benjamin: "Electric Arc Welding," American Technical Society, Chicago, 1952.
Morris, Joe Lawrence: "Welding Principles for Engineers," Prentice-Hall, Inc., Englewood Cliffs, N.J., 1951.
"Welding Handbook," American Welding Society, Miami, 1968.
American Welding Society Publications, 2501 Northwest 7th Street, Miami, Fla. 33125.
  Safe Practices for Welding and Cutting Containers That Have Held Combustibles, A6.0–64.
  Recommended Safe Practices for Inert-Gas Metal-Arc Welding, A6.1–66.
  Safety in Electric and Gas Welding and Cutting Operations—ANSI Standard, Z49.1-67.
Code of Minimum Requirements for Instruction of Welding Operators: Part A—Arc Welding of Steel, B2.1-45.
  A Test Program on Welding Iron Castings, D11.1-65.
  Recommended Practices for Repair Welding of Cast-Iron Pipe, Valves, and Fittings, D10.2-54.
Henry, O. H., G. E. Claussen, and G. E. Linnert, "Welding Metallurgy," American Welding Society, New York, 1949.
Sosnin, H. A.: "Arc Welding Instructions for the Beginner," The James F. Lincoln Arc Welding Foundation, Cleveland.
Linnert, G. E.: "Welding Metallurgy," American Welding Society, Miami, 1966.

Chapter **6**

Gas Welding in Maintenance

**Engineers of Union Carbide Corporation, Linde Division,
New York, N.Y.**

AIR-ACETYLENE SOLDERING, HEATING, AND BRAZING

An air-acetylene appliance produces a flame with a temperature of approximately 4000°F by mixing acetylene with atmospheric air in much the same way that air is mixed with city gas in a kitchen range. The correct mixture produces a pale blue flame with a bright, sharp inner cone that is hot enough for light silver soldering (brazing), for most soft soldering, and for hundreds of heating jobs. Air-acetylene appliances are used throughout industry as companion equipment to the oxyacetylene blowpipe for applications requiring clean, ready-to-use heat but not the extremely high temperatures of the oxyacetylene flame.

An air-acetylene outfit consists of torch handle, a torch stem or tip, a pressure-reducing regulator, a cylinder of acetylene, and a hose for connecting the torch to the regulator and tank. Interchangeable stems or tips that give various sizes and types of flames are available (see Fig. 6-1). The acetylene cylinders themselves come in all sizes, including small portable units (see Fig. 6-2).

Soldering The air-acetylene torch is extensively used for all kinds of soldering with both soft and silver (hard) solder. Although soft soldering is more widely used, silver soldering (also referred to as brazing) is sometimes used for soldering sweat-type fittings in addition to the more precise soldering associated with jewelry and instrument manufacturing. With an air-acetylene torch, the silver solder used must have a melting point lower than 1500°F. If sweat-type fittings being silver-soldered are larger than 1½ in. in diameter, or if a great number of joints are being made, an oxyacetylene torch is recommended, since its greater flame temperature speeds the work. Silver soldering commercial metals over $\frac{1}{32}$ in. thick is also best done with an oxyacetylene torch. In contrast to silver soldering, practically all soft soldering can be done with an air-acetylene torch.

Caution: Silver soldering requires a special rod and a special flux, usually in paste form. Care should be taken to follow the manufacturer's directions. The fumes from some silver solders are toxic; therefore, special ventilating precautions are necessary.

When using air-acetylene appliances, you have a choice of two soldering methods:

    1. The open (direct) flame method. The flame heats the workpiece, and the workpiece melts the solder in conjunction with the flame. The advantages of the open (direct) flame method include

        a. Speed (no copper intermediary to be heated).

        b. Greater diversity in the uses to which the flame can be put.

        c. Greater efficiency in the use of fuel (the gas goes further because it is applied directly to the workpiece).

        d. More heat because of direct application of the flame.

2. The enclosed (indirect) flame method. The flame is applied to the soldering copper. The copper in turn heats the workpiece. The workpiece, in conjunction with the soldering copper, melts the solder where it is needed. The advantages of the enclosed (indirect) flame method include
   - a. Heat is better controlled.
   - b. Less experience is needed on the operator's part.
   - c. More delicate work is possible, especially where damage to the adjacent materials might result from the use of an open flame.

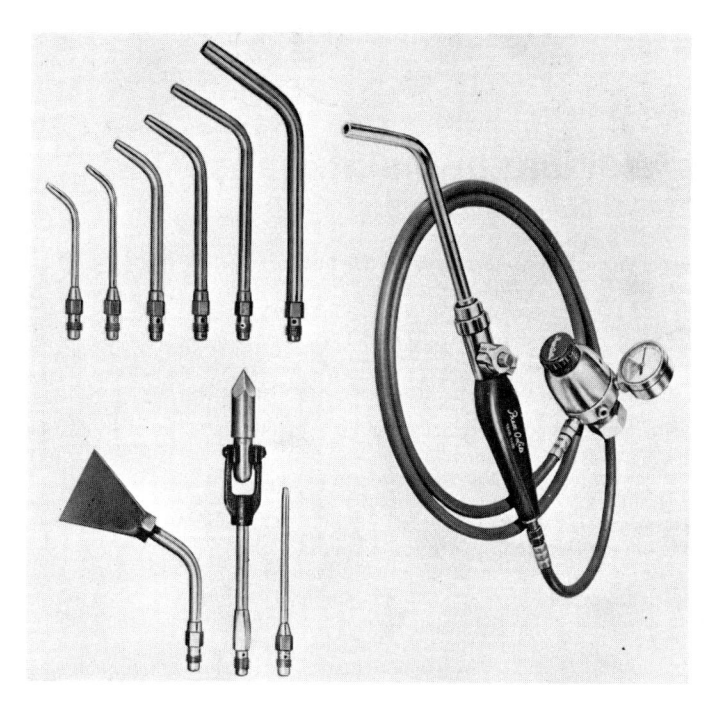

Fig. 6-1 A typical air-acetylene outfit consisting of a regulator, torch handle, and attached torch stem with interconnecting hose. Also shown are some of the typical interchangeable torch stems available. Notice the special-purpose stems: a hatchet-shaped paint-burner stem and a soldering-iron stem.

See Table 6-1 for commonly used soft solders and Table 6-2 for soldering fluxes.

Sheet-metal working Sheet-metal soldering can be done with either the enclosed (indirect) flame method or the open (direct) flame method depending on the choice of the operator. Many types of joints can be made in sheet metal. Joints described on the following pages are most widely used.
   1. The lap joint (see Fig. 6-3):
      - a. Thoroughly clean the edges to be joined.
      - b. Flux the edges by dipping them in a bath of hydrochloric (muriatic) acid, or using a brush, paint them with it.
      - c. If you are using a soldering iron, tin the iron first and then tin the edges. If you are using a soldering torch, tin the edges. The edges should be tinned along their entire length and then placed so that the tinned edges overlap. Use C-clamps to hold them together if you have them.
      - d. Next, pressing down on the soldering iron, run it up and down over the seam until a fillet of solder is visible. If you are using a soldering torch, move it back and forth with the flame touching the work until the fillet appears. In both cases, where no fillet appears, add more solder.

e. When making a long seam with a plain lap joint, it is best to tack the seam first. Tacking means applying drops or spots of solder at intervals along a seam to hold it in place. Clean, flux, and tin the entire job. Heat the seam, and apply solder spot by spot. Then do the regular soldering job on the whole seam. If the tacks tend to melt or the seams to pull apart when you near them with the torch, proceed as follows:
      (1) Press the pieces of metal together at the trouble spot with a stick.
      (2) Reheat the tacks and the solder that has been previously applied as tinning. Keep pressing the heated area together with the stick until the solder has cooled and formed a bond. Proceed with the soldering job.
   f. When the joint is finished, wipe off all excess solder with a stiff bristle brush and wash off the excess flux with hot water.

TABLE 6-1 Commonly Used Soft Solders

Composition, %	Melting range, °F	Gives best results when used for
Tin, 63; lead, 37	361	Critical electronic work, coatings for printed-circuit boards
Tin, 60; lead, 40	361–374	
Tin, 50; lead, 50	361–420	General purposes
Tin, 40; lead, 60	361–460	Automobile radiators, roofing seams, wiped joints in plumbing, dip coatings
Tin, 35; lead, 65	361–478	
Tin, 30; lead, 70	361–496	Filling dents in automobile bodies
Tin, 20; lead, 80	361–534	Apply by wiping; some dip coating
Tin, 15; lead, 85	438–553	
Tin, 10; lead, 90	514–574	Where higher-melting-point solders are necessary
Tin, 5; lead, 95	574–596	
Tin, 96; silver, 4	430	Food-handling equipment, plumbing, heating, refrigeration tube joints where higher-temperature or higher-strength solders are necessary
Tin, 95; antimony, 5	450–464	Some electrical and copper-tubing joints. Do not use on zinc or galvanized sheet

TABLE 6-2 Soldering Fluxes

Metal	Flux to use*
Aluminum	Aluminum application flux
Block tin	Rosin or zinc chloride
Brass	Rosin, zinc chloride, or muriatic acid†
Cast iron	Zinc chloride or muriatic acid
Chromium	Muriatic acid
Copper	Rosin, zinc chloride, or muriatic acid
Gun metal	Rosin, zinc chloride, or sal ammoniac
Inconel	Strong zinc chloride
Iron (galvanized)	Muriatic acid
Iron (tin-coated)	Rosin or zinc chloride
Lead	Rosin, zinc chloride, or muriatic acid
Monel	Zinc chloride or muriatic acid
Nickel	Zinc chloride or muriatic acid
Pewter	Rosin, pewter application flux
Stainless steel	Strong zinc chloride
Steel (plain)	Zinc chloride
Steel (galvanized)	Muriatic acid
Steel (tin-coated)	Rosin or zinc chloride
Terne plate	Rosin or zinc chloride
Tin	Rosin or zinc chloride
Zinc	Strong zinc chloride or muriatic acid

*Nearly all these fluxes are available commercially in paste form. Pastes are usually preferred because they give excellent results on most jobs and are easy to use.
†Muriatic acid is a mild form of hydrochloric acid.

2. Lock joint (see Fig. 6-4):
   a. Thoroughly clean surfaces that will form the joint.
   b. Form the lock joint between the two sheets.
   c. Pound the joint tight with a composition mallet, or use a block of wood between the sheets and a steel hammer. Try to get the joint as flat and tight as possible.
   d. Apply acid flux along the seam, and heat the seam.
   e. Apply just enough solder to seal the seam. (You have already made the seam mechanically strong by hammering and forming the lock joint.)
   f. If the seam is fairly long, you can run the flame a few inches ahead of the solder instead of heating and soldering a section at a time.
   g. Remove all excess solder with a stiff bristle brush, and wash off excess flux with hot water.
3. Flange joint:
   a. A flange joint is generally used in conjunction with rivets or spot welds. The solder is used to make the seam tight to air, gas, or water.
   b. Before the joint is formed, the area to which the solder will be applied must be thoroughly cleaned and must remain clean until the seam is finished.
   c. A tinning coat of solder can be applied to the seam before it is riveted or spot-welded.
   d. Either use acid core solder, or flux the joint with hydrochloric (muriatic) acid.
   e. Heat the joint with either a torch or soldering iron. Capillary attraction will draw the solder into the seam. Fill the joint with the desired amount of solder.
   f. Remove all excess solder with a stiff bristle brush; wash off excess flux with hot water.

Automobile-body soldering Automobile-body soldering is done to fill in dents that cannot be hammered out completely, rough spots, and welded seams. Either soldering method can be used, direct (open) flame or indirect (closed) flame. Where, the deposits of solder to be made are considerable or in places where an open flame would not damage chrome finishes or glass, we recommend the open (direct) flame method because of its speed and the rapidity with which the solder can be deposited. For the places adjacent to glass or chrome finishes, use the enclosed (indirect) flame method. When you have decided which method to use, proceed as follows:

1. Grind away the paint from the dented area, and polish with steel wool or emery cloth.
2. Flux thoroughly and, after heating, apply enough solder to tin the dent.
3. Fill in the dent by adding solder from a bar and smoothing with a maple paddle. Take care not to melt the solder until it runs. Melt it just enough to make it pasty; then smooth with the paddle.
4. When the dent is filled in, heat the solder slightly and smooth it again before letting it cool.
5. Finish the job with rasps, body files, and emery cloth. Clean, prime, and paint.

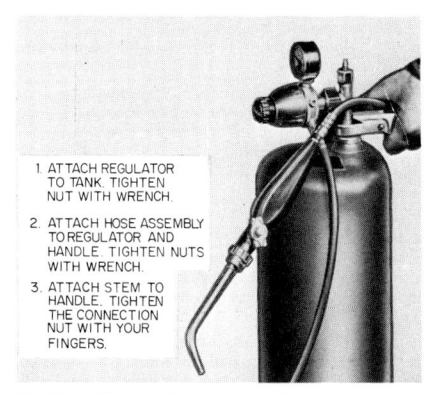

Fig. 6-2 Connecting a typical air-acetylene outfit. The standard type of portable outfit is shown. Smaller tanks also are available.

Electrical connections For soldering electrical connections (Fig. 6-5) the enclosed (indirect) flame method is preferred. Prepare the electrical connections the way you usually do, and proceed as follows:
1. Thoroughly clean the connections.
2. Apply a noncorrosive flux paste.
3. Tin the soldering iron with a thin coat of solder.
4. Tin the wires, and melt enough solder onto them to be sure you have a good electrical connection.

NOTE: Where very large connections are to be made, an open-flame stem can be used.

Installing sweat-type fittings The following is the most efficient method for making sweat-type joints as recommended by two of the leading copper-tube manufacturers. The air-acetylene torch with a direct (open) flame is used by literally thousands of plumbers and is universally recognized as the best means of making these joints. The torch saves time and money, and a relatively inexperienced worker can do a good job with very little training and practice.

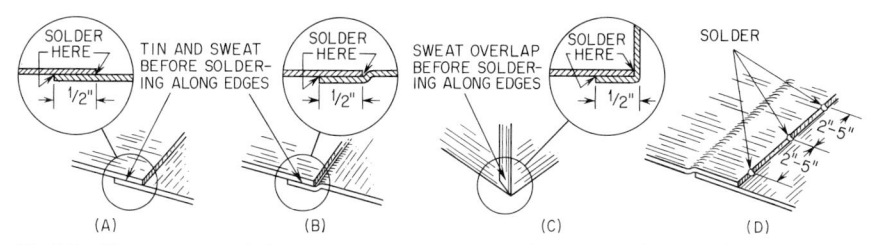

Fig. 6-3 Three variations of a lap joint (A, B, and C); at the right (D), notice how to tack a long seam by applying spots of solder at intervals.

There are two basic types of sweat-type fitting: the plain type and cast type. With the plain type the solder is fed at the point where the fitting and the tube join. With the cast type the solder is fed through precast holes in the fitting itself. The instructions below will work equally well with both types (see Figs. 6-6, and 6-7).

1. Cut the tube to the length required with a hacksaw (32 teeth to the inch), or a disk cutter. Make certain that the tube ends are cut square. Special vises which hold the tube securely and guide the saw blade are furnished by a number of manufacturers.

2. Ream the tube, and remove burrs on the outside. Use a sizing tool if necessary to correct any possible distortion of the tube from handling. The point of a sizing tool is inserted in the end of the tube and is hammered until the tube is again round.

3. Clean the outside surface of the tube and the inside surface of the fitting until the metal is bright. All traces of discoloration must be removed. This must be done even though the tube may appear to be perfectly clean, and it is particularly important when soldering larger-size joints. No. 00 steel wool is very satisfactory for cleaning tubes and fittings. Do not use files or rough sandpaper, as they score the surface and may result in a poor joint.

4. Apply a thin, uniform, and complete coating of a reliable brand of soldering flux or paste to the cleaned portion of both tube and fitting. Do not apply the flux too thickly, as excess flux may form bubbles when heated and prevent the solder from creeping into the joint. After the tube has been inserted into the fitting as far as it will go, revolve the fitting once or twice to spread the flux evenly.

5. Apply the flame evenly all around the circumference of the fitting, and as it

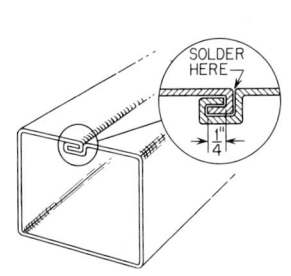

Fig. 6-4 A lock joint where mechanical strength is provided by the joint rather than by the solder bond.

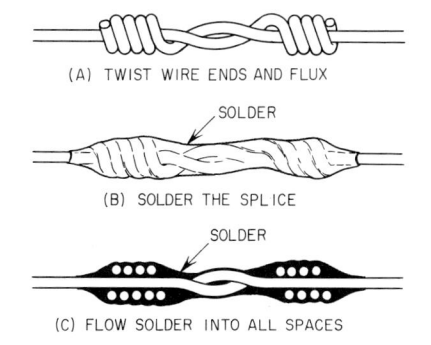

Fig. 6-5 Soldering electrical connections.

becomes heated, move the flame back and forth to prevent overheating. Occasionally test the heat by touching the fitting with solder where the tube and fitting join. Do not let the flame touch the solder while testing the temperature of the joint.

It is important not to overheat the joint. If the connection is heated too much, the flux may be burned out from inside the joint and the solder will not spread properly. An overheated joint causes the solder to seep through the joint and run away.

During the heating operation, adjacent wood surfaces should be protected from the heat by means of sheet asbestos. Because of its narrow, concentrated flame, the air-acetylene torch can be used very close to wood surfaces without scorching them.

6. Remove the flame, and apply solder to the edge of the fitting where it comes in contact with the tube as soon as the fitting has reached the correct temperature to melt the solder. Be sure that enough solder is used.

Enough solder to make an efficient joint will be automatically sucked in by capillary attraction. When a line of solder shows completely around the fitting a fillet of solder appears in the chamfer at the end

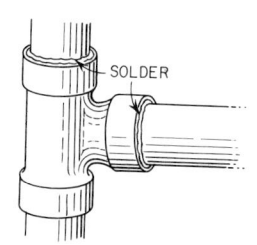

Fig. 6-6 Where to solder plain-type sweat fittings.

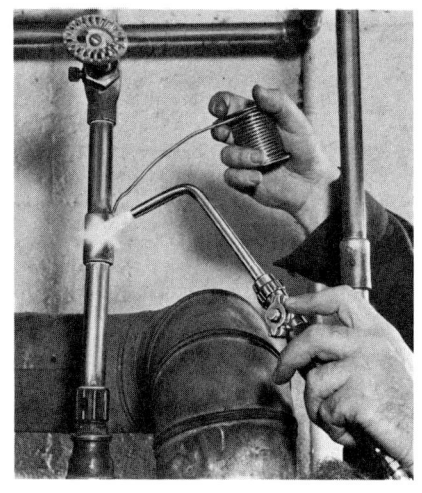

Fig. 6-7 Soldering 1-in. copper tubing and fitting with precast holes.

of the fitting, and the joint has all the solder it will take. Wipe off any excess solder or flux.

7. Slightly reheat the connection in order to help the solder permeate the metal. Remove the flame, and continue to feed solder to make certain the joint is filled.

8. Permit the connection to cool for a fraction of a minute. A rag or wad of waste saturated with water will hasten the cooling. Remove all surplus solder from around the edges with a brush. This operation will show whether or not the solder has filled the joint.

9. When disconnecting a soldered tube from a fitting on which other soldered connections are to be left intact, the application of wet cloths to the parts which are not to be disconnected will prevent melting of the solder at such connections.

10. More than ordinary care should be exercised in soldering fittings $2\frac{1}{2}$ in. in diameter and larger. It is essential that the heat be uniformly distributed around the entire circumference of the fitting and not concentrated in one spot.

When making large-diameter joints, a tip producing a large flame should be used. The flame should be directed on the fitting to avoid any unnecessary annealing of the tube.

For assembling lines 3 in. in diameter or over, it may be advisable to use two or three torches. Solder should then be applied simultaneously at two or more points.

11. In applying solder to a tee, feed solder from both ends of the fitting.

12. Solder when confined between two surfaces will run uphill (by capillary attraction), and joints can be made in almost any position.

13. In sweating male and female adapters, care should be taken to allow more time for the solder to set, as these heavier fittings do not cool so quickly.

Paint burning An air-acetylene torch with a paint-burning stem is a quick, easy, and economical means of removing old, cracked, and checked paint from a surface that can stand a moderate amount of heat. The number of coats of paint is not important; it just takes a little more time to remove them. Paint can be removed from wood, canvas, brick, stone, or metal.

Caution: Avoid inhaling any dust or fumes that may be given off in the paint-burning operation. Such dust and fumes may be toxic, particularly if the paint being removed contains lead or cadmium compounds.

There are two methods of removing paint. They are listed below as Method A and Method B. We suggest you try both methods. You can then use the one that suits your particular type of work. Once the old paint is removed and rough spots smoothed, the surface is ready for a new coat of paint.

Method A (Fig. 6-8):

1. Hold the paint burner in your left hand. Hold the putty knife (with a stiff blade about 3 in. wide) in your right hand.

2. Move the torch backward and forward 1 in. from the painted surface about 6 in. at a stroke. Follow the movements of the torch with a steady forward movement of the putty knife, keeping the putty knife hot with the flame.

NOTE: You will find it advisable to wear asbestos or other heavy flame-resistant gloves when burning paint. The putty knife gets very hot after a while; so you should protect your right hand. Cloth (cotton) gloves are not satisfactory.

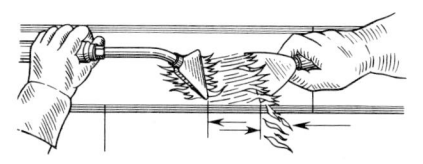

Fig. 6-8 Paint burning (Method A).

Fig. 6-9 Paint burning (Method B).

3. Moving the torch back and forth changes the paint to a plastic state and keeps the putty-knife blade hot. A hot blade reduces the tendency of the paint to stick to it.

Method B (Fig. 6-9):

1. Move the torch more or less steadily from right to left over the painted surface. Bring the paint to a bubbly plastic state. Scrape off the paint as soon as it bubbles. Do not let the flame touch the blade of the putty knife.

2. The putty knife should have a back-and-forth motion which will intermittently expose the scraped area to the flame. This method is recommended for particularly heavy or stubborn paint.

Miscellaneous air-acetylene applications There are many repair and maintenance jobs aside from soldering and brazing that can be done efficiently with an air-acetylene torch. A few of these applications are given below.

Loosening Nuts and Bolts. Frequently you come across a bolt that resists all attempts to loosen it with a wrench. Heat the nut for several minutes and let it cool; then try the wrench again. Generally, you will now find the nut ready to turn.

Freeing Frozen Shafts. A frozen shaft of small diameter can be freed by heating the collar that holds it. Heat the collar, not the shaft. You will find that you can separate the parts quite quickly no matter how tightly they are frozen together.

Lead Working. The air-acetylene torch can be used to build up lead battery terminals. Any of the standard stems can be used depending on the amount of work to be done and the speed with which you want to do it.

It is recommended that you use a form, where possible, to keep the lead in the shape of a battery terminal and to prevent it from running on the battery. Put the form over the old terminal and keep adding melted lead until the desired height and shape of the terminal are attained.

The air-acetylene torch can be used to repair lead-lined vats, wipe joints in lead pipe and lead-covered cable, and solder battery-cable lugs.

Caution: When working with lead in a confined space, be very sure of your ventilation and, if possible, use a suitable air line mask.

Anchoring Bolts in Concrete or Stone. Firmly anchoring a large bolt in concrete or stone can be solved as follows:

1. Drill a hole in the concrete or stone with a star or other type drill. It is best to dish or widen the bottom of the hole slightly to increase the stability of the bolt after the solder sets. Make certain all free moisture or water is removed from the cavity.

2. Heat the solder (bar solder is best) in a ladle with an air-acetylene torch until the solder is molten.

3. Place the bolt in the hole thread-end up, and pour molten solder around it until the solder is level with the floor.

4. This type of mounting will give years of satisfactory service; if it should become loose, just reheat the solder with the torch and it will be as tight as ever.

Cutting Asphalt Tile. The air-acetylene torch has been used with good success by asphalt-tile contractors for heating tiles that have to be bent, formed, or cut. After a few seconds of heating, the tile can be shaped or cut with great ease.

Cutting Safety Glass. Using an air-acetylene torch with a medium-sized stem, the following procedure can be used when cutting safety glass:

1. Score both sides of the glass with a glass cutter and break the glass.

2. Soften the plastic filler by running the torch back and forth along the lines of the cut.

3. Wobble the glass from side to side several times. Then hold the glass to one side while you cut the heat-softened plastic filler with a razor blade.

Precautions and safe practices

1. Do not let acetylene escape near any possible source of ignition. Accumulations of acetylene in certain proportions may explode if ignited.

2. *Never* store acetylene tanks in a closed or confined space, such as a closet.

3. *Never* solder a container that contains or has contained flammable liquids or vapors (including gasoline, benzene, solvents, and other similar or dissimilar materials) unless the container has been thoroughly purged of all traces of flammable material and vapors. Be sure that any container you work on is vented. We urge that before you do work of this kind, you get Booklet A-6.0.40 from the American Welding Society, 2501 Northwest 7th St., Miami, Fla. 33125.

4. *Never* use a tank with a leaking valve.

5. *Do not* make any repairs to an acetylene tank, except to tighten the packing-gland nut on the valve.

6. *Do not* abuse or drop tanks or handle them roughly.

7. *Never* use a tank as a roller. Never use a wrench or pliers on the tank valve. Always use a valve key.

8. *Never* allow full tank pressure to enter a stopped hose. Always use a regulator when there is a needle valve on the torch handle.

9. Examine your hose for leaks frequently. Dipping it in a bucket of clean water, with the pressure in the hose, is the quickest and easiest way.

10. *Do not* use hose that is worn or any equipment that is in need of repair.

OXYACETYLENE WELDING, CUTTING, GOUGING, AND HARD-FACING

Metal production, fabrication, and repair, as they are known today, would be impossible without the oxyacetylene process and its flame of approximately 6000°F. The oxyacetylene process is built on two principles: (1) acetylene burned with an equal amount of oxygen produces an intensely hot flame that will melt and fuse most metals, and (2) a jet of oxygen striking a piece of ferrous metal that has been heated to its kindling temperature will rapidly burn the metal away.

Welding and brazing Welding with an oxyacetylene blowpipe is simple. You put two pieces of metal together, then melt the edges with an oxyacetylene flame. The molten metal flows together and forms a single, solid piece of metal. Welding rod similar to the base metal is usually added to strengthen the joint. This is known as *fusion welding*. If you use a steel rod, the process is sometimes called *steel welding*.

Braze welding is another method. In braze welding, the two pieces being joined are heated to a dull red. They are not melted. A flux is added to clean the metal and protect it from the air. When the pieces are dull red, molten-bronze welding rod is added to form a strong bond. This bronze weld is generally as strong as the base metal.

Building up worn parts with bronze or steel welding rod, heating and forming of metals, gouging, hard facing, and soldering are other jobs done by the oxyacetylene flame.

Cutting Oxygen cutting is similar to the eating away of steel by ordinary rusting, only it is very much faster. In rusting, the oxygen—in the air or in water—affects the metal

Gas Welding in Maintenance 4-87

slowly. Directing a jet of pure oxygen at metal heated almost to the melting point actually speeds up chemical reaction of rusting to such an extent that the metal ignites and burns away. The iron oxide melts and runs off as molten slag to expose more iron to the action of the oxygen jet. This makes is possible to cut iron and steel leaving a smooth, narrow cut. An oxyacetylene outfit is shown in Fig. 6-10.

Fig. 6-10 A complete oxyacetylene outfit. The welding blowpipe shown may be adapted for cutting by exchanging the welding tip with a cutting attachment. Blowpipes designed especially for cutting only are also available.

Setting up an outfit Suggestions and recommendations for safe handling of oxyacetylene equipment have been set forth by the International Acetylene Association. They are included here in brief form.
1. Fasten your cylinders.
    a. Use a cylinder truck, or tie them to a post or bench with a chain, wire, or strap iron.
2. "Crack" the cylinder valves.
    a. Stand behind them; open each a fraction of a turn and close immediately. This blows out any dirt that may have collected in the valve.
3. Connect the regulators.
    a. Oxygen connections have right-hand threads.
    b. Acetylene connections have left-hand threads.
4. Loosen the pressure-adjusting screws on the regulators.
5. Open the oxygen cylinder valve slowly, then as far as you can.
6. Open the acetylene cylinder valve only 1½ turns.
    a. Stand to one side of the regulator gauges when opening these valves.
    b. Leave the T wrench in the acetylene cylinder in case you have to shut it off quickly.

7. Attach the hoses to the regulators, then to the blowpipe.
   a. Use oxygen to blow out new hoses.
8. Attach a welding head to the blowpipe.
   a. Tighten the connection nut.
9. Turn off valve at blowpipe, admit pressure to hose, and test for leaks.
10. Adjust the oxygen pressure.
    a. Open the blowpipe oxygen valve.
    b. Turn in the regulator pressure-adjusting screw until the regulator gauge shows desired pressure.
    c. Close the blowpipe valve.
11. Adjust the acetylene pressure.
    a. Same procedure as for oxygen.

Flame adjustment   The three basic types of flames for an oxyacetylene blowpipe are shown in Fig. 6-11.

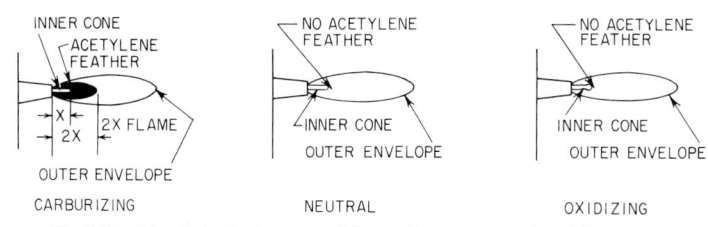

Fig. 6-11   The three basic types of flames for an oxyacetylene blowpipe.

1. To adjust to an excess acetylene (carburizing) flame, start with both blowpipe valves closed. Then
   a. "Crack" the blowpipe oxygen valve; open the acetylene valve about a full turn. Light the blowpipe.
   b. Increase the oxygen supply until you see three distinct parts to the flame: a brilliant inner cone, a whitish acetylene feather, and a bluish outer envelope. This is a carburizing flame.
   c. The amount of excess acetylene in the flame is expressed as a ratio of the total length of the feather to the total length of the inner cone. Thus, in a 2X flame, the acetylene feather is twice as long as the inner cone.
2. To adjust to a neutral flame
   a. Proceed as above, but keep adding oxygen until the acetylene feather just disappears.
   b. This leaves two parts to the flame: a brilliant white inner cone and a bluish outer envelope.
   c. At this point, the blowpipe is burning equal amounts of oxygen and acetylene. This is a neutral flame.
3. To adjust to an oxidizing flame
   a. Proceed as above until a neutral flame is obtained.
   b. Keep adding oxygen beyond the point where the acetylene feather disappears until the inner cone shortens (about 20 percent shorter than a neutral inner cone) and becomes "necked-in."
   c. A harsh sound also characterizes this oxidizing flame unless a very low flow is used.
4. Carburizing, neutral, and oxidizing flames can be harsh or soft. You get a harsh flame when using almost the maximum flow through a tip; you get a soft flame when using less than normal flow.
   a. In a harsh flame, the pressures approach blowoff; that is, a slight increase in pressure causes a gap to appear between the flame and the tip.
   b. In a soft flame, the gas flow is reduced with the blowpipe valves. The inner cone is about half as long as that in a harsh flame.

Braze welding   (see Fig. 6-12) Braze welding is a process which enables you to weld various metals and alloys without melting the base metal. Using a bronze rod (which melts between 1500 and 1650°F) as a filler metal, you can make strong joints in many metals and alloys. The process is similar to soldering, the difference being that solder melts at a much lower temperature than the bronze rod does and is of much lower strength.

In braze welding, a slightly oxidizing flame is generally used, since a carburizing flame gives off certain gases that dissolve in the molten puddle and leave weak, porous spots in the weld.

You can braze-weld cast, malleable, wrought, and galvanized iron; carbon steels and alloy steels, copper, brass and bronze; nickel; Monel; Inconel; and other metals. The following are some of the features and advantages of braze welding:

1. Braze welding can be used for many repair and fabrication jobs.
2. It is faster than fusion welding because less heat is required to melt the filler metal.
   a. This means that you use less gas; so costs are lower.
   b. Less heat means less distortion in the piece being braze-welded.
   c. More work can be done in less time with this fast process.

Fig. 6-12 A braze weld.

Fig. 6-13 A fusion weld.

3. Braze welding produces good strong joints.
   a. Bronze rod, properly deposited, can have a tensile strength up to 56,000 psi.
   b. Tensile strength of plain low-carbon steel is about 52,000 psi.
4. Braze welding can be used for joining dissimilar metals: cast iron to steel, iron to copper.
5. Braze welding can be used to join malleable iron parts and to repair large castings.

Fusion welding   (see Fig. 6-13) Fusion welding is the joining of metal by melting and fusing the edges together. The joint is a thorough mixture of the base metal and the welding or filler rod used to build up the seam. There is no sharp line of demarcation as with a braze weld. The filler rod is used in all cases, except when you are welding sheet metal, and should be about the same composition as the base metal. For example, you use steel rods when welding various plain-carbon or alloy steels while cast-iron rod is used for welding iron castings.

Fusion welding is used mainly where you cannot use braze welding. It has a wide appeal to small users for light-gauge, mostly nonproduction work and for maintenance jobs, although it has largely been displaced for large-scale production work by electric-arc welding. Fusion welding, rather than braze welding, is necessary for parts that will be in use at high temperatures. As a braze-welded joint becomes heated, it loses its strength rapidly, since the filler metal will melt at about 1650°F.

Parts subjected to great tensile stresses, i.e., great pulling loads, should be fusion-welded. For example, some steels have tensile strengths up to 90,000 psi and more. These exceed the tensile strength of a braze weld (up to 56,000 psi) and must be fusion-welded with a special steel rod if the joint strength is to equal or exceed the strength of the base metal.

Fusion welding can also be used where an approximate color match between welded parts is necessary.

Fusion welding uses a neutral or slightly oxidizing flame, since a carburizing flame can cause entrapments in the filler metal.

Weld preparation
1. As a part of your preparation for welding, you should select and prepare the proper joint design for your work (see Fig. 6-14).
   a. Square-edge butt and flange welds are commonly used in sheet-metal work. In the latter, the edges are turned up and melted to form the joint and no filler rod is needed.
   b. Lap joints are rarely used except where one cylindrical section fits inside another.

c. The butt joint with beveled edges (the V) is most widely used. For plate over 3/16 in., bevel the edges by oxygen cutting, grinding, or machining to an included angle of 90°.
d. The corner fillet and double-V joints can also be used, depending on the demands of the job.
e. In the double-V joint, the plates are welded from both sides. It is generally used for work thicker than 1/2 in.

2. The second step in preparation is to clean the edges of any oil or grease, dirt, scale, or rust with steel wool, a wire brush, or some other means.

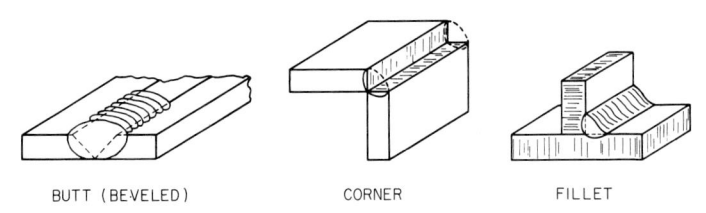

BUTT (BEVELED)   CORNER   FILLET

Fig. 6-14 Butt, corner, and fillet joints.

3. Before welding, the pieces must be properly spaced, since they will tend to expand during welding. Two types of spacing are used to counteract expansion. On material 1/8 in. thick or less, the edges are generally placed parallel to each other about 1/8 in. apart. They should then be tack-welded every 6 in. or so to prevent undue distortion. This space between the plates allows the molten filler metal to flow to the bottom of the joint. Good penetration is thus assured. For material over 1/8 in. thick, a progressive method of spacing is used (see Fig. 6-15). For every foot of weld length, the pieces should be spread apart about 1/8 in. For example, welding pieces 2 ft long, you would leave the pieces approximately 1/4 in. apart at the finishing end of the weld. At the starting end, the pieces should be spaced about 1/16 in. Slightly more spacing is required for unbeveled edges.

4. Tack-weld the pieces at start and finish. In sheet metal, tack-weld about every 6 in. This keeps the pieces in alignment and prevents them from drawing too close together as they are heated during welding. There is a great deal of strain on a tack weld as a result of these internal expansion and contraction stresses. So make your tack welds carefully, and make them strong.

Blowpipe motion   There is no hard-and-fast rule which will tell anyone exactly how to move a blowpipe when welding. Each welder develops a natural motion after a little practice. Figure 6-16 gives a suggested pattern for moving the blowpipe while welding. The motion is effectively a series of semicircles, wide enough to ensure heating beyond the limits of beveling and moving forward to a slight extent in each blowpipe swing. For fusion welding, it is important to try to move the flame around in front of the rod at the end of each sweep so that complete melting of the edges is obtained. With braze welding, it is not necessary to melt the edges of the joint; so the blowpipe motion is generally faster.

Making a weld   The procedures for making a braze or fusion weld are essentially the same except that, for fusion welding, the edges of pieces being joined are actually melted, while for braze welding, the edges of the pieces being joined are heated to a dull red. In both fusion and braze welding, the filler rod is melted to furnish the filler metal for the seam. The following brief discussion applies to both fusion and braze welding. Nevertheless, the difference between the two welding methods should always be remembered.

1. Steel thicker than 5/16 in. can be welded in two or more passes, while material over 1/2 in. should always be multipass welded (see Fig. 6-17).

2. Lay in the first pass or root weld from 2 to 3 in. long. After making this beginning section of the root weld, go back and build up the finishing weld to the desired reinforcement.

3. There are two reasons for making a root weld for about 2 1/2 in. and then returning for the second pass or finishing weld before continuing with the root weld (see Fig. 6-18):

a. You take advantage of the heat left in the plate when you made the beginning section of the root weld. If you continued all the way across the plate, this heat would be lost when you returned for a second pass.
   b. Experience shows that best results in strength, uniformity, and appearance are achieved when this system is used.
4. The following points are important in making a good weld:
   a. Do not add filler rod until you have formed a molten puddle (fusion welding) or heated to a dull red (braze welding).
   b. Keep the rod in the puddle.

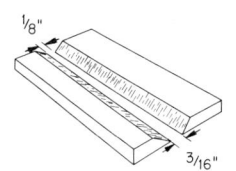

Fig. 6-15 When making a butt weld of pieces over $1/8$ in. thick, progressive spacing will counteract distortion caused by expansion.

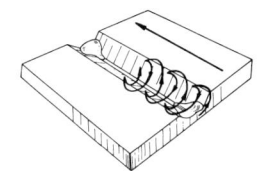

Fig. 6-16 Blowpipe motion for fusion and braze welding.

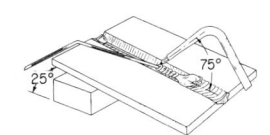

Fig. 6-17 Braze or fusion welding of two pieces thicker than $3/16$ in. is best done with two passes. Notice that the weld proceeds in stages.

   c. Keep your eye on the leading edge of the puddle to ensure that you always have thorough fusion or heating.
   d. Remember that the rod is deposited evenly by constantly melting it into the molten puddle, not by applying the flame directly to the rod.
5. The blowpipe should be tilted slightly to an angle of 75° with the plate surface, to ensure a certain amount of preheat as the weld proceeds. The plate may be tilted upward to an angle of perhaps 25° to aid in even buildup.
6. The blowpipe should be directed squarely into the V between the plates so that both sides will be heated evenly.
7. Proper weld sequence is shown below. The first root weld is made for about $2\frac{1}{2}$ in.; the first finishing weld is about half this length. Each successive pass, both root and finishing, is about the same length as the first section. The final section of finishing weld will be a bit longer than any other part to make up for the shortness of the first finishing weld.
8. Never make a flush weld if maximum strength is desired. Always provide reinforcement; that is, make sure that the lowest ripple is $1/16$ in. above the surface of the plate.

Heavy braze welding The following points apply to such heavy jobs as repairing heavy steel or iron castings:
   1. Preparation of the work is important. First, vee out the crack with your gouging nozzle on steel parts or by chipping, grinding, or machining if the piece is cast iron. Be sure to clean thoroughly a generous space on each side of the V to permit the crown of the weld to lap over and give additional strength.

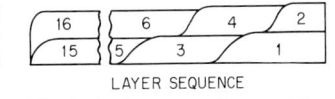

LAYER SEQUENCE

Fig. 6-18 Sequence of root and finishing welds for a weld made in two passes (layers).

   2. On cast-iron pieces, it is fairly certain that graphite (pure carbon) flakes are embedded in the surface and have been smeared by machining or grinding. In the presence of an oxidizing flame, this carbon will unite with oxygen and burn off as a gas. Use steel wool or a wire brush on the surface to complete the cleaning job.
   3. Choose a location where it will be possible to set up a temporary preheat furnace. The reasons for preheating are:
      a. It is easier to braze-weld if heat is stored up in heavy pieces. A fairly small and convenient welding flame can be used if the pieces are at about 500°F.
      b. If the pieces are cold, heat from the welding flame would be rapidly drawn away.

c. Preheating will help prevent excessive internal stresses from occurring as the piece cools.

   4. Depending on the shape of the piece, it may be possible to make a double V and have a welder work on each side of the joint.

   5. Since it is easier to build up a weld in successive horizontal layers, position the work if possible so that your weld line is flat.

   a. If possible, it is desirable to support your starting weld on a carbon block or piece of firebrick. If this cannot be done, use a piece of 10-gauge sheet or carbon plate fitted to the bottom of the abutting pieces. When the weld is finished and the casting is cool, you can remove the sheet if necessary by chipping.

   b. Use plenty of flux or flux-coated rod so the tinning action will take place automatically and stay well ahead of the weld itself.

   6. When completed, the weld can be cleaned by starting at the top and working downward with a large oxidizing flame, melting the runovers.

   7. In cases where the castings are spread out, be sure that they are well supported, since cast iron is weakened when heated to a high temperature.

Fusion-welding cast iron

   1. Cast iron does not have the strength and ductility of steel. Without careful cleaning before and proper cooling after welding, a casting may become hard and brittle and possibly crack.

   a. Clean off any dirt, scale, and grease that might weaken the final weld with a wire brush, a grinder, or a file.

   b. In order to equalize internal expansion and contraction stresses introduced during welding, preheat small castings locally with your blowpipe. Large castings should be placed entirely in a preheat furnace and raised to a temperature of approximately 500°F. The stresses of concentrated welding heat might crack the casting without this preheating.

   2. Molten cast iron is very fluid and may tend to fall through. It is also a good idea to weld "in the flat" with some sort of backup where possible. Carbon blocks may be removed after the weld has cooled.

   3. Bevel the edges, by chipping or grinding, to an included angle of about 90°.

   4. To help further in cleaning the edges so that a clean, sound weld will be obtained, use a flux that will chemically float out dirt, slag, and oxide inclusions.

   5. Add just enough flux so that all the impurities are cleaned and fluxed out of the weld zone.

   6. Use only one pass. It is not necessary to fill in a root weld and then a finishing weld as was the case with steel.

   7. Cast iron must be cooled slowly after welding. Sudden chilling of a recently welded cast-iron part can cause it to crack. Fast cooling also makes a casting hard, brittle, and subject to being cracked easily; slow cooling imparts softness and ease of machinability. Small parts can be placed into a can of lime or cement or some similar material so that they will cool properly. Larger castings can be left in the preheat furnace for slow, even cooling.

For recommended welding methods see Tables 6-3 and 6-4.

Oxygen cutting Iron burns (all burning is an oxidation process) as readily as wood or paper if it is heated to the right temperature and is exposed to a large amount of pure oxygen. Metals like aluminum, stainless steel, and magnesium also oxidize, but it takes even more heat to melt their oxides than it does for iron oxide. Other means must be used to cut them. Oxygen cutting is primarily intended for cutting ferrous metals (iron or steel).

The first step in oxygen cutting ferrous metals is to preheat the metal until it is red hot. At this point, the metal is said to be at its kindling or ignition temperature—it is ready to burn away. The actual cut in the metal is started by directing the pure oxygen stream from a cutting blowpipe at the preheated metal. The hot iron and the oxygen react instantly, producing so much heat that the oxide formed melts and flows or is blown away. As the oxide flows away, the cut progresses through the metal as the next layer of metal is exposed to the oxygen. When the blowpipe is moved along the line of cut, the heat of the reaction between the iron and oxygen raises the temperature of these successive layers of metal.

TABLE 6-3 Recommended Welding Methods (Ferrous)

Metal	Welding method	Flame adjustment	Recommended welding rod	Flux
Steel, cast.........	Fusion weld	Neutral	High-test steel	None
Steel pipe	Fusion weld	Neutral	High-test steel	None
	Steel welding	Carburizing	CMS steel	
Steel plate	Fusion weld	Neutral	Drawn iron	None
	Steel welding	Carburizing	High-test steel	
			CMS steel	
Steel sheet	Fusion weld	Neutral	Drawn iron	None
	Bronze weld	Slightly oxidizing	High-test steel Bronze	Brazing None
High-carbon steel .	Fusion weld	Carburizing	High-test steel CMS steel	None
Manganese steel ...	Fusion weld	Slightly oxidizing	Same composition as base metal	None
Cromansil steel....	Fusion weld	Neutral	High-test steel CMS steel	None
Wrought iron	Fusion weld	Neutral	High-test steel	None
Galvanized iron ...	Fusion weld	Neutral	Drawn iron	None
	Fusion weld	Neutral	High-test steel	None
	Bronze weld	Slightly oxidizing	Bronze	Brazing
Cast iron, gray.....	Fusion weld	Neutral	Cast iron	Ferrous
	Bronze weld	Slightly oxidizing	Bronze	Brazing
Cast iron, malleable	Bronze weld	Slightly oxidizing	Bronze	Brazing
Cast-iron pipe, gray	Fusion weld	Neutral	Cast iron	Oxweld ferrous
	Bronze weld	Slightly oxidizing	Bronze	Brazing
Cast-iron pipe	Fusion weld	Neutral	Cast iron Same composition as base metal	Ferrous
Chromium-nickel ..	Bronze weld	Slightly oxidizing	Bronze	Brazing
Chromium-nickel steel castings	Fusion weld	Neutral	Same composition as base metal 25-12 chromium-nickel steel Columbium-bearing 18-8 Stainless steel	Stainless steel
Chromium-nickel steel (18-8)	Fusion weld	Neutral	Columbium-bearing 18-8 Stainless steel	Stainless steel
Chromium-nickel steel (25-12)	Fusion weld	Neutral	Same composition as base metal	Stainless steel
Chromium steel ...	Fusion weld	Neutral	25-12 chromium-nickel steel Columbium-bearing 18-8 Stainless steel	Stainless steel
Chromium steel (4–6 percent)	Fusion weld	Neutral	Columbium-bearing 18-8 Stainless steel	Stainless steel
Chromium iron	Fusion weld	Neutral	25-12 chromium-nickel steel Columbium-bearing 18-8 Stainless steel Same composition as base metal	Stainless steel

4-94 Basic Maintenance Technology

Oxygen cutting is used almost everywhere—for cutting straight lines and circles in plate, for cutting shapes to accurate dimensions in single pieces and in stacks, for trimming plate to size and beveling it for welding, for piercing holes, for cutting I beams and other structural members to size, and for many other uses. The oxygen-cutting blowpipe is also a prime fabricating tool in industry for preparing plates and cutting

TABLE 6-4 Recommended Welding Methods (Nonferrous)

Metal	Welding method	Flame adjustment	Recommended welding rod	Flux
Aluminum	Fusion weld	Slightly carburizing	Aluminum	Aluminum
Brass	Fusion weld	Oxidizing	Bronze	Brazing
	Bronze weld	Slightly oxidizing	Bronze	
Bronze	Fusion weld	Neutral	Bronze	Brazing
	Bronze weld	Slightly oxidizing	Bronze	
Copper (deoxidized)	Fusion weld	Neutral	Deoxidized copper	None
	Bronze weld	Slightly oxidizing	Bronze	Brazing
Copper (electrolytic)	Fusion weld	Neutral	Cupro	None
	Bronze weld	Slightly oxidizing	Bronze	Brazing
Everdur bronze	Fusion weld	Slightly oxidizing	Everdur bronze	Silicon bronze
Nickel	Fusion weld	Slightly carburizing	Same composition as base metal	None
Monel metal	Fusion weld	Slightly carburizing	Same composition as base metal	Monel
Inconel	Fusion weld	Slightly carburizing	Same composition as base metal	Inconel
Lead	Fusion weld	Slightly carburizing	Same composition as base metal	None

structural members in the shipbuilding, heavy-machinery, and building-construction industries. Oxygen cutting is also extensively used for demolishing and scrapping of machinery, obsolete equipment, unsafe or unwanted structures; for cutting heavy scrap to smaller size; for removing bolts and rivets; and for similar work (see Fig. 6-19).

Oxygen cutting is very versatile in that steel, wrought iron, and cast iron can be cut in almost any form, of almost any thickness. Hand cutting is restricted to thicknesses of about 1 in. Machine cuts have been made, however, in material of about 6 in. in thickness.

The process is inexpensive. Initial equipment cost and subsequent upkeep costs are very low compared with other means of doing the same job. The gas costs are almost negligible when you consider the variety and quality of the work done. The equipment needed for oxygen cutting is easily portable and can be taken almost anywhere for on-the-job use. The process is very fast. Depending on the thickness of the material, speeds up to 500 fph can be attained. The process is easily learned. The correct techniques can be studied and picked up in a few minutes.

Oxygen-cutting equipment (See Fig. 6-20) Oxygen cutting requires the same equipment needed for welding, including a welding blowpipe fitted with a cutting attachment and a special nozzle. Where you are going to do oxygen cutting for long periods, a cutting blowpipe is more desirable than a cutting attachment. The cutting nozzles come in various sizes. The thickness of the metal and its surface condition determine the size of the nozzle needed. For example, five different-sized nozzles handle all thicknesses up to 12 in.

Various accessories, which supplement basic equipment, are available for making special types of cuts. In freehand guided cutting, the blowpipe head can be drawn along a bar or straightedge. This will assure an accurate square or beveled straight-line cut. Circles or disks with 2 in. or greater diameters can be accurately made with circle-cutting attachments. Where high accuracy is required in cutting straight lines, circles, or shapes, special machines are available which mechanically hold, guide, and advance the blowpipe over the work. Little or no finishing is required on these high-quality machine cuts.

Preparation for cutting

1. First, select a suitable place for working—make sure there is no combustible material at hand. Use asbestos or sheet-metal shields to protect floors of wood or other materials, where necessary. Protect your legs and feet from sparks and slag.

2. A clean metal surface means lower gas consumption and a good-quality cut. So remove all the dirt and paint you can by scraping or wire brushing.

3. Look at the instruction sheet for your cutting attachment or blowpipe to find out what size nozzle to use for the thickness of metal you are cutting.

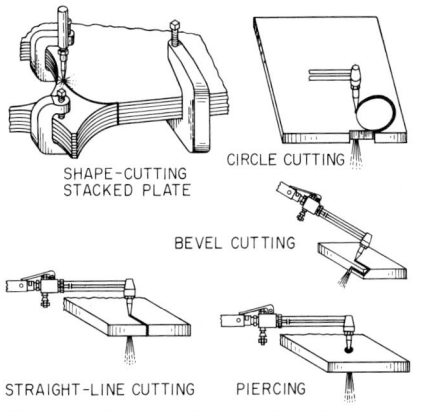

Fig. 6-19 Some of the jobs done by oxygen cutting.

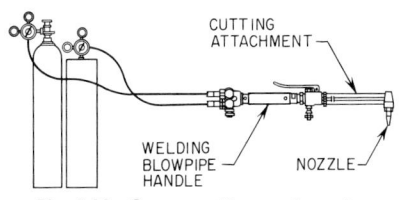

Fig. 6-20 Oxygen-cutting equipment.

4. The adjustment of the flame for a cutting attachment or blowpipe is different from that for a welding blowpipe because the latter has no cutting oxygen stream.
   a. If the cutting oxygen valve is opened after the preheat flames are adjusted to neutral, the preheat flames will lack oxygen. This is because both preheat and cutting oxygen come from the same source and part of the preheat supply has been diverted to form the cutting-oxygen stream.
   b. To correct this, the preheat flames should be adjusted with the cutting oxygen level down.
   c. Also, oxygen-flow adjustments must always be made with the needle valve on the cutting attachment. Open the blowpipe oxygen valve wide, leave it that way while cutting, and adjust the flame with the other valve.

Making the cut

1. During cutting, hold the blowpipe in one hand and guide the blowpipe by resting it on your other hand.
   a. A piece of firebrick on the plate will provide a rest for your hand as well as indicate the proper spacing of the blowpipe from the work.
   b. Make sure nothing will prevent you from finishing the cut without interruption.
2. Hold the blowpipe so that the preheat cones just lick the work surface. Preheat the starting point on the edge to a bright red (see Fig. 6-21).
3. Start the cut by slowly pressing down the cutting-oxygen lever (see Fig. 6-22).
   a. Keep the tip vertical and always the same height above the work.
   b. Do not advance the blowpipe until the cut is completely through the metal.
   c. Continue the cutting action by moving the blowpipe along the line of cut at a uniform rate (see Fig. 6-23).
4. If you move the blowpipe too slowly, you will melt over the edges of the cut and give it a ragged appearance.
5. If you move the blowpipe too fast, the cutting jet will not penetrate the metal completely and you will lose the cut. In this case, release the cutting-oxygen lever, go back to where you lost the cut, and start over again.

a. Experience is the only way to learn exactly how fast to move the blowpipe.

b. When the cut is finished, release the cutting-oxygen lever and turn off the preheat flames.

Gouging Gouging or grooving is merely a special type of oxygen cutting. It is a means of removing a narrow strip of metal from the surface of a plate. You use the same equipment for gouging that you use for cutting, except that you must have a large-bore,

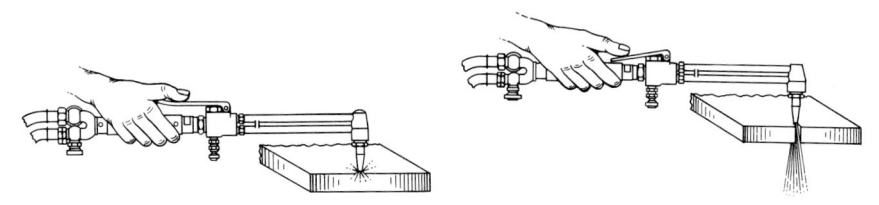

Fig. 6-21 Preheating.

Fig. 6-22 Beginning a cut.

low-velocity nozzle. As in cutting, the operation centers around three main steps: preheating, starting the groove, and progressing. Other things to be watched during gouging include

1. Pulling the nozzle back along the plate surface after preheating, then opening the cutting-oxygen lever. This ensures that the stream will fall on hot metal, not on relatively cold metal ahead of the preheated spot.

2. Keeping the flames low. If the inner cones of the preheat flames on the lower side of the nozzle are just barely touching the work, you will get maximum efficiency from the preheat flames.

3. Keeping the blowpipe moving in a straight line. When making a long groove, there is a tendency to move the blowpipe toward you as the groove proceeds and describe a long arc instead of a straight line in the plate.

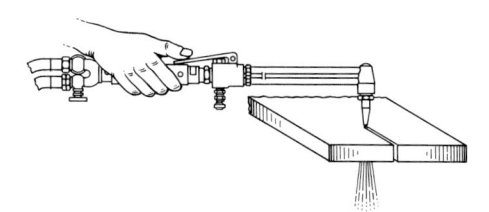

Fig. 6-23 Cutting.

With the step-back method of gouging, you will have less tendency to lose the cut or swing out of line than if you gouge in one continuous pass.

1. The groove is carried progressively across the plate in a series of short gouges.

2. Start the groove, then continue it for about 3 in. Lift up the blowpipe, bring it back about ½ in., and restart the groove.

3. As each short pass is completed, the nozzle is drawn back slightly to restart the groove.

4. Repeat these steps until you have reached the full length of the desired groove.

Gouging is used in three main applications:

1. Removing defective welds. When a weld does not have a good appearance or is not as strong as it should be, it can be removed by gouging and replaced. You can also remove the old weld and have the piece ready to be rewelded all in one operation by gouging.

2. Opening up cracks in castings so that sound repairs can be made by welding.

3. Dismantling welded structures to permit reuse of most of the parts, thus obtaining maximum salvage.

Using a special gouging nozzle, you can cut grooves from ⅜ to ½ in. wide by ⅛ to ⁷⁄₁₆ in. deep. These variations in groove dimensions are controlled by three factors:

  1. By the angle of the nozzle with respect to the work (see Fig. 6-24). A flat angle gives a shallow groove, and a steeper angle a deep groove.
  2. By the speed of travel of the blowpipe. The faster you move, the shallower the gouge becomes.
  3. By the oxygen pressure. High pressures wash a bit more metal out of the groove than lower pressures.

Hard-facing Hard-facing is the process of applying a layer of special alloy on a metal part or surface to protect it from wear. The big difference between hard-facing and the fusion welding is that the hard-facing alloy does not mix with the base metal to any extent. In fusion welding, complete penetration is necessary, but in hard-facing, it should be avoided. This is important because mixing of the base metal with the hard-facing alloy would dilute and soften the deposit. In hard-facing, the surface of the steel picks up carbon from an excess acetylene flame. The carbon lowers the melting point of the steel and causes it to melt quickly to a depth of only a few thousandths of an inch. This very thin film of melted steel fuses with the hard-face deposit to make a strong bond between the deposit and the steel.

Metals that can be hard-faced include carbon and low-alloy steels (covering 95 percent of the wear problems you will normally encounter), all forms of cast iron (except chilled), and many other special alloys.

With the longer life of hard-faced parts (2 to 25 times longer), the reduction of maintenance labor and of replacement parts used is dollars saved. Here are a few typical examples of how hard-facing increases the life of parts:

Part Hard-Faced	Times Longer Life
Pump shaft	3
Clutch plate	7
Valves, valve seats	7
Valve-seat inserts	15
Hand shovels	3
Spray nozzle disks	12
Cams	6
Shear blades	10
Mill hammers	5
Punches	13

Hard-facing rods There are a number of hard-facing rods available to help you solve particular wear problems resulting from such factors as abrasion, impact, corrosion, and heat. Very often more than one cause of wear is present. Your problem then is to choose the hard-facing alloy best suited to combat the combination of factors. You should consider every job as a special problem. The same rod used for one job will not necessarily work on the same or similar part in another instance. If you are in doubt about which rod to select, test several under actual conditions. Manufacturer's data will usually help you select the proper rod, but often you must make the final decision in the light of what you can find out about the wear conditions involved. Tables 6-5 and 6-6 show the particular characteristics of some of the hard-facing rods available.

How to hard-face steel

  1. Clean the surface to be hard-faced by filing, wire brushing, or grinding. Edges or corners that might become overheated during hard-facing should be grooved out as shown and filled with hard-facing deposit. Use your cutting blowpipe or attachment and grooving nozzle (see Fig. 6-25). If an edge or corner of the part takes a lot of pounding or impact in use, machine the corner or edge as shown in Fig. 6-26. The dotted lines in the illustration show how the hard-face deposit should be built up to the original contour of the part.
  2. Parts more than 2 in. in thickness should be preheated throughout to prevent the deposit or the part itself from cracking when it cools. You can preheat medium-sized parts with your blowpipe. Use a neutral flame. Move the flame in a wide circle over the part. Gradually make the circles smaller and smaller until the part turns a dull red color. Large surfaces or bulky parts should be preheated in a furnace. Heat the part until it turns a dull red color.
  3. Deposits up to ⅛ in. in thickness can be made in one pass. Best impact resistance is obtained from deposits ¹⁄₁₆ in. in thickness, never over ⅛ in. If you want to build up a

Codes, Specifications, and Welding Standards*

Title	Published by	Field of application	Source
	General		
AWS Terms and Definitions, A3.0	AWS,‡ 1976, 80 pp.	Welding, cutting, brazing	AWS‡
Resistance Welding Equipment, ANSI C88.2	ANSI,¶ 1969	General welding	ANSI¶
AWS Welding Symbols, A2.0, ANSI Y32.3	AWS,‡ 1968, 90 pp.	Engineering-shop drawings	AWS‡
Master Chart of Welding and Allied Processes	AWS,‡ 1976	Wall size 22 by 28 in. Desk size 8½ by 11 in.	AWS‡ AWS‡
Electric Arc-Welding Apparatus, ANSI C87.1	National Electrical Manufacturers Association, 1970	General welding	ANSI¶
Safety in Welding and Cutting, ANSI Z49.1	ANSI,¶ 1973	General welding	ANSI¶
Safety Standard for Transformer-Type Arc-Welding Machines, ANSI C33.2	Underwriters' Laboratories, 1972	General Welding	ANSI¶
Welding Symbols Chart	AWS,‡ 1976	Wall size 22 by 28 in. Desk size 11 by 17 in.	AWS‡ AWS‡

Boilers and pressure vessels		
ASME Boiler and Pressure Vessel Code		
Sec. I, Power Boilers	ASME,§ 1977	Power boilers in stationary service	ASME§
Sec. III, Div. 2, Code for Concrete Reactor Vessels and Containments	ASME,§ 1977	Nuclear power plants	ASME§
Sec. IV, Heating Boilers	ASME,§ 1977	Boilers in operation at less than 15 psig and for hot-water heating and supply	ASME§
Sec. VIII, Pressure Vessels	ASME,§ 1977	Pressure vessels	ASME§
Inspection and testing		
Appendix, Inspection of Welding, 3d ed.	API,† 1978	Refinery equipment	API†
Standard Methods for Mechanical Testing of Welds	AWS,‡ 1974	Welding shops and fabricators of welded structures	AWS‡

*All are available from sponsoring organization. For convenience, AWS is given as source when possible.
†American Petroleum Institute, 2101 L St., N.W., Washington, D.C. 20037.
‡American Welding Society, 2501 N.W. 7th St., Miami, Fla. 33125.
§The American Society of Mechanical Engineers, Order Department, United Engineering Center, 345 East 47th St., New York, N.Y. 10017.
¶American National Standards Institute, 1430 Broadway, New York, N.Y. 10018.

Codes, Specifications, and Welding Standards* *(Continued)*

Title	Published by	Field of application	Source
Piping			
Power Piping, ANSI B31.1	ANSI,¶ 1977	Pressure piping systems	ANSI¶
Standard for Welding Pipelines and Related Facilities (Std 1104)	API,‡ 1977	Cross-country petroleum and natural-gas pipelines	API†
Standard for Qualification of Welding Procedures and Welders for Piping and Tubing, D10.9	AWS,‡ 1969, 72 pp.	All piping systems	AWS‡
Structural (building)			
Rules for Arc and Gas Welding and Oxygen Cutting of Steel Covering the Specifications for Design, Fabrication, and Inspection of Arc and Gas Welded Steel Structures and Qualifications of Welders and Supervisors. Cal. 1-38-SR.	Board of Standards and Appeals, New York, 1968, V. 53, Bull. 51, p. 1338; Amendments: 1974, V. 59, Bull. 4, p. 34; V. 59, Bull. 13, p. 259; 1977, V. 62, Bull. 22, p. 500	Buildings in New York	Board of Standards and Appeals, City of New York, 80 Lafayette St., New York, N.Y. 10013
Structural Welding Code, D1.1	AWS,‡ 1975, 166 pp.	Highway, railway, bridges, buildings, and tubular structures	AWS‡
Safe Practices for Welding and Cutting Containers That Have Held Combustibles, A6.0	AWS,‡ 1965, 16 pp.	Shops engaged in welding or cutting operations on combustible solids, liquids, or gases	AWS‡

Specifications for Field-Welded Tanks for Storage of Production Liquids, 8th ed. (Spec. 12D)	API,† 1977	Oil-field service—capacities over 500 bbl	API†
Specifications for Shop-Welded Tanks for Storage of Production Liquids, 7th ed. (Spec. 12F)	API,‡ 1977	Oil-field service—capacities to 440 bbl	API†
Recommended Rules for Design and Construction of Large, Welded, Low-Pressure Storage Tanks, 6th ed. (Std. 620)	API,† 1977	Petroleum products storage—for internal pressures of 15 psig or less	API†
Welded Steel Tanks for Oil Storage, 6th ed. (Std. 650)	API,† 1977	Oil storage at atmospheric pressure	API†
Standard for Welded Steel Elevated Tanks, Standpipes, and Reservoirs for Water Storage, D5.3	American Water Works Association and AWS,‡ 1973	Elevated steel water tanks, standpipes, and reservoirs	AWS‡
Water Tanks for Private Fire Protection, NFPA Std. No. 22	National Fire Protection Association, 1976	Field-welded tanks, gravity and pressure towers, etc.	National Fire Protection Association, 470 Atlantic Ave., Boston, Mass. 02210

*All are available from sponsoring organization. For convenience, AWS is given as source when possible.
†American Petroleum Institute, 2101 L St., N.W., Washington, D.C. 20037.
‡American Welding Society, 2501 N.W. 7th St., Miami, Fla. 33125.
§The American Society of Mechanical Engineers, Order Department, United Engineering Center, 345 East 47th St., New York, N.Y. 10017.
¶American National Standards Institute, 1430 Broadway, New York, N.Y. 10018.

Codes, Specifications, and Welding Standards* (Continued)

Title	Published by	Field of application	Source
Qualifications			
Qualifications Test for Gas Welders (General Specifications for Inspection of Material, Appendix VII, Welding, Part E, Sec. E-2)	Bureau of Supplies and Accounts, U.S. Dept. of the Navy	All gas welding done for the Navy Dept.	Bureau of Supplies and Accounts, U.S. Dept. of the Navy, Washington, D.C. 20350
Welding Procedure and Performance Qualification, B3.0	AWS,‡ 1977, 97 pp.	Industry, welding instructors, and code-writing bodies wishing to prescribe methods	AWS‡
Welding and Brazing Qualifications, Boiler and Pressure Vessel Code, Sec. IX	ASME,§ 1977	Boilers and pressure vessels	ASME§

*All are available from sponsoring organization. For convenience, AWS is given as source when possible.
†American Petroleum Institute, 2101 L St., N.W., Washington, D.C. 20037.
‡American Welding Society, 2501 N.W. 7th St., Miami, Fla. 33125.
§The American Society of Mechanical Engineers, Order Department, United Engineering Center, 345 East 47th St., New York, N.Y. 10017.
¶American National Standards Institute, 1430 Broadway, New York, N.Y. 10018.

Fig. 6-24 The angle of the nozzle with respect to the work controls the depth of the groove.

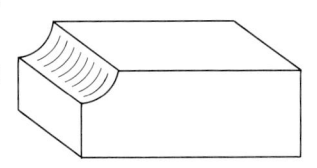

Fig. 6-25 A grooved edge for hard-facing.

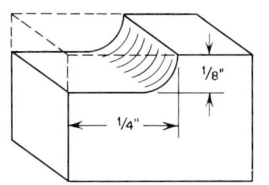

Fig. 6-26 A machined edge for hard-facing.

TABLE 6-5 Characteristics of Hard-Facing Rods

Hard-facing rod	Tensile strength, psi	Hardness on Rockwell C scale	Melting point, °F	Contains
Haynes:				
90	63,000	45–55	2390	Iron, chromium, carbon
92	25,000	64	2012	Iron, molybdenum, carbon
93	43,000	57–62	2225	Iron, chromium, molybdenum, cobalt, vanadium, carbon
94	60,000	50–61	....	Iron, chromium, boron, cobalt carbon
Haynes Stellite:				
1	47,000	46–54	2828	Cobalt
6	105,000	33–44	2327	Chromium
12	76,000	37–47	2306	Tungsten
1016	....	58	....	Cobalt, chromium, tungsten, carbon
Hascrome	40,000	28–43	2500	Iron, chromium, manganese

TABLE 6-6 Characteristics of Hard-Facing Rods

Hard-facing rod	Resistance to			
	Abrasion	Impact	Corrosion	Hot abrasion
Haynes:				
90	A	C	C	NR
92	A	NR	NR	NR
93	A	D	C	NR
94	A	D	D	NR
Haynes Stellite:				
1	A	NR	A	A
6	B	A	A	B
12	A	B	A	A
1016	A	NR	C	A
Hascrome	C	A	D	NR

A, excellent; B, high; C, good; D, fair; NR, not recommended.

badly worn surface with hard facing to a depth greater than 1/8 in., you should use more than one deposit.

4. Hard-facing rods are applied with a carburizing flame—a flame using more acetylene than oxygen. The extra acetylene shows up as a whitish feather around the inner cone. Use the amount of excess acetylene specified by the rod manufacturer.

5. Low- and medium-carbon steels are the most widely used metals and are the easiest to hard-face. The following instructions are for the hard-facing of these steels:

6. If you have selected a rod, prepared the part, and set up your welding outfit, you are ready to start hard facing. Begin by heating the part (see Fig. 6-27).

4-104 Basic Maintenance Technology

7. Now adjust to a carburizing flame. Reduce the amount of oxygen until you have the proper flame, depending on the rod you are using.

8. Hold the carburizing flame over the heated area. The tip of the inner cone should be just off the steel surface—about ⅛ in. as shown. Hold the flame there until the metal under the flame starts to sweat.

9. Next lift the welding blowpipe a little and put the rod into the flame so that it just touches the sweating surface. Lower the blowpipe until the inner cone of the flame just

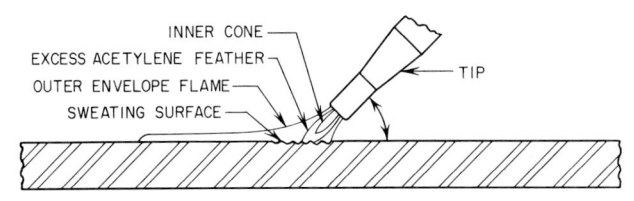

Fig. 6-27 Heating the surface before depositing the hard-facing rod.

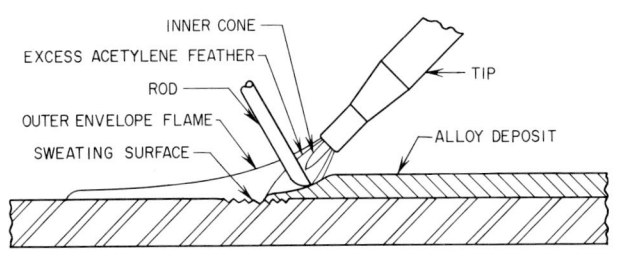

Fig. 6-28 Depositing the hard-facing rod.

touches the rod and is about ⅛ in. from the steel surface, as shown in Fig. 6-28. A small puddle of melted rod will form on the sweating surface. If the first few drips of the melted rod foam or bubble or do not spread evenly, the surface is too cold. Take the rod away and start over again.

10. Next take the rod out of the puddle. Spread the puddle over the sweating surface by pointing the flame into it—do not use the rod to spread it. If there is not enough hard-facing deposit to cover the wearing surface, continue the process.

11. Point the flame so that it touches the forward end of the puddle and the steel surface.

12. When the surface sweats, add more metal to the puddle from the rod. Then, as you did before, remove the rod and spread the puddle with the flame. Repeat until the entire surface is covered.

13. Allow the part to cool slowly to prevent cracks and stresses in the hard face. Small- and medium-sized parts can be cooled in air. Large or bulky parts should be wrapped in asbestos paper or buried in asbestos, slaked lime, wood ashes, or another insulating material until they cool. Parts that are liable to crack should be put in the preheating furnace while they are still hot from hard-facing. Then they should be brought to an even red heat and, with the heat turned off, allowed to cool overnight in the furnace with the door closed.

Hard-facing cast iron

1. Cast iron does not sweat like steel, and it melts at about the same temperature as the rod. *So be careful*—do not melt the base metal too deeply.

2. Use a little less acetylene in the flame than you would for steel.

3. Use cast-iron brazing flux when you apply the rod.

4. A crust will form over the surface of the cast iron when it is heated. To get a good bond, you will have to break the crust with the end of your rod.

5. Very thin cast-iron parts should be backed up with wet asbestos or carbon paste to keep them from melting.

Finishing the hard-facing deposit

1. *Heat treating* of the hard-faced parts is *usually not necessary.* The only time you will heat-treat a part after hard facing is when you want to toughen the base metal. To do that, heat the whole part to a dull red heat. Then dip it in oil. Do not use water for the quench because it may crack the deposit and base metal.

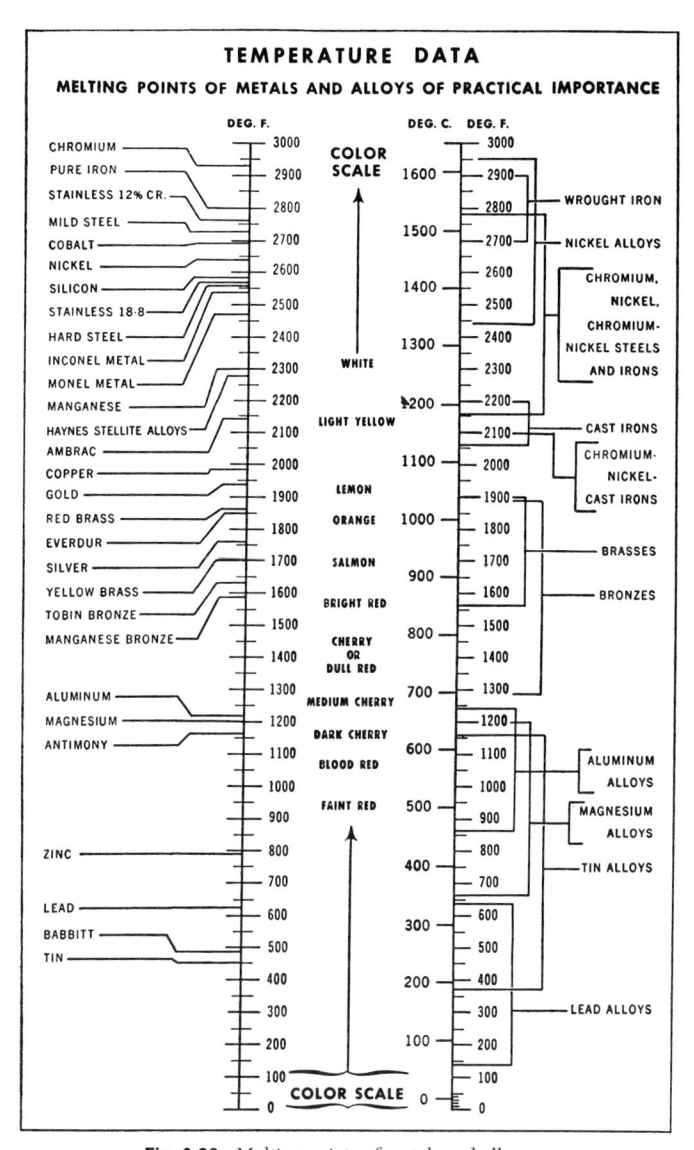

Fig. 6-29 Melting points of metals and alloys.

2. *Surface cracks* are usually caused by insufficient preheat or by cooling the part too quickly. You will find, however, that a surface crack will not harm the properties of the hard face or the strength of the part. If you want to repair a cracked surface

   a. Preheat the piece as for hard facing.
   b. Heat the metal around the crack to a dull red.
   c. Then melt the edges down into the crack.
   d. Add a little metal from the rod.
   e. Now slowly move the flame away from the hot spot to prevent quick cooling.

3. You can grind a hard-faced part to exact size or remove high spots on the surface. Use a grinding wheel not coarser than 46 or finer than 60 in Grade I or J of the Norton system. The speed of the wheel should be between 2800 and 4200 sfpm. Higher speeds might crack the hard-face surface.

See Fig. 6-29 for melting points of metals and alloys.

REFERENCES

"The Oxy-Acetylene Handbook," Union Carbide Corporation, Linde Division, 3d ed., 1976.
Linnert, G. E.: "Welding Metallurgy," American Welding Society, 3d ed., vol. 1, *Fundamentals*, 1965; vol. 2, *Technology*, 1967.
"Brazing Alcoa Aluminum," Aluminum Company of America, 1967.
"Welding Alcoa Aluminum," Aluminum Company of America, 1972.
"Stellite Hard-Facing Products," Stellite Division, Cabot Corporation, 1975.
Jefferson, T. B.: "The Welding Encyclopedia," Jefferson Publications, 17th ed., 1974.
"Welding Handbook," American Welding Society, 7th ed., vol. 1, *Fundamentals of Welding*, 373 pp.; vol. 2, *Welding Processes: Arc and Gas Welding and Cutting, Brazing, and Soldering*, 570 pp.

Chapter **7**

Metal Resurfacing

K. N. KELLEY[1]
Manager, Marketing Communications, Stoody Company, Industry, Calif.

Construction equipment is subjected to constant battering from wear-inducing forces to a degree perhaps unequaled in any other industry. Minimizing or circumventing these forces through wear-avoidance practices ranging from design factors through preventive maintenance is of primary concern to both the design and the maintenance engineer.

Yet it is not possible to insulate many working parts from the destructive influences which cause metal wear. Recourse in these cases is to increase the inherent wear resistance of the part as much as possible. Manufacturers have a number of options at their disposal, including a variety of hardening and coating techniques. At the maintenance level the most widely used and effective procedure is surfacing by welding processes.

INTRODUCTION

These introductory paragraphs will briefly review several techniques for increasing wear resistance that are commonly employed in the manufacture of metal parts. Note that these procedures are usually not feasible in the typical maintenance shop nor are they suitable for restoring parts where significant amounts of wear have already occurred; the remainder of this chapter will deal more fully with the subject of rebuilding and hard-facing as practiced generally in construction machinery maintenance shops.

Manufacturing techniques include through-hardening by conventional heat-treating methods and the following surface-modification procedures:

Flame hardening The flame-hardening process hardens the surface or specific portions of a hardenable steel part. It is a selective heat treatment accomplished by the direct impingement of a flame on the surface, followed by an air or water quench. The chemistry of the surface is not changed; wear resistance is enhanced simply by increased hardness resulting from a metallurgical transformation.

Hardness can be increased to ¼ in. in depth. Control factors in flame hardening are of critical importance since unequal stress leading to fracture or warpage can easily be generated; however, some maintenance shops have mastered flame-hardening techniques for specialized applications.

Diffusion alloying Carburizing and nitriding are well-known examples of a procedure for surface chemistry modification called *diffusion alloying*. Alloying of the surface with carbon or nitrogen is accomplished in the presence of a reactive gas or salt at relatively high temperature. Increased wear resistance is a result of a carbide or nitride constituent at the surface—usually about 1 percent—in combination with elements such as chromium or molybdenum which may be present in the base metal.

Carburizing and nitriding are highly suitable processes for the mass production of small

---

[1]The author is now retired.

parts, and they are reasonably inexpensive. However, the treatment is essentially superficial, seldom exceeding 0.020 to 0.025 in. in depth, and its benefits are limited in service under normal loads to resisting the effects of sliding or rolling friction.

Other examples of diffusion alloying include the addition in a similar manner of chromium or aluminum to steel surfaces. The presence of chromium in the first few thousandths of an inch of the surface will increase wear resistance, provided sufficient carbon is present in the base metal.

Hard chrome plating Hard chrome plating is an effective means of protecting steel parts from both abrasive wear and corrosion, provided the allowable limits of wear do not exceed the depth of the relatively thin layer deposited. The process is also frequently used to add stock to parts which have been mismachined undersize.

Hard chrome plating is an electrolytic procedure generally available from specialty shops. It differs from decorative plating chiefly in that the deposited surface layer is essentially pure chromium metal as opposed to the copper-nickel-chrome combinations normally used for decorative work. Substantially greater amounts of buildup are feasible in hard chrome plating, up to as much as 0.040 in., or even more. However, because the deposit is relatively brittle and subject to fracture or peeling under certain types of stress loading, thickness is normally limited to 0.020 in. or less.

Unlike decorative plating, hard-chroming requires finishing by grinding and therefore the deposit is made oversize to provide grinding stock. The thickness of the chromium layer is quite uniform except at sharp corners such as the crest of threads where excessive buildup can occur; platers use special techniques to equalize thickness at such points when this is necessary.

WELDED OVERLAYS

The process of depositing a layer of metal on a steel part by welding is commonly called *surfacing*. When the weld metal is added simply for the purpose of restoring a worn part to its original configuration, the term correctly applied is *rebuilding*. Hard-facing (or hard-surfacing) is defined as the addition of an overlay of wear-resistant alloy to worn, partially rebuilt or new parts.

Rebuilding and hard-facing is a distinct branch of the welding art. It had its origins in the early 1920s and from the beginning has been closely associated with the construction industry. It is probable that the total usage of welding alloys to protect and restore equipment used in earth-moving, digging, and crushing operations exceeds that of all other industries combined.

While the basic equipment, processes, and skills required for weld-surfacing are essentially the same as those used in metal-joining, the materials and many of the techniques are quite specialized. A competent welder will have no difficulty in adapting his skills to the requirements of rebuilding and hard-surfacing; however, in the choice of alloy and procedures, a thorough understanding of the basic principles involved is essential (Fig. 7-1).

FUNCTIONS OF REBUILDING AND HARD-FACING

While rebuilding and hard-facing are typically classified as maintenance functions, the benefits of the latter process often clearly surpass those that might be expected of a simple maintenance procedure, although this is often not fully understood. By extending part life or by modifying or improving part performance, hard-facing can result in economic rewards considerably greater than those related merely to the salvage of worn parts.

Obviously, the reduction of replacement part costs through the restoration of used parts is the principal goal of rebuilding applications where the deposited metal has a wear resistance approximately equal to that of the original part. Here the economic criteria for the application simply involve balancing the cost of rebuilding versus the replacement cost. Frequently the cost saving is significant, especially where large parts are no longer usable after relatively small amounts of metal loss have occurred or where the rebuilding procedure can be automated for minimum application cost.

Where the use of hard-facing, which typically may extend the wear life of parts from 3 to 10 times, is feasible, the added economic advantages become very significant. First, hard-faced parts can generally be refaced and put back in service, compounding the initial cost benefit. Further, by extending part life hard-facing can eliminate one or more downtime periods. In some applications the cost of unscheduled equipment shutdowns for part replacement or repair can be extremely costly.

Because hard-faced parts maintain their original size and shape longer, equipment operates with greater efficiency. The economic benefit which can be gained with increased production from crusher hammers or rolls that stay close to size may sometimes totally overshadow any other cost considerations which might be associated with a decision concerning hard-facing as opposed to replacement.

The performance of parts with a cutting edge, such as shovel teeth, blades, or tillage-type tools, can be enhanced by a technique which takes advantage of the bimetallic nature of hard-faced surfaces. By placing the alloy on only one side of the cutting edge, selective

Fig. 7-1 Typical hard-facing activity shows group 3 alloy weld-cast into lengths of pipe driven into worn sheepsfoot tamps.

wear is encouraged; as the unprotected side wears away, a fresh sharp edge of hard-facing material is continuously exposed.

Finally, hard-facing can be used to modify the surface of parts in other ways which may be desirable, and the nature of the alloys is such that the surface will tend to retain the modified contours in service. For example, weld beads running across the surface of a roll can improve its gripping action; beads running parallel to material flow can help a part slide more freely. Tungsten carbide alloys, which produce a heterogeneous deposit, can be used to provide a nonslip surface or a serrated edge to improve the cutting action of various parts.

REBUILDING AND HARD-FACING PROCESSES

Almost all the welding methods commonly used for joining metal parts are also used in hard-facing and rebuilding operations. These include manual gas and electric arc, semiautomatic and fully automatic processes. Thermal spraying (flame spraying and plasma arc) is primarily a surfacing process.

The selection of the most appropriate method is limited to some extent by the equipment available but is primarily dictated by the requirements of the application. Labor costs, the type of alloy to be used and its cost, and part configuration must all be considered in making this decision.

Oxyacetylene welding Oxyacetylene or gas welding is the slowest and therefore often the most costly method of depositing overlay material. It is, however, a preferred method for certain types of parts and for one important group of hard-facing alloys (tungsten carbide).

Use of the torch permits welding on thin-edge tools without burning through and is sometimes the only practical method for parts of this type. Gas welding also is used when dilution of the overlay alloy with the base metal must be held to a minimum. Finally, it is a desirable method for depositing tungsten-carbide-type rods since the low heat input and fast-freezing puddle minimize the tendency of the carbide particles to dissolve in the molten steel.

Flame spraying, which utilizes alloy wires or powders instead of welding rods, is essentially a variation of oxyacetylene welding and offers the same advantages but is limited in both deposit thickness and the variety of alloys available.

Manual arc welding Manual electric-arc welding is still the most widely used method for rebuilding and hard-facing applications (although the semiautomatic process is increasingly popular, especially in the construction industry). Almost all types of the commonly specified alloys are available as coated electrodes, including tungsten carbide and nonferrous cast rods.

Manual arc welding is relatively fast and many of the electrodes available are quite inexpensive. It is an excellent procedure for general-purpose use on a variety of parts.

Semiautomatic welding Semiautomatic welding is a superior choice for a wide variety of applications where part size and shape will permit its use. Deposition rate is high and the process utilizes continuous wires which eliminate the waste of stub ends and lost time for electrode changes.

Semiautomatic welding equipment is commonly available. Both open-arc and gas-shielded arc (GMAW) methods are used but open-arc is common for hard-facing because most of the materials do not require external shielding.

Automatic welding Automatic welding equipment, which eliminates the need for manual control of the weld bead, is available in considerable variety; several types are especially designed for construction equipment—specifically for tractor undercarriage components and for various types of crushers.

Automatic systems facilitate the use of the submerged-arc welding method wherein the molten weld puddle is shielded from the atmosphere by a layer of granular flux. This makes possible the use of relatively low alloy and inexpensive materials which are not suitable for open-arc welding. The equipment will also handle open-arc wires and may be adapted to gas shielding where required.

Conventional machines are most commonly used for cylindrical parts which are rotated in a positioner to advance the weld bead. Equipment is also available for flatwork where the welding head travels while the work remains stationary. General-purpose equipment, which can manipulate the arc in two directions to provide a variety of overlay patterns on flat surfaces, has been introduced in recent years and is especially suited to maintenance shop applications.

The advantages of automatic welding are high deposition rates and precise control of deposit appearance and chemistry. Its use is limited, however, to applications where part size and shape are appropriate, where the deposit can be programmed in advance and, except for several applications using specialized equipment, to instances where the work can be brought to the equipment.

SELECTION OF REBUILDING AND HARD-FACING ALLOYS

A large variety of rebuilding and hard-facing alloys are commonly available. To select an alloy for a specific application it is necessary to recognize the wear factors present and the way in which they operate under the particular service conditions involved.

Types of wear Materials engineers recognize many types of wear. For a practical understanding of wear as it relates to the selection of surfacing materials the following five broad categories will be useful:

1. *Abrasion.* Abrasion is defined as the grinding or scratching action of hard particles, such as sand or rocks, rubbing or sliding against a surface. When combined with heavy loading the condition is often called gouging.

2. *Impact.* Surface deterioration which results from a blow or series of blows is classified as impact-type wear. It can take the form of deformation, fracture, or spalling (spalling means that the surface peels or breaks off in pieces).

3. *Corrosion.* Corrosion is a type of wear associated with chemical or electrochemical attack; rust is a common example. Corrosion problems can often be anticipated and avoided by the use of appropriate materials or by painting, plating, or rubber-coating steel

parts, but where corrosive attack is combined with other wear factors such as abrasion the use of special wear and corrosion-resistant alloys is indicated.

4. *Heat.* Heat affects wear resistance by softening metal surfaces, making them more susceptible to abrasive and corrosive attack. Where heat is the only problem it can be dealt with, as in the case of corrosion, by more direct means than hard-facing. When combined with abrasion, the use of specialized hard-facing alloys is required.

5. *Stress-related wear.* Stress related wear is a result of metal fatigue. Adhesion, or galling, between two metallic surfaces is an example. Fretting, caused by vibration between parts, is another type of wear generated by stress.

Most wear problems encountered under actual service conditions involve more than one of these factors. A careful analysis should be made of the degree to which each type of wear is present in a particular application, and the selection of material should be predicated on that determination.

Wear typically starts at a specific location on a part and its progressive effects may be observed throughout the part's service life. A study of this process will provide useful guidance in establishing hard-facing procedures and material choices. Sometimes a small amount of alloy applied at the point where wear begins will effectively deter its further progress. In many instances a combination of two or more surfacing materials is indicated, depending on the severity of the wear attack on various areas of the part.

Types of rebuilding and hard-facing alloys Despite occasional claims to the contrary there is no universal hard-facing alloy. The many products on the market are each designed to satisfy certain specific and varied requirements of the many applications for which they are used. Cost, weldability, soundness of deposit—in addition to the basic requirement of resisting one or more of the various wear types—are all of concern to the user. The enhancement of one feature is generally a trade-off accomplished at the expense of another.

Therefore, the choice of the correct material for a particular job is a matter of considerable importance and is as vital to the success of the application as the skill of the welder. The recommendations provided by the various manufacturers are of assistance in making the most advantageous choice, but they are sometimes quite contradictory. It will be helpful if the basic principles of the metallurgy of wear-resistant materials and the categories into which they can be grouped are understood.

Almost any hard material, if it is compatible with the base metal, can be used for surfacing and will provide some added wear resistance; cast iron and certain tool steels are examples. But most alloys classified as hard-facing materials have one characteristic in common: Their microstructure consists of a network of metal carbides evenly distributed throughout a matrix of softer material. These carbides are precipitated as the molten metal in the weld deposit cools. They are much harder than steel and harder than the apparent hardness of the deposit itself. They carry the wear load and, as the matrix wears away between them, fresh surfaces are constantly exposed. Thus a hard-facing deposit with an apparent hardness of less than 50 Rockwell C will outwear hardened steel with a hardness over 60 Rockwell C; the hardness reading reflects the matrix hardness primarily and the carbides to only a small extent.

Hard-facing alloys typically consist of iron—or in some cases cobalt or nickel—with varying amounts of carbon plus chromium, tungsten, molybdenum, or other carbide-forming elements added. As a rule, more carbides are formed and the abrasion resistance of the weld deposit is increased as the percentage of these added elements is increased.

Alloys used for rebuilding and hard-facing can be conveniently classified into five major groups. The appropriate category into which a particular alloy should be placed can usually be determined from the manufacturer's literature, and the material's classification provides guidelines for its use.

Group 1. Build-Up Materials. The alloys in this category are not hard-facing materials but are important in the rebuilding of worn parts. They are designed to provide a tough underbase for hard-facing deposits in areas where severe wear has occurred.

Materials used for rebuilding carbon-steel parts contain up to 6 percent alloys (other than iron). They combine strength and toughness, can be deposited crack-free in multiple layers, and are usually readily machinable. The value of these alloys as an underbase is that they will not readily deform under the more brittle hard-facing layers. This can occur if ordinary steel welding rods are used for buildup, with the result that the hard-facing cracks and spalls.

These materials are often used as the final overlay on parts subject to impact, metal-to-metal sliding or rolling—shafts, gears, and tractor rails are typical examples—or where a machined finish is required. Also included in this category are materials designed for buildup on austenitic manganese steel parts (also called Hadfield's manganese or, simply, manganese). These alloys typically contain about 14 percent manganese with smaller amounts of nickel or molybdenum and sometimes chromium.

Like the base metal itself, manganese alloys are tough and work-harden in service under impact. They are used both for buildup prior to hard-facing and as an overlay on parts subject to impact such as shovel bucket and rock-crusher components.

Group 2. Low-Alloy Ferrous Materials. Most authorities classify iron-base materials up to approximately 20 percent alloy content as low alloy. They have greater shock resistance but less abrasion resistance than materials with alloy contents above 20 percent.

Alloys in this group form relatively few, if any, carbides but owe their wear resistance to the special characteristics of their various chemistries. Their properties are inherent, as deposited, and not developed as a result of heat treatment as in the case of machinery-grade steels. Group 2 alloys are appropriate for hard-facing nonlubricated metal-to-metal rolling or sliding parts and parts subject to considerable impact and low-to-moderate abrasion. Examples include sprockets and drive tumblers plus a variety of machinery parts.

Low-alloy ferrous materials are generally the least expensive hard-facing materials and are commonly available as manual (stick) electrodes, often with iron-powder coatings for high deposition rates. They are also widely used in wire form in submerged-arc automatic applications on parts such as tractor rolls but are seldom applied by the oxyacetylene method.

Group 3. High-Alloy Ferrous Materials. These are the most common hard-facing alloys, containing over 20 percent alloying elements. They form carbides and, as a rule, the higher the alloy content within this group, the greater the abrasion resistance and the lower the impact strength.

High-alloy deposits typically cross-check upon cooling; cross-checks are normal and generally desirable. Deposits are usually not machinable but can be ground if necessary—though sometimes with difficulty. In most applications, deposits should be limited to two layers. Where more buildup is required to reach finish size, it is advisable to rebuild to within two layers with a group 1 material.

Applications for group 3 alloys include those involving wear from abrasion by rocks, sand, ore, and cement and bulk materials which contain dirt or other gritty particles. Typical parts hard-faced with these materials are ripper teeth, dozer end bits, shovel teeth and buckets, draglines, conveyor parts, buckets and mixer paddles. Group 3 alloys are available as bare rods, coated electrodes, and wires for semiautomatic and fully automatic applications (Figs 7-2 and 7-3).

Group 4. Tungsten-Carbide Materials. This special group includes the most wear-resistant hard-facing alloys. These are steel tubes which contain crushed particles of tungsten carbide in various mesh-size ranges. In depositing these materials, only the matrix material is melted; the carbide particles are not (they will, however, *dissolve* in the molten weld puddle and care must be taken to minimize this effect). The method by which the deposit resists wear is the same as that of the group 3 alloys, but in this case the carbides are of visible size (ranging from a fine powder to almost ⅛ in. in diameter) and are pure tungsten carbide, substantially harder than the microscopic carbides formed as a group 3 alloy deposit cools. Tungsten-carbide deposits are unique in that the deposit is itself abrasive as well as abrasion-resistant. Deposits of all but the finest mesh sizes are sandpaperlike on the surface. They can provide a nonslip surface that won't wear out and can improve the cutting action of some parts such as posthole augers. Group 4 alloys provide maximum abrasion resistance for parts in contact with the earth, including digging tools and most tillage implements.

Mesh-size ranges (these are expressed as 5–8, 8–10, 10–20 and so on—the higher the number, the smaller the particle size) are determined by the application. Deposits made with the larger particles provide better cutting action and more impact strength; deposits made with the smaller particles are more wear-resistant. If the application involves abrasive materials which are very fine, tungsten-carbide materials may be unsuitable since it is possible to cut between the carbide particles and undermine them. Group 4 alloys should *never* be used on metal-to-metal applications.

Tungsten-carbide materials are available as bare rods, coated electrodes, wires, and powders. They are expensive, so the economics of the application must be carefully analyzed, but for jobs where extreme abrasion is present they are often the best possible choice.

Group 5. Nonferrous Materials. The nonferrous alloys are those in which iron is not a principal ingredient. They are most often cobalt base or nickel base and, like the group 3 alloys, precipitate carbides in the weld deposit which are responsible for their wear resistance.

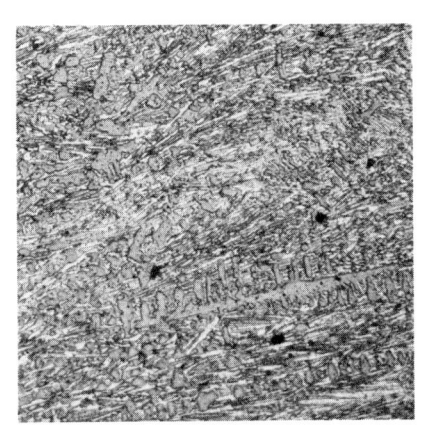

Fig. 7-2 Micrograph of medium-alloy group 3 material shows small eutectic carbides (white areas).

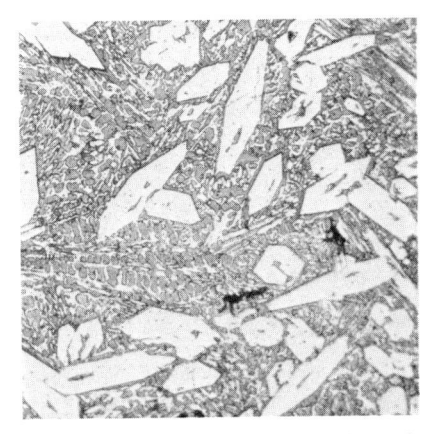

Fig. 7-3 High-alloy group 3 material reveals both primary (large white areas) and eutectic carbides.

Cobalt-base alloys are extremely versatile hard-facing materials. Depending on the amount of carbide-forming elements contained (principally, chromium and tungsten, with silicon and sometimes nickel, boron, or molybdenum), the abrasion resistance of the various grades is comparable with the group 3 alloys from the low to the high end of the scale. Deposits are resistant to corrosion and retain useful hardness at temperatures up to 1500°F. Cobalt-base alloys are resistant to galling and are excellent for metal-to-metal applications such as bearings, seals, and valves.

The nickel-base alloys are similar to the cobalt base and are useful in certain corrosion situations; no single alloy or class of alloys is resistant to all corrosive media. Nickel-base alloys are the predominant type used for thermal spraying.

Nonferrous hard-facing alloys are relatively expensive (though not as expensive as the group 4 alloys) and principally available as bare or coated rods or powders. Several types can be machined with carbide tools; the harder ones must be ground if a finish is required.

Not all materials sold for wear resistance will fit precisely into the groups above. There are copper aluminum and copper nickel brazing alloys with good wear resistance which are used for bearings and similar parts. Certain tool steel rods are quite similar to group 2 alloys. High-carbon 420 stainless steel is a useful hard-facing material in some applications. The best guidelines for selecting the right material are those outlined in the preceding paragraphs, plus manufacturers' recommendations and experience gained by observing the performance of parts that have been rebuilt.

Forms of rebuilding and hard-facing alloys Rebuilding and hard-facing alloys are available as bare and coated rods, continuous wires, and powders. They often differ in form from conventional welding materials. Because they are inherently hard they cannot be pulled through drawing dies to form wire, as is ordinary carbon steel; therefore, manufacturers use several other methods to produce them. One is to melt the alloy in a furnace and cast it into rod form. Cobalt- and nickel-base alloys are commonly made in this manner, as are several high-alloy iron-base products.

Low- to medium-alloy iron-base electrodes are often produced by adding the carbon and alloy requirements to the coating applied to a carbon steel core. The coating thus serves a dual purpose, shielding the arc and contributing to the analysis of the weld deposit.

Many hard-facing electrodes and almost all hard-facing wires for semiautomatic and automatic application are produced by a method called *fabrication*. Here the alloying ingredients are contained in a tube formed of carbon steel, often with shielding or fluxing materials included. As with the alloy-coated electrodes, the welder actually creates the alloy during the welding process.

REBUILDING AND HARD-FACING METHODS

Base metals In all types of welding, it is important to be aware of the chemical composition of the base metal. This is essential in hard-facing because parts intended for severe service are frequently made of high-carbon or -alloy steels, cast iron, or Hadfield's manganese; these can present welding problems.

Cast iron is extremely crack-sensitive and not all cast-iron parts can be hard-faced. High preheat temperatures and slow cooling are necessary and the deposit should be peened to help relieve stresses. Many hard-facing alloys are not recommended for cast iron; manufacturers' literature will usually specify those which are.

Since austenitic manganese steel is widely used for construction machinery parts, it is most important to identify this material which is significantly affected by welding temperatures. Hadfield's manganese steel derives its unique toughness and work-hardening characteristics not only from its composition—12 to 14 percent manganese and 1.2 to 1.4 percent carbon—but also from a rapid water quench from 1800 to 2000°F which makes it soft and ductile. The manganese-alloy electrodes used for rebuilding this material contain additional alloying elements which compensate for the fact that the weld deposit will not have the benefit of this special processing, but the properties of the base metal itself can be damaged or destroyed if sufficient heat input occurs during welding to alter the effects of the original heat treatment.

In addition, intermediate alloys formed from a mixture of Hadfield's manganese and ordinary carbon- or low-alloy steels generally have extremely poor physical properties. These will occur in the fusion zone between the deposit and the base metal. In this case and in the case of excessive heat input, spalling of the welded overlay will occur in service.

Austenitic manganese steel can be identified with a pocket magnet. Parts are nonmagnetic when new, weakly magnetic after work hardening. Heat should always be held to a minimum when welding this base metal, and only an alloy recommended for manganese welding should be used.

An ideal base metal for hard-facing is a medium-carbon steel, 0.30 to 0.50 percent. Manufacturers who design parts with hard-facing in mind frequently use medium-carbon steel and equipment owners ordering replacement parts often specify this base metal to facilitate rebuilding.

Preparation for welding Surface preparation is especially important in rebuilding and hard-facing because worn parts may be contaminated with a variety of foreign substances. Dirt, oil, and rust should be removed, as failure to clean the surface to be welded can result in porosity in the deposit and spalling.

Deep cracks should be gouged out to their full depth with a torch by grinding or machining and the metal replaced with a compatible welding material. Fatigued or rolled-over metal should also be removed. Heat-checks (alligator hide) are not necessarily harmful. Stress cracks, however, which run in fairly straight lines at right angles to the direction of the load, should be repaired.

When rebuilding parts which have been previously hard-faced, the alloy applied should be compatible with the previous deposit. If some spalling of the old deposit has occurred, make sure that portions of the remainder have not loosened. Tap the part with a hammer to locate these areas, and remove them. If only small amounts of the old hard-facing deposit remain and if they are sound, it's fairly safe to proceed with reapplication. If more than 20 percent of the original thickness remains, a good rule is to remove it or return the part to further service until more has worn away.

Preheat Preheat is an important element in many rebuilding and hard-facing jobs. It is necessary in some instances to avoid damage to the base metal or distortion of the part. Inadequate preheat can also result in deposits that will crack or spall.

As the carbon content of carbon-steel parts increases, the preheat requirements increase. Alloy steels also generally need preheat in proportion to the amount of alloy present; chrome nickel stainless steel alloys are an exception. Cast iron requires a very high preheat, preferably to a dull red color. Preheating of austenitic manganese steel should generally be avoided, although in very cold weather parts may be warmed to no more than 200°F.

Temperature-indicating crayons may be used to determine when proper preheat temperatures are reached. The manufacturers of these materials provide charts which show the precise temperature ranges recommended for welding various base metals. When parts require preheat, they should be held at the specified temperature a sufficient length of time for it to reach the core. The part should be held at or near this temperature throughout the welding process. On large parts the use of a supplementary heating torch may be necessary. All parts that require preheat should be slow-cooled.

The use of some hard-facing materials makes it necessary to preheat parts even when the base-metal composition does not. Metals expand when heated and contract as they cool. When a weld bead is first deposited it is fully expanded; the base metal, if not adequately preheated, is not. As the weld cools it will contract and must either stretch, crack, or distort the part. Most hard-facing alloys, especially those in groups 3 and 5, have almost no ductility (ability to stretch); so preheat is especially important. When adequate preheat cannot be accomplished—because the part is austenitic manganese, because it is too large, or because time or equipment is not available—special attention must be given to the hard-facing alloy selected and to welding procedures.

Fig. 7-4 Deposit cross sections (magnified) compare penetration of straight versus reversed polarity.

Welding procedures Rebuilding and hard-facing are fusion welding processes. Fusion between the deposit and the base metal is of concern regardless of the method of welding used because it affects both the soundness of the bond and the chemistry of the deposit. Insufficient penetration can result in spalling; too much will diminish the wear-resistant properties of the overlay because of dilution. Hard-facing alloys are designed to tolerate some dilution with the base metal, but penetration should be held to the minimum amount consistent with a strong bond (Fig. 7-4).

Oxyacetylene Hard-Facing. Oxyacetylene hard-facing is most suitable for thin-edge tools and for applications where dissolving tungsten-carbide particles or dilution must be held to a minimum. It is not recommended for cast iron or large parts where a high preheat is required.

Surface impurities will adversely affect oxyacetylene deposits, as will oxides formed during welding. Therefore it is necessary to clean worn parts thoroughly by grinding, and it is advisable to lightly grind new parts prior to surfacing. The area ground should be slightly larger than the area to be surfaced. The part should be positioned for downhand welding and a jig should be used if it will have to be turned during the surfacing operation. If preheating is required, it is helpful to group small parts to simplify this operation.

The diameter of the rod used is determined by the size and thickness of the part; small-diameter rods are required for delicate work but larger diameters will reduce both time and material costs. A tip size about three times larger than normally used for mild-steel welding with the same-diameter rod is recommended.

A carburizing flame should be used to reduce oxides; it will also add carbon to the surface of the part which lowers its melting point and aids fusion. A 3X flame (the feather is three times as long as the inner cone) is recommended for cobalt-base alloys and fine-mesh tungsten-carbide rods. A 4X flame is used for larger-mesh tungsten-carbide rods and most iron-base hard-facing alloys (Fig. 7-5). Nickel-base alloys are an exception to the rule; most should be applied with a neutral flame.

The entire part should be preheated with the torch and then the flame should be concentrated on the area to be hard-faced until it reaches red heat. The flame should then be directed on the point where the deposit will start, until the surface appears shiny and

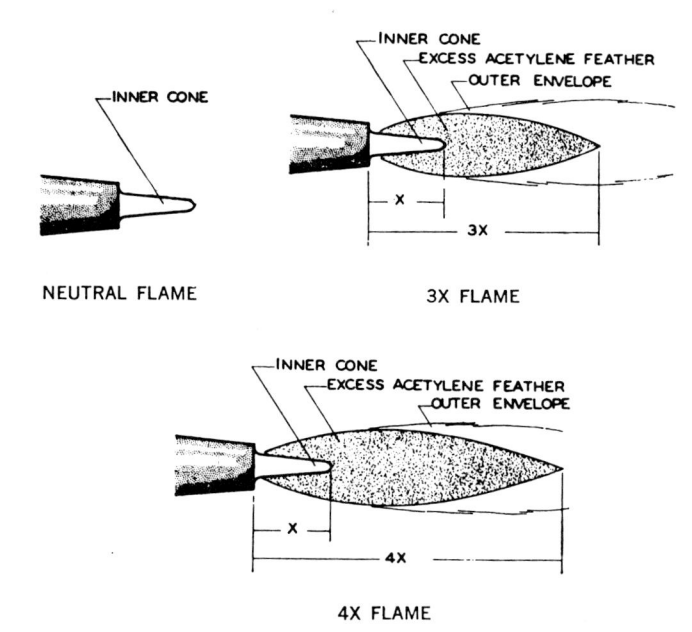

Fig. 7-5 Comparison of 3X and 4X flames with neutral flame.

watery. This is the sweating temperature—it indicates that the surface is beginning to melt. The tip of the rod should be moved under the edge of the flame so that a drop will melt and fall into the sweating zone. As the torch is moved forward, the drop will spread and follow the heat of the flame. As more alloy is added in this manner, the heat should be directed more at the tip of the rod than on the work, but enough heat should be kept on the base metal that it remains at sweating temperature ahead of the alloy puddle.

Different alloys vary substantially in their ability to flow or to stack, and either characteristic may be desirable in a particular application. The skillful oxyacetylene welder, however, can exert considerable control over penetration, deposition rate, and the shape of the deposit.

Manual Arc Hard-Facing. Penetration is of primary concern in arc welding because it's easy to overdo it. It is generally desirable to hold penetration to the minimum amount that will ensure a sound bond. The use of straight polarity will result in less penetration than reverse polarity; however, reverse polarity must be used when recommended for a particular electrode. The lowest possible amperage will minimize penetration, but amperage should not be less than the lowest recommended for the electrode. The use of the largest-diameter electrode compatible with the part will also minimize penetration.

The arc should be directed toward the pool of molten alloy rather than the base metal to limit penetration. A tight overlap of adjacent stringer beads will significantly reduce penetration, but don't exceed 50 percent—too much overlap can reduce penetration to a point where the bond is inadequate and spalling will occur.

The use of two layers is the most effective way to reduce the effect of dilution and is recommended for all alloys except those in group 4 (tungsten carbide) which should be limited to a single pass. However, no *more* than two passes are recommended for most group 3 and group 5 alloys.

Stringer beads, weave beads (Fig. 7-6) and horseshoe-shaped beads are all used in manual arc surfacing. In general, an oscillated bead is less likely to spall than stringer beads, especially when depositing hard-facing alloys over areas rebuilt with group 1 alloys. The manufacturer's literature for a particular rod will usually suggest the best technique.

Semiautomatic Hard-Facing. The semiautomatic method is increasingly popular for rebuilding and hard-facing, particularly in the construction field where large amounts of metal must be deposited on large parts. Most of the rules for manual electric surfacing apply to semiautomatic operation.

Wires for semiautomatic hard-facing are of the tubular (fabricated) type. Because they will crush under pressure, special care must be given to the feed mechanism of the semiautomatic machine. Feed rolls must be carefully aligned and, if the unit is not already so equipped, should be replaced with a special geared type which grips without excessive pressure. Most semiautomatic hard-facing wires run open arc. Welding should be done in the flat (downhand) position whenever possible. Some wires can be welded vertically or semivertically if necessary.

The manufacturer's recommendations for voltage and amperage settings should be followed; most wires operate dc reverse polarity within a range of 200 to 400 amp. The shortest possible arc is usually recommended to minimize spatter.

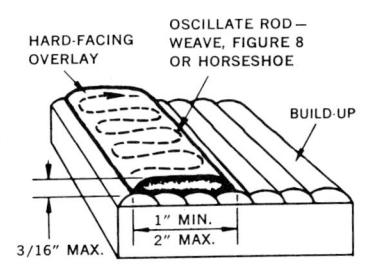

- Use weaving bead instead of stringer bead when applying hard-facing.
- Limit single pass bead thickness to 3/16".
- Use same technique for second layer.
- Avoid severe quench.

Fig. 7-6 Proper process for applying hard-face overlay.

Automatic Rebuilding and Hard-Facing. A great deal of construction equipment rebuilding and hard-facing is done on automatic welding machines, in dealer shops, job shops, some large contractor-owner maintenance shops and, with specialized equipment designed for specific applications, in the field. Where a considerable amount of welding is to be done on a regular basis on a particular part, the investment in automatic equipment is worthy of consideration.

In addition to the usual welding considerations of voltage, amperage, and polarity, the operator of automatic equipment is concerned with such variables as travel speed, stepover, wire extension, and, when welding round parts, the "lead." Each of these factors is important to the success of the application.

Wire manufacturers' recommendations should be heeded with regard to the welding parameters. In general, amperage requirements are in the range from 300 to 400 amp, though some wires can be deposited at 500 amp or more; best results will usually be obtained at the low end of the recommended range. Voltage influences bead shape because the width will increase as the voltage increases; the normal range for most wires is 28 to 32 volts. Straight polarity results in less dilution, higher hardness, and faster deposition; reverse polarity yields a sounder and smoother deposit.

Once the welding parameters have been set, the speed at which the arc travels is a primary determinant of bead shape. About 20 to 30 in./min is normal for most wires. When welding rotating round parts, the lead—the distance between top center and the point at which the arc contacts the work—also influences bead shape. It is normally set at 1½ to 2 in., depending on part diameter.

As in manual arc welding, step-over affects penetration. A tighter step-over will reduce penetration, but step-over should not exceed 50 percent.

Wire extension—the distance between the contact tip and the work—is normally set at 1½ to 2 in. Deposition rate can sometimes be increased by increasing the extension but this can also result in some loss of precision in the placement of the weld bead.

Automatic welding machines designed for such parts as tractor rolls and rails, which are rebuilt with low-alloy materials, are equipped with flux hoppers which supply a constant

4-118 Basic Maintenance Technology

layer of neutral submerging flux to the surface of the part for arc shielding. Machines designed for mounting directly onto roll or impact crushers use small-diameter high-alloy wires and will operate open arc (Figs. 7-7 and 7-8).

Thermal Spraying. Thermal spraying is used not only for hard-facing but for a variety of surfacing operations including the buildup and repair of flawed or mismachined parts. Several distinct processes are included in this classification. Many of the alloys are

Fig. 7-7 Automatic impactor-bar rebuilding machine rebuilds parts in place. Similar equipment is designed for roll crushers.

actually brazing materials; the procedures used produce an overlay with a bond that ranges from surface adhesion to true fusion (Fig. 7-9).

In flame spraying, the alloy in powder form is injected into the gas stream of an oxyacetylene torch from a hopper connected to the torch. The flame is directed on the work and the particles pass through it and adhere to the surface. This type of spraying is often done with a mounted gun on round parts in a lathe. The as-sprayed deposit is

Fig. 7-8 Submerged-arc automatic welding system is designed specifically for rebuilding tractor undercarriage rollers.

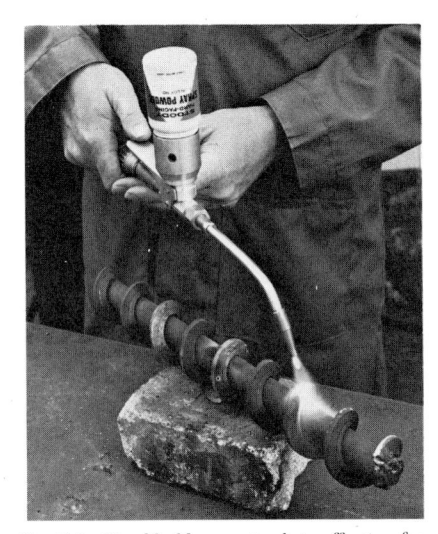

Fig. 7-9 Hand-held spray torch is effective for thin deposits on small parts.

essentially held by a mechanical bond, and it is then fused with the torch or in a furnace. Hand-held spray torches, with a powder hopper mounted on the torch, are commonly used for small or thin parts. Here the procedure is very similar to oxyacetylene welding although fusion with the base metal is minimal.

Most spray powders used for manual torch application are nickel base and should be applied with a neutral flame (manufacturer's instructions should be followed for other types). The torch should be adjusted for a slightly reducing flame; it will change toward neutral when the trigger is pulled to inject the powder. Preheat the part lightly with the torch 1½ to 3 in. away before starting to apply the powder. When it reaches sweating temperature and appears shiny, the trigger should be pulled. The powder should flow on and fuse. It may be necessary, from time to time, to release the trigger and fuse the overlay with the torch.

Fig. 7-10 A one-bead-high, 1 in. sq. waffle pattern on a crusher roll produces 7/16-in. chips.

In torch spraying and all thermal spraying, parts must be very clean—free of oxides, grease, moisture, or other contaminants. Spray parts as soon as possible after cleaning; even an 8-hr delay can allow an undesirable surface film to build up.

Plasma-arc spraying makes possible the use of a wider variety of powders as it results in true fusion; penetration can be easily controlled and overlays from as little as 0.010 in. to as much as ¼-in. thick can be deposited.

Metallizing, on the other hand, produces a deposit which is essentially mechanically bonded. It is used to apply both powders and wires. The wire melts in the flame of the special gun and is carried in droplets onto the work by compressed air. Any metal that can be drawn into wire form can be applied with this process.

SURFACE-CHECKS IN HARD-FACING

Surface-checks, also called cross-checks or relief-checks, are not only normal for most medium- and high-alloy materials but must occur to relieve stresses which would otherwise be locked in and result in major cracks or spalling, either immediately or in service (Fig. 7-10). Manufacturers' literature should identify those materials which are supposed to surface check. A good pattern will exhibit fine tight cracks across the weld beads at intervals of ½ to ¾ in. They should not open up and must not extend into the base metal.

If an adequate pattern does not occur naturally, it should be induced by sponging the deposit with a wet cloth or spraying it with a fine water mist. It can also be accelerated during the cooling period by occasionally striking the deposit with a hammer.

HARD-FACING PATTERNS

While a smooth surface is sometimes required for hard-facing deposits and some parts are machined or ground after surfacing, it may not be necessary or even desirable. Placement of the weld beads in the proper locations and in the proper pattern can increase wear resistance beyond that provided by the alloy itself; correct placement can save material and even improve performance of the part (Fig. 7-11).

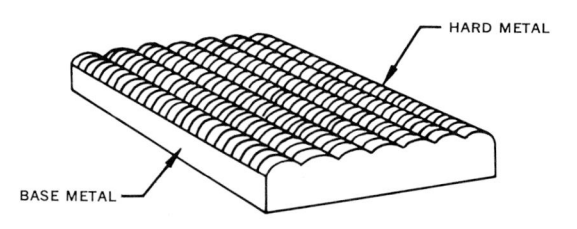

CROSS-CHECKS ARE NUMEROUS, TIGHT, CLOSELY SPACED AND DO NOT EXTEND INTO THE BASE METAL.

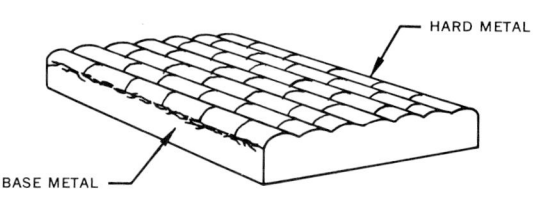

IF CROSS-CHECKING DOES NOT OCCUR IN HIGH ALLOY DEPOSITS, CRACKS WILL PROBABLY APPEAR. HAIRLINE CRACKS AT THE INTERFACE CAN FOLLOW AND CAUSE SPALLING.

Fig. 7-11 Proper and improper cross-checking.

Often it is desirable to put less hard-facing material on a part. In rocky soil, parallel stringer beads, spaced some distance apart, act as rails to carry the material if they run in the same direction as the load. Parts working in fine sand, however, can be improved by a pattern of spaced beads at right angles to the flow of the material. Here the material will build up between the beads and thereafter wear against itself as long as the beads remain. A cross-hatched or waffle pattern is often effective for the same reason and is especially suitable for working in slightly damp soil. A pattern of regularly spaced dots is useful for protection against solid surfaces and saves substantially in both material and application time (Figs. 7-12 to 7-14).

Applying hard-facing to both sides of a tooth or cutting edge will result in the part becoming more and more blunt as it wears. Applying it to one side only will produce a part which self-sharpens as the unfaced side wears away. Experience will show which side is best for surfacing.

The best guide for establishing the pattern for hard-facing any part is to observe, at intervals during its operation, the wear pattern that develops. Thicker deposits can be applied at the points of first and severest wear and lesser amounts at subsequent points. Some parts can be adequately protected by as little as a single drop of hard-facing applied to the precise point where wear begins.

HARD-FACING APPLICATION EXAMPLES

There are hundreds of construction machinery parts that can be salvaged, protected, or improved by rebuilding and hard-facing. The typical examples that follow illustrate basic techniques that can be applied to a variety of parts.

Cable Sheaves. Most cable sheaves should be preheated and slow cooled. Sheaves should be placed on a jig that can be turned or on a rotating positioner for downhand welding (Fig. 7-15). Badly worn areas should be built up with group 1 alloys. They should then be finished with circumferential beads (around the part) using a slight weave and either a group 1 or tough group 2 alloy. Cable sheaves are good parts for semiautomatic application and best for full automatic.

Fig. 7-12 Waffle pattern on shovel bucket traps soil; compacted soil provides added protection from wear.

Fig. 7-13 Dozer-blade resurfacing shows intricate pattern of checkerboard beads.

Chutes (Baffle Plates). Stringer beads should be applied ½ to 1½ in. apart in the direction of material flow; spacing will depend on size of the material being processed. An intermediate group 3 material should be used and the application repeated before the deposit is entirely worn away.

Shafts. Shafts are generally made of medium- or low-alloy steels and can be rebuilt manually with group 1 alloys. Thorough preheat and slow cooling is required; the need for preheat increases with larger parts and care should be taken to maintain temperature during welding. Grinding or machining of worn areas prior to welding is advisable. Weave beads about 1 in. wide running longitudinally should be used and the part should be built oversize about ⅛ in. to allow for finish machining.

Large shafts can often be rebuilt most economically by the automatic submerged-arc process. Shafts subject to severe wear, as in a packing gland, are often rebuilt manually with a machinable group 5 (cobalt-base or nickel-base) alloy. Where minimum buildup is required, flame spraying is a simple and effective procedure.

Swing Hammers. Swing hammers come in many shapes and sizes. Almost all are subject to severe wear and lose efficiency in the process. Many small hammers may often be hard-faced when new, before being put into service. A small amount of group 4 alloy on the striking edge will keep them out to size for full working efficiency even though unprotected areas may wear away behind the deposit (Fig. 7-16).

Large hammers have often lost a considerable portion of their original size before

Fig. 7-14 Dot pattern saves alloy and welding time, limits heat input on manganese shovel.

being removed from service. Copper plates clamped around the part to form a cavity the shape of the required finish size can be used to simplify the rebuilding process. The process of filling the cavity with semiautomatic wire is sometimes called *weld casting*.

Where substantial buildup is required, a group 1 alloy that is compatible with the base metal should be used. Choice of overlay alloy will generally be from groups 3 or 4.

Engine Valves. Almost all internal combustion engine valves can be hard-faced; the procedure is economically most feasible with the large diesel types (many are hard-faced by manufacturers in original equipment).

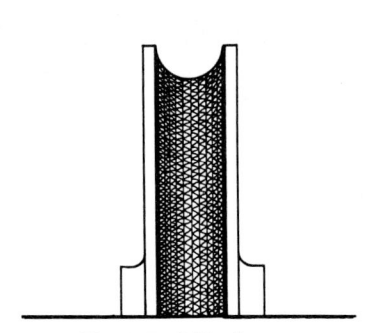

Fig. 7-15 Cable sheave.

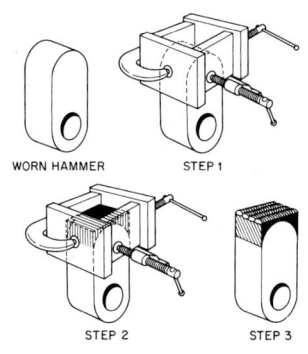

Fig. 7-16 Swing crusher hammers.

The area to be hard-faced should be undercut ⅛ in. by machining or grinding. Use a jig so that the part can be positioned for downhand welding and rotated during the process. Apply a nonchecking grade of cobalt-base (group 5) alloy by the oxyacetylene method. The seating area can be built up sufficiently oversize to allow for finishing by grinding or machining.

Teeth. Teeth at the leading edges of shovel buckets, ditch diggers, dredge cutters, and similar equipment are generally hard-faced, often even though they may be of a throwaway, replaceable type or made of special wear-resistant alloys. One reason is that hardfacing can be applied in a manner that will improve their operating characteristics while increasing service life. The hard-facing is placed chiefly on one side of the part. The unprotected side will wear away more rapidly, producing a continuously self-sharpening point (Fig. 7-17). Care should be taken to place the deposit on the leading edge of the tooth so that it will be supported as wear occurs.

The type of material to be handled by the equipment should be considered. Beads running the length of the tooth, parallel to material flow, will act as runners for large particles such as rock or slag, but fine materials like sand or dirt can work

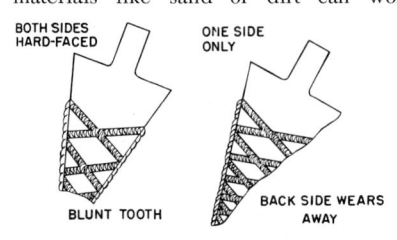

Fig. 7-17 Hard-facing for self-sharpening.

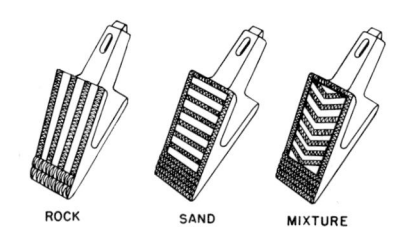

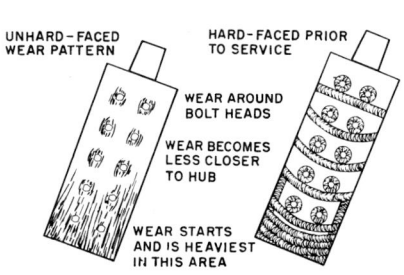

Fig. 7-18 Applying hard-facing to match wear pattern. Note (top) how material flow is controlled.

between them and erode the tooth itself. Spaced beads across the tooth or a waffle (cross-hatch) pattern will result in fine materials becoming packed between the beads, giving additional protection to the tooth (Fig. 7-18).

Groups 3 and 4 alloys, singly or in combination, are generally used for teeth; tungsten-carbide materials are often applied just at the leading edge with the remainder of the surface faced with a group 3 material. Many large teeth are made of austenitic manganese steel-base metal and proper welding precautions should be observed.

Augers and Conveyor Screws. Augers and conveyor screws should be positioned vertically in a rotating fixture for hard-facing the flight faces, horizontally for hard-facing

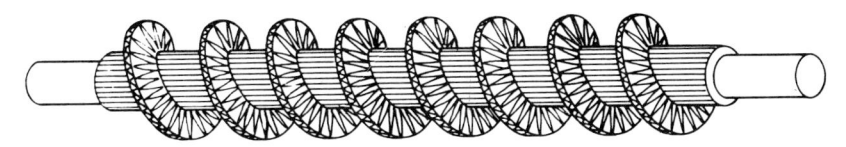

Fig. 7-19 Curbing-machine augers. Hard-face flight faces and peripheries and shaft with coated rod. Reapply hard-facing as necessary.

the periphery. The wear pattern on worn parts should be observed as a guide to material placement. A fine-mesh (40-Down) group 4 rod is a good choice for severe wear applications. A group 3 material may be used when the edge must be ground or in lesser wear situations.

Hard-face posthole auger cutter teeth on the top side only for self-sharpening effect (Fig. 7-19).

Crusher Roll Shells. Semiautomatic or automatic welding is recommended for rebuilding or hard-facing rock-crusher roll shells whenever possible. If the crusher is used regularly and is subject to fairly severe wear, an automatic unit which may be temporarily mounted to rebuild the rolls during normal downtime periods is a good investment. These machines use semiautomatic-type open-arc wires and are capable of both circumferential welding and a variety of transverse or cross-hatch patterns.

Most crusher roll's shells are made of austenitic manganese steel. Normal precautions for rebuilding this material should be observed and special care taken that excessive heat doesn't build up in the roll during the welding process. The temperature of the roll should never exceed 500°F nor should even moderate temperatures be maintained for extended periods of time. High temperatures in localized areas must also be avoided.

Size is a factor to be considered in relation to heat input; most crusher roll shells are large enough to provide sufficient mass for uniform heat distribution and dissipation. However, water cooling may be necessary on some smaller rolls. High-arc travel speeds achieved with automatic equipment or skip welding techniques with manual or semiautomatic wires are the best ways to avoid localized heating.

New shells should be put into service for a short time—about one shift—to remove residual stresses before welding. They may then be hard-faced with minimum buildup required and maintained on a regular schedule.

In rebuilding roll shells, the surface must be thoroughly cleaned prior to welding, and cracks repaired. Wedge bolts should be loosened if extensive rebuilding is required and retightened after the roll has cooled.

Where substantial buildup is required, a compatible buildup alloy (manganese alloy for Hadfield's manganese shells) should be used to bring the roll to within two layers of finish size. Transverse crescent-shaped weave beads, ½ to ¾ in. wide, are generally deposited when using manual or semiautomatic welding. Three or four beads are deposited in the area of deepest wear, then the roll is rotated to a new position. This process is repeated

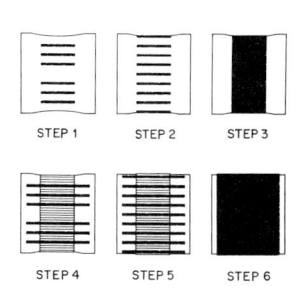

Fig. 7-20 Skip weld process used in restoring worn crusher roll shell to size and contour. Skip welding provides proper heat distribution, minimizing thermal stress.

until the pattern covers the entire circumference, then continued until the spaces between the beads are filled in. Successive layers of transverse beads are added until the roll is built up to the required diameter for hard-facing (Fig. 7-20). Circumferential beads are usually applied when building up worn rolls automatically.

Two layers of a group 3 alloy are applied for hard-facing. In both rebuilding and hard-facing the deposit should not come closer than 1 in. to the extreme edge as cracking can start at this point. A variety of patterns may be applied for the final overlay and will affect the size and shape of the crusher's output.

Chapter **8**

Maintenance of Plain Bearings

W. N. GOLDENBOGEN
Supervisor, Material Quality, Gould, Inc., Clevite Engine Parts Division, Cleveland, Ohio

DESIGN

Materials Plain bearings, or sleeve bearings, are designed to support rotating or oscillating shafts and at the same time protect them from damage. The ideal bearing offers low friction, low journal wear, conformability, and embeddability. During abnormal operating conditions the bearing metal should yield rather than damage or distort the shaft. Observation of this yielding should indicate to the operator or mechanic that unusual conditions exist. This should be a warning sign to investigate thoroughly and make minor repairs or adjustments before costly maintenance becomes necessary.

Designers, machinists, and metallurgists have combined their efforts to improve bearing performance. Refinements in materials, manufacturing methods, tolerances, and lubricants have increased the life and capacity of bearings.

Chief among design factors are materials, loads, size tolerances, temperature, and lubrication. Except for load-carrying capacity, the well-known white metals, tin-base and lead-base babbitt, are still the best all-around bearing metals. Where size is no limitation and unit loads can be kept low (2000 psi maximum), these metals have operated on shafts with hardnesses as low as 170 Brinell. With increased power ratings from smaller machines, the increased loading has demanded bearings with greater load capacity. Bronzes, copper-leads, cadmium base, aluminum base, and silver have been developed to meet this need. Some surface qualities have been sacrificed to boost the fatigue resistance (load capacity), and a corresponding increase in shaft hardness has been required. Plated overlays of lead- or tin-base material have been used in the range of 0.001 in. thick to improve the surface action of the bronzes and silver. The fatigue life of babbitt increases as the thickness drops below 0.008 in. (see Fig. 8-1). Automotive-bearing thicknesses are 0.002 to 0.004 in. today. The use of these thin MICRO[1] layers is not practical where conditions of dirt or wear may expose the steel backing.

Loads Modern bearing design demands that bearing loads be accurately determined. After due consideration of inertia, deflection, distortion, shaft whip, radial loads, and shaft speeds, it is possible to select a bearing material and a suitable bearing area that can be expected to perform satisfactorily. Figure 8-2 gives some idea of the relative load-carrying capacity of several typical bearing materials.

Tolerances After selection of a suitable bearing material, attention is given to establishing dimensional tolerances. Permissible variations are listed in Tables 8-1 to 8-5.

[1]Registered trademark of Clevite Corp.

TABLE 8-1 Case Tolerances

Bore tolerance:
   0.001 in. up to 10-in. bore
   0.002 in. over 10-in. bore

Taper tolerance:	Normal service	Heavy-duty service
1-in. length	0.0002 in.	0.0001 in.
1- to 2-in. length	0.0004 in.	0.0002 in.
Over 2-in. length	0.0005 in.	0.0003 in.

Out-of-round tolerance:
   0.001 in. max allowed if horizontal is larger than vertical

Alignment:
   Alignment bar with diameter 0.0005 to 0.00075 in. under low limit of case bore should turn freely with the use of small lever when cap bolts are properly torqued

TABLE 8-2 Shaft Tolerances

	Automotive	Heavy duty
Diameter tolerance:		
Up to 1½-in. journal	0.0005 in.	0.0005 in.
1-up to 10-in. journal	0.001 in.	0.001 in.
Over 10-in. journal	0.002 in.	0.002 in.
Diametral-taper tolerance:		
Up to 1 in. of length	0.0002 in.	0.0001 in.
1 up to 2 in. of length	0.0004 in.	0.0002 in.
Over 2 in. of length	0.0005 in.	0.0003 in
Out-of-round condition:		
Up to 3-in. diameter	0.0003 in	0.0002 in.
3- to 5-in. diameter	0.0005 in.	0.0003 in.
Over 5-in. diameter	0.001 in	0.0004 in.
Maximum misalignment:		
Adjacent main journals	0.001 in.	0.0005 in.
Crankpin parallel with main journals	0.001 in	0.0005 in.
End clearances:		
Shaft diameter		
2–2¾ in.	0.003–0.007 in.	0.003–0.007 in.
2¾–3½ in.	0.005–0.009 in.	0.005–0.009 in.
3½–5 in.	0.007–0.011 in.	0.007–0.011 in.
Over 5 in.	0.009–0.013 in.	0.009–0.013 in.
Shaft hardness:		
Brinell	200	300
Shaft-journal finish (all applications):		
Microinches	15 max	
Waviness	0.0001 in. max	
Lobing	0.0001 in. max	
Chatter	0.00005 in. max	

TABLE 8-3 Connecting-Rod Tolerances

	Automotive	Heavy duty
Diameter:		
Up to 3¼-in. diameter	0.0005 in.	0.0005 in.
3¼ to 10-in. diameter	0.001 in.	0.001 in.
Over 10-in. diameter	0.002 in.	0.002 in.
Taper, hourglass, or barrel shape:		
1-in. length	0.0002 in.	0.0001 in.
1- to 2-in. length	0.0004 in.	0.0002 in.
Over 2-in. length	0.0005 in.	0.0003 in.

Out-of-round:
   0.001 in. max if rod is larger horizontally than vertically

Parallelism and twist between rod bore and wrist-pin bore when measured 6 in. from the end of wrist-pin bushing 0.001 in. max

TABLE 8-4 Spread

Free spread (width across the open ends):
   Main bearings—crankcase bore plus 0.005–0.020 in. depending on the thickness and structural stiffness of the bearing

TABLE 8-5 Recommended Oil Clearances

Bearing-oil clearances:
   The general rule for the size of the oil clearance for pressure-lubricated bearings is to allow 0.001 in. for each inch of journal diameter, subject to modification according to the bearing metal alloy used

Bearing alloy	Shaft diameters	
	2–2¾ in.	2¹³⁄₁₆–3½ in.
Lead- and tin-base babbitts	0.0015–0.0025 in.	0.0025–0.0035 in.
Cadmium	0.002–0.003 in.	0.003–0.004 in.
Copper-lead	0.0025–0.0035 in.	0.0035–0.0045 in.

Lubrication To make any bearing perform with maximum life requires adequate lubrication. Not only should the supply of oil be maintained, but also an oil of proper viscosity should be used. A weight of oil that will provide a liquid film and consume the least amount of power is normally selected. Usually the equipment manufacturer makes a recommendation for the grade of lubricant to be used. The ideal design will provide for a large enough volume of oil to keep the bearing reasonably cool. The volume of oil to be pumped is a function of the temperature rise and will determine the size of the grooves and clearances. A temperature rise of 60°F is considered a safe figure for satisfactory operation, with maximum of 290°F.

Grooving Grooving design is important to the operation of plain bearings because it distributes the lubricant from the point of entry to the places where it is needed. The edges of the axial grooves must be well blended to avoid shearing of the oil film on the sharp corners. Properly designed grooves prevent lubricant leakage from the ends of the bearing. Introduction of the oil ahead of the highly loaded area is essential to the development and maintenance of a hydraulic film to separate moving surfaces. Figures 8-3 to 8-7 show examples of grooving design.

CARE AND MAINTENANCE

Lubricant Selection of lubricants for modern complex machinery should not be done entirely on a dollars-and-cents basis. The least expensive oils may eventually cost the most. Nor will high-priced oils provide the greatest protection against breakdown. Usually the equipment manufacturer has decided which of the many additives are required for any particular condition. For maintenance purposes it is necessary only to see that the recommended lubricants are properly applied.

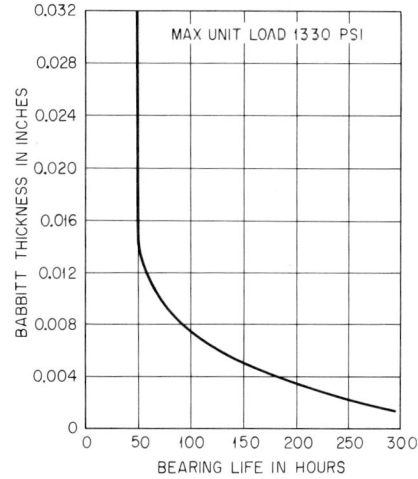

Fig. 8-1 Bearing life versus babbitt thickness.

By this we mean that drain and replacement schedules are observed, filtration is provided if necessary, contamination is minimized, and operating temperatures are kept as low as possible.

The statement that a lubricant never wears out may be true. However, in many cases, before the lubricant has a chance to wear out, it has been contaminated with foreign material (water, metal chips, abrasive particles, or acidic organic compounds) which will cause wear or loss of bearing material. Changing the lubricant becomes necessary to remove these potential sources of bearing wear and destruction.

Cleanliness The life of a lubricant can be extended by continuous removal of contaminants (filtration) and good housekeeping to exclude dirt and water. Such items as air filters, clean lubricant containers or transfer equipment, and tightly fitting covers on oil reservoirs are important.

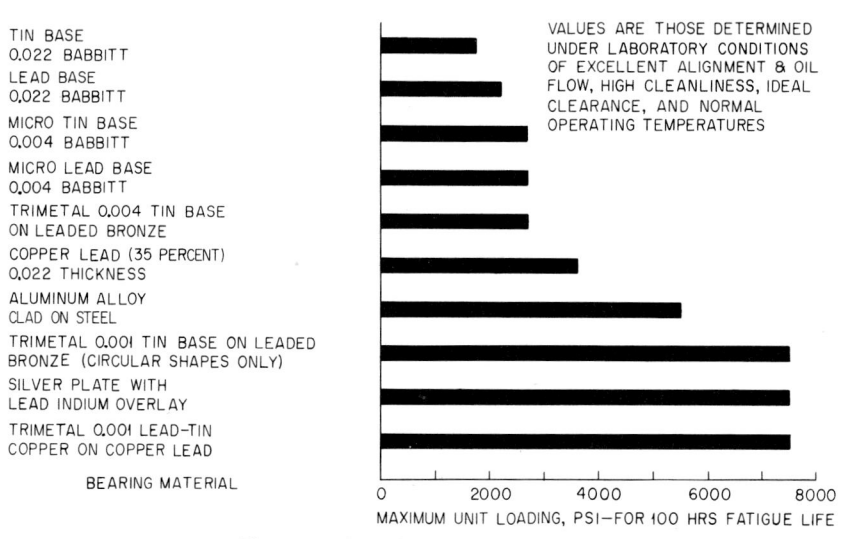

Fig. 8-2 Relative load-carrying capacity.

Temperature In addition to the maintenance of the lubricant, there are at least two other precautions which can be observed in bearing operations. They are temperature and pressure. High temperatures rob bearing materials of their strength (Fig. 8-8). High temperatures (above 275°F) also promote rapid breakdown of lubricants to form sludges or corrosive compounds. Temperature control of bearings therefore becomes an important factor in their life. By suitable means, either a temperature gauge at the bearing or in the oil or by feeling, the temperature of a bearing should be determined at frequent intervals. If an abnormally hot bearing is discovered, some means must be found to cool it immediately. If necessary, shut down the machinery.

Pressure When pressurized lubrication systems are used, a normal operating pressure is established. A frequent look at the pressure gauge will reveal any great variation from this normal value. Either an abnormally high or low pressure can be a danger signal.

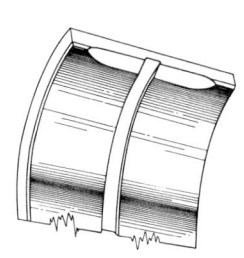

Fig. 8-3 Spreader groove and chamfer at parting line.

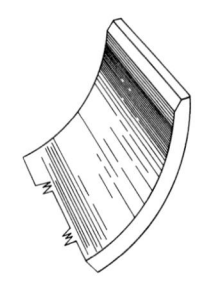

Fig. 8-4 Chamfer at parting line.

If a pump or line fails, lubrication of a part will be impaired and dangerous conditions may develop. On the other hand, the oil flow may be restricted. Although the gauge reading may be high, the flow of lubricant may be inadequate to prevent trouble. Investigate these conditions, and take corrective measures immediately.

INSPECTION AND RECONDITIONING

When it becomes necessary to dismantle a bearing, certain precautions should be observed. It is extremely important to mark or identify each part so that when the

Fig. 8-5 Main-bearing oil groove and hole.

Fig. 8-6 Grooving in main bearing.

machinery is reassembled, it can be installed in its original position. The matching of parts during the original (break-in) period makes them unfit for operation in other positions.

Journal Examination of the parts of a bearing can be a valuable source of information for the immediate trouble and for averting possible future difficulty. If the journal surfaces are excessively worn, ridged, or scored, the shaft must be reground for further service. For minimum specifications of a shaft to continue in service see Table 8-2.

If the shaft is to be used without reconditioning, it should be thoroughly cleaned, including oil passages. Also, carefully measure the diameters of the journals, because these values will be used to select the size of the replacement bearings that will provide the proper oil clearance.

Surface areas of a reground shaft are made up of a series of tiny sharp ridges. These are created by the cutting action of the abrasive grains on the face of the grinding wheel. Although these ridges are scarcely detectable, they present an unsatisfactory surface which will cause excessive bearing wear unless the roughness is reduced by a finishing operation. For final polishing, set the shaft in V blocks and polish off the ridges with a fine emery cloth and light machine oil, using a reciprocating motion. Some prefer to place the shaft in a lathe and polish while the shaft rotates in the *same direction* as it would *in an engine*.

All fillets should be checked to ensure against interference with the ends of the bearings. The conditioned shaft should be thoroughly washed, and all oil passages cleaned.

Connecting rods If connecting rods are involved, two conditions require checking: parallelism and twist (see Table 8-3). The rod bore and piston pin should be parallel within 0.001 in. in 6 in. A bent rod will cause an uneven distribution of load

Fig. 8-7 Thrust-face tapered lands.

on the bearing area, forcing the piston skirt out of parallel with the cylinder bore. This will result in uneven and excessive bearing wear, piston-ring wear, and out-of-round cylinder wear.

Twist in a rod should also be limited to 0.001 in. in 6 in. Out-of-roundness beyond 0.001 in. for rod and main bores causes variation in oil clearance. A maximum out-of-round rod should never be matched with a maximum out-of-round shaft. Bore finish should be coarse enough to ensure proper bearing contact, yet smooth enough to promote good heat transfer; 60 to 90 microinches is recommended.

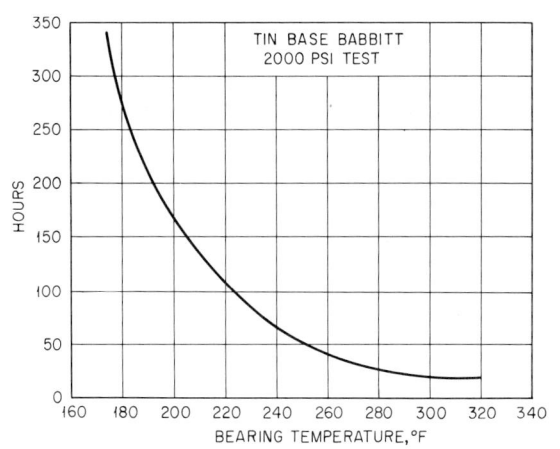

Fig. 8-8 Temperature versus fatigue life.

Bearing replacement The usual practice is to replace or renew the bearing surfaces at the time of overhaul. However, these may be times and conditions under which it is not necessary to do so. The condition of the bearings must be evaluated and compared with the cost of a subsequent overhaul. If the bearings are not worn, fatigued, or damaged in any way and all the applicable tolerances are within established limits (consult manufacturer's service manual), it is not necessary to replace the parts.

Many teardowns reveal the bearings to be ready for renewal or replacement. Such evidence as wear, edge loading (Fig. 8-9), fatigue (Fig. 8-10), embedment (Fig. 8-11), scoring (Fig. 8-12), lack of clearance (Fig. 8-13), and hourglass journal damage (Fig. 8-14) without a doubt calls for replacement or renewal of the bearings. After the journals have been examined or repolished, the job of selecting or fitting the bearings to the shaft is begun. It is advisable to follow the manufacturer's original equipment specifications for bearing materials and running clearances.

Fig. 8-9 Edge loading has caused fatigue at upper edge of bearing.

Fig. 8-10 Condition of bearing surface resulting from fatigue.

REASSEMBLY

Crush It is of utmost importance that the bearing inserts have good contact with the housing or seat. To assure this, the diameter of the two inserts at right angles to the parting line when placed together is slightly greater than the diameter across the parting surface when the bearing is in place, thus requiring this amount to be compressed when the bearing is drawn up tight. For example, each half shell is made slightly in excess of an exact half circle. The excess is called crush (see Fig. 8-15), and its purpose is to permit the shell to be firmly clamped in the bearing seat. If the bearing does not have the proper amount of crush, it will not be held securely and will have a slight degree of movement during operation.

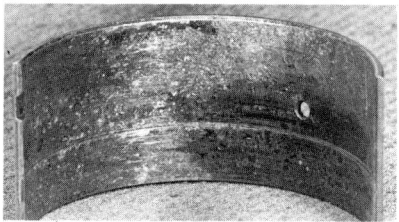

Fig. 8-11 Result of high temperatures caused by dirt.

Fig. 8-12 Scoring caused by circulating hard particles.

Loose inserts also will allow oil to work in between the back of the bearing and the housing. This cuts down the heat conductivity and tends to raise the bearing temperature. Also, a certain amount of flexing of the insert will take place, which adds to the normal friction heat, and under a retarded rate of heat transfer, will lead to a premature bearing failure due to overheating.

Insufficient crush can be due either to filing of the parting surfaces of the shells or to the presence of dirt and foreign matter between the parting faces of the bearings and bearing caps. The dirt will act as shims to prevent the faces from coming together as they should. Under no circumstances are the parting surfaces of the bearing insert, the cap or shank, or the saddle to be filed. In assembling the bearing, be absolutely sure that no dirt, nicks, or burrs remain upon the parting faces of either the cap or saddle.

Do not attempt any operation on the bearing insert other than correcting the spread, and this only when necessary. The spread (see Table 8-4) is built into the bearings so that the inserts have to be lightly pressed into place. If the parts have excessive spread, they can be tapped gently on the end to cause close-in. If they are too loose, the insert can be opened by placing it on a wooden block with convex side up and tapping with a mallet.

Bolt torque On all service installations it is an absolute necessity to use recommended bolt torque values and a torque wrench (Fig. 8-16) when tightening the bearing nuts or cap screws. Almost all the engine builders perform their boring operations with

Fig. 8-13 Wiping caused by insufficient clearance.

the bolts torqued to the same specifications as recommended in their service manuals. It is well to remember that any variation in bolt torque may seriously affect the crankcase or rod-bore sizes, bearing crush, clearances, and resulting bearing performance.

Oil clearance The various bearing metals have individual requirements for oil clearances. A general rule for the amount of oil clearance for pressure-lubricated bearings is to allow 0.001 in. for each inch of shaft-journal diameter. Table 8-5 lists recommended oil clearances for bearing alloys and shaft sizes.

Measure oil clearance The inside diameter of the bearing, in assembly, can be measured with inside micrometers if dial indicator bore gauges are not available. The journal diameter is best measured with micrometers that read in ten-thousandths of an inch.

There are various practical methods for determining oil clearance. A material which will deform can be squeezed between the journal and the bearing with the cap bolts properly torqued (see Fig. 8-17). After removal of the cap, the flattened material is compared with a prepared chart and the tolerance can be read in thousandths directly from the chart. Several commercial items of this nature are available from automotive-parts manufacturers. An alternative method is the use of lead or brass shims whose thickness ranges are standardized. A shim of suitable thickness, shorter than the bearing and about ¼ in. wide, should, when clamped between the shaft and bearing, allow the shaft to turn easily. A shim 0.001 in. heavier than the required clearance should lock the shaft from rotation. This check requires experience and care to avoid damaging the bearing inserts, and it is made with all bearing caps loose except for the position under consideration. It is necessary to apply the correct torque to the clamping bolts. Extreme care must be used to eliminate false readings that can be caused by housing bore or journal misalignment. If out-of-roundness is found to exist, use the largest journal diameter, because minimum clearance is the critical condition.

Fig. 8-14 Fatigue caused by hourglass journal.

End clearance Table 8-2 lists the recommended values of end clearances for various sizes of shafts. Checking this dimension is absolutely necessary, since the lack of clearance can easily cause thrust-bearing failure. End movement can be measured by forcing the shaft in each direction and checking either with shims at the thrust faces or with an indicator on the flywheel face.

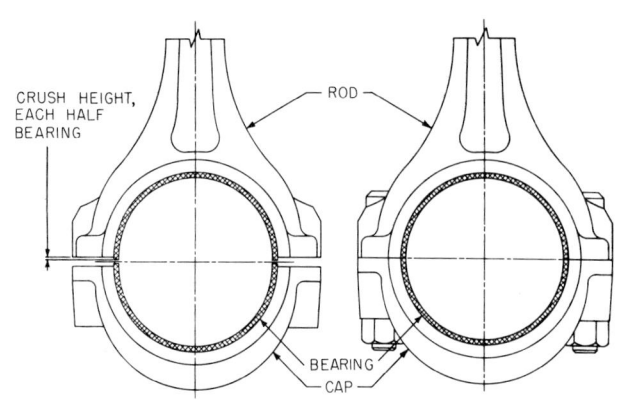

Fig. 8-15 Diagrammatic illustration of bearing crush.

Final checking In all final assembly operations and after finished surfaces have been prepared, use the utmost care to exclude dirt, chips, and all foreign matter from bearing surfaces. If included in the assembly, these particles will damage the bearing and the journal by scoring and wearing during the initial revolutions of the engine. Excessive conditions of dirt have been known to cause almost immediate failure through wear and high temperatures due to friction.

Preliminary lubrication Another precaution is to flood the unit whenever possible with clean oil just prior to initial start-up after the final assembly. This will provide temporary lubrication until the normal supply of oil is available through the lubrication system.

Free rotation If there is any doubt about the clearance and contact area between journals and bearings, it can be checked by manual rotation of the

Fig. 8-16 Torque wrench.

Fig. 8-17 Gauging oil clearance by means of plastic material.

unloaded shaft. It must rotate freely. Any indication of binding must be traced, and its cause removed. The best way to make this check is with the block mounted on a stand. Assemble the shaft with a light, uniform coating of bluing on the journals, and after rotating the shaft a couple of revolutions, turn the block over and rotate the shaft again. The transfer of blue to the bearing shells will indicate alignment condition and causes for binding. Lack of contact on a bearing also means trouble. It could be excessive clearance, or it could be shaft and/or case bore misalignment. A suitable blue pattern covering 45 to 90° at the center of all bearings will predict good bearing performance. This check is also a good place to pick up evidence of hourglass journals, taper, shaft burrs, and fillet ride.

RENEWING CAST-BABBITT LINERS

Rebabbitting large cast-in-place bearings may be not only desirable but practical in maintaining bearings used in heavy stationary machinery. During the teardown, one should look for the usual signs of misalignment, uneven wear, lack of lubrication, excessive shaft wear, excessive dirt, and high temperatures so that an attempt can be made to correct these conditions during reassembly or in the method of operation after overhaul.

Reclamation of housing Removal of the old babbitt liner can be done by heating the inside surface with a "buffalo" torch. Use as low a temperature as possible to avoid distortion of the steel back. Wiping the molten surface with a dry cloth will effectively remove all but a thin layer of the babbitt. If the removal of the old babbitt can be accomplished without severe oxidation, the residual layer will serve as the bonding layer for the new babbitt. The surface should be light golden or straw colored; otherwise, do not attempt to flux and rebabbitt without further preparation. If the remaining metal is brown or black colored or if there appear to be cracks in the surface, a light machine cut should be taken to expose sound metal of the steel shell.

For proper tinning, the steel surface must be chemically clean and slightly etched. Dip the whole shell into a hot (180°F) alkaline metal cleaner until no water breaks appear, which will indicate a clean surface. After rinsing in clean running water, pickle the shell in 1 part water and 1 part muriatic acid at 160°F for 2 to 4 min or until the surface has a gray matted finish. Remove from the acid, and rinse in clean water. Immerse the shell in soldering flux (commercial brands available) kept above 150°F. After removal from the flux, dip the shell into molten pure tin at approximately 550°F. Use caution when dipping the wet shell in the molten tin to avoid spattering of hot metal because of steam generation. Wear safety clothing to protect from possible burns. Allow the shell to remain in the tin pot long enough to approach the temperature of the tin. The length of time required for this will depend on the mass of metal present.

To pour the babbitt, remove the shell from the tin bath and attach the heated core and end plates. Immediately pour the heated babbitt (700 to 800°F) into the annular space between the core and shell. Pour sufficient metal at one time to fill the entire space. This will prevent lamination and segregation. If at all possible, it is desirable to cool the shell quickly by quenching with water applied on the bottom side. As soon as the babbitt has set, the assembly can be thoroughly cooled and knocked apart.

Determine the dimensions to which the bearing is to be machined by checking the journal size and making sure that its condition is satisfactory or by reconditioning it. Apply the manufacturer's recommendations for tolerances and grooving. Use sound machining, locating, and measuring techniques. The final cut should be made in such a manner that the best possible surface finish is obtained.

Precautions in checking clearances, alignment, lubrication, and cleanliness as described earlier are recommended.

Chapter **9**

Maintenance of Rolling Bearings

W. RALPH GOOD
SKF Industries, Inc., King of Prussia, Pa.

GENERAL

Reliable bearing performance is a key factor in optimizing maintenance costs. If this goal is to be attained, it is necessary to follow the equipment builder's operating and maintenance recommendations. For special cases, most bearing manufacturers maintain service departments to render technical assistance. Also, most areas of the world are serviced by distributors, who usually represent more than one bearing manufacturer. As with all mechanical equipment, bearings should be handled carefully and sensibly to prevent damage from mechanical abuse and contamination.

BEARING DESIGNS AND NOMENCLATURE

Nine basic types of rolling bearings and their standard nomenclature are shown in Fig. 9-1. Some of these basic types are available in many variations; for instance, single-row deep-groove ball bearings are generally available in nine different configurations as shown in Fig. 9-2. Tapered roller bearings can come in more than 20 different configurations, some of which are shown in Fig. 9-3. Cylindrical roller bearings may be obtained with one, two, or four rows of rollers. The other basic types do not come in large numbers of configurations, but it should be noted that all types of rolling bearings may vary greatly in internal design, depending on the manufacturer. It is not within the scope of this book to describe all the various designs of rolling bearings used in machinery. Rather it alerts maintenance personnel to their variety. Details are given in manufacturers' catalogs.

BOUNDARY DIMENSIONS

In general, ball, spherical roller, and some cylindrical roller bearings are made to metric boundary dimensions and tolerances which have been standardized by ISO (International Standards Organization). Therefore, bearings from all manufacturers throughout the world are physically interchangeable. Most tapered roller bearings are made to inch dimensions. However, the AFBMA (the U.S. Anti-Friction Bearing Manufacturers Association) has recently proposed to ISO several new series of metric-dimensioned bearings which in all probability will become available in the near future. Also, some metric tapered roller bearings to the present ISO boundary plans are made in Europe and Asia. Interchangeable units are thus available from several manufacturers. In most cases identical numbers are used.

Inch-dimensioned cylindrical roller bearings do not conform to an ISO standard and will vary depending on the manufacturer.

BEARING SERIES

For any given bore size all types of rolling bearings are manufactured in several series for different severity of service. For instance, most ball bearings are made in three series for light, medium, and heavy duty. These are designated as the 2-, 3-, and 4-diameter series. Spherical roller bearings are normally available in eight different series as shown in Fig. 9-4. Tapered roller bearings have a larger number of series or duty classifications, but all series are not necessarily available in every bore size.

BEARING NOMENCLATURE

The illustrations below identify the bearing parts of the nine SKF basic bearing types. The terms used conform with the terminology section of the AFBMA* Standards—and are mutually accepted by the anti-friction bearing manufacturers.

*Anti-Friction Bearing Manufacturers Association, Inc.

Self-Aligning Ball Bearing

Single Row Deep Groove Ball Bearing

Angular Contact Ball Bearing

Double Row Deep Groove Ball Bearing

Spherical Roller Bearing

Cylindrical Roller Bearing

Tapered Roller Bearing

Ball Thrust Bearing

Spherical Roller Thrust Bearing

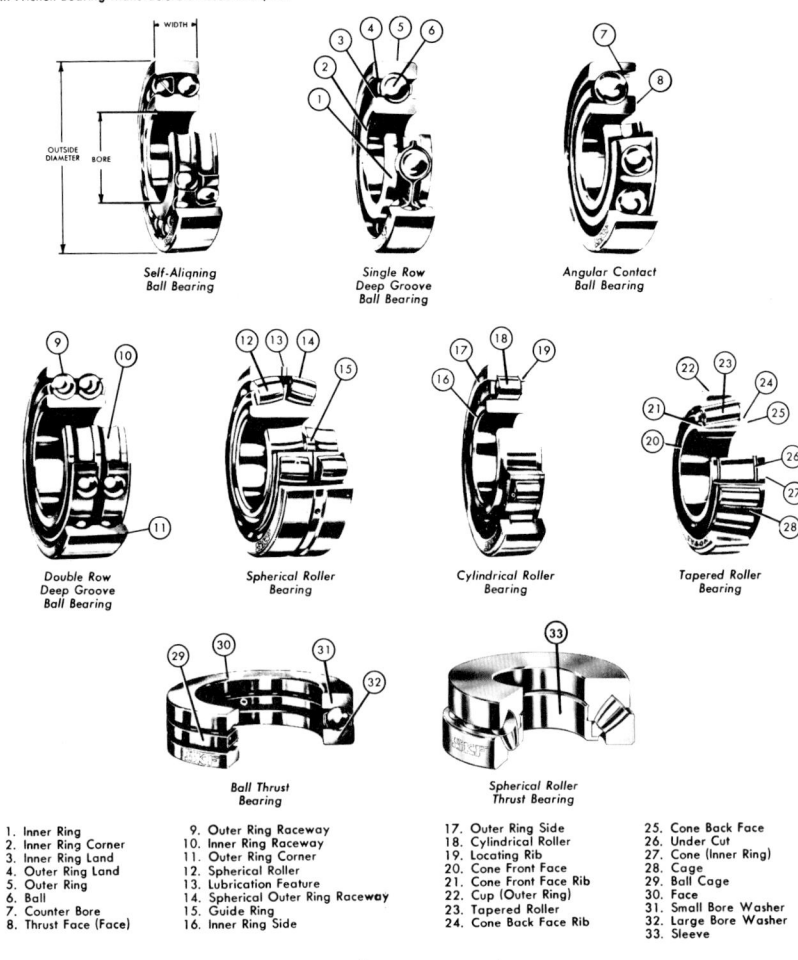

1. Inner Ring
2. Inner Ring Corner
3. Inner Ring Land
4. Outer Ring Land
5. Outer Ring
6. Ball
7. Counter Bore
8. Thrust Face (Face)
9. Outer Ring Raceway
10. Inner Ring Raceway
11. Outer Ring Corner
12. Spherical Roller
13. Lubrication Feature
14. Spherical Outer Ring Raceway
15. Guide Ring
16. Inner Ring Side
17. Outer Ring Side
18. Cylindrical Roller
19. Locating Rib
20. Cone Front Face
21. Cone Front Face Rib
22. Cup (Outer Ring)
23. Tapered Roller
24. Cone Back Face Rib
25. Cone Back Face
26. Under Cut
27. Cone (Inner Ring)
28. Cage
29. Ball Cage
30. Face
31. Small Bore Washer
32. Large Bore Washer
33. Sleeve

Fig. 9-1 Basic bearing nomenclature.

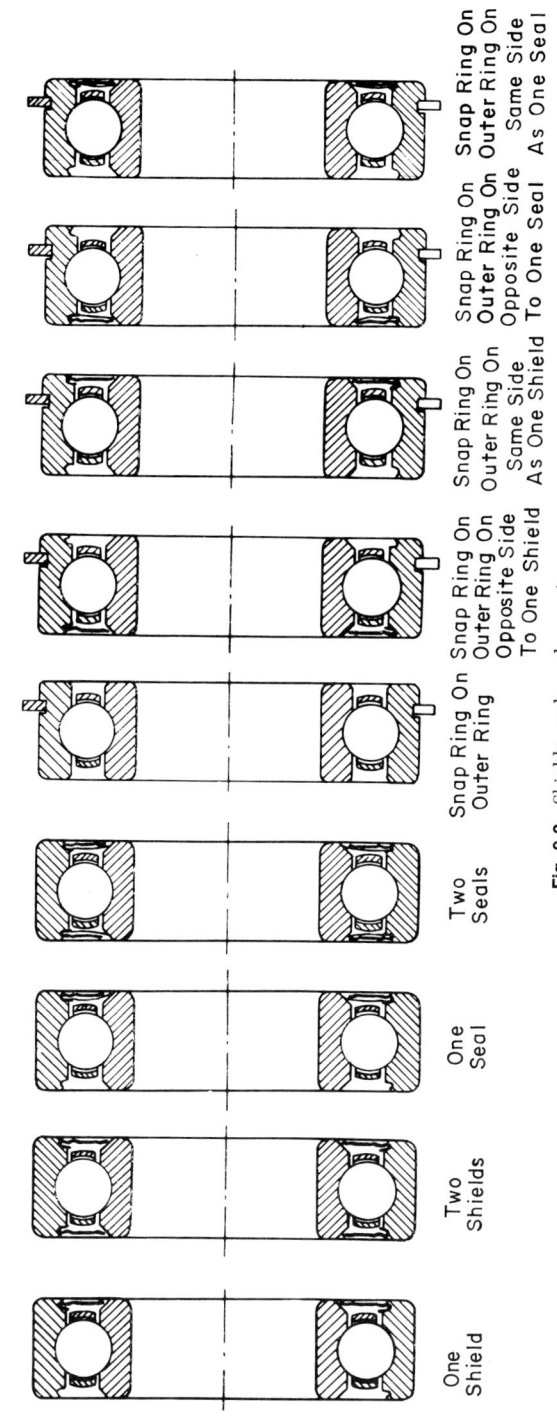

Fig. 9-2 Shields, seals, and snap rings.

LOAD RATINGS

All manufacturers of rolling bearings establish a load rating for each bearing produced. An industry-approved method for calculating this rating exists, but not all manufacturers use the method. The unfortunate situation therefore exists that two almost identical bearings produced by different manufacturers can have vastly different ratings.

Ratings are expressed as the load which will give a rating life of a certain number of revolutions. Rating life is defined as the number of millions of revolutions at a given constant speed that 90 percent of the bearings will complete or exceed before first

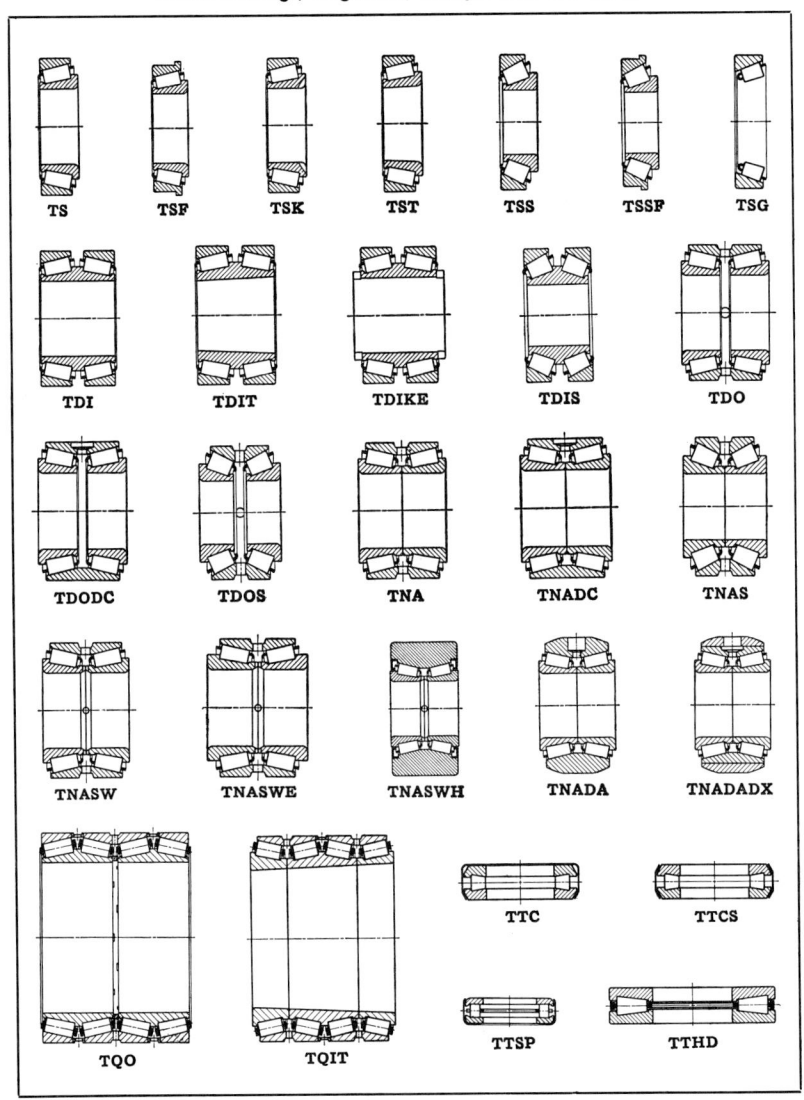

Fig. 9-3 Types of tapered roller bearings.

evidence of fatigue develops. In other words, this is a reliability or statistical rating, the only mechanical component so rated. The most common basis of rating is 1 million revolutions, but tapered roller bearings are rated on the basis of 90 million revolutions, usually expressed as 500 rpm and 3000 hr. Hence it can easily be seen that comparing manufacturers' ratings as published in their catalogs can be misleading if appropriate adjustments are not made to published values.

SHAFT AND HOUSING FITS

It is a basic rule of design that one ring of a rolling bearing must be assembled with its shaft or housing with an interference fit, since it is virtually impossible to prevent creep by clamping the rotating ring axially. Generally it is the rotating ring that is tight, but more

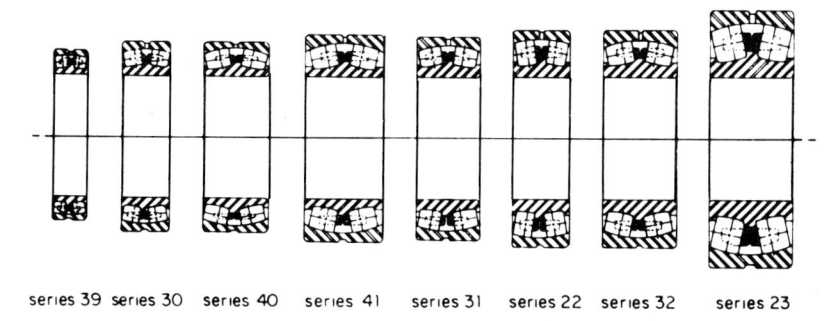

series 39 series 30 series 40 series 41 series 31 series 22 series 32 series 23

Fig. 9-4 Roller bearing series.

correctly stated it is the ring which rotates relative to the load. In some special cases this is not the rotating ring; for instance, in a vibrating unit where vibration is produced by eccentric weights, the load rotates with the rotating ring and it is best to have the stationary ring have the tight fit.

Except for special cases as illustrated above, the stationary ring can normally be assembled with shaft or housing with a slip fit.

The magnitude of the interference fit will vary with the severity of duty and type of bearing. Ball bearings under normal-load conditions will have approximately 0.00025 in. interference per inch of shaft when the inner ring is the tight fit. Roller bearings will have fits of approximately 0.0005 in. per inch of shaft. Fits will be increased for heavy-duty service and decreased for light duty. In general, when the outer ring is the tight fit, the interference is less than a corresponding shaft fit.

All bearing manufacturers show recommended fitting practices for their bearings in their general catalogs. With the exception of tapered roller bearings, the recommendations are normally expressed in ISO standards. ISO standards designate the fit between the bearing outside diameter and the housing by a capital letter and a number such as H7, J6, or P6. Fits between the shaft and bore of the bearing are designated by lowercase letter and number such as g6, m5, or r7. In the ISO system the letter indicates the class or type of fit and the number the tolerance range.

BEARING MOUNTINGS

When a shaft is mounted on rolling bearings, some provision must be made for thermal expansion and/or contraction of the shaft. Also, the shaft must be located and held axially so that all machine parts remain in the proper relationship dimensionally. When the inner ring has the tight fit, it is usually locked axially relative to the shaft by locating it between a shaft shoulder and some type of removable locking device. A specially designed nut as shown in Fig. 9-5 is normal for a through shaft. A clamp plate as shown in Fig. 9-6 is normally used when the bearing is mounted on the end of the shaft. For the locating or held bearing of the shaft, the outer ring is clamped axially, usually between housing shoulders or end-cap pilots. This type of mounting restricts axial movement in the shaft to the end movement resulting from the internal clearance of the bearing. If required, this

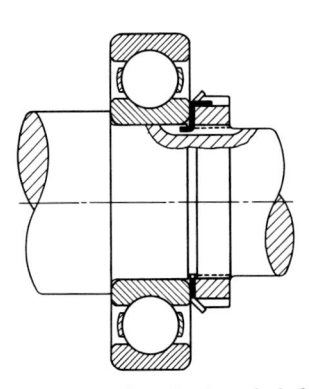

Fig. 9-5 Special nut for through shaft.

Fig. 9-6 Cone-spacer adjusting device.

can be zero if the appropriate bearing type is used. The outer rings on all other bearings on the shaft should not be secured axially, and enough clearance should be provided between the side face of the stationary ring and the nearest housing shoulders to allow for anticipated expansion or contraction. A typical held-free mounting is shown in Fig. 9-7.

Certain types of cylindrical roller bearings are capable of absorbing shaft expansion internally simply by moving one ring relative to the other as shown in Fig. 9-8. The advantage to this type of mounting is that both inner and outer rings may have a tight fit. This may be desirable or even mandatory if significant vibration and/or unbalance exists in addition to the applied load.

Where bearing centers are short and minimum thermal expansion is expected, an opposed mounting as shown in Fig. 9-9 may be used. In addition to its simplicity, this mounting has the advantage that thrust in one direction will be taken on one bearing and thrust in the other direction on the other bearing. Obviously, the clearance between the side face of the bearing and the housing shoulder must be carefully controlled or the shaft will shift excessively in an axial direction.

Single-row tapered roller bearings and angular-contact ball bearings require special consideration. For example, if a radial load is applied to a single-row tapered roller

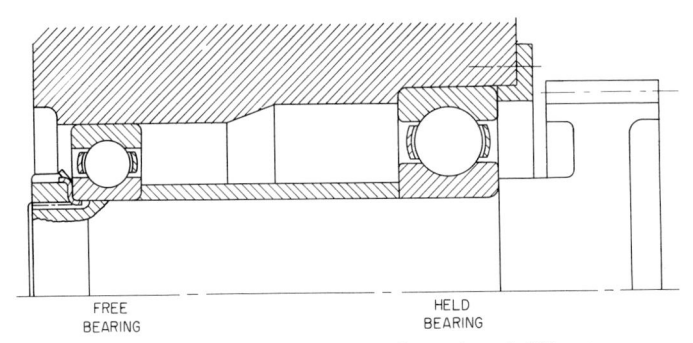

Fig. 9-7 Axial shaft locations, one free and one held bearing.

bearing, an axial component of the load is generated by the angle of the roller set which tends to separate the bearing unless this induced thrust is resisted by another bearing properly mounted to resist the movement. The other bearing is normally another single-row tapered roller bearing. A mounting of this type may have the bearings arranged in one of two ways as shown in Fig. 9-10. The upper portion of Fig. 9-10 shows the included angles of the conical portions of the bearing, or the cup roller track, open away from each other. This is known as an *indirect* mounting. The lower half shows the included angles of the conical portions of the bearings open toward each other. This is known as a *direct*

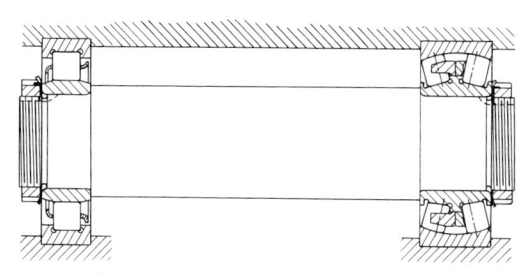

Fig. 9-8 Axial shaft locations, roller bearing used for floating location.

mounting. It should be noted before progressing further with a description of this type of mounting that the point of reaction of the load on the centerline of the shaft, or the effective center, is not at the geometric center of a single-row tapered roller bearing but at some point "O" as determined by the angle of the roller relative to the centerline of the shaft.

Therefore, if the bearings of two different mountings are physically located the same distance apart with an indirect mounting, the effective centers of the bearings are farther apart than with the direct mounting, a more desirable arrangement when an overturning load exists.

With either a direct or an indirect tapered roller bearing mounting it is necessary to set the running clearance of the bearings when they are assembled. This is done by adjusting the cones in an indirect mounting and the cups for a direct mounting. Figures 9-11 and 9-12 show two ways of adjusting cones by nuts, and Fig. 9-13 shows a method of shimming for cone adjustment. Figures 9-14 to 9-16 show three ways of shimming cups in a direct mounting. Proper running clearance is controlled by measuring the end movement, or end lateral, of the shaft. The machine builder's recommendation for proper end lateral should be strictly followed. It will usually be indicated on the drawing of the particular part or given in the maintenance manual for the equipment.

Obviously, the only provision for thermal expansion in either of these mountings is the end lateral of the assembly. For that reason they should be used only where bearing centers are relatively short or where little temperature variation is anticipated.

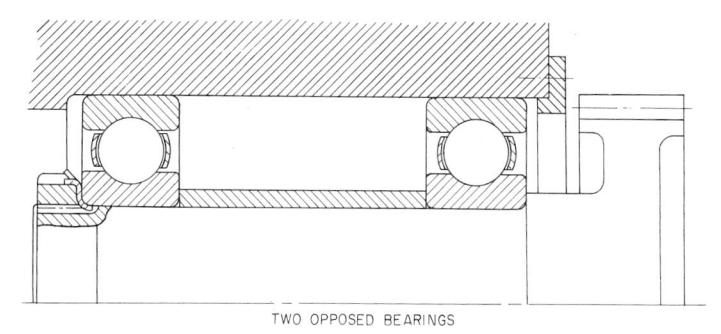

TWO OPPOSED BEARINGS

Fig. 9-9 Axial shaft locations, opposed mounting of ball bearings.

4-142   Basic Maintenance Technology

Two-row tapered roller bearings are mounted the same as other types of bearings. Proper end lateral is established in the factory.

Angular-contact ball bearings are rarely used singly. However, if they are, they must be mounted in a similar manner to single-row tapered roller bearings. Thus much smaller running clearances used in ball bearings make a mounting of single angular-contact ball

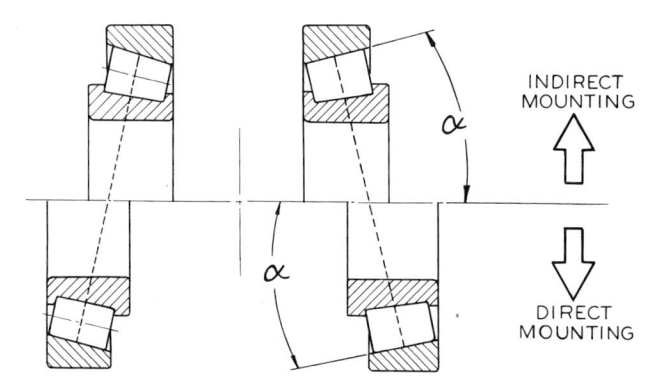

Fig. 9-10   Indirect and direct mounting.

bearings very difficult to adjust properly. Angular-contact bearings could be substituted for the tapered roller bearings of Fig. 9-10, and the same comments and nomenclature would apply for single-bearing mountings.

However, angular-contact ball bearings are normally used in pairs as shown in Fig. 9-17. The side faces of these bearings are especially ground in the factory to permit them to be mounted side by side as shown in Fig. 9-17. Face-to-face (Fig. 9-17A), back-to-back (Fig. 9-17B), and tandem (Fig. 9-17C) is the common terminology for these mountings. When two or more bearings are stacked in tandem for high thrust loads, usually another bearing in the assembly is mounted face-to-face or back-to-back with the tandem stack. When mounted in any of these arrangements, they may be considered as one multiple-row bearing. Because methods of face grinding may differ from one manufacturer to

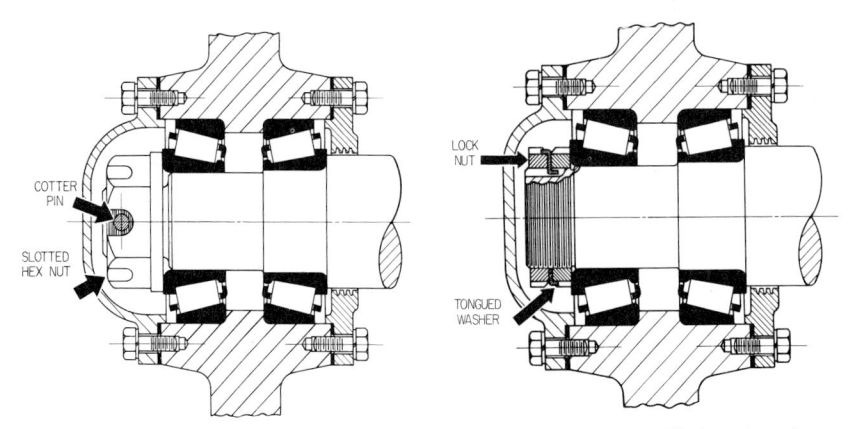

Fig. 9-11   Slotted-nut adjusting device.

Fig. 9-12   Double-nut and lock-washer adjusting device.

Maintenance of Rolling Bearings **4-143**

another, it is advisable not to mix brands in a pair of tandem-group bearings. The bearing number should indicate in some way that the bearings have been properly ground for mounting in pairs. Bearings for single mounting are available and should not be used as part of a pair.

A large percentage of spherical roller bearings are made with tapered bores. Some ball,

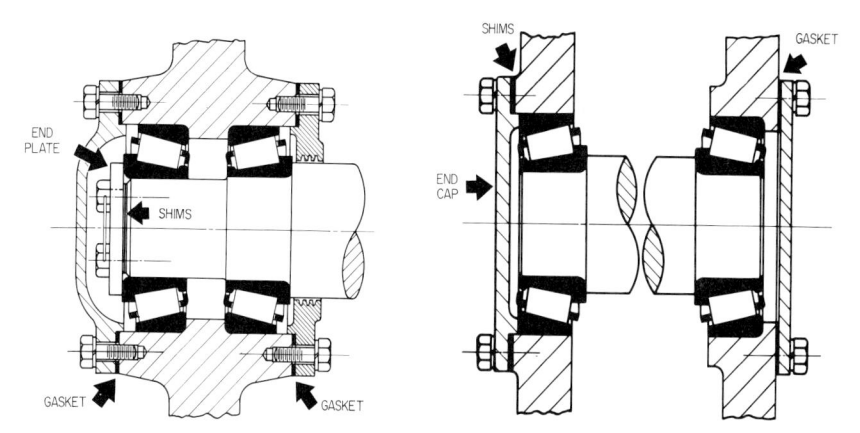

Fig. 9-13 End-plate and shims adjusting device.

Fig. 9-14 End-cup and shims adjusting device.

tapered roller, and cylindrical roller bearings are also available with tapered bores. These bearings may be mounted directly on the shaft as shown in Fig. 9-18. However, many tapered-bore bearings are mounted on one of two types of sleeves as shown in Figs. 9-19 and 9-20. European machinery builders are particularly partial to use of sleeve mountings.

The adapter sleeve may be mounted as shown in Fig. 9-19 or with a shaft shoulder ring as shown in Fig. 9-21. With a removable type of sleeve as shown in Fig. 9-20, the bearing must always be against a shaft shoulder.

The taper is 1 to 12 on diameter in all but the widest series of spherical roller bearings when a flatter 1 to 30 taper is used. Some four-row cylindrical-roller rolling-mill bearings will also use a 1 to 30 taper in the bore of the inner ring.

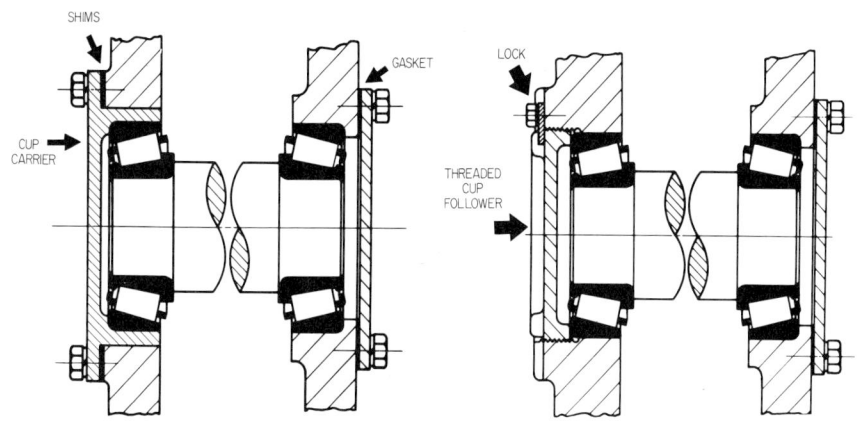

Fig. 9-15 Cup-carrier and shims adjusting device.

Fig. 9-16 Threaded cup-follower adjusting device.

MOUNTING AND DISMOUNTING OF ROLLING BEARINGS

General The most important thing to remember when mounting or dismounting a rolling bearing of any type is to apply the mounting or dismounting force to the side face of the ring with the interference fit. Keep this force from passing from one ring to the other through the ball or roller set. This is particularly important during mounting. Cleanliness is, of course, extremely important. Not only the bearing but also the shaft and housing must be free from chips, burrs, and dirt.

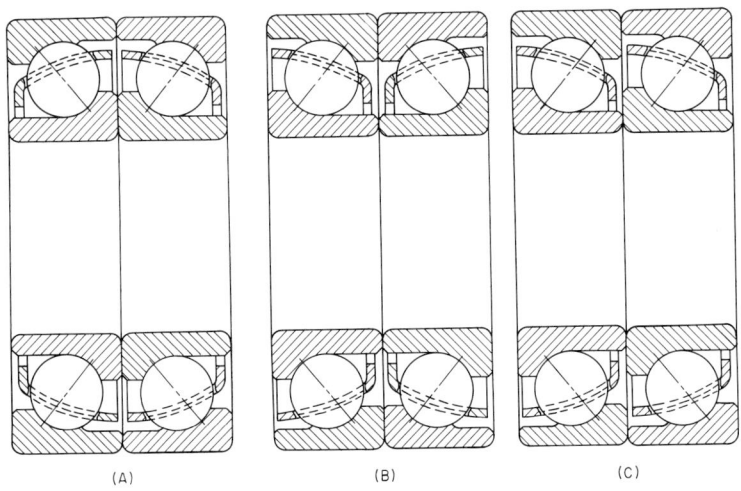

Fig. 9-17 Methods of mounting two flush-ground angular-contact bearings. (A) Face-to-face. (B) Back-to-back. (C) Tandem.

Bearings should be kept wrapped until the last possible moment. Since most modern rust preventives used by bearing manufacturers are compatible with petroleum-base lubricants, the slushing compound is normally not removed. However, there are exceptions to this rule. If oil-mist lubrication is to be used and the slushing compound has hardened in storage or is blocking lubrication holes, it is best to clean the bearing with kerosene or other appropriate solvent. Obviously, the other exception would be if the slushing compound has been contaminated with dirt or foreign matter before mounting. It is also permissible and sometimes desirable to wipe the rust preventive from the bore or outside diameter of the bearing, depending on which surface will have the tight fit. Before mounting or dismounting a bearing, always take the time to collect the proper tools and accessories. The use of inappropriate tools is a major cause of bearing damage. Also, remember, never strike a bearing directly with a hammer, sledge, or mallet.

Cold mountings All small bearings (4-in. bore and smaller) may and sometimes must be mounted cold by simply forcing them on the shaft or into the housing. However, it is important that this force be applied as uniformly as possible around the side face of the bearing and to the ring to be press-fitted. Mounting fixtures should be used. These can be a simple piece of tubing of appropriate size and a flat plate as shown in Fig. 9-22. Do not try to use a drift and hammer, because the bearing will become cocked on the shaft. Force may be applied to the simple fixture described above by striking the plate with a hammer or by an arbor press as shown in Fig. 9-23. It is a good idea to apply a coat of light oil to the bearing seat on the shaft and bore of the bearing itself before forcing on the shaft. It should be noted that all sealed and shielded ball bearings must be mounted cold in this manner.

Temperature mountings The simplest way to mount any open straight-bore bearing, no matter what size, is to heat the entire bearing and simply push it on its seat and hold in place until it cools enough to start gripping the shaft. For tight outside-diameter fits the housing may be heated if practical; if not, the bearing may be cooled by dry ice. However, if the ambient conditions are humid, cooling the bearing introduces the possibility of condensation on the bearing which will induce corrosion later.

There are several acceptable ways of heating bearings. Some of these are as follows:

1. Hot plate: a bearing is simply laid on an ordinary hot plate until it reaches the approved temperature. The disadvantage of this method is that the temperature is difficult to control. A Tempilstik or pyrometer should be used to make certain the bearing is not overheated.

2. The temperature-controlled oven: This method needs little comment. The bearings should be left in the oven long enough to heat thoroughly. However, never leave bearings in a hot oven overnight or over a holiday or weekend.

3. Induction heaters are available which can be used to heat bearings for mounting. One of these is shown in Fig. 9-24. It must be remembered that this is a very quick method of heating and that some method of sensing the ring temperature must be used or the bearing may be damaged. A Tempilstik or pyrometer can serve this purpose.

4. A hot-oil bath may also be used to heat the bearing and, in fact, is the most practical means to heat larger bearings. This method has some drawbacks, as the temperature of the oil is difficult to control and may overheat the bearing or even become a fire hazard. A mixture of soluble oil and water can eliminate both these disadvantages. Make the mixture 10 to 15 percent soluble oil. This solution will boil at approximately 210°F, which is hot enough for most bearing fits.

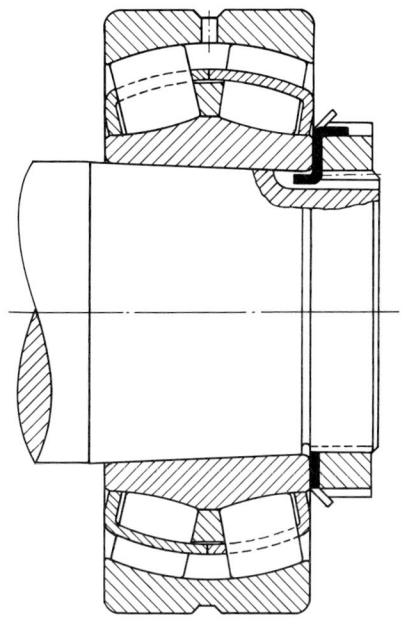

Fig. 9-18 Direct mounting.

The heating solution should be placed in a tank or container which has a grate several inches off the bottom, as shown in Fig. 9-25. This will allow any contaminants to sink to the bottom and keeps the bearings off the bottom of the container.

As mentioned above, 210°F is hot enough to mount most bearings. If using one of the other methods of heating or another solution, 250°F will do the bearing no harm. However, this temperature should not be exceeded for small ball bearings (2-in. bore and smaller). Larger bearings can be heated somewhat higher than this without harm, but metallurgical damage will occur at over 300°F.

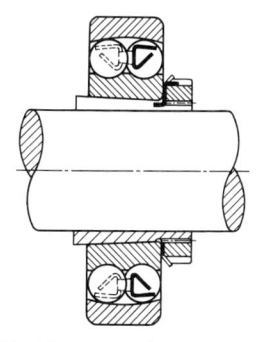

Fig. 9-19 Mounting with an adapter sleeve.

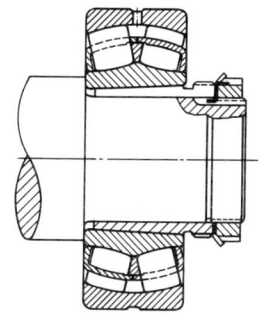

Fig. 9-20 Mounting with removable type of sleeve.

4-146 Basic Maintenance Technology

Mounting tapered-bore bearings Tapered-bore bearings can be mounted simply by tightening the locknut or clamping plate, which will locate it on the shaft until the bearing has been forced up the taper the proper distance. However, especially for large bearings, this technique will require a good amount of brute force. There are special techniques that may be used to reduce the amount of force required.

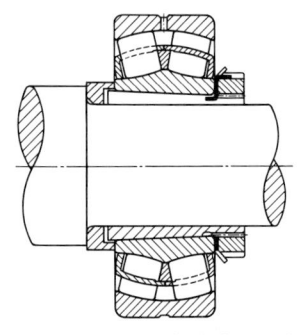

Fig. 9-21 Mounting with shaft shouldering.

Fig. 9-22 Mounting using flat plate.

Before reviewing the mounting techniques for tapered-bore roller bearings, we will discuss the special case of self-aligning ball bearings. The bearing should be put on its tapered seat and the locknut hand tightened until all looseness is removed between adjacent parts. Then using a spanner wrench, not a drift and a hammer, tighten the nut one-eighth turn farther. Bend a lockwasher tab into the nut slot nearest to a washer tab in a tightened direction. At this point, the outer ring should rotate as well as swivel freely.

Tapered-bore spherical roller bearings can be mounted a bit more scientifically. Since the internal clearance in a roller bearing is significantly larger than in a ball bearing, this clearance can be measured with a thickness gauge. As the bearing inner ring is pushed up the tapered seat, the inner ring expands, thereby reducing the internal clearance. Hence the amount of this reduction is a direct function of the interference fit between the bore of the bearing and the shaft. Therefore, if we measure the internal clearance of the bearing unmounted and control the amount the clearance is reduced during mounting, we control the shaft fit within very close limits.

The internal clearance of a spherical roller bearing is measured as follows: The bearing is unwrapped and placed on a table so that it can be easily handled. With one hand grasping the lower portion of the inner ring, oscillate the inner ring and roller set in a circumferential direction to seat the lower rollers properly in the sphere of the outer ring, on the roller paths of the inner ring and against the separate guide ring between the two rows of rollers. Select a gauge blade of perhaps 0.003- or

Fig. 9-23 Mounting using an arbor press.

0.004-in. thickness or less for small bearings. The usable length of the blade should be somewhat longer than the length of a roller. It should not be equal to or greater than the width of the bearing. While pushing the top roller against its guiding surface, insert the blade between two rollers and the outer ring and slide the blade circumferentially toward the roller at the top of the bearing, as shown in Fig. 9-26. The blade should pass between

Fig. 9-24 Induction heater, as used to heat bearing prior to mounting. *(Reed Electric Sales & Supply Inc., Portland, Ore.)*

the uppermost roller and the inside of the outer ring. Do this with successively thicker feeler blades until a blade will not pass. Move it so that it approaches the bite between a roller and the outer ring sphere; then with one hand grasping the inner ring as described earlier, slowly roll the uppermost roller under the feeler blade. With the blade between the uppermost roller and the sphere, attempt to swivel the blade and withdraw it axially. The swiveling motion helps to center the roller in its proper operating position, and withdrawing it with the characteristic wiping feel of a line-to-line contact will show that thickness to be the looseness over that roller. If the blade becomes much looser during the swiveling and withdrawing process, attempt the same procedure with a blade 0.001 in. thicker and continue until a blade cannot be swiveled or withdrawn. The internal clearance over that roller will be the blade that can be swiveled and withdrawn after a thicker one has jammed.

Repeat this procedure in two or three other locations by resting the bearing on a different spot on its outside diameter and measuring over different rollers in one row. Either repeat the above procedure for the other row of rollers or measure each row alternately in the procedure described above. Make a note of this unmounted internal clearance.

After the unmounted radial clearance is measured, the bearing is placed on its tapered seat. If the shaft provides for a locknut, it is then assembled, but the lock washer is left off the shaft at this point. The locknut should then be tightened against the bearing, pushing it up the taper until the internal clearance is reduced by the specified amount as shown in Table 9-1. An impact-type spanner wrench as shown in Fig. 9-27 is ideal for tightening the nut.

Fig. 9-25 Hot-oil bath for bearing.

The amount of force required to drive a tapered-bore bearing can be greatly reduced if the shaft is drilled and grooved as shown in Fig. 9-28. If these fittings are available, attach a hydraulic pump to the connection at the end of the shaft. Drive the bearing on the taper just enough so there is some interference; then build up hydraulic pressure under the bore of the bearing. A

TABLE 9-1 Recommendation for Driving a Spherical Roller Bearing on a Tapered Seat
(Values in inches)

Bearing bore, mm		Reduction in radial internal clearance		Min permissible final clearance after mounting bearings with clearance		
Over	Incl.	Min	Max	Normal	C3	C4
40	50	0.0010	0.0012	0.0008	0.0012	0.0020
50	65	0.0012	0.0015	0.0010	0.0014	0.0025
65	80	0.0015	0.0020	0.0010	0.0016	0.0030
80	100	0.0018	0.0025	0.0014	0.0020	0.0030
100	120	0.0020	0.0028	0.0020	0.0025	0.0040
120	140	0.0025	0.0035	0.0022	0.0030	0.0045
140	160	0.0030	0.0040	0.0022	0.0035	0.0050
160	180	0.0030	0.0045	0.0024	0.0040	0.0060
180	200	0.0035	0.0050	0.0028	0.0040	0.0065
200	225	0.0040	0.0055	0.0030	0.0045	0.0070
225	250	0.0045	0.0060	0.0035	0.0050	0.0080
250	280	0.0045	0.0065	0.0040	0.0055	0.0085
280	315	0.0050	0.0075	0.0043	0.0060	0.0095
315	355	0.0060	0.0085	0.0047	0.0065	0.0100
355	400	0.0065	0.0090	0.0050	0.0075	0.0115
400	450	0.0080	0.0105	0.0050	0.0080	0.0120
450	500	0.0085	0.0110	0.0065	0.0090	0.0135

NOTE: The axial displacement of the bearing is approximately 16 times the clearance reduction.

pressure of 3000 to 6000 psi will be needed, but with this pressure between the bore of the bearing and the shaft it is possible to float the bearing up the taper with much less torque applied to the locknut or clamp plate than in a dry mounting.

Another convenient way to mount a tapered-bore bearing is to use a hydraulic nut or mounting tool as shown in Fig. 9-29. This technique can also be adapted to sleeve mountings that are large enough to be drilled and grooved.

Cylindrical and tapered roller bearings with tapered bores are not as common as their spherical counterparts, and the manufacturer will have specific mounting instructions for each application.

Fig. 9-26 Determining internal bearing clearance.

Dismounting of bearings A wide variety of tools are available commercially which are designed to remove a rolling bearing from its seat without damage. Typical bearing pullers are shown in Fig. 9-30. In removal we should again keep in mind the basic rule to apply force to the ring with the tight fit. Pullers can normally be applied to bearings so that

Fig. 9-27 Use of impact-type spanner wrench.

this rule is observed. However, sometimes supplementary plates or fixtures may be required.

For smaller bearings, an arbor press is equally effective at removing as well as mounting bearings. Also techniques such as the one shown in Fig. 9-31 may be used where size permits.

Hydraulic removal Where shafts have been designed to apply hydraulic pressure to the fit between shaft and bearing, removal is quite simple. First the locking device,

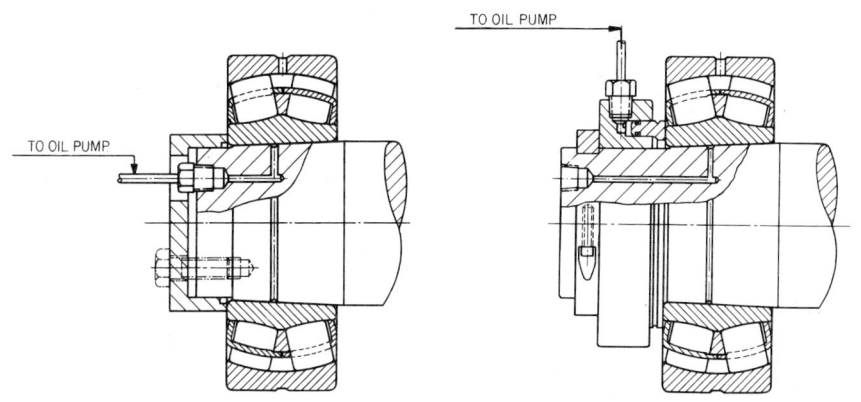

Fig. 9-28 Drilling and grooving to reduce driving force.

Fig. 9-29 Use of hydraulic nut.

whatever it is, should be backed off a distance greater than the axial movement of the mounting; ¼ in. will be sufficient in virtually every case. Then connect a hydraulic pump to the fitting provided at the end of the shaft as shown in Fig. 9-32A and start building up pressure. When the pressure becomes great enough to break the fit, usually about 3000 to 6000 psi, the bearing will literally jump off the taper with a sharp bang. The retaining

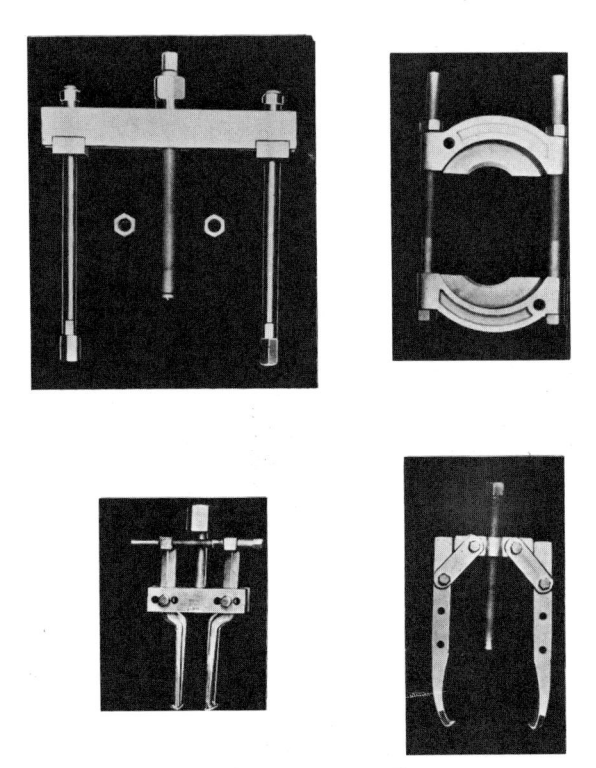

Fig. 9-30 Typical bearing pullers.

device, still being loosely connected, will prevent the bearing from coming off the end of the shaft.

Hydraulic pressure may be used with straight-bore bearings, but a puller must be used in conjunction with the hydraulic pump, as there will be no axial component of the hydraulic pressure to blow the bearing off its seat. See Fig. 9-32B.

Larger sleeve mountings may also be designed to utilize hydraulic pressure for dismounting. If this feature is available, follow the same procedure as outlined above. However, if the sleeve mounting does not have this feature, other techniques such as shown in Fig. 9-33 must be used. For withdrawal sleeves a special nut must be used as shown in Fig. 9-34. For larger withdrawal sleeves a hydraulic nut is desirable for dismounting.

LUBRICATION

The primary purpose of lubrication in a rolling bearing is to separate the contacting surfaces, both rolling and sliding. This purpose is rarely achieved 100 percent, and boundary lubrication or partial metal-to-metal contact frequently occurs. By far the most common lubricants are petroleum products in the form of grease or liquid oil.

Generally the machine builder decides whether a bearing will be grease- or oil-lubricated and normally will recommend the basic specifications of the required lubricant. However, because the machine designer cannot foresee all the variable conditions under which his equipment will operate, some judgment is required on the part of maintenance personnel. Some knowledge of the lubricants' specifications is therefore useful.

Oil lubrication For oil lubrication, the Annular Bearing Engineers Committee (ABEC) has issued the following recommendations:

The friction torque in a ball bearing lubricated with oil consists essentially of two components. One of these is a function of the bearing design and the load imposed on the bearing, and the other is a function of the viscosity and quantity of the oil and the speed of the bearing.

It has been found that the friction torque in a bearing is lowest with a very small quantity of oil, just sufficient to form a thin film over the contacting surfaces, and that the friction will increase with greater quantity and with higher viscosity of the oil. With more oil than just enough to make a film, the friction torque will also increase with the speed.

The energy loss in a bearing is proportional to the product of torque and speed, and this energy loss will be dissipated as heat, and cause a rise in the temperature of the bearing and its housing. This temperature rise will be checked by radiation, convection and conduction of the heat generated to an extent depending upon the construction of the housing and the influence of the surrounding atmosphere. The rise in temperature, due to operation of the bearing, will result in a decrease in viscosity of the oil, and therefore, a decrease in friction torque compared with the friction of starting, but soon a balanced condition will be reached.

With so many factors influencing the friction torque, energy loss, and temperature rise in a bearing lubricated with oil, it is evidently not possible to give definite recommendations for selection of oil for all bearing applications, but two general considerations are dominant:

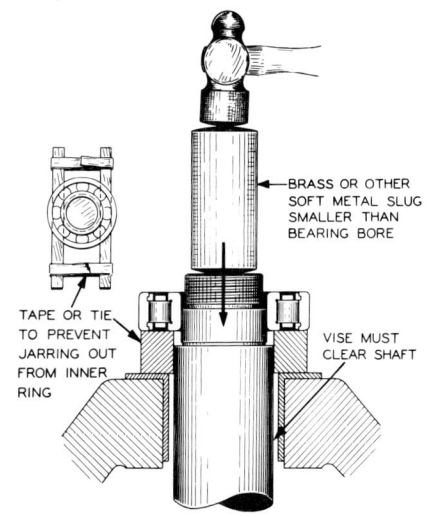

Fig. 9-31 Use of an arbor press to remove small bearings.

1. The desire to reduce friction to a minimum, which requires a small quantity of oil of low viscosity.

2. The desire to maintain lubrication safely without much regard for friction losses, which results in using larger quantities of oil and usually of somewhat greater viscosity in order to reduce losses from evaporation or leakage.

This second condition is most frequently met when bearings have to operate in a wide range of temperatures. An oil that has the least changes with changes in temperature, i.e., an oil with high viscosity index, should be selected.

In the great majority of applications pure mineral oils are most satisfactory, but they should, of course, be free from contamination that may cause wear in the bearing, and they should show high resistance to oxidation, gumming, and to deterioration by evaporation of light distillates and they must not cause corrosion of any parts of the bearing during standing or operation.

It is self-evident that for very low starting temperatures an oil must be selected that has sufficiently low pour-point, so that the bearing will not be locked by oil frozen solid.

In special applications, various compounded oils may be preferred, and in such cases the recommendation of the lubricant manufacturer should be obtained.

For cases where the bearing load is not known, it is a good rule to select an oil that will have at least the following viscosities at the operating temperature:

For ball bearings and cylindrical roller bearings .... 70 SSU
For spherical roller bearings ....100 SSU
For spherical roller thrust bearings ....150 SSU

It should be kept in mind that the temperature of the oil is usually 5 to 10°F higher than the bearing housing. For example, assuming that the temperature of the bearing housing is 170°F, the temperature of the oil will usually be 175 to 180°F. The viscosity of the oil at this temperature should therefore be 70, 100, or 150 Saybolt Seconds Universal (SSU), depending on the bearing type as indicated above. To obtain the approximate viscosity at any other temperature, the diagram in Fig. 9-35 may be used as a guide. For example, refer to the diagram: If the oil viscosity at 180°F should be 70 SSU, read up from 180 on the temperature coordinate to the intersection of 70 on the viscosity coordinate. From this point, follow a line parallel to one of the oblique lines to point 100 on the temperature coordinate. Then, reading over to the viscosity coordinate, find that an oil having a viscosity of about 360 SSU at 100°F is required. This is the customary way of specifying

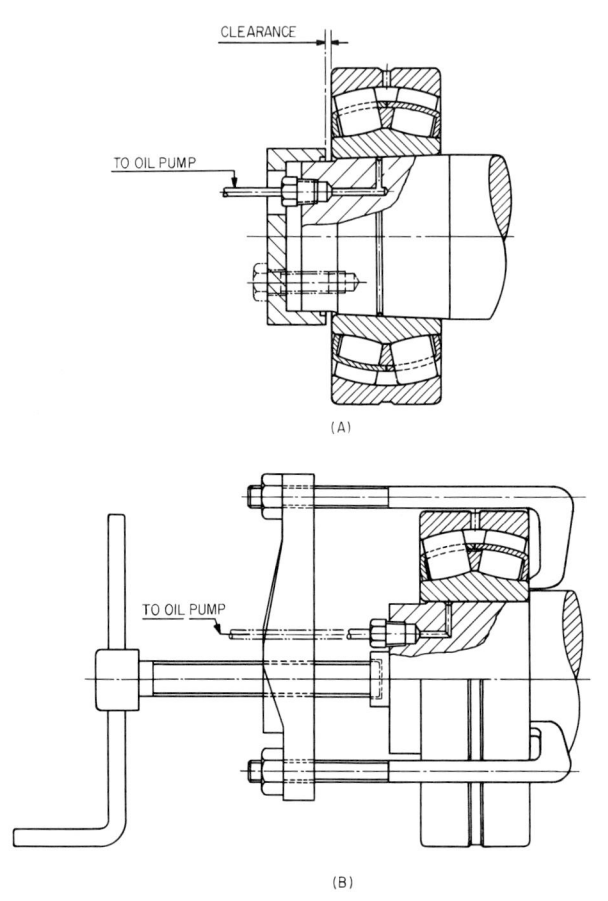

Fig. 9-32 Hydraulic removal. (A) By connection to pump. (B) In conjunction with a puller.

oil viscosities, since the viscosity ratings are usually given at temperatures of 100 and 210°F, the latter being for very heavy oils. Information regarding change in viscosity with change in temperature for a particular oil should, however, be obtained from the oil supplier.

For bearings operating at extremely slow speed, that is, 10 rpm or less, heavy oils which

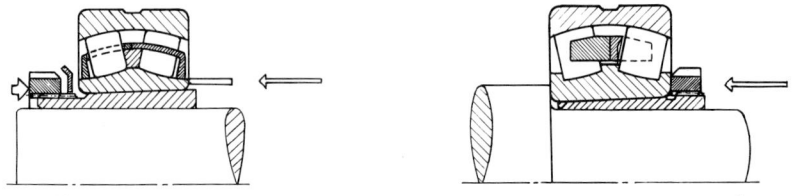

Fig. 9-33 Bearing removal.

Fig. 9-34 Bearing removal.

have high viscosities at operating temperatures are required. For these applications, it is best to consult the bearing manufacturer.

Grease lubrication Where grease lubrication is used, we need to consider a few of the basic physical and chemical characteristics of the lubricant. Greases are a mixture of lubricating oil and usually a soap base. The base merely acts to keep the oil in suspension.

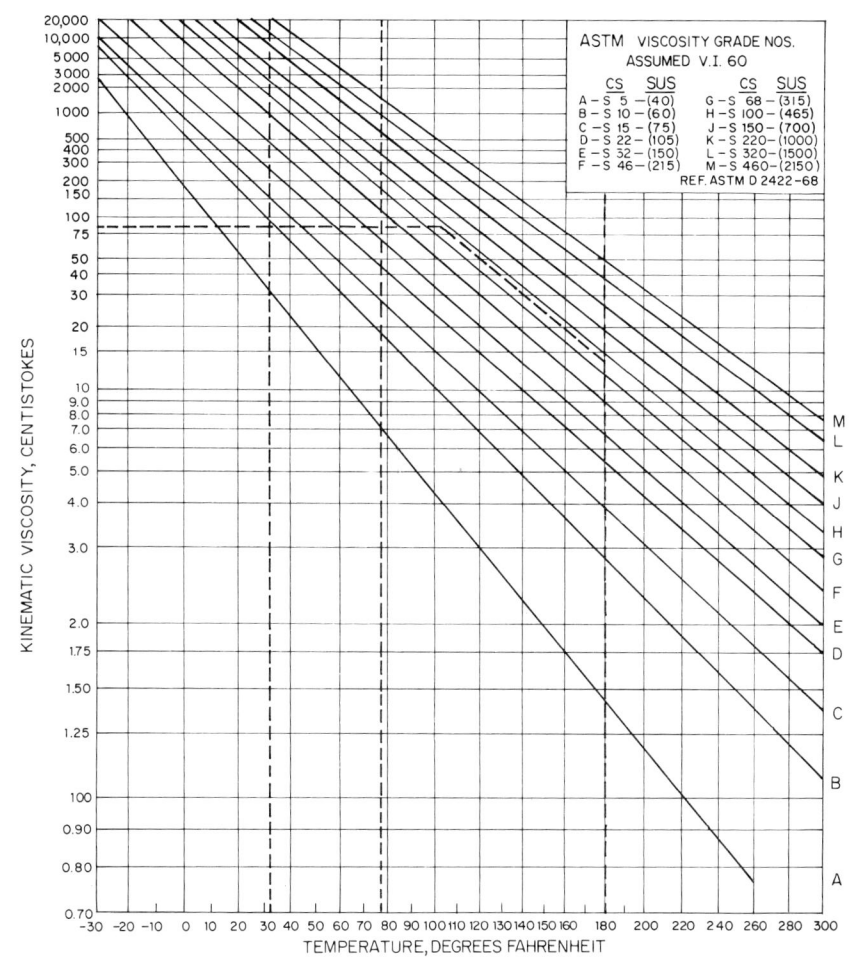

Fig. 9-35 Temperature-viscosity diagram.

When moving parts of a bearing come in contact with the grease, a small quantity of oil will adhere to the bearing surfaces. Oil is therefore removed from the grease near the rotating parts. Bleeding of the grease obviously cannot go on indefinitely; so new grease must come in contact with the moving part or a lubrication failure will result.

Many maintenance departments want to use one grease to lubricate all bearings in the plant. Some lubricant suppliers even advocate this technique. However, it is a risky procedure at best, since there is no true universal ball- and roller-bearing grease. A ball bearing is best lubricated with a fairly stiff grease which will channel. On the National Lubricating Grease Institute (NLGI) code, greases of the number 2 consistency, or 265 to 295 worked penetration, are normally recommended. For roller bearings a grease stiff enough to channel is not desirable, since the full width of the roller track would soon be starved for lubricant if the grease is not soft enough to slump back into the bearing when it is pushed aside. This generally means greases in the number 0 or 1 consistency class with worked-penetration numbers of 355 to 380 for grade 0 and 310 to 340 for a number 1 grease.

Whatever the consistency of the grease, it is still the properties of the oil compounded in the grease that determine if the bearing will be satisfactorily lubricated. All statements and guidelines outlined above in the discussion of oil lubrication also apply to grease-lubricated bearings.

Another characteristic of a grease that must be considered is its drop point. This is the temperature at which the grease passes from a semisolid to a liquid. Typical dropping points are as follows:

                Calcium .... 160–210°F
                Sodium .... 275–350°F
                Lithium .... 350–400°F
                Bentone .... 500°F plus
                Silicone .... 500°F plus
                Calcium complex .... 500°F plus
                Aluminum complex .... 450°F plus

The drop point is the characteristic referred to when a grease is advertised as being good up to 400°F. Whether it will lubricate a bearing or not is still a function of the viscosity of the lubricating oil, not of the drop point of the base. In fact, common industrial bearings made of standard through-hardened or case-hardened materials have temperature limitations of 200 to 300°F depending on the material and how it was heat-treated. The bearing manufacturer should be consulted for specific information.

Chapter **10**

Maintenance of Mechanical Power Transmission Equipment

A. BISHOP
Product Manager, Mounted Bearings, Dodge Division, Reliance Electric Company, Mishawaka, Ind.

R. ELSON
Product Specialist, Mounted Bearings, Dodge Division, Reliance Electric Company, Mishawaka, Ind.

M. D'HOORE
Product Specialist, Babbitted and Bronze Sleeve Type Bearings, Dodge Division, Reliance Electric Company, Mishawaka, Ind.

C. HUEMMER
Product Manager, V-Belt Drives, Tapered Bushings, Dodge Division, Reliance Electric Company, Mishawaka, Ind.

G. BOEYER
Product Manager, Shaft Mounted Speed Reducers, Flexible Couplings, Dodge Division, Reliance Electric Company, Mishawaka, Ind.

W. STREJC
Product Specialist, Chain Drives, Dodge Division, Reliance Electric Company, Mishawaka, Ind.

H. BURTON
Product Manager, Dry Fluid Drives and Couplings, Dodge Division, Reliance Electric Company, Mishawaka, Ind.

A. SAREEN
Product Specialist, Dry Fluid Drives and Couplings, Dodge Division, Reliance Electric Company, Mishawaka, Ind.

MAINTENANCE: TAKE CARE OF IT NOW—OR PAY LATER!

Too often, maintenance is done only *after* the damage is done. That's exactly when it's too late. An expensive and probably critical piece of equipment is down and you're stuck with costly repairs, to say nothing of even costlier downtime losses.

4-156   Basic Maintenance Technology

The right time to consider maintenance is *before* equipment failure. There is absolutely no substitute for a program of planned preventive maintenance, especially for mechanical power transmission components, to prevent the hair-raising problems of catastrophic equipment failure and the ensuing cost-raising problems of expensive replacement, installation, and lost production time.

As a guide to proper maintenance of the many mechanical power transmission components used in construction equipment, this chapter concentrates on various facets of installation—as a function of maintenance—and on maintenance procedures themselves. Its objective is to help you get all the life and performance capabilities out of your equipment the manufacturers have built into it.

MOUNTED BEARINGS

Bearings are used to support rotating shafts carrying loads of various types. Mounted bearings (Fig. 10-1) such as pillow blocks, flange bearings, and take-ups are frequently used since they offer additional advantages over the use of bare bearings:
1. Their integral housings make them easier and quicker to install.
2. Bearings are protected by seals to give them longer service life.
3. They are available in a wide variety of housing configurations, types of seals, mounting devices, rolling elements, and a range of sizes to match most application requirements.
4. Manufacturers such as Dodge® supply most of their mounted bearings completely assembled, adjusted, and lubricated, saving installation time and eliminating the possibility of contamination during installation.

Because these bearings are often used on equipment subject to shock loads, vibration, and operation in dirty, wet, or abrasive environments, proper selection, installation, and maintenance are critical to obtaining maximum performance and service life.

Selection of the correct mounted bearings is primarily a function of the equipment manufacturer. However, the user should also be concerned since he will be operating the equipment. Proper selection, installation, and maintenance are cheaper than repeated bearing failures and will result in reduced operating costs.

All manufacturers of mounted bearings supply installation manuals, which should be followed for correct installation and maintenance procedures. The following recommendations will help you know what to look for and what to avoid in order to assure long life and trouble-free performance.

Fig. 10-1 Special-duty pillow-block mounted bearing.

Mounting There are three basic types of arrangements used to mount bearings to a shaft: setscrew, eccentric collar, and adapter. The setscrew and eccentric collar types are used for small to medium sizes of light to moderately loaded bearings. Bearings having these types of mountings are positioned on the shaft and then secured to the shaft.

Setscrew Types. Setscrews are mounted either in the inner race or in overcollars. They should be properly tightened to keep the inner race of the bearing secured to the shaft.

Eccentric Collar Types. These consist of cam locking of collar and inner race of bearing to shaft. Eccentric collar mounts must be tightened in direction of rotation only.

Adapter Types. These result in the equivalent of a press fit yet can be used with economical commercial shafting. Adapter type mountings are recommended for applications involving heavy loads with shock and/or vibrations.

Shafting Proper shafting is a critical factor of mounted bearing installations. Industry standards for shafting tolerances, such as those listed in Table 10-1, should be consulted to make sure that the shafting is within recommended values on critical applications. Shaft fits for other than adapter-type mount bearings should be as snug as practical.

TABLE 10-1 Recommended Shaft Tolerances

Nominal shaft size, in.	Commercial shaft tolerance*	Recommended Shaft Tolerances		
		Setscrews in inner race or collar	Eccentric lock collar	Adapter mounting
Up to 1½	+.000−.002	+.0000−0005	+.0000−.0005	+000−.002
Over 1½ to 2½	+.000−.003	+.0000−.0010	+.0000−.0010	+.000−.003
Over 2½ to 4	+.000−.004	+.0000−.0010	....	+.000−.004
Over 4 to 6	+.000−.005	+.0000−.0015	....	+.000−.005
Over 6 to 8	+.000−.006	+.0000−.0020	....	+.000−.006

*Cold-finished low-carbon bars (Ref. A.I.S.I. Tables 5-1 and 5-2).

Alignment Bearing alignment is another installation factor to be considered. Bearings should be carefully aligned at installation. Careful alignment procedures optimize bearing and sealing capabilities.

Load direction Preferred direction of load is from the shaft into the base. If this is not practical, housing, cap and base bolt strengths, and bolt-tightening procedure, must be considered.

Seals Mounted bearings probably fail more from contamination and lack of proper lubrication than from fatigue of the bearing elements. Since the bearing seals keep contamination out and lubricant in, selection of the correct seal will affect overall bearing performance. Clearance seals operate with small clearance between seal and shaft or inner race. Misalignment will cause a seal to close up on one side and open up on the opposite side.

Labyrinth Seals. These are a type of clearance seal in that they have no rubbing parts. Labyrinth seals have radial passages generally requiring at least reversal in directions through which the grease passes to keep contaminants out of the bearing.

Lip Seals. These seals offer more initial resistance to entrance of foreign material; however, the seal and the surface against which it rubs are subject to damage from abrasive contaminants. The seal can also be damaged by heat conducted through the shaft or generated by the friction between the seal and the rubbing surface. Replacement should be made when loss of seal flexibility impairs its capability.

Combination Seals. These incorporate both labyrinth and rubbing-seal action. The rubbing-seal portion is generally protected by a metallic labyrinth seal and is within the confines of the seal.

Flingers Locking collars and adapter nuts used to lock a bearing to a shaft also act as flingers, adding to sealing efficiency and protecting the seal. While some bearings have no device other than a seal to act as flingers, many types of bearings are furnished with collars at either one or both ends of the bearing.

Lubrication Bearing manufacturers' installation and operating manuals provide specific information regarding lubrication schedules for applications at various operating speeds. Relubrication schedules established by experience are often required because of the wide range of operating conditions involving varying amounts of dirt, moisture, and heat.

The type of grease is important since grease not only acts as a lubricant but also provides a barrier against contamination of rolling elements. Lithium-base NLGI #2 greases are commonly used because of their water-resistant qualities and wide operating-temperature range. Other greases may be better suited for special application requirements, such as extreme temperatures, extreme loads, or chemical compatibility. Static oil, circulating oil, or oil mist are used for higher speed applications.

Too much grease can cause a bearing to run hot, particularly at higher operating speeds. Removal of a grease fitting until excess grease purges from a bearing will generally relieve this situation.

Troubleshooting Discovering and correcting operational problems in time will result in continued operation instead of unexpected failure. Careful observation generally pinpoints problems. Tips on what to look for as part of a regular maintenance program include:

1. Check seals. Look for signs of wear and heat. Fresh grease showing at seals is desired.
2. Check lubricants. There should be no pronounced changes in consistency or color and a minimum of contaminants should be present.
3. Check for heat, noise, or vibration. All of these can be signs of lack of lubrication. If the addition of a small amount of grease does not eliminate these symptoms, check with vibration analysis equipment.
4. Check mountings. Examine all fastening devices periodically for proper tightness.

After an initial short run-in period, also check base bolts, cap bolts, setscrews or collars, and adapter nut for proper tightness. Check bearings for freedom of operation and operating temperatures. It is not uncommon for a bearing operating temperature to be 70° to 80°F above the ambient temperature at higher operating speeds.

BABBITTED AND BRONZE SLEEVE-TYPE BEARINGS

Sleeve-type bearings have stood the test of time in literally thousands of construction applications where they have been proved for stock product economy, long life, and minimum maintenance characteristics.

The sleeve bearing can be simply described as a bearing housing which holds a sleeve of soft material—normally, babbitt or bronze—with a lubricant between the sleeve and the shaft. Theoretically, the bearing can deliver infinite life if the lubricant film is properly maintained.

The following guidelines are suggested for proper installation and maintenance of sleeve bearings to get dependable performance and long service life.

Maintain adequate grease lubrication High starting friction is a fact of life with sleeve bearings because of their large shaft and sleeve contact area. Yet, it is this large contact area that gives the bearing its heavy-load-carrying capabilities, quiet operation, and vibration- and shock-dampening characteristics.

In repetitive start and stop operation, lubrication tends to break down, resulting in metal-to-metal contact and destructive wear of the soft metal sleeve. Care should be taken to see that adequate lubrication is maintained under all operating conditions and especially when a bearing is operating under 10 rpm.

Maintain proper bearing alignment With non-self-aligning bearings, proper alignment is critical for uniform distribution of load under all operating conditions. If misalignment occurs, the shaft will tend to wear the cap (top) and base (bottom) on the opposite sides of the same sleeve bearing. Misalignment can be corrected by shimming the bearing housing.

Do not exceed prescribed rating loads Any good power transmission engineering catalog contains load rating tables for sleeve bearings. These ratings are the industrial standards established by the Mechanical Power Transmission Association (Standard No. 401). Starting and occasional peak loads may be allowed to exceed these ratings but never by more than 100 percent.

A general rule of thumb is that babbitted sleeve bearings will take approximately 250 psi. To find an approximate load rating at slow speed, use the formula

$$\text{psi} = \text{load} \div \text{shaft diameter} \times \text{length through bore}$$

Load ratings are slightly reduced with an increase in speed.

Maintain correct load direction Direction of load should not be closer than 30 deg to the grease groove. If a bearing has a cap which is not gibbed or doweled to its base, the load should be on the base and not closer than 30 deg to the joint between cap and base. When the direction of load is in the joint area between cap and base, angle babbitted and bronze pillow blocks are recommended. Typical applications are conveyor head shafts or shafts carrying heavy gear or chain drives.

Maintain shaft journal surface The finish on a shaft journal must be equal to that of commercial steel shafting (about 32 microinches) and the diameter must be within the tolerance of commercial steel shafting. A shaft must also be clean and free from rust, nicks, or burrs which would hamper operation and reduce bearing life. The better the shaft finish, the longer the sleeve bearing will operate with normal maintenance.

Inspect for wear periodically Soft bushing material is used in sleeve bearings so that wear is on the bearing and not on the shaft. By visually inspecting the ends of sleeve bearings periodically, replacement can be scheduled to eliminate downtime and catastrophic failure.

Do not operate at temperatures higher than recommended Ambient operating temperatures should not exceed 130°F for babbitted sleeve bearings and 300°F for bronze sleeve bearings. These temperature extremes are industry standards. If a shaft transmits heat from any source, shaft temperature at the bearing should not exceed these extremes.

Protect against adverse operating conditions The soft bushing material used in sleeve bearings does have self-healing characteristics. However, it is best to keep contaminants out or to remove them as soon as they get in. This is especially important in the dirt and dust of construction service. Greasing bearings to purge contaminants and to maintain adequate lubrication cannot be done too often in the effort to ensure long bearing life and dependable performance.

TAPERED BUSHINGS

Tapered bushings, such as the Taper-Lock® manufactured by Dodge, offer the most practical alternative to the restrictions of shrink-fit or bored-to-size bushings (Fig. 10-2).

The advantages of tapered bushings over these other types are numerous. They allow you to meet a maximum of application requirements with a minimum inventory of bushings. They provide easy-on, easy-off installation, good gripping strength, and excellent concentricity, and they are readily available in a wide range of sizes without special machining.

Selection of the tapered bushing is based on the torque capacity of the component—sheave, sprocket, clutch, etc.—and the standard commercial shafting being used. Incorrect selection or installation can result in the components flying off while in use, causing costly equipment damage and personal injury. The following suggestions will help you get dependable performance and long life from tapered bushing installations.

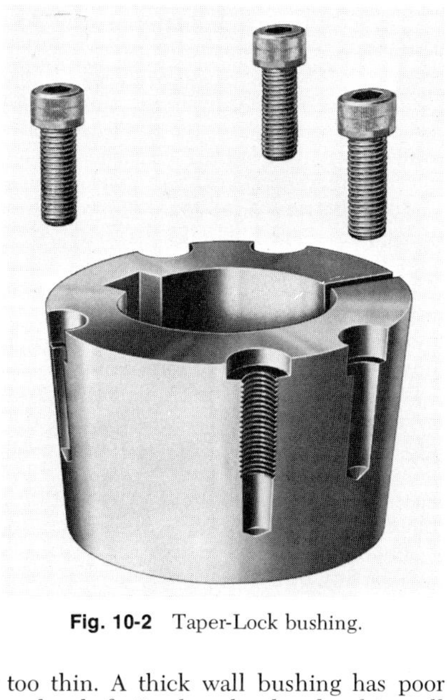

Stick to manufacturer's size recommendations Generally, the component manufacturer will indicate the correct bushing size to meet torque and shaft requirements. For best results, stick to those recommendations.

Fig. 10-2 Taper-Lock bushing.

Consider wall thickness Avoid bushings whose walls are relatively too thick or too thin. A thick wall bushing has poor flexibility and will resist good gripping action on the shaft. On the other hand, a thin wall bushing is too fragile and could easily be broken, especially in the removal process.

Clean shaft bushing and component To take maximum advantage of a bushing's capabilities, be sure that the shaft is free of nicks and burrs and that no dirt or chips are on the bushing and component taper bore. Such contamination will impair the precision-machined fit of the tapered bore, the bushing, and the bushing shaft. The result will be lack of concentricity, an imbalanced component, or gripping strength on the shaft that is less than optimal.

Tighten mounting screws Follow the bushing manufacturer's recommended torque values in tightening mounting screws. To assure proper seating, tighten screws alternately, using a torque wrench and cheater bar to get sufficient torque values.

Instructions for Dodge Taper-Lock bushings recommend that a bushing be rapped in with a hammer and drift as part of the installation sequence. This procedure supplements the mounting screws in assuring proper seating of the bushing in the component. Flanged bushings should *not* be rapped in.

When removing the component from a shaft, the opposite technique is often helpful. In other words, after the mounting screws are loosened, the hub is rapped rather than the housing. This tends to back the bushing out of the component and makes disassembly easier than with flanged bushings.

V-BELT DRIVES

V-belt drive performance depends on four basic factors: drive design, component selection, installation, and maintenance. If any of these are neglected, the system cannot function properly and premature failure is likely.

Generally, the user of construction equipment has little control over drive design and component selection. These are functions of the equipment manufacturer and the equipment in which the drives are installed. However, installation as it relates to component or belt replacement and maintenance as it relates to prevention of trouble are factors which the user can control for optimum return on the investment in V-belt drive equipment.

The following is a quick guide to proper installation and maintenance procedures.

Proper Installation Minimizes Future Maintenance

Inspecting sheaves Sheaves that are worn or in less than perfect condition can seriously shorten belt life. Using a sheave groove gauge, check for groove wear or distortion. If grooves are worn, the belt will bottom out, resulting in slippage and belt charring or burning. If sidewalls are dished, the bottom shoulder of the sheave will wear the bottom corner of the belt, causing premature failure. Also check sheave grooves for rust and wipe clean of oil and grease. Inspect grooves for cracks, chips, or burrs which could damage the belts. Don't use a sheave that has been cracked or damaged in transit; it could fly apart and cause serious damage to equipment and injury to operating personnel.

Mounting new sheaves Three types of sheave bushings are in general use today: the split-taper-keyed bushing; the Q-D, or quick-disconnect; and the Taper-Lock bushing manufactured by Dodge. For best results, bushed sheaves should be mounted and tightened according to specific instructions supplied by the manufacturer. Refer to the instruction/installation manual. Always be sure that the mating surfaces of the sheave and hub are free of foreign material—dirt, grease, paint, or burrs. Sheave wobble could indicate the presence of foreign material and should be corrected.

Checking alignment V-belt drives do not require as tight alignment tolerances as most other drives. However, unless the belts run through the sheave grooves in a relatively straight line, wear will be accelerated. Using a straightedge, steel tape, or even a piece of string, see if both the driver and driven sheaves are properly aligned to ensure maximum operating efficiency. Another common cause for misalignment is incorrectly mounted sheaves. Rotate drive slowly to check for wobble, which indicates improper installation.

Selecting belts Always select belts to match sheave grooves. The instruction/installation manual will usually include selection tables and conversion charts. A sheave-groove gauge will also be useful in this procedure to determine the proper cross section of belt required. Belt brands should never be mixed on a multibelt drive. Always use a matched set from a single manufacturer; otherwise belt and drive life will be substantially shortened because of differing belt-performance characteristics.

Installing a belt Prior to installing belts, be sure that power is turned and locked off so there is no chance of accidental or inadvertent start-up while work is being done. Never pry or roll belts into the sheaves. It is not only dangerous to an installer's hands, but it also tends to shorten belt life even if there is no visible damage. Instead, use the drive take-up and drop belts into the grooves. Use a bar if necessary but only to move the motor, never to pry the belts.

Tensioning belts properly Proper tensioning is probably the most critical factor contributing to long V-belt drive life with minimum maintenance. Too much tension shortens belt and bearing life, while too little can cause a belt to slip under load. Generally, the best tension is the least tension at which belts will not slip under full load.

Refer to manufacturer's specifications for proper tensioning and use a tension tester to meet recommended values.

Proper Maintenance Minimizes Costly Failures

Prevent oil and grease buildup Standard service belts exposed to excessive oil and grease contamination are likely to fail prematurely. Check and replace bearings and/or bearing seals if necessary to remove the most common source of oil and grease leakage. If the operating environment is naturally oily or greasy, special oil-resistant belts should be used. Drip shields should be provided where necessary to prevent oil or grease leakage onto the belts from equipment above.

Prevent dirt accumulation Dirt buildup on belts or in sheave grooves accelerates wear and impairs traction. Check and clean when necessary for optimum performance and long belt life.

Never use belt dressing Belt dressing of any kind is unnecessary and actually harmful to V-belts which operate from the wedging action of the belt in the sheave groove. Belt dressing causes belt slippage and attracts dirt and grit which accelerate wear of both the sheave and the belt.

Troubleshooting V-belt drives are an extremely reliable and efficient means of power transmission that require few special tools or procedures for good maintenance. The main ingredients of a successful preventive maintenance program are visually inspecting, listening, and then correcting apparent drive problems. An easy-to-use table of common problems, symptoms, and required corrective action is included for convenience in troubleshooting V-belt drives on a regular basis (see Table 10-2).

SHAFT-MOUNTED SPEED REDUCERS

A shaft-mounted speed reducer is an expensive piece of machinery usually performing a critical function in construction equipment (Fig. 10-3). Install and maintain it properly and it will give you years of dependable service. Abuse or ignore it and you will pay a high price in lost time and emergency repairs.

Maintenance Begins with Proper Installation

The first step in maintaining a shaft-mounted reducer is to install it properly. Begin by obtaining the correct reducer. This may sound elementary, but, on a job site where several reducers are being installed, it can save embarrassing switching later.

Check the driven shaft The driven shaft should be examined for burrs and to make sure it is straight. Bent shafts can cause excessive reducer movement and will eventually loosen the torque arm bolts.

Position and tighten the reducer Position the reducer on the driven shaft as close as possible to its supporting bearing. Then, tighten it to the shaft, using the procedure recommended by the manufacturer. Most manufacturers use either setscrew or tapered bushings to secure the reducer against axial movement on the driven shaft. Rotate the reducer to its running position and secure the torque arm to a rigid support.

Remove shipping tape With the reducer installed in its running position, locate the magnetic drain plug in the bottom hole and the vent plug in the top hole. It is extremely important that shipping tape covering the vent hole be removed. Failure to do this will result in excessive internal pressure which can blow the reducer's oil seals or cause oil to leak past the seals.

Fig. 10-3 Shaft-mounted speed reducer.

TABLE 10-2 V-Belt Drive Failures and Remedies

V-belt drive failures and remedies

Problem	Probable causes	Solution

(Table 10-2 cross-references V-belt drive problems with their probable causes and solutions in a matrix format.)

Problems (rows): Loose Cover & Swell; Weathering or "Craze" Cracks; Gouges; Spin Burn; Envelope Wear; Uneven Envelope Wear; Ply Separation; Side Split; Broken Belts; Belts Turn Over; Hardening & Premature Cracking; Belt Squeal; Excessive Stretch; Excessive Vibration; Belts Too Long At Installation; Belts Too Short At Installation; Mismatched Belts At Installation; Cut Thru on Top (Joined Belts); Flange Wear (Synchronous Belts); Web Fabric Wear (Synchronous Belts); Tooth Shear (Synchronous Belts).

Probable causes (columns): Excessive Oil; Exposure to Elements; Pried Over Sheaves; Contact w/Obstruction; Insufficient Tension; Stalled Driven Sheave; Constant Slippage; Rough Sheaves; Substandard Sheaves; Excessive Tension; Shock Load; Foreign Material; Excessive Dust; Drive Misalignment; Worn Sheaves; Excessive Vibration; High Ambient Temperature; Excessive Tension; Drive Underbelted; Inadequate Tension; Damaged Tensile Member; Incorrect Belts; Incorrect Drive Setup; Insufficient Take Up; Improper Matching; Mixed Old & New Belts; Non Parallel Shafts; Different Manufacturers; Belt/Pulley Incompatible.

Solutions (columns): Lubricate Properly; Clean Sheaves & Belt; Replace Belts; Provide Protection; Install Properly; Check for Belt Length; Remove Obstruction; Tension Properly; Free Sheaves; Replace Sheaves; File Smooth; Redesign Drive; Operate Properly; Align Drive; Provide Ventilation; Check for Proper Belt; Check Machinery; Use Only New Belts; Use Single Source; Check Fit; Replace Pulleys.

4-162

Maintenance of Mechanical Power Transmission Equipment   4-163

Add proper lubrication   Use only the recommended quantity and grade of oil specified on the name plate and installation instructions. As a safety check, most manufacturers provide an oil-level plug which should be used during filling and subsequent level checks. For reducers equipped with backstops, do not use oils that contain graphite or molybdenum disulfide because these oils can cause sprags to malfunction.

Mount input sheave   Proper mounting is important because it affects the life of the input shaft bearings. The sheave should be as large or larger than the minimum recommended by the manufacturer. It should be mounted as close to the reducer bearing retainer as possible to minimize the overhung load.

Check alignment and adjust V-belts   Both the motor and reducer sheave should be checked for alignment with a straightedge. Realign, if necessary, and adjust V-belt tensions. The use of a tension tester will prevent overtightening the belts, which could shorten the life of the motor and reducer bearings.

Continued Maintenance Prolongs Life

If all the reducers that failed prematurely were broken down by categories, the number one culprit would be poor maintenance. This covers a variety of sins, but first and foremost among them is failure to change oil as recommended.

Change oil frequently   Most manufacturers state the maximum operating period between oil changes should be 6 months or 2500 hr of operation, whichever occurs first. Such figures are fine for normal installations, but, under the extreme operating conditions of the construction industry where dirt, dust, and occasional high temperatures are common, the oil should be changed every 1 to 3 months.

Consider seasonal changes   Where seasonal changes are extreme, grades of oil should be changed to suit warm or cold weather conditions. When changing oil, a careful inspection of the magnetic drain plug can provide valuable information about the condition of the gears and bearings and thereby possibly prevent costly unscheduled downtime.

Operating inspection   In addition to regular oil changes, proper maintenance should also include periodic inspections of the reducer while it is operating. In general, such inspection will provide numerous clues that will prevent premature failure and expensive equipment shutdowns.

Troubleshooting   The following list enumerates symptoms of trouble to look for and what each may indicate:

1. Excessively high running temperatures can indicate overload or too high an oil level.
2. Excessively loud noise levels can indicate impending bearing or gear failure.
3. Oil on the reducer can indicate a damaged seal or loose oil plugs.
4. Loose or turned V-belts can indicate that the torque arm may not be properly fastened to a rigid support.

If any of these conditions exists, stop the machinery and provide whatever service is necessary to correct the malfunction before it does excessive or permanent damage.

CHAIN DRIVES

Since the early 1890s when chain drives were developed to power bicycles and the first automobiles, they have emerged as one of the most economical methods of power transmission available to industry. In the face of rising costs, chain drives offer one of the best horsepower-per-dollar packages for meeting numerous application demands in construction equipment. However, to take full advantage of the economy and savings inherent in chain drives, correct lubrication and maintenance are vital. Without these, an investment in chain-driven equipment can become a burdensome expense because of early failure and performance problems.

The following suggestions are recommended to ensure that a chain drive delivers the life and performance capability built into it by the component manufacturer.

Installation Recommendations

Machinery Preparation   Prior to mounting drive components, be sure that rigid support members are present to oppose vibration and drive movement during operation. Shafting and shaft-support bearings must also be able to withstand the bending movement

4-164 Basic Maintenance Technology

imposed by drive operation. Nonrigid support, excessive shaft displacement, or improper bearings can all lead to premature drive failure.

Alignment Both shafts must be aligned parallel to each other (Fig. 10-4), so that they rotate in the same plane of operation. It is also necessary to make sure that shafts are level. These alignment factors can be verified by using a machinist's feeler bar and a spirit level.

On portable machinery, where misalignment could be a frequent problem, the use of a wide-clearance- or offset-sidebar-type chain might be desirable. Such chains can take 4 in. of lateral displacement or 8 deg of twist per 4-ft length.

Cleaning components All shafts, sprockets, bushings, and keys, must be thoroughly cleaned to remove chips, dirt, and burrs that could interfere with drive operation and wear-drive components.

Mounting sprockets Two mounting methods are generally in use: setscrew with key, and tapered bushing. Both offer the user certain advantages and restrictions within the capabilities of their design. But industry experience tends to prove that the taper bushed sprocket—because of its excellent concentricity—leads to maximum drive life.

Installing chain Any length of chain with an even number of pitches requires only one connecting link for assembly. Chain with an odd number of pitches requires one connecting link plus an offset link. Because odd-pitch chain and offset links tend to weaken a chain assembly, their use should be avoided whenever possible.

Proper tensioning Do not install chain too tightly or it will impose bearing loads greater than normal. On the other hand, loose installation creates drive noise and chain-speed pulsations. Either way, the result is abnormal chain and sprocket wear. To tension

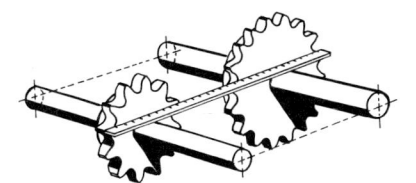

Fig. 10-4 Sprocket alignment.

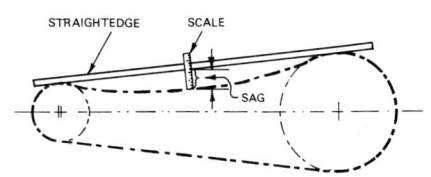

Fig. 10-5 Chain-sag measurement—horizontal drive.

chain properly, turn the free sprocket until lower strand is tight. This creates a sag in the upper strand. With the use of a straightedge and scale (Fig. 10-5), adjust the sag so that it does not exceed 2 or 3 percent of the horizontal length between sprockets or the center distance of the total chain.

Maintenance Recommendations

Inspection As with all forms of power transmission equipment, there is no substitute for a systematic, periodic inspection of chain drives to prolong service life. Because the greatest amount of adjustment is usually necessary during the first few hours of run-in after drive installation, inspection should occur more frequently during the first 8 hr of operation but can become less frequent after that period.

Lubrication The importance of maintaining adequate lubrication of chain-drive components can't be emphasized enough. Most chain-drive failures result from a lack of lubrication that can cause destructive metal-to-metal contact of moving parts. Most chain-drive-component suppliers such as Dodge provide factory prelubrication. However, it is critical that some method of regular lubrication maintenance be provided by the chain-drive user—either manual, drip, bath, or oil stream. Consult the manufacturers' instructions for their recommendation for a particular drive speed and application.

Cleaning Chain should be cleaned at least once a year or immediately if the presence of corrosion or foreign material is detected. Proper procedure for cleaning roller chain is as follows:
 1. Completely remove the chain strand from the sprockets.
 2. Thoroughly brush-clean chain in kerosene.
 3. Inspect the chain strand for signs of wear or corrosion.
 4. Soak the chain in an oil bath so lubricant thoroughly penetrates components.
 5. Allow chain to hang freely to drain off excess lubricant.

6. Clean and inspect sprockets and shafts.
7. Reassemble chain strand around the sprockets.

Replacement Should excessive wear of either the chain or the sprockets be detected, they should be replaced immediately and simultaneously. Placing good chain on badly worn sprockets or worn chain on good sprockets tends to shorten drive life. When replacing chain, length should be determined by the number of pitches rather than by the measured length, which might actually be longer than necessary because of some chain elongation.

Troubleshooting Periodic inspection can reveal several drive problems which, if caught and corrected soon enough, can save the drive, your temper, and catastrophic failure with its ensuing downtime losses. For convenience and easy reference, these problems, their causes, and the corrective actions necessary are listed in Table 10-3.

FLEXIBLE COUPLINGS

Flexible couplings such as the Para-Flex® by Dodge are turning out to be the most maintenance-free available to users of construction equipment (Fig. 10-6). Properly installed, Para-Flex will provide years of trouble-free service and will even give an early warning that element failure is approaching.

Proper alignment is vital As in all elastomeric shear-type couplings, service life is a function of alignment. The closer the coupling is aligned, the longer its service life will be. Check carefully during installation.

Tightening the clamping bolts One factor that more than any other can cause premature failure and operating problems is insufficient tightening torque on clamping bolts. Bolts supplied are grade 8, with a minimum tensile strength of 150,000 psi, so there is little likelihood of stripping threads or snapping off the bolt head in tightening a bolt properly. Use of a torque wrench is recommended.

Periodic visual inspection The coupling element should be checked periodically for signs of cracking. These cracks indicate that the element is reaching the end of its useful life and should be replaced in the next 3 to 6 months.

Fig. 10-6 Para-Flex coupling.

The element will not fail catastrophically except under extreme overloads, in which case it acts like a mechanical fuse to protect the driven machinery. Under normal operating conditions, it will give adequate warning before failure and replacement can be scheduled during routine maintenance shutdown.

DRY FLUID DRIVES AND COUPLINGS

The Flexidyne® dry fluid drive and coupling manufactured by Dodge is extensively used in construction equipment to provide smoother starts, prevent breakage, and reduce maintenance on motors, gears, bearings, and driven machinery (Fig. 10-7).

The fluid is heat-treated steel shot, a measured amount of which—called the *flow charge*—is contained in a housing keyed to the motor shaft. When the motor is started centrifugal force throws the flow charge out to the perimeter of the housing, packing it between the housing and rotor. In this operating mode, power is transmitted to the load. After the starting period of slippage between the housing and rotor, the two become locked together and full-load speed is achieved with operation continuing without slip and at 100 percent efficiency.

TABLE 10-3 Solutions for Chain and Sprocket Failures

Common symptoms	Probable cause	Corrective action
Nonsymmetrical wear on sprockets or rollers	Shafts out of parallel or not in same plane	Realign shafts
Wear on inside of roller plates or side tooth form of sprocket teeth	Sprockets offset on shafts (misaligned), or out of parallel	Realign sprockets
Wear on tips of sprocket teeth	Chain elongated excessively	Replace chain
	Improperly cut sprockets	Replace with correct sprocket
Worn or hooked sprocket teeth	Unhardened sprockets	Replace with hardened sprockets
Wear on edges or sides of link plates	Chain contacting case or fixed object	Increase case clearance or move fixed object
Excessive vibration	Excessive eccentricity or face run out in sprocket	Replace with properly machined sprocket
	Broken or missing roller	Repair or replace chain
Premature elongation	Inadequate or contaminated lubrication, or underchaining	Increase oil flow or redesign
Brown-red oxide in chain joints and oil	Inadequate lubrication	Improve lubrication

Chain jumps sprocket teeth	Wear to vertical limit or excess initial slack	Adjust centers or idler
	Wear to mismesh on large sprocket	Replace chain
Broken chain parts	Drive overloaded	Redesign or avoid
	Excessive slack causing chain to jump teeth	Periodically adjust center distance
	Foreign object	Prevent entry
	Excess chain speed	Redesign or avoid
	Poorly fitting sprockets	Replace
	Inadequate lubrication	Proper lubrication
	Corrosion	Prevent or use noncorrosive chain
Excessive noise	Chain contacting fixed objects	Remove objects
	Inadequate lubrication	Improve lubrication
	Broken or missing rollers	Repair or replace chain
	Misalignment	Check shaft and sprocket and realign
	Chain jumping sprocket teeth	Adjust center distance

Properly installed, the Flexidyne requires only minimum maintenance to deliver years of economical service. An instruction/maintenance manual is supplied with each unit and should be consulted for installation and maintenance procedures. However, the following guidelines are suggested to provide dependable, long-term performance.

Placing the unit If the output shaft or motor shaft extends too far into a driven hub, the Flexidyne will not slip—no matter how much flow charge is used. Excessive vibration may also be caused by this type of improper installation. A quick inspection through the output hub will determine whether this is the reason for no slippage.

Mounting and aligning Such factors as flexible mounting structure and misalignment of other components are major causes of vibration and shortened service life. Dodge

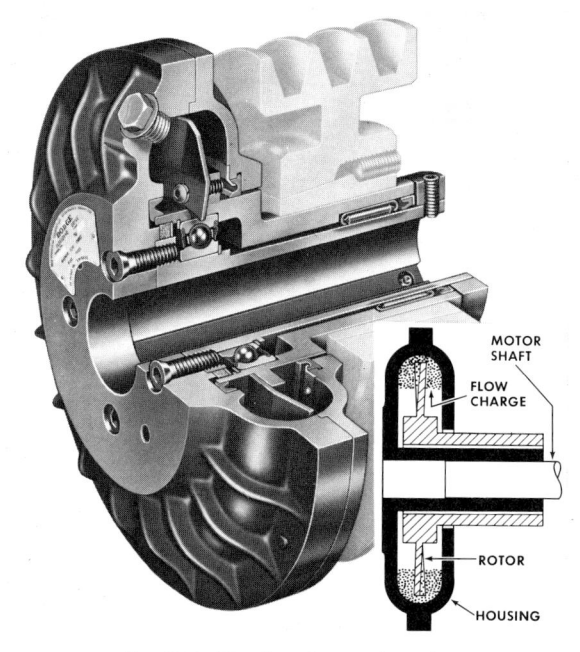

Fig. 10-7 Flexidyne drive and coupling.

recommends that units be aligned with a dial indicator and that motor bases be doweled to assure positive alignment.

Adverse operating conditions In applications where a Flexidyne is subjected to unusually corrosive elements or a very humid condition, stainless steel flow charge will provide better service than the standard cast-steel charge. Another use of stainless steel flow charge is in installations subjected to long periods of idleness where the drive is exposed to wide ranges of temperature or humidity.

High-speed operation Generally, considerably higher-than-normal motor speeds should be avoided when using the Flexidyne. However, where high speeds are unavoidable, it is important that the Flexidyne be connected to the driven machinery or that some resistance be offered to the rotation of the rotor. The rotor must slip for a minimum of 1 to 1½ sec in the flow charge so that the charge will be evenly distributed in the housing. Otherwise, an undesirable out-of-balance condition is likely to result.

Changing operating characteristics Acceleration time and starting torque provided by the Flexidyne can be changed with the addition or removal of flow charge. Consult the instruction/maintenance manual for amounts of flow charge to be used. The charge can be poured in or out of the housing through an easily accessible filler plug.

Overload protection A Dodge speed-drop cutout is included routinely with large Flexidyne units and is available as an option on smaller units. It must be installed to protect against heat damage in applications where overload conditions are anticipated.

The cutout can be connected to interrupt current or to activate a bell, light, or other type of warning device. For hazardous atmospheres, a special explosion-proof cutout is available.

Frequent starting For normal service, involving three or four starts a day of not over 6-sec acceleration time each, the flow charge should be changed after 10,000 hr of operation. Under adverse conditions, such as more frequent starting or extended acceleration time, flow charge should be changed more often. For very extreme conditions, it may be necessary to change the flow charge every 3 to 4 months. If in doubt, consult the instruction/maintenance manual included with each unit.

Lubrication The Flexidyne is lubricated at the factory by Dodge and requires no further lubrication. Never apply oil or grease to the flow charge.

Erratic acceleration If a Flexidyne unit is experiencing erratic or nonrepetitive acceleration, the reason is usually a breakdown of the flow charge into a powdery form. This deterioration is caused by prolonged use without adding new shot. The old flow charge needs to be removed and replaced with new shot.

Slippage while running If the Flexidyne slips while running, it is an indication that the flow charge has worn out and needs replacement. Also, under long service, the rotor will eventually wear and replacement will be necessary. The need for new shot or a new rotor will usually be signaled by a loss in the power transmitting capacity of the Flexidyne.

Chapter **11**

Maintenance of Scaffolds and Ladders

LEONARD SAFIER
President, Patent Scaffolding Company, A Division of Harsco Corporation, Fort Lee, N.J.

Major strides have been made in recent years in the availability of different types and wider choices of scaffolds and ladders for above-the-ground work. Today's manager has more varied and versatile equipment available than ever before; consequently, without realistic guidance, it is easier to make the wrong choice of equipment.

Additionally, more consideration has to be given to employee safety, since current OSHA regulations require safety and protective components which in preceding years were never a major issue. In pre-OSHA days, it was important only to supply a guardrail; under today's working conditions, this guardrail must be accompanied by a midrail and toeboard plus, under certain circumstances, 19-gauge wire mesh installed between guardrail and toeboard.

In general, the wise manager should insist upon having a variety of access equipment at his disposal; the days have passed when you could make do with one old weatherbeaten stepladder and one extension ladder in poor condition. Certain types of ladders and scaffolds are especially suited to certain specific tasks but can often be most unsuitable for other work of another nature requiring a scaffold more specifically suited for the work. Consideration must always be given to the type of equipment which will produce the highest specific work output and at the same time accomplish this in a safe manner with safe equipment. Sturdy work platforms of high stability and ample size result in freely moving, confident workers; the solidity of the work platform is a very important psychological factor and one which is commonly overlooked. Many people are unable to climb more than a few feet above the ground without feeling the necessity of having to hang onto something tightly with at least one hand; consequently, a man assigned to work on a ladder or at a precarious perch could produce only a small fraction of his work output doing the same work at ground level.

In considering the type of equipment for specific tasks, the following typical questions should be asked:

1. How high is the work?
2. Is it spot work or is it continuous in a horizontal- or vertical-pass direction?
3. Is it work which can be most easily done from top to bottom over a length of 10, 20, or 30 ft?
4. Is it a large area which will require a combination of frequent horizontal and vertical passes?
5. Is the area below the work suitable for support from the ground?
6. Is there a wall or an unobstructed space to support a ladder or scaffold?

7. Must people or conveyances pass unobstructed beneath the work?
8. Depending on question 2, how frequently is it necessary to move the work support?
9. How many men and how much equipment must be supported on the scaffold?

The following is a list of scaffolds and ladders, followed by basic descriptions to enable you to determine the best equipment to use for individual circumstances:

Stepladders and extension ladders—wood, aluminum, and fiberglass
Welded aluminum folding or sectional scaffolding (rolling towers)
Welded sectional steel scaffolds and rolling towers
Safety swinging scaffolds, one- and two-point suspension
Tube and coupler scaffolds (steel and aluminum)
Special-design scaffolds

STEPLADDERS AND EXTENSION LADDERS

Stepladders and extension ladders are preferably used for spot work at relatively low heights with no obstructions. The size and weight of a ladder are important; there is room for only one man on a ladder, but if two men are required to carry and lift the ladder into position, then one man is working only a small part of the time.

Choice of materials for ladders—wood, metal, fiberglass.

1. Wood is preferred for regular use in relatively low-height operations and is the workhorse of the industry. However, a wood ladder (or any other type for that matter) can be abused through overfamiliarity. Frequent inspection, properly documented, is essential. Wood is often abused because of its ability to withstand high excessive loads for short periods of time, such as walking or working on it while it is in a *flat* position. Wood is organic and therefore biodegradable; therefore, wood-ladder life can be extended by water-repellent and antifungicide treatments such as pentachlorophenol.

2. Aluminum ladders are preferred by many for their lighter weight. However, Underwriters' Laboratories, Inc., issue the following CAUTION printed on their listing label:

ELECTRICAL SHOCK HAZARD—METAL LADDERS SHOULD NOT BE USED WHERE CONTACT MAY BE MADE WITH ELECTRICAL CIRCUITS. REFER TO INSTRUCTION LABEL.

3. Fiber-glass ladders are preferred for use in workplaces involving electrical hazards, are about the same weight as wood ladders, and have the additional advantage of insulation properties. Fiber glass is less subject to the effects of abuse (resulting from being dropped or otherwise mishandled) than either wood or metal ladders, since fiberglass side rails have superb recovery from bending or distortion. For proper maintenance, fiber-glass ladders should be coated with a hard floor wax or car wax to reduce the tendency of glass fibers to bloom at areas of friction and scraping.

Some things are so simple, both in the way they are made and in the ways they are used, that the scientific principles upon which they are based are barely visible. With all the simplicity it has constantly retained, the ladder as we know it in its best form today represents a high degree of engineering skill, scientific accuracy, and, most important, dependable safety.

Vital differences in ladders may not be detected except by an expert; hence industry finds it profitable to seek expert ladder advice. Proper weight, exact balance, scientific proportions, dependable quality of materials, character of workmanship, and, of utmost importance, adaptability of a certain type of ladder to the particular kind of service for which it is intended—these factors are essential in the modern ladder even though they may not always be visible to the uninitiated.

Most of the states and principal industrial groups take the ladder question seriously and have acknowledged the importance of ladder efficiency and safety by establishing rigid codes designed to ensure the use of ladders built for the special kind of service required.

A program of ladder upkeep and care should be as much a part of any company's safety program as its maintenance of equipment or any other efficiency or safety devices. The correct use of ladders must be considered no less than the matter of choice when purchases are made.

Three ladder groups Generally, ladders can be classified in three groups: extension ladders, single (straight) ladders, and stepladders. Also, there are special-purpose ladders, which may constitute a fourth classification.

Extension ladders are used in building construction, painting, plastering, maintenance, and almost everywhere that an adjustable ladder is needed.

Single ladders, which of course are not adjustable for height, are generally used when one type of work is to be done at a more or less standard height.

Both extension and single ladders should be fitted with ladder feet to prevent slipping.

Stepladders are what the name implies. They also are self-supporting. Selection for both efficiency and safety is of prime importance. The use of the proper type of stepladder, whether standard classification or special purpose, is receiving more and more attention from industrial executives responsible for their selection.

For design and construction details, the "Safety Code for Portable Wood Ladders" (A14.1-1968), published by the American National Standards Institute (ANSI) is a reliable guide. Since not all ladders meet the standards of safety necessary in industrial plants, the following condensed information on the most important types of ladders should be helpful to those responsible for procurement and factory maintenance. In addition, Underwriters' Laboratories, Inc., approval of a ladder indicates it meets basic standards of design and construction for safety for various categories of use.

Extension ladders Extension ladders (Fig. 11-1) consist of two sections with three-section ladders also available for longer lengths. The strength and safety of such ladders come from the type and quality of wood used in the side rails, as well as their size. Rung size also is important. The required thickness of the side rails depends on the length of the ladder and the type of wood used. The distance between the side rails of the bottom section of a parallel-side ladder should be at least 14½ in. inside to inside for ladders of extended lengths up to 28 ft. Between 28 and 40 ft it should be a minimum of 16 in., and for ladders over 40 ft it should be 18 in. Two-section extension ladders longer than 60 ft are not allowed under the ANSI code.

Good-grade rope (5/16 in.-diameter minimum) and pulley (1¼-in. diameter minimum) for raising the upper section are essential features of well-built extension ladders, because weakness at this point may result in serious injuries. All holes to hold the wood rungs must either extend through the side rails or be bored to give at least ⅞-in. length of bearing to the rung tenon.

Extension ladders should be equipped with automatic spring locks, which enable the worker who is on the ground or floor to raise and lower the upper section. No manual adjustment is necessary. This automatic spring lock, like all the other ladder hardware, is better if plated to resist rust.

Single ladders Single ladders longer than 30 ft are not permitted under the ANSI code. Rungs should be not more than 12 in. apart, and all holes for rungs should be drilled in the same manner as for extension ladders. Rungs must be tight in the hole and secured in place with nails to prevent turning. Pressed-steel rung braces under several rungs is one earmark of a good single ladder.

Fig. 11-1 Extension ladder meeting ANSI Safety Standard A14.1 and listed by UL, Inc.

The width between side rails at the base, inside to inside, must be at least 11½ in. for all ladders up to and including 10 ft. This minimum width increases 14 in. for each additional 2 ft of length.

Because ladders are subjected to rough usage in such trades as masonry and building construction, the extra-heavy-duty single ladder is preferred. Both the side rails and the rungs are heavier than in the standard single ladder, although the length is limited to 30 ft. Mason's ladders should measure at least 12 in. between side rails (inside to inside) up to 10 ft, with ¼ in. for each additional 2 ft of height. Rungs must be between 8 and 12 in. apart.

Stepladders Stepladders should be made so that the treads will be level in the open position. Good-quality stepladders are designed so that, when open, the slope of the front section is at least 3½ in./ft and the slope of the back section at least 2 in./ft for each 12 in. of side rail. Stepladders, in accordance with ANSI and UL code requirements, should never be furnished in lengths greater than 20 ft, and steps should be uniformly spaced not more than 12 in. apart. Good-quality ladders are equipped with steel safety spreaders, so designed that they will not injure hands when opening and closing. The spreaders also act as braces between the front and rear side rails.

The inside-to-inside measurement of side rails at the top should be at least 11½ in., with an increase of at least 1 in. for each foot of ladder length. This assures a safe, wide base.

Fig. 11-2 Platform stepladder. Preferred because it permits worker to have both hands free.

Safe footing is assured in high-quality ladders by reinforcing steps by means of trussing and bracing, substantially attached by rivets, bolts, or screws. Ladders should be checked frequently to be sure that steps are securely fastened.

Stepladders for heavy-duty use are usually identified by a rung-type back, this construction being more rigid than the slatted back. It permits working from either side.

The minimum dimensions of parts of the heavy-duty ladder (or equivalent cross section developing an actual working stress per square inch as required by the ANSI Code) are shown in the table on the following pages.

The platform stepladder (Fig. 11-2) is by far the most popular, having proved itself by reducing the number of accidents resulting from falls and dropped objects. The 14- by 19-in. platform gives the worker a a firm footing and a guard on three sides, at the same time permitting him to work with both hands.

In the safety platform ladder, steps are truss-rodded and also knee-braced. The steel spreader is of the safety type, a shield over the joint preventing injury to the worker's hands. Good spreaders do not permit the ladder to fold up accidentally. Holes in the top are used as a tool rack, thus reducing danger from falling tools.

Special-purpose ladders There are many types of special-purpose ladders, such as shelf, fruit pickers', trolley, decorators, and paperhangers'. One of the more familiar types is the sectional which, as the name implies, is made in interlocking sections, either

Maintenance of Scaffolds and Ladders 4-175

continuous taper or interchangeable. It is used widely by window cleaners. A big advantage is portability, since it can be knocked down into small units. Sectional ladders should not be longer than 31 ft and should have an overlap of at least 1 ft.

All special-purpose ladders should conform to the ANSI standards.

Metal ladders In recent years, single, extension, and stepladders made of aluminum or magnesium alloys have come on the market. They are light in weight and resist climatic conditions. However, because they are conductors of electricity, they should not be used around electrical equipment. It is well to tag or paint instructions to this effect on the ladder.

Upon receipt, metal ladders should be examined for sharp edges and burrs on the side rails, tops, and bottoms; such defects can cause painful cuts. The bottoms should be protected to prevent the marring of floors. The best method is to use safety shoes, which also help to prevent slipping.

Metal ladders are now covered by ANSI code A14.2, effective Feb. 1, 1972.

Precautionary measures Where special groups are using ladders, such as plumbers, electricians, and millwrights, the ladders should be properly identified, with the members of each craft held responsible for their particular ladders. The use of just any ladder the worker comes across is likely to lead to costly disaster. Instructing workers as to ladder usage is extremely important.

LADDER SAFETY RULES

LADDER SAFETY RULES #402W FOR WOOD
SINGLE & EXTENSION LADDERS

FOLLOW THESE INSTRUCTIONS FOR YOUR SAFETY
AND THAT OF OTHERS

1. Inspect ladder carefully on receipt and before EACH use. Test all working parts for proper attachment and operation. Ladders found to be damaged, defective or with missing parts should be withdrawn from use and marked "DO NOT USE." Never use a ladder known to have been dropped until it has been carefully re-inspected for damage of any nature.

2. Install and use this ladder in compliance with the Regulations of the Occupational Safety and Health Act – 1970, and with all other applicable governmental regulations, codes and ordinances. Ladder usage must be restricted to the purpose for which the ladder is designed.

3. Keep nuts, bolts, and other fastenings tight. Oil moving metal parts frequently. Obtain replacement parts from the manufacturer. Do not allow makeshift repairs. Replace frayed or badly worn rope promptly. Keep rungs free of grease, oil, paint, snow, ice or other slippery substances.

4. Ladders must stand on a firm level surface. Always use safety feet and other suitable precautions if ladder is to be used on a slippery surface. Never use an unstable ladder. Ladders should not be placed on temporary supports to increase the working length or adjust for uneven surfaces.

5. Face ladder when ascending or descending. Always place ladder close enough to work to avoid dangerous over-reaching. Keep work centered between side rails. Side loading should be avoided.

6. Sectional (Window Cleaners) Ladders must be assembled in proper sequence with a base section at the bottom, and equipped with safety feet where slippery conditions exist. Maximum assembled length must not exceed 21 ft. for Standard Sectional Ladders or 31 ft. for Heavy Duty. Do not intermingle Sectional Ladders of different types or strength. All Safety Rules printed herein apply to Sectional Ladders except nos. 10 & 11.

7. Never place ladders in front of doors or openings unless appropriate precautions are taken.

8. Before installing an extension or single ladder always insure that working length of ladder will reach support height required. It should be lashed or otherwise secured at top to prevent slipping and should extend at least 3 feet above a roof or other elevated platform. Never stand on top three rungs of an extension or single ladder.

9. Install a single or extension ladder so that the horizontal distance of that ladder foot from the top support is $1/4$ of the effective extended length of the ladder ($75\frac{1}{2}°$ angle). Always insure that both side rails are fully supported top and bottom. Never support ladder by top rung.

10. Overlap extension ladder sections by at least:
    3' each overlap for total nominal lengths up to & including 36'; 4' each overlap for total nominal lengths over 36', up to and including 48'; 5' each overlap for total nominal lengths over 48', up to & including 60'.

At overlaps, fly or upper sections must always be outermost so as to rest on lower section(s).

11. Be sure all locks on extension ladders are securely hooked over rungs before climbing. Make adjustments of extension ladder heights only when standing at the base of the ladder. Never

extend a ladder while standing on it. For 3-section ladders, always fully extend top section first. Ladders must not be tied or fastened together to provide longer sections other than manufactured for.

12. Water conducts electricity. Do not use wet ladders where direct contact with a live power source is possible. Use extreme caution around electrical wires, services and equipment. Provide for temporary insulation of any exposed electrical conductors near place of work.

13. A ladder is intended to carry only one person at a time. Do not overload. For support of 2 persons special ladders are available. NEVER use a ladder in a horizontal position, never sit on a ladder when it is on edge and never use a ladder in a flat position as a scaffold plank.

14. Store ladders on edge in such a manner to provide easy access for inspection. Provide sufficient supports to prevent sagging. Never use ladders after prolonged immersion in water, or exposure to fire, chemicals, fumes or other conditions that could affect their strength.

15. Only premium grade extension ladders should be used in conjunction with ladder jacks and stages or planks.

16. For further instruction on the care of Wood Single and Extension Ladders refer to the American National Standard, Safety Code for Portable Wood Ladders, ANSI A14.1 – 1968.

LADDER SAFETY RULES #402A FOR ALUMINUM SINGLE & EXTENSION LADDERS

FOLLOW THESE INSTRUCTIONS FOR YOUR SAFETY AND THAT OF OTHERS

1. Inspect ladder carefully on receipt and before EACH use. Test all working parts for proper attachment and operation. Ladders found to be damaged, defective or with missing parts should be withdrawn from use and marked "DO NOT USE." Never use a ladder which has been dropped until it has been carefully re-inspected for damage of any nature.

2. Install and use this ladder in compliance with the Regulations of the Occupational Safety and Health Act – 1970, and with all other applicable governmental regulations, codes and ordinances. Ladder usage must be restricted to the purpose for which the ladder is designed.

3. Keep nuts, bolts and other fastenings tight. Oil moving metal parts frequently. Obtain replacement parts from manufacturer. Do not allow makeshift repairs. Never straighten or use a bent ladder. Replace frayed or badly worn rope promptly.

4. Ladders must stand on a firm, level surface. Always use safety feet. If ladder is to be used on a slippery surface take additional precautions. Never use an unstable ladder. Ladders should not be placed on temporary supports to increase the working length or adjust for uneven surfaces.

5. Face ladder when ascending or descending. Always place ladder close enough to work to avoid dangerous over-reaching. Keep work centered between side rails. Side loading should be avoided.

6. Keep rungs free of grease, oil, paint, snow, ice, or other slippery substances.

7. Never place ladders in front of doors or openings unless appropriate precautions are taken.

8. Before installing an extension or single ladder always insure that working length of ladder will reach support height required. It should be lashed or otherwise secured at top to prevent slipping and should extend at least three feet above a roof or other elevated platform. Never stand on top 3 rungs of an extension or single ladder.

9. Install a single or extension ladder so that the horizontal distance of that ladder foot from the top support is $1/4$ of the effective working length of the ladder ($75\frac{1}{2}°$ angle). Always insure that both side rails are fully supported top and bottom. Never support ladder by top rung.

10. Overlap extension ladder sections by at least:
    3′ each overlap for total nominal lengths up to & including 36′.
    4′ each overlap for total nominal lengths over 36′ up to & including 48′.
    5′ each overlap for total nominal lengths over 48′, up to and including 60′.
At overlaps, fly or upper sections must always be outermost so as to rest on lower section(s).

11. Be sure all locks on extension ladders are securely hooked over rungs before climbing. Make adjustments of extension ladder heights only when standing at the base of the ladder. Never extend a ladder while standing on it. For 3-section ladders, always fully extend top section first. Ladders must not be tied or fastened together to provide longer sections than manufactured for.

12. Metal and water conduct electricity. Do not use metal, metal reinforced or wet ladders where direct contact with a live power source is possible. Use extreme caution around electrical wires, services and equipment. Provide for temporary insulation of any exposed electrical conductors near place of work.

13. A ladder is intended to carry only one person at a time. Do not overload. For support of 2 persons special ladders are available. NEVER use a ladder in a horizontal position, never sit on a ladder when it is on edge and never use a ladder in a flat position as a scaffold plank.

14. Store ladders on edge in such a manner to provide easy access for inspection. Provide sufficient supports to prevent sagging. Never use ladders after exposure to fire, chemicals, fumes or other conditions which could affect their strength.

15. Portable ladders are designed as one-man working ladders, including any material supported by the ladder. There are three classifications:

Maintenance of Scaffolds and Ladders 4-177

Type I – Heavy Duty for users requiring not more than a 250 pound load capacity for maintenance, construction or heavy duty work.

Type II – Medium Duty for users requiring not more than a 225 pound load capacity for painting, or other medium duty work.

Type III – Light Duty for users requiring not more than a 200 pound load capacity for service requirements such as general household use. Not for use with stages or planks.

16. Only Type I and Type II extension ladders should be used in conjunction with ladder jacks and stages or planks.

17. For further instructions on the care of Aluminum Single and Extension Ladders refer to the American National Standard, Safety Code for Portable Metal Ladders, ANSI A14.2 – 1972.

LADDER SAFETY RULES #403W
FOR WOOD STEPLADDERS

FOLLOW THESE INSTRUCTIONS FOR YOUR SAFETY
AND THAT OF OTHERS

1. Inspect ladder carefully on receipt and before EACH use. Test all working parts for proper attachment and operation. Ladders found to be damaged, defective or with missing parts should be withdrawn from use and marked "DO NOT USE." Never use a ladder that has been dropped or tipped over until it has been re-inspected for damage of any nature.

2. Install and use this ladder in compliance with the Regulations of the Occupational Safety and Health Act – 1970, and with all other applicable governmental regulations, codes and ordinances. Ladder usage must be restricted to the purpose for which the ladder is designed.

3. Keep nuts, bolts, and other fastenings tight. Oil moving metal parts frequently. Obtain replacement parts from the manufacturer. Do not allow makeshift repairs.

4. Ladders must stand on a firm level surface. Always use safety feet and other suitable precautions if ladder is to be used on a slippery surface. Never use an unstable ladder. Never "walk" a stepladder while on it. Ladders should not be placed on temporary supports to increase the working length or to adjust for uneven surfaces.

5. Face ladder when ascending or descending. Always place ladder close enough to work to avoid dangerous over-reaching. Keep work centered between side rails. Side loading should be avoided.

6. Keep steps free of grease, oil, paint, snow, ice, or other slippery substances.

7. Insure that stepladders are fully opened with spreaders locked. Do not stand on top, pail rest or rear rungs of stepladders.

8. Never place ladders in front of doors or openings unless appropriate precautions are taken.

9. Water conducts electricity. Do not use wet ladders where direct contact with a live power source is possible. Use extreme caution around electrical wires, services and equipment. Provide for temporary insulation of any exposed electrical conductors near place of work.

10. A ladder is intended to carry only one person at a time. Do not overload. For support of 2 persons special ladders are available. NEVER use a stepladder in a closed or horizontal position, never sit on a ladder when it is on edge and never use a ladder in a flat position as a scaffold plank.

11. Store ladders on edge in such a manner as to provide easy access for inspection. Provide sufficient supports to prevent sagging. Never use ladders after exposure to fire, chemicals, fumes or other conditions which could affect their strength.

12. Portable ladders are designed as one-man working ladders, including any material supported by the ladder. There are three classifications:

Type I – Industrial – for Heavy Duty and Industrial use.

Type II – Commercial – for Medium Duty and Light Industrial use.

Type III – Household – for Light Duty such as light household use.

13. For further instructions on the use and care of Wood Stepladders refer to the American National Standard, Safety Code for Portable Wood Ladders, ANSI A14.1 – 1968.

LADDER SAFETY RULES #403A
FOR ALUMINUM STEPLADDERS

FOLLOW THESE INSTRUCTIONS FOR YOUR SAFETY
AND THAT OF OTHERS

1. Inspect ladder carefully on receipt and before EACH use. Test all working parts for proper attachment and operation. Ladders found to be damaged, deformed, defective or with missing parts should be withdrawn from use and marked "DO NOT USE." Never use a ladder that has been dropped until it has been carefully re-inspected for damage of any nature.

2. Install and use this ladder in compliance with the Regulations of the Occupational Safety and Health Act – 1970, and with all other applicable governmental regulations, codes and ordinances. Ladder usage must be restricted to the purpose for which the ladder is designed.

3. Keep nuts, bolts, and other fastenings tight. Oil moving metal parts frequently. Obtain re-

placement parts from manufacturer. Do not allow makeshift repairs. Never straighten or use a bent ladder.

4. Ladders must stand on a firm, level surface. Always use safety feet. If ladder is to be used on a slippery surface, take additional precautions. Never use an unstable ladder. Never "walk" a stepladder while on it. Ladders should not be placed on temporary supports to increase the working length or to adjust for uneven surfaces.

5. Face ladder when ascending or descending. Always place ladder close enough to work to avoid dangerous overreaching. Keep work centered between side rails. Side loading should be avoided.

6. Keep steps free of grease, oil, paint, snow, ice, or other slippery substances.

7. Insure that stepladders are fully opened with spreaders locked. Do not stand on top, pail rest or rear rungs of stepladders.

8. Never place ladders in front of doors or openings unless appropriate precautions are taken.

9. Metal and water conduct electricity. Do not use metal, metal reinforced or wet ladders where direct contact with a live power source is possible. Use extreme caution around electrical wires, services and equipment. Provide for temporary insulation of any exposed electrical conductors near place of work.

10. A ladder is intended to carry only one person at a time. Do Not overload. For support of 2 persons special ladders are available. NEVER use a stepladder in a closed or horizontal position, never sit on a ladder when it is on edge and never use a ladder in a flat position as a scaffold plank.

11. Store ladders on edge in such a manner as to provide easy access for inspection. Provide sufficient supports to prevent sagging. Never use ladders after exposure to fire, chemicals, fumes or other conditions which could affect their strength.

12. Portable ladders are designed as one-man-working ladders, including any material supported by the ladder. There are 3 duty classifications:

> Type I – Heavy Duty – for users requiring not more than 250 pound load capacity for maintenance, construction or heavy duty work.
>
> Type II – Medium Duty – for users requiring not more than a 225 pound capacity for painting, or other medium duty work.
>
> Type III – Light Duty – for users requiring not more than a 200 pound load capacity or service requirements such as general household use. Light duty ladders should not be used with scaffold planks.

13. For further instructions on the care of Aluminum Stepladders refer to the American National Standard, Safety Code for Portable Metal Ladders, ANSI A14.2 – 1972.

WELDED ALUMINUM SCAFFOLDS

This type of scaffold affords firm, solid work platforms for use by one or more men. Because of their lightness, they are fast and easy to erect and are therefore preferred where a number of off-the-floor jobs are required to be done in a large number of positions. Their lightness, mobility, and ease of erection make them most suitable for light-duty work, especially where the equipment requires frequent erection and dismantling.

These scaffolds are prefabricated from high-strength aluminum-alloy tubing and are equipped with casters as necessary for easy mobility of the erected scaffold. The types usually most practical for maintenance work are aluminum rolling scaffolds with internal stairways, and aluminum ladder scaffolds.

Folding ladder scaffolds (Fig. 11-3) are built in one-piece base sections which speed the erection and dismantling process. The ladder-type base sections are 29 in. or 4 ft 6 in. wide, with spans of 6, 8, or 10 ft between frames. In this type of unit, the two diagonal braces and one horizontal brace are integral parts of the folding unit. Intermediate, extension, and guardrail sections can be placed atop the folding unit, using individual end frames and braces.

A larger folding-type scaffold has base dimensions of 4 ft 6 in. by 6 ft. This unit has an internal stairway, and the upper sections as well as the base section are one-piece folding units. When the scaffold must be erected higher than recommended for a base of this size, outriggers can be used. They clamp to the legs of the base section. Means for leveling, to compensate for uneven ground, are part of the leg equipment. The casters on the legs are locked at both wheel and swivel. Folding scaffold sections are, of necessity, heavier than individual components of demountable sectional scaffolding.

Sectional aluminum stairway scaffolds are designed with end frames of various heights to provide different working levels, adjustable bottom sections with casters but without the folding feature, intermediate sections, half sections, and guardrail sections. All components are demountable so as to be light and easy to handle for erection and dismantling. Internal stairways are used. Outriggers may be used to increase the base area (see Fig. 11-4).

The folding, sectional-stairway, and ladder scaffolds are used for outdoor cleaning and maintenance work—ladder scaffolds for low to medium height and one-man jobs, and

Fig. 11-3 Folding aluminum ladder scaffolds are designed so that the end frames will not fall over at any point during the erection or dismantling process. It is a completely freestanding unit at all times.

folding or sectional-stairway types for higher or heavier work. They are especially suitable when the work is horizontal. Indoors, they simplify work on walls and ceilings, and often are suitable for group lamp replacement.

WELDED SECTIONAL STEEL SCAFFOLDS

Used in situations similar to those of aluminum scaffolds, welded sectional steel scaffolds are heavier and therefore more suitable for heavy-duty work requiring relatively infrequent erection and dismantling. They can be assembled as a rolling scaffold and have mobility similar to aluminum scaffolds but are heavier and more cumbersome to handle. Some end frames have integral exterior ladders. Adjustable extension legs may be used for leveling. Casters lock at the wheel and swivel.

Steel ladder scaffolds, similar to the aluminum ones, are used for heavier-duty work in restricted spaces. Steel-pivoted diagonal cross bracing is used as with larger steel frames. An often useful accessory for the steel ladder scaffold is the bridging trestle. This replaces the diagonal cross braces at the bottom level and permits the scaffold to clear obstructions or permits the passage of traffic beneath the scaffold without interference.

For access to high work areas where relatively heavy work has to be done, such as the replacement of a crane motor or large heating unit, the steel scaffold is unsurpassed in

Basic Maintenance Technology

Fig. 11-4 Outrigger supports should be used when working at platform heights greater than four times the smallest dimension. Shown is an aluminum sectional scaffold 4 ft 6 in. wide.

Fig. 11-5 Sectional rolling scaffold. Steel frames are 5 ft wide and joined by pivoted diagonal braces of lengths from 4 to 10 ft. A hook-on type of access ladder is used with a 3-ft-high grabrail at top, guarded by steel-wire rope with snap hooks. The complete lift of the ladder is set back 7½ in. from the scaffold frames.

Fig. 11-6 Safety swinging scaffold (two-point suspension). Utilizes steel-wire rope with ratchet-action raising and crank-handle lowering. Note use of guardrails, midrails, and toeboards. Similar scaffolds also available with power machines and aluminum platforms.

strength and versatility. It is now available with OSHA-complying external stand-off access ladders (Fig. 11-5), or even *internal* stair-type systems called step units.

SAFETY SWINGING SCAFFOLDS

These are generally two-men work platforms, either wood or aluminum, suspended from roof supports either inside or outside buildings (see Fig. 11-6). They are most suited for successive work operations vertically above each other. Swing-stage platforms are generally available in lengths from 8 to 32 ft, and the industry norm is 28 in. in width. They can be used in conjunction with hoisting apparatus consisting of rope blocks and falls, manually operated steel-wire-rope hoisting mechanisms, and air or electrically operated machines at the platform level. All hanging equipment requires most extreme care in safe rigging procedures by experienced personnel. In general, they are seldom economical for work at heights of less than 30 ft, *unless* access to such work is impractical by other types of ladders or scaffolds.

Swinging scaffolds are particularly suitable for cleaning and painting, tuck pointing, window washing, and similar jobs on exterior walls or tanks where large vertical range and quick up-down mobility are required. They are used also where the surface below the work is crowded, or where conditions are unsuitable for support of ground-based scaffolding. They are recommended for light and medium loads.

Safety belts and separately attached lifelines are essential for proper worker safety as well as insistence on the installation and use of guardrails, midrails, and toeboards. Wire mesh between guardrail and toeboard is required by OSHA *only* when employees are required to work under the scaffold.

As well as the conventional two-point suspension swing-stage platforms, additional items for specialized work are available in the form of one-man work cages and bosun's chairs. The cages are often available with extensions attached to each side of the cage for use by two men. Generally, the cages are used with power-operated winches and the bosun's chairs with powered winches or blocks and falls.

TUBE AND COUPLER SCAFFOLDS (STEEL AND ALUMINUM)

These scaffolds provide the greatest versatility in scaffolding odd shapes such as processing works and refineries and in erecting to extreme heights. They are erected from four basic components: baseplates, interlocking tubing or pipe, bolt-activated couplers for making right-angle connections, and adjustable couplers for making connections at other than right angles.

Horizontal runners can be placed at any point on the vertical posts and, in turn, bearers at any point on the runner, thereby obtaining maximum versatility. This type of scaffolding is unsurpassed in providing work platforms for spheres, cylinders, and other odd-shaped vessels such as those in refineries. It also can be used to build storage racks of virtually any size and capacity, and is even the most suitable type of equipment to scaffold certain buildings having uneven exteriors and projections (Fig. 11-7).

This scaffold also is available in all-aluminum components (Fig. 11-8), making it particularly useful in corrosive atmospheres. Both types can be made into rolling scaffolds with the addition of special casters at the base.

SPECIAL-DESIGN SCAFFOLDS

Special-purpose scaffolds are available specifically designed for access purposes where standard scaffolding components cannot be used easily or are even impossible to use. Such scaffolds, stationary and mobile, are frequently used in the aircraft and aerospace industries, as well as for special requirements on many projects.

Safety requirements Almost all present-day steel and aluminum scaffolds are listed under the Reexamination Service of Underwriters' Laboratories, Inc. The UL seal on maintenance scaffolds means not only that the product is properly designed to sustain the loads for which it was intended but also that certain manufacturing standards are included in the design to assure maximum strength. One of the first things a maintenance department should look for is a UL listing sticker or label. After purchase of the equipment, it becomes the maintenance department's responsibility to make sure not only that the

Fig. 11-7 This shows Tubelox scaffolding used to scaffold building surfaces with a difficult roof-overhang condition. The versatility of Tubelox is similarly utilized for unusually shaped structures.

Fig. 11-8 Aluminum ladder scaffold. Basic unit consists of 6-ft ladder frames, diagonal braces, platform, and adjustable casters. Note use of guardrails, midrails, and toeboards.

scaffold is used properly but also that it is maintained properly. Of course, all equipment must comply with the appropriate OSHA regulations concerning its manufacture, installation, and use. Near the end of this section an OSHA scaffolding checklist for scaffolds of various types is reproduced which can be used to assist in assuring that OSHA compliance is being maintained.

Safe use and safety rules Any reputable manufacturer of ladders and scaffolds will furnish (with or attached to their equipment) information on specific items and include a set of safety rules which, if followed, can drastically reduce many industrial accidents associated with the use of this equipment. Copies of safety rules should be freely available from the manufacturer of the product—not only at the time of first delivery, but cheerfully and freely in later years when requested. Printed safety rules are *not* intended to be retained with the delivery slip (to which they are frequently attached) and sent into the office for billing purposes and thence forgotten. Neither are they for the purpose of padding out a foreman's hip pocket while the men involved in doing the work have never seen or read them. Safety rules must be read and clearly understood by the men *doing* the work; it is up to the judgment of the workman's immediate superior as to whether best results are obtained by having the workman read the safety rules or by having them read to him along with an explanation of the reasons why certain safety precautions are vital to freedom from injury.

Recently, in the promulgations of OSHA, two sets of regulations for scaffolding were made. One set is for use in construction (1926.451) and the other set is for use in general industry (1910.28), although both can apply in certain circumstances. The need for maximum employee safety in the erection and use of scaffolds is strongly emphasized. The OSHA regulations should be thoroughly read and observed, especially where the use of certain safety components has traditionally *not* been customary. Briefly, the OSHA regulations merely describe in detail the safe use of scaffolds, and therefore it is to the benefit of employer and employee to be completely familiar with such requirements pertinent to the work to be done. There is no shortcut to OSHA compliance; however, it need not be dwelled on at length in this chapter, since the general requirements and principles are familiar.

OSHA SCAFFOLDING CHECKLIST

This highlights certain basic safety precautions concerning scaffolds and can be used as a base for expansion to cover your own particular maintenance requirements. This list is basic and does not purport to be all-inclusive or to encompass all circumstances. Such determinations must be made by the employer in his own individual circumstances.

OSHA SCAFFOLDING CHECKLIST

Project _____
Inspection area _____ Area supervisor _____
Inspected by _____ Date _____

Ladders and scaffolding	Yes	No	Action/comments
Manually propelled mobile scaffolds			
1. When using freestanding mobile scaffold towers, do you restrict tower heights to four times the minimum base dimension? (3 and 3½ in some states)			
2. Do casters have a positive locking device that will hold the scaffold?			
3. Are mobile scaffolds properly braced by cross and horizontal bracing?			
4. Is the cross bracing of such length as will automatically square and align vertical members so that the erected scaffold is plumb, square, and rigid?			
5. Are all brace connections secured?			
6. Are platforms tightly planked for the full width of such scaffolds, except for the necessary entrance opening?			
7. Do you provide a ladder or stairway for proper access and exit?			

4-184 Basic Maintenance Technology

OSHA SCAFFOLDING CHECKLIST (*Continued*)

Project _____
Inspection area _____ Area supervisor _____
Inspected by _____ Date _____

Ladders and scaffolding	Yes	No	Action/comments
8. Is such a ladder or stairway affixed to or built into the scaffold?			
9. Is it so located that when in use it will tend to tip the scaffold?			
10. Are landing platforms provided at intervals not greater than 35 ft? 30 to 35 ft?			
11. Is the force necessary to move your mobile scaffold applied near or as close to base as practicable?			
12. Do you make adequate provision to stabilize the tower during movement from one location to another?			
13. Do you permit scaffolds to be moved only on level floors, free of obstructions and openings?			
General scaffolds			
14. When scaffolds are in use, do they rest upon a suitable footing?			
15. Are they plumb?			
16. Have you installed guardrails at all open sides and ends of the scaffolds?			
17. Are such guardrails made of not less than 2- by 4-in. lumber (or other material providing equivalent protection)?			
18. Are they approximately 42 in. high, with a midrail of 1- by 6-in. lumber (or material providing equivalent protection), and toeboards 4 in. high?			
19. Is wire mesh installed between the toeboard and the guardrail, extending along the entire opening?°			
20. Does such mesh consist of No. 19 gage U.S. Standard wire ½-in. mesh, or the equivalent?			
21. Is the planking used of scaffold-plank quality, even if not officially "graded" as scaffold plank?			
22. Does the span of the planks exceed the maximum allowable depending on the designation light duty, medium duty, or heavy duty? (1926.451(a)(10)?			
23. Are your men instructed always to replace guardrails, midrails, and bracing if they have had to be temporarily removed for passing materials and equipment?			
24. Do your sectional scaffolds over 125 ft in height have drawings designed by a registered professional engineer? They must.			
25. Do you obtain safety rules and instructions on use of scaffolds from your scaffolding supplier? You should— they are free.			
Ladders			
26. Are both legs of rung ladders supported at top and bottom?			
27. Is the footing slippery, and if so are safety feet installed?			
28. Can the work height be reached without standing on the top of stepladders or the top three rungs of rung ladders?			
29. Does the extension ladder "fly" section rest on top of the base section and not hang underneath?			
30. Does the wood look crumbly? If so, check by inserting a sharp knife end under a sliver of wood and pry up. If it results in a long splinter, it is O.K.; if it results in "crumbling" of the wood, it is decayed.			
31. Never drop ladders or allow them to fall unless they are thoroughly and minutely inspected for damage before reuse.			

Maintenance of Scaffolds and Ladders 4-185

OSHA SCAFFOLDING CHECKLIST (*Continued*)

Project _____
Inspection area _____ Area supervisor _____
Inspected by _____ Date _____

Ladders and scaffolding	Yes	No	Action/comments
32. Never use a bent aluminum ladder or one which has been bent and restraightened; the material has been overstressed on both occasions and is no longer reliable.			

° Wire mesh required only when employees are required to work or pass underneath the scaffold.

APPLICATION

Let us examine a typical example, and first assume that the exterior windows of a 400-ft-long by 24-ft-high building are to be cleaned. The building front is brick up to 6 ft, with standard glass panes from 8 to 24 ft. One man is assigned to do the job, with occasional additional help if he needs it. Six factors are:
1. The work is within ladder range.
2. The window frames afford support.
3. The area below is asphalted, with some grass sections.
4. No traffic need pass below—wheeled or pedestrian.
5. The work ranges both horizontally and vertically.
6. There is no chance of electrical contact.

At first consideration, most factors indicate this work could be done by one man with a 28-ft aluminum extension ladder, which has a maximum extended height of 25 ft; if the man does not use the upper three rungs (safe practice), he will stand on a rung at 22 ft and be able to reach 28 ft high. He can move the ladder easily by himself. He can reach 2 ft either side of the ladder and can cover a 5-ft-wide strip of windows in a vertical pass. This is obviously the correct choice, yes? No!

The man will spend excessive time climbing up and down and relocating the ladder. His *productive* working time will be not more than 50 percent. Remember—a ladder must be installed one-fourth of its extended length away from the building at base; therefore, although he can reach the upper-level windows easily, he will have to lower and reposition the ladder frequently to reach the lower levels. Resorting to overreaching sideways and behind the ladder is unsafe and is the initiating factor in many ladder accidents. A platform stepladder with the platform 18 ft high is a better possibility. The best choice, however, is a 29-in.-wide by 10-ft-long aluminum ladder scaffold, about 19 ft high to top working level, with outboard safety supports and a so-called climb-through wood platform which has a hatch which can be slid out of the way to climb through and thence replaced to make a full platform cover. The reasons for the choice of this equipment are

    1. The man can achieve all heights by varying the levels of the plywood platform for a full working width of 10 ft; from one work level plus his own reach he can clean, say, 60 sq ft, compared with a typical 6 sq ft from one extension-ladder position.

    2. He can reach all heights with one initial placement plus only *two* platform repositionings.

    3. He can cover 10 ft by 24 ft, less 6 ft brick, that is, 180 sq ft, with minor downtime for repositioning.

    4. When a 10-ft vertical pass is completed, he can *roll* the scaffold himself to the next position. With certain youth and agility he could even erect it himself; the aluminum components are extremely light; otherwise, he would require help for initial installation.

    5. For the grass areas (or even at the rear of ornamental-shrub beds), the scaffold can be rolled on leveled planks. Four-inch steel channels can also be used as wheel guides and bridges over minor humps and deviations from level. Level such earth or grass with sand so that there are no "holes" *under* the planks.

NOTE: Aluminum scaffolds are standardly equipped with 5-in.-diameter casters but are optionally available with 8-in. ones; both have a 24-in. range of screw-leg adjustment for support from different levels. The 8-in. wheels provide easier rolling over rough surfaces such as old asphalt.

Maintenance Aluminum scaffolds of all types require minimum maintenance. Stairways, ladders, and platforms should be inspected frequently, and any grease or oil should be removed immediately. Make sure the plywood platforms or platform planks are solid, with no splits. Do not store platforms near excessive heat, to avoid drying and warping. Casters should be cleaned and lubricated, and brakes should be checked for satisfactory operation. Threads on extension legs should be cleaned and lubricated periodically for smooth operation. Coupling pins used to join frames vertically should be kept clean so that upper frames slip over the pins easily and freely.

With aluminum scaffolds, slight bends in the tubing due to severe impact or mishandling should be straightened. The spring-lock devices used to fasten the braces should be kept free from dirt to ensure proper operation of the lock. On the more popular types of aluminum scaffolding, this mechanism is exposed and can be cleaned easily with a wire brush.

The steel types of scaffolding should be kept clean by scraping or wire brushing; any rusted spots on frames or braces should be scraped and touched up with quick-drying enamel. Stud threads should be lubricated, and wing nuts run off and on to ensure fast, secure fastening during erection. All frames should be checked frequently for missing vertical-coupling sprockets and the pins that lock them in place, as well as wing nuts. Cross braces should be straightened if bent, and the alignment of the tops of the frames should be checked and braces realigned if necessary.

For maximum safety, swinging scaffolds must be properly maintained. The operating mechanism, or winch, should be kept free of dirt and grit at all times. A wire brush usually is satisfactory for this work. All operating parts should be properly lubricated as outlined in the manufacturer's instructions. The safety devices in the winches should be inspected frequently. Pawls and pawl springs should be checked for proper working condition. Teeth on the drum casting should be inspected, and if broken or worn, the manufacturer should be consulted about replacement. The steel cable should be run off and checked for excessive damage and kinks and then rewound through an oily rag to clean the cable and give it a thin coat of oil. Worm and gear mechanisms should be checked for excessive play, cleaned, and repacked with fiber grease. The stirrup should be checked for alignment and straightened, and all painted surfaces of the machine should be recoated where necessary for rust prevention.

Wooden platforms require careful inspection and maintenance. Grease or oil spilled on the platform should be removed immediately. After a job, the platform should be placed across horses for inspection, overhaul, and repair. Mortar, concrete, and paint should be removed with a wire brush and scraper. Broken rungs, slats, and damaged or missing toeboards should be replaced, as should missing hinges and hooks and eyes. After necessary repairs have been made, the platform should be given a thick coat of quick-drying paint. Finally, clean and examine the S and L hooks from which the scaffold is hung, clean and inspect center stanchions, and replace missing or defective wing nuts and bolts.

Tube and coupler scaffolding should undergo a systematic inspection. Bent tubes should be strengthened and, in case of seriously damaged tubes, discarded or cut into shorter lengths for short bearers. Very dirty or rusty tubes should be cleaned with a wire brush. During this operation, inspect the male and female fittings for damage which would affect their safety and then clean with a wire brush. Remove damaged couplers from stock. Studs should be kept covered with a light film of oil, and catch bolts should be checked for stripped threads.

Ladders and scaffolds are vital accessories for all "off-the-ground" work. It behooves all persons, employers and employees alike, to keep their equipment in first-class operating condition and always use it with safety as the maximum prerequisite.

Chapter **12**

Chain Hoists and Chain Slings

H. E. HOLLIDAY*
Service Manager, CM Hoist Division,
Columbus McKinnon Corporation,
Tonawanda, N.Y.

J. D. ZAJAC
Product Standards and Service, CM Hoist Division,
Columbus McKinnon Corporation,
Tonawanda, N.Y.

GENERAL

Chain hoists, both hand and electric, are a widely used type of material-handling equipment. Their simplicity, dependability, and relatively low cost have made them standard material-handling equipment in manufacturing plants, foundries, mills, repair shops, and garages and in practically every phase of the construction field.

This chapter describes the various types of chain hoists and their relative advantages and usual applications. It provides information on preventive maintenance, inspection, and trouble shooting.

TYPES OF CHAIN HOISTS

Rigger Ratchet Hoist. This is the simplest and least expensive type of chain hoist, with approximately 15:1 mechanical advantage. This lightweight portable tool can be used for pulling horizontally, vertically, or at any angle. A directional level mechanism determines if the load is being applied to or released from the tool by the up-and-down movement of the handle. When loading, a driving pawl engages the ratchet and, by turning the liftwheel, causes tension on the chain. By shifting the direction lever to the unload position, the tension can be released one tooth at a time (Fig. 12-1).

Cyclone and Satellite Spur-Geared Hoists. This type of hoist is more efficient than the rigger ratchet hoist, with mechanical advantages of approximately 22:1 for low-capacity units varying upward with capacity increase, usually obtained by handwheel size and a set of reduction gearing (Fig. 12-2 to 12-4). The initial cost is higher than the ratchet hoist. A self-energizing Weston-type brake is incorporated, which enables the load to be lowered with comparatively little effort, at a very slow rate, and with precise positioning if desired. This type of hoist is an accepted standard for industrial applications requiring high speed and efficiency. The spur-geared hoist is generally used for vertical loading and is available in several model variations. An overload protection device is often incorporated.

*Deceased.

4-188   Basic Maintenance Technology

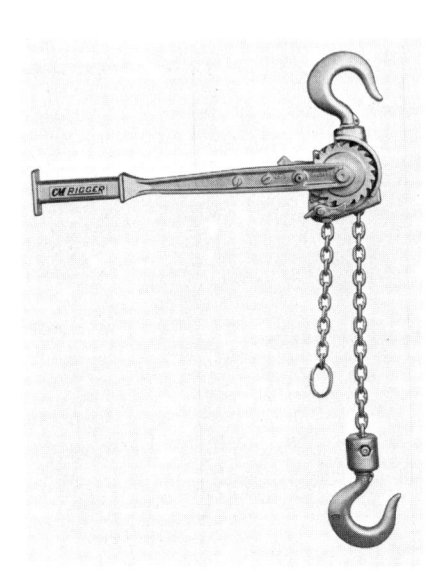

Fig. 12-1 Rigger ratchet hoist.

Fig. 12-2 Cyclone spur-geared hoist.

Fig. 12-3 Satellite spur-geared hoist.

Fig. 12-4 Multiple-reeved cyclone spur-geared hoist.

Modern Spur-Geared Hoists. These hoists utilize more compact design, lightweight alloys, and more antifriction bearings to achieve greater portability through a weight reduction.

Cyclone Spur-Geared Low-Headroom Trolley Hoist. This hoist is built integrally with a trolley for installations where headroom is limited. The 1-ton model requires approximately 13-in. headroom. In heavier capacities, the headroom saving is correspondingly greater. This type of hoist is available with plain or geared trolleys. The geared trolley is moved along the beam by a hand chain drive (Fig. 12-5).

CM Model B Puller (Lever-Operated Hoist). These hoists are lightweight portable tools which are designed for pulling, lifting, or dragging with capacities from ¾ ton to 6 tons. (Figure 12-6 illustrates a ¾-ton model.) A reversing mechanism located in the lever permits short-stroke operation for loading and unloading. This tool has a mechanical advantage of approximately 25:1. An automatic friction-type load brake holds the load

Fig. 12-5 Cyclone spur-geared low-headroom trolley hoist.

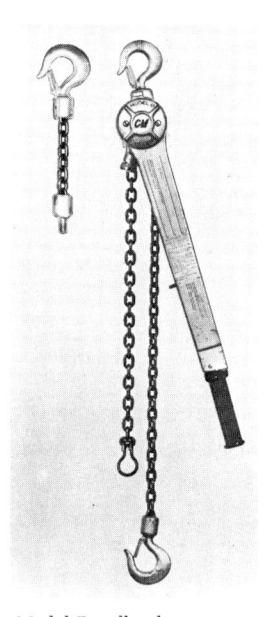

Fig. 12-6 Model B puller, lever-operated hoist. At left, an anchor sling attachment.

securely. These pullers are available with a load-sensing device to warn the operator the tool is being loaded beyond its rated capacity. In close headroom operation, an anchor sling is available (shown on left of puller photo, Fig. 12-6).

Lodestar Electric Hoists. These hoists are used for high-speed repetitive duty. They are equipped with either pushbutton or pendant rope controls. Electric hoists are usually equipped with safety limit switches to control both up and down travel. This control prevents the load hook from jamming against the bottom of the hoist or the chain from running out of the hoist (Figs. 12-7 and 12-8).

Electric hoists are available for use with different power supplies. Many are equipped with single-phase 115-volt motors which can be plugged into a conventional three-prong receptacle. Some manufacturers offer three-phase dual-voltage single-speed models. Three-phase single-voltage two-speed models are also available. Both types are designed to operate on 50 or 60 Hz ac 230 or 460 volt power for dual voltage and 230 or 460 volts for two-speed models. Recent electric hoist models are equipped with load protectors to prevent overloading of the hoist (Fig. 12-9).

Electric chain hoists are available in capacities from ⅛ to 5 tons. With lifting speeds

Fig. 12-7 Electric hoist with pendant control station.

Fig. 12-8 Electric hoist with motor-driven trolley and pendant control.

from 2½ to 64 fpm, these are widely used throughout industry because of their convenience, low cost, and durability.

SELECTION OF A CHAIN HOIST

In selecting either hand-operated or electric-powered hoists, certain considerations are basic and common to both types. Figure 12-10 illustrates the importance of performance and physical characteristics of both hand and electric types which must be considered in selecting a hoist for a given use and installation.

Initial cost, frequency of use, labor savings, safety, portability, and maintenance requirements are overall factors in selection of a hoist. Of prime importance is the capacity of the heaviest load to be lifted. Environmental conditions such as moisture, heat, chemicals, and foreign material in the atmosphere, must also be considered. These conditions may require weatherproofing or special protection. Under normal conditions, standard hoists are satisfactory. Other factors which affect the selections of a hoist are headroom, height of lift, location, height of hand chain, type of suspension, and trolley clearances.

PREVENTIVE MAINTENANCE

The design of modern hand and electrical hoists is such that maintenance has been reduced to a minimum. However, correct and adequate maintenance of all types of lifting equipment is of the utmost importance. Many older models still in active service will be included in the maintenance instructions published by the manufac-

Fig. 12-9 Electric-hoist load protector.

turer. Hoist operating mechanisms are subject to wear and should be inspected. A regular periodic inspection should be initiated to ensure that worn or damaged parts are removed from service before they become unsafe.

Individual applications will determine the frequency of inspection. However, the following formula can be used as a guide to the amount of time which should be allowed for general maintenance. The times obtained from this formula do not refer to major items such as gears, frames, and liftwheels but rather to inspection of chain, hooks, limits of travel, brakes, and electrical controls.

$$Hm = \frac{(W \times A) + K + M}{90}$$

where Hm = maintenance hours per month required
 W = working hours of hoist per week
 A = age of hoist in years
 K = humidity
  40% = 10   80% = 52
  50% = 13   90% = 80
  60% = 20   100% = 120
  70% = 38
 M = Atmospheric conditions (dust)
  Clean = 10
  Medium dusty = 20
  Foundry dusty = 40

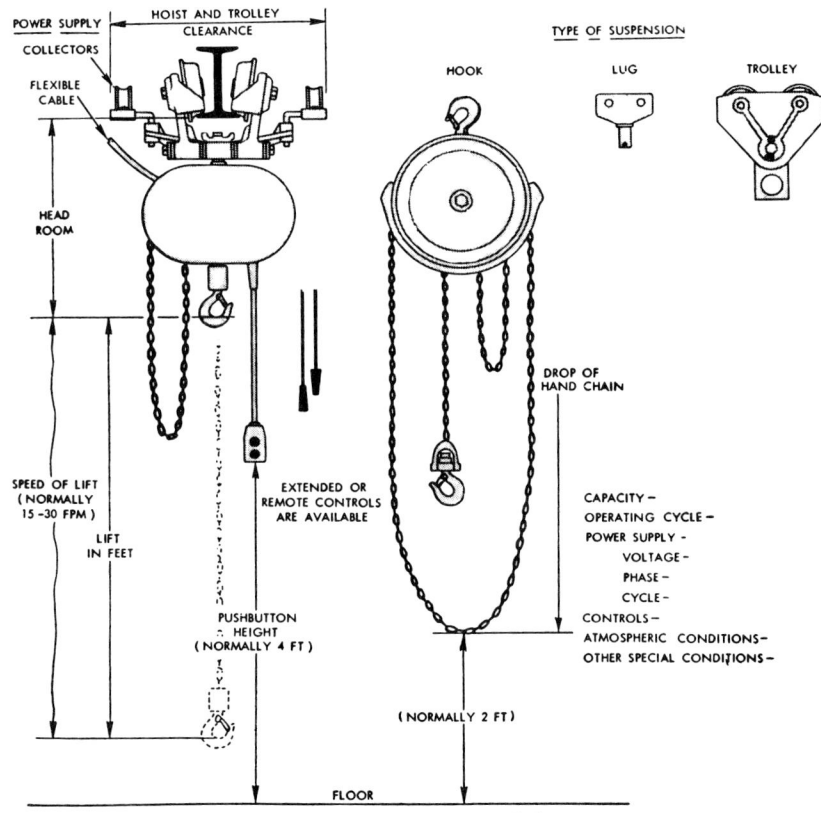

Fig. 12-10 Hoist-instruction check diagram.

Preventive maintenance takes into account the type of service for which equipment is used and whether equipment is in regular use or frequently idle. Inspection and testing procedures depend on these factors.

Service

Types of Service for hand- and electric-operated chain hoists are defined as follows:
   1. *Normal service* that service which involves operation with randomly distributed loads within capacity or uniform loads up to 65 percent capacity for not more than 15 percent for hand-operated hoists and 25 percent for electric- or air-operated hoists during a single work shift.
   2. *Heavy service* that service within rated capacity which exceeds normal service.
   3. *Severe service* that service which involves normal or heavy service in adverse environmental conditions.

Inspection

Inspection of hand- and electric-operated chain hoists is a critical part of preventive maintenance.

Initial inspection Prior to use, all new and altered hoists shall be inspected by the user to ensure compliance with the provisions outlined.

Procedure Inspection procedure for hoists in *regular service* is classified as frequent or periodic based on the intervals at which inspection should be performed. The intervals depend on the nature of the critical components of the hoist and the degree of their exposure to wear, deterioration, or malfunction. Maximum intervals between frequent and periodic inspections are outlined:

1. *Frequent inspection*—visual inspection by the operator or other designated personnel with records not required.
   a. Normal service—monthly.
   b. Heavy service—weekly to monthly.
   c. Severe service—daily to weekly.
   d. Special or infrequent service as authorized by a qualified individual—before and after such special or infrequent period of service with records of operation.
2. *Periodic inspection*—visual inspections by appointed person making records of apparent external conditions to provide the basis for a continuing evaluation.
   a. Normal service—equipment in place; yearly.
   b. Heavy service—as in (*a*) unless external conditions indicate that disassembly should be done to permit detailed inspection; semiannually.
   c. Severe service—as in (*b*); quarterly.
   d. Special or infrequent service as authorized by a qualified individual—before the first such period of service and as directed by the qualified individual for any subsequent special or infrequent period of service.

Frequent inspection Inspect items such as those in the following list for damage at the intervals outlined (under the preceding heading Procedure) or as specifically indicated. Be sure to include observation during operation for any damage that might appear between the regular inspections. Carefully examine deficiencies and determine whether they constitute a safety hazard. Inspect:
   1. Braking mechanism for evidence of slippage under load.
   2. Load chain for lubrication, wear, or twists and broken, cracked, or otherwise damaged links. Check chain also for deposits of foreign material that might be carried into the hoist mechanism.
   3. Hooks for deformation, chemical damage, or cracks. Hooks damaged from chemicals, deformations, and cracks, or hooks having more than 15 percent in excess of normal throat opening or more than 10° twist from the plane of the unbent hook must be replaced.
   NOTE: Any hook that is twisted or has a throat opening in excess of normal indicates abuse or overloading of the unit. Other load-bearing components of the hoist should be inspected for damage.
   4. Hooks for proper operation of latch if one is used.

Periodic inspection Make a complete inspection of the hoist at the intervals outlined (under the previous heading Procedure) depending upon its activity, severity of service, and environment, or as specifically indicated below. Include the requirements of frequent inspection and, in addition, items such as those in the following list. Carefully examine any deficiencies and determine whether they constitute a safety hazard.

1. Check for external evidence of wear of chain, load sprockets, idler sprockets, and handwheel pockets or for chain stretch.

2. Inspect hook-retaining nuts, collars and pins, welds, or riveting used to secure the retaining members.

3. Examine brake mechanism for worn, glazed, or oil-contaminated friction disks, worn pawls, cams, or ratchets. Watch for corroded, stretched, or broken pawl springs.

4. Look for worn, cracked, or distorted parts, such as hook blocks, suspension housing, outriggers, hand chain wheels, chain attachments, clevises, yokes, suspension bolts, shafts, gears, and bearings.

5. Check for loose bolts, nuts, or rivets.

6. Inspect supporting structure and trolley for continued ability to support the imposed loads.

7. Make sure the warning label on proper use of the hoist is attached and legible.

Idle-hoist inspection Hoists not in regular use must be checked carefully before they are put back into service.

1. A hoist that has been idle for a period of 1 month or more but less than 6 months must be inspected by, or under the direction of, a designated person. It must meet the requirements of frequent and periodic inspections before placing it in service.

2. A hoist that has been idle for a period of 6 months shall be given a complete inspection to conform with the requirements of frequent and periodic inspection.

Inspection records Written, dated, and signed inspection reports and records shall be made on critical items such as brakes, hooks, and chains and the time intervals specified in the inspection classification. Records should be readily available.

Testing

Operation tests All new hoists shall be tested by the hoist manufacturer. All altered or repaired hoists or hoists that have not been used within the preceding 12 months shall be tested before use by, or under the direction of, a designated person to ensure compliance with the standard, including the following:

1. All functions of the hoist including hoisting and lowering shall be checked with the hoist suspended in the unloaded state.

2. After testing in the unloaded state, a load of 50 lb times the number of load-supporting parts of the chain shall be applied to the hoist in order to check proper load control.

Load tests All new hoists shall be tested with at least 125 percent rated load. If testing by the manufacturer is impractical, the user shall be notified and the test shall be accomplished at the job site by or under the direction of an appointed person.

All hoists in which load-sustaining parts have been altered, replaced, or repaired shall be tested statically or dynamically by, or under the direction of, an appointed person and a written report prepared and made readily available. The applied test load shall be 125 percent of the rated load. The replacement of normal maintenance items such as chain is specifically excluded from this hoist load test. A functional test of the hoist under a load of at least 50 lb times the number of load-supporting parts of the chain shall be made.

On hoists incorporating load-limiting devices which prevent the application of 125 percent of rated load, a load test shall be conducted with at least 100 percent rated load, after which the function of a load-limiting device shall be tested. The hoist manufacturer should be consulted for the value of the upper load limit of the load-limiting device.

HOIST OPERATOR INSTRUCTIONS

Prior to the start of normal operations, the hoist operator should test all limit switches, brakes, and other safety devices. Any failure should be reported immediately and the equipment should be removed from service until repairs can be made. It should be the responsibility of the operator to perform the frequent inspection.

The hoist operator should be responsible for the safe operation of the equipment. At no time should he leave the control position with a load suspended. Limit switches are for emergency use and should not be used for ordinary operation. During normal operation, if the equipment fails to respond properly, the failure should be reported immediately.

TABLE 12-1 Troubleshooting Guide for Spur-Geared Hoists

Problem	Probable cause	Check/remedy
Hoist is hard to operate in either direction.	Load chain worn long to gauge, thus binding between liftwheel and chain guide.	Check gauge of chain. Replace if worn excessively.
	Load chain rusty, corroded, or clogged up with foreign matter such as cement or mud.	Clean by tumble polishing or solvent. Lubricate with penetrating oil and graphite. (SC-46 and SC-146)
	Load chain damaged.	Check chain for gouges, nicks, bent or twisted links. Replace if damaged.
	Liftwheel clogged with foreign matter or worn excessively, causing binding between the liftwheel and chain guide.	Clean out pockets. Replace if worn excessively.
	Hand-chain worn long to gauge, thus binding between handwheel and cover.	Check gauge of chain.
	Handwheel clogged with foreign matter or worn excessively, causing binding of chain between the handwheel and cover.	Clean out pockets. Replace if worn excessively.
	Liftwheel or gear teeth deformed.	Excessive overload has been applied. Replace damaged parts.
Hoist is hard to operate in the lowering direction.	Brake parts corroded or coated with foreign matter.	Disassemble brake and clean thoroughly. (By wiping with a cloth—not by washing in a solvent.) Replace washers if gummy, visibly worn, or coated with foreign matter. Keep washers and brake surfaces clean and dry.
	Chain binding.	See first three check/remedy items in this table.
Hoist is hard to operate in the hoisting direction.	Chain binding.	See first three check/remedy items in this table.
	Chain twisted. (3-ton capacity and larger)	Rereeve chain, or on 3- and 4-ton unit, if both chains are twisted, capsize hook block through loop in chain until twists are removed. Caution—do not operate unit in hoisting direction with twisted chain or serious damage will result.
	Overload.	Reduce load or use correct capacity unit.
Hoist will not operate in either direction.	Liftwheel gear key or friction hub key missing or sheared.	Install or replace key.
	Gears jammed.	Inspect for foreign material in gear teeth.

TABLE 12-1 Troubleshooting Guide for Spur-Geared Hoists (*Continued*)

Problem	Probable cause	Check/remedy
Hoist will not operate in the lowering direction.	Locked brake due to a suddenly applied load, shock load, or load removed by other means than by operating unit in the lowering direction.	With hoist under load to keep chain taut, pull sharply on hand chain in the lowering direction to loosen brake.
	Chain binding.	See first three check/remedy items in this table.
	Lower hook all the way out. Load chain fully extended.	Chain taut between the lift wheel and loose end screw. Operate unit in hoisting direction only.
Hoist will not operate in the hoisting direction.	Chain binding.	See first three check/remedy items in this table.
Hoist will not hold load in suspension.	Lower hook or load side of chain on wrong side of liftwheel.	Lower hook must be on same side of liftwheel as upper hook. Refer to assembly. Rereeve chain.
	Ratchet assembled in reverse.	Ratchet must be assembled as shown.
	Pawl not engaging with ratchet.	Pawl spring missing or broken pawl binding on pawl stud. Replace spring and clean so pawl operates freely and engages properly with ratchet. Do not oil.
	Ratchet teeth or pawl worn or broken.	Replace pawl and/or ratchet.
	Worn brake parts.	Replace brake parts which are worn.
	Oily, dirty, or corroded brake friction surfaces.	Disassemble brake. Clean thoroughly. (By wiping with a cloth—not by washing in a solvent.) Replace washers if gummy, visibly, worn, or coated with foreign matter. Keep washers and brake surfaces clean and dry.

When operating hoisting equipment, the operator must always use safe material-handling methods. He is responsible for the safe slinging of loads and must know the location and operation of all main electric-power feeder switches and emergency stop buttons.

REPAIRS

Hoist manufacturers have available maintenance and parts manuals for use when repairs become necessary. These manuals give assembly and disassembly instructions, location, and identification of all parts. When replacement parts are needed, name plate information, especially the serial number, should be included with correspondence. Table 12-1 is a troubleshooting guide for the spur-geared hoist. Tables 12-2 and 12-3 are troubleshooting guides for the electric hoist.

TABLE 12-2 Troubleshooting Guide for All Electric Hoists

Problem	Probable cause	Check/remedy
Hook does not respond to the control station.	No voltage at hoist—main line or branch circuit switch open; branch line fuse blown or circuit breaker tripped.	Close switch, replace fuse or reset breaker.
	Phase failure (single phasing, three-phase unit only)—open circuit, grounded or faulty connection in one line of supply system, hoist wiring, reversing contactor, motor leads or windings.	Check for electrical continuity and repair or replace defective part.
	Upper or lower limit switch has opened the motor circuit.	Press the "other" control and the hook should respond. Adjust limit switches.
	Open control circuit—open or shorted winding in transformer, reversing contactor coil or speed selecting relay coil; loose connection or broken wire in circuit; mechanical binding in contactor or relay; control station contacts not closing or opening.	Check electrical continuity and repair or replace defective part.
	Wrong voltage or frequency.	Use the voltage and frequency indicated on hoist identification plate. For three-phase dual-voltage unit, make sure the connections at the conversion terminal board are for the proper voltage.
	Low voltage	Correct low-voltage condition.
	Brake not releasing—open or shorted coil winding; armature binding.	Check electrical continuity and connections. Check that correct coil has been installed. The coil for three-phase dual-voltage unit operates at 230 volts when the hoist is connected for either 230-volt or 460-volt operation. Check brake adjustment.
	Excessive load.	Reduce loading to the capacity limit of hoist as indicated on the identification plate.
Hook moves in the wrong direction.	Wiring connections reversed at either the control station or terminal board (single-phase unit only).	Check connections with the wiring diagram.
	Failure of the motor reversing switch to effect dynamic braking at time of reversal (single-phase unit only).	Check connections to switch. Replace a damaged switch or a faulty capacitor.
	Phase reversal (three-phase unit only).	Refer to installation instructions.

Chain Hoists and Chain Slings 4-197

TABLE 12-2 Troubleshooting Guide for All Electric Hoists (*Continued*)

Problem	Probable cause	Check/remedy
Hook lowers but will not raise.	Excessive load.	Reduce loading to capacity limit of hoist as indicated on the identification plate.
	Open hoisting circuit—open or shorted winding in reversing contactor coil or speed selecting relay coil; loose connection or broken wire in circuit; control station contacts not making; upper limit switch contacts open.	Check electrical continuity and repair or replace defective part. Check operation of limit switch.
	Motor reversing switch not operating (single-phase unit only).	Check switch connections and actuating finger and contacts for sticking or damage. Check centrifugal mechanism for loose or damaged components. Replace defective part.
	Phase failure (three-phase unit only).	Check for electrical continuity and repair or replace defective part.
Hook raises but will not lower.	Open lowering circuit—open or shorted winding in reversing contactor coil or speed selecting relay coil; loose connection or broken wire in circuit; control station contacts not making; lower limit switch contacts open.	Check electrical continuity and repair or replace defective part. Check operation of limit switch.
	Motor reversing switch not operating (single-phase unit only).	Check switch connections and actuating finger and contacts for sticking or damage. Check centrifugal mechanism for loose or damaged components. Replace defective part.
Hook lowers when hoisting control is operated.	Phase failure (three-phase unit only).	Check for electrical continuity and repair or replace defective part.
Hook does not stop promptly.	Brake slipping.	Check brake adjustment.
	Excessive load.	Reduce loading to the capacity limit of hoist as indicated on the identification plate.
Hoist operates sluggishly.	Excessive load.	Reduce loading to the capacity limit of hoist as indicated on the identification plate.
	Low voltage.	Correct low voltage condition.
	Phase failure or unbalanced current in the phases (three-phase unit only).	Check for electrical continuity and repair or replace defective part.
	Brake dragging.	Check brake adjustment.

TABLE 12-2 Troubleshooting Guide for All Electric Hoists (*Continued*)

Problem	Probable cause	Check/remedy
Motor overheats.	Excessive load.	Reduce loading to the capacity limit of hoist as shown on the identification plate.
	Low voltage.	Correct low-voltage condition.
	Extreme external heating.	Above an ambient temperature of 104°F, the frequency of hoist operation must be limited to avoid overheating of motor. Special provisions should be made to ventilate the space or shield the hoist from radiation.
	Frequent starting or reversing.	Avoid excessive inching, jogging, or plugging. This type of operation drastically shortens the motor and contactor life and causes excessive brake wear.
	Phase failure or unbalanced current in the phases (three-phase unit only).	Check for electrical continuity and repair or replace defective part.
	Brake dragging.	Check brake adjustment.
	Motor reversing switch not opening start winding circuit. (Single-phase unit only.)	Check switch connections and actuating finger and contacts for sticking or damage. Check centrifugal mechanism for loose or damaged components. Replace defective part.
Hook fails to stop at either or both ends of travel.	Limit switches not opening circuits.	Check switch connections, electrical continuity, and mechanical operation. Check the switch adjustment. Check for a pinched wire.
	Shaft not rotating.	Check for damaged gears.
	Traveling nuts not moving along shaft—guide plate loose; shaft or nut threads damaged.	Tighten guide plate screws. Replace damaged part.
Hook stopping point varies.	Limit switch not holding adjustment.	Check switch connections, electrical continuity, and mechanical operation. Check switch adjustment. Check for pinched wire. Check for damaged gears. Tighten guide plate screws. Replace damaged part.
	Brake not holding.	Check the brake adjustment.

LUBRICATION

All moving parts of a hoist for which lubrication is specified shall be regularly lubricated. Lubricating methods shall be checked for proper delivery of lubricant. Particular care should be taken to follow the manufacturer's recommendations as to points and frequency of lubrication, as well as to quantity and type of lubricant to be used. Be cautious when substituting lubricant since most manufacturers recommend a specific lubricant only.

Chain Hoists and Chain Slings 4-199

TABLE 12-3 Troubleshooting Guide for Two-Speed Electric Hoists

Problem	Probable cause	Check/remedy
Hoist will not operate at slow speed in either direction.	Open circuit.	Open or shorted motor winding loose or broken wire in circuit, speed-selecting contactor stuck in opposite speed mode. Replace motor, repair wire, and/or repair speed-selecting contactor.
	Phase failure.	Check for electrical continuity and repair or replace defective part.
Hoist will not operate at fast speed in either direction.	Open circuit.	Open or shorted motor winding, loose or broken wire in circuit, speed-selecting contactor stuck in opposite speed mode. Replace motor, repair wire, and/or repair speed-selecting contactor.
Hook will not raise at slow speed.	Open speed-selecting circuit.	Open or shorted winding in speed-selecting contactor coil. Loose connection or broken wire in circuit. Mechanical binding in contactor. Control station contacts not making or opening. Replace coil; repair connection, contactor, or control station.
Hook will not lower at slow speed.	Phase failure.	Check for electrical continuity and repair or replace defective part.
Hook will not raise at fast speed.	Excessive load.	Reduce loading to capacity limit of hoist as indicated on the identification plate.
	Phase failure.	Check for electrical continuity and repair or replace defective part.
	Brake not releasing.	Check electrical continuity and connections. Check that correct coil has been installed. The coil for three-phase dual-voltage unit operates at 230 volts when the hoist is connected for either 230- or 460-volt operation. Check brake adjustment.
Hook will not lower at fast speed.	Phase failure.	Check for electrical continuity and repair or replace defective part.
	Brake not releasing.	Check electrical continuity and connections. Check that correct coil has been installed. The coil for three-phase dual-voltage unit operates at 230 volts when the hoist is connected for either 230- or 460-volt operation. Check brake adjustment.
Hook moves in proper direction at one speed—wrong direction at other speed.	Phase reversal.	Wiring reconnected improperly. Interchange two leads of motor winding that is out of phase at the speed-selecting relay.

Load chain articulates slowly under high bearing pressures and should be lubricated as specified by the hoist manufacturer. In the absence of recommendations, use antiweld or EQ-type lubricant, applied sparingly but frequently as it dissipates during use. Hand chain is lightly loaded and normally needs no lubrication.

WELDED LINK LOAD AND HAND CHAIN

Hoist hand chains and load chains are carefully manufactured. The size of the links and pitch are held to very close tolerances. Pockets of liftwheels, sheaves, and handwheels are accurately formed to ensure a close fit. Hoist manufacturers have charts available for the user of various sizes of chain pitch. Often, this includes information on permissible wear beyond which a chain is no longer safe and should be replaced. Figure 12-11 is an

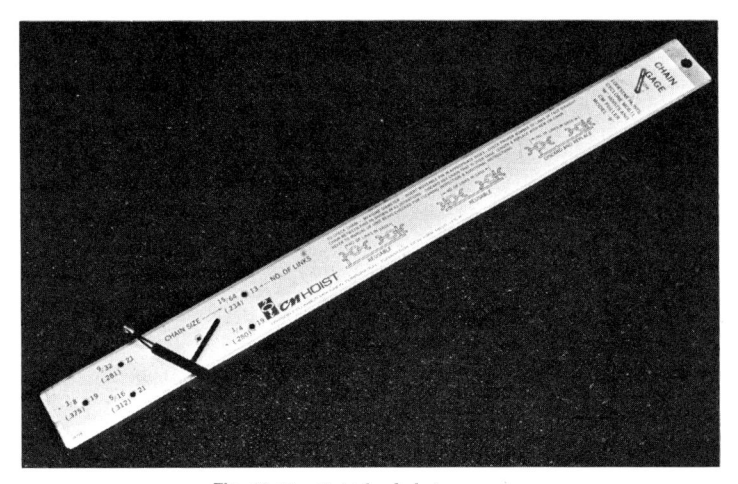

Fig. 12-11 Hoist-load chain gauge.

example of a chain gauge used for determining if the chain is within allowable limits. Lacking manufacturer's data, a fairly accurate check on wear can be made by comparing dimensions of a worn section versus an unused section, such as the loose end. If a used chain is 1½ percent longer than an unused chain for electric and air hoists (2½ percent for hand hoists), replace the chain. Before gauging, the chain must be cleaned in a solvent. After gauging, check for nicks, gouges, and twisted links by a link-by-link inspection. If any of these conditions exists, the load chain must be replaced.

PREVENTIVE MAINTENANCE VERSUS BREAKDOWN MAINTENANCE

Preventive maintenance may be defined differently in scope and intensity in various industries. It has been recognized as extremely important in the reduction of maintenance costs and improvement in the reliability of equipment and production output. Preventive maintenance consists of inspecting equipment and keeping records showing wear or other deterioration of parts. Such records may include repetitive servicing, lubricating, painting, and cleaning. They are the basis for routine inspection and point to the need for major overhaul before breakdown.

Obviously, preventive maintenance reduces the corrective maintenance workload. As preventive maintenance takes over, the timing of the corrective workload is shifted from when you *must* do it back to when you *want* to do it. Thus work can be done more efficiently and at lower cost. However, not every plant can expect to derive equal benefits. For this reason, a preventive maintenance program should be approached with caution.

The cost of a preventive maintenance program is initially high. But the shutdown of equipment for no reason other than periodic inspection may be intolerable from a production standpoint. To help in cost reduction, a good recordkeeping system is essential in planning, scheduling, and training of maintenance personnel. Anyone who expects quick, full benefits from a preventive maintenance program will be disappointed, since a program such as this may take several months before much progress is shown. Breakdown in equipment could result in severe damage and, thus, be far more costly to repair. After inspection, an inspector can report worn parts prior to failure, allowing lead time to order replacement parts. A preventive program must be tailored to fit a company's needs. Such industries as plating companies, chemical plants, and heavy manufacturing plants may require more frequent maintenance.

Some companies prefer a program of replacement units whereby an entire unit can be removed from service and replaced with an identical unit. Such a philosophy can be used for certain types of equipment in a plant where standardization exists.

Sometimes when a centrally administered preventive maintenance program cannot be achieved, qualified mechanics can be assigned to individual pieces of equipment. Because of their familiarity with the equipment, mechanics can effectively reduce cost and breakdown in plants where equipment is not used continually. A complete overhaul of production equipment can be accomplished during a shutdown period similar to the automobile industry changeover for a model year.

In a continuous manufacturing plant, vital inspection and replacement is difficult. Often, these inspections and replacements can be accomplished during the same time it takes to perform primary repairs. However, such methods require reporting and recordkeeping of deficiencies during normal plant operation. Parts must be available and personnel must be familiar with the equipment. A sound preventive maintenance program undoubtedly has advantages over the breakdown approach. Although some plants would prefer breakdown to deliberate shutdown preventive maintenance, we might expect newly created state and federal laws to insist on preventive maintenance.

Chapter **13**

Maintenance of Belt Conveyors and Conveying Equipment

R. A. HOPPENRATH
Product Application Manager—Conveyors
Barber-Greene Company, Aurora, Ill.

Belt conveyors are the key elements in nearly every system for moving bulk materials. They will operate for many years with a minimum of attention but they do require care.

Conveyors appear to be relatively simple mechanisms; therefore they are often neglected. Yet, regardless of the care and ingenuity which have gone into their manufacture, conveyors can serve well only when operated correctly, maintained properly, and lubricated regularly. Maintenance of conveyors is just as important as taking proper care of more complex pieces of equipment. Failure of just one conveyor can shut down an entire operation.

GENERAL

Typically, a conveyor consists of the belting itself, a structure to support it, a drive mechanism, idlers to hold the belt, a method of tensioning the belt, feeding and discharge hoppers, and a variety of accessories which perform such special functions as clearing the belt where sticky or wet material is being handled.

Other accessories include covers that prevent material from blowing off the belt in high winds. Other covers prevent the belt from being blown off the idlers or protect the belt and its contents from sun and weather. Then there are special idlers which cushion the load and keep the belt aligned.

Portable and permanent conveyors Portable conveyors are equipped with pneumatic-tired wheels so they can be moved about readily. Some models have hinged booms to reduce their length for transport. Permanent conveyors have a pair of A-frame supports instead of wheels. Both portable and permanent conveyors are frequently used in the same plant (Figs. 13-1 and 13-2).

Portable conveyors are supplied with screw take-up tail ends (Fig. 13-3) for belt tensioning. Permanent conveyors are provided with either screw take-ups or fixed bearings if the conveyor was ordered with automatic gravity take-up.

Fixed tail ends (Fig. 13-4) are assembled and properly aligned at the factory. No further pulley alignment is necessary. Gravity take-ups are either the vertical or the horizontal type. The latter are generally used where space is limited.

Horizontal gravity take-ups (Fig. 13-5) require a support frame for a movable carriage and pulley. A cable is attached to the movable carriage and to a counterweight to keep the belt automatically at proper tension.

4-204 Basic Maintenance Technology

The maintenance instructions outlined in this chapter apply to both portable and permanent conveyors. Most components are similar on both types of machines.

To neglect maintenance of any conveyor component is to invite trouble. Operating costs will soar, conveyor life will be shortened, and the whole plant operation will be made inefficient by frequent conveyor breakdowns.

Trouble-free belt conveyor operation results from:
1. Selection of quality components
2. Careful installation and initial start-up
3. Proper operating, lubrication, and maintenance procedures
4. Routine inspection

Working around any kind of machinery may be dangerous. Mechanics and operators must be extremely careful while performing maintenance on belt conveyors.

Safety There is no set of safety rules which can cover every conceivable danger. Safety is basically common sense. One simple rule which could prevent thousands of

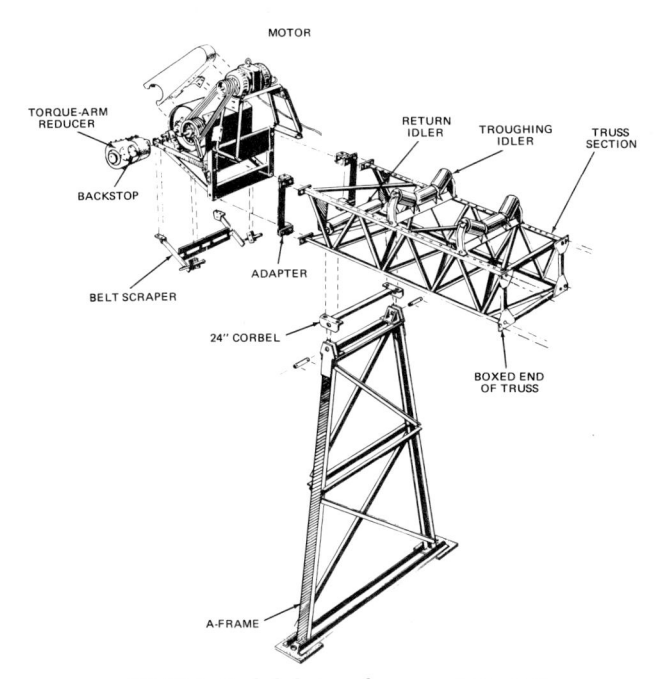

Fig. 13-1 Exploded view of permanent conveyor.

serious accidents every year is: *Never attempt to clean, lubricate, or adjust any machine while it is in motion.*

Owners are advised to consider the use of safety switches and sturdy guards. All personnel who work near equipment powered by electric motors should observe these safety rules:
1. All electrical equipment must be grounded.
2. Always pull the main switch before disconnecting a plug from a receptacle.
3. Never grease, adjust, or repair a machine without removing the plug from its receptacle or pulling the main switch.
4. Use only the proper size and style of fuses; keep a supply on hand and never substitute pieces of metal for proper fuses.
5. Keep electrical cables in good condition at all times and keep them out of water.

Professionalism A true professional is the operator or mechanic who knows the value of proper lubrication. This cuts down lost time. Lost time means lost wages for everyone. Correct lubrication also increases the life of a machine, thus ensuring job security.

The real professional
 1. Keeps the grease gun clean and wipes each grease fitting with a clean cloth to prevent grit from being pumped into the bearing
 2. Keeps grease and oil containers clean and their covers in place to keep dust and dirt out of the lubricant
 3. Keeps each lubricant container well labeled and always uses the correct lubricant in the correct place
 4. Studies lubrication charts and does not guess when lubricating a machine
 5. Keeps machines as clean as possible and removes excess grease and oil which may accumulate during the day
 6. *Lubricates ball or roller bearings carefully so as not to blow seals through overgreasing*

MOST VULNERABLE COMPONENT—THE BELT

The conveyor belt is the most expensive and vulnerable part of the machine and should be cared for accordingly. Most conveyors are designed to provide maximum belt life, provided that proper attention is given to operation and maintenance.

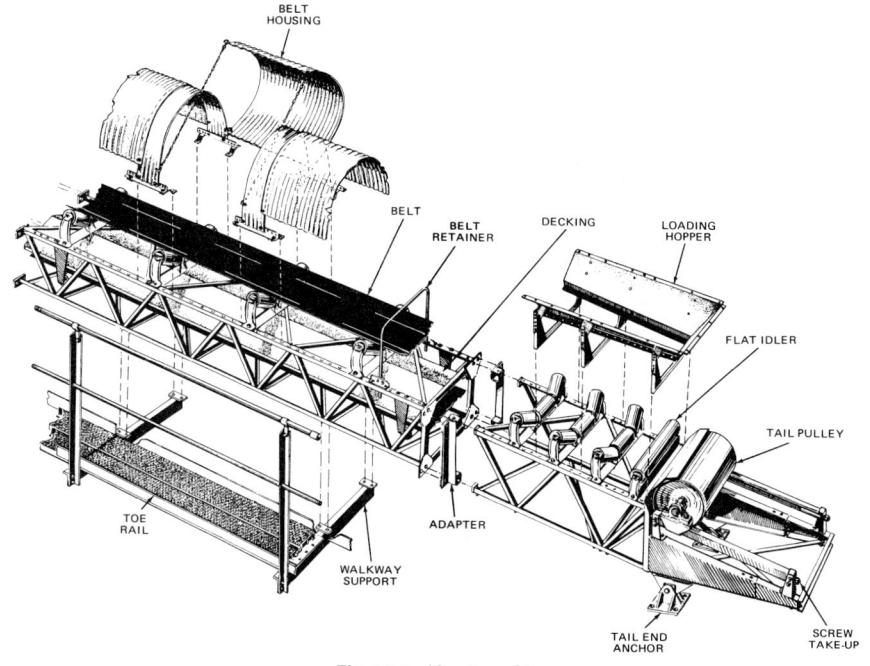

Fig. 13-1 (Continued.)

Good-quality conveyor belting is generally constructed of a high-tensile synthetic carcass with square edges and an extra-heavy cushion layer of rubber between all plies.
 1. It has high strength, impact resistance, and great flexibility.
 2. It has no shrinkage or excessive elongation.
 3. It is rot-proof and mildew-proof.
 4. It is resistant to chemicals, alkalis, minerals, and organic acids.

Belt wear The following items are common causes of excess conveyor belt wear and suggestions for their correction:
 1. *Improper installation.* The belt must line up so that it runs squarely on the center of pulleys and idlers. If not properly aligned, belts will not give maximum service. Obstructions will damage the belt or permit abrasive material to rub against the moving belt, causing excessive wear and premature failure.

Fig. 13-2 Portable conveyor.

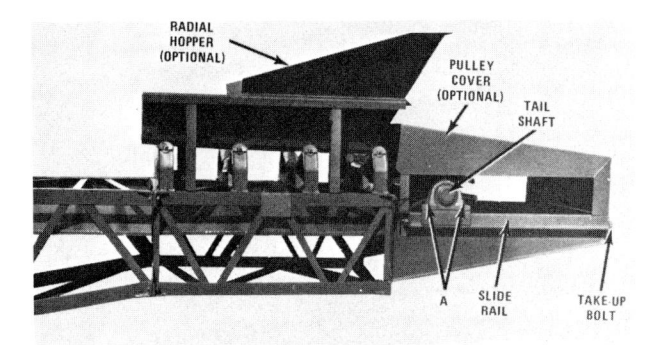

Fig. 13-3 Tail end with take-up screw.

Fig. 13-4 Fixed tail end.

Maintenance of Belt Conveyors and Conveying Equipment 4-207

2. *Belt tension.* Conveyor take-ups should be adjusted to provide just enough tension to operate the belt under load, without slipping. Excess tension places the belt under strain. Slippage between drive pulley and belt obviously will wear the belt.

3. *Starting under load.* The conveyor should never be started under load if it can be avoided. If possible, empty the belt before stopping and start the conveyor before loading.

4. *Moisture.* Cuts or tears should be cleaned with gasoline and filled in with rubber cement. Otherwise, moisture and abrasive particles can enter and start breaking down the carcass of the belt.

5. *Oil and grease.* Petroleum oil or grease is extremely harmful to rubber. Care should be taken while lubricating the conveyor to prevent lubricant from contacting the belt.

6. *Damaged idlers.* Idlers that do not turn freely because of improper lubrication or damaged bearings cause drag and wear on the belt. Hard-turning idlers should be repaired or replaced. Do not allow material to build up on idlers.

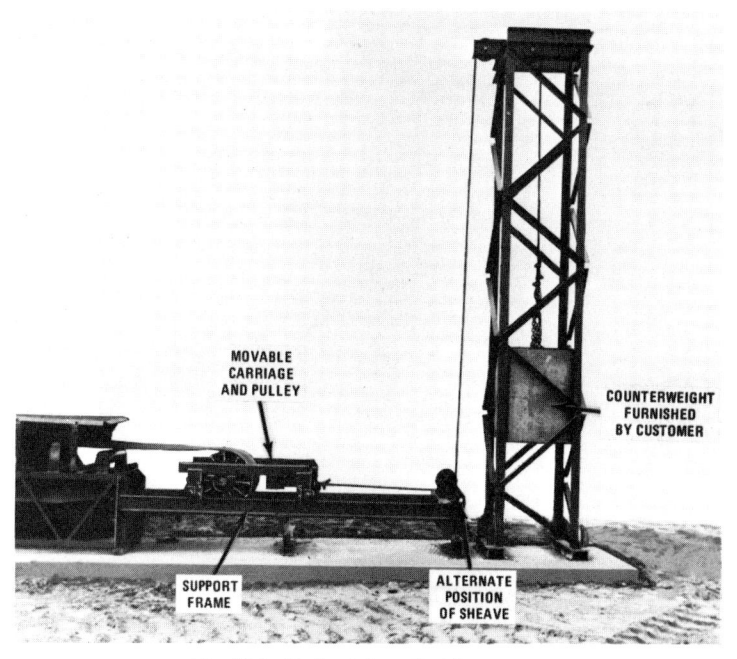

Fig. 13-5 Horizontal gravity take-up.

7. *Buildup on idlers.* Some materials tend to build up on idler rollers—especially return idlers and snub pulleys where the carrying side of the belt contacts them.

Buildup will cause difficulty in training a new belt. Damage to the belt can occur when a material builds up to such an extent that it will cause a belt to rub against obstructions. Buildup will be worse in freezing weather.

8. *Friction from external causes.* Hoppers, skirtboards, and chutes should be adjusted so that they do not touch the belt. Flashing on hoppers and skirts should be kept in proper alignment.

9. *Decking.* Decking over the return belt should be kept clean so that material is not carried between belt and pulley.

Inspection Inspect the entire length of the belt as installed on the conveyor to be sure that there are no obstructions which will rub, tear, or cut the belt while it is running. Be sure that no particles of the material handled can catch at some point and damage the belt. Make certain that no stones can drop or bounce onto the return run of the belt and get between the belt and foot pulley. Check belt lacings periodically and replace if they are defective.

Belt Tension Belt tension should be checked at frequent intervals and adjusted when necessary. Adjustment of the conveyor take-up (Fig. 13-3) should provide just enough tension to operate the belt under load without slipping. Excess tension places the belt and drive machinery under strain. Slippage between drive pulley and belt will cause serious belt wear.

If the conveyor is to be shut down for a length of time, it is advisable to remove the belt and store it until ready to resume operations. This is particularly true if the conveyor is exposed to severe weather.

Store the belt in a cool, dry place. Do not lay the roll so that the belt rests on edge. Table 13-1 shows common problems that occur with conveyor belts and recommendations for solutions.

BELT-CUTTING PROCEDURE

Cutting a belt and fitting it to a conveyor require special care. Improperly cut belts cause a number of conveyor problems (as outlined in the preceding section) and can shorten belt lift drastically. The correct procedure is listed:
 1. Measure 8 ft from the end of the belt as shown in Fig. 13-6.
 2. In four places, measure from the outside edge of the belt to the center.

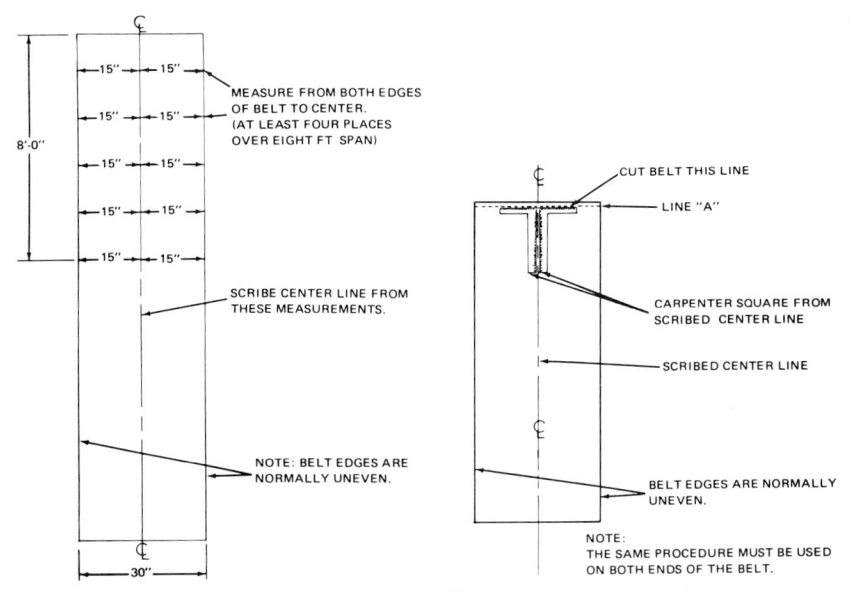

Fig. 13-6 Measurements for cutting conveyor belt.

Fig. 13-7 Scribing a centerline in the conveyor belt.

 3. Scribe the centerline through points determined when measuring from the belt edge (Fig. 13-7).
 4. Locate a carpenter's square on scribed centerline. Scribe a line from center of belt to outside edge in both directions. This line is designated as line A in Fig. 13-7.
 5. Cut the belt along line A.
 6. The same procedure should be used on the other end of the belt.
  NOTE: Because of the unevenness of belt edges, it is extremely important to mark off points in four places over an 8-ft span.

A belt tightener (Fig. 13-8) should be used to bring the two ends of the belt together for splicing.
  NOTE: It is extremely important that conveyor belting be cut square before beginning installation of lacing.

TABLE 13-1 Troubleshooting Chart

Problem	Probable cause	Recommendation
\<Making conveyor belts run straight\>		
Belt runs true when empty, crooked when loaded.	Off-center loading or variations in nature or formation of load.	Adjust chute and other loading devices so that load is delivered to center of belt and in line with direction of belt travel.
Belt climbs sideways on some idlers.	Loose idler.	Return idler to proper position; fasten securely.
	Idler sticks or jams.	Lubricate properly. Replace any sticking idlers having worn spots.
	Idlers out of alignment.	Realign idlers while belt is unloaded.
Part of belt running off idlers is in vicinity of splice.	Improper splice; ends not cut squarely for a mechanical splice.	Resplice. Make sure ends are square for mechanical-type splice.
Same section of belt repeatedly runs off idlers along entire conveyor.	Crooked belt.	Replace with straight belt.
Belt with worn edge becomes crooked.	Worn edge becomes stretched because of high-friction pull or shrinks from moisture absorption.	Eliminate cause of wear. Repair damaged edge.
Top cover gouged or grooved.	Material trapped under skirts.	Prevent jamming by providing an increasing gap under skirts in direction of belt travel. Increase belt tension or space loading-point idlers more closely.
Blisters in cover.	Fine material working into cuts or punctures.	Make spot repair.
Lengthwise strip swelling of bottom cover.	Oil.	Avoid over lubrication and spillage of oil and grease. Use oil resistant belt if necessary.
Bottom cover wear.	Drive pulley.	Increase belt tension if belt rating permits. Lag drive pulley.
	Sticking rollers.	Lubricate properly and replace idlers as required.
	Excessive troughing idler tilt.	Reset not more than 2 deg from upright.
	Corroded troughing idler rolls.	Replace.
\<Carcass damage\>		
Carcass breaks.	Impact.	Load lumps between idlers. Decrease height of free fall of lumps.
	Material trapped between belt and pulley.	Use plows ahead of tail pulley on return side. Use deflector over take-up pulley.
	Material buildup on pulleys	Use scrapers.

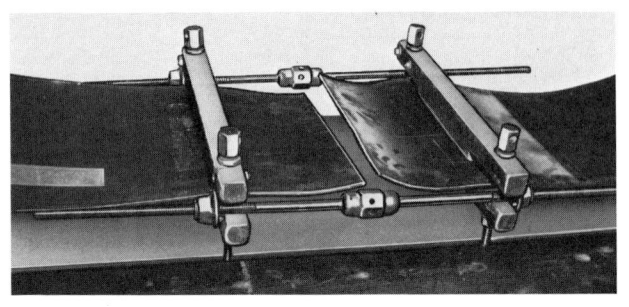

Fig. 13-8 Belt tightening for splicing.

1. Square belt ends and cut to length. To simplify the cutting job, use an Alligator Wide Belt Cutter.

2. Support belt ends with wood plank. Nail Flexco Templet in position with belt ends tight against lugs. Punch or bore bolt holes.

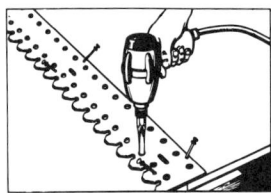

3. An impact tool with Flexco Power Punch or Flexco Power Boring Bit speeds hole boring operation. Remove templet. Leave plank under belt ends for a work surface. All work can be done from the top of the belt.

4. Assemble bolts in bottom plates. Fold one belt end back out of the way. Then insert bolts from under side along one row of holes.

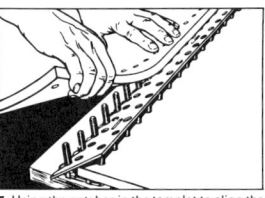

5. Using the notches in the templet to align the opposite row of bolts, place the other end of the belt over the bolts. Press belt onto bolts with hands. Remove templet. Continue to press belt until it is in place.

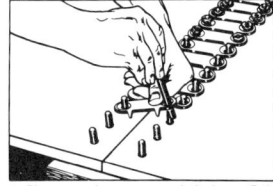

6. Place top plate over one bolt. Insert Bolt horn Tool through the other plate hole and over the second bolt to pry it into place.

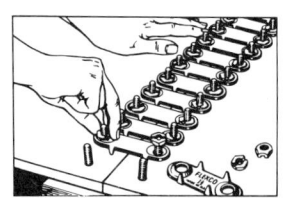

7. Assemble all top plates same way as in Direction No. 6. Start nuts down by hand far enough so that wrench will engage bolts.

8. Before tightening fasteners, cut a piece of Flexco-Lok® Tape three times the width of the belt plus six inches and cut a point on one end. Thread pointed tape between fastener teeth on top of belt, back through the bottom plates, and across the top again.

9. Pull tape tight and hold in position by tightening a fastener at each end of the splice. Then snug down all other plates.

10. Tighten all fasteners from edges to center. Tighten all nuts uniformly. A Flexco Power Tool Wrench used with an impact tool will speed this step considerably.

11. Hammer plates in belt with metal or hard wood block as illustrated. Then retighten nuts.

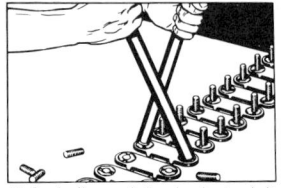

12. Break off excess bolt ends using two bolt breakers. On belts with thick rubber covers, retighten all nuts after a few hours running.

Fig. 13-9 Procedure for applying conveyor belt fasteners.

4-210

APPLYING CONVEYOR BELT FASTENERS

Fasteners should be retightened at least once after the first 24 hr of service, especially on belts with thick rubber covers (Fig. 13-9).

INSTALLATION OF BELT

Alignment of the whole conveyor should be checked by instrument or stringline. The alignment must be almost perfect to ensure proper belt operation.

Check all flashings on hoppers and skirt boards. Where necessary, make adjustments so that they do not put too much pressure on the belt. Take-ups, such as screws at tail ends and gravity take-ups, should be brought to minimum position and secured there.

When the truss section is lined up and the other adjustments have been made, the belt may be installed. The roll of belting should be kept level and placed in alignment with the installation. Placing a shaft through the center of the roll and supporting it at each end will allow the belt to feed easily and smoothly off the roll. However, control the roll so that it does not unwind and telescope. Throughout installation the belt must be kept taut.

CAUTION: Most conveyor belts have a thicker rubber cover on the carrying side than on the pulley, or underside. Be sure that this carrying side is uppermost when the belt is placed on the troughing carriers. A clamp and cable may be attached to the end of the belt and the cable pulled over the head pulley. Fasten the belt and thread the cable through the return. Bring the ends of the belt together at a convenient place for splicing. In cases where the belt is too long for one roll, of course, splices must be made at all connecting points.

TRAINING OF BELT

A newly installed belt must be lined up to run on the center of the pulleys and idlers. Misalignment can cause belt training problems. The tail-end take-up pulley and all idlers should be squared up so they are at 90 deg with the centerline of the conveyor. The belt can be aligned only by adjusting the troughing and return idlers. Self-aligning idlers should be adjusted so that the belt will touch the guide rollers before it hits the troughing or return idlers.

If the belt is a reversing type and self-aligners are used, they will be supplied with an actuating shoe. The latter will be bolted to the outside of the idler frame and will tip the idler in the proper attitude to center the belt.

When starting a new belt for the first time, run it slowly and intermittently. Check the entire length to be certain it is not running off at any point to an extent that will damage the belt.

Do *not* attempt to train the belt by adjusting the head-end drive pulley. Drive shaft bearings are set at the factory before shipment. If they are moved, it will throw the drive belts out of line.

The return run should be checked first for belt alignment. When a belt runs off a return roll, the edge may be damaged. To train the return belt, start at the head end. In general, adjustments are made 15 to 20 ft behind the point where the belt runs off.

All troughing idlers are provided with slotted holes for the hold-down bolts to permit shifting the ends for belt alignment. To adjust idlers loosen the hanger bracket bolts slightly and move the idler slightly toward the head shaft on the side opposite the point where the belt runs off. The belt will shift to the side where it touches the roll first.

Do not tighten hold-down bolts for troughing and return idlers until the belt has been trained. Adjust idlers (Fig. 13-10) with the belt running. The effect of idler shifting is not immediate. Wait a few minutes to see if the belt trains properly before making further adjustments.

NOTE: When proper training has been finished, tighten all mounting bolts to prevent shifting of idlers.

When training the belt on the carrying side, start at the tail pulley. Follow the same adjustment procedure on troughing rollers as described previously for return rolls. To repeat, make adjustments behind the point where the belt runs off and wait a few minutes for the belt to train. Tighten all hold-down bolts after the belt is properly trained.

Do not try to correct belt alignment by extreme shifting of one idler but rather by slightly changing a number of idlers. Proper alignment of idlers at the head end will train a belt to run on the center of the head pulley.

If the belt does not run central after adjustment of idlers and take-up screws, check to be certain the belt lacing is square.

Permanent stretch of a conveyor belt which has a synthetic carcass will happen fast

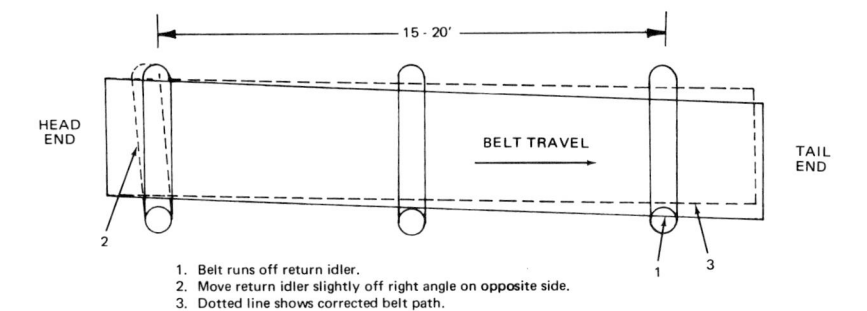

1. Belt runs off return idler.
2. Move return idler slightly off right angle on opposite side.
3. Dotted line shows corrected belt path.

Fig. 13-10 Belt training by idler adjustment (return idler shown).

during the first few hours of operation. It should be checked frequently during the first week. Although permanent stretch occurs more rapidly with synthetic-carcass belts, the total stretch will be no greater than for other types of belting materials.

Conveyor take-up (Fig. 13-11) should be adjusted to provide enough tension for the belt to operate under load without slipping. Correct belt alignment must be maintained while tension adjustments are made. When correctly aligned and adjusted, the belt will be centered on the tail pulley while the conveyor is operating. The amount of belt tension is correct when the belt sags slightly between the return rolls.

Do not apply excessive tension because damage to the belt and to pulley bearings may result. After the belt has been run in, tension should be checked at frequent intervals and take-ups adjusted when necessary.

Fig. 13-11 To adjust conveyor belt, turn take-up screws (A) clockwise until desired belt tension is obtained.

Fig. 13-12 Keep the rubber flashing of the loading hopper (shown here) adjusted.

OPERATING PRECAUTIONS

Certain precautions must be observed to ensure safe operation of conveying equipment.

1. Always check belt alignment when starting the conveyor for the first time on a new setup. Be sure that the belt does not rub or catch on any obstruction.

2. When starting operation, be sure that the belt is started and moving freely before

any material is discharged to the hopper. Also, be sure that the discharge to the hopper is such that the hopper is not buried in spilled material.

3. Clean out the hopper end at frequent intervals to remove any accumulation of material. Keep the belt, pulleys, and return idlers clean and free of accumulated material or sticky material.

4. Keep hopper flashing (Fig. 13-12) in proper adjustment.

5. Unless absolutely necessary, the conveyor should not be started under load. If possible, empty the belt before stopping and start the conveyor before loading.

6. If material is fed onto the belt from gravity chutes, it is often possible to overload the conveyor and cause serious damage to the power unit, drive machinery, or belt. The feed of material onto a belt must be regulated not to exceed the conveyor's rated capacity in tons per hour.

7. When handling oily, abrasive materials, such as metal chips, turnings, or oiled coal, use a special Barprene belt. This belt is designed to withstand the corrosive action of oil and grease.

8. To handle hot material, use a high-termpearture belt and high-temperature grease for lubrication.

9. Check motor rotation every time electrical leads are changed after moving a conveyor.

Overloading a conveyor As already described, every conveyor is designed to handle a given maximum capacity. If material is fed onto the belt from gravity chutes, it is often possible to overload the conveyor and cause serious damage to motor, drive machinery, or belt. Feed of material onto the belt must be regulated not to exceed the rated capacity in tons per hour for which it was designed.

It is also possible to overload a conveyor by changing operating conditions. For example, if a conveyor is designed to handle a specified capacity in a horizontal position and later is elevated to operate at an 18-deg angle, capacity must be reduced to avoid overloading the conveyor.

Normally, rated capacity is for a conveyor at an 18-deg incline. A belt will not be fully loaded at the rated capacity. Capacity of a conveyor is decreased when the angle of operation is increased. It is advisable to stay below the maximum angles given by the manufacturer. A conveyor loaded with more than rated capacity has extra wear, which reduces the life of the mechanical components and may burn out the motor.

Capacity measurement in the field To determine the actual capacity of a belt, it is necessary to know belt speed. Then, the capacity for any belt width may be determined in the following manner:

1. Measure a length on the loaded belt equal to the length given in Table 13-2 for the corresponding belt speed.

TABLE 13-2 Belt Conveyor Capacity Measurement

Belt speed, fpm	Belt length ft	in.	Belt speed, fpm	Belt length ft	in.	Belt speed, fpm	Belt length ft	in.
200	6	0	300	9	0	400	12	0
210	6	3½	310	9	3½	410	12	3½
220	6	7¼	320	9	7¼	420	12	7¼
230	6	10¾	330	9	10¾	430	12	10¾
240	7	2½	340	10	2½	440	13	2½
250	7	6	350	10	6	450	13	6
260	7	9½	360	10	9½	460	13	9½
270	8	1¼	370	11	1¼	470	14	1¼
280	8	4¾	380	11	4¾	480	14	4¾
290	8	8½	390	11	8½	490	14	8½

2. Weigh the material contained in this length of belt.

Each pound of material on this length of belt represents 1 ton per hr that the conveyor is handling.

Feeding Material should be delivered to the hopper with as little drop and impact as possible. Large-size material must be carefully fed to the hopper to prevent damage to the belt.

It is desirable to feed at a uniform rate of speed so that the load on the belt will be evenly distributed along the entire conveyor length. The feed should also be such that the material is carried on the center of the belt so the belt will run true.

Adjusting V-belt tension When they are being mounted, V-belts must not be stretched over the rims of sheave grooves. Reduce the span between sheaves enough to permit the belts to be assembled without stretching.

Tighten new belts about 2 times normal tension. There will be a rapid drop in tension during the run-in period (the first 24 to 48 hr) while the belts seat themselves in grooves (Table 13-3).

TABLE 13-3 V-Belt Tension Adjustment

V-belt cross section	Average Small sheave diam. range, in.	Drive ranges Small sheave rpm range	Speed Ratio range
3V	2.65–3.35	1200–3600	2.00–4.00
3V	4.75–6	900–1800	2.00–4.00
5V	7.1–9	600–1500	2.00–4.00
5V	12.5–16	400–800	2.00–4.00
8V	18–22.4	200–700	2.00–4.00

After the first day or two, check for the correct amount of tension in each belt. If the belt deflection force is over 1.5 times normal, the belts are too tight. If the force is below normal belt tension, they are too loose (Table 13-4).

Adjustment Procedure

STEP 1. Measure span length t of drive (Fig. 13-13).

STEP 2. At center of span t, apply a force perpendicular to the span large enough to deflect one belt on the drive 1/64 in. per inch of span length from its normal position.

TABLE 13-4 Recommended Belt Deflection Force in Pounds per Belt

For normal tension, lb	For 1.5 times normal tension, lb	For 2 times normal tension, lb
3	4½	6
4	6	8
8	12	16
10	15	20
20	30	40

Conveyor hopper flashing The rubber hopper flashing (Fig. 13-12) on conveyors is adjustable. It should be kept adjusted so that it contacts the belt evenly but lightly. Contact that is too hard will cause rapid belt wear.

Hydraulic hoist assembly Whether hand- (Fig. 13-14) or power-actuated (Fig. 13-15), a hydraulic hoist will operate smoothly and with little trouble, free of dirt and foreign matter. All hose connections must be kept tight.

Hand hydraulic pump The hand hydraulic pump used in conjunction with a hydraulic ram for raising a boom is a double-action piston pump. It forces oil from a reservoir into the ram.

If the hand pump fails to raise the boom, either the ram is leaking or the check ball is not seating properly. If the fault is with the check ball, follow this procedure:

1. Remove the valve plug, the spring, and, finally, the ball.
2. Use a magnet to clean out any chips that might be on the valve seat.
3. Replace the check ball and tap lightly with a small hammer and brass rod.
4. Reassemble the valve plug.

In some instances air will get into the line. Sometimes this air will get into the oil passages in the pump, causing it to function only on one side. To expel this air
1. Open the release valve.
2. Give the pump handle a few quick strokes, thus expelling air from oil passages back into the oil reservoir where it will no longer be harmful.

Fig. 13-13 Conveyor-drive V-belts are adjusted by the head shaft torque arm reducer tie-rod turnbuckle.

Fig. 13-14 Hand-operated hydraulic boom hoist.

Hydraulic ram The hydraulic ram used on either hand- or power-actuated boom hoists will require little or no maintenance, provided the correct oil is used and it is kept clean and free of dirt. Inspect the ram periodically for leakage around the packing. If leakage is excessive, install a new packing kit.

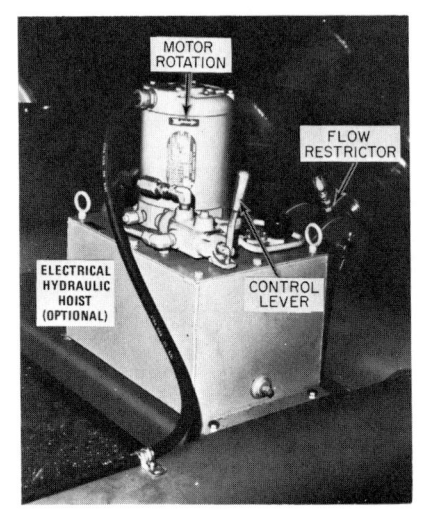

Fig. 13-15 Electrical hydraulic boom hoist.

Tires and wheels Periodically, check tire pressure and keep inflated to correct pressure.

Belt scraper The function of a belt scraper (Fig. 13-16) is to wipe the belt clean of wet, sticky materials. The scraper prevents material buildup on return rolls, eliminating excessive belt tension or failure as well as bearing failure. Return rollers must be kept clean if they are to effectively train a belt onto the belt pulley.

A scraper is mounted so that the scraper blade contacts the belt prior to leaving the head pulley. Compression springs maintain an even scraper pressure against the carrying surface of the belt.

Spring tension is adjustable through use of adjusting nuts. Scraper blades are reversible for maximum wear.

Adjusting belt-scraper tension Tighten nuts increasing the spring tension on the belt by the scraper blade (Fig. 13–17). Proper adjustment is obtained when the rubber blade contacts the belt firmly and evenly. Do not use too much tension because it will cause excessive wear and premature breakdown of the rubber scraper blade.

ADDED MAINTENANCE AND ADJUSTMENTS FOR GASOLINE ENGINE DRIVEN CONVEYORS

Engine Refer to the engine manufacturer's manual for specific maintenance instructions.

Head shaft drive chain adjustment
1. Loosen the drive shaft bearing hold-down bolts (A), four on each side (Fig. 13-18).
2. Loosen lock nuts (B) on retainer clips (C), two on each side.
3. Move shaft to obtain proper chain adjustment. Be sure to move each side the same amount to maintain proper sprocket and sheave alignment.
4. Hold clips (C) firm against bearings and tighten nuts (B).
5. Tighten bolts (A).

Drive shaft V-belt adjustment After adjusting the drive chain, it is necessary to adjust the drive shaft drive belts.
1. Loosen bolts (A) on the countershaft bearing support plates on each side (Fig. 13-19).
2. Loosen lock nuts (B) on adjusting bolts (C), one on each side of the conveyor.
3. Take up on nut (D) to remove excess slack in the V-belts. Be sure to take up equally on both take-up bolts so sheave alignment will remain parallel.
4. Tighten lock nuts (B) and bearing support plate bolts (A). After making this adjustment, it will be necessary to adjust the countershaft drive belts.

Countershaft drive belt adjustment
1. Loosen the two bolts (A) in each of the four-corner power unit sill brackets (Fig. 13-20).
2. Loosen lock nuts (B) on take-up bolts (C).
3. Tighten adjustment nuts on take-up bolts (C) until the drive belts are snug. (Do not run with belts too tight.) Be sure to take up equally on each take-up bolt so sheave alignment will remain parallel.
4. Tighten lock nuts (B) and bolts (A).

LUBRICATION

Nothing can add to the life of a conveyor more than thorough and proper lubrication at correct intervals. When time and availability of a machine are at a premium, a breakdown caused by improper lubrication is absolutely inexcusable since this can so easily be avoided (Fig. 13-21).

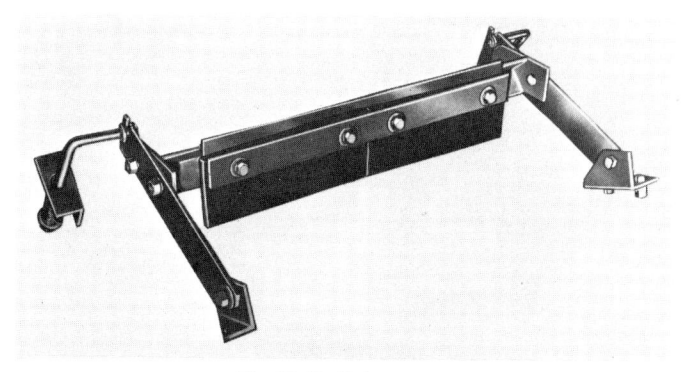

Maintenance of Belt Conveyors and Conveying Equipment 4-217

Fig. 13-16 Belt scraper.

Fig. 13-17 Belt scraper adjustment.

Fig. 13-18 Head shaft drive chain adjustment.

Fig. 13-19 Drive shaft V-belt adjustment.

Fig. 13-20 Countershaft drive belt adjustment.

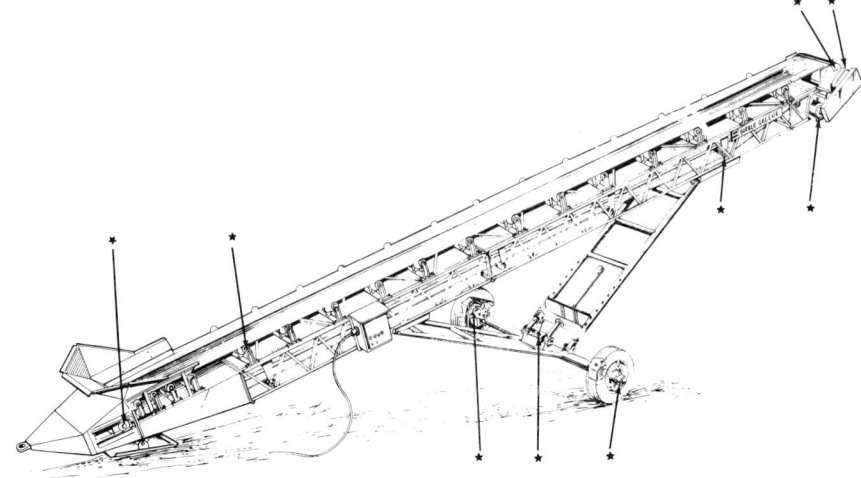

Fig. 13-21 Typical lubrication points.

caused by improper lubrication is absolutely inexcusable since this can so easily be avoided (Fig. 13-21).

NOTE: The number of shots from a grease gun called for in lubrication instructions is based on greasing with a standard 13-oz grease gun. One shot with such a gun equals 1/54 oz. In other words, it takes 54 shots to deliver 1 oz of grease.

The following greases (or others equal to the characteristics of those listed) are recommended for bearings. These greases are suitable for temperature ranges of -10 to $+200°F$ under normal speeds and conditions. For continuous cold weather operation a lighter-grade is advisable.

Consult the factory for recommendations for other temperatures or abnormal conditions. It is important that a grease be used that is suitable for prevailing conditions. Apply lubricant on the basis of lubrication character and temperature.

Characteristics	Temperature, °F	Shots
Penetration unworked	0	165–200
Penetration unworked	77	220–290
Penetration worked	77	235–290
Dropping point	0	300 min.
Mineral oil gravity	—	24–29
Pour	0	0–+15
SSU	100	275–335
SSU	210	50–60

Place safety first Observe the following safety precautions before servicing and lubricating idlers:

1. Wear safety glasses when inspecting idlers while a belt is carrying material.
2. Do not lubricate idlers while a belt is running.
3. Be sure that conveyor drive is locked out before removing an idler or component part.
4. Do not leave tools or parts lying on a belt.

Idlers Keeping idler bearings properly greased is important for the following reasons:

1. To minimize bearing rolling friction and the wear it causes.
2. To avoid the tendency of old grease to thicken with time and lose some of its lubricating quality.
3. To assist seals in preventing dust and moisture from entering bearings and causing wear or corrosion. A good packing of grease not only lubricates but also serves as a barrier against foreign matter.

Usually, a manufacturer will indicate what normal lubrication intervals should be for idler bearings. For some ball or roller bearing idlers, for example, the lubrication interval is 800 to 1200 operating hours under average operating conditions. It is important to note, however, that this is only a guide. The proper lubrication interval may vary from job to job, from conveyor to conveyor, and even from one group of idlers to another on the same

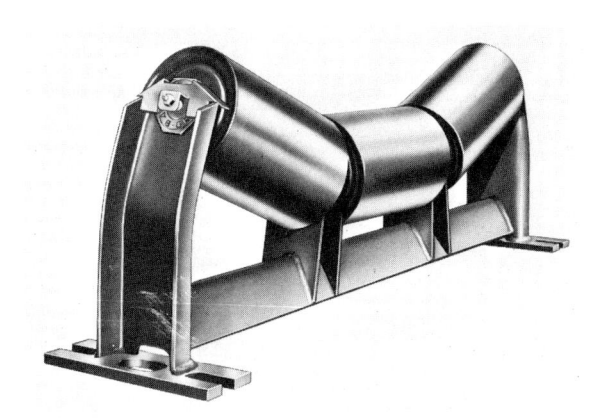

Fig. 13-22 Troughing idler.

4-220 Basic Maintenance Technology

conveyor. The operator must therefore determine the correct lubrication interval for a particular application (Fig. 13-22).

Most conveyor idlers have lubrication fittings on each side. On some types all three rolls of a troughing idler can be lubricated from either side of the belt from either one of the fittings (Fig. 13-23).

Idlers should be lubricated with a high-quality No. 2 or No. 1 grease, either calcium or lithium base. The operating range for this lubricant should be from -10 to $+200°F$. Using a high-quality grease of this type exclusively will help ensure efficient, trouble-free performance and the longest possible service life.

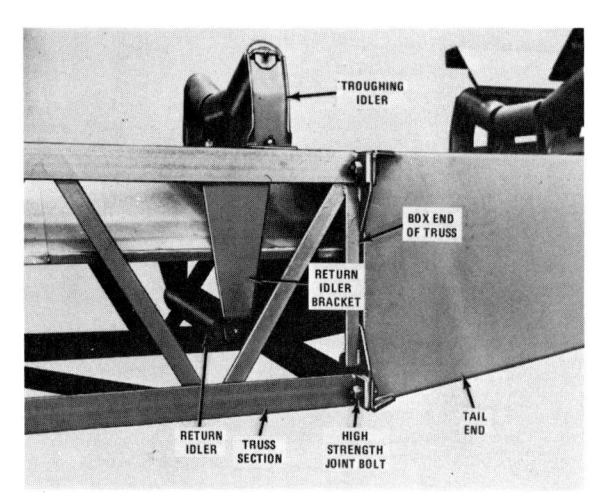

Fig. 13-23 Troughing and return idlers shown in relative working positions.

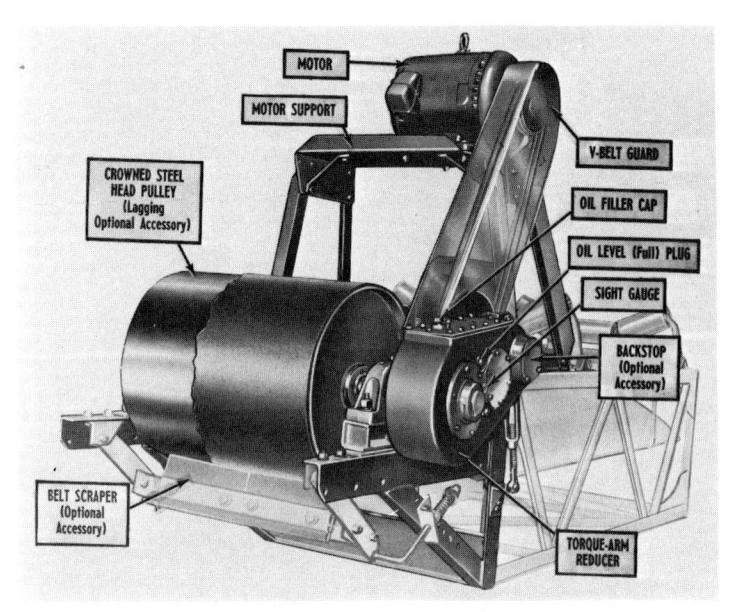

Fig. 13-24 Permanent conveyor head and drive.

Antifriction bearings One factor that determines the lubrication interval is the effectiveness of bearing seals. Another is the set of conditions that prevails at the individual conveyor. If a belt handles much fine, corrosive, or wet material, or if the atmosphere contains a great deal of dust or moisture, idlers should be greased more often than they would be under normal conditions.

With some systems of positive lubrication it is possible to change grease or flush out old grease while replacing it with fresh, clean lubricant. Position a container on the far side to catch old grease so that it does not come in contact with the belt. For normal lubrication, however, the appearance of additional grease at the fitting on the far side of the idler is assurance that all bearings in the rolls have been lubricated. Note that some bearings can be damaged by overgreasing, as explained.

Antifriction bearings used on some conveyors have been lubricated at the factory. The grease cannot be seen since it is concealed within the bearing by the grease retainer seals. Overgreasing distorts and damages these seals, allowing dirt to enter and greatly shorten the life of the bearing.

CAUTION: Never use a power-operated grease gun on antifriction bearings. Lubricate only as directed by the lubrication chart.

Drive Check the oil level in a gear reducer box every 40 to 50 hr of operation. Fill box (Fig. 13-24) to oil level plug, using a good grade of gear lubricant. For temperatures above 32° use SAE 90 gear lubricant. For temperatures below 32° use SAE 80 gear lubricant. Drain, flush out, and refill gear box after 300 hr of operation.

For reducers with bearings requiring external lubrication, use a good grade of grease. For specific grade and type, refer to the operator's manual for the manufacturer's recommendation.

Oil-tight chain guards Fill to oil level plug with a good-quality clean SAE 30 motor oil. SAE 20 motor oil may be used in cold weather. Check oil level once a week *when drive is not operating.* Drain and refill every 3 months.

Plain bearings Plain bearings should be greased every day the conveyor is in operation. Use a good-quality bearing grease.

Electric motors Unless otherwise specified, electric motors are equipped with ball bearings and should be greased only once or twice a year, depending on service. The bearings should be taken out, washed with gasoline, and repacked with electric motor grease, a process that should be performed only by a competent electrician.

Operation of electrically powered equipment can be as safe or as hazardous as you make it. Follow these simple rules for your safety and the safety of others:

1. All electrical equipment must be grounded.
2. Always pull the main switch before disconnecting a plug from its receptacle.
3. Never grease, adjust, or repair a machine without removing the plug from its receptacle or pulling the main switch.
4. Use only the proper size and style of fuse, keep a supply on hand, and never substitute pieces of metal.
5. Keep electrical cables in good condition at all times and keep them out of water.

Gasoline engines Gasoline engines should be lubricated as specified in the operator's manual supplied by the manufacturer.

Chapter **14**

Maintenance of Hydraulic Hose and Fittings

ROBERT L. OLSON, P.E.
Product Application Department, The Gates Rubber Company,
Englewood, Colo.

Power is transmitted through the medium of hydraulics by pressurizing and pumping a liquid at one point and converting it to work at another point. Hydraulic liquids are routed from point to point under pressure through a number of conveyance systems. Usually a combination of conveyance systems is used in a single hydraulic circuit. Pipe, tubing, sandwich channels, drilled holes, hose, couplings, and adapters are among the most popular. When a flexible conveyor is required between two points in a circuit, such as when one point moves relative to the other point, hose is usually used as the conveyor. Wherever hose is used in this type of circuit, couplings are required. This chapter is primarily intended to discuss hose and couplings, but, since couplings and adapters are so closely related, adapters will be included.

HOSE

Generally, hose is designed into a circuit as a flexible connection or vibration dampener. It consists of three major parts: tube, reinforcement, and cover (Fig. 14-1).

Tube

The innermost part of the hose, the tube, is the only part of the hose in contact with the liquid being conveyed. Its purpose is to contain the fluid and keep it from passing into the more porous reinforcement.

There are many tube materials in the hose types available (Table 14-1), having particularly strong resistance to certain degenerative characteristics. For instance, a tube material that is ideal for phosphate ester fluids is not suitable for use with petroleum-base hydraulic oils and vice versa. Most hydraulic-tube materials are compounded using acrylonitrile and butadiene as base polymers. This combination is known in the rubber industry as Buna-N. The chemical and physical properties of this synthetic rubber are subject to a considerable amount of control through compounding with a number of other ingredients necessary to make the material useful.

Chemical compatibility between the tube and the hydraulic fluid is very important. Another factor that can be equally

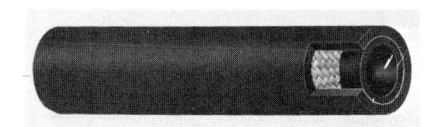

Fig. 14-1 Hose consists of three major parts: tube, reinforcement, and cover.

4-224 Basic Maintenance Technology

important is temperature. Almost any hydraulic tube stock is compounded to function properly between −40 to +200°F (−40 to +93°C). At temperatures under −40°F, normal tube material will stiffen and tend to become brittle. At temperatures over 200°F it will harden and sometimes crack. Specially compounded tube material is available for either of these extremes. Certainly there are many other factors a compounder keeps in mind when designing tube materials, but, for a user, the above are the important factors.

Reinforcement

Many reinforcement materials and configurations are used in the manufacture of hydraulic hose. Their purpose is to keep the tube from expanding when its contents are pressurized. Naturally, the greater the pressure requirements of a hose, the greater the strength requirements of the reinforcement.

TABLE 14-1 Characteristics of Hose Stock Types
Choose a hose stock type that best suits application. Chemical and physical properties are subject to a considerable amount of control through compounding; therefore, the characteristics shown for each stock type are generalized to some degree. Tube and cover stocks occasionally may be upgraded to take advantage of improved materials and technology.

Chemical name	Polychloroprene (Neoprene)	Acrylonitrile and butadiene (Buna-N)	Isobutylene and isoprene (Butyl)	Chlorosulfonated polyethylene (Hypalon)	Ethylene propylene diene (EPDM)
ASTM-SAE designation SAE J14 SAE J200	SC BC	SB BG	R AA	TB CE	R AA
Tensile strength	Good	Fair to good	Fair to good	Good	Good
Tearing resistance	Good	Fair to good	Good	Fair	Good
Abrasion resistance	Good to excellent	Fair to good	Fair to good	Good	Good
Flame resistance	Very good	Poor	Poor	Good	Poor
Petroleum oil and commercial gasoline	Good	Good to excellent	Poor	Good	Poor
Resistance to gas permeation	Good	Good	Outstanding	Good to excellent	Fair to good
Weathering	Good to excellent	Poor	Excellent	Very good	Excellent
Ozone	Good to excellent	Poor	Excellent	Very good	Outstanding
Heat	Good	Good	Good to Excellent	Very good	Excellent
Low temperature	Fair to good	Poor to fair	Very good	Poor	Good to excellent
General chemical resistance	Good	Fair to good	Good	Good	Good

Maintenance of Hydraulic Hose and Fittings   4-225

OPEN BRAID

TWO WIRE BRAID

COMPACT BRAID

THREE WIRE BRAID

TWO BRAID

FOUR SPIRAL BRAID

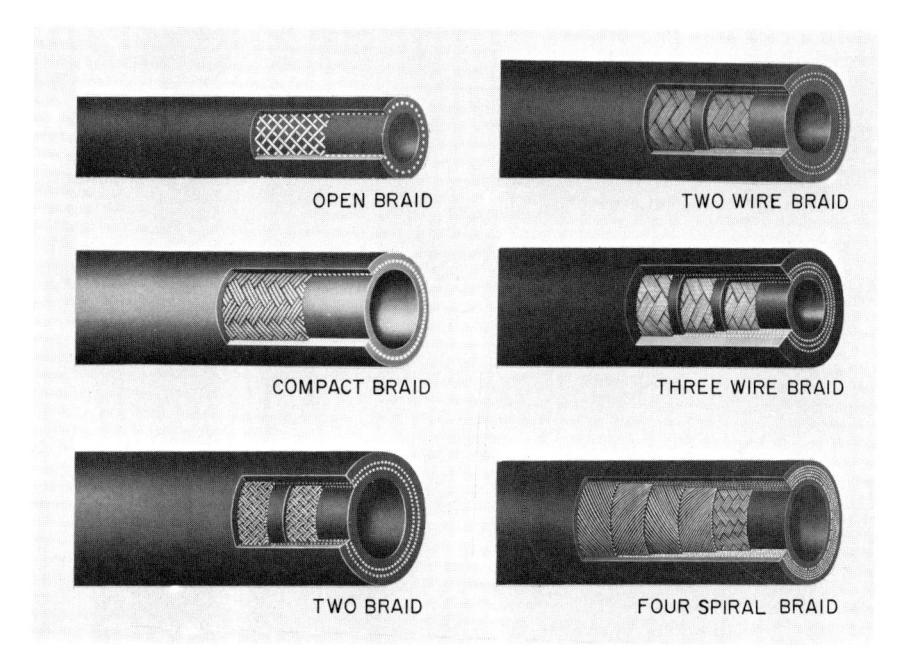

Fig. 14-2 Cutaway drawings of hydraulic hose show reinforcement materials and configurations used to keep a tube from expanding when its contents are pressurized.

The lowest-pressure hydraulic hose generally will have one layer of open-fiber braid. In recent years, the fiber used is synthetic material such as polyester which, because of its added strength over natural fibers, has allowed less dense braiding in some cases or higher pressure ratings for the hose in other cases. As pressure requirements increase, the density of the braid must be increased until a point is reached where additional pack is impractical. Then the designer will use two braids for extra strength. As pressure requirements increase beyond the capability of two-fiber braid, one-wire braid is used. Again, densities of braid and wire diameter may be varied by a designer to meet strength requirements and second- and third-wire braids may be added. Spiral-wire hose has become popular for extremely high pressure applications. It consists of wires lying side by side in a spiral configuration around the hose, each layer spiraling in the opposite direction from that of the preceding layer. The most common spiral hose is *four spiral*; it has four layers or plies of wire. However, six spiral hose is also available (Fig. 14-2).

In designing hose, it has been found that combinations of the above are sometimes advantageous. Fiber braid and wire braid on the same hose and wire braid with spiral wire offer some advantages. The possibilities become limitless for the hose designer, whose knowledge can produce an ideal hose for almost any application.

An interesting observation at this point is that the larger the hose diameter, the less pressure it will accommodate using the same reinforcement construction. For instance, a one-wire-braid ¼-in.-ID hose is rated to burst at 12,000 psi, and a 1-in.-ID hose of the same construction is rated to burst at 3000 psi. This is evident in hose catalogs. The smaller the hose, the higher the pressure rating. The ratio will not be the same as above because hose manufacturers tend to use heavier reinforcement in the larger sizes but the difference is still quite noticeable.

To understand this, visualize a 1-in. length of ¼- and 1-in. hose (Fig. 14-3). Slit each sample on a side parallel to its axis and flatten it; there will be a piece about 1 × ¾ in. or ¾ sq in. from the ¼-in. hose sample and a piece about 1 × 3 in. or 3 sq in. from the 1-in. sample. Now, the pressure in pounds per square inch against ¾ sq in. will apply only about one-quarter of the stress on the wall—where the ¼-in. hose was slit—that it will on the wall of the 1-in. hose of 3 sq in. In other words, the force trying to tear the 1-in.-long section of wall on the ¼-in. hose is only about one-quarter the force trying to tear the 1-in.-long section of wall on the 1-in. hose because the pressure per square inch is pressing against about one-quarter of the square inches in the ¼-in. hose.

Cover

The purpose of cover is to protect reinforcement from damage caused by abrasion or moisture. Hydraulic hose cover is compounded to be resistant to tearing, abrasion, oil,

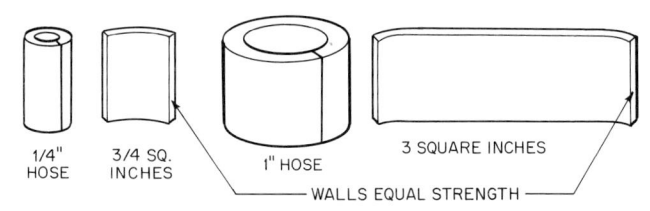

Fig. 14-3 The larger the hose diameter, the less pressure it will accommodate using the same reinforcement construction. In this illustration, the force trying to tear the section of wall on the ¼-in. hose is only about one-quarter the force trying to tear the wall on the 1-in. hose because the pressure per square inch is pressing against about one-quarter the square inches in the ¼-in. hose.

weathering, and ozone. The base polymer is generally polychloroprene (Neoprene). Special compounds are used for extreme temperature, flame resistance, nonconductivity, or other special hose applications. Another variation is called a *cotton cover*. This is an exposed synthetic rubber-impregnated cotton braid which is desirable when the hose is required to slip on a smooth surface or in other applications where the hose is not subject to damage through abrasion. The real bonus in this kind of construction is that the cover contributes to the strength of the hose.

While on the subject of cover, reference to thin-cover wire hose is frequently made. It is necessary to go back in history to fully explain the term. In the early days of wire-braid hose a reasonably heavy cover became standard to protect the costly wire braid. It was quickly determined that for a coupling to stay on the ends of a hose under the pressures the hose was capable of withstanding, the cover had to be removed in the area of the coupling so that the coupling could bite into the steel wire. Later, in an attempt to eliminate this operation of skiving (or buffing) the cover off, a coupling system was developed that would work without removing the cover, provided the cover was not too thick. At this point, thin-cover hydraulic hose was developed, and it has become very popular. Today, regular-cover wire hose is used only where extreme abrasion is evident. Thin-cover hose is easier to couple and is used extensively.

Selecting Hydraulic Hose

To determine the correct hose for an application it is necessary to know the flow velocity of the liquid in the circuit, the maximum operating pressure of the circuit including surges, the temperature range of the fluid, the type of fluid used in the system, and any unusual requirements of a hose such as its resistance to flame or its conductivity.

A nomographic chart (Table 14-2) may be used to select the correct hose size for a given hydraulic system. The velocity of the hydraulic fluid should not exceed the range shown in the right-hand column. When oil velocities are higher than recommended in this chart, the results are turbulent flow with loss of pressure and excessive heating. Higher velocities may be used if the flow of hydraulic fluid is intermittent or only for short periods of time.

The velocity of hydraulic fluid in suction lines should always fall within the range recommended to ensure efficient operation of the pump. The following is an example of the use of this chart: What size hose assembly is recommended to carry 10 gal of oil per minute, and what will be the velocity of the oil through the hose assembly? The hose assembly is to be used in a pressure line and flow is to be continuous.

Locate the flow, 10 gpm (left-hand column), and a velocity, 10 ft/sec (right-hand column), since it is near the center of the recommended range. Lay a straightedge across

Maintenance of Hydraulic Hose and Fittings 4-227

TABLE 14-2 Flow Capacity of Hose Assembly at Recommended Flow Velocities

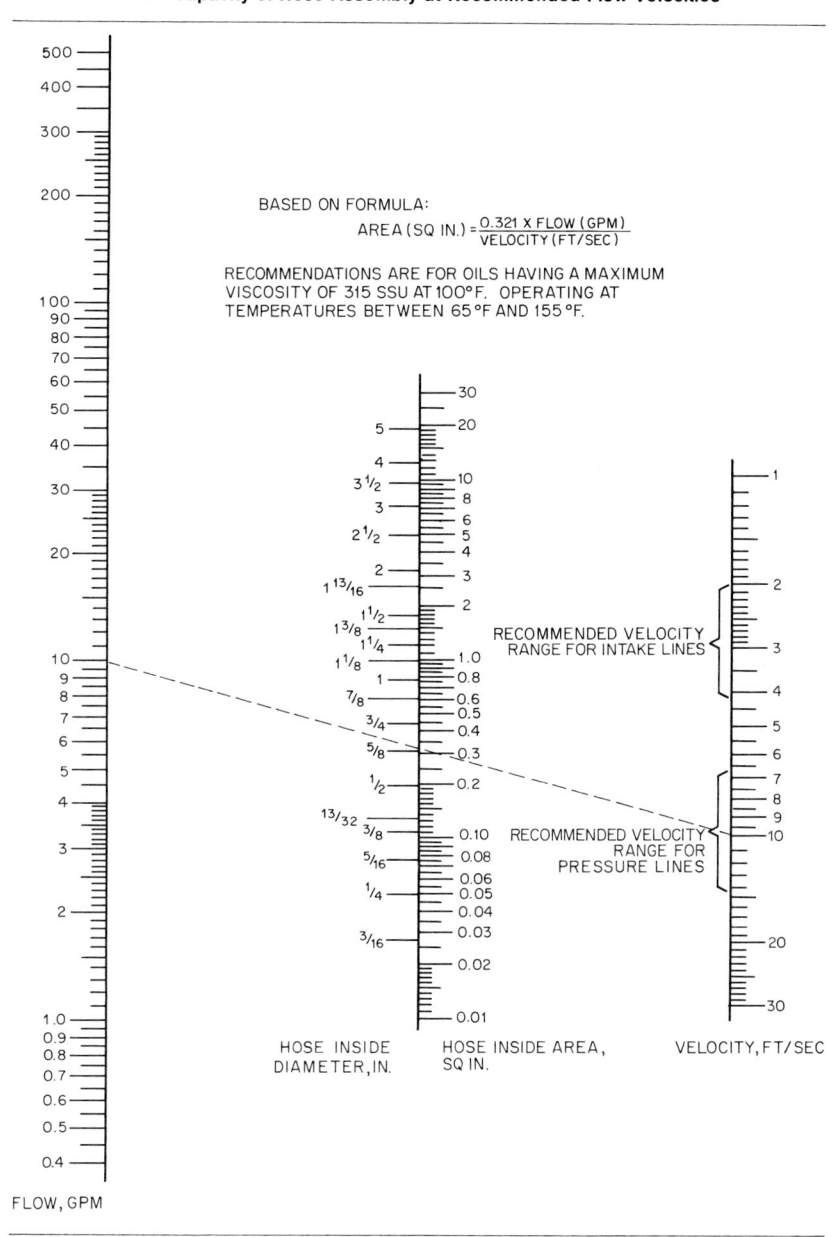

BASED ON FORMULA:
$$\text{AREA (SQ IN.)} = \frac{0.321 \times \text{FLOW (GPM)}}{\text{VELOCITY (FT/SEC)}}$$

RECOMMENDATIONS ARE FOR OILS HAVING A MAXIMUM VISCOSITY OF 315 SSU AT 100°F. OPERATING AT TEMPERATURES BETWEEN 65°F AND 155°F.

RECOMMENDED VELOCITY RANGE FOR INTAKE LINES

RECOMMENDED VELOCITY RANGE FOR PRESSURE LINES

FLOW, GPM | HOSE INSIDE DIAMETER, IN. | HOSE INSIDE AREA, SQ IN. | VELOCITY, FT/SEC these two points. The straightedge crosses the center column nearest the ⅝-in. hose assembly. Keeping the straightedge on 10 gpm in the left-hand column, cross the center column at the ¾-in. assembly, the ⅝-in. assembly, and the ½-in. assembly. Reading the right-hand column, the straightedge crosses it at 7.5, 10.3, and 16 ft/sec, respectively. Since 7.5 and 10.3 are within the recommended velocity range for pressure lines, either a ¾- or ⅝-in. hose assembly may be used.

Problems concerning suction hoses are solved in a similar manner, except that the recommended velocity range for intake line (right-hand column) and the values for suction hose assemblies (center column) are used. When the hose size has been determined, it is necessary to consult a manufacturer's catalog or the SAE handbook to select the correct hose. These references list the characteristics and dimensions of the most commonly used hose. A complete SAE handbook or individual specifications may be obtained by contacting the Society of Automotive Engineers, Inc., 400 Commonwealth Drive, Warrendale, PA 15096.

If individual specifications are preferred, the following list may aid in a selection:

SAE J30	Fuel and Oil Hoses
SAE J343	Tests and Procedures for SAE 100R Series Hydraulic Hose and Hose Assemblies
SAE J514	Hydraulic Tube Fittings
SAE J515	Hydraulic O-Ring
SAE J516	Hydraulic Hose Fittings
SAE J517	Hydraulic Hose
SAE J518	Hydraulic Flanged Tube Pipe, and Hose Connections, 4-bolt Split Flange Type
SAE J533	Flares for Tubing
SAE J926	Hydraulic Pipe Fittings
SAE J1402	Air Brake Hose—Automotive

A common method of selecting hose for maintenance of existing equipment is to remove the failed hose and judge what the hose type is either by reading the lettering (if it has not been obliterated) or by examining the hose construction and dimensions and duplicating the hose as closely as possible. This procedure is not recommended, but it is done every day and it keeps equipment working.

COUPLINGS

Hydraulic hose assemblies generally consist of a length of hose and two couplings (Fig. 14-4). There are literally thousands of couplings on the market and not too many chances to substitute one for another. Be cautious in selecting the correct stem and ferrule for the hose, and where threads are involved, be sure they are the correct threads, especially in high-pressure applications.

A question often is: Can one manufacturer's coupling successfully be applied to another manufacturer's hose? This is done frequently, but always with some degree of risk. Each manufacturer designs couplings to fit its own hose properly. Even though hose from two manufacturers may comply completely with the same SAE specification, there can be differences in tube hardness and other factors that cause differences in stem and ferrule configurations and dimensions. Each manufacturer tests extensively to be certain its hose and couplings are well matched and compatible. Minute design changes take place in the final design of a coupling. The benefits from this are lost when the coupling is applied to another hose. As mentioned, it is done and may be good first aid in an emergency, but is not recommended and manufacturers will not guarantee it.

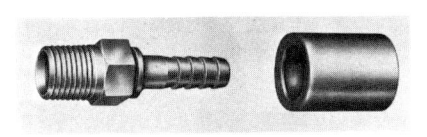

Fig. 14-4 Be sure to select the correct stem (left) and ferrule (right) for hydraulic hose. Correct choice of threads is particularly important in high-pressure applications.

With this knowledge, selecting couplings for an application becomes easy. A hose manufacturer's catalog always recommends couplings for use with the hose. For most hose there is a selection of permanent or reusable couplings. Both have advantages and disadvantages, but the selection is generally determined by the coupling installation equipment the installer has available.

Reusable couplings, as the name implies, can be removed from a worn-out hose and placed on a new hose. This flexibility saves paying for new couplings when a hose is replaced and ensures having the correct couplings on hand. Usually these couplings can be changed without special tools. They are available for most hydraulic hose. The disadvantages of reusable couplings are that generally the initial cost is higher and that the time required to install them is greater. Whereas their length of service and reliability have been more than satisfactory, it is generally understood in the industry that permanent couplings are superior. This is substantiated by the fact that five out of seven of the largest hose manufacturers do not list reusable couplings for their highest-pressure hose in their catalogs. Another disadvantage is that reusable couplings are generally larger,

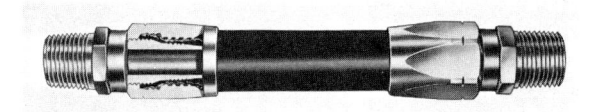

Fig. 14-5 Screw-type reusable coupling has deep, coarse convolutions inside the ferrule that form left-hand threads to assist in getting it over the end of hose. *(Anchor Coupling Co.)*

heavier, and bulkier than the permanent type. Mentioning these disadvantages is not meant to imply reusable couplings do not have a place. They certainly do—in fact, there are a large number of low-pressure hose types for which only reusable couplings are specified.

The most common reusable coupling is the screw type; it has deep, coarse convolutions inside the ferrule that form left-hand threads to assist in getting it over the end of the hose (Fig. 14-5). For thin-cover no-skive wire hose these convolutions are made rather sharp to cut through the cover and grip on the wire. The alternative has flatter-shaped convolutions to avoid cutting the wire on hose when the cover has been skived away in the area of the ferrule. One end of the ferrule has a reduced diameter that does not fit over the hose but is threaded to receive the threads provided on the stem. The end of the stem is tapered ahead of the threads and, as it is rotated into the end of the ferrule, it expands the diameter of the hose, forcing a tight fit onto the convolutions inside the ferrule (Fig. 14-6).

Other reusables are mostly stems similar to crimp-type stems with devices such as two- and four-bolt clamps for higher-pressure applications or stressed steel bands in various configurations for lower-pressure applications to compress the hose onto the stem (Fig. 14-7).

For low-pressure applications, a push-on hose and coupling system is available that

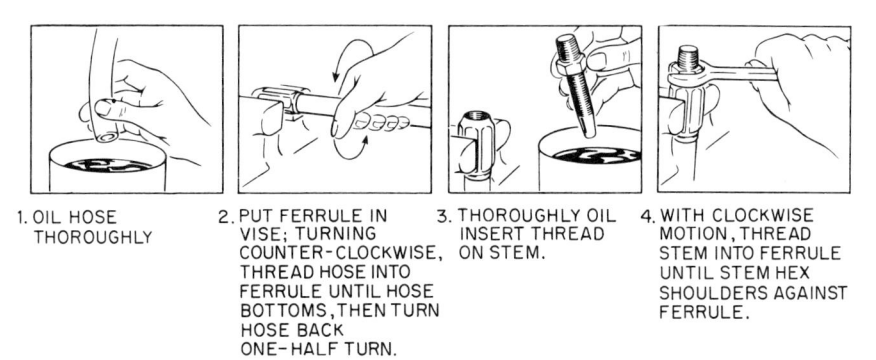

1. OIL HOSE THOROUGHLY
2. PUT FERRULE IN VISE; TURNING COUNTER-CLOCKWISE, THREAD HOSE INTO FERRULE UNTIL HOSE BOTTOMS, THEN TURN HOSE BACK ONE-HALF TURN.
3. THOROUGHLY OIL INSERT THREAD ON STEM.
4. WITH CLOCKWISE MOTION, THREAD STEM INTO FERRULE UNTIL STEM HEX SHOULDERS AGAINST FERRULE.

Fig. 14-6 How to assemble reusable couplings in four easy steps.

simply involves pushing the stem into the end of the hose for coupling. The hose is designed to grip the stem in Chinese finger-puzzle fashion when pressurized.

Permanent Couplings

The term *permanent* is used to indicate that once this coupling is attached to the hose it is not designed to be removed without destroying it (Fig. 14-8). Usually a ferrule of the crimp type is used with a stemmed fitting to constitute a coupling. The primary advantage of the permanent coupling is its reliability; a second advantage is that usually less time is required for installation; third, the permanent coupling and ferrule cost is about one-third to one-half the cost of a reusable; fourth, it is lighter in weight and less bulky.

Crimp-type ferrules vary from thin metal as seen on garden hose to heavy machined metal. Most hydraulic ferrules are machined and have circumferential serrations

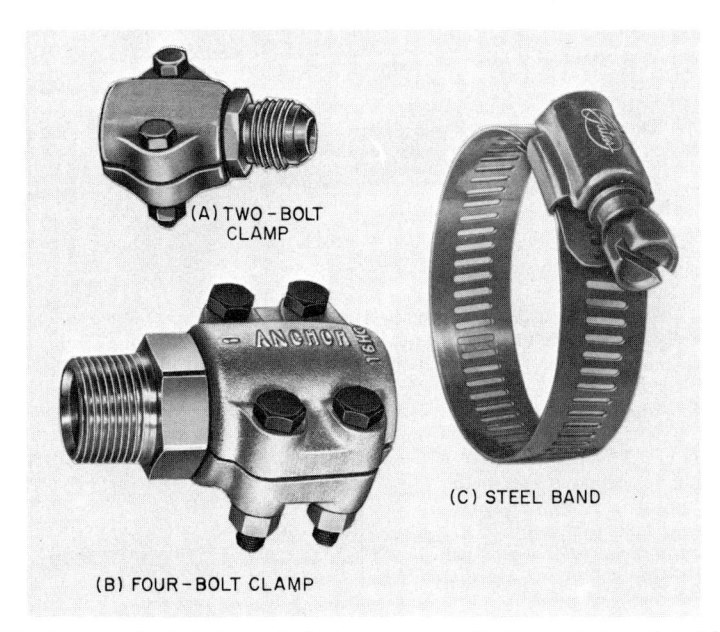

Fig. 14-7 Two- and four-bolt clamps for higher-pressure applications and stressed steel bands in various configurations for lower-pressure applications are other types of reusable stems for hydraulic hose. *(Anchor Coupling Co.)*

machined on the inside. Again, for no-skive hose these serrations are left sharp to cut through a cover to grip the reinforcement or left flat for skived ends (Fig. 14-9). Stems are designed with saw-toothed circumferential serrations for ease of insertion into the tube and to grip the tube in the other direction. The serrations in the ferrule and stem are designed together so that the hills in the ferrules lie in the valleys of the stems and vice versa to provide optimum gripping onto the hose end.

Crimping the ferrules on permanent couplings is the one thing that cannot be done with tools common around a construction project, although in recent years crimping machines are becoming a part of large contractors' equipment. In addition to having the equipment, it is important to have

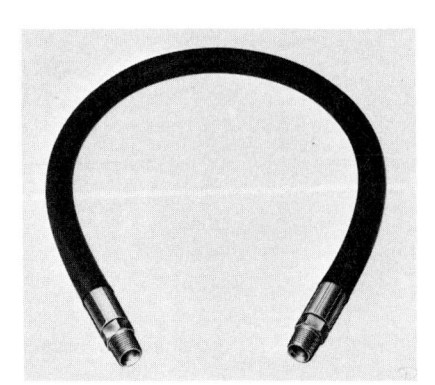

Fig. 14-8 Permanent coupling, once attached, cannot be removed without destroying hose. These couplings are very reliable and require less time for installation.

Fig. 14-9 Most hydraulic ferrules have circumferential serrations machined on the inside. Stems are designed with saw-toothed circumferential serrations for ease of insertion into tube and to grip tube in other direction.

someone available with adequate knowledge to select correctly the hose and fittings for an application and to couple the hose correctly (Fig. 14-10).

A ferrule must be crimped to a specified diameter, usually ±0.005 in. Crimping is usually accomplished with a small hydraulic press having a ram capable of generating from 25 to 100 tons of force. It is not meant to be implied here that all this is in any way difficult to learn; rather, it should not be attempted by an individual who is insufficiently trained. Improperly coupled hose under high pressure can fail, causing undue downtime and possible injury or worse in some cases. Any good hose supplier has the capability and will be more than willing to spend the necessary time to train a customer or his personnel to couple hose properly. In cases where this is not practical, it is urged here that coupled assemblies be purchased complete from a competent hose supplier.

FITTINGS AND ADAPTERS

It was stated earlier that hose is used as a conveyor between two points. The configuration of the connector at each of these points determines what is required on the ends of the

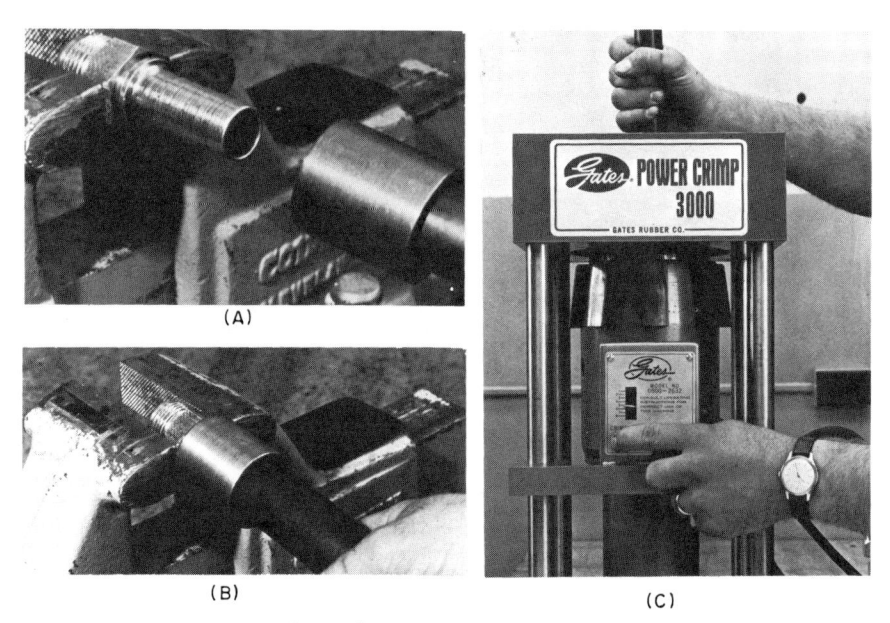

Fig. 14-10 Power crimping machine makes easy crimping of ferrules on permanent couplings. (A) Put stem in vise and place ferrule on hose. (B) Push hose into stem. (C) Insert into machine and push button until pressure is applied and coupling is crimped.

coupling stems. Again, the selection of the correct fitting is extremely important. The following has been prepared to assist you in verbally describing or selecting from a catalog the most frequently used hydraulic fittings. The term *fittings* is being used because it includes couplings and adapters. Frequently both are required to make a hose connection.

Threaded Fittings

There are basically two thread systems used in hydraulics: Iron Pipe Thread and SAE Standard Screw Thread.

The Iron Pipe Thread system generally employs a sealing fit of the mating male and female threads to obtain a leak-proof joint. The thread forms used in the Iron Pipe system are designated by symbols such as NPTF, NPSF, and other combinations using the letters shown below.

Definitions

N	National	F	Fuels
P	Pipe	S	Straight Thread
T	Tapered Thread	M	Mechanical Joint

Descriptions of iron pipe threads

NPTF This is a dryseal thread; the National pipe tapered thread for fuels. This is used for both male and female ends. The sealing fit of the mating threads produces the leak-proof joint.

NPSF The National pipe straight thread for fuels. This is sometimes used for female ends and properly mates with the NPTF male end. However, the SAE recommends the NPTF thread in preference to the NPSF for female ends.

NPSM National pipe straight thread for mechanical joint. This is used on the female swivel nut of iron pipe swivel adapters. The leak-proof joint is not made by the sealing fit of threads, but by a tapered seat in the coupling end.

Descriptions of SAE standard screw threads

Couplings having the SAE standard screw thread usually employ such things as "O" rings, compression sleeves, or tapered seats to obtain a seal.

JIC The Joint Industry Conference specifies a 37-deg angle flare or seat to be used with high-pressure hydraulic tubing. Couplings specified as JIC have 37-deg tapered seats.

SAE A term usually applied to fittings having a 45-deg angle flare or seat. Soft copper tubing is generally used in such applications as it is easily flared to the 45-deg angle. These are for low-pressure applications—such as for fuel lines and refrigerant lines.

Identifying threads A caliper, a thread gauge, and a seat gauge are required to accurately identify threads.

A *caliper* (Fig. 14-11A) measures outside diameter at the largest point on the threads of a male fitting or the inside diameter of a female fitting.

A *thread gauge* (Fig. 14-11B) determines threads per inch. A gauge and fitting should be held toward a light to ensure accuracy.

A *seat gauge* (Fig. 14-12) determines whether seat angles are 37-deg Joint Industry Conference (JIC) or 45-deg SAE. The angles of a coupling and a gauge are matched with the gauge held parallel to the centerline of the coupling.

With the information given above, the charts (Fig. 14-13) may be used to identify the fitting type and nominal size.

Fitting shapes The next thing to determine is the shape of a fitting and, in some cases, whether a swivel is required. The threaded portion of a solid fitting is firmly attached to the hose and cannot rotate unless the hose rotates. When installing an assembly, often the first end can be screwed into place and tightened if it has a solid fitting. The hose then cannot be rotated to attach the second end, so there must be a swivel fitting either on the second end of a hose or on the point to which it is being connected. A swivel fitting allows the thread portion to be rotated independently with respect to a hose or other mounting (Figs. 14-14 and 14-15).

Straight fittings are the most common; however, angle fittings are frequently used. The majority of angle fittings are bent-tube type; however, the block type is not uncommon. Both are used as space savers, and they help to avoid bending the hose too tightly. Angle fittings are found commonly in 45-deg and 90-deg angles (Fig. 14–16 to 14–18).

Flange Fittings

The popular alternatives to threaded fittings are flange fittings. They are primarily used on the larger ID hydraulic hoses because of the difficulties encountered in installing larger-diameter threaded fittings. Flange fittings are identified by a nominal size which is related to the tube diameter on which the flange is brazed.

From the above it can be seen that a nominal 1-in. "O" ring flange is 1¾ in. in diameter and is available for ¾ and 1 in. ID hose, but, in either case, it is called a 1-in. "O" ring flange (Fig. 14-19). "O" ring flange fittings form a seal by compressing the "O" ring against a flat or otherwise prepared surface; generally, two flange halves and four small

(A)

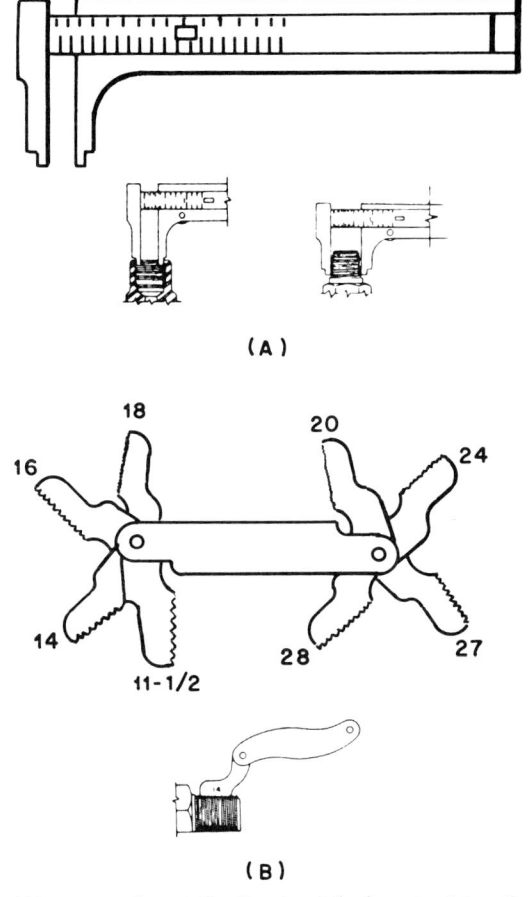

(B)

Fig. 14-11 Caliper (A) measures the outside diameter at the largest point on the threads of a male fitting or the inside diameter of a female fitting. Thread gauge (B) is used to determine threads per inch.

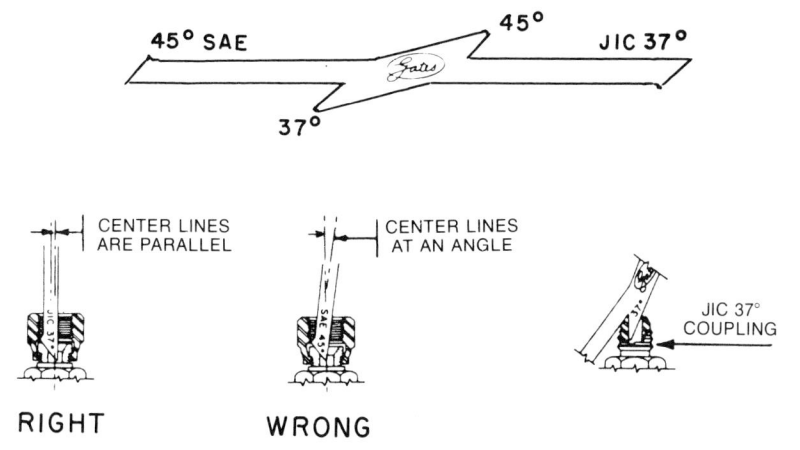

RIGHT　　WRONG

Fig. 14-12 Seat gauge is used to determine whether the seat angles are 37° JIC or 45° SAE.

4-233

IRON PIPE THREAD FITTINGS

MALE THREAD			FEMALE THREAD		
NOMINAL THREAD SIZE, IN.	No. THRDS. PER IN.	THREAD O.D., IN.	NOMINAL THREAD SIZE, IN.	No. THRDS. PER IN.	THREAD I.D., IN.
1/8	27	13/32	1/8	27	23/64
1/4	18	35/64	1/4	18	15/32
3/8	18	43/64	3/8	18	19/32
1/2	14	27/32	1/2	14	3/4
3/4	14	1 1/16	3/4	14	61/64
1	11 1/2	1 5/16	1	11 1/2	1 13/64

NPTF SOLID MALE | NPTF or NPSF SOLID FEMALE | NPTF SWIVEL MALE | NPSM SWIVEL FEMALE

JIC FLARED TYPE FITTINGS (37°)

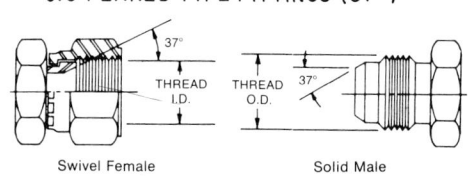

TUBE O.D., IN.	NOMINAL THREAD SIZE, IN.	No. THRDS. PER IN.	THREAD O.D., IN.	NOMINAL THREAD SIZE, IN.	No. THRDS. PER IN.	THREAD I.D., IN.
1/8	5/16	24	5/16	5/16	24	17/64
3/16	3/8	24	3/8	3/8	24	21/64
1/4	7/16	20	7/16	7/16	20	25/64
5/16	1/2	20	1/2	1/2	20	29/64
3/8	9/16	18	9/16	9/16	18	1/2
1/2	3/4	16	3/4	3/4	16	11/16
5/8	7/8	14	7/8	7/8	14	13/16
3/4	1 1/16	12	1 1/16	1 1/16	12	31/32
7/8	1 3/16	12	1 3/16	1 3/16	12	1 7/64
1	1 5/16	12	1 5/16	1 5/16	12	1 15/64
1 1/4	1 5/8	12	1 5/8	1 5/8	12	1 35/64
1 1/2	1 7/8	12	1 7/8	1 7/8	12	1 51/64
2	2 1/2	12	2 1/2	2 1/2	12	2 27/64

Swivel Female | Solid Male

SAE FLARED TYPE FITTINGS (45°)

NOMINAL THREAD SIZE, IN.	No. THRDS. PER IN.	THREAD O.D., IN.	NOMINAL THREAD SIZE, IN.	No. THRDS. PER IN.	THREAD I.D., IN.
5/16	24	5/16	5/16	24	17/64
3/8	24	3/8	3/8	24	21/64
7/16	20	7/16	7/16	20	25/64
1/2	20	1/2	1/2	20	29/64
5/8	18	5/8	5/8	18	9/16
11/16	16	11/16	11/16	16	5/8
3/4	16	3/4	3/4	16	11/16
7/8	14	7/8	7/8	14	13/16
1 1/16	14	1 1/16	1 1/16	14	63/64

Swivel Female | Solid Male

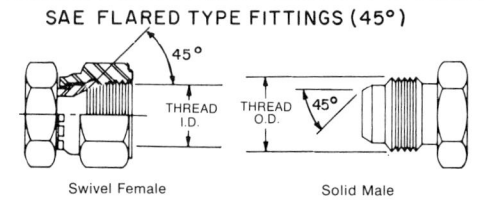

Fig. 14-13 Charts used to identify fitting type and nominal size.

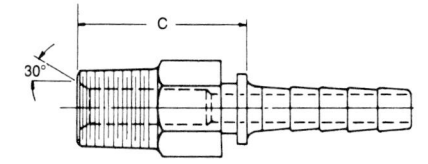

Fig. 14-14 NPTF solid male fitting.

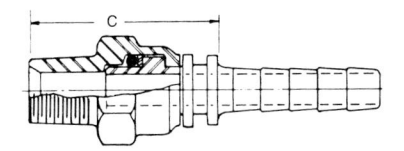

Fig. 14-15 NPTF swivel male fitting.

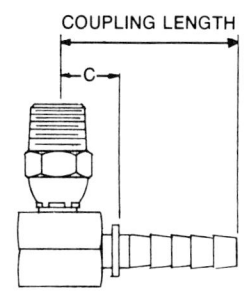

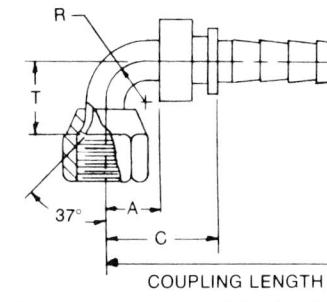

Fig. 14-16 NPTF swivel male—90° (block type).

Fig. 14-17 45° JIC swivel female—bent tube.

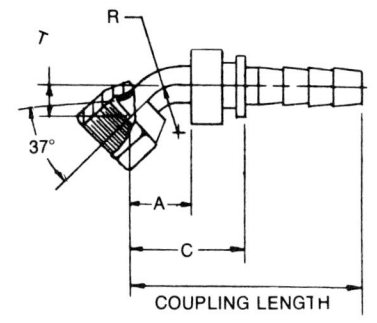

Fig. 14-18 90° JIC swivel female—bent tube.

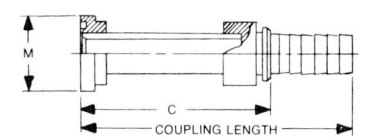

HOSE I.D., IN.	NOM. FLG. SIZE, IN.	M FLG. DIA., IN.	COUPLING LENGTH, IN.	C CUT-OFF, IN.
1/2	1/2	1-3/16	3.29	1.86
1/2	3/4	1-1/2	3.29	1.86
3/4	3/4	1-1/2	3.74	2.04
3/4	1	1-3/4	3.74	2.04
1	1	1-3/4	3.62	1.91
1	1-1/4	2	3.62	1.91
1-1/4	1-1/4	2	4.63	2.33
1-1/4	1-1/2	2-3/8	4.63	2.33
1-1/2	1-1/2	2-3/8	4.94	2.30
1-1/2	2	2-13/16	4.94	2.30
2	2	2-13/16	6.08	2.54

Fig. 14-19 "O" ring flange—straight.

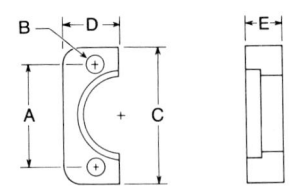

NOM. FLG. SIZE, IN.	A DIM., IN.	B DIM., IN.	C DIM., IN.	D DIM., IN.	E DIM., IN.
1/2	1.50	0.34	2.12	0.91	0.63
3/4	1.88	0.41	2.56	1.03	0.75
1	2.06	0.41	2.75	1.16	0.75
1-1/4	2.31	0.47	3.13	1.44	0.75
1-1/2	2.75	0.53	3.69	1.62	0.81
2	3.06	0.53	4.00	1.91	0.81

Fig. 14-20 "O" ring flange fittings are available with tubes bent to a variety of angles.

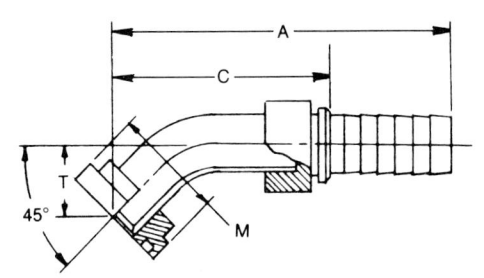

Fig. 14-21 "O" ring flange—45°.

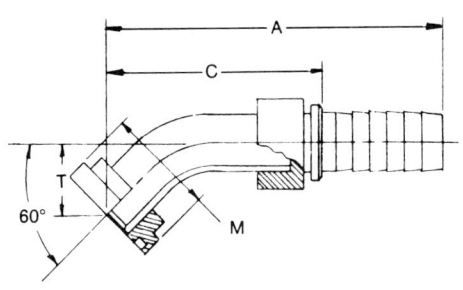

Fig. 14-22 "O" ring flange—60°.

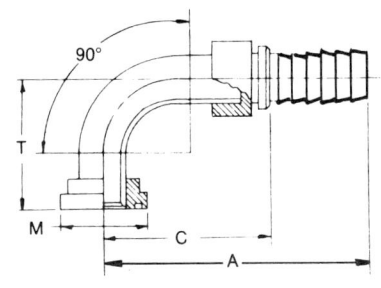

Fig. 14-23 "O" ring flange—90°.

Fig. 14-24 On straight fittings the overall assembly length is measured end to end.

bolts and washers are used. These fittings are available with tubes bent to a variety of angles (Figs. 14-20 to 14-23).

Overall Assembly Length

The method of measuring and identifying the overall assembly length (OAL) has been standardized and, when used properly, can be understood accurately by anyone using it. On straight fittings, OAL is measured from end to end as shown (Fig. 14-24). Angular fittings are measured to the centers of the ends (Fig. 14-25).

Fig. 14-25 Angular fittings are measured to the centers of the ends.

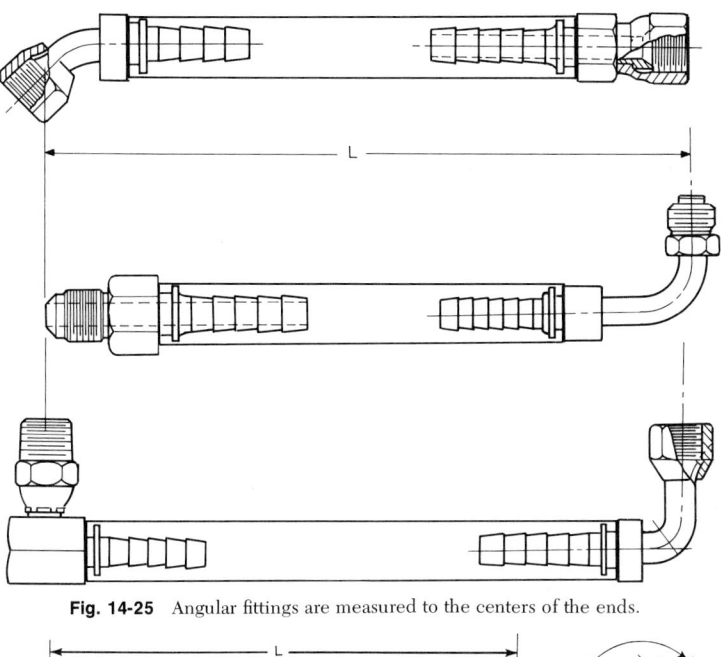

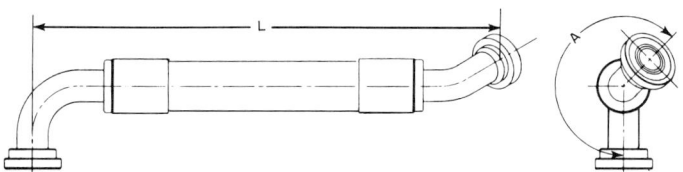

Fig. 14-26 Offset measured with first coupling vertically downward and measured clockwise. Measure from centerline of flanged head to centerline of flanged head for length of assembly.

Another measurement is necessary when two angular couplings are required on the same assembly. It is called the *orientation* and is measured with the first coupling pointing vertically downward; looking at the assembly from the second coupling end, the angle in degrees is measured in the clockwise direction. The length is again measured from centerline to centerline (Fig. 14-26).

Determining the assembly length Hydraulic hose will lengthen or shorten when pressurized and it is important to consider this when determining an assembly length. A hose can shrink up to 4 percent or elongate up to 2 percent, according to SAE standards for higher-pressure hoses, and to 3 percent either way for lower-pressure hoses. Sufficient allowance should be made to permit such changes in length (Fig. 14-27).

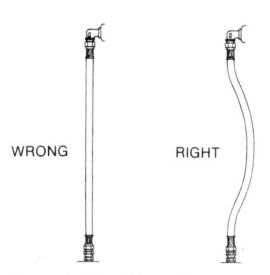

Fig. 14-27 Allow for some slack in hose when determining assembly length.

To cut a hose to the correct length (the so-called cutoff length) to meet an assembly-length specification, coupling lengths must be taken into consideration. In most cases the manufacturer's catalog indicates the amount of hose the coupling replaces. These dimensions must be acquired for each coupling and deducted from the overall length to determine the cutoff length (Figs. 14-28 and 14-29).

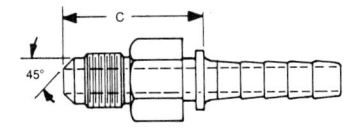

HOSE I.D., IN	THREAD SIZE	HEX SIZE, IN.	BORE SIZE, IN.	COUPLING LENGTH, IN.	C CUT-OFF IN.	STD. PACK (UNITS PER PACK)
1/4	7/16-20	0.50	0.156	2.17	1.08	5
1/4	1/2 -20	0.56	0.156	2.23	1.14	5
3/8	5/8 -18	0.69	0.232	2.21	1.20	10
3/8	3/4 -16	0.75	0.265	2.40	1.39	5
1/2	3/4 -16	0.81	0.342	2.88	1.44	5
1/2	7/8 -14	0.88	0.406	3.02	1.58	5
3/4	1-1/16 -14	1.13	0.625	3.53	1.82	5

Fig. 14-28 SAE solid male. Manufacturer's catalog indicates amount of hose the coupling replaces.

In summary, using the information acquired, a hose assembly may be completely identified as in the following example:

Hose	½ in. 100R1
First coupling	½ in. NPTF solid male
Second coupling	¾-16 JIC swivel female
OAL	9.5 in.

The assembly in the example would resemble Fig. 14-30, which shows permanent couplings, although the example didn't specify them. Either permanent or reusable could have been used.

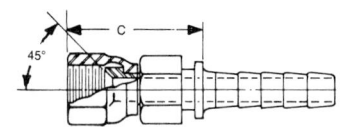

HOSE I.D., IN.	THREAD SIZE	HEX SIZE, IN.	BORE SIZE, IN.	COUPLING LENGTH, IN.	C CUT-OFF IN.	STD. PACK (UNITS PER PACK)
1/4	7/16-20	0.56	0.156	2.31	1.29	5
1/4	1/2 -20	0.69	0.156	2.38	1.29	5
1/4	5/8 -18	0.75	0.156	2.47	1.38	10
3/8	5/8 -18	0.75	0.265	2.39	1.38	10
3/8	3/4 -16	0.88	0.265	2.47	1.46	5
1/2	3/4 -16	0.88	0.406	2.95	1.52	5
1/2	7/8 -14	1.06	0.406	3.04	1.61	5
3/4	1-1/16- 14	1.25	0.625	3.39	1.68	5

Fig. 14-29 SAE swivel female.

Maintenance of Hydraulic Hose and Fittings 4-239

Coupling installation equipment Hose and coupling manufacturers provide a variety of tools for coupling installations ranging from small fixtures to high-pressure crimping machines. The reason for the variety is that no one system is right for everyone. A relatively small user might elect to keep an inventory of reusables as opposed to a large user who might find that the saving in permanent coupling inventory cost or value more than offset the cost of the crimping equipment. This can only be determined by working with a hydraulic hose supplier and comparing actual costs.

Cutters. Hydraulic hose can be cut with a standard fine-tooth hacksaw but the high-tensile wire used in medium- and high-pressure hose is tough and hard to cut. The next step up is a converted power handsaw which is generally equipped with a sharp circular blade without teeth. The mounting provides pins to arc the hose, which greatly reduces friction and makes a fast cut possible. Cutters are available specially manufactured for hose; they can get

Fig. 14-30 This completed assembly uses a ½-in. 100R1 hose, with the first coupling a ½-in. NPTF solid male and the second coupling a ¾-16 JIC swivel female. Overall assembly length is 9.5 in.

Fig. 14-31 Hose-cutting saw has sharp circular blade without teeth.

more elaborate, such as the combination cutter and skiver shown in Fig. 14-31. Both cutter and skiver are available as separate machines, and there are skivers available that are less elaborate and less expensive (Fig. 14-32).

Machines are also available to reduce the time and work required to install reusable

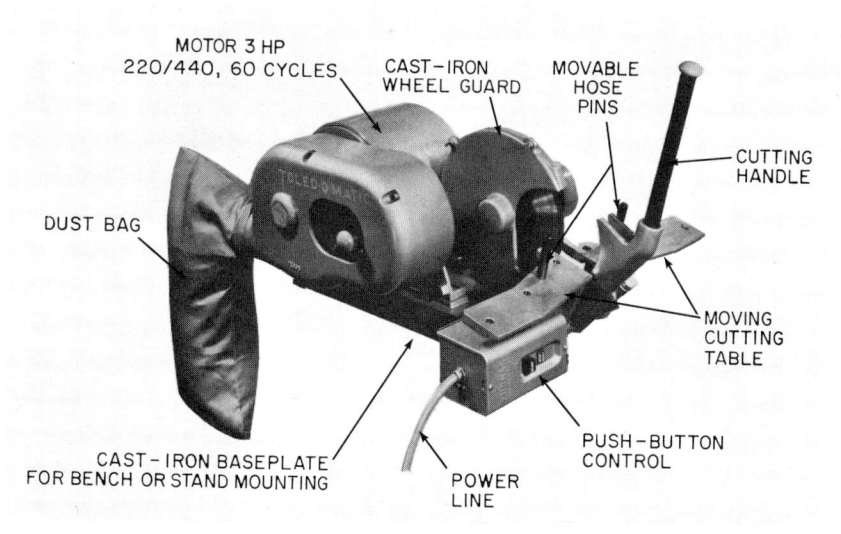

Fig. 14-32 Skiving brush housing equipped with dust opening for connection to exhaust system or dust bag furnished.

4-240　Basic Maintenance Technology

couplings. Some are simply rotating chucks for screw-type couplings. The one shown in Fig. 14-33 has a vise to hold the hose. These machines are a great asset if many reusable assemblies are to be made because so many revolutions are required to complete an assembly and considerable torque is required to make the assembly, particularly on the larger-diameter hoses.

Crimping Machines. For permanent assemblies, a crimping machine is usually required. Crimpers are available from hand-lever-operated types for low-pressure hose to 100-ton hydraulic presses powered by 10,000-psi pumps.

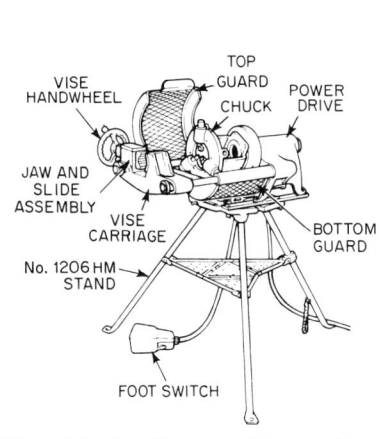

Fig. 14-33 Installing reusable couplings is made easier with machine like this.

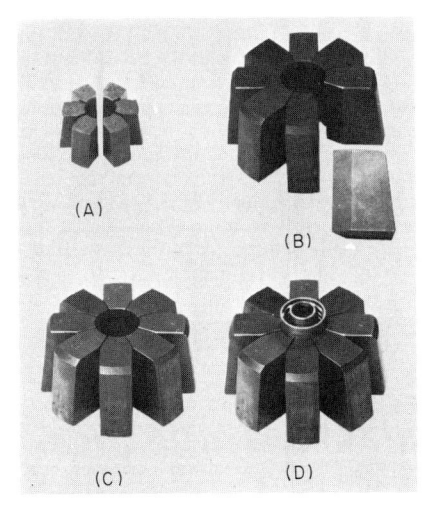

Fig. 14-34 Die fingers for use in crimping machines come in various sizes. (*A*) Small six-finger set. (*B*) and (*C*) Large eight-finger set. (*D*) Large eight-finger set showing coupling seated inside fingers.

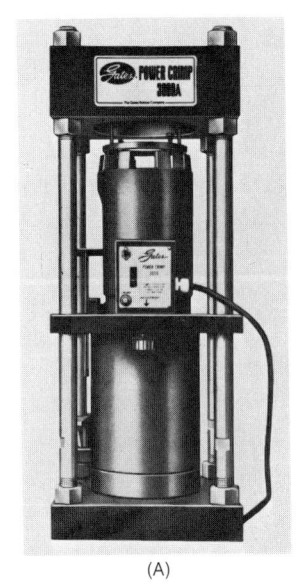

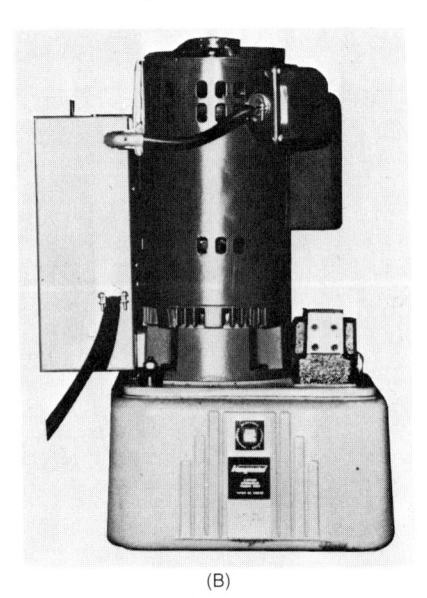

Fig. 14-35　(*A*) 100-ton power crimper and (*B*) 10,000-psi pump.

Maintenance of Hydraulic Hose and Fittings 4-241

Crimping machines usually come equipped with a number of sets of crimping fingers or dies to accommodate a range of hose sizes. Generally, there are six or eight fingers to a set (Fig. 14-34). The crimping faces of the fingers contact the ferrule and squeeze it onto the hose. The surface opposite the finger face is rounded and tapered to fit inside a tapered cone. As the fingers are forced farther into the cone they are forced together uniformly, causing the crimp. The small conical angle provides great mechanical advantage from the press ram through the fingers to the ferrule.

The larger crimping machines shown will crimp hose sizes from $3/16$ to 2 in. ID. These machines are generally powered by an electric-driven hydraulic pump. Smaller crimpers are available that will crimp $3/16$ to 1 in. ID hose that can be operated using a hand pump, air pump, or electro-hydraulic pump (Fig. 14-35).

A hand pump is often installed with a small crimper on a piece of mobile equipment such as a pickup truck for field crimping. In cases where the truck is equipped with compressed air, an air pump is sometimes used (Fig. 14-36).

(A)   (B)

(C)

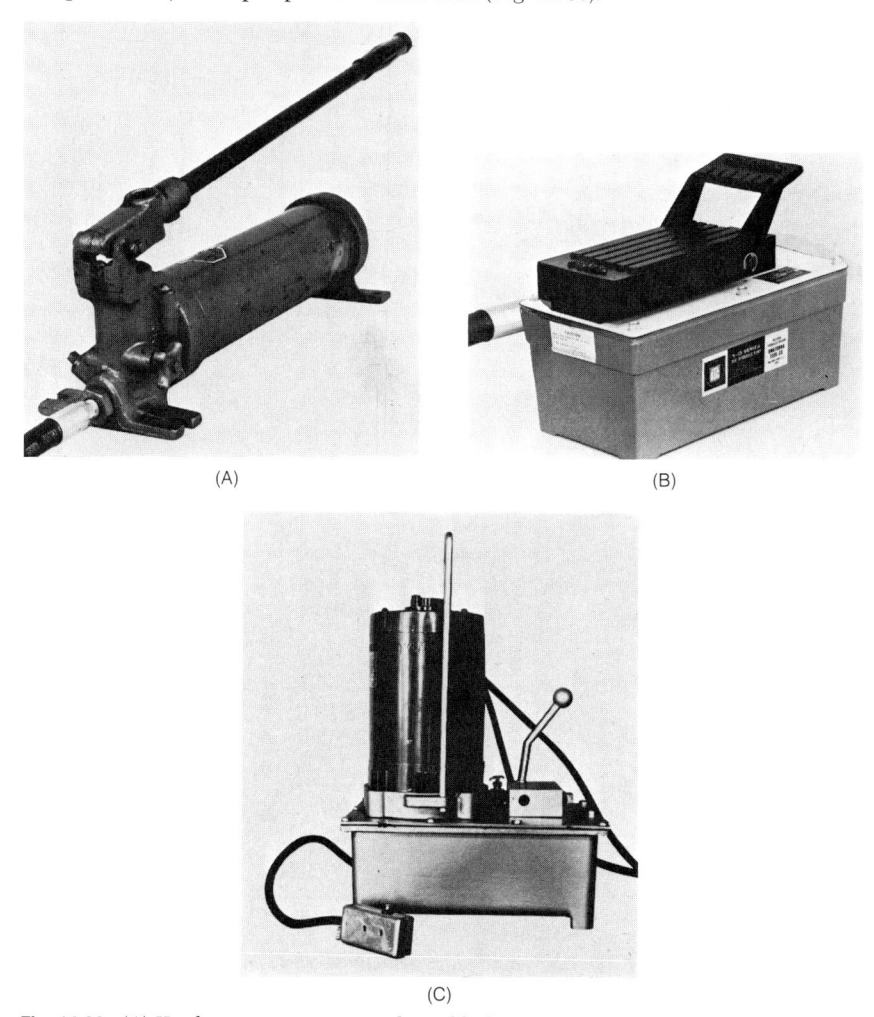

Fig. 14-36 (A) Hand pump—convenient and portable for use anywhere. Makes original equipment manufacturer factory-type assemblies on the spot to minimize downtime. (B) Air hydraulic pump—uses standard air compressor power (90 psi). Convenient and versatile for in-shop use. (C) Electro-hydraulic pump—delivers high oil volume for fast coupling speeds. Perfect for shops doing volume assembly work. Operates on 115 volt ac single-phase current.

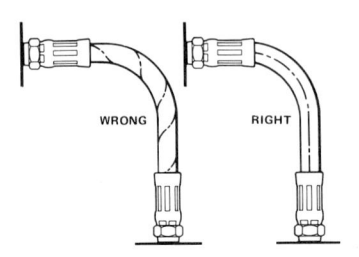

Fig. 14-37 Take care not to twist hose during installation. Correct position can be determined by the printed layline on the hose. Pressure applied to a twisted hose can cause hose failure or loosening of connections. *(Weatherhead Co.)*

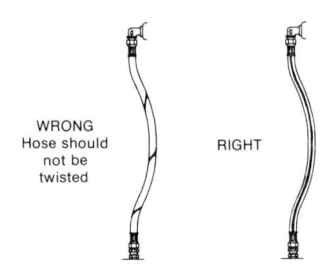

Fig. 14-38 Twisting hose will cause threaded fitting to loosen or cause undue stress within hose.

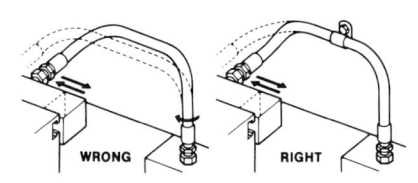

Fig. 14-39 To avoid twisting in hose lines bent in two planes, clamp hose at change of plane (as shown). *(Dayco Corp.)*

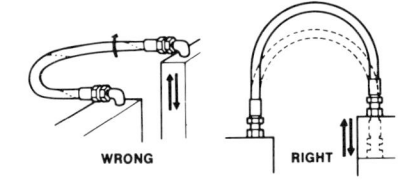

Fig. 14-40 To prevent twisting and distortion, hose should be bent in the same plane as the motion of the boss to which the hose is connected. *(Dayco Corp.)*

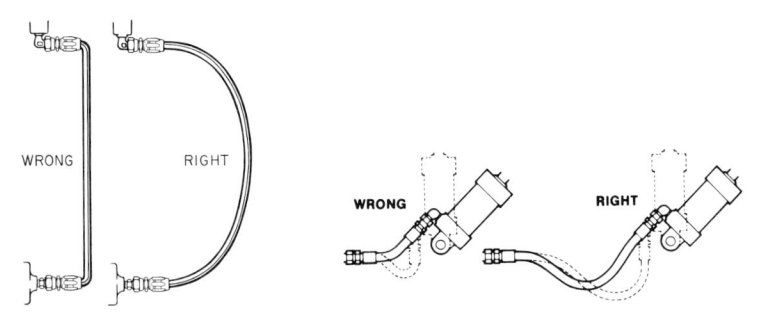

Fig. 14-41 Always avoid bending hose more sharply than the prescribed minimum-bend radius (left). *(Aeroquip Corp.)* Adequate hose length is most important to distribute movement on flexing applications and to avoid abrasion (right). *(Dayco Corp.)*

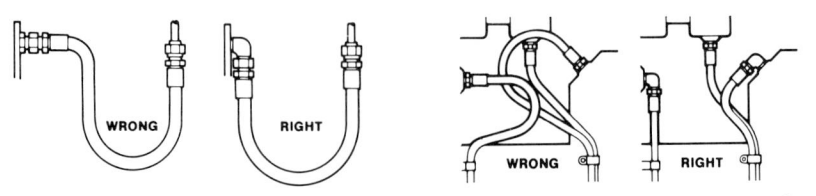

Fig. 14-42 Avoid sharp twist or bend in hose by using proper angle adapters (left). Obtain direct routing of hose through use of 45° and 90° adapters and fittings. Improve appearance by avoiding excessive hose length (right). *(Dayco Corp.)*

Installation of assemblies There are right and wrong ways to install hydraulic assemblies—a wrong installation can loosen threaded fittings and greatly reduce the life of an assembly. The difference between bending and twisting should be recognized, namely, that hydraulic hose can be bent but definitely should not be twisted.

Two basic situations can cause hose to twist: (1) twisting the hose with a wrench during installation, and (2) attaching an assembly end to a moving part in a manner that would cause the hose to twist rather than to bend. Either of these will cause a threaded fitting to loosen or cause undue stress within the hose (Figs. 14-37 and 14-38).

Bending hose into too small a radius is definitely discouraged. Every hydraulic hose has a manufacturer's and/or an SAE minimum-bend radius. This radius is measured to the inside surface of a hose. As an example, a hose having a minimum-bend radius of 5 in. should be bent no tighter than it would be bent if it were wrapped on a drum or pole with a diameter of 10 in. Angle fittings and adapters can frequently be used to reduce assembly length and to avoid sharp bends. Such fittings can also be used to dress up the appearance of an installation (Figs. 14-39 to 14-42).

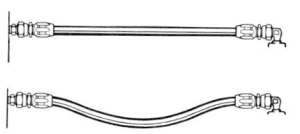

Fig. 14-43 Compensate for length change. *(Aeroquip Corp.)*

As mentioned earlier, but important enough to repeat, always be sure to leave enough slack in an assembly to allow for the possible +2 to −4 percent change in length when the hose is pressured (Fig. 14-43).

Obviously, good judgment should be used in any installation. Hose must be routed where it will be protected against outside damage as from snagging, rubbing, or contacting hot objects. Aside from this, hose requires little care and no maintenance.

Adapters

Adapters are available for several purposes. The ends of adapters can readily be identified, using the fitting-identification section presented earlier. Most hydraulic hose and coupling catalogs include a selection of the most commonly used adapters and most hose and coupling suppliers stock them.

A common use for adapters is to connect mismatched connections. For example, a hose

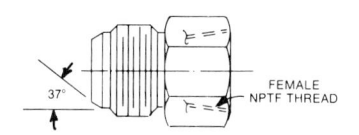

TUBE O.D., IN.	JIC MALE THREAD	NPTF FEMALE THREAD	APPROX. LGTH. HOSE DISPLACED, IN.
1/4	7/16 - 20	1/8 - 27	23/32
1/4	7/16 - 20	1/4 - 18	15/16
5/16	1/2 - 20	1/8 - 27	23/32
5/16	1/2 - 20	1/4 - 18	7/8
3/8	9/16 - 18	1/4 - 18	27/32
3/8	9/16 - 18	3/8 - 18	29/32
1/2	3/4 - 16	3/8 - 18	1
1/2	3/4 - 16	1/2 - 14	1-7/32
5/8	7/8 - 14	1/2 - 14	1-1/8
3/4	1-1/16 - 12	3/4 - 14	1-5/16
1	1-5/16 - 12	1 - 11-1/2	1-15/32

Fig. 14-44 JIC solid male to NPTF solid female.

with ¾-in. male NPTF thread is to be connected to a tube with ½-in. female JIC thread. An adapter with ¾-in. female NPTF thread on one end and a ½-in. male JIC on the other end would be required (Fig. 14-44).

Another common use is to change the angle of a connection. 45 deg and 90 deg adapters are available and commonly used (Figs. 14-45 and 14-46). A T adapter is used to

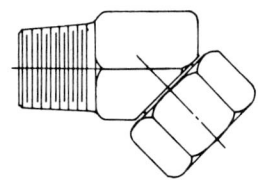

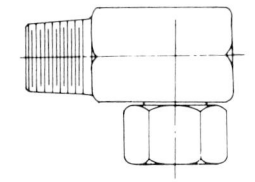

Fig. 14-45 NPTF solid male to NPSM swivel female (45°).

Fig. 14-46 NPTF solid male to NPSM swivel female (90°).

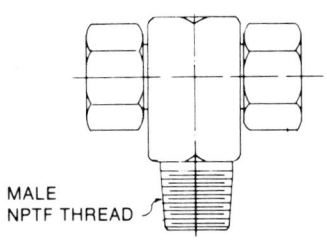

Fig. 14-47 T—NPTF solid male lower end, NPSM swivel female branches.

interconnect an additional line to a circuit. These are manufactured with a variety of end combinations and, usually, the combination required is available (Fig. 14-47).

Innumerable special-purpose adapters can be seen on equipment. They are usually identified by their odd shape, and they are mentioned here only to provide the knowledge that there are nonstandard devices that can be purchased only from an equipment manufacturer or his dealers.

Conclusion

Having read to this point, you know that lack of information can be costly. Fortunately, hose and fitting suppliers have knowledgeable people who are more than willing to assist with any related problem. These people are backed up by factory specialists. Why not take advantage of them?

Chapter **15**

Steam and Hot Water Cleaning

WILLIAM AXELSON
Vice President, The Hotsey Corporation, Englewood, Colo.

The old adage "cleanliness is next to godliness" is an essential principle in the maintenance of construction equipment. Machinery involved in earthmoving, whether building up or tearing down, excavating or lifting, is being used for one of the dirtiest jobs machinery is called on to handle.

Because of this, construction equipment owners and operators have placed tremendous emphasis on proper equipment cleanup. They know that dirt, grime, and grease are some of the tougher opponents of a successful equipment maintenance program.

Experience has shown that
- Dirt can add more than 1000 lb of unneeded weight to an earthmover.
- Dirt and dust can clog an air-cooled power shovel to the point of collapse.
- Dirt can foul a diesel tractor engine into failure.
- Dirt can bring bulldozers to a grinding halt.
- Dirt can ruin hydraulic cylinders, corrode wiring, and destroy components.
- Dirt can make any engine run at excessive temperatures, increasing both fuel consumption and the possibility of cracked blocks.
- Dirt can be harmful to employee morale and can damage a company's reputation.

All these factors contribute to the fact that dirt, grime, and grease can cripple construction equipment performance and dramatically increase operating costs.

HOW TO CLEAN EQUIPMENT

Economics dictates that a construction equipment supervisor should utilize a regularly scheduled machinery cleaning program to spot potential problem areas before they become serious. Also, machinery must be clean before servicing, repairing, or painting.

Statistical evidence has shown that cleaning equipment is a chore that costs time and money. Research indicates that a mechanic in an equipment repair shop spends 20 percent of his time cleaning parts at his work bench or cleaning so that parts can be either removed or replaced. That is 1 hr out of every 5 spent on the job. For this reason, construction equipment operators should implement some type of routine machinery cleaning program.

Pressure Washer and Steam Cleaner

The use of a high-pressure hot water washer or steam cleaner will quickly and efficiently remove crusty, gritty, or corroded buildups of dirt, grease, and grime from transmissions, track and roller assemblies, shovel hoists and crowds, engine blocks, drive trains, and dozer buckets. Once the equipment is clean, the proper repairing, welding, and painting

Basic Maintenance Technology

can take place immediately. Hours of wiping, swabbing, and wirebrushing can be reduced to minutes of cleaning when a pressure washer or steam cleaner is used.

In actuality, cleaning construction equipment is similar to washing your hands. You need four basic elements:

- Water
- Pressure
- A chemical agent (if the surface is greasy)
- Heat (150 to 210°F to activate the chemical and lower the viscosity of the oil)

The more there is of each element, the more effective the cleaning job will be. And the pressure washer and steam cleaner provide these basic elements.

There are several important factors to be taken into consideration when setting up a routine cleaning program.

Portability versus permanence   An equipment superintendent must evaluate carefully *where* the majority of cleaning will be done. If most of the cleaning will be done in the shop (or adjacent to it), a permanent installation will do the best job. In this situation, the pressure washer's heat and pressure units can be located remotely and multiple cleaning stations located in areas of the yard or shop where cleaning will be done.

If the majority of the cleaning will be done in the field and the construction site, then a portable cleaner mounted on a pickup truck would be preferable.

Ideally, a construction equipment concern should have both systems. As explained below, on-site cleaning is just as important as cleaning at the shop prior to repair.

Size of washer or cleaner   Cleaning energy is determined by a simple formula.

$$\text{Amount of water discharged} \times \text{pressure} = \text{cleaning energy}$$

Industry agrees that a cleaner that discharges 120 gph at 500 psi is fine for light-duty cleaning. A machine that delivers 210 gph at 650 psi offers more impact for medium-duty cleaning such as degreasing engines or cleaning farm equipment.

Heavy-duty equipment requires a heavy-duty washer, one that can deliver 240 to 540 gph at 1000 to 2000 psi. A heavy-duty unit that produces 540 gph at 2000 psi generates an astonishing 1 million units of cleaning power.

Experience shows that construction equipment maintenance supervisors should consider cleaning equipment that can deliver at least 240 gph and 1000 psi. This will provide enough cleaning power to blast loose the heavy accumulations of dirt, grime, and grease that can rob a machine of its usefulness.

ESTABLISH A CLEANING PROGRAM

Heavy construction equipment operators and maintenance personnel should adopt a regular cleaning program that will, over the long term, save machinery repair costs, reduce labor time, and save on fuel costs.

Based on in-the-field experience, a guideline to follow includes daily on-site cleaning, on-site inspection, and cleanup before maintenance.

Daily, on-site preventive cleaning   When a truck-mounted washer or cleaner is used, equipment should be cleaned each day. A few minutes' cleaning will greatly extend machinery life. It will rinse radiators and oil coolers, free crawler tracks, clear air intakes, unclog power trains, clean cement batchers—in short, cleaning will help assure optimum service from equipment that takes terrible punishment. Daily cleaning also makes it easier for an inspector to spot the more visible trouble areas before they occur.

On-site inspection of machinery   Daily cleaning of construction equipment as a part of preventive maintenance will help catch major problems. But all heavy construction equipment should be thoroughly inspected in the field after each 600-hr operational cycle. A few minutes' cleaning of the forklift, backhoe, earthmover, front end loader, crane, or dump truck will prepare it for a careful going over.

Such a cleaning-inspection routine, in the field, can mean the difference between downtime and optimum performance. And it can help keep equipment where it belongs—in the field. Field inspection saves time because the equipment need not be taken back to the shop for inspection. And, using a pressure washer or steam cleaner, an inspector need not fear overlooking a metal fissure, cracked block, hydraulic or oil leak, worn V-belt, frayed cable, corroded battery, or a bent tie rod.

Cleanup before maintenance Should the routine daily cleaning procedure or on-site inspection reveal a problem area, the machinery should be taken to the shop for repairs. Most companies have developed shop procedures to a point where successful repairs are virtually assured. One thing these shops all have in common is a clean machine before repair gets under way.

Pressure washers and steam cleaners quickly and efficiently remove crusty, gritty, and corroded buildups of dirt, grease, and grime from transmissions, track and roller assemblies, shovel hoists and crowds, engine blocks, and drive trains. Once dirt and grime have been removed, repairing, welding, and painting can take place immediately.

Because of the unique flexibility of a cleaning system, an equipment maintenance supervisor can also
1. Add acid to phosphatize parts
2. Clean metal chips out of crevices in parts
3. Sanitize machinery
4. Deodorize equipment
5. Spray insecticides to control insect populations

WASHING TIPS AND TACTICS

As the cleaning program is implemented, several helpful hints can make machinery cleaning more effective.

1. Take heavily greased equipment to a grease pit for washing so that effluent can be properly contained according to environmental considerations.
2. Begin cleaning machinery from the bottom up. Work progressively upward so that soap will not run down dry surfaces.
3. Always rinse from the top down so upper surfaces stay clean.
4. Use a water-soluble oil when rinsing sensitive machine surfaces that are susceptible to rust. As the water evaporates, a thin film of oil is deposited which inhibits the rusting process.
5. Make sure electrical outlets stay dry. The same goes for switch boxes or terminals. Spray engine components such as distributor caps, generator and battery terminals, and spark plug connectors with silicone to promote water repellency.
6. When cleaning parts or assemblies in the shop, spray toward drains. When drains are far apart, remove flushed water with a squeegee.
7. Don't use hot water to remove earth, rock dust, or attached soil—use cold water to save heat and conserve fuel.
8. Clean greasy, oily surfaces with hot water and a detergent.
9. Various cleaning gun nozzles provide different spray patterns for specific cleaning applications. A 0-deg nozzle provides a solid, forceful stream of high-pressure water. When it is necessary to blast loose tough accumulations of dirt and grime, 0 deg is most effective. A broader range (up to 40 deg) is most advantageous for cleaning large, flat surfaces.
10. Hold the cleaning wand 6 to 12 in. from the surface for best results.
11. The trigger shutoff on the pistol grip of the cleaning gun allows for fast water cutoff, thus saving water and energy.
12. A so-called plumbed-in central cleaning system for in-shop use has several advantages. First, the heat and water pressure are generated in a central location. Second, simple hose-connection cleaning stations can be positioned in the most appropriate locations or equipment bays. This will enable the construction equipment operator to fashion a system of the exact specifications required, including volume, temperature, pressure, and additional chemicals.
13. Always remember: Use a simple two-step process when cleaning with a pressure washer or steam cleaner—apply water (and chemicals or detergents if necessary) and rinse the surface clean.

HOW HEAVY-DUTY CLEANERS OPERATE

There are two primary methods of cleaning construction equipment: the steam cleaner and the hot water high-pressure washer.

Steam cleaner This machine produces very wet or saturated steam, with or without chemicals, at the nozzle of a discharge gun. Saturated steam is produced by superheating water to develop pressure in a partially restricted water-heating coil. As the heated, pressurized water leaves the discharge nozzle and enters the atmosphere, it explodes into tiny droplets of saturated steam (Fig. 15-1).

The steam cleaner operates in the following way: Water is pulled from a float tank and is pumped into the water-heating coil. Detergents may be added to the water if desired. A restricting nozzle at the end of the steam gun maintains a small internal back pressure and keeps the coil full of water. When the volume of water being pumped is known, a certain amount of heat can be added to superheat the water in the coil. As heat is applied, the water temperature in the coil rises and the water expands, causing an increase in internal pressure. This expansion of water develops pressure in steam cleaners, not a water pump.

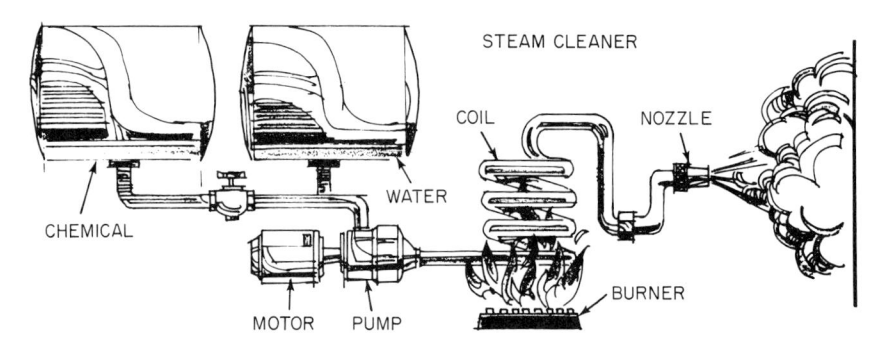

Fig. 15-1 Arrangement of a saturated steam cleaning system.

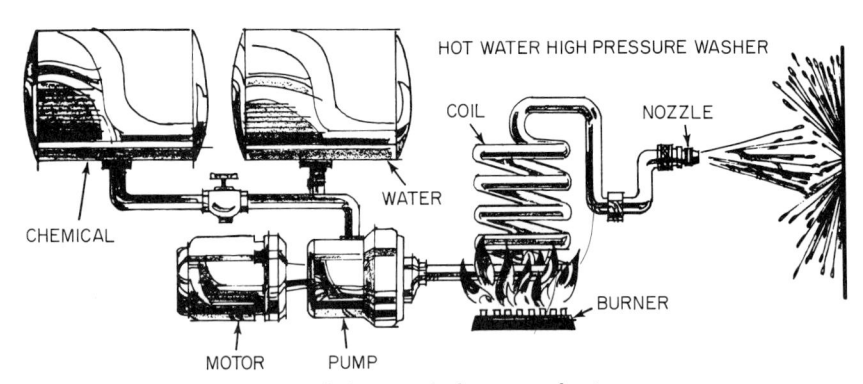

Fig. 15-2 Arrangement of a hot water high-pressure cleaning system.

The boiling point of water rises under pressure, and, when the water is heated to 330°F, the operating pressure is 100 psi. But, as the superheated water hits the atmosphere, it instantly dissipates a tremendous volume of heat or energy. This rapid loss of heat yields clouds of steam vapors (which can inhibit vision in cool weather).

Hot water high-pressure washer This washer uses a water pump to increase pressure tremendously. The pump, the heart of the pressure washer, pushes the volume of water and creates high pressure. The water in the pressure system need not be superheated, but is controlled at temperatures less than 212°F. The discharged water remains in a solid state and flows in a stream with very little surface area. The water temperature leaving the nozzles of both the steam cleaner and the high-pressure washer is less than 212°F since boiling water will not exceed 212°F at atmospheric pressure (Fig. 15-2).

ADDITIONAL CLEANING EQUIPMENT

In addition to a basic cleaning system, one should consider three additional pieces of equipment, the parts washer, the SandJet and the ChemJet, which can further reduce labor costs and save time. Tank cleaners and degassers should be considered if bulk tanks need frequent cleaning.

Parts washer This piece of equipment can greatly ease the job of construction equipment mechanics and repair specialists who deal with dirty or greasy parts. Washers speed the cleaning job and are easy, clean, and safe to use. They make fast work of hard-to-clean parts such as pistons, bearings, carburetors, gears, valves, and seals.

A parts washer should feature large cleaning tanks (18 × 36 in.) and a special safety lid, which will close automatically and shut off the pump in case of fire.

The washer should be operated automatically by a switch which turns on the pump when the lid is raised. The pump should be located in the fluid reservoir to assure a constant supply of solvent.

When the lid is closed, the pump will shut off automatically, assuring that no solvent fumes can escape when the cleaner is not in use. A drain shut-off valve allows the tank to be filled for soaking extra-grimy parts.

Fig. 15-3 Steam cleaning proves useful in removing dirt or grime from dozer buckets. This cleaning reveals cracks or metal fissures.

SandJet This machine hooks up to a pressure washer and offers an excellent alternative to sandblasting. Because the attachment works with the washer, the sand runs off the surface being cleaned for easy recovery and reuse. There is no dust pollution because of the combination of sand and water and users do not have to screen areas to keep airborne sand under control.

One person can operate the SandJet, and the only protective clothing needed is a pair of goggles. The SandJet can be used to clean steelwork, crane booms, dippers and buckets, masonry, tanks, boilers, engine parts; it can also be used to get any metal ready for painting such as heavy machinery, trucks, tractors, fork lifts, and draglines.

ChemJet This machine also easily attaches to the wand of a pressure washer. A vacuum is created by the washer and it draws any nonviscous fluid and mixes it with the pressure washer's water flow. This chemical applicator has many uses, including phosphatizing and etching raw steel, brightening raw aluminum, stripping paint, and cleaning timbers and supports.

Tank cleaners and degassers These pieces of equipment are useful to construction equipment operators who often need to clean tanks thoroughly (especially fuel cells and bulk storage tanks) before repairs are made. This is an especially important safety procedure when tanks are welded or braised. Use of a tank degasser cuts cleaning time and eliminates a great deal of downtime.

Using a five-phase program involving steam, heat, rinsing, heating, and flushing, a cleaner can completely remove tank contamination as determined by a zero reading on a so-called sniffer instrument.

This specialized cleaning system can be used in degassing tanks for welding and repair

Fig. 15-4 Job-site earth, entrapped in treads and undercarriages of tracked vehicles, adds hundreds of pounds of useless weight to equipment, increasing fuel consumption and reducing engine life. A sound pressure wash relieves this problem.

Fig. 15-5 Tires, like tracks, need to be dirt- and grime-free to enhance life. Consider this in relation to current tire replacement costs.

purposes and in rendering a tank clean for storage or when tank fuel loads must be changed (diesel and gasoline, for instance).

All construction equipment maintenance programs should include a cleaning procedure. This will be an investment with wide-ranging benefits for companies that have to service and maintain expensive equipment. Overwhelming evidence supports the contention that adherence to a cleaning program will help assure long, economical machinery life (Figs. 15-3 to 15-5).

Section **5**

Maintenance of Power Systems

Chapter **1**

Maintenance of Electrical Power Systems

JESS JOHNSON
Kohler Co., Kohler, Wis.

Popular, independent electrical power supply systems at construction sites today are the small, portable gasoline-engine-driven generators in the 500- to 5000-watt range. These lightweight air-cooled units provide ac and/or dc electrical energy for power tools, lighting, floodlighting, warning flashers, and welders at sites remote from commercial power sources. This chapter will be devoted almost entirely to the maintenance of these small portable generator sets.

Generator sets are built to withstand the harsh environment prevalent at most construction sites, but to remain reliable, they require a specific amount of attention at the intervals recommended by their manufacturers (Fig. 1-1). All too often, the unit that performs most reliably does so unnoticed and becomes, unfortunately, the unit most likely to be neglected in terms of service until it's too late to prevent major problems. For example, a simple task such as checking oil level before each start-up may not be done and costly internal damage to the engine may result. Such damage might occur because no one in the crew has been assigned direct responsibility for servicing the unit. A good maintenance program costs very little in time and material, yet lack of proper service leads to costly repairs and, perhaps most important to a contractor, costly downtime.

Maintenance of engine-driven generator sets can be divided into two categories: general routine service and preventive maintenance service.

Routine Service

As can be seen in Table 1-1, the routine service requirements for a typical portable engine-driven generator set are minimal and not difficult to perform. They are important, however, and the person assigned responsibility for servicing a unit should keep an accurate operating hour-service log to record when required services were performed. A sample of a typical service log is shown in Fig. 1-2.

Under the dusty, dirty operating conditions normally encountered at most construction sites, the importance of changing oil and servicing the air cleaner at the prescribed intervals cannot be overemphasized. A dirt-clogged air cleaner contributes to an overrich fuel mixture, leading to formation of harmful sludge in the crankcase of the engine. Unfiltered air entering through loose, missing, or improperly installed air cleaner components quickly causes internal damage—as little as one small teaspoonful of dirt thus introduced can ruin piston rings and the cylinder bore in just a few minutes.

5-2 Maintenance of Power Systems

Even when not contaminated by dirt, lubricating oil absorbs the contaminants which are normal by-products of combustion and which can eventually deteriorate and harm the engine. The best protection is to change oil at specified intervals.

Dust and dirt can also accumulate on air intake screens, cooling fins, and cooling air outlets so that damage could occur because of overheating. To avoid this, a generator set should be operated in an area free of chips, dust, and dirt. As a further precaution, intake screens should be cleaned daily before each start-up, and an operating unit should be stopped and cleaned whenever buildup is noted. Accumulation on brushes and commutator should be blown out at frequent intervals with dry compressed air to allow full output of the generator.

Fig. 1-1 Typical application of engine generator on construction site.

Electrical connections on a generator set should be checked and tightened periodically to prevent shorting and possible damage to electrical devices inside the control box. At the same time, safety guards should be checked to assure that all are in place and securely tightened.

TABLE 1-1 Routine Service Schedule

Frequency	Service
Every day	Check oil level Clean air inlets and outlets Check fuel filter Replenish fuel supply Service air cleaner (if dirty)
Every 25 operating hr	Change crankcase oil Service air cleaner Service fuel filter
Every 50 operating hr	Clean external surfaces Clean commutator Retighten electrical connections
Every 100 operating hr	Service spark plugs Check breaker points Replace air cleaner

Preventive Maintenance

In addition to routine services, a generator set should receive preventive maintenance of tuneup services at periodic intervals. Benefits of such services will not only be noted immediately in improved performance but in continued satisfactory performance throughout an extended service life. Such services are often best performed at service centers authorized by a generator manufacturer, but they are not too numerous or too difficult to

OPERATING HOUR - SERVICE LOG

The following is provided to help you keep an accumulative record of operating hours on your generator set and the dates required services were performed. Enter hours to the nearest quarter hour.

DATE RUN	OPERATING HOURS		SERVICE RECORD		DATE RUN	OPERATING HOURS		SERVICE RECORD	
	HOURS RUN	ACCUMULATIVE	DATE	SERVICE		HOURS RUN	ACCUMULATIVE	DATE	SERVICE

Fig. 1-2 Typical service log.

be performed by an experienced member of a maintenance crew. Under normal conditions, a generator set should be tuned up about every 500 hr of operation. A typical tuneup on a small air-cooled engine-powered generator follows.

TUNEUP RECOMMENDATIONS

Exterior: Thoroughly clean, especially cooling fin and air intake areas
Test: Crankcase vacuum and/or compression
Cylinder head: Remove and clean combustion chamber
Air cleaner: Service oil bath or replace element on a dry type
Spark plug: Replace
Breaker points: Replace
Condenser: Replace
Valves: Check valve to tappet clearance and adjust as needed
Breather: Service
Ignition: Check and adjust timing
Carburetor: Restart engine and adjust under load

When a generator set is in for tuneup, this is a good time to look it over closely; there are usually a number of indicators that tell you what condition it's in. Service hints can often be given to the operators as a result of such observations. For example, the condition of the electrodes and spark plugs is an excellent indicator of operating conditions.

5-4 Maintenance of Power Systems

Spark plug analysis When removing spark plugs, always check the firing end as the appearance here gives a very good indication of operating conditions. If abnormal conditions are indicated, always check the number of the removed plug—it may be of the wrong heat range for the engine. If the center electrode is worn round, don't try to square it with a file for reuse—replace the plug to prevent the misfiring often encountered when using worn plugs. Some common firing-end indicators are listed in the following guides for spark plug analysis.

Normal. A plug taken from an engine operating under good conditions will have light- or gray-colored deposits. If the center electrode is not rounded off, a plug in this condition could be regapped and reused.

Worn-Out. On a plug which has been in service too long, the center electrode will be rounded off and the gap will be worn 0.010 in. more than the original setting. Replace worn plugs since they require excessive voltage to fire properly.

Carbon-Fouled. Soft, sooty black deposits indicate incomplete combustion from rich carburetion, weak ignition, retarded timing, or poor compression.

4-CYCLE AIR COOLED ENGINES-TROUBLE SHOOTING GUIDE

PROBLEM	FUEL RELATED CAUSES			IGNITION CAUSES		OTHER CAUSES						
	NO FUEL	IMPROPER FUEL	FUEL MIX. WRONG	NO SPARK	POOR IGNITION	IMPROPER COOLING	IMPROPER LUBRICATION	POOR COMPRESSION	VALVE PROBLEMS	CARBON BUILD-UP	GOVERNOR FAULTY	ENGINE OVERLOADED
WILL NOT START	X			X				X	X			
HARD STARTING		X	X	X	X			X	X			
STOPS SUDDENLY	X			X			X	X				
LACKS POWER		X	X	X	X			X	X	X		X
OPERATES ERRATICALLY		X	X	X							X	
KNOCKS OR PINGS		X	X		X				X			X
"SKIPS" OR MISFIRES			X	X								
BACKFIRES			X	X					X			
OVERHEATS			X	X	X	X			X			X
IDLES POORLY		X			X							

Fig. 1-3 Typical engine troubleshooting guide.

Wet-Fouled. A wet-fouled plug could be caused by drowning with raw fuel or oil in the combustion chamber. The raw fuel problem may be caused by operating with too much choke. Oil in the combustion chamber area is usually caused by worn rings or valve guides.

Overheated. Overheating is indicated by chalk-white-colored deposits, not burned black as might be expected. This condition is also usually accompanied by excessive gap erosion. Overadvanced timing, lean carburetion, clogged air intake, or blocked cooling fins are some of the causes of overheating.

Tests Crankcase vacuum and compression tests should be made on engines brought in for tuneup (Fig. 1-3). These tests and checks are described as follows:

Crankcase Vacuum Test. A partial vacuum should be present in the crankcase when an engine is operating at normal temperatures. An engine in good condition will have a crankcase vacuum of 5- to 10-in. water column as read on a U-tube water manometer or ½ to 1 in. Hg as calibrated on a mercury vacuum gauge. A crankcase vacuum check is best accomplished with a U-tube manometer. If the vacuum is not in the specified range, consider one or more of the following factors—the condition easiest to remedy should be checked first:

1. Clogged crankcase breather can cause pressures to build up in the crankcase. Disassemble breather assembly and thoroughly clean; then, recheck pressure after reinstalling.

2. Worn oil seals can cause lack of vacuum. Oil leakage is usually evident around worn oil seals.

3. Blow-by, leaky valves can also cause positive pressures. These conditions can be confirmed by making a compression test on an engine.

When using a manometer, place the cork end into an oil fill hole (the other end is open to the atmosphere) and measure the difference between columns. If the water column is higher in the tube connected to the engine, vaccum or negative pressure is indicated. If the higher column is on the atmospheric side of the manometer, positive pressure is present.

Compression Test. The results of a compression check can be used to determine if an engine is in good operating condition or if reconditioning is needed. Low readings can indicate several conditions or a combination of different conditions (Table 1-2).

TABLE 1-2 Low Compression Troubleshooting Guide

Possible cause	Remedy
Cylinder head gasket blown	Remove head, replace gasket, reinstall head, recheck compression
Cylinder head warped or loose	Remove head, check for flatness (see cylinder head service), reinstall and secure in proper sequence to specified torque value
Piston rings worn—blow-by occurring	Recondition engine
Valves leaking	Recondition engine

Higher-than-normal compression can indicate that excessive carbon deposits have built up in the combustion chamber. To check compression, remove spark plugs and run engine up to a speed of at least 800 rpm. Be sure the air cleaner is clean and exhaust is not restricted before checking compression. Set throttle and choke wide open, insert gauge in spark plug hole, and take several readings. Consistent readings in the 110- to 120-psi range indicate good compression. Reconditioning is indicated if readings fall below 100 psi. On two-cylinder engines, take readings on both cylinders.

Cylinder head service To maintain top operating efficiency and performance, cylinder heads should be removed and carbon cleaned out about every 500 hr of operation. This service should be included in every tuneup. Carbon buildup can be especially heavy in larger-engine models which are often run at reduced load. Constant speed operation also seems to cause increased accumulation of carbon in the combustion chamber. When removing carbon, use a piece of wood or plastic to avoid scratching the aluminum, particularly in the gasket seat area. Always use a new cylinder head gasket and tighten capscrews in the proper sequence and to the torque valve specified in the manufacturers' service manuals when reinstalling cylinder heads.

Air cleaner Check the condition of the air cleaner on units brought in for tuneup. If poor service is indicated by clogged, dirty elements or improper installation, look for worn-out piston rings or sludge deposits in the oil pan. Service oil bath cleaners or replace dry-type air cleaner elements as part of the tuneup. The service recommendations on some typical air cleaners are as follows:

Dry Air Cleaner. Elements should be replaced when power loss is noted or after 100 to 200 hr if engine is operated under good clean air conditions—service and replace element more frequently under extremely dusty or dirty conditions. Dry elements should be cleaned after about 50 hr of operating—remove the element and tap lightly on a flat surface to loosen dirt. Replace the element if dirt does not drop off easily. Do not wash dry elements in any liquid or attempt to blow dirt off with an air hose as this will puncture the filter element. Carefully handle a new element—do not use if gasket surfaces are bent or twisted (Fig. 1-4).

Oil Bath Air Cleaner. If operating under extremely dusty conditions, it may be advantageous to install an oil bath air cleaner in place of a standard cleaner, thus eliminating the need for frequent replacement of the dry element. Normally converting to an oil bath cleaner involves removal of the dry-type cleaner and installation of an elbow and the oil bath unit in its place. The oil bath cleaner should be serviced after every 25 hr of operation; however, if extremely dusty or dirty conditions exist, service the cleaner

5-6 Maintenance of Power Systems

TABLE 1-3 Generator Troubleshooting Guide

Problem	Possible cause	Suggested remedy
No output	Loose terminal connections	Check for loose or bad connections and tighten all brush connections.
	Brushes not seated	Check for loose springs or brushes sticking in holder. Correct any cause for brushes not riding properly.
	Dirty commutator	Poor contact caused by buildup of dirt or oily film on commutator collector rings. Clean with coarse cloth or fine sandpaper or stone (don't use emery paper).
	Residual magnetism lost	After long periods of storage, it may be necessary to "flash" generator to restore magnetism. To do this, lift all brushes off commutator, connect positive (+) battery to positive generator terminal, then momentarily touch negative (−) battery to negative generator brush.
	Short in ac circuit	If engine labors while running or jerks while being cranked, check for short circuit in ac line. If short develops in ac armature, the armature will get very hot.
Low output or excessive drop in voltage	Engine speed too low	Check with tachometer. No-load speed should be about 3750 rpm; speed under load, 3600 rpm. Readjust governor speed.
	Overload	Make sure plant capacity is not being exceeded. Reduce load to 1500 watts; or, at extreme altitudes, derate plant about 3% per 1000 feet of elevation.
	Wrong ac lines	If external leads are too small or too long, excessive resistance is created which reduces output. Shorten lines or use larger-gauge wire.
	Engine in poor condition	Poor compression, excessive carbon, faulty ignition, wrong polarity, or any other condition causing poor performance may show up in reduced output.
	Brush angle wrong	If brush ring shifts or is positioned wrong when installed, brushes will be out of neutral zone, resulting in low output and/or excessive arcing. Position notch in brush ring 57° above horizontal plane (about 5/16 in. above end bracket leg).
Excessive arcing	Brushes sticking	If brushes are too large or too small, they may stick or cock in holder and chatter. Use proper size.
	Brush tension wrong	If spring tension is wrong, brushes may chatter. Adjust tension.
	Wrong brushes	Brush grade and material must be correct—use only specified brushes.

more frequently—even every 8 hr or twice daily if conditions warrant. To service normal-capacity oil bath air cleaners, remove the wing nut and remove the air cleaner components as a unit (Fig. 1-5).

Fuel system services Tuneups should include a complete check of the fuel system, including reconditioning of the carburetor and/or fuel pump if needed. Check fuel lines; replace hose in bad condition. Service the fuel filter if an engine is so equipped. Test the

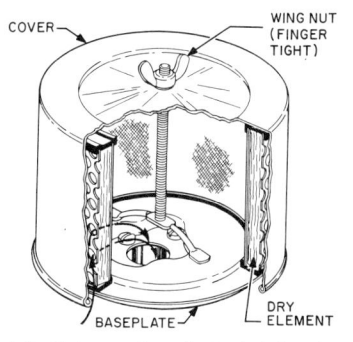

Fig. 1-4 Cutaway view of a typical dry-element air cleaner.

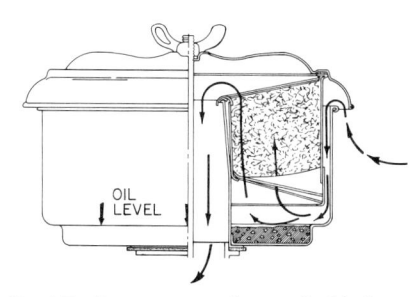

Fig. 1-5 Cutaway view of a typical oil bath air cleaner.

engine under load to check carburetor and pump. If readjustment of carburetor does not restore proper idle or operation, it should be reconditioned.

Exhaust smoke indicators Many people do not realize that even an engine in good condition will consume a certain amount of oil. As engine hours accumulate, ring and bore wear will result in even higher consumption. While blue exhaust smoke could indicate excessive consumption due to wear, it can also be caused by diluted oil or by operating with too much oil in the crankcase.

Generator service Generators do not normally require service on a regular basis. However, it is a good idea to remove the end cover and check the commutator and brushes at least every 6 months or every 50 hr or more often under dusty, dirty conditions. Visually check the commutator first—if a thin skinlike film of uniform thickness is evident on the surface, it is an indication of normal operation. The film acts as a lubricant and promotes longer brush life. If the surface is streaked or has ridges of dirt, clean it with a coarse cloth or, if this doesn't work, use fine sandpaper or a

Fig. 1-6 Checking generator brushes.

commutator stone—do not use emery cloth. Lift brushes and check their surfaces—replace them if unevenly worn or if worn down to about one-half their original length. Other common causes for rapid brush wear are wrong brush tension, rough commutator surface, high mica on commutator, or brush chatter. Blow dust out with dry compressed air after servicing the generator (Fig. 1-6).

Some common generator problems that may be easily detected and often corrected without test instruments are stated in Table 1-3, which is a troubleshooting chart. If routine service and suggested corrective action fail to solve a problem, contact a qualified technician to locate and correct the generator malfunction.

Chapter **2**

Maintenance of Diesel Power Systems

VERNON D. HAGELIN[1]
Technical Services, Deere & Company, Moline, Ill.

Need for preventive maintenance of diesel engines cannot be too highly stressed. Regular servicing and prompt attention to warnings of trouble will help prevent costly major repairs, keep necessary repairs from cutting into prime working time, improve operating efficiency, and reduce fuel costs.

Adequate records of filter and oil changes and of other services which should be performed periodically are an important aid to proper servicing and preventive maintenance. Such records should be kept with the machine on which the engine is being used or in the service department.

Many companies make diesel engines, but no two use exactly the same combination or design of cooling and fuel-injection systems, filters, governors, and other components.

This chapter concentrates on the fundamentals of preventive maintenance, common to most diesel engines, which the engine owner or operator should understand. Repairs which involve major engine or component disassembly should be performed only by expert technicians.

Manufacturers' manuals are a must for proper maintenance. If these manuals specify measures more stringent than those recommended here, *follow the manuals*.

The chapter concludes with a general troubleshooting chart. Some sections on various systems also give suggestions for finding malfunctions which are obviously in those systems.

The following are areas of preventive maintenance to be discussed:
- Visual inspection
- Starting engine
- Stopping engine
- Fuel system
- Cooling system
- Lubrication system
- Air intake and exhaust
- Electrical system
- Diagnosing malfunctions

VISUAL INSPECTION

Daily visual inspection of an engine is an important part of adequate preventive maintenance. It is also recommended before any tuneup procedure is started.

[1]The author is now retired.

Oil leaks

An external leak could be the cause of using too much oil.
    Check oil level.
    Look for leaks at the oil pan, drain plugs, front and rear seals, and gaskets.
    Check the coolant for oil contamination.
    Check the oil-cooler top and bottom covers.
    Check oil-filter housings.

Coolant leaks

    Check coolant level.
    Look for coolant leaks at the radiator, water pump, hoses, and oil-cooler inlet and outlet.
    Periodically examine all water hoses for softening, swelling, hardening, and cracking, any of which indicate need for immediate replacement.

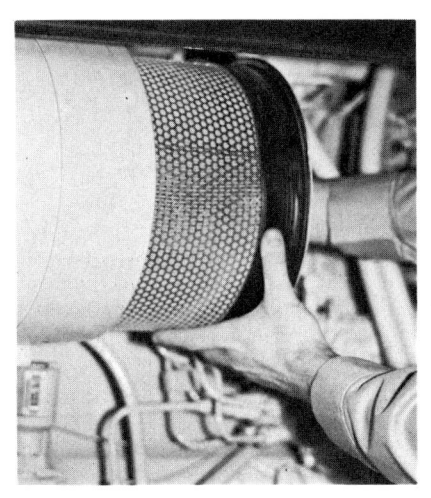

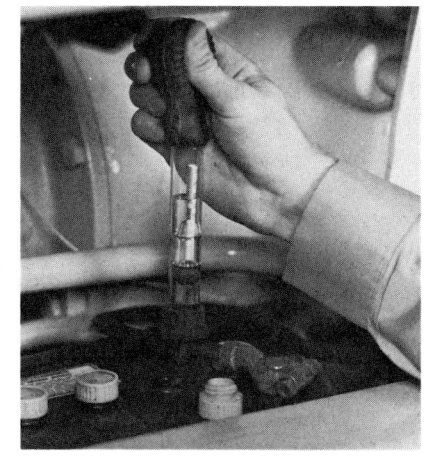

Fig. 2-1 Check the air filter.

Fig. 2-2 Check battery electrolyte level.

    Check the radiator for trash buildup, bent fins, kinks, dents, fractured seams, and cracked tubes.
    Be sure fan blades are straight and far enough from the radiator to prevent striking the core.
    Be sure the fan belt is in good condition and under proper tension.

Fuel system

    Inspect fuel-tank seams and fuel-pump inlet and outlet connections for leaks. If the fuel-supply pump has a primer level, be sure it is in the lowest position.
    Examine fuel-filter inlet and outlet connections.
    Check high-pressure fuel-supply connections to make sure none is twisted, kinked, broken, or leaking.
    Inspect fuel-delivery and leak-off lines of fuel-injection nozzles for leakage.

Air supply

    Check air filters (Fig. 2-1), air hoses, air-cleaner intake and outlet connections, and the air restriction indicator.

Electrical system

    Check battery electrolyte level and specific gravity (Fig. 2-2); be sure cap vents are open.
    Inspect all connection, especially for corrosion, at battery terminals.
    Look for bare wires which could cause shorts.
    Check for overheated parts (these often have an odor like burned insulation).
    Check alternator drive-belt tension (Fig. 2-3).

Turbocharged engines
Check air inlet and outlet connections.
Be sure oil inlet and drain lines are not twisted, kinked, or broken.
Be sure there is no strain on connector between turbocharger and manifold.

STARTING ENGINE

Before starting an engine, make the visual inspection already outlined. Though some of these checks may be made during the engine warmup period, many should not be attempted with the engine running.

Crank the engine no longer than 20 or 30 sec at a time. If it does not start, allow the cranking motor to cool for 2 or 3 min before cranking again. If the engine will not start after several attempts, refer to the trouble-shooting section (Diagnosing Malfunctions) at the end of this chapter.

After starting the engine, wait until

Fig. 2-3 Check alternator drive-belt tension and fan-belt tension.

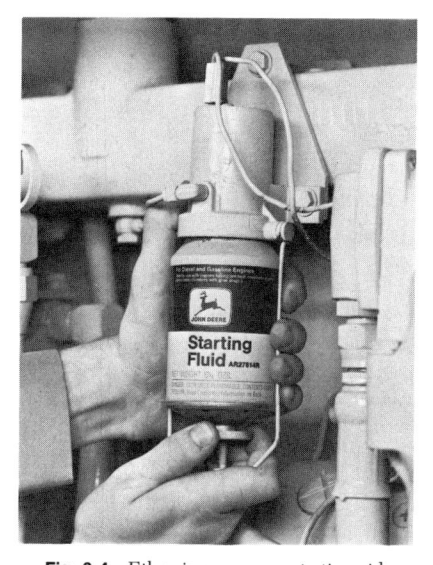

Fig. 2-4 Ether is a common starting aid.

gauges show proper oil-pressure and operating temperature before placing it under load. Run the engine at part load in midspeed range for 5 to 10 min [longer if ambient temperature is near or below 0°F (−18°C)] or until the gauges show proper readings; then apply partial load for a few minutes before going to full load. Do not race the engine during warmup.

Blocking air flow through the radiator will speed warmup, but be sure to remove the blocking material immediately after operating temperature is reached. (If the engine does not warm up properly, check the thermostats.) Partial blocking may also be used if the engine must be kept running in extremely cold weather with no load; however, regardless of ambient temperature, long idling periods must be avoided. An idling engine will not reach optimum operating temperature, resulting in less efficient combustion. Also, oil pressure may not reach its most effective level, preventing some engine parts from receiving adequate lubrication.

To prevent possible damage to an electrical system, the master switch must be in the ON position when the engine is running. The switch should be in the OFF position only when the engine is shut down.

Cold-weather starting Diesel engines depend on heat in combustion chambers for ignition, so starting them in cold weather can be a problem. Some common starting aids are ether (direct-injection systems only; Fig. 2-4), electrical glow plugs which preheat combustion chambers, oil heaters, coolant immersion heaters, battery warmers, and booster batteries.

Ether. Ether has a much-lower ignition point than diesel fuel, and some engines have adapters for injecting it. Heat from this initial ignition warms the fuel-air mixture and normal combustion follows.

Inject ether *only* while the engine is being cranked. If ether or other starting fluid is to be sprayed into the precleaner, use it sparingly and *only* after engine cranking begins. Excessive or improper use of ether can damage an engine.

Never use ether or other starting fluid in conjunction with glow plugs.

Glow Plugs. Glow plugs, which contain heating elements, fit into small turbulence chambers in the engine head. When turned on before using the starting motor, they warm the fuel-air mixture, leading to normal combustion.

Some manufacturers recommend glow-plug heat times of up to 3 min, depending on ambient temperature, before using the starter.

Immersion Heaters. Coolant immersion heaters are available for some diesel engines; they are particularlyuseful if the ambient temperature is expected to drop to $-10°F$ ($-23°C$). These heaters plug into 110-volt outlets (sometimes 220; see instructions), and they usually are left connected for at least 8 hr.

Some diesel engines have excess-fuel buttons to increase fuel delivery when starting; their use is explained in the manufacturer's manual.

Follow the manufacturer's instructions when using oil heaters or battery warmers.

STOPPING ENGINE

Diesel engines develop very high temperatures when working under load; they must be cooled gradually before being stopped.

Operate the engine at half-throttle (no load) for 3 to 5 min before shutting down and then at low idle for 1 or 2 min, to permit the engine and turbocharger to cool from their operating temperatures.

During hot weather, a longer cooling-off period may be desirable.

After stopping the engine, fill the fuel tank to prevent overnight condensation of moisture, and clean trash and dirt from the engine and radiator.

FUEL SYSTEM

Diesel engines burn a wide variety of fuels, but use of clean fuel is imperative, and selection of the right type is important. Check recommendations in the operator's manual. Do not use furnace fuel in modern diesel engines.

The two major grades of diesel fuel are 1-D and 2-D.

Grade 1-D is more volatile and is recommended for modern high-speed engines with variable loads and speeds, as well as for operation in extremely cold weather or at altitudes above 5000 ft.

Grade 2-D is recommended for high-speed engines with relatively high loads and more uniform speeds, as well as for engines not requiring the higher volatility of 1-D.

Ignition qualities of diesel fuels are measured by the cetane number, which is roughly comparable to the octane rating of gasoline. Cetane ratings vary from 33 to 64.

High-cetane fuels permit engines to be started at lower air temperatures and higher altitudes, provide faster engine warmup without misfiring, reduce varnish and carbon deposits, and help eliminate knock caused by slow ignition. However, a too-high cetane rating can cause incomplete combustion.

Ether has a cetane rating of 85 to 96, is highly volatile, and often is used as a cold-weather starting aid. However, it is highly explosive and can damage an engine if too much is sprayed in before cranking. CAUTION: Spray ether into the engine *only* while operating the starter.

Troubleshooting Elsewhere in this chapter is a section devoted to diagnosing most diesel-engine malfunctions. However, the following list may be of value if the malfunction is obviously in the fuel system:

Engine will not start, starts hard, or misfires
  Fuel tank empty
  Water in fuel
  Clogged supply line or filter

Faulty transfer pump
Air lock in injection pump
Fuel-cap vent plugged
Governor linkage to pump loose or broken
Drive shaft or key sheared
Pump plunger or distributor seized
Stuck valves or plugged orifices in nozzles
Excess-fuel control (if any) not activated
Shut-off control not deactivated
Engine will not idle smoothly
Faulty governor idling adjustment
Worn throttle linkage
Stuck plunger or pump rack or sticky or stuck control
Stuck nozzle valve or faulty nozzle-opening pressure
Leaky delivery valve
Pump out of time or calibration; loose control sleeves
Dirty filter
Engine smokes and knocks
Improper fuel
Excessive fuel delivery—faulty fuel-stop setting
Pump out of time
Dirty or fouled nozzles
Nozzle-opening pressure too low
Valves stuck open
Faulty turbocharger
Engine lacks power
Clogged filters
Pump timing retarded
Pump plungers or distributor worn
Faulty nozzles
Governor out of adjustment
Faulty aneroid to control fuel-air mixture
Faulty turbocharger

Filters Diesel fuels tend to be impure, while injection parts are precision made, emphasizing the need for frequent filter attention. Further, filters and their frequent inspection are not intended to compensate for a contaminated fuel supply.

Many diesel engines have three stages of progressive filters: a screen at the tank or transfer pump to remove large particles, a primary filter to remove small particles, and a secondary filter to remove tiny particles. Some filters not only remove suspended matter from fuel but also soluble impurities. Most filters have a water trap where water or heavy sediment can settle to be drained later.

Check all filters frequently. Change filter elements according to instructions in the manufacturer's manual, and keep in mind that it's better to change them too often than not often enough. Change the elements more often when operating in extreme dust or dirt. Check water traps at frequent intervals and drain out water and sediment.

Fuel-transfer pump Analyzers are available for checking fuel-transfer pump delivery and pressure. Usually, however, visual inspection is sufficient.

For a visual check, disconnect the pump-to-filter line at the filter (Fig. 2-5). Set the throttle so the engine will not start, and turn the engine over several times. If fuel spurts from the line, the pump is operating properly.

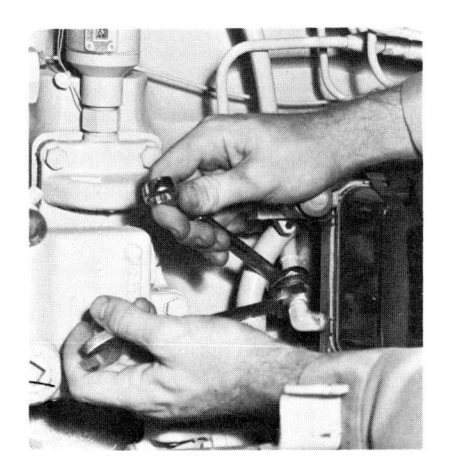

Fig. 2-5 Disconnect the pump-to-pump filter line at filter.

5-14   Maintenance of Power Systems

If little or no fuel flows, check for the following:
- Primer lever (if so equipped) left in upward position
- Leaking sediment-bowl gasket
- Plugged sediment-bowl screen
- Loose or damaged connections
- Air leak in inlet line
- Clogged fuel lines
- Loose cover screws on pump

If the trouble is not in these areas, repair or replace the pump. Most pump manufacturers sell repair kits, but it may be more economical to replace the entire unit. Also, some pumps are sealed units and so must be replaced if defective.

If repair is needed, disassemble the pump as outlined in the manufacturer's manual, inspect as follows, and replace parts as necessary:
- If the pump is a diaphragm type, look for punctures or leaks in the diaphragm. Check the slot in the diaphragm pull rod for wear.
- Examine the cover and body assembly for cracked or warped gasket surfaces.
- Examine valve and cage assemblies for worn valves or broken springs.
- Check the diaphragm and rocker-arm spring for proper tension.
- Inspect the rocker-arm link and pin for wear or damage.
- Inspect the filter screen for punctures and clogging.

Fuel lines   Diesel engines have three types of fuel lines:

Heavyweight.   These are high-pressure lines between injection pump and nozzles.

Medium-Weight.   These lines can handle light to medium pressures between tank and injection pump.

Lightweight.   These lines are used when there is low or no pressure of leak-off fuel from nozzles to tank or pump.

Inspect fuel lines periodically for leaks which indicate loose connections, breaks, or flaws. Connections should be snug, but not overtightened. (Overtightening could strip threads or damage sealing surfaces.)

Injection pumps   Injection pumps are precision instruments, critical in adjustment, and easily damaged by careless handling. Only qualified technicians should disassemble or service these pumps, closely following the manufacturer's manual and using the special tools the manual specifies.

Never steam-clean an injection pump during engine operation, and do not spray a warm pump with cold water; pump seizure could result.

Fuel-injection nozzles   Fuel-injection nozzles are comparatively simple devices but so important to proper diesel performance that their faulty operation can cause a variety of engine problems, ranging from hard starting to lack of power.

Various manufacturers use several different types of nozzles. While common features of these nozzles will be discussed here, each operator should consult the manual for his engine to secure specific information. Table 2-1 lists some problems and their possible causes.

The operator or owner should not attempt to remove or test nozzles unless he has the proper equipment and technical skill.

Before removing nozzles for testing and servicing, clean the area around them and remove and cap the injection and leak-off lines. Soak the nozzle assemblies in clean solvent or fuel after discarding the outer seals. Clean carbon and dirt deposits from spray tips and nozzle bodies with a soft-bristle brush (Fig. 2-6); *never* use emery cloth or a steel-wire brush.

When testing nozzles, follow the manufacturer's manual. Use a nozzle tester (Fig. 2-7), a high-pressure hand pump which forces fuel through the nozzle. CAUTION: Fuel comes out of the nozzles at extremely high pressure and can penetrate clothing and skin; point the nozzle away from yourself and any bystander.

Test for three things: spray pattern, opening pressure, and valve leakage.

Spray Pattern.   Fuel should be finely atomized and evenly distributed. There should be no stream or large visible drops. If the spray pattern is poor, look for a clogged or eroded orifice or a bent valve.

Opening Pressure.   Pressure needed to open the nozzle is checked by pumping the tester steadily until the gauge needle falls rapidly. Check the pressure reading against the manufacturer's manual. If pressure is too low, a weak or broken spring may be the cause,

Maintenance of Diesel Power Systems   5-15

or spring pressure may need adjustment. If pressure is too high, the tip may be plugged, or the valve may be binding in the valve guide.

Valve Leakage. This is the third possible malfunction to check. If these tests show the nozzle is not working properly, it must be disassembled, inspected, cleaned, and, if necessary, repaired. This requires a special tool kit and probably should be entrusted to an experienced technician.

CAUTION: When working on several nozzles, do not mix parts.

TABLE 2-1   Troubleshooting Fuel-Injection Nozzles

Problem	Possible cause
Nozzle opens at wrong pressure	Faulty spring-pressure adjustment
Nozzle will not close	Broken spring
Poor spray pattern	Plugged or chipped orifices (spray tip only); chipped or broken pintle end; deposits on pintle seat; chipped pintle seat
Poor misting of fuel	Plugged or chipped orifices (spray tip only); valve not free; cracked tip
Valve operates erratically	Valve spring misaligned; spring broken; deposits on pintle seat; bent valve; distorted body
Valve will not operate	Valve spring misaligned; spring broken; varnish on valve; deposits in seat area (pintle tip only); bent valve; valve seat eroded or pitted; distorted body
Too much fuel leaks off	Valve guide worn
Too little fuel leaks off	Varnish on valve; insufficient clearance between valve and guide

Inspect the parts for wear, chipped edges, scratches, misalignment, and breakage. Chipped, broken, or bent parts should be replaced, and if damage is extensive, replace the entire nozzle. Some nozzle parts are sold in matched sets, and should be so installed.

If the valve on some types of nozzles is sticking but not bent, a little polishing or use of injector lapping compound around the valve-guide area may be all that is needed. Other nozzle parts often may be salvaged by lapping their surfaces to remove tiny scratches or burrs—improper lapping can result in excess wear.

Before reassembling nozzles, flush each part in clean oil. Then place the entire nozzle in the tester for retesting.

Fig. 2-6   Clean nozzle with soft bristle brush.

Fig. 2-7   Test nozzles with nozzle tester.

Fuel storage The importance of proper fuel storage cannot be stressed too highly. Many diesel-engine difficulties can be traced to dirty fuel or fuel that has been in storage too long.

Keep all dirt, scale, water, and other foreign matter out of the fuel, and avoid storing fuel for a long period of time. Flush fuel-storage tanks periodically.

Water can condense in partly filled fuel containers. The containers should have provision for draining water which settles in their bottoms.

Governors Diesel-engine governors are speed-sensitive devices which act on the engine throttle to maintain a selected speed, limit slow and fast speeds, and shut down the engine if it threatens to overspeed.

Governors are relatively trouble-free but do need occasional adjustment and servicing. Two principal requirements for proper functioning are a vibration-free drive and clean moving parts. Manufacturer's instructions should be followed in servicing governors.

Here are some suggestions for diagnosing governor malfunctions.

Erratic engine operation, hunting, or misfiring
   Idle spring missing, broken, or wrongly adjusted
   Wrong control spring, or worn or broken spring
   Parts worn, sticking, binding, or improperly assembled
   Faulty high-idle adjustment
   Adjusting screw needs adjustment

Engine idles erratically
   Idle spring missing, broken, or improperly adjusted
   Parts worn, sticking, binding, or improperly adjusted
   Wrong governor spring

Engine not receiving fuel
   Shut-off control not deactivated
   Parts worn, sticking, binding, or improperly adjusted

Engine does not develop full power or speed
   Wrong governor spring
   Faulty fast-idle adjustment
   Adjusting screw needs adjustment
   Parts worn, sticking, binding, or improperly adjusted

Bleeding If the fuel system has been opened (e.g., to replace a fuel filter), air may get in to form an air lock that would keep fuel from reaching or going through the injection pump. Then the engine will not start or will run poorly.

The following procedure is recommended for bleeding the system:

1. Fill the fuel tank with the correct fuel.
2. Loosen the bleed plug on the fuel filter and open the fuel shut-off valve at the tank. If there are dual filters, bleed the filter nearest the tank first.
3. Open the fuel-supply valve. If the system has a manual priming pump or lever, pump the lever until a solid, bubble-free stream of fuel flows from the opening. If the lever will not pump fuel and no resistance is felt at the upper end of the stroke, turn the engine with the starter to change the position of the fuel-pump cam.
4. Tighten the bleed plug.
5. If the engine has dual filters, repeat the bleeding process on the other filter (Fig.2-8).
6. When bleeding is completed, leave the primer lever at the lowest point of its stroke.
7. If an airlock is still present, bleed the injection lines.
8. Using wrenches (two required), loosen the injection-line nuts on at least two lines. CAUTION: Loosen injection-line connectors only one turn to avoid excessive spray.
9. Bleeding half the injection lines usually is sufficient; the others will bleed themselves when the engine starts running.

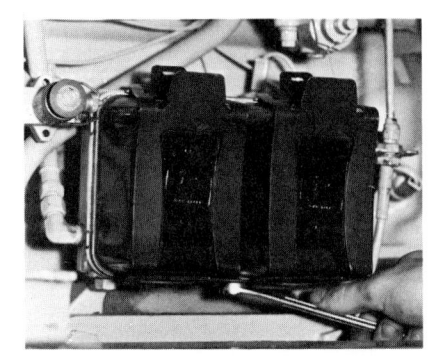

Fig. 2-8 Bleed air from fuel line at filters.

10. Crank the engine until fuel without foam flows around the connectors. Then tighten the connections carefully until snug and free of leaks. Do not bend the line connections.

COOLING SYSTEM

The intense heat of combustion in modern high-performance engines may cause such components as valves, pistons, and rings to operate near critical temperature limits even when the cooling system is operating normally.

Overheating seriously affects engine lubrication. High metal temperatures may destroy the lubricating film, accelerate oil breakdown, and cause formation of varnish. Cylinder heads and engine blocks often are warped and cracked by the terrific strains set up in the metal by overheating, especially when followed by rapid cooling.

Common causes of cooling malfunctions are clogged systems, low coolant level, and defective water pumps and thermostats.

Clogging Rust clogging, perhaps the most common cause of cooling-system trouble, can be avoided by periodic rustproofing and cleaning when necessary.

Rust forms on walls of the engine water jacket and other metal parts. Particles settle in the jacket and radiator water tubes, cutting down heat transfer until the engine overheats. Overheating stirs up more rust in the block and forces it into the radiator, which eventually gets clogged.

Rust and scale can also build up on the water side of the combustion chambers, causing overheating and eventual engine damage. Rust, scale, and grease can be removed by double-action cleaners, which are harmless to cooling-system metals and connections if used according to directions.

If rust and grease are not completely neutralized and flushed out, they can destroy the corrosion inhibitors in later fills of antifreeze and antirust solutions.

Drain the entire system at least once a year. If the liquid is rusty, use a cooling-system cleaner. Otherwise, use a radiator flush, or flush with plain water. Corrosion inhibitors will not clean out rust already formed.

It is desirable to maintain at least a 25 percent solution of antifreeze containing dependable inhibitors, even during the warm season.

Coolant Even distilled water can cause rust in a system, so if water alone is used as a coolant add a can of rust inhibitor. Keep the system filled to a level midway between the radiator core and the bottom of the filler neck.

If antifreeze coolant is used, ethylene glycol types are recommended, because most diesel engines develop temperatures above the boiling point of alcohol. Add rust inhibitor if the antifreeze does not already contain it.

Use only as much antifreeze as expected temperature extremes require. Because of the nature of ethylene glycol, too strong a cooling-system solution can actually reduce protection against freezing.

Never pour hot water into a cold engine or cold water into a hot engine; a cracked cylinder head or block could result.

Remember that the term *permanent antifreeze* does not mean the solution is good for more than one season; it means the solution will not boil away at normal engine operating temperatures. Adding rust inhibitors or fresh antifreeze to used solutions will not restore full-strength corrosion protection.

Repair radiator or other cooling-system leaks before installing antifreeze coolant. Follow instructions when adding sealing solutions to correct minor leaks. Some of these solutions will react with antifreeze and rust inhibitors and seriously affect coolant performance.

Radiator Though such cooling-system malfunctions as external leaks may be obvious, check the entire system before servicing. Use a pressure tester according to the manufacturer's instructions. Check the radiator (Fig. 2-9), water pump, hoses, drain cocks, and cylinder block for leakage.

Inspect the radiator for bent fins and tubes with cracks, kinks, dents, and fractured seams. Only experienced radiator technicians should make needed repairs.

Radiator cap Most radiator caps have a pressure valve to vent coolant or steam if pressure reaches a certain point and a vacuum valve which opens to prevent vacuum in the cooling system.

5-18 Maintenance of Power Systems

Using a tester available from the dealer, check both valves periodically for proper opening and closing pressures. If either valve malfunctions, use a new radiator cap.

External leaks At best, external leaks will cause loss of costly antifreeze. But they also can cause insufficient engine cooling with possible damage.

Most radiator leakage is due to mechanical failure of soldered joints caused by cooling-system pressure or engine or frame vibration. Other possible sites for external leakage are hose connections, core-hole plugs, gaskets, and stud bolts and capscrews.

Check the cylinder block for coolant leakage before and after it gets hot and while the engine is running.

Internal leaks Coolant may leak into an engine because of a loose cylinder head or sleeve joint, defective gaskets, a cracked or porous casting, or malfunction in the push-rod compartment.

Water or antifreeze will form sludge when mixed with engine oil, possibly causing lubrication failure (Fig. 2-10), sticking piston rings and pins, sticking valves and valve lifters, and extensive engine damage. If crankcase oil looks milky, the cause may be antifreeze contamination.

If a coolant leak exists in the push-rod compartment, it may be necessary to pressurize the cooling system and tear down part of the upper part of the engine.

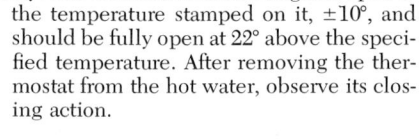

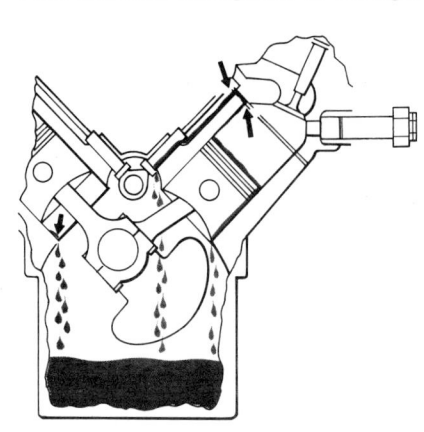

Fig. 2-9 Use pressure tester to check radiator.

When replacing cylinder-head gaskets, use only new gaskets designed for the engine in question. Be sure the head and block surfaces are clean, even, and smooth.

Follow the manufacturer's torque specifications and the sequence for tightening cylinder-head bolts.

Thermostats Faulty or improper thermostats can cause engines to warm up too slowly or to operate at the wrong temperatures. Use only the type specified by the engine manufacturer. Never run an engine without thermostat protection (some engines have two or more).

Discard broken, faulty, or corroded thermostats. Do not use bellow-type thermostats in high-pressure cooling systems.

To check a thermostat (Fig. 2-11), suspend the unit and a thermometer in a container of water and, while stirring, heat the water gradually. The thermostat should begin to open at the temperature stamped on it, ±10°, and should be fully open at 22° above the specified temperature. After removing the thermostat from the hot water, observe its closing action.

Fig. 2-10 Coolant mixes with engine oil through internal leaks.

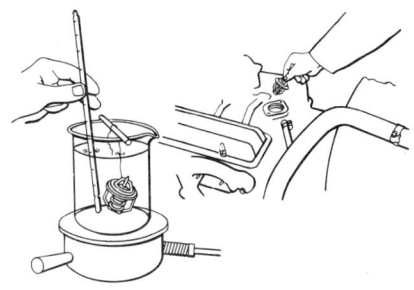

Fig. 2-11 Check condition of thermostat.

When installing a thermostat, clean the gasket surfaces and use a new gasket. Position the thermostat with the expansion element toward the engine (the frame must not block water flow).

Hoses Periodically check all hoses for hardening, cracking, softening, and swelling (Fig. 2-12). If hoses must be removed, check any inside reinforcing springs for corrosion.

Fig. 2-12 Inspect hoses for wear.

Replace hoses often enough to be sure they are always pliable and able to pass coolant without leaking or shedding small particles of rubber which could clog the radiator.

Use only the best available hoses, and coat connections with nonhardening sealing compound when installing. Tighten hose clamps securely. A pressurized cooling system can blow off an improperly installed hose.

Water pump Some diesel-engine water pumps turn at 4000 rpm and pump as much as 125 gal (479 liters) of coolant per minute. Overheating occurs quickly if a pump malfunctions.

Pump malfunctions may be caused by leaks in the housing, broken or bent vanes on the impeller, and damaged seals and bearings. If a pump must be removed and disassembled for inspection, replace all damaged or worn parts and use new seals and gaskets when reassembling. Follow the manufacturer's instructions.

Filters Some engines have filters in the cooling system. The filter element and resistor plates often contain chemicals which remove or neutralize corrosives, alkalize the coolant enough to prevent corrosion of metal parts, and form rustproof films on metal surfaces.

Periodic filter servicing should include draining sediment from the lower sump and replacing the filter element when necessary.

Fan and fan belt Effective preventive maintenance must include regular checking of fan-belt condition and tension, similar to the procedure shown in Fig. 2-3. Replace belts when wear warns of early failure.

Adjust fan-belt tension as specified by the manufacturer (Fig. 2-13). Too much tension causes premature failure of belt and fan bearings. Too little permits belt slippage and causes insufficient cooling and excessive belt wear.

Fan service usually consists of making sure the blades are straight and far enough from the radiator so they will not strike the core.

Aeration Aeration in a cooling system, caused by air mixing with the coolant, can speed formation of rust and corrosion. It can also cause foaming, overheating, and loss of coolant through the overflow pipe.

Aeration may result from a leak in the system, turbulence in the top tank, and too-low coolant level.

Use the following steps to check for aeration in the cooling system:
1. Adjust coolant to correct level.
2. Replace pressure cap with plain but airtight cap.

5-20 Maintenance of Power Systems

3. Attach rubber tube to lower end of overflow pipe.

4. With transmission in neutral, run engine at high speed until temperature gauge stops rising and stabilizes.

5. Without changing engine speed or temperature, place end of rubber tube in container of water.

6. A continuous stream of bubbles from the tube will show that air is being drawn into the cooling system.

Exhaust-gas leakage A cracked head or loose cylinder-head joint can allow hot exhaust gas to be blown into the cooling system under combustion pressures even though the joint may be tight enough to keep liquid from leaking into a cylinder.

Exhaust gases dissolved in the coolant will destroy the inhibitors and form acids which cause corrosion, rust, and clogging. The cylinder-head gasket may burn or corrode because of the gases, and excess pressure also may force coolant out of the overflow pipe.

Fig. 2-13 Adjust fan-belt tension.

Check the cooling system for exhaust-gas leakage if the coolant is rusty or if there are severe rust clogging, corrosion, or overflow losses.

1. Warm up the engine and place it under load.

2. Remove the radiator cap and look for excessive bubbles or an oil film in the coolant.

3. Make this test quickly, before boiling starts, for steam bubbles will be misleading.

Flushing cooling system Cooling systems should be flushed and thoroughly checked at least once a year and always before installing new antifreeze solution.

Incomplete flushing, such as hosing out the radiator, will close the thermostat and prevent thorough flushing of the water jacket. Either remove the thermostat or, after filling the system with clean water, run the engine long enough to open the thermostat.

After thorough flushing, open all drain points to drain the system completely. Clean out the overflow pipe; remove insects and dirt from radiator air passages, the radiator grille, and screens. Check the thermostat, radiator pressure cap, and the cap seat for dirt and corrosion.

LUBRICATION SYSTEM

Faulty lubrication is perhaps the greatest single cause of premature engine failure. It can result from using the wrong grade or weight of lubricating oil, insufficient oil in the crankcase, deficient oil pressure, lack of proper oil additives, or contaminated oil, any of which can lead to a major and costly engine overhaul long before normal wear would make it necessary.

Oil contamination The major causes of lubricating-oil contamination include the following.

Maintenance of Diesel Power Systems 5-21

Improper Storage and Handling. Store lubricants in a clean, enclosed area, and keep all covers and spouts on oil containers when not in use. These practices not only keep dirt out of the lubricant but reduce condensation of water in the containers.

Dust Breathed into Engine with Combustion Air. Regularly clean or replace air filters and the breather on the oil filler.

Cold Engine. A cold engine greatly reduces fuel-burning efficiency and partially burned fuel blows by the piston rings into the crankcase. Oxidation of this fuel in the oil forms a harmful varnish which deposits on engine parts. A misfiring engine will also cause this contamination.

Cold engines also contaminate lubricating oil when water vapor, a normal product of combustion, condenses on cold cylinder walls and is blown past the rings into the crankcase. It then combines with oxidized oil and carbon particles to form sludge, which can effectively block oil screens or passages. Warm up the engine properly before applying full load, making sure the engine reaches operating temperature each time it is started. Use the proper thermostat to warm up the engine as quickly as possible.

Oxidation. This occurs when hydrocarbons in the oil combine with oxygen in the air to produce acids which are highly corrosive and create harmful sludges and varnish deposits.

Carbon Particles. A product of normal engine operation, these particles are created when oil on the upper cylinder walls is burned during combustion. In addition to contaminating the oil, excessive carbon deposits can cause piston rings to stick in their grooves.

Engine Wear. Tiny metal particles are constantly being worn from bearings and other parts. They tend to oxidize and contaminate the oil.

Antifreeze. Should it leak from cooling into lubrication systems, antifreeze can cause sludge formation. Leaking can occur if head gaskets are damaged by improper use of starting fluids, or if head bolts are not torqued to specifications when the head is removed and replaced.

Additives One important way to combat oil pollution is to start with good-quality oil containing proper additives. Remember that these additives eventually wear out, so change oil before they are exhausted.

Anticorrosion additives protect metal surfaces from corrosive attack.

Oxidation-inhibitor additives keep oil from absorbing oxygen, preventing oxidation and varnish and sludge formation.

Antirust additives prevent rusting of metal parts during storage periods, downtime, and even overnight. They also help neutralize harmful acids, and they form a protective coating which repels water droplets and protects the metal.

Detergent additives prevent deposits and help keep metal surfaces clean. They hold carbon particles and oxidized oil in suspension, so these will be eliminated when oil is drained. Black oil is evidence that these detergents are keeping such contaminants in suspension instead of letting them accumulate as sludge.

Oil filters Proper attention to oil filters is essential in guarding against oil contamination, and it is one of the easiest and most important forms of preventive maintenance.

Two types of oil filters are in general use: *Surface filters* have a single surface which stops or removes dirt particles larger than holes or openings in the filter; *depth filters* contain a large volume of filter material which removes particles suspended in the oil as well as some water and water-soluble impurities (Fig. 2-14).

Many depth filters have a relief or bypass valve, which opens as the filter becomes clogged allowing oil to bypass to the bearings. Otherwise, pressure would build up on the inlet side of the filter, causing the pressure-regulating valve in the engine to open completely, sending all oil back to the crankcase, and

SURFACE FILTER   DEPTH FILTER

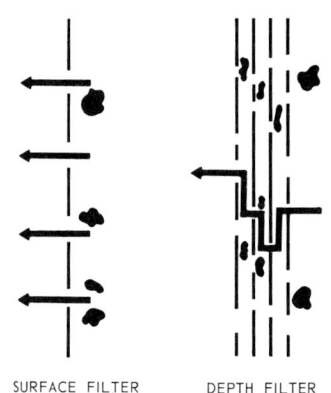

Fig. 2-14 Filtering methods: surface versus depth.

seriously damaging the engine. Replace such filters only with an exact duplicate of the original.

Follow the manufacturer's instructions for checking filters. When servicing any filter, use new gaskets and seal rings. Tighten the housing firmly but not too tightly. Then run the engine until the oil pressure registers and check for leaks.

Pressure-regulating valves These valves maintain correct oil pressure in the lubrication system regardless of engine speed or oil temperature. They also bypass oil at filters and oil coolers. Most are adjustable.

Servicing usually is confined to cleaning parts and the valve bore in which the valve slides with a proper solvent. Also check the valve poppet for wear or nicks which might cause it to hang up in the bore.

Check engine oil pressure after servicing any pressure-regulating valve. Causes of too-low oil pressure include low crankcase-oil level, too-thin oil in the crankcase, worn engine bearings, worn oil pump, filter or pump leaks, faulty regulating-valve spring, and improperly adjusted regulating valve.

Causes of too-high oil pressure include too-heavy oil in the crankcase, stuck regulating valve, and improperly adjusted regulating valve. A defective gauge may show deficient or excessive oil pressure.

Oil coolers Many lubrication systems have oil coolers which use engine coolant to dissipate heat from the oil. These coolers may be mounted internally (Fig. 2-15) in the crankcase or externally on the outside of the engine block.

Normal maintenance of the cooling system usually will keep the oil cooler clean. When cleaning the lubrication system, remove the cooler and use solvent to clean the oil passages.

When replacing or installing an external oil cooler, use new gaskets and be sure the capscrews are properly tightened.

Ventilation All diesel engines have ventilation systems to carry away fuel vapor and water vapor so vapors will not condense into liquids which drain into the crankcase. Manufacturers use a variety of ventilation systems, but these general suggestions will aid in proper servicing.

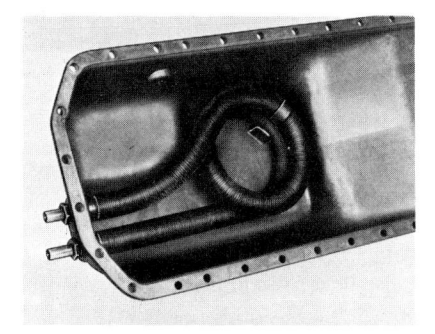

Fig. 2-15 Internal oil cooler.

Begin with the air inlet, which may be the oil-filler cap or the main air cleaner. In either case, service regularly according to instructions in the manufacturer's manual.

If the system has a regulating valve, either clean it with a solvent or replace it, according to the manufacturer's instructions.

If the system has a vent tube, periodically remove it and clean it with a solvent. If the vent tube has a filter, be sure to clean it also.

Oil consumption Some oil consumption is normal during diesel-engine operation. However, if oil consumption seems excessive, check the following possible causes.

Wrong Weight or Grade of Oil. Follow manufacturer's instructions, giving proper consideration to type of engine service and prevailing ambient temperature.

Engine Not Run Long Enough under Load to Seat Rings. Some variation in oil consumption can be expected during break-in, but the level should be stabilized before 250 hr of operation if the engine was broken in properly. If not, assume that a problem exists and must be corrected.

Pressure-Regulating Valve Improperly Adjusted. Rings and valves are flooded with oil.

Crankcase Breather. If plugged, the breather can cause increased crankcase oil pressure.

External Oil Leaks. Even if small, these can add up to loss of quarts of lubricant between changes. Check front and rear seals, all gaskets, and filter-attaching points.

Engine Blow-by. Fumes from the crankcase vent should be barely visible with the engine at fast idle under no load. Excessive blow-by may indicate that piston rings and cylinder liners have worn to the point where the rings cannot seal off the combustion chambers.

Valve-Guide Seals. If defective, these seals permit an excessive amount of oil to enter combustion chambers. To check, warm up the engine and let it idle slowly for 10 min. Then remove the exhaust and intake manifolds. If valve ports and undersides of valve heads are wet with oil, it has been drawn through the valve guides.

If the cylinder head is removed, check the piston heads. If they are wet with oil, it may have been drawn past the piston rings; if excessive blow-by has indicated worn rings, this will serve to confirm it.

Excessive Engine Speeds. Excessive speed is another and common cause of excessive oil consumption. Observe fast-idle limits specified by the manufacturer.

AIR INTAKE AND EXHAUST

Improper or inadequate intake air filtering can lead to early engine failure and a costly overhaul. Enough dust-laden air can pass through an almost invisible crack or leak over a period of time to damage an engine severely.

Intake system Diesel-engine manufacturers use one or more of several types of air filters, but one thing is necessary for all—periodic checking (Fig. 2-1), daily for some types and more often in dusty or dirty conditions.

Rules for servicing air filters depend on type of filter, contamination of air, and type of application. Normal service intervals are specified by the manufacturer, but frequent inspection tells whether they are adequate for conditions under which an engine is operating. Some units have indicator lights which simplify checking.

General Maintenance Rules.
- Keep filter-to-engine connections tight.
- Keep air cleaner properly assembled so all joints are oil- and air-tight.
- Periodically inspect entire air-intake system, including any hoses and the intake manifold.
- Service oil-bath cleaners often enough to prevent oil from becoming thick with sludge. Use specified grade of oil and keep at proper level in cup; oil from an overfilled cup can be drawn into the engine, where it becomes fuel. The engine may overspeed and be damaged by this uncontrolled additional fuel.
- Never wash dry filter elements in fuel oil, gasoline, or solvents; use methods recommended by the manufacturer.
- Periodically inspect the rubber dust-unloading valve (if used), squeezing the rubber end of the valve to be sure it is not clogged.

Servicing Precleaners.
1. Remove and empty bowl (Fig. 2-16).
2. If unit has prescreener, blow or brush off chaff or other foreign matter.

Servicing Oil-Bath Cleaners.
1. Stop engine and remove oil cup.
2. Clean cup if more than ¼ in. of sediment has accumulated. Clean cup if oil has thickened or contains water.

Fig. 2-16 Periodically service the precleaner.

3. Remove caked dirt from bottom of cup, then wash with clean diesel fuel—*never* with gasoline, naphtha, benzene, or other highly flammable solvent.
4. Clean dirt tray (if used).
5. Refill cup to oil-level mark (never above) and replace.

Servicing Dry-Type Cleaners.
1. If cleaner has dust cap, empty it daily. If it has an automatic dust-unloading valve, check this daily to detect any clogging.

Maintenance of Power Systems

2. If unit has restriction indicator, clean whenever the indicator signals a restriction. Some indicators may stick, so watch conditions and total hours of operation.

3. If unit does not have restriction indicator, clean at recommended intervals and more often in dusty conditions.

4. Remove dusty elements and tap gently on heel of hand as element is being rotated. Do not tap on hard surface.

5. If tapping does not remove dust, use a compressed-air cleaning gun (pressure not over 30 psi), blowing up and down pleats and from inside to outside.

6. If element is oily or sooty, use compressed air as described above; then soak and gently agitate element in solution of lukewarm water and commercial filter-element cleaner or equivalent nonsudsy detergent.

7. Rinse element in clean water, shake off excess water, and allow element to dry thoroughly. Do not use compressed air to dry because the wet element can be ruptured easily by air pressure.

8. Inspect element for damage by placing light inside; discard filter if even slight damage is apparent.

9. Discard element after recommended service period (such as 1 year or six washings), if cleaning attempts fail, or if gasket is damaged or missing.

10. Clean the inside of the filter housing with a damp cloth, install the element in the housing with gasket and fin end first, and draw cover tight. Reset restriction indicator if one is used.

Testing Intake System. If air flow into an engine is restricted, the vacuum in the cylinders increases, possibly causing oil to be drawn into the combustion chambers and thereby increasing oil consumption. A vacuum can be detected by an auxiliary vacuum gauge available from the dealer or manufacturer. Proceed as follows:

Warm up the engine.

On engines with restriction indicators, remove the indicator, install a pipe-tee fitting, reinstall the indicator, and connect the gauge to the fitting. (Be sure to remove the pipe-tee after the test.)

On engines without restriction indicators, connect the gauge to the intake manifold.

Run engine at fast idle. Check the gauge reading against the manufacturer's specifications. Too high a reading indicates restriction in the air-intake system.

On engines with restriction indicators, check the indicator by gradually closing the air-intake opening with a board or metal plate, meanwhile watching the gauge. If the indicator is defective, replace it.

Exhaust system Service consists of cleaning carbon buildup from inner passages of the exhaust manifold, replacing defective mufflers, and keeping the entire system free from leaks—especially if the engine is in a machine with a cab or other operator enclosure into which deadly carbon monoxide gas might penetrate.

Turbocharger Turbochargers are exhaust-driven turbines which operate centrifugal compressors to increase air supply to combustion chambers. They run at very high speeds, which can range from 40,000 to more than 100,000 rpm, yet they are relatively simple devices and, if properly serviced, will operate with little or no attention.

Some troubleshooting hints are the following:

Noisy operation or vibration
  Bearings not being lubricated
  Leak in intake or exhaust manifolds
  Improper clearance between turbine wheel and housing
  Rotary vanes out of balance

Engine will not deliver rated power
  Clogged or leaking manifold system
  Foreign matter lodged in compressor, impeller, or turbine
  Excessive carbon buildup behind turbine wheel
  Seized bearing in rotating assembly

Oil-seal leakage
  Faulty seal
  Restriction in air cleaner or air intake creating suction

Inspecting turbochargers

▪ Regularly check mounting and connections for prompt detection of oil or air leakage.

- Check crankcase vent to be sure there is no restriction in air flow.
- Operate engine at approximate rated output and listen for unusual turbocharger noise. Abnormal shrill whine could mean bearings are about to fail, but do not confuse this with normal whine heard during so-called rundown as engine speed is reduced.
- Other unusual turbocharger noises could mean improper clearance between turbine wheel and housing.
- Check for unusual turbocharger vibration while engine is operating at rated output; dirty air may have pitted the rotary vanes and caused an imbalance.
- Check exhaust smoke under engine load conditions. Excessive smoke may indicate incorrect fuel-air mixture, which could be due to engine overload, engine malfunction, or turbocharger malfunction. Excessive smoke during acceleration is normal.

ELECTRICAL SYSTEM

Diesel-engine electrical-system maintenance consists largely of battery care, though attention should be given to such components as gauges, the drive belt for the alternator or generator, and the starting motor.

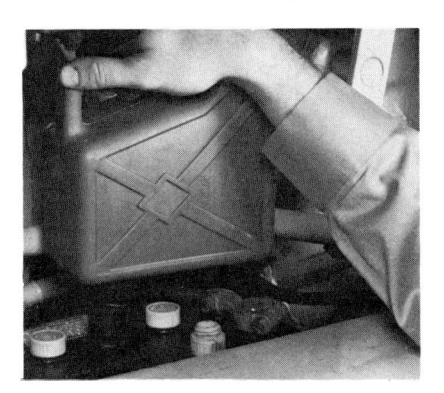

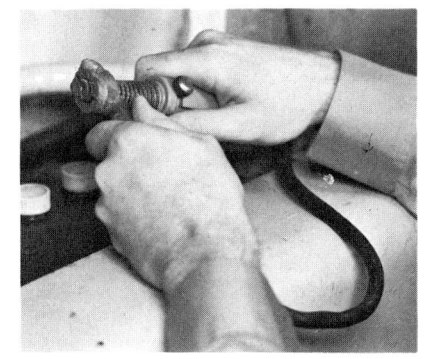

Fig. 2-17 Fill battery with distilled water.

Fig. 2-18 Clean battery terminals.

Batteries

Check electrolyte level every 50 hr, also inspecting vent holes in caps to be sure they are open. Electrolyte level should be at the bottoms of the filler necks and above the tops of the battery plates.

Use distilled water to bring the electrolyte to proper level (Fig. 2-17). Do not use hard water; dissolved minerals will leave deposits on the plates which interfere with chemical action.

Do not overfill. Excess electrolyte will escape through the vent holes and leave deposits on the tops of batteries.

Add only water unless electrolyte has been lost by spilling.

If water must be added to the battery in freezing weather, immediately run the engine long enough to assure proper mixing.

Specific gravity

Before adding water, check electrolyte specific gravity (Fig. 2-2), which should be between 1.215 and 1.270 when electrolyte temperature is 80°F (27°C).

If the hydrometer does not have temperature correction, add four gravity points (0.004) for each 10° above 80, and subtract four points for each 10° under 80.

Battery cells that differ more than 50 specific gravity points indicate an unsatisfactory battery condition caused by an internal defect, short circuit, or deterioration from extended use. The battery should be replaced.

Cleaning batteries

Clean batteries every 250 hr, more often if there are heavy deposits on the case and terminals.

Remove the cable terminals carefully; never pry against the battery case. Clean the terminals (Fig. 2-18) both inside and out, and dip them in a solution of 2 tablespoons of baking soda to 1 pt of water.

Clean the posts (Fig. 2-19) and top of the battery, and brush on a fresh mixture of baking soda and water. Apply this solution until foaming stops, then flush with clean water, making sure neither the solution or flushing water gets through vents in the caps (the electrolyte could be neutralized or contaminated). Dry the battery and posts with a clean cloth.

After replacing the cable terminals, apply a coating of petroleum jelly or light grease to the terminals and clamps to protect them from corrosion.

Be sure each cable terminal is on the proper post, or the generator or alternator may be damaged. Do not pound or force the clamps into place.

Recharging batteries

Cell caps should be removed when recharging batteries, and be sure to follow the charger manufacturer's instructions (Fig. 2-20).

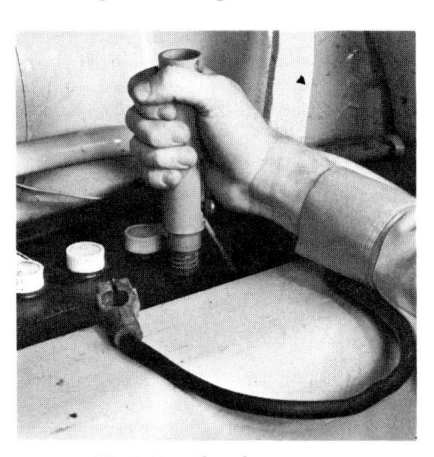

Fig. 2-19 Clean battery posts.

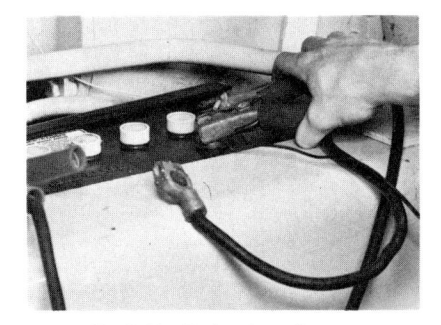

Fig. 2-20 Recharging a battery.

Badly sulfated batteries will not accept fast charging without danger of damage; they must be recharged slowly. The normal slow-charging period is 12 to 24 hr, but badly sulfated batteries may require 60 to 100 hr for complete recharge.

Booster batteries

If booster batteries are needed for cold-weather starting, be sure all electrical switches and accessories are turned off.

Connect cables first to the machine battery and then to the booster battery. Remove cables from the booster battery first. This guards against explosion of gas from the machine battery.

When attaching booster-battery cables, be sure to connect positive (+) to positive and negative (−) to a grounding point away from the battery. Do not use a 12-volt booster battery with a 6-volt machine battery.

Replacing batteries

Troubleshoot the battery before you replace it; many batteries returned for warranty claim or trade-in have nothing wrong except they are completely discharged.

Choose a replacement battery of an ampere-hour rating at least equal to the original. If accessories have been added, a larger size may be needed.

Remember the cheapest replacement battery is not always the most economical. Consider such factors as ampere-hour rating, construction, and length of warranty.

Safety

Always disconnect the battery ground strap or cable before working on any part of the electrical system of an engine to prevent injury from sparks, short circuits, or the engine accidentally starting.

Do not lay metal tools or other objects across the battery.

Hydrogen gas, released from the electrolyte (more is generated while charging), is extremely flammable, so keep sparks and fires away. Be sure the room where batteries are being charged is well ventilated.

Battery acid is harmful to skin and to most materials. Immediately remove any clothing on which acid is spilled.

If acid contacts skin, rinse the affected area 10 to 15 min with running water.

If acid splashes into the eyes, flush with running water 10 to 15 min. Also be sure to force the lids open while washing the eyes. Then, the victim should *immediately go to a doctor* for further treatment.

Acid spilled on the floor or on paint or metal surfaces of a machine can be neutralized by using a mixture of 1 lb of baking soda to 1 gal of water or 1 pt of household ammonia to 1 gal of water.

Alternator Alternators seldom give trouble unless abused, though bearings may fail and insufficient belt tension may reduce output. Check belt condition and tension occasionally, following the manufacturer's belt-tension specifications. If a new belt is installed, check tension after a few hours of operation and compensate for any stretch.

Disconnect the battery ground cable before working on or near the alternator or regulator. If alternator wiring is disconnected, be sure it is properly replaced before the battery is reconnected.

Failure to observe any of the following precautions may result in serious damage to the alternator, regulator, or electrical system:

Never ground a terminal or connect a jumper wire to any alternator terminal.

Never ground the alternator output terminal.

Never ground the alternator field terminal or the field circuit between the alternator and regulator.

Never disconnect or connect any alternator wires with batteries connected or the alternator in operation.

Never attempt to polarize the alternator or regulator.

Generator and starting motor If the engine has a generator instead of an alternator, its service is much the same as for the starting motor.

Most generators and starting motors have sealed bearings which require no lubrication, but some have oil cups for lubrication every 250 hr. Do not overlubricate; excessive amounts of oil can get on the brushes and cause failure.

Occasionally check the condition of the generator belt, and check its tension against manufacturer's specifications every 250 hr. If a new belt is needed, use an exact duplicate and recheck tension after a few hours of operation.

When checking generator brushes (Fig. 2-21), do not pull the brush-connector wire while the brush is under spring tension or allow the tension arm to snap down on a brush. Replace the brushes with new ones if they are worn so tension arms are against the brush holders instead of the brushes or if the brushes are worn to half their original length.

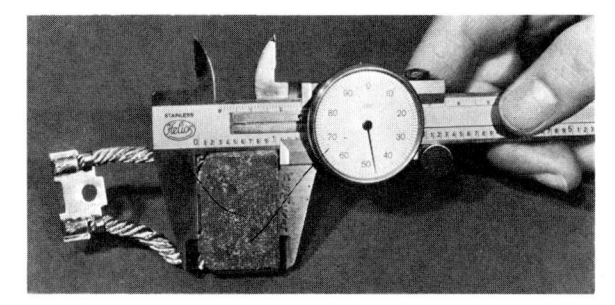

Fig. 2-21 Check generator brushes.

Check new brushes for binding action in the holders. If the brush binds, clean the holder with a cloth or sandpaper and wipe off dirt and grit; do not use a solvent, which could soften insulation on the wires.

Inspect the generator or starting motor for signs of overheating. If solder has been thrown against the cover band or inside of the housing, the unit requires expert care by an experienced technician.

If the commutator is scored or scratched, it must be machined by a competent technician. If merely glazed or dirty, lightly sand it with No. 00 sandpaper, then blow away dust and dirt. Do not use emery cloth, for remaining particles will cause arcing and rapid wear.

5-28   Maintenance of Power Systems

If any wire leads were disconnected from the starter or generator while servicing, polarize the generator before starting the engine, using the procedure outlined in the manufacturer's manual.

Gauges Frequently check all gauges when operating a diesel engine. If readings are not normal, stop operation until the cause is found. Keep in mind when looking for the cause that malfunction of the sending or receiving unit of the gauge itself is a possible cause for improper readings.

DIAGNOSING MALFUNCTIONS

Prompt diagnosis and correction of even minor engine malfunctions can, at least, improve operating efficiency and reduce downtime during prime working periods. At best, it may prevent costly major overhauls and permit scheduling corrective action until more convenient times.

Many malfunctions can be corrected in a relatively short time, perhaps during or after the regular work day. Others may require partial engine disassembly or attention from a competent technician, who should check the entire engine as he corrects the immediate trouble.

Here are some symptoms and the malfunctions they suggest.

Engine starts hard or will not start
Fuel system
  Fuel tank empty
  Wrong type of fuel
  Water, dirt, or air in fuel system
  Fuel lines clogged or restricted
  Fuel filter restricted
  Faulty fuel-transfer pump
  Faulty injection pump
  Faulty injection nozzles
  Fuel shut-off is engaged
  Air-intake system restricted
  Air leak in suction side of fuel system
Lubrication
  Wrong oil viscosity for ambient temperature
Electrical system
  Battery weak or dead
  Corroded or loose battery cables
  Cranking speed too slow

Uneven running
Engine
  Faulty valve clearance
  Stuck or burned valves
  Leaking cylinder-head gasket
  Low compression
  Worn or broken compression rings
  Faulty timing
  Engine overheating

Frequent stalling
Engine
  Valves sticking or burned
  Incorrect timing
  Engine overheating

Uneven running or frequent stalling
Fuel system
  Wrong type of fuel
  Water, dirt, or air in fuel system
  Air leak in suction side of fuel system
  Fuel line clogged or restricted
  Fuel filter restricted
  Faulty fuel-transfer pump
  Faulty injection pump
  Faulty injection nozzles
  Injection-nozzle leak-off lines clogged
  Injection pump out of time

Engine misses
Engine
  Weak valve springs
  Faulty valve clearance
  Burned, warped, pitted, or sticking valves
  Low compression
  Worn camshaft lobes
  Engine overheating
Fuel system
  Water, dirt, or air in fuel
  Wrong type of fuel
  Faulty injection nozzles
  Faulty injection pump
  Faulty fuel-transfer pump

Engine lacks power
Engine
  Dirty air-intake system
  Blown cylinder-head gasket
  Worn camshaft lobes
  Burned, warped, pitted, or sticking valves
  Faulty valve clearance
  Faulty valve timing
  Weak valve springs
  Low compression
  Engine overheating
Fuel system
  Wrong type of fuel
  Water, dirt, or air in fuel system
  Air leak in suction side of fuel system
  Fuel line clogged or restricted
  Fuel filter restricted

Speed-control linkage improperly
   adjusted
Faulty fuel-transfer pump
Faulty fuel-injection pump
Faulty fuel-injection nozzles
Injection-nozzle leak-off line clogged
Injection pump out of time
Clogged manifold system

Black or gray exhaust smoke
Engine
   Engine overloaded
   Faulty engine timing
   Restricted air cleaner
   Dirty air-intake system
   Faulty turbocharger
Fuel system
   Wrong type of fuel
   Excessive fuel delivery
   Faulty injection nozzles
   Faulty or improperly adjusted aneroid
   Injection-nozzle leak-off line clogged
   Injection pump out of time

White exhaust smoke
Engine
   Low compression
Fuel system
   Improper fuel
   Faulty injection nozzle
   Injection pump out of time

Slow acceleration
Engine
   Components worn
   Sticky governor
Fuel system
   Improper fuel
   Faulty injection nozzle

Abnormal engine noise
Engine
   Excessive valve clearance
   Worn cam followers
   Bent push rods
   Worn rocker-arm shafts
   Worn main or connecting-rod bearings
   Foreign material in combustion chamber
   Worn piston-pin bushing and pins
   Scored pistons
   Faulty engine timing
   Excessive crankshaft-end play
   Loose main-bearing caps
   Worn timing gears
   Worn oil-pump gears
   Broken oil-pump shaft
   Engine oil level low
   Camshaft oil-pump gear worn or broken
   Gears worn or broken
   Fan striking radiator
   Faulty turbocharger bearings

Excessive oil consumption
Engine
   External oil leaks
   Excessive engine speeds
   Piston rings not seated
   Piston rings worn or broken
   Piston rings sticking in grooves
   Scored pistons
   Faulty piston-ring tension
   Piston-ring gaps not staggered
   Excessive ring-groove wear
   Oil-return slots in pistons clogged
   Worn valve guides or valve stems
   Restricted crankcase breather
   Restricted air-intake system
   Excessive main or connecting-rod
      bearing clearance
   Worn crankshaft thrust bearing
      (misaligned piston and rod)
   Faulty front or rear crankshaft seal
   Excessive loss through vent tube
   Faulty turbocharger oil seal
   Crankcase oil too thin
   Oil level too high

Low oil pressure
Engine
   Oil level low
   Clogged oil-pump inlet screen
   Faulty oil pump
   Oil too thin
   Internal oil leakage
   Faulty regulating-valve operation
   Faulty oil-pressure indicator

High oil pressure
   Oil too heavy
   Faulty regulating-valve operation
   Faulty oil-pressure indicator

Engine overheats
Engine
   Defective head gasket
   Incorrect engine timing
   Scored pistons
   Engine oil level low
   Engine overloaded
Cooling system
   Low coolant level
   Air in coolant
   Faulty thermostat
   Cooling system limed
   Faulty water pump
   Radiator dirty or clogged
   Loose or broken fan belt
   Bent or broken fan blade
   Faulty radiator pressure cap
Fuel system
   Improper fuel
   Excessive fuel delivery
   Faulty injection-pump timing

Engine runs cold
Faulty thermostat
Faulty temperature gauge

Water pump leaks or is noisy
Worn seal and/or shaft or bearing
Worn or broken gasket
Faulty impeller

Oil in coolant or coolant in crankcase
Leaking head gasket
Cracked cylinder-block water jacket
Cracked cylinder liner
Cylinder-liner packings leaking

Turbocharger malfunctions
Abnormal whine
  Faulty bearings
Other noise; vibration
  Defective bearing lubrication
  Leak in intake or exhaust manifolds
  Improper clearance between turbine wheel and housing
Engine does not deliver rated power
  Clogged manifold system
  Foreign matter in compressor, impeller, or turbine
  Excessive buildup in compressor
  Leak in engine intake or exhaust manifolds
  Bearing seizure in rotating assembly
Oil-seal leakage
  Faulty seal
  Suction caused by restricted air cleaner or air intake

Battery malfunctions
Undercharged battery
  Excessive load from added accessories
  Excessive engine idling
  Low charging-system output
  Current leakage
Low battery output
  High resistance in circuit
  Low electrolyte level
  Low specific gravity
  Defective cell
  Cracked or broken case
  Faulty terminal connections
  Battery too small
Battery uses too much water
  Cracked case
  Battery being overcharged
  Defective battery

Starting circuit If starting motor does not operate, connect voltmeter to solenoid S terminal and a good ground. With key switch on, press start switch.
Voltmeter registers battery voltage
  Defective starting motor
  Defective starter solenoid switch

Voltmeter does not register
  Defective key switch or start switch
  Faulty starting-circuit relay
  Faulty wiring between key switch and starting-circuit relay
  Faulty wiring between battery and solenoid S terminal
  Faulty start safety switch
Starting motor will not spin or engine will not crank
  Faulty battery
  Burned or faulty solenoid switch contacts
  Open, shorted, or grounded solenoid-switch pull-in windings
  Poor brush contact or worn-out brushes
  Burned commutator
  Commutator mica too high
  Open or grounded field winding
  Faulty brush-spring tension
  Grounded brush holder
Solenoid switch chatters
  Low battery
  Poor connection
  Open solenoid hold-in circuit
Starting motor spins but does not crank engine
  Faulty overrunning clutch pinion
  Broken drive lever
  Broken drive-lever pivot bolt
  Broken magnetic-switch plunger hook
  Faulty overrunning clutch
Engine cranks slowly
  Too-heavy transmission oil
  Low battery
  High resistance in battery cables
  Faulty battery-terminal connections
  Burned or faulty solenoid-switch contacts
  Faulty brush contact or worn-out brushes
  Burned commutator
  Commutator mica too high
  Shorted or grounded armature coil
  Faulty brush-spring tension
  Armature rubbing pole core
Starting motor keeps running
  Defective solenoid
  Defective start switch
  Shorted wiring
Abnormal noise while cranking
  Armature interfering with stationary components
  Starting-motor drive gear worn

Charging circuit
Low charging-system voltage
Faulty wiring
Low alternator amperage output
Defective regulator
Defective battery
Slipping drive belt

Low charging-system output
  Slipping drive belt
  Defective battery
  Grounded, shorted, or open field windings
  Defective rectifier bridge
  Defective diode trio
  Defective stator
  Defective regulator
High charging-system voltage
  High resistance at regulator connections
  Defective regulator
  Shorted field windings
  Grounded brush-sleeve clip
Noisy alternator
  Worn or defective bearings
  Defective drive belt
  Loose mounting or drive belt
  Pulley not properly aligned

Gauges
A gauge does not register
  No current to gauge
  Faulty sending or receiving unit
  Poor ground connection
  Connecting wire grounded to unit

A gauge consistently registers too high
  Faulty connection
  Broken connecting wire
  Poor ground at sending unit
  Failure of gauge or sender (usually sender)
Fuel gauge always shows empty
  Poor ground at receiver
  No current to receiver
  Grounded wire between sender and receiver
  Hole in sender float
Fuel gauge always shows full
  Poor ground at sender
  High resistance or open circuit between sender and receiver
  Defective sender or receiver

Fuel shut-off solenoid
High current draw
  Shorted windings
Low or no current draw
  High resistance in internal connection or wire
  Open circuit windings

Chapter **3**

Maintenance of Gasoline Power Systems

C. L. FRICKE
Engine Service Manager, Briggs & Stratton Corp., Milwaukee, Wis.

At almost every engine service meeting, someone asks, "How long should an engine last?" In all probability, the speaker answers by talking about the importance of preventive maintenance. Most small air-cooled engines are capable of operating 1000 hr with only routine maintenance and occasional repair as a result of normal wear. However, only a small percentage of air-cooled engines operate for more than 500 hr, because maintenance is marginal.

No contractor would consider paying two times the regular price of a tool at the time of purchase. Yet, this is exactly what happens when improper maintenance robs half the expected life of that tool. Some years ago, a member of the Outdoor Power Equipment Institute said that the equipment common to his industry was built to deliver 10 years of dependable service. He added that few owners will receive more than 4 years of service because their maintenance procedures were incomplete, improper, or nonexistent.

In the construction field, contractors must be concerned about the cost of equipment downtime. Lost time, resulting from equipment which does not operate, is intangible. The cost of lost time is not easily measured. But it is generally conceded that the cost of lost time far exceeds the cost of a preventive maintenance program.

Manufacturers of small air-cooled engines are particularly concerned about their product, since they realize that the small engine is usually an orphan on construction job sites. Everybody operates the small engines, but it seems no one is really responsible for maintenance.

In the construction industry, a small engine is rarely operated by only one man. Instead, engine-powered portable tools are moved from one job site to another. During any given day, several men may operate the same engine. For this reason, it is usually impractical to attempt to keep a log of the number of hours the engine is operated. Therefore, most contractors find it best to have one man perform preventive maintenance on all small engines on a calendar basis. In this way, preventive maintenance is performed every third or fourth day or perhaps on a certain day of each week.

In discussing the matter of engine life and preventive maintenance with contractors throughout the country, it became obvious that contractors who are pleased with the performance of their small air-cooled engines are those who assigned the responsibility of preventive maintenance to a single person. They charged one individual with performing preventive maintenance as recommended by the engine manufacturer.

Most small engine downtime results from one or more of the following conditions:

    1. Wear, caused by dirt and abrasives which enter the engine because of improper air-cleaner maintenance

2. Scoring, seizures, or parts breakage which occurs when an engine is operated without a sufficient amount of oil in the crankcase or with a lubricant of improper viscosity

3. Damage from abrasives which enter the engine when a spark plug has been improperly cleaned on an abrasive blast-cleaning machine

4. An excessive accumulation of combustion deposits which rob power and shorten valve life

5. Overheating, resulting from plugged cooling fins on the cylinder and cylinder head

6. Damage caused by improper storage during periods when the engine is not being operated

The following pages discuss these areas in greater detail.

AIR CLEANER—FUNCTION AND DESIGN

The average small engine uses about 10,000 gal of air for each gallon of fuel. It is obvious that the air which an engine breathes must be clean. If the air is contaminated with dust and dirt, abrasives eventually find their way into the engine crankcase and contaminate the engine oil. Since most small air-cooled engines do not include an oil filter, abrasives will circulate to internal moving parts again and again.

Briggs & Stratton engines, from 2 to 9 hp, are equipped with an Oil-Foam air cleaner, utilizing a reusable polyurethane foam element housed in a metal cannister. The foam element is oiled so that the millions of microfine cells trap dirt before it can enter the engine. Oil-Foam air cleaners are designed in such a way that sealing lips of the foam element actually become a gasket when the air cleaner is assembled. This design prevents dirt from bypassing the oiled form element.

SERVICING THE OIL-FOAM AIR CLEANER

Oil-Foam air cleaners should be cleaned and reoiled after every 25 hr of operation under normal conditions—more frequently under extremely dusty conditions. Engines of 2 to 5 hp use a rectangular-shaped Oil-Foam air cleaner of the type shown in Fig. 3-1. Service is performed as follows:

1. Remove mounting screw.
2. Remove air cleaner carefully to prevent dirt from entering carburetor.
3. Take air cleaner apart.
4. *a.* Wash foam element in kerosene or liquid detergent and water to remove dirt.
   b. Wrap foam in cloth and squeeze dry.
   c. Saturate foam in engine oil. Squeeze to remove excess oil.
   d. Assemble air-cleaner parts—fasten to carburetor with screw.

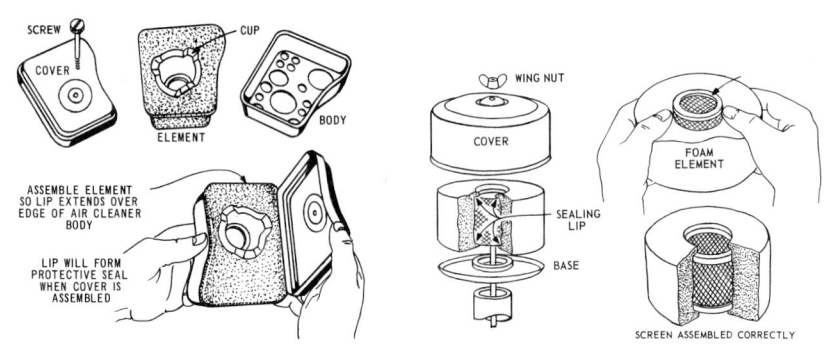

Fig. 3-1 Air-cleaner maintenance procedure, 2 to 5 hp engines.

Fig. 3-2 Air-cleaner maintenance procedure, 6 to 9 hp engines.

Engines of 6 to 9 hp use a cylindrical-shaped Oil-Foam air cleaner of the type shown in Fig. 3-2. Service is performed as follows:

1. Remove wing nut and cover.
2. Lift foam element from base.

3. Push down foam element as shown and pull out screen.
4. *a.* Wash foam element in kerosene or liquid detergent and water to remove dirt.
   b. Wrap foam in cloth and squeeze dry.
   c. Saturate foam in engine oil. Squeeze to remove excess oil.
   d. Put screen inside element. Be sure sealing lip is over end of screen (top and bottom).
5. Reassemble parts as shown. Assemble wing nut finger tight.

Many engines from 10 to 16 hp are equipped with a heavy-duty dual-element air cleaner, utilizing both a washable paper cartridge and an oil-foam precleaner sleeve.

SERVICING THE DUAL-ELEMENT AIR CLEANER

Clean and reoil the foam precleaner sleeve at 3-month intervals or every 25 hr, whichever occurs first (Fig. 3-3).
1. Remove wing nut and cover.
2. Remove foam precleaner sleeve by sliding it off the paper cartridge.
3. *a.* Wash foam in kerosene or liquid detergent and water to remove dirt.
   b. Squeeze foam dry.
   c. Oil foam with 1 oz of engine oil. Squeeze to distribute oil evenly.
4. Install foam precleaner sleeve over paper cartridge.
5. Reassemble cover and wing nut. Assemble wing nut finger tight.

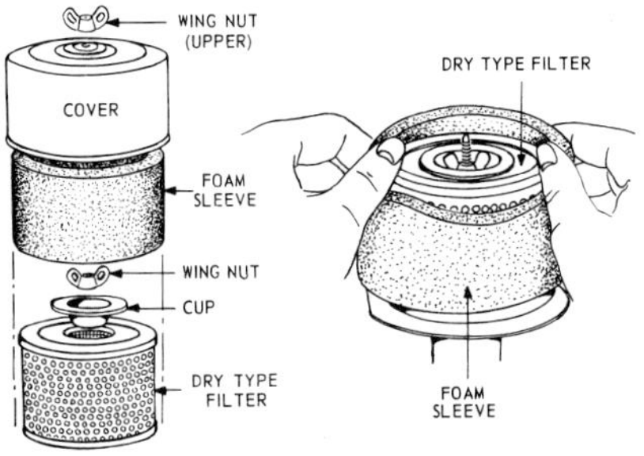

Fig. 3-3 Dual-element air-cleaner maintenance procedure.

Yearly or every 100 hr, whichever occurs first, remove the paper cartridge. Clean the cartridge by tapping gently on a flat surface. If the cartridge is very dirty, replace the cartridge or wash it in liquid detergent and water. Rinse the cartridge until the water remains clear. The cartridge must be air dried *thoroughly* before using. NOTE: Service the air-cleaner assembly more frequently under dusty conditions.

LUBRICATION AND FUEL RECOMMENDATIONS

Many engines are damaged because they are operated without a sufficient amount of oil in the crankcase or because the oil is not of the proper viscosity. The problem generally occurs because the crankcase oil level is not checked at reasonable intervals. It is good practice to check the oil level each time an engine is refueled.

Use a high-quality oil classified "For Service SC, SD, SE, or MS." This is the same type of oil usually recommended for use in truck, automotive, and large stationary engines. Such oils do include a detergent, which helps to keep internal parts cleaner by retarding the formation of gum and varnish deposits. Nothing should be added to the recommended oil.

Select oil viscosity for the season in which an engine will be operated. Briggs & Stratton recommend the following: In summer (above 40°F), use SAE 30; if not available, use SAE 10W-30 or SAE 10-W 40. In winter (under 40°F), use SAE 5W-20 or SAE 5W-30; if not available, use SAE 10W or SAE 10W-30. Below 0°F, use SAE 10W or SAE 10W-30 diluted 10 percent with kerosene.

Use clean, fresh, lead-free or leaded regular-grade automotive gasoline. Purchase gasoline in relatively small quantities, so it can be used within 60 days. In this way, the fuel will remain fresh and the volatility will be correct for the season.

BLAST CLEANING SPARK PLUGS

Cleaning spark plugs on an abrasive blast-cleaning machine has long been an accepted field practice. Many conscientious repairmen and engine owners blast-clean spark plugs frequently, simply as a matter of routine maintenance. They believe that blast cleaning spark plugs is a way to salvage plugs, decrease repair costs, and improve engine performance. Unfortunately, this is often false.

We investigated complaints on premature wear and short engine life and found that spark plugs had been cleaned on an abrasive blast-cleaning machine. We then used an abrasive blast-cleaning machine to clean spark plugs on several new engines. After only a few hours of operation, extreme wear had occurred on internal engine parts. The wear had been caused by abrasive grit which remained within the plug after blast cleaning. During operation, grit had fallen into the combustion chamber, then worked its way past the piston rings toward the crankcase.

When a spark plug has been cleaned by the abrasive blast-cleaning method, some of the abrasive grit remains tightly packed within the plug after the cleaning operation. Although instructions with spark plug cleaning machines call for solvent cleaning to remove oil deposits before blasting, contacts with local automotive service stations indicate that few, if any, cleaning machine operators degrease the plugs. Further, simply blowing out the plugs with a forceful blast of air after blast cleaning is not adequate to remove all the tightly packed grit. In tests, spark plugs that contained even a small amount of grit caused considerable wear after a few hours of operation.

Fig. 3-4 About 200 mg of abrasive blast-cleaning grit will fill a circle approximately the size of a nickel. This amount can easily remain within a spark plug after blast cleaning.

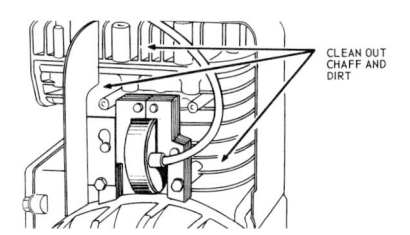

Fig. 3-5 Area for blower housing cleaning.

To get an idea of the way in which spark plugs are being blast-cleaned in the field, we had two plugs cleaned at each of five local automotive service stations. We then washed them out in gasoline to remove any abrasive grit which remained after cleaning. All the spark plugs contained enough grit to cause engine wear. The residue of grit in the 10 plugs ranged between 32 to 372 mg. To give some idea of the quantities involved, 200 mg of grit will fill a circle the size of a nickel (Fig. 3-4).

Engines that have been damaged by abrasive blast-cleaning grit usually have very high piston ring wear, scratched or scored cylinder bores, and extreme amounts of smooth wear on the crankshaft crankpin and bearing journals. The grit, which can be introduced into an engine as a result of improperly cleaned spark plugs, may well account for cases of unexpected wear in which there is no evidence of abuse or neglect by the operator.

In view of our findings, we most urgently recommend that dirty spark plugs be cleaned only in a solvent, or by scraping the electrodes and then regapping. If a dirty spark plug cannot be cleaned in this manner, it should be discarded and a new plug used.

CLEAN OUT COMBUSTION CHAMBER

Industrial engines are often operated at constant speed and at relatively constant load. Under these conditions, the use of leaded automotive fuels results in a gradual buildup of deposits in combustion chambers. When combustion deposits accumulate in excessive amounts, engines lose power. In addition, particles from the deposits may lodge between exhaust valves and seats, causing valves to burn. Removing combustion deposits, after every 100 to 300 hr of operation, will restore power and increase valve life.

Clean out the combustion chamber as follows:

1. Remove cylinder head screws. Be sure to note if the screws are of different lengths and have steel washers because they must be replaced in their original positions.

2. Turn the crankshaft until the piston is at the top of the cylinder bore and both valves are closed. Scrape and wire-brush the combustion deposits from the cylinder head and combustion chamber.

3. Reuse the cylinder head gasket only if it is in good condition. Reassemble the cylinder head. Turn each screw with a wrench until the screw head is lightly seated.

4. Use a socket wrench with a 6-in. handle and turn all the screws one-quarter turn. Run the engine for about 5 min and retighten all the screws approximately one-quarter turn.

CLEAN COOLING SYSTEM

Air-cooled engines require a generous flow of cooling air, which is directed across cooling fins on the cylinder and cylinder head. Over an extended period of time, dirt and chaff may clog cooling fins, interrupting the flow of cooling air. When this occurs, the engine may overheat severely. Periodically, remove the blower housing and clean out dirt and debris (Fig. 3-5).

STORAGE INSTRUCTIONS

Engines to be stored over 30 days should be completely drained of fuel to prevent gum deposits from forming on essential carburetor parts, the fuel filter, the fuel lines, or the tank.

1. Remove all fuel from the fuel tank. Run the engine until it stops from lack of fuel. Remove the small amount of fuel that remains in the sump of the tank by absorbing it with a clean dry cloth.

2. While the engine is still warm, drain the oil from the crankcase. Refill with fresh oil.

3. Remove a spark plug, pour 1 oz (2 or 3 tablespoons) of engine oil into the cylinder, and crank slowly to distribute the oil. Replace the spark plug.

4. Clean dirt and chaff from the cylinder, the cylinder head fins, and the blower housing.

There is no magic royal road to success with small engines. Engine life, performance, and dependability will be in proportion to maintenance. A preventive maintenance program is a must for contractors who cannot afford unnecessary engine downtime.

Chapter 4

Principles and Maintenance of Hydraulic Systems

MARVIN SHERMAN AND LINCOLN BURROWS
Sperry Vickers, Division of Sperry Rand Corporation, Troy, Mich.

Hydraulic power has long been one of the most useful and versatile methods of actuating and controlling construction machinery. The brute force and very precise control provided by hydraulic systems have taken much of the work out of operating earthmoving and material-handling equipment and self-propelled vehicles of all kinds. As applications of this form of power grow, so does the need for a better understanding of hydraulic system operation and maintenance.

FUNDAMENTALS

A hydraulic system is used to transfer mechanical energy from one place to another, using pressure energy as the medium. Mechanical energy, driving the hydraulic pump, is converted to pressure energy and kinetic energy in the fluid. This is reconverted to mechanical energy to move a load, such as a bucket, boom, or blade.

The power of the fluid flowing under pressure can be applied directly to the load to produce rotary or linear motion without gears, chains, belts, magnets, or commutators. This direct application of force by an incompressible, nearly frictionless fluid is one of the main factors contributing to the high efficiency of a hydraulic system. Since the fluid also serves as a lubricant, it helps reduce wear and extend the life of hydraulic components.

When hydraulic fluid is confined in a pipe or hose it acts as if it were a steel rod. Power applied at one end of the line causes equal power to be applied at the other end. The power source, that is, the pump, can be connected to the load with relative ease. Hydraulic pipe and hose can be installed around corners, along booms, and through panels. Hydraulic drive costs are competitive with those of mechanical systems and are lower than those for electrical drives on a cost-per-horsepower basis.

The high power-to-weight ratio of hydraulic components is one factor. The economic advantage of hydraulics is particularly evident where a single hydraulic pump is used to drive several hydraulic motors at various locations on a construction vehicle. The high power-to-weight ratio of hydraulic pumps and motors often permits them to be installed in places on a vehicle that would be too small for electrical or mechanical devices having equal capabilities. For example, the cubic space required for a 100-hp electric motor is almost 14 times that required for a 100-hp hydraulic motor.

All hydraulic systems are basically the same regardless of the application. A typical circuit (Fig. 4-1) includes four basic components: (1) a tank (reservoir) to hold the fluid, (2) a pump to force the fluid through the system, (3) valves to control fluid pressure, flow rate, and direction, and (4) one or more actuators to convert the energy of fluid movement into mechanical force to perform the work. The actuators are either cylinders for linear motion or motors for rotary motion.

These basic components can be combined in a variety of ways to achieve various motions and, if necessary, at various rates and various locations on a machine. Circuit drawings or diagrams show how the hydraulically operated machine works. These diagrams are also essential for installation of the hydraulic components, for general maintenance work, and for troubleshooting the system.

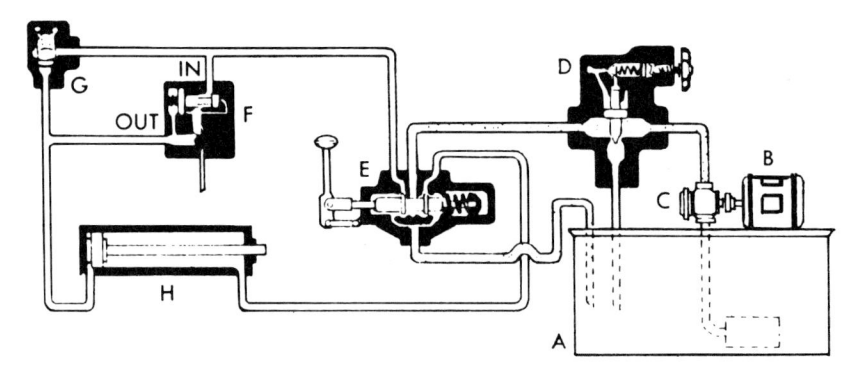

List of Components

A—Reservoir
B—Electric Motor
C—Pump
D—Maximum Pressure (Relief) Valve
E—Directional Valve
F—Flow Control Valve
G—Right-Angle Check Valve
H—Cylinder

Fig. 4-1 Typical hydraulic circuit is shown here in both pictorial forms and as it would appear using ANSI symbols.

Figure 4-2 shows how graphic symbols are used in most circuit diagrams. These symbols, specified by the American National Standards Institute (ANSI), show the function of system components rather than their shape, internal construction, or operation. The same type of component is often represented by somewhat different symbols, depending on its use in the system.

PUMPS

The pump is the heart of a hydraulic system and, right or wrong, gets most of the attention when a problem develops in a system. Of all the components, it is the most susceptible to failure. It often runs continuously through a work shift, and it is the first component exposed to fluid contaminants when they leave the reservoir. Because of its high performance and relatively small size, a hydraulic pump is more vulnerable to damage from improper operation or maintenance. Still, with proper operation and maintenance practices, the pump can be a most reliable component.

Some equipment operators who stress good maintenance can predict within a few hours of operation when a pump will need rebuilding. They include rebuilding as part of a maintenance program and have very few actual failures.

Classification and rating Practically all pumps used in today's hydraulic systems are positive displacement units. That is, they provide a given amount of fluid for every revolution, stroke, or cycle. Fluid output, except for internal leakage losses, is independent of system pressure.

Pumps are classified as fixed- or variable-displacement units, with displacement depending on the working relationship of internal operating elements. In a *fixed-displacement pump*, the relationship cannot be changed, and pump delivery can be varied only by changing pump speed. The amount of flow going to various parts of the hydraulic circuit is controlled with various valves. In a *variable-displacement pump*, displacement can be varied by an integral controlling device which adjusts the physical relationship of the pump's operating parts.

Pumps are generally rated on the basis of pressure and volumetric output. The *pressure rating* indicates how much pressure the pump can withstand safely for a given time without damage to its parts. This, in turn, determines how much load a system can handle.

Volumetric output, or flow, is given in gallons per minute (gpm) at a designated drive shaft speed. The SAE ratings for vane pumps on construction machinery are designated at 1200 rpm and 100 psi outlet pressure. Other frequently used terms for pump flow rating include *delivery rate, capacity,* and *pump size*. A third means of rating is on the basis of pump *displacement*. This refers to the amount of fluid delivered in one revolution of a pump's rotating members. Displacement is expressed in cubic inches per revolution.

Most hydraulic pumps are the rotary type. That is, a rotating assembly carries the fluid from the pump inlet to the outlet. The type of element that transfers the fluid further establishes another basic approach to pump classification. These elements can be in the form of *vanes, pistons,* or *gears*, which are the three most common types of pumps.

Figure 4-3 shows the configuration of a simple vane pump. A cylindrical rotor with movable vanes in radial slots rotates in a circular housing. As the rotor turns, centrifugal force drives the vanes outward so that they are always in contact with the inner surface of the housing. This action promotes both long life and efficiency since the vanes move automatically to compensate for vane tip wear.

Pumping chambers are formed between succeeding vanes carrying fluid from the inlet to the outlet. A partial vacuum is created at the inlet as the space between vanes increases. Then, fluid is squeezed out at the outlet as the pumping chamber size decreases.

The pump shown in Fig. 4-3 is referred to as *unbalanced* because high pressure is generated on only one side of the rotor shaft. Most of today's vane pumps are of the *balanced* design shown in Fig. 4-4. In this case, an elliptical housing forms two separate pumping chambers on opposite sides of a rotor so that the side loads cancel out, increasing bearing life and permitting higher operating pressures.

Piston pumps A piston-type pump is really a rotary-reciprocating pump. Several pistons—usually seven or nine—reciprocate in rotating cylinder barrels. The pistons retract while passing the inlet port to create a vacuum and permit oil to flow into the pumping chambers. They extend at the outlet to push the oil into the system.

Most piston pumps are classified as *radial* and *axial*. In a radial piston pump (Fig. 4-5) the pistons are arranged like wheel spokes in a short cylindrical block. This cylinder block is rotated by the drive shaft inside a circular housing. The block turns on a stationary pintle that contains the inlet and outlet ports. As the cylinder block turns, centrifugal force slings the pistons outward and they follow the circular housing. The housing centerline is offset from the cylinder block centerline. The amount of eccentricity between the two determines the piston stroke and therefore the pump displacement. Controls are applied to change the housing location and thereby vary the pump delivery from zero to maximum.

In axial piston pumps, the pistons stroke axially, or in the same direction as the cylinder block centerline. Axial piston pumps can be of in-line or angle design.

Gear pumps External and internal gear pumps make up most of the types of pumps classified under gear pumps. *External gear pumps* have the largest application in power transmission. *Internal gear pumps* are used for automatic transmissions and power steering units in automobiles.

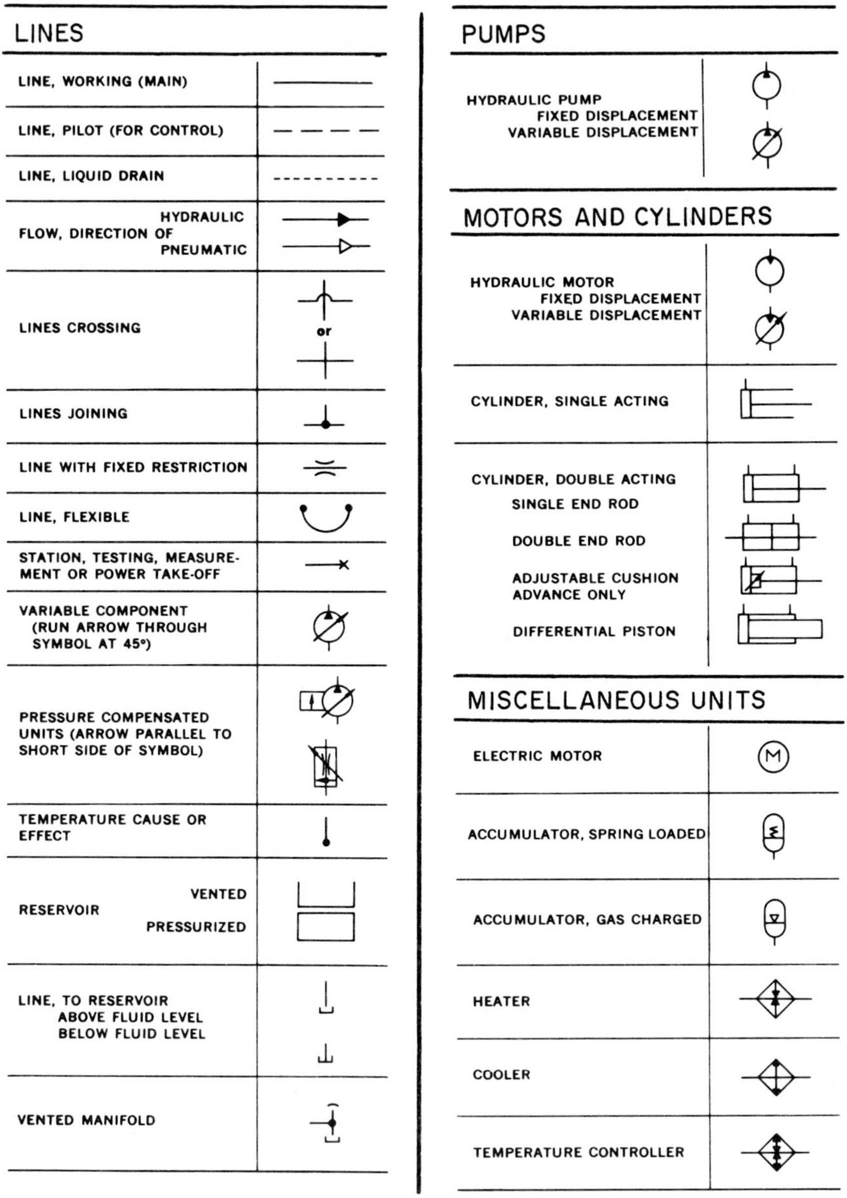

LINES

LINE, WORKING (MAIN)	————
LINE, PILOT (FOR CONTROL)	— — — —
LINE, LIQUID DRAIN	- - - - - -
FLOW, DIRECTION OF HYDRAULIC PNEUMATIC	—▶ —▷
LINES CROSSING	─┼─ or ─┬─
LINES JOINING	─┴─
LINE WITH FIXED RESTRICTION	─⋈─
LINE, FLEXIBLE	⌣
STATION, TESTING, MEASUREMENT OR POWER TAKE-OFF	—✕
VARIABLE COMPONENT (RUN ARROW THROUGH SYMBOL AT 45°)	⌀
PRESSURE COMPENSATED UNITS (ARROW PARALLEL TO SHORT SIDE OF SYMBOL)	
TEMPERATURE CAUSE OR EFFECT	│•
RESERVOIR VENTED PRESSURIZED	⊔ ⊟
LINE, TO RESERVOIR ABOVE FLUID LEVEL BELOW FLUID LEVEL	
VENTED MANIFOLD	

PUMPS

HYDRAULIC PUMP FIXED DISPLACEMENT VARIABLE DISPLACEMENT	◯ ⌀

MOTORS AND CYLINDERS

HYDRAULIC MOTOR FIXED DISPLACEMENT VARIABLE DISPLACEMENT	◯ ⌀
CYLINDER, SINGLE ACTING	
CYLINDER, DOUBLE ACTING SINGLE END ROD DOUBLE END ROD ADJUSTABLE CUSHION ADVANCE ONLY DIFFERENTIAL PISTON	

MISCELLANEOUS UNITS

ELECTRIC MOTOR	Ⓜ
ACCUMULATOR, SPRING LOADED	
ACCUMULATOR, GAS CHARGED	
HEATER	◇
COOLER	◇
TEMPERATURE CONTROLLER	◇

Fig. 4-2 Symbols for hydraulic-system representation as set down by ANSI.

MISCELLANEOUS UNITS (Cont.)

FILTER, STRAINER	
PRESSURE SWITCH	
PRESSURE INDICATOR	
TEMPERATURE INDICATOR	
COMPONENT ENCLOSURE	
DIRECTION OF SHAFT ROTATION (ASSUME ARROW ON NEAR SIDE OF SHAFT)	

METHODS OF OPERATION

SPRING	
MANUAL	
PUSH BUTTON	
PUSH-PULL LEVER	
PEDAL OR TREADLE	
MECHANICAL	
DETENT	
PRESSURE COMPENSATED	
SOLENOID, SINGLE WINDING	
REVERSING MOTOR	

PILOT PRESSURE REMOTE SUPPLY	
INTERNAL SUPPLY	

VALVES

CHECK	
ON-OFF (MANUAL SHUT-OFF)	
PRESSURE RELIEF	
PRESSURE REDUCING	
FLOW CONTROL, ADJUSTABLE— NON-COMPENSATED	
FLOW CONTROL, ADJUSTABLE (TEMPERATURE AND PRESSURE COMPENSATED)	
TWO POSITION TWO CONNECTION	
TWO POSITION THREE CONNECTION	
TWO POSITION FOUR CONNECTION	
THREE POSITION FOUR CONNECTION	
TWO POSITION IN TRANSITION	
VALVES CAPABLE OF INFINITE POSITIONING (HORIZONTAL BARS INDICATE INFINITE POSITIONING ABILITY)	

5-44 Maintenance of Power Systems

The external gear pump (Fig. 4-6) consists of two meshed gears in a closely fitted housing. The inlet and outlet ports are opposite each other. One gear is powered and, in turning, drives the other. In moving past the inlet, gear teeth create a partial vacuum. Oil is drawn from the inlet and carried to the outlet in the pumping chambers formed between the gear teeth and the housing. The internal gear pump (Fig. 4-7) has an inner

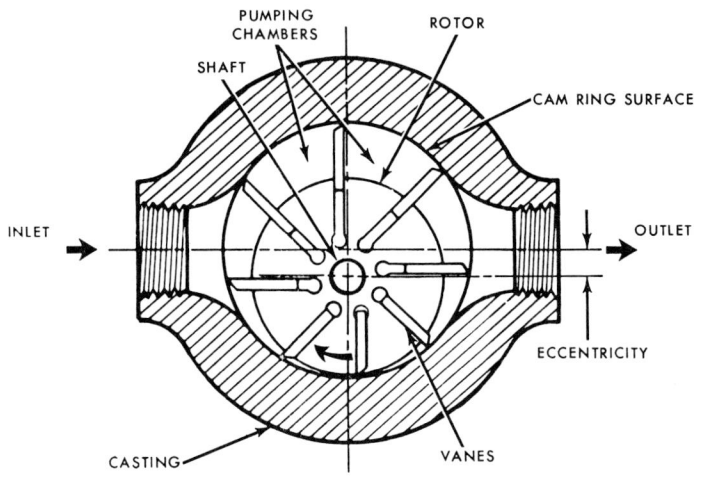

Fig. 4-3 Vane pump—unbalanced design.

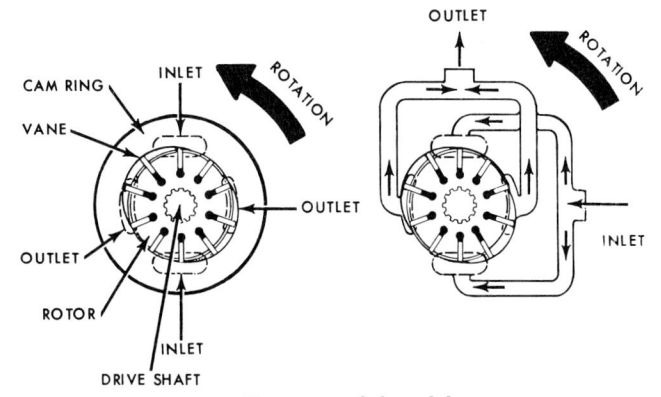

Fig. 4-4 Vane pump—balanced design.

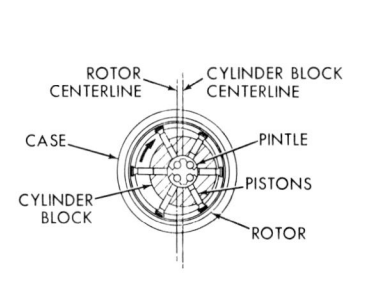

Fig. 4-5 Radial piston pump.

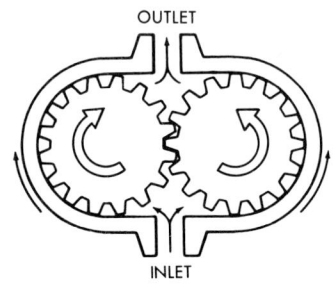

Fig. 4-6 External gear pump.

gear keyed to the drive shaft, a larger external gear, a crescent seal, and a closely fitted housing. The two gears are not concentric; as they rotate, pumping chambers open up between them at the inlet and close off at the outlet. The crescent seals the inlet port from the outlet and both gears carry oil past it.

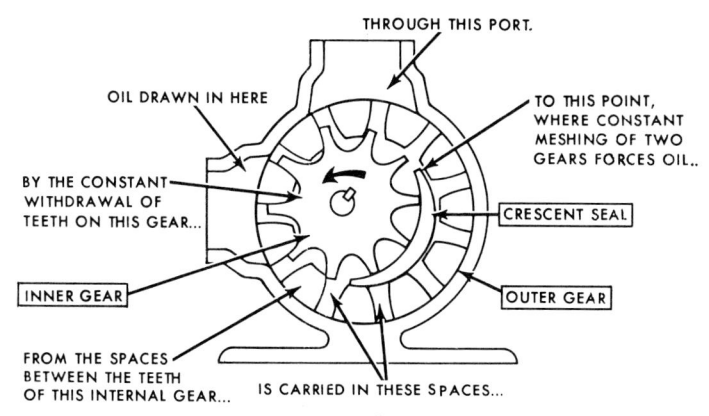

Fig. 4-7 Internal gear pump.

VALVES

Hydraulic valves, which are used to control actuators, fall into three categories: pressure, flow, and directional. As might be expected, some valves have multiple functions that cover more than one category.

All pure *pressure-control* valves operate in a condition approaching hydraulic balance; that is, pressure is effective on one side or end of a ball, poppet, or spool and is opposed by a spring. This is shown in Fig. 4-8, which details a simple check valve of ball and seat placed between two ports. Flow through the seat pushes the ball away to permit free flow, and flow in the other direction pushes the ball against the seat to seal the passage. A check valve could be either a pressure or a directional-control valve or both.

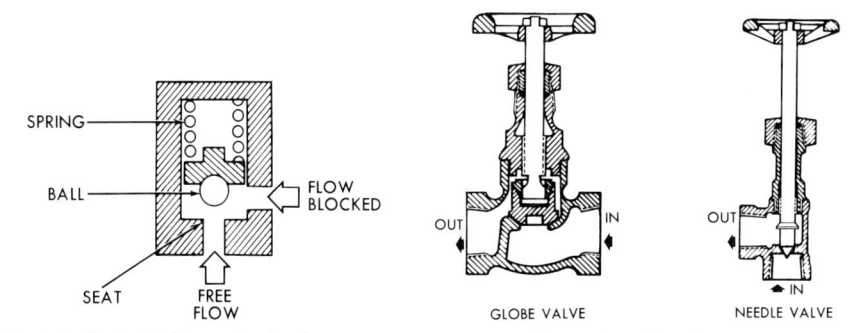

Fig. 4-8 Typical ball-type check-valve arrangement.

Fig. 4-9 Globe- and needle-valve principles and flow pattern.

A *flow-control* valve controls actuator speed by regulating flow. Flow-control valves are rated according to gallon-per-minute capacity and operating pressure. They are classified as *adjustable* or *nonadjustable*. Fixed-size orifices are one form of nonadjustable flow-control valve, but two of the most common are *needle* and *globe* valves. Figure 4-9 provides a schematic of these two valves.

While the basic check valve can and does serve as a direction-control valve, the most common is the reversing directional (or four-way) valve (Fig. 4-10). This valve has at least two finite positions with two possible flowpaths in each extreme position. It must have four ports: P (pump or pressure), T (tank or reservoir), and actuator ports A and B. In one extreme position, the valve has the pump port connected to the B actuator port and the A port to the tank. In the opposite position, flow is reversed; the pump is connected to the A port and the B port to the tank.

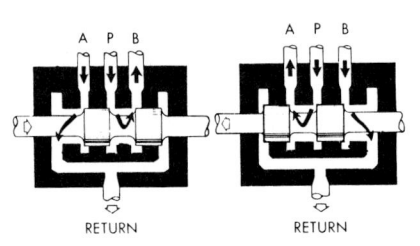

Fig. 4-10 Reversing directional valves have at least two finite positions, with two possible paths in each extreme position.

ACTUATORS

Actuators in a hydraulic system are devices that convert pressure energy into mechanical force and motion. They are either *linear* or *rotary*. A linear actuator (cylinder or ram) can give force and motion outputs in a straight line. A rotary actuator (or motor) produces torque and rotating motion.

The *cylinder* is a piston or plunger operating in a cylindrical housing and the ram is a single-acting plunger-type cylinder. The basic parts of a cylinder are identified in Fig. 4-11. The rotary actuator motor is a pump that is being pushed instead of doing the pushing. The main types of motors are identical to the pumps—vane, piston, and gear. Nearly all are reversible-type actuator motors.

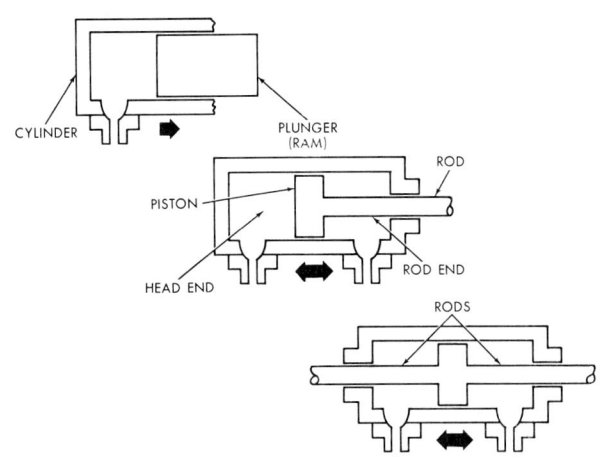

Fig. 4-11 Basic parts and nomenclature of hydraulic cylinders.

MAINTENANCE MORE IMPORTANT

The importance of hydraulic maintenance has gone through a step change. Today's sophisticated construction machinery is designed to meet the demand for increased productivity, meaning bigger payloads delivered faster under more precise control techniques and at higher operating temperatures. Good maintenance practices, most of which are simply a matter of commonsense, can bring out the best from hydraulic components.

Although most of the recommendations made here may have a familiar ring, their importance justifies repetition. The following are problems handled all too frequently by service personnel in working with construction equipment users:

    Wrong grade or type of hydraulic fluid
    Dirty or clogged fluid filters
    Insufficient fluid in the reservoir
    Pump driven in wrong direction
    Loose pump inlet lines

Any of these situations could be avoided through relatively simple maintenance procedures.

Fluid factors As indicated in the preceding list, and as verified by maintenance records, about three-quarters of all hydraulic service problems can be traced directly or indirectly to hydraulic fluid. It might be a matter of the wrong fluid, or dirty fluid, or perhaps not enough fluid.

The importance of keeping things right with hydraulic fluid cannot be overly stressed. It is the lifeblood of a system, serving not only as the means of transmitting power but also as a lubricant and coolant. Such things as cleanliness, viscosity, and temperature have a great effect on overall system performance.

In the first place, the correct fluid must be used. Selection depends on such characteristics as viscosity index, pour point, acidity, lubricity, usable temperature range, and cost. Operators should keep in mind, however, that the best approach to fluid selection is to follow the recommendations of the machinery manufacturer.

Most fluids have a petroleum base with special additives to provide the qualities required for hydraulic system operation. For severe service such as construction work, antiwear additives help provide reasonable operating life for hydraulic components. Several years ago hydraulic equipment manufacturers were recommending the use of regular crankcase oils with SAE classifications such as SC, SD, and SE. Now, the major oil companies offer specially formulated hydraulic fluids with the antiwear qualities of these crankcase oils.

Viscosity. Fluid viscosity is extremely important. Hydraulic fluid must have sufficient body to provide adequate sealing between working parts of pumps, valves, cylinders, and motors, but not enough to cause pump cavitation or sluggish valve operation. Viscosity is especially critical during start-up at low temperatures. If viscosity is too high, the pump may not get enough fluid and could actually seize up for lack of lubrication.

For extremely wide seasonal variations, the viscosity grade of the fluid should be changed in the spring and fall, as with automobile engines. Original equipment manufacturers' (OEM) operation and maintenance manuals should be checked for the proper grades. Where hydrostatic transmissions are involved, their control mechanisms might require still different viscosity fluids.

Petroleum-base fluids tend to thin out with increasing temperature and to thicken with decreasing temperature. This characteristic is expressed as *viscosity index* (V.I.). A high V.I. number means that viscosity changes at a slow rate with respect to temperature changes.

In general, the V.I. of mobile hydraulic fluids should not be less than 90. Multiple viscosity oils like SAE 10W-30 include additives to improve V.I. These oils generally exhibit both a temporary and a permanent decrease in viscosity due to oil shear that occurs in the operating hydraulic system. Actual operating viscosity can be far less than that shown in the oil specification data. For this reason, such oils should have high shear stability to assure that viscosity remains within recommended limits (Table 4-1).

TABLE 4-1 Typical Oil-Viscosity Recommendations for Mobile Hydraulic Systems

Hydraulic-system operating temperature range (min* to max)	SAE viscosity designation
−10°F to 130°F (−23°C to 54°C)	5W 5W-20 5W-30
0°F to 180°F (−18°C to 83°C)	10W
0°F to 210°F (−18°C to 99°C)	10W-30**
50°F to 210°F (10°C to 99°C)	20-20W

*Ambient start-up temperature.

Hydraulic components have closely fitted mating parts, and fluid lubricity is a must for reasonably long operating life. Crude oil and animal or vegetable oils should never be used. They do not lubricate properly and do not resist rust, corrosion, or foaming.

5-48 Maintenance of Power Systems

Oxidation Resistance. Oxidation resistance has a major effect on the life of hydraulic fluids. Air, heat, and contamination tend to promote oxidation.

All hydraulic fluids will combine with air to a certain degree. High operating temperatures speed up this action. For every 18° rise in temperature the oxidation rate doubles, creating contaminants which then tend to accelerate further contamination. Better grades of fluids usually include inhibitors to retard this action, but operating temperatures should still be watched closely.

To help avoid trouble, bulk fluid temperature measured at the reservoir should be kept in the 100 to 130°F range because localized hot spots may be 100° hotter than the bulk fluid temperature. A thermometer attached to the end of a wire and placed in the reservoir fill pipe can be used to measure fluid temperature. Also, temperature stickers are available for attaching to the side of a reservoir or suspected hot spot. Spots on the sticker change color permanently to indicate the maximum temperature.

When draining a system, try to remove all the used fluid so that none remains to mix with the new fluid. In most cases, bleeding at the lowest point in the system will help. It's also a good idea to drain only after the fluid is warmed up. This helps keep any contaminants in suspension so that they will be removed with the drained fluid. After draining, accumulated deposits should be flushed out with light viscosity fluid. This fluid should have a rust inhibitor to protect metal surfaces against rust after removing the hot fluid.

When a hydraulic pump or motor fails, the system should be considered contaminated, and the unit should be removed for repair. After removal of the pump or motor, the reservoir should be drained and the fluid discarded. Then the reservoir should be flushed to remove all contamination. Lines, cylinders, and valves should also be flushed.

Filtration and handling Fluid contamination must be avoided at every point, both in and out of the system. This includes storage and handling. Drums should be stored on their sides and, if possible, covered to prevent accumulation of dirt and water. Also, they should be kept full to reduce the chance for condensation to build up. Special care is a good idea when transferring fluid from drums into the hydraulic reservoir, and vice versa.

Many equipment builders recommend using a fluid transfer pump equipped with a

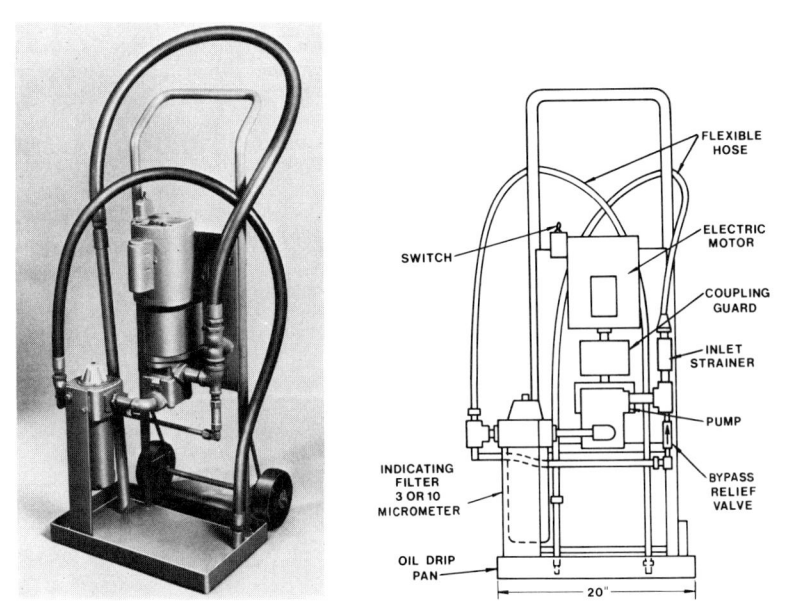

Fig. 4-12 *A* and *B* Portable hydraulic filters can be very helpful in keeping hydraulic systems clean when used on a regularly scheduled basis, following component repair or replacement, or during break-in of new machines.

filter of 25-micron rating, or finer. Portable filters are available for this and other applications (Fig. 4-12A and B). These portable units can be used for periodic filtering of system fluid while the machinery is in operation or for cleaning up the system following repair or replacement of hydraulic components. They are also useful during the initial break-in period on a new machine when so-called built-in contaminants such as metal chips, lint, or paint chips can be dislodged and circulated through the system. At times, service and maintenance procedures require that fluid be drained from a system. Here the portable filter can be used as a unit to transfer fluid into a drum and then subsequently back to the machine.

Establishing a schedule for using this type of auxiliary filtration must be based on experience—preferably documented with good records. The many variables to which each machine is exposed create system-contaminant levels which build up at different time rates in the individual machines. Retaining and analyzing maintenance and production records on these machines provide guidance in setting up schedules for using portable filters.

Hand-carried kits are available which provide on-the-spot fluid-contaminant-level determinations. A sample of fluid is drawn from a system and filtered through a very fine membrane disk. Discoloration of the disk can be compared with that of disks obtained previously. Tying these results into other maintenance records gives a higher degree of confidence that the machine will not have trouble from system contamination.

Typical of filters installed as an integral part of a hydraulic system are micronic-particle units with a replaceable element made of alpha cellulose (Fig. 4-13). The element normally recommended provides two-stage filtration to trap and retain a majority of particles more than 8 to 10 microns in size. For comparison, the smallest particle a sharp eye can see is about 40 microns.

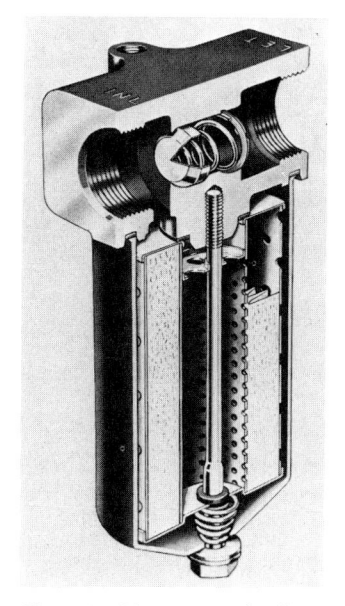

Fig. 4-13 Micronic-particle filters are often used as an integral part of a hydraulic system. The spring-loaded bypass valve at the top opens if the filter element becomes so loaded that it restricts flow excessively.

The main system filter should include a nominal-pressure bypass check valve to protect the system during cold starts and whenever the filter element becomes clogged. When possible, filters should be located in the tank return line where they can trap contaminants before the fluid reenters the reservoir. This location also permits using a low-pressure-type filter. Since filter elements must be changed at regular intervals, the filter should be located in an accessible area and not inside the reservoir.

Reservoir Airborne contaminants—and they're abundant around construction sites—are only one reason for good filtration practice. Another factor is that hydraulic reservoirs have tended to get smaller. This limits the opportunity for dirt particles to settle out before the fluid reaches the pump. It also means that there is a smaller amount of fluid in today's systems so that a given quantity of fluid makes more passes through the system in any given period of time. Contaminants thus have more chance to cause damage.

Inside the reservoir, a strainer is often used on the end of the pump inlet line (Fig. 4-14); this is in addition to the main system filter. This strainer should be cleaned periodically to help assure that sufficient fluid reaches the pump at all times.

The reservoir air breather vent usually includes a filter which also needs attention. This vent provides a means of maintaining atmospheric pressure to force fluid into the pump. A dirty vent filter could be plugged enough to reduce atmospheric pressure at the pump inlet, which, in turn, could produce a vacuum condition that is high enough to cause pump cavitation and, eventually, pump failure.

Cylinder rods and seals are also of concern. A worn or damaged piston rod seal or wiper seal can permit dirt and air to enter a hydraulic system.

Maintenance of Power Systems

Troubleshooting Improving the service life of hydraulic equipment is basically a matter of maintaining proper conditions within the system. The user should start with a clean fluid, keep it clean, and be on the alert for those symptoms which indicate possible trouble. Routine checks and simple corrective action can solve such difficulties when they

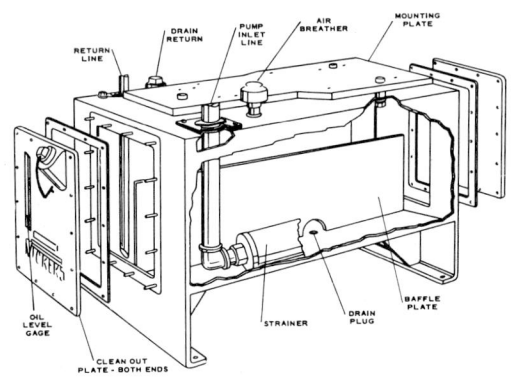

Fig. 4-14 Typical reservoir. The unit here conforms to ANSI specifications.

begin and long before they lead to major problems. The following list provides some helpful troubleshooting pointers:

Noisy Pump.
   Cavitation (pump starving)
   - Clean inlet strainer
   - Check inlet piping for obstruction
   - Fluid viscosity too high
   - Operating temperature too low
   Pump picking up air
   - Oil level low
   - Loose or damaged intake pipe
   - Worn or damaged shaft seal
   - Aeration of fluid in reservoir (return lines above fluid level)

Other.
   - Worn or sticking vanes
   - Worn ring
   - Worn or damaged gears and housings
   - Shaft misalignment
   - Worn or faulty bearings

Low or Erratic Pressure.
   - Contaminants in fluid
   - Worn or sticking relief valve
   - Dirt or chip holding valve partially open
   - Pressure control setting too low

No Pressure.
   - Oil level low
   - Pump drive reversed or not running
   - Pump shaft broken
   - Relief valve stuck open
   - Full pump volume bypassing through faulty valve or actuator

Actuator Fails to Move.
   Faulty pump operation (see Noisy Pump)
   Directional control not shifting
   - Electrical failure, solenoid, limit switches, etc.

- Insufficient pilot pressure
- Interlock device not actuated

Mechanical bind
Operating pressure too low
Worn or damaged cylinder or hydraulic motor

Slow or Erratic Operation.
- Air in fluid
- Fluid level low
- Viscosity of fluid too high
- Internal leakage through actuators or valving
- Worn pump

Overheating of System.
Continuous operation of relief setting
- Stalling under load, etc.
- Fluid viscosity too high

Excessive slippage or internal leakage
- Check stall leakage past motors and cylinders
- Fluid viscosity too low

Start-up procedures The high speeds and high pressures of hydraulic systems require careful start-up procedures. This is necessary for increased service life, reduction of vehicle downtime, and improved operating efficiencies. By following these recommended practices, you will be taking a big step toward trouble-free hydraulic system operation.

There are three steps that should be followed before the initial vehicle start-up. First, always be sure the system is filled with an oil meeting the vehicle manufacturer's recommendation. Second, check all lines, fittings, and components in the system. Keep them as clean as possible. Third, vent the air from the system. A good job of eliminating air will pay big dividends in efficient operation and longer equipment service life.

1. Fill all pump and motor housings with hydraulic fluid through the inlet ports. If the unit is mounted so that the inlet port cannot be used, fill through the outlet port or drain port. Obviously, this will have to be done prior to installing inlet and outlet connections. The inlet and outlet connections should remain plugged until the connections are ready to be made.

2. When the system is completely plumbed up, fill the reservoir.

3. Break the outlet connection loose on all pumps or use the air bleed valve and hold cracked open until air in the outlet line is purged and solid oil runs out. Bleed air from the highest point in the system and tighten the connection.

4. Loosen the connection at check valves, motor outlet, and heat exchanger outlet (provided the heat exchanger is above the oil level in the reservoir). Turn the engine over on the starter—do not start the engine. Tighten each connection in turn as a solid stream of fluid starts to be expelled.

5. When all connections are tight and the system has expelled as much air as possible, start the engine. Engine speed should be in the low rpm range for the next few minutes when each component in turn is operated.

6. Check fluid level in reservoir and add if necessary.

7. Check system for leaks. Be sure inlets to pumps are tight and not leaking air into the system. Air leaks into the system will cause the pump to be noisy and will damage the unit.

8. Examine the fluid in the reservoir. On initial start-up there may be some air which should come out as bubbles on the surface of the fluid. The fluid in the reservoir should clear up in a short time. (Length of time is dependent on how much air was removed from the system prior to start-up.)

Chapter **5**

Maintenance of Electric Motors

R. E. ARNOLD[1]
Former Manager, Product Engineering, General Electric Company, Industrial Motor Division, Schenectady, N.Y.

GENERAL

Electrical equipment will operate better, last longer, and require less repair if it is kept clean and dry and is lubricated properly. This is particularly true of electric motors, which are the driving forces for many operations. The failure of windings and/or bearings means delays from hours to days on vital work.

The electric motor, however, is only one component in an electric drive, and it is extremely important that all system components be correct and properly maintained if unplanned downtime is to be avoided. The power supply must be correctly matched; the control must function well; and particularly, overload, overcurrent, and overtemperature protective devices must be coordinated with the motor and function properly; and the driven equipment must be maintained.

PLAN

Perhaps the most important part of maintenance is to have a well-documented plan that is faithfully carried out. Any good plan will be based on the economics.

If the motor drive is vital to the job, maintenance of the highest order is indicated, but depending upon the physical size, the value, and the complexity of motor, different approaches can be taken. For example, on small, low-cost, more or less standard induction motors, probably the best solution would be to use motor spares or spare parts and keep the ordinary maintenance practices to a minimum.

However, if the motor is a large and/or a high-cost unit, the spare-motor philosophy is not practical because of the investment cost; a high degree of maintenance is justified, with the possibility of carrying spare parts, such as bearings, and form-wound coils if used.

If the motor is small with special features, such as high-temperature bearings to operate in a high-temperature environment, again a high degree of maintenance is indicated with a definite need of carrying critical parts, particularly the bearings.

Motors used in noncritical applications within a relatively dry, clean, normal-temperature environment require only minimum maintenance with the motors lubricated on a schedule suggested by the motor manufacturer. AC wound-rotor and dc armatures (rotors) using commutators or collector rings should have a brush-replacement schedule.

On large motors, of the order of 500 hp and larger for alternating current, and 300 hp and larger for direct current, the minimum maintenance schedule should be based on the manufacturer's recommendations. The value of the equipment is such that it is very

[1]The author is now retired.

important to follow the motor manufacturer's recommendations. There may be other maintenance procedures that can be used, but even these should be checked with the motor manufacturer if not obviously correct. A suggested maintenance schedule for large machines is shown in Table 5-1. Such a schedule would apply to dc motors and ac motors and generators by using the appropriate parts. This schedule could be adjusted to meet the needs of particular installations. Records are important for guidance and should show when more or less attention is justified on particular items.

Advancements have been made in motor inherent protection devices which will protect motor windings from overtemperature as well as overload in most cases and thus prevent motor burnouts.

There are now on the market metal-oxide variable resistors which, when properly placed in the motor circuit, will prevent high-voltage surges from overstressing the insulation. Where repetitive frequent starting, plugging, reversing, and jogging are encountered, such devices should be considered.

The above two items are mentioned because good planning includes upgrading of the equipment and system to prevent failures and in case of failure to take advantage of available improvements in the components and systems. A typical example is the use of a higher-temperature class of insulation for rewind on the stators of ac machines.

The economics of the maintenance plan must include the cost of the loss of production and work-force efficiency as well as the cost of repairs balanced against the cost of the maintenance of the equipment.

In some large operations where many machines are involved, maintenance scheduling can be computerized.

PART I. ALTERNATING-CURRENT INDUCTION MOTORS

The induction motor is one of the oldest types of motors and in its commonest form, the squirrel-cage motor, the simplest. The first commercial installation was in 1889, and the induction motor with distributed primary and secondary windings was developed in 1892. Since the first commercial installation, many designs of polyphase induction motors have been developed and two types have become the recognized standard: (1) the squirrel-cage rotor and (2) the wound-rotor construction.

The characteristic features of the stationary primary element—distributed windings and comparatively small air gaps—are common to both types. The squirrel-cage motor has no external secondary or rotating connection, while the secondary or rotor windings of the wound-rotor motor usually are connected through slip rings and brushes to some form of adjustable resistance.

The modern induction motor, especially the squirrel-cage type, is undoubtedly the most rugged rotating electrical apparatus ever developed. The majority of maintenance requirements, outages, and repair costs would therefore depend to a large extent on the correctness of application. However, the cardinal principle of any electrical maintenance is *keep the apparatus clean and dry*. This immediately points to periodic inspections, which are a highly desirable check on operating conditions.

Before details dealing primarily with maintenance are discussed, a few of the fundamental characteristics of the induction motor which may aid in solving some of the maintenance problems are presented.

Slip At no load an induction motor will run at practically synchronous speed, but when it is loaded, the speed at which the motor will run is below synchronous speed by an amount known as the slip. Thus, if the synchronous speed of a given motor is 1800 rpm and the full-load speed is 1750 rpm, the slip at full load is 50 rpm, or 50/1800, or 2¾ percent. The slip of any induction motor is a function of the losses in the secondary (resistance times the current squared in watts). The higher the secondary resistance, the greater will be the starting torque with any given current; the higher the slip, the higher also are the losses and thus the lower the efficiency.

Torque In analyzing some of the maintenance problems, a consideration of torque characteristics is sometimes important. There are two torque characteristics to consider: starting torque and breakdown torque. Motor torque, in the design state, must be balanced against efficiency and power factor. High starting torque results in lower efficiency and power factor as well as poorer speed regulation. However, if the starting torque is too low,

TABLE 5-1 Suggested Maintenance Schedule

Component	Monthly Inspection or maintenance operation
BEARINGS	Make sure that grease or oil is not leaking out of the bearing housings. If any leakage is present, correct the condition before continuing to operate
Ball and roller	Listen to a few bearings on a sampling basis. Bearings that get progressively noisier will need replacement at next shutdown
Sleeve	Check the oil level. Check the oil color through the sight gauge. Slightly cloudy oil is all right. Black oil is a danger sign
BRUSHES	Check the brush length. Replace when rivet or clip will rub commutator before next inspection. Inspect for worn or shiny brush clips, frayed or loose pigtails, and clipped or broken brushes. However, many dc brushes have no clips. Pigtails are tamped. There is danger that the pigtail of a worn-out brush will cut the commutator. Such brushes have a wear marker on the pigtail. When it gets below the top of the box, the brush should be discarded. Remove a few brushes to check the brush-commutator contact face in the case of dc motors and the brush-collector contact face in the case of ac motors and generators. Burned areas indicate commutation or sliding contact troubles. WARNING: HIGH VOLTAGE. ELECTRIC SHOCK AND ROTATING PARTS CAN CAUSE SERIOUS OR FATAL INJURY. AVOID CONTACT WITH LIVE ELECTRICAL PARTS AND MOVING MECHANICAL PARTS. (IT IS WELL TO NOTE THAT SILICON-CONTROLLED RECTIFIER DRIVES MAY HAVE HIGH VOLTAGES AT THE BRUSHES EVEN WITH THE ARMATURE STATIONARY. LINE SWITCHES SHOULD BE OPEN)
COMMUTATORS	Check the commutator for roughness by carefully feeling the brushes with a fiber stick. Also occasional wiping is recommended using a piece of coarse or nonlinting cloth. Jumping brushes give advance warning that a commutator is going rough. Observe the commutator for signs of threading. If threading is getting worse—take action; threading healed over—all right. Check for excessive commutator wear rate, streaking, copper drag, pitch bar marking, and heavy-slot bar marking. Commutator should have not more than 0.0025 in. total indicator runout or 0.0002 bar-to-bar steps. For high-speed commutators surfacing should meet 0.0010 and 0.0001 limits, respectively
COLLECTORS	Check the collector for roughness, dust, and wear. Ordinarily the rings will require only occasional wiping using a piece of coarse or nonlinting cloth. If the brushes are bouncing up and down on a cycle basis, check for collector-ring concentricity
MECHANICAL	
Air filters	Replace when necessary. Clogged filters cause overheating and lead to premature insulation failure
Bolts	Perform visual observation for loose bolts, loose parts, or loose electrical connections
Noise and vibration	Check for any unusual noise, vibration, or change from previous observations. Loose pole bolts often are the source of magnetic noise when motors are fed from rectifiers

TABLE 5-1 Suggested Maintenance Schedule (*Continued*)

Component	Every 6 Months Inspection or maintenance operation
BEARINGS	
Ball and roller	Listen to all bearings. Pull back bearing cap to inspect grease condition on a few representative machines
Sleeve	Take samples of oil from representative bearings for acidity (neutralization number) test. See manufacturer's sleeve-bearing relubrication recommendations. (Oil acidity is affected most by atmospheric contaminants and temperature—take one sample in each different area.) Change oil if required
COMMUTATORS	Check risers for cracks. If there are cracks, also check end of shaft keyway and shaft fan. (Cracks here mean extreme torsional vibration in system)
INSULATION	Measure 1- and 10-min insulation resistance and calculate the polarization index. Compare with records. Wipe deposits from brush-holder stud insulation and commutator creepage path or collector creepage path. Remove heavy deposits from around field-coil connections on dc machines where grounding might occur. Blow deposits out of the commutator rise area or the collector area with clean dry air. Blow out any blocked ventilation openings in windings. Make visual inspection for signs of overheating (dry, cracked, "roasted-out" insulation and varnish)
MECHANICAL	
Bolts	Check all electrical connections for tightness. Look for signs of poor connections (arching, discoloration, heat). Adjust inspection period to suit experience. Inspect foundation for signs of cracking, displaced foot shims; check foot bolts for tightness. Check frame-split bolts, brush holders, brush-holder studs, and bracket bolts, etc., on sampling basis. Check all coupling bolts
Shaft	Check corners of exposed end of shaft keyway for cracks (due to extreme torsional vibration). If there are cracks, check fan commutator risers and dc machines and the fan and character of the applied load on ac machines
Ventilation	Check for clogged screens, louvers, filters, etc.
Vibration	Check for excessive vibration (more than 0.002 to 0.003 in.) that will indicate change in balance or alignment. If actual operation cannot be seen, check for other signs of vibration (loose parts, chafing, shiny spots, rust deposits)
Yearly	
BEARINGS	
Sleeve bearings	Drain housings. Remove top half of housing. Lift top half of bearing. Inspect bearing surface and rings for wear. If excessive wear or sludge is found in bottom of housing, roll out bottom half of bearing for inspection. Clean if necessary and refill. SEE SLEEVE-BEARING RELUBRICATION RECOMMENDATIONS OF MANUFACTURER

the motor may not be able to start the load. High breakdown or high maximum running torque also results in low power factor and high starting current. If the breakdown torque is too low, the motor may stall on ordinary overloads, which can result in so-called roasted insulation if adequate protection is missing.

Voltage and frequency To obtain optimum results, induction motors should be operated at their normal rated frequency and voltage. Of course, some variation can be tolerated, voltage limits being approximately plus or minus 10 percent from nameplate and frequency plus or minus 5 percent. Both should never be varied at the same time to the extreme allowable limits; neither should they be varied at the same time in opposite direction. The following tabulation shows the effect of variation of voltage and frequency on performance of two-, four-, and six-pole motors of normal design.

	Power factor	Torque	Slip	Full-load efficiency
Voltage high	Decreased	Increased	Decreased	Approx. same
Voltage low	Increased	Decreased	Increased	Slightly lower
Frequency high	Increased	Decreased	Same	Approx. same
Frequency low	Decreased	Increased	Same	Slightly lower

A good rule to follow is that if normal frequency is changed by no more than 10 percent, a corresponding change in voltage should be made proportional to the square root of the ratio of the frequencies. It is not desirable to operate at decreased frequency with less than normal voltage because of increase in current input and temperature unless the load is reduced correspondingly.

Stator windings At first glance, the stator of an induction motor appears to be so rugged and simple that the necessity for maintenance is frequently overlooked. However, an inspection of any electrical-repair-shop records certainly indicates that the stator of the induction motor is a vulnerable item.

Trouble with stators usually can be pinned to one of the following causes: overloading, single-phase operation, moisture, bearing trouble, or insulation failure.

Main contributing factors to stator failures usually are dust and dirt. Some forms of dust or dirt are highly conductive and contribute to insulation breakdown. A good example of this is found in rubber-mill operations where a large amount of lampblack is used in processing the rubber. The type of lampblack used in processing synthetic rubber is more conductive than that used in processing natural rubber, so that failure rates increased when synthetic rubber came in. This is but one of many examples of conductive dirt and dust hazards. Certain steel-mill applications are others.

In addition to insulation failures due to conducting dust, restriction of ventilation can result from dust clogging ventilating ducts, thus causing overheating with possible resultant insulation failure due to excessive temperatures. With the loss of the cooling system or the restriction of ventilation, the failure occurs in many motors because they are inadequately protected by overload relays. Since the heaters are selected on the basis of current, they will not protect the motor when the cooling means fails because the current is affected only slightly to the higher temperatures. Inherent protection based on the temperature of the winding is generally safer. Periodic cleaning with clean, dry compressed air usually will suffice to keep dust accumulations to a minimum. However, some types of dust or dirt have a tendency to stick to windings, and blowing with air does not give a completely satisfactory cleaning. Other methods of cleaning are covered later.

One of the natural enemies of insulation is moisture. Some types of modern insulation have reasonable resistance to moisture, but in general, all types of windings should be kept dry. Many applications make it almost impossible to accomplish this unless special enclosures or other means to keep out moisture are employed. Some success for motors operating in damp locations has been obtained by using special treatment of windings. This will be discussed later.

Vibration frequently hastens winding failures. Vibration during operation may cause coil movement which eventually breaks or wears through insulation. As the motor becomes older, the insulation dries out and loses its flexibility. The mechanical stresses resulting during starting, plugging, and reversing, as well as natural stresses occurring under normal operation, may precipitate short circuits in coils or failures to ground. Periodic varnish and drying treatments properly performed tend to maintain a solid winding, thus minimizing coil movement.

Rotor windings General comments on stator windings here and in subsequent discussion apply equally to windings of wound-rotor motors. However, since the rotor is a moving part, additional maintenance problems are introduced.

Practically all wound rotors have three-phase windings and can therefore have trouble due to single-phase operation. An open circuit in a rotor shows up in lack of torque and slowing down in speed. It usually is accompanied by a grumbling noise and sometimes by failure to start the load. The most logical place to look for an open rotor circuit is in the secondary resistor or the circuit external to the rotor. The stud connections to the slip rings should be checked also, as the open circuit may be found there.

On rotors of greater capacity, where the coils are made of copper strap, clips are used to connect the bottom and top halves of the coil. These connectors should be checked for signs of heating, indicating a partial open. Such end connections, if faulty or improperly made, are a common source of open rotor circuits. Most manufacturers now braze these end connections instead of soldering, which minimizes faulty connections.

A ground in the rotor circuit will not affect the motor performance unless a second ground develops, which may cause the equivalent of a short circuit. This unbalances the rotor electrically and causes reduced torque, excessive vibration, sparking at the collector ring or brushes, or uneven wear of collector-ring brushes.

A reasonably successful method for checking for short circuits in rotor windings is to raise the brushes and energize the stator. If the rotor is free of short circuits, there should be little or no tendency to rotate even when the motor is disconnected from the load. If there is evidence of considerable torque, or if there is a tendency to come up to speed, the rotor should be removed and the winding opened to determine where the fault exists.

Another check which can be made with the rotor in place and with stator energized and brushes raised is to check the voltage across the rings to determine if they are balanced. When this check is made, the rotor should be rotated to several positions and readings taken at each position to be sure that inequality in voltage readings is not due to the relative positions of stator and rotor phases.

Squirrel-cage rotors with die-cast or pressure-cast aluminum rotors comprise the large majority of induction motors. They are very rugged and generally require little maintenance because of the absence of joints and connections. However, they may give trouble because of open circuits (bars) or high-resistance points between the rotor and the end rings. The symptoms of such trouble are reduced torques, higher slip under load, increased heating, and sometimes a noticeable noise.

Open bars associated with cast rotors are seldom visible. Therefore, if open bars are suspected, a method of checking is to apply 10 to 25 percent single-phase voltage to the stator with an ammeter in one line, slowly rotate the shaft, and watch the ammeter for significant changes in current. A significant current change with rotor position indicates a defect in the cast winding cage. Usually in this case, a pulsing sound related to slip under load is the first observation. Of course, if this occurs, a new replacement rotor is indicated.

Large induction motors and some high-resistance rotors are made with the bars brazed to the end rings for technical or economic reasons. A cracked rotor bar on a brazed cage with bar extensions between the iron core and the end ring can usually be seen if one has a reason to look for the defect. If there is trouble in the brazed joints, generally localized heating will cause discoloration. Cracked bars should be replaced, and with high-resistance joints further brazing is necessary, generally in a repair shop.

If cracked rotor bars or die-cast rotors are visible where the bars and end rings meet, the rotor should be replaced, but the manufacturer should be informed so that he can determine if it is a manufacturing defect, a design problem, or an application problem.

Repairs to squirrel-cage rotors, such as replacing bars or brazing broken bars, should be attempted only to competent personnel. Considerable skill is required, and proper care should be exercised in such repairs.

As stated previously, a small air gap is a characteristic of an induction motor. The size of the air gap has a direct bearing on the power factor of the motor, and any alterations that affect the air gap, such as grinding the rotor body, increase magnetizing current and lower the power factor.

Good maintenance on sleeve bearing motors includes a periodic check of motor air gap with feeler gauges to ensure against a worn bearing that would permit a rotor rub. Measurements should be taken on the coupling or driving end of the motor. Four measurements of air gap should be made approximately 90° apart, and one of the points should be on the load side, that is, the point on the rotor corresponding to the load side of the bearing.

On larger motors, a record of air-gap measurements should be kept so that a comparison can be made with previous checks to determine the amount of bearing wear. A rub (rotor on stator) resulting from bearing wear can generate enough heat to cause insulation failure.

Overloading of motors because of increased demands on the driven machine increases the operating temperature, resulting in shortened life of insulation. Momentary overloads within reasonable values usually do no damage; consequently, a thermal-overload device offers best protection. Since the best place to measure the thermal effect of overloads is on the motor, a thermal device can be applied directly to the motor winding. Most manufacturers can supply such a device, which provides effective protection against sustained overtemperature.

The polyphase induction motor is beyond doubt the simplest and most foolproof piece of rotating electrical apparatus. Experience indicates that the most frequent cause of winding failures is probably moisture associated with dirt, dust, chemicals, etc. Since bearings and lubrication of bearings are common to all types of motors, they will be treated in a separate section.

PART II. THE DIRECT-CURRENT MOTOR

The dc motor is more likely to be damaged in operation than the ac motor because a number of current-carrying parts are exposed. This motor comprises two parts: the stationary part, or field, and the rotating part, or armature.

The field of a dc motor consists of a frame and field poles which are fastened to the inner circumference of the frame. The poles are steel, usually laminated, and have mounted on them the field windings which furnish the excitation for the motor. Field windings, in general, are subject to the usual failings of electrical equipment. They can become dirty or oil-soaked, which can interfere with heat being dissipated, causing eventual burnout. Excessive field current caused by malfunctioning of control will cause excessive heating and failure. Field heating can be caused by high voltage, too low speed, brushes being off neutral, overloads, or a partial short in one field coil. An open circuit in a field coil can result in failure to start or excessive speeds at light loads and heavy sparking of the commutator.

The armature of a dc motor consists of two main parts: the windings and the commutator. Cleanliness, while important on all electrical equipment, is particularly so on the dc commutator and brush rigging. Oil, dust, grease, moisture, and corrosive gases should be kept away from commutators and brushes, as they cannot give good performance under adverse conditions such as these.

The armature is the heart, so to speak, of the dc motor. The main line current flows through it, and if the machine is overloaded, the armature shows the first signs of distress. Reasonable attention from the standpoint of cleaning should result in little or no trouble with the armature under normal operating conditions. Repairs to armatures should be done only by competent personnel. Care should be exercised in handling an armature, some of the more important items being:

1. Never roll an armature on the floor; a coil may be injured or a binding band nicked.
2. Support or lift an armature only by its shaft, if possible; otherwise use a wide lifting belt under the core.
3. Never allow the weight of the armature to rest on the commutator or coils.

Windings should be preserved with periodic varnish treatment followed by baking where possible. This will be treated later.

If necessary to renew the banding on an armature, duplicate the banding originally supplied by the manufacturer; in other words, do not change material, diameter, width of band, or location. Band width should not be increased, as to do so may cause restriction of ventilation and also cause heavy current in the bands sufficient to overheat and melt the band solder.

The commutator is probably the most vulnerable part of a dc motor, as it is an exposed current-carrying part rotating at relatively high speeds. The success or failure in operation of a dc machine depends to a large extent on commutation. Regardless of any other excellent feature a dc machine has, if the commutation is unsatisfactory, the machine has little commercial value. It is the intent at this time not to go into the details of manufacture or assembly of commutators but to assume that the manufacturer has produced a device which, under proper conditions, will give trouble-free operation.

Assuming that the design of the machine is such that good commutation may be expected, continued satisfactory operation depends on maintaining the commutator surface in good condition. In general, this means that the surface should be smooth, concentric, and properly undercut. The brush holders should work smoothly and be free of dust and dirt; the brushes should be of the proper grade and manufactured to correct size and tolerance. Brush holders must be spaced equally around the circumference and have correct spacing from the commutator surface (usually $\frac{1}{16}$ to $\frac{3}{32}$ in.). While conditions under which dc machines are operated vary widely, there are some basic conditions which must be maintained on all such machines to ensure satisfactory commutation, brush life, and a minimum of commutator wear. These conditions and recommended maintenance methods are listed below.

1. The commutator must be concentric. On high-speed machines with peripheral speeds of 9000 fpm or above, the commutator should be concentric within 0.0005 in., which is the practical limit in grinding. For peripheral speeds of between 5000 and 9000 fpm, the concentricity should be within 0.001 in. On slow-speed, large-diameter commutators, this figure can be 0.003 in.

There should be no abrupt change in surface from bar to bar. Variations of over 0.0005 in. are enough to cause bad commutator performance. This bar-to-bar roughness can be detected by using a stick sharpened like a pencil and held on the revolving surface at a slightly inclined angle. If the commutator has negligible bar-to-bar roughness, the stick, so used, will feel as if it is moving over a smooth glass surface.

2. To get a commutator concentric, a grinding rig should be used in all but a few special cases. The grinding rig consists of an abrasive-stone setup similar to a lathe tool in a carriage which can be moved back and forth along the commutator and equipped with a feed to advance the stone into the commutator surface (see Fig. 5-1). The support must be rigged so that the grinding stone is subjected to a minimum of vibration. On most dc machines such a rig can be mounted on a brush arm by removing the brush holders. To be sure of obtaining maximum rigidity, it may be desirable to brace the brush-holder bracket arm during the grinding operation. In some cases, it is possible to support the grinder on parallels fastened to the bedplate, in which case the entire brush rigging can be removed.

Fig. 5-1 Commutator grinding rig in position.

Grinding should be done, if possible, with the armature in its own bearings and, in the case of a constant-speed machine, at rated speed. Low-speed grinding, if there is any evidence of unbalance, will cause the commutator to run eccentrically at rated speed. Care should be exercised to prevent copper and stone dust from getting into the windings. It is important to provide a vacuum cleaner on the grinding rig. In extreme cases where a cleaner is not available, the commutator necks and risers and the coil ends can be covered with paper or cloth to keep the grinding dust out of the machine.

To grind a commutator, the armature must be rotated. Each case must be treated separately, as local conditions will govern. At times it is possible to run the motor with half the brushes out, grinding half the commutator and then repeating with the other half. A driving motor can be used, coupled or belted to the armature on which the work is being performed. On some types of equipment, it is difficult because of space or other considerations to grind the commutator in place. If the machine speed is relatively low, the work should be done in a lathe by taking a very fine cut off the surface and then polishing with a stone. On high-speed machines, a special rig should be used equipped with a grinding

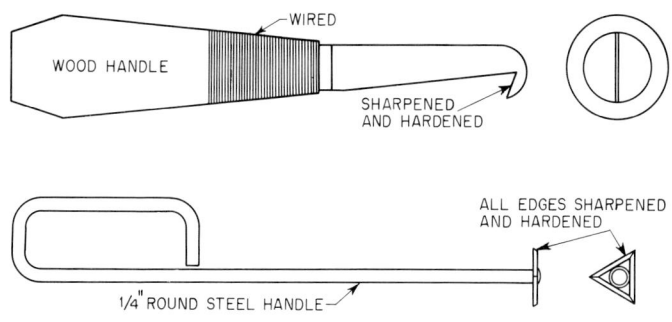

Fig. 5-2 Tools for commutator undercutting.

device, a driving motor, and an adapter for bearings, so that the armature can be run in its own bearings. Grinding should be done at a speed as near to rated maximum speed as possible and still not have the grinding stone affected by vibration.

Three grades of stones are used in grinding commutators: coarse, medium, and fine. The coarse stone has a grit of approximately 80 mesh and takes off large amounts of copper. In general, the use of the coarse stone is not too desirable, because if a great deal of copper has to be removed, it is better to use a lathe tool which can be set up in the grinding rig. The medium stone has a grit of approximately 120 mesh and is used for the bulk of grinding work, finishing up with the fine stone with a grit of approximately 200 mesh for the fine finish. The stone should be contoured or block, and sandpaper should be used to avoid "crowning" the bar.

3. After grinding is completed, all commutator slots should be cleaned out thoroughly and bar edges beveled. Beveling accomplishes two things: burrs, caused by the stone dragging copper over the slots, are removed, and the sharp edge is eliminated at the entering side of the bar under the brush. Bevel as little as possible to just remove "wire edge." This is accomplished with a special beveling tool which should have about $\frac{1}{32}$-in. chamfer at 45° for bars of medium thickness. For thinner or wider bars, the beveling can be changed accordingly.

4. Almost all modern dc machines have undercut mica in the commutator slots. This undercutting should be kept at $\frac{1}{32}$ to $\frac{3}{64}$ in. If, when a commutator is to be ground, sufficient copper is removed so that the undercutting will be shallow, then the commutator should be reundercut before grinding is started. This is done by using a small, circular, high-speed saw about 0.003 in. thicker than the actual thickness of the mica. Where the extent of undercutting is not great, a hacksaw blade mounted in a wooden handle is a popular tool. It is sketched, along with another tool, in Fig. 5-2. Otherwise, one of the undercutting tools put out by any one of a number of manufacturers can be used. Care should be taken when undercutting that a thin sliver of mica is not left against one side of the slot. Slots should be checked, and if slivers are present, hand scarfing will be necessary. If these slivers are left, subsequent operation of the machine raises the mica above the edge of the bar and one of the worst conditions for good commutation develops. As explained previously, the bars should be scarfed after the undercutting is completed. Final polish can best be obtained by using white brush-seater stone on the commutator surface after sanding in the brushes.

5. The bottom of the brush-holder box should be set at the correct angle and the correct distance from the commutator surface. The distance from the bottom of the brush box to the commutator surface on most machines is $1/16$ to $3/32$ in. Failure to maintain the proper spacing results in brushes riding the commutator surface poorly and also has the effect of shifting neutral on machines using inclined brush holders. This shift in neutral contributes to poor commutation. To check this spacing and to ensure uniformity, a piece of hard fiber or similar material of the proper thickness should be used as a gauge.

Spacing of brush holders around the commutator should be maintained evenly, and the spacings should not deviate more than plus or minus $1/32$ in.; also, the holders should be aligned on the brush arm in a straight line parallel to the length of the commutator bars. A popular method of checking brush-arm spacing is to stretch a piece of paper or adding-machine tape around the commutator under or near one brush path and mark the brush position on the paper with a pencil. The paper is then removed, and spacing is checked for uniformity. The brush arms can be shifted to obtain equal spacing. The check should be taken several times to average out any errors. After this work has been done, the brushes should be reinstalled or replaced with the brush holders staggered properly as per manufacturer's recommendations.

Brush and brush-holder tolerances Brushes should slide freely in the brush boxes. The standard tolerances for brush width and thickness are plus 0.000 to minus 0.004 in. up to and including a brush width of $3/4$ in. For brushes over $3/4$ in. wide the tolerance is plus 0.000 to minus 0.015 in. The average industrial-type brush box maintains a tolerance of plus 0.003 to 0.006 in. on the thickness and plus 0.002 to 0.010 in. on the width.

With these close tolerances in the brush boxes and brushes, it can be readily seen that not much dust or dirt is required between brushes and boxes to cause sticking brushes. For this reason a clean machine is essential to good commutator performance. At times brushes will appear to be tight in the holder even after a thorough cleaning. The boxes should then be checked to determine if they have been warped. This occurs at times if the machine has been run with excessive load which caused heating or if the machine has had a severe flash which caused heating of the brush holders. The reaction surface in the box must be flat.

Pressure Brush pressure springs should be set at the manufacturer's recommended values and should be adjusted to be as uniform as possible. Uniform pressures prevent selective action whereby certain brushes tend to take more than their proper share of the load. Tension can be measured with a straight spring scale and should be taken just as a new brush on being lowered touches the current-collecting surface. The pull should be taken in the direction of normal brush motion to avoid setting up friction values differing from those affecting the brush in operation. Dividing the pull obtained with the scale by the cross-sectional area of the brush gives the pressure in pounds per square inch.

Variables affecting commutation The maintenance suggestions outlined above should result in satisfactory commutation, brush life, and commutator conditions. However, there are a large number of variables which can upset the delicate balance between brush and commutator. The resulting unbalance is indicated in a number of ways. Following are listed several of the more common adverse conditions encountered and their remedies, together with what experience indicates as possible causes. These should serve as a guide to the maintenance man in solving his commutation problems but should not be taken as the final analysis on any particular trouble job.

Brush chatter is caused by high friction between the brush and the commutator or by a poor commutator surface. A common cause of high friction between brush and commutator is light-load running. On machines which operate for extended periods with brush densities in the order of 25 amp/sq in. and below, the brushes have a tendency to develop a highly polished glaze on the commutator which invariably causes brush chatter. In order to correct this condition, the brushes in one or more paths around the periphery should be lifted in order to raise the current density in the remaining brushes to approximately 55 amp/sq in. It is always better to operate a brush overloaded for a short period than to operate it lightly loaded for an extended period. If it is not possible to raise the brushes as recommended above, a grade of brush with a slight amount of cleaning action should be used to prevent the formation of the high-friction film, a grade of brush more suitable for low-current density should be installed, or a combination of both can be used.

High friction can also be caused when a commutator is run hot for an extended period.

This condition usually does not develop unless the machine is run well beyond rated capacity and cannot be corrected by any maintenance program. It may be necessary to remove the high-friction film with white seater stone before a normal commutator film can be restored.

Poor commutator surface and high mica also cause brush chatter, which, however, can be distinguished from chatter caused by light load or overload conditions by its lower frequency. It can be corrected by grinding the commutator, trimming out mica fins, scarfing, and polishing.

Brush chipping can result from brush chatter as described above. It can also result from extremely light or extremely heavy spring pressure, high mica, improper fit of brushes in the holders, or severe shock or vibrations set up outside the machine. Correction can be accomplished as described previously.

Threading or Streaking of Commutator Surface. This is caused by the breakdown, in the brush paths, of the film on the commutator surface. The tendency then is for the current to pass through the area where the film has been broken. The surface condition then is further aggravated, and finally threads or streaks are worn around the periphery of the commutator. This condition is caused by particles of copper being embedded in the face of the brush. These particles work-harden and become tools, cutting through the commutator film. Brush faces should be inspected periodically and corrected if necessary.

Selective action, the tendency for one brush or group of brushes to carry more than its share of the load, is also a prime cause for threading or streaking.

Flat spots on the commutator sometimes develop which, if permitted to go uncorrected, will become as much as 0.030 in. deep. The spots may begin singly or at one or two pole pitches apart. Often the flat spots will multiply, with a second, third, and fourth set following the initial set, displaced by several bars. The cause is always a cyclical disturbance, either electrical or mechanical. Electrical sources might be an open or partially shorted armature coil. Mechanical sources would include mechanical unbalance, coupling misalignment, and high or low commutator bars. Flexible coupling should be lined up to about 0.002 in. both transversely and angularly. Atmospheric contamination often aggravates this condition.

Threading or streaking can be reduced by raising the current density above 30 amp/sq in. or by using a brush with some cleaning action. Electrographitic brushes with a mild polishing action are most suitable for this type of situation.

If selective action is causing the streaking or threading, it is necessary to check conditions which could cause unbalance in electrical paths. Electrical resistance in one path different from that in parallel paths causes selective action. Terminal connections, spring pressure, shunt-to-brush connections, brush size, brush-holder spacing, and brush material should be checked. Hardness of brushes ordinarily does not contribute to threading. However, the ash content of a brush may do this, especially if the ash particles are hard, as they can be if ash content is improperly controlled in manufacture.

Sparking. Practically every abnormal condition of the commutator, brush holders, brushes, fields, or armature results in sparking. If the maintenance practices, as outlined previously, have been followed and the machine was correct originally, we must then look for the cause of injurious sparking somewhere in the electrical or magnetic circuit of the machine or in the brush grade. The following are conditions to be looked for: bad connections between armature coils and commutator bars, particularly those which are bad only when the armature is rotating; short-circuited field coils; open-circuited or short-circuited armature coils; unequal air gaps due to worn bearings; improper brush grade; partial short in shunt field; or neutral set-off. The following is a convenient checklist on the most general causes of sparking:

Rough, eccentric, or dirty commutator
Brushes sticking in holders
Incorrect brush tension
Poor brush fit
Brushes not parallel with commutator bars
Unequal brush-holder spacing
Vibration
Brushes off neutral
Unequal spacing of main or commutating poles

Reversed or short-circuited commutating-pole coils
Reversed compensating winding
Open circuit in armature winding
Short circuit in armature winding
Unequal air gap
Grounds
Atmosphere or other commutator contamination (high friction)

Checking Neutral. When a machine is reassembled after it has been dismantled for cleaning or repairs, it is frequently necessary to check and set neutral on the machine. The following is an outline of the kick-neutral method.

This method is based on measurement of voltages induced in the armature coils as the current in the main field on the machine is interrupted. Voltages induced in the conductors located at equal distances to the right and left of the pole centers are equal in magnitude and opposite in direction. If the terminals of a low-range voltmeter are connected to commutator bars corresponding to conductors located midway between poles, no deflection will be caused by breaking the field current. When the brushes are set so that the centerlines of their faces correspond with the centerlines of the commutator bars between which there is no induced voltage, they are on neutral (see Fig. 5-3). A much simpler method is to tie the voltmeter to two studs of opposite polarity.

If the number of commutator bars is not evenly divisible by the number of poles, use the following method: With the machine at standstill, raise all brushes. Replace one of them on each arm by a special brush of the same thickness. This special brush should be beveled to a knife-edge parallel with its longer side and in the center of its face. Connect leads from adjacent brush arms to a dc voltmeter, preferably one having 0.5-, 1.5-, and 15-volt scales. Separately excite the shunt field from a dc source through a quick-break switch. Insert enough external resistance in the excitation circuit to keep the field current small at the beginning. Use the smallest field current that gives a good deflection on the low scale of the voltmeter. When "kick" voltage is read for the first time, begin with the 15-volt scale and change to lower scales only when it is certain that the voltage is within their respective ranges. Before the switch is opened for each reading, wait long enough for the induced voltage caused by closing the circuit to decay. Shift the rocker ring to the point at which the voltage is minimum when the field circuit is opened. If the machine has double brush holders, the center of the brush holder is placed on the neutral mark instead of either of the double holder brushes.

If the number of bars between centerlines of brushes on adjacent arms results in half a bar being included in the commutator pitch (such as 20½ bars between centerlines), this alternative method is used: Raise all brushes. With the voltmeter points on bars 1 and 21 in the approximate neutral zone, open the field circuit as described in the paragraph above and read the deflection. Move the voltmeter points to bars 1 and 22, and read the deflection as the field circuit is opened. Rotate the armature slightly until the two readings are equal but opposite in polarity. This indicates that the correct neutral is exactly on the centerline of bar 1 and on the mica between bars 21 and 22. The rocker ring is shifted until the centerlines of the arc of the brush surfaces are exactly over these positions. The same procedure applies here for double brush holders.

If the number of bars is evenly divisible by the number of poles, it is possible to be off neutral one-half bar with the method just described. If the armature can be rotated, use this alternative method: Raise all brushes. Determine the commutator pitch. Full-pitch windings are *now very rare.* Coil throw (back pitch) is stated in slots. For simplex lap windings, commutator pitch is one segment, or for a simplex wave winding is the number of segments per pairs of poles. With the voltmeter points on bars 1 and 21 in the approximate neutral zone, open the field circuit as described in the above paragraph and read the deflection. Move the voltmeter points to bars 2 and 22, and read the deflection as the field is opened. Rotate the armature in either direction, and repeat these operations until the two readings are equal but opposite. This indicates that the correct neutral is exactly on the mica between bars 1 and 2 or between bars 21 and 22. The rocker ring is shifted to these points, as explained in the preceding paragraph.

If the armature cannot be rotated, the neutral is located by the use of a curve or a calculation (see Fig. 5-4). If the number of bars is divisible by the number of poles, proceed as outlined in the above paragraph. Read the induced voltages on bars 1 and 21, 2 and 22, 3 and 23, etc., until a point is reached at which the polarity of the induced voltage reverses. Then record four readings, two on either side of the reversing point; plot the induced voltages as ordinates and the number of commutator bars as abscissas. Keep in mind that the number indicates the centerline of the end of the bar. After the exact point of reversal has been determined from the curve, mark the relative position on the commutator. This is the correct neutral. Shift the rocker ring as described in one of the above paragraphs. It is possible to calculate the distance from the centerline of a bar on either side of the point of reversal to the neutral without plotting a curve. Only two readings are necessary, one on either side of the point of reversal. Measure the distance between the centerlines of two adjacent bars. The distance from the centerline of one bar to neutral is found by dividing the reading on that bar by the sum of the two readings. This quotient is expressed in percentage of the total distance between centerlines.

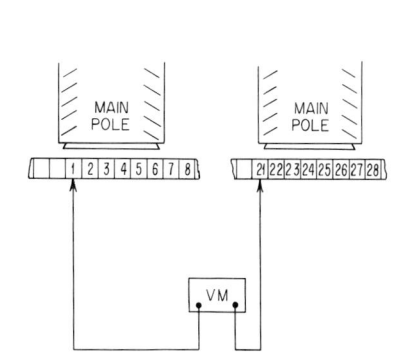

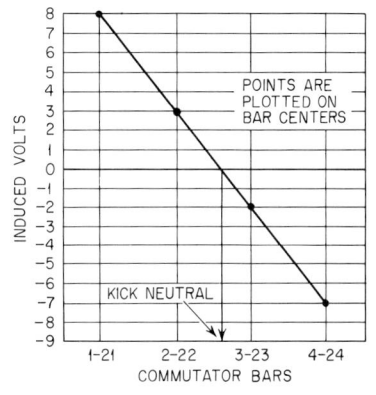

Fig. 5-3 Instrument connections for reading voltage induced in armature.

Fig. 5-4 Curve method of determining neutral.

Effect of Setting Brushes off Neutral. At times it is advisable to shift brushes off neutral. This may be done to obtain results that are too expensive to get otherwise or to get better performance curves. When brushes are moved in a direction opposite rotation, the motor becomes undercompounded and compensation increases. Moving brushes in the direction of rotation overcompounds the machine and undercompensates it. Should there be a lack of compensation in a motor which is to run in one direction only and the additional amount of compensation needed is not too great, the machine may be made to commutate properly by moving the brushes slightly against rotation. If the machine is of a reversing type, this cannot be done because poor commutation would be obtained in one direction of rotation. Sometimes a motor-speed curve has a hump in it. Often this can be corrected by moving the brushes off neutral a slight amount as long as commutation is not affected. The effect of shifting brushes off neutral must be observed over an extended period of time. The limit of good commutation determines how far the brushes may be moved. If good commutation is not obtained by moving the brushes, a change should be made in the commutating-pole air gap. This is seldom required but probably should be done by the manufacturer because of the equipment needs and the skills required.

Flashing. A question which has been argued pro and con for years is "Can the brush in itself be the cause for a machine to flash over?" A great deal of study and investigation has been devoted to this. A flash usually is the result of a suddenly charged condition in current or voltage or in the field strength of the machine. Disregarding control failures, the flashing results from a short circuit imposed by brush shunts breaking and touching on the opposite polarity, brushes breaking and causing shorting of commutator bars at the commutator surface, a piece of carbon from broken brushes becoming lodged between opposite polarities of the wiring around frame of the motor circuit, or bar burning on flat spots on the commutator. The large factor of safety in today's standard of commutation is such that a given grade of brush is rarely the cause of flashing.

Commutator Films. The most satisfactory commutator film is evenly colored and is between a light brown or straw color and dark brown. The most important point is that it be a uniform color and not a highly glazed or extremely dull finish.

Undesirable films due to atmosphere conditions are usually quite dark, black or gray. These films can be detected by the fact that the discoloration of the copper is seen on the surface, not in the brush paths.

The amount of contamination in the air that can be injurious to a commutator surface by causing undesirable film is extremely low. Conditions which cannot be detected by smell often cause deleterious results. For example, it is known that 3 parts of sulfur in 1 million parts of air are sufficient to cause heavy formation of undesirable film. When such conditions are found, a cleaner brush should be used with enough abrasive action to prevent the formation of the heavy film or else a brush whose film-forming qualities may develop a satisfactory film of copper oxide on the commutator surface before the injurious film can be established.

Copper Pickup. This undesirable feature is the transfer of copper particles from the commutator surface to the brush faces where they become embedded. Once started, copper pickup is usually progressive to the point where commutation and current collection is impaired. When copper becomes embedded in a brush face, the contact drop is decreased to the point where commutation becomes impossible. The rubbing effect of the copper particles against the commutator surface causes threading. Rubbing work-hardens the particles so that they act as tiny cutting tools. In order to remedy this condition, the proper bar-edge beveling should be maintained, the machine must be kept clean by blowing out thoroughly after working on the commutator, and brush-seating stones should be used sparingly. Also, the brush faces should be sanded to clean out the copper particles.

Brush Wear. This is one of the controversial subjects with operators of machines using brushes. All operators naturally desire longest possible brush life. However, compromises which result in increased commutator wear or poor commutation must not be resorted to in order to obtain long brush wear. Brushes wear away both electrically and mechanically, and these two have a complicated and strange relationship.

Brush life (for the most common electromagnetic brushes) is at its best when the load on the brush is around 55 amp/sq in. Lightly loaded brushes set up friction and chatter conditions, as explained previously.

A newly installed machine frequently has short brush life, caused usually by building and cement dust from the new construction. This condition gradually improves with time, although sometimes over a year may be required to overcome this trouble.

All the possible adverse conditions discussed above have an effect on brush life. Correction of the particular trouble also tends to correct short brush life.

PART III. INSULATION AND ITS CARE

The most important item in the maintenance of electrical equipment is taking care of the insulation. To take care of the insulation, keep it clean, keep it dry, and keep it cool. These three musts are interrelated; for example, if the external surfaces are kept clean, the motor will run cooler because of better heat transfer. A fourth point would be to make sure the power supply and the motor control are basically sound.

In many cases this means that good preventive maintenance can give added years of trouble-free life. It is extremely important to prevent insulation failure, because if the insulation fails, the motor or generator winding most generally has to be removed from the electromagnetic core and a new winding must be inserted and varnish-treated with the proper procedures before that particular motor can be used. This rewinding is time-consuming and varies from 2 days to several weeks depending upon the distance and the availability of the service facilities, the size of the equipment, and the availability of the insulating material as well as the conductors.

Such failures then cannot be considered as temporary shutdown failures, and it is important to have a maintenance plan and program. If the electrical rotating machine is a key to a continuous process, a spare motor is indicated or at least a stator if the machine is an ac squirrel-cage induction motor, or a rotor if the machine is a dc motor.

If the motor is not critical, the rewind time is possibly satisfactory, but in any case, a maintenance plan and program is justifiable.

Let us examine the conditions for long, trouble-free insulation life.

Keep it clean This means the surface of the insulating material shall be free of dirt, dust, salts, oil, cement dust, carbonaceous materials, chemicals, lint, fibers, etc. Quite often, the basic way to do this is to select the right kind of equipment enclosure, for example, a totally enclosed, fan-cooled motor instead of an open, open-dripproof, or splashproof motor, or separate clean air ventilation provided by a blower rather than some form of open motor using the surrounding contaminated air. Many failures would be eliminated, particularly where abrasive dusts such as cement dust, fine conducting dusts such as carbon dust, and strong chemical atmospheres are present.

However, even some dusts that seem innocuous will absorb moisture easily and cause the winding to become wet, which will reduce insulation resistance and dielectric strength; and the dirt, with the absorption of moisture, will affect surface tracking.

Chemical atmospheres can be very bad, because they may react with the insulation materials, causing the insulation to deteriorate.

Even if you enclose the motor, it is necessary to keep the external metal surfaces clean for good heat transfer and the external ventilation paths clean to obtain the maximum air flow for cooling purposes.

For good maintenance practices to give long insulation life, keep the motor clean by (1) removing the source of the contaminant, if possible; (2) selecting the proper enclosure; (3) cleaning the windings; and (4) cleaning the surfaces and keeping the ventilating paths open.

Keep it dry It is extremely important for long life to keep electrical equipment dry. Moisture and humidity can be particularly harmful to insulation, because if the insulation is porous, moisture will be absorbed and the insulation resistance can be reduced drastically. Also the surface of the motor in locations where it is very humid or cold can be completely covered with a thin film of water, which would make the external surface of the winding reasonably conductive. With this condition any pinhole in the insulation can cause a ground or a short between turns because the insulation resistance is lowered drastically, sometimes in the order of several magnitudes, like 1000 to 1.

Dust, dirt, salt, oil, chemicals, and other contaminants generally greatly accelerate the wetting of the surface, emphasizing the fact that the machine should be kept clean as well as dry.

It is important to remember that all insulation is affected to some degree by absorption of moisture or by collection on the surface, but the degree is widely different with various insulations. The higher-temperature insulations are more nonhygroscopic (moisture-resistant) than the lower-temperature insulations, and vacuum-pressure impregnation (not as important on random-wound as on form-wound) is much more moisture-resistant than the normal varnish dips and bakes. Also, more dips and bakes are helpful in decreasing moisture-absorption characteristics of the windings because of the added thickness of varnish and the reduction of pinholes that reduce surface tracking when wet.

In the selection of equipment, proper choice of enclosure can ensure against water in the solid form leaking, dripping, and being blown into the motor when such hazards are present. If recurring flooding of an induction motor is known to happen, encapsulation of the motor windings might be an answer to keep the motor running during such periods and thus keep a continuous process rolling.

If motors are used in very humid or very cold atmospheres, the use of space heaters might be indicated when the motors and generators are not operating (not energized). Space heaters that keep the air temperature higher around the windings by about 3 to 6°C (37½ to 50°F) will prevent surfaces from becoming wet. Such devices are very important in such areas as the Gulf of Mexico coast and Alaska, for standby, storage, and very intermittently operated machines. When machines are in operation, they generate enough heat to drive off moisture.

Keep it cool This means the machine temperatures should be kept within the temperature rating specified on the machine nameplate. Machines are generally designed to meet the ANSI (American National Standards Institute) standards, IEEE (Institute of Electrical and Electronics Engineers) standards, and NEMA (National Electrical Manufacturers Association) standards for insulation and systems. The insulation systems are classified as O, A, B, F, and H with their hottest-spot temperatures of 90, 105, 130, 155, and 180°C, respectively. At this time, O is used only and rarely in small devices; A insulation is used in small motors, particularly fractional-horsepower where flux limitations are met before temperature-rise limitations; B and F insulation are used on integral-horsepower ratings, with B insulation being the NEMA standard on ratings 1- to 200-hp polyphase ac induction motors; and H insulations are used for special motors and special applications, particularly high ambients like 50 and 65°C. The pertinent data on these standards are shown below.

Insulation class	Hottest-spot temp, °C	Permissible rise, °C			
		By thermometer* method	By resistance or embedded detector*		
			TEFC	Max at 1.15 SF	TENV
O	90	35	45		
A	105	50	60	70	65
B	130	70	80	90	85
F	155	105	105	115	110
H	180	130	125		135

*Based on 40°C ambient and suitable hot-spot allowances for the methods of measuring temperatures.

The above table is associated with maximum temperatures, but in actual practice the operating temperatures are generally well below the limits. For example, the ambient temperature of most applications is in the 24 to 30°C range, and usually machines are not fully loaded on a continuous basis.

It is advisable to keep the operating temperature of the machine as low as possible because the thermal aging of the machine is dependent upon the average temperature and the insulation life is generally based on an approximate rule of thumb that the thermal life of insulation is doubled for each 10°C decrease in temperature. Although this number can be different and can vary from 8 to 15°C, it does give a reasonable concept of life within the usual operating-temperature ranges. Therefore, to emphasize our first two points, keeping the motor clean tends to keep it cool, resulting in longer life, and keeping the machine dry makes it easier to keep the motor clean.

In the last few years, in the interest of reducing overall noise, enclosures have been placed around equipment with little regard to the effect of the enclosure on ventilation and the adverse thermal effects on the equipment. Therefore, any time an enclosure is used to surround equipment or the electrical machine, the thermal effects on the equipment and on the machine and the ambient must be checked. The ambient is important because of its relation to overload relays in the control and the ambient of the control center. When the motor ambient is very hot or above normal and the ambient of the control elements, including the overload relay heaters, is normal, the control will not protect the motor from overload. If the ambients are colder than normal, nuisance tripping will occur and control should be ambient-compensated.

Proper power supplies It is important to have the proper power supply to avoid troubles. Low voltage generally results in higher conductor losses and makes the machine run hotter. High voltage results in higher electromagnetic-core losses and, depending upon saturation, can cause the machine to run hotter. Care should be taken to keep the voltages within the standard tolerances allowed in NEMA and ANSI standards.

Many direct-current motors are operated from rectifier power supplies, and the pulsating voltage and current waveforms affect the performance by increasing motor heating and degrading commutation. Because of these effects, it is necessary that the motors be designed or specially selected to suit this type of operation. Since there are many different power supplies, it is important that the proper power supply be matched with the motor. The motor and power-supply manufacturer(s) should be contacted if there is any question of compatibility. The motor nameplate normally specifies by code the type of power supply for which the dc motor is intended.

This is also true of ac adjustable-speed motors supplied by inverters, converters, primary-voltage static power supplies, etc., and it is very important that the motors and power supplies match to prevent undue heating.

Hermetic motors are a special case because they are designed to operate in a hermetically sealed enclosure with one of the freons, gas, or liquids being used as the cooling medium. Such motors generally require special insulating materials to be compatible with a specific freon gas or liquid, and it is therefore important to rewind with the special materials in case of a winding breakdown. Generally, the safest way is to have the rewinding done under the direction of the motor manufacturer or the equipment manufacturer.

Cleaning and Drying Insulation

The operating instructions provided by machine manufacturers always emphasize the importance of keeping electrical apparatus clean, dry, and cool, as mentioned above. Favorable locations, adequate ventilation, application of heaters to prevent condensation of moisture on the motor when the machine is out of service for any length of time, and suitable covers all help to reduce the number of outages and to lessen maintenance costs.

However, motors do get dirty, and when they do, the insulation must be cleaned and restored to as near its original condition as possible and as soon as possible. The amount of cleaning and the method of cleaning are dependent upon the type of dirt, the amount of moisture, how soon the machine is required to go back in service, the availability of a spare machine or component, etc. After cleaning, it is important that the machine is dry. Visual examination is not sufficient, and tests should be made to make sure that the motor windings have an insulation resistance of not less than

$$\frac{\text{Rated voltage}}{(0.075 \times \text{hp rating}) + 1000} \text{ megohms}$$

which would be 0.46 megohm for a 10-hp 460-volt motor and 2.14 megohms for a 1000-hp 2300-volt motor, as two examples. However, a minimum of 1.0 megohm is preferable.

If an electrical machine has been stored in a damp location or has been in a humid environment without being operated for a long time, it should be dried out.

Windings can be dried in a number of ways: (1) Bake the motor in an oven, preferably a circulating-air oven, at a temperature not exceeding 90°C, until the insulation resistance becomes practically constant. (2) Enclose the motor with canvas or similar covering, leaving a hole at the top for moisture to escape, and insert heating units or lamps. (3) Pass a current at one-tenth rated voltage (with rotor locked) through the stator windings, and increase the current gradually until the winding temperature reaches 90°C. Do not exceed this temperature. In a dc machine pass current through the exciting field coils.

As stated above, when harmful dirt accumulations are present, a variety of cleaning techniques are available. First dry dust, dirt, or carbon should be vacuumed without disturbing adjacent areas or redistributing the contamination. Use a small nozzle or tube connected to the vacuum cleaner to enter into narrow openings, as between commutator risers on dc machines. A soft brush on the vacuum nozzle will loosen and allow removal of dirt more firmly attached.

After the initial cleaning with vacuum, high-velocity air may be used to remove the remaining dust and dirt. It is important to vacuum first so that extraneous material such as conducting or other harmful particles will not be driven into the windings or other critical areas. Note the highest pressure for compressed air should be 30 psi, which is the legal limit (OSHA), and during the operation, safety glasses and/or other protective equipment should be used to prevent possible eye injury. Care must be taken to make sure that the air supply is dry.

The presence of oil makes thorough, effective cleaning of machines in service virtually impossible, and repair-shop or service-shop conditioning is recommended. Oil on a surface forms a flypaper which attracts and holds firmly any entrained dust. Neither suction nor compressed air is effective; consequently only accessible areas may be cleaned. First, remove as much of the dirt as possible by wiping with clean, dry rags. For areas not readily accessible, a clean rag is drawn through an opening by means of a hooked wire and the rag is drawn alternately back and forth. This process is continued until a clean cloth thus applied will stay clean. Cloths should be changed frequently in any wiping operation; otherwise, the dirt or contamination picked up by the cloth may simply be transferred to another previously uncontaminated area.

To simplify removal of oily dirt, solvents are commonly prescribed, but in the field on assembled machines, liquid solvents are strongly discouraged, particularly on dc machines and ac machines with commutators or collectors. Generally the solvent application in the field should be by a wiping rag barely moistened (not wet). Where dirt is heavy, repeated wipings may be required. Liquid solvent can carry conducting contaminants (metal dust, carbon, etc.) deep into hidden but critical areas to produce shorts and grounds, thus causing machine failure. Without special testing equipment such as surge testers, weaknesses that could result in shorts in an armature, wound rotor, or rotating field cannot be exposed. Grounding weakness may be studied with a megger, but even here, acceptability of the test is questionable because the megger does not develop enough electrical energy to expose all the weak grounding paths.

Solvents may be toxic or flammable. Therefore, adequate ventilation must be provided to minimize fire hazards and health hazards caused by the use of solvents for cleaning purposes. Keep away from sparks, heat, or flame to prevent fire or explosion hazard.

Prior to cleaning and other maintenance activities electric circuits must be deenergized. Also electric circuits should be grounded to discharge capacitors prior to cleaning or maintenance. These precautions are necessary to avoid electric shock which can cause serious or fatal injury.

Certain freons are recommended for cleaning because they are not flammable, have good solvency for grease and oil, are considered safe with most varnishes and insulations, and have a low order of toxicity. Inhibited methyl chloroform is also acceptable. Carbon tetrachloride is effective and nonflammable but is very toxic in confined spaces with repeated usage and is therefore not recommended. The chlorinated solvent is sometimes used in very stubborn cases, but all the safety precautions must be used. Toluene and xylene and Stoddard solvent (hydrocarbon vents) possess good solvency but are not recommended, since they are flammable and attack varnishes quite readily. Steam cleaning is not recommended because, as with liquid solvents, conducting contaminants may be carried deep into inaccessible areas, resulting in shorts or grounds.

Carbon-brush performance may be ruined by absorbed solvents; so brushes should be removed on dc machines, ac wound rotors, and synchronous generators (where used) before solvent wiping.

The above really refers to field-service cleaning. In a repair shop or service shop with a disassembled machine, there are many more options, and generally the shop has more equipment, more instrumentation, and more know-how to perform the necessary operations.

The first step is to get an initial insulation-resistance reading on each machine component. Low readings would be expected with badly contaminated machines, but failure would indicate electrical damage calling for repair, not just cleaning.

Cleaning with water and detergent is a very effective method of cleaning windings when used with a low-pressure steam jenny (maximum steam flow 30 psi and 90°C), but to minimize possible damage to varnish and insulation, a fairly neutral nonconducting type of detergent should be used. A pint of detergent to 20 gal of water is recommended. If a steam jenny (steam-spray machine) is not available, the cleaning solution may be applied with warm water by a spray gun. Tightly adhering dirt will require additional agitation by gentle brushing or wiping. After the cleaning operation, the windings should be rinsed with water or low-pressure steam.

Even plain water can be used to wash out motors that have been plugged with mud or other inert foreign matter by plant operations or floods. The motors can be washed with water from a hose and disassembled so that all parts can be thoroughly cleaned. When water is applied to insulated parts, the pressure should not exceed 25 psi.

After either cleaning operation described above, the surface moisture should be wiped off with a clean cloth and the insulation dried promptly to keep the penetration of water as low as possible. It is advisable to dry the windings further, preferably in a circulating-air oven at a temperature not exceeding 90°C until the insulation resistance becomes practically constant. Other methods such as hot-air heaters are also satisfactory, but care must be taken to keep the temperature within limits and to make sure that local hot spots are not developed.

If an oven cannot be used, the machine's own frame with the addition of some covers will usually make an effective enclosure to contain heat. Some flow of air is desirable to allow the moisture to be carried away. Methods of generating heat include blowing hot air through the machine, heating with heat lamps, or passing a current at low rated voltage with the rotor locked for ac machines and passing a low current through the main-field-coil windings for dc machines.

Solvents can be used in the repair shop or service shops under controlled conditions. If the dirt incrustations are not removable by wiping, blowing, suction, or steam cleaning and the electric-circuit insulation is not restored, solvent cleaning is employed. The two types of solvents are petroleum distillates and other solvents, comprising chlorinated solvents, mixtures of chlorinated and petroleum solvents, and coal-tar solvents. Preferences have been shown for both types.

Actual immersion in trichlorethane or perchlorethylene with air agitation or suspension of the part in a vapor degreaser will soften dirt accumulations, permitting compressed-air removal. Repeated cleanings in this manner may restore the insulated circuits. Under no conditions should carbon tetrachloride or any *chloride* solvent be used because of the highly toxic effect. However, taking the necessary precautions, the *chlorinated* solvents are used effectively.

The petroleum distillates (hydrocarbon solvents) are classed as safety-type solvents and have flash points at about 100°F (38°C). They can be supplied by practically all oil companies under various trade names. Stoddard solution or solvent as described in the National Bureau of Standards Commercial Standard CS 3-40 is a good solvent of this type. It is sometimes modified with perchlorethylene and methylene chloride to give better results. However, all such solvents are flammable, and the vapors form explosive mixtures with air if used at temperatures above their flash point. Since their flash point is higher than that of other types of petroleum distillates, such as gasoline, however, the so-called safety type presents less of a fire hazard. Complete safety precautions should be taken.

An appreciable percent of complex structures (such as dc armatures) may never recover with submerged solvent cleaning, because conducting contaminants become trapped in critical fissures. Also commutators, collectors, etc., should not be cleaned by submerging in a solvent.

While electric-circuit insulation is restored by submerged solvent cleaning, revarnish treatment is required. This also applies to water and detergent cleaning and is recommended for steam cleaning. The varnish used should be what the manufacturer recommends.

One other method of cleaning should be mentioned. Ground-up corncobs, peanut husks, and the like have been employed for mild abrasion and absorption of oil and grease of contaminated electrical components. Where visual contamination is removable in this manner, its use is recommended, as the meal has an affinity for oil and grease and will clean a machine in excellent fashion. When this method is used, the machine should be covered and the dirty meal drawn off by a vacuum. The operators should use a dust mask or respirator.

Silicone high-temperature insulation represents a special case, and the manufacturer should be consulted.

Testing Insulation

The main test in maintenance work is the measurement of insulation resistance. The best guide in making these tests is IEEE 43, "Recommended Guide for Testing Insulation Resistance of Rotating Machiner," published by the Institute of Electrical and Electronics Engineers, 345 East 47th Street, New York, N.Y. 10017. This serves as probably the best indication of whether the machine is in condition for operation and dielectric testing.

The insulation-resistance test gives a good indication of the condition of the insulation, particularly from the standpoint of moisture and dirt. The value of resistance depends upon the type size, voltage rating, etc., of the machine. A good general minimum value is

$$\frac{\text{Rated voltage}}{(0.75 \times \text{hp rating}) + 1000} \text{ megohms}$$

or approximately 1 megohm for each 1000 volts of operating voltage with a minimum value of 1 megohm. Preferably, the resistance corrected to 40°C should measure at least 1.5 megohms. However, don't be fooled; high-insulation-resistance values do not necessarily assure high dielectric strength, although low-insulation resistance may indicate low dielectric strength. The importance of the measured value lies in the relative readings of insulation values under similar conditions at various times. They usually indicate under these conditions the effectiveness of the maintenance work.

A sudden drop or a consistent trend toward low values of insulation resistance gives evidence that the insulation system is deteriorating and that failure may be imminent.

Tests to obtain values of insulation resistance test only the value of resistance to ground and in addition, winding to winding in the case of dc fields. Disconnect all external leads for the test.

The armature and individual field coils in a dc machine can be tested only to ground and not from one circuit to another or along creepage paths from one exposed voltage to another. Superficial cleaning methods may greatly improve resistance to ground but may actually further deteriorate the insulation condition across shorting paths (by washing contaminants into the machine). A machine whose insulation resistance is low could be given an in-place cleaning and later fail because cleaning was not complete and damaging amounts of contamination were left that could not be detected by testing insulation resistance to ground.

Insulation resistance can be measured by a self-contained instrument such as the familiar megger, either hand- or motor-operated; by the electronic type, with a resistance bridge; or with a milliammeter, a voltmeter, and a dc supply. Any of these instruments used in insulation testing must be well maintained and calibrated to be sure that the readings are factual.

As stated above, the insulation resistance of apparatus in service should be checked periodically at approximately the same temperature and under similar conditions of humidity to determine possible insulation deterioration. If such measurements show wide variations, the cause should be determined and corrective measures taken to forestall an insulation failure. On large machines a further refinement is used by obtaining the polarization index. The polarization index is obtained by dividing the 10-min insulation-resistance value by the 1-min insulation-resistance value. The index is a useful means of determining if a machine is suitable for overpotential testing or for further operation. A history of this index is valuable in deciding whether or not the insulation system is deteriorating. If insulation-resistance values are taken for use in calculating a polarization index only, they do not have to be corrected for temperature, since this does not affect the ratio.

The recommended minimum value of polarization index is 2.0 for large form-wound machines. This is subject to somewhat the same limitations that are applied to insulation-resistance measurements.

Dielectric Tests

The purpose of dielectric tests is to determine if the insulation on the machine can withstand the voltage stresses set up during normal and possible abnormal conditions during operation.

The application of the ac high voltage necessary to make a dielectric test presents hazards in that not only can the ac voltage used puncture or break down the insulation, but often severe burning of the machine laminations occurs because the capacity needed to test larger machines is such that in case of breakdown, a large amount of power follows in the arc established. However, in many cases, the risk involved does not outweigh the possible long outage that might occur if the insulation failed while in an operation driving an important load. To diminish that risk, it is important that a satisfactory insulation resistance is obtained because a lot of damage can be done by applying high voltage to a wet and/or dirty winding.

The test voltage applied to new machines or to the winding of machines completely rewound with new coils and insulating material is specified by IEEE and ANSI standards as twice rated voltage plus 1000 volts held for 60 sec, with the exception of field windings of synchronous motors, which are given a test voltage of 10 times the exciter voltage but not less than 1500 volts. For machines in commercial operation or for repaired machines, no standards have been set, but established practice is to use an ac test voltage between 65 and 75 percent of the test voltage for new windings. The lower value should be used for older windings.

Within recent years, high-voltage dc testing has become more and more accepted. It has numerous advantages over ac testing. The capacity used is small, and the test effect in searching out weak insulation is comparable with ac testing. The unit used is considerably smaller physically than the test transformer, the equipment needed to test the largest machine being easily transported in a car, whereas the transfomer requires a large truck. The device is electronic and consists essentially of a high-voltage rectification circuit. Instruments measure the current and voltage. Another advantage, which is of prime importance, is that in case of an insulation failure during test, no iron burning results because of the small amount of power used. The test equipment operates from the 60-cycle lighting circuit. Test values have been established whereby a dc test voltage is applied 60 percent greater than the ac test voltage ordinarily used. The cost of the dc test outfits is considerably less, especially when compared with the cost of a test transformer with capacity to test the windings of large machines. The same dc test outfit can be used to test the windings of machines, from the smallest to the largest. This is generally used for large machines 6000 volts or over. See IEEE 95, "Insulation Testing of Large A-C Rotating Machines with High Direct Voltage."

A further test is being used for dielectric testing of high-voltage machines using ac high voltage at very low frequencies (about 0.1 Hz). A working group in IEEE is developing a standard. There are several advantages: the newer equipment is much lighter and more portable than the established ac equipment; it gives a better ac search than high-voltage dc testing; and it does not generate corona during the testing period.

Turn-to-turn insulation can be checked with a surge-comparison tester and is used to locate insulation faults and winding dissymmetries in all classes of equipment regardless of size. It is an electronic device and is portable, so that it can be used for maintenance work as well as shop work. High turn-to-turn voltages are applied without excessive winding to ground stresses, and the testing is nondestructive. Both the surge test and the high-voltage dc testing should be done by experienced, trained operators.

PART IV. BEARINGS AND LUBRICATION

Proper care of bearings, which includes lubrication, is one of the more important maintenance items pertaining to motors. The rotor of a motor, which transfers the electrical energy from the power supply into mechanical energy through the rotating shaft to drive the load, is supported on bearings.

The designer has a choice between sleeve bearings and antifriction ball or roller bearings, but for normal applications, economics will decide the choice of the bearing and lubrication system. Despite the trend in integral-horsepower motors from sleeve to ball bearings in the range of 1 to 500 hp, millions of fractional-horsepower motors are made with wick-oiled sleeve bearings.

In the integral-horsepower range of 1 to 500 hp, almost all normal motors are made with ball bearings or roller bearings except for some of the larger, higher-speed motors. The trend to ball bearings is such that 100 hp and smaller sleeve-bearing motors are considered special.

Over 500 hp, ac and dc motors are normally made with oil-lubricated sleeve bearings for the high surface speeds with large shaft diameters and usually have split bearings for ease in replacement. However, some dc machines as large as 2000 hp are equipped with antifriction bearings.

An analysis of induction-motor failures over a long period shows that bearings are one of the principal causes of trouble and the successful operation of a motor depends on a good preventive-maintenance program. Not only will loss of lubrication fail the bearings, but the failure might result in other problems such as winding or stator-rotor rubbing problems on sleeve-bearing motors.

On small motors with wick-oil sleeve bearings, good maintenance requires that the bearings be reoiled once a year or every 2000 operating hours, whichever comes first. Approximate amounts of oil to be added vary from 30 drops for a 3-in.-diameter rotor to 100 drops for a 9-in. motor. If light turbine oil is unavailable, SAE 10 automotive oil will suffice. During any other motor maintenance work, the yarn or felt packing should be replaced. After the bearing housing is washed, the wick cavity is repacked with clean wool yarn saturated with oil.

For ball bearings and roller bearings, good maintenance requires that the bearings be regreased on a schedule which will vary with motor size, speed, duty, and environment. The larger the motor, the higher the speed, the hotter the temperature or ambient, the greater the vibration, the more severe the duty, and the higher the loading, the more frequent should be the regreasing. The interval of times for relubrication vary from 10 years to 1 month depending upon the above factors. For example, a 1-hp motor used in infrequent operations, such as portable tools, normally used in a clean and reasonable temperature ambient of 10 to 30°C (50 to 86°F), would need to be regreased only once in

10 years while a 200-hp motor used continuously in a dirty, hot environment with severe duty and heavy vibration would need to be regreased every month. This constitutes a range of 84 to 1.

Table 5-2 is an approximate guide for relubrication of electric motors. Note that some manufacturers recommend longer periods and some shorter periods for regreasing. These periods are based on deep-groove conrad ball bearings or roller bearings with open, single-shielded, or double-shielded bearings with a grease reservoir adjacent to the bearings.

TABLE 5-2 Guide for Maximum Relubrication Periods*

Service	Motor horsepower				
	¼–10	15–40	50–150	200–250	Over 250
Easy: infrequent operation, valve operators, door openers, portable tools	7 years	5 years	3 years	2 years	9 months
Standard: one or two shifts, machine tools—air-conditioning conveyors, compressor refrigeration, laundry, textile machinery, woodworking, water pumping, generally Class B insulation	5 years	3 years	1 year	9 months	6 months
Severe: continuous running. Fans, pumps, motor-generator sets, coal and mining machinery, steel mills, some Class F insulation	3 years	1¼ years	6 months	4 months	3 months
Very severe: dirty and vibrating applications, hot pumps and fans, high-ambient Class H insulation	9 months	4 months	3 months	2 months	1½ months

*Some manufacturers recommend longer periods and some recommend shorter periods, based on open and shielded bearings with grease reservoirs adjacent to bearings.

Sealed bearings are not used generally because the grease, and therefore the oil, for lubrication is limited to what is inside the bearing proper and then the life of the bearing becomes the life of the grease. However, on some applications, particularly some in the military field where only 4000 to 5000 hr of life is needed, they are extremely satisfactory because once installed, there is no question concerning the proper grease; there is almost no chance of getting dirt in the bearings; and there is no chance of overgreasing. Sealed and shielded bearings require a stiffer grease to provide greater mechanical stability to minimize churning while closely confined near the rotating balls.

Shielded bearings do allow some feeding of oil from the grease in the housing reservoirs, and open bearings allow the most.

The greases used for ball and roller bearings have constantly been improved over the years, and the modern ball-bearing greases are very stable chemically.

If kept clean and at the intended operating temperatures, these modern greases will lubricate satisfactorily for many years. As a result, some smaller motors are not provided with grease fittings or plugs and are intended to operate without greasing maintenance.

Starting in 1925, the sodium-calcium mixed-base grease replaced calcium cup grease; in 1940, oxidation inhibitors were introduced for general use; in 1953, lithium-soap grease was introduced; and in 1968, the synthetic thickeners in petroleum grease appeared for use in higher bearing temperatures that might be encountered with Class F insulation motors. (The much higher temperatures associated with Class H insulated motors used silicone grease brought out in 1945 or some of the recently developed greases using synthetic thickeners, such as polyurea.) On high-temperature applications always use a grease the same as or compatible with the original grease.

A conscientiously applied program of preventive maintenance will add years of useful life to bearings. Machines with antifriction bearings are shipped from the factory with the bearings packed with grease. If the machines have been stored for a long period of time, it is advisable to regrease the bearings before operation, especially for motors of 150-hp size and larger. The grease used as a lubricant in grease-lubricated antifriction bearings does not lose its lubrication ability suddenly, but the oil bleeds out of the grease over a period of time. For a given bearing and assembly, the loss of lubricating ability of a grease with age depends primarily on the type of grease, the size of the bearing, the speed at which the bearing operates, the temperature and the severity of the loading, and operating conditions. Although it is not possible to predetermine accurately when new grease must be added, good results can be obtained with the following procedure:

1. Wipe all lubrication fillings clean.
2. Remove the relief plug and free the hole of hardened grease.
3. With the machine running, add grease slowly with a hand-operated pressure gun until new grease is expelled through the relief hole. (If the fittings are not safely accessible with the machine running, grease may be added sparingly with the machine at rest.)
4. It is recommended that the motor be regreased at standstill for safety reasons as well as to prevent grease leakage along the shaft-end shield fit and into the internal end shield cavity and the motor windings.
5. Run the machine from 10 to 20 min with the relief plug removed to expel excess grease.
6. Clean and replace the relief (outlet) plug.
7. Make sure the new grease is clean. Dirt and contaminants introduced when adding grease can cause bearing failure, and some foreign particles such as silica or cast-iron dust are almost sure to cause bearing problems.
8. Use the grease (or an equivalent grease) recommended by the motor manufacturer. Some very good greases are not compatible with each other, and if there is any doubt about compatibility with mixed greases, the motor manufacturer should be consulted.
9. Do not overgrease, because this can cause bearings to heat up considerably and cause failure. Sometimes motors are furnished with grease plugs instead of fittings to prevent casual overgreasing.

If a motor has to be disassembled for any reason, such as cleaning or rotor or stator repairs, the bearings and housings should be cleaned of the old grease by washing with a grease-dissolving solvent and the bearing should be repacked at about 50 percent full with the manufacturer's recommended grease or equivalent. Again, cleanliness is the critical word because of the small clearances in ball and roller bearings. Compressive stresses commonly range well over 1,000,000 psi in the minute contact area in the heavily loaded ball-race contacts. It is easily seen that a piece of grit or cast-iron dust could destroy the balls and races.

If a bearing has to be replaced, care must be taken in removing the bearing and reinstalling either the old bearing or a new one. Care must be taken to avoid scratching or nicking critical surfaces. A bearing puller should be used to remove the bearing. It is similar in looks and works like a gear puller or pulley puller. Wherever possible, apply the pulling force on the inner ring or race which is on the shaft. Pulling on the outer race transmits the force through the balls in the bearings, which presents the possibility of damage by brinelling. In case you have to pull on the outer race, the bearing should be replaced. As a matter of fact, any pulled bearing is generally replaced unless it is hard to procure a new one.

Installation of a replacement bearing starts with making sure that it is the correct type. The bearings in many cases are identified on the motor nameplate. If the original bearing called for is not available, it might be wise to check the manufacturer for a suitable substitute. Before reassembling a bearing, all bearing and machined surfaces should be thoroughly cleaned with a suitable solvent. Examine the machined fits of the end shield, cartridge, slinger, and grease caps for burrs. It is important that these surfaces be smooth.

A typical way to reassemble a bearing is as follows:

1. Inspect the bearing housing and related parts for foreign material. Clean, if necessary.
2. The machined fits and critical surfaces of the end shield, bearing cartridge, grease cap, shaft, and bearings should be free of all nicks, scratches, or burrs. If any polishing is

done, care should be taken to avoid a deposit of metal dust in and around the bearing assembly.

3. The internal surface of the bearing cartridge or housing should be coated with a thin film of recommended grease as well as the shaft and shaft fit of the grease cartridge and grease cap. These precautions, although not absolutely essential, will guard against corrosion of the critical surfaces.

4. On smaller bearings, heat the bearing in an oven to 100 to 125°C (but not higher) and place on the shaft while making sure that it is against the locating shoulder. The cooling of the bearing will give a tight fit with the shaft. If no suitable heat is available, a piece of pipe can be used as a pressing fixture to press the bearing tightly against the locating shoulder. The bearing should be gently tapped into place, with pressure applied to the inner race only. Any pressure applied to the outer race is transmitted through the balls to the inner race. Small indentations or impressions in the races (brinelling) will result in making the bearings noisy and will also cause premature bearing failure.

On larger motors with larger bearings, the bearing can be heated in hot oil 50 to 125°C (122 to 257°F) and placed on the shaft. Hold the bearing against the shaft shoulder until the bearing cools.

5. Replace the bearing nut and washer if used.
6. Secure the bearing cap and housing and make sure the shaft turns freely.

If a bearing failure in a specific installation or application occurs too frequently to be considered the result of normal fatigue, then something is wrong and bearings will fail repeatedly until the "something wrong" is corrected. If possible, a bearing should be removed at the first sign of noise or heating so that a bearing specialist or the motor manufacturer can examine the bearing. Much can then be determined by a knowledgeable expert or the manufacturer's bearing specialist. Complete failures, on the other hand, destroy most evidence of the cause of trouble.

Sleeve bearings with oil lubrication are used for larger motors, particularly ones with a large shaft, to avoid the fatigue limits with rolling bearings, to operate above the speed limits for grease lubrication, and for easier bearing replacement. They are commonly used with split-end shields.

Motors with sleeve bearings are normally lubricated with turbine-type mineral oils, which offer the longest life of any petroleum lubricants available for self-contained or circulating oil systems. Their rust inhibitors also minimize corrosion on the shaft, bearing, and housing surfaces.

As a guide for selecting the proper grade of turbine oil, Tables 5-3 and 5-4 show the viscosities for several classes of motor operation. A light oil of 150 SSU (Saybolt Seconds Universal) viscosity at 100°F is recommended for fractional-horsepower motors and large units at 1500 rpm and faster. Lower speeds, high temperatures, and high loads normally call for the more severe requirements of some rolling-contact thrust bearings (such as for vertical pumps) call for heavier oils up to 600 SSU. It can be seen that manufacturers' recommendations should be followed for the best operation.

TABLE 5-3 Suggested Viscosities for Electric-Motor Oils

	Oil-viscosity grade, SSU at 100°F	
Type of bearing system	Below 1500 rpm	1500 rpm and up
Sliding contact		
Pressure-lubricated	300	150
Ring-oiled	300	150
Disk	300	150
Wick-oiled (fractional-hp)	150–300	150
Plate-type, thrust	300	300
Rolling contact		
Ball and cylindrical roller	150	150
Spherical roller	600	600

However, other grades can often be substituted successfully. For example, a 300 SSU oil can give satisfactory overall performance in a high-speed motor, even though the bearing temperatures are higher than desired and power losses slightly greater. On the other hand, motors running below 1500 rpm start more easily with a light oil.

Oils other than the turbine type are normally avoided. Automotive oils not only are more expensive, but their detergents may cause foam, emulsions, and other operating difficulties. Gear oils generally give shorter life, except where their ability to withstand extreme pressures is required, as in gear motors.

TABLE 5-4 Typical Motor-Oil Properties

Characteristic	Grade		
	Light	Medium	Heavy
Viscosity, SSU at 100°F	140–170	270–325	540–700
Viscosity index, min	85	85	85
Pour point, °F, max	0	5	25
Flash point, °F, min	380	400	420
Neutralization number, mg of KOH per gram, max	0.15	0.15	0.8
Aniline point, °F, min	195	195	
Turbine-oil oxidation test, hr, min (ASTM D943)	1000	1000	1000
Rust-prevention test (ASTM D 665)	Pass	Pass	Pass

Synthetic oils intended for aircraft, fire resistance, or other special applications are normally unsuited to electric motors, since most of them attack the paint, rubber, and insulations usually used. If extreme temperatures suggest using a synthetic oil, the complete motor should be considered for compatibility with all components. Under such conditions the best answer might be a circulating oil system, which has supplemental heating or cooling to keep oil temperature at a level which permits use of a turbine oil and avoids the compatibilty problem.

Maintenance with oil lubrication consists mainly of checking the oil every 3 or 4 months for level and appearance. Like ball-bearing relubrication, the ideal interval varies depending on ambient temperature, cleanliness, and severity of service. Level is usually checked at standstill, since rotation of an oil ring or other components may distort the reading. However, the manufacturer sometimes indicates that the oil level should be set while running. If the oil looks clean and not discolored, simply add enough to replenish the level.

Yearly oil changes are common unless ambients or operating conditions call for other intervals. The acidity of the oil can serve as a limiting guide for oil changes. When oxidation raises the neutralization number above 1.0 (ASTM D 974), oil gets darker, varnish deposits can be expected, and bearing corrosion rate increases. Oil should be changed promptly. If an acidity check is impractical, a visual observation is usually adequate. Do not add oil until it drops below full level and flooding should be avoided.

Any cleaning that is needed usually occurs when the oil is changed. This generally flushes water, dirt, and sediment from the bearing housing or oil reservoir. During disassembly for a general motor cleaning, the bearing and housing should be washed with a solvent. Some manufacturers recommend hot kerosene. Coating the drain plug with a sealer such as alkyd-resin compound or No. 3 Permatex before replacing it will prevent leakage.

Extreme care is required in the disassembly of a bearing to prevent nicking or burring of the bearing or machined running surfaces. In addition, the surfaces of the journal and the bearing must be protected from rust and damage when exposed during the process of disassembly and reassembly.

The sleeve bearing, if it must be replaced or rebabbitted, should be ordered from the motor manufacturer or a reliable supplier. If rebabbitting is indicated, the babbitt recommended by the motor manufacturer should be used and no changes should be made except by the motor manufacturer or a knowledgeable expert. There are several types of babbitt, but only two general classes are in use. The tin-base type contains 80 to 90 percent tin with the remainder divided equally between copper and antimony, and the lead-base type runs 75 to 85 percent lead with 5 to 10 percent tin and 5 to 10 percent antimony. Tin-base babbitt is generally used in corrosive atmospheres.

Clearance between the bearing and the journal is important, and the following general rules apply:

1. For shafts 1 in. in diameter and less, the diametrical clearance should be 0.001 in./in. of shaft diameter.

2. For shaft larger than 1 in. in diameter the diametrical clearance should be 0.002 in./in. of shaft diameter.

Since cleanliness is always important, all bearing and machine surfaces should be thoroughly cleaned with a suitable solvent before reassembly. Examine all machine fits for burrs, and remove if present. Remove all oil compound from sealing surfaces.

Prior to actual reassembly, the following precautions should be observed:

1. Inspect the bearing housing and related parts for foreign matter. Clean, if necessary.

2. Inspect the journals and polish them with crocus cloth if any scratches are detected. Do not allow any metal dust to fall into the housing when polishing the journals.

3. Spread a thin coat of oil over the journal and bearing surface before reassembling.

4. Sealing surface of the end shield should be coated with a sealing compound.

A suggested maintenance schedule on sleeve includes a monthly check of the oil level, a check that the oil rings that carry the lubricant to the journal and the bearing are rotating, and a check of the oil color through a sight gauge. Slightly cloudy oil is all right. Black oil is a danger sign. On a 6-month basis, check the oil for acidity and change it if required.

A good maintenance schedule, conscientiously followed with the proper materials and proper procedures, will give long life and service (Table 5-5).

TABLE 5-5 Good Maintenance Practices

Keep motor off line when not needed	Saves unnecesary wear of brushes, commutator, and bearings, saves lubrication
Do not leave field circuit unless motor has been especially designed for this type of duty	Check temperature of shunt fields with themometer to see that it does not exceed 90°C. When field must be excited, caution maintenance men to be sure field circuit is opened before working on the motor. On ac check stator windings
Keep motor clear of metal dust or cuttings that can be drawn into windings and pole pieces	Magnetic attraction will draw metal parts into the air gap and damage windings. Cast-iron dust particulary damaging
Reassembling of motor	Be sure to retain proper air gaps in motor by checking bore of pole faces before removing poles from the frame. Mark shims and poles. Reassemble, replacing poles and liners in original position on dc motors
Note wearing parts and parts frequently replaced to determine anticipated repairs	Carry in proper storeroom stock of replacement parts. Make survey of standard repair parts to avoid duplication of parts to be carried

The main points to remember about bearings are to handle them carefully, remove and install them properly, and lubricate them with the proper lubricants. Keep them clean.

Tables 5-6 to 5-8 summarize maintenance procedures and provide a ready reference.

If the maintenance man or inspector is to do a satisfactory job, proper tools and instruments are necessary. Also, he should have a good knowledge of the electrical and mechanical characteristics of the equipment under his care, together with an understanding of the correct operation of the equipment. The following, as a minimum, should be made available:

1. Tools necessary to disassemble apparatus
2. Extension cords, safety type
3. Flashlights, rubber or molded cases
4. Air-gap gauges
5. Micrometers, inside and outside
6. Dial indicator with assortment of brackets

Maintenance of Electric Motors 5-79

7. Megger, 500-volt
8. Volt-amp-ohmmeter tester such as Simpson, Triplett, or equivalent
9. Thermometers and levels
10. Portable instruments such as ammeters, voltmeters, and graphic meters

Access to instruction books furnished by the manufacturer with equipment purchased should be a must.

Only by having suitably instructed personnel with adequate equipment can motors be given the attention they need for long, trouble-free operation.

TABLE 5-6 AC and DC Motor Check Chart

Trouble	Cause	What to do
Hot bearings—general	Bent or sprung shaft	Straighten or replace shaft
	Excessive belt pull	Decrease belt tension
	Pulley too far away	Move pulley closer to bearing
	Pulley diameter too small	Use larger pulleys
	Misalignment	Correct by realignment of drive
Hot bearings—sleeve	Oil grooving in bearing obstructed by dirt	Remove bracket or pedestal with bearing and clean oil grooves and bearing housing; renew oil
	Bent or damaged oil rings	Repair or replace oil rings
	Oil too heavy	Use a recommended lighter oil
	Oil too light	Use a recommended heavier oil
	Insufficient oil	Fill reservoir to proper level in overflow plug with motor at rest
	Too much end thrust	Reduce thrust induced by driven machine or supply external means to carry thrust
	Badly worn bearing	Replace bearing
Hot bearings—ball	Insufficient grease	Maintain proper quantity of grease in bearing
	Deterioration of grease or lubricant contaminated	Remove old grease; wash bearings thoroughly in kerosene and replace with new grease
	Excess lubricant	Reduce quantity of grease. Bearing should not be more than half filled
	Heat from hot motor or external source	Protect bearing by reducing motor temperature
	Overloaded bearing	Check alignment, side thrust, and end thrust
	Broken ball or rough races	Replace bearing; first clean housing thoroughly
Oil leakage from overflow plugs	Stem of overflow plug not tight	Remove; recement threads; replace and tighten
	Cracked or broken overflow plug	Replace the plug
	Plug cover not tight	Requires cork gasket, or if screw type, may be tightened

TABLE 5-6 AC and DC Motor Check Chart (*Continued*)

Trouble	Cause	What to do
Motor dirty	Ventilation blocked, end windings filled with fine dust or lint	Clean motor will run 10 to 30°C cooler. Dust may be cement, sawdust, rock dust, grain dust, coal dust, and the like. Dismantle entire motor and clean all windings and parts
	Rotor winding clogged	Clean, grind and undercut commutator, or clean and polish collector. Clean and treat windings with good insulating varnish
	Bearing and brackets coated inside	Dust and wash with cleaning solvent
Motor wet	Subject to dripping	Wipe motor and dry by circulating heated air through motor. Install drip- or canopy-type covers over motor for protection
	Drenched condition	Motor should be covered to retain heat and the rotor position shifted frequently
	Submerged in flood waters	Dismantle and clean parts. Bake windings in oven at 105°C for 24 hr or until resistance to ground is sufficient. First make sure commutator bushing is drained of water and completely dry

TABLE 5-7 DC Motor Check Chart

Trouble	Cause	What to do
Fails to start	Circuit not complete	Switch open, leads broken
	Brushes not down on commutator	Held up by brush springs; need replacement. Brushes worn out
	Brushes stuck in holders	Remove and sand; clean up brush boxes
	Armature locked by frozen bearings in motor or main drive	Remove brackets and replace bearings or recondition old bearings if inspection makes possible
	Power may be off	Check line connections to starter with light. Check contacts in starter
Motor starts, then stops and reverses direction of rotation	Reverse polarity of generator that supplies power	Check generating unit for cause of changing polarity
	Shunt and series fields are bucking each other	Reconnect either the shunt or series field so as to correct the polarity. Then connect armature leads for desired direction of rotation. The fields can be tried separately to determine the direction of rotation individually and connected so both give same rotation
Motor does not come up to rated speed	Overload	Check bearing to see if in first-class condition with correct lubrication. Check driven load for excessive load of friction
	Starting resistance not all out	Check starter to see if mechanically and electrically in correct condition

TABLE 5-7 DC Motor Check Chart *(Continued)*

Trouble	Cause	What to do
	Voltage low	Measure voltage with meter and check with motor nameplate
	Short circuit in armature windings or between bars	For shorted armature inspect commutator for blackened bars and burned adjacent bars. Inspect windings for burned coils or wedges
	Starting heavy load with very weak field	Check full field relay and possibilities of full field setting of the field rheostat
	Motor off neutral	Check for factory setting of brush rigging or test motor for true neutral setting
	Motor cold	Increase load on motor so as to increase its temperature, or add field rheostat to set speed
Motor runs too fast	Voltage above rated	Correct voltage or get recommended change in air gap from manufacturer
	Load too light	Increase load or install fixed resistance in armature circuit
	Shunt field coil shorted	Install new coil
	Shunt field coil reversed	Reconnect coil leads in reverse
	Series coil reversed	Reconnect coil leads in reverse
	Series field coil shorted	Install new or repaired coil
	Neutral setting shifted off neutral	Reset neutral by checking factory setting mark or testing for neutral
	Part of shunt field rheostat or unnecessary resistance in field circuit	Measure voltage across field and check with nameplate rating
	Motor ventilation restricted, causing hot shunt field	Hot field is high in resistance, check causes for hot field, in order restore normal shunt field current. Restore ventilation
Motor gaining speed steadily and increasing load does not slow it down	Unstable speed load regulation	Inspect motor to see if off neutral. Check series field to determine shorted turns. If series field has a shunt around the series circuit, that can be removed
	Reversed field coil shunt or series	Test with compass and reconnect coil
	Too strong a commutating pole or commutating-pole air gap too small	Check with factory for recommended change in coils or air gap
Motor runs too slow continuously	Voltage below rated	Measure voltage and try to correct to value on motor nameplate
	Overload	Check bearings of motors and the drive to see if in first-class condition. Check for excessive friction in drive

TABLE 5-7 DC Motor Check Chart (*Continued*)

Trouble	Cause	What to do
	Motor operates cold	Motor may run 20 percent slow owing to light load. Install smaller motor, increase load, or install partial covers to increase heating
	Neutral setting shifted	Check for factory setting of brush rigging or test for true neutral setting
	Armature has shorted coils or commutator bars	Remove armature to repair shop and put in first-class condition
Motor overheats or runs hot	Overloaded and draws 25 to 50 percent more current than rated	Reduce load by reducing speed or gearing in the drive or loading in the drive
	Voltage above rated	Motor runs drive above rated speed requiring excessive horsepower. Reduce voltage to nameplate rating
	Inadequately ventilated	Location of motor should be changed, or restricted surroundings removed. Covers used for protection are too restricting of ventilating air and should be modified or removed. Open motors cannot be totally enclosed for continuous operation
	Draws excessive current owing to shorted coil	Repair armature coils or install new coil
	Grounds in armature such as two grounds which constitute a short	Locate grounds and repair or rewind with new set of coils
	Armature rubs pole faces owing to off-center rotor causing friction and excessive current	Check brackets or pedestals to center rotor, and determine condition of bearing wear for bearing replacement. Check pole bolts
Hot armature	Core hot in one spot, indicating shorted punchings and high iron loss	Sometimes full slot metal wedges have been used for balancing. These should be removed and other means of balancing be investigated
	Punchings uninsulated Punchings have been turned or band grooves machined in the core Machined slots	No-load running of motor will indicate hot core and drawing high no-load armature current. Replace core and rewind armature. If necessary to add band grooves, grind into core. However, treated glass roving properly varnished and properly processed is the more common method of banding windings in the small and medium sizes. Check temperature on core with thermometer not to exceed 90°C
Hot commutator	Brush tension too high	Limit pressure to 5 psi. Check brush density and limit to density recommended by the brush manufacturer
	High brush friction caused by atmospheric contaminants	Remove cause

TABLE 5-7 DC Motor Check Chart (*Continued*)

Trouble	Cause	What to do
	Brushes off neutral	Reset neutral
	Brush grade too abrasive	Get recommendation from manufacturer
	Shorted bars	Investigate commutator mica and undercutting, and repair
	Hot core and coils that transmit heat to commutator	Check temperature of commutator with thermometer to see that total temperature does not exceed ambient plus 55°C rise, total not to exceed 105°C. Class F and H will be hotter
	Inadequate ventilation	Check as for hot motor
Hot fields	Voltage too high	Check with meter and thermometer and correct voltage to nameplate value
	Shorted turns or grounded turns	Repair, or replace with new coil
	Resistance of each coil not the same	Check each individual coil for equal resistance within 10 percent, and if one coil is too low, replace coil
	Inadequate ventilation	Check as for hot motor
	Coil not large enough to radiate its loss wattage	New coils should replace all coils if room is available in motor
Motor vibrates and indicates unbalance	Armature out of balance	Remove and statically balance, or balance in dynamic balancing machine
	Misalignment	Realign. Alignment of flexible couplings generally must be closer than coupling manufacturer recommends
	Loose or eccentric pulley	Tighten pulley on shaft or correct eccentric pulley
	Belt or chain whip	Adjust belt tension
	Mismating of gear and pinion	Recut, realign, or repair parts
	Unbalance in coupling	Rebalance coupling
	Bent shaft	Replace or straighten shaft
	Foundation inadequate	Stiffen mounting place members
	Motor loosely mounted	Tighten holding-down bolts
	Motor feet uneven	Adds shims under foot pads to mount each foot tight
Motor sparks at brushes or does not commutate well	Brush setting not true neutral	Check and set on factory setting or test for true neutral
	Commutator rough	Grind and roll edge of each bar
	Commutator eccentric	Turn and grind commutator

TABLE 5-7 DC Motor Check Chart (*Continued*)

Trouble	Cause	What to do
	Mica high-hot undercut	Undercut mica
	Commutating pole strength too great, causing overcompensation, or strength too weak, indicating undercompensation	Check with manufacturer for correct change in air gap or new coils for the commutating coils
	Shorted commutating pole turns	Repair coils or install new coils
	Shorted armature coils on commutator bars	Repair armature by putting into first-class condition
	Open-circuited coils	Same as above
	Poor soldered connection to commutator bars	Resolder with proper alloy of tin solder. Current motors use mostly TIG welded
	High bar or loose bar in commutator at high speeds	Inspect commutator nut or bolts and retighten and grind commutator face
	Brush grade wrong type. Brush pressure too light, current density excessive, brushes stuck in holders. Brush shunts loose	See brushes
	Brushes chatter owing to dirty film on commutator	Resurface commutator face and check for change in brushes
	Vibration	Eliminate cause of vibration by checking mounting and balance of rotor
Brush wear excessive	Brushes too soft	Blow dust from motor and replace brushes with a changed grade as recommended by manufacturer
	Commutator rough	Grind commutator face
	Abrasive dust in ventilating air	Reface brushes and correct condition by protecting motor
	Off-neutral setting	Recheck factory neutral or test for true neutral
	Bad commutation	See corrections for commutation
	High, low, or loose bar	Retighten commutator motor bolts and resurface commutator
	Brush tension excessive	Adjust spring pressure not to exceed 2 to 2½ psi
	Electrical wear due to loss of film on commutator face	Resurface brush faces and commutator face
	Threading and grooving	Same as above
	Oil or grease from atmosphere or bearings	Correct oil condition and surface brush faces and commutator
	Weak-acid- and moisture-laden atmosphere	Protect motor by changing ventilating air, or change to enclosed motor

TABLE 5-7 DC Motor Check Chart (*Continued*)

Trouble	Cause	What to do
Motor noisy	Brush singing	Check brush angle and commutator coating; resurface commutator
	Brush chatter	Resurface commutator and brush face
	Motor loosely mounted	Tighten foundation bolts
	Foundation hollow and acts as sounding board	Coat underside with soundproofing material
	Strained frame	Shim motor feet for equal mounting
	Armature punching loose	Replace core on armature
	Armature rubs pole faces	Recenter by replacing bearings or relocating brackets or pedestals
	Magnetic hum	Refer to manufacturer
	Belt slap or pounding	Check condition of belt and change belt tension
	Excessive current load	May not cause overheating, but check chart for correction of shorted or grounded coils
	Mechanical vibration	Check chart for causes of vibration
	Noisy bearings	Check alignment, loading of bearings, lubrication, and get recommendation of manufacturer
	Magnetic noise	Tighten pole bolts. Motors supplied with power from an SCR source will generally have more noise caused by SCR ripple in armature current usually 120, 180, and 360 Hz with harmonics
Wrong rotation	Wrong connection	Consult connection diagram

TABLE 5-8 AC Motor Check Chart

Trouble	Cause	What to do
Motor stalls	Wrong application	Change type or size of motor. Consult manufacturer of driven equipment
	Overloaded motor	Reduce load
	Low motor voltage	See that nameplate voltage is maintained within standard tolerances
	Open circuit	Fuses blown; check overload relay, starter, and push button
	Incorrect control resistance of wound rotor	Check control sequence. Replace broken resistors. Repair open circuits

TABLE 5-8 AC Motor Check Chart (*Continued*)

Trouble	Cause	What to do
Motor connected but does not start	One phase open. Motor may be overloaded	See that no phase is open. Reduce load
	Rotor defective	Look for broken bars or rings
	Poor stator-coil connection	Remove end bells, locate with test lamp, and repair
Motor runs and then dies down	Power failure	Check for loose connections to line, to fuses, and to control
Motor does not come up to speed	Not applied properly	Consult supplier for proper type and size
	Voltage too low at motor terminals because of line drop	Use higher voltage on transformer terminals or reduce load
	If wound rotor, improper control operation of secondary resistance	Correct secondary control
	Starting load too high	Check load motor is supposed to carry at start
	Low pull-in torque of synchronous motor	Change rotor starting resistance or change rotor design. Consult manufacturer
	Check that all brushes are riding on ring	Check secondary connections. Leave no leads poorly connected
	Broken rotor bars	Look for cracks near the rings. A new rotor may be required, as repairs are usually temporary
	Open primary circuit	Locate fault with testing device and repair
Motor takes too long to accelerate	High WK^2. Excess loading	Reduce load. Check moment of inertia with equipment manufacturer
	Poor circuit	Check for high resistance
	Defective squirrel-cage rotor	Replace with new rotor
	Applied voltage too low	Get power company to increase voltage tap
Wrong rotation	Wrong sequence of phases	Reverse connections of motor or at switchboard
Motor overheats while running under load	Check for overload	Reduce load
	Wrong blowers or air shields may be clogged with dirt and prevent proper ventilation of motor	Good ventilation is manifest when a continuous stream of air leaves the motor. If not, check with manufacturer
	Motor may have one phase open	Check to make sure that all leads are well connected

TABLE 5-8 AC Motor Check Chart (*Continued*)

Trouble	Cause	What to do
	Grounded coil	Locate and repair
	Unbalanced terminal voltage	Check for faulty leads, connections, and transformers
	Shorted stator coil	Repair and then check wattmeter reading (form coils)
	Faulty connection	Indicate by high resistance
	High voltage, low voltage	Check terminals of motor with voltmeter
	Rotor rubs stator bore	If not poor machining, replace worn bearings (sleeve bearings)
Motor vibrates after corrections have been made	Motor misaligned	Realign
	Weak foundations	Strengthen base
	Coupling out of balance	Balance coupling
	Driven equipment unbalanced	Balance driven equipment
	Defective ball bearing	Replace bearing
	Bearings not in line	Line up properly
	Balancing weights shifted	Rebalance rotor. Securely fasten blancing means
	Wound rotor coils replaced	Rebalance rotor
	Polyphase motor running single-phase	Check for open circuit in one line or phase
	Excessive end play	Adjust bearing or add washer
Unbalanced line current on polyphase motors during normal operation	Unequal terminal volts	Check leads and connections
	Single-phase operation	Check for open contacts
	Poor rotor contacts to control wound-rotor resistance	Check control devices
	Brushes not in proper position in wound rotor	See that brushes are properly seated and shunts in good condition
Scraping noise	Fan rubbing air shield	Remove interference
	Fan striking insulation	Clear fan
	Loose on bedplate	Tighten holding bolts
Magnetic noise	Air gap not uniform	Check and correct bracket fits or bearing
	Loose bearings	Correct or renew
	Rotor unbalance	Rebalance

REFERENCES

Acknowledgment is made to the following books, pamphlets, and articles:
"Productive Maintenance," General Electric Company.
"Maintenance Hints," Westinghouse Electric Corporation.
Various maintenance articles published in *Power Engineering*.
Booser, E. R.: "Lubrication Plan for Rerated Electric Motors," General Electric Company, June 6, 1967.
Various maintenance and instruction bulletins of motor manufacturers.

Section **6**

Maintenance of Ground Contact Elements

Chapter **1**

Upkeep and Maintenance of Tires for Construction Equipment

LOUIS A. ARBORE, MANAGER

GEORGE ZAMBELAS, ENGINEER
Michelin Tire Corporation, Technical Group, Earthmoving Dept.,
Lake Success, N.Y.

INTRODUCTION

This chapter was composed to help you properly select, use, and maintain tires for construction equipment. Tire costs are a significant part of overall equipment maintenance costs, so it is very important that you know how to get the most out of your tire dollar.

To select a tire properly, you must know as much as possible about the specific application you are considering since each application, even on the same job site, can be very different. Specific job conditions, previous tire performance, and vehicle type and usage are to be thoroughly evaluated and matched to the capabilities of new tires that are being considered.

Tire cost is another factor that must enter into the tire selection procedure. Initial cost as well as expected performance must be considered. Choosing the least expensive tire may not be the answer. Consider also the increased productivity and lower downtime when using tires of higher quality.

After choosing tires, use and maintain them properly to get maximum performance. By familiarizing yourself with the information in this chapter and by contacting tire and vehicle manufacturers for additional information, you will be one step closer to realizing lower tire costs (Fig. 1-1).

SELECTING AN EARTHMOVER TIRE

The following is a list of the main factors to consider before selecting a tire:

Vehicle
- Its original equipment (tire size)
- Its loaded axle weight

Site
- Type of surface, condition of haul road
- Type and condition of loading and dumping areas

Use of Vehicle on Site
- Length of round trip (cycle)
- Number of runs (cycles) covered in 1 day's work
- Number of working hours in a 24-hr day
- Average number of miles covered in 1 hr (average speed taken from whole day's travels)
- Maximum loaded speed

Other Considerations
- Behavior of vehicle/tire combination (traction, flotation)
- Previous tire performance
- How tires wear and reasons for final removal
- Any sidewall, tread, bead problems
- The tire
- Cost (original purchase price and cost per mile based on expected performance)

Fig. 1-1 Tire maintenance in construction begins with an evaluation of tires on the scrap pile. Experience in this area reveals the main causes of failure and suggests ways the job site can be changed to improve tire life.

Size and Load Index
1. The size will normally be the same as that of the tire originally fitted to the vehicle. However, optional sizes available for the vehicle should also be considered.
2. Load index.

NOTE: It sometimes may be advisable to recommend tire- and rim-size changes to match job conditions.

Type. In general, the choice of tread design and type of tire is essentially determined by
1. The type of surface the tires are to work on and the problems to be resolved (present and/or future) including muddy conditions, posing grip problems; rocky conditions, posing problems of possible cuts, tread hacking, and shock ruptures (tread or wall); and conditions where both of these problems are encountered (need for grip and resistance to rocky terrain).

NOTE: Each of these problems can be encountered to varying degrees. It is necessary to evaluate them.

2. The average number of miles covered per hour. When this is determined it should be compared with the tire manufacturer's maximum limitations established for each tread/type of tire.

With the above determinations in hand, the proper inflation pressures may now be ascertained (Fig. 1-2).

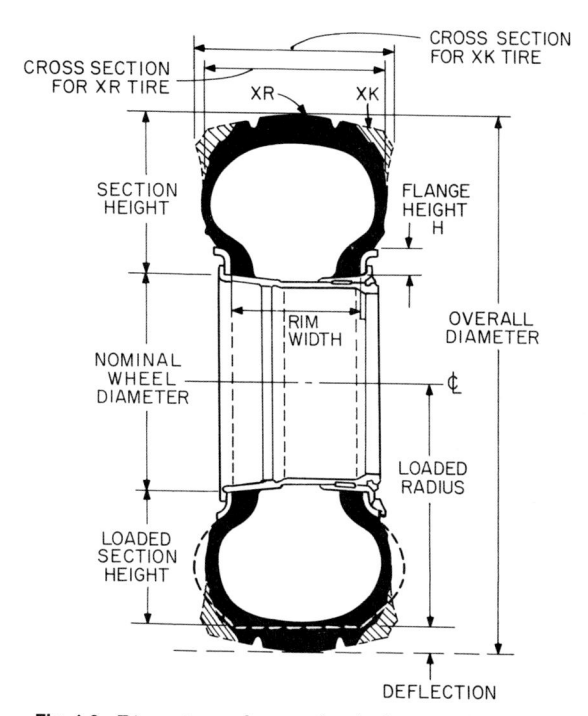

Fig. 1-2 Dimensions to be considered when specifying a tire.

AIDS IN TIRE SELECTION

One help to the earthmover industry is selecting transport tires through the use of the ton-mile-per-hour (tmph) system. Briefly, it consists of establishing the job requirements in terms of mean tire load (tons) and workday average speed (mph). The product of these two figures gives a job tmph value. A tire is then selected which has a tmph value near or above the job value (the tire tmph value or rating is determined by test).

The tmph system was designed for use as a tool to help select the proper tire for any given job. It is important to understand that proper tire application depends on many important factors, some of which are not taken into consideration with the present tmph system. Some of these factors are excessive grades or curves, poor haul-road conditions, wheel position (drive, steer, trail), and heat generation from brakes and planetaries.

A better alternative to the tmph system is to consider loads and speeds (both maximum and average) separately, as well as other important job conditions. First, the tire's load-carrying capacity at various maximum speeds listed in the tire manufacturer's load and inflation tables is compared with the actual tire loads found (or estimated) for a particular job site.

Then, the required work day average speed (WDAS) is determined for each application. (WDAS equals the number of miles traveled in a day, divided by the total shift hours including breaks). This value is then compared to the allowable WDAS listed for each tire as determined by the tire manufacturer.

After a tire is found which has load and speed capabilities that meet or exceed expected job conditions, other job-site factors must also be considered to assure that tire and job are properly mated. As mentioned before, some factors are the existence of severe grades or curves or the conditions of haul roads. Tire manufacturers or their representatives should be consulted to assure that all these conditions are properly evaluated.

CONTACT AREA

On occasion, users or equipment manufacturers request ground-contact-area information from tire manufacturers. This information is used to help predict a tire's flotation and/or ground-contact-pressure characteristics. It must be remembered, however, that flotation/mobility is dependent on more than just contact area. Nevertheless, it is important.

When contact area values are given one must know the specific conditions for which they are valid, especially if they are to be used in comparing tire brands. Actual contact area, that is, rubber-to-ground interface, is a function of many variables, in particular, tire load, inflation pressure, ground penetration, tire casing construction, and tire tread pattern.

Usually, the area is specified as a *projected* area on a steel plate for a given load and inflation pressure. (The projected area includes the actual rubber in contact plus the voids within the outline of the contact rubber.)

Although contact area is important in evaluating a tire's flotation/mobility capability, there are other factors to consider. To achieve good flotation, the ground must be disturbed as little as possible. The lower the ground pressure the less the ground will be upset. Theoretically, then, the greater the contact area, the lower the ground pressure (ground pressure equals tire load divided by contact area) and, consequently, the better the flotation. But ground-contact pressure is not uniform throughout the contact patch. Both tire construction and tread design affect ground-pressure uniformity. The radial tire, for example, has a more uniform distribution of ground pressure from the center of the patch to the edge than a bias tire. Under deflection, the center of the patch of a bias tire has relatively low ground pressure, but the edge pressures are relatively high. Therefore, the tire has a tendency to dig in at the edges, thus disturbing the ground. The radial patch, owing in part to its belt-stabilized tread, has more uniform ground pressure, which results in better flotation characteristics.

Fig. 1-3 Track print left behind in soft ground reveals how radial design and belt-stabilized tread provide uniform ground contact with minimum disturbance of the earth, assuring good traction and flotation.

Moreover, in order to improve flotation, in the field the inflation pressure can be lowered with the approval of the tire manufacturer, who will usually require a reduction of speed. (The contact area is then increased.) This can be done with a radial tire with more success than with a bias tire. Lowering the pressure in a bias tire results in significant tread distortion (hence, ground disturbance), whereas the radial tire with its stabilized tread is less affected and consequently its flotation/mobility capability can be improved (Fig. 1-3).

Of course, in any discussion of flotation, traction, and mobility, there are many other considerations (particularly, those pertaining to the ground or soil itself), which are not included in the discussion above. These, however, are beyond the scope of this chapter.

STORAGE AND HANDLING OF TIRES

Proper tire storage and handling are items which are frequently ignored by people in every phase of the earthmoving industry. To preserve the inherent quality of tires and the

Upkeep and Maintenance of Tires for Construction Equipment 6-5

initial investment in them, there are certain basic procedures to follow when storing and handling tires.

Storage

To avoid all premature aging and degradation of tires during storage, it is necessary to protect them from

    1. Inclemencies, such as changes of temperature, drafts, extreme heat (more than 100°F), and humidity.

    2. Ozone sources, such as arc welders, spark-producing motors, mercury vapor quartz bulbs, battery chargers with mercury rectifiers, and direct exposure to the sun.

    3. Distortion, such as from stacking. Upright positioning of earthmover tires is preferred over stacking to minimize distortion and stress.

If tires are to be stored for any significant length of time, it is recommended that they be placed in closed storage areas which do not contain the conditions just described. In addition, items should be stored to allow the oldest tires to be shipped first. This of course applies to all rubber products, such as tubes, valves, and flaps.

Handling

To eliminate deterioration of beads and the subsequent consequences, follow these rules regarding tire handling:

    1. Do not lift a tire by the beads directly with a crane's hook.

    2. Use flat straps or webbing, not metal slings or chains to lift tires.

    3. Pick tires up under the tread and not by the beads when using a forklift truck.

    4. For all tires shipped with bead protectors, leave this protection on until mounting time. Also, save the protectors since they can be used to cover a tire's beads when it is dismounted for retreading or repair.

The aim of the rules of storage and handling, as outlined here, is to preserve the quality of new tires right up to the time of mounting. Strict adherence to these rules will keep your tires factory fresh and ready for use. Individual tire manufacturers should be consulted for their specific recommendations on storage and handling.

MOUNTING INSTRUCTIONS

    1. Tubeless-type rim:
        Examination of Fig. 1-4 reveals key areas for attention on the tubeless-type rim.
    2. Valve installation:
        a. The valve consists of:
            (1) A sealed base, consisting of a body, a sealing gasket, and a tightening nut (except when the valve hole is in the side of the rim; see Fig. 1-5).
            (2) A large-bore valve stem complete with sealing ring, large-bore (jumbo) core, and a cap.

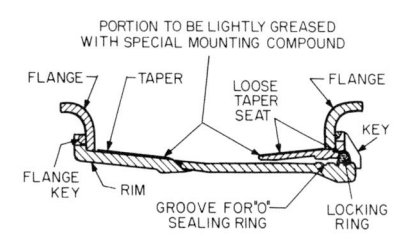

Fig. 1-4 Tubeless-type rim.

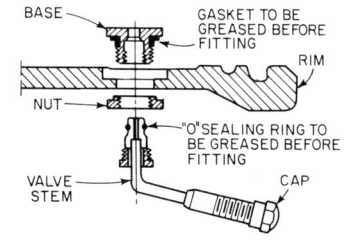

Fig. 1-5 Valve installation procedure.

        b. Before installation of the valve, it is imperative:
            (1) That the parts of the rim and the valve base which come in contact with the gasket be absolutely clean and free of scratches or other defects which might cause a leak.

6-6 Maintenance of Ground Contact Elements

  (2) That the seating surfaces of the "O" sealing ring in the valve stem and the base are clean and free from cracks, scratches, or burns which could cause a leak (Fig. 1-6).
 c. Installing the valve:
  (1) *Thoroughly clean and lightly grease the part of the rim on which the gasket will be placed.*
  (2) Grease the gasket and place it on the valve base. Place the assembly in the valve hole, screw the nut up *Tightly*.
  (3) Clean the valve stem and grease its "O" sealing ring; then, with the stem pointing in the right direction, screw it into the base and tighten, making sure that the stem does not change its position (Fig. 1-7).
   NOTE: When removing the base, *use a new gasket on reinstallation.*

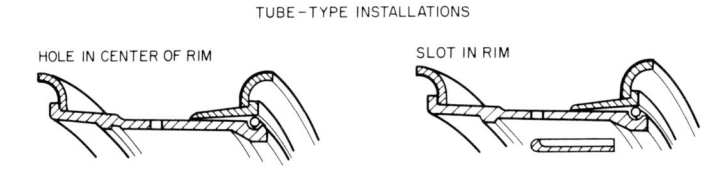

TUBE-TYPE INSTALLATIONS

HOLE IN CENTER OF RIM   SLOT IN RIM

TUBELESS INSTALLATIONS

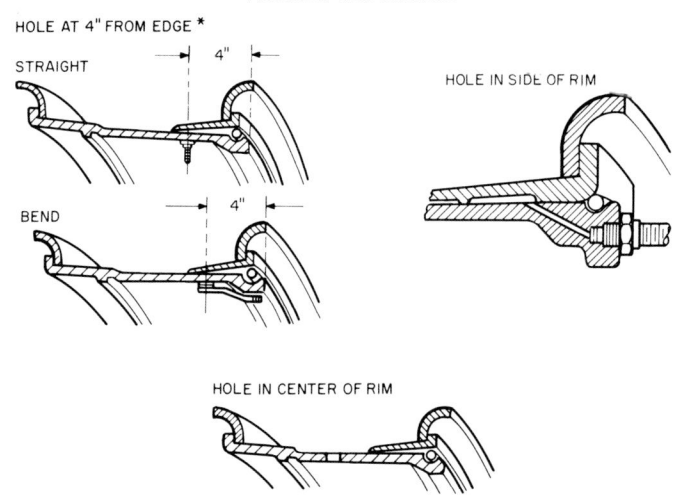

HOLE AT 4" FROM EDGE *
STRAIGHT   4"   HOLE IN SIDE OF RIM

BEND   4"

HOLE IN CENTER OF RIM

*VERIFY MEASUREMENT, EXCEPTIONS EXIST.

Fig. 1-6 Different valve arrangements are necessary depending on the rim or wheel and on tube-type or tubeless mounting. When selecting valves, make sure the rim or wheel matches the tire (tube type or tubeless) and flap combination, where applicable.

 3. Tire mounting:
  a. General:
   As the tire has to form an airtight assembly with the rim, it is necessary
  (1) To fit it on the correct rim and flange. They must be clean and in good condition.
  (2) To carefully follow the instructions and methods described below (Fig. 1-8).
  b. Preparation of the rim:
  (1) If necessary, using a file and wire brush, remove any rough pieces of metal or old rubber which might affect air sealing. Then, wipe down the rim and loose bead seat with a dry rag.

Upkeep and Maintenance of Tires for Construction Equipment   6-7

(2) Those parts of the rim and of the loose taper bead seat on which the tire beads will rest, as well as the groove for the "O" sealing ring, must be perfectly clean and free from burrs.

c. Preparation of the tire:
  (1) If necessary, remove any accumulation of rubber or grease which might be stuck to the beads. In doing this, be careful not to damage the tire beads.
  (2) Wipe the tire beads with a dry rag.

d. Mounting operation:
  (1) Put the inner flange into position and check that it is keyed to the rim. (Some rims are manufactured without flange keys.)
  (2) Fit the first flange seal correctly on the rim and push it until it rests against the flange. (Only use flange seals if mounting a used tire and a sealing problem exists.)
  (3) Lightly lubricate the taper portion as indicated under paragraph 1 (Fig. 1-4).
  (4) Place the tire on the rim and push it back as far as possible.
  (5) Center the tire correctly on the rim.
    NOTE: Bad centering of the tire can make
    (a) Placing of the loose taper seat difficult
    (b) Inflation difficult
    (c) "O" sealing ring come out of its groove during inflation
  (6) Fit the second flange on the loose taper seat; check the keying (Fig. 1-9).
  (7) Fit the second flange seal on the loose taper; push it until it rests against the flange, *if applicable*.

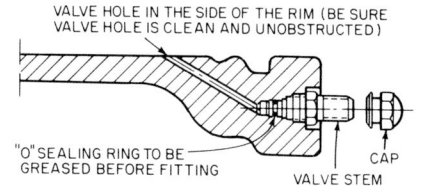

Fig. 1-7 Valve installation when the valve hole is in the side of the rim.

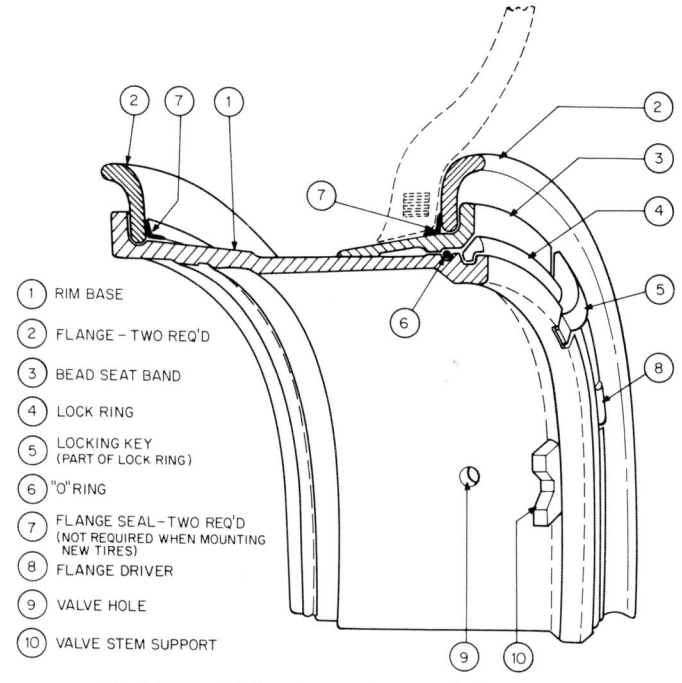

1. RIM BASE
2. FLANGE – TWO REQ'D
3. BEAD SEAT BAND
4. LOCK RING
5. LOCKING KEY (PART OF LOCK RING)
6. "O" RING
7. FLANGE SEAL – TWO REQ'D (NOT REQUIRED WHEN MOUNTING NEW TIRES)
8. FLANGE DRIVER
9. VALVE HOLE
10. VALVE STEM SUPPORT

Fig. 1-8 Typical five-piece earthmover tubeless-type rim.

6-8 Maintenance of Ground Contact Elements

(8) Lightly lubricate, with mounting compound, the two portions as indicated under paragraph 1 (Fig. 1-4).
(9) Fit the assembly of taper seat on the rim inward so that it clears the groove of the "O" sealing ring (Fig. 1-10).
(10) Place "O" ring in groove, being careful that it is not twisted, and lightly lubricate the exposed portion of the "O" ring. The "O" ring should be clean and dry before placing in the groove.
(11) Fit the locking ring.
   NOTE: Before inflation:
   (a) Check the keying.
   (b) Make sure that the "O" ring is correctly positioned in the groove.
   (c) Correct centering of the tire on the rim.

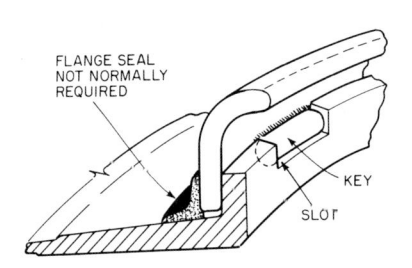

Fig. 1-9 Flange and key seated on taper seat.

Fig. 1-10 Fitting the taper seat assembly to clear the "O" ring.

e. Inflation:
Remove the valve core and inflate to 90 psi, observing the normal safety precautions for inflation. (Avoid standing in front of the tire.) At this pressure, all parts of the rim assembly should be in place. Check this. Bring the pressure to its working value and screw in the core and cap.

f. Checking for leaks:
Apply soap solution
(1) To valve and valve stem.
(2) To "O" ring (where accessible).
(3) Between tire beads and flanges.
(4) Between flange/bead seat band and lock ring.

Check proper bead seating of earthmover tires. Improper earthmover tire mountings can be the cause of many tire problems. It is important that the mounting instructions on these pages should be followed closely. Flange bead seals should not be used with new tires. They are usually used because of a leak. Air leaks are frequently the result of improper mounting or faulty rim components which may be masked by the use of bead seals. Therefore, before using bead seals, all rim components and the seating of the bead should be checked. Failure to follow rules can prove disastrous. (See Fig. 1-11.)

It is also important to have the tire as concentric to the rim as possible before inflating. Therefore, the tire should not be resting solely on the top of the rim. It should be resting on the ground with a minimum of weight on it; the vehicle may be partially jacked up or the hoist supporting the tire may be adjusted. If

Fig. 1-11 An example of shoddy workmanship by the tire mounter. Mounting studs are broken or missing. Nuts are missing and the assembly is improperly torqued. All add up to an undesirable and unsafe situation.

possible, after inflating to the seating pressure (90 psi), the tire should be run for about a half-hour. After this time, the pressure can be adjusted to the recommended working pressure (after the tire cools down). See Fig. 1-12 for examples of proper and improper bead seating.

Keep in mind that although it is easier to mount a radial earthmover tire, compared with a conventional tire, care must be taken to ensure that the radial is mounted properly. By following the above recommendations and instructions, many tire problems can be avoided.

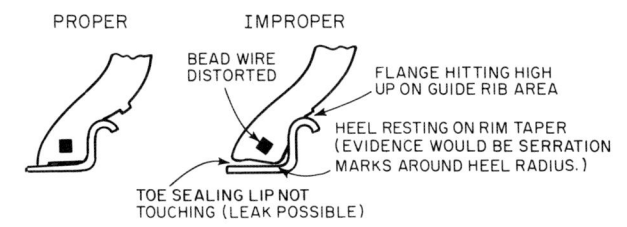

Fig. 1-12 Examples of proper and improper bead seating.

4. Tire demounting:
   After the vehicle has been blocked up to a suitable height
   a. Deflate the tire by unscrewing the valve core.
   b. Push the tire bead away from the seat.
   c. Push the loose seat inward far enough to clear the "O" sealing ring. Remove the locking ring and the "O" sealing ring.
   d. Withdraw the flange and loose seat.
   e. Free the second bead from the rim.
   f. Remove the tire from the rim.
   g. Remove the flange seals from the loose seat and the rim (if applicable).
   CAUTION:
   (1) Never apply heat to the rim when the tire is mounted. Always demount the tire from the rim when heat treatment is required (welding, cleaning).
   (2) With dual fitment, *both* inner and outer tires must be deflated before removing *either* tire from the vehicle.

DETERMINING INFLATION PRESSURE

Any determination of proper tire inflation pressure requires assessment of axle loads. Ideally, equipment should be weighed, both loaded and unloaded. In lieu of this, axle loads can be estimated. Some tire manufacturers have equipment charts which give recommended inflation pressures for most of the more popular earthmover machines. The following paragraphs outline the procedures for determining pressures for different types of machines.

Dumpers and scrapers
- Estimate material density.
- Estimate equipment load capacity (use equipment charts as a guide, but consider any deviation from the standard capacities, sideboards, e.g.).
- Calculate payload.
- Using gross vehicle weight (GVW) distribution as listed in equipment charts, calculate axle loads. Add any significant weight contributed by optional equipment. (Divide the axle load by the number of tires per axle to give the tire load.)
- Determine the top speed of a vehicle loaded (generally, 30 to 49 mph speeds are used).
- Knowing load and speed, use the carrying capacity chart for the tire under consideration to determine the proper inflation pressure.

EXAMPLE:
An empty six-wheel end-dump truck has the following axle weights:

 Front (2 tires)   = 30,000 lb
 Rear (4 tires)   = 31,000 lb
      Tire size, front and rear: 18.00–33

6-10   Maintenance of Ground Contact Elements

Operating, it was observed to be carrying heaped loads of earth and rock (approximately 75/25). Its volumetric capacity according to the manufacturer's specifications[1] is 30 cu yd (heaped). The density of material is 2650 lb/cu yd. Therefore, payload is

$$30 \text{ cu yd} \times 2650 \text{ lb/cu yd} = 79{,}500 \text{ lb}$$

The gross vehicle weight is then

Empty weight—front	30,000 lb
Empty weight—rear	31,000 lb
Payload	79,500 lb
Total GVW	140,500 lb

Knowing the GVW distribution[1] to be 33 percent front and 67 percent rear, the loaded axle weights may be calculated

$$\text{Front} = 0.33 \, (140{,}500 \text{ lb}) = \frac{46{,}365}{2} = 23{,}183 \text{ lb per tire}$$

$$\text{Rear} = 0.67 \, (140{,}500 \text{ lb}) = \frac{94{,}135}{4} = 23{,}534 \text{ lb per tire}$$

The loaded machine was observed to be hitting a top speed of 30 mph. With the foregoing information we can now determine the recommended inflation pressure from the carrying capacity chart. Refer to the chart for the 18.00–33X tire in the *Michelin Earthmover Tire Data Book.*

Front tires   At 30 mph we see that the recommended inflation pressure for the load of 23,183 lb would be 90 psi.[2]

Rear tires   At 30 mph and 23,534 lb, inflation pressure would also be 90 psi.[2]

At this point other factors which may affect the final recommended pressure should be reviewed. For example, what is the ambient temperature? (See the temperature correction chart.) What are the flotation requirements? What is the condition of the haul road? These factors could require adjusting the inflation pressure.

Loaders

- Estimate material density (refer to Table 1-1, covering load material weights).
- Estimate equipment load capacity. Use equipment charts as a guide, but consider other-than-standard-size buckets.
- Calculate payload.
- Determine axleloads (Fig. 1-13). Of course, this figure will vary with the position of the bucket. Nevertheless, an intermediate position may be used in calculations. The position generally used when determining inflation pressures for normal loading operations is one somewhat between the carry position and the extended position. To complete this axleload calculation, the vehicle wheelbase and empty weights must be known. The equipment charts give the empty axle weights. The wheelbase must be measured. (In lieu of this information, the axle loads may be estimated in a different manner as described below.)
- After axle loads are found, the inflation pressures are obtained from the carrying capacity charts.

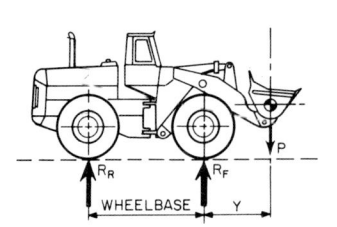

Fig. 1-13   Basic dimensions for axle-load calculations.

Front Axle

A load in the bucket results in a weight increase on the front axle and a decrease on the rear axle. This increase/decrease is a function of the payload, the bucket position, and the wheelbase.

For the front axle the loaded weight could be expressed as

Front axle loaded = unladen front weight + payload + weight transferred from rear

---

[1] This information may be found in equipment charts.

[2] NOTE: The two-star (**) tire meets this pressure requirement.

An illustration will show how the axle loads may be calculated with the above information.

Material density	2500 lb/cu yd
Bucket capacity	4 cu yd
Weights empty	Front: 20,000 lb
	Rear: 26,000 lb
Wheelbase	80 in.
Distance of bucket CG to front axle centerline (Y):	50 in.
Tire size	23.5–25

The basic formula (see diagram) is

$$\text{Front axle load } (R_F) = \text{empty load} + \text{bucket load} \ \text{ or } \ \frac{WB + Y}{WB}$$

TABLE 1-1 Load Material Weights (Approximate)

Material	Weight, lb/cu yd (loose)
Basalt	3300
Bauxite	2400–3200
Caliche	2100–2500
Clay	
Natural bed	2200–2800
Dry	1850–2500
Wet	2500–2900
Clay and Gravel	
Dry	2000
Wet	2200–2800
Coal	
Anthracite	1450–2000
Bituminous	1350–1600
Lignite	1225
Copper ore	2800
Earth	
Dry loam	1550–2100
Moist	2080–2600
Wet	2700–2900
Earth, sand, and gravel	2650
Earth and rock	
25/75	3300
50/50	2900
75/25	2650
Gravel	
Dry	2500
Wet	3400
Iron ore	4150–5500
Limestone	2400–2600
Sand	
Dry	2400–2950
Moist	2800–3100
Wet	3100–3250
Sand and gravel	
Dry	2900–3100
Wet	3400
Sandstone	2600–2950
Shale	2100
Slag	3000
Stone, crushed	2400–2900
Taconite	2900–3850

6-12   Maintenance of Ground Contact Elements

Therefore, for this example

$$\text{Front axle load} = 20{,}000 \text{ lb} + (2{,}500 \text{ lb/cu yd})(4 \text{ cu yd}) \frac{80 \text{ in.} + 50 \text{ in.}}{80 \text{ in.}}$$
$$= 20{,}000 \text{ lb} + 16{,}250 \text{ lb}$$
$$= 36{,}250 \text{ lb}$$

Divide this figure by the number of tires per axle to derive the tire load (18,125 lb).

Using the Michelin carrying capacity chart for our 23.5-25X tire, the following inflation pressure is required for the front axle:

$$\text{Front} \quad \frac{36{,}250}{2} = 18{,}125 \text{ lb} \quad \text{per tire at 40 psi}$$

IMPORTANT: When the Y dimension and the wheel base are not known, their effect on the front axle loads as given in the formula can be estimated. The value for the $(WB + Y)/WB$ for most loaders falls between 1.6 and 1.8. This we will call the *leverage factor*. Therefore, in the above example, inserting a leverage factor of 1.6 results in

$$\text{Front axle load } 20{,}000 + (2{,}500)(4)(1.6) = 20{,}000 + 16{,}000$$
$$= 36{,}000 \text{ lb per axle or } 18{,}000 \text{ lb per tire}$$

Rear Axle

The basic formula is

$$\text{Rear axle load } (R_R) = \text{empty load} - \text{bucket load} \frac{Y}{WB}$$

In determining inflation pressures for the rear tires, however, the rear axle load with a loaded bucket is *not* required. The rear axle load is greater with an empty bucket. Therefore, this figure is used when determining inflation pressure.

Again, using the Michelin carrying capacity chart for the 23.5–25X tire, the following inflation pressure is required for the rear axle:

$$\text{Rear} \quad \frac{26{,}000}{2} = 13{,}000 \text{ lb per tire at 30 psi}$$

As with other earthmoving vehicles, other factors affecting the final pressure recommendation should be reviewed. Traction and flotation requirements, for instance, may dictate some pressure adjustments.

Lift and carry   If the loader is to be operated in a lift-and-carry application, the recommended inflation pressures could be somewhat different. In general, they are higher. For these applications the individual tire manufacturers should be consulted.

Graders and dozers   Inflation pressures are functions of the vehicle axle field weight and vehicle speed. The axle weights for some of these machines may be found in the equipment charts. The speed schedule generally used is 30 mph.

PRESSURE CORRECTIONS FOR HIGH AMBIENT TEMPERATURES

The abundant air mass contained in an earthmover tire is significantly affected by ambient temperatures, and adjustments to the basic inflated pressures are therefore required to avoid harmful underinflation.

To assist in obtaining maximum possible service life, Table 1-2 has been prepared to show pressure corrections for given ambient temperatures.

EXAMPLE: The basic inflation pressure for a Michelin 18.00–25XRB** tire on the rear of an R30 Euclid end-pump truck (maximum speed 30 mph) is 100 psi.

If the tire is to be inflated where the ambient temperature is 100°F, it must be inflated to 108 psi. This is found in Table 1-2 by looking down the left-hand column to the recommended book pressure of 100 psi, then reading across to the 100°F temperature column. At this intersection we read 108 psi.

This chart should also be referred to when the tire's cold inflation pressure is being checked. For instance, if the same machine in the example above was checked early in the morning (after it had been idle all night) when the ambient temperature was 80°F, then, according to the chart, we should find a tire pressure of 104 psi. (Look down the left-hand column to 100 psi, then across to the 80°F temperature column; at the intersection read 104 psi.)

Cold inflation pressure is defined as that pressure at which the temperature of the air inside the tire is equal to the temperature of the air surrounding the tire. This condition is

TABLE 1-2 Adjusted Inflation Pressure (psi)

Recommended inflation pressure, psi	Ambient temperature, °F										
	65	70	75	80	85	90	95	100	105	110	115
30	30	31	31	31	32	32	33	33	34	34	35
35	35	36	36	37	37	37	38	38	39	39	40
40	40	41	41	42	42	43	43	44	44	45	45
45	45	46	46	47	47	48	49	49	50	50	51
50	50	51	51	52	53	53	54	55	55	56	56
55	55	56	57	57	58	58	59	60	61	61	62
60	60	61	62	62	63	64	64	65	66	67	68
65	65	66	67	67	68	69	70	71	72	72	73
70	70	71	72	73	73	74	75	76	77	78	78
75	75	76	77	78	79	79	80	81	82	83	84
80	80	81	82	83	84	85	86	87	88	89	89
85	85	86	87	88	89	90	91	92	93	94	95
90	90	91	92	93	94	95	96	97	99	100	100
95	95	96	97	98	100	100	102	103	104	105	106
100	100	101	103	104	105	106	107	108	109	110	111
105	106	107	108	109	110	111	112	113	115	116	117
110	111	112	113	114	115	116	118	119	120	121	122

found when the tire has been idle for approximately 8 hr. Please note that when a tire has been exposed to the sun for a while its inflation air temperature will be significantly higher than the weather bureau ambient air temperature even though the tire has been idle. This results in a higher pressure reading and, therefore, cannot be considered as the cold pressure.

NOTE: The goal of this adjustment is twofold:
1. To minimize the tire deflection at high ambient temperatures and thereby control excessive temperature/pressure buildup
2. To decrease the chances of operating underinflated owing to a drop in the high ambient temperature

Important

1. Michelin's carrying capacity charts are based on an ambient temperature of 65°F.
2. The foregoing corrections pertain to cold inflation pressures only. Adjustment should not be made on a working tire.

When considering the values of inflation pressures checked while working, anything in excess of 15 percent pressure buildup from cold pressure due to working alone (i.e., excluding the pressure buildup due to ambient temperature increase which could be another 10 percent) would require investigation.

SOME CAUSES OF PREMATURE DETERIORATION IN EARTHMOVER TIRES

A large number of earthmover tires retire from service prematurely because of:
- Incorrect inflation
- Overloading
- Excessive speed
- Severe shocks
- A combination of the above factors

One particular damage is the separation of certain elements in a tire's construction. This is usually the result of excessive heat generation due to
- Speeds higher than those recommended for the loads and pressure concerned
- Underinflation or overloading
- Heat generated by parts of the vehicle such as brake drums and planetary gears

Separation can be aggravated or caused by mechanical forces such as
- Shocks from badly maintained road surfaces
- Hammering due to road surfacing
- Lateral forces occurring in tight bends

To help avoid the above damage, watch the maintenance of roads and, if possible during their construction, insist on large-radius bends. Table 1-3 gives maximum recommended speeds for flat (unbanked) turns of different radii.

TABLE 1-3   Speed Restrictions in Flat (Unbanked) Curves*

Speed, mph	Minimum turning radius, ft
5	50
6	70
7	90
8	120
9	150
10	190
11	230
12	270
13	320
14	370
15	420
16	480
17	540
18	610
19	680
20	750
21	830
22	910
23	990
24	1080
25	1170
26	1270
27	1370
28	1470
29	1580
30	1690

*EXAMPLE: The radius of a curve in a haul road is 750 ft. Speed in negotiating turn should not exceed 20 mph.

Ply separation can be produced by lateral forces occurring in tight turns. It is important to bank haul-road curves properly and to limit speed in relation to the curve. If sharp bends cannot be avoided, take them at the lowest possible speed.

Eliminating these causes of premature tire deterioration can minimize downtime (loss of production) and the risk of vehicle damage and/or personal injury.

Chapter **2**

Wear and Maintenance of the Undercarriage

K. D. ANDERSON
Terex Division, General Motors Corporation, Hudson, Ohio

Owners of crawler tractors have the right to expect optimum performance. New units will deliver that optimum capability, but no make or model can continue such performance without timely and effective inspection, maintenance, and service.

At the outset, the components of a crawler undercarriage are perfectly matched according to dimension. But, as the components work together, they wear. Wear causes mismatch of parts and ultimately accelerates the wear rate of the entire system. This deterioration rate and general repair costs can be controlled. The following should help to lower maintenance costs, increase undercarriage life, and improve job performance and profitability.

WEAR

The undercarriage, the major wear item of a track-laying vehicle, does have a normal wear pattern. Understanding how wear occurs and the wear characteristics of parts can aid in avoiding unscheduled downtime.

Basic to a normal wear pattern is internal pin and bushing wear. Link side wear, accelerated sprocket tooth wear, too much track sag—all these may occur as a result of internal pin and bushing wear.

While a tractor is operating, the chain is in tension. Because the bushing is pressed into one link and the pin is pressed into the adjacent link, the tension in the chain causes contact on only one side of the pin OD and the bushing ID. This contact occurs on the same side whether the tractor is going forward or in reverse. Wear at the pin OD and bushing ID occurs because there is relative motion between the pin and bushing when the chain bends around the sprocket and idler. This internal pin and bushing wear is called *pitch wear* or *track stretch* because it causes the track pitch to increase and the chain actually gets longer (Fig. 2-1).

Normal wear of the sprocket occurs along with pitch wear because the elongation of the chain causes the sprocket to pick up the bushing closer to the tooth tip. Wear occurs as the bushing slides down to the tooth root. This is also called *forward drive side wear*.

In reverse operation the opposite side of the sprocket tooth, usually called the *reverse drive side*, contacts the bushing with the teeth at the top of the sprocket. The effect of reverse operation is more severe to both sprocket and bushing because there is scrubbing action between the sprocket and the bushing as the bushing enters the sprocket. This scrubbing action, or relative motion, between sprocket and bushing does not take place during forward motion of the tractor.

6-16 Maintenance of Ground Contact Elements

In addition to forward and reverse drive side wear, there may be some degree of tip wear on the sprocket. Since the tips of sprocket teeth are left in an as-cast condition it is possible that a tip will get scuffed when the bushings enter the sprocket. This is normal wear and will not affect the performance of the sprocket.

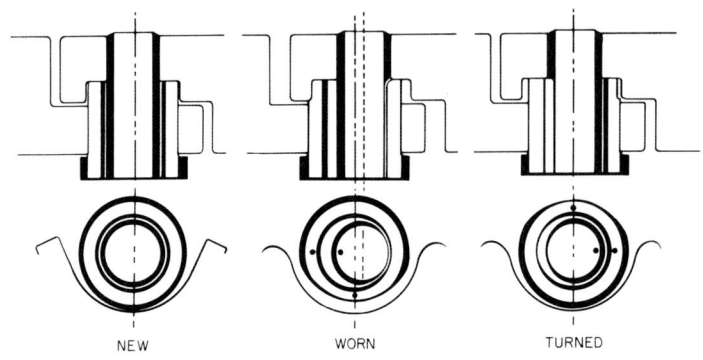

NEW   WORN   TURNED

Fig. 2-1 Pin and bushing wear for new and worn conditions and the results of turning.

The rail surface of the link is the side opposite the surface to which the track shoes are bolted. This is the wearing side of the link. It comes in contact with the roller and idler tread surfaces. The rail gets narrower over the pin and bushing bores and consequently wears faster at this point. When the chain travels around the idler, the center of the rail is the only point of contact. This causes the center of the rail to wear faster also. As a result, normal link rail wear becomes wavy as wear progresses (Fig. 2-2).

The normal wear pattern for a roller is directly related to link rail wear. As a link rides over the rollers, the roller tread wears down. Even though this is normal wear, there is a limit which can be tolerated. If the tread wears down too far, the roller flange will hit the pin boss, causing wear of the roller flange and the links. The links may be damaged to the extent that they won't hold the pins (Fig. 2-3).

Fig. 2-2 Normal wavy link wear—shaded portion—for the link rail.

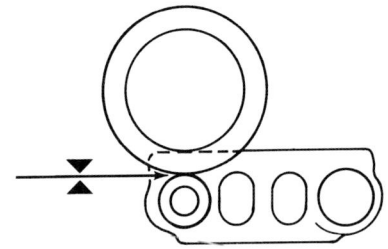

Fig. 2-3 Excessive damage means that links cannot hold the pins.

Another basic contributor to wear is front idler and/or track frame misalignment. Track frame mounting-point wear can allow toe-out while the unit is working in the forward direction. This type of misalignment may cause excessive end wear of pins, off-center external wear of bushings, rail side wear and sprocket tooth gouging of the inside of the links, side wear of the sprocket and sprocket teeth, and flange wear of rear rollers.

The front idler assembly may also be misaligned. This misalignment can affect the wear of the front idler flange, the link side rails, and the front track roller flanges. The idler is guided on the roller frame with a series of wear bars and plates which are shimmed to align the idler with the track rollers. There are also side wear plates which serve to guide the idler as it recoils back and forth. Improperly shimmed or unequally worn wear plates will cause the idler to run off center or out of line with the front track rollers. This will allow the links to ride against one flange of the idler and interfere with the front track roller flanges.

The entire alignment problem can be thought of as that of a pulley system with the idler and sprocket as the pulleys. If any part is out of line, the chain will interfere with it and cause wear.

How materials affect wear Nothing affects the life or wear rate of a crawler undercarriage like nature. The normal wear rate of all undercarriage components, especially the chain, is greatly affected by soil conditions. For example, sand greatly accelerates the wear rate of the entire undercarriage but the pins and bushings suffer the most. Sand works into the link counterbore and then into the bushing where it speeds up wear of pins and bushings. Link counterbores can be so badly worn that the links will not be rebuildable. The life of the pin and bushing can be shortened to a fraction of the life of other components.

Rock has its greatest effect on track shoes. Grousers wear fastest. The plate wears thin and beam strength is significantly reduced. If subjected to impact, shoes can bend or crack as they approach the rebuild limit. Track links are subjected to impact and twisting when working in rocky ground, sometimes causing fatigue cracking. These heavy impact loads, when concentrated in one area such as the front or rear rollers, may cause accelerated wear of these parts but normal wear on the rest of the undercarriage.

Wet clay-based soil poses a particular problem to sprockets. The wet clay may pack in the roots of the teeth. As a result, a mismatch of track chain pitch and sprocket pitch can occur which can produce severe wear of the tooth tip on the reverse drive side when the tractor is moving forward. Bushing OD wear is accelerated and in severe cases bushings may even crack.

Generally, coal serves as a lubricant to an undercarriage. The life of an undercarriage working in this material will be much better and longer than that of a comparable unit doing the same work in sandy soil.

The moisture content in soil increases its abrasive effect. A moderately abrasive, dry soil can become very abrasive when water is applied. Tractors working in a moist riverbed, for example, will have shorter pin and bushing life than identical machines a few hundred yards away on higher, drier ground.

The general relationship between different materials and corresponding undercarriage life appears graphically in Fig. 2-4.

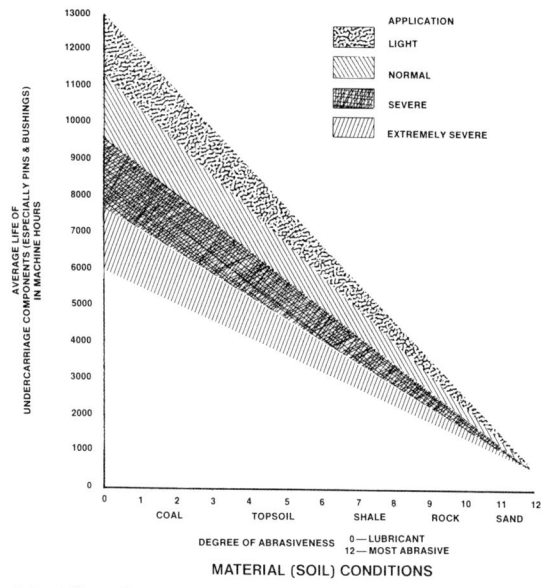

Fig. 2-4 Effect of applications and soil conditions on undercarriage life.

TABLE 2-1 Relationship of Applications to Wear Patterns

Application	Material	Load	Travel pattern	Probable wear characteristics
Ripping or quarrying	Rock, concrete, ore, shale (hard and rough)	High impact, heavy side loads, concentrated, constant	Heavy and frequent cycling and maneuvering	Heavy tread wear on rear rollers
Heavy digging and dozing	Rock, concrete, ore, shale (hard and rough)	High impact, heavy side loads, concentrated, constant	Heavy and frequent cycling and maneuvering	Heavy tread wear on front rollers
Landclearing Logging Digging Dozing	Sand, gravel, stumps (rough and abrasive)	Frequent high impact and side loads, packing	Heavy cycling and maneuvering	Accelerated pin and bushing OD wear; sprocket tip wear due to packing; roller tread wear proportional to soil abrasiveness; link side wear
Leveling Plowing Pushing Pulling Loading	Clay, gravel (rather rough)	Moderate loads—some impact; side loads	Moderate cycling and maneuvering	Pin and bushing wear; sprocket root wear; roller tread wear in proportion to soil abrasiveness
Mining Ditching Grading	Coal, clay, brush (loose and soft)	Uniform loads and little impact	Little maneuvering	Pin and bushing OD wear but not severe; even wear on entire undercarriage
Finish grading Stockpiling Spreading Side sloping	Clay, light gravel, firm earth (smooth and clean)	Little or no impact loads	Very little cycling or maneuvering	Link side wear and retention guard wear; idler wear plate; roller flange wear
Landfill	Trash, garbage, fill dirt	Little impact loads, packing	High cycling and maneuvering	Pin and bushing wear; sprocket tip wear; idler wear plates

Applications and wear The type of question in which a unit is used will also affect the wear rate and wear pattern of the undercarriage components. The wear chart (Table 2-1) contrasts results for various applications and environments.

Working requirements are closely related to applications. For example, applications which require consistent turning to the right or left will cause faster wear on one track. Side hill work speeds wear of roller flanges and link sides. High speed accelerates undercarriage wear. Over rough ground, high speeds create heavy loads which damage track alignment.

OPERATOR TECHNIQUES AND WEAR

In the same way that the driver of a car affects tire life, the crawler tractor operator can control the wear rate of undercarriage components. Abusive operating techniques can be expected to reduce overall track life, compared with the life conscientious operation would yield. The following are some dos and don'ts for operators:

Don't ...

... go too fast unless productivity is worth the increased wear. This is especially true for reverse operation.
... spin the tracks. When the track slips, the undercarriage components wear faster and no work gets done.
... take such deep cuts with the blade or overuse down pressure.
... park the machine in mud, water, or corrosive environments.

Do ...

... ease up on load when the track begins to slip.

... change operating direction often enough to balance wear.

The operator can also enhance maintenance of an undercarriage with the following helpful hints:

1. Make daily visual inspections of equipment. Check for loose bolts, leaking seals, and abnormal wear. Report all items that need attention to the maintenance team so needed adjustments can be made before extensive damage occurs.

2. Clean mud and debris from undercarriage so rollers can turn properly. Do this as required but always at the end of the day.

3. Do not allow either obvious or subtle problems to go without alerting the maintenance crew.

MAINTENANCE PRACTICES

Without doubt, components of an undercarriage do wear. The rate of wear, however, can be controlled through timely, proper inspection and maintenance. The maintenance cost for a crawler tractor's undercarriage can be as much as half the cost of maintaining the entire unit. Properly maintaining an undercarriage can mean cost savings and less unscheduled downtime.

For maximum service life for all track components, keep the track properly adjusted. The track should be adjusted so that there is about 1½ in. of sag between the front carrier roller and the idler. If the track is adjusted too tightly, a great amount of friction will exist between the pins and bushings as they hinge and travel around the sprocket and idler. This friction causes accelerated wear of pins, bushings, sprocket, idler, and rollers. Severely tight tracks absorb a great amount of horsepower and reduce the amount of power available for work. Also, an extremely tight track can cause severe damage to the final drive hubs, bearings, and gears.

However, if the track is too loose, service life is likewise reduced. Loose tracks fail to stay properly aligned and tend to come off when the tractor is turned. This causes wear to the idler center flanges, the roller flanges, and the sides of the sprocket teeth. A loose track will whip at high speeds, resulting in impact loads on carrier rollers.

It is important that tracks be adjusted under actual working conditions of the machine. If the machine works in material which has a packing characteristic, then the material should not be removed when adjusting track for slack.

6-20 Maintenance of Ground Contact Elements

Links normally wear on the rail surface. The case depth of this heat-treated surface will diminish and approach a value beyond which the link is not rebuildable. Beyond this limit the link must be replaced in order to maintain adequate clearance between roller flanges and pin boss. The link is rebuilt by passing layers of weld over the rail surface.

Track rollers and carrier rollers should be checked to ensure free rotation and no leakage of lubricant. All rollers do not wear at the same rate. Uneven wear can be balanced by changing roller positions in the group, by switching them from one side to the other, or by turning them end for end. Changing the position of the rollers, in some cases, distributes the wear and extends the service life of the roller group. If the position of rollers is to be changed it should be done when they are about 50 percent worn.

When rollers are worn up to 75 percent, the maintenance supervisor has to consider the remaining service life of the rollers. They may be used until they are worn out, but this may cause pin boss damage to links and accelerated wear to other parts.

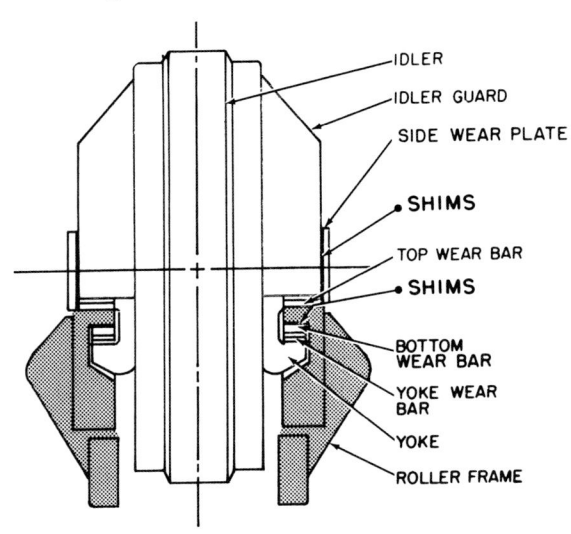

Fig. 2-5 Idler assembly cross section reveals the intricacy of this undercarriage element.

The other alternative is to rebuild the rollers by building up the worn treads and flanges with weld. It is very important that the flanges of the front and rear rollers be in good enough condition to properly guide the track under the middle rollers.

The greatest wear problem with idlers is misalignment. Shims can be added or removed to make up for wear and maintain alignment. The side wear plates which guide the idler as it recoils can be turned end for end to allow the unworn top edge to be used after shimming becomes impractical. The bottom wear bars provide replaceable wear surfaces on which the idler rides during recoil. Worn idlers can be reconditioned by building up the worn rim surface and the sides of the center section with weld (Fig. 2-5).

Sprockets should be checked regularly with a wear gauge to determine when to replace them. If they show more wear on one side of the tooth than on the other, they can be switched from side to side to prolong life and balance wear. Turning pins and bushings will also increase sprocket life. It is not good practice to rebuild worn sprockets. If sprocket teeth are filled with weld, the sprocket will be out-of-round and bushings will wear at a more rapid rate.

Optional equipment helps There are working conditions in which it is not economically advantageous to allow parts to wear normally. Normal wear in such conditions is too fast, too destructive, and too expensive to be compensated by replacement or rebuilding. In such cases, there are usually optional parts available which will prove to wear at a much slower rate and to be more productive than the standard part.

In clay-based mud, a normally equipped tractor will have problems with mud packing between sprocket teeth and causing abnormal, accelerated sprocket wear. Use of a relieved tooth sprocket can materially reduce the mud packing but may also result in accelerated bushing wear caused by the narrow tooth root. These wear rates should be carefully analyzed to produce the most economical balance.

In severe job conditions, extreme-service track shoes usually prove to be tougher and more durable than the standard shoe. They're specially made and proven for high-impact jobs such as rock work.

Track roller guards are optional items which prevent rocks and other debris from entering the track roller area. They, in conjunction with retention guards in the front and rear, also serve to prevent so-called snaking of the tracks. Snaking may occur on uneven ground, side slopes, or in constant maneuvering (turning) applications.

All of these items can affect the life of undercarriage components. It's important to equip a tractor properly to achieve maximum production and service life.

Rebuilding versus replacement As the components of an undercarriage wear, the owner should be concerned about the economic decisions that must be made: Should the parts be replaced, rebuilt, or run to destruction?

Running a component until it is destroyed is most likely to have an adverse affect on the entire undercarriage. For example, if the chain is run to destruction it will affect the sprocket pitch, the rollers, and the idlers by accelerating their wear rates. Normally, when any component is run to destruction, the complete undercarriage will be affected and will have to be replaced.

A more difficult decision is choosing between rebuild and replacement. The decision will probably be based upon the cost savings afforded by each. Rebuilding is the process of welding, or more specifically hard facing, the wear surfaces of eligible parts. Weld is deposited in layers until a wear surface comparable to the new part dimension is acquired. If rebuilding is delayed, the amount of weld that must be applied and the additional labor involved will make the cost of the rebuilt part approach the price of a new part. Conversely, if a part is rebuilt too early, any case hardness remaining will be affected by welding such that, after rebuild, the life of the part may actually be shortened. Therefore, the most important consideration with rebuilding is to rebuild at the proper time. The cost of rebuilding will vary from one area to another. Local sources should be contacted to determine the cost of rebuilding components in a specific area.

When replacing components on an undercarriage, new parts should not be matched with badly worn parts. Mating a worn surface with a new part will only result in accelerated wear of the new part. For example, when a new sprocket is used with a worn chain, the sprocket will wear very rapidly until it matches the chain pitch. Likewise, when new and used links and track rollers are mismatched, the new components have accelerated wear rates.

In deciding whether to rebuild or to replace worn components, the total cost of each alternative and the corresponding cost per hour should be calculated. The following is the cost-per-hour calculation for rebuilding:

Cost of part, new $_____
Rebuild cost $_____
Estimated downtime cost $_____
Removal replacement cost $_____
Other related costs $_____
A = total cost with rebuild $_____
Hours on part before rebuild _____ hr
Estimated life after rebuild _____ hr
B = total hours with rebuild _____ hr

$$\text{Total cost per hour (with rebuild)} = \frac{A}{B}$$

The cost per hour should be calculated because generally rebuilding will give a part additional life. But the additional cost incurred may not be justified by this increased life. The total cost per hour with rebuild should be compared with the cost per hour without rebuild.

$$\frac{\text{Cost of new part}}{\text{Estimated life}}$$

The lower of the two cost-per-hour figures may be the most economical.

SUMMARY

This chapter emphasizes the normal wear patterns that can be expected and the wear characteristics produced by various working conditions and styles of operation. It includes maintenance hints for critical components of an undercarriage. The most important consideration in maintenance efforts, including rebuilding and replacement, is timing. Keen observation and analysis of wear characteristics and rates, coupled with timely, effective maintenance, can make the difference between profitable and unprofitable operations.

A crawler tractor owner needs a reliable source for analysis of wear and potential undercarriage life to eliminate unnecessary costs. The original equipment dealer can assist in cutting costs to a minimum and realizing the best possible life for the undercarriage.

Chapter **3**

Dozer Moldboard and Ripper Tooth Maintenance

HERMAN A. HULLMANN
Technical Services Representative, Fleet Services,
Fiat-Allis Construction Machinery Inc.,
Springfield, Ill.

Dozer and ripper manufacturers provide the construction machinery industry with several types of dozers and rippers to suit various job requirements. The steel industry provides alloy steels for cutting edges, end bits, ripper shanks, and points to withstand abrasion and/or the shock loads they are subjected to. It is therefore the responsibility of the user, with assistance from the manufactuer, to select the dozer and ripper best suited for a particular job application. Select the dozer cutting edges, ripper shanks, and points that give optimum performance.

Types of Dozers

There are basically four types of dozer moldboards: the angle moldboard, the straight moldboard, the semi-U moldboard, and the full-U moldboard.

Angle dozer The angle dozer utilizes a C frame and has three positions of the moldboard: straight and 25° angle to the right and to the left. It is a lighter-weight dozer and should be used primarily in dirt or loose-material applications (Fig. 3-1).

Straight moldboard The straight moldboard is the most rugged of the moldboards and is the one best suited for rock or severe applications (Fig. 3-2).

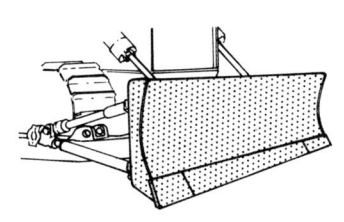

Fig. 3-1 Angle dozer.

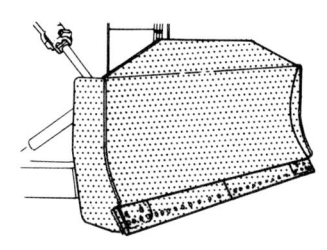

Fig. 3-2 Straight dozer.

Semi-U moldboard The semi-U moldboard, which utilizes the characteristics of both the straight and full-U moldboard, is a rugged, all-purpose moldboard. It works well in all applications and is now the most popular moldboard in the construction machinery industry (Fig. 3-3).

Full-U moldboard The full-U moldboard is longer than the straight blade with approximately one-quarter of each end of the blade angled forward. The U shape of the moldboard will hold and carry more material in front of the blade, making it ideal for moving loose or ripped material in land reclamation and similar applications. It is also used in severe applications, but is not recommended as the primary rock moldboard. (Fig. 3-4).

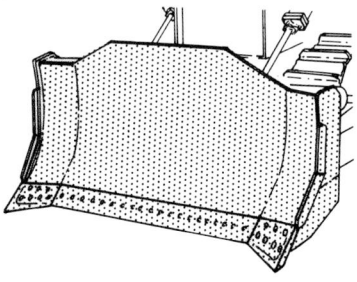

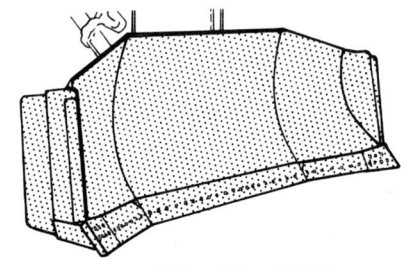

Fig. 3-3 Semi-U moldboard blade.

Fig. 3-4 Full-U moldboard blade.

Types of Rippers

Basically, there are only two types of rippers: the pull or tow type and the integral tractor-mounted ripper. The towed ripper is rapidly disappearing and is no longer considered in most construction applications. The integral mounted ripper, which is mounted directly to the rear housing of the tractor, provides more maneuverability and tractor balance. It is more compact and utilizes hydraulic cylinders and tractor weight for better penetration and ripper depth control. With today's bigger tractors, such as the Fiat-Allis 41B, ripping that previously required blasting is being done economically.

Variations in lifting and control arrangements provide three classes within the integral mounted rippers: the straight bar, the parallelogram, and the radial ripper.

Straight bar ripper The straight bar ripper, which is primarily used on smaller tractors, is a lightweight ripper that utilizes one to five shanks (Fig. 3-5).

Multishank operation is suitable for relatively easy soils, that is, top soils, glacial till (without boulders), chalk, or weak sandstones. It produces relatively high volumes in good conditions but is unsuitable for slabby material or boulders as the distance between the tractor and shank is usually 36 in. or less. Single-shank operation is suitable up to medium-strength or broken limestones or glacial till with small boulders.

Parallelogram ripper The parallelogram ripper is the most practical ripper for all applications. It is heavily constructed and utilizes one to three shanks. Most parallelogram designs have fixed shanks and a constant point angle of penetration as the ripper beam is lowered to provide positive control over ripping depth. Variations in the parallelogram design offer free-swinging or swivel-action shank brackets and a hydraulic pitch adjustment. In difficult materials, swivel-action brackets permit the shank to face into the loads to prevent slewing and to reduce stresses on ripper and tractor. Swivel-action brackets

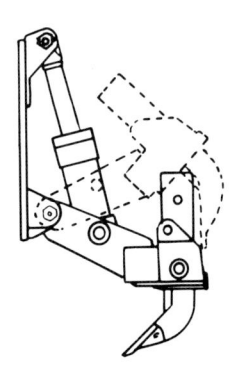

Fig. 3-5 Straight bar ripper.

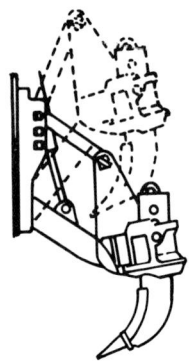

Fig. 3-6 Parallelogram ripper.

also permit steering a tractor during ripping operations without generating side-thrust loads. The advantage of the hydraulic pitch is the ability to change the point angle quickly and easily from the seat of the tractor. A steep point of entry for fast, easy penetration and then a flatter point angle when ripping depth is obtained result in longer point life and increased production (Fig. 3-6).

Multishank operation is widely used for glacial till (without large boulders), medium or broken limestones, or coal, but it is not suitable for work in slabby materials where raking may occur. Single-shank operation of a parallelogram ripper is suitable for the most difficult materials.

Radial ripper The radial ripper acts in an arc, the beam pivoting to raise and lower the shank. The angle of the point, in relation to the ground, changes as the point is lowered. Some radial ripper designs have one to three shanks on an offset pattern; other designs have only a single shank (Fig. 3-7).

Multishank operation is very much the same as with the multishank parallelogram rippers; but owing to the shank configuration design, it is not as susceptible to raking. Single-shank operation of a radial ripper is suitable for the most difficult materials, including some igneous rocks. The radial shank is considerably longer than the parallelogram shank and will attain a deeper ripping depth. For example, the Fiat-Allis 41B parallelogram ripper has a maximum ripping depth of 42 in., while the 41B radial ripper has a maximum ripping depth of 84 in.

Dozer and Ripper Installation

1. When installing a dozer or ripper on a tractor, it is vitally important that all paint and high spots be removed from any matting surfaces to prevent loosening of the attaching part and to obtain proper adjustments.

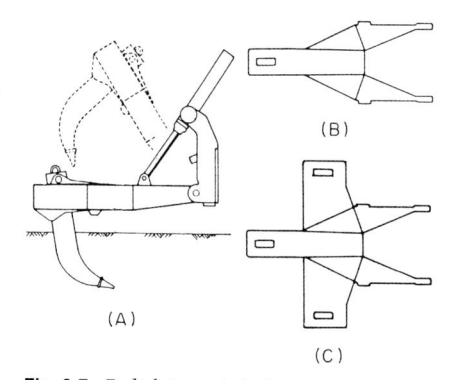

Fig. 3-7 Radial ripper (A), showing a single (B) and multishank (C) arrangement.

2. All bolts and capscrews should be torqued to the manufacturer's specifications and not overtorqued. Overtorquing stretches the capscrews beyond the yield point, resulting in broken capscrews when shock loads are applied. Frequent inspections should be maintained to ensure loosening does not occur.

3. If a dozer is to be used to push load scrapers, a push plate should be installed on the face of the moldboard to prevent damage to the moldboard.

Dozer Cutting Edges and End Bits

Construction machinery life is based on hours of operation or use. Hour life of ground-engaging tools is too variable to list in specific hours. Hour life of cutting edges may vary from a few hundred to several thousand hours, depending on the following elements:
1. Component material and design
   a. Each manufacturer of cutting edges and end bits may use a different alloy steel and heat-treat procedure to obtain optimum performance for their design.
   b. There are usually two classes of cutting edges: standard service and severe service.
      (1) A standard edge is thinner, with a harder heat treatment to resist abrasion, but it will fail in severe shock or rock applications. The thinner edge also has better digging ability in tightly compacted materials.
      (2) A severe service edge is thicker with less brittle heat treatment to withstand the shock loads in severe rock applications, but it is subject to a faster wear rate in highly abrasive materials.
   c. Some of the options for end bits are (Fig. 3-8)
      (1) Standard end bits for average dozing and abrasive materials.
      (2) Severe service end bits for high-impact applications.
      (3) Adapters for end bits to prolong the life of end bits in corner digging applications.

6-26 Maintenance of Ground Contact Elements

    d. The width of the backup plate or cutting-edge bolting plate on the moldboard is vitally important to the support of the cutting edge and end bits. All cutting edges are reversible, to double the life of a cutting edge before replacement is necessary. Inspect frequently for worn or broken edges and end bits, and turn or replace them before the bolting and support area of the moldboard is damaged. Any repair in this area is difficult and costly. End bits usually wear more rapidly and will require replacement before cutting edges.

    e. When turning or replacing cutting edges and end bits, be sure the support area is clean and without rough or high spots before installation. Edges bolted to uneven surfaces will be in stress and will break when shock loads are applied. Plow bolts should be properly seated in the edge and torqued to the correct specification (Fig. 3-9).

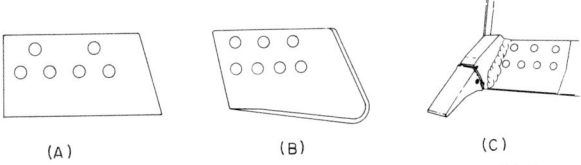

Fig. 3-8 Optional end bits include (A) standard, (B) severe service, and (C) special adapters.

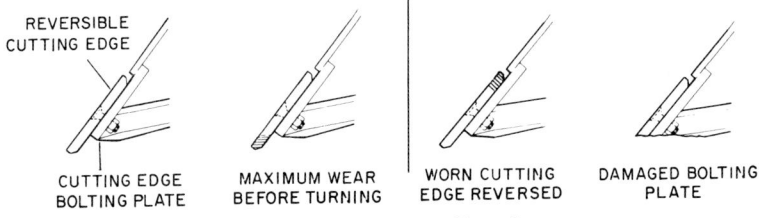

Fig. 3-9 Reversible cutting edges and how they operate.

2. Operator experience
    a. Operator experience is a vital factor in prolonging the hour-life of ground-engaging tools. An experienced operator will move more material with less downtime and fewer costly repairs than the inexperienced operator. Remember operators can make or break a project.
    b. Conduct schools. Let experienced operators share their skills and technology with newer and less experienced operators.
3. Abrasiveness of soil or rock
    a. Material abrasiveness is an uncontrollable but major factor in determining hour life of cutting edges and end bits. Hour-life may vary from a few hundred hours in very highly abrasive material to several thousand hours in less abrasive material.
    b. The highest or hardest heat-treated edges, without excessive breakage, should be used in any application. Keeping proper records is the best method for determining which type is suited for the application to obtain optimum performance.
4. Size of ripped or blasted material
    a. The fragmentation of ripped material is much smaller and more uniform than blasted material and is easier to doze.
    b. Standard cutting edges and end bits work well in ripped material, where as blasted material requires the severe service.
5. Down pressure applied to component. Constant or prolonged down pressure and the abrasiveness of the material are definite factors in the wear rate of cutting edges and end bits. Down pressure can be reduced in most applications by ripping.
6. Cycle time
    a. Hour-life is based on tractor operating hours, and the number of hours the cutting edge is actually engaged in the ground determines its end life.

b. Long or slow returns have a longer cycle time, less actual engaged time, and/or longer overall operating hours on a cutting edge.
   c. Short, fast cycle time results in increased production and also in fewer hours of life on the cutting edge. Study of production will show correct cycles for application.

Ripper Shanks

Even though there is a wide range of ripper shanks available, there are only three basic designs: the straight shank, the curved shank, and the double-offset shank. Matching the shank design to the duty is of vital importance for economic operation, greater production, better control of fragmentation, and longer component life. Most manufacturers will advise on the correct shank for a specific application.

Straight shank Straight shanks are particularly suitable for slabby and blocking materials.

Curved shank Curved shanks give a lifting action, resulting in good fracture characteristics in fine-grain, unbroken materials.

Double-offset shank Double-offset shanks are most effective in nonabrasive coal and other light, easily fractured, easily penetrated materials. With wear plates, these shanks

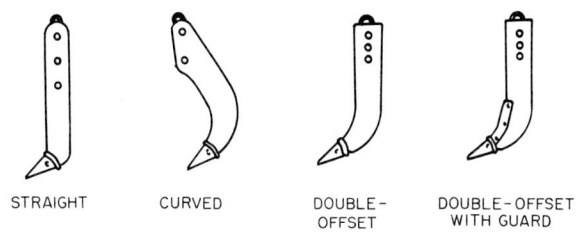

STRAIGHT   CURVED   DOUBLE-OFFSET   DOUBLE-OFFSET WITH GUARD

Fig. 3-10 Typical ripper shank forms.

give maximum ripping production in the widest variety of materials. They are the shanks best suited for most general construction and quarry applications. Wear plates, in split segments or one-piece wraparounds, greatly prolong shank life (Fig. 3-10).

Shank positions Shank position combinations in the tool bar should be as shown in Fig. 3-11. Improper shank positions cause twisting and side thrusts, resulting in broken ripper frames and shanks.

Shank repair kits Worn or broken shank ends can be repaired with a weld-on shank end. Most shank manufacturers offer a shank repair kit with complete installation instructions. As the shank material may vary from one manufacturer to another, it is vitally important to read and follow the instructions provided by each manufacturer.

Ripper Points

Ripper points may also be referred to as ripper teeth or tips, and they constitute the part of the ripper that takes the brunt of the punishment. Different point manufacturers may use different alloy steels and heat-treat procedures to meet specific requirements for their design.

Ripping capability is determined by the amount of pressure applied per square inch of point surface. A given size tractor can only produce a given amount of pressure on a point. Therefore, the sharper the point, the greater the amount of pressure per square inch of point surface and the greater the ripping capability. The applied pressure per square inch of point surface decreases at a rapid rate as the point wears and becomes dull. In some tough applications, it is

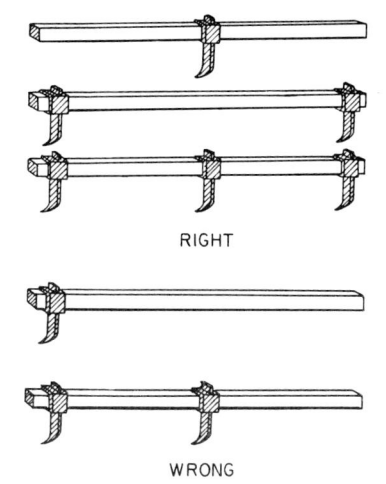

RIGHT

WRONG

Fig. 3-11 Proper shank position combinations.

6-28 Maintenance of Ground Contact Elements

more economical to change points with only 50 percent wear to maintain a sharp point and an economical ripping operation (Fig. 3-12).

Point selection There are a variety of points to meet every ripping requirement. Selecting the right point for your specific job is vitally important to assure optimum performance. Manufacturers provide points of different lengths and designs, each with a specific alloy steel and heat treatment to meet these requirements. Each manufacturer has a method of identifying various points and will assist in making the right selection. As an example, one manufacturer uses red or blue paint in the point pocket combined with the length of the point as follows:

1. Long *blue* point for normal application with resistance to normal impact and abrasion.

2. Short, sharp *blue* for high impact in severe rocky applications.

3. Long *red* for abrasion and moderate impact.

4. Extra-long *red* for extreme abrasion.

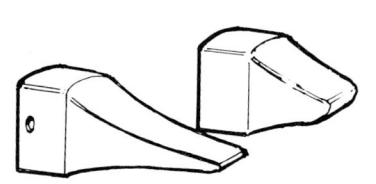

Fig. 3-12 New and half-worn ripper points.

Knowing the different types of points and keeping a proper record is the best method for determining the right point for the application.

Point angle The point angle of penetration is the angle formed by the front or top face of the point and the ground level at point of entry into a material (Fig. 3-13). Before the material can be ripped, it must be penetrated, and the point angle of penetration is an important factor in penetration. Tighter or harder materials require a steeper angle of penetration. In certain rock formations, angles of 45° or more are necessary to obtain penetration. Steeper angles may cause higher-than-average point breakage, but steeper angles may make the difference in whether a material can be ripped. Even with higher point costs, ripping is usually more economical than drilling and blasting.

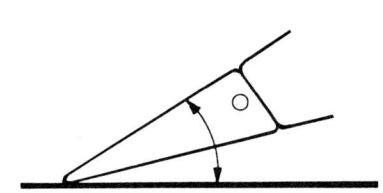

Fig. 3-13 Ripper tooth point of penetration.

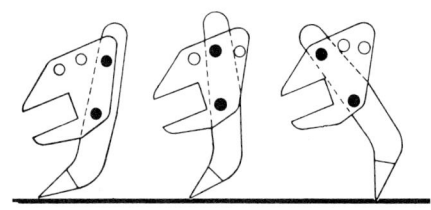

Fig. 3-14 Swivel-action bracket rippers.

Changing the point angle of penetration may be accomplished in three ways:

1. Some swivel-action bracket rippers have a combination of holes in the shank pocket to change the point angle of penetration (Fig. 3-14).

2. With variable parallelogram rippers, the point angle of penetration is changed by hydraulic cylinders or a combination of holes for the top parallel bar (Fig. 3-15).

3. On radial rippers, the angle of point penetration can easily be changed by simply raising or lowering the shank in its retainer (Fig. 3-16).

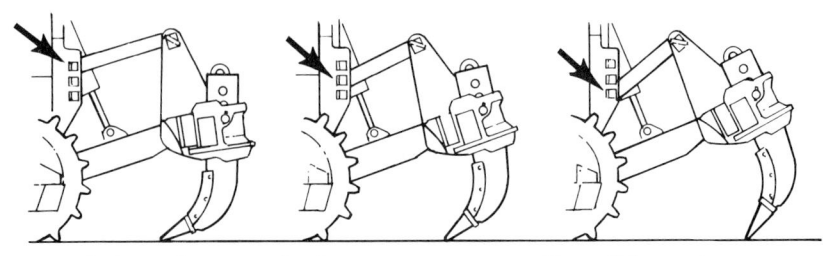

Fig. 3-15 Changing angle of point penetration on variable parallelogram rippers.

Excessive point failures

Some common causes of excessive point failures are
1. Operator inexperience
2. Incorrect point material or point length for an application
3. Worn or broken shank nose
4. Wrong shank angle
5. Backing with point in ground
    a. Operating a ripper is an art requiring the best manpower. Ripping should never be faster than first gear and reduced throttle. If ripping is easy, rip deeper or add more shanks. Speed increases wear and the possibilities of impact damage.
    b. A long and hard point for abrasion, working in rocky material, will result in broken points. A shorter and less brittle point, for impact application and working in abrasive material, will result in points wearing out prematurely. Select the correct point for the application.
    c. Fit of a point on a shank nose is vital to support of the point. A worn or broken shank nose will allow the point to move, and either breakage will occur from a low-support contact area or the point will have too steep a penetration angle. Repair the shank nose by installing a shank repair kit.
    d. Too flat a point penetration angle will result in excessive point wear and lack of penetration. Too steep a point penetration angle will result in excessive point breakage. Adjust the point angle of penetration to obtain a quick penetration and ripping depth without excessive point breakage.

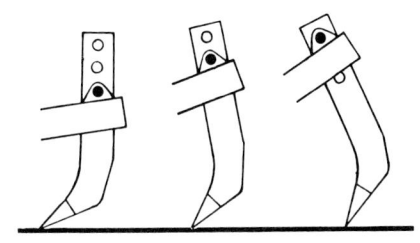

Fig. 3-16 Changing angle of point penetration on radial rippers.

    e. A tractor should never be backed with the point in or near the ground—a rock could catch the back side of the point, resulting in a broken point or loss of the point.

Hard-Facing

Replaceable items such as dozer cutting edges, end bits, ripper points, and shank guards are heat-treated to maximum hardness for best resistance to wear. Rebuilding or hard-facing of these items is seldom economical and can often result in failure. It is therefore not recommended. However, those who still want to rebuild or to hard-face should obtain and follow the instructions of the manufacturer to assure the best possible job and to minimize the danger of failure.

In most cases, the dot method of hard-facing is recommended over the stringer beads since cracking of the base metal adjacent to the stringer beads may result. The dot method consists of a pattern of dots $3/8$ to $1/2$ in. in diameter located on approximately 1 to $1\,1/2$ in. centers. The dot method is less expensive because less labor and materials are required, and it is considered as effective as continuous stringer beads. The dots are excellent indicators of extent of wear and should be rewelded before they are completely worn off.

Chapter **4**

Maintenance of Earthmoving Buckets and Bucket Teeth

ENGINEERING STAFF
ESCO Corporation, Portland, Oreg.

Proper selection from the many different types and sizes of earthmoving bucket teeth available and careful maintenance can help increase productivity and at the same time hold down operating costs.

Most equipment owners see to it that expensive machinery is regularly maintained, with engines, hydraulic systems, drive trains, tracks, and tires getting periodic attention. But too often the digging end of that equipment is neglected. And when the bucket doesn't dig or load to capacity, the total machine doesn't do its designed job. Proper teeth can reduce tire wear on front end loaders. And, with the right teeth, regularly maintained, all equipment can operate with less fuel consumption.

The many varieties of earthmoving equipment teeth dictate that each manufacturer's instructions be followed. Tooth locking systems vary; steels used in different points and adapters require different welding techniques and materials. Even digging angles vary for different types of buckets, and teeth must be installed correctly to operate efficiently.

The instructions in this chapter relate to tooth installation and maintenance for the more commonly used buckets in the construction and mining industries: front end loaders, hoe dippers, shovel dippers, clamshell buckets, and dragline buckets. By following these instructions, better performance can be obtained from equipment at lower cost.

Front End Loader Buckets and Teeth

These buckets need teeth to dig into material. They increase penetration of a lip by concentrating force on tooth points rather than on a large lip area.

Some buckets still use solid one-piece weld-on teeth. But most use two-piece (point and adapter) teeth. Solid teeth may originally be inexpensive but maintenance is expensive. When worn, they must be burned off and new ones welded onto the lip. Thus, they are usually run while dull in an effort to get longest life from them.

Two-piece teeth are recommended. They are more economical and versatile. Adapters are bolted or welded to the bucket lip and the points are mechanically attached. When points are worn, they are changed quickly and easily. Also, different point styles can be used to match the material being handled.

Welded adapters are preferred because they are stronger than bolted types. The advantage of bolted adapters is that they can be removed when working, whereas teeth could damage a structure such as in unloading barges.

How to space teeth Too many or too few teeth can adversely affect loader performance. This typical rule of thumb is often used to obtain the correct number of teeth: Adapter nose width, multiplied by 4 and divided into inside lip width of bucket. For example, adapter nose is 3½ in., multiplied by 4 to give 14. Divide this into a lip width of 123 in. to give 8.8 teeth. Round off to 9 teeth. Place a tooth at each end of the bucket and divide the rest evenly along the lip. Divide the remaining 7 teeth into the lip width of 123 in. to get 17.6 in. spacing between teeth.

Adapter types Some manufacturers offer different styles of adapters. A single-leg adapter welds to the top of a lip, leaving the bucket bottom flush for cleanup work.

Others have 1½ or 2 legs which slip over the lip and are welded or bolted to it. These are stronger and provide a wear-resistant runner for the bucket bottom.

Nose types and locking devices Most adapters have a flat wedge-shaped nose which mates with the box section of the replaceable point. These accept head-on loads quite well, but they require a heavy box section on the point and a large pin to resist loads from the side, top, and bottom. They are subject to breakage when vertical or side loads become more than they can bear.

One manufacturer uses a conical nose. It resists loads equally from all directions. A flat rim on the tip of the nose provides holding power so only a small vertical pin is needed to retain the point on the adapter.

A rubber lock fits into a keyslot which squeezes against a corrugated pin, wedging the point onto the nose. The pin can be driven out from top or bottom, making it easy to change points. It is safer to replace points with this system than with side-locked points. Usually there isn't much room to swing a hammer to remove side-locked points.

Point selection Points are offered in a variety of sizes and styles. Some manufacturers provide thicker points for loading rock where penetration is no problem. These give more wear metal to resist abrasion. Long, sharp points are used for penetrating less abrasive material. There are broader points for cutting into coal, hardpan, and clay.

Welding points Points should not be hard-faced or weld-repaired. The alloys and heat treatment make them extremely hard and welding heat can destroy their metallurgical qualities.

Welding two-leg adapters Follow the detailed welding instructions given in "How to Weld Alloy-Steel Bucket Components," later in this section.

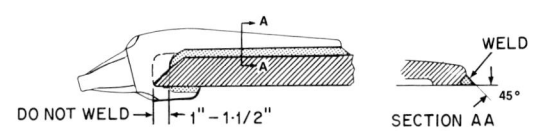

Fig. 4-1 Front end loader two-leg adapters. Fillet weld should cover J groove for effective weld strength through fillet throat. Hold weld angle close to 45°.

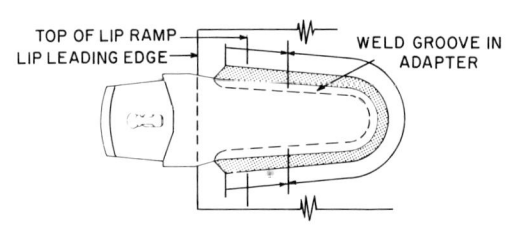

Fig. 4-2 Top view, welding adapters on front end loader buckets.

Use E7016 or E7018 electrodes or E70 wire, all of low-hydrogen content.
1. Locate adapters on bucket lip.
2. Preheat and tack adapters to lip. Use a Tempilstik to check temperature.
3. Weld around both legs but not within 1 to 1½ in. (25 to 38 mm) of the leading edge, as shown in Fig. 4-1. Start weld as shown in Fig. 4-2 at approximate center of leg but behind lip ramp on top leg. Weld one pass around back of leg to center of opposite side. Then backstep with one pass on each side, starting 1 to 1½ in. (25 to 38 mm) back from lip edge to center of leg.

Repeat until weld is complete. Maintain full fillet size to stop off point and tie in the ends.

4. Grind front of weld to smooth contour to relieve stress concentrations (Fig. 4-3).

Welding one-leg adapters

1. Locate adapters on lip.
2. Preheat and tack adapters to lip.
3. Weld completely around adapter legs and along loading edge of lip, as shown in Fig. 4-4.

CAUTION: Welding to lip leading edge may cause cracks in lip if these instructions and good welding techniques are not carefully followed.

Start weld at approximate center of leg behind top lip ramp and weld one pass around back of leg to center on opposite side. Then backstep with one pass from lip leading edge to center of leg, using starter plates, as shown in Fig. 4-5. Do this to both sides. Repeat until weld is finished.

4. Burn off starter plates. Finish welding to lip leading edge, carefully smoothing contours at weld junctions. This weld can be either of two types shown in Fig. 4-6, J groove or fillet. If lightener cavity extends forward of lip leading edge, as shown in Fig. 4-7, weld to edge of cavity. Otherwise, weld across entire width of adapter. Maintain full fillet size over entire weld.

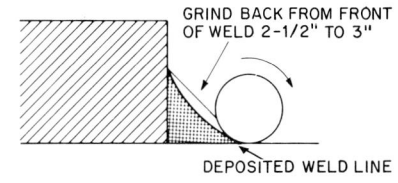

Fig. 4-3 Grind weld back from 2½ to 3 in. to reduce stress concentrations.

5. Grind off front of weld to maintain smooth contour and avoid stress concentration.

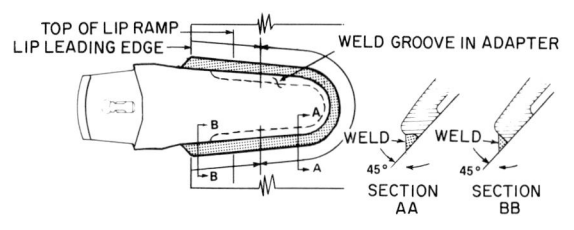

Fig. 4-4 When welding one-leg adapters on front end loader buckets, fillet weld must cover J groove (section *AA*) for effective weld strength through fillet throat. Use fillet weld only in forward area (section *BB*) where there is no J groove.

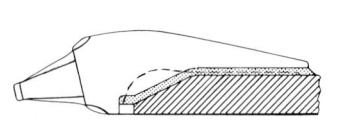

Fig. 4-5 When using a starter plate, backstep from the starter plate. Then remove before welding to lip of leading edge.

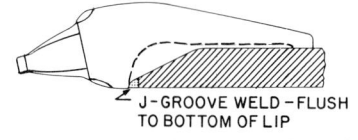

Fig. 4-6 J-groove weld at leading edge is flush with bottom of lip. (Bevel or fillet weld would be equal to lip leading-edge thickness.)

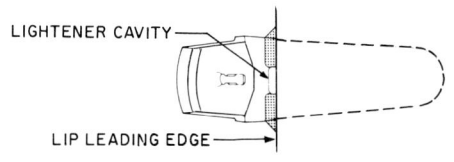

Fig. 4-7 Top view of one-leg adapter, showing lightener cavity.

Weld sizes

Bucket size		Weld size	
Cubic yards	Cubic meters	Inches	Millimeters
2	1.8	½	13
2¼–5	2–4.6	⅝	16
5½–12	5–11	¾	19
13 and up	12 and up	1	25

Teeth for Shovel Dippers, Hoe Dippers, and Dragline Buckets

There are many types of teeth used on earthmoving buckets. Designs from different manufacturers have similarities as well as differences. Instructions given here are from ESCO. Even so, these instructions do not cover every design variation. Check with the supplier to be sure of proper instructions for maintaining teeth on equipment.

How to install tooth horn adapters Used on hoe dippers, shovel dippers, and dragline buckets, these adapters fit on integrally cast horns on a bucket lip. They are clocked onto the horns by two methods: either fluted spools and wedges or key locks.

Fluted Spool and Wedge

1. Before installing, clean all adapter and lip mating surfaces.

2. Install new adapter and check clearances, as shown in Fig. 4-8. There must be clearances at A, AA, B, and CC. Bearing surfaces must be uniform at D and DD. If there are no clearances, rebuild the horns, as shown under "Rebuilding Tooth Horns."

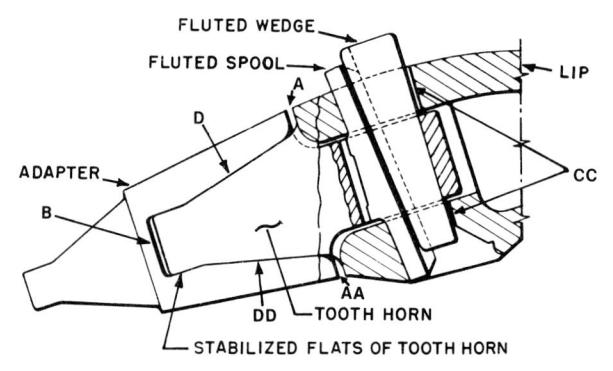

Fig. 4-8 Clearances and bearing areas of tooth horn adapter with fluted spool and wedge lock.

3. Insert fluted spool into cavity with large lug at bottom and facing forward.

4. Insert wedge behind spool, small end down. Drive in with several heavy blows of a hammer until there is no apparent movement—but not to the point of refusal. Use a 4-lb (2-k) hammer.

NOTE: Spool and wedge must be installed properly or it will cause breakage or loss of adapter or breakage of lip tooth horn.

5. Check tightness of wedge regularly. Loose wedges can lead to adapter loss or premature failure. After initial installation, tighten wedges every 2 to 4 hr until no longer needed to prevent looseness. Burn off any part of wedge extending from lip bottom.

6. If wedge drives flush with top of horn, use next oversize wedge.

7. To remove, drive out wedge from bottom upward with drift pin and hammer.

Key Lock

1. Clean adapter and lip mating surfaces.

2. Install new adapter and check clearances A, AA, B, and CC. There must be uniform bearing surfaces at D and DD. If no clearances, rebuild tooth horns (see Fig. 4-9).

3. Insert key into cavity and drive with a 4-lb (2-kg) hammer until there is no apparent movement with several heavy blows—but not to point of refusal.

4. Heat and bend tang at an angle (see Fig. 4-10).

5. Check tightness of key regularly. Loose keys can lead to adapter loss or premature failure. After initial installation, tighten keys every 2 to 4 hr until no longer needed to prevent looseness. Burn off key extending past lip bottom. Heat and bend tang as required.

6. If key drives flush with top of horn, use next oversize key.

7. To remove, straighten tang and drive out key from bottom upward with drift pin and hammer.

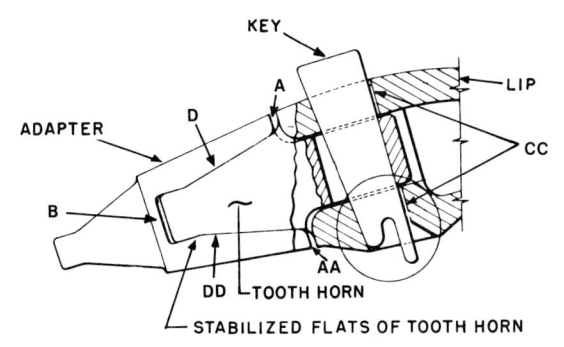

Fig. 4-9 Clearances and bearing areas of tooth horn adapter with key lock.

Maintaining Whistler adapters Whistler adapters are used on shovel dippers and front end loader zipper lips.

1. When lip is new, make a template to locate the leading edge of the keyslot. This will aid in rebuilding the lip and keyslot when they are worn. Cut template to approximate shape of lip bearing pad. Slip over lip and against its leading edge. Mark location of keyslot front edge, both inside and outside the lip.

2. Before installing new adapter, check to see if lip needs rebuilding.
   a. Clean bearing surfaces of lip, adapter, C-clamp, and wedge.
   b. Place new lip template on each bearing pad so it touches lip leading edge. Forward edge of the keyslot must line up with marks on template within ½ in. (13 mm).
   c. Install new adapters but not the clamps or wedges.

3. New adapter must bear against the lip leading edge and both top and bottom legs must contact lip for 3 to 4 in. (76 to 102 mm) back from leading edge (see Fig. 4-11).

4. The rear end of the adapter legs must either contact the lip bearing pad, or the total gap (gaps of both legs added) must not exceed 3/16 in. (5 mm) (Fig. 4-11). If any of these conditions are not met, the lip bearing pads are too worn and must be rebuilt.

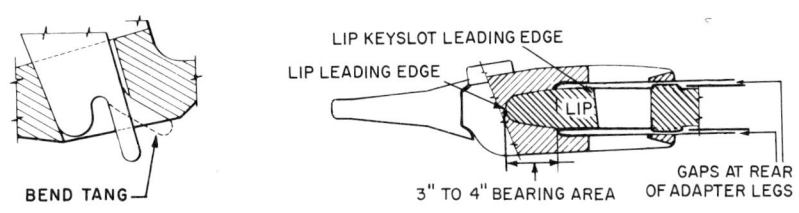

Fig. 4-10 Bend tang on key lock used on tooth horn adapters.

Fig. 4-11 Bearing surfaces and clearances of Whistler adapter.

5. With new adapters in place on lip, install C-clamp, large end first, from inside bucket or dipper.

6. Insert wedge forward of C-clamp, small end first, with flutes engaging C-clamp flutes. Drive wedge with 10 to 20 lb (4.5 to 9 kg) hammer until wedge moves only ⅛ in. (3 mm) per blow, with several successive blows—but not to point of refusal (see Fig. 4-12).

6-36 Maintenance of Ground Contact Elements

7. Check wedge tightness regularly. A loose wedge can lead to adapter loss or premature failure. Tighten every 2 to 4 hr until no longer needed to prevent looseness. Burn off any part of wedge extending below adapter bottom to prevent it from being knocked loose when digging (Fig. 4-12).
8. If wedge drives flush with top of adapter, use next oversize wedge.
9. To remove, drive out wedge from bottom upward with drift pin and hammer.

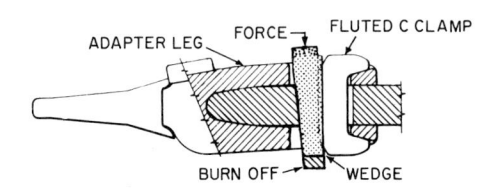

Fig. 4-12 Installing fluted C-clamp and wedge on Whistler adapter.

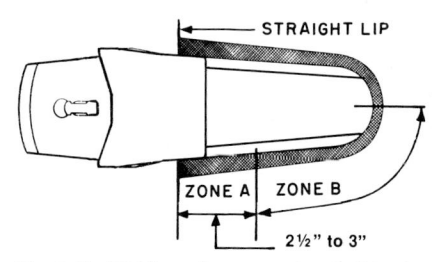

Fig. 4-13 Welding adapters to clamshell buckets; top view shows two welding zones.

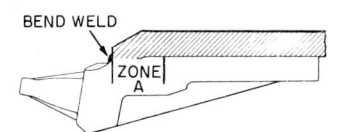

Fig. 4-14 Side view of clamshell bucket adapter. Bead weld to leading edge and grind to allow lips to close.

Installing Adapters on Clamshell Buckets

Use E7016 or E7018 electrodes or E70 wire, both of low-hydrogen content. Follow welding details under "Welding ESCO Alloy-12 Series" (see Figs. 4-13 to 4-15).
1. Position adapters on lip.
2. Preheat and tack adapters to lip.
3. Use fillet weld in the two zones as follows:

Bucket size		Fillet weld size			
		Zone A, reduce gradually to zone B			
Cubic yards	Cubic meters	Inches	Millimeters	Inches	Millimeters
½–5	0.5–4.6	⅝	(16)	⅜	(10)
5½–10	5–9	1	(25)	½	(13)
10¼ and up	9.5 and up	1½	(38)	¾	(19)

4. Bead weld to lip leading edge and grind flush to allow lips to close.

Installing Adapters on Hoe Dippers and Dragline Buckets

Use low-hydrogen content E7016 or E7018 electrodes or E70 wire. Follow welding details given under "Welding Structural Components."
1. Position adapters on lip.
2. Preheat and tack adapters to lip.
3. Use same weld sizes in zones *A* and *B* as given in "Clamshell Buckets" above (see Figs. 4-15 and 4-16).
4. Weld from rear forward with rotated weld patterns top and bottom and side to side to minimize weld stresses.
5. Grind front ends of welds at least 1½ to 2 in. back to reduce weld stresses.

Maintenance of Earthmoving Buckets and Bucket Teeth 6-37

How to Install Adapters and Teeth on Integral Nose Tooth Bases

There are two types of noses cast into the lips of hoe dippers: shovel dippers and dragline buckets. One has a flat wedge nose with a rounded tip. The other is a patented design with a double conical nose and a flat rim at the tip. These bearing surfaces are different, but the teeth are locked to the noses by the same method.

1. Make a template of the tooth base when it is new to aid in rebuilding when it is worn. Template must fit exactly the tapered bearing surfaces where the adapter or tooth contacts the base. Cut it to fit the base halfway between the edge of the keyslot and the side of the base. Mark on the template the distance from the edge of the keyslot. Cut a clearance over the front face of the base 7/16 in. (11 mm).

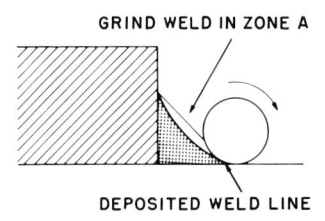

Fig. 4-15 Grind weld in zone A to reduce stress concentration.

4-16 Hoe and dragline adapters; side view shows weld zones A and B.

2. When installing new box teeth or Kwik Tip adapters, check the fit to see if tooth base needs rebuilding.
   a. Clean all bearing surfaces of tooth base, adapter, spool and wedge, and keyslots of base and adapter (see Fig. 4-17).
   b. Visually check both sides of each base for wear. Severely peened or rounded areas are signs of wear which cause adapters to rock on base. This can cause breakage. Build up these areas as described under "Rebuilding Lip Tooth Bases."
   c. If template of new tooth base is available, place it on each base, both sides of keyslot, at location marked on template.
   d. If template contacts forward edge of base but doesn't contact both tapered surfaces at the same time, tooth base needs rebuilding.
   e. Rebuild any area on tapered bearing surfaces and stabilizing flats on conical tooth bases that is 1/8 in. (3 mm) or more from contacting template when it is located at the marked distance from the keyslot.
   f. Finally, place new adapters on tooth base and insert spool (see Fig. 4-18). Check for bearing of spool against back of adapter keyslot. If there is clearance, tooth bases must be rebuilt. NOTE: Some spools are straight and can be inserted either end first. Most spools are tapered and must be inserted large end first from top of tooth base. Tapered spools are marked on small end with Up or at center with Up and an arrow (see Fig. 4-19).
   g. If spool properly bears against back of adapter keyslot, install wedge. Check for clearance of wedge with front of adapter keyslot and for distance of top of

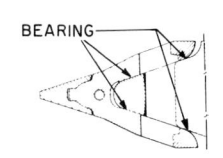

Fig. 4-17 Bearing areas of adapter and integral nose tooth base.

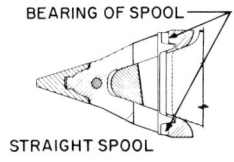

Fig. 4-18 Checking bearing area of integral nose with spool.

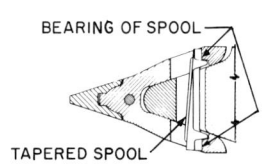

Fig. 4-19 Checking bearing of tapered spool on integral nose and adapter.

6-38 Maintenance of Ground Contact Elements

wedge above top of adapter. If standard wedge drives flush with top of adapter, use oversize wedge. If oversize wedge drives flush with top of adapter or touches front edge of keyslot, rebuild tooth bases.

NOTE: Insert small end of wedge first from top of tooth base. Spool and wedge must be installed properly or adapters will be lost or broken or tooth base broken.

3. If tooth bases are not worn, install adapters and spools and wedges as described in f and g above.

4. Drive wedge with 10 to 20 lb (4.5 to 9 kg) hammer until wedge moves only ⅛ in. (3 mm) per blow with successive blows. Do not drive to point of refusal.

5. Check wedge tightness regularly. Loose wedges can lead to adapter loss or failure. After initial installation, tighten wedge every 2 to 4 hr until no longer needed to prevent looseness. Burn off wedge extending below adapter bottom to prevent it from being knocked loose when digging (see Fig. 4-20).

6. If wedge drives flush with top of adapter, use next oversize wedge.

7. To remove, drive out wedge from bottom upward with drift pin and hammer.

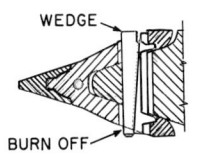

Fig. 4-20 Integral nose with adapter, spool, and wedge.

How to Weld Alloy-Steel Bucket Components

While buckets made of plate steel present no special welding problems, many buckets and components, including tooth and lip mounts, require special techniques. There are many types of alloys used in buckets, adapters, and points.

Varying amounts and combinations of carbon, nickel, chrome, silicon, molybdenum, and manganese give toughness and hardness to bucket parts. Some alloys are weldable, some are not. Some are heat-treated to obtain maximum hardness and welding heat can ruin them.

It is best, of course, to obtain specific welding instructions from the manufacturers, as there is no way to tell what alloy a particular part is made of.

Welding bucket teeth Most manufacturers do not recommend welding or hard-facing replaceable points on two-part teeth. They are cast of alloys and heat-treated to obtain optimum hardness to resist abrasion and breakage.

Welding ESCO alloy-12 series These alloys are used by one manufacturer for bucket parts.

Alloy	Bucket part	Weldability
12C	Front end loader adapters	Excellent
12E; 12F low hardness	Bucket lips and other structural components	Excellent
12 high hardness	Medium-size bolt-on or shank adapters.	Fair
12H	Large bucket adapters, shank, Whistler, and Kwik Tip types	Not recommended
12M low hardness	Bucket components of small to moderate thickness	Excellent
12M high hardness	Bolt-on adapters and wear-protection parts	Not recommended
12S	Bucket and dredge points, ripper points, and bits, and blades	Not recommended
12T, 12F	Large bucket structural components	Excellent

Welding structural components ESCO alloys 12C, 12E, 12F, and 12M low hardness and 12T have excellent weldability. They are used for weld-on adapters and for structural components on dragline bucket lips, cheeks, and arches; shovel dipper beams and most dipper fronts and lips; dredge cutter arms, rings, and hubs.

Structural components of these alloys may be hard-faced. However, since replaceable wear-resistant shrouds are available, hard-facing is not usually required.

These alloys are magnetic and can be distinguished from nonmagnetic manganese steel with a magnet. Since the two different alloys require different welding materials and techniques, it is important to check for magnetism before welding.

Welding electrodes Use E7016 or E7018 low-hydrogen electrodes to weld these alloys to each other or to other low-alloy or carbon-steel castings or forgings.

Controlling moisture in low-hydrogen electrodes Purchase low-hydrogen electrodes in 10-lb hermetically sealed containers. Larger sizes may have too much moisture which can cause underbead cracking. Electrodes pick up moisture when exposed to air. After opening container, keep electrodes dry.

1. Remove only a ½ hr supply a a time. Keep the remainder in a ventilated holding oven at 150°F (66°C).

2. If a partially used container is stored for later use, bake in a 500°F (260°C) ventilated oven for 2 hr before using.

3. If electrodes are taken from a cardboard box or other nonairtight container, do not use until they have been baked for 2 hr in a 500°F (260°C) ventilated oven. Then place in a ventilated holding oven while still warm, removing only a ½-hr supply at a time. Keep holding oven at 150°F (66°C).

4. Rebake electrodes exposed to air for more than ½ hr as in step 2.

Welding techniques Preheat heavy sections to 350 to 400°F (177 to 204°C) with torch. Check temperature with Tempilstik. Remove chill from light sections in cold weather.

Bevel the joints for 100 percent weld penetration. Rebuild worn areas to original contours with E7016 or E7018 before hard-facing. Build up excessively worn areas with low-alloy steel plate to restore original contour.

Weld stringer beads with a slight weave that is not more than three times rod diameter. Remove slag after each pass. Peen each bead to reduce stress concentrations. Maintain interweld temperature of less than 500°F (260°C).

After welding or hard-facing, postheat weld and areas around weld uniformly to 350 to 400°F (177 to 204°C), then air cool.

Dot hard-facing This hard-facing method is preferred to any other. It consists of spots or dots about ⅜ to ½ in. (10 to 13 mm) diameter, welded in rows about 1 to 1½ in. (25 to 40 mm) apart.

It is faster, uses less material, and costs less than continuous stringer bead or solid overlay. It also reduces cracking and weld failure of other methods.

Dots are also good wear indicators. Reweld them before they are completely worn off. Use Stoody 31 or equivalent electrodes for hard facing.

Welding manganese steel For rebuilding structural components such as tooth bases, use lip bearing pads made of ESCO alloy-14 manganese steel.

Check with a magnet to distinguish nonmagnetic manganese from magnetic alloys as it requires different materials and techniques.

Do not preheat—it will destroy toughness and make steel brittle.

Preparation Bevel the joints with a torch, grinder, or arc air to obtain 100 percent weld penetration. Avoid overheating when gouging.

Grind or torch-cut work-hardened surfaces at least ¹⁄₃₂ in. (1 mm) deep before welding. Check with prick punch to be sure all work-hardened surface is removed. Scarf out cracks completely before welding.

Welding Skip weld and avoid wide weaving beads to distribute heat evenly along joint. Make beads less than 5 in. (125 mm) long. Weld with arc at low current setting (cold arc).

You should be able to put your bare hand on the part within 6 in. (150 mm) of the weld without burning your hand.

Peen each bead immediately to reduce contraction stress.

Rebuild worn areas to original contour with austenitic manganese nickel electrodes before hard facing.

Use these rods Use austenitic chromium nickel stainless-steel electrodes to weld manganese steel.

6-40 Maintenance of Ground Contact Elements

To weld manganese to ESCO 12T, 12M, 12E, and other low-alloy or carbon steel, use stainless-steel electrodes types 307, 308, 309, 310, 312, or 316.

Austenitic manganese nickel rods are sometimes used, but welds are not as durable as stainless welds.

Do not use mild steel or other electrodes to weld manganese to other materials.

Rebuilding tooth horns

1. When horn is new make a template to aid in rebuilding. Scribe it as shown in Fig. 4-21. Use to check keyslot leading-edge bevel and width.
2. If tooth horn does not have stabilizing flats, build them up as shown in Fig. 4-22. Rebuild nose as shown in Fig. 4-23.
3. Grind off work-hardened surfaces if tooth horns are of manganese steel.
4. Use a new adapter, fluted spool and wedge, or key as a gauge.
5. Build up tooth horn with weld until there is bearing at surfaces D and DD with clearance of $1/64$ to $1/32$ in. (0.4 to 0.8 mm) at E and EE. With adapter bearing on surfaces CaD and DD, B must be $1/8$ to $3/8$ in. (3 to 10 mm) maximum. Clearance A must be at least twice that of B. Top of wedge should protrude $3/4$ in. (19 mm) above tooth horn. Use standard feeler gauges (see Fig. 4-24).

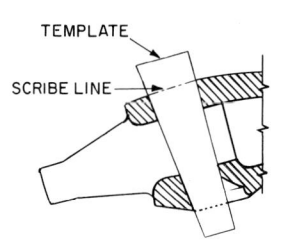

Fig. 4-21 Template for rebuilding tooth horns.

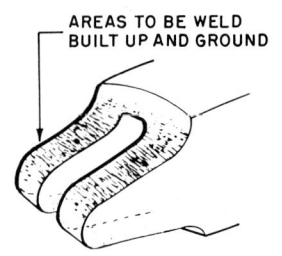

Fig. 4-22 Areas of tooth horn to weld built up and ground.

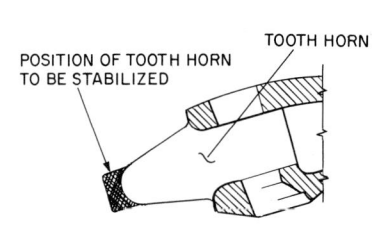

Fig. 4-23 Adding stabilized flats to tooth horn.

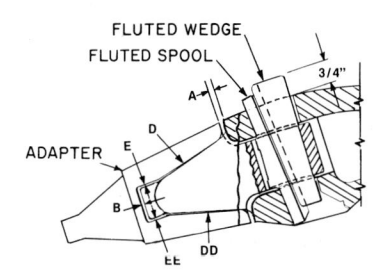

Fig. 4-24 Bearing areas and clearances of tooth horn and adapter.

Rebuilding Whistler lip adapter bearing pads

1. Use template to check wear. If it shows keyslot in bearing pad is too close to lip leading edge, examine keyslot by placing straightedge against its forward edge (see Figs. 4-25 and 4-26).
    a. If forward edge is rounded so straightedge rocks when placed against it, rebuild forward edge and grind smooth so straightedge will not rock. Rebuilt forward edge should line up with marks on template when it is properly positioned in contact with lip leading edge (see Fig. 4-27).
    b. If forward edge of keyslot is straight (the straightedge will not rock more than $3/32$ in. (2.5 mm), then the lip bearing pad can be rebuilt without welding keyslot.
2. Weld buildup and grind leading edge of lip back 3 to 4 in. (76 to 102 mm) to provide bearing against new adapter. Use template as a guide for amount to be added to leading edge by positioning template so marks for keyslot line up with forward edge of keyslot.

3. Weld buildup and grind rear adapter bearing pads so the total gap with a new adapter (measuring gap under each leg and adding them) is not more than 1/16 in. (1.5 mm).
4. Check fit. Use new Whistler adapter, C-clamp, and wedge as gauges.
   a. W-1 wedge should insert through lip and just into lower-leg keyslot of adapter with moderate pounding.
   b. If wedge does not insert far enough, grind buildup from leading edge. If it inserts too far, weld more buildup at lip leading edge (see Fig. 4-27).

Rebuilding lip tooth bases (integral noses) These tooth bases are cast as part of the lip on certain types of hoe dippers, shovel dippers, and dragline buckets. They are also called *integral noses*.

There are two types: conical nose and nonconical nose. Instructions for both are given.

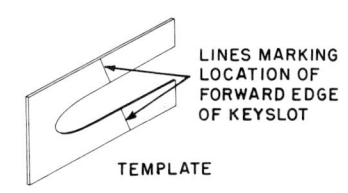

Fig. 4-25 Make a template for rebuilding Whistler lip bearing pads. Scribe lines to mark forward edge of keyslot.

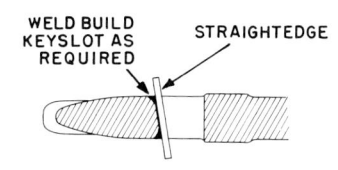

Fig. 4-26 Check keyslot with straightedge. Rebuild keyslot as required.

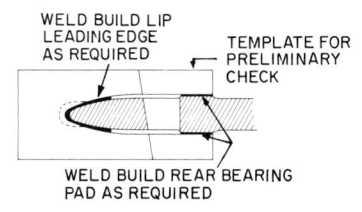

Fig. 4-27 Areas to rebuild on lip leading edge and rear bearing pads; solid areas show weld buildup needed.

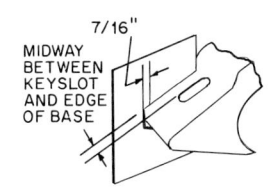

Fig. 4-28 Use a template to check wear on lip tooth base.

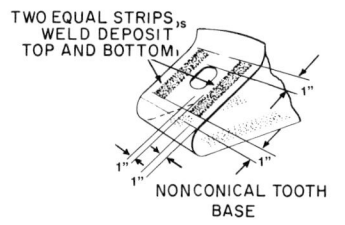

Fig. 4-29 Buildup on worn nose with two equal strips of weld (top and bottom) on nonconical tooth base.

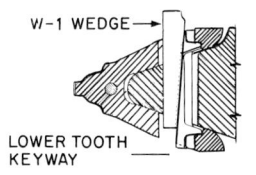

Fig. 4-30 Check the fit of rebuilt tooth base with adapter and wedge.

Nonconical lip tooth bases

1. Make a template as described in "How to Install Adapters and Teeth on Integral Nose Tooth Bases." Use template and an adapter to check for wear (see Fig. 4-28).

2. Rebuild rounded or worn areas back to original contour, using a Kwik Tip adapter or solid tooth as a gauge. Weld two equal strips about 1/8 in. (3 mm) thick on top and bottom of base. Strips should extend from 1 in. (25 mm) behind keyslot to within 1 in. (25 mm) of leading-edge curve or flat of nose. Keep weld 1 in. (25 mm) away from keyslot and 1 in. (25 mm) away from outside edges (see Fig. 4-29).

3. Check fit, using new Kwik Tip adapter or solid tooth and fluted spool and wedge as gauges.
   a. Fit tooth to base until W-1 wedge will insert through base and adjust into lower tooth keyway with moderate pounding (see Fig. 4-30).

6-42 Maintenance of Ground Contact Elements

b. Check to see if four-point bearing is achieved. Chalk, set tooth or adapter, and grind high spots. Repeat to reach four-point bearing.

Rebuilding conical lip tooth bases Follow above instructions with these exceptions.

1. Use rounded buildup strips to follow contour or conical Kwik Tip adapter or tooth (see Figs. 4-31 and 4-32).

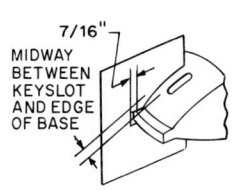

Fig. 4-31 Use a template to check wear on a conical nose lip tooth base.

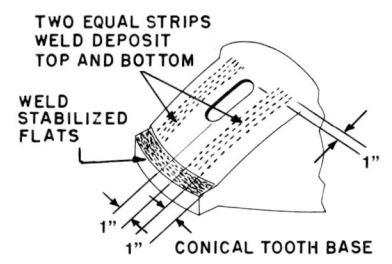

Fig. 4-32 Areas to build up on a conical lip tooth base. Weld stabilized flats and apply two strips. Weld deposit on top and bottom of nose.

2. Build up stabilizing flats as close as possible to forward end of adapter or tooth cavity. Top and bottom surfaces must be parallel.

Chapter 5

Maintenance of Vibratory Compactor Drums

ARNOLD DEICHEL
Manager, Product Services
Ingersoll-Rand Company, Compactor Division
Shippensburg, Pa.

Vibratory roller compactors have proven to be profitable in site preparation for a broad range of construction projects ranging from atomic generating sites and airports to shopping areas.

As a result of vibratory compaction of soils ranging from Georgia red clay to Mojave desert sand, contractors have realized site-preparation cost efficiencies and increased profits. The key to the successful operation of the vibratory rollers is a systematic preventive maintenance program as well as routine servicing consistent with the manufacturer's recommendations. Field maintenance starts on receipt of the equipment.

Every piece of construction equipment should have a receipt-condition report attached to record the actual condition of the unit when it arrives at its destination. Fleet managers should check all pertinent data to be certain it corresponds with the equipment ordered—model and serial numbers must match; all keys, parts, service manuals, and auxiliary equipment should arrive at the same time. Visual inspection will determine the condition of tires, gauges, seat, glass, battery and accessories, and the overall appearance of the equipment.

Claims for any transit damage should be immediately filed with the delivering carrier—with a copy of the report sent to the sales representative of the manufacturer. A typical report lists all pertinent condition received data that should be recorded and returned to the factory. The receiver should make a copy for his files and send an additional photofax copy to the carrier (Fig. 5-1).

A predelivery inspection and warranty registration form will help assure trouble-free starting and operation. This comprehensive checklist serves a dual purpose for start-up and routine equipment maintenance. Each unit has such a plaque; it spells out clearly what field maintenance steps must be taken at specific operating intervals.

Periodic service bulletins are issued to advise the latest tips on maintenance as well as auxiliary and product improvement equipment available. Field reports frequently provide hints for service expediency that can save the contractor both time and money.

Maintain master records A master maintenance record card (Fig. 5-2) can be kept up-to-date with all pertinent information included relating to each machine in the field. This file should be maintained throughout the life of the equipment. If any unusual activity is noted on a particular compactor by maintenance mechanics, it signals a need for action.

Spare parts availability is the key to an effective construction equipment support program. Most manufacturers maintain parts inventories throughout the United States

6-44 Maintenance of Ground Contact Elements

with levels based on experienced need. Naturally, the user should stock certain recommended spare parts for quick field replacement to minimize equipment downtime.

Each unit is supplied with a permanently affixed lubrication schematic diagram (Fig. 5-3) which lists recommended lubrication and maintenance schedules. This same chart usually appears in the compactor's operator manual.

Ingersoll-Rand
Machine Receipt Condition

| MODEL NUMBER |
| SERIAL NUMBER |
| DEALER NAME |
| LOCATION |
| PARTS & SERVICE MANUALS ☐ |
| IGNITION KEYS ☐ |

VISUALLY INSPECT

TIRES ☐	SEAT ☐
GAUGES ☐	GLASS ☐
BATTERY ☐	CONTROLS ☐
ACCESSORIES ☐	PAINT ☐

COMMENTS

File Claim with Carrier for Transit Damage   Form 8100-10

Fig. 5-1 Maintenance begins with a record of machine condition at the time of receipt.

Field Troubleshooting A Key

To keep downtime to a minimum, a systematic, easy-to-read troubleshooting manual is available to locate malfunctions and, where practical, perform field corrections. Simplified fault-logic diagrams help to quickly pinpoint and correct problems so that they can be corrected quickly and efficiently. For example, if vehicle response is sluggish, maintenance personnel should check for a plugged filter, suction line clogged, parking brake engaged, air in the system or, possibly, even a worn pump or motor in the drive circuit. Once the problem is found, its correction is normally simple.

Plugged Filter Change filter.

Suction Line Clogged Inspect suction line from the reservoir to the transmission pump; maximum suction should not be greater than 10 in. of mercury.

Parking Brake Engaged Disengage parking brake lock usually located under the driver's seat by lowering lever to the release position. If brakes fail to release, check the master cylinder for clearance between the brake pedal linkage and the cylinder on air brake systems. Verify gauge pressure and whether the parking control valve is pushed in.

Air in System Check for low fluid level; check inlet filter and suction line for leaks allowing air to enter the system.

Broken or Crimped Control Cable Inspect the control cable from the console to the transmission for movement and wear, and replace if necessary. Do not attempt to move control unless engine is running.

Pump Drive Disconnected (on Clutch-Equipped Machines Only) Inspect and adjust clutch assembly between engine and pump drive for proper engagement. Clutch requires tension of 110 ft-lb to operate—measured at the extreme end of the handle.

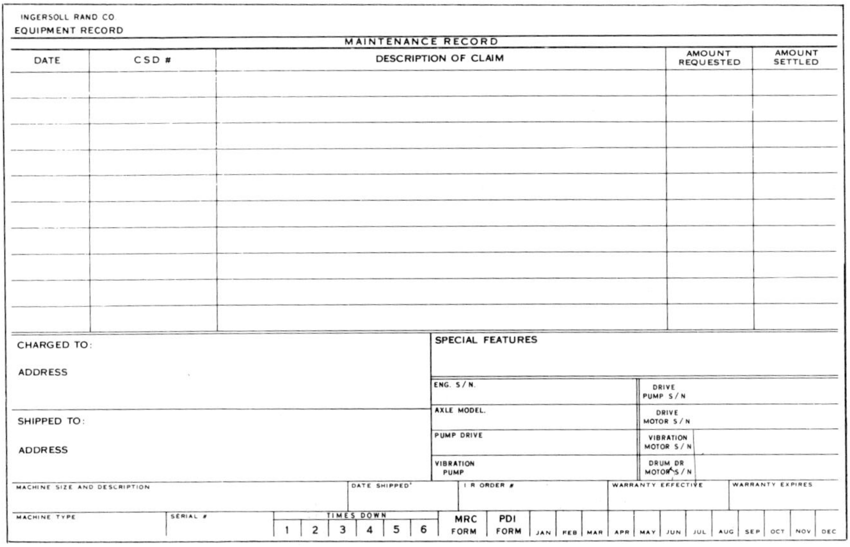
INGERSOLL RAND CO
EQUIPMENT RECORD

MAINTENANCE RECORD

DATE	CSD #	DESCRIPTION OF CLAIM	AMOUNT REQUESTED	AMOUNT SETTLED

CHARGED TO:
ADDRESS
SHIPPED TO:
ADDRESS
SPECIAL FEATURES
ENG. S/N. | DRIVE PUMP S/N
AXLE MODEL. | DRIVE MOTOR S/N
PUMP DRIVE | VIBRATION MOTOR S/N
VIBRATION PUMP | DRUM DR MOTORS S/N
MACHINE SIZE AND DESCRIPTION | DATE SHIPPED | I R ORDER # | WARRANTY EFFECTIVE | WARRANTY EXPIRES
MACHINE TYPE | SERIAL # | TIMES DOWN 1 2 3 4 5 6 | MRC FORM | PDI FORM | JAN | FEB | MAR | APR | MAY | JUN | JUL | AUG | SEP | OCT | NOV | DEC

Fig. 5-2 Typical master maintenance record card.

Do Not Disturb Pressure Levels

Before beginning additional troubleshooting, it is important to understand that hydrostatic transmissions must maintain certain pressures to function properly. Variation from the proper pressure levels will damage or render the transmission inoperable.

For hydrostatic heavy-duty transmission, four pressures must be monitored to accurately diagnose a malfunction in the transmission (Fig. 5-4). Pressure gauges should be installed to permit proper troubleshooting and diagnosis (Fig. 5-5).

1. *Charge pump inlet suction.* The maximum vacuum at the charge pump inlet should not exceed 10 in. of mercury under normal operating conditions. It is normal for the vacuum to be higher during cold start up.

2. *Charge pressure.* The minimum allowable charge pressure is 130 psi above case pressure. Normal charge pressure is 190 to 210 psi above case pressure when pump is in neutral and 160 to 180 psi above case pressure when pump is in stroke position.

3. *System or high pressure.* The maximum system pressure obtainable is controlled by the high-pressure relief valves located in the motor manifold. Relief valves have a coded number stamped on the exposed end, stating the valve setting, for example, Sundstrand 50 equals 5000 psi; Eaton 500 equals 5000 psi.

4. *Case pressure.* Transmission case pressures should not exceed 40 psi under normal operating conditions, except during cold start-up.

Vibration and Steering Start-up

Good start-up procedures can help keep equipment working efficiently and economically. We all require a high degree of reliability, increased service life, and elimination of

6-46 Maintenance of Ground Contact Elements

Model SP-54
Ingersoll-Rand
Compaction Division

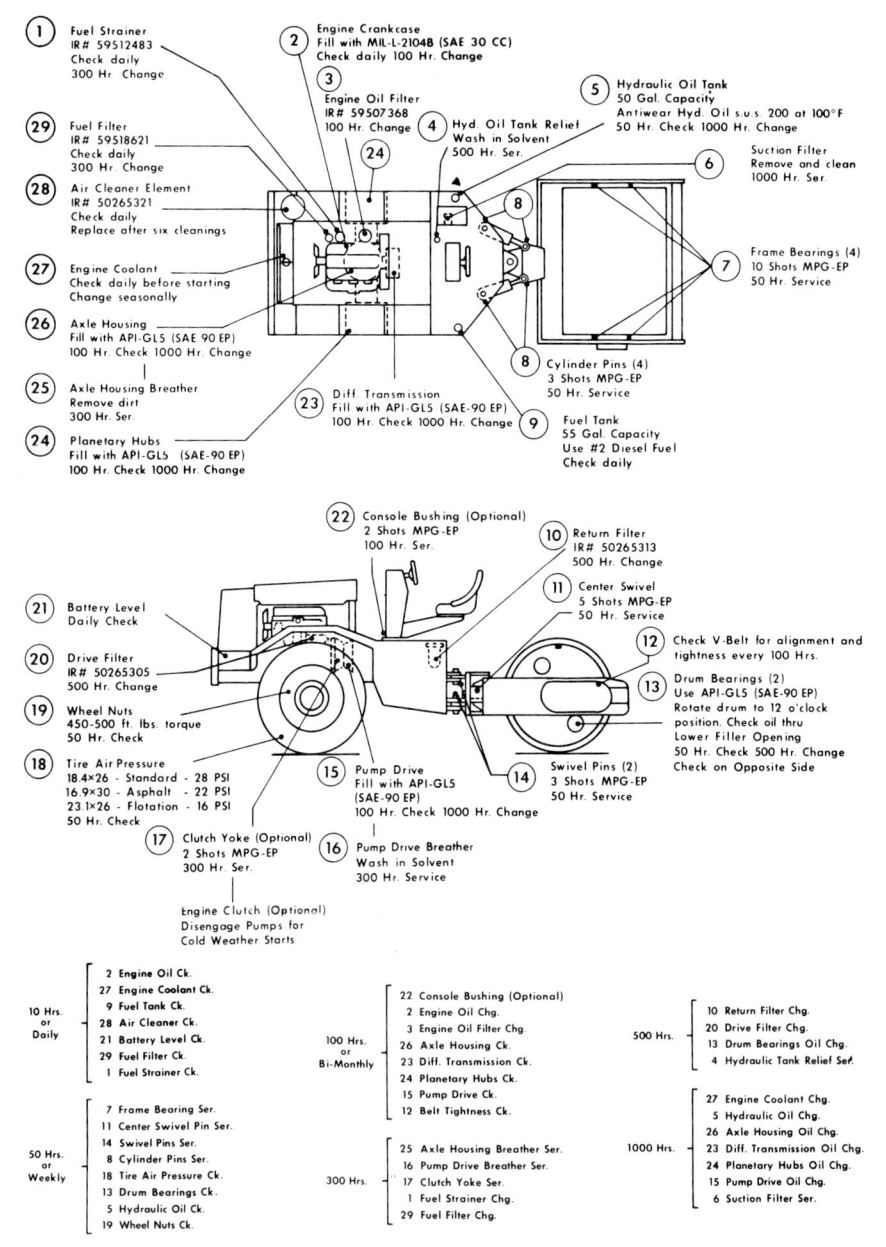

(1) Fuel Strainer
IR# 59512483
Check daily
300 Hr. Change

(2) Engine Crankcase
Fill with MIL-L-2104B (SAE 30 CC)
Check daily 100 Hr. Change

(3) Engine Oil Filter
IR# 59507368
100 Hr. Change

(4) Hyd. Oil Tank Relief
Wash in Solvent
500 Hr. Ser.

(5) Hydraulic Oil Tank
50 Gal. Capacity
Antiwear Hyd. Oil s.u.s. 200 at 100°F
50 Hr. Check 1000 Hr. Change

(6) Suction Filter
Remove and clean
1000 Hr. Ser.

(7) Frame Bearings (4)
10 Shots MPG-EP
50 Hr. Service

(8) Cylinder Pins (4)
3 Shots MPG-EP
50 Hr. Service

(9) Fuel Tank
55 Gal. Capacity
Use #2 Diesel Fuel
Check daily

(29) Fuel Filter
IR# 59518621
Check daily
300 Hr. Change

(28) Air Cleaner Element
IR# 50265321
Check daily
Replace after six cleanings

(27) Engine Coolant
Check daily before starting
Change seasonally

(26) Axle Housing
Fill with API-GL5 (SAE 90 EP)
100 Hr. Check 1000 Hr. Change

(25) Axle Housing Breather
Remove dirt
300 Hr. Ser.

(24) Planetary Hubs
Fill with API-GL5 (SAE-90 EP)
100 Hr. Check 1000 Hr. Change

(23) Diff. Transmission
Fill with API-GL5 (SAE-90 EP)
100 Hr. Check 1000 Hr. Change

(22) Console Bushing (Optional)
2 Shots MPG-EP
100 Hr. Ser.

(21) Battery Level
Daily Check

(20) Drive Filter
IR# 50265305
500 Hr. Change

(19) Wheel Nuts
450-500 ft. lbs. torque
50 Hr. Check

(18) Tire Air Pressure
18.4×26 - Standard - 28 PSI
16.9×30 - Asphalt - 22 PSI
23.1×26 - Flotation - 16 PSI
50 Hr. Check

(17) Clutch Yoke (Optional)
2 Shots MPG-EP
300 Hr. Ser.

(16) Pump Drive Breather
Wash in Solvent
300 Hr. Service

(15) Pump Drive
Fill with API-GL5
(SAE-90 EP)
100 Hr. Check 1000 Hr. Change

(14) Swivel Pins (2)
3 Shots MPG-EP
50 Hr. Service

(13) Drum Bearings (2)
Use API-GL5 (SAE-90 EP)
Rotate drum to 12 o'clock
position. Check oil thru
Lower Filler Opening
50 Hr. Check 500 Hr. Change
Check on Opposite Side

(12) Check V-Belt for alignment and
tightness every 100 Hrs.

(11) Center Swivel
5 Shots MPG-EP
50 Hr. Service

(10) Return Filter
IR# 50265313
500 Hr. Change

Engine Clutch (Optional)
Disengage Pumps for
Cold Weather Starts

10 Hrs. or Daily
- 2 Engine Oil Ck.
- 27 Engine Coolant Ck.
- 9 Fuel Tank Ck.
- 28 Air Cleaner Ck.
- 21 Battery Level Ck.
- 29 Fuel Filter Ck.
- 1 Fuel Strainer Ck.

50 Hrs. or Weekly
- 7 Frame Bearing Ser.
- 11 Center Swivel Pin Ser.
- 14 Swivel Pins Ser.
- 8 Cylinder Pins Ser.
- 18 Tire Air Pressure Ck.
- 13 Drum Bearings Ck.
- 5 Hydraulic Oil Ck.
- 19 Wheel Nuts Ck.

100 Hrs. or Bi-Monthly
- 22 Console Bushing (Optional)
- 2 Engine Oil Chg.
- 3 Engine Oil Filter Chg.
- 26 Axle Housing Ck.
- 23 Diff. Transmission Ck.
- 24 Planetary Hubs Ck.
- 15 Pump Drive Ck.
- 12 Belt Tightness Ck.

300 Hrs.
- 25 Axle Housing Breather Ser.
- 16 Pump Drive Breather Ser.
- 17 Clutch Yoke Ser.
- 1 Fuel Strainer Chg.
- 29 Fuel Filter Chg.

500 Hrs.
- 10 Return Filter Chg.
- 20 Drive Filter Chg.
- 13 Drum Bearings Oil Chg.
- 4 Hydraulic Tank Relief Ser.

1000 Hrs.
- 27 Engine Coolant Chg.
- 5 Hydraulic Oil Chg.
- 26 Axle Housing Oil Chg.
- 23 Diff. Transmission Oil Chg.
- 24 Planetary Hubs Oil Chg.
- 15 Pump Drive Oil Chg.
- 6 Suction Filter Ser.

Fig. 5-3 This schematic diagram becomes a sort of portable maintenance manual since it can be affixed like a decal to the side of the unit as a permanent source of information.

downtime for our equipment. Combining good start-up procedures with a good preventive maintenance program, users can be assured of efficient, economical equipment operation resulting in greater profits from each project completed.

Air must be expelled from a hydraulic pump at start-up or it will not prime. Air in the pumping chamber can cause cavitation with resultant pump failure. The following practices are a sure first step toward trouble-free pump operation.

| CASE-PRESSURE GAUGE | HIGH-PRESSURE GAUGE | LOW-PRESSURE GAUGE | VACUUM GAUGE |

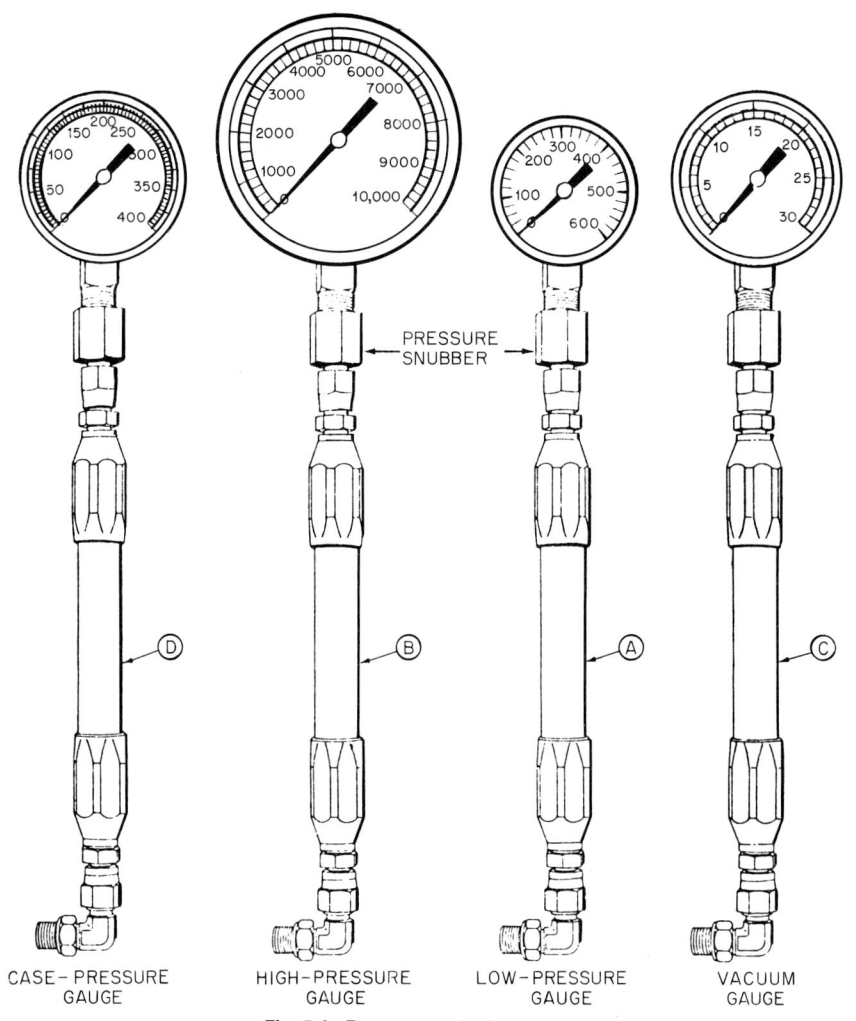

Fig. 5-4 Pressure monitoring gauges.

Remember it only takes 20 sec to burn out a system; therefore, be certain that all components are filled with clean fluid from clean containers before hydraulic lines are installed. It doesn't pay to economize with used fluids; it may cause an extra maintenance job before the next scheduled time. Always refer to the manufacturer's specifications when changing or adding oils.

Flush and clean the reservoir and lines before mounting the system. Fill the reservoir with new, clean fluid as recommended by the manufacturer. With caution, crack open a

6-48   Maintenance of Ground Contact Elements

pressure line fitting at the pump outlet port to bleed air from the system and to prime all newly installed components to assure lubrication at start-up. Now the engine is ready to start.

Turn over the engine by rotating the starter several times—for about 1 min—then start the engine and set the speed between 800 and 900 rpm (avoid high-speed start-up). With

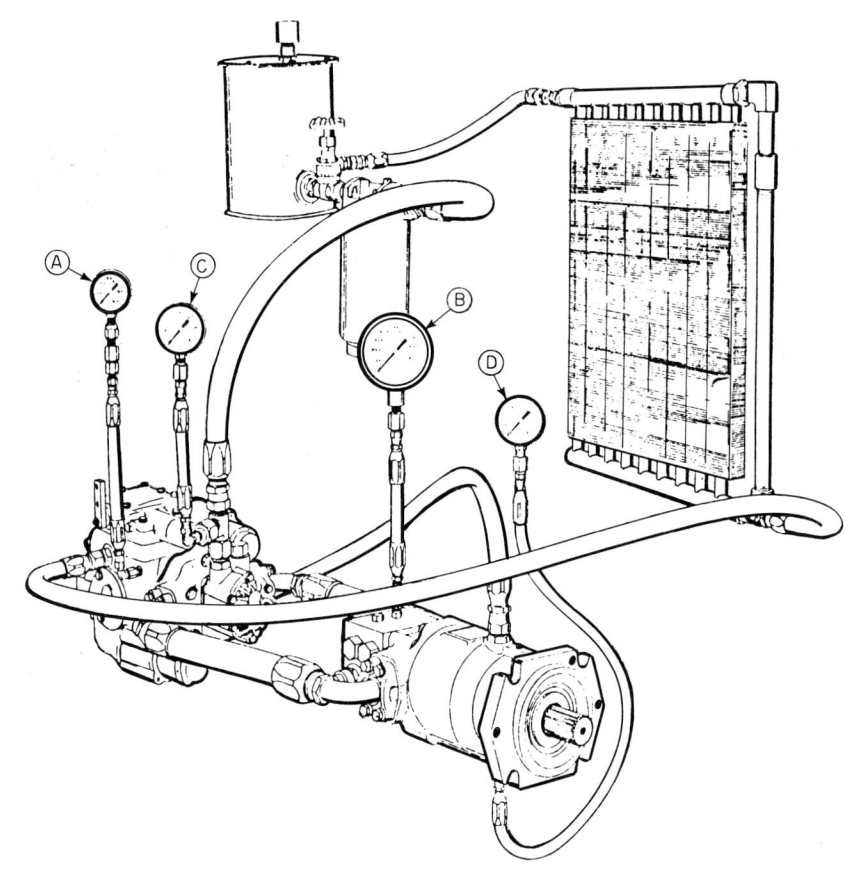

Fig. 5-5 Proper pressure gauge installation. At points A and B, the guage connection is 7/16 in. × 20 SAE "O" ring for all series. At C the gauge connection may also be connected on the suction side of the inlet filter; at D there is a reducer fitting from case port to gauge hose assembly.

the pump primed, air and fluid will bleed through the loose fitting. When a solid stream of fluid begins to flow, all air is bled from the system and the fitting should then be retightened.

Finally, examine the fluid in the reservoir. There may be some bubbles at the top of the fluid—that is the residue of any air which may have been left in the system. This will clear up shortly; then fill the hydraulic oil reservoir to the recommended level.

Cold Weather Start-up Procedures

Consistent daily start-up procedures assure maximum efficiency and service from vibratory compaction equipment.

1. Ensure that oils used in the system conform to manufacturer's specifications for local temperature conditions.

2. Start the engine and allow to idle at a speed between 800 and 900 rpm.

Maintenance of Vibratory Compactor Drums 6-49

3. Do not allow cylinders to travel to the end of their stroke cycle, or permit pressure to build up to the setting on the relief valve. Cold fluid makes relief valves sluggish and can add from 500 to 1000 psi to the maximum setting on the valve.

4. When all components are warm to the touch, the vehicle can be safely placed into service.

Troubleshooting the Vibration Circuit

Whenever difficulties are experienced in the vibration circuit—low frequency, erratic vibration, or no vibration—the following series of checks can be easily performed in the field to ascertain how to correct most malfunctions. Since these are general suggestions, it is always best to refer to the operator's manual for the individual unit for specific guidance before attempting repairs.

Low RPM or No RPM You might find low engine rpm or an inoperative relief valve, worn pump and/or motor, crimped suction line or clogged filter, inoperative control valve, pump not engaged to engine, eccentric shaft unable to rotate, or loose drive belts. Correction for these malfunctions can normally be made in the field.

1. *Relief valve inoperative.* Clean, check, and replace as necessary.

2. *Pump or motor worn.* Disassemble pump assembly and inspect internal components for excessive wear. Where necessary, replace defective components and check for possible contamination of fluid. Always replace contaminated fluid in accordance with recommended procedures.

3. *Inoperative control valve.* Remove control valve. Check valve for internal leakage. Repair or replace valve assembly.

4. *Pump not engaged to engine (where applicable).* Adjust engine clutch assembly to required 110 ft-lb to engage clutch.

5. *Eccentric shaft not rotating.* Check eccentric shaft bearings for resistance to rotation. Replace bearings if necessary, following procedure outlined in service bulletin.

6. *Loose drive belts, where applicable.* Tension as per operator's manual.

Erratic RPM

Erratic rpm can have several causes—air in fluid, relief valve defective, loose vibration drive belt (where applicable), damaged eccentric shaft bearing, or clogged suction line or filter. Follow procedures for insufficient or no rpm for all appropriate causes except:

1. *Air in fluid.* Tighten leaky inlet connections. Fill reservoir to correct level and bleed air from system.

Fig. 5-6 In a typical application, this vibratory drum compactor achieves about 67 percent density while eliminating air pockets and eventually reaching a 75 percent density compaction level through compaction.

2. *Relief valve setting below operating pressure.* Reset relief valve to setting as outlined in Ingersoll-Rand service bulletin 8040-25.

3. *Loose vibration drive belts (where applicable).* Adjust V-belt and realign pulleys. Consult operator's manual for proper tension for individual unit.

Major Repairs or Overhauls

Many major repairs can be accomplished in the field. Remember that there is no substitute for consistent service and preventive maintenance to assure maximum machine performance with minimum downtime.

When troubleshooting a vibratory compactor, remember to check for a simple solution first. Often, the difficulty may be solved simply, for example, by replacing a clogged filter.

Should any unusual operational malfunctions occur outside the scope of routine troubleshooting, it is always best to contact the service representative. He should know the equipment and be able to provide the technical backup needed to get the compactor back into operation quickly (Fig. 5-6).

Section 7
Maintenance Programs for Equipment Entities

Chapter 1
Diesel Tractor Maintenance

CATERPILLAR TRACTOR COMPANY
Peoria, Ill.

THE MAINTENANCE OBJECTIVE

The ultimate maintenance objective for the heavy equipment owner or manager is *to control* the machine element of his business—to be able to plan his jobs, because he can schedule machine availability and know in advance what his operating costs will be.

Making machinery a *controllable* factor in the operation enables the contractor to bid and schedule a job with greater knowledge and confidence about earthmoving costs and ability to finish the job on time. In other words, the maintenance program contributes to his ability to operate a business profitably. Most decisions regarding the type and degree of a tractor maintenance program depend on its contribution to the profitability of the business.

In times of material and energy shortages, decisions regarding amounts and type of maintenance practices may depend on conservation considerations or the necessity to preplan replacement of parts and machinery. Still, profit is the ultimate motive.

In any case, the construction equipment owner should get a clear idea of how much control over machine availability and cost is needed to meet his business objectives. Today he has a wide variety of elements—tools, training, inspection techniques, failure forecasting services, component replacement programs, communication, and record keeping methods—from which to fashion a product support program that fits his needs.

Routine maintenance Virtually all tractor manufacturers provide a book of specific instructions for routine lubrication and maintenance of the machine. These recommendations, based on research, testing, and years of experience with a particular design, are considered to be practices which give optimum machine performance for the expense and effort involved. These recommended practices represent what the manufacturer considers necessary to maintain the performance standard expected from his product.

Standard lubrication and maintenance procedures are, of course, set for typical or average working conditions. Applications significantly lighter or more severe than what is typical may mean that the procedure should be modified.

Fuel system maintenance Proper fuel system maintenance has a great influence on the life and operability of the diesel tractor. The correct grade and quality of fuel, for instance, contributes not only to the productivity of the machine through the amount of engine power available but also affects engine life.

Fuels with high sulfur content cause rapid deterioration of the engine lubricating oils that protect and cool engine parts, necessitating more frequent oil change. If fuels have a high moisture or sediment content, regularly draining condensate from tanks is necessary to prevent engine stalling and damage to fuel injection system. If fuel shortages mean lower-quality fuels will be in use, then vigilance in fuel system maintenance will be necessary to compensate.

Fuel should be kept clean during storage, transportation, and refueling. The best fuel, if contaminated by dust, can damage an engine. Fueling should never take place in a dusty locale. Dust should be wiped away from the tractor's fuel tank neck and cap before opening.

Fuel tanks should be filled at the end of the work day to prevent overnight water condensation in empty tanks. Condensate should not only be drained from tractor fuel tanks but also from storage tanks.

Lubrication

Use Recommended Type of Lubricant. Lubricants must clean and cool, as well as lubricate, parts. They prevent metal-to-metal contact, thereby reducing friction and wear.

Roller bearings, if in preloaded applications, must have a special type of lubricant with adequate film strength. For wheel bearings, a correct viscosity grease is needed; if too heavy, it will squeeze out or channel in cold weather, leaving dry areas on the bearing which can be damaged by condensed moisture.

The wrong viscosity oil in the hydraulic system causes cavitation damage that can damage hydraulic pumps or overheat the system.

The wrong engine oil can cause sticking of piston rings with loss of engine compression and sizing, carbon buildup in the cylinder, and possible breakage of piston rings.

Lubricating oil of improper type or viscosity in transmission or final drives can cause pitting or scoring damage that weakens gears.

Use Right Amounts of Lubricant. Check Level of Oil Compartments. Inadequate oil to roller bearings can cause overheating that damages the bearing. Inadequate oil in a gear compartment can cause severe scoring of gears. (See Fig. 1-1.)

Fig. 1-1 Daily maintenance requires checking levels in oil-filled components.

Careful filling of the hydraulic system is important; measure the amount and follow procedure. Low oil level can cause cavitation damage to the hydraulic pump (noisy pump is a symptom) or overheating of the system, which causes even more severe pump damage.

Inadequate engine oil can cause crankshaft bearings to overheat and seize. Insufficient oil to a turbocharger for a period as short as 5 sec can cause damage.

Change Oil and Filters at Proper Intervals. Should the oil filter become plugged, most diesel engines have an oil filter bypass valve to forestall immediate damage to

Fig. 1-2 By inspecting contents of used filters, the maintenance man gets an idea of oil and machine condition.

engine and turbocharger from lack of lubricant. But if oil filter change is neglected, damage will eventually occur since contaminated oil is being provided to bearings, etc. Oil should be changed at recommended intervals to remove contaminants and renew oil-lubricating qualities. (See Fig. 1-2.)

Abrasives in oil can scuff pistons, wear main and rod bearings, wear piston rings, scratch turbocharger bearings, and cause short engine life. Dirt is the main contributor to short engine life.

Clean practices in filling the hydraulic system are necessary because high pressures make these systems sensitive to dirty oil damage. Never open a hydraulic tank in dusty conditions. Use the correct hydraulic filters. Abrasive wear to hydraulic pumps will cause loss of power.

Dirty oil will also cause wear to roller bearings and pitting of gears that leads to tooth breaking.

Service Air Cleaner When Needed. Do not reuse a damaged air cleaner element. Inspect air inlet system for leaks since dirt entering can drastically reduce engine life. Abrasive material or dirt coming through a damaged air cleaner element can cause damage to the turbocharger and abrasive wear of piston rings and liners with resulting poor compression. Restrictions cause high exhaust temperatures producing carbon deposits in the turbocharger and shortening valve, piston, and manifold life.

Keep Cooling System Radiators or Heat Exchangers Clean. Use proper coolant, check coolant level, check hose condition, and look for leaks.

Inadequate engine coolant can cause overheating and piston seizure. Impure water in the cooling system, with chloride content too high, can cause cylinder liner pitting. A plugged radiator core can cause engine overheating; a plugged aftercooler core can cause high exhaust temperature and damage to the engine components.

Inspect Seals and Lines for Leaks, Crimps, etc. Leaky fuel lines are a fire hazard.

Water leaks through seals or gaskets into compartments cause rust or etching of roller bearings. Abrasive materials such as dust or grit cause wear to bearings.

Maintenance Programs for Equipment Entities

Air leaks in hydraulic system suction lines can cause aeration of oil and erosion of metal in pump. Abrasives getting into the hydraulic system through defective rod wiper seals or on dented or scored cylinder rods can severely wear a hydraulic pump. Rod wiper seals should be checked with grease or with a feeler gauge.

Breaks, leaks, or restrictions in engine oil lines can cause inadequate oil flow and consequent damage to engine and turbocharger. Be on the lookout for oil leaks.

Change or Service Compartment Breathers at Recommended Intervals. Do clean and careful work.

If breathers to an oil-lubricated housing plug up, seals can be blown out allowing oil leakage or dirt entry. Carelessly serviced breathers can let dirt into oil compartments, causing rapid wear to bearings, gears, etc.

Battery Care. Battery care determines the length of battery life. Lack of water, insecure fastening of the battery, undercharging, and overcharging can cause problems.

Engine inspections and adjustments Incorrect rack settings can cause scuffed rings, eroded pistons, high exhaust temperature, and damaged turbocharger. Overheating and turbocharger damage can also result from restrictions in the exhaust system.

Timing and valve adjustment (Fig. 1-3) are important to avoid rough combustion that damages pistons and scuffs rings; burnt valves; and subsequent engine failure.

Neglecting to adjust drive belts can cause overheating of an engine with a belt-driven water pump or fan, or inadequate battery charging with a belt-driven alternator or generator.

Drive train maintenance Maintenance of correct transmission oil level is necessary to prevent torque-converter overheating that breaks down oil-lubricating ability.

Brakes and Drive Bearings. Brake bands of track-type tractor steering clutches must be properly adjusted, or hard steering and poor operation of the tractor will result. Brake systems of wheel-type tractors must be kept adjusted for safety reasons and to prolong brake life. Maintenance, depending on the system, will include fluid-level checks for leaks, draining moisture from air accumulator systems, as well as mechanical adjustments.

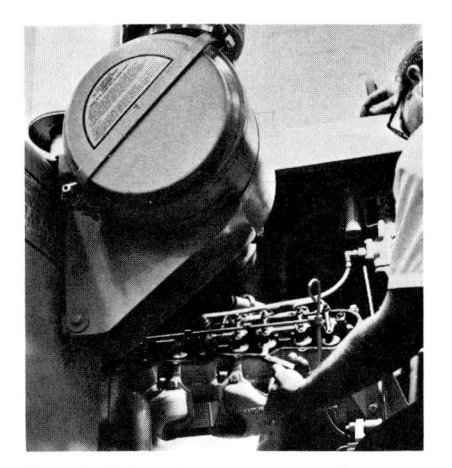

Fig. 1-3 Valve adjustment is a simple maintenance procedure that can prevent expensive engine damage.

Final drive bearings on crawler tractors must be adjusted to prevent bearing damage and misalignment of gears that can cause gear tooth fractures.

Linkages and Joints. Control linkages should be greased, adjusted, and inspected for wear. Failure to do so can cause rapid wear and damage to the components they control. Joints inadequately lubricated can cause poor dozer control.

Report all suspected problems The benefits of preventive maintenance will be ineffective if this is not done promptly. Problems with the machinery must not be ignored by anyone who works with the equipment.

Poor tractor performance, leaks, or broken or cracked parts observed by operating or service personnel should be promptly reported and, in many cases, remedied before further operation of the machine.

Minor problems discovered in maintaining or operating the tractor can be corrected in the field. These are preventive repairs such as replacing hoses, shimming ball joints, or welding cracked metal parts or loose wear places. Replacing track shoes, lost bolts, straps, or clips is the type of thing which should be done to prevent compounding damage.

Declining performance of major tractor components such as loss of engine power, smoky exhaust, unusual noises, sluggish hydraulics or jerky shifting, overheating, or poor starting should be reported for examination by skilled service people.

Plans for tractor maintenance Putting the manufacturers' recommendations into effect requires a system heavily reliant on good communication.

To organize a maintenance program, responsibilities must be assigned and a communication system of input and output devised. No matter who has the responsibility for performing lubrication and maintenance tasks—the operators, maintenance specialists, or a combination of both—the responsible people must be part of a workable communication network.

Since most lubrication and maintenance recommendations are scheduled on the basis of the tractor's time in service (calculated either by engine service meter or by calendar time), this information must be part of the input that sets the lubrication and maintenance function into operation. Figure 1-4 shows an information flow that starts with the combined input of the manufacturer's recommendations and the tractor's service time. Such a constant information flow is necessary for wise purchasing and operation of machines. It gives the lubrication, maintenance, and service people the necessary information to keep the tractor repaired and productive. It also lets the contractor management know what is being done and what the tractor is costing.

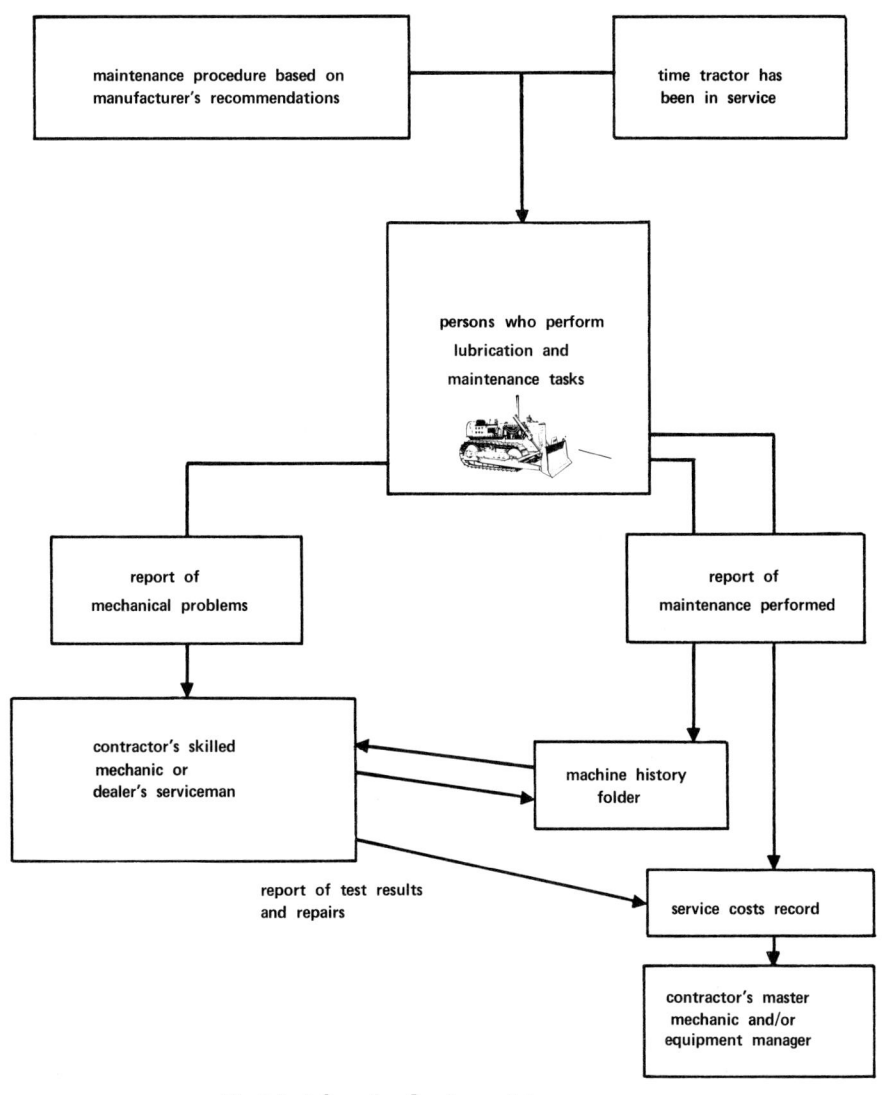

Fig. 1-4 Information flow in a maintenance program.

7-6 Maintenance Programs for Equipment Entities

To accomplish this information flow, there are many different methods, some quite elaborate. However, since any maintenance system consumes manpower and material, it should be kept as uncomplicated as possible. The need for simplicity is second only to the need for effectiveness. Any preventive maintenance system should have built-in checks to see that the work actually gets done and then is reported back in a form which makes the information usable.

MAINTENANCE PROGRAMS

Following are two types of maintenance programs. They by no means cover all types of plans in use today but give an idea of the variety available.

Packaged maintenance A new idea in maintenance makes scheduling, reporting, and record keeping remarkably simple for the contractor who owns just a few machines. The equipment dealer keeps track of the service intervals for each tractor enrolled in the program. When each succeeding service interval approaches, the contractor receives from the dealer a package of the materials to be used—the filters, gaskets, and oil sample bottles—everything except the hand tools and the lubricants (Fig. 1-5).

The tractor operator or maintenance man can simply be handed the package when it's time to service the machine. The contents of the box initiate the recommended tasks. As the maintenance is performed, the activity is reported on a short machine inspection form (Fig. 1-6A) which encourages the maintenance man to observe any abnormalities in machine performance and draw attention to them.

When the maintenance is completed, the oil samples plus the inspection report are sent back to the equipment dealer. The oil is analyzed for metal content which would show abnormal wear. The oil analysis is coupled with information from the maintenance man's inspection and a report made by the dealer to the tractor owner.

Each kit contains precisely the items needed for the particular inspection interval at hand. This not only saves a small contractor the trouble of ordering and stocking filters and gaskets, but receipt of the package initiates the maintenance work, and filing the report provides a check on whether or not it's been done. In addition, it places the burden of

Fig. 1-5 In a packaged maintenance program, receipt of a box containing all needed filters, gaskets, and the like initiates the contractor's performance of periodic maintenance.

consolidating and interpreting inspection reports with the equipment dealer, though many customers keep duplicate copies of the maintenance inspection reports.

Another form (Fig. 1-6B) enclosed in the kit is one to order parts. With the completion of the inspection report, the owner learns whether or not he needs new parts. If he does, he uses the form supplied for ordering replacements recommended in the inspection report.

As for the maintenance intervals, these are outlined in advance by the owner and dealer

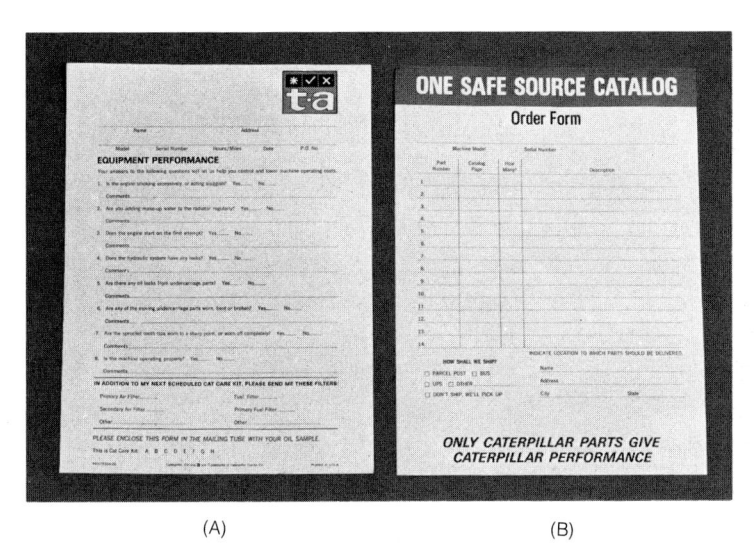

Fig. 1-6 Packaged maintenance includes a short inspection by the contractor's maintenance mechanic or supervisor who refers his results to the equipment dealer, using (A) a set checklist and, where needed, (B) a parts order form.

at the time the tractor is enrolled in the program. What items come with each package vary according to what maintenance is scheduled.

Program tailored for a specific job A maintenance program can be tailored to serve an individual business. One example of an owner-developed program is that of a California company engaged in agriculture and land development work.

Because it operates on a large scale and has many years' experience in one geographic area, this company has developed a maintenance regimen which truly fits its business and brings outstanding results. It operates about 100 crawler tractors in the 140- to 180-hp class.

The company operates its crawler tractors in two 10-hr shifts per day during 9 months of the year. At some time during the 3-month slack period, each tractor is brought in to a central shop for 1500-hr (annual) maintenance. The rest of the year the tractors are serviced from maintenance trucks which daily come to the work sites.

The maintenance trucks are supplied with 1000 gal of diesel fuel, engine oil, hydraulic oil, water, transmission oil, gasoline, compressed air, and grease. They are equipped for fast fueling and vacuum draining of oil compartments. Used lubricants are stored on the truck and hauled away for disposal.

Each tractor receives fuel and maintenance once per 20-hr period. The machines are equipped with auxiliary fuel tanks to make this interval possible.

A simple maintenance record card (Fig. 1-7) is kept on the tractor in a box attached to the side. It is updated each time service is performed. Remaining with the machine, it gives operators, and field mechanics called in for inspections or repairs, a recent history—up to 2 weeks—of maintenance performed.

The precise tasks to be done each visit are scheduled from the central office. Routinely, a maintenance crew will fuel the tractor, clean the air conditioner filters, and blow dust out of the tractor cab with high-pressure air. The air cleaner indicator and the oil level in all compartments will be checked. If 240-hr service is due, the engine oil and filters will be changed and the fuel pump housing, pilot bearing, and U joints greased. With fast fueling and vacuum oil-draining equipment, all this can be accomplished in about 10 min (Fig. 1-8).

A permanent maintenance log on each machine is kept at the central office. From there, maintenance is initiated, recorded, and monitored. All records are kept by service meter hour. Forms in the logbook indicate the regular hourly intervals the company has determined best fit its operation.

The dispatcher handles the mathematical calculations that determine when the next service is scheduled (Fig. 1-9). Whenever a certain service procedure is due within 80 hr, the dispatcher puts it on the day's schedule sheet which indicates when 240- and 500-hr maintenance is due. For each machine due within a week, the dispatcher lists the last registered service hour reading and the hour reading when maintenance is due and also calculates how many hours until maintenance or how long it is past due and which maintenance man is responsible. Each member of the maintenance crew gets a copy and can tell at a glance which machines need service in addition to regular fueling and checks.

As he works, each maintenance man fills out a daily report form for all the tractors serviced, including reports of oil consumption. The reports are submitted and logged each day, determining what appears on the following day's schedule sheet. Whenever any scheduled maintenance goes unperformed for 20 hr, the dispatcher contacts the maintenance man by radio.

Fig. 1-7 A 2-week record of daily maintenance is kept on the machine for quick inspection by maintenance and service people.

The company has two field mechanics and three more in the shop. The field service trucks have beds large enough to accommodate small tractor components. Each is equipped with a 1000-lb crane and air compressor. When not making repairs, the field men are constantly inspecting machines and troubleshooting, often at an operator's or maintenance man's request.

When repairs are needed, they are usually performed in the company shop. In extremely busy times, rebuilt exchange components may be used.

Annual cleaning and inspections—the 1500-hr maintenance—is done in the shop, if possible, and includes service on air cleaner, final drives, track rollers, steering clutches, and transmission. Undercarriage wear is calculated at this time, under the equipment dealer's track inspection program.

This company prefers to turn pins and bushings at 10,000 service hours and to buy new undercarriages at 16,000 to 18,000 hr on 140-hp tractors and at 23,000 to 25,000 hr on 180-hp tractors. It usually meets this goal.

Most contractors will not find their working materials so ideal as a clay loam nor their working conditions so uniform year to year that they can so precisely predict and schedule repairs. But this company's experience with a large fleet of crawler tractors indicates that regular, carefully applied maintenance practices can bring predictable results, making a business more manageable.

THE MAINTENANCE ORGANIZATION

The types of personnel and equipment employed for tractor maintenance should be approached with as much care as selection of the earthmovers in the construction business.

If decisions regarding either service personnel or equipment are made haphazardly, under-utilization of time and talent and over-investment in equipment are definite risks.

Skill levels In planning for maintenance personnel and equipment, a concept of *levels* or *classes* can be used. To develop this concept, consider listing all the activities involved in a complete overhaul of a tractor engine. The number of tasks are arranged in

Fig. 1-8 Daily servicing can be as little as a 10-min stop.

Fig. 1-9 All periodic maintenance is scheduled and recorded by the dispatcher.

7-10 Maintenance Programs for Equipment Entities

columns according to the amount of expertise and knowledge of engines needed to perform them:

Skill level 1	*Skill level 2*	*Skill level 3*
Engine removal	Job planning	Failure analysis
Disassembly	Evaluation	
Use of hand tools	Inspecting	
Parts cleaning	Testing	
Use of service literature	Reporting	
Parts ordering		
Storing, moving parts		
Reassembly		
Hoisting		
Engine installation		
Lubrication		

Shop studies, not only for engine rebuild but for tractor service work in general, have shown that at least 70 percent of the time needed for service work is spent on the activities which fall under skill level 1; 20 to 30 percent of the time is spent on skill level 2 tasks and under 10 percent on the analysis type of work that requires a high degree of knowledge about machinery.

With the skill levels proportionately charted on a triangle, we can see that a service shop needs more people at the lower skill levels and fewer at the high levels. For people with higher-level skills to spend time routinely on lower-level tasks would be inefficient use of their training (Fig. 1-10).

Use of this concept permits the service shop owner to practice economy in terms of training time and the level of skill he pays for. The largest number of employees do not need great training—they need manual skills best taught on the job. The few employees who need higher skills can be sent to formal schools to become specialists, and, of course, the shop owner will have an investment in their training as well as in their higher salaries. Sources for training highly skilled service people will be discussed later in this chapter.

If this pyramid concept applies to the repair shop or tractor dealer's service department, where do the contractor's maintenance personnel needs fit in? Depending on union regulations or local custom, some portion of the routine maintenance of tractors may be done by operators. As for the contractor's personnel devoted primarily to maintenance and repair, the same pyramid concept of skill levels can apply. His personnel pyramid will overlap that of the dealer service organization or repair shop he relies on routinely.

The contractor's field maintenance people who perform the lubrication, fueling, and adjustments can also be trained to help out with level 1 disassembly and assembly tasks, hoisting, cleaning, etc., in the contractor's service shop, if he chooses to operate one. In addition, the contractor may have a couple of servicemen at skill level 2 to plan repair work, examine parts, and do preliminary testing or inspecting of machines that are not running correctly.

For the analyses of complete tractor systems and machine applications which require level 3 skills and training, the contractor may turn to the dealer's service organization since it is uneconomical for all but the largest construction companies to train and pay for skills needed only occasionally.

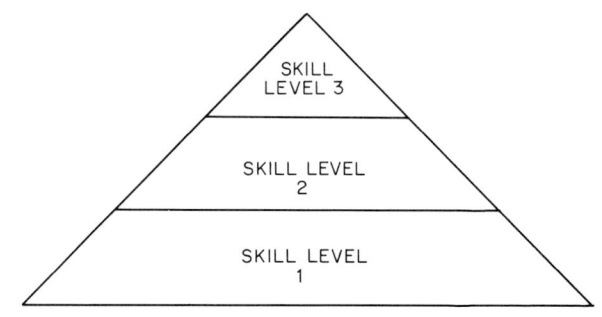

Fig. 1-10 Skill levels of personnel needed for tractor repair organization.

Under this concept, the contractor's service personnel needs would also form a triangle which would overlap (in part) the dealer's service organization, as shown in Fig. 1-11.

Whenever the contractor's needs temporarily exceed, by volume or skill needed, the capability of his own maintenance organization, he can turn to the tractor dealer for whom it is more economical to maintain a larger, more highly trained, service organization. In

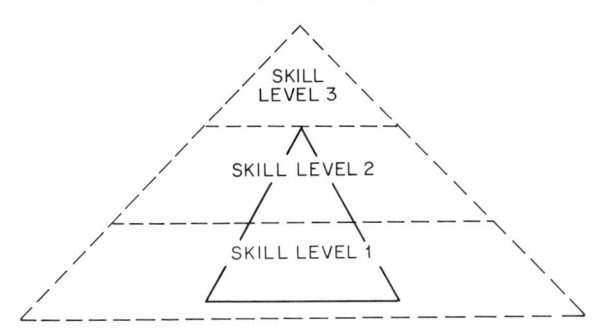

Fig. 1-11 Contractor's needs in lubrication and maintenance personnel overlap (small triangle) those of the repair organization and take the same form with the need for fewer people with higher skill levels.

addition, the dealer will have the latest technical information and diagnostic tools available.

Specific maintenance skills needed To perform recommended lubrication and maintenance of tractors, plus function as level 1 mechanics in a contractor's service shop, the following technical skills will be needed:

1. Care and use of hand tools: box end, open-end wrenches; socket wrenches; screwdrivers; hammers; pliers, diagonals (side cutters); chisels, punches; files; hacksaws.

2. Use of shop tools and equipment: hoists, lifts, and capacities; spreader bars, chains, and slings; jacks and capacities; blocking and jack stands; storage fixtures and racks; air and electric wrenches; electric drill; bench grinder; thread taps; bearing heaters.

3. Use of shop cleaning equipment: low- and high-pressure water cleaning; steam cleaning or heated high-pressure washing; parts cleaning-washing (hand wash); scraping.

4. Storage of components and parts.

5. Operation of tractors: moves machines into and out of shop; machine operating safety; machine safety devices (brakes, locks, lift arm supports); reads machine gauges, indicators.

6. Operation of material-handling equipment.

7. Operation of vehicles: operation and safety; preventive maintenance; servicing fuel and oil.

8. Use, care, and safety needs of hand-welding equipment.

9. Use and care of precision measuring devices: torque wrenches; micrometers; dial indicators; feeler (clearance) gauge.

10. Disposal of contaminated fuels and lubricants and used coolants: disposal equipment and methods; storage.

11. Disassembly/assembly of lines, hoses, electrical connectors, and wires: bending and flaring tubing; inspecting for wear; inspecting for leaks (air, coolant, fuel, oil, and sources); installing batteries and connectors; servicing batteries (clean, fill, hydrometer check).

12. Installation and position of gaskets, "O" rings and seals.

13. Use of miscellaneous materials: sealing compounds, liquid gasket, and special lubricants.

14. Knowledge of fasteners: types and grades of fasteners (bolts, nuts, washers); fastener cleanliness and lubrication.

15. Personal safety: safety glasses; safety shoes; danger of loose clothing.

16. Shop safety: lifting correctly; climbing on or around machines; greasy and wet floors; parts on floor; flammables; battery acid, explosive hydrogen gas; cleaning pressure, caustic burns; welding-acetylene precautions, sparks, splatter burns, clothing; hoisting,

7-12 Maintenance Programs for Equipment Entities

lifting materials; raised machine implements-block, support; all safety rules and regulations.

Most construction companies, except the very smallest ones, will need at least one level 2 serviceman who, when called to the field or assigned to a machine in the shop, can make a basic machine inspection. He should be able to recommend whether to take a machine out of service or leave it running to the end of the shift. He should have basic knowledge of the scope of repair work needed, and his opinion of the equipment failure will help the equipment manager decide whether repair should be done in the field, in the contractor's shop, or by another repair source such as the tractor dealer.

Once a tractor is brought to the shop, the level 2 serviceman must be able to perform further tests and inspections, plan the repair job, guide the work of other level 1 technicians, evaluate reusability of parts, and report the parts, supplies, and time used repairing the tractor.

Maintenance equipment and facilities A triangle also could represent the frequency of different types of maintenance and repair jobs—the routine maintenance tasks are by far the majority. Preventive repairs may be next; in-shop repairs are less frequent; and major overhauls and rebuilds (the minority) are at the apex of the triangle (Fig. 1-12).

However, the *investment* in facilities and tools for repair could be represented by an overlapping, inverted triangle, the overhauls and rebuilds requiring far more in tools and facilities than the more routine service work (Fig. 1-13).

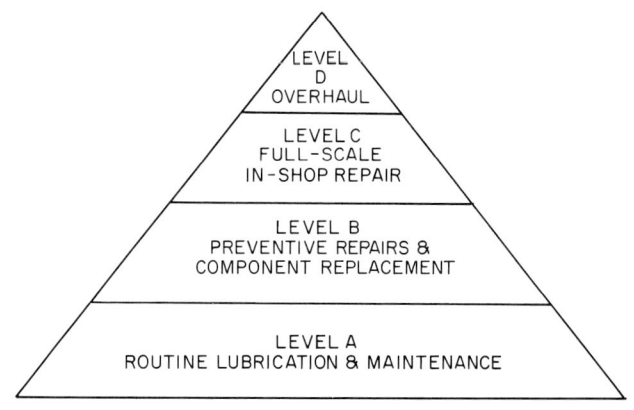

Fig. 1-12 Frequency of different types of maintenance and repair jobs.

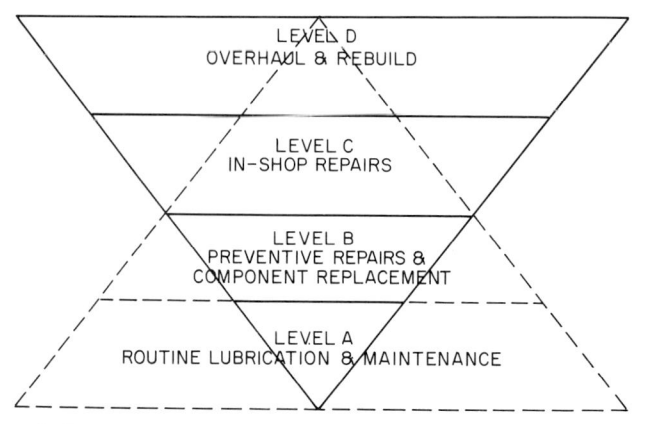

Fig. 1-13 Level of investment in equipment and facilities for different types of repair and maintenance jobs is superimposed over pyramid showing the frequency of different types of jobs.

For the lubrication and maintenance tasks, a minimum of hand tools are used for checks and adjustments. Most routine lubrication is performed at the job site, so a fuel and lubrication truck is part of the level A maintenance equipment. Oils and greases, filter elements, cooling water, and antifreeze solutions should be carried on it. Wrenches, grease guns, a torque wrench, etc., can be carried on this vehicle or on a small service

Fig. 1-14 A ¾-ton pickup equipped with hand tools and special selected tools is adequate for troubleshooting and minor repair.

Fig. 1-15 A truck-mounted crane with a minimum capacity of 4000 lb is helpful in making field repairs.

truck. Pumps or fuel and lubrication trucks are usually air-operated. The compressed air is also used for tires and cleaning dirt off machines.

In some cases a ½- or ¾-ton pickup truck equipped with hand tools and selected special tools will be adequate for field troubleshooting and minor repairs (Fig. 1-14). If there is no shop on the job site, or if some equipment is located too far away from the home-base service shop, better equipped service trucks will be required for preventive repairs (level B).

A 1- or 1½-ton truck with field-service body and truck-mounted crane may be needed. A 4000-lb capacity crane that will lift smaller engines and transmissions is the largest crane that can be used on this size of truck (Fig. 1-15).

A large assortment of hand tools and common pullers will be needed for level B service. For field use, oxyacetylene cutting and welding equipment and electric generators to power drills, impact wrenches, and lights could be needed. Commonly used parts such as nuts, bolts, washers, etc., must be included.

A level B repair shop can either be in a small, permanent building or, at a remote job site, in a temporary building or specially designed trailer which has workbenches and job cranes, air compressor, and generator. If permanent, it may also include fuel storage tanks, wash rack, grease pit, and tire changing equipment.

The level C and D repair facilities provide the ability to completely recondition the tractor through component exchange (level C) and/or to practice specialty rebuilding of major components such as engine and transmissions (level D). The amount of specialized equipment needed for ascending levels of service increases geometrically. Most smaller companies will want to rely on an outside service source such as the equipment dealer for repairs of this type.

Because the volume of service work increases with fleet size, big companies frequently have level C repair facilities and staffs. The types of equipment needed for such facilities are discussed under Repair Alternatives in the next section of this chapter.

Maintenance record keeping No business, nor even a single piece of equipment, can be wisely managed without good records. Each tractor should have a history file into which records of lubrication and maintenance, repairs, inspections, uptime, downtime, and cost information are logged. This will give factual performance data to help an equipment manager or a service manager evaluate a single tractor.

But to be truly useful as a management tool, composite records must be kept, records that give an overview of the contractor's machine fleet and that give the history of a single tractor's performance at a glance.

Composite records of fuels, lubricants, maintenance, repairs, working time and conditions, downtime, and inspection results can do several things for the contractor:

1. Serve as a control to assure the recommended maintenance job is being done
2. Provide a tractor history to help in identifying and diagnosing machine problems
3. Call attention to excessive costs or downtime
4. Help to make repair versus replacement decisions
5. Help to evalute the company's maintenance program, including its costs

When combined with data on labor costs, machine depreciation, and productivity, this information can be used to determine such necessary items as:

1. How low to bid a job, yet do it profitably
2. Optimum time to trade machine—before excessive repair costs, yet not so early as to take a loss on the investment
3. Evaluate tractor obsolescence in terms of other units available
4. Help determine if larger-sized or higher-priced machine is more economical in the long run
5. Which type of machine (crawler tractor versus wheel tractor, for instance) does certain jobs most economically
6. Which manufacturer's equipment is most economical

Good machinery records are necessary for knowledgeable management of the business. Keeping cost records gives the contractor a clearer profit picture.

There are several methods of keeping records: Most begin with daily logging of fuel, oil, and grease used, location, hours worked, job type. This information can be logged in the pocket-size book by the operator or maintenance person responsible (Fig. 1-16). Maintenance times, parts used, and costs can be reported by the serviceman.

A set of records is kept for each machine. These can be compiled on a monthly basis by a bookkeeper. At year's end, a service review record can be filled out to give a year's service record at a glance. This record can remain in the machine history folder.

Computerized record keeping has increased the usefulness of machine records to the point of making them a sophisticated management tool. An example of what's available in the computerized time and cost record service is what an earthmoving equipment dealer in the Midwest offers machine owners.

The contractor keeps cost records on fuel, oil, grease, hydraulic oil, filters, repair parts, repair labor; hours of machine availability, total hours worked, hours not worked due to downtime, hours not worked for other reasons, and an estimate of the hourly cost of downtime. At the beginning of every month, the dealer sends the contractor a data-processing card for each machine in the program. The customer inserts monthly total information on the card and returns it to the dealer where the card is keypunched and fed into the computer.

By midmonth, the contractor has a printout containing monthly and year-to-date cost figures. At the end of his fiscal year, the contractor also receives a summary of the cumulative cost per hour for the current and preceding years. Cumulative costs reported include depreciation and replacement, investment based on current yield of his yearly average investment, maintenance and repair, and downtime and lost production due to obsolescence.

The most popular use of the service is to signal when certain costs suddenly go out of line, or to realize more fully the cost of downtime, or to determine the most economical

WORK AND COST RECORD

Service Meter Reading   Date

ENGINE

Diesel/Gas Added (Gals.)	Fuel Filter Changed	Air Filter Changed	Battery	
			Checked	Replaced

Machine Lubed	Oil Added or Changed (Gals.)	Oil Sampled	Oil Filter Changed

TRANSMISSION AND CONVERTER

Oil Added or Changed (Gals.)	Oil Sampled	Filter Changed

FINAL DRIVE/DIFFERENTIAL

Oil Added or Changed (Gals.)	Oil Sampled	Filter Changed

HYDRAULIC CONTROL

Oil Added or Changed (Gals.)	Oil Sampled	Filter Changed

AVAILABILITY

Scheduled Clock Hours	Clock Hours Not Worked Due to On-Shift		Total Clock Hours Worked
	Repairs	Maintenance	

REMARKS

WORK AND COST RECORD

Service Meter Reading   Date

ENGINE

Diesel/Gas Added (Gals.)	Fuel Filter Changed	Air Filter Changed	Battery	
			Checked	Replaced

Machine Lubed	Oil Added or Changed (Gals.)	Oil Sampled	Oil Filter Changed

TRANSMISSION AND CONVERTER

Oil Added or Changed (Gals.)	Oil Sampled	Filter Changed

FINAL DRIVE/DIFFERENTIAL

Oil Added or Changed (Gals.)	Oil Sampled	Filter Changed

HYDRAULIC CONTROL

Oil Added or Changed (Gals.)	Oil Sampled	Filter Changed

AVAILABILITY

Scheduled Clock Hours	Clock Hours Not Worked Due to On-Shift		Total Clock Hours Worked
	Repairs	Maintenance	

REMARKS

Fig. 1-16 A daily record book is a sound aid to logging maintenance costs regularly. This is a typical page from such a booklet.

time to update equipment. As an additional service, the dealer can use the accumulated data to prepare a more sophisticated financial analysis for equipment investment and replacement. The report covers fixed costs based on market values and investment costs; variable operating costs; net operating revenue which takes into account gross income versus cash outflow; and lastly, rate of return on investment.

MAINTENANCE PLANNING

The maintenance plan of the future should be accomplished more easily and quickly than it is today. Recognizing that maintenance creates a high labor cost, tractor manufacturers compete to make their products easy to maintain.

Tractor design and maintenance trends Caterpillar, for instance, challenges its tractor designers to simplify maintenance through a numerical rating system. By assigning points for the frequency of tasks and the number of operations, the difficulty of access, and the hazards to performing all required maintenance activity, they come up with a cumulative total for each machine design. The goal is a design with a low maintenance index score.

Adoption of the airlines' fast-fill fuel systems and vacuum oil change equipment (Fig.

1-17) are two improvements which will directly cut maintenance time for the earthmoving industry. Development of multiple-specification oils and greases will simplify maintenance by reducing the number of different lubricants needed. Development of better lubricants will extend oil change periods. And there are many improvements which are features of the tractor itself.

A major time-saver is the replacement of dipsticks with exterior sight gauges for checking oil compartment levels. Another is the elimination of grease points with sealed, lifetime-lubricated linkage pins and joints.

Just as sealed track eliminated the daily job of greasing rollers, sealed and lubricated track will extend the interval at which it is necessary to turn pins and bushings.

An innovation that improves maintenance may be as simple as replacing oil pan drain plugs with spigots so the oil doesn't run down the mechanic's arm. Or the innovation may be a change in an entire machine system: Replacement of cable and mechanical controls by hydraulics has eliminated cable maintenance and reduced linkage adjustments. Automatic transmissions have eliminated clutch service.

On wheeled tractors, the growing use of disk brakes has reduced brake adjustments. On crawler tractors with segmented sprockets, sections of teeth can be replaced without removing the track (Fig. 1-18).

Fan drive belts are self-adjusting, and some have been replaced entirely by thermostatically controlled viscous couplers.

Spin-on, throwaway oil filters and compartment breathers eliminate the need to clean filter housings. Daily air cleaner service is not necessary on self-cleaning dust-ejector-type models. Indicators tell when maintenance is required.

At the same time, some trends point toward more maintenance: The concern for controlling emissions and economizing fuel use calls for more frequent engine adjustments. Another factor is the move toward operator enclosure. This often requires cab air

Fig. 1-17 Vacuum oil-changing equipment helps drain and refill the crankcase in just a few seconds.

conditioning with filters which must be regularly cleaned. Frequent window washing is needed to improve visibility.

Because the operator of the future will be encased, away from the sound of his machine, he probably will have more dash gauges, warning horns, etc., to keep him aware of how the machine is operating and to help him identify problems.

Fig. 1-18 With bolt-on sprocket segments it is unnecessary to break track.

Tractor inspection: failure forecasting Specialized diagnostic tools are moving inspection and failure forecasting far ahead of where they were just a few years ago. Use of atomic absorption spectrophotometry for oil analysis has perhaps made the greatest change in the state of the art, but there are a variety of special tools for investigating the tractor's mechanical systems.

Unfortunately, most of these are expensive, so their employment must be handled wisely. For many contractors it is more economical to purchase an inspection service than to invest in a lot of equipment.

In the ideal inspection program, specialized tools are no replacement for a common-sense approach to machine operation. They work best when integrated into a system that relies on alertness to changes in tractor performance.

For an efficient inspection, the easiest checks are made first. The first level of inspection activity should not require taking a single wrench to the machine, but it should be made by a knowledgeable inspector. Only if abnormalities appear is it prudent to go further and apply diagnostic equipment to the tractor system.

The inspection program Caterpillar recommends has three parts to the first-level (initial) inspection.

Oil Analysis. An oil sample is sent to the equipment dealer for an analysis which can detect wear of internal components that cannot be inspected without taking the machine apart.

At oil change periods, a small sample, drawn from the tractor's oil-filled compartments, is mailed to the dealer who analyzes its microscopic metal content with special atomic absorption spectrophotometry equipment (Fig. 1-19). When this is done regularly, a record or wear trend is established for that specific machine. If the amount of a certain metal or other foreign matter rises suddenly above the normal wear trend, it can indicate that a failure is imminent.

The type of metal or material found in the oil (copper, iron, aluminum, silica, or chromium) indicates what part—piston or bearing, gear, clutch plate, etc.—is wearing. The dealer, knowing his products and their wear trends, can forecast a failure and immediately contact the tractor owner.

While oil analysis is available from several sources, to be truly effective at predicting failures the program must have the tractor manufacturer's and dealer's input. Establishing

7-18 Maintenance Programs for Equipment Entities

the metal content of oil is not really meaningful for forecasting failures unless the party which interprets the data knows what is normal or abnormal for the model of tractor. Moreover, two to three samples must be taken and recorded at regular intervals to establish a wear trend which reflects the conditions in which a particular machine is working.

Oil analysis, if a hit or miss proposition, is not nearly so effective at gauging internal machine wear and predicting wear failures as it can be unless it is done regularly by people who know the machinery they are testing.

Fig. 1-19 Atomic absorption spectrophotometry detects traces of metal in oil samples, thus indicating parts wear.

Visual Inspection and Operating Checks. By first looking over and then operating all parts of a tractor, a knowledgeable serviceman can do an effective job judging the condition of a machine.

Before starting the tractor, the inspector looks over about 70 parts for exterior damage, dirt, or leaks that would indicate need for service. He must rate them either as up to standard or needing attention.

He then makes operating checks, relying on the sound and feel of tractor systems, to rate about 50 items as up to standard or needing attention. Among these are 10 *critical indicators:* engine starting ability, engine smoke, radiator pressure, electrical system output, torque-converter temperature, track pitch, link height, pin and bushing, track roller tread wear, and hydraulic implement cycle time. If any of these is substandard, he initiates a second-level inspection of the particular machine system involved. This investigation, to be done efficiently, utilizes specialized testing tools.

System Diagnoses. Critical indicators of trouble with the electrical system indicate a second-level inspection that includes a cell-by-cell complete test and inspection of batteries. Alternator or generator output, regulator performance, and amperage draw and glow-plug draw can be checked by an electrical test instrument.

If engine performance is substandard, an engine performance test plus oil analysis can point out the problem. The serviceman checks low and high idle and the full-load balance-point rpm with a tachometer. A manifold pressure sensor tests turbocharger boost. The rack setting and compression are checked. Fuel pressure and oil pressure can be tested through the full operating range by use of a tetragauge. An injection timing light indicates the exact moment of fuel injection and checks the accuracy of timing advance. Fuel flow can be measured accurately with a horsepower meter, and the horsepower this fuel will produce can also be visually displayed.

For troubleshooting cooling system problems, a variety of diagnostic instruments can be used. A phototachometer will measure fan speed (Fig. 1-20). An air flow meter can be used to check flow at various places on the radiator core. A thermister-thermometer arrangement permits simultaneous temperature readings to be taken at different points in the cooling system (Fig. 1-21). Pressures are measured by a special radiator gauge.

For checking transmission performance, a combined tachometer and hydraulic test box are used. A hydraulic flow meter allows an operating check of the implement system efficiency (Fig. 1-22). If this is not normal, then the pump and individual cylinder circuits can be tested one by one to pinpoint the trouble.

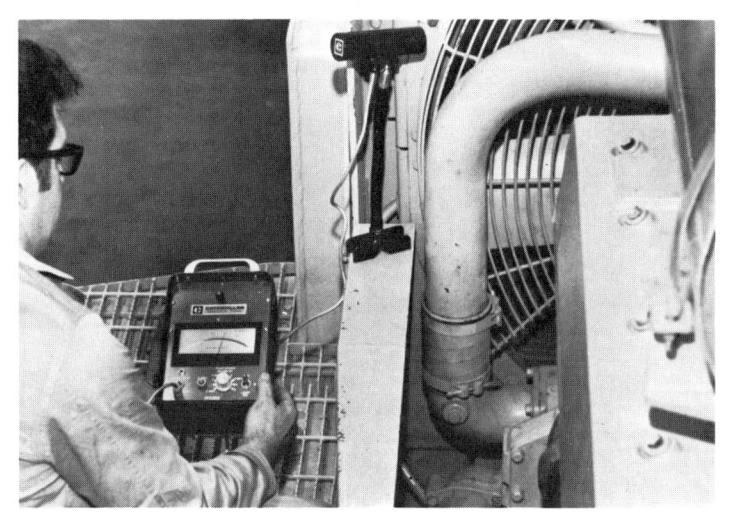

Fig. 1-20 Phototachometer measures fan speed.

These special instruments permit inspection and isolation of a problem with minimum disassembly of the tractor (Fig. 1-23). Inspection with these can be economical in that time and money are not wasted tearing down a tractor and inspecting parts to try to find the cause of inadequate performance.

Organized approach Inspections are too often overlooked or neglected when machinery appears to be running correctly. A formal procedure and regular scheduling are often necessary to overcome resistance to spending time on a tractor about which there have been no complaints.

However, the inspection and failure forecasting services can pay for themselves if they help the contractor decide whether to start a new job or new season with equipment in its present condition; detect shortcomings in tractor maintenance; increase the service life of components by detecting the first signs of unusual wear; permit downtime to be scheduled, thus increasing machine availability; reduce owning and operating costs by getting problems repaired before they cause further damage.

Replace-before-failure decision When a part is showing wear or damage, the decision when and whether to replace it before failure will depend on several things.

One major consideration is whether failure or deterioration of the part will cause damage to other parts. The tractor owner weighs the cost of immediately taking a machine out of production versus possibly higher costs of repair if service is delayed to the most convenient time.

Running to failure in certain tractor systems such as hydraulics, turbocharger, and transmission can cause contamination that will be very expensive to repair. Bearing damage in final drives, engine, turbocharger, or transmission, for instance, can lead to misalignment of parts that will create severe damage if the situation is not quickly

7-20 Maintenance Programs for Equipment Entities

corrected. A simple-to-fix air inlet leak, if given immediate attention, can prevent very extensive engine damage.

Other items such as a leaking cylinder rod or radiator can be temporarily compensated for until there is time for repair.

Does the failing component contribute directly to production? Engine, transmission, water pump, and hydraulic systems are necessary to keep a tractor at work. But repair of a gauge or alternator could possibly be delayed if the production schedule is tight.

Fig. 1-21 The thermister thermometer can give continuous temperature readings from various locations in the cooling system.

Distance from a service source is another consideration. If equipment is working at a remote site, it will be more economical to consolidate repairs so that all major or minor problems are fixed at one time. A very thorough inspection and full maintenance are advisable before a tractor is sent to a remote place.

The importance of machine availability or the necessity of planning downtime are two other factors. When a tractor is in a key job, such as pushloading scrapers on a big construction project, to take an ailing machine out of production for repairs or let it run until it breaks down is a hard decision that is always a bit of a gamble. It will be necessary to weigh the probable time loss for immediate repair versus the potential time loss if run to failure causes more extensive repairs to be needed. One compromise solution to this dilemma is off-shift exchange for a rebuilt component (to be discussed under Planned Component Replacement).

The value of inspection is that it makes a repair-before-failure decision possible. This is bound to help the entire business process. It permits the tractor owner to take advantage of cost savings on repairs—flat rate or trade-in of a component for rebuilding. It also helps with production scheduling and budget planning. In short, it improves the climate for decision making by providing the time to weigh alternatives; after a breakdown, the manager has only the ability to react.

Under most conditions the repair versus run-to-failure decision will be based on economics, the cost of repair versus the cost of lost production time. In times of short supply on replacement parts, the emphasis may be altered. Repair may be done earlier and at less convenient times to preserve the machine and get longer life than would be economical when parts are in free supply. An example could be undercarriage wear. When pins and bushings reach the wear limit for turning, the tractor owner must turn them regardless of convenience, or the ability to reuse their other side will be lost. Likewise grousers can be permitted to wear only so far before their potential for rebuilding is lost.

Planned component replacement Selected machine components can be replaced before accelerated wear causes them to fail or be costly to rebuild. Because more parts of the old component will be reusable before heavy wear has taken place, rebuilt exchange components can be gotten for a lower price. However, the biggest advantage of a component exchange program is the ability to plan downtime far in advance.

The time selected for component exchange is based on wear data from similar machines in similar working conditions. These are calculated by the equipment manufacturer. A wear curve is plotted on a graph which represents averages of time in

Fig. 1-22 Hydraulic flow meter is used to measure hydraulic pump outlet or efficiency of individual circuits in the implement system.

Fig. 1-23 The specialized instruments for testing tractor systems with minimum machine disassembly.

service for a component versus the cost to rebuild it. The point on the curve which represents the longest in-service time before rebuild costs sharply increase is the optimum time for component replacement (Fig. 1-24).

Repair alternatives Speed and economy are the goals for repair when it is needed. Whether or not an equipment owner should have his own tractor repair shop will depend on the size of his operation. Building, equipping, and maintaining a facility plus employing full-time mechanics amount to a sizable fixed cost. Unless the contractor has a large enough repair volume to justify this constant expense, he should consider outside sources for repairs.

Maintaining a repair facility and organization versus using an equipment dealer's services, for instance, is a judgment that can be based on facts and figures. By using the shop cost analysis chart (Fig. 1-25) the tractor owner can calculate his own costs versus the charges made by an outside service source. In operating one's own service shop there are 12 possible types of direct overhead expense plus four categories of allocated overhead. These can be divided by annual hours of service work performed to give an hourly average for overhead expense. Added to the mechanic's wages, it indicates a true cost per hour that can be compared to the hourly charge of an outside repair organization.

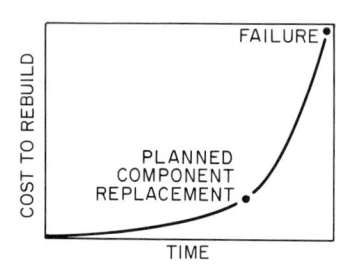

Fig. 1-24 For each component, the cost to rebuild will increase as the component works, slowly at first and then more rapidly. At a given point, normal wear accelerates into abnormal wear. With planned component replacement, you replace a component before this acceleration begins. You get a substantial portion of the life of a component while enabling it to be rebuilt at lower cost. Once this replacement point is passed, the cost to rebuild climbs more quickly for each hour of use, ending in component failure.

Parts inventory The size of inventory will depend on who is doing repairs. If the contractor's own service organization does this, parts inventory may be more complex and require a larger on-hand parts stock so as not to disrupt the efficiency of his shop operation.

To know when and how many parts to order, it is necessary to make a parts forecast and determine lead time (how long it will take to get a part after it is ordered).

A simple method of forecasting parts usage is to average historical usage over a certain period of time. Experience has shown that using a *moving average* based on the last four quarters (12 months) of usage works well.

Lead time, too, will be based on experience. The distance from the parts source, availability of people to go after parts, plus the equipment dealer's normal inventory and response to emergency orders will have to be considered in calculating lead time (Fig. 1-26). If lead time is short—the tractor owner is close to a dependable parts supply—fewer parts will have to be kept in stock.

For each of those items which must be stocked, an *order point* should be established. The order point should be set at the number of parts which the forecast shows will be needed during the lead time (between the time the order is placed and the new stock arrives). Many users set the order point slightly higher than the forecasted usage. Thus they add a margin of safety in case an unusual number of parts are needed.

Repair facility For fully servicing diesel tractors (major component rebuilding not included), a contractor will need a facility which includes repair bays equipped with air, water, and electrical outlets. For larger tractors, these bays should be about 30 × 60 ft. They should be served by an overhead 7½-ton traveling crane. In addition to normal workbenches in repair bays, a heavy-duty metal bench is desirable as a platform for heavy hammering and as a central location for tools including a bench grinder, arbor press, and heavy-duty vise.

For track-type equipment, special hard-surface floors will be needed. Steel rails embedded in the concrete are another possibility. In addition to the shop proper, several support areas are needed. These include an outdoor machine cleaning area plus an indoor cleaning area for tractor components, equipped with a 2½- to 3-ton jib crane and a drainage system.

Also needed are a tool crib, parts storeroom, machine shop, and welding booth.

Shop tools If equipped as follows, the facility described above can be used to (1) completely recondition machines by installing rebuilt components, (2) clean radiators, (3) rebuild brake bands, and (4) recondition final drives, steering clutches, and hydraulic cylinders.

In addition to numerous hand tools, many special tools and major shop tools needed include central compressed air. Steam or heated high-pressure washing equipment is recommended for the machine cleaning area. Parts cleaning tanks of one or two sizes are needed. Hydraulic cylinder repair can be simplified with a special holding stand (Fig. 1-27).

Also necessary are welding equipment and a radiator cleaning tank. A grinder, a lathe, and glass bead blasting equipment are used for reconditioning parts. A portable hydraulic pin press for removing the master track pin will dramatically cut the time spent removing track from a crawler tractor.

Parts and component exchange Components such as alternators/generators, transmissions, turbochargers, torque dividers, engine, hydraulic hose couplings, brake bands, and undercarriage can be purchased from equipment dealers rebuilt to factory specifica-

SHOP COST ANALYSIS

Servicemen Salaries: Average Hourly Wage paid Servicemen $________

OVERHEAD COSTS

*Direct Overhead Expense

Management Salaries: Salaries and bonuses of master mechanics and shop foremen $________

Other Shop Personnel: Cost of auxiliary shop workers (mechanics' helpers, the fellow on the steam cleaner, etc.) $________

Employee Benefits: Life & Health Insurance, pensions, state unemployment, FICA $________

Allowed Time: Wages paid for vacations, holidays and sick leave $________

Truck Expense: Operating expense, including insurance and taxes of trucks and other transportation equipment engaged in repair work $________

Shop Re-work and Warranty: Value of parts and labor invested in doing a repair job over $________

Repair and Maintenance of Shop Equipment: Cost to repair and maintain welding equipment, air compressor and steam cleaner $________

Small Tools and Supplies: Uniforms, gasoline, oil, welding rods, drills, grinders, etc. Expenses of maintaining tool cribs should be included $________

Lost and Clean-up Time: All labor expense for non-productive time $________

Service Training: Wages paid and the expenses of training meetings $________

Equipment Depreciation: Depreciation on transportation and other shop equipment $________

Miscellaneous Direct Expense: All other expense of a direct nature $________

**Allocated Overhead Expense

Administrative: Allocated % x top management salaries and bonuses. (Equipment superintendent) $________

Office: Allocated % x office salaries, supplies and postage $________

Occupancy: Allocated % x real estate, insurance, taxes, utilities, repairs, maintenance, rent, depreciation and amortization of leasehold improvements $________

Other Allocated Expense: Allocated % x insurance and taxes (other than real estate), telephone, janitors, watchmen and other miscellaneous allocated expense $________

Total Overhead $________

Total Overhead / ***Total Service Hours = Hourly Overhead Rate

Hourly Rate of Servicemen $________

Hourly Overhead Rate +$________

Cost of Your Shop Per Hour $________

*Direct Expenses: Costs incurred directly by the Maintenance and Repair Department

***Total Service Hours: Number of Man Hours charged to repair jobs during the accounting period

**Allocated Expenses: Expenses incurred by the entire organization and whose services the Maintenance and Repair Dept. uses. e.g. Payroll office. These expenses are allocated to the Maintenance and Repair Department in proportion to the usage of the services. The usage is the allocated percentage figure.

PEG07066-00 Caterpillar, Cat and ⊞ are Trademarks of Caterpillar Tractor Co. PRINTED IN U.S.A.

Fig. 1-25 Shop cost analysis chart.

7-24 Maintenance Programs for Equipment Entities

tions and pretested (Fig. 1-28). This saves production time in getting the machine back to work sooner. It also vastly reduces the need for specialized repair equipment and highly trained shop personnel. In many cases exchange components are available for a flat-rate price.

Component rebuilding The most basic requirement for rebuilding tractor components is having properly trained service personnel who perform many functions required

Fig. 1-26 Level of parts inventory and lead time to get a part will depend on the contractor's proximity to the equipment dealer's inventory and the size of this inventory.

Fig. 1-27 Demonstration of a special stand for holding hydraulic cylinders that are being repaired.

in reconditioning many types of equipment. The second requirement is specialized tooling. Engine reconditioning requres many special tools including valve grinding equipment and specific tools for working on fuel systems. An engine dynamometer, fuel injection test bench, and electrical testing unit are large investments that enable quality checks to be made before the unit is reinstalled (Fig. 1-29).

Hydraulic test benches are required to test power shift transmissions (Fig. 1-30).

Undercarriage rebuilding requires large investments in track press, track wrenches, and rebuilding equipment.

The specialized tools represent a big investment but give a very high-volume service shop the efficiency to keep its costs, per service hour, competitive with a smaller, less equipped shop. This is one reason that practically all but the largest construction companies turn to an outside service such as the equipment dealer to do the more sophisticated repair work.

In-shop inspection A part of all repair time is spent getting the tractor to the shop and then disassembling some parts to reach the damaged components. To reduce the incidence of repair, it is wise to inspect all machine systems while the tractor is in for repairs. With the specialized test equipment mentioned earlier in "Tractor Inspection: failure forecasting," this can be accomplished with minimum time expenditure. Correcting all detected problems at that time will forestall the day the tractor will again come in for repair.

Repair versus replacement decisions At some point the tractor owner faces the decision of whether to continue repairing or to replace his machine. If he bases this decision on a guess, he either may keep the machine too long or trade too soon, risk downtime and repairs that cut deep into profits, or not get the full value from the original investment.

There is a fairly simple way to figure the optimum replacement time, if the owner has kept accurate cost records. By adding five kinds of costs to get a cumulative cost per hour, he can tell whether his costs are decreasing or rising as the tractor gets older. The optimum time to trade is when the cost per hour starts going up. The five costs to be added are (1) depreciation and replacement costs, (2) investment costs, (3) maintenance and repair costs, (4) downtime costs, and (5) obsolescence costs.

Fig. 1-28 Rebuilt turbochargers are common items in parts exchange programs.

Depreciation and Replacement Cost. The difference between what is paid for a machine and what is redeemed on a sale or trade-in is depreciation, an out-of-pocket cost. Add to it the replacement cost. New tractors invariably cost more due to price increases, some of which is offset by higher productivity.

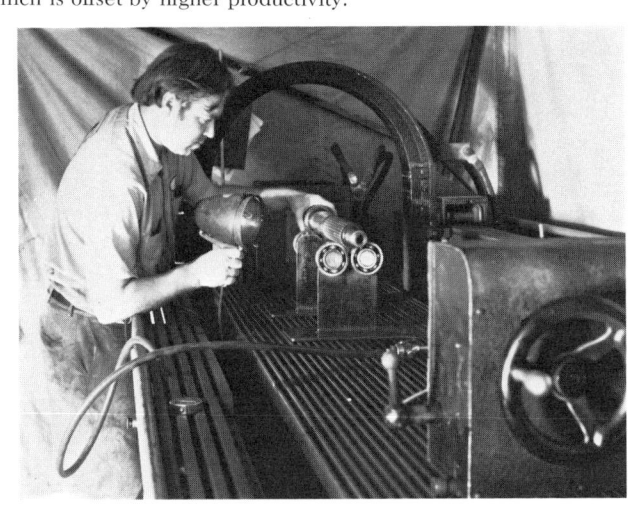

Fig. 1-29 Service mechanic uses strobe equipment to check a splined shaft for cracks before using it in tractor rebuilding.

7-26 Maintenance Programs for Equipment Entities

Investment Costs. Whether purchased for cash, by installment, or leased, some type of finance charge or investment expense has to be charged against the machine.

Maintenance and Repair Costs. These tend to grow as a machine ages.

Downtime Costs. Lost production also increases with machine age. Accurate records are needed to compare downtime with work time and to estimate the financial burden it creates.

Obsolescence Costs. As tractor designs improve, newer models are frequently capable

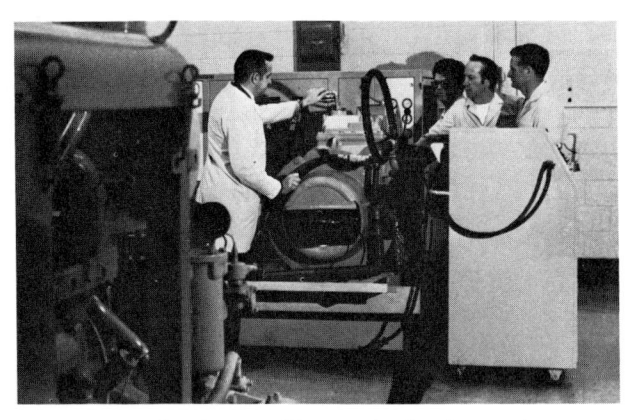

Fig. 1-30 The transmission test bench is one of the specialized sophisticated instruments that enables maintenance people to perform sound repair work.

of doing more work in the same time. An equipment dealer can tell how much productivity is lost by continuing to operate an old machine instead of replacing it with a new one. This can be converted to a cost figure.

Annually these costs can be figured and divided by the number of hours the machine has operated, thus giving a cost per hour. Figure 1-31 shows a chart and graph representing a hypothetical tractor operated 1500 hr/year. Both show how cumulative cost per hour dropped for 3 years then began to rise. For this particular unit, ideal trading time would occur at the end of 3 years. However, each machine will differ according to working conditions. So, only the owner's records provide the accuracy necessary to manage repair versus replacement decisions wisely.

Diesel Tractor Maintenance 7-27

CUMULATIVE COSTS PER HOUR YEAR

FACTORS	1	2	3	4	5	6	7	8	DECISION
Depreciation & Replacement Costs	7.47	6.03	5.40	5.10	4.67	4.40	4.14	3.92	Keep
Investment Costs	2.36	2.13	1.95	1.80	1.66	1.55	1.46	1.38	Keep
Maintenance & Repair Costs	0.93	1.33	1.69	2.00	2.32	2.62	2.97	3.33	Trade
Downtime Costs	0.24	0.36	0.45	0.54	0.62	0.71	0.80	0.90	Trade
Obsolescence Costs				0.30	0.48	0.60	0.86	1.05	Trade
Total Cumulative Costs per hr	11.00	9.85	9.49	9.74	9.75	9.88	10.23	10.58	Keep or Trade
Cumulative Hours	1500	3000	4500	6000	7500	9000	10,500	12,000	

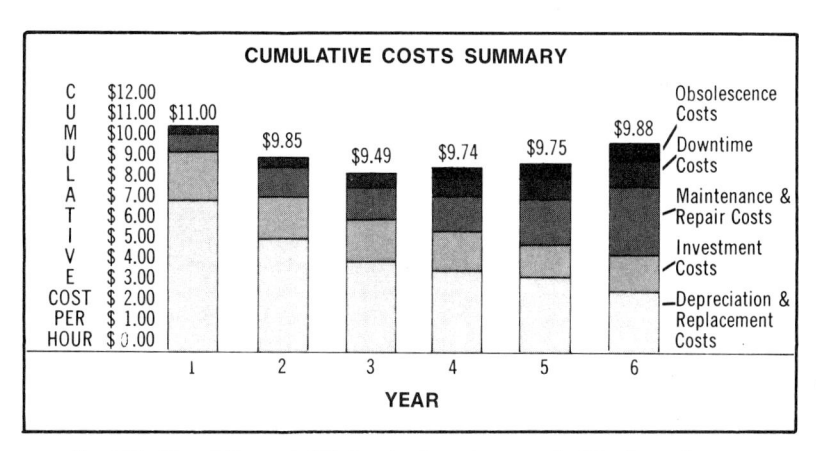

Fig. 1-31 Cumulative costs fall, then begin to rise, over the life of a machine.

Chapter **2**

Scraper and Scraper Blade Maintenance

TEREX DIVISION
General Motors Corporation, Hudson, Ohio

Good scraper maintenance includes everything from wiping the windshield to overhauling the engine. (See Fig. 2-1.) Although these two extremes are widely separated levels of scraper maintenance, both are equally important to the primary objective of good maintenance: keeping the machine producing on the job.

The following scraper maintenance procedures are divided into two general categories: mechanical maintenance and structural maintenance. Mechanical maintenance coverage includes only the general service/maintenance procedures scheduled on a regular basis. Detailed overhaul procedures for the scraper components is left to the service/overhaul manuals provided by the scraper manufacturer. Scraper structural maintenance covers typical service and maintenance of the apron, elevator, bowl, rolling floor, tailgate, and other standard scraper operating structures requiring periodic service and adjustments.

ENGINE

Proper scraper engine maintenance requires attention to each of the four major engine systems: intake air, fuel, cooling, and lubrication.

Air intake system

1. Check the engine intake air restriction daily. An air restriction gauge that continually monitors engine air cleaner air flow restriction is usually found attached to the air cleaner in many late-model scrapers. It indicates the degree of air cleaner element plugging as the red band rises in the gauge window. When the red band fills the window, the air cleaner has reached the maximum allowable air flow restriction and must be serviced to restore normal engine efficiency. Air cleaners on scrapers not equipped with these restriction gauges should be checked for the degree of air cleaner restriction with a manometer or other gauge as specified in the scraper manual.

2. Keep the air cleaner dust cup clean (Fig. 2-2). Air cleaners equipped with dust cup vacuator valves automatically allow the dust cup to empty, but the dust cup and valve should be checked regularly and unplugged, if needed, to keep the dirt from building up to the point of choking the air precleaner tubes. Some dirt buildup in the dust cup is not harmful provided that it is not allowed to get high enough to plug the air precleaner section, if the air cleaner is so equipped (Fig. 2-3). To prevent rapid air cleaner plugging, keep the dust cup clean and the dust valves, if installed, open.

3. When the daily inspection of the air cleaner dust cup indicates dirt buildup in the precleaner tubes, use a soft-bristled brush to clean out the dust and loose dirt. The

7-30 Maintenance Programs for Equipment Entities

precleaner section will have to be removed from the air cleaner assembly and washed to remove soot or other dirt that cannot be brushed from the tubes.

4. Service the air cleaner secondary filter element as soon as the air restriction gauge or manometer reading indicates the need as recommended by the scraper manufacturer. If the air cleaner is equipped with an inner safety element, clean or replace this element

Fig. 2-1 Full view of scraper reveals complexity of maintenance problem.

Fig. 2-2 Keep the air cleaner dust cup and vacuator valve clean.

also, according to the manual instructions, to restore the air cleaner assembly to maximum efficiency.

5. Eliminate all air leaks in the air-cleaner-to-engine piping.

Fuel system The clean fuel required for trouble-free engine performance starts with the fuel supply. Make sure that fuel storage tanks and drums are protected from dirt and water contamination. Clean the fuel hose nozzle and fuel tank cap before filling the tank. Fill the tank as often as possible to reduce moisture condensation inside the tank to a minimum. Always try to fill the fuel tank before parking the scraper overnight or longer.

1. Drain the fuel filters daily by opening the filter housing drain cocks to drain water

Scraper and Scraper Blade Maintenance 7-31

and sediment from the fuel system. Drain the filters only until the fuel runs clear; don't drain excessive fuel.

2. Check the fuel tank level and fill the tank, if low, before starting the engine. Open tank fuel line valves, if closed (Fig. 2-4).

3. Be sure to install new fuel filter elements and clean the filter housings at the intervals recommended in the scraper maintenance manual (Fig. 2-5).

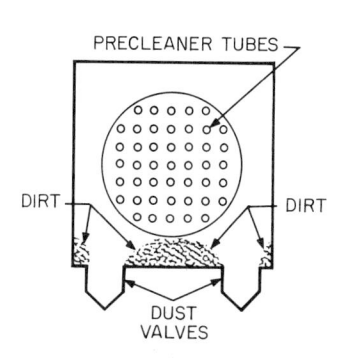

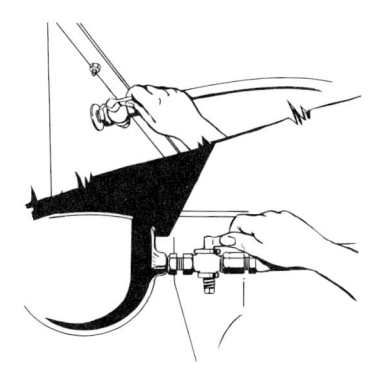

Fig. 2-3 Dirt buildup in the air cleaner.

Fig. 2-4 Proper method for opening fuel tank valves.

Fig. 2-5 Be sure to install a new filter element.

4. The fuel tank breather should be cleaned as required to ensure proper tank venting. Clean the breather exterior before removing it to keep dirt from falling into the tank (Fig. 2-6).

5. Don't overlook the fuel tank filler tube screen (Fig. 2-7). This screen, if installed, should be cleaned periodically to eliminate it as a potential source of fuel contamination.

Cooling system

1. Check the engine coolant daily and add coolant to the recommended level, if low. Always use care when removing the radiator cap. Some caps have vent stops, holes, or other features to allow any pressure in the cooling system to be dissipated before the cap is completely removed. Be sure to follow the recommendations in the scraper maintenance manual for proper venting of the cooling system pressure.

2. While the radiator cap is removed, check the condition of the cap sealing gasket

7-32 Maintenance Programs for Equipment Entities

and seat and the mating flange of the radiator filler neck. Clean up nicks and replace the cap gasket as required to ensure a pressure-tight seal.

3. If antifreeze solution is installed in the cooling system, check the concentration of the antifreeze with a reliable tester. Follow maintenance manual recommendations for adding antifreeze and/or corrosion inhibitors to the cooling system to provide needed protection.

4. Clean the exterior of radiator core fins of dust, dirt, or other debris with compressed air and/or water. Direct the air or water stream through the core in the opposite

Fig. 2-6 Proper method for cleaning fuel tank cap breather.

direction to the normal air flow. When cleaned, check for bent or damaged tubes and fins and for coolant leaks. Repair or replace the radiator, as required, to ensure trouble-free engine cooling.

5. Check fan belt tension and pulley alignment as specified in the scraper manual (Fig. 2-8). The maximum pulley misalignment allowable is usually $\frac{1}{16}$ in. When checking belts make sure that they do not bottom in the pulley grooves or project more than $\frac{1}{16}$ in. above the grooves. Check also for glazing, cracking, or fraying of the belt(s). In multibelt installations, make sure that any variation in circumference between belts does not exceed $\frac{1}{16}$ in. Replace belts for any of these conditions. Remember, all belts in a multi-belt installation must be replaced as a set to ensure that all belts are properly matched to distribute the load equally. When replacing fan belts, do not pry or otherwise force replacement belts onto the pulleys. Release the belt-adjusting pulley, install the belt(s), then tighten the belt with the adjusting pulley. Recheck all new belts at short intervals until they take their initial set, then follow the scraper manual instructions for the normal inspection and adjustment intervals.

6. Lubricate all fan belt pulleys as recommended in the scraper maintenance manual.

7. With the engine running, check for air bubbles in the coolant flow through the radiator top tank. Air bubbles indicate a suction leak in the water pump or pump inlet line or fittings. Tighten or replace hoses and fittings as required to stop suction leaks.

Fig. 2-7 Clean the fuel tank filler tube screen.

8. Check for proper thermostat operation on the engine temperature gauge. Thermostats should maintain coolant temperature within the operating limits specified in the scraper manual. Replace thermostats that allow engine overheating or overcooling.

9. Drain, flush, and refill the cooling system as specified in the scraper manual to ensure clean, rust-free coolant with the required antifreeze and/or corrosion inhibitor concentrations.

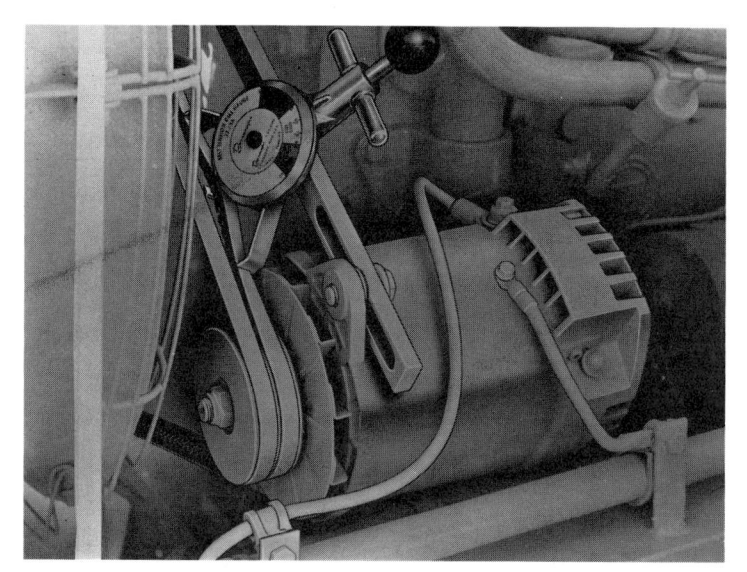

Fig. 2-8 Fan belt tension checks are an important part of routine maintenance.

Lubrication system

1. Check the crankcase oil level before starting the engine at the beginning of each shift, or as otherwise specified in the scraper manual (Fig. 2-9). Add the specified oil, as required, to the proper operating level. Do not overfill the crankcase. Excess oil can cause oil foaming and engine overheating, to say nothing of being wasteful.

2. Drain and refill the crankcase, replace the oil filters, and clean the filter housings as specified in the scraper manual. Be sure to use the recommended oil filter elements.

Fig. 2-9 Check oil level in the crankcase.

3. Clean the engine crankcase breather filter as required. The recommended cleaning interval should be specified for average job operating conditions but can vary widely between jobs, seasons, etc.

4. Check the engine oil pressure gauge reading regularly. Make sure that the lubricating system maintains the oil pressure required at both idle and maximum rated engine speeds. This reading is a constant indication of both the condition of the engine lubrication system and the general mechanical condition of the engine.

GENERAL ENGINE MAINTENANCE

Keep the engine well-tuned and the throttle and shut-off linkage well-lubricated. A periodic engine stall speed check, if recommended in the scraper manual, will help indicate the loss of engine efficiency and the need for maintenance or repair.

Transmission The following transmission maintenance practices apply generally to converter-planetary gearing-type transmissions. They are only general guides, however, and should be used in conjunction with the service/maintenance instructions in the manual for the transmission.

1. Check the transmission oil level. The level should be checked before starting the engine, particularly after the scraper has been parked for an extended period, to ensure that it contains adequate oil for initial transmission operation. The operating oil level should then be checked when the oil has warmed to normal operating temperature.

    a. Before starting the engine, remove the transmission dipstick and make sure the oil level is high enough for initial transmission operation. On transmissions equipped with check cocks or check plugs, open the upper check cock or remove the upper check plug and add oil, if low, to the level of the upper check cock or check plug.

    b. Start the engine and allow the transmission oil to warm to normal operating temperature. With the bowl resting on the floor or ground, apron and ejector completely lowered, and the wheels completely blocked, shift the transmission momentarily through all gear ranges and back to neutral. Adjust engine speed to 1000 rpm and remove the dipstick or the lower check plug, or open the lower check cock. Add oil, if low. Do not overfill as excess oil causes oil foaming which results in transmission overheating.

2. Lubricate the transmission shift linkage, if mechanical, and check the shift lever operation for control valve detent feel. Adjust the linkage for positive detenting action in each transmission range according to instructions in the scraper service/maintenance manual.

3. Clean transmission breather and oil filler cap. Wash the breather in a suitable solvent and let dry to ensure proper transmission housing ventilation.

4. Drain and refill the transmission oil sump, clean the filter housings, and install new elements as recommended in the scraper manual. Check the used filter elements for evidence of metal particles trapped in the filter medium. Steel chips in the element indicate wear or damage to gears, bearings, or shafts. Aluminum chips (nonmagnetic) in the filter element are signs of erosion or damage to torque converter elements.

5. Check the scraper instrument panel gauges for transmission oil-operating temperature and pressure in all operating ranges at various engine and travel speeds. Excessively high or low readings or erratic readings indicate either mechanical or hydraulic trouble in the transmission.

6. Carefully inspect the transmission housing, the oil filter and oil cooler, and all external hoses and fittings for oil leaks. Repair or replace any leaking components.

7. A good test of the condition of a transmission equipped with a torque convertor is the stall speed test. Be sure to follow the scraper manual or transmission manufacturer's instructions for checking stall speeds to avoid damage to the converter and transmission. This test can be helpful in determining whether the cause of poor scraper performance is the engine, the transmission, or both. A below-normal stall speed indicates that the engine is not delivering its normal rated output to the transmission. An above-normal stall speed usually is an indication that the transmission is not capable of absorbing the required amount of engine power and is, therefore, not performing to specifications.

Scraper and Scraper Blade Maintenance   7-35

Transmission air shift system

1. Check the air shift system controls and fittings for air leakage. Charge the scraper air system to operating pressure, shut off the engine, and observe the air pressure gauge for a loss of pressure over a reasonable period. Listen for audible signs of air leakage at all shift controls, hoses, and fittings. Check suspected leak points with soap solution. Repair

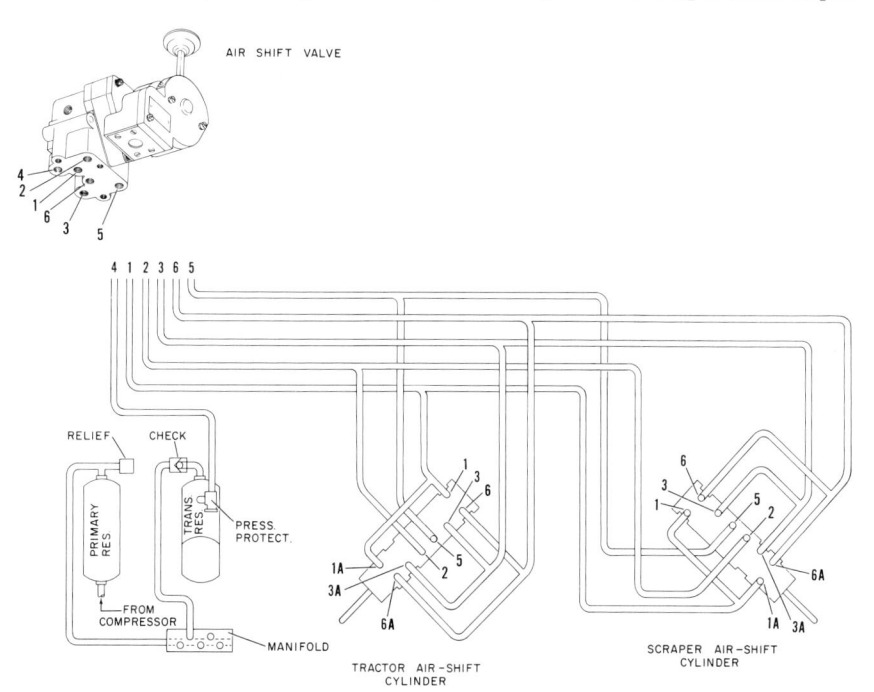

Fig. 2-10 Transmission air shift system.

or replace any leaking components. A loss of 20 psi air pressure in 30 min in the shift system air supply is a sufficient leakage rate to prevent proper transmission shifting.

2. Check the available air-shifting pressure at the transmission shift cylinder with an accurate air pressure gauge. If the available air pressure is below the specified pressure, the transmission cannot be properly shifted. Find and correct the cause of the inadequate air pressure to ensure proper transmission operation.

3. Excessive scraper brake chamber push rod travel can use sufficient air, when brakes must be frequently applied, to prevent positive shifting of air-shifted transmissions. Keep the scraper brake chamber slack adjusters properly adjusted to minimize brake chamber push rod travel and avoid transmission air-shifting problems.

DRIVELINES

1. Check the driveline regularly for loose bolts, loose flanges, and backlash. Use a small prybar to check the driveline flanges at the transmission and differential for looseness. If loose, drop that end of the driveline and test the flange for backlash between the splines and flange. Replace flanges that do not fit snugly.

2. Check the driveline universal joints for worn bearings, crosses, and seals. If bearings and crosses are worn, both must normally be replaced as an assembly.

3. Check for excessive wear of the slip-joint splines. Replace the driveline if badly worn.

4. Grease the driveline universal joints at the recommended intervals (Fig. 2-11). If the driveline crosses are equipped with relief-type bearings, apply the lubricant until it

flows through the needle bearings and past the seals to ensure complete bearing lubrication and to clean out old lubricant. If the universal joint crosses do not have relief-type seals, lubricate the driveline only to the degree specified in the scraper manual, to prevent popping the seals with lubricant applied under pressure.

5. Tighten all driveline bolts and the slip-joint retainer to the recommended torque values at the specified intervals.

Fig. 2-11 Be sure to thoroughly grease the drive line, including universals and splined shaft.

Fig. 2-12 Carefully check the oil level in the differential.

DIFFERENTIAL, AXLE, AND FINAL DRIVE

Maintenance of the standard and No-Spin type differentials, axles, and final drive gearing requires regular checking and refilling of their lubricant reservoirs, and replacing lubricants at specified intervals. Differentials equipped with air-powered lock-up clutches require additional periodic maintenance of the differential air circuit.

1. Check the differential housing lubricant level at the specified interval (Fig. 2-12). Add lubricant, if low.

2. Check the final drive planetary housing lubricant level and add lubricant as required.

3. Clean the differential housing breather as frequently as required by the job operating conditions (Fig. 2-13).

4. Drain the differential and final drive gearing lubricant and refill with fresh lubricant at the specified interval. When a differential is rebuilt, the initial lubricant

should be drained and replaced with fresh lubricant after a short operating period to flush away any foreign matter produced during the run-in period.

Differentials equipped with air-powered lock-up clutches require additional periodic maintenance of the air supply system.

1. Check the air regulator-lubricator oil reservoir daily. Drain moisture and sediment, as required, and fill the reservoir with recommended lubricant.

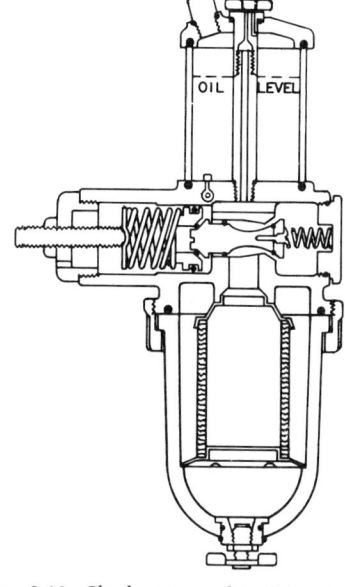

Fig. 2-13 Clean out the differential breather.

Fig. 2-14 Check air regulator-lubricator oil level.

2. Drain the air regulator lubricator and replace the air filter at the specified interval.
3. Check the air regulator-lubricator oil level (Fig. 2-14) daily and refill as required with the specified lubricant.

POWER TAKE-OFF

1. Check the power take-off (PTO) oil level at the intervals specified in the scraper maintenance manual. Remove the oil level check plug, or open the oil level check cock and add the recommended oil until it flows from the check cock or check plug hole, if low.
2. Drain and refill the PTO at the specified interval. Remove the lubricant drain plug, drain the oil completely, and refill the housing to the normal operating level with fresh lubricant. If possible, check the drained lubricant for debris and metal chips that indicate damage to gears, bearings, seals, or other PTO internal components.

STEERING SYSTEM

Keeping the steering system oil clean and all steering components properly lubricated and adjusted are the basic requirements for scraper steering system maintenance.

1. Check the steering tank oil level and refill as required (Fig. 2-15).
2. If the scraper has a vented steering oil tank, clean the tank air breather filter regularly (Fig. 2-16). A well-ventilated tank allows proper oil flow through the outlet line to prevent pump cavitation. Remove and clean the tank breather element as often as required by job conditions. Light dust can be shaken, vacuumed, or blown from the filter element with compressed air. Heavier sooty dirt should be washed from the breather

7-38 Maintenance Programs for Equipment Entities

element in a detergent solution. After washing, rinse the element in clear water and dry thoroughly.

3. Check the steering oil filter restriction gauge or warning light regularly, if the scraper is so equipped, to determine the degree of filter element plugging. Change the filter element when the gauge or warning light so indicates, or at the interval specified in the scraper manual. A dirt-plugged filter forces the oil through the filter bypass valve. This rapidly increases the dirt pickup rate of the steering oil which, in turn, increases the wear rate of all steering system components.

4. Check the steering system operating pressure as specified in the manual to determine the condition of the system components. A steering system that cannot develop the operating pressure required for adequate steering rates indicates a severely worn pump, valve, or cylinders.

5. Check the steering gear oil level every 100 hr or as specified in the scraper maintenance manual (Fig. 2-17). Drain and refill the oil reservoir with fresh lubricant every 1000 hr.

6. Drain and refill the steering tank and clean the filler neck screen as specified in the scraper manual.

Fig. 2-15 Decals remind operator or maintenance mechanic to check the steering tank oil level.

Fig. 2-16 Remove and examine the steering oil tank breather.

Scraper and Scraper Blade Maintenance 7-39

7. Lubricate the steering column, steering cylinder mounting pins, reversing valve rollers, and steering frame pins with the specified lubricants at the intervals required by job conditions or by the maintenance manual schedule.

8. Check all steering components, hoses, and fittings for wear, damage, and oil leaks, particularly at maximum pressures, and tighten fittings to stop any leaks. Replace hoses or fittings and seals that cannot be tightened adequately to stop leaks.

9. In nonpressurized steering systems, remove the oil tank cap and observe the oil flow through the tank for air bubbles during high-speed steering pump operation. Air bubbles indicate air suction leaks through oil seals and hose fittings on the low-pressure side of the steering hydraulic system. Other symptoms of air leakage into the oil are a noisy pump, erratic steering, and oil overheating. An air suction leak at the inlet fitting of a

Fig. 2-17 Check the oil level in the steering gear.

noisy steering pump can often be detected by coating the fitting with oil to momentarily stop the leak and quiet the pump.

10. Steering time from side to side should be specified in the scraper manual and is the normal test for steering system condition. Slow steering, particularly when coupled with oil overheating, is an indication of worn steering system components, low relief valve setting, the wrong oil viscosity, low oil level, or air suction leaks.

HYDRAULIC SYSTEM

The oil tank, pump, valve, cylinders, and connecting hoses that operate the scraper bowl, ejector, apron, elevator, rolling floor, etc., should be serviced regularly to maintain them in peak operating condition.

1. Check the hydraulic tank oil level at the specified interval. Make sure that all oil lines and cylinders are fully charged by operating all cylinders. Retract all cylinders completely before checking the tank oil level since, if the oil is added to the system with the cylinders extended, the system will be overfilled and the tank will overflow when the cylinders are fully retracted. Fill the oil tank, if low, to the proper operating level indicated by the sight gauge or check cocks.

2. On scrapers equipped with oil filter restriction gauges, check the restriction gauge reading at the specified interval with the engine idling and the oil at normal operating temperature. If the gauge reads above the maximum allowable pressure, replace the oil filter element. A higher reading indicates the oil filter is plugged thereby allowing oil to bypass the filter. Under this condition, the oil dirt pickup rate will increase and accelerate hydraulic component wear rates. The oil filter element should be replaced, regardless of the restriction gauge reading, at the interval specified in the scraper manual to avoid a deceptive reading produced by a defective filter element or filter bypass valve. After a

hydraulic component failure, the oil filter element should be replaced on an accelerated schedule to ensure that all debris from the failed component is cleaned from the system; for example,

Oil change	Hours
First	4
Second	10
Third	40
Fourth	80
Fifth	160

Subsequent changes should be made at the normal specified interval.

3. Clean the hydraulic oil tank breather as specified, or according to the job operating conditions. Wipe the breather exterior and remove it from the tank. Then vacuum or use compressed air to blow the dirt from the breather element. Soot-type deposits that can be cleaned in this manner should be washed off in a detergent solution. Be sure to thoroughly rinse and dry the breather element before reinstalling it on the hydraulic tank.

4. Lubricate all bowl, pull yoke, and cylinder mounting pins and linkage with the required lubricants at the specified intervals, or as determined by job conditions (Figs. 2-18 to 2-20). Be sure that lubricant applied through connecting tubing from remote-mounted lube fittings reaches the component(s). Plugged component lube passages, hoses, fittings, or loose connections can prevent adequate lubrication and shorten service life of the component(s).

5. Drain the oil tank, clean the filler screen, and refill with new oil at 1000 hr or as specified in the scraper manual. Be sure to use the recommended hydraulic fluid of the required viscosity for the operating conditions (ambient temperature, etc.). Be sure to install new oil filter elements and gaskets before refilling the oil tank.

6. Test the operation of the hydraulic system to determine the condition of the system components. Slow, sluggish, or erratic operation of the bowl, ejector, apron, elevator, and rolling floor should be diagnosed and corrected without delay to prevent breakdowns on the job. Consult the scraper maintenance manual for diagnosis procedures and required corrective actions to keep the hydraulic system at peak efficiency.

Fig. 2-18 Grease upper bowl cylinder pins.

AIR SYSTEM

Proper operation of the scraper air system is critical for the safety of both the operator and the scraper. Hence, the importance of regular maintenance of the air system cannot be overemphasized.

1. Drain all air reservoirs and, if so equipped, air system moisture ejectors daily. All air tank drain cocks should be opened fully whenever the scraper is parked at the end of the shift and left open to allow the tanks to drain completely, until it is to be restarted.

2. Check the air compressor drive and mounting at the specified interval. Tighten all drive and mounting bolts to the required torque, if loose. Check compressor drive belts and replace if worn, glazed, or cracked. Be sure to replace *all* belts of a multibelt set even if only one looks worn or defective.

3. If the compressor oil supply is self-contained, check the oil level and fill, if low. Wash the compressor air inlet filter element in solvent at the interval specified in the scraper manual, saturate with engine oil, and squeeze dry. Install the cleaned filter with a new gasket.

4. Remove and clean or replace the compressor governor air filter.

5. Check the cut-in and cut-out settings of the governor and adjust, if necessary, to the scraper manual specifications. Be sure that the air system safety valve will open at the maximum allowable pressure.

6. Make sure that all pressure protection valves in the air system are adjusted to the required opening and closing pressures.

7. Check all air hoses, fittings, and components for damage and deterioration. Repair or replace all suspected parts.

8. With the air system charged to maximum operating pressure and the engine off, test all air system connections for leaks. Apply a soap solution to all connections and tighten all leaking connections to stop air leaks. Replace seals, fittings, or components in which leaks cannot be stopped. Pay particular attention to the air compressor discharge line for signs of excess carbon. If sooty, remove the line and check the compressor unloading valves for carbon. Clean all carbon from the compressor cylinder head and unloading valves. Listen at the compressor for air leaks that indicate leaking unloader piston seals. Replace leaking seals.

Fig. 2-19 Grease lower bowl cylinder pins.

Fig. 2-20 Grease ejector linkage pins.

BRAKES

Badly worn or out-of-adjustment brakes prevent normal scraper operating efficiency and create a constant operating hazard. Excessive brake drag, on the other hand, robs scraper power and produces brake overheating and fading. Worn or poorly adjusted brakes also

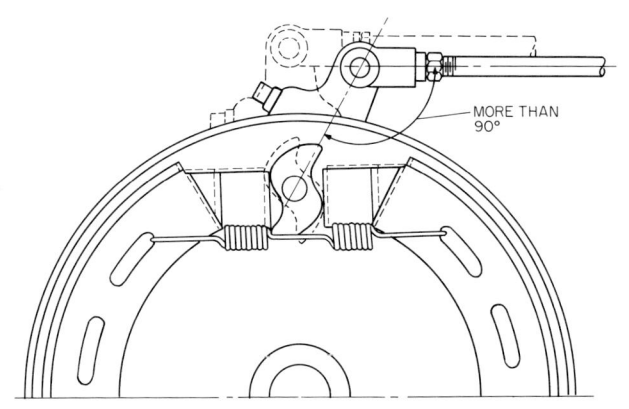

Fig. 2-21 Proper procedure for checking slack adjusters.

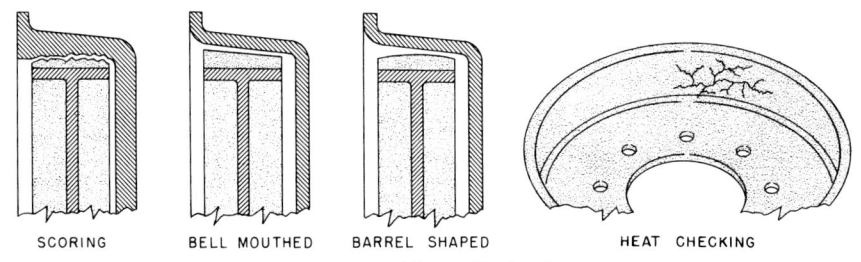

Fig. 2-22 Typical forms of brake shoe wear.

require excessive brake chamber push rod travel. Under conditions requiring frequent braking, the excessive air volume needed to apply badly worn brakes can deplete the available air supply below that required for normal operation of air-powered accessories, such as transmission-shifting air cylinders. Hence, brakes must be properly maintained for both safety and maximum production.

1. Check slack adjusters for slightly more than a 90° angle between the brake chamber push rod and the slack adjuster arm. The angle should be approximately the same at all brake chambers (Fig. 2-21). If necessary, adjust the brakes:
    a. Back out the slack adjuster lock screw enough to turn the adjusting shaft.
    b. Turn the adjusting shaft tight; then back off one-third turn and tighten the lock screw.
    c. Recheck all slack adjusters to ensure that the brake chamber push rod travel will be about the same at all wheels for equal application of all brake shoes.

2. Remove the wheel and brake drum assemblies and check brake parts at the intervals specified in the scraper manual (Fig. 2-22).
    a. Check for cracked and bent brake backing plates.
    b. Check cams for flat spots that cause wheel pulling.
    c. Check for bent camshafts and worn bushings and pins that cause camshaft binding.

d. Clean linings and bolt and rivet holes of rust and corrosion and check brake linings for excessive wear and flat spots.
   e. Check shoe rollers for binding and flat spots.
   f. Check brake drums for excessive wear, cracks, and distortion.

Remove worn or damaged parts and repair or replace as required. Turn brake drums and install oversized linings as specified in the scraper manual to return brakes to normal efficiency.

WHEELS AND TIRES

All loads placed on the scraper while loading and hauling in all operating conditions are concentrated on the wheels and tires. Hence, lack of proper maintenance of these components can quickly result in scraper downtime. To prevent such needless downtime, all tires and wheels should be checked regularly and serviced as required.

   1. Tire inflation and condition should be carefully checked daily. Use an accurate tire gauge and check inflation while the tires are *cold*. Tire pressure increases as tires warm up during operation. *This is normal*. Do not bleed air from tires that are checked while warm to reduce them to the specified air pressure. Tires bled in this manner will be dangerously underinflated when they cool down.

   2. Inspect all tires thoroughly for damage and deterioration before operating the scraper. Remove all stones or other objects embedded in the tire treads or carcase. Sand, gravel, and water can work into a tire carcase exposed by damage or extreme wear and separate the plies. This will quickly ruin a tire that might otherwise have been returned to service if repaired in time.

   3. Always clean all oils, greases, or other such materials from tires as soon as possible. These materials are highly damaging to tire treads.

   4. Make sure that all tires have valve caps tightened securely. These caps keep valves airtight and prevent dirt from plugging or damaging valve stems.

   5. Check all rim parts carefully for distortion, cracks, or other damage that can produce air leaks and operating hazards. Check tire rims and tighten all wheel nuts as required. Keep all exposed wheel surfaces well-painted. Replace all wheel and rim parts that are in questionable condition.

   6. Grease-lubricated wheels should be removed to repack the bearings as specified in the scraper manual. Check all internal parts while the wheels are dismounted and repair or replace as required to prevent unexpected downtime from faulty wheels and tires.

ELECTRICAL SYSTEM

The condition of the scraper electrical system is easily determined by the operation of the electrical components. Batteries that remain fully charged, or are quickly recharged after extensive use, yet do not use up excessive cell water are in good condition and indicate a sound altenator/generator and connecting wiring system. A cranking motor that turns the engine flywheel at normal engine speeds means that the batteries and cranking motor circuit are also performing satisfactorily. If the remaining electrically powered accessories also operate normally, the entire scraper electrical system is sound. To keep it in good condition, the electrical system should be maintained as specified in the scraper manual, with emphasis on the following areas:

   1. Keep the exterior of the batteries clean, and the connecting cables well-secured to the battery posts. Corrosion buildup on the cable clamps and battery hold-downs should be neutralized with a baking soda solution and rinsed clean. Be sure to prevent the neutralizing solution and rinse water from entering the battery cells.

   2. After cleaning the battery exterior, check the electrolyte level and specific gravity. Electrolyte should be above the tops of the plates in each cell. If low, fill to the specified level with demineralized water. Use an accurate hydrometer to check the electrolyte specific gravity and be sure to compensate for the electrolyte temperature when reading the specific gravity. By keeping records of battery specific gravity readings and cell water usage, the voltage regulator can be adjusted as required to avoid overcharging or undercharging the batteries.

3. Check the generator/alternator drive belt tension and adjust to the scraper manual specifications. Do not overtighten the belt.

4. Test the cranking motor mounting bolts and tighten securely, if needed.

5. Check the instrument panel ammeter reading with all lights turned on and the engine operating at substantial speed to make sure that the generator/alternator is delivering sufficient output to meet all electrical loads. Adjust the voltage regulator or repair the generator/alternator if required to meet the electrical loads.

6. Test the operation of all electrically powered gauges and accessories. Repair or replace any gauge, accessory, or connecting wiring to the inoperative component(s).

7. Carefully inspect all wiring and terminals. Replace broken, frayed, or deteriorated wiring, cables, connectors, and terminals. Be sure that all cables, harnesses, clamps, straps, and component mountings are clear of moving parts to prevent chafing during scraper operation.

SCRAPER BOWL MAINTENANCE

Proper maintenance of the scraper bowl should include everything from simple paint touch-up to straightening and reinforcement of the complete bowl structure as needed. The ejector, apron, or elevator assembly, rolling floor, and other assemblies that make up the bowl should also have periodic attention to keep them in top working order. Maintenance of the hydraulic cylinders, engine, transmission, and other mechanical components that are mounted on the scraper bowl is covered in previous sections.

Lubrication All rolling, rotating, sliding, or other working parts of scraper bowl components should be lubricated at the intervals and with the lubricants specified in the scraper maintenance manual. Typical grease points to be lubricated on a scraper are:
1. Steering frame upper hitch pin
2. Steering frame lower king pin
3. Bowl hoist linkage pins (Fig. 2-23)
4. Bowl apron cylinder and linkage pins
5. Bowl ejector cylinder pins
6. Steering cylinder mounting pins
7. Steering cylinder reversing valve pins (Fig. 2-24)
8. Pull yoke mounting pins.

Typical grease points to be lubricated on elevating-type scraper bowls include:
1. Elevator stub spindle (Fig. 2-25)
2. Elevator idler roller (Fig. 2-26)
3. Upper elevator stub spindle

Fig. 2-23 Bowl hoist linkage pins are an important lubrication point.

Scraper and Scraper Blade Maintenance   7-45

    4. Rolling floor rollers, pull arms, and elevator arms
    5. Endgate stabilizer rollers (Fig. 2-27)
    6. Centralized lube fittings (Fig. 2-28)

Scraper blade arrangement Proper scraper cutting blade arrangement for the type of job upon which the scraper is to be used is a normal maintenance requirement for

Fig. 2-24 Steering reversing valve pins demand lubrication attention.

Fig. 2-25 On elevating scraper bowls, elevator stub spindles are key grease points.

scrapers. The scraper cutting blades should be arranged in one of the following patterns (Fig. 2-29) to provide the best scraper performance for the applications described.
    1. Four-inch drop center: This arrangement is recommended in rock-free soils when maximum center-heaped loads are desired.
    2. Straight edge: Use this arrangement for making level cuts and fills.
    3. Maximum blade overhang: Recommended for fast loading and maximum blade wear life in loose, rock-free soils.
    4. Minimum blade overhang: Recommended for rocky soils.

Fig. 2-26 Note contacting surfaces on elevator idler rollers and their need for lubrication.

Fig. 2-27 Be sure to grease endgate stabilizer rollers.

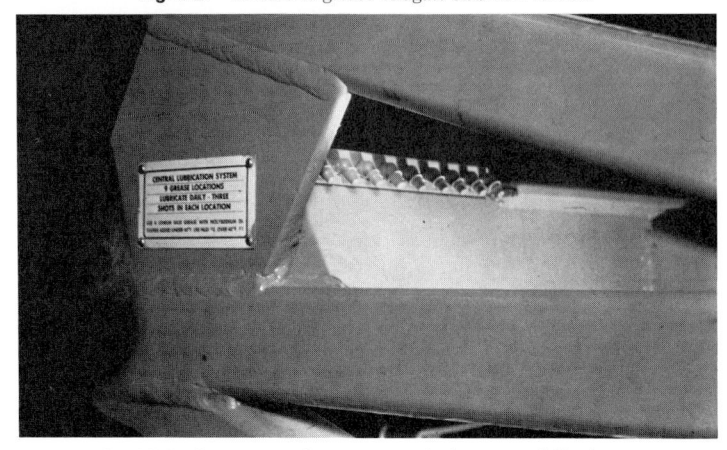

Fig. 2-28 Centralized lube fittings provide an easy method to cover difficult access scraper areas.

Scraper and Scraper Blade Maintenance 7-47

5. One-side cutting pattern: Recommended for beginning cuts on slopes, or for crowning cuts.

Scraper blade installation Most scraper cutting blades are identical, reversible, and, therefore, interchangeable. Thus worn blades can be replaced or rearranged as required for maximum performance or to provide new cutting edges. Blades are normally attached to the blade base with hardened plow bolts that must be tightened securely (Fig. 2-30).

FOUR INCH DROP CENTER

MAXIMUM OVERHANG

STRAIGHT EDGE

MINIMUM OVERHANG

ONE SIDE CUTTING EDGE

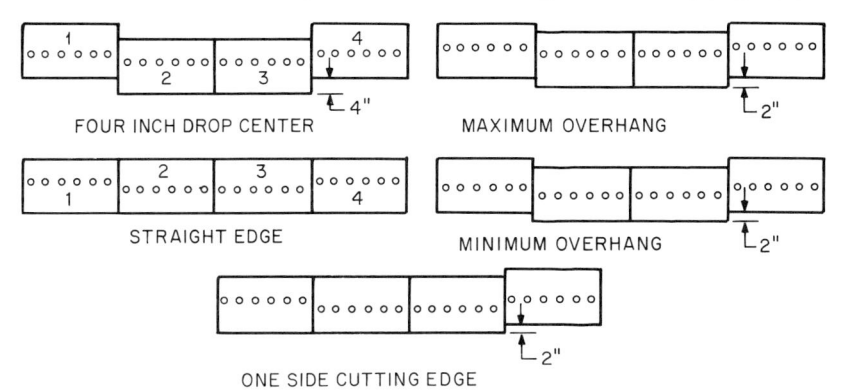

Fig. 2-29 Typical cutting blade arrangement.

Fig. 2-30 Scraper blade installation.

1. Install the blade mounting bolts and tighten to the specified torque.
2. Pound the bolt heads and retighten as required. Alternate tightening and pounding until the bolt torque remains constant under additional pounding.

Bowl side wear blades The blow side wear blades cannot be interchanged between the right and left sides of the bowl but can be reversed on the same side when worn (Fig. 2-31). Tightening side wear blade attaching bolts is the same as the procedure listed above under Scraper Blade Installation.

ELEVATOR MAINTENANCE

Chain slack adjustment Check the elevator chain slack and adjust as required to manual specifications. Measure the slack from the bottom of the elevator frame perpendicularly to the top of the chain midway between the chain idler rollers (Fig. 2-32). The chain is adjusted, if required, by extending or retracting the chain tail rollers. Add or release grease from the tail roller grease ram to increase or decrease the elevator chain slack to specifications as described in the scraper maintenance manual. Be sure to tighten all lock bolts to maintain the proper adjustment.

Chain centering Check the elevator frame for rectangularity and the chain for centering on the frame during operation. The diagonal distance between opposite corners of the chain assembly should be equal to ensure that the chain is properly aligned (Fig. 2-33). Adjust the tail roller hangers as required for proper chain centering during operation. Be sure to maintain the required chain slack. Too little slack will overload the spindle bearings and accelerate bearing wear. Too much slack will permit the chains to run off the chain rollers and damage the entire assembly.

Elevator gearbox Check the elevator gearbox oil level as specified in the scraper manual. The elevator should be stopped, the bowl level, the blade lowered, and the fill plug wiped clean. If the gearbox is to be drained and refilled, drain the lubricant only while at operating temperature. Remove and clean the gearbox breather, blow dry, and reinstall. Clean the magnetic drain plug and install securely in the gearbox drain hole. Refill the gearbox with the recommended lubricant to the fill plug level and replace the plug securely.

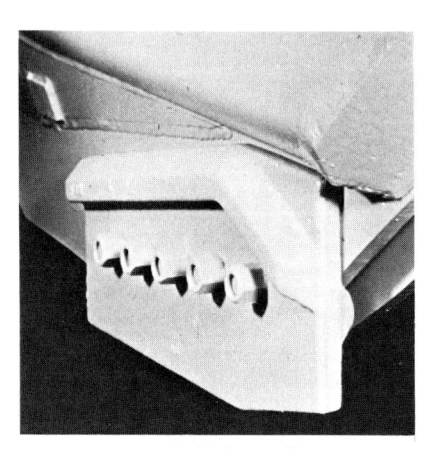

Fig. 2-31 Side blade installation.

Rolling floor Clearance between the strike-off blade and moldboard should be checked and adjusted to manual specifications as required (Fig. 2-34). Remove any rocks or other obstructions holding the rolling floor from its foremost position. If the sliding blocks are worn so that the strike-off blade is not parallel to the rolling floor, install shims between the blocks to provide the required clearance.

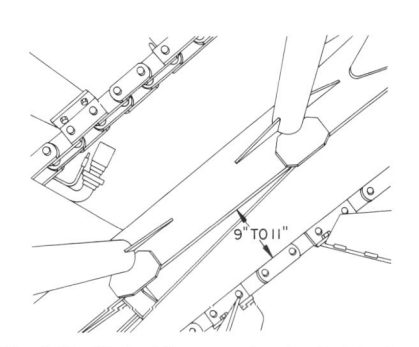

Fig. 2-32 Method for measuring the slack in the elevator chain.

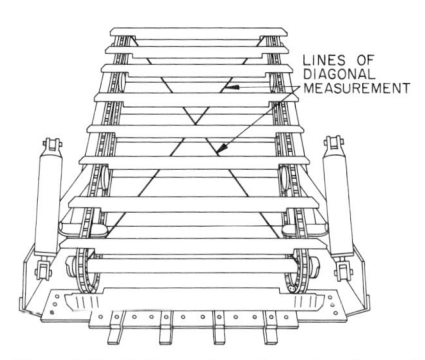

Fig. 2-33 Taking the elevator's diagonal measurements.

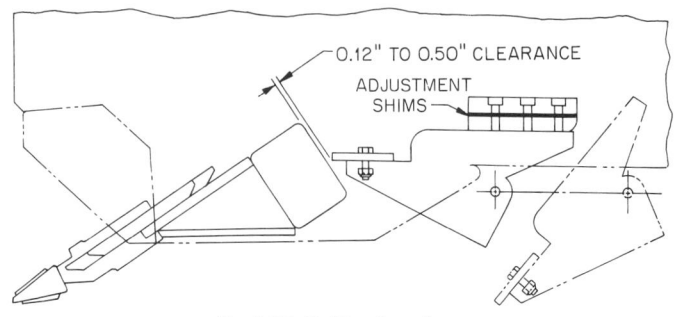

Fig. 2-34 Rolling floor clearance.

Check the rolling floor cylinder mounting clearance and adjust the cylinder hangers as required (Fig. 2-35). Add or remove shims from the rolling floor cylinder hangers to provide the specified rolling floor clearance with the cylinders fully extended, and tighten the hanger mounting bolts securely to maintain the adjustment (Fig. 2-36).

Check the clearance between the inside of the rolling floor roller flanges and the outside of the roller track and adjust as follows, if necessary, for proper operation:

Remove the rolling floor mounting bolts from the rolling floor and separate the hanger and adjusting shims.

Add shims to increase or remove shims to decrease the clearance between the roller flanges and the outside of the roller track.

Replace the roller hanger assembly and tighten the mounting bolts securely.

Endgate The elevating scraper endgate must have equal operating clearance at each end between the frame walls. Check carefully for endgate clearance and center the endgate, if necessary, as follows:

1. With the endgate fully retracted, check the clearance between the endgate and inside frame walls on both sides of the endgate (Fig. 2-37).

2. If clearance is unequal, loosen the endgate stabilizer track hanger bracket mounting bolts and pry the stabilizer track bracket left or right to equalize the clearance between the endgate and frame walls. Tighten the bracket mounting bolts securely (Fig. 2-38).

Required clearance between the endgate and frame crossmember with the endgate fully retracted is established by adding or removing shims between the endgate cylinder mounting bracket and endgate (Fig. 2-39). Loosen the mounting bolts in the bracket slotted holes and remove the bracket bolts securing the adjusting shims and tailgate. Add shims to increase the endgate clearance, or remove shims to decrease the clearance, as required, and tighten all mounting bolts securely (Fig. 2-40).

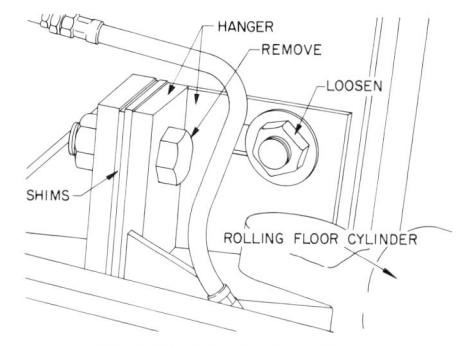

Fig. 2-35 Adjusting the rolling floor.

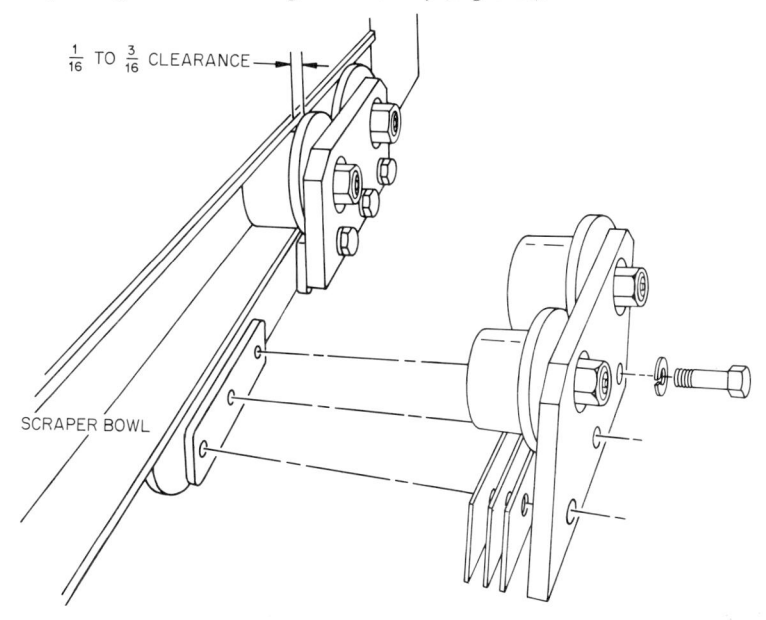

Fig. 2-36 Floor roller adjustment.

7-50 Maintenance Programs for Equipment Entities

   Structural maintenance Good scraper bowl structural maintenance begins with a daily walkaround inspection of the complete bowl. Any visual damage or defects to the bowl structures and working parts should be corrected. Pay particular attention to the apron or elevator, the ejector, tailgate, and rolling floor and to their operating levers and linkages. Obvious interference between these components and frame or chassis members during scraper operation must be corrected.

   Ejector stops Ejector stops are welded inside the bowl to limit ejector travel and prevent overstroking the ejector cylinder. To repair or replace a damaged or worn upper ejector stop (Fig. 2-41), extend the ejector cylinder fully and back off 1 in. Mark the ejector stop location and lower the ejector completely. Then weld the stop in place. If a worn stop is to be rebuilt, lower the ejector and apply the weld, as required, to the proper dimensions for a sound stop installation.

   Lower ejector stops (Fig. 2-42) are used to prevent bottoming the cylinder when the ejector is completely retracted. To repair or replace the lower stops, lower the ejector fully, then raise it 1 in. and mark the ejector location. Raise the ejector completely and block securely with the proper lock pins, etc. Then weld the lower stop blocks in location on the bowl.

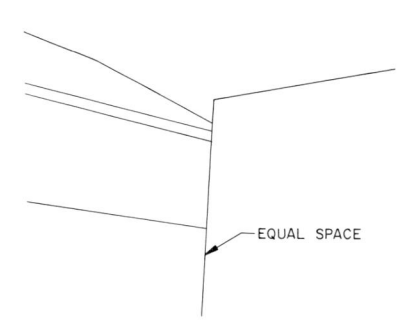

Fig. 2-37 Endgate clearance measurement.

   NOTE: Always preheat the area to be welded to ensure sound welds and prevent underbead cracking. Also, double-ground the welder near the weld area to prevent electrical system damage from stray weld currents.

PATCHING, STRAIGHTENING, AND REINFORCING STRUCTURES

Periodically inspect the bowl, tail, and attached parts for bends and for cracked and broken welds. Any damage found should be repaired as soon as possible to prevent major structural failures. All threaded parts should be inspected for stripping.

   Patching There are two methods to be used when patching a hole in a scraper structural plate.

   1. On an outside surface where no moving parts will come into contact with the patch, trim off the curved edges of the hole and place a patch of the same thickness and type of steel over the hole. This patch should overlap the hole at least 2 in. all around. Tack weld the patch in a few places to hold it in position, then weld a fillet all around the edge of the patch.

   2. When patching a hole where moving parts must pass over the patch, such as on the rolling floor, the hole should be trimmed with an air arc and a patch of the same shape and contour as the panel placed in the hole. Before welding the patch, tack weld a piece of

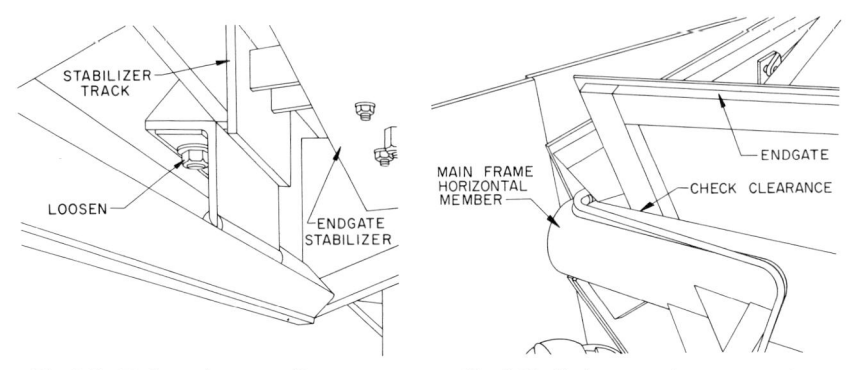

Fig. 2-38 Endgate clearance adjustment.   **Fig. 2-39** Endgate travel measurement.

strip steel across it to provide a grip and to keep the patch in the proper position. Next, weld around the patch enough to hold it firmly in place and remove the steel strap. After removing the strap, complete the weld and grind it down to finish the job. If the hole is completely through a double-paneled wall, each side of the wall can be repaired in the

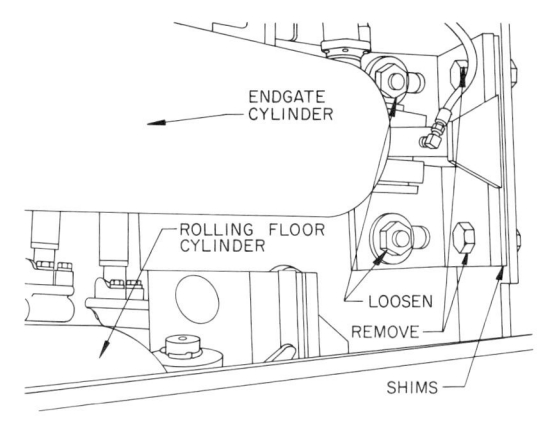

Fig. 2-40 Endgate travel adjustment.

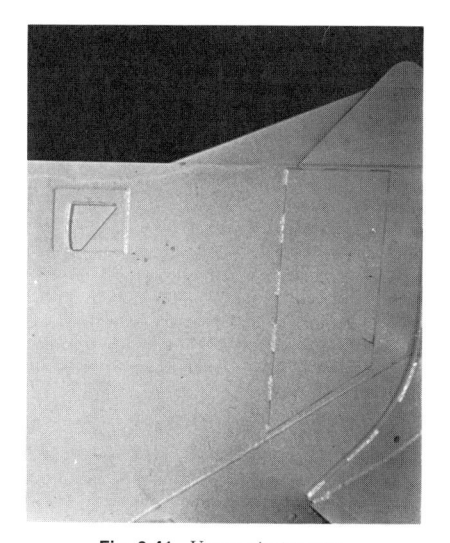

Fig. 2-41 Upper ejector stop.

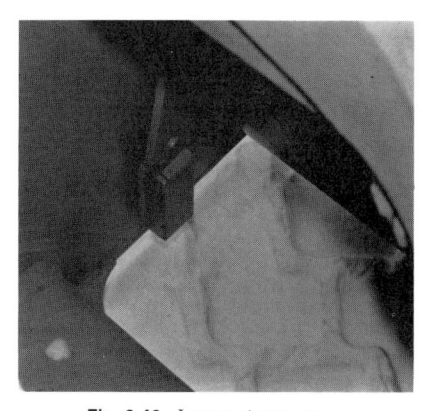

Fig. 2-42 Lower ejector stop.

manner suggested. To obtain the original strength, if the center frame is damaged, the frame member between the panels must also be repaired.

Reinforcing Structural reinforcement can be made with channel, angle, or flat steel stock. Whenever possible, the reinforcement should be extended well beyond the bent, broken, or cracked area. The reinforcement stock thickness should not exceed that of the base stock, and the material should be of the same strength.

Straightening Hydraulic straightening and aligning equipment should be used to straighten bent or twisted parts. However, if heat must be applied, never heat the metal beyond a dull, cherry red color as too much heat will weaken the metal. When it is necessary to heat the metal, apply the heat uniformly over the area to be straightened and protect the heated surface from sudden cooling. Structural parts that cannot be straightened should be replaced.

7-52   Maintenance Programs for Equipment Entities

Welding   Electric arc welding is recommended for all welded chassis repairs. Bent, twisted, and misaligned parts should be straightened and realigned before welding repairs are attempted. Successful weld repairs will depend, in large degree, upon the use of proper welding equipment, materials, and the ability of the welder.

Always fasten the welding machine's ground cable to the piece being welded, if possible. The current from the welding electrode to the ground cable always follows the path of least resistance. If the ground clamp is attached to the tractor when the scraper bowl is being welded, for example, welding current must pass through the hitch to return to the welding machine. Small electric arcs can then be produced across the hitch connecting parts which may cause weld blotches on their wearing surfaces and increase the wear rates of these moving parts.

Always disconnect the battery cables to isolate the batteries from the welding current passing through the scraper structure and to prevent possible battery damage.

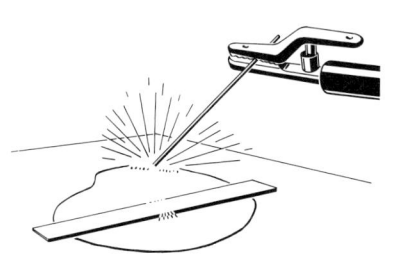

Fig. 2-43 Proper method for applying a patch.

Never alter a roll-over protection system (ROPS) structure or mounting plates in any way. Any alteration made may void the SAE certification for these structures. Personal injury or damage may result if such alterations are made.

Painting   Thoroughly clean all areas to be painted and remove rust. Apply a suitable primer coat and the finish specified.

Chapter **3**

Motor Grader Maintenance

JIM D. SMITH
General Service Manager, Austin-Western Division, Clark Equipment Company, Aurora, Ill.

RICHARD A. MARTIN
Service Training Supervisor, Austin-Western Division, Clark Equipment Company, Aurora, Ill.

INTRODUCTION

A motor grader is one of the most versatile tools in the arsenal of earthmoving machinery. Whether as an all-around utility tool or as a principal piece of equipment on a given job, the motor grader can perform more different kinds of tasks than most other pieces of equipment.

The reason for this machine's multiduty role is that grader owners—contractors, highway road departments, producers, and miners alike—look upon their graders as both a production and a finishing tool. Graders are used for heavy grading, spreading, bluetopping, building roads, preparing land for development, bulldozing, scarifying, and even land cleaning.

Graders are used to fight snow, open new roads, and maintain existing thoroughfares, haul roads, and natural road drainage systems. Each year, more than 5000 new units with sophisticated features are purchased in the United States alone. An average highway maintenance grader performs 3500 hr/year. Construction-applied graders accumulate a little less.

Graders see action at construction materials plants and at coal and other mineral mines. They work on city streets, on rural roads, in mountainous terrain, at beaches, and at parks.

Small wonder, then, why a motor grader is rated one of the most versatile tools in contracting and highway maintenance. Like other pieces of mechanized equipment, a motor grader requires an on-going maintenance effort for it to perform the job it was designed to do. Yet because of the machine's lightness and precision capabilities, a motor grader demands stringent maintenance checks.

A machine performs only as well as its operator. But its ability to perform is directly related to how well the machine has been maintained. The more efficient the maintenance, the better and longer the machine will be able to perform. A motor grader, perhaps more than any other piece of earthmoving equipment, needs to be finely tuned to perform its function efficiently.

A look at the basic design and function of the machine tells why.

Design and function Fundamental design components of a motor grader include a power plant and drive train, a moldboard and blade, a power transfer system (mechanical or hydraulic) that delivers energy to the machine's working parts, and a control system to refine power to do the job.

One main frame supports the engine housing, drive train, operator's station, operating components, and power demand system. The frame runs on four or six wheels.

While tractor shovels and scrapers can be judged by how much material they'll load, haul, and dispose of within a set distance and given amount of time, a motor grader cannot.

A grader's function is not to carry material but to move it. The machine does so only short distances at a time and cycles in repetitive passes. Depending on the underfoot, a motor grader could require various accessories to assist the blading. For example, scarifying attachments help loosen and lift tough underfoot to be moved. If obstructions need be cleared, or if the assignment is snow removal, then a dozer blade or snowplow can be fitted on many a grader's front end. Other functional attachments include snow blowers, power tilt blades, V-type and reversible snowplows, snow wings, winches, and more.

Not only do the main components and power and control systems require maintenance effort, but any attachments or accessories added to the machine will demand their own service and repair efforts.

Depth of owner maintenance The key to any effective service and repair program starts with preventive maintenance (PM). It is the considered belief of most construction equipment manufacturers that maintenance impact within any organization begins at the top and depends on the willing cooperation of management, operators, and service personnel.

The extent to which service and repair of motor graders is undertaken within an organization depends on several factors. First, the ability and facilities of the company readily indicates how much field and shop service can be performed. Self-service and repair determination requires analyses of dealer or distributor support and such an organization's service capabilities.

It may be less expensive for a contractor, highway department, materials producer, and/or miner to perform total service and repair in-shop than it would be, say, to ship components or full machines for dealer service. Some grader owners perform service and repair on every component of a motor grader. Others, depending on their organizational capabilities, perform limited repair. Still other owners rely totally on dealer service contracts to keep machines in operational conditions. Cost studies must be undertaken to make such determination.

Service starts with PM A preventive maintenance program relies heavily on the machine operator. The extent of operator involvement, however, will again depend on the company or department organization, on local union requirements, and machine application.

Operator involvement in an organization's maintenance effort is crucial to effective service, repair, and the efficient working life of the machine. For it is the operator who is the first to know and help identify mechanical disorders in machinery performance.

All PM programs consist of daily, weekly, and monthly checks. In most cases, grader manufacturers prefer checks to be done on an hours-worked schedule based on normal operating conditions. Certain work conditions require stepping up maintenance checks. The time and effort spent on checking for proper running order pays off where it counts: in extended useful machine life and efficient on-the-job performance.

Like any part of a maintenance program, PM must be stressed from upper management down to the operator level. Sophisticated programs feature administrative procedures requiring participation from operator, oiler (if, and where necessary), job superintendent, servicemen, service management, and equipment superintendent. But any program will only be as good as management insists.

Stress cleanliness to operator Knowing a machine and how it works is the first step to planned maintenance and extended useful life. All operators should thoroughly familiarize themselves with the essential points of operating and servicing a grader before engine ignition. Service and operating manuals are included with machines for that purpose. It wouldn't hurt to have experienced operators, even with the equipment superintendent present, review in detail with a new operator the operating and service features peculiar to different makes of graders.

The first step to keep a motor grader running efficiently is to keep the machine clean—both inside and out. For example, operators should never remove inspection cover plugs or breathers without first removing all dirt from around them to prevent contamination of the engine, gear cases, and other internal parts.

Simply inspecting the outside of the machine and wiping off dirt with *clean* rags will help put sound maintenance in action. By frequently cleaning a grader an operator will first remove harmful dirt and then discover possible problem areas that may be starting. Loose connections, nuts, and bolts will be discovered before they develop into major problems or component failures.

Fuel should always be handled carefully to prevent dirt and water from entering the fuel system and causing a failure. Oil, grease, and fuel containers at a job site and in fueling areas should be kept clean and well-covered.

An often overlooked or carelessly checked component on a motor grader is the engine's air cleaner. A dirty air cleaner lets dirt enter the engine, and dirt is the cause of most engine wear. In dusty operating conditions, it may be necessary to clean the filter several times a day. But when servicing a dry-type air cleaner, always be sure to clean from the inside using an air line with less than 100 psi or with sudsy warm water. After washing, the cleaner should be allowed 24 hr to dry.

It's also important that instructions pertaining to the engine's cooling system be followed carefully, and in most cases, manufacturers recommend using permanent-type antifreeze. Also, lubricants specified by manufacturers are recommended according to the requirements of their grader. It's best not to second-guess the people who put the machine together.

When oil levels on a new machine first indicate that lubricant should be added, original oil should be drained and an approved lubricant available locally should be installed. All manufacturers leave the owner responsible for changing lubricants in the field and using lubricants that suit local climatic conditions. So, it's better to play it safe than to mix unknown brands of oil that could contain unknown additives. When flushing housings and gearboxes, mineral spirits or regular flushing oils should be used instead of gasoline, kerosene, or other thin fuels.

All fluid levels should be checked before starting the engine. And before applying a load, the engine should have plenty of time to warm up to the operating temperature specified by the manufacturer.

Visual checks by the operator can save trouble on future work days. Nuts, bolts, hydraulic fittings, and connections should be visually checked at least weekly.

Typical recommended lubrication schedules take shape after 8 operating hours and extend to semiannual (approximately every 1000 hr) to annual and periodic lubrication.

Preventive maintenance extends beyond normal component checks for lubrication, nuts and bolts, and hose connections. Most grader manufacturers describe general service of engines, drive trains, differentials, braking systems, axles, tires, hydraulics, blade, and moldboard care in their operator's manuals. Detailed servicing and repair on these components and hydraulic valving accompany grader manufacturers' manuals, courtesy of original equipment manufacturer (OEM) suppliers. The main parts of a grader are shown in Fig. 3-1.

Maintenance of hydraulic and other power transfer systems, blades, engine, and power train and steering will be discussed in greater detail later in this chapter.

GENERAL PM CHECKS

Engines, steering For maximum engine efficiency and long service life, the engine manufacturer's recommendations as outlined in applicable engine operator's and maintenance manuals should be followed at all times. General cleanliness and daily visual inspections will detect minor adjustments and needed maintenance.

Steering mechanisms for all motor graders should be checked daily before putting the machine into operation. There are three different steering systems available on today's motor graders: an all-wheel-steer (AWS) grader (in both four- and six-wheel models), an articulated frame grader, and a straight frame grader. Mechanical linkages and/or hydraulic connections should be checked daily on all models.

The steering angle of the rear axle of a four-wheeled AWS grader and the front axles of all AWS models should be equal in both directions. The front axle steering angle should measure 25° each way; rear axles should provide 15° each way.

7-56 Maintenance Programs for Equipment Entities

Tire pressure Most graders are shipped from the factory with 60 to 70 psi of air in the tires. Tire pressure should be brought to the manufacturer's recommendations, and the rolling radius of each tire should be checked daily to ensure the machine will be level and perform to specifications. Rolling radius is the distance in inches from the ground to the center of each driving axle at the wheel. In matching new tires, the grader should be on a

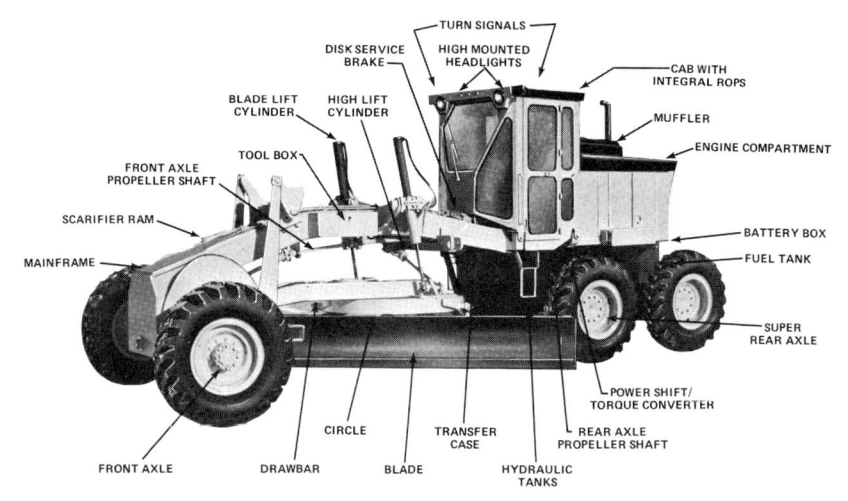

Fig. 3-1 Location of the major components of a motor grader.

hard, level surface and the rolling radius should be brought to within $\frac{1}{4}$ in. of each tire. Overinflation of each tire of 2 to 4 lb, *as required*, will result in matched or uniform rolling radii at all tires, assuming that all tires are of the same name brand, size, type of tread, and number of plies.

Matched rolling radii provide equal tire-to-ground traveling speed on all tires. Higher tire mileage, satisfactory axle gear lubricant temperatures, and maximum work factor from fuel as it's burned will result from properly matching rolling radii.

As tire treads wear, it may be necessary to overinflate certain tires as much as 5 lb to maintain uniform rolling radii within $\frac{1}{8}$ in.

Retreaded tires should be installed in complete sets, and all used tires should be the same size, have the same number of plies, bead seat, and nearly the same type of and height of lugs at the treads.

Even though some graders may have compensating differentials on the axles, it is important that the same rolling radii be maintained.

Inflation pressures are established to fit a combination of three factors: load, job, and speed. With or without overload, overinflation causes cord stress which reduces resistance to blowouts from impacts and increases the danger of rock cuts. On the other hand, underinflated tires are subject to an increased percentage of deflection and excessive flexing. Operating a motor grader with underinflated tires will result in uneven or spotty tread wear, radial sidewall cracks, ply separation, and loose or broken cords inside the tire.

When tires operate in soft soil or sand, such as in a typical cut-and-fill road construction job or in rough grading, inflation requirements are appreciably lower than those required for pavement, packed dirt, or gravel surfaces. The reason is that the tire makes an impression in the soft surface which cradles the tire, preventing extreme deflection. Better traction, lower rolling resistance, less cutting, and impact breaks are indirect but worthwhile benefits derived by lowering inflation under such circumstances.

Do not bleed tires to correct buildup of air pressure. A certain air pressure rise during grader operation is normal because of heat building up in the tire. Hot air expands inside the tire and the tire's shell restricts the expansion. Pressure will stabilize as the temperature balance is reached between internal heating and external cooling.

Tire bleeding will also aggravate rather than cure a high-pressure reading. Reduced pressure will cause increased flexing or bulging, creating more heat and more pressure. Air pressure should be checked only when the tires are considered to be at normal pressures. During a 24-hr operation, as during snow clearance, for example, a correction factor can easily be determined by experiment. By checking as many tires as possible when they are cold and again after at least 2 hr of operation, a factor can be judged to help determine how hot your machine's tires really are. If necessary, reduce speed to cool overpressure tires.

Overloading the tires with ballast or other techniques on road graders is not recommended by most manufacturers.

Blade and circle field service Most new graders feature blade side-shifting and full circling from the cab without requiring the operator to leave his station. In most cases, the moldboard can be fully side-shifted right or left and revolved while the blade is in the ground and while the grader is moving. The moldboard should be raised out of the ground, however, when it is being completely reversed.

Owners should caution new operators to not abuse the moldboard. An operator should stop the grader and back away from immovable objects like stumps and rocks. Operators should also use caution around street manholes and bridge abutments.

When the machine is operated under muddy conditions or during freezing weather, always ensure that a coating of antirust or water-resistant grease is applied to all shafts on the moldboard. And to ensure normal life of the moldboard, have your operators fully side-shift to the right and left sides after each day's work. By doing so, all debris will be removed from both shafts. Water-resistant grease should then be applied by hand over the entire exposed length of both upper and lower moldboard shafts. When the machine is idle for an extended period, coat the shafts and moldboard with an antirust or water-resistant grease to keep them from rusting. *But* in dry or dusty conditions do not grease the lower side-shift shaft.

The circle should be rotated at least halfway around every day so that high-pressure grease may be evenly spread by hand over the inside bore and bottom flange of the circle. The top side of the circle should not be lubricated. See Fig. 3-2.

On all motor graders not equipped with power tilt-type moldboards, adjust blade tilting links by loosening link bolts on either side of the circle. Resetting will generally require the following procedures on hydraulic graders:

1. For general grading, set both tilting links in central position.
2. For cutting hardpan, clay, or caliche, tilt moldboard to suit local conditions by lowering blade lift rams with cutting edge on ground.

Fig. 3-2 When performing preventive maintenance on the moldboard, a grader's circle should be rotated at least halfway around every day so that high-pressure grease may be evenly applied by hand or brush over the inside bore and bottom flange of the circle.

3. For light maintenance or cutting ice, tilt moldboard ahead the full distance by lifting blade rams. Moldboard will then fall forward.

When a machine is equipped with a power tilt moldboard, adjustments can be made from the cab.

HYDRAULICS

Two types of systems Today's motor graders transfer engine power to working functions through three different types of systems: hydraulic, mechanical, and hydraulically assisted mechanical.

Two types of hydraulic systems are common to motor graders: closed- and open-pressure systems. Closed-pressure systems use pressure-compensated piston pumps to regulate the flow of hydraulic fluid. Open-pressure systems, on the other hand, employ gear and vane pumps.

Hydraulic oils are required to perform the dual functions of lubrication and transmission of power. Oil should therefore be selected with care and preferably with the assistance of an oil supplier. Generally, a high-quality oil of proper viscosity (measure of fluidity) meeting the API classification of MS or DS may be used in motor graders. ML type is not recommended.

Check your grader's manual for oil recommendations under normal operating temperatures of the machine's hydraulic system.

Handling oil properly, therefore, is the start to maintaining an efficient power transmission system. Before opening an oil drum, carefully wipe the top of the drum to prevent dirt from falling into the oil. If it does, ensure that the oil is cleaned before using. Most large particles can be removed by straining through a 100-mesh screen. Remove any remaining dirt by letting it settle.

When, for any reason, the hydraulic oil in the grader's system needs to be removed, the service department must drain the entire tank reservoir. Oil should necessarily be replaced after 2000 hr, either by disconnecting hoses and draining each line and ram separately or by flushing the system briefly with fuel oil.

Connections should be checked concurrent with oil changes, and tightened if necessary. Loose connections will permit air to be drawn into the system, causing noisy or erratic operation, or will permit hydraulic oil to leak. Filters and the system's reservoir should also be checked for contamination while the system has been drained. Replace dirty filters and clean the reservoir before adding new oil.

To refill the system, first check all lines, couplings, and plugs to ensure there will be no leaks in the system once it has been filled. Be especially careful of joints that were disconnected for draining.

Before removing the filler cap to add oil to the system, wipe off both the filler plug and the filler nozzle with a clean, lint-free cloth. It's especially important to watch for metallic chips, bits of waste, and other contaminants that may cause damage to the system. The reservoir should be tightly closed after filling.

Begin by filling the tank itself to the full mark on the sight gauge. Proceed then, one function at a time, to fill each line by gradually working each control lever to its extreme positions in a deliberate back-and-forth movement until no further sound of cavitation is heard. Do that under a no-load circumstance, or damage will result. Check oil reservoir level after each mode is refilled.

When all lines have been refilled, add more fluid as may be necessary to again bring the level in the reservoir back to the full mark.

Troubleshooting and PM A closed hydraulic system is totally sealed, except for its breather adjacent to the filler cap. The system consists of a reservoir, filter, filler screen, pump, control valve bank, rams, oil motors, priority and divider valves, and required piping. All grader hydraulic systems require periodic attention to keep them trouble-free.

Early care of trouble signs can produce significant savings in hydraulic maintenance. The first step to system maintenance is knowing how to read warning signs. Leaking oil deposits on the grader or on the ground are the first signs of trouble. Dirty, discolored oil or low oil in the system's reservoir can also be a bad indication.

The main symptoms to watch for are excessive temperatures and noise. These usually indicate a defective pump. Monitor the system's overall performance. If the grader functions slower than normally, less powerfully, or less responsively, then it's time to check the system for flow rate and pressure. The check should be made at engine speed with the blade under load.

An operator can normally hear any unusual sounds, malfunctioning valves, or a noisy pump. Investigate immediately because noise generally indicates cavitation, excessive fluid aeration, or worn internal parts.

Should a pump stop working, the prognosis usually shows that a major internal pump component has failed. The diagnosis is to remove the pump, examine it, and rebuild or replace as necessary.

During daily inspection, do not be concerned with the cause and cure–only the symptoms. An operator should watch for and report any indications which may lead to future operational trouble. Some items that may cause poor performance or eventual failure are listed below and should be checked periodically.

    1. *Reservoir.* Check the fluid level at reasonably close intervals to be sure it is maintained at the full mark. If the level goes down faster than usual, there is a leak somewhere in the system.

    2. *Return line filter.* The return line filter should be checked frequently and replaced as necessary.

    3. *Oil viscosity.* Never use fluid that is too thick. Oil viscosity should be determined by local temperature levels and operating conditions. Fluid may thicken through oxidation or contamination. Oxidation results in gum, sludge, plugged valves, and excessive wear not only on the pump but also on the cylinder and valving.

High-pressure hydraulic systems demand close tolerance on machined parts. Contamination can destroy finishes and machined fits. Thorough precautions against foreign particles entering the system should always be practiced. If cleanliness of the system is at all questionable, drain and flush it.

The system's reservoir should be checked periodically for dirt, metal particles, or other contaminants. Clean fluid is the best insurance for long hydraulic system life. It's best to maintain the in-line filter element in your grader's hydraulic system in the best possible state of repair. Don't be satisfied with just *good* repair.

Hydraulic connections must be kept tight to prevent fluid leaks and contaminants from entering the system. Line leaks may also permit air to enter the system, which will cause aeration and ruin the system.

No manufacturer can set hard-and-fast rules for making oil changes because of the variety of service conditions. Only experience can dictate how long the service period must be for a particular application. Normally, a grader can operate safely without change of fluid for about 1000 hr, providing the system is kept clean and filters are cleaned or replaced on schedule. Rougher-than-normal or extended-hours usage demand more frequent servicing. The harder the use, the more frequent the care.

Failure in a hydraulic system can also be due to inadequate maintenance or a system component malfunction. A thorough understanding of the system and its components is the first step for reliable troubleshooting. In fact, the best aid to troubleshooting is confidence of knowing the system. Every component has a purpose, and the construction and operating characteristics of each part should be understood.

Service personnel should know the capabilities of individual grader hydraulic systems. Each component of a peculiar system has its own maximum rated speed, torque, or pressure. Loading the system beyond specification invites the possibility of failure.

When checking pressures, use a gauge at indicated checkpoints.

The ability to recognize hydraulic trouble indications normally improves with experience. Three general procedures have the greatest effect in maintaining efficiency, performance, and life of the system:

    1. Cleaning or replacing filters and strainers

    2. Maintaining a clean, sufficient quantity of hydraulic fluid in the reservoir of the proper type and viscosity

    3. Keeping oil connections tight, but not to the point of distortion so that air is excluded from the system

Tables 3-1 to 3-5 contain five main categories of troubleshooting effects and maintenance techniques. Each heading is an effect which indicates a malfunction in the system. For example, if a pump is exceptionally noisy, refer to Table 3-1 Excessive Noise. The noisy pump appears in Col. A under the main heading. In Col. A there are four probable causes for a noisy pump. The causes are sequenced according to the likelihood of happening or the ease of checking the problem. Cause 1 is cavitation, and the remedy is "a." If the first cause does not exist, check cause 2 and continue accordingly until the correct cause is identified.

Aeration Aeration and cavitation are two distinct phenomena that can occur in a hydraulic system, causing noisy operation, erosion of metal, and accelerated wear. While their effects are similar, they have different causes and require corrective action. Because

TABLE 3-1 Troubleshooting Based on Excessive Noise

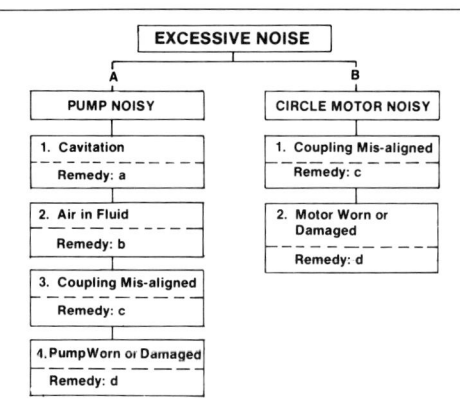

```
                        EXCESSIVE NOISE
                       ┌────────┴────────┐
                       A                 B
                  PUMP NOISY        CIRCLE MOTOR NOISY

                1. Cavitation       1. Coupling Mis-aligned
                  Remedy: a           Remedy: c

                2. Air in Fluid     2. Motor Worn or
                  Remedy: b            Damaged
                                      Remedy: d
                3. Coupling Mis-aligned
                  Remedy: c

                4. Pump Worn or Damaged
                  Remedy: d
```

REMEDIES:

a. Any or all of the following: Replace dirty filters—Clean clogged inlet line—Clean reservoir breather vent—Change system fluid—Change to proper pump drive motor speed—Overhaul or replace supercharge pump

b. Any or all of the following: Tighten leaky inlet connections—Fill reservoir to proper level (with rare exception all return lines should be below fluid level in reservoir)—Bleed air from system —Replace pump shaft seal

c. Align unit and check condition of seals and bearings

d. Overhaul or replace

of their distinctness and commonness of occurrence, they are discussed here as separate maintenance items.

All hydraulic fluid contains some dissolved air, usually about 10 percent by volume. Under increased pressure, hydraulic fluid will absorb much more air. Aeration in a hydraulic circuit is the presence of free air in places where there ought to be only fluid.

Free air in the hydraulic system can make the fluid spongy, causing erratic pressure and reduced effectiveness of the fluid as a lubricant. Free air may accelerate breakdown of the fluid as the bubbles implode with great force. Difficulties with aeration will occur more frequently as flow velocities increase in hydraulic components.

Aeration may be accompanied by foaming in the reservoir. But aeration damage will be most severe in the pump. It can cause erosion marks on the end plates, between the ports. Aeration bubbles may cause gear teeth to bounce as they turn, resulting in rippling. Aeration can sometimes be diagnosed by characteristic sounds—as if the pump is pumping marbles—and by the presence of foam in the reservoir.

Excess air can be taken into the fluid in several ways. The most common is when the fluid level in the reservoir is too low to cover the intake opening. Any time that a

Motor Grader Maintenance 7-61

whirlpool is seen at the intake, air is being pulled into the pump along with the fluid. Restrictions or obstructions in inlet piping can create pressure drops that allow free air to form and be taken into the pump inlet. The most common places for air to be introduced into a hydraulic system or for aeration to occur are:
1. Damaged inlet line: loose or defective fittings or seals at any component.
2. Damaged return line: loose or defective fittings at any component.
3. Damaged or worn cylinder rod, packing, or seals.
4. Cracked junction blocks, tees, or piping.
5. Fluid level too low.
6. Return fluid discharged above fluid level in reservoir.
7. Flow in reservoir is turbulent.
8. Air trapped in system during original filling or when adding fluid or makeup.

Aeration causes lack of lubricity and overheating. The phenomenon will cause hydraulic pump failure and eventual system breakdown. It could cause jerky and uneven

TABLE 3-2 Troubleshooting Based on Excessive Heat

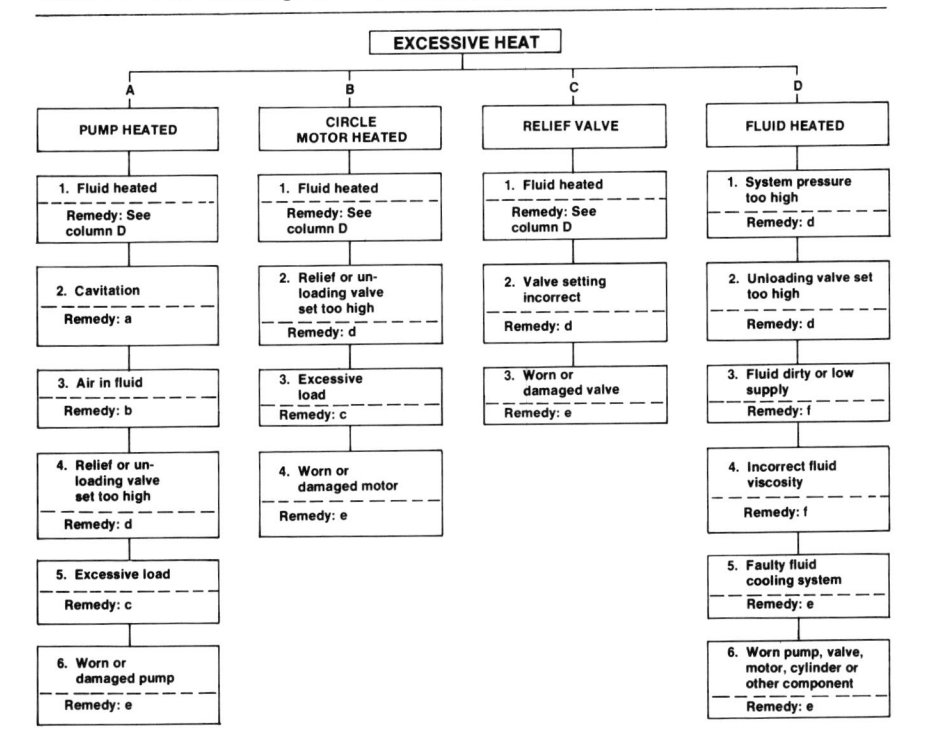

EXCESSIVE HEAT

A: PUMP HEATED	B: CIRCLE MOTOR HEATED	C: RELIEF VALVE	D: FLUID HEATED
1. Fluid heated — Remedy: See column D	1. Fluid heated — Remedy: See column D	1. Fluid heated — Remedy: See column D	1. System pressure too high — Remedy: d
2. Cavitation — Remedy: a	2. Relief or unloading valve set too high — Remedy: d	2. Valve setting incorrect — Remedy: d	2. Unloading valve set too high — Remedy: d
3. Air in fluid — Remedy: b	3. Excessive load — Remedy: c	3. Worn or damaged valve — Remedy: e	3. Fluid dirty or low supply — Remedy: f
4. Relief or unloading valve set too high — Remedy: d	4. Worn or damaged motor — Remedy: e		4. Incorrect fluid viscosity — Remedy: f
5. Excessive load — Remedy: c			5. Faulty fluid cooling system — Remedy: e
6. Worn or damaged pump — Remedy: e			6. Worn pump, valve, motor, cylinder or other component — Remedy: e

REMEDIES:

a. Any or all of the following: Replace dirty filters—Clean clogged inlet line—Clean reservoir breather vent—Change system fluid—Change to proper pump drive motor speed—Overhaul or replace supercharge pump

b. Any or all of the following: Tighten leaky inlet connections—Fill reservoir to proper level (with rare exception all return lines should be below fluid level in reservoir)—Bleed air from system—Replace pump shaft seal

c. Align unit and check condition of seals and bearings—Locate and correct mechanical binding—Check for work load in excess of circuit design

d. Install pressure gauge and replace if necessary to correct pressure

e. Overhaul or replace

f. Change filters and also system fluid if of improper viscosity—Fill reservoir to proper level

7-62 Maintenance Programs for Equipment Entities

TABLE 3-3 Troubleshooting Based on Incorrect Flow

```
                        INCORRECT FLOW
                              |
        ┌─────────────────────┼─────────────────────┐
        A                     B                     C
    NO FLOW              LOW FLOW             EXCESSIVE FLOW
```

A. NO FLOW
1. Pump not receiving fluid — Remedy: a
2. Pump drive motor not operating — Remedy: e
3. Pump to drive coupling sheared — Remedy: c
4. Pump drive motor turning in wrong direction — Remedy: g
5. Directional control set in wrong position — Remedy: f
6. Entire flow passing over relief valve — Remedy: d
7. Damaged pump — Remedy: c

B. LOW FLOW
1. Flow control set too low — Remedy: d
2. Relief or unloading valve set too low — Remedy: d
3. External leak in system — Remedy: b
4. Worn pump valve, motor, cylinder, or other component — Remedy: e

C. EXCESSIVE FLOW
1. Flow control set too high — Remedy: d

REMEDIES:

a. Any or all of the following: Replace dirty filters—Clean clogged inlet line—Clean reservoir breather vent—Fill reservoir to proper level—Overhaul or replace pump
b. Tighten leaky connections—Bleed air from system
c. Check for damaged pump or pump drive—Replace and align coupling gear
d. Replace
e. Overhaul or replace
f. Check position of manually operated controls—Check electrical circuit on solenoid operated controls
g. Reverse rotation

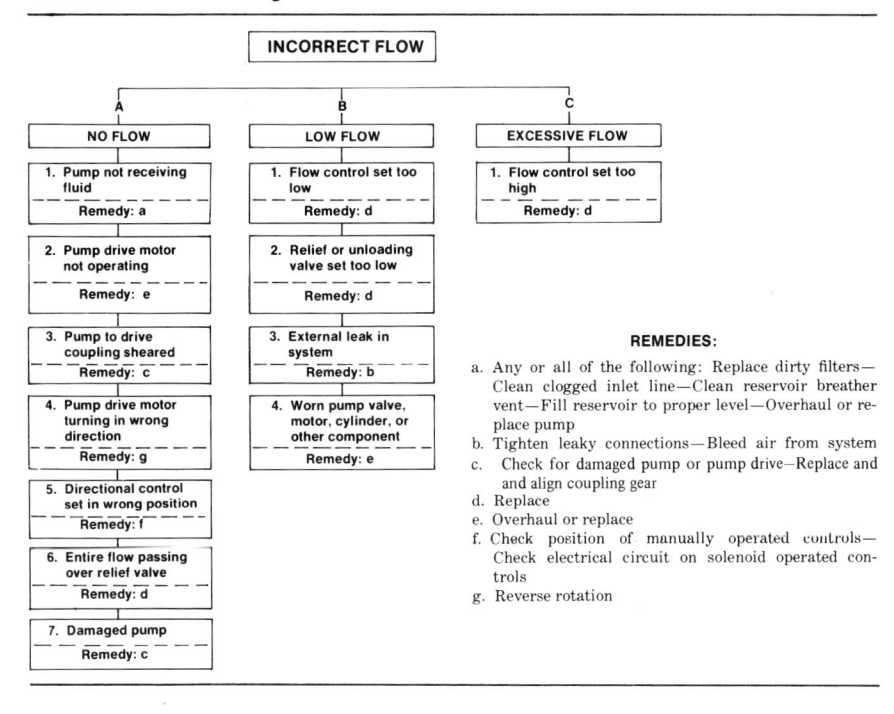

TABLE 3-4 Troubleshooting Based on Incorrect Pressure

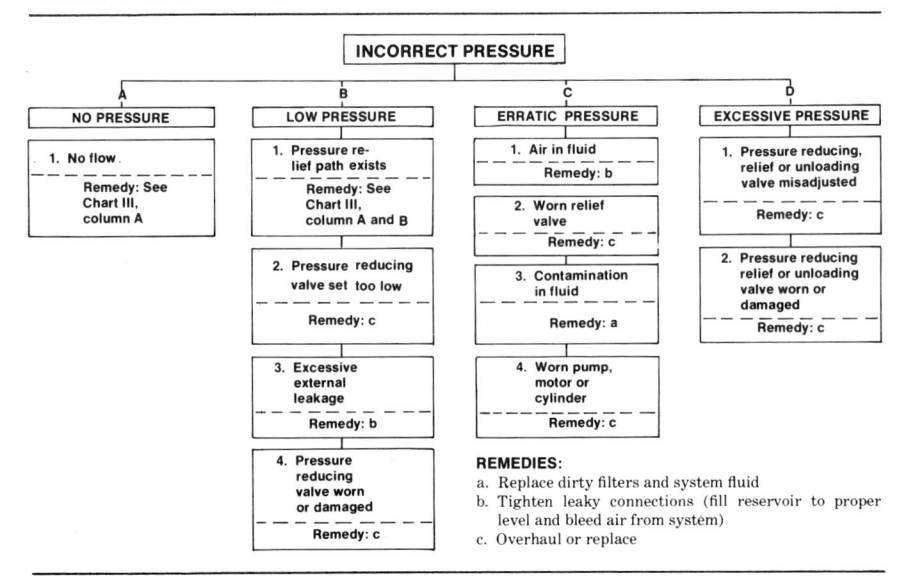

```
                        INCORRECT PRESSURE
                              |
    ┌──────────────┬──────────┴──────────┬──────────────┐
    A              B                     C              D
NO PRESSURE   LOW PRESSURE         ERRATIC PRESSURE  EXCESSIVE PRESSURE
```

A. NO PRESSURE
1. No flow. — Remedy: See Chart III, column A

B. LOW PRESSURE
1. Pressure relief path exists — Remedy: See Chart III, column A and B
2. Pressure reducing valve set too low — Remedy: c
3. Excessive external leakage — Remedy: b
4. Pressure reducing valve worn or damaged — Remedy: c

C. ERRATIC PRESSURE
1. Air in fluid — Remedy: b
2. Worn relief valve — Remedy: c
3. Contamination in fluid — Remedy: a
4. Worn pump, motor or cylinder — Remedy: c

D. EXCESSIVE PRESSURE
1. Pressure reducing, relief or unloading valve misadjusted — Remedy: c
2. Pressure reducing relief or unloading valve worn or damaged — Remedy: c

REMEDIES:

a. Replace dirty filters and system fluid
b. Tighten leaky connections (fill reservoir to proper level and bleed air from system)
c. Overhaul or replace

movement in pumps and motors which, in combination with the causes just mentioned, will cause eventual failure.

Loss of lubricity in a hydraulic component will eventually result in seizure and subsequent pump failure. Overheating is caused by a breakdown of hydraulic fluid as a result of oxidation. Oxidation of fluid leads to sludging and varnish formation. Operating a

TABLE 3-5 Troubleshooting Based on Faulty Operation

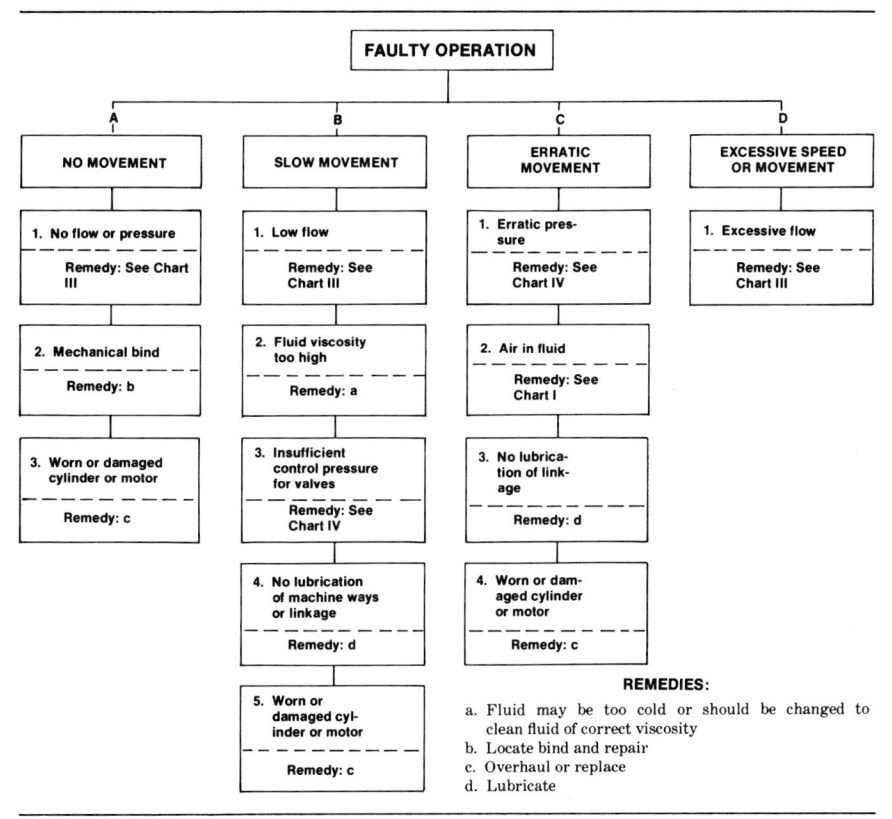

A NO MOVEMENT	B SLOW MOVEMENT	C ERRATIC MOVEMENT	D EXCESSIVE SPEED OR MOVEMENT
1. No flow or pressure Remedy: See Chart III	1. Low flow Remedy: See Chart III	1. Erratic pressure Remedy: See Chart IV	1. Excessive flow Remedy: See Chart III
2. Mechanical bind Remedy: b	2. Fluid viscosity too high Remedy: a	2. Air in fluid Remedy: See Chart I	
3. Worn or damaged cylinder or motor Remedy: c	3. Insufficient control pressure for valves Remedy: See Chart IV	3. No lubrication of linkage Remedy: d	
	4. No lubrication of machine ways or linkage Remedy: d	4. Worn or damaged cylinder or motor Remedy: c	
	5. Worn or damaged cylinder or motor Remedy: c		

All items share headings FAULTY OPERATION.

REMEDIES:
a. Fluid may be too cold or should be changed to clean fluid of correct viscosity
b. Locate bind and repair
c. Overhaul or replace
d. Lubricate

system with aeration can oxidize the whole charge of the fluid, and eventually sludge or varnish will cause the motor or pump to overheat excessively and cause failure.

Regular inspection and PM are the best ways to prevent air from being introduced into a system. Obviously, the cause should be found and corrected when present to prevent premature failure of the pump or breakdown of the lubricating ability of the fluid.

Return fluid entering the reservoir will create aeration if it is discharged above the main body of fluid in the tank. To prevent this condition, maintain sufficient fluid in the tank to keep the return line submerged. The pump intake line should always be below the surface for the same reason.

After the system has been completely drained and flushed, a tendency for the fluid to aerate may exist until all air is purged from all lines and components. To correct this condition, a thorough purging of the entire system should be performed.

Cavitation Cavitation is a vacuum in the fluid, and it occurs when components do not completely fill. It can also occur in motors or cylinders where the load overruns the delivery from the pump.

The characteristic sound of cavitation is a high-pitched scream which increases with the degree of cavitation and with increased operating pressure.

Pump cavitation may be caused by restricted inlet line, by a clogged inlet filter, by fluid that is too high in viscosity, or by excessive length in the inlet line. Also, if the pump is too high above the fluid level in the reservoir, the lift may be too much to allow the pump to fill. Cavitation from fluid that is too thick sometimes is avoided by operating at reduced engine speed until the fluid is warmed up and becomes less viscous.

Typical effects of cavitation are eroded end plates characterized by pitting between inlet and outlet ports, or rippled signs with gear teeth ends worn flat on the end and pitting around the inlet ports. Effects are much the same as aeration.

Whenever cavitation is detected, it should be promptly corrected or the life of the pump will be shortened. When it is suspected, check hydraulic components and lines on the entire machine to determine if corrective action is required.

Hydraulic pumps service As previously mentioned, two types of hydraulic systems are common in today's motor graders: a closed-center hydraulic system and an open-center hydraulic system. A closed-center system will consist of pressure-compensated piston pumps to deliver hydraulic fluid to working parts. In such a system, the hydraulic fluid flows to the valve bank and stops, thus maintaining the amount of pressure required to operate the system's functions on demand. But because no valve is 100 percent efficient, a small amount of leakage oil—no more than 0.52 qt/min—will flow through the valve bank to the circuits' return line.

Then, when a valve is actuated in a closed-center system, the waiting fluid, with pressure built up to the system's designed psi, will flow through the valve bank to a directed function. As it does, the decrease of pressure in the line causes the compensator on the system's pump to activate the pump's swash plate (piston pivot block) in the unit to begin charging oil to the system.

In an open-center system, on the other hand, the hydraulic fluid flows at all times at the same rate and is diverted only when the pumps are activated by the valving. In an open system, gear- and vane-type pumps are used to supply power on demand to the grader's functions.

A grader's main hydraulic system could encompass all standard functions on the machine except those involved with the control of blade functions or those specifically added for certain operational accessories. The main system would include pumps, valve banks, filters, and tanks plus control valves, rams, and other accessories used for steering and braking. It would not, however, include the filter and combination valve used in a torque converter and transmission units as they, along with the torque-converter charging pump, constitute an independent, closed system within themselves.

Inspection and repair of most hydraulic *piston pumps* are similar, but recommendations from the manufacturer should be followed. General guidelines can be followed, however.

After disassembly and before inspection, all parts should be washed in a clean mineral spirit solution and air-dried. Inspect the flat surface of the back plate; the finish on the pump's piston block side should be smooth and free of grooves or metal buildup. The back plate should be replaced if it shows any signs of wear. The piston guide should be tight in the housing. Refer to Fig. 3-3.

Next, inspect the piston block. The surface that contacts the back plate should be smooth and free of grooves and metal buildup.

The pump's pistons should move freely in the piston block bore. If pistons are sticky in the bore, examine the bore for scoring or contamination.

Examine the outside diameter of the pistons for finish condition. They should not show wear or deep scratches. The shoes should fit snuggly on the ball end of the pistons. The flat surface of the shoes should be smooth, flat, and show no signs of metal flaking or buildup. Do not lap piston shoes. See Fig. 3-4.

Examine the spider. It should be flat and have no cracks or signs of wear in the pivot area. The pivot should also be smooth and show no signs of wear.

Inspect the cam plate for condition of the finish of the polished shoe surface. It should show no signs of scoring or flaking.

Inspect the shaft for fretting in the bearing and spline areas. Then, inspect thrust bearing and washers for wear.

Inspect the needle bearings in the housing assembly. If they are free of excessive play and remain in the bearing cage, there is no need to replace the bearing. Inspect the compensator springs for breakage or weakness and the spools for scoring. "O" rings, snap rings, gaskets, and shaft seal should be replaced, prior to reassembly.

To service, inspect, and repair a main hydraulic *gear pump*, the following general guidelines should be followed. All parts should be inspected for wear and replaced if necessary. If wear is the result of abnormal conditions such as contamination or cavitation, the situation should be corrected before returning the pump to service. Refer to Fig. 3-4 for parts of a main hydraulic gear pump.

A hydraulic motor used in conjunction with open-circuit systems consists of as few as three moving parts that are lubricated by the hydraulic fluid which drives the motor. The

(A)

(B)

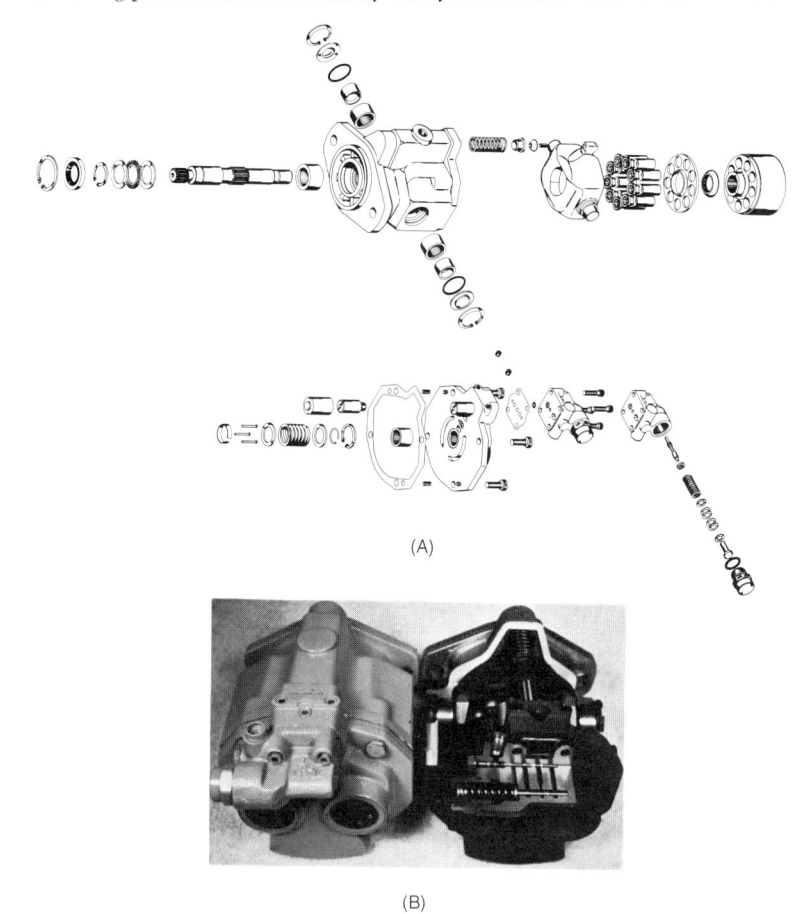

Fig. 3-3 Hydraulic piston pump shown in (A) exploded and (B) assembled and cutaway views.

parts consist of matched gerolor sets, spline drive, and the output drive. Motors are instantly reversible by merely reversing the flow of hydraulic fluid.

To remove a hydraulic motor, first disconnect all hydraulic lines at the motor fittings. Ends of lines and motor ports should be plugged to prevent any foreign matter from entering the hydraulic system. Next, remove nuts and lock washers from the bolts securing the motor to the gear housing adapter. Work the motor outward until the splined end of the output shaft clears the splined coupling on the worm shaft, and remove the motor. Thoroughly clean the exterior of the motor before starting disassembly. The work area where the motor is disassembled should also be thoroughly clean.

The matched gerolor sets can be removed and replaced without completely disassembling the motor. The difference between motors, if any, would be in the thickness of the gerolor sets. During reassembly make certain that the gerolor set being installed is the same thickness as the one which was removed.

Disassembly of the actual gerolor is not recommended. If this unit is scored or worn, the entire gerolor assembly must be replaced. Light nicks on the edge or corner of one of the parts may be removed with an india stone. But extreme care should be taken when removing any nick and when cleaning so as not to lodge any foreign objects between moving surfaces of the gerolor assembly.

Remove the splined drive from housing. Inspect the splines on the drive shaft for nicks,

Fig. 3-4 When making a piston pump inspection, make sure that shoes fit snugly and that all surfaces are smooth.

cracks, and burrs. Small nicks and burrs can be removed with No. 600 grit abrasive paper or with india stone. If cracks or excessive wear are found, the drive must be replaced.

Hydraulic valve banks All hydraulic graders are equipped with a set of main valve banks to control fluid pressure to the machine's working functions. Arrangement of the valve bank will vary among manufacturers. Some models have one bank with eight valves; others have 2 four-bank valves on each side of the cab. Valves direct power to side shift, scarifier, right- and left-hand blade lifts, blade power tilt, high lift, circle rotation, and, for AWS machines, to rear steer.

Some blade lift circuits also incorporate an integral float position, engaged by pushing the control lever forward past the DOWN position, which allows the blade to follow the contour of the road or land surface for such operations as snow removal.

To service after disassembly, inspect cleaned parts and identify and remove nicks and burrs from all parts with fine emery cloth. Each spool should be inspected for damage and freedom of movement in the spool bore. If the spool or spool bore is damaged, the complete working section assembly will have to be replaced.

The system relief valve as shown in Fig. 3-5 is not classified by most grader manufacturers as a repairable item. If a thorough cleaning in solvent does not correct relief valve difficulty, the part should be replaced.

Inspect condition of the flat surface of seats. The flat surface should be in perfect condition without scratches, burrs, or surface defect of any type on the sealing surface because control of work port leakage depends on the condition of these parts.

Fig. 3-5 In the case of the hydraulic system relief valve, if thorough cleaning in solvent does not correct the difficulty, the part must be replaced.

Inspect detent pawls in the detent cap and screw for wear. It is not necessary to inspect "O" rings and backup washers, as these parts should be replaced as new items. All parts should be thoroughly cleaned and dried, and metal parts should be oiled prior to reassembly.

Hydraulic rams Servicing procedures for the hydraulic cylinders used on most graders are basically identical with a few exceptions. Using the proper-type spanner wrench, the cylinder gland should be unscrewed from the housing. Pull piston and rod assembly from the cylinder and remove the locknut from the rod. Remove the piston from the rod and remove piston ring bearing, ring seal, and "O" ring from the piston. Remove the gland from the rod, and remove the backup ring, "O" ring, rod seal, and wiper from the gland. Refer to Fig. 3-6 as an example.

Once disassembled, inspect the finish in the housing bore for scratches and scores. Imperfections will reduce the efficiency of the cylinder and measurably shorten the life of

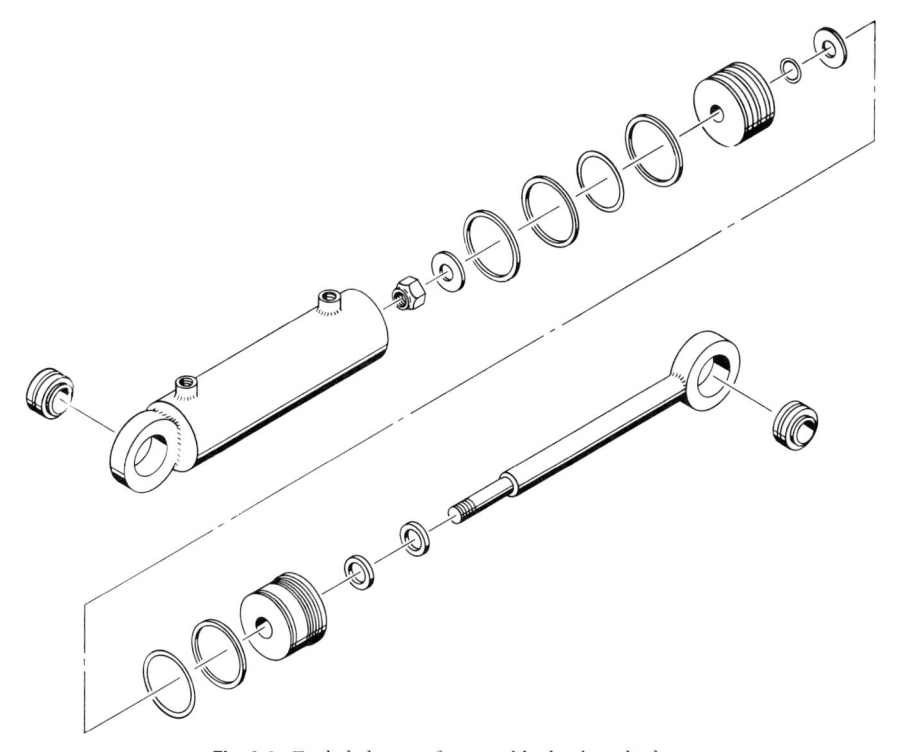

Fig. 3-6 Exploded view of a typical hydraulic cylinder.

the piston ring bearing and seal. Look for fretting or other signs of movement between the piston and the shoulder of the rod. Movement would indicate improper torque on the locknut and can cause internal leakage. Check for uneven wear patterns on the housing bore, indicating nonconcentric bore wear which would result in internal leakage.

Scratches and scores on the rod will ruin the rod seal and wiper, initially allowing external leakage of hydraulic fluid, attended by contamination of the entire hydraulic system. Chronic loss of rod seal and wiper could indicate a bent rod. The rod is best checked between V blocks with a dial indicator but may be checked when assembled in the housing. At no point in the rod and piston assembly travel should more than 50 lb·ft torque be required to rotate the rod 360°.

All nonmetallic wearing parts such as "O" rings should be replaced during reassembly. All parts should be thoroughly cleaned with a high-detergent soap or hydraulic oil, not with solvent-type cleaners.

All hydraulic hoses, piping, fittings, and connections in main and ancillary systems should be checked daily for leaks and tightness. Be sure that there is no frayed metal on steel tubing and fittings and that rubber hoses are not rubbing against each other or any part of the grader.

Hydraulic service-brake service Some hydraulic-powered graders are equipped with a hydraulic brake system. The design of one such system on Clark graders, as an example, includes two identical brake application valves in the system. Each valve features an inlet port, two application ports, and a return port. When the service brake is applied, when force is applied against the push rod, its movement compresses the metering spring which, in turn, forces the metering piston down. Piston movement then opens the inlet valve by unseating the metering valve, permitting fluid from brake line accumulator to pass through the inlet port and out through the application ports.

As the fluid pressure on the metering piston balances the applied force exerted by the metering spring, the valve automatically returns to a holding position and the spring-loaded inlet valve closes.

When applied force from the brake pedal is either reduced or eliminated, the hydraulic pressure in the application section of the valve will force the metering piston upward, away from the exhaust valve. This allows the fluid to pass through the return port to the pump until the force on the metering piston balances the force from the metering spring. Throughout the operating cycle, a pressure-balancing piston within the metering valve provides a smooth hydraulic balance.

To service the system's master cylinder, disassemble and clean all parts according to the manufacturer's service manual. Take special care to see that no residue is left in ring and cup seating grooves. Check interior bores of the valve body and sliding surfaces of the pistons for nicks and scratches. Buff with a crocus cloth to remove minor marks. More severe damage would call for replacement. Check spring and actuating spring and replace all "O" rings and U cups, regardless of apparent good condition. Coat new "O" rings with a thin layer of petroleum jelly before installation. Replace boot and reassemble according to manufacturer's directions.

A shuttle valve is normally placed where the brake lines from both valves join to form a single line to the service brake unit. It is used with the Clark system so that only the left-hand brake valve will activate the clutch disconnect unit. Except for new "O" rings, replacement of this unit is recommended over repair.

STEERING CIRCUIT SERVICE

Three types of steering systems are incorporated in today's motor graders: hydraulic-over-mechanical front wheel steer on a straight frame grader, hydraulically controlled articulated frame, and all-wheel-steer (AWS) hydraulic.

For a mechanically steered grader, adjustment of worm gear drives and linkages from the steering column are the servicing requirements. Hydraulic-over-mechanical graders call for maintenance and adjustment of worm drive gearing and U-joints connecting the linkage from the wheels to the steering column.

Articulated frame graders require maintenance of steering rams and the frame's hinge point, plus checks on flexible hoses.

An AWS power grader incorporates an open hydraulic system on its front steering with a pump located on the rear of the engine block. That pump is gear-driven off the engine camshaft. It is a single-chamber, one-direction gear pump which pressurizes and directs the flow of hydraulic oil to the grader's Orbitrol front steer unit exclusively. Oil gallonage pumped in excess of that required for the steer system is diverted back to the tank through a priority flow divider which is an integral part of the pump.

The flow divider is built into the back plate of the pump assembly and is designed to divide the flow of oil from the pump into two separate outlets, a priority port and a secondary port. The priority port is the fixed flow port and is used to operate the power steering system.

The secondary port is used to drain any excess gpm than the power system requires. The priority port is set at 6 gpm. *Total* output of the pump should never be more than 17.6 gpm.

The working parts of the flow divider consist of a spool, spool spring, and dampening disk. The spool is drilled inside to provide an orifice, which regulates the 6 gpm flow rate to the priority port. The oil from the pump directed through the orifice creates a pressure drop as it passes through. The pressure differential increases as the pump output is increased and causes the spool to compress the spring.

As that happens, the holes to the priority port begin closing, and flow to the secondary port begins. The dampening disk between the spool and spool spring acts as a shock absorber to dampen the movement of the spool.

A relief valve in the back plate controls the pressure in the priority circuit. It has nothing to do with oil pressure of the secondary port. No attempt should be made to adjust the flow divider or replace internal parts. It is recommended that all troubleshooting techniques be used before any attempt to service the internal parts of the pump be made.

To service the pump, first clean all disassembled parts in a clean mineral spirit solution and dry by air or lint-free cloth.

On mechanically steered graders, inspect drive gear shaft for broken keyway. Inspect both the drive gear and idler gear shafts at bearing points and seal areas for rough surfaces and excessive wear. If the shafts measure less than 0.6850 in. in bearing area, the gear assembly should be replaced. Inspect gear face for scoring and excessive wear. If the gear width is below specifications, replace the assembly. Assure that snap rings are in grooves on either side of the drive and idler gears. If the gear teeth edges are sharp, break the edges with emery cloth.

Oil grooves in bearings of both front and back plates should be in line with dowel pin holes and 180° apart. If the inside diameter of bearings in the front plate exceeds 0.691 in., the front and back plates should be removed. Bearings in the front plate should be flush with islands in the groove pattern. Check for scoring on the face of back plate, and if wear exceeds 0.0015 in., replace it.

Check inside gear pockets for excessive scoring or wear. Body should be replaced if inside diameter of gear pocket exceeds 1.719 in.

If a shop test stand is available, the following procedures are recommended for testing rebuilt pumps.

First, mount the pump on the stand, making sure that the proper level of clean oil is available in the reservoir. Check suction line for leaks and obstructions. Next, start the pump and run for 3 min at zero pressure. Intermittently load the pump to 500 psi for 3 min. Step up loading to 1000, then 2000 psi at 3-min intervals. Remove the pump from the test stand and check for freeness of the drive shaft. Check also for leaks.

If a shop stand is not available, mount the pump on the grader and run pump at one-half engine speed and zero pressure. By operating the steer valve, build pressure intermittently for 3 min. Increase engine speed to full throttle and build pressure for 3 min. Then, idle engine and inspect for leaks.

Power steering maintenance One of the most useful features on modern motor graders is the power steering system. The system could consist of essentially a gerolor motor which is substituted for a spool valve used with tiller bar steer or a complicated linkage frequently found with mechanical steer graders. An example of such a system is the Orbitrol power steering used on Clark power graders.

The effect of power steering is that the operator may steer the grader like a car or truck, instead of controlling direction by slapping as must be done with tiller bar operation, or by brute force as may often be required with mechanical steer.

The Orbitrol steer unit receives hydraulic fluid directly from a steer pump which contains its own valve. Refer to Fig. 3-7.

Likewise the Orbitrol unit has its own direct return to the main hydraulic drain line. Piping connections to and from the unit to the front steering ram itself are direct along the grader frame.

A normal periodic functional check of the entire grader's power steering system will generally be adequate to ensure satisfactory service. The oil level of the reservoir that supplies the system is most important. If the oil level drops appreciably over short periods of use, it is wise to search for a leak in the system.

A black accumulation of dirt at a fitting can indicate a leakage point. Clean the area completely with a solvent-wetted cloth, steam clean, or otherwise clean off any debris from the immediate area and any dirt accumulation above the area so that contamination will not enter the system while the connection is open. Then be extremely careful to apply compound sparingly to the male fitting only. Do not let any compound enter an area where it may be washed into the oil stream.

To continue the functional check of the system, turn the steering wheel through the full

7-70 Maintenance Programs for Equipment Entities

travel with the vehicle power on. Do this at engine idle and full throttle, while the machine is standing still with the steered wheels on dry concrete and while the machine is rolling slowly.

Any speed irregularities and sticky sensations may indicate dirt in the fluid. If under any of these conditions the steering wheel continues to rotate when started and released, a condition known as *motoring* exists. This may also indicate dirty fluid in the system.

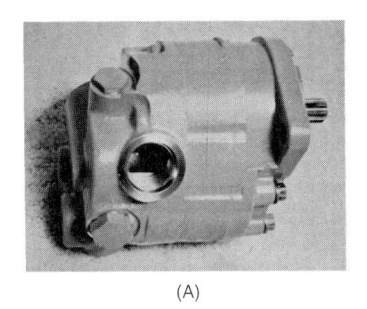

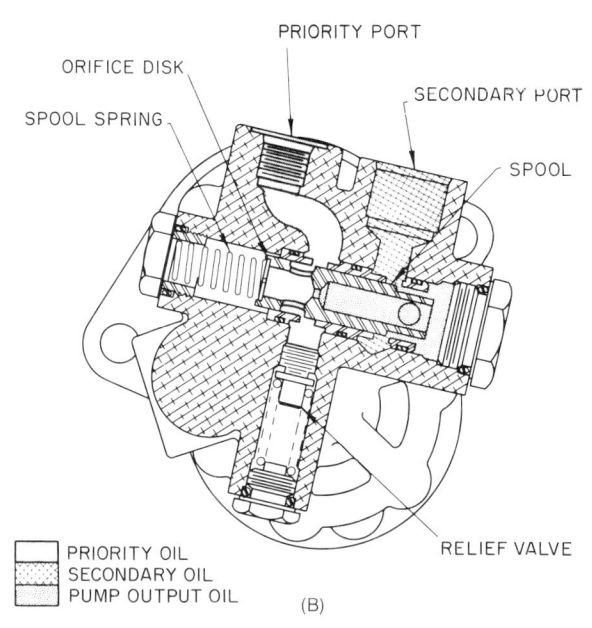

Fig. 3-7 (A) Orbitrol power steering pump and (B) cutaway of its priority flow divider.

If a dirty fluid is suspected, clean or replace the filter element in the system. This is located in the return line. Drain and replace as much of the oil as possible; crank the pump over by hand to exhaust oil from it and swing the cylinder through a full travel. *But do not forcibly rotate the Orbitrol steering wheel if a dirty fluid is suspected.*

Refill the system with clean oil, run the system briefly, recheck, and refill as necessary to obtain proper fluid level. Operate the system for a short time to determine whether a correction has resulted. It is sometimes less costly to rinse and reclean the system twice than to completely tear down and reassemble a unit, and the clean fluid will definitely protect all the components of the system.

In the functional check, determine also that the actuating cylinder achieves full travel without hesitation. If the cylinder seems to pause in its travel while it should be moving smoothly, this may indicate that it contains trapped air. In filling and refilling a system, it is sometimes necessary to lift the vehicle weight off the steered axle or to remove the cylinder and hold it in a position such that the ports are uppermost so that air will be bled back to the system reservoir and effectively exhausted from the system at the reservoir vent.

Check the left and right degrees of travel and check for toe-in and toe-out also. Proper wheel alignment is every bit as important on a power-steered vehicle as on any other to ensure satisfactory tire life and geometrically true steering.

Inspect to ensure that the system has adequate power. The grader will steer completely while standing still on a smooth hard surface. Any evidence of hard steering can indicate either reduced oil flow to the control or reduced system relief pressure. Adequate oil flow under all conditions can best be checked by timing the full travel of the cylinder with the steered axle unloaded and loaded. A great difference at low engine and a slight difference at high engine speed may indicate a defective pump drive or priority valve.

Adequate oil pressure can only be determined by connecting a pressure gauge (2000 psi full scale recommended) at the pump outlet port or at the in port of the Orbitrol. With the engine running at a medium speed, turn the steering wheel to one end of the travel and hold the cylinder at the travel limit briefly, just long enough to read the pressure gauge. Never hold a system at a relief pressure for more than a few seconds at a time. Longer operation at relief pressure can overheat most systems quite rapidly. The pressure relief valve is a protection for all the various parts of the steering system. There is no pressure relief in the Orbitrol. The power steering pressure relief valve is located in the steer pump.

If the system is reported to operate extremely hot, connect a pressure gauge as above and operate the engine at near full throttle. Rotate the steering wheel slowly in each direction and bring the wheel to the position that shows the lowest pressure reading. This places the control section of the unit in neutral.

Now, turn the steering wheel to a limit stop and hold it there for 1 or 2 sec. Release the steering wheel gently and watch the gauge. If the pressure does not drop to very nearly the same neutral pressure as measured when placing the control in neutral deliberately, a binding control shaft or dirt between the spool and sleeve of the control valve can be the cause or difficulty.

If the recentering characteristic as measured above is erratic and if the control feels slightly sticky through most of the travel, apply the pressure gauge in the out line of the Orbitrol. This return-line pressure should be below 30 psi during all periods of normal operation. Check this downstream line to ensure that no fittings are obstructed.

If you need or wish to accomplish repairs within the Orbitrol unit itself, use the procedures outlined in applicable service manuals for dismantling, inspection, and reassembly.

Steering rams adjustment To inspect and repair steering rams on all hydraulically powered steering systems, check all sliding surfaces—piston, cylinder, inside walls, piston rod—for nicks, scratches, or other damage. Minor nicks or scratches may be smoothed off with a crocus cloth. Larger marks would suggest replacement of the defective part.

Wash all parts except seals and "O" rings in a clean mineral spirit solution and air-dry. "O" rings, poly cup, and wiper should be discarded and replaced each time ram is disassembled, regardless of apparent condition. Teflon seal should also be examined and replaced if nicked, stretched, or damaged. Coat all new "O" rings with a light coating of petroleum jelly before reassembly, and soak new seal in clean hydraulic fluid for a minimum of 5 min before reinstallation.

The rear steer ram on four-wheel AWS machines is located on the left side of the machine, parallel to the rear axle and just to the rear of it. The ram is connected at the body end by a cap-and-ball joint at the rear of the gear carrier, and at the piston rod end by a pin and fork connected to the left hub. This ram is controlled through the main valve bank and is customarily used in conjunction with the steering wheel to provide Clark's standard AWS feature.

The ram is removed from the machine by unbolting the ball cap, removing the fork pin, and disconnecting the two hydraulic hoses. It may then be lifted out to a suitable work location.

Internal disassembly, repair, inspection, and reassembly is identical to that for the front steer ram outlined over the last few pages, except that the rear ram is not equipped with a small "O" ring in the piston. Refer to sequence for front ram.

After the ram has been reassembled at the work location, remount on machine by inserting fork pin, reattaching ball cap and required shims, and reconnecting hydraulic lines. Test machine before returning to regular service.

The pair of rams utilized for rear steer function on six-wheel Clark graders are located beneath and extending to the rear of the fifth wheel. They are connected on the piston end to the back side of the rear axle housing, adjacent to the pivot box. The cylinder end of the ram is connected to a lower crossmember at the rear of the grader frame through a simple pin-and-keeper arrangement.

After disassembling rear steer rams on tandem units, and smoothing small nicks and scratches, wash all parts except seals and "O" rings in a clean mineral spirit solution and air-dry. "O" rings, poly cup, and wiper should be discarded and replaced each time ram is disassembled, regardless of apparent condition. Teflon seal should also be examined and replaced if nicked, stretched, or damaged. Coat all new "O" rings with a light coating of petroleum jelly before reassembly, and soak new seal in clean hydraulic fluid for a minimum of 5 min before reinstallation.

BLADE MAINTENANCE

The following part of the chapter on grader maintenance deals with the maintenance of all components and functions which are in some way associated with the operation of the grader blade itself.

This section examines the mechanical parts in detail, discussing their location, operation, servicing, and repair. Adjustment procedures for both mechanical blade movements and hydraulic power tilt and side shift are detailed to explain most field maintenance requirements.

Location, inspection, and repair of circle drives and blade lifting rams are also discussed.

Drawbar and circle A grader's drawbar (refer to Fig. 3-8) forms the backbone of its blade. With its companion circle, the two assemblies support the full thrust of the moldboard and provide the framework through which the machine's functional tool is raised, lowered, and rotated.

Certain adjustments must be made and should be checked each time the circle is lubricated. For example, after heavy operation, the drawbar may develop some play. To adjust the play at the forward ball joint on the drawbar, set the moldboard against the ground and rock the machine forward and back slightly. Disassemble the socket of the ball joint and adjust the numbers or thicknesses of shims as necessary to achieve zero movement of the drawbar's stem after reassembly.

Most drawbar and circle assemblies are equipped with three sets of shims on the right and left extremes of the grader's drawbar. Vertical adjustment of all three sets is performed as follows:

1. Set the moldboard against the ground and run the machine ahead enough so that there is tension on the blade.

2. Measure the points on all shim retainers to ensure that the clearances are at least between 0.015 and 0.062 in. Add or remove the shims as necessary to maintain the limits.

3. After reshimming, rotate the circle 360° to check for binding and again add or subtract shims as necessary to correct.

Horizontal circle play is held in adjustment by setscrews which pass through threaded holes in an extension on the drawbar into each of the side shim retainers. By adjusting these screws, the retainers are moved toward or away from the circle teeth, thereby limiting or permitting horizontal play of the circle. Ideally, zero clearance is best, although older machines can be adjusted to within 0.062 in.

Ram ball joints located at the rear of the drawbar should be checked for excessive play each 50 hr for the first 250 hr of operation and every 250 hr thereafter. If properly adjusted, joints should be tight, but not to the point of binding. Adjustment is maintained by removing end caps from rod pistons and adding or subtracting shims as necessary.

Worm gear circle motors A grader's worm gear circle drive mechanism is usually located on top of the machine's drawbar. The circle is driven by a hydraulic motor through a worm and gear reduction unit.

Most worm gear circle motors are equipped with bolted flange-type mounts. To service the motor, first disconnect hydraulic lines from the motor and remove the corner mounting bolts to permit removal of motor to a clean work area. Place the motor in a vise, taking care to clamp across the flange instead of the housing, as the latter will cause distortion. Always use a protective device such as special soft jaws or hard rubber. Next, remove the

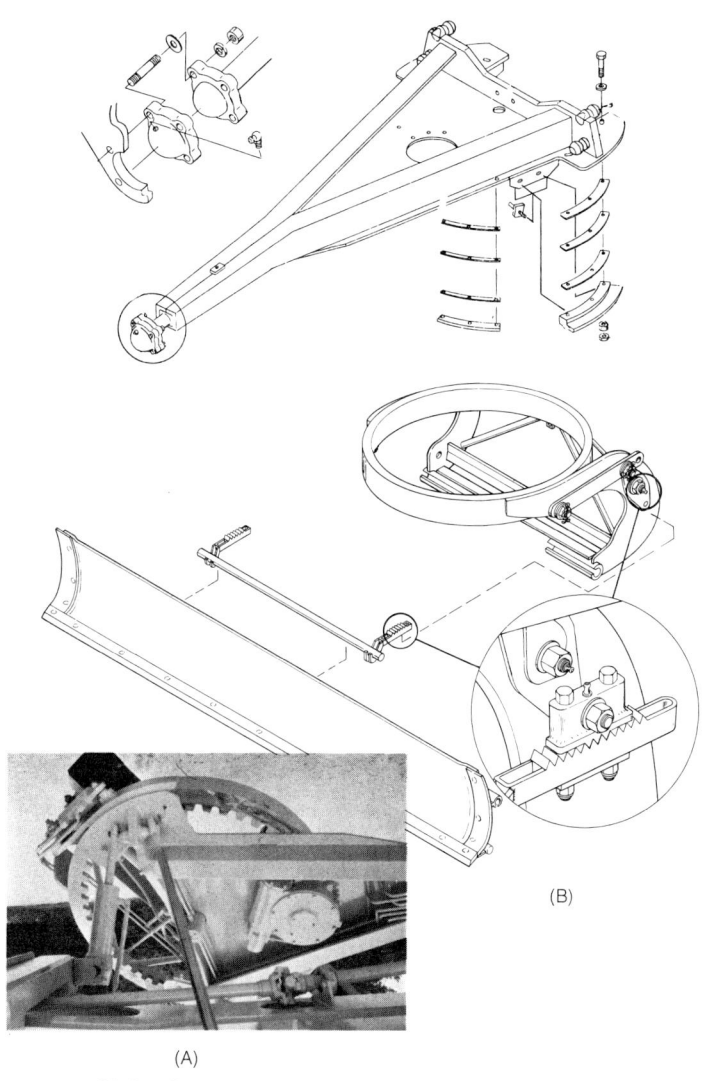

Fig. 3-8 (*A*) Assembled and (*B*) exploded views of drawbar and circle showing relationship of moldboard.

cap screws from the end cap, exercising care to avoid scratching any of the mating surfaces.

Remove the gerolor assembly, spacer, and splined drive shaft, then the cap screws holding the wear plate. Remove the cap, bearing races, thrust bearing, and radial bushing.

Proceed to remove the motor and other internal parts according to the manufacturer's recommendations found in your service manual. Check all parts for scratches and burrs that could cause leakage. Replace all parts on which defects will be too severe. Then, to ensure that all mating surfaces are perfectly smooth, place a piece of at least No. 600 grit paper on a smooth surface. Place the parts on the grit paper and stroke gently in a figure-eight motion several times. Wash all parts in a clean mineral spirits and allow to air-dry, or blow dry with an air hose. *Do not wipe dry*, as lint from the cloth deposited on critical internal parts can cause leakage or malfunction. Discard and replace all old seals and "O" rings with new ones coated with petroleum jelly. Dip new seals into a clean hydraulic oil solution before installing and reassembly.

Prior to starting a worm gear circle motor after repair or replacement and installation, if necessary to do so, disconnect one of the hydraulic lines to the motor and manually add hydraulic fluid until it begins to flow back out of the port. Reconnect the hydraulic line and tighten.

It is especially important that hydraulic fluid used to drive this motor is filtered and clean. Contaminated fluids cause wear on the bearings, gerolor star and rolls, and drive splines, reducing the motor's operating life and increasing the likelihood of breakdown.

Poor installation practices which may lead to misaligned shafts should be avoided. Such motors are adequately designed to withstand sufficient side and thrust loads, but to impose an unnecessary continuous force on the motor shaft can accelerate wear and reduce the work life of the motor.

In the event that difficulties arise which affect the operation of the worm gear circle motor, possible causes and corrective measures are listed in Table 3-6.

Worm gear circle drives Most worm gear circle drives are factory adjusted and should not be tampered with unless major repairs or complete overhaul are undertaken. When new components are required, the worm and gear should be replaced as a matched set.

In most cases, the unit can be serviced while either mounted on the machine's drawbar or disconnected by removing the hydraulic lines to the motor, moving the circle to clear pinion teeth, then unbolting housing and removing the gear drive as a unit. Care should be taken to cap and seal all hydraulic lines to prevent contamination when the motor is disconnected.

Disassembly procedures will vary for each manufacturer's gear. Care should always be taken, however, to properly identify each part during disassembly. Refer to Fig. 3-9 for an exploded view of a typical circle drive worm gear.

After disassembly, all parts should be thoroughly cleaned in mineral solvent or other suitable cleaning solution. Discard and replace all gaskets and seals removed during teardown. Shim groups should be kept together as units and not separated. All bearings removed should be replaced. Any bearings not removed should be inspected for dirt and foreign matter in the grease. Remove and replace all grease if dirt or material is found.

Finally, inspect gearing pinions for broken or missing teeth and replace damaged parts. Shims may be replaced by installing identical sizes and thicknesses.

The worm gear circle drive should be inspected monthly or every 200 hr of operation to add lubricant if needed. The drive's case should be emptied of all grease annually and refilled with new 80–90 EP multipurpose gear lubricant.

When reassembling the worm gear drive, proper tooth contact pattern must be realized. Add or deduct from shims to achieve a ratio according to manufacturer specifications.

The contact pattern for the worm gear can be manually turned when assembling. Proper tooth contact during operation under load is approximately 80 percent of the tooth length starting at the leaving toe and progressing toward the entering heel of the tooth. Idler bearing should be shim-adjusted without drag from a mating part.

Manual blade shift and tilt Graders equipped with manual side shift and tilt moldboards will be similarly equipped as shown in Fig. 3-10; the exploded view of a typical right-hand tilt plate is shown in Fig. 3-11.

On such equipment, remove the manual shift and tilt by first lowering the moldboard to the ground, crossways to the frame. Remove cotter pins and slotted nuts from the left- and right-hand tilt plates. Remove right- and left-hand shoulder bolts and nuts to dismantle the pitch-adjusting lock. Then remove the nut from the pivot bolt and raise the moldboard slightly to allow it to tilt forward. Lower it again to take weight off the bolts. With the top of the moldboard leaning forward, carefully block it to keep it from falling. Blocks should also be applied under the shifter rail.

Next, pry the left- and right-hand tilt plates away from the circle leg enough to clear the

Motor Grader Maintenance 7-75

bolts. Move the grader back from the moldboard and remove the left-hand flange socket from the shifter rail and slide the left-hand tilt plate from the rail. To remove the right-hand tilt plate assembly, turn the lock control shaft until the lockpin is in a raised position. Then, slide it off the left-hand end of the shifter rail and prepare for inspection.

Inspect all parts for wear or other damage. Working surfaces should be lubricated before reassembly.

To reassemble, first install the plate's spring onto its lockpin and insert the lockpin into

TABLE 3-6 Troubleshooting the Worm Gear Circle Motor

TROUBLE	CAUSE	CORRECTION
NO PRESSURE	1. Check pump.	1. Repair
FLUCTUATING PRESSURE	1. Check fluid level. 2. Check for broken lines. 3. Broken pump—worn pump 4. Relief valve stuck.	1. Refill if needed. 2. Repair 3. Disassemble, inspect pump parts. 4. Disassemble valve, clean, replace damaged parts.
RELIEF VALVE CHATTER	1. Damaged valve. 2. Dirt between piston and seat in relief valve control head.	1. Repair 2. Disassemble, clean.

NOISY PUMP	1. Restricted intake.
	2. Cavitation at pump inlet.
	3. Pump picking up air
	   a. around shaft or head packing.
	   b. at loose or broken intake pipe.
	4. Worn pump.
	5. Excessive pressure.
		1. Clean intake strainer; check intake piping for obstruction.
		2. Fluid viscosity too high, intake partially restricted.
		3. Replace packing, grease pump fitting Repair or replace pipe.
		4. Disassemble pump, inspect internal parts for wear.
		5. Check relief valve setting for line restriction (clogged or undersize lines).

FOAMING OIL	1. Wrong grade oil. 2. Inadequate baffling. 3. Return line above fluid surface.	1. Replace. 2. Correct. 3. Extend line or add fluid.
SLOW OPERATION OF MOTOR	1. Worn pump. 2. Worn motor. 3. Extremely high fluid temperature causing pump and motor to slip (temperatures increase as pump and motor wear.) 4. Inadequate size oil lines. 5. Pump cavitation. 6. Plugged filter. 7. Relief setting too low.	1. Repair or replace pump. 2. Replace worn parts or motor. 3. Increase reservoir size and use a high viscosity index oil. 4. Increase oil line size. 5. Increase oil line size to pump. 6. Replace filter element or clean filter. 7. Set relief valve for proper psi.
MOTOR WILL NOT TURN	1. Shaft seized in housing due to excessive side load or misalignment. (Note maximum radial loading on shaft.) 2. Large contaminating particles in fluid such as machining chips or sand. Very dirty fluid. 3. Broken shaft from extreme side loads or misalignment.	1. Replace housing assembly if damaged 2. Flush new systems—use better filtration. 3. Correct and Replace.
MOTOR RUNS WITHOUT TURNING SHAFT	1. Broken shaft.	1. Replace shaft assembly. Check housing for wear and replace if necessary. Check for misalignment.
MOTOR TURNS IN WRONG DIRECTION	1. Hose connections wrong. 2. Wrong timing.	1. Reverse connections. 2. Retime. See Reassembly Step 8.
LEAK AT SHAFT	1. Worn or cut shaft seal.	1. Replace seal, polish shaft at seal area with a No. 400 wet or dry sanding cloth. Check for misalignment.
LEAK BETWEEN FLANGE AND HOUSING	1. Loose flange. 2. Damaged seal between housing and flange. 3. Leak in body plug seal.	1. Tighten. 2. Replace seal. Check housing surface at seal for sharp nicks or deep scratches. 3. Replace faulty o-ring.
LEAK BETWEEN HOUSING AND SPACER PLATES OR BETWEEN SPACER PLATES AND GEROLER	1. End cap bolts loose. (NOTE: All motors are tested and rated at a maximum back pressure of 1000 psi.)	1. Tighten the 7 cap screws at geroler end of motor See Reassembly Step 10 for torque.
LEAK AT OIL PORTS	1. Poor fittings. 2. Damaged threads.	1. Replace fittings carefully. 2. Replace housing or use nut such as "true seal".

*Wherever difficulties have been caused by dirt in the system, check fluid supply and fluid filter, change if necessary. Fluid must be kept free of water and foreign material. Continuing trouble may indicate a dirty new fluid supply, or that the fluid is breaking down under operating conditions. Use high quality fluid and operate system within recommended temperature limits.

the tilt plate's box with the pin slot toward the tilt plate. Install end plate onto the box and install and tighten cap screws and lock washers.

Then install the locknut onto the lockpin and tighten until the boss on the lock cam enters the slot in the lockpin's side. Install the side plate with cap screws and lock washers and tighten. Install the control shaft into the lock cam; turn the control shaft to align pinholes in shaft and drive pin into plate. Check lock and unlocking action by turning shaft. Should any tightness appear, loosen locknut until it is eliminated.

Coat slide rail and underside of slide angle with grease before installing the plates. Install the right-hand tilt plate onto the moldboard, and move grader into position until right-hand circle leg enters the space between the tilt plate and lock housing. Install the left-hand tilt plate onto the moldboard shifter rail and onto left-hand circle leg bolts.

Again move the grader forward until the desired moldboard pitch or angle is obtained. Install and tighten nuts on pivot bolts and put on the cotter pins. Install pitch-adjusting locks, making sure that the grooves in the locks mesh with those on circle legs. Install and tighten tilt locknuts. Turn lock control shaft until the lockpin rests on top of the slide rail. Move moldboard or circle until the pin drops into the desired notch on rail.

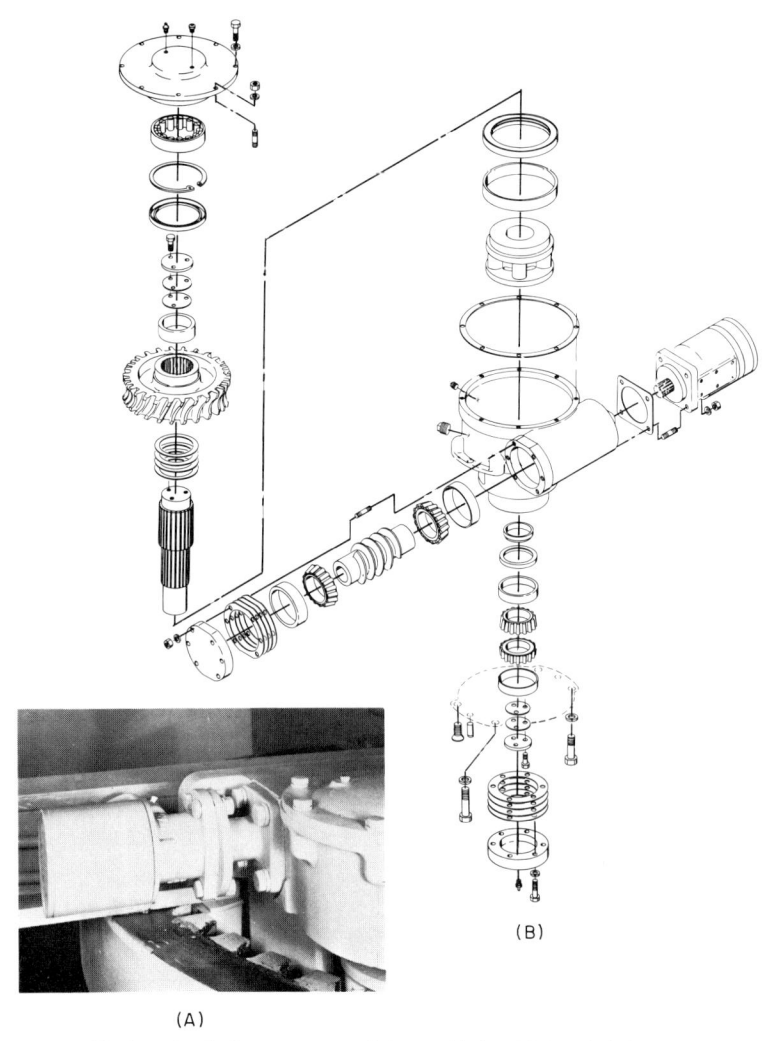

Fig. 3-9 Circle drive worm gear (A) assembled and (B) exploded view.

Blade hydraulic rams A swivel valve is necessary on hydraulic-powered graders to provide the power and to drain the lines to hydraulic side shift ram and, when used, power tilt moldboard rams. The swivel valve is required on those graders having 360° circle rotation and moldboard hydraulic side shift.

Maintenance and repair of both a grader's side shift and scarifier rams are normally the same. *Side shift rams* are located behind and mounted to the moldboard assembly of the grader. Its piston end connects to the left rear of the moldboard, while its cylinder end connects to the right leg of the circle.

The *scarifier ram*, depending on the type of grader, will be located where the scarifier

EXPLODED VEIW OF R.H. TILT PLATE
1. Tilt Plate   4. Control Shaft   7. Pin
2. Lock Pin   5. Lock Cam   8. End Plate
3. Spring   6. Side Plate   9. Nut

Fig. 3-10 Manual blade shift and tilt mechanism.

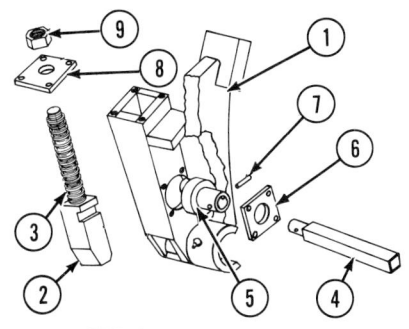

Fig. 3-11 Right-hand manual blade shift and tilt mechanism, exploded view.

is mounted. Both the side shift and scarifier rams normally consist of cap-and-ball-type connections at the piston end.

To service, disconnect both hydraulic hoses from the rams. Support the ram with blocks, chains, or slings, then remove ball cap nut, ball, and both halves of the socket. (Refer to Fig. 3-12 for an exploded diagram.)

Do not lose shims which may be of different thicknesses and place them aside for identification and replacement at the same points. Take the ram to a suitably clean work area for disassembly, inspection, and servicing. Ram will be remounted by reversing the procedure, taking care to include the proper number and correct-size shims when installing the ball cap.

Optional *power tilt rams* available on current model graders are normally located on the side arms of the grader circle and replace standard equipment supplied for manual tilt. In most cases, they are held in place with the same cap nuts and on the same studs as are used on machines not equipped with this option.

Before removing it from the machine, be sure that the moldboard is set firmly against the ground. Just as when making manual adjustments, the moldboard could shift when the rams are removed.

Blade lift rams are located in vertical positions on both sides of the grader's main frame, just ahead of the operator's station. These rams control the height of the cutting edge off the ground.

To service, the rams can be left in place

Fig. 3-12 Location of scarifier and dozer blade rams.

or removed from the machine by, first, lowering the moldboard to the ground, shutting down the engine, and then disconnecting the hoses at tops and bases of the rams. Then, disconnect the rams' end caps and swing the unit free from the drawbar.

If dismounting is preferred, do so by removing the bolts anchoring the ball socket housings. Watch, identify, and care for the shims! Another, more expedient, method of removing the ram from the machine is to remove setscrews and pull both pivot pins from the socket housings.

For graders equipped with a *high-lift ram,* as shown in Fig. 3-13, it will come factory mounted for right-hand operation and be located on a special pivot unit found on the right-hand side of the grader frame, behind the blade lift ram bracket as shown in Fig. 3-18. This ram controls the angle of the blade relative to horizontal and when used in conjunction with other blade controls can swing the blade up to a maximum of 90°, or to full vertical for rough sloping or other similar applications.

Like the blade lift rams, it's possible to service the high-lift ram without removing it

Fig. 3-13 Location of high-lift ram.

from the grader. To do so, merely disconnect the ram at the ball cap end, removing ball cap and shims. Retract the ram enough to permit it to clear the circle and be swung to vertical. Disconnect both hydraulic hoses to relieve suction in ram; disassembly for servicing may then proceed.

For complete removal from the machine, continue by shifting support of the ram's weight to a sling or chain hoist. Loosen and remove lower and upper pivot caps and shims. The ram may now be pulled out of the brackets and moved to a suitable work location.

Disassembly, inspection, and maintenance of most hydraulic rams are similar. The procedure for specific rams used on different model graders are detailed in the operator's and service manuals supplied with the machines. However, a general discussion earlier in this chapter on ram servicing, under the heading Hydraulic Rams, can be reviewed.

Cutting edges and moldboards The best tip for PM of moldboard and blade is an operational necessity: Keep the cutting edges full and replace when they show signs of wear; before grading starts to damage the moldboard, keep the moldboard slide rail clean and lubricated by wiping clean with a cloth soaked in diesel fuel.

ENGINE AND POWER TRAIN MAINTENANCE

Full operating and service information for any engine of a motor grader may be obtained by consulting the engine manufacturer's operating and service manuals supplied with the machine.

It is not the scope of this chapter to discuss in detail engine maintenance and engine servicing. Parts of the engine that will be reviewed in this chapter are the air cleaner and cooling system.

Motor Grader Maintenance 7-79

Generally, when servicing an engine or any part of the engine, be sure it is stopped. Before cleaning, servicing, lubricating, checking belt tension, adjusting brakes or clutch, removing housing covers, working on the hydraulic system or making repair—shut down the engine. When making adjustments which require a running engine, set the parking brakes, place the transmission in neutral, and block the wheels. Always have two men working, one at the controls and one at the engine.

Air cleaner The dry type, the most commonly used on current graders, is similar in many respects to the air cleaner used in an average automobile. Both filter air through coated paper cartridges and, under normal operating circumstances, neither requires as frequent a schedule of major maintenance as does an oil bath air cleaner.

Frequency of service with a dry-type cleaner can range from 4 to 120 hr. Daily inspection, however, is recommended; more often if dust conditions warrant, as the dry-type air cleaner will not function properly when dust deposits build up past a 2-in. level in the dust cap. See Fig. 3-14.

A tight positive seal *must* be maintained between cap and cleaner body after every servicing, as leakage at this point will increase dust loading and make it necessary to service the filter cartridge more often. Because of the normal frequency of dust cup

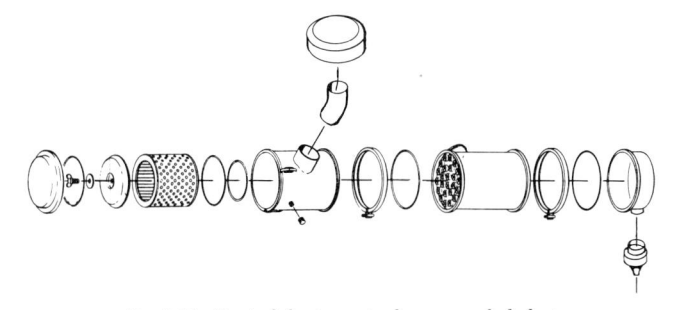

Fig. 3-14 Typical dry-type air cleaner, exploded view.

servicing and the possibility of "O" ring damage, it is recommended that several replacement "O" rings be kept in stock at all times. Excessive smoke and/or loss of power may also indicate the need for filter cartridge servicing.

To clean, stop the engine, then wipe the cover and upper portion of the air cleaner. Loosen clamps and remove cover to unscrew the wing bolt which holds the cartridge in position. Then, lift the cartridge from the cleaner. Considerable dust can be dislodged by slapping the side or bottom rim of the cartridge with the palm of your hand. But, do not bang the bottom rim of the cartridge against a hard surface.

If compressed air at 100 psi or less is available, blow out the cartridge from its clean air side. An even, bright red pattern seen through the cartridge when a light is held inside means that the cartridge is clean. Have several spare cartridges on hand to minimize downtime.

With both filter cartridge and dust cup removed, inspect tubes by looking through them with a bright light. Remove deposit with a stiff fabric brush. It may also be practical to wash the filter with any good, nonsudsing household detergent.

Before replacing the cartridge, wipe out any dust which may be at the bottom of the cartridge chamber. Check top and bottom gaskets, filter cartridge gasket, and restriction tap plug to ensure that all are seated, undamaged, and leak-free. Slip cartridge into position and secure with wing nut so that the unit cannot be rotated. Be sure that the rubber-metal washer is in place under the wing nut and that it is in good condition. After ensuring cover gaskets are in good condition, the cover should fit into position smoothly without forcing. If not, it may not be properly seated. Check and try again.

As an added precaution against dirt getting into the engine, frequently inspect the connections between the air cleaner and cleaner pipe and between the air cleaner and engine. Also check the clamps periodically and tighten when necessary.

Failure to perform maintenance as outlined above may result in unnecessary dust entering the engine, causing rapid premature wear to such internal engine parts as piston and sleeve groups, valves, and crankshaft assembly. The cost from such negligence must be borne by the owner, including labor for parts installation. For operation's sake and your own peace of mind, replace the filter after each 1000 hr of use. While the cartridge may function sufficiently for a longer period, the small savings gained does not justify the risk.

Cooling system Most graders are shipped from manufacturers complete with a fill of ethylene glycol permanent antifreeze, unless specified by the owner. When it becomes necessary to add or to drain and refill the cooling system with fresh antifreeze, ensure that the added or replacement solution does *not* have antileak additives. Such an additive is incompatible with corrosion-resistant components in the engines used on these machines.

The cooling system in most graders automatically maintains the most desirable engine temperature under normal operating conditions. When the engine is started cold, a bypass thermostat prevents circulation to the radiator and permits coolant to circulate only through the engine block. When an efficient operating temperature has been reached, the thermostat opens, allowing the coolant to circulate throughout the engine and into the radiator.

Also, to deliver maximum cooling air, a fan must have all blades fully functioning. Check the fan periodically for damaged blades and make sure blades revolve without interference. Play at the fan hub should also be checked as excessive movement is an indication of loose or worn bearings, a situation that should be promptly corrected.

Fan belt tension should also be checked frequently, and the radiator should be kept filled with either permanent antifreeze or clean, soft fresh water, weather permitting. Keep the front of the radiator clean, and if spaces between the fins become clogged, clean with an air or water hose in the direction opposite the normal flow of air.

Generally speaking, coolants will protect against corrosion and scale but will not prevent deterioration of the hoses in use. Hoses will harden and crack with age, causing coolant leaks. Hardened hoses will transmit engine vibrations and rocking action to radiator components, placing a stress on soldered seams. The inside lining of hoses may also deteriorate, permitting particles of rubber to flake off and clog narrow passages of the radiator core.

To drain the cooling system, open the drain cocks at the bottom of the engine block and radiator. Remove radiator filler cap to allow air to displace the coolant. Check to see that the drain points are not plugged and that the solution drains out completely.

Clean and flush the system before refilling. Check radiator, water pump, all gaskets and hose connections and repair leaks and worn hoses and defective parts as may be necessary.

Operation of power shift transmission and torque converter The transmission and hydraulic torque portion of the power train enacts an important role in transmitting engine power to the driving wheels. To properly maintain and service these units, it is important to first understand their function and how they operate.

The transmission and torque converter function together and operate through a common hydraulic system. It is necessary to consider both units in the study of their function and operation.

A shift control valve assembly is mounted directly on the side of the converter housing. The function of the control valve assembly is to direct oil under pressure to the desired directional and speed clutch. A provision is made to neutralize the transmission when the brakes are applied with the left foot; this is accomplished through a brake-actuated shut-off valve. The speed and direction clutch assemblies are mounted inside the transmission case and are connected to the output shaft of the converter by either direct gearing or drive shaft. The purpose of the speed or directional clutches is to direct the power flow through the gear train to provide the desired speed range and direction.

With the engine running, the converter charging pump draws oil from the transmission sump through the removable oil suction screen and directs it through the pressure-regulating valve and oil filter.

The pressure-regulating valve maintains pressure to the transmission control cover for actuating the direction and speed clutches. This requires a small portion of the total volume of oil used in the system. The remaining volume of oil is directed through the torque converter circuit to the oil cooler and returns to the transmission for positive lubrication. This regulator valve consists of a hardened valve spool operating in a closely fitted bore. The valve spool is spring-loaded to hold the valve in a closed position. When a specific pressure is achieved, the valve spool works against the spring until a port is exposed along the side of the bore. This sequence of events provides the proper system pressure.

After entering the converter housing, the oil is directed through the stator support to the converter blade cavity and exits in the passage between the turbine shaft and converter support. The oil then flows out of the converter to the oil cooler. After leaving the cooler, the oil is directed to a fitting on the transmission and then, through a series of tubes and passages, lubricates the transmission bearings and clutches. The oil then gravity drains to the transmission sump.

The hydraulic torque converter consists basically of three elements and their related parts to multiply engine torque. The engine power is transmitted from the engine flywheel to the impeller element through the impeller cover. This element is the pump portion of the hydraulic torque converter and is the primary component that starts the oil flowing to the other components which results in torque multiplication. This element can be compared to a centrifugal pump in that it picks up fluid at its center and discharges at its outer diameter.

The torque-converter turbine is mounted opposite the impeller and is connected to the output shaft of the torque converter. This element receives fluid at its outer diameter and discharges at its center. Fluid directed by the impeller out into the particular design of blading in the turbine and reaction member is the means by which the hydraulic torque converter multiplies torque.

The reaction member of the torque converter is located between and at the center or inner diameters of the impeller and turbine elements. Its function is to take the fluid which is exhausting from the inner portion of the turbine and change its direction to allow correct entry for recirculation into the impeller element.

The torque converter will multiply engine torque to its designed maximum multiplication ratio when the output shaft is at zero rpm. Therefore, as the output shaft is decreasing in speed, the torque multiplication is increasing.

The shift control valve assembly consists of a valve body with selector valve spools. A detent ball and spring in the selector spool provides one position for each speed range. A detent ball and spring in the direction spool provides three positions, one each for forward, neutral, and reverse.

With the engine running and the directional control lever in neutral position, oil pressure from the regulating valve is blocked at the control valve, and the transmission is in neutral. Movement of the forward and reverse spool will direct oil under pressure to either the forward or reverse direction clutch as desired.

When either directional clutch is selected, the opposite clutch is relieved of pressure and vents back through the direction selector spool. The same procedure is used in the speed selector.

The director or speed clutch assembly consists of a drum with internal splines and a bore to receive a hydraulically actuated piston. The piston is made oil-tight by sealing rings. A steel disk with external splines is inserted into the drum and rests against the piston. Next, a bronze disk with splines at the inner diameter is inserted. Disks are alternated until the required total is achieved. A heavy backup plate is then inserted and secured with the snap ring. A hub with OD splines is inserted into the splines of disks with teeth on the inner diameter. The disks and hub are free to increase in speed or rotate in the opposite direction as long as no pressure is present in that specific clutch.

To engage the clutch, as previously stated, the control valve is placed in the desired position. This allows oil under pressure to flow from the control valve, through a tube, to a chosen clutch shaft. This shaft has a drilled passageway for oil under pressure to enter the shaft. Oil pressure sealing rings are located on the clutch shaft. These rings direct oil under pressure to a desired clutch. Pressure of the oil forces the piston and disks against the heavy backup plate. The disks, with teeth on the outer diameter, clamping against disks with teeth on the inner diameter, enable the hub and clutch shaft to be locked together and allow them to drive as a unit.

There are bleed balls in the clutch piston which allow quick escape for oil when the pressure to the piston is released. See Fig. 3-15 for a diagramed system.

Transmission and torque-converter troubleshooting The following data are presented to help locate the source of difficulty in a malfunctioning powershift transmission and hydraulic torque converter. It's necessary to consider the torque-converter charging pump, transmission, oil cooler, and connecting lines as a complete system when troubleshooting the source of a problem. The proper operation of any of these units depends significantly on the condition and operation of the others.

By studying the principles of operation and referring to data in service manuals, it may be possible to correct any malfunction that may occur in your grader's system.

Troubleshooting consists of two classifications: mechanical and hydraulic.

Mechanical. Prior to checking any part of the system from a hydraulic standpoint, the following checks should be made:

1. Ensure all control lever linkage is properly connected and adjusted at all connecting points.

2. Shift levers and rods should be checked for binding or restrictions in travel that would prevent full engagement. Shift levers first by hand at the control valve, and if full engagement cannot be obtained, the difficulty may be in the control cover and valve assembly.

Hydraulic. Before checking the torque converter, transmission, and allied hydraulic system for pressures and rate of oil flow, the following preliminary checks should be made: The transmission's oil level should be checked with the oil temperature between

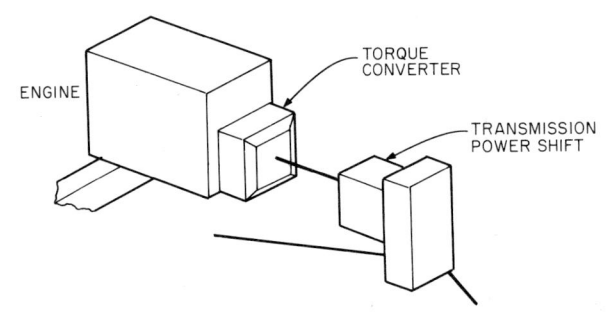

Fig. 3-15 Schematic of the relationship of the torque converter to the transmission and engine.

180 and 200°F. To raise oil temperature, either work machine or stall it out by engaging shift levers in forward and high speed, applying brakes. Accelerate engine half to three-quarters throttle. Hold the stall until desired converter outlet temperature is reached. *Be careful!* Full-throttle stall speeds for an excessive period will overheat the converter.

If the mechanical and hydraulic prechecks do not reveal a trouble area, proceed to find the problem.

If, for example, the system is *overheating*, the cause could be a low oil level, a plugged suction screen, or a defective oil pump in the system.

Causes for a *noisy converter* could be any of three possibilities: worn coupling gears, worn oil pump, worn or damaged bearings. Poor gears and oil pump mean replacement. And a complete disassembly of the system will be necessary to determine what bearing is faulty.

If the system is producing *low clutch pressure,* it could mean that oil levels are low or that the clutch pressure-regulating valve spool or piston bleed valve are stuck in their open positions. Remedies for those three causes are fairly simple. However, if the problem is a faulty charging pump or broken or worn clutch shaft or piston sealing rings, the remedies are more complex and call for replacement parts.

Lack of power could be due to low engine rpm at converter stall or overheating in the transmission/converter system. If low engine rpm is the suspected cause, tune the engine and check the governor.

Contaminants in the system The presence of water and/or ethylene glycol coolant mixtures in the transmission or torque-converter oil can be detrimental to the reliability and durability of the system's internal components.

Such foreign substances deteriorate nonmetallic parts and highly loaded steel parts such as bearings and gears because it reduces lubricity. Additionally, the frictional capacity of the drive clutch plates can be impaired as a result of surface film or impregnation. Glycol will deteriorate clutch plate material.

Should you suspect contamination, a sample of the oil should be obtained when the transmission is at normal operating temperature to ensure that any contaminant is thoroughly dispersed. Some of the conditions that indicate an excessive amount of water and/ or glycol include rusted or pitted transmission parts, oil spewing from the transmission breather, oil in the radiator, blistered or wrinkled gaskets in uncompressed areas, and cloudy or gray appearance of the transmission oil.

A qualified laboratory or the oil supplier will provide an analysis of the degree of contaminant and possibly a clue to its source.

Because systems vary so greatly, any maintenance of the powershift transmission and torque converter should adhere to manufacturer recommendations that are part of the technical service and operator's manual furnished with your motor grader. Consult a factory service representative or qualified dealer service department for advice, assistance, or actual repair.

Transfer cases If internal parts of a grader's transfer case are required, it's best to remove such units as a single assembly. Before beginning removal, determine if noise is actually generating from the transfer case.

Operators usually insist that operating noise comes from the transfer case. In numerous instances, however, investigation revealed that noisy operation was due to any of the following: fan out of balance or bent fan blades, crankshafts out of balance, rough engine idle producing a rattle in the gear train, clutch assembly out of balance, worn-out universal joints, loose or broken engine mounts, wheels out of balance, tire treads humming or vibrating at certain road speeds.

Try to locate and eliminate noise by means other than transfer case removal. If, however, the noise is found to definitely be in the transfer case, categorize the noises and, if possible, determine what position the shift unit is in when the noise occurs:

1. *Growling and humming.* These noises are caused by worn, chipped, rough, or cracked gears. As gears continue to wear, the grinding noise will become more noticeable, particularly in the gear position that throws the greatest load on the worm gear.

2. *Hissing or thumping.* Hissing noises can be caused by bad bearings. As bearings wear and retainers start to break up, the noise could change to a thumping or bumping.

3. *Gear whine.* This noise is usually caused by lack of backlash between mating gears. Improper shimming of unit is the big offender in this case.

4. *Improper lubricant.* Transmissions with improper lubricant or lack of lubricant can cause a unit to run hotter than normal, causing insufficient cooling and lubricant to cover the gears.

If the noise occurs when the machine is in neutral gear, possible causes include use of incorrect grade of lubricant, scuffed gear tooth contact surface, insufficient lubricant, excessive backlash in gears.

Before removing, drain the case of gear oil, flush twice with diesel fuel or mineral spirits, and run the machine slowly back and forth for several minutes with the cleaning solution.

Follow instructions in your service manual carefully and use a chain or cable hoist having a minimum of $1/2$-ton capacity along with sufficient wood blocking. A hydraulic jack may also be useful.

After following disassembly instructions and guides furnished in the grader's manual, thoroughly clean the transfer case housing and all parts in the solution. Inspect all pieces, paying close attention to gear teeth and shaft splines, ensuring that they are free of metal chips and that tips are not chipped or broken. Check bearing surfaces for pitting and scoring and replace all questionable or damaged parts. Ball and roller bearings should be checked closely for contaminated grease and should be replaced if any doubt exists.

Check all spacers for damage or deterioration and replace if any is found. All seals and gaskets should be replaced, regardless of apparent condition. All snap rings should be concentric, and the circular tension should lock the fings into the snap ring grooves. All metal faces at gasketed or sealed joints should be flat, smooth, and free from nicks, burrs, and scratches.

Propeller shafts should be cleaned and repacked with fresh lubricant every 5000 hr, concurrent with disassembly, cleaning, and inspection of U joints. Grease fittings are provided on each bearing group for weekly lubrication of bearings between cleanings. Cork washers should be changed each time U joints are disassembled and inspected.

Follow directions in your grader's service manual and be sure that yokes are properly timed and that yoke arms are on identical planes. Where a flat surface is not available, the planing can be checked by sighting through the cap screw holes of one of the yokes along the shaft. If the yokes are properly timed, it should be possible to sight down along the shaft directly through the cap screw holes of the yoke on the opposite end. See Fig. 3-16.

AXLES AND WHEELS

Maintenance service of wheels and axle assemblies will vary with the type of system on each different model motor grader. In this portion of the chapter, discussion will be focused on the maintenance of axles and wheels used in all-wheel-drive motor graders simply because of their relative complexity to a grader driven by its rear axles only.

Axle assemblies on such graders are hydraulically steered, double-reduction units, deriving power first through spiral pinions and bevel gears, then through a fully enclosed pinion and spur gear running in oil. They are equipped with dust-proof oil seals and antifriction ball and tapered roller bearings throughout, with the axle drive shafts being connected through a common differential unit.

To repair parts located at either end of the axle housings on most graders, it isn't necessary to remove the axle housing from the grader frame. Where gear carrier or axle housing repairs are necessary, however, the entire axle assembly should be removed from the frame.

No matter what type of motor grader you may have, it is highly recommended that

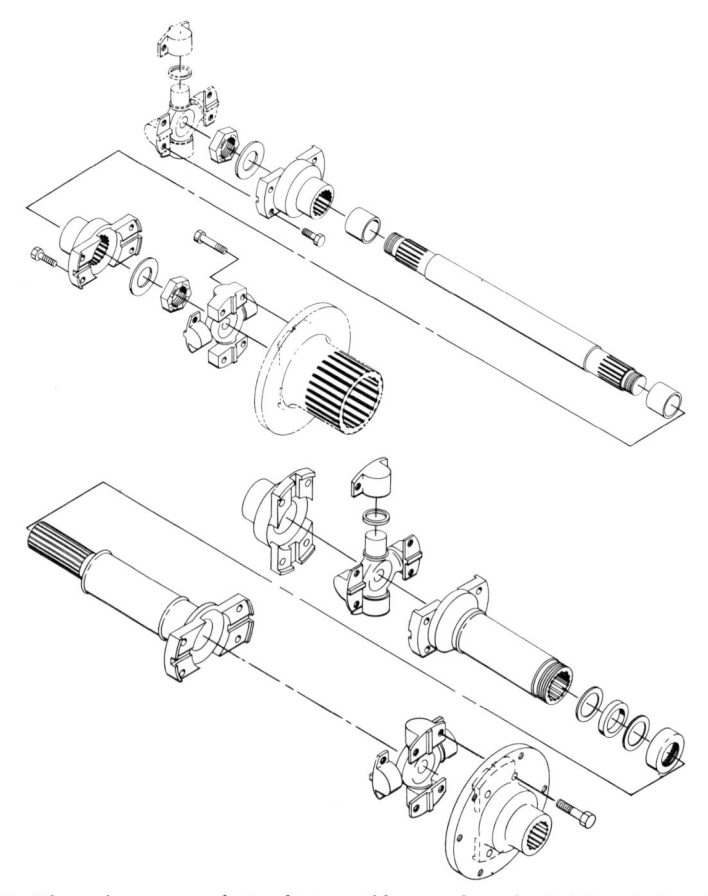

Fig. 3-16 When yokes are properly timed it is possible to see down the shaft directly through the cap screw holes of the yoke on the opposite end.

directions in the service manual provided with the machine be followed to the letter. Disassembly of right- and left-hand wheels will be the same.

Once removal and disassembly of steering joints has been completed, steps should be taken to inspect all roller bearings for chipping, flat spots, or excessive wear. Refer to Fig. 3-17 and replace both cone and cup if necessary. Bearings should be inspected carefully for contaminated grease and throughly cleaned or replaced if contamination is found.

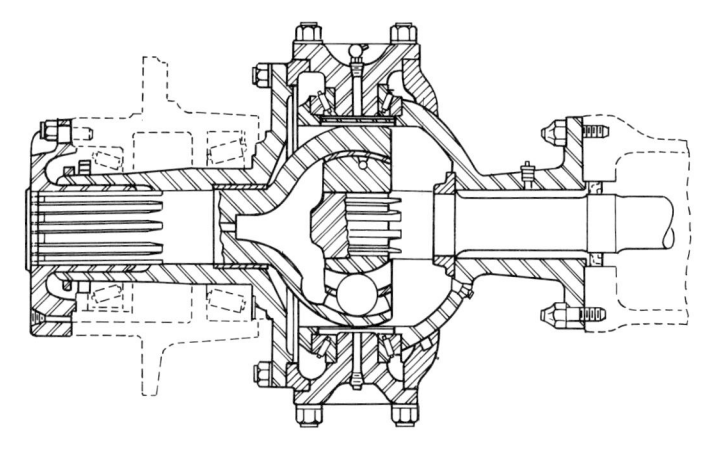

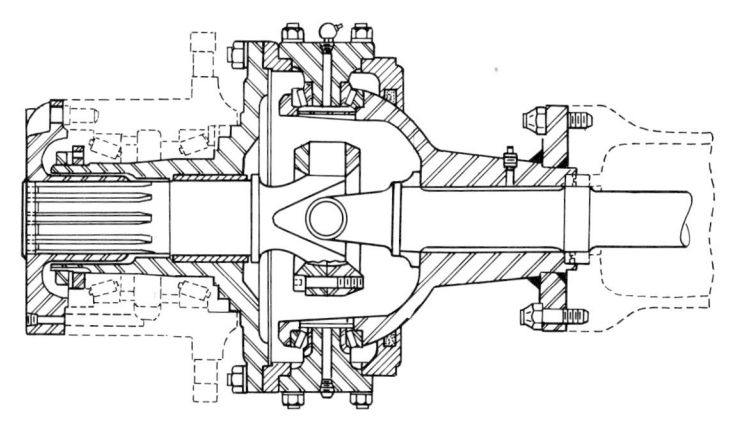

Fig. 3-17 Two different types of steering joints.

Wash all old grease from the inside wheels and from axle end and bearings. Repack bearings with wheel-bearing grease when reassembling.

If the hub felt is worn or damaged, replace it by prying off the inner bearing and removing the washer and felt. The gasket should be discarded and replaced regardless of apparent conditions.

Whenever a *steering joint* is removed from its housing, it's a good idea to wash and inspect it. If disassembly is required, particularly in a constant-velocity joint, the following procedures, as applicable, are recommended.

Using a bronze plug and hammer, tap the steering joint group off the end of the axle shaft. By pushing down on one side of the inner race, the opposite side will automatically come up, and one ball can then be removed. By pressing down on another point, another ball can be removed, and so on until all the balls are removed.

Roll both the cage and the inner race until they are standing on edge. If the rectangular slots are not aligned with the outer race, revolve the cage slightly to align. Then lift out the inner race and the cage.

Next roll the inner race at right angles to the cage so that one lug projects through one rectangular slot in the cast. Remove the inner race from the cage.

If your grader is equipped with Cardan joints on rear axles, separate the Cardan rings by removing cap screws. Pull short and long axle segments apart by hand, then clean and

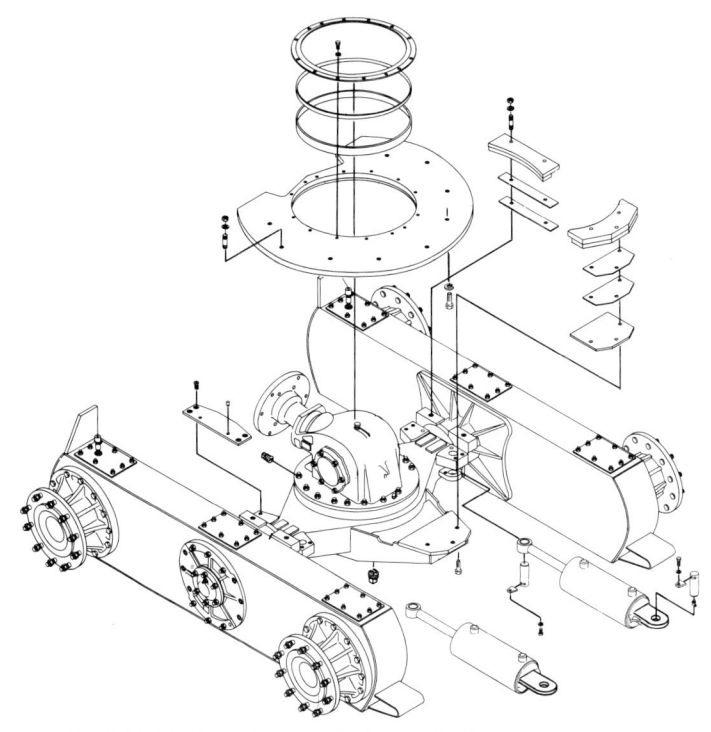

Fig. 3-18 Tandem axle assembly of a six-wheel motor grader.

inspect all parts. Follow timing and reassembly instructions carefully to place units back in service.

Wheel steering knuckle assemblies on front axles and on rear axles of four-wheel graders must be removed and disassembled according to procedures outlined in service manuals. Inspection and repair that follows consists of, first, discarding and replacing felt ring and seal, soaking new seal in oil before reassembly. All parts should be washed in mineral spirits solution and carefully inspected before reassembly. Bearings should be checked especially closely for chips, flat spots, or wear; both cone and cup should be replaced if replacement of either is required.

Carefully clean and wash bearing caps and cups and repack with lithium-lead EP wheel bearing grease after old grease has been flushed.

Pivot bearings must be adjusted with steering knuckles installed on the axle and the wheel hubs removed. The steering joint may be removed, but its removal is not mandatory for servicing.

Tandem units on six-wheel machines work by transmitting power through the drive axle to the dual drive socket. The front sprocket chain (see Fig. 3-18) drives the wheelset to the front, while the rear sprocket chain drives the wheel set to the rear on either side of the machine.

Chains and sprockets may be inspected by removing desired inspection covers from the top of the pivot case. To remove a chain, block up both pivot cases so wheels can turn freely and remove inspection cover over affected outer sprocket.

Then, rotate the wheels until the master link on the chain is located at the top of the sprocket. The master link can be easily spotted as it is the only link fastened with pins and cotter keys instead of rivets. Remove the keys, pins, and link from the chain, and the remainder of the chain may then be threaded out of the pivot box. Check condition of gaskets when inspection plates are removed and replace if necessary.

To adjust a drive chain, select one wheel from either pivot case, jack up the case at that point, and remove inspection cover over the outer sprocket for that wheel. Turn the top of

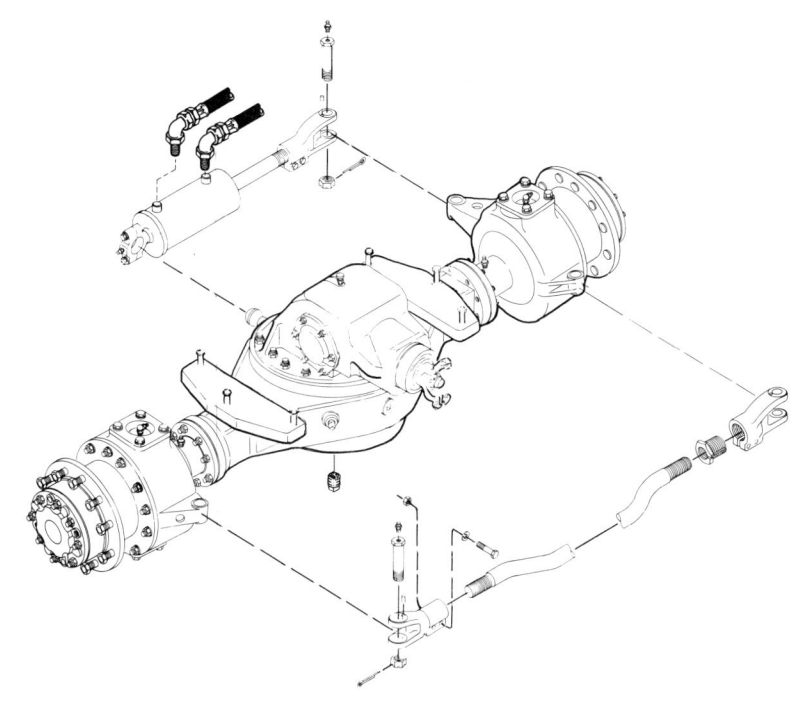

Fig. 3-19 To remove the drive axle, first drain pivot case of all oil and disconnect both drive chains.

the wheel toward the front of the grader and hold while inserting a short bar through the inspection hole and under the chain. The chain should lift about ½ in. If too little or too much slack, readjust by removing tire and rim. Because hub centers are slightly eccentric, chain slack can be added or removed by removing cap screws from the hub and rotating the hub clockwise or counterclockwise as needed to attain proper slack.

To remove the *drive axle,* refer to Fig. 3-19 as a guide and first drain pivot case of all oil and disconnect both drive chains as outlined above. Remove bearing cap and cap screw and washer assembly from front end of shaft.

Remove bearing support plate. If the going gets difficult, remove setscrews after the studs and cap screws have been withdrawn and replace with ½-13 × 1 in. or longer cap screws. Draw up those cap screws evenly until plate pops loose. The sprocket and shaft may then be drawn out of the axle housing with a suitable lifting device.

Check ball bearing for wear or contamination and replace if necessary. Check bushing for pitting or scoring and sprocket teeth for chipping or wear. Discard and replace seal and gaskets, soaking new seal in lubricating oil before reinstallation. Sprockets are interchangeable.

It is easier to wash out a main *axle housing* before disassembly is started. Remove the drain plug and allow oil to drain from the case. Replace plug and refill with mineral spirit solution and run the machine forward and backward several times and drain again. Repeat to ensure parts are thoroughly cleaned.

To remove *front axles* from the grader's frame, place the blade until it's square across the frame, and block both rear wheels securely to prevent possible machine movement. Remove nuts from pivot pins and setscrews, then disconnect steering ram hydraulic hoses at their union ends. Unlock and remove the capscrews from front propeller shaft unit and universal joint fitting yoke.

Next, start the engine and raise the blade as high as possible and place blocks under each end of the blade approximately 1 ft from the ends. Force the blade downward to lift the front of the frame up. Lift front only enough so that axle pivot pins are free of excessive binding.

Remove the pivot pins and again force the blade downward, this time enough to lift frame completely free of axle. Then, roll axle out from under the front end of the frame. The reassembly procedure is the reverse of the above.

On four-wheel graders it is preferred that the rear axle be rolled from under the grader to remove the gear carrier.

Start by placing the machine on level ground and revolve the blade until it's square across the frame. Block each end of the blade in a way similar to what is required for front axle removal.

Then, block the front wheels securely and move the blade downward to stabilize the grader. Referring to Fig. 3-19, remove cap screws and nuts from right and left axlemounts. Disconnect and cap hydraulic hoses at their union ends, then remove cap screws from the yoke while allowing propeller shaft to fall free. Disconnect parking brake release cable and again force the blade downward to lift the front of the grader.

It'll be necessary to jack up under the rear bumper plate to lift the frame off its rear axle, but by lifting the front end first, lifting will be much easier and tipping less likely.

Roll the rear axle back toward the rear and block under the axle bracket on the grader's frame. Then remove the jack from under the bumper plate.

To *remove rear axle and tandem units* on six-wheel machines proceed as outlined above, stabilizing the machine for safety. Unbolt six studs as shown in Fig. 3-18. The studs attach the support plate to the main frame.

Then remove the cap screws at the yoke and disconnect propeller shaft and parking brake release cable. Remove pins with keepers to disconnect the rear steering rams (if part of the machine on all-wheel-drive units) from the rear axle housing.

Start engine and slowly and evenly move blade downward onto blocking to raise rear end of main frame sufficiently to allow the rear axle (with wheels and tires) to roll away from under the frame. Place a jack under the rear bumper as a safety measure.

Remove the support plate from axle housing after removing six nuts from studs. Three guides should be marked to be reinstalled in the identical positions. The shims should be wired to each guide during dismantling.

Inspect the support plate bushing and replace it if damaged or worn. Manually lubricate the lower side of the plate near guides with lithium-lead EP gun grease.

To remove gear carriers from axle housing, refer to procedures listed in appropriate servicing manuals. Gear carriers should be inspected to check that all roller bearings are not chipped, have no flat spots, or show signs of excessive wear. Ball bearings should likewise be inspected for wear and replaced if necessary. All bearings should be cleaned and lubricated prior to reassembly.

Check all pinions and gears, paying particular attention to shaft splines and gear teeth to ensure that they are free of metal chips and that tips are not chipped or broken. Pinion and gear should be replaced only as matched sets. Seals and gaskets should be discarded and replaced, regardless of apparent condition.

Bevel gear and pinion adjustment No other phase of axle maintenance is more important, yet less understood, than adjustment of bevel drive gears and pinions to secure proper tooth contact.

When the tooth contact area is toward either the heel or toe, for example, teeth will be overloaded at the ends and chipping may occur. On the other hand, when the contact area is too high, then load application is farther from the base of the teeth, thus increasing bending load and possibly resulting in breaking teeth out at the base. Or, when the contact area is less than specified, a higher localized load results in squeezing out a film of lubricant. A metal-to-metal contact such as this causes scoring, scuffing, or galling, which contributes to noisy operation, your first sign of rapid wear.

Factory adjustment of matched bevel drive gear and pinion sets is as correct as special equipment and years of manufacturing experience can make it. Adjustment should not be disturbed except when absolutely necessary. For example, should it be required to dismantle and reassemble a gear carrier group to replace some part other than gear and pinion, do not disturb the shim adjustment. Always replace worn bearings.

Additionally, backlash should be checked and noted before disassembling the unit so that the gear may be adjusted to original backlash when making final adjustments. Backlash should measure 0.008 in.

Checking for tooth contact area and location is accomplished by applying oiled red lead

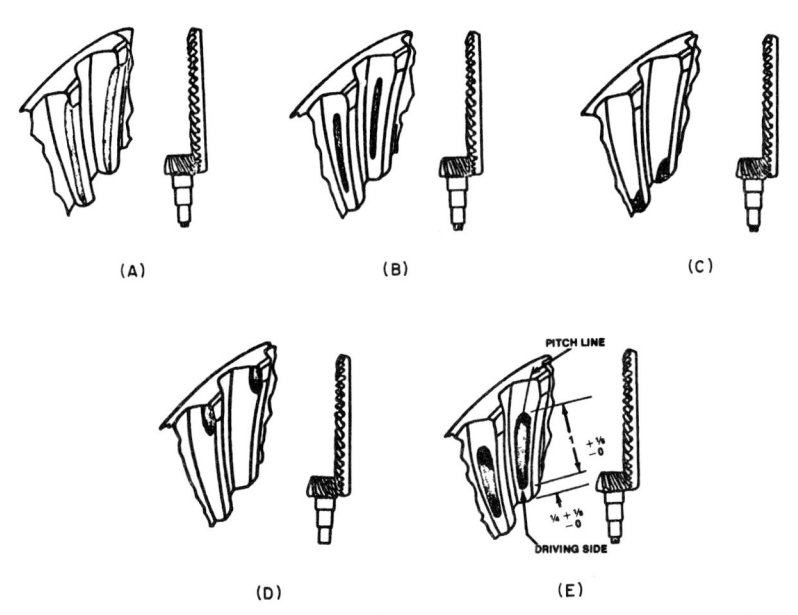

Fig. 3-20 (A) High, narrow tooth contact; (B) low, narrow tooth contact; (C) short toe tooth contact; (D) short heel tooth contact; (E) proper contact.

lightly to the drive side of the bevel gear teeth with no load on the gear set. When the pinion is rotated, then, red lead is squeezed away by contact of the teeth, leaving bare areas the exact size, shape, and location of the contacts. Sharper contact impressions can be obtained by applying a small amount of resistance to the gear when rotating the pinion.

When making adjustments, always check the drive side of bevel gear teeth. Coast-side contact will automatically correct when drive-side contact is correct. As a rule, coating about 12 gear teeth with red lead is sufficient for checking purposes.

Figure 3-20A shows *high, narrow contact,* which is undesirable and will result in noisy operation, galling, and overloading the teeth. To correct contact, add shims to move pinion in toward the toe of gear teeth (toward the center of the axle) to lower contact area to proper location. This adjustment will decrease backlash between pinion and gear teeth, which can be corrected by moving the gear away from the pinion. Several adjustments of both pinion and gear may be required before correct contact and backlash are secured. Correct backlash is 0.008 in. for the gear illustrated.

Figure 3-20B shows *low, narrow contact,* which will result in galling, grooving, and noise. To secure correct contact, remove shims to move pinion out from gear (away from center of axle) a sufficient amount to move contact area to proper location.

Figure 3-20C shows *short toe contact,* which, because contact area overlaps the toe of bevel gear teeth, will result in chipping of tooth edges and excessive wear at this point. To secure correct contact, move the gear away from the pinion. Doing so will increase lengthwise contact and move the contact area toward the heel of the gear teeth. Correct backlash is obtained by moving pinion in toward center of axle.

Figure 3-20D shows *short heel contact,* a condition which produces approximately the same results as short toe contact. To correct this condition, the gear must be moved in toward the pinion to increase lengthwise contact and move contact toward the toe of teeth. Proper backlash is obtained by moving pinion out toward the heel of gear teeth, away from the center of the axle.

Figure 3-20E shows correct tooth contact when adjustments have been properly made, which will result in efficient, long life of machine parts.

Chapter **4**

Static Weight Compactor Maintenance

ROGER ADE
Marketing Services, Hyster Company,
Construction Equipment Division, Kewanee, Ill.

Static weight rollers, better known as steel wheel rollers, are built for one purpose only—compaction. Contractors who build highways, airport runways, and parking lots know compaction is an important necessity in building these load-bearing surfaces. As a result, static weight rollers have become a very important tool in the contractor's arsenal of equipment.

Since the very early 1900s when steam rollers were used, the static weight roller has evolved into three basic categories. The *three-wheel roller*, with its two drive wheels and guide roll in a tricycle arrangement, is used primarily for breakdown rolling of asphaltic mixtures. The weight of this type of roller ranges from 5 to 14 tons.

The second category of static weight roller is the *three-axle tandem roller*, seldom used in today's road building. Weights range from 10 to 20 tons or more.

The *two-axle tandem roller* is the most popular of the three categories. Weights range from 3 to 14 tons or more (Fig. 4-1). The weight of each roller can be varied within a few tons by ballasting. Smaller rollers in this class may have auxiliary pneumatic tires to facilitate transportation from one job site to another. Tandem rollers are used extensively for asphalt finish rolling.

Modern static weight rollers are relatively easy to operate and maintain compared to other types of construction equipment. However, most tandem rollers are cumbersome to service because of their boxy design. Rollers with full-width hoods exposing the engine compartment aid in servicing the machine.

PREVENTIVE MAINTENANCE PROGRAMS

A main concern for any contractor on a paving job is preventing downtime. Good coordination between the asphalt plant, haul trucks, and paver is essential for a smooth-flowing paving operation. When asphalt is laid, the static weight roller must be ready to work. If not, the entire operation including paver, haul trucks, and men could be brought to a standstill.

The contractor must compact asphalt when it is hot to meet compaction densities required by the job specifications. As the asphalt cools, compaction becomes more difficult. A roller malfunction can be damaging to the contractor's timetable, especially if another roller is not on hand. This is one reason why a strong preventive maintenance program is essential to reduce costly downtime.

7-92 Maintenance Programs for Equipment Entities

There are other reasons for a preventive maintenance program (Fig. 4-2). For example, there are limited working days each year when a contractor can lay asphalt. For best results, weather conditions should be warm and dry. Contractors strive to get the most out of all their equipment on the days they can work. Good preventive maintenance is required if production is to be maximized.

Higher prices for fuel, parts, and labor are always on the contractor's mind. Longevity of

Fig. 4-1 The two-axle tandem roller, ranging from 3 to 14 tons or more, is the most popular type of static weight roller for asphalt paving work.

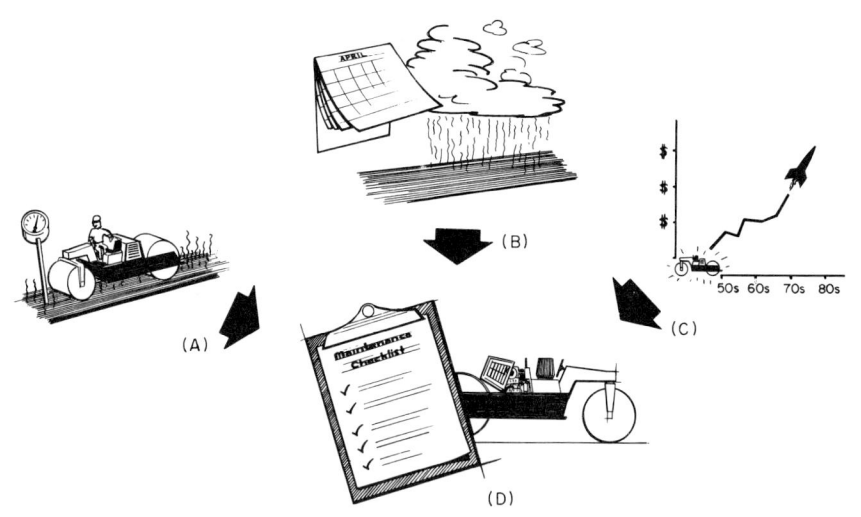

Fig. 4-2 Three major factors for keeping a roller in good running condition, (A) newly laid asphalt must be compacted while it is still hot, (B) working days are limited, and (C) costs are skyrocketing for fuel, spare parts, labor, and new equipment, are considered in a strong preventive maintenance program (D).

a piece of machinery is becoming much more important. Good preventive maintenance will prolong the life of a roller, allowing the contractor to get the most from his investment.

For these reasons—working when the asphalt is hot, limited working days, and prolonging the life of a unit—sound preventive maintenance is required. These items, coupled with increasing costs, quickly convince a contractor that a strong preventive maintenance program is a must.

PREVENTIVE MAINTENANCE REQUIREMENTS

All construction equipment manufacturers strongly recommend a preventive maintenance program. The owner's and operator's manual supplied with each unit provides a recommended maintenance schedule.

A strong maintenance program has three requirements (Fig. 4-3). The first and most important is the people who actually *perform* daily preventive maintenance.

The operator should accept the responsibility of walking around the machine to inspect daily checkpoints. The owner's and operator's manual, which locates all checkpoints, should be read, understood, and kept handy for further reference. There may also be a

Fig. 4-3 Successful preventive maintenance programs have three basic elements that make them work: (A) an observant operator who regularly inspects machines and reports current or potential problems, (B) quick responsive action on these reports, and (C) accurate records.

maintenance decal on the roller to direct the operator to maintenance checkpoints (see Fig. 4-4).

Any failure to report or repair current or potential problems could cause serious problems later. A good reporting system should be initiated to quickly and accurately communicate problems to the proper persons.

The second requirement in a good preventive maintenance program is quick and positive response to a known or potential problem. If the operator indicates a potential problem, take action *immediately*—even if it is a minor adjustment! Any problem has a potential of causing a major breakdown.

The third factor in a good maintenance program is a system for recording costs and hours. Well-kept records can serve as guides indicating when to replace or rejuvenate an old roller. Keeping good records of maintenance performed on the unit also provides a basis for planning efficient maintenance schedules. For example, it may be determined through records that the air filter should be cleaned more periodically than the manufacturer recommends. This action may be small, but in the long run, engine wear will be reduced, prolonging the life of an engine.

Records may also indicate the type of maintenance program for the unit. Most maintenance schedules are based on hours. However, because of certain applications or varying conditions, the need for maintenance cannot always be measured in hours. In these cases, a weekly, biweekly, or monthly schedule may be more applicable.

There are other factors, of course, which contribute to a good maintenance program.

One is the operator's attitude toward the machinery he works with. An operator will have a better attitude toward his roller if it is clean and in good working condition. Regular cleaning improves the appearance and removes harmful dirt from the roller. Painting and repairing visible dents each year will keep the roller looking like new, and the operator will care for it as if it were new.

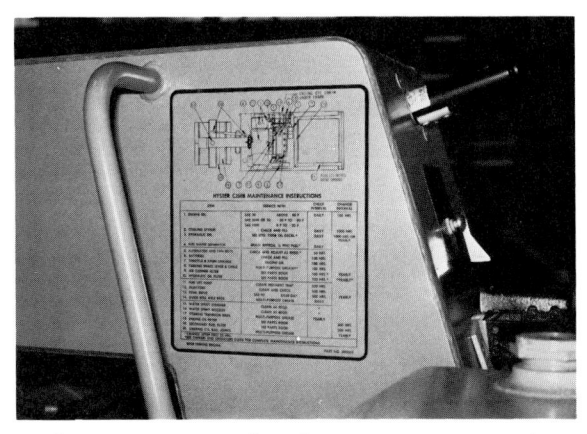

Fig. 4-4 Most modern compactors carry along their own maintenance markings as a guide to both operators and mechanics. Get both these groups to learn where these references are and how to use them.

PREVENTIVE MAINTENANCE PROCEDURES

Daily Maintenance Checkpoints

Static weight rollers begin their work day well after other units on a paving job. The asphaltic mix must be produced, transported, and laid down before the roller can begin compaction. During this time, preventive maintenance should be performed.

It doesn't take long to walk around the roller to check daily maintenance points. Daily checkpoints are all listed in the manufacturer's owner's and operator's manual (Fig. 4-5). Efforts to follow these recommendations will, in the long run, prolong the life of the roller and, more importantly, prevent a major breakdown on the job.

As the unit is visually inspected, check for liquid running down the rolls or frame, indicating a leak. Puddles on the ground may also tip off a fluid loss. Leaks should be traced immediately to determine their origin. Usually a loose fitting is the cause. However, by careful examination, a blown gasket or seal may be discovered which would require immediate attention.

Engine Lubrication Oil. This is a daily checkpoint not to be overlooked. The oil level should be up to the *full* mark on the dipstick. To accurately check the oil level, start and warm the engine. Stop the engine and allow about 5 min for the oil to drain off internal parts. This eliminates the possibility of overfilling. The oil should generally be changed every 100 hr or 2 weeks, even more often where adverse conditions prevail. When checking the oil level, clean around the dipstick area to prevent contamination.

If additional oil is required, use only approved oils. It is important to wipe off all oil that spills on the engine. Loose oil mixed with dirt can hide engine leaks caused by blown gaskets or seals. Such leaks can cause major failures if left undetected.

Cooling System. Checkpoints in the cooling system include coolant, radiator cap, and drive belts. Coolant level should be above the top of the coils. Always fill with clean water to prevent radiator clogging. And make sure the radiator cap seals the system properly. A worn cap will not hold pressure and may cause the system to overheat. Antifreeze levels should be checked during cold weather. An overnight drop in temperature could damage an unprotected cooling system. Drive belts should be examined for cracking and water.

Fuel System. Contamination of the fuel system can be kept to a minimum by observing a few simple daily rules. Always fill the fuel tanks at night when the unit is shut down to prevent water condensation on the interior walls of the tank from contaminating the

Fig. 4-5 Daily maintenance checkpoints include (A) engine oil, (B) cooling system, (C) fuel system, (D) hydrostatic system, (E) air cleaner.

fuel. The primary fuel strainer on diesel engines should be drained to remove dirt and water accumulations. Foreign particles or water in the fuel system can cause rough running, abnormal wear, and premature failure.

Hydraulic System. The hydraulic system is probably the most complex and most expensive system on modern static weight rollers. Special attention given to daily checkpoints helps ensure detection of potential breakdowns that could be very costly.

Check the oil level when the hydraulic system is cold. If the steering system and hydrostatic transmission share a common reservoir, fully extend the steering cylinder before checking the hydraulic oil level. Most manufacturers provide a sight gauge at the hydraulic tank, as shown in Fig. 4-5. The hydraulic tank oil should be visible or slightly above gauge level when oil is cold. Make absolutely sure that the oil level can be seen in the sight gauge before starting the engine and operating its hydraulic systems.

If additional oil is required, use only the manufacturer's recommended oil. Incorrect oil can cause abnormal wear and shortened component life. Do not completely fill the tank with oil. Air space is designed into the tank for expansion at warm temperatures. Pressurization will occur in the tank at operating temperature on machines using a nonvented hydraulic reservoir.

Air Cleaner. The air cleaner is an important part of a well-running engine. It may not be necessary to remove the element daily if there is an air restriction indicator as shown in Fig. 4-5.

Water Spray System. This system requires some attention in combating contamination. It is a good practice to flush water spray manifolds periodically and to fill only with *clean* water.

The manufacturer's owner's and operator's manual will indicate daily grease and lubrication points such as guide rolls, steering trunnions, and certain drive train components. Failure to properly lubricate these points can cause excessive wear and premature failure.

100-hr or 2-week Maintenance Checkpoints

Air Cleaner(s). These should be checked every 100 hr or 2 weeks (see Fig. 4-6). Reusable elements can be cleaned periodically, but after 10 cleanings they should be replaced. Air restriction indicators on some models signal when the filter needs cleaning or changing.

Engine Lubricating Oil. Change the engine lubricating oil and filters at intervals of 100 to 200 operating hours. Refer to the manufacturer's owner's and operator's manual for specifications on the type of oil to be used as well as change intervals.

Batteries. Batteries need attention periodically to ensure a normal life. Make sure batteries are secured at all times. Vibrations and jolts can damage internal plates and cause premature failure. The water level in each cell should never be below the top of the plates. Keep terminals clean and tight.

If batteries are continuously losing charge, it may be time to replace them. *But*, do not overlook the possibility of an electrical system failure. A defective alternator or regulator, or an external short could continually drain battery power.

Hydraulic System. Examine this system regularly to check for contamination. A vacuum gauge provided by the manufacturers of rollers with hydrostatic-type transmissions will indicate if a filter needs replacing. A reading above 10 in. Hg indicates the need for a filter change.

Check for loose hose fittings and worn hoses. Tighten or replace fittings and hoses which may fail under high pressures.

Transmission, Throttle, and Park Brake Adjustments. These should always be adjusted properly, not only as a good maintenance procedure but for personal safety as well. Periodic checks and adjustments keep the roller fine-tuned for the operator. Oil sprayed on push-pull cable ends will facilitate smooth operation and rust prevention.

After these checkpoints have been reviewed, the roller is ready to work. Before putting the roller on the job, however, allow it to warm up thoroughly. Periodically scan the gauges to detect any abnormalities.

While running the unit, watch for trails of liquid indicating leaks. Any sluggishness or abnormal behavior should be looked after as soon as possible. Again, there is a possibility of a minor problem developing into a major breakdown if it is not detected and corrected.

Preventive maintenance doesn't stop when the engine is shut down at the end of the

Static Weight Compactor Maintenance   7-97

day. That is the time to check into problems and report them to the proper persons. Any adjustment or repair requirements can be made before the next work shift.

To have a successful preventive maintenance program, the daily and periodical checkpoints must be inspected regularly. Preventive maintenance can be performed after the work shift. Contractors who have a second shift often perform maintenance during the evening, allowing operators to start the next day without delay.

It doesn't matter when preventive maintenance is performed—just so it *is* performed.

MAINTENANCE AND TROUBLESHOOTING OF MAJOR SYSTEMS

This section is divided into the major systems common in modern static weight rollers. Although daily maintenance can provide a constant watch on the roller, there are many elements of each system that require special attention. This section deals with good maintenance and troubleshooting procedures for each system.

All static weight rollers have power trains consisting of an engine, transmission, and final drive assembly. These will be the first items discussed. Related systems such as cooling, fuel, and steering follow. By inspecting each system periodically, you can eliminate unnecessary downtime.

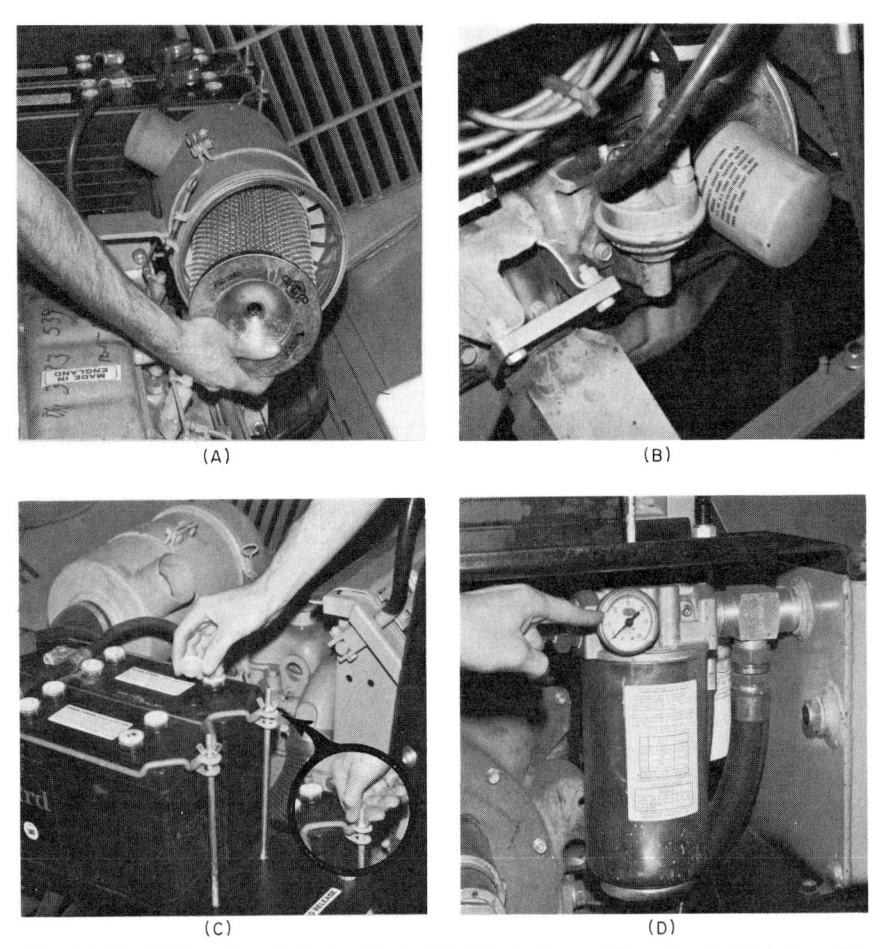

Fig. 4-6 The 100-hr or 2-week checkpoints include (A) air cleaner, (B) engine oil, (C) battery, (D) hydrostatic systems, plus various controls and adjustment points.

Engine

Lubricating Oil. The lifeblood of any engine is the lubricating oil. Special additives such as detergents and antifoam and corrosion inhibitors work to clean, lubricate, and reduce engine wear. Proper oil selection is critical to maximize fuel economy and minimize friction and wear.

Common oils are SAE 30, SAE 20, and SAE 10W. Refer to the manufacturer's manual for specific recommendations. One oil may be preferred over another because of ambient temperatures or working conditions.

Oil viscosity is based on thickness. Essentially, the higher the number, the thicker the oil. The letter W following an oil grade (e.g., SAE 10W) indicates that it is a good cool weather oil.

Ambient temperature is the biggest factor in selecting an oil recommended by the maufacturer. Of course, in cold weather a thicker oil such as SAE 30 will require more effort in starting the engine than, say, a SAE 10W grade. Thicker oils are recommended in warm weather operations.

Each additive in an oil performs a particular function in prolonging engine life. Detergents and dispersants reduce sludge and grime suspended in oil. Antioxidants and corrosion inhibitors serve to protect oil consistency. Foam inhibitors deter air bubbles which tend to reduce the ability of oil to perform effectively.

If regular inspections of the oil are made, any sudden rise in the usage rate can be discovered quickly. This may give an early warning of a blown gasket or seal before major damage occurs.

Oil Analysis Services. These services are extremely effective in detecting internal engine problems before failure occurs. Scientific analysis of oil samplings can detect minute changes in the oil consistency caused by unusual wear or damaged parts.

For example, an increase of copper particles suspended in oil may indicate flaking or unusual wear in a thrust washer. Early detection of such a problem could allow enough time to pull the roller off the job when the work schedule is light and repair the problem. This prevents the inevitable failure on the job site.

Oil analysis is invaluable when weighed against the detrimental effects of downtime. Oil companies, private institutes, and some local dealerships provide this service. Take advantage of it!

Engine Crankcase Breather, Blower Screen, and Air Box Drain. These maintenance checkpoints are often neglected. They should be free of excessive dirt. The air box drain removes contaminants from the system. Thorough cleaning each year is recommended even though encrusted dirt may not be visible.

There is no substitute for a well-tuned engine. Smooth-running engines save fuel and reduce wear. Also, keep a watchful eye on the electrical system. Excessive dirt and loose connections can cause deterioration and failure of wires.

Hydrostatic Transmissions

Modern static weight rollers utilize hydrostatic transmissions because of their smooth operation. On asphalt, where the rollers are constantly moving forward and backward, hydrostatic transmissions offer smooth directional change without scuffing the mat and damaging the surface. Infinitely variable changes in speed and roller direction can then be controlled through one lever.

Hydrostatic transmissions work in a closed-loop hydraulic circuit (Fig. 4-7). The variable-displacement hydrostatic pump mounted to the engine converts mechanical rotary energy into hydraulic flow. When the variable-displacement pump swashplate is tilted, by the speed/directional control, a positive stroke to the pistons is created. The pistons displace oil, creating a flow. High-pressure lines carry the flow to the motor which is coupled to a final drive, where it is then converted back to rotary energy. Speed is controlled by adjusting the displacement (flow) of the transmission.

Other components working in the circuit include the hydraulic reservoir, filters, and oil cooler. Regular maintenance checks performed on these components can prevent damage to the transmission caused by excessive contamination, heat, or pressure.

Maintenance on any hydraulic system starts with cleanliness. Abrasion, caused by contamination, not only wears a transmission out quickly but can ultimately cause irreparable damage. Contamination may be classified in three categories: soft, abrasive, and chemical. Bits of rubber hose or seals are forms of soft contamination. Abrasives include lint and dirt. Chemical contamination may include unapproved oils, additives, or water. When contamination is found in one part of the system, the whole system must be flushed and new hydraulic oil added.

Using the wrong type of oil can be just as damaging to a transmission as dirt. Use *only* approved oils specified in the owner's and operator's manual supplied with the roller.

Always keep the oil in the hydraulic reservoir at the proper level. Too little oil causes

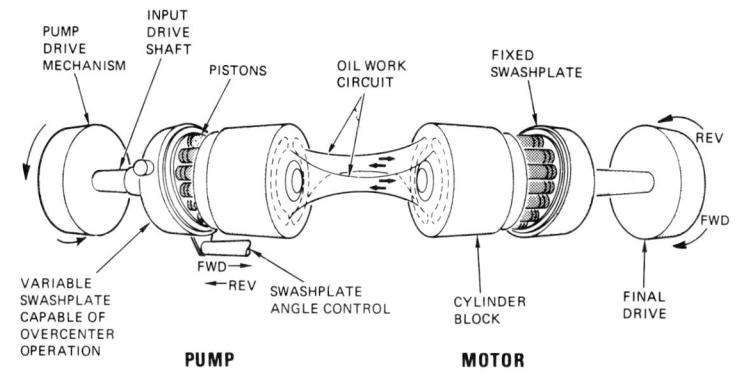

Fig. 4-7 Hydrostatic transmissions convert mechanical rotary energy from the engine into hydraulic flow and back to mechanical energy in the final drive.

Fig. 4-8 Vacuum pump readings should not exceed 10 in. Hg at normal operating temperatures.

overheating. Low oil levels can also cause aeration which works as an abrasive in the form of air bubbles. And, of course, adequate oil is required to produce power.

In Hyster rollers and possibly in other rollers, all the oil from the reservoir is passed through a 10-micron suction filter before it reaches the pump. A vacuum gauge mounted on the suction filter indicates the flow at which the transmission is pulling oil through the filter (Fig. 4-8). The maximum vacuum should not exceed 10 in. Hg at normal operation temperatures. During cold weather start-up it is acceptable to have a higher vacuum because of the thicker, cold oil. However, the roller should *never* be operated until the oil is at normal operating temperature and vacuum is below 10 in. Hg.

Hydraulic hoses need attention periodically to detect unusual wear. Vibration and constant rubbing against the frame causes outer hose coverings to wear. Cut or blistered hoses should be replaced before they have a chance to fail.

The most important factor in hose maintenance and replacement is selecting the proper hose. When selecting a hose, make sure it has the proper pressure rating. Allow for the manufacturer's recommended bend radius to prevent unnecessary stress on the hose and couplings. A short hose will put undue stresses on both hose and couplings. Also allow for expansion and contraction of hoses under varying pressures. Of course, the best way to ensure selection of an approved hose is to order approved replacements from your roller dealer.

When installing a hose, refrain from twisting it. Also, double-check the fittings to insure proper coupling. Loose connections can allow air to enter the system or hydraulic oil to leave the system. New "O" rings in high-pressure couplings will provide a good seal.

Troubleshooting a hydraulic system Troubleshooting the hydraulic system should be done with a systematic procedure. Start by first inspecting those possible problem areas which can easily be reached. Eliminate each possible cause involving no disassembly before tearing down transmission components.

Of course, before one can troubleshoot the hydrostatic system, he must know the system

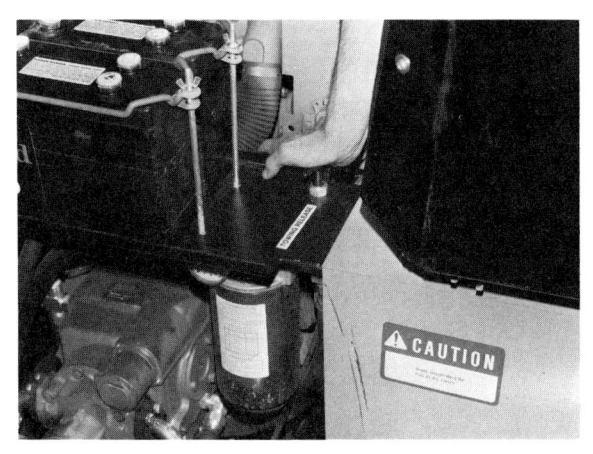

Fig. 4-9 A partially opened towing release valve allows hot oil to bypass the oil cooler, thus overheating the system.

thoroughly. Many procedures require the use of pressure gauges. So, know the system's pressure requirements as listed in the manufacturer's service manual.

Following is a list of the most common malfunctions associated with a hydrostatic transmission. Before disassembling any transmission components, check to make sure you are not voiding the warranty.

Neutral Difficult or Impossible To Find. The control linkage should always be adjusted properly, not only as a good maintenance practice but also for personal safety. However, roller vibrations and wear can spur periodical adjustment.

If the linkage is properly adjusted, the displacement control valve mounted on the pump may be out of adjustment or have internal broken linkage. This valve controls the angle of the pump's swashplate, which in turn, ultimately controls oil displacement. Adjustment of this valve is very critical and should be done with caution. Block the drive roll off the ground and adjust as recommended in the manufacturer's service manual.

System Operating Hot. Low oil level or a dirty suction filter are two possible causes of overheating. Both will result in an inadequate amount of lubrication and cause heat buildup. Next, feel the oil cooler with your hand to make sure it is dissipating heat. Make sure the towing bypass valve (if one is provided) is completely closed (Fig. 4-9). A partially opened valve will allow hot oil to return to the low-pressure side of the system, bypassing the oil cooler.

Some consideration must be given to working conditions. Are you working the system beyond its designed capabilities? Check the system's high pressure to ensure it is within the specified limits.

Transmission Operates in One Direction Only. If it is determined that the control linkage is adjusted properly, check the system's high-pressure relief valves. There are two of these valves (one for each direction) in the motor manifold assembly mounted on the hydrostatic motor. They serve to prevent abnormal pressure surges, caused by rapid acceleration or sudden braking, from entering the motor. A defective valve can be detected by merely swapping valves (Fig. 4-10). If the unit does not operate in the other direction, you may suspect a defective valve.

Defective charge check valves may cause operation in one direction only. Check valves serve to admit oil from the charge pump to the low-pressure side, maintaining adequate lubrication oil. This oil is also used to hydraulically control the displacement control valve. It controls the output of the transmission. Of course, a faulty check valve could prevent the oil from reaching the control valve, thus not actuating the valve.

To determine if a check valve is faulty, remove the valve and pour a small amount of oil in the bottom. Let it sit for a while. A good check valve will hold this oil. Replace the valve if it does not.

If pressurized oil reaches the control valve, check the valve for broken or damaged linkage.

The shuttle valve, located in the motor manifold, should be checked if the cause has not yet been determined. The shuttle valve simply allows high-pressure oil to flow in the right direction. A broken spring in the valve or contamination causing the spool to stick can be determined by disassembling the valve.

Fig. 4-10 A defective high-pressure relief valve can be detected by swapping it with the other high-pressure relief valve.

Unit Will Not Operate in Either Direction. The first three items to check are the amount of oil in the reservoir, the control linkage, and the towing release valve. Next, it is necessary to check the charge-pressure reading from the hydrostatic pump (Fig. 4-11). The pressure specifications for the neutral and foward/reverse directions will be given in the manufacturer's service manual. If pressure readings do not match specifications, the charge pump, charge relief valves, or inlet filter may be the cause.

If it is determined that the charge pressure is as specified, it will be necessary to check the system's high pressure (Fig. 4-12). The high-pressure relief valves may be malfunctioning. Check for contamination or wear and replace if necessary. Make sure you are not working the system beyond its designed capabilities.

Inspect the displacement control valve for broken linkage or damage. Replace if necessary.

System Response Sluggish. Low charge pressure reaching the displacement control valve can cause sluggish response. If the pressure readings are not within manufacturer's limits, the problem may be faulty charge relief valves in the pump or motor. Low charge pressure may also be caused by a plugged inlet filter. Inspect the displacement control valve for malfunctioning if the charge pressure is within limits.

Contamination partially plugging the control office in the pump may cause sluggish response. If this is the case, the inlet filter, and possibly hydraulic oil, should be changed to prevent damage to the transmission.

Of course, an old transmission system will eventually develop enough internal wear or damage that it may cause slower response. This would indicate the need for replacement or overhaul of the transmission.

System Operates Noisily. A noisy system usually indicates air in the hydraulic system. Air can enter from loose hose couplings or filter connections or low oil level in the reservoir. Considerable foaming in the reservoir indicates aeration. Of course, aeration of the hydraulic oil will reduce its lubricating ability.

Cavitation, or a vacuum of fluid, will have much the same effect as aeration. It may be caused by the motor load overrunning the delivery of the pump. Restricted inlet lines or a clogged inlet filter will, of course, prevent adequate oil flow.

Refer to manufacturer's service manual if questions arise concerning troubleshooting

Fig. 4-11 High-pressure check ports in the hydrostatic motor include neutral and forward and reverse.

Fig. 4-12 Charge-pressure check ports in the hydrostatic pump and motor.

and repair procedures. The service manual will also specify what you can and cannot disassemble within the warranty policies when repairing the roller.

Torque Converter, Transmission, and Clutch

Before the advent of the hydrostatic transmission, powertrains consisted of a clutch, torque converter, and mechanical-type transmission. Although they are not as smooth as hydrostatic transmissions, proper adjustment will promote smooth and positive operation.

Check the oil level in the transmission every 50 hr of operation. Do not overfill a transmission when replenishing oil. Excessive oil develops foaming and high oil temperatures. Change transmission oil and filters once yearly.

Maintain the proper lubricant level in the torque converter as well. Avoid excessive dirt buildup around check and fill ports as excessive dirt entering the torque converter or transmission can reduce their life and efficiency.

Final Drive

There are two basic types of final drives used in modern static weight rollers. The first is the pinion and spur gear combination, a long-time workhorse in the industry. However, this type is gradually being replaced by the planetary-type final drive. Planetary final drives deliver a smoother power transfer, have better load distribution, and are more compact in design.

Pinion and spur gear final drives require daily and sometimes twice-daily lubrication. Bearings must be greased regularly and gears lubricated by brushing or pouring oil on their contact surfaces. Refer to the manufacturer's owner's and operator's manual for special maintenance schedules.

In sharp contrast, planetary-type final drives require little daily maintenance. Periodic checks of the oil level should be made. Drain and flush the oil once a year and examine the lubricant for unusual wear.

A daily walk around the roller should include inspection of the final drive seals. If leakage is detected, the roller should be pulled off the job as soon as possible before major and expensive failure occurs.

Common causes for final drive failures are lack of lubrication, bearing or seal damage, and overloading. Excessive heat buildup is a sign of inadequate lubrication or the wrong type of lubrication. Whenever disassembling a final drive, replace all seals and "O" rings with new ones. Final drives and downtime are too expensive to trust reused seals and "O" rings.

Cooling System

The most common cause of overheating problems is lack of coolant. Restricted coolant and air flow also has an effect on the cooling efficiency.

Dirt and corrosion will eventually clog a radiator's tubing and restrict coolant flow and heat distribution. Coolant additives, such as rust inhibitors, work to cut corrosion; however, the system should be flushed each year to ensure cleanliness.

Restricted air flow may stem from several areas. Excessive dirt in the cooling fins will reduce air flow. Clogged or bent fins reduce the area in which the radiator can disperse heat (see Fig. 4-13).

Shrouds and baffles mounted between the fan and radiator direct air flow, increasing the fan's efficiency. Both must be in place and properly adjusted. Too much clearance between the blade tips and shroud will allow the fan to draw air from around the shroud and not through the radiator as intended. Baffles must be in place to prevent warm engine compartment air to recirculate through the radiator.

Fan belt adjustment has a definite effect

Fig. 4-13 Bent or clogged radiator fins reduce air flow and can cause overheating.

on the air flow. Fan speed is reduced if the belts or pulley grooves are worn enough to cause slippage. Loosely adjusted belts cause excessive belt vibration and wear. Belt life can be reduced if adjusted too tightly, causing undue stress.

The cooling system is pressurized to raise the coolant boiling point. Badly worn or damaged radiator caps allow pressure leaks, causing the liquid to boil quicker. When inspecting the cap, check the sealing surfaces. Also, inspect the radiator neck for damage. The cap should seal at the minimum recommended pressure rating (usually 7 psi).

Fuel System

All through this chapter *cleanliness* has been stressed as a main factor in roller maintenance. The fuel system is no exception. It too must be kept clean for maximum performance. Association with a reputable fuel dealer who provides consistently clean fuel is to your advantage.

Gasoline fuel systems consist of a tank, carburetor, and fuel line connecting the two. The fuel filter, located in the inlet line to the carburetor, requires changing at least every 300 hr.

To reduce overnight water condensation, fill the fuel tank after the last work shift. No matter how careful you are, however, the tank will accumulate water and dirt sediments. Periodically drain the tank to remove this contamination.

With ever-increasing fuel bills, it is very important to keep the carburetor on gasoline engines properly adjusted. It should be inspected periodically to detect leakage and maladjusted linkage.

Diesel fuel systems are much more complex than gasoline systems. The precision and high-quality materials used by diesel engine manufacturers, however, can also be harmed by contamination. To prevent this, most diesel systems provide two fuel filters which require daily draining. Replaceable elements should be changed at least as often as the manufacturer's recommended maintenance schedule, more often in dusty applications.

Drain the tank periodically to remove water and dirt sediments. Check fuel lines and fittings for fuel leakage or air infiltration.

Troubleshooting a diesel fuel system

Clogged or Restricted Fuel Lines and Filters. These problems cause hard starting, rough idle, and erratic operation. If you find excessive contamination in the system, determine whether the contamination is from prolonged buildup or from newly pumped fuel. After detecting excessive contamination, it would be wise to replace fuel filters also.

Rough or Erratic Engine Performance. This problem may be caused by air leaks in the line. Loose or broken connections on the suction lines reduce the amount of fuel reaching the cylinders and allow air to enter the system.

Excessive Black Smoke. Excessive smoke is often the symptom of an inferior grade of fuel. After tanks and lines have been drained of such fuel, replace with a higher grade of fuel. The excessive smoke will disappear after a short time.

Improper Timing. Excessive black, white, or blue smoke can also be caused by improper timing. Improper injection pump timing will also cause a rough running engine. Follow the manufacturer's recommended troubleshooting procedure when timing problems occur.

Before disassembling components of the fuel system, know the manufacturer's warranty policy. An authorized engine dealer will be able to answer your questions or repair a malfunctioning system.

Steering System

Hydraulic steering systems are widely used on static weight rollers. Components in such systems include a hydraulic oil reservoir, hydraulic pump, control valve, and hydraulic cylinder(s).

Most hydraulic steering systems share the oil reservoir with the hydrostatic transmission. The precautions taken to ensure clean hydraulic fluid in the transmission will serve to provide clean fluid in the steering system. The steering system has a relief valve for protection from sudden or excessive pressure surges.

Hydraulic pumps, mounted on the engine, may be either gear or vane type and driven by a camshaft, pump drive mechanism, or auxiliary belt off the engine. The control valve is usually mounted just below the steering wheel. It serves to meter the proper amount of oil to the steering cylinder mounted to the guide roll's kingpin.

Maintaining the steering system requires regular inspection of the pump, control valve, and cylinder for leakage or damage. Check hydraulic oil daily to assure proper level.

Troubleshooting the steering system

Low Oil Level or Contamination. These problems cause slow or sluggish steering response. If the transmission shares the hydraulic oil, it too will show these symptoms. If you do find contamination in the system, clean or replace the filters and change the oil.

Spongy or Jerky Actions. These actions and a noisy pump indicate air filtration, usually from the suction side of the system. Inspect connections, hoses, and seals to find the source of air leaks. After correction, bleed the system to eliminate air bubbles.

Slow or Hard Steering. These symptoms may indicate several problems. First of all, make sure the pressure relief valve is adjusted properly. Refer to the manufacturer's service manual for proper pressure settings. If the pressure relief valve is not functioning, remove and clean or replace as necessary.

Built-up Dirt and Sludge. Dirt and sludge decrease the system's performance and increase wear. Needless to say, this is not good for the systems. Eventually, worn parts will cause excessive internal leakage causing slow or hard steering.

If there is no steering response, check for a broken belt, shaft, or coupling at the power transfer point between the engine and pump or for internal damage in the pump, control valve, or cylinder.

WINTER STORAGE AND REPAIRS

In many areas of the country, contractors cannot lay asphalt surfaces in the winter months because of inclement weather. The static weight roller, along with the other asphalt paving equipment, remains idle until the following spring. This is an excellent time to repair and rejuvenate the unit for the next season's work.

Before putting a roller in winter storage, certain precautions should be taken to prevent damage and corrosion caused by the changing weather conditions (Fig. 4-14).

First of all, *drain all water ballast in the drums!* Water expands when it turns to ice, and it doesn't let a steel drum stop it. Warping, cracking, and bending will ruin the drum.

A mixture of fuel oil and 10 percent preservative oil (such as Shell Ensis 30) should be run in a diesel engine for a period of 15 to 20 min before engine shutdown. After allowing the engine to cool, turn the engine over several times without starting. This sprays the mixture into the combustion area.

The fuel tank should be filled to the top. The air cleaner or carburetor intake, exhaust, and crankcase breather openings must be taped or otherwise sealed to prevent the entrance of moisture. Winterize the cooling system to withstand the lowest temperature the unit will be exposed to during storage. Change engine oil and filter. Cover the engine with a tarp or cover the hood openings with plywood or hardboard.

Remove batteries and store in a cool dry place above freezing. Check and maintain water level and specific gravity while in storage.

Drain the water spray tank and pump of all water. It is especially important to drain the pump when it is mounted under the tank. Drain plugs or drain cocks are located in the bottom of most water spray pumps. Cap or tape the spray nozzles and coat the tank fill cap thread with grease if it is metal.

Coat hydraulic cylinder rods and exposed control valve spools with heavy grease. All grease-packed bearings should be repacked and bearing preload adjusted. Spray all linkage, control cable ends, hinges, and other such moving parts with a rust-preventative oil.

Of course, there's nothing like a new paint job during the winter to improve the looks of the roller. However, good looks isn't the only advantage of a new paint job.

Repainting requires that the roller be thoroughly cleaned. Harmful dirt otherwise not detected in normal maintenance procedures is usually removed. This prevents a certain amount of corrosion caused by chemicals in the dirt. Rust is removed by sanding or

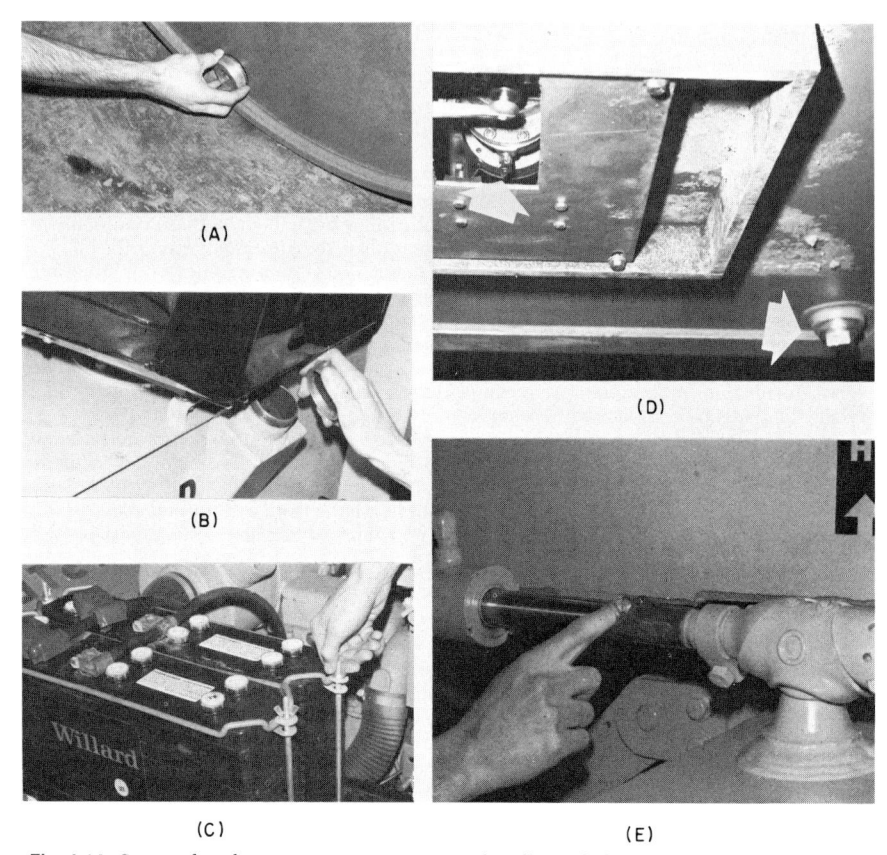

Fig. 4-14 Steps to be taken to winterize a static weight roller include (A) drain ballast, (B) fill fuel tank, (C) remove battery, (D) drain water spray system, and (E) coat exposed surfaces.

blasting, allowing the new paint to seal the surface. Repair of dents and blemishes checks the spread of corrosion. Repainting should be considered as much a part of good maintenance as regular preventive maintenance procedures. An operator will tend to take better care of the unit if it looks and runs like new. In turn, he'll maintain it better, thus reducing downtime and costs. And that's what maintenance is all about.

Chapter 5

Hydraulic Excavator and Backhoe Maintenance

ROBERT W. WOLTERSDORF
Technical Services Manager, Crane and Excavation Group,
Koehring Corporation, Milwaukee, Wisc.

INTRODUCTION

Hydraulic excavators and backhoes (Figs. 5-1 and 5-2) will perform only as well as they are maintained. Neglect of this piece of equipment is a very costly mistake made by many contractors.

Minutes spent on daily maintenance could very conceivably save many hours or days of downtime. What is the cost of one man's wages for 30 min each of preoperation and postoperation maintenance? Compare this to the cost of downtime on the job. On most jobs, the hydraulic excavator or backhoe is the main piece of equipment. All others depend upon this piece being in constant operation. Downtime for the hydraulic excavator or backhoe means trucks waiting, workers idle, and costly time in job completion. You cannot replace lost time!

There is *no substitute* for proper maintenance. Inspection, lubrication, and cleanliness are the three points of a successful operation. *Keep your machine productive—keep your machine safe!*

MACHINE LOGBOOK

A definite scheduled maintenance procedure should be planned and carried out regularly. Perhaps the best method to assure proper maintenance is to keep a daily log (Fig. 5-3) for each piece of equipment. Some manufacturers provide a logbook for their equipment. If this is not available, some type of maintenance checklist should be made up for use. Be sure this checklist is initialed by the person responsible for performing these duties to encourage conscientiousness.

DAILY INSPECTION

Daily inspection of the complete machine should be as much an item of maintenance as lubricating or adjusting the machine (Fig. 5-4). At the end of the work day, *inspect the entire machine* for leaks, loose connections, signs of wear, cracks, etc. Report any signs of trouble discovered during this inspection. Remember to check nuts and bolts. A loose connection quickly pounds itself into a dangerous connection. Take special care of all heat-treated bolts. They were installed because they were required. *Never substitute a lesser-grade bolt when replacing hardware!*

7-108　Maintenance Programs for Equipment Entities

Never take anything for granted. Don't assume that everything is all right at the start of the work day just because everything seemed all right at the end of work yesterday. *Before* beginning operation, *thoroughly inspect* the *entire machine* for signs of vandalism. An isolated machine arouses curiosity. Check all cab doors for signs of tampering. Check for loose connections.

Fig. 5-1 Hydraulic excavator with backhoe attachment.

Fig. 5-2 Hydraulic excavator with front shovel attachment.

GOOD HOUSEKEEPING

Cleanliness should not be overlooked as an item of maintenance. A clean machine is a safe machine. In the process of cleaning, future trouble may be discovered and avoided.

Cleanliness and inspection go hand in hand. While inspecting the machine, the operator should also be cleaning areas and observing if there has been leakage.

Hydraulic system　Keep the areas around the hydraulic control valve, hydraulic pumps, and hydraulic motors clean. In this way, seal or ring leakage in its early stages can be discovered (Fig. 5-5).

Hydraulic Excavator and Backhoe Maintenance   7-109

Fig. 5-3 Daily logbook is a definite asset in encouraging proper maintenance.

Fig. 5-4 Daily inspection. Trouble in the making might be discovered during this *before* and *after* work inspection.

Fig. 5-5 Hydraulic system cleanliness. Leakage can more easily be detected if the machine is kept clean.

Operator's cab The inside of the cab should be kept clean at all times (Fig. 5-6). Dirt or debris under the foot pedals could be a potential hazard. Do not keep tools, rags, or grease guns on the floor. Keep the windows clean and the windshield washer reservoir filled with an ample amount of fluid.

Undercarriage A regular cleanout in this area will help in preventing and detecting early problems. This equipment is designed for job applications which create debris. When traveling, this debris will have a tendency to try to build up in the track rollers and tumblers. Provide the operator with a stout shovel and assign him the responsibility of keeping the undercarriage clean of all debris (Fig. 5-7).

Alert the crew on the job site to visually inspect the undercarriage when walking by the machine during the work shift. Watch for rocks or debris that may become wedged between the rollers. In this instance, alert the operator so that he can dislodge the wedged material before any harm can be done to the rollers or tumblers.

When working in areas of continuous freezing and thawing, mud can create a problem (Fig. 5-8). Mud, allowed to harden overnight, will restrict the move-

Fig. 5-6 Good housekeeping. Keep your own working area clean and safe.

Fig. 5-7 Debris in the undercarriage can be the start of future problems.

ment of the rollers and tumblers. This can cause premature wear and improper operation. At the end of the day's shift, back out of the excavation area to a dry position and clean this mud from between the rollers and tumblers.

By keeping the undercarriage clean, oil leaks in the rollers and tumblers can be detected at an early stage.

LUBRICATION

The importance of lubrication cannot be overemphasized. Every manufacturer provides a lubrication chart permanently fixed to the machine. The operator or oiler should study the chart carefully and set up a pattern to follow so that no point of lubrication is overlooked during his regular lubrication schedule (Fig. 5-9). This is especially important in reference to the points that require greasing daily or even more often (Fig. 5-10).

Lubrication, done properly, depends on four essentials: the correct lubricant, the right quantity, the right time, and the proper application. All four essentials must be followed to guarantee that equipment is serviced correctly. Let's break the four essentials down and find out why they are necessary.

The correct lubricant As in all products, lubricants have certain properties to perform a function. Be especially conscious of ambient temperatures. A grease or oil that flows at 50°F may become solid at −10°F. If any special grease is recommended, it is because of the specific requirement of this application. Never substitute a lesser-grade or lesser-value lubricant. Always use at least the equivalent of the manufacturer's recommendations. This is especially true for hydraulic oils used in the system. (Refer to the next section, Hydraulic Fluids.)

The right quantity All drive cases should be kept to the level specified by the manufacturer (Fig. 5-11). Too low a level results in bearing or gear failure. Too high a level might result in too great a pressure buildup at higher operating temperatures or leakage into other areas.

Usually, the manufacturer will stipulate the quantity of lubricant to be used, especially on specific applications such as ring gear bearing. Generally speaking, four to six shots from the grease gun will take care of lubrication points on the base machine. For the attachment bushings, pump until clean grease is visible.

Fig. 5-8 Mud, if allowed to dry overnight (or to freeze overnight in winter) can stop movement and cause premature wear.

Fig. 5-9 Checking the lubrication chart. Know the lubrication points and follow the manufacturer's recommendations.

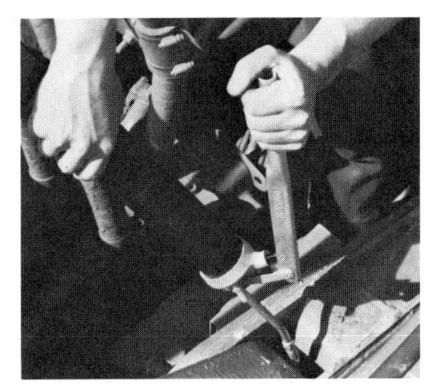

Fig. 5-10 Applying lubricant for bearings and bushings. Regular lubrication is a *must* in maintaining hydraulic excavators.

The right time Lubrication frequency as given on the lubrication chart should never be extended beyond the manufacturer's recommendations. This chart is to be used as a guide. In many instances, under unusual or severe operating conditions. (Fig. 5-12) the frequency of lubrication may be cut in half. (For example, when digging in water or under extreme abrasive conditions, lubricate the attachment fittings every 2 or 4 hr as conditions warrant.) Use good judgment and common sense when determining the frequency of lubrication for your equipment, dependent upon job conditions.

The proper application Before attempting any lubrication, be sure of cleanliness for the lubricants (Fig. 5-13). Clean the outside of the grease gun before using and wipe off the outside of the fittings before applying the lubricant to prevent grit from being forced into the fitting. Keep all oil containers, funnels, and oil spouts clean.

Drain all oil cases when the oil is hot so that the draining oil can carry off the sludge.

Automatic lubricators Some machines are equipped with automatic lubricating systems. The word *automatic* means self-acting, but it does not mean self-inspecting or self-filling! The oiler or operator must inspect the automatic lubricator for the quantity of lubricant in the reservoir (Fig. 5-14). He must also inspect the terminal points of the grease lines to make sure the proper amount of grease is getting to the bearings. The injectors should also be checked periodically for the manufacturer's recommended adjustment.

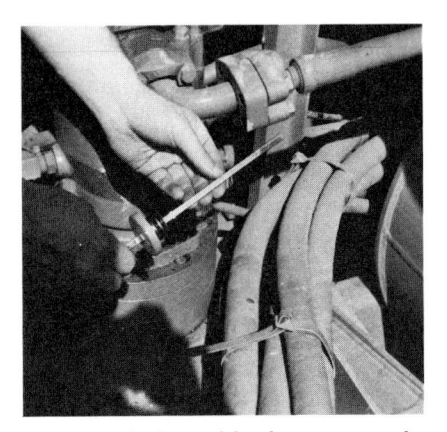

Fig. 5-11 Checking oil levels to maintain the proper quantity of oil at the specified level.

HYDRAULIC FLUIDS

Hydraulic excavator manufacturers do a great deal of research and testing of hydraulic oils to select an oil best-suited for their equipment. They recommend an oil suitable to the demands of the job applications of a hydraulic excavator.

There are several different types of fluids generally used in hydraulic systems: engine oils, automatic transmission fluids, and hydraulic oils. While engine oils and automatic transmission fluids are used generally for smaller construction equipment, they are not recommended for hydraulic excavators.

Hydraulic excavator manufacturers recommend hydraulic oils be used in their equipment because their qualities are most suitable. Hydraulic oils are antidetergent, thus permitting water, etc., to settle at the bottom of the reservoir where it can be drained off.

Fig. 5-12 Job conditions. Severe service dictates a more rigid scheduling of maintenance intervals.

These oils also contain antiwear additives such as zinc diophosphate for lubricating in the critical tolerance areas. They also contain antioxidants, corrosion inhibitors, and antifoam agents.

Use the manufacturers' recommendations for the hydraulic oil to be used in their system (Fig. 5-15). Most manufacturers provide a list of oil companies and their hydraulic fluid comparable to the fluid used in the excavator at time of shipment from the factory.

Changing from one oil manufacturer to another is permissible when the complete system is drained and refilled; however, mixing oils on an add-on basis is not necessarily recommended. Unless it is a physical impossibility to stay with the same oil, do not mix brands of oil in the system. Various brands could have a chemical reaction between themselves. Do this only on an emergency basis.

HYDRAULIC SYSTEM

Maintenance of the hydraulic system is a relatively simple matter because it is an entirely closed system. Cleanliness is the key factor in hydraulic system performance, efficiency, and life. Remember this *before* attempting *any* service operations. All the surrounding area of the part to be serviced should be thoroughly cleaned and free of all contaminants.

Hydraulic tank Be sure that the fluid level in the reservoir is kept at the proper level at all times (Fig. 5-16). When adding oil or changing oil, it is desirable to filter the oil with

Fig. 5-13 Zerk fittings. Clean fittings *before* applying the lubricants to prevent pumping dirt or grit into the bearings.

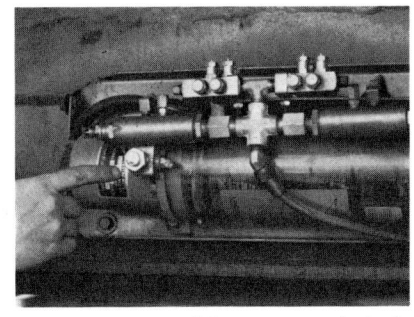

Fig. 5-14 Automatic lubricators can only do the job if the reservoir has a supply of lubricant.

Fig. 5-15 Checking the manual to find which hydraulic fluid the machine is designed to use. Use the fluid recommended by the manufacturer.

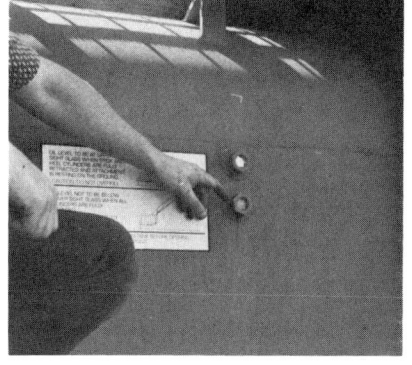

Fig. 5-16 Hydraulic tank. Fluid level in the reservoir *must* be kept at the proper level for maximum operating efficiency.

Maintenance Programs for Equipment Entities

a 10-micrometer filter *before* it enters the tank because the new oil is probably dirtier than the oil that has been circulated and filtered through the system. New, uncontaminated oil should look clear, not hazy or cloudy.

Weekly, before the day's shift starts, crack open the drain cock at the bottom of the reservoir. Screen the fluid through a cloth to catch the sediment deposited at the bottom (Fig. 5-17). Do not become alarmed if some unusual sediment is found in the hydraulic system. Following is a list and explanation of various sediments that may possibly be found when cracking the drain on the tank:

Condition	Cause
Cloudy or milky fluid	There is water in the oil. Continue to drain until the cloudiness is gone.
Lint	Sometimes is present in some oils due to refining methods. Be aware that filters may clog more readily if this condition exists.
Carbon (looks like cinders, but crumbles between fingers)	Oil is oxidizing. Some carbon particles usually form in the oil cooler. When these particles are excessive, the oil is worn out and must be changed.
Rubber (or synthetic)	Hose deteriorating or being cut by dirt or chips. "O" rings wearing or being pinched by loose flanges or fittings. Packing or "O" rings being scuffed by roughness or dirt on rams or valve spools.
Metal flakes	Cavitation of pump. Inspect for the cause.

Hydraulic filters A good recommendation for new equipment is to change or clean (depending upon the type used) the hydraulic filter elements after the initial 50 hr of operation (Fig. 5-18). Thereafter, change or clean the elements at the manufacturer's recommended interval. If filter indicator gauges are provided with the equipment, be very observant as the gauge needle approaches the danger level (Fig. 5-19). It may rise very quickly once the element is beginning to get clogged. The safest measure is to change elements earlier than recommended.

CAUTION: Long periods of operation at the borderline of the danger level could affect the life of the pumps!

Hydraulic lines Be sure that all connections are kept tight, but not to the point of distortion, so that no oil is allowed to leak out nor air to be drawn into the system (Fig. 5-20). Air in the system can cause damage to the components.

Hydraulic components All hydraulic components must be disassembled and assembled in spotlessly clean surroundings (Fig. 5-21). Clean all metal parts with mineral oil solvent. After the parts have been dried thoroughly, lay them on a clean, lint-free surface. Whenever any component is disassembled, replace all "O" rings, backup rings, and seals. Lubricate all parts with clean hydraulic oil before reassembly. Use small amounts of petroleum jelly to retain the "O" rings in place during assembly.

Priming and venting Some manufacturers utilize an air pressurization system on the fluid within the hydraulic tank. In this instance, the hydraulic system must be vented and

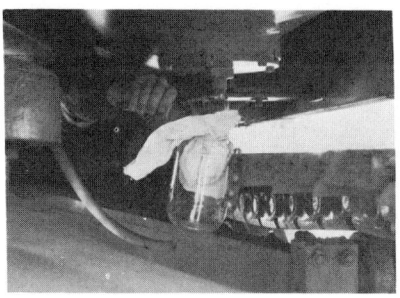

Fig. 5-17 Hydraulic tank drain, Drain the water settled at the bottom of the hydraulic reservoir. Catch the sediment and examine.

Fig. 5-18 Hydraulic filters. Change or clean elements as required.

Hydraulic Excavator and Backhoe Maintenance 7-115

primed each time the system is depressurized (Fig. 5-22). Be sure to follow the manufacturer's recommendations for this procedure to prevent pump cavitation.

AIR SYSTEM

Air systems generally consist of an air compressor, governor, reservoir, filter, lubricator, alcohol evaporator, moisture ejector, pressure regulators, relay valves, and air actuators.

Air reservoir The air reservoir should be drained completely every day when the machine is shut down (Fig. 5-23). Most manufacturers provide a moisture ejector in the system, but draining the tank daily assures removal of all contaminants.

Fig. 5-19 Filter indicator gauges. Be observant of the condition of the filter elements.

Fig. 5-20 Hydraulic connections keep the oil in and air out of the hydraulic lines.

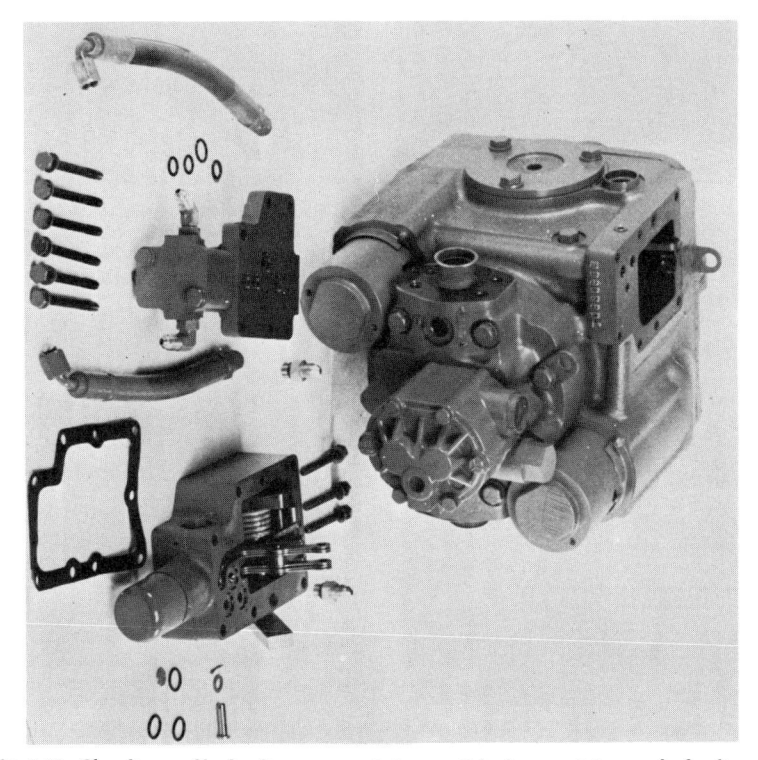

Fig. 5-21 Cleanliness of hydraulic components is essential when servicing any hydraulic parts.

Maintenance Programs for Equipment Entities

Moisture ejector If the machine is equipped with a moisture ejector, it should be checked periodically for proper seating and operation (Fig. 5-24). Manually depress the pin in the exhaust port. This lifts the exhaust valve off its seat, thereby draining the sump area. If the valve is inoperative or becomes sluggish, remove the complete valve assembly and check the filter. Clean if necessary.

Air filter The air filter should be checked twice weekly (Fig. 5-25). Open the drain cock to remove any liquid accumulated in the bottom of the bowl.

Air line lubricator Drain off the sediment collected in the bowl by means of the petcock (Fig. 5-26). Clean the bowl occasionally.

Alcohol evaporator Every 3 months, remove the packing in the bottom of the body near the mounting flange and clean thoroughly with a cleaning solvent (Fig. 5-27).

The rest of the components should be observed regularly to be sure they are functioning properly. They should be relatively trouble-free. If a malfunction occurs to any of these components, disassemble and clean with a solvent.

PUMP DRIVE CASE

Hydraulic excavators demand severe service from the pump drive cases; therefore, it is extremely important that the pump drive case be properly maintained after the initial

Fig. 5-22 Priming and venting hydraulic system. Prevent the possibility of pump cavitation by following the manufacturer's recommendations.

Fig. 5-23 Drain air reservoir completely at the end of the day's work for removal of contaminants and for safety when the machine is left unattended.

Fig. 5-24 Testing the air system moisture ejector for operation.

Fig. 5-25 Draining contaminants from air system filter.

Hydraulic Excavator and Backhoe Maintenance 7-117

hours of operation and also throughout the life of the equipment. Check the oil level of the pump drive case daily and keep at the specified level (Fig. 5-28).

Changing the oil in the pump drive case is much less expensive than replacing component parts and having unnecessary downtime.

Fig. 5-26 Draining sediment from air line lubricator.

Fig. 5-27 Be sure alcohol evaporator is functioning properly in freezing climate.

Fig. 5-28 Pump drive case. Keep the oil clean and at the specified level.

A good rule of thumb to follow for pump drive case maintenance is as follows:

1. Drain and flush the hot gear case (at the end of the work shift) after the initial 100 hr of operation.

2. After the initial oil change, all other oil changes are to be done whenever the engine oil is changed, with a *maximum* of 500 hr between changes.

3. Clean and inspect the pump drive case breather every oil change.

RING GEAR AND BEARING

To ensure uniform distribution of grease throughout the bearing, the machine should be rotated at least two complete revolutions while greasing (Fig. 5-29).

Inasmuch as the ball bearings are under heavy service, it is recommended that they be lubricated each week or every 40 operating hours. Use shorter intervals between greasings in tropical areas or where there is high humidity, dust, or wide ranges in temperature.

Under extremely dusty or dirty conditions, sufficient grease should be added to flush out contaminated grease. Under less severe conditions, add grease until it appears at the seal.

Before storing a new or used machine, thoroughly lubricate the bearing so that grease can be seen venting at the seals. This procedure should be repeated at least twice a year or in line with climatic conditions.

Fig. 5-29 Proper lubrication of ring gear and bearing is critical.

Fig. 5-30 Bathe track drive chains with oil every day.

TRACK DRIVE CHAINS

Some hydraulic excavators make use of large drive chains to propel the tracks. These chains will operate much longer before replacement if they are lubricated daily (Fig. 5-30).

Even though the chain is subject to abrasive material such as dirt and sand, it will give many more hours of operating life if *once daily* a small stream of oil is poured on the full length of the chain. Allow the lubricant to reach the pin bushing and roller bushing contact areas.

The lubricant recommended is determined largely by the temperature in which the machine is being operated. If the temperature ranges from 20 to 40°F, use SAE 20 oil. When the temperature ranges from 40 to 100°F, use SAE 30 or 40 oil. (Even drained crankcase oil performs a good lubricating job.)

Since some machinery is washed down in the evening and the lubricant may be washed from the chain, lubrication is best accomplished in the morning before the machine begins working.

TRACK DRIVE SPROCKETS

If the sprocket teeth are badly worn, the sprocket rim (if available) or the complete sprocket should be replaced before new chain is installed (Fig. 5-31). If this is not done, the new chain will wear rapidly.

Never run badly worn chains over new sprockets!
Never run new chains over badly worn sprockets!

Building up the teeth of the sprockets by welding should be discouraged, as the correct tooth contour cannot be accurately formed in the field.

TORQUE RATINGS

Heat-treated bolts are used almost exclusively in hydraulic excavators. When disassembling or servicing the various assemblies, special attention must be taken to the bolt heads (marked). When reassembling, to avoid future trouble, be sure the nuts are pulled up to the proper torque (Fig. 5-32).

Excavator manufacturers usually supply a chart in their service manuals listing the torque ratings for the heat-treated bolts used in their equipment. Some manufacturers also list the torque values for special applications where the part is illustrated in the parts book.

Fig. 5-31 Track drive sprocket. Tooth and chain contour must be matched.

SPECIAL COMPOUND USAGE

The use of degreasers, sealing compounds, and nonseizing compounds is very predominant in hydraulic excavators. Be aware of this when servicing the equipment, especially when servicing the attachment cylinders, pivot links, bushings, etc.

The manufacturer generally lists the areas and compounds used. If not in a listing, it may be found on the parts pages where the parts are illustrated.

PARTS REPLACEMENT

Whenever service is required on a hydraulic excavator, *new* seals, "O" rings, and backup rings should always be installed if they have been removed while gaining access to the part to be serviced. *Never* reuse seals or rings. It will only create an area susceptible to leakage.

Another thing to remember when replacing parts is *never mismatch parts*. Mating a worn part with a new surface will only result in rapid wear of the new item.

Fig. 5-32 The proper torque for heat-treated bolts must be maintained.

Fig. 5-33 Hydraulic pressure relief valves should be kept at factory settings. *Never* overpressure the system.

HYDRAULIC PRESSURE RELIEF VALVES

The hydraulic pressure relief valves mounted in the hydraulic control valves are factory-set units. Hydraulic excavator manufacturers, as a rule, recommend that the end users *do not* attempt to *reset* them. The machine is designed to perform most efficently at the factory-set pressure setting (Fig. 5-33).

Raising the pressure settings of the control valve reliefs will only do harm to the system and create a potential hazard. Only a qualified factory-trained serviceman should check the pressure settings of these.

CONCLUSION

The owner of a hydraulic excavator or backhoe is entitled to the best performance and the lowest cost operation that is obtainable from the equipment.

Most hydraulic excavator manufacturers provide in their machines a high standard of long-life performance. The degree to which this performance is obtained depends largely on the maintenance and care provided by the owner. Therefore, end users must realize that a neglected machine is very susceptible to breakdowns. The manufacturer has no direct control over machine application, operation, inspection, lubrication, or maintenance.

To keep the machine operating at maximum efficiency, responsible personnel must know and understand the equipment. The owner must *insist* that both the operator and those responsible for maintenance read the manufacturer's manual. Each manufacturer's machine has its own design peculiarities that may or may not mean an unusual maintenance factor in one area or another. *Know the equipment you work with!*

Chapter **6**

Power Shovel Maintenance

GEORGE O. FORSYTHE
General Service Manager,
Northwest Engineering Company,
Green Bay, Wisc.

MAINTAINING A POWER SHOVEL

In the arsenal of excavators available for modern-day earthmovers, one machine stands as the tantamount replica to the first of all tools used for moving material: the shovel.

Designed to do just what it is called, today's power shovel remains among the most popular production tools for repetitive, continuous excavation.

But unlike its ancestors, the power shovel incorporates an extent of technological sophistication that requires a corresponding degree of care and treatment other than just cleaning a cutting edge. Power shovels are designed for cycle duty and therefore are subjected to the rigors of continuous operation.

In this chapter, we discuss the principal areas of power shovel maintenance. Every effort has been made to provide as thorough a general discussion as necessary for owner levels of repair and service. Yet, because of varying user requirements and maintenance capabilities, equipment and control variations will exist from machine to machine, by manufacturer and model.[1]

Specific information for your machine not contained in this discussion should therefore be obtained directly from the shovel's manufacturer or distributor. At the very least, operator and service manuals should be referred to as guides.

Power shovels are used in mining both metallic and nonmetallic minerals. They see widespread duties in crushed stone plants, sand and gravel pits, and quarries. They're used by cement companies, road builders, and general excavators.

More than 3000 units have been purchased for the surface mining of bituminous coal and lignite since the Power Crane and Shovel Association and the Bureau of Mines began keeping records of manufacturers' shipments. At least 7500 power shovels are used in multiple roles by the construction minerals industries, where, it has been reported, 44 percent of the plants deploy an average of 1.53 units each.

Available with bucket capacities that range from a utility size of ¾ cu yd to the jumbo-yard class used in stripping operations, power shovels are built with three different types of power plants: diesel, electric, and diesel-electric. Many of the machines are convertible to dragline, crane, or clamshell operation by changing front end attachments.

Diesel-powered shovels continue to be the most popular throughout all industries. As much as 95 percent of all units in the field are diesel-powered. Because of that, discussion of the maintenance procedures for power shovels will be limited to the diesel engine type. Maintenance procedures for the rotating base, crawler base, and front end attachments can generally apply to all machines.

[1]The model used for discussion in this chapter is a Northwest 80-D power shovel.

7-122 Maintenance Programs for Equipment Entities

Like other pieces of equipment, power shovels require ongoing maintenance programs for them to perform the job that they're designed to do. While a machine will only produce as well as its operator, its productive capability is directly related to how well the machine has been maintained. The more thorough its upkeep, the longer and more efficient the machine will perform. Some power shovels have been serving owners for more than 40 years. A look at the designed functions of the power shovel will help illustrate key maintenance areas.

Function and design Unlike draglines, hydralic excavators, and clamshells, a power shovel is designed to operate directly against a face or a bank, which is displaced as this type of excavator travels (Fig. 6-1). The shovel can exert sustained, simultaneous crowd-

Fig. 6-1 A power shovel is designed to operate directly against a face or a bank, which is displaced as the excavator travels.

ing and hoisting forces. A shovel has the ability to dig below grade and scale material from high on a face. In usual operation, the power shovel is used to load a hopper or truck or to blend material mix on a high face.

A shovel is positive in action and capable of digging the hardest types of material because the machine's design permits power to be applied to force its dipper into a bank.

Power shovels consist of a crawler mounting, an upper structure which rests on a rotating base and houses the excavator's main machinery (see Fig. 6-2). A shovel's front end attachment includes a shovel boom, dipper stick, and dipper. The dipper stick, dipper, and boom are equipped with sheaves and wire ropes for moving the dipper to its work functions of cutting, loading, and dumping. The dipper stick acts as the arm for the dipper, moving back and forth along a pivot point in the boom called the *shipper shaft*. Power for these motions normally comes from the main machinery through wire rope.

Since the shipper shaft acts as a fulcrum or pivot point, both dipper and stick can be raised or lowered by the machine's hoist line as well as pushed out or retracted by the shovel's crowd mechanism. Motions can be performed simultaneously, permitting operational flexibility.

Where a dragline offers the ability to reach for material, drag it in, lift, and then cast it to another location on a work site, a shovel works within the reach of its boom and dipper stick. The machine rotates and either loads a hauler or stockpiles the material for later rehandling.

Where a backhoe will dig in a direction toward its base, a backward motion, a shovel's dipper breaks out and digs away from its main machinery. A shovel loads material by gravity through its dipper's door, letting material fall out from the rear of the dipper. A backhoe, on the other hand, must rotate its bucket down and forward for the material to fall out.

Not only do the main components and power system of a shovel require maintenance effort, but its attachments and accessories will demand their own service and repair.

Depth of operator maintenance The depth and extent of owner maintenance programs will depend on several factors. An owner can contract for full or partial service from an equipment distributor or perform all or partial servicing internally. A principal consideration is to determine internal ability and facilities to indicate the extent of field and shop maintenance that can be performed. Self-service and repair determination also requires a qualitative analysis of distributor support, location, and service capabilities.

Some power shovel owners perform total service on their machines. Others totally rely on distributor and manufacturer field service representatives. Whatever the case, the *bare minimum* standard of owner service is a preventive maintenance program performed by operator, oiler, and/or qualified service mechanic.

A well-maintained machine will achieve greater production and last longer than one

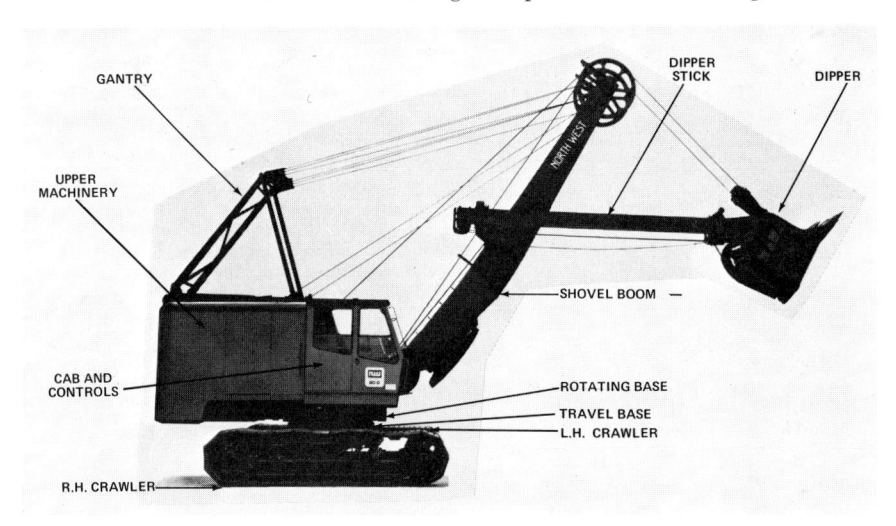

Fig. 6-2 External view of a power shovel and its principal components.

accustomed to haphazard service and operation. It follows then, that the better an operator knows his machine and how to properly operate it, the longer the machine will sustain a useful life.

Effective maintenance begins with the operator, and proper care for equipment will only result from management influence. When management involvement in a maintenance program is understood throughout all levels, the importance of a conscientious effort will underscore the productivity of the machinery in use.

Preventive maintenance begins with proper operation, and all operators should be trained on the characteristics of different models of power shovels. Superintendents should insist on every operator becoming thoroughly familiar with his machine. The best place to start is with the operator's manual, supplied with every machine from all manufacturers. Such a manual outlines and discusses not only the operational characteristics peculiar to the make and model but also discusses and lists service techniques and requirements. Surprisingly, many times such manuals are misplaced in the field or simply overlooked.

PREVENTIVE MAINTENANCE

A thorough preventive maintenance (PM) program will extend beyond normal checks for lubrication, fittings, nuts, and bolts. Preventive maintenance starts with cleanliness, and manufacturers stress the importance of keeping machinery clean for inspection and early identification of abnormal wear and tear. General field service of a power shovel's main components—its power train, drive gears, shafts, hoists, travel bases, and attachments—are detailed in individual service and operator's manuals.

Maintenance of the main components along with schedules for periodic checks will be discussed in individual sections of this chapter: Upper Machinery and Rotating Base, Travel Base, Attachments.

It is widely agreed that the most critical part of PM, however, is *lubrication.* Regular lubrication is essential to the operation and life of any power shovel. The time intervals for lubrication specified in operator and service manuals of particular machines must be followed. Lubricants specified by manufacturers or their approved equivalents should be used. And, machines operated in severe conditions like mud, dirt, and dust should be serviced at shorter intervals than those outlined by manufacturers for normal operating conditions.

TABLE 6-1   List of Typical Lubricants for a Power Shovel

Application	Lubricant
Engine crankcase	Follow engine builder's recommendations.
Engine drive gear case Boom hoist worm gear case Travel gear case	Above 32°F use an extreme pressure No. 140 transmission gear oil. Below 32°F use an extreme pressure No. 80 transmission gear oil with a viscosity index of not less than 100, or a minimum viscosity of 210°F of 75 SSU.
Ring gear and pinion Dipper sticks Planetary gears All open, exposed gears	Use a good quality, sticky open-gear compound that either requires heating to apply or can be applied cold. *Do not use a spray type lubricant.* If the gear compound which can be applied cold does not form a tough, long-lasting lubricating film under severe operating conditions, then change to a gear compound which requires heating before it can be applied. Pour or brush on when the lubricant is hot enough to flow on the teeth.
Crawler drive chains Clevises, pins, levers All hand-lubricated oiling points Shovel crowd drive chain	Use same oil as used for engine.
Wire rope	Refer to the wire rope manual supplied with your machine.
Anti friction bearings—grease lubricated Sleeve bearings (bushings)— grease lubricated	Use an extreme pressure, heavy-duty, high dropping-point grease (min. 375°F).
Air system oilers	Use a lightweight spindle oil or transformer oil having a pour point of −25°F or less and a viscosity 50 SSU or less at 100°F and also a 165°F minimum aniline point. An oil with rust and oxidation inhibitors is preferred.
Air compressors Air compressor filters	Use a heavy-duty, nondetergent engine oil having viscosity as follows: Below 0°F: SAE No. 10W 0 to 60°F: SAE No. 20 Above 60°F: SAE No. 30
Centralized oiling system Pressurized and drop feed	Use an extreme pressure, transmission gear oil as follows: Above 32°F: Viscosity No. 90 Below 32°F: Viscosity No. 80 With a viscosity index of NOT less than 100 or a minimum viscosity of 210°F of 75 SSU. In general, use an extreme pressure transmission gear oil having as high a viscosity as weather conditions will permit.
Roller bearing dipper trip	Use a general-purpose grease with a low freezing point (−40 to +250°F range).

Always refer to the list of lubricants specified in your machine's service and/or operator's manual. A list of typical lubricants used for power shovels by Northwest Engineering Company is provided as Table 6-1.

Throughout this chapter, reference to *heavy-duty grease* means a *number one, heavy-duty, extreme pressure, high dropping-point grease* (minimum 375°F).

Adjustments to a power shovel's components must be made only when required and only by an experienced operator or mechanic. When adjustments are required, the procedures in service manuals should be followed as stated to avoid premature wear or damage to other components within the system.

When performing maintenance, it is highly recommended that records be kept on the general condition of the components and what maintenance was performed.

UPPER MACHINERY AND ROTATING BASE

Lubrication and Adjustments

The upper machinery and rotating base of a power shovel consist of the following components and are illustrated in Fig. 6-3: engine power train, engine driving gears, horizontal reverse shaft, drum shaft assemblies, air supply system, centralized oiling

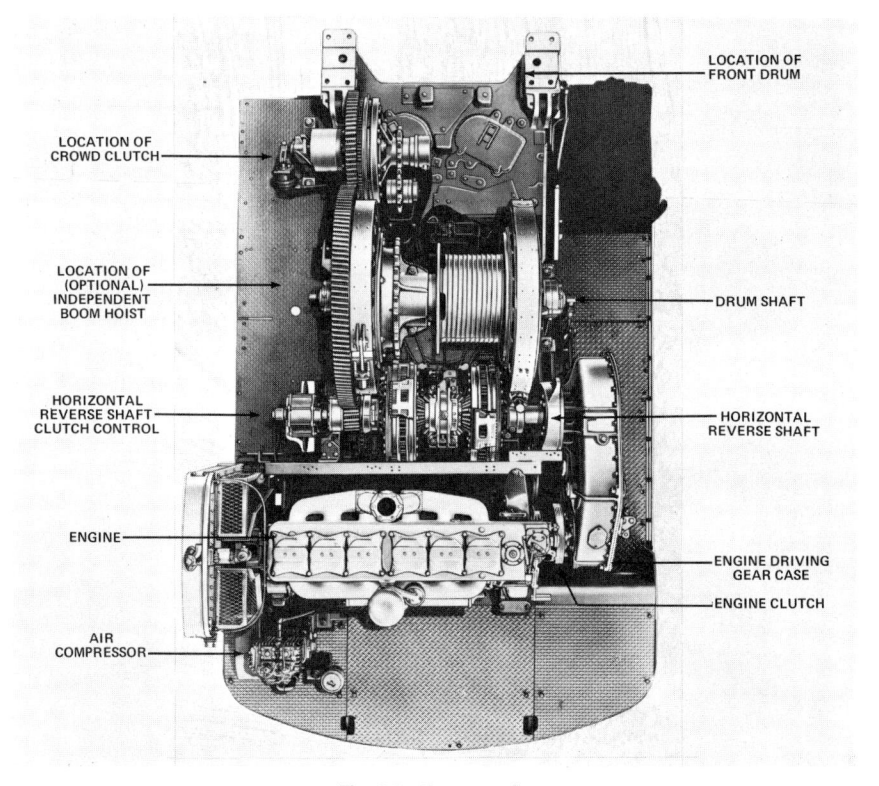

Fig. 6-3 Upper machinery.

system. Beneath the upper machinery are the power shovel's rotating rollers and rotating base. A front drum attachment and independent boom hoist offered by some manufacturers on certain models are also mounted in the upper machinery.

Engine power train Lubrication of the engine power train should be performed according to a planned maintenance schedule. Daily, a power shovel's engine, clutch, and gear case should be inspected for cleanliness, proper adjustments, and lubrication.

Before operating the machine, the operator should ensure that fuel, oil, and engine coolant levels are within manufacturer's specifications. Before starting and during opera-

7-126 Maintenance Programs for Equipment Entities

tion, checks should be made for leaks and broken or frayed lines and hoses. Fuel, oil, and air filters should be clean and serviceable. Gear oil level should continually be checked as well as the clutch lining for wear. Proper engine adjustments should be made according to procedures outlined in the engine manufacturer's maintenance manual provided with the machine. Detailed engine care steps are provided for that purpose.

Daily Checklist. A general guide for daily PM checks on a machine's engine is provided below:

Engine and radiator	Check for leaks
Fluid levels	Check coolant, engine oil, and transmission
Fan belt	Check for wear and proper adjustment
Fan	Tighten any loose mounting bolts
Exhaust system	Check that drains are clear and that exhaust functions properly
General	Tighten loose hoses, lines, and fittings

All adjustments to the engine should be performed according to the recommendations outlined in respective manuals.

Engine Clutch. Sleeve bushings (item 3, Fig. 6-4) for the engine clutch should be lubricated at least once every 40 hr of operation. To lubricate, remove small pipe plug in the rim of the outer disk and pour 20 to 30 drops of light engine oil into the opening.

The clutch's shifter rod bearing (item 2) should be lubricated at least every 20 hr of operation with a heavy-duty grease. To do so, first shut off the engine and engage the clutch. Too much grease could, however, prevent the clutch from engaging. Pins, toggle, shifter shaft, and upper and lower bushings (items marked 3, Fig. 6-4) should be lubricated with engine oil every 40 hr to prevent friction wear.

Adjustments of a typical engine clutch can be explained by referring to Fig. 6-5. The engine clutch should be adjusted when necessary to prevent it from slipping when under full load and engine power. To adjust the clutch, loosen the locking screw and turn the adjusting collar clockwise one or two notches at a time until properly adjusted. The locking screw may then be tightened in one of the notches of the adjusting collar closest to the correct adjusting point.

Engine Driving Gear. Engine driving gears (Fig. 6-6) should be lubricated with an EP transmission gear oil. Different types of transmission oils should never be mixed. Since oil circulates among the gears and splashes over the components as the gears rotate, mixing oils could cause an adverse chemical reaction, breaking down the protective qualities of the oil and thus causing damage to the gears and bearings.

Fig. 6-4 Engine clutch components.

Fig. 6-5 Engine clutch adjustment.

Operators should check the gear case oil level at least once every 300 hr by removing a level plug usually located on the sides of gear cases. The power shovel should be on level ground when the check is made.

No adjustments need be performed to engine driving gears because they are factory-set.

Proper Care of Shafts

Each shaft in a power shovel's upper machinery calls for different types of lubrication at varying intervals, and each will require periodic adjustment. Shafts included as part of the upper machinery are the horizontal reverse, vertical reverse, drum, rotation, and boom. The boom hoist shaft can be either independent or nonindependent. Main components that require attention include bearings, clutches, brakes, and gears.

Horizontal reverse shaft Once the engine clutch is engaged, the horizontal reverse shaft (Fig. 6-7) will automatically be set in motion. Because of its constant rotation, the bearings of this shaft require frequent lubrication.

The shaft's helical pinion (item 8, Fig. 6-7C) and bevel pinions (item 3, Fig. 6-7A) call for lubrication at least once every 8 hr of operation with a sticky open-gear compound.

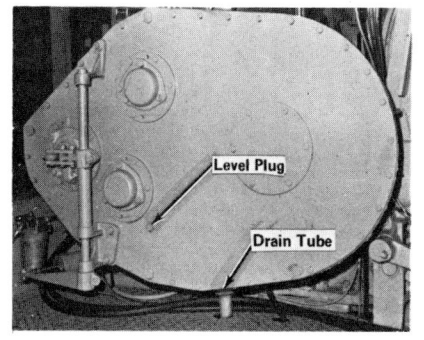

Fig. 6-6 Engine driving gears.

The shifter key slots (item 4, Fig. 6-7A), located on the adjustment side of the clutch, and the clutch driver keys in the hub of each outer driver also require lubrication each 8 hr of operation. Use engine oil to lubricate these parts.

On some power shovels, an adjustable swing pressure regulator is provided in the air line at the inlet side of the horizontal reverse shaft clutch control valve. This regulator limits the maximum pressure available to the control valve when the clutches are being used to swing the machine. The pressure regulator is not intended to correct faulty clutch operation caused by improper adjustment.

Use a heavy-duty grease every 300 operating hours to lubricate the center bearing of the shaft (item 1, Fig. 6-7A) and the bevel pinion outer and inner bearings (item 2, Fig. 6-7A). The left-hand bearing will require lubrication with the same type grease every 500 hr (item 7, Fig. 6-7C). The right-hand bearing (item 6, Fig. 6-7B) is lubricated from the EP transmission gear oil by the rotation of the engine driving gears.

The only adjustments that need be made to the horizontal reverse shaft are those for the left- and/or right-hand clutches and clutch control. To adjust the left-hand clutch, loosen the Allen head cap screws on the clutch's adjusting collar. Rotate the collar to obtain the correct adjustment. (A clockwise rotation will tighten the collar, counterclockwise will loosen.) After tightening the Allen head cap screws to 130 lb · ft (lubricated) torque, move the clutch control forward to toggle. Check the distance the clutch shifter key moves. If

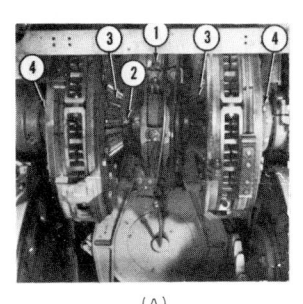

Fig. 6-7 Horizontal reverse shaft, showing (A) shaft assembly, (B) right-hand side, (C) left-hand side.

the key moves at least ³⁄₁₆ in. and not more than ³⁄₈ in., the clutch has been adjusted properly. Finally, move the control lever to the rear toggle position and check the right-hand clutch adjustment in the same manner.

Be careful not to overtighten the cap screws in the clutch's adjusting collar. Overtightening will cause distortion and binding of the outer driver hubs. The collars' clearances around the shaft should be equal.

Vertical reverse shaft The bevel gear and spur pinion of a power shovel's vertical reverse shaft (Fig. 6-8) should be lubricated at least once every 8 hr of operation. When lubricating, use a sticky, open-gear compound.

The upper and lower bearings of the vertical reverse shaft require lubrication every 40 hr of operation with a heavy-duty grease. Fittings are located on the right-hand A-frame member. No adjustments are required for this shaft.

Drum shaft assembly Areas of concern for maintaining and servicing the drum shaft assembly (Figs. 6-9 to 6-11) include its bearings, clutches, and brakes. Always avoid excessive lubrication of the drum shaft and use a heavy-duty grease. Fittings are located on the drum shaft for lubricating each of the sleeve bearings (item 1, Fig. 6-9). When a machine is equipped with an independent high-speed boom hoist, a fitting is provided to lubricate the left-hand drum shaft bearing at 300-hr intervals. Sleeve bearings and both right- and left-hand shaft bearings should be lubricated every 1000 hr.

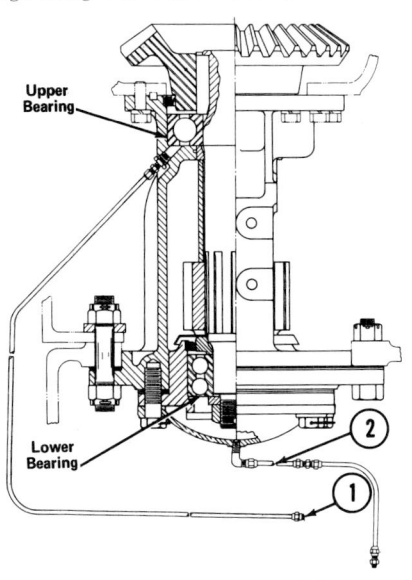

Fig. 6-8 Vertical reverse shaft.

Lubricate the drum clutch gear teeth after every 8 hr operation (item 4, Fig. 6-9) with a sticky, open-gear compound to prevent friction from metal to metal contact.

Lubricate the right- and left-hand drum clutch's cylinder pivot pins, toggle pins and link bushings, bell crank, pivot pin and bushings, live-end strut bushings, rod end pins, and dead end link pin and bushings with a few drops of engine oil at least once every 8 operating hours.

The right- and left-hand drum brakes require lubrication with engine oil of their toggle pins, yoke end pins, rod yoke end pins, live-end pins, and connection pins at a similar frequency with a light engine oil.

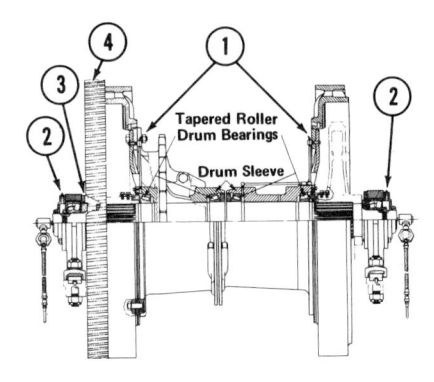

Fig. 6-9 Drum shaft assembly; right- and left-hand view of drum

Outer and inner bushings of the drum brakes should be lubricated after every 8 operating hours. Right- and left-hand brake shaft outer bushings are lubricated with a heavy-duty grease through fittings located in the toggle support. The left-hand inner bushing is lubricated with a heavy-duty grease through a fitting in the left end of the shaft. The right-hand inner bushing is lubricated with engine oil at the right-hand side frame.

As with the horizontal reverse shaft, once the engine clutch is engaged, the drum shaft is in continuous rotation. To adjust the drum's clutches, adjust them one at a time. First, disengage the engine clutch and stop the engine. With the drum shaft clutch disengaged, add or remove shims from the adjusting nuts and center guide to create approximately a $\frac{1}{16}$-in. uniform clearance between the friction blocks and the inner clutch surface of the friction flange (see Fig. 6-10).

Next, release the locknut on the turnbuckle rod end of the clutch and turn the turnbuckle so that the rod ends are pulled closer together. Doing so will expand the clutch

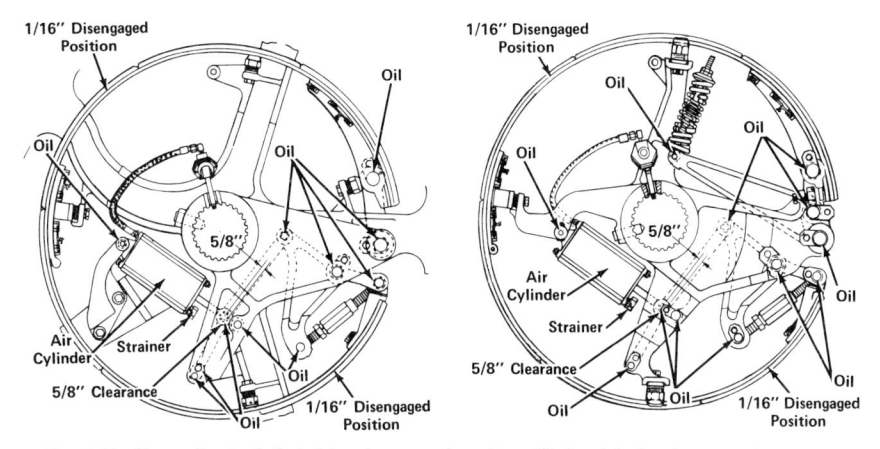

Fig. 6-10 Drum clutch. *Left*, left-hand, engaged position. *Right*, right-hand, engaged position.

band until approximately $\frac{5}{8}$-in. clearance develops between the toggle stop pin and piston rod end when the clutch band is fully engaged without drum rotation. This should happen when the air system pressure is at 90 psi in a normal operating temperature. Then, tighten the locknut on the turnbuckle rod end.

Clutches assembled with cushion springs have been compressed to preset lengths. Do not decrease the length under normal operation. In abnormal conditions, however, it may be necessary to increase their tension. If the clutch band requires frequent adjustment or the band is slipping and overheating the friction flange, the cushion clutch springs may not be handling the load. When such conditions develop and it is impossible to eliminate the trouble by proper adjustment of the clutch control parts, the springs may have to be replaced. It is suggested that a factory service representative be contacted for advice.

Linings of the drum brakes should never be cleaned with a cleaning fluid or other liquid. Always sand or lightly file the lining. Liquid can cause the lining and rivet holes to accumulate dust, dirt, and grease.

When the brake linings begin to wear, adjustments should be made on the brakes' dead end eyebolt to compensate for wear. The following adjustment procedures generally apply to most drum shaft brakes. Refer to Fig. 6-11, which illustrates left- and right-hand brakes, and turn the brake adjusting nut clockwise one-half revolution at a time until the brake will hold the load when the brake pedal is pushed forward and latched. Keep the tension in the release springs adjusted to prevent the brake band from dragging on the outer surface of the friction flange when the brake pedal is released.

Nonindependent boom hoist The nonindependent boom hoist (Fig. 6-12) is normally driven by a worm gear whose parts are lubricated with transmission gear oil contained in the gear's housing. The housing is filled by removing the pipe plug located alongside the boom hoist forked lever on the top side and rear left corner of the rotating base casting. (Refer to Fig. 6-12A.) The gear's worm, bushing thrust washers, worm

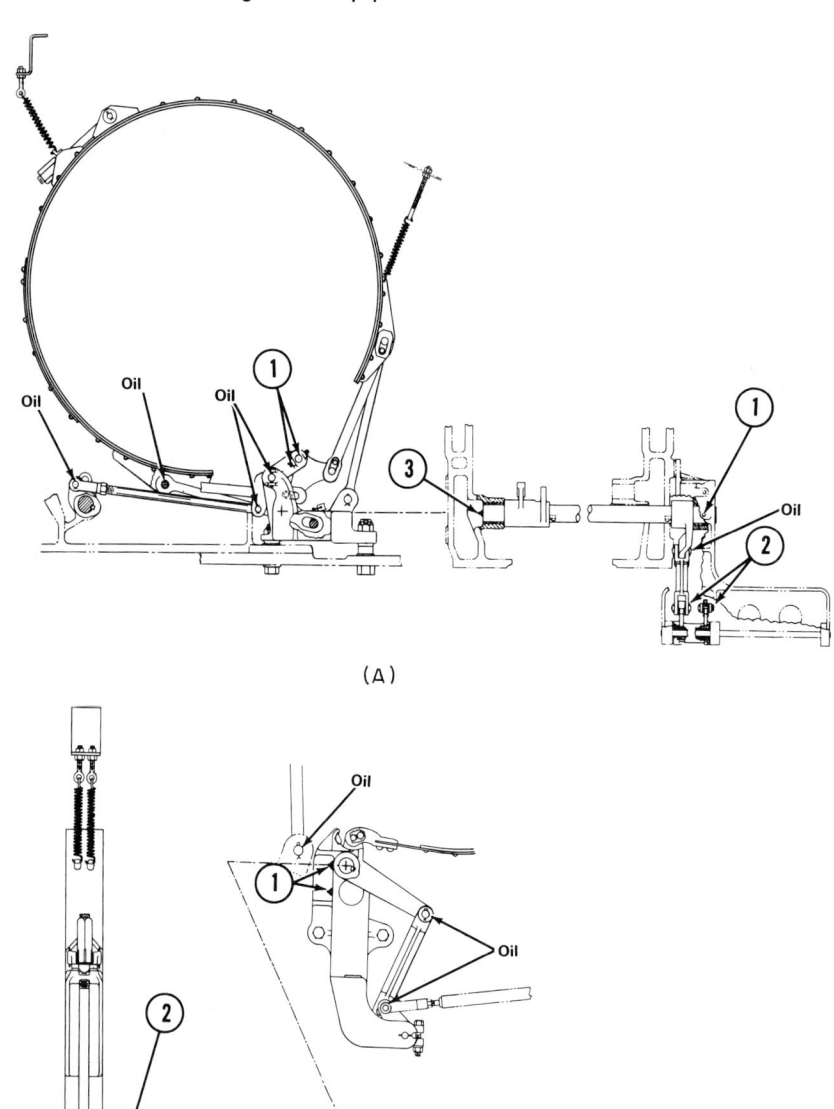

Fig. 6-11 Drum brake.

Power Shovel Maintenance 7-131

wheel, cone brake wheel, floating brake cone ring, and floating brake cone are lubricated in this manner.

The housing should be drained, flushed, and refilled twice a year under normal operating conditions. Check the shovel manufacturer's specifications for the right viscosity oil and volume. When adding oil, always use the same type. Mixing two types of oil

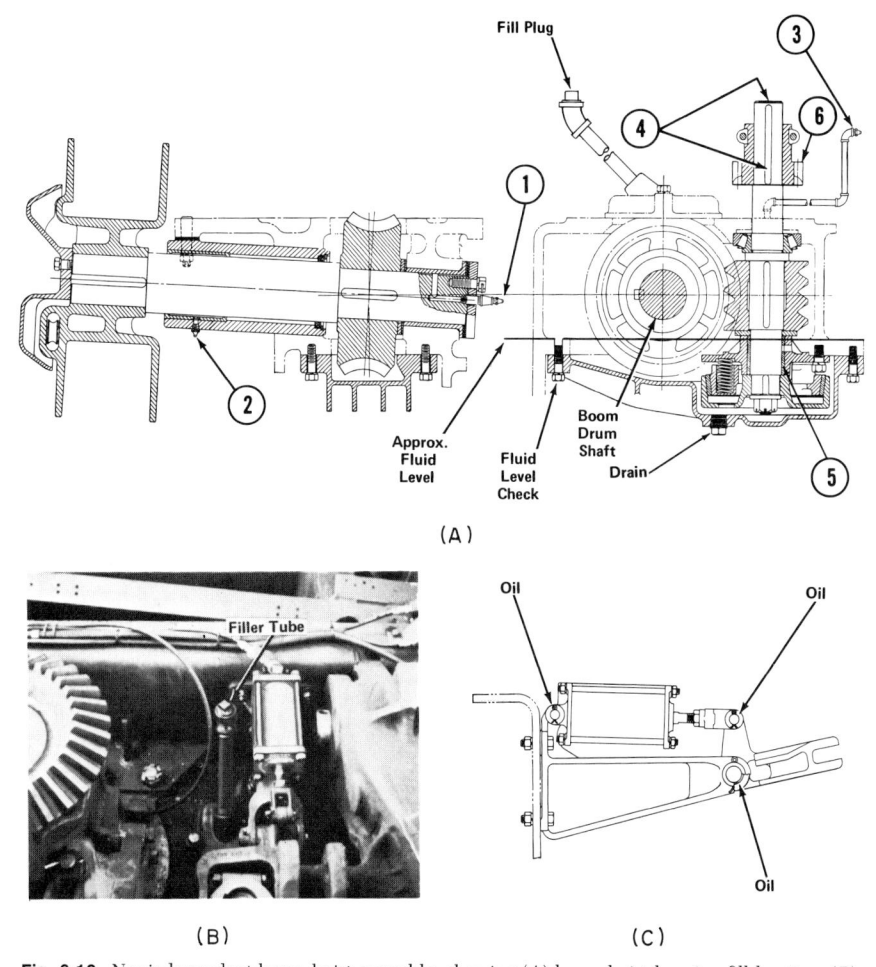

Fig. 6-12 Nonindependent boom hoist assembly, showing (A) boom hoist housing fill location, (B) boom hoist drum and worm shafts, (C) boom hoist control.

may cause a chemical reaction which could break down protective qualities of the oil and damage the parts it would normally lubricate.

When the boom is used frequently, lubricate at least once every 8 operating hours.

Check the oil level at least once after every 40 hr of operation by removing the cap screw in the center of the bottom cover. Keep filled to the lower side of the boom drum shaft. Grease the inner bushing and worm wheel thrust washer (item 1, Fig. 6-12*B*), outer bushing (item 2), and worm shaft bearing (item 3) with heavy-duty grease every 40 hr. Splines on which the pinion slides and the shifting collar (item 4) should be lubricated with engine oil every 8 hr. The teeth of the sliding pinion (item 6) are to be lubricated with a good, sticky open-gear compound applied to the teeth of the rotating gear.

7-132 Maintenance Programs for Equipment Entities

The only lubrication required for the worm gear control is oiling with engine oil of the cylinder anchor pin at least once every 40 hr to eliminate wear and binding. (See Fig. 6-12C).

Springs in the nonindependent boom hoist floating brake must exert sufficient pressure against the brake cone to hold the brake wheel stationary at any time when the sliding pinion is disengaged from the rotating gear.

The floating brake cone has six holes which hold the brake springs. When the boom hoist is assembled at the factory, the two brake springs will exert enough pressure to hold the brake wheel stationary. However, the pressure between the floating brake cone and ring and the cone brake wheel can be increased by adding extra brake springs to this assembly—installing three, four, or up to six springs.

Independent boom hoist Always avoid excessive lubrication and keep the linings of the independent boom hoist (Fig. 6-13) free from grease and dirt.

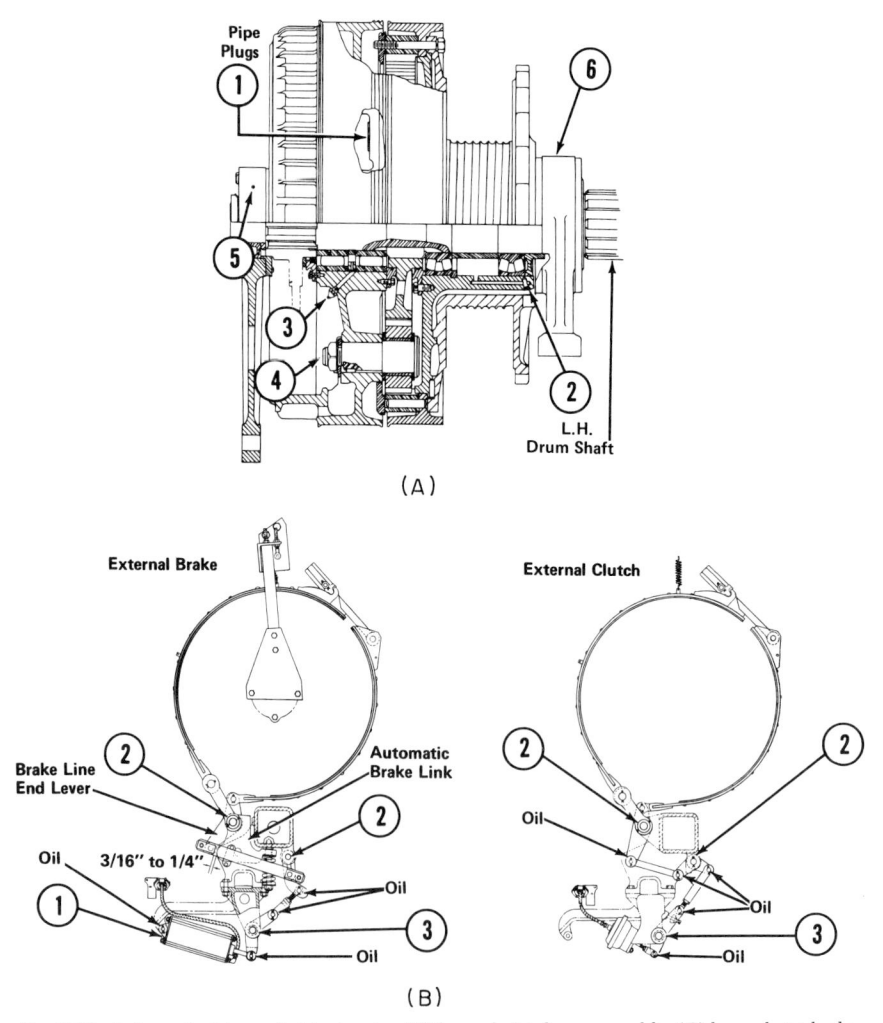

Fig. 6-13 Independent boom hoist, showing (A) boom hoist drum assembly, (B) boom hoist brake and clutch controls, and, opposite page, (C) boom hoist control internal band, (D) boom hoist control support, (E) boom hoist sheave assembly, (F) boom hoist clutch throwout.

Power Shovel Maintenance 7-133

The teeth of the planetary pinions, planetary gear, and sun pinion should normally be lubricated at least once every 40 operating hours. Refer to Fig. 6-13A and remove the pipe plugs (item 1) and lubricate using a good, sticky open-gear compound. If the gears are well-lubricated and the lubricant is clean, inspect for wear and reschedule lubrication for another time.

Rope drum sleeve bearings (item 2) should be lubricated each 300 operating hours with a heavy-duty grease through fittings located in the rope drum sleeve hub. Clutch roller bearings (item 3) should also be lubricated each 300 hr using a hand-held grease gun through fittings located in the clutch drum hub.

Lubricate the independent boom hoist's planetary pinion bushings and washers (item 4) at intervals of every 4 hr with heavy-duty grease through fittings located in the outer end of each planetary pinion pin. The outboard strut bearing (item 5) is lubricated with a heavy-duty grease every 300 hr through fittings at the top of the outboard strut. The inboard bearing (the left-hand drum shaft bearing, item 6) should be lubricated at 300-hr intervals through heavy-duty grease fittings located in the drum shaft bearing housing.

Clean and lubricate the air cylinder strainer (item 1, Fig. 6-13B) with light engine oil once every 150 hr, or more frequently under dusty conditions. The air cylinder, clutch,

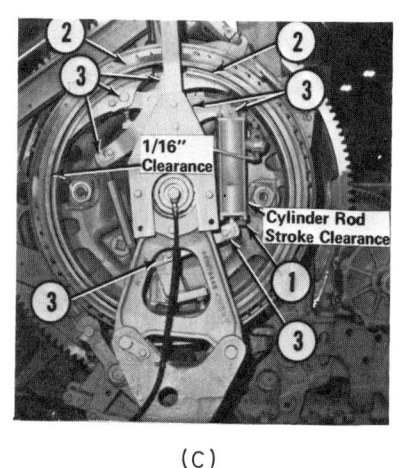

(C)

(D)

(E)

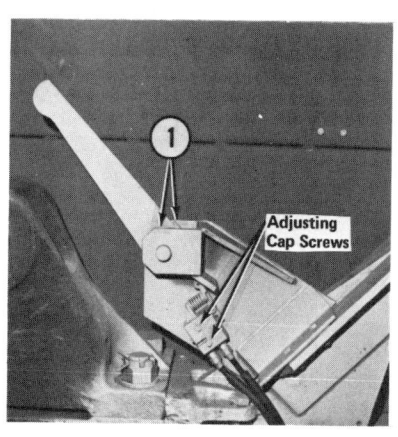

(F)

and brake pins are located at the lower right-hand side of the drum and clutch assembly (items 1 and 2). Each should be lubricated with engine oil at the end of each 8 operating hours. Rocker lever bushings (item 3), located in the brake and clutch rocker levers, should be lubricated with a heavy-duty grease every 8 hr.

Also clean and lightly oil the boom hoist's air cylinder breather (item 1, Fig. 6-13C) at least once every 150 hr, or more frequently when in dusty conditions. If the clutch lining (item 2, Fig. 6-13C) becomes sticky, grabs when applied, or fails to hold when properly adjusted, remove the glaze from the lining with an emery cloth.

All the pins and bushings (item 3, Fig. 6-13C) connecting the air cylinder and toggle linkage to the clutch should be lubricated with engine oil at least once every 40 hr of normal operation. If the boom is operated continuously, lubricate as frequently as every 8 hr. But avoid excessive lubrication.

The independent boom hoist's ratchet pawl control (Fig. 6-13D) contains a pawl spacer (item 1), wire rope socket (item 2), and the pulley pin (item 3), which should be lubricated with engine oil every 40 operating hours. The control support's air cylinder breather (item 4) should be cleaned and oiled with engine oil at least once every 150 hr.

The deflecting sheave assembly's bushing and thrust washer should be lubricated every 4 operating hours with a heavy-duty grease through a fitting located in the hub of the deflecting sheave (item 1, Fig. 6-13E). Lubricate the inner and outer shaft bushings through fittings located at the bottom of the drum shaft sheave and in the end of the shaft retaining washer every 150 hr with a heavy-duty grease.

The hoist's clutch throwout should be lubricated after every 8 hr with a heavy-duty grease through fittings located at the end of the throwout lever shaft (item 1, Fig. 6-13E).

The only adjustments required on an independent boom hoist are adjustments to the brake and clutch controls and to the internal band control. *Before beginning checks and adjustments, lower the boom onto a support.*

Refer to Fig. 6-13B and follow this procedure:

1. Adjustment of the external clutch band is controlled by an adjusting nut. When the control valve lever is pulled backward as far as possible, the external clutch band must be adjusted tight enough to hold the clutch drum from turning, thus raising the boom. The length of the piston rod and brake and clutch control linkage are factory-set and should never be changed.

2. The toggle adjusting screw is properly set at the factory and should not be adjusted.

3. The adjusting nut, located near the top of the external brake band, is used to adjust the brake band. When the control valve lever is in its neutral or vertical position, the brake must be adjusted tight enough to keep the boom from moving downward.

4. The brake band release springs should be adjusted with a slight tension when the brake is applied. Too much tension may cause the band to chatter when lowering the boom.

To adjust the internal clutch control, refer to Fig. 6-13C and follow these steps:

1. Lower the boom to a suitable support, and disengage the engine clutch.

2. When the internal clutch band is fully released, there should be $1/16$-in. uniform clearance between the clutch band lining and the drum. Band release adjusting screws are provided to control this clearance.

3. Adjust the clutch band by loosening the turnbuckle locknut and turning the turnbuckle to push the rod ends farther apart, thereby expanding the clutch band.

The length of the wire control rope on the ratchet pawl control (Fig. 6-13D) should be adjusted so that the ratchet pawl is released from the rope drum when the control handle is moved to the OFF position. It should be engaged with the rope drum when the control handle is moved to the SET position.

No adjustments are required for the deflecting sheave assembly.

Before adjusting the clutch throwout (Fig. 6-13F), move the machine to firm level ground with no load suspended from the boom.

Then, loosen the cap screws and move the cam *up* to decrease the boom angle. Move the cam *down* to increase the boom angle. When the desired angle is attained, tighten the cap screws.

Rotating shaft Because of its importance to the function of a power shovel and its continuous use, the rotating shaft assembly (Fig. 6-14) requires close attention during operation. At least five of the six major components of this shaft demand daily or twice-daily lubrication. On some models, the shaft's upper bushing thrust collar support rods and jaw clutch are lubricated automatically through a centralized oiling system. The gear thrust washer (item 3, Fig. 6-14A) is lubricated from the same EP transmission gear oil which enters the oiling collar from the centralized oiler. Teeth of the shaft's rotating pinion (item 5, Fig. 6-14A) and the rotating gear (item 6) should be lubricated with a

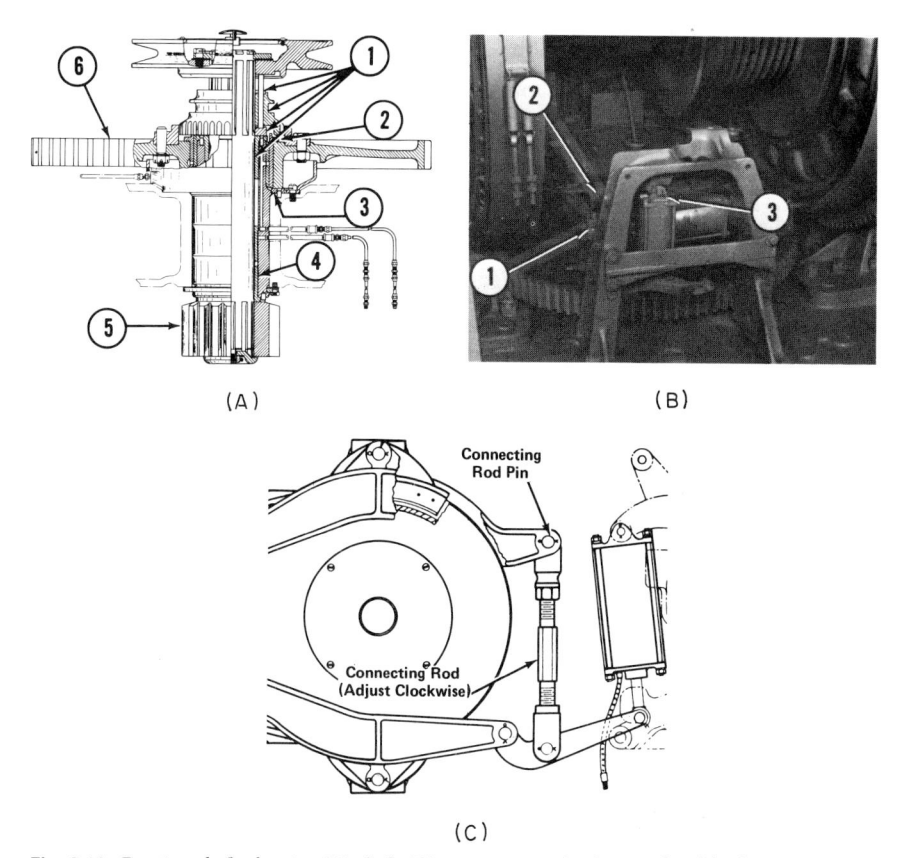

Fig. 6-14 Rotating shaft, showing (A) shaft, (B) rotating gear clutch control and brake, (C) rotating shaft brake adjustment.

sticky open-gear compound. The lubrication should be applied evenly to all teeth at least once every 8 operating hours to avoid friction and wear from metal-to-metal contact. The rotating shaft's gear bushing (item 2) and lower bushing (item 4) should be lubricated through a fitting located on the right-hand A-frame member lubrication plate every 4 hr of operation with a heavy-duty grease.

Lubrication points for a typical rotating shaft gear clutch control and rotating brake are illustrated in Fig. 6-14B. Rod end bushings (item 1) should be lubricated with engine oil every 20 hr. The shifting fork control shaft and bushings (item 2) should be greased every 8 hr with a heavy-duty grease. Movable parts of the cylinder anchor pin, rod end pins and bushings, toggle lever, connecting rod pin, and brake lever hinge pin (item 3) should be lubricated with engine oil at least every 8 hr.

The only adjustment area called for on the rotating shaft assembly is the rotating brake. If the brake fails to hold when traveling, use the following procedures for adjustment. Lower the power shovel's boom to the ground on a suitable support to prevent damage to it and to ensure that the rotating base will not rotate. Next, move the swing brake valve to the OFF position, making sure that the brake is completely released. Adjust the rotating brake when air pressure has been built up to working pressure. Once it has, shut off the engine and disengage the engine clutch. (Refer to Fig. 6-14C, a schematic of a rotating shaft brake.) Remove the connecting rod pin from the left-hand brake lever and turn clockwise one-half turn at a time until the brake is properly adjusted to hold the upper structure from rotating. Do not overadjust the brake too tightly. Doing so will cause the brake lining to wear more rapidly and prevent the upper structure from rotating freely.

Air supply system   Most air supply systems (Figs. 6-15 to 6-17) consist of an air compressor, air tank receivers, and the necessary valves, filters, and lines. A typical air supply system, one used on Northwest Engineering Company power shovels, is diagramed in Fig. 6-15. The entire system should be checked periodically for loose fittings and lines, condensation, and for dirty filters.

Manuals for various air compressors used in power shovels are supplied with the power shovel upon delivery and should be referred to for the manufacturer's recommended maintenance procedures. The first four items in the checklist that follows are provided as an aid in PM; the remaining three supersede information in compressor manufacturers' manuals:

Air Compressor Checks (see Fig. 6-16).

1. Compressor .... Check for leaks.
2. Drive belts .... Check for proper tension.
3. Fan .... Check guard for proper installation and tightness.
4. Air cleaner ..... Check for cleanliness and serviceability. Service at least once a week under normal operating conditions, daily under dusty conditions.
5. Crankcase ..... Fill with a heavy-duty, nondetergent engine oil. Change oil at least every 150 hr, more often under adverse conditions. Recommended viscosity:

Below 0°F:	SAE 10W
Between 0 and 60°F	SAE 20W
Above 60°F:	SAE 30W

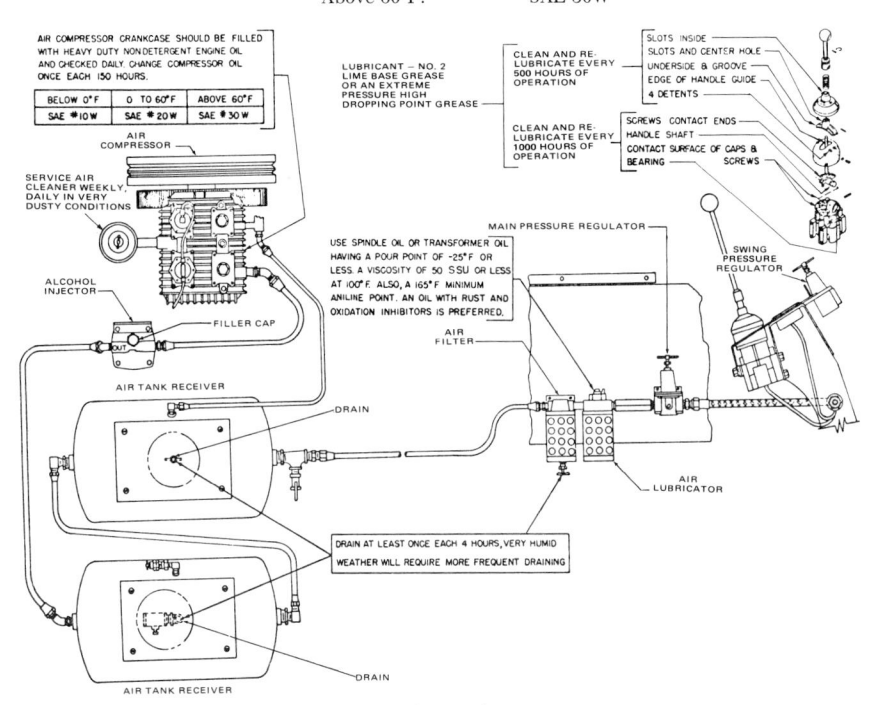

Fig. 6-15   Air supply system.

Each air tank should be drained once every 4 hr to remove condensation. Drain the tanks more often in humid environments.

During cold weather, keep the system's alcohol injector (refer to Fig. 6-17) filled with methyl alcohol. The air line lubricator bowl should be filled with a lightweight spindle or transformer oil having rust and oxidation inhibitors. The oil should have a pour point of $-25°F$ or less and a viscosity of 50 SSU or less at $100°F$.

The system's air filter should be drained at least every 4 hr, more frequently under humid conditions. Clean and service the filter element every 300 hours. Refer to the manufacturer's service bulletin for cleaning and maintenance instructions.

The air line lubricator lubricates the internal parts of the system's control valves,

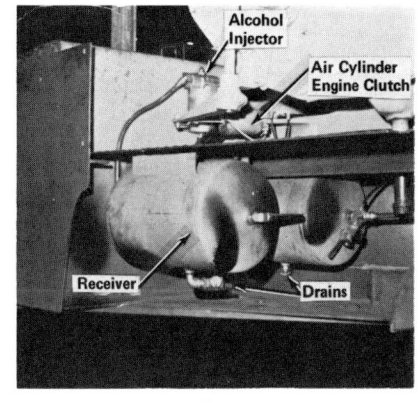

Fig. 6-16 Air compressor.   **Fig. 6-17** Air supply system components.

cylinders, brake chambers, and all other components. To maintain proper lubrication of these components, it is imperative that they are operated at least once every 8 hr.

When the lubricator is properly adjusted, one drop of oil will automatically be delivered with each rapid engagement of the right drum clutch cylinder. Check oil flow by repeatedly engaging the drum clutch at least four times in rapid succession. Under continuous operation, the air system lubricator should use one bowl of lubricant every 40 hr.

With the engine running, engine clutch disengaged, and the system at 90 psi, adjust the air line lubricator needle valve clockwise to *reduce* the oil flow and counterclockwise to *increase* the oil flow.

The air supply pressure regulating valve is preset for 90 lb working pressure and should not be changed. Compressor drive belts should be checked and adjusted at least once every 300 hr.

Centralized oil system Two types of centralized oil systems are used with power shovels: a drop feed and a pressurized system (Figs. 6-18 and 6-19). Each type of system lubricates the following components through a metered multiple oiler located on the right-hand side A-frame member of most models: splines at the upper end of the vertical travel shaft, the gear clutch, spur pinion bushing, splines at the upper end of the rotating shaft, sliding gear clutch, support rods, split ring, upper rotating shaft bushing, rotating gear bushing, and thrust washer.

The pressurized centralized oiling system is activated by pressure from the air system. At the start of a work day, turn the system's shut-off cock to a horizontal feed position. At the end of the work day or when the machine will be idle for a period of time, turn the cock to the vertical position to stop the oil flow. A centralized oiling system of the pressurized type is illustrated in Fig. 6-18.

A drop feed centralized oiling system is illustrated in Fig. 6-19. The system's transparent oil reservoir cap, usually located on the right-hand side of the power shovel's A-frame, should be filled with an extreme pressure transmission gear oil. The oil should have as heavy a viscosity as weather conditions will permit. Do not allow the oil level to go below 1 in. from the bottom of the cup. Each feeder should be adjusted to feed the proper amount of lubricant to the part it is to lubricate.

7-138 Maintenance Programs for Equipment Entities

In a pressurized system, individual feed valve toggle assemblies are provided for metered lubrication. To adjust, raise each valve toggle lever to a vertical position. Adjust the knurled knob that is located below each valve toggle until the desired flow rate is visually established in the sight glass. The approximate drop rate of oil is 12 drips per minute.

In a drop feed system, check each feeder periodically by watching the oil dropping past the transparent window, just below the adjusting screw in the feeder body. The transpar-

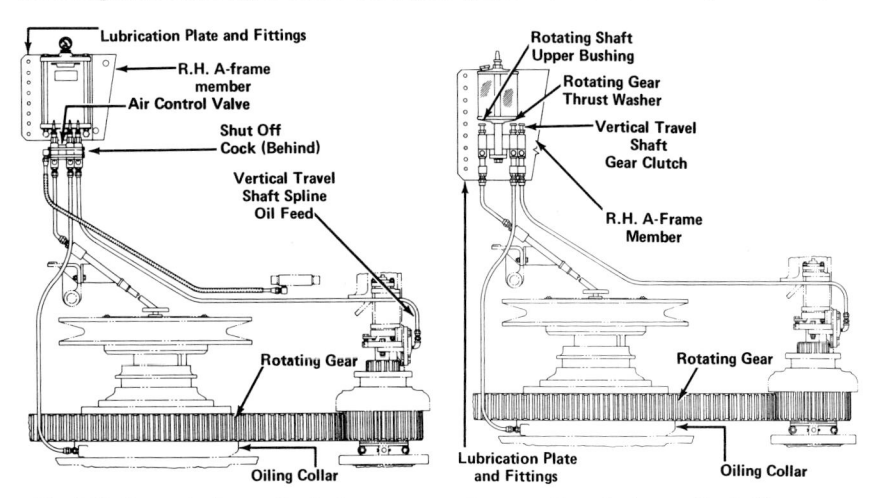

Fig. 6-18 Pressurized centralized oil system. **Fig. 6-19** Drop feed centralized oiling system.

ent oil reservoir cup has a lever shut-off valve located on top of the oil cup that should be raised to a vertical or feed position at the start of a day's operation. Lower shut-off valve to a horizontal position to stop the flow of oil at the end of a work session or when the machine will be idle for extended periods of time.

Rotating and equalizer rollers The surface of the equalizer rollers and the ring gear flange above the rollers should be kept free from grease and oil. These rollers must rotate and never slide against the ring gear flange.

Lubrication of the rotating rollers consists of greasing the front rotating roller bushings and washers (item 1, Fig. 6-20A), the front rotating roller support pins and washers (item 2), and the rear rotating roller bushings, washers, and pins (item 3) every 8 hr with a heavy-duty grease.

Adjust the equalizers (Fig. 6-20B) so that neither rear nor front rollers can lift up more than $1/8$ in. when the machine is operating. To adjust, balance the rotating base so that all the rotating rollers rest evenly on the ring gear. Next, check both the front and rear equalizer assemblies by rotating the equalizer bracket until one roller is in contact with the roller path of the ring gear. If more than $1/8$-in. clearance exists between the top of the loose equalizer roller and the flange of the ring gear, that roller should be adjusted.

To adjust the front equalizer roller, refer to a schematic as illustrated in Fig. 6-20B and remove the cap screw from the adjusting lever. Turn the adjusting lever until proper $1/16$-in. clearance is obtained between the roller path of the ring gear and the loose roller. Reposition the adjusting lever on the splines of the equalizer shaft so that one of the two holes lines up with the tapped hole in the rotating base. Choose the position on the spline so that it will allow the holes to line up with the least amount of back-off on the adjusting lever. When proper clearance is established and adjustment completed, install and tighten cap screws to a torque of 200 lb·ft (lubricated).

To adjust rear equalizer, loosen rear equalizer shaft clamp screws. Rotate rear equalizer shaft until there should be approximately $1/16$-in. but never less than $3/64$-in. clearance between the top of the loose equalizer roller and the flange of the ring gear. Then tighten clamp's screws to 300–400 lb·ft torque (lubricated). Keep clearance between the ring gear flange and the two rear sets of equalizer rollers as equal as possible.

Batteries Check each battery for proper water level once a week. When required, fill with distilled water to the full mark indicated on the battery. Using an alkaline solution of bicarbonate of soda and water, wash batteries when required to remove battery acid from the top of the battery and corrosion from the terminals. In cold weather, always recharge the batteries after adding water to prevent freezing.

CAUTION: Batteries give off hydrogen gas which can be ignited by sparks, flame or smoking, causing the battery to explode. Extreme care should be exercised when removing cables or performing maintenance.

CAUTION: When cleaning terminals, disconnect the ground cable first. Injury could result if the ungrounded cable contacts the ground, causing electrical sparks.

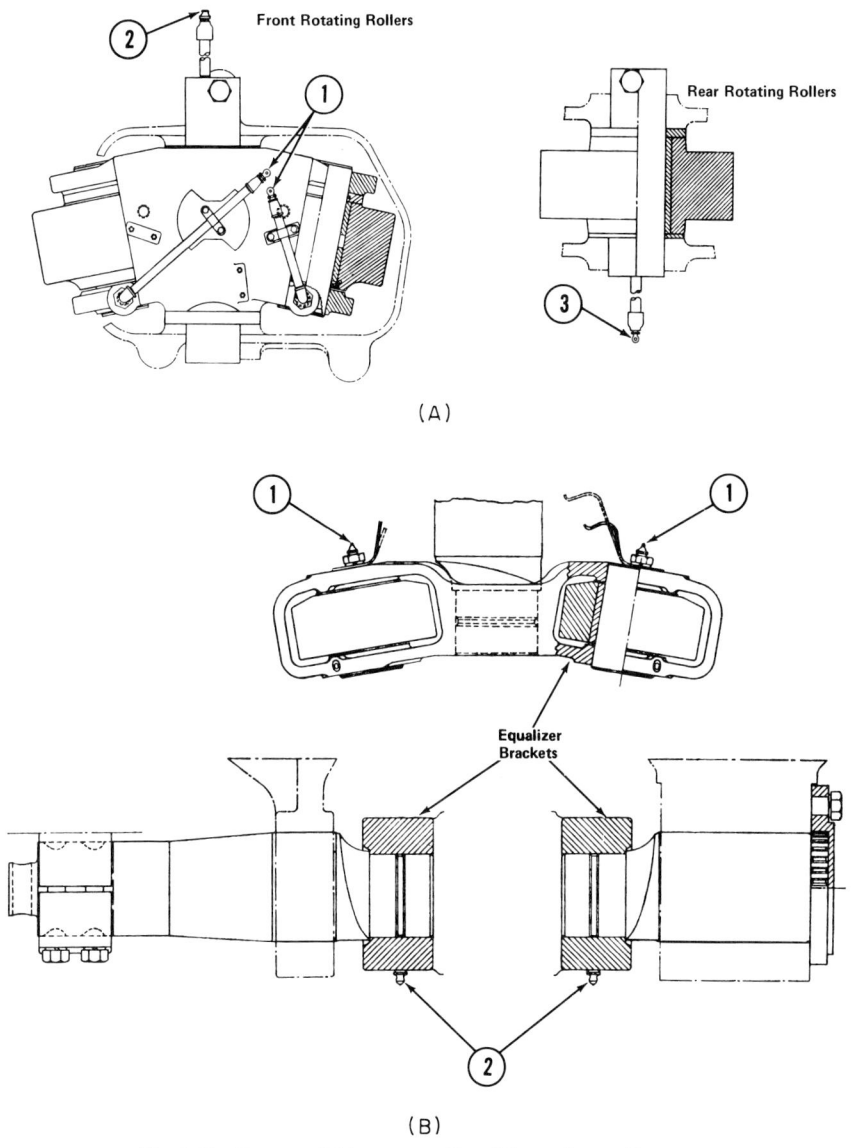

Fig. 6-20 Diagram of (A) rotating rollers, (B) equalizer rollers.

7-140　Maintenance Programs for Equipment Entities

Excessive heat or high charging rate are causes of water evaporation in batteries. Always check for proper ventilation to the battery area, correct belt tension, and charging rate. Refer to the battery manufacturer's service manual for charging system specifications.

Troubleshooting and lubrication charts for the upper machinery and rotating base assembly follow as Table 6-2 and as Fig. 6-21 and Table 6-3 working together.

TABLE 6-2　Troubleshooting Charts—Upper Machinery

Malfunction	Probable cause	Corrective action
	Engine	
1. Engine fails to crank when starter switch is pushed in.	a. Defective starter switch.	a. Replace starter switch.
	b. Defective cables.	b. Replace cables.
	c. Defective batteries.	c. Charge or replace batteries.
	d. Defective starting system.	d. Refer to the engine manufacturer's maintenance manual.
	e. Mechanical seizure of engine parts.	e. Refer to the engine manufacturer's maintenance manual.
2. Engine cranks but fails to start.	a. Improper or contaminated fuel in fuel tanks.	a. Drain and refill.
	b. Defective fuel or air inlet system.	b. Check and replace filters.
	c. Restricted or leaking fuel lines or fittings.	c. Check fuel lines, tighten or replace if necessary.
	d. Incorrect fuel injector or rack timing.	d. Refer to the engine manufacturer's maintenance manual.
3. Engine fails to develop full power.	a. Throttle not adjusted properly.	a. Check and adjust linkage if required.
	b. Restricted fuel filter or line.	b. Check, clean, or replace.
	c. Incorrect governor adjustment.	c. Check governor adjustment.
	d. Faulty injectors or incorrect fuel injector and rack timing.	d. Refer to the engine manufacturer's maintenance manual.
4. Engine overheats as indicated by temperature gauge.	a. Low coolant level in radiator.	a. Check and refill.
	b. Defective temperature gauge.	b. Replace gauge.
	c. Engine coolant circulating system defective.	c. Refer to the engine manufacturer's maintenance manual.
5. Engine oil pressure low as indicated by oil gauge	a. Low oil level.	a. Check and refill.
	b. Clogged oil filter.	b. Replace oil filter.
	c. Defective oil gauge.	c. Check and replace.

Malfunction	Probable cause	Corrective action
	Engine clutch	
1. Clutch will not engage.	a. Linkage out of adjustment.	a. Check and adjust if necessary.
	b. Lack of air pressure.	b. Check air supply.
	c. Air cylinder defective.	c. Repair or replace air cylinder.
	d. Excessive grease on clutch linings or driving disk.	d. Remove linings, clean, and replace.
	f. Engine clutch locknut valve in wrong position.	f. Change to correct position.
	g. Insufficient or excessive lubrication on sliding sleeve.	g. Refer to engine lubrication instructions in text.
2. Clutch will not disengage.	a. Loss of air pressure.	a. Check air supply.
	b. Broken linkage.	b. Replace linkage.
	c. Defective air cylinder.	c. Repair or replace air cylinder.
	d. Defective control switch.	d. Check and replace control switch.
	Engine driving gears	
1. Excessive noise in gear case.	a. Lack of gear oil.	a. Check level and refill.
	b. Worn bearing.	b. Check shaft end play. If worn, contact manufacturer.
	c. Lack of lubrication on horizontal reverse shaft spur pinion and bevel pinion.	c. Apply open gear compound to horizontal reverse shaft spur pinion and bevel pinion.
	d. Moisture—ice.	d. Drain and flush gear case and refill with proper grade oil.
	e. Loose horizontal reverse shaft gear.	e. Tighten gear washer cap screw.
	Horizontal reverse shaft	
1. Clutch will not engage.	a. Loss of air.	a. Check air supply system.
2. Clutch drags.	a. Friction facings dirty or have excessive grease.	a. Remove and clean.
	b. Improper adjustment.	b. Adjust clutch.
3. Clutch fails to hold.	a. Clutch improperly adjusted.	a. Adjust clutch.
	b. Clutch facings are glazed.	b. Remove and clean glaze from lining.

(continued)

TABLE 6-2 Troubleshooting Charts–Upper Machinery (*Continued*)

Malfunction	Probable cause	Corrective action
Vertical reverse shaft		
1. No backlash in the bevel gear.	a. Excessive gear compound.	a. Remove excessive gear compound.
2. Air chamber piston rod does not engage the clutch wheel to friction block.	a. Excessive friction block wear.	a. Adjust shifter screw if blocks are to correct size. If blocks are worn, replace blocks.
Rotating brake		
1. Brake will not engage or disengage.	a. Loss of air supply to air cylinder.	a. Check air supply system. Locate and repair line or fitting.
	b. Defective air cylinder.	b. Replace air cylinder or contact your dealer's service department.
	c. Toggle out of adjustment.	c. Check toggle adjustment and correct.
2. Brake will not hold.	a. Brake lining worn.	a. Check brake adjustment and replace lining if necessary.
3. Clutch control will not operate.	a. Defective linkage or out of adjustment.	a. Replace linkage or adjust as required.
	b. Defective air cylinder.	b. Check air cylinder and replace if necessary.
Drum shaft		
1. Clutch will not engage or disengage.	a. Lack of air in system.	a. Repair or check air supply.
	b. Defective air cylinder.	b. Repair or replace air cylinder.
2. Clutch grabs or chatters.	a. Clutch linings glazed or greasy.	a. Remove lining and clean.
	b. Check for rough friction flange.	b. Clean area of foreign material.
3. Clutch fails to hold.	a. Clutch out of adjustment.	a. Adjust clutch.
	b. Lining is glazed.	b. Remove and file glaze from lining.
4. Clutch requires frequent adjustment, slips, or overheats.	a. Clutch out of adjustment.	a. Adjust clutch.
	b. Cracked or broken linkage.	b. Replace defective part as required.
	c. Linkage binding.	c. Lubricate linkage.
5. Brakes grab or chatter.	a. Brake linings oily or greasy.	a. Remove lining and clean.
	b. Check for rough friction flange.	b. Clean area of foreign material.
6. Brakes fail to hold.	a. Brake incorrectly adjusted.	a. Adjust brake.

Malfunction	Probable cause	Corrective action
	Drum shaft	
	b. Brake lining glazed.	b. Remove lining and clean.
	Independent boom hoist	
1. Brake or clutch linings grab or chatter.	a. Linings are dirty or greasy.	a. Remove linings and clean.
2. Brake or clutch will not hold.	a. Improperly adjusted.	a. Check and readjust as required.
	b. Linings glazed.	b. Remove and clean glaze from linings.
3. Internal clutch will not engage.	a. Clutch improperly adjusted.	a. Adjust clutch and check air system.
4. Boom will not raise.	a. External clutch band improperly adjusted.	a. Adjust clutch band as recommended.
5. Boom hoist stops before correct boom angle is obtained.	a. Boom hoist clutch throw-out improperly adjusted.	a. Adjust angle as recommended by moving Pilotair valve up to decrease angle or down to increase angle.
6. Ratchet pawl will not disengage.	a. Control rope broken.	a. Replace control rope.
	b. Air cylinder assembly defective.	b. Check and replace air cylinder assembly if required.
	c. Loss of air to air cylinder.	c. Check and repair air supply system.
	d. Defective air supply valve.	d. Refer to manufacturer's service manual.
7. Ratchet pawl will not engage.	a. Inoperative air cylinder.	a. Check air supply and air cylinder, replace if required.
	b. Broken ratchet pawl return spring.	b. Replace return spring.
	Nonindependent (worm gear) boom hoist	
1. Boom won't raise or lower.	a. Sliding pinion won't engage with rotating gear. Rust or open-gear compound on pinion.	a. Sliding pinion binding, clean and lubricate worm shaft and keys with oil.
	b. Air cylinder won't extend.	b. Check air system. Repair or replace air cylinder.
2. Boom drifts after raising or lowering.	a. Disengagement of the sliding pinion from the rotating gear while the worm shaft is still rotating.	a. Stop the worm shaft from rotating before disengaging the sliding pinion.
	b. Brake springs in the floating brake cone not	b. Add additional brake springs, three, four, or

(continued)

TABLE 6-2 Troubleshooting Charts–Upper Machinery (*Continued*)

Malfunction	Probable cause	Corrective action
	Nonindependent (worm gear) boom hoist	
	exerting sufficient pressure against the brake cone to hold the brake wheel stationary at anytime when the sliding pinion is disengaged from the rotating gear.	six in the floating brake cone (refer to worm gear boom hoist adjustments).
	c. Worn parts in the floating brake cone assembly.	c. Disassemble floating brake cone assembly and replace worn parts.
	Air supply system	
1. No pressure in system.	a. Reservoir drains open.	a. Close drains.
	b. Open line.	b. Check and repair line fitting or valves.
	c. Compressor drive belts broken or loose.	c. Check, adjust, or replace drive belts.
	d. Compressor malfunction.	d. Refer to the compressor manufacturer's service manual.
2. Compressor will not build up to full pressure.	a. Open line fitting or valve.	a. Check lines, fittings, and valves. Replace as required.
	b. Safety discharge valve defective.	b. Replace discharge valve.
	c. Defective compressor and improperly adjusted unloader and/or governor.	c. Refer to compressor manufacturer's service manual.
3. Compressor builds up full pressure ahead of reservoir—no pressure to system.	a. Restriction in line.	a. Remove and clean restriction from line or replace if necessary.
4. Full pressure at reservoir but insufficient pressure at operator's console.	a. Improperly adjusted or defective air supply regulating valve.	a. Inspect, adjust, or replace.
	Centralized oil system—pressurized (opt.)	
1. Oiling system will not operate.	a. Air control valve OFF.	a. Turn valve ON.
	b. Incorrect oil used in system.	b. Drain, clean, and refill oil reservoir. Readjust valves.
	c. No air from air supply system.	c. Check air supply system.

Malfunction	Probable cause	Corrective action
	Centralized oil system—pressurized (opt.)	
	d. Shut off cock in OFF position.	d. Turn to ON position.
	e. Central feed valve clogged or defective.	e. Clean, adjust, or replace valve.

TABLE 6-3 Lubrication Chart (Upper Machinery), for Fig. 6-21

Key no.	Lubrication points	Frequency, hr	Grease	Oil	Gear compound	Check	Change
1	Engine (refer to engine service manual)						
2	Engine clutch sleeve bushings, pins, and toggle linkage	40		X			
3	Clutch linkage	20	X				
4	Engine driving gear case	300		X		X	
4	Engine driving gear case	1000					X
5	Horizontal reverse shaft right-hand bearing	See 4		X			X
6	Right-hand brake shaft inner bushings	8		X			
7	Clutch driver keys and shifter rod key shots	8		X			
8	Bevel pinion outer bearing	300	X				
9	Bevel pinion inner bearing—center bearing	300	X				
10	Horizontal reverse shaft bevel pinions	8			X		
11	Horizontal reverse shaft helical pinion	8			X		
12	Horizontal reverse shaft left-hand bearing	1000	X				
13	Left-hand drum brake cross shaft bushing	8	X				
14	Vertical reverse shaft upper bearing	40	X				
15	Vertical reverse shaft lower bearing	40	X				
16	Vertical reverse shaft—spur pinion—bevel pinion	8			X		
17	Air cylinder rod pins—rod end bushings	20		X			
18	Rotating shaft shifting collar	8	X				
19	Nonindependent (worm gear) boom hoist thrust washer inner bearing	8	X				
20	Nonindependent (worm gear) boom hoist outer bearing	8	X				
21	Nonindependent (worm gear) boom hoist housing	40				X	
22	Nonindependent (worm gear) housing	Every 6 mo.					
23	Nonindependent (worm gear) boom hoist spur pinion	8			X		
24	Nonindependent (worm gear) sliding pinion	8		X			
25	Nonindependent (worm gear) continuous cylinder anchor pin—rod end pin	8		X			
26	Boom hoist cross shaft	40		X			
27	Right-hand drum shaft bearing	1000	X				

TABLE 6-3 Lubrication Chart (Upper Machinery), for Fig. 6-21 (*Continued*)

Key no.	Lubrication points	Frequency, hr	Grease	Oil	Gear compound	Check	Change
28	Right-hand drum clutch cyl., toggle, pivot, pins and bushings	8		X			
29	Drum shaft sleeve bearings, inner and outer	1000	X				
30	Drum clutch gear teeth	8			X		
31	Left-hand drum shaft bearing	1000	X				
32	Independent boom hoist rope drum sleeve bearings	300	X				
33	Independent boom hoist rocker lever bushings	8	X				
34	Independent boom hoist air cylinder strainer	150		X		X	
35	Independent boom hoist air cylinder brake and clutch linkage	8		X			
36	Independent boom hoist planetary pinion bushings and washers	4	X				
37	Independent boom hoist clutch roller bearings	300	X				
38	Independent boom hoist outboard strut bearing	300	X				
39	Independent boom hoist planetary pinions, gear, and sun pinion	40			X		
40	Rotating gear bushing—thrust washer	4	X	X			
41	Rotating gear—spur pinion teeth	8			X		
42	Air cylinder pins—shifting fork bushings	8		X			
43	Cylinder anchor pins—rod end bushings—pins	8		X			
44	Shifting fork control shaft bushing and washer	8	X				
45	Rotating shaft lower bushing	4	X				
46	Equalizer roller bushings—thrust washers	8	X				
47	Front and rear equalizer roller brackets—shafts	8	X				
48	Front rotating roller support pins and washers	8	X				
49	Front rotating roller bushings and washers	8	X				
50	Rear roller shaft (not shown)	8	X				
51	Right-hand cross shaft bushings	8	X				
52	Air compressor (refer to engine service manual)						
53	Independent boom hoist clutch throwout	8	X				
53	Independent boom hoist clutch throwout pin	8		X			
54	Deflecting sheave bushings and thrust washers	4	X				
55	Deflecting sheave outer drum shaft	150	X				

Power Shovel Maintenance 7-147

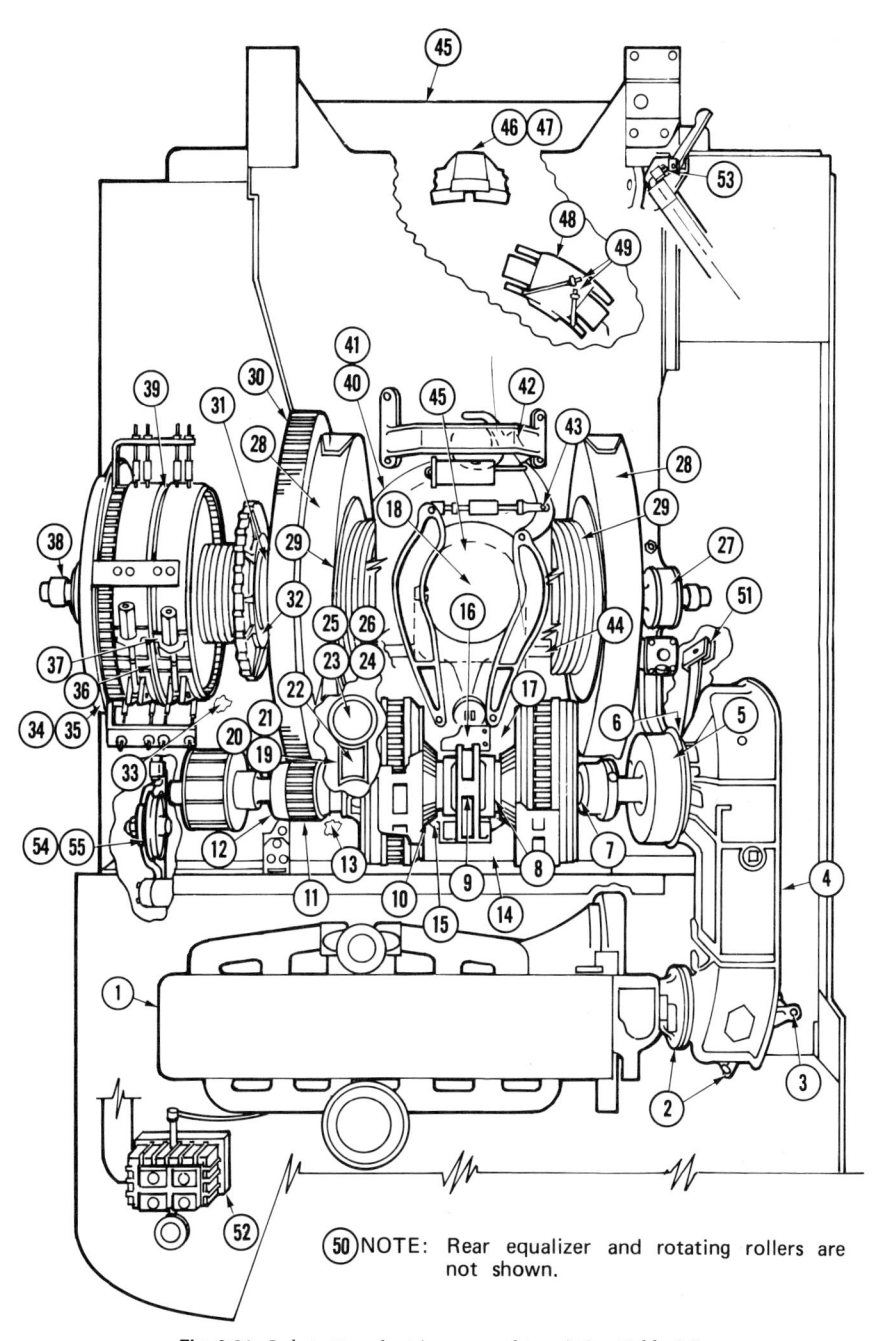

Fig. 6-21 Lubrication chart (upper machinery). See Table 6-3.

50 NOTE: Rear equalizer and rotating rollers are not shown.

TRAVEL BASE

Lubrication and Adjustments

Major components of a power shovel's travel base consist of the following, illustrated in Fig. 6-22: ring gear, left- and right-hand crawlers, idler sprockets, idler rollers, crawler rollers, drive chains and sprockets, outboard bearing supports.

CAUTION: Never attempt to lubricate or adjust the crawler while the machine is in motion. If movement of the track is necessary to reach a component, be sure that all personnel are clear of the track.

Lubrication should always be performed according to a preplanned maintenance sched-

Fig. 6-22 Travel base assembly.

ule to eliminate the possibility of premature wear or damage. When the machine is working under dusty conditions or in water or mud, lubrication should be scheduled more frequently to force out any harmful accumulation of foreign matter.

When the machine is traveling continuously, lubricate all roller bushings, drive shaft bushings, idler shaft bushings, idler roller bushings, outside bearing bushings, and drive chains at least once every 30 min.

For ease of maintenance checks, service, and adjustments, discussion of the travel base assembly is broken down into two sub-assemblies: the crawler and the travel base.

Crawler assembly The number of lubrication points for a crawler will vary depending on the length of the crawler side frame (Fig. 6-23).

The following items requiring lubrication with a heavy-duty grease, at least once every 4 hr or under adverse conditions once every 30 min, are keyed as callouts to Fig. 6-23:
1. Drive sprocket shaft bearing bushings and thrust washers
2. Crawler roller bushings, shafts, and washers
3. Idler roller shaft bushings and wearing washers
4. Chain deflecting roller bushings, washers, and pins
5. Idler sprocket shaft bearing, bushings, and thrust washers
6. Sprocket shaft outside bearing bushings

The crawler's drive chain will require a liberal quantity of light oil between the sides of its side bars for oil to penetrate the outside surfaces of the bushings, thus lubricating both pins and bushings. Never apply a heavy gear oil to these surfaces. Such a lubricant will coat only the outside and not reach the chain pins and bushings. Also apply a liberal amount of light oil between the side bars and rollers. Then, travel the machine to work in the oil (refer to Fig. 6-24). Under normal operating conditions, lubricate the chain every 20 hr.

Use a heavy-duty grease to lubricate the outboard bearing bushing, thrust washers, and bearing spacers through fittings located on each drive sprocket's outboard bearing housing (Fig. 6-25) at least once every 4 hr.

Tread link pins are normally run dry without lubrication. Yet, if the power shovel is working on pavement or on a timber runway continuously, the tread link pins should be lubricated with engine oil at least every 20 hr of normal operation.

Power Shovel Maintenance 7-149

Adjustments to the crawler should be made as required from daily checks. The first area of concern is the crawler tracks. Tracks should be adjusted so that a limited amount of slack or sag exists on their top sides to prevent excessive tension or wear. When checking the track for adjustment, always travel the machine forward a short distance with the drive chain to the rear, thus allowing slack at the top of the tracks (see Fig. 6-26).

Stop the machine; place a straight edge parallel to and on top of the tread belt to determine the amount of slack in the belt between the idler roller and drive sprocket. Slack should never be more than 3 in. for normal operating conditions (Fig. 6-26).

Operating the machine with too much slack will cause excessive wear not only on the track but also on the drive sprockets and track links.

Adjust the tread belt slack at either the drive sprocket or idler sprocket end. Follow these steps for adjusting both the tread belt and the drive chain:

Crawler Track Adjustment

1. Loosen the crawler-bearing U bolts (item 1, Fig. 6-26) for both the inner and outer drive shaft bearings until the rear U-bolt spacers (item 5) are free. Remove the take-up bolt keeper pin from the outboard bearing take-up bolt socket before adjustment is made.

2. Remove cotter key (item 3) and loosen the tension bolts (item 4).

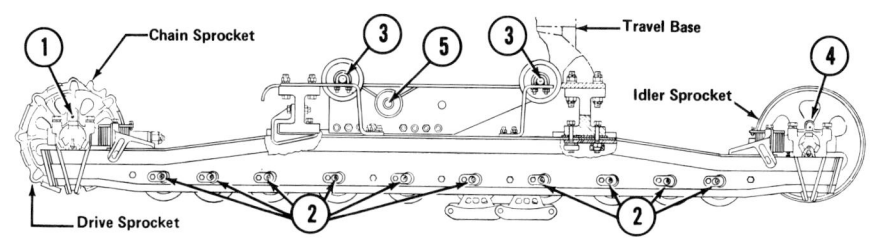

Fig. 6-23 Crawler side frame.

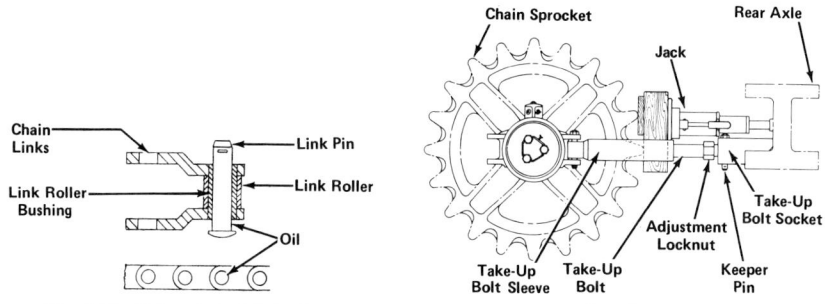

Fig. 6-24 Drive chain lubrication.

Fig. 6-25 Outboard bearing support.

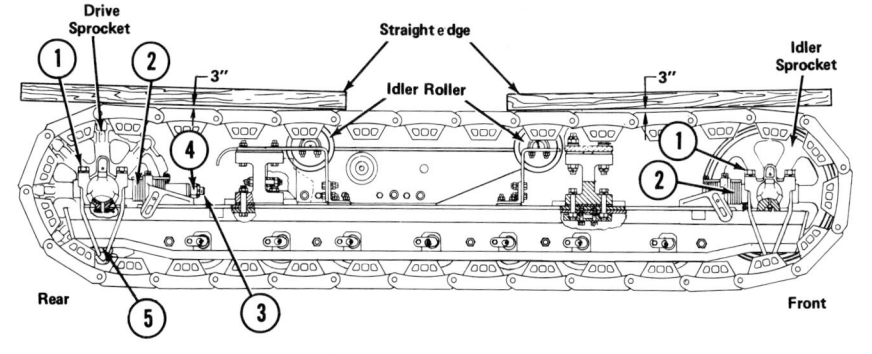

Fig. 6-26 Track adjustment.

Maintenance Programs for Equipment Entities

3. Place a suitable jack between the H-beam axle and the rim of the drive sprocket (Fig. 6-25), as low as it can conveniently be arranged. Then, push the drive shaft bearings toward the rear of the side frame. Do not use the outboard bearing take-up bolt sleeve as a jack or to move the track back.

4. Install take-up shims (item 2) between the drive shaft bearings and the retaining bearings as required.

5. When adjustment for the chain tension is completed, measure the distance from the center of the drive sprocket shaft to the axle on each track to determine if the shafts and axles are parallel.

6. Tighten all U bolts, tension bolts, and shim keeper bolts.

7. If equipped with side bearing supports, loosen the adjusting nut and adjust the take-up bolt so the keeper pins can be installed by hand into the take-up bolt sockets.

If the drive chains are properly adjusted and have no appreciable amount of slack and only the tread belt requires adjustment, make the adjustment at the idler sprocket only. When making the adjustment to the belt, keep the crawler drive sprocket shaft parallel to the H-beam axle and look for any noticeable side wear on sprockets which would indicate misalignment.

Travel base Travel base components are lubricated by different methods: splash lubrication from the gear case, centralized oiling from the metered oiler mounted on the right-hand A-frame member, grease through a high-pressure grease gun, engine oil, and an open-gear compound.

Travel base gears, pinions, shaft, bushings, washers, ratchet travel brake parts, and jaw clutches are lubricated with EP transmission gear oil splashed over these parts while the machine is traveling. The travel base gear guards will require approximately 12 ½ gal of transmission gear oil for both sections. Refer to Fig. 6-27 and pour about 5 gal in the right-hand gear guard and the balance in the left-hand gear guard by lifting the fill caps located in the rear of the travel base casting. Check the oil level at least once every 300 hr by removing the level test plug.

At least twice a year, gear guards should be drained and flushed with fuel oil. If the machine is driven through water or under conditions which might force water into the gear guards, draining should be performed and the guards refilled more frequently.

Do not allow water accumulation to freeze because ice formed on the gears and jaw clutches will cause steering to become inoperative.

Air cylinder pins and the shifting fork bushings of the vertical travel shaft gear clutch control (Fig. 6-28) should be lubricated with engine oil. A few drops of oil applied to these parts every 8 hr will sufficiently lubricate the components for normal operating conditions.

The upper bushings, split collar, and oiling washer on the vertical travel shaft are to be

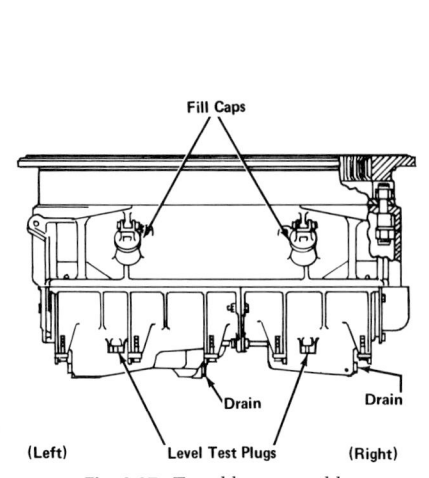

Fig. 6-27 Travel base assembly.

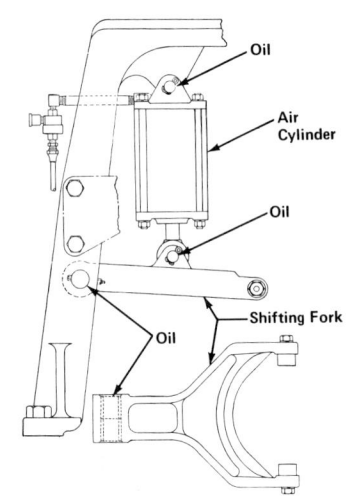

Fig. 6-28 Travel shaft gear clutch control.

lubricated with oil from the oiling washer (see Fig. 6-29). Fill the oiling washer every 4 hr under normal operation, every 30 min for continuous operation.

A gear compound must be brushed or poured on the ring gear, rotating gear, and vertical travel shaft spur pinion at least once every 8 hr. Apply the compound to all the teeth of each gear to prevent friction wear from metal-to-metal contact (see Fig. 6-30).

CAUTION: Be sure engine clutch lockout switch is activated and engine is shut off when applying gear lubrication. Because of limited accessibility to these gears, do not apply gear compound and rotate at the same time.

Adjustments to the travel base should be

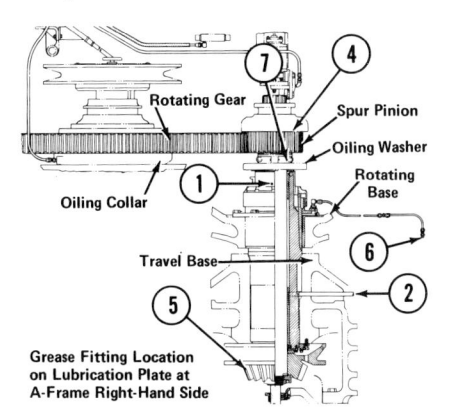

Fig. 6-29 Center pin and vertical travel shaft.

Fig. 6-30 Gearing diagram.

made only when necessary and then only as directed to reduce premature wear and damage. Most linkages, air cylinders, and bearings are preset and do not require periodic adjustments.

For example, no adjustments are required for the shifting fork in the travel shaft gear clutch control. The air cylinder piston rod is properly adjusted to the correct length at the factory and should not be changed.

The travel brake adjustment is located on the front side of the travel base casting (Fig. 6-31) and is used only to compensate for brake lining wear. When adjustment is needed, first set the brake valve to the OFF position. Tighten the brake by turning adjustment screw clockwise. Do not overtighten because if the adjustment is too tight, the lining will wear more rapidly and the machine will fail to travel freely.

NOTE: 1. A ratchet-type travel brake (digging locks) on a power shovel prevents crawler from traveling either forward or backward. Use rear ratchet travel brake when traveling uphill and use front and/or rear brakes during a digging operation.

2. A ratchet travel brake control valve (digging lock) must be in OFF position to unlock brakes when steering the machine.

3. Do not engage ratchet travel brake when machine is traveling. No adjustments for ratchet travel brake assembly are required.

Troubleshooting and lubrication charts for the crawler and travel base assemblies follow as Table 6-4, working with Fig. 6-32.

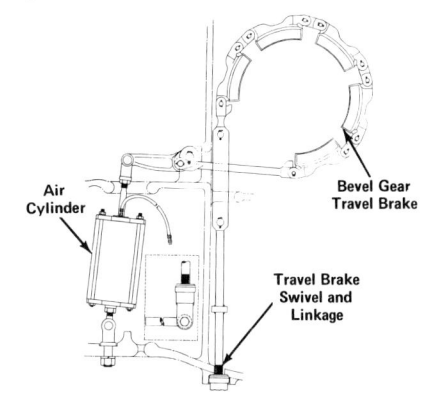

Fig. 6-31 Travel brake adjustment.

TABLE 6-4 Troubleshooting Charts—Travel Base

Malfunction	Probable cause	Corrective action
Crawler		
1. Machine pulls to one side.	a. Worn or damaged track components.	a. Check track components, replace as required.
2. Crawler throws tracks.	a. Incorrect adjustment.	a. Adjust track to compensate for slack.
	b. Worn-out sprockets.	b. Check sprockets, replace if required.
	c. Worn-out tread links.	c. Check tread links, replace if required.
3. Excessive noise in crawler assembly.	a. Worn drive sprockets.	a. Check sprockets and replace if required.
	b. Worn track pins.	b. Check and replace track pins if required.
	c. Worn track tread links.	c. Repair or replace.
4. Crawler drive chain throws off during operation.	a. Incorrect adjustment.	a. Check and adjust.
	b. Worn drive sprocket.	b. Check sprocket and replace if required.
5. Excessive crawler drive chain noise.	a. Lack of lubricant.	a. Lubricate pins, links, and bars with oil.
	b. Defective pin, link, or bar.	b. Check and replace as required.
Travel base		
1. Machine will not steer right or left.	a. Winter operation—gear guard filled with ice, preventing operation of steering controls.	a. Melt ice, drain and refill with oil. CAUTION: Nylon lines in travel base.
	b. Defective air cylinders.	b. Check air cylinder, replace if required.
	c. Linkage broken.	c. Check and replace linkage as required.
	d. Horizontal reverse shaft clutches need adjustment or worn out.	d. Adjust clutches or replace if required.
	e. Ratchet travel brake will not disengage.	e. Check for the air cylinders possibly binding or loss of air pressure.
2. Machine will not travel freely (refer to text).	a. Bevel gear travel brake too tight.	a. Adjust brake (refer to text).
3. Excessive noise in gear case.	a. Lack of lubricant.	a. Check and fill.
	b. Wrong type of oil used.	b. Drain and refill.
	c. Excessive water in gear case—diluted oil.	c. Drain and refill.

Power Shovel Maintenance 7-153

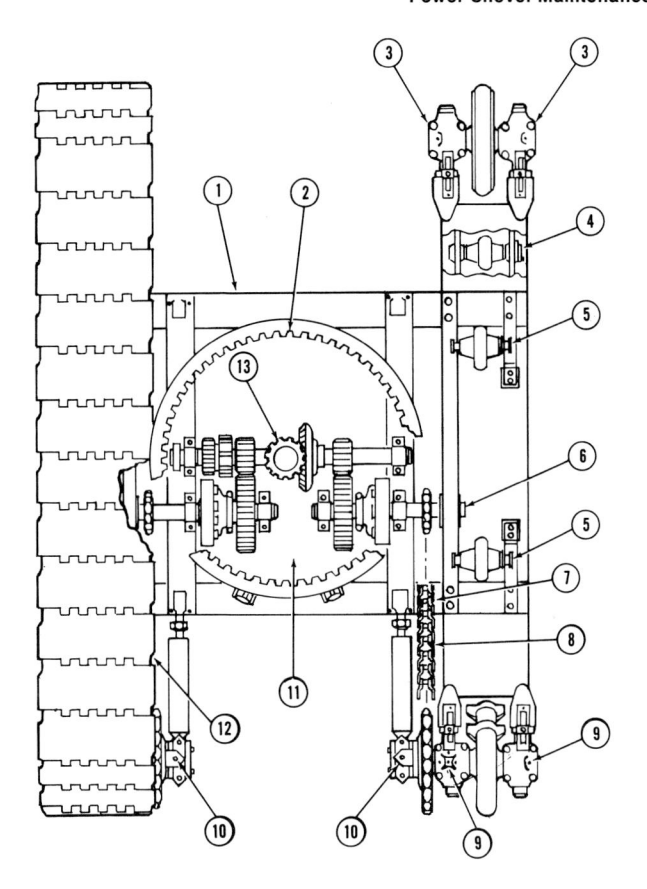

Key No.	Lubrication Points	Frequency	Grease*	Oil*	Gear Compound*	Check	Change
1	Vertical Travel Shaft Lower Bushing (Fitting in front axle)	4	X				
2	Ring Gear	8			X		
3	Idler Sprocket Shaft Bearing Bushings and Washers	4	X				
4	Crawler Roller Bushings, Shafts and Washers	4	X				
5	Idler Roller Shaft Bushings and Wearing Washers	4	X				
6	Sprocket Shaft Outside Bearing Bushings	4	X				
7	Chain Deflecting Roller Bushings, Washers and Pins	4	X				
8	Chain Links, Pins, Rollers and Bushings	20		X			
9	Drive Sprocket Shaft Bearing Bushings and Washers	4	X				
10	Outboard Bearing Bushings, Washers and Spacers	4	X				
11	Travel Base Gear Case and Components	300		X		X	
11	Travel Base Gear Case and Components	1000		X			X
12	Tread Link Pins	20		X			
13	Vertical Travel Shaft Spur Pinion	8			X		

Fig. 6-32 Lubrication chart.

ATTACHMENTS

Lubrication and Adjustments

Components of a power shovel's attachments, illustrated in Fig. 6-33, consist of the shovel's front drum, boom, dipper stick, dipper, shipper shaft assembly, boom head shaft, gantry, gantry bridge sheaves, and crowd bracket sheave.

Lubrication of attachments, like other components of a power shovel, should always be performed according to a preplanned maintenance schedule. Doing so will eliminate the possibilities of premature wear and damage. Records of service should be kept on the general condition of the assemblies and what maintenance was required.

CAUTION: Never attempt to lubricate or service any part of attachment assemblies while they are moving or when machine is working.

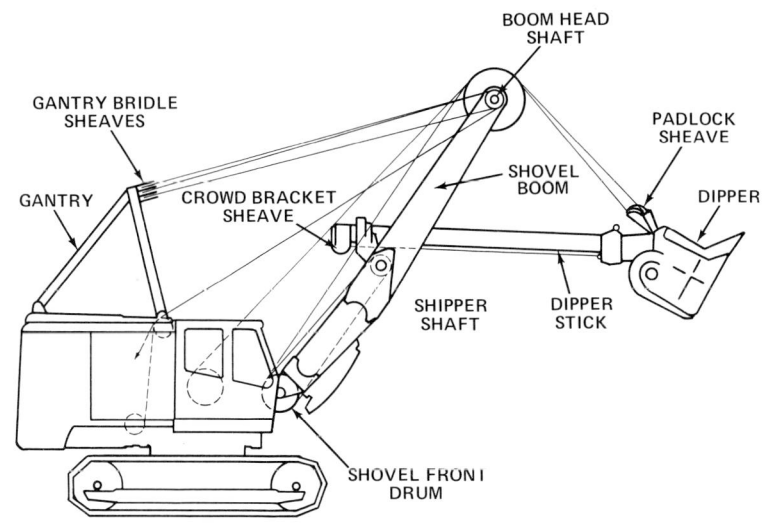

Fig. 6-33 Front end attachment components.

Shovel crowd clutch Twelve components of the shovel's crowd clutch require service and PM. The gear teeth should be lubricated with an open-gear compound at least once every 8 hr. The crowd drive chain should be lubricated with a light engine oil. Apply a liberal amount of oil at least once every 8 hr to penetrate between the side bars to the bushings, and run the chain after each lubrication. Do not apply open-gear compound or grease on the chain because it will not penetrate between pins and bushings. Table 6-5 lists typical lubrication points and schedules for a shovel crowd clutch which are keyed to Fig. 6-34.

The crowd's shifter screw lever compensates for clutch friction block wear. When the friction blocks have worn so that the air cylinder piston no longer engages the clutch wheel with the friction blocks, and enough material still exists on the blocks, then the shifter lever may be retarded one spline on the shifter screw (Fig. 6-34).

The chain should be adjusted periodically so that no more than $1\frac{1}{4}$-in. slack or less than $\frac{7}{8}$-in. slack exists. If the chain is adjusted too tightly, the front drum bushings and left-hand main drum bearings will wear more rapidly.

To adjust the crowd chain, refer to Fig. 6-35 and remove the $\frac{3}{4}$-in. pipe plug from the special pipe coupling, exposing the adjusting, pressure grease gun fitting. Tighten the crowd chain by slowly pumping heavy-duty grease into the fitting to obtain the $\frac{7}{8}$- to $1\frac{1}{4}$-in. slack. Use a hand grease gun. To loosen the chain, remove the pressure grease gun fitting and push down on the top strand of the crowd chain.

The shovel dipper trip is located on top of the crowd clutch gear teeth and requires lubrication of the driving gear teeth with a quality, sticky open-gear compound which can be poured or brushed on the gears. End bearings, thrust sleeve, lever and cylinder anchor

TABLE 6-5 Lubrication Schedule–Shovel Crowd Clutch

					Lubrication interval	
No.	Item	Lubricant type	By means of	Location	Normal operation, at least once	Adverse conditions
1	Crowd clutch gear teeth	Open-gear compound	Refer to text for shovel crowd clutch lubrication.			
2	Shifter screw and nut assembly	Heavy-duty, extreme pressure, high dropping-point grease	Pressure grease gun and fitting	In the shifter screw nut housing	Each 4 hr	As required
3	Shifter key plunger and ball	Heavy-duty, extreme pressure, high dropping-point grease	Pressure grease gun and fitting	In the left-hand bearing housing	Each 4 hr	As required
4	Clutch wheel and shifter collar thrust washers	Heavy-duty, extreme pressure, high dropping-point grease	Pressure grease gun and fitting	In the hub of the clutch wheel	Each 4 hr	As required
5	Clutch wheel bushings NOTE: *Be sure to lubricate each of the three fittings, item 5, when performing maintenance.*	Heavy-duty, extreme pressure, high dropping-point grease	Pressure grease gun and fitting	In the hub of the clutch wheel	Each 4 hr	As required
6	Left-hand crowd shaft bushing and spacing collar	Heavy-duty, extreme pressure, high dropping-point grease	Pressure grease gun and fitting	In the hub of the drive sprocket	Each 4 hr	As required
7	Right-hand crowd shaft bushing and spacing collar	Heavy-duty, extreme pressure, high dropping-point grease	Pressure grease gun and fitting	At the right-hand end of the drive sprocket	Each 4 hr	As required
8	Drive sprocket thrust washer	Heavy-duty, extreme pressure, high dropping-point grease	Pressure grease gun and fitting	In the right-hand outer end of the crowd shaft bracket	Each 4 hr	As required
9	Air chamber rod end pin	Engine oil	Oil can	Shifter screw lever	Each 20 hr	As required
10	Idler sprocket bushings and washers (chain tightener sprocket)	Heavy duty, extreme pressure, high dropping-point grease	Pressure grease gun and fitting	In the end of the chain tightener pin	Each 4 hr	As required
11	Air cylinder adjusting rod	Heavy-duty, extreme pressure, high dropping-point grease	Pressure grease gun and fitting	In the adjustment end of air cylinder rod	Each 4 hr	As required

pins, the rod end pin (items marked 4, Fig. 6-36), and air cylinder (item 5) require lubrication every 8 hr with a light engine oil. After 40 hr of operation, the dipper trip's friction flange, inner and outer bearings, thrust sleeve, and spacing washer (item 2, Fig. 6-36) should be lubricated with a heavy-duty grease. The rope drum, its inner and outer bearings, and related parts (item 3) should also be greased at least every 40 operating hours with a heavy-duty grease.

Fig. 6-34 Shovel crowd clutch. Numbers refer to Table 6-5.

To compensate for clutch friction disk wear, adjust the shaft spring on the end of the dipper trip rope taut for all operating conditions. If the spring is adjusted too tightly it will cause the dipper trip to overheat.

The yoke end of the swivel lever spring eyebolt, located at the lower end of the dipper stick, should be tight against the lug on the backhaul rope bracket before the dipper trip shaft spring is adjusted. The adjusting screw in the short lever is properly adjusted when sufficient pressure is applied to the dipper trip to open the dipper door, and the long lever does not interfere with the left end bearing. The stop screw, located in the left end bearing is factory-set and should not be changed.

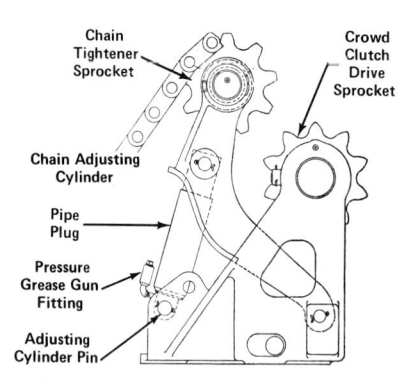

Fig. 6-35 Crowd chain adjusting cylinder.

Front drum The front drum sprocket and crowd chain are lubricated as required with a liberal quantity of light engine oil at least once each 8 hr to penetrate between the side bars to the bushings. Run the chain after each lubrication to work in the oil between the pins and bushings and between the rollers and bushings. Do not apply an open-gear compound or grease on the chain because this will prevent the oil from penetrating between the pins and bushings.

Front drum bushings (item 1, Fig. 6-37) located in the hub of the front drum should be lubricated with a heavy-duty grease every 4 operating hours. The take-up drum bushing should be lubricated with a heavy-duty grease also in its fitting located in the rim of the take-up drum (item 2) every 40 hr. Clamping collar plungers (item 3) should be lubricated with engine oil every 20 hr. Adjustments are not required for the front drum.

A-frame and gantry The gantry tension member bushing and the horizontal strut pins (item 1, Fig. 6-38), guide roller bushings (item 4), vertical strut upper pins, removable gantry strut pins (item 5), and the right- and left-hand A-frame sheaves (items 2 and 3) should be lubricated each 20 hr with a heavy-duty grease or engine oil as specified by the manufacturer. No adjustments to the gantry are required.

Fig. 6-36 Shovel dipper trip located on top of the crowd clutch gear teeth.

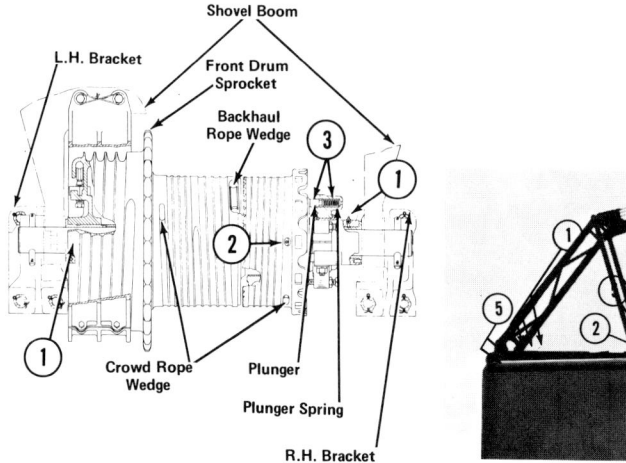

Fig. 6-37 Front drum.

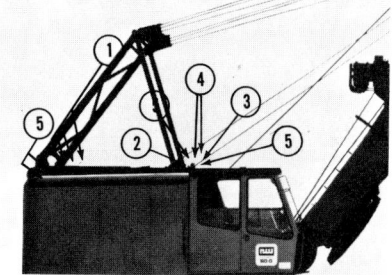

Fig. 6-38 A-frame and gantry.

No adjustments are required for a power shovel's bridle assembly. However, the bridle's yoke hanger bushings (item 1, Fig. 6-39) and sheave bushings and washers (item 2) should be lubricated with a heavy-duty grease at least once every 40 hr through grease fittings provided.

Boom assembly A power shovel's boom assembly has three areas of lubrication: hinge pin, shipper shaft, and boom head assembly (Fig. 6-40). Use a heavy-duty grease to lubricate the boom hinge bushings (item 1, Fig. 6-41) at least every 8 operating hours. If the boom is raised or lowered continuously, always lubricate more frequently. Inspect the bushings for wear daily.

The dipper stick saddle liners (item 1, Fig. 6-42) should be lubricated with a quality, sticky open-gear compound at least once every 8 hr. The shipper shaft's saddle bushings and washers (item 2), its crowd sheave bushings and backhaul sheave bushings (item 3), plus the fairlead sheaves (item 4) should be lubricated with a heavy-duty grease after every 4 hr of operation.

Each sheave in the boom head assembly has its own grease fitting and should be lubricated according to manufacturer specifications. A typical schedule calls for lubricating the boom point sheave bushings and washers (item 1, Fig. 6-43) with heavy-duty grease every 20 operating hours. Boom hoist sheave bushings, washers, and shaft (item 2) should be greased every 8 hr.

The only adjustments required to the shovel's boom are to its saddle liners. An approximate clearance of 1/8 in. should be maintained between the dipper stick and the liners. To adjust, add saddle liner shims between top and bottom liners and saddle to decrease the clearance between the dipper stick and top and bottom liners. Add saddle liner shims between side liners and saddle to decrease clearance between sides and dipper stick.

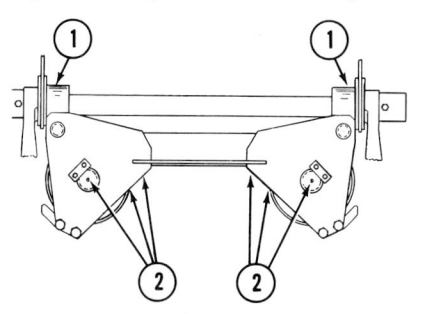

Fig. 6-39 Bridle: folding gantry.

After you have adjusted the saddle liners, change the quantity of spring adjusting washers. To determine the quantity of washers, (1) install saddle liner bolt spring on the saddle liner bolt, (2) insert saddle liner bolt into the dowel locking bar and turn the bolt one or two threads into the saddle liner, (3) slide the spring against the dowel locking bar, and (4) check the gap between the spring and head of the saddle liner bolt.

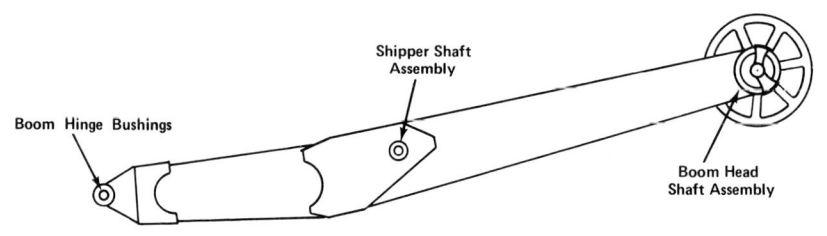

Fig. 6-40 Boom assembly.

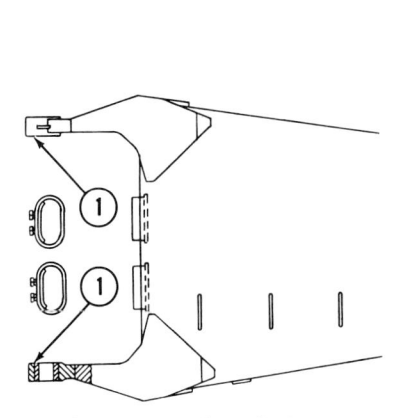

Fig. 6-41 Boom hinge bushings.

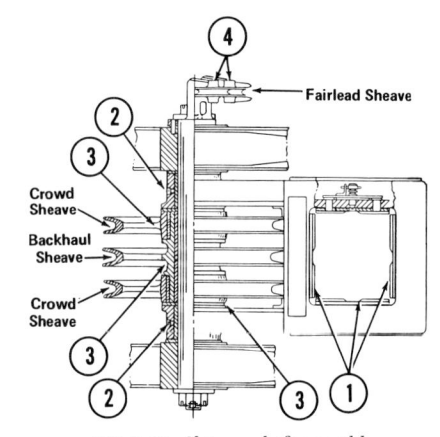

Fig. 6-42 Shipper shaft assembly.

Such a procedure will determine the number of washers required. To install the washers, remove the saddle liner bolt, and place the washers on the bolt and tighten until the shoulder of the bolt bottoms against the saddle liner.

Dipper and stick For all parts of the dipper stick, use a good, sticky open-gear compound. Lubricate the saddle liners at least once every 8 hr. Figure 6-44 shows lubrication points for the padlock sheave (item 2) that requires lubrication with a heavy-duty grease every 4 operating hours. Sheave bushings (item 3) should also be lubricated every 4 hr with a heavy-duty grease. The dipper door hinge bushings should be lubricated every 8 hr at the outer end of each hingepin (item 4) with a heavy-duty grease.

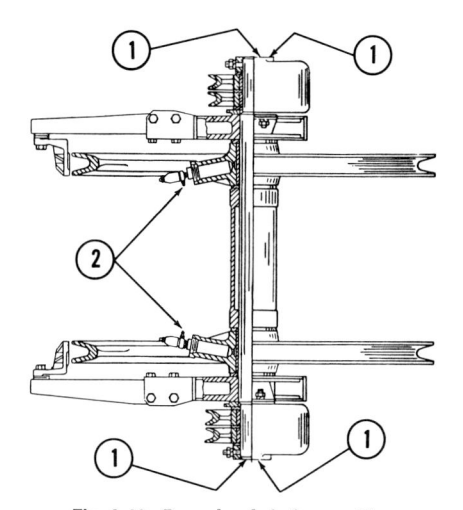

Fig. 6-43 Boom head shaft assembly.

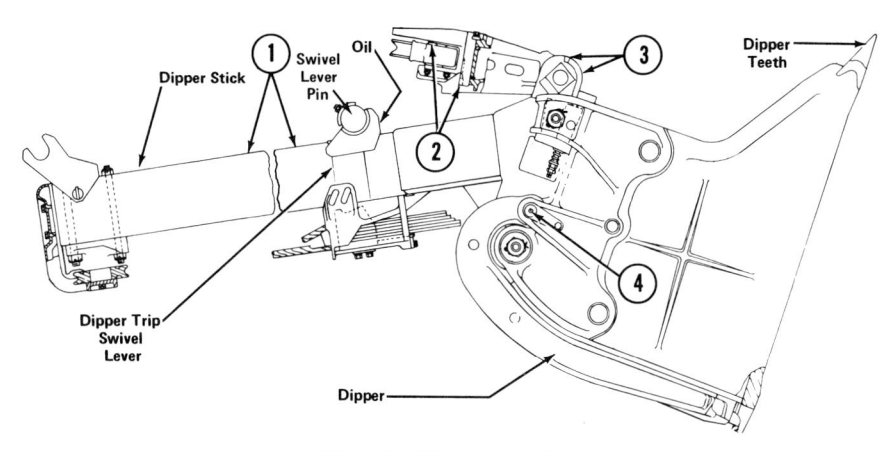

Fig. 6-44 Dipper and stick.

The dipper latch lever pin, dipper latch, latch lever, roller, and pins (Fig. 6-45) are all lubricated with engine oil. A good schedule to follow is every 4 hr.

The dipper body is connected to the front end of the dipper back by means of the tie bolt to permit the so-called slope or rake of the dipper to be set in either of two positions.

When digging conditions require a change to prevent excessive wear at the heel or rear part of the dipper front, remove the front tie bolt and dipper connection wedges. Adjust the locknuts at the lower end of each connection wedge until there is approximately $1/32$-in. clearance between the coils in the dipper wedge springs.

The dipper latch stop washers are used to control the distance the dipper latch bar enters to hold in the dipper front. Keep the bar adjusted so the dipper trip can easily and quickly open the dipper door.

Wire rope inspection and lubrication The following inspection of wire rope is given as basic reference.

All wire ropes should be visually inspected daily. A thorough inspection should be made at least once a month. Records should be kept as to rope condition, date inspected, and when lubricated. Notations should also be kept covering any damage, corrosion or rust, broken wires, and the fitting condition of the wire rope to the drums, sheaves, and pulleys.

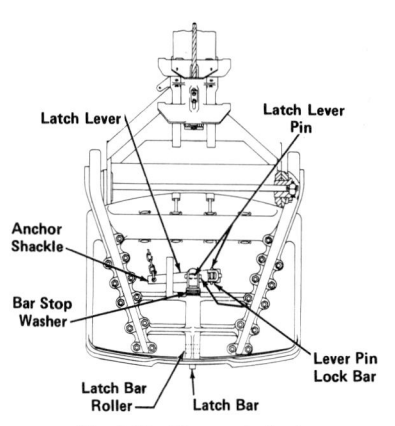

Fig. 6-45 Dipper trip latch.

The following conditions are provided to help determine the safety of the wire rope, and if replacement is required.

1. Corrosion causes deterioration.

2. Broken wires that occur in the valleys between strands are caused by fatigue and breakage of internal wires not visible during inspection.

3. Extensive wear, abrasion, scrubbing, and peening of outside wires cause distortion of rope structure.

4. Extensive reduction in wire rope diameter indicates deterioration of internal core or wire strands.

5. Running ropes: six randomly distributed broken wires in one rope lay or three broken wires in one strand of a rope lay.

6. Pendants or standing ropes: more than one broken wire in one rope lay.

7. Severe kinking or crushing in any wire rope causing distortion.

8. Any excessive rusting or development of broken wires near an attachment.

All surfaces that the wire rope comes in contact with should be inspected. Any condition which may cause damage to the wire rope should be noted and corrective action taken to repair or replace.

Lubricate the rope regularly with a good wire rope lubricant which may be sprayed or brushed on. If rope is lubricated regularly, the lubricant will penetrate throughout each strand including inner wires to reduce wear.

CAUTION: Never attempt to lubricate wire rope while it is moving. To effectively lubricate the entire rope length changes of position must be made but only when all personnel are clear of the rope and attachments.

Experience has proved that changing or reversing a wire rope end for end distributes, over different sections of the rope, the wear and fatigue from bending, thus increasing the service received from the wire rope.

Breaking in a new rope with a light load, or with no load, for a short period of time gives the rope an opportunity to adjust itself to the operating conditions, thus increasing the service received from the rope.

Sudden application of loads or jerking a wire rope may not break the rope, but this type of abuse does cause rapid deterioration and shortens rope life. Some other forms of abuse that reduce rope service are kinking, dragging over obstacles, crosswinding on drums, failing to properly lubricate, overloading, and using the wrong type or construction of rope.

Chapter 7

Asphalt Paving Equipment Maintenance

BARGER K. MACY
Technical Publications Supervisor
Blaw-Knox Construction Equipment, Inc.
Mattoon, Ill.

Although a good maintenance program is the key to lower operating costs with any piece of construction equipment, it becomes even more important with asphalt paving equipment, because so many other operations on the construction job are affected if the paving machine is down or not operating properly. This machine must be available to efficiently lay all the bituminous material being produced by the asphalt plant and delivered hot by the trucks. If problems develop during the paving operation, the asphalt plant, hauling trucks, and the paving and rolling operations are all affected.

Equipment types covered in this chapter include self-propelled asphalt pavers-finishers, sometimes referred to as spreaders; self-propelled power boxes; towed tail gate spreaders; and self-propelled road wideners. Although they may be designed for various sizes and types of jobs, they incorporate the same basic operating principles and contain the same major components, no matter who manufactured the machine. All have a hopper, which receives material, and conveyors and/or augers, which deliver and distribute material to a screed or strike-off. The screed levels and compacts, whereas the strike-off only levels material on the grade. Most use an engine for either traction and/or material handling drives. Many of the self-propelled units have hydrostatic traction drive and material feed systems. Some still use mechanical drives. All use hydraulic power to control and operate the machine. Many of these machines incorporate sophisticated electronic controls to maintain finished surface specifications.

The recommendations for maintenance of asphalt paving equipment covered in this chapter are general enough to pertain to all types described with a few exceptions. For example, some maintenance procedures to be performed on a track-driven machine would not apply to a wheel-driven machine (Fig. 7-2).

It is very important to consult the service or maintenance manual furnished with the machine for specific maintenance information, service points, and adjustments of any piece of construction equipment (Fig. 7-3). Engine manuals supplied by the engine manufacturer should also be referred to for routine servicing information. Also, remember the operating environment affects the frequency of normal scheduled service. Climate, altitude, dust conditions, and even types of paving materials can create the need for more intensive preventive maintenance than is normally required. Good judgment, common sense, and a watchful eye coupled with a conscientious maintenance program can prevent costly situations from occurring.

Finally, keep a set of records to indicate what service has been performed on the paving machine, whether maintaining a single tow box or a yard full of boxes, pavers, and wideners.

PREOPERATION CHECKOUT PROCEDURE

Before starting to pave, the entire machine should be inspected, using the following checklist as a guide. It is recommended that this procedure be performed at least once a week to prevent component failure and unnecessary downtime.
1. Check engine cooling system level.
2. Check engine lubricating oil level.
3. Check pump drive box oil level.
4. Check traction drive transmission oil level.
5. Check traction drive axle oil level.
6. Check hydraulic reservoir fluid level.
7. Check battery water level.
8. Check brake master cylinder fluid level.
9. Check drive tire pressure.
10. Conveyors adjusted for proper tension.

Fig. 7-1 The asphalt paving train in action.

Fig. 7-2 Maintenance operations on wheel-driven asphalt pavers are sometimes different from those for track-driven vehicles.

11. Check conveyors for excessive wear of drag bars, chains, sprockets, or conveyor belt.
12. Augers adjusted for proper amount of play.
13. Check augers for excessive wear.
14. Entire machine lubricated per instructions furnished by the manufacturer.

Fig. 7-3 Service manuals form the cornerstone of any sound maintenance program.

15. All machine functions operating properly.

Also check the following items periodically.
16. Engine clutch properly adjusted.
17. Engine rpm (no load—idle and full).
18. Engine properly tuned and operating smoothly.
19. Engine fan belt tight.
20. Engine mounting bolts tight.
21. Transmissions, axles, and other mechanical drive components mounting bolts tight.
22. Wheel lug nuts tight.
23. All hydraulic lines and fittings tight.
24. Hand brakes properly adjusted.

DAILY WASHDOWN PROCEDURE

Daily performance of the following washdown procedure with *kerosene* or *diesel fuel* will prevent operational difficulties and premature component failure caused by a buildup of paving material on machine components as well as contribute toward the overall operational safety and appearance of the unit (Fig. 7-5).

Move the machine away from the freshly laid mat and position so the solvent residue will not affect continuation of the paving operation. Raise the screed or strike-off. Clean the machine with a spray applicator and scraper starting at the front and working rearward.

1. *Push rollers:* Clean so they turn freely.
2. *Hopper and flow gates:* Clean all sides from top down. CAUTION: Do not operate conveyor while cleaning the hopper. Do not stand in the hopper if the engine is running.
3. *Conveyors—pavers:* Thoroughly spray each chain and drag bar several times with conveyors running for adequate cleaning. *Road wideners:* Scrape off excess material from

the running belt. CAUTION: Do not soak belt. Clean rollers beneath the belt as soon as material starts to build up. Material buildup will appear as lumps in the belt at the rollers. Dirty rollers will cause rapid belt wear. The rollers are difficult to clean, but every effort must be made to do so.

4. *Exterior:* Raise hoppers and remove all excess material.

5. *Rear conveyor drive sprockets—pavers:* Wash thoroughly while running to maintain self-cleaning action of sprockets.

6. *Auger and auger bearings—pavers:* Scrape out around bearings. Scrape augers and wash down entire auger area while running augers.

7. *Screed—Pavers:* Scrape off any material caught on the frame and screed plate. Thoroughly clean the face and bottom of the screed. Grease vibrator bearings.

8. *Edger plates:* Remove and clean so they slide up and down properly. Inside surfaces should be smooth.

9. *Hydraulic screed extensions—pavers:* Clean thoroughly, especially slide surfaces which mate. Grease slides.

10. *Strike-off—road wideners:* Thoroughly clean the face and bottom edge of the strike-off.

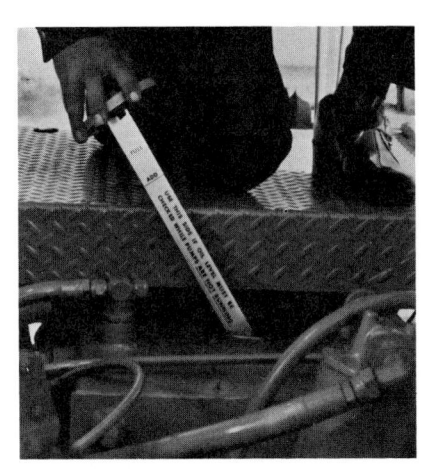

Fig. 7-4 Oil level checks form an important aspect of preventive maintenance.

PREVENTIVE MAINTENANCE TIPS

1. Establish a regular maintenance program, including records of service performed, to prevent normal wear from becoming a major expense.
2. Always lubricate a paving machine while it is warm to ensure that lubrication reaches all critical points. All bearings in the material handling areas should be *flush-greased daily while still warm* to extrude any material which has worked into the bearings during operation.

Fig. 7-5 Washdown procedure proves a vital key to paver performance.

3. Periodically inspect the grease tubes which run to bearings to be sure they are not damaged.
4. Stock known high-wear repair parts for quick routine or emergency replacement.
5. Conveyor chains are one of the most expensive repair items on a paver. Keep properly adjusted to obtain maximum life. (See section on Paver Conveyor Adjustment in this chapter.) On machines which permit it, reverse the entire

Fig. 7-6 Be sure asphalt pavers receive regular lubrication.

conveyor chain assembly to obtain many additional hours of paving from the existing chains. Most wear on the pins, links, and bars only occurs on the load-bearing side.
6. Conveyor belts on road wideners should be maintained for proper tension and tracking to obtain maximum belt life. (See section on Road Widener Conveyor Adjustments in this chapter for correct adjusting procedure.)
7. Check conveyor, auger, and traction drive chains and any drive belts for proper tension. See appropriate Adjustments sections.
8. Augers should be re-hard-surfaced *before* wear reaches the main auger flight material. Replace auger wear plates if badly worn on machines so equipped.
9. Check hopper hinges for wear. Repair if necessary.
    a. Ideally, if the screed is properly used, the screed plate should wear evenly over the entire surface. However, it probably will show slightly more wear toward the center.
    b. If paving materials are found sticking to the screed plate, improper paving procedures have been used.
        (1) Screed has not been preheated to near paving material temperature.
        (2) Screed has been used paving on its strike-off.
    c. If the screed plate is excessively worn near the front, screed has been used paving on its strike-off.
    d. If the screed plate is excessively worn near the rear, screed has been used paving on its tail.
    e. If screed plate is badly warped, screed has been excessively preheated.
    f. Lubricate the vibrator shaft bearings immediately after preheating the screed to compensate for grease which flows out of the bearings.
10. Check the screed regularly for worn parts. (Refer to the Screed Adjustments

section of this chapter.) Also consult the manufacturer's paving manual for correct use of the screed to obtain maximum life of the screed plate.
11. Electronic components do not normally require scheduled maintenance. However, they should be periodically checked to see if they are functioning properly. When not in use, demountable components such as control boxes, sensors, and reference beams should be carefully stored where they will not be damaged.
12. Track systems should be inspected regularly for worn components. Service the track system at a convenient time and location away from the job site. Replace all questionably worn parts which might fail during paving.
13. Take special care to prevent contamination from entering the hydraulic system when servicing. Dirt or moisture in the hydraulic system is the major cause of hydraulic component failure.
14. Do not overgrease clutches.
15. On machines equipped with kerosene screed heaters and washdown systems, drain water and dirt traps regularly.
16. On mechanical drive machines equipped with compressed air systems, drain water and dirt traps regularly.

OFF-SEASON STORAGE

After the paving season is over or when the machine is to be idle for a period of time, the following procedure is recommended to protect the paving machine and ensure its availability when paving is to be resumed.
1. Clean the machine thoroughly, removing *all* paving material.
2. Fill the fuel tank with fuel.
3. Drain the engine crankcase, change engine oil filter, and refill engine crankcase with recommended lubricant. Refer to engine manual for further recommendations concerning off-season storage.
4. Drain the hydraulic reservoir, change main hydraulic filters, and refill reservoir with recommended fluid.
5. Remove battery and place on trickle charge.
6. Remove seat and store in a dry place.
7. Lubricate entire machine per the instructions given in the machine service manual.
8. Check the machine thoroughly for worn or damaged components and repair or replace as required. Perform needed overhaul work as required.

PAVER CONVEYOR ADJUSTMENT

Maintaining the paver's two conveyors in proper adjustment will ensure maximum service life from the conveyors, sprockets, and shafts. A properly adjusted conveyor clears the ground by a minimum of 5 in., as shown in Fig. 7-8, so it will not hang up on obstructions. Do not overtighten chains. They should contain enough slack to pass aggregate through the drive sprockets without binding up. Adjustment of the conveyors is accomplished by moving the conveyor idler shaft and bearing arrangements located under the front of the paver hopper.

To adjust the conveyors, raise the front hopper apron [Fig. 7-7(1)] and raise the hopper sides. Loosen the cap screws securing the conveyor idler shaft pillow block bearings [Fig. 7-7(2)] to the paver frame. Loosen the jam nuts securing the adjusting screws [Fig. 7-7(3)] and using the adjusting screws, move the idler shaft bearings forward or rearward as required to obtain the 5 in. minimum ground clearance of the conveyors (Fig. 7-8). Care must be taken to adjust both sides of a conveyor equally to produce a true running nonbinding conveyor. With adjustment completed, retighten the bearing cap screws and lock the jam nuts on the adjusting screws.

On new machines, it may be necessary to adjust the conveyors several times during the paver's break-in period to compensate for chain run-in. This condition will diminish after the break-in period, and only periodic observation will then be required.

Maintaining the conveyors in proper adjustment may use up all the adjustment travel after many hours of operation. However, adjustment capability can be regained by removing either the offset half links or full links from the conveyor chains as required.

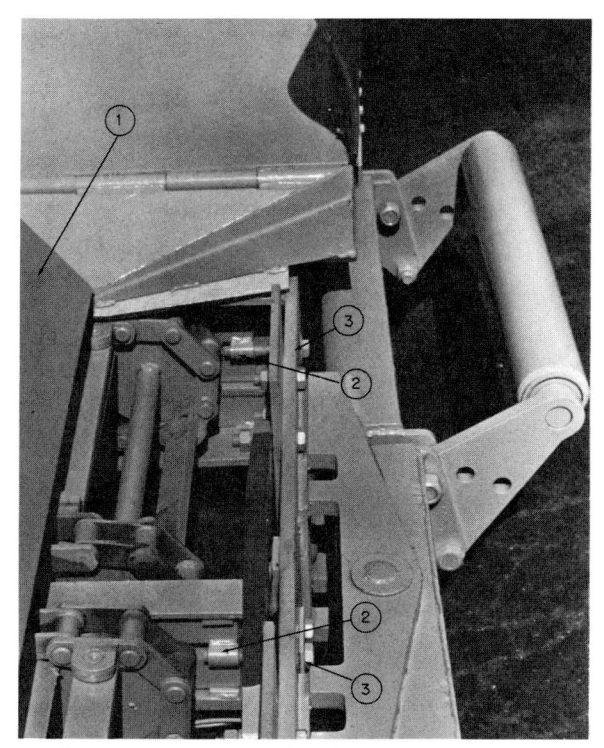

Fig. 7-7 Conveyor adjustment must be done in logical sequential steps and involve the (1) apron, (2) pillow block bearings, and (3) adjusting screws.

Fig. 7-8 Conveyor chain should be set for a minimum ground clearance of 5 in.

NOTE: If the offset half links are removed, an additional half link adjustment can be obtained at a later date by removing a full link and reinstalling the half link.

Extended service life of a conveyor assembly can be obtained by reversing the complete assembly before it is completely worn out.

AUGER AND CONVEYOR DRIVE CHAIN ADJUSTMENT

There should be a slight amount of play in the augers to avoid shocks and binding as they stop and start (Fig. 7-9). Approximately ¾ in. of free movement is satisfactory. This free movement is checked by manually rotating each auger by hand (Fig. 7-10).

Fig. 7-9 Area where auger adjustment is made.

The auger drive chains are normally totally enclosed in a case at the rear center of the machine. Consult the operators manual for adjustment procedure on your particular paver or tow box.

The conveyor drive chains should contain very little slack. On many pavers the conveyors are driven by the same chains as the augers. On others a separate set of conveyor drive chains are located on the outer rear of each conveyor. Consult the operators manual for adjustment procedure on your particular paver.

It is extremely important to maintain proper adjustment of these drive chains to ensure smooth operation of the paver's feeder systems and to prevent damage to the chains.

ROAD WIDENER CONVEYOR ADJUSTMENTS

The conveyor belt must be kept in proper adjustment to operate acceptably and to prevent rapid wear. Proper tension and alignment of the conveyor belt is maintained by adjusting the four tension bolts [Fig. 7-11(1),(2)] on each end of the conveyor assembly. When adjusted properly, the conveyor belt should not slip under load. However, do not tighten to the point where the joint fitting pulls out of the belt. A properly adjusted belt will tend to crawl forward under no load and crawl rearward loaded. It should never crawl far enough to wear at either edge.

NOTE: The conveyor belt will crawl toward the tight side of the shafts. To keep centered, tighten tension bolts evenly. When installing a new belt, take up most of the tension at the discharge end so that final adjustment can be made at the tail shaft end

Fig. 7-10 Manual rotation of augers is important to upkeep.

while paving. Most of the stretch in the conveyor belt will occur after running the first couple of loads of hot mix material. After the initial adjustment to compensate for stretching, little stretching should occur. However, the belt should be checked regularly for proper tension, alignment and wear or damage.

New belt After installing a new belt, run it briefly and adjust the four tension bolts to produce a *flat but unstretched* belt. Lay out and mark 50 in. on the belt in the hopper area. The joint should not be located inside the 50-in. dimension. Tighten the belt until the distance between the marks is 50 $\frac{3}{4}$ in. $+ \frac{1}{4}$ or $- \frac{1}{8}$ in.

Retensioning After approximately 80 hr of operation, loosen the belt to remove tension, but maintain a *flat unstretched belt*. Again, mark a 50-in. dimension on the belt and stretch to 50 $\frac{3}{4}$ in. $+ \frac{1}{4}$ or $- \frac{1}{8}$ in. If additional tension is required, tighten in $\frac{1}{8}$-in. increments until the belt no longer slips under proper loading conditions.

A cocoa mat cleaner [Fig. 7-11(3)] and neoprene wiper [Fig. 7-11(4)] minimize material leakage and prevent material sticking to the conveyor belt. These belt wipers are adjustable and should be kept in contact with the conveyor belt but should not exert excessive pressure on the belt.

The service life of a conveyor belt is significantly shortened when it is not properly adjusted and cleaned. Daily cleaning is very important to obtain maximum conveyor belt life. (See Daily Washdown Procedure.)

TRACTION DRIVE CHAIN ADJUSTMENT

Proper adjustment of the pavers traction drive chains is essential to obtain a smooth finished mat and to protect the various drive train components from shock load damage. Adjust for approximately $\frac{3}{4}$ in. of slack as illustrated (Fig. 7-12).

On a new machine, it may be necessary to adjust the traction drive chains during the

7-170 Maintenance Programs for Equipment Entities

paver's break-in period to compensate for chain run-in. Occasional observation should be made thereafter.

Maintaining the traction drive chains in proper adjustment may use up all the adjustment travel after many hours of operation. However, adjustment capability can be regained by removing either the offset half link or a full link from the drive chains as required.

Fig. 7-11 In making road widening adjustments, the key areas for consideration include the tension bolts (1) and (2), the cocoa mat cleaner (3), and the neoprene wiper (4).

SCREED ADJUSTMENTS

The amount of wear on a screed plate depends on whether the screed is properly adjusted during paving (Fig. 7-13). Those adjustments which produce a high-quality mat also produce maximum screed plate life. Conversely, a screed which is producing a poor mat

Fig. 7-12 Traction drive chain adjustment includes a ¾-in. slack allowance.

Fig. 7-13 Wear on the screed is a function of the care taken in its adjustment.

7-172 Maintenance Programs for Equipment Entities

can be expected to show early signs of wear. So it is important to know how to properly adjust the screed to obtain the best possible mat and, as a by-product, lower screed plate repair costs.

Mat or depth or thickness is established by the screed floating up or down until the bottom of the screed is parallel with the grade over which it is traveling. Controlling mat thickness is accomplished by controlling three forces that act upon the screed.

1. Force 1: paving speed
2. Force 2: angle of attack
3. Force 3: head of material

If one of these three forces is varied during the paving operation, it will affect the mat thickness, quality, and texture (Fig. 7-14).

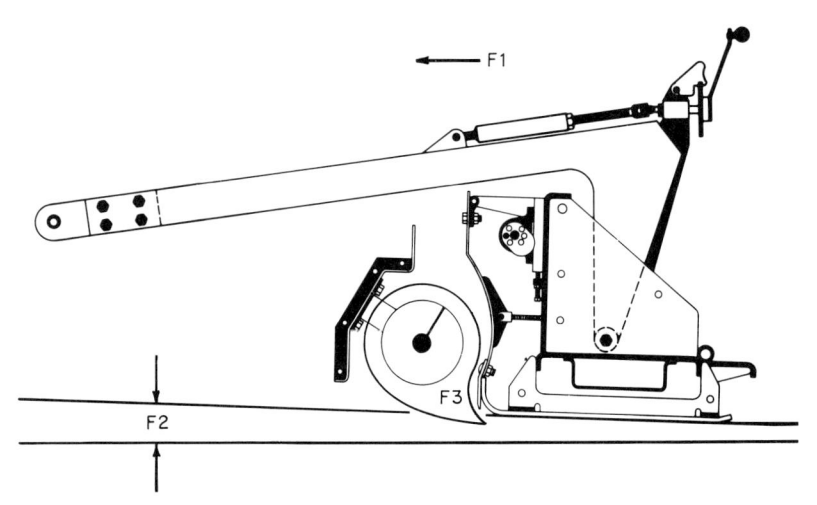

Fig. 7-14 Forces acting on the paver screed.

Force F 1, paving speed Any disturbance in the paving speed will be reflected in the riding surface. Frequent speed changes, speeding up and slowing down, without changing the angle of attack can result in a mat deficiency. Increased speeds will decrease mat thickness, and decreased speeds will increase mat thickness.

Paving speed should be sufficient to use the asphalt being produced, not allowing a buildup of trucks at the paver nor allowing the paver to sit idle waiting for trucks. The best paving speed for any paving operation is to match material usage with asphalt plant production, striving to run the paver as close to 100 percent of the time possible.

Force F 2, angle of attack This is the angle between the screed bottom and the grade. The free-floating screed raises or lowers itself until its bottom is parallel to the grade over which it travels. Controlling this angle controls the mat depth or thickness. For example, the thickness of the mat is 1 in. and the bottom of the screed is set 1 in. above and parallel to the grade. Once the angle of attack is set, *and other forces remain unchanged,* the screed will maintain this average depth. Scratching over high spots and filling in low spots produces a smooth riding surface. Increasing this angle will thicken the mat, while decreasing the angle will thin the mat.

Manually operating the depth cranks will cause the screed assembly to pivot, thus changing the angle of the screed bottom in relation to the side arms and grade. On pavers equipped with automatic joint matcher or grade and slope control, mat depth is controlled during the paving operation by raising or lowering the screed pull point. Depth cranks should not be adjusted after initial settings when using automatic controls. The mat thickness does not change immediately but takes approximately the length of the paver to be realized in the finished surface.

It is very important to maintain the correct angle of attack across the entire screed. A twisted screed plate cannot produce a quality mat and obviously will wear rapidly. A screed riding on its nose will produce little compaction and will wear the screed plate very quickly in the front area. A screed continually riding on its tail will wear the screed plate thin in the rear area.

Force F 3, head of material The head is the amount of material that lies directly in front of the screed. This material is distributed along the full width of the screed by the augers. A *constant* head of material is the most important and the most difficult of the three forces to control. Approximately 90 *percent of all paving deficiencies* are a direct result of not maintaining a constant head of material.

A constant head of material is not only an even depth of material along the entire length of the screed, but is consistent in temperature, aggregate blend, and mixing procedures.

Figure 7-15 shows how the head of material reacts or flows during the paving operation. The line of shear occurs at the point where the material either goes under the screed to become part of the finished mat or goes up and back into the augers for later placement. Disturbing this smooth transition of material will definitely result in laying an inferior mat.

Overfeeding the augers, as illustrated in Fig. 7-16, will result in a variety of mat deficiencies such as ripples, auger shadows, short and long waves, etc. Not only will paving quality suffer but excessive wear will occur as the augers bore through this large head of compacting material.

Underfeeding the augers, as illustrated in Fig. 7-17, or fluctuating the head of material in front of the screed (Fig. 7-18) will also result in an extremely poor riding surface.

Properly setting the automatic auger-conveyor shut-off in conjunction with properly

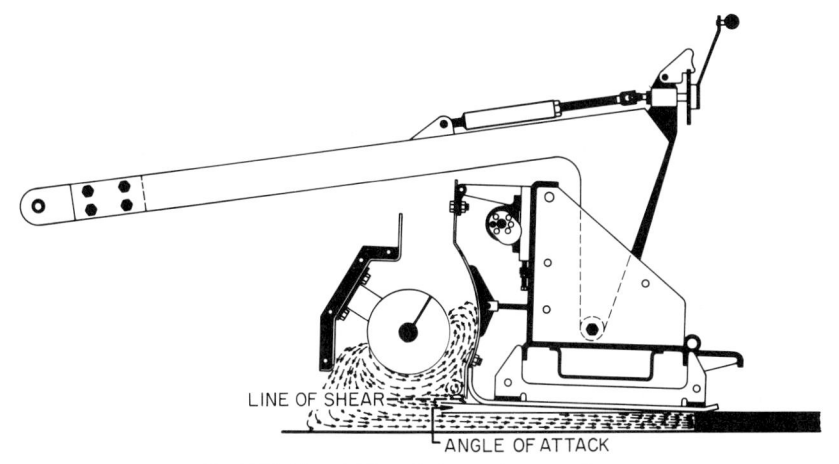

Fig. 7-15 Material flow during paving operation.

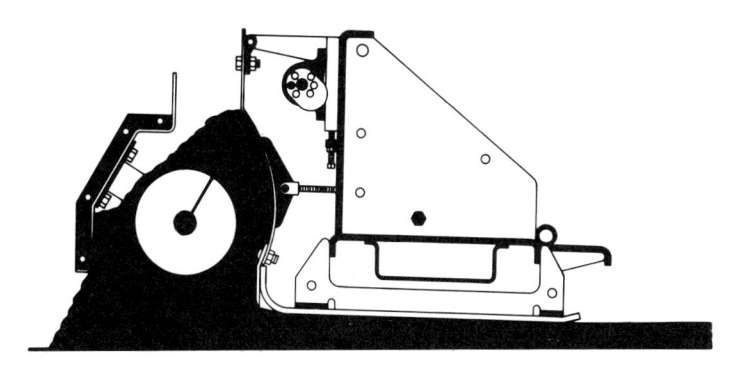

Fig. 7-16 Results of overfeeding the augers.

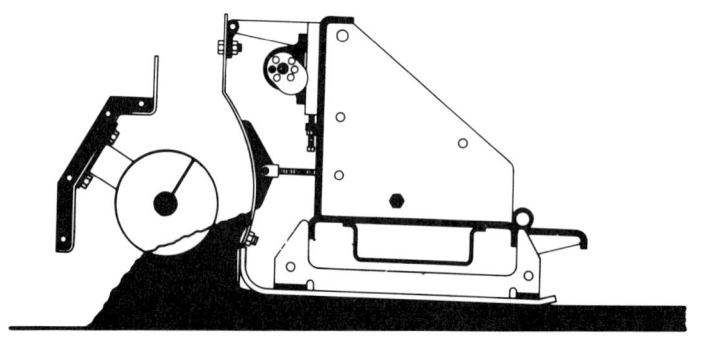

7-174　Maintenance Programs for Equipment Entities

Fig. 7-17　Results of underfeeding the augers.

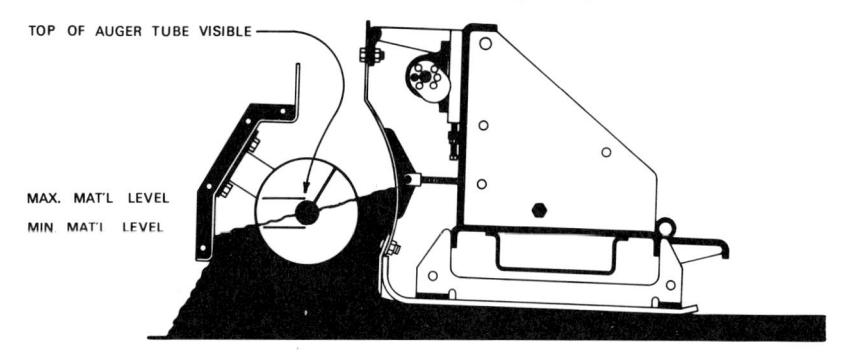

Fig. 7-18　A fluctuating head of material in front of the screed causes a poor riding surface.

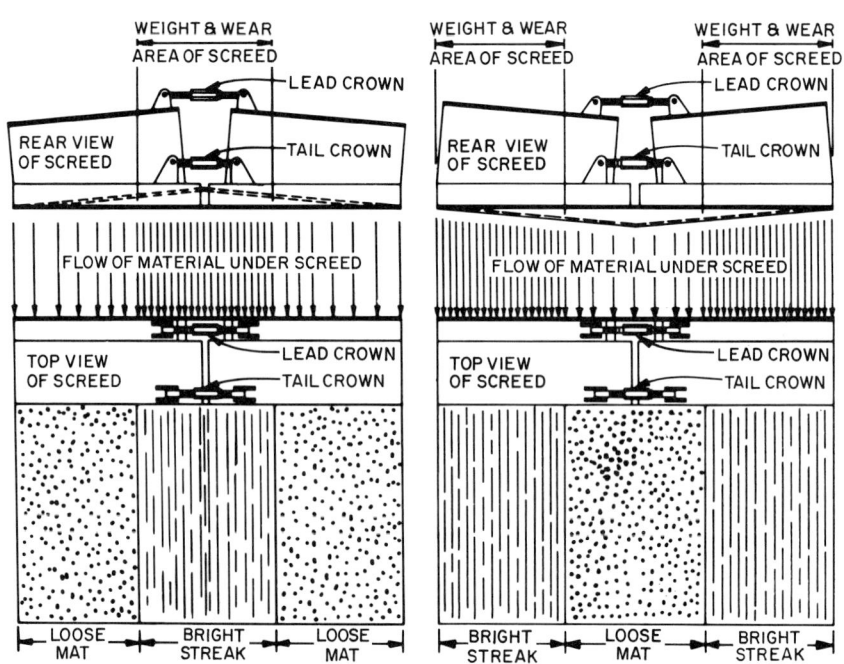

Fig. 7-19　Improper lead crown adjustment produces wear at the center of the screed.

Fig. 7-20　Improper lead crown adjustment can also produce wear at the screed's outer edges.

setting the height of the flow gates and selecting the correct auger-conveyor speed for your particular paving operation will ensure quality paving, production, and maximum life from your paver's auger and screed assemblies. Flow gates should normally be set about half open for basic screed width paving. As screed width is increased, the flow gates should be closed down even more to prevent the piling of excess material in the center of the screed while the screed extensions remain starved for material. Some experience will be required to obtain the best flow gate settings. Failure to make the adjustments will result in either overfeeding or underfeeding the augers.

Crown adjustment Many paving jobs require a specified crown to be formed into the riding surface between lanes and/or between main surface and shoulders. Crowning lugs are provided in the center of the basic screed and on special screed extensions for this purpose. The lower rear lug may be adjusted to produce the desired crown in the finished surface. This adjustment is called *tail crown*. The upper forward lug must be properly adjusted in relation to the rear lug to produce consistent material flow under the screed, tight mat texture over the entire width of the screed, and even weight distribution over the entire screed. This adjustment is called *lead crown*. Improper lead crown adjustment will result in various mat faults and excessive wear on the screed plate in localized areas.

Figures 7-19 and 7-20 show the relationship in the screed between lead crown and tail crown and also locate the turnbuckle assemblies which control both these adjustments. As indicated in Fig. 7-21, a good starting point when beginning to pave is to induce $\frac{1}{8}$ to $\frac{3}{16}$ in. more lead crown than tail crown into the screed. As the paving operation progresses, further adjustment of lead crown can be made to produce a tight, evenly textured mat along the entire width of the screed, as illustrated in Fig. 7-21.

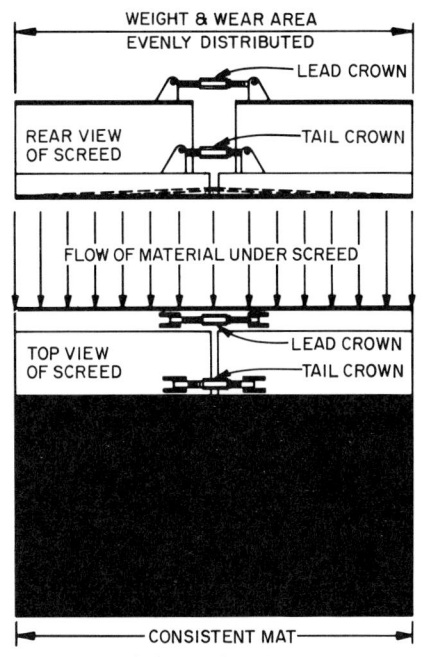

Fig. 7-21 With the weight and wear areas of the screed evenly distributed, wear is minimized and a consistent mat is obtained.

Figures 7-19 and 7-20 illustrate the effects produced from improperly adjusting lead crown. Both conditions illustrated will not only produce a mat of poor texture and rough riding qualities but will also cause excessive wear on the screed plate due to the uneven weight distribution of the screed.

When making lead crown adjustments rotate the turnbuckle only a quarter turn at a time and observe the texture of the mat as the paver moves forward 10 to 20 ft. Continue this procedure until optimum mat texture is achieved.

CAUTION: A considerable change in lead crown can cause a change in tail or road crown. Therefore, it is good practice to recheck road crown after obtaining optimum mat quality.

SUMMARY

Proper machine adjustments and correct paving procedures are as important a part of good maintenance on paving equipment as servicing the machine according to the manufacturer's specifications. Following the preceding recommendations will result in extended paving equipment life, reduced operating costs, and improved machine availability.

Chapter **8**
Hydraulic Crane Maintenance

CHARLES ISENBERGER
Grove Manufacturing Company
Shady Grove, Pa.

The term *hydraulic crane* as used in this chapter implies the type of mobile crane equipped with a hydraulically extendable and retractable telescoping boom.

In some instances, the term *hydraulic* is applied to cable-operated or fixed-boom cranes equipped with hydraulically actuated controls. The material that follows deals solely with mobile cranes that have telescoping booms powered via a closed-circuit hydraulic system.

Telescoping-boom cranes arrived in earnest and in ever-increasing numbers immediately following World War II. They are popular in the construction equipment industry because a telescoping boom is self-contained, can be quickly extended and retracted, and needs no auxiliary transport or assembly/disassembly time or labor. Contractors discovered that this capability of rapid job-to-job mobility and job site readiness represented substantial savings in time, manpower, and equipment rental costs.

Consequently, telescoping-boom cranes have been taking over applications normally covered by fixed-boom cranes, first in the lower-capacity, 100-ft reach class, and increasingly with the advent of more powerful, longer-reach hydraulic cranes, in the class above 100 tons and with reaches approaching the 300-ft range.

It is expected that with advanced technology this trend will continue and larger hydraulic cranes will be introduced.

By the nature of the machine, it follows that a sound maintenance program for a hydraulic crane is markedly different from that traditionally followed for a mechanical or cable-operated crane.

An analogy that will be helpful in developing a preventive maintenance program for any hydraulic crane is to think of the crane's hydraulic pump as its heart, its hydraulic oil as its blood, and its valving, reservoir, and hydraulic lines as the crane's circulatory system. Each of the foregoing components is *that* important to the healthy, normal functioning and life expectancy of the unit.

Thus it becomes vitally necessary to deal with hydraulic system maintenance as the most important consideration in caring for the telescoping-boom crane as a whole.

All hydraulic crane functions are either hydraulically or electrically actuated and fully hydraulic in operation.

MAIN HYDRAULIC CRANE FUNCTIONS

The main functions are, of course, elevating and lowering the boom, extending or retracting the boom, rotating the crane turntable, raising or lowering the hook on the hoist line, extending or retracting the hydraulic outriggers, and, on cranes so equipped, extending or retracting the counterweight.

Boom elevation, telescoping, and outrigger functions are all activated by hydraulic double-acting cylinders. The hoist and turntable swing systems are driven by hydraulic motors.

MAJOR HYDRAULIC CRANE COMPONENTS

Superstructure In most hydraulic cranes, the superstructure consists of the turntable assembly with its 360° swing mechanism, the electrohydraulic swivel (or equivalent system) cab assembly, the boom assembly, the hoist(s) and mounting assembly, the counterweight, and associated crane control componentry and hydraulic system piping. Most hydraulic crane superstructures are also equipped with gas or diesel engines. Other units are powered by an engine in the carrier as in single-engine truck cranes or in self-propelled rough-terrain cranes.

Boom swing system A large-diameter roller (or ball) bearing joining the superstructure and carrier provides a continuous 360° swing in either direction. The swing motor is hydraulically driven, providing low-speed, high-torque power for swinging the turntable through a gearbox power transfer system. The latter is generally a sun-and-planet-type gear arrangement, operating the turntable swing at just a few revolutions per minute.

Boom elevating system The crane boom is typically elevated by dual (sometimes single) double-acting hydraulic cylinders. On the most popular makes of cranes the boom can be lowered below horizontal to facilitate cable reeving, jib erection, and boom maintenance.

Boom telescope system On most hydraulic cranes, double-acting hydraulic cylinders are installed within the telescoping sections to achieve boom extension or retraction (some are externally mounted). The fly or outer section is sometimes powered, sometimes power-pinned, and sometimes manually extended and pinned.

Hydraulic hoist system The main (and auxiliary, if so equipped) hydraulic hoist provides power for load-raising or lowering operations. A hydraulic motor (or motors as in some dual-speed hoists) drives the hoist drum through a planetary gear-reduction system.

Hydraulic outrigger system For stability during craning work, hydraulically powered double-acting cylinders extend and retract the crane's horizontal outrigger beams and raise or lower the vertical stabilizing jacks.

HYDRAULIC CIRCULATORY SYSTEM

As indicated earlier, hydraulic oil is the lifeblood of any telescoping-boom crane's circulatory system since it is the medium for transmitting power—for lubricating the system—and serves as the system's coolant.

Therefore, selection of the proper oil is a vital requirement for satisfactory performance and service life of the crane circulatory system. It should be selected with care and with the assistance of a reputable supplier. Attention should be given to two important oil factors:

1. *Antiwear additives.* The hydraulic oil selected must contain the necessary additives to ensure high antiwear characteristics and excellent chemical stability.

2. *Viscosity.* The hydraulic oil selected must have proper viscosity to maintain a lubricating film at the operating temperature of the system.

For optimum performance, suitable types of oil are:

1. *Crankcase oil.* If it meets the prevailing API (American Petroleum Institute) service classification of MIL-L-2104C, crankcase oil is suitable.

2. *Antiwear-type hydraulic oil.* There is no common designation for oils of this type; however, they are produced by all major oil suppliers and provide the required antiwear qualities of MIL-L-2104C crankcase oils.

Hydraulic oil specifications The following table summarizes by viscosity oil types recommended for use in mobile hydraulic crane systems. All must meet the current API service classification of MIL-L-2104C:

Operating temperature range of hydraulic system, min–max°	SAE* viscosity designation
0–180°F (−18–83°C)	10W
0–210°F (−18–99°C)	10W–30
50–210°F (10–99°C)	20–20W

*Society of Automotive Engineers.

Temperatures shown in the above table are cold start-up minimum-to-maximum operating limits. Suitable start-up procedures must be followed to ensure adequate lubrication during system warm-up.

Arctic conditions When operating in subzero temperatures, common practice is to use special heating equipment before starting. However, owing to variables not only in temperatures but in available equipment, the best advice is to consult the equipment manufacturer's factory for specific recommendations.

During cold weather starting, high-speed operation of system components must be avoided until the entire hydraulic system is warmed up to provide adequate lubrication. Start-up of each operation (boom elevation, extension, swing, hoist, etc.) must be performed so all trapped oil returns to the system's temperature.

Draining and flushing Whenever it becomes necessary to change hydraulic oil, it is important that whoever is performing the task knows the reason for the change. Just draining the hydraulic reservoir is not enough, even if it is only for viscosity reasons. Remember that oil trapped in the system, especially when hydraulic cylinders are involved, accounts for a large percentage of the total volume in the system.

If a component has been changed because of a failure that might allow metal or abrasive particles to enter the system, all systems must be thoroughly checked, drained, and flushed individually by removing lines and/or components that may have become contaminated.

Special cleaning oils containing compounds that remove gum and sludge and also pick up loose rust are available in a wide range of viscosities. These oils can be put to work while accomplishing a thorough cleaning. They are usually recommended for a period of up to 50 hr operation before changing to regular oils. When the cleaning oil is in the system, check and clean filters and strainers frequently. When change is made to regular oil, flush the system to prevent intermixing with cleaning oil.

CAUTION: When flushing cylinders, disconnect return lines only. When changing hydraulic oils, it will always be necessary to recheck the reservoir oil level after a brief operation and to add oil to compensate for that which was pumped into the lines and components.

Flushing Procedures

1. Remove oil reservoir drain plug. Allow approximately 3 min after oil stops flowing from drain port for side walls to drain.

2. Install drain plug. Fill reservoir with a 50-50 mixture of fuel oil and clean hydraulic oil.

3. Cycle crane through *all* hydraulic functions several times and return to stowed-boom position.

4. Remove drain plug, completely drain, flush, and reinstall drain plug, and refill reservoir with clean hydraulic oil. (Refer to the previous table under Hydraulic Oil Specifications.)

HYDRAULIC PUMPS

The required flow of hydraulic oil (expressed in gallons per minute, or gpm) to supply all crane functions is provided by the crane's hydraulic pumps. Generally the pumps are driven by a power take-off (PTO) from the engine.

Each hydraulic pump incorporates the following major components: shaft end cover, matched gears, bearings, and gear housing, bearing carrier, connecting shaft, and port end cover. The number of gear housings and bearing carriers is dependent on the number of pump sections incorporated in the pump assembly.

Inspection Visually inspect pumps for damage, corrosion, loose or missing parts, and evidence of leakage between gear housings, bearing carriers, and/or port end cover. Check that pump is properly mounted on pump drive and that all lines are attached securely.

Removal

1. Disconnect suction lines from pump; cap lines and openings.

2. Tag and disconnect pump pressure lines; cap lines and openings. CAUTION: Keep pump as level as possible when removing from pump drive to avoid damaging the spline.

3. Remove bolts and washers securing pump to drive housing; remove pump and

cover opening. *Do not grip on or near any machined surfaces during removal, disassembly, or reassembly.*

To avoid possible damage to a new or rebuilt pump, reduce main relief valve pressure (psi) before operating pump. After pump has run in for about 5 min at zero pressure (all controls in neutral), adjust relief valve pressure setting. Failure to observe this precaution can result in almost immediate pump failure should relief pressure setting be excessive.

TABLE 8-1 Hydraulic Pump Troubleshooting

Trouble	Probable cause	Remedy
Pump not delivering fluid.	Coupling or shaft sheared or disengaged.	Check pump drive is properly engaged. If drive shaft or coupling is damaged or sheared, remove and repair or replace as necessary.
	Reservoir-to-pump supply line broken or restricted. Air entering at suction manifold. Pump not priming.	Clean, repair, or replace line as necessary. Check all lines for security, manifold for cracks and proper attachment. Tighten, repair, or replace components as necessary.
	Internal contamination.	Repair or replace pump. Drain, flush, and refill system with recommended oil.
Excessive pressure buildup.	System relief valve set too high.	Use adequate pressure gauge and adjust system relief valve, as necessary.
	Restricted pump-to-control valve supply line.	Clean, repair, or replace line as necessary.
Pump noise (accompanied by oil foaming in reservoir).	Air entering at suction manifold.	Check all lines for security, manifold for cracks and proper attachment. Tighten, repair, or replace components as necessary. Assure that oil level in reservoir is adequate. (Fill to HIGH mark on dipstick.)

Pump installation

1. Engage pump spline in pump; secure with washers and bolts.
2. Connect suction hoses to pump.
3. Connect pressure hoses to pump, ensuring that hoses are connected to correct port (as marked prior to pump removal).

Functional check of hydraulic pump

1. Start crane engine and allow for warm-up.
2. Shut down engine and engage pump drive.
3. Restart engine and adjust throttle to one-half governed rpm. Allow engine to run with no load applied.
4. Using control lever, build up pressure intermittently for about 3 min.
5. Increase engine speed to recommended operating rpm and again build up pressure intermittently for about 3 min.
6. Reduce engine speed to idle rpm and allow to run for approximately 5 min.
7. Shut down engine and check pump for leakage.
8. Check pump oil delivery in accordance with following:
    a. Disconnect pressure (circuit supply) line from applicable pump section.
    b. Connect flow meter between pump outlet and circuit supply line.
    c. Start crane engine and adjust speed to recommended rpm.
    d. Check reading on flowmeter against gpm level given in pump specifications.

e. Reduce engine speed to idle rpm; shut down engine. Repeat above steps for remaining pump sections.

HYDRAULIC CRANE CONTROL VALVES

Hydraulic valves, arranged in banks, are connected to the control levers in the crane operator's cab by mechanical linkage and control the amount of oil volume acting on the various crane functions.

Inspection Inspect control valves for visible damage, binding spools, and evidence of leakage. If excessive internal leakage is suspected during operation with spool in center position, it is possible that the area between the spool and working section bore of the valve body is worn beyond serviceable limits. If this condition exists, the spool and body must be replaced as an assembly.

Main relief valves are incorporated in the control valve assemblies. The compression of the pilot spring within the relief valve body controls the relief opening in the pilot poppet valve. When the pilot poppet valve opens, hydraulic oil enters under pressure and opens the large poppet in the valve, thereby protecting the system components from hydraulic pressures over and above their design ratings.

Adjustment of the main relief valves should be made only by qualified personnel using the proper equipment.

Valve leakage checks If a hydraulic valve leaks enough to drip oil externally, it is leaking badly enough to take the crane out of service for immediate repairs.

External leaks sometimes develop at fittings and seals. Spool seals are susceptible since they are subject to wear. Seals may be damaged by too-high temperatures or by dirt or paint accumulation in the spool. Damaged seals *must* be replaced.

Warped mounting surfaces can distort the assembly and cause leakage. To check for valve distortion, loosen the mounting bolts lightly. If the leakage is reduced when the bolts have been backed off slightly, you know that distortion was the problem. To correct this condition, shim the valve assembly level and retighten the mounting bolts.

A component functioning at reduced efficiency may indicate the relief valve for that component is malfunctioning. Assuming preliminary checkout reveals that adequate volume is being supplied to the affected valve bank, relief valves are properly adjusted, and the component is not at fault, the next step would be to check the valve for scored or worn parts.

Scoring from contamination Contamination is the number one problem in hydraulic crane systems, and evidence of it is in scoring of valve components. External contamination from dust or internal contamination from deteriorating components or oxidized oil can score or severely wear on valve components. If scoring occurs, the components *must* be replaced.

Valves should also be checked for rust. Rust or dirt collecting on the linkages can prevent free movement of the spool and keep it out of true center position.

Excessive system pressure can create both internal and external leaks in valves that are otherwise in sound condition. Therefore, it is extremely important that relief valves be adjusted only by qualified personnel using the proper equipment.

Causes of sticking valve spools Some of the most common causes for stiff spool movement or jammed spool action are system overheating, excessive pressure, contaminated or deteriorated oil, or warped mountings. When deteriorated or contaminated oil is the cause, flushing the system and replenishing with clean oil may solve the problem. If the spool bores are badly scored or galled, the valve must be removed for servicing. If the oil is scorched or deteriorated, similar treatment is required.

Warpage also occurs when mounting plates are not level or become distorted from machine damage. As mentioned previously, the valve can be shimmed level.

HYDRAULIC OIL BREAKDOWN

Oil breakdown will occur if the oil becomes contaminated with air, water, or dirt or if the oil is exposed to excessively high temperatures or pressures. Even with normal usage, oils deteriorate and the additives that inhibit rust, oxidation, and foaming lose their power.

Follow the recommended oil change intervals and procedures. They are intended to get the oil out of the system before it starts breaking down and before deterioration harms the system.

Good oil is always a good investment. The money saved by switching from the recommended grade to a cheaper grade will probably be spent repairing or replacing prematurely worn components, or cleaning sludge out of the system.

Cause of faulty check valves Most control valve assemblies have integral relief check valves (Fig. 8-1).

CAUTION: Do not adjust any relief valve unless a preliminary pressure check reveals

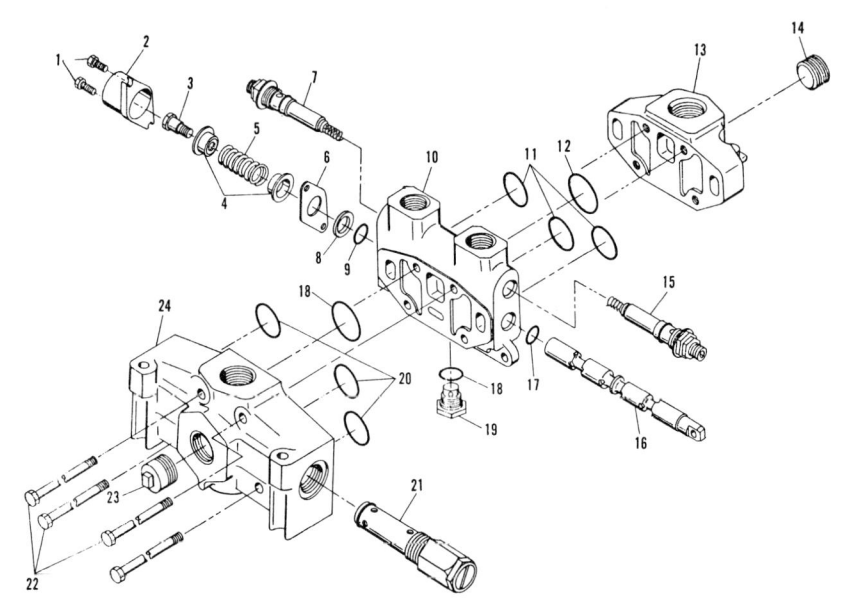

Fig. 8-1 Typical control valve assembly: (1) cap screws; (2) cap; (3) bolt; (4) spring guides; (5) spring; (6) retaining plate; (7) circuit relief valve; (8) backup ring; (9) ring seal; (10) housing; (11) "O" ring; (12) "O" ring; (13) outlet relief; (14) plug; (15) circuit relief valve; (16) spool; (17) ring seal; (18) "O" ring; (19) load check valve; (20) "O" ring; (21) main relief valve; (22) tie bolts; (23) plug; (24) inlet section.

that the pressure setting does not agree with that specified for the valve. Only qualified service personnel should attempt adjustment of hydraulic check valves.

HYDRAULIC SYSTEM FILTER

The hydraulic crane's filter for its hydraulic system is typically installed in the reservoir return line. The element is of the replaceable cartridge type, and it is recommended that the original element be replaced upon completion of the first 50 hr of crane operation. Subsequent element replacement should be governed by the atmospheric conditions under which the hydraulic crane is working. In an average climate the element should be replaced after 200 hr of service. Should the machine be subjected to an unusually contaminated atmosphere, it is recommended that the element be replaced more frequently.

To avoid unnecessary loss of hydraulic oil from the reservoir when replacing the element, extend and set the outriggers and position the boom at maximum elevation and extension.

1. Remove drain plug from bottom of filter, allow residue oil to drain, and replace plug.

2. Remove bolts and washers securing housing assembly to base section; remove housing.

If filter element is found to be excessively dirty, drain, flush, and refill system with clean hydraulic oil.

TABLE 8-2 Crane Control Valve Troubleshooting

Trouble	Probable cause	Remedy
Sticking spool.	Excessively high oil temperature.	Eliminate any restriction in pipe line or filtering system
	Dirt in oil.	Change oil and flush system.
	Pipe fittings too tight.	Check torque. Retorque as necessary.
	Valve warped from mounting.	Loosen valve and check.
	Excessively high pressure in valve. (Relief valves not working properly.)	Check pressure at inlet and at working ports.
	Handle or linkage binding.	Free linkage.
	Spacer bent.	Replace valve.
	Return spring damaged.	Replace faulty parts.
	Spring or valve cap binding.	Loosen cap, recenter, and retighten.
	Valve not thoroughly warmed up.	Allow time for system warm-up.
Leaking seals.	Paint on or under seal.	Remove and clean, as necessary.
	Excessive back pressure.	Open or enlarge line to reservoir.
	Dirt under seal.	Remove and clean, as necessary.
	Scored spool.	Replace valve.
	Loose seal plates.	Clean and tighten plates.
	Cut or scored seal.	Replace faulty parts.
Unable to move spool in or out.	Dirt in valve.	Clean and flush out valve assembly.
	Spool cap full of oil.	Replace seals.
	Bind in linkage.	Free linkage.
Load drops when spool moved from neutral.	Dirt in check valve.	Disassemble and clean check valve.
	Scored check valve poppet or seat.	Replace poppet or lap poppet to seat.
Poor hydraulic system performance or failure.	Damaged pump.	Check pressure or replace pump.
	Dirt in relief valve.	Disassemble and clean relief valve.
	Relief valve damaged.	Replace relief valve.

(continued)

TABLE 8-2 Crane Control Valve Troubleshooting (*Continued*)

Trouble	Probable cause	Remedy
	Worn cylinder(s) or motor(s).	Repair or replace damaged components.
	Load too heavy.	Reduce load. (Refer to load chart for rated capacities.)
	Internal valve crack.	Replace valve.
	Spool not at full stroke.	Check movement and linkage.
	Oil low in reservoir.	Add oil. Fill to FULL mark on dipstick.
	System filter clogged.	Clean or replace filter element.
	Line restricted.	Check lines. Clean or repair as necessary.

3. Remove and discard housing seal and filter element.
4. Clean housing with soft bristle brush and 50-50 mixture of clean hydraulic oil and fuel oil. Assure housing seal is properly positioned in housing groove.
5. Install new housing seal and filter element.

HYDRAULIC CRANE SWING SYSTEM

The swing system of the typical hydraulic crane consists of the swing motor, hydraulic swing and mechanical brake, swing gearbox, and swing bearing (Fig. 8-2). The system is powered by the swing motor mounted on top of the swing box assembly and permits a continuous 360° rotation of the crane superstructure via the crane turntable. A control lever (or foot pedal) actuates swing to left or right, and a pedal controls operation of the swing brake.

Swing box The turntable swingbox is serviced with SAE 90 gear oil. The gear case should be drained, flushed, and refilled every 2500 hr of operation, or every 12 months, whichever occurs first.

Servicing procedure

1. Remove drain plug and allow approximately 3 min for sidewalls to drain after oil stops flowing from drain port.
2. Flush case with light flushing oil.
3. Install drain plug and refill case with oil specified in lubrication chart for crane or with SAE 90 gear oil.

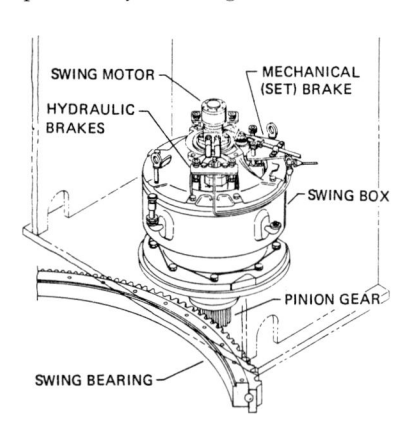

Fig. 8-2 Crane swing system components.

HYDRAULIC CRANE SWING BEARING

The swing bearing is the most critical maintenance point of a hydraulic crane. It is here, at the exact centerline of rotation, that the stresses of loads are concentrated. In addition, the bearing provides the only attachment between the crane carrier (or frame of a self-propelled or rough-terrain-type crane) and the crane superstructure. Therefore, proper care of the bearing and periodic maintenance of the turntable-to-bearing attach bolts is a *must* to ensure safe and efficient crane operation.

TABLE 8-3 Crane Swing System Troubleshooting

Trouble	Probable cause	Remedy
Boom swing operation erratic in either direction.	Damaged relief valve.	Replace relief valve.
	Swing brake dragging (not releasing properly).	Readjust and/or replace necessary parts.
	Low engine rpm.	Increase engine rpm to obtain smooth swing operation.
	Low hydraulic oil.	Increase hydraulic oil to proper level.
	Improper movement of control lever to neutral. Insufficient lubricant on swing bearing.	Feather controls to neutral to maintain smooth stopping action. Lubricate bearing properly. (Refer to Fig. 8-12.)
	Machine not level.	Level machine using outriggers.
	Damaged swing motor.	Replace swing motor.
	Excessive overload.	Reduce load. (Refer to load capacity chart.)
	Restricted or partly clogged hydraulic hose or fittings. Pump cavitation in swing section.	Replace hose or fittings. (Refer to manufacturer's specifications.) Retighten suction hose or replace any damaged fitting.
	Improperly torqued turntable bolts.	Retorque turntable bolts evenly.
	Excessive preload on upper and lower pinion shaft bearing.	Adjust as necessary.
	Improperly torqued swing motor attachment bolts.	Retorque swing motor attachment bolts.
	Malfunction of the swing box.	Remove swing box and make necessary repairs.
	Worn or damaged pump section.	Repair or replace damaged section.
Boom swing operation erratic in one direction only.	Relief valve inoperative.	Clean and readjust and/or replace relief valve.
	Machine not level.	Level machine.
	Turntable bearing binding due to continuous limited swing. (Example: concrete pourer.)	Rotate machine in both directions several times 360° and lubricate turntable bearing.
	Restricted hose or fitting.	Replace hose or fitting.
	Remote mounted crossover relief valve malfunctioning.	Replace crossover relief valve.
	Damaged swing pinion.	Replace pinion.
Boom will not swing in either direction.	Damaged relief valve.	Remove, clean, and repair or replace relief valve.

(continued)

TABLE 8-3 Crane Swing System Troubleshooting (*Continued*)

Trouble	Probable cause	Remedy
	Damaged swing motor.	Repair or replace swing motor.
	Broken swing motor drive coupling.	Replace drive coupling.
	Swing brake not releasing properly.	Repair as necessary.
	Completely inoperative crossover relief valve.	Replace crossover relief valve.
	Internal damage to swing box.	Remove swing box and repair.
	Worn or damaged hydraulic pump section.	Replace pump section.
Swing operation slow in either direction.	Damaged relief valve.	Adjust, repair, or replace valve.
	Improperly adjusted swing brake.	Readjust.
	Improperly adjusted crossover relief valve.	Adjust valve.
	Improperly lubricated swing bearing.	Lubricate bearing per recommendations.
	Improper size hose and/or fittings installed.	Refer to manufacturer's specifications.
	Clogged or restricted hydraulic hoses or fittings.	Clean or replace damaged parts.
	Worn or damaged output shaft bearing.	Replace bearings.
	Damaged outrigger selector switch.	Repair or replace switch.
	Worn or damaged swing motor.	Repair or replace motor.
	Worn or damaged hydraulic pump section.	Repair or replace pump section.
Swing motor continues to operate when swing control is in neutral position.	Control valve spool sticking or valve otherwise damaged.	Repair or replace valve.
Swing motor turning in wrong direction.	Improper port connections.	Reverse port connections.
Swing motor noisy.	Air in system.	Bleed air from highest point in circuit and replenish fluid.
	Motor binding.	Repair or replace motor.

The inner race of the swing bearing may be secured to the mounting plate on the carrier by a specified number of grade 8 bolts. Generally, an equal number of those bolts secure the bearing outer race to the superstructure.

It is recommended that the attach bolts be inspected and retorqued as required after the first 300 hr of crane operation. The bolts may loosen in service due to vibration, shock

Hydraulic Crane Maintenance 7-187

loads, and temperature changes. Thereafter, periodic inspection of the bolts should be made every 500 hr to be sure they are secure.

WARNING: Failure to maintain proper tightness of all turntable attach bolts could result in damage to the crane and possible injury to personnel.

Repeated retorquing may cause mounting bolts to stretch. If bolts keep working loose, they should be replaced with new bolts of the applicable grade and size.

HYDRAULIC SWIVEL COUPLING

Removal

1. Remove electrical swivel coupling per instructions in section on Electrical System.

2. Tag and disconnect hydraulic lines to swivel coupling barrel; cap lines and openings.

3. Remove bolts and washers securing swivel to turntable.

4. Withdraw swivel coupling sufficiently to remove hydraulic lines from bottom of swivel coupling. Tag and remove lines; cap lines and openings.

5. Withdraw swivel coupling from the crane.

NOTE: Any maintenance requiring disassembly of the hydraulic swivel coupling should also include replacement of all seals and rings. Aligning discarded seals and rings

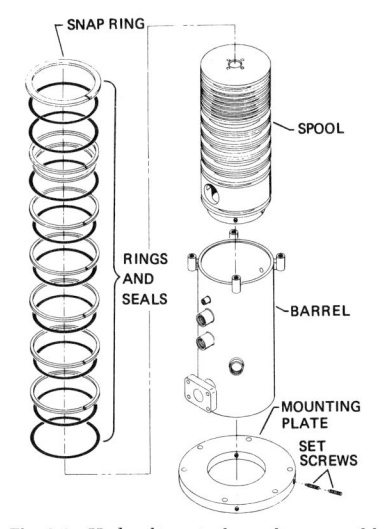

Fig. 8-3 Hydraulic swivel coupling assembly.

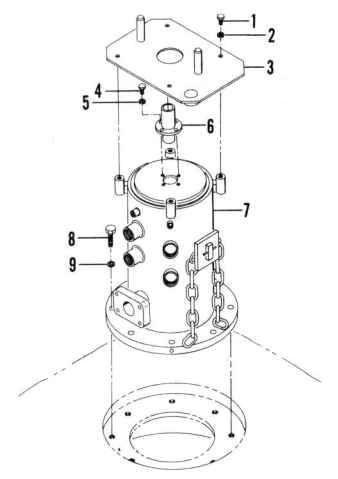

Fig. 8-4 Removing and installing hydraulic swivel coupling: (1) bolt; (2) washer; (3) mounting plate; (4) bolt; (5) washer; (6) coupling; (7) hydraulic swivel coupling; (8) bolt; (9) washer.

in the order of disassembly will facilitate the installation of new seals and rings. When installing, avoid stretching the seals or scratching grooved and gland surfaces.

Inspection Clean spool and housing with suitable solvent and dry with compressed air. Check spool and inside of barrel for scratches, grooves, scoring, etc. If any grooves have developed with a depth exceeding 0.005 in., the unit should be replaced.

Reassembly of hydraulic swivel coupling

1. Lubricate spool.
2. Install new seals and rings on spool.
3. Insert but do not force spool into barrel; secure with snap ring.
4. Install top plate (mounting plate).

Installation
 1. Connect hydraulic lines to bottom of swivel coupling as marked prior to removal.
 2. Secure mounting plate to turntable with attaching bolts and washers.
 3. Connect hydraulic lines to swivel coupling barrel.
 4. Install electrical swivel coupling in accordance with instructions (see Electrical System section of this chapter).

Functional check Activate hydraulic system. Rotate superstructure and cycle the outriggers. Observe outriggers for proper operation. Check hydraulic swivel and connections for security and evidence of leaks.

HYDRAULIC TELESCOPING-BOOM ASSEMBLY

The multisection telescoping-boom assembly is mounted on the turntable and consists of the individual nested sections of the boom and the internal hydraulic cylinders which extend and retract the telescoping sections.

Graphite-impregnated wear pads reduce friction during extension and retraction, and adjustable wear pads provide for proper boom alignment.

Hydraulic boom inspection Visually inspect each telescoping section for adequate lubrication. Observe extended sections for evidence of cracks, warping, or other damage. Periodically check security of boom wear pads. Check boom nose sheaves for security and freedom of movement.

Boom alignment and servicing Lubrication of the boom is limited to the sides and bottom of the bottom plates of the telescoping sections. Extreme pressure (EP) grease is recommended.

Boom alignment is achieved by adjusting the wear pads located at various points in the boom assembly. Adjustment procedure is as follows:
 CAUTION: Be sure outriggers are properly extended.
 1. Lubricate boom completely.
 2. Adjust side wear pads snug against boom section, and back off one-quarter turn.
NOTE: When extending and retracting boom during alignment, movement should be stopped if a restriction is encountered, and wear pads readjusted as necessary to provide free travel of affected boom section(s).
 3. Retract and extend boom. Check for high point where boom has brushed wear pads at widest point.
 4. Retract boom sections to align high point on boom section with adjacent wear pads.
 5. Turn adjusting screws snug against boom section, then adjust out one-eighth turn.
 6. Attach a weight (do not exceed lifting capacity chart) to hook and extend boom full length. Check for side deflection. If the boom deflects to the left, the forward left adjustable wear pad would be adjusted in and the rear left pad would be adjusted out, away from the internal boom section in a similar manner. The forward right adjustable wear pad will be adjusted out, and the right rear pad adjusted in.

Boom removal procedure
 1. Extend and set crane outriggers.
 2. Remove hook block or headache ball and wind remaining wire rope onto hoist drum.
 3. Elevate boom slightly to expose lift cylinder rod end anchor shafts. Assure that boom is fully retracted. *Check with manufacturer to be sure of boom weight. Be sure that blocking or lifting device is capable of supporting boom assembly.*
 4. Block or suspend boom in position.
 5. *Attach lifting devices fore and aft of lift cylinder attach fittings* (Fig. 8-5A).
 6. Engage hose reel retaining pin.
 7. Tag and disconnect lines to telescope cylinders; cap lines and openings.
 8. Safety-block the boom lift cylinders. Failure to do this properly could result in injury to personnel.
 9. Remove cap screws and washers from anchor shaft retainer bolt. Remove anchor shafts (Fig. 8-5B and C).
 10. Activate hydraulic system and withdraw lift cylinder rod ends from attach fittings.
 11. Remove lift cylinder safety blocks and lower lift cylinders (Fig. 8-5D).

12. Disconnect boom angle indicator. CAUTION: Remove all hydraulic power before proceeding.

13. Remove pivot shaft (Fig. 8-5*E*).

14. Raise boom clear of crane (Fig. 8-5*F*) and lower boom assembly for repairs. Follow the reverse procedure for hydraulic boom installation.

NOTE: Failure to reinstall any shims removed during disassembly of trunnion mount style could result in binding and shearing of the boom pivot shaft.

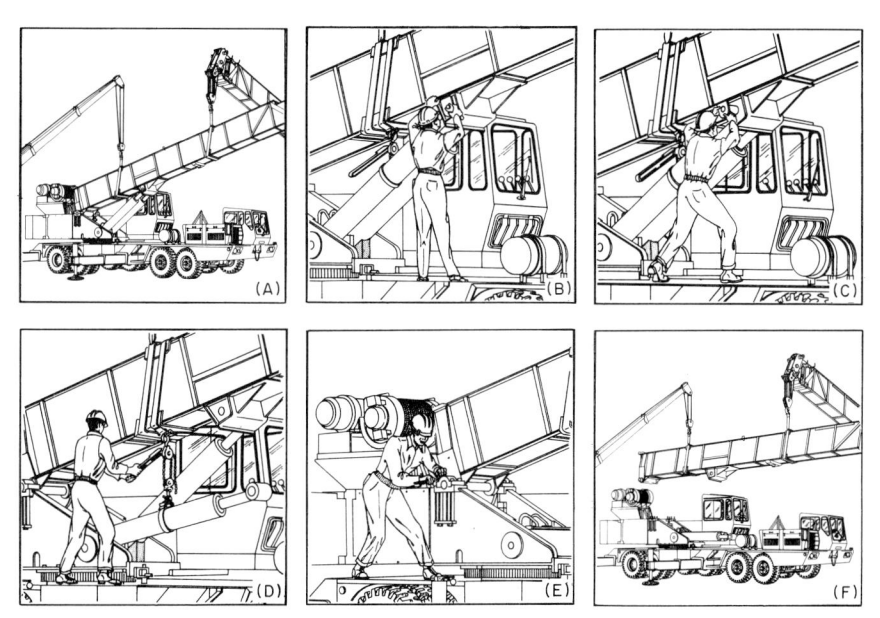

Fig. 8-5 Steps in removing and installing boom assembly (see text).

Checking drift of boom elevation cylinders If the hydraulic crane boom has a tendency to drift down when elevated, the following procedure should be followed to locate the malfunction:

A. Check for Holding Valve Leakage

1. Elevate boom approximately 8 in. from horizontal.
2. Disconnect lowering circuit line from holding valves. If valve leaks, reconnect hydraulic line, lower boom, and shut down operation.
3. Check seals for damage and inspect valve for scoring, wear, and evidence of foreign material; repair or replace as required.

B. When Boom Drifts at One Elevation Only. If boom drifts down consistently only at one specific boom angle, this indicates the possibility of a damaged area in one of the cylinder barrels.

1. Elevate boom to position where drift occurs.
2. Disconnect lowering circuit line from each cylinder.
3. Inspect for leakage to determine damaged cylinder(s); repair or replace as necessary.

C. When Boom Drifts at All Angles. If the boom drifts down at all angles, proceed as follows:

1. Elevate boom approximately 8 in. from horizontal.
2. Shut down all operations and disconnect the lowering circuit line from each lift cylinder.
3. Allow excess oil to drain from fittings. If oil continues to drain from a cylinder port, the cylinder is damaged. Repair or replace as necessary.

HYDRAULIC HOIST (WINCH)

The typical hydraulic hoist consists of a base, the drive mechanism, and the hoist drum. The primary drive and final drive housings are fastened to the hoist base. The drum rotates on antifriction bearings which extend from either side of the drum into the drive housings. The brake assembly and the hydraulic motors are mounted onto the primary drive housing.

Inspection Visually inspect the hydraulic connections and piping for evidence of leaks indicating parts that may be loose or damaged. Check general condition of hoist assembly and security of hoist mounting bolts.

Alignment Correct hoist mounting alignment with the crane boom is important to maintain level cable wrap, optimum cable life, and avoidance of unequal loading of the hoist bearings and gear case. Lineup is a relatively simple procedure requiring a minimum of equipment. If misalignment exceeds $\frac{1}{2}°$, proper adjustment should be made immediately. This can only be accomplished with the cable removed from the drum.

Alignment procedure
1. Lower boom to level position.
2. Find a line on top of the hoist drum which is parallel to the drum axis as follows:
    a. Use a Miracle Point Gage and find a 0° dial point next to each flange on top of the drum. NOTE: If this special equipment is not available, sufficient accuracy in locating a centerline may be obtained by using a steel square against the machined inner surfaces of both drum flanges. It is advisable to avoid using any cast surface in this procedure unless a check from both flanges indicates that the resultant line is straight.
    b. Draw a line between the hoist drum flanges, passing through both points as located above, and determine the midpoint.
3. Find the midpoint of lines drawn across the top of the boom section, perpendicular to its length, at both ends of the boom section.
4. Check as follows to see if the hoist is aligned perpendicular to the boom:
    a. String a chalk line from the point on the outer boom centerline across the base section centerline and the drum midpoint.
    b. Pull line taut, aligning it directly over the top of the hoist drum midpoint.
    c. With a protractor, measure the angle between the chalk line and the cross line drawn on hoist. If measurement of angle exceeds tolerance of $90° \pm \frac{1}{2}°$, realignment will be necessary.
5. If realignment is necessary, remove hoist mounting bolts and shift hoist as necessary to achieve minimum angular tolerance. Trial-and-error location may be necessary for proper lineup of bolt holes and stop blocks. If all bolts cannot be reinserted, or stop blocks interfere with lineup, slight elongation of hoist bolt holes and/or grinding of the mounting lugs might be necessary. *Extreme care should be taken to avoid overcorrection since very small adjustments result in large angular changes.*

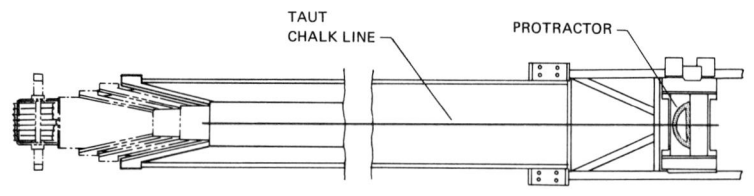

Fig. 8-6 Hoist-to-boom alignment.

Removal of hoist (steps)
1. Prior to initiating hoist removal, let out all cable and disconnect from hoist drum.
2. Attach suitable lifting device to hoist.
3. Tag and disconnect hydraulic lines from hoist and hoist motors. Lift hoist free of pad and remove.

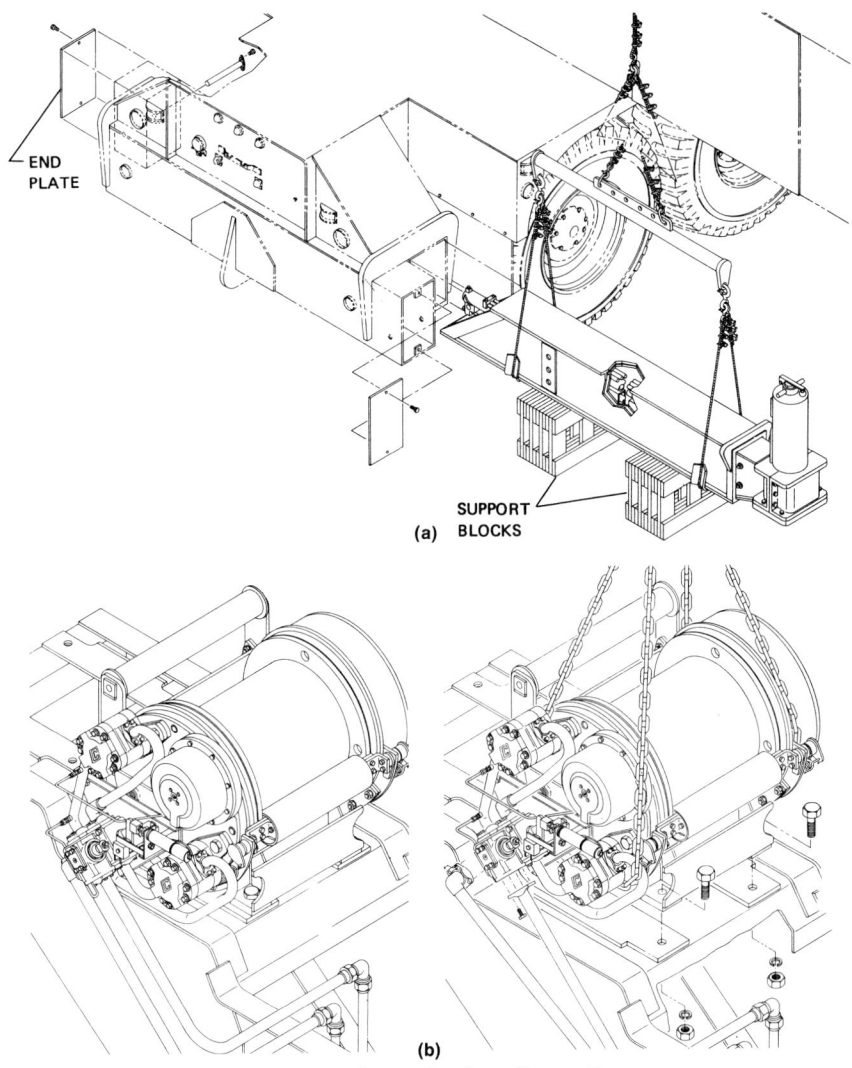

Fig. 8-7 Typical removal and installation of hoist.

NOTE: Any maintenance requiring disassembly of the hydraulic hoist should also be accompanied by replacement of all seals and rings.

Installation of hoist After necessary maintenance of hoist component has been performed, follow these steps for proper reinstallation:

1. With hoist supported in suitable lifting sling, position hoist on mounting plate.
2. Secure hoist to mount with attaching hardware at required points.
3. Check clearance between mount and hoist with feeler gauge at final attach point. If necessary, shim as required to provide level mounting surface.
4. Connect hydraulic lines to hoist and hoist motors, assuring that proper lines are connected to correct ports (as marked prior to removal).
5. Remove lifting sling from hoist.
6. Reinstall cable on hoist.
7. Adjust hoist mounting and alignment as required.

7-192 Maintenance Programs for Equipment Entities

Functional check of hoist
1. Attach a load and functionally check hoist by raising and lowering load on line.
2. Check for smooth and efficient operation of brake system.
3. Observe hoist hydraulic conditions and connections for security and freedom of leaks.

REEL FOR HYDRAULIC HOSES

The hydraulic crane's hose reel provides automatic stowing of the boom telescope cylinder hydraulic hoses. The basic components of the hose reel assembly are reel, stand, hub, spring, and spring housing. The reel consists of an inner and outer flange with a divider through the center. A housing mounted on the outside of the inner flange contains the spring which maintains a constant tension on the hose reel.

Inspection Visually inspect the hose reel mount for security. Observe for complete rewinding of hydraulic lines onto hose reel with boom fully retracted.

The hose reel spring is lubricated through an access hole located on the left side of the hose reel, above the center shaft. Lubrication every 50 operating hours is recommended.

With the boom fully retracted, spray motor oil under pressure into the hose reel spring housing. The spring should be saturated with oil. If air pressure lubricating equipment is not available, use a pump oilcan to ensure complete saturation of the spring.

Hose reel removal procedure
1. Tag and disconnect hydraulic hoses from hose reel. Cap hoses and openings.
2. Engage hose reel retaining pin.
3. *Tag and disconnect hydraulic hoses from telescope cylinders. Cap hoses and openings.*
4. Remove bolts and washers securing hose reel to frame. Remove hose reel.

CAUTION: Use care when removing bearing from hose reel to avoid damaging inner race. *Use extreme caution* when lifting reel from spring cover. If retaining band is broken and spring has expanded, make no attempt to remove spring by hand!

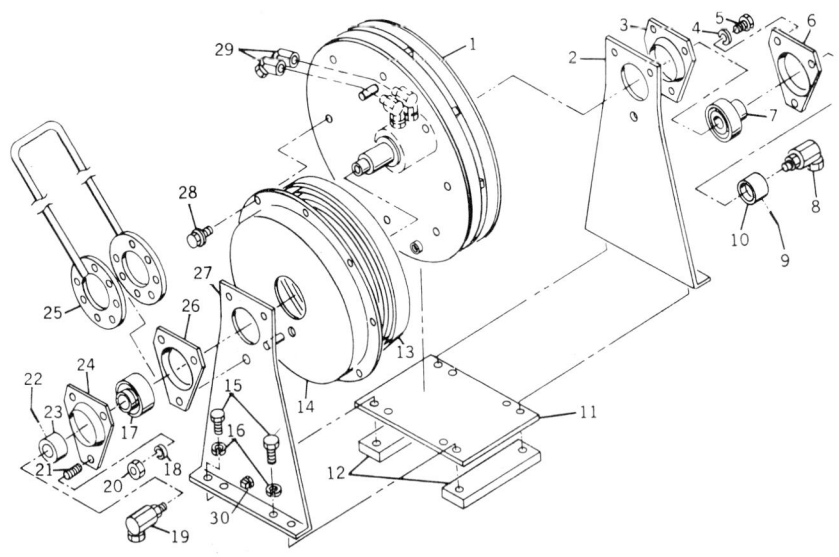

Fig. 8-8 Hydraulic hose reel components: (1) reel assembly; (2) side plate; (3) bearing retainer; (4) lockwasher; (5) bolt; (6) bearing retainer; (7) bearing; (8) 90° elbow; (9) set screw; (10) bearing cover; (11) baseplate; (12) mounts; (13) spring assembly; (14) spring cover; (15) four bolts; (16) four lock washers; (17) bearing; (18) lock washer; (19) 90° elbow; (20) nut; (21) stud; (22) setscrew; (23) bearing cover; (24) bearing retainer; (25) hose guard; (26) bearing retainer; (27) side plate; (28) bolt; (29) two 45° elbows; (30) retaining pin.

HYDRAULIC OUTRIGGERS

The hydraulic crane's outrigger system typically consists primarily of a control panel, double-box housing, telescoping beams, extension cylinders, jack cylinders, or perhaps hydraulic cantilever-type outriggers that flop down into position.

Again, typically, electrical switches on the crane's control panel determine the mode of operation (extension or retraction) and the selection of the desired cylinders.

Inspection Visually inspect outrigger beams for excess dirt accumulation. Check hydraulic connections for security. Observe fully extended outrigger for straightness of beam and general condition of outrigger assembly.

In most cases, it is not necessary to remove outrigger beams for removal of extension cylinders, but if a beam must be removed, this is the recommended procedure:
1. Activate hydraulic system; extent subject outrigger slightly to facilitate attaching of lifting device to beams.
2. On opposite side of outrigger housing:
    a. Remove end cover.
    b. Remove cap screws securing shaft to outrigger housing; remove shaft.
    c. Tag and disconnect hydraulic lines to extension cylinder. Cap lines and openings.
3. Attach suitable lifting device to outrigger beams.
4. Position supporting blocks beneath beams.
5. Withdraw outrigger beams and position on supporting blocks.

When trouble is located and corrected, essentially follow the reverse of the above procedure for reinstallation of hydraulic beams.

Removal of extension cylinder
1. Activate hydraulic system; extend subject outrigger slightly to gain access to rod and retaining pin. (Refer to Fig. 8-9A and B.)
2. Extend adjacent outrigger slightly to gain additional access to hydraulic connections of subject outrigger. CAUTION: Be sure hydraulic power is removed before proceeding.
3. Remove end plate from outrigger housing.
4. Tag and disconnect hydraulic lines to cylinder; cap all lines and openings.
5. Remove extension cylinder anchor shaft.
6. Remove extension cylinder from outrigger beam.

Disassembly of Extension/Retraction Cylinder
1. Remove oil from extension cylinder.
2. Remove bolts securing headplate to cylinder barrel. Use extreme care in handling or laying down cylinder rod. Damage to rod surface may cause undue maintenance and expense.
3. Withdraw cylinder rod assembly from cylinder barrel. Cover cylinder barrel opening to prevent contamination from dust and dirt.
4. Secure cylinder rod from moving at rod end; remove piston locknut from rod.
5. Remove piston, spacer, head, and headplate from cylinder rod.
6. Remove all seals and rings from piston, head, and headplate. *Aligning discarded seals and rings in order of disassembly will facilitate installation of new seals and rings.*
7. *Clean all parts with solvent and dry with compressed air. Inspect all parts for serviceability.*
8. Stone out minor blemishes and polish with a fine crocus cloth.
9. Clean with solvent and dry with compressed air any parts that have been stoned and polished.

ELECTRICAL SYSTEM

In general, the hydraulic crane's electrical system is powered from the crane carrier's engine-driven alternator and two lead-acid storage batteries. The system is the single-wire, negative ground-return type utilizing the machine's superstructure as a ground. Typically, electric power activates or assists the following crane systems: ignition and

7-194 Maintenance Programs for Equipment Entities

starting, instruments and indicator lights, windshield wiper, heater, outriggers, and load moment indicator or anti-two-block device.

Maintenance of the electrical system primarily consists of maintaining the electrical components of the engine, for example, keeping batteries charged, making periodic output checks of the alternator, making voltage regulator adjustments (as necessary), replacing diodes in the alternator. Standard repair and adjustment procedures for engines with alternators incorporated in the electrical system would be applicable.

Minor electrical system maintenance includes repair or replacement of damaged wiring, switches, indicator and panel lamps, etc. Standard wiring practices should be observed when replacement is necessary.

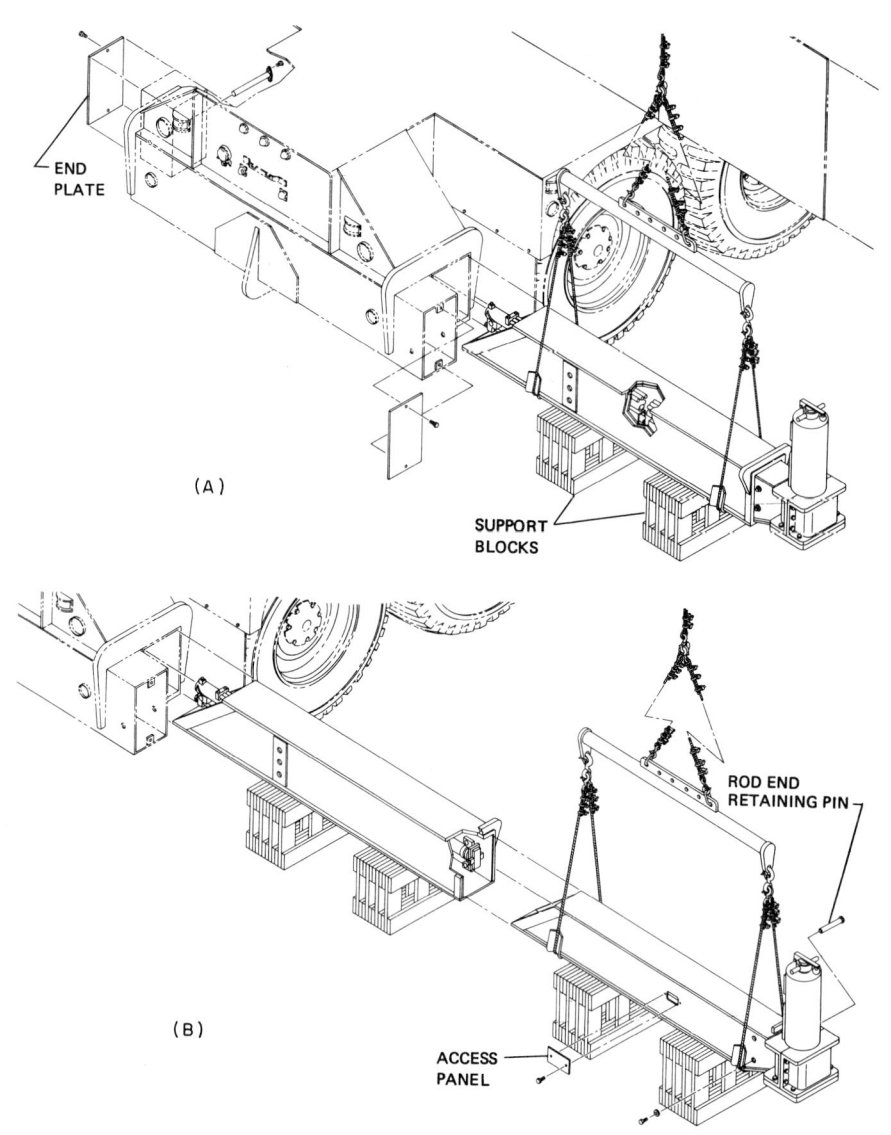

Fig. 8-9 Steps in (A) removing and (B) installing outrigger beam.

TABLE 8-4 Outrigger Troubleshooting

Trouble	Probable cause	Remedy
Slow or erratic operation of outrigger extension cylinders.	Damaged relief valves.	Remove relief valve; clean or replace.
	Low hydraulic oil.	Replenish oil to proper level.
	Sticking solenoid valve spool.	Repair or replace valve spool.
	Improper ground to base of solenoid.	Ground properly.
	Damaged "O" rings and swivel.	Remove swivel and replace "O" rings.
	Directional selector switch sticking (if so equipped).	Clean or replace switch.
	Collector ring dirty or glazed.	Clean and deglaze collector ring.
	Damaged wiring to solenoid switch.	Replace wiring.
	Weak brush springs on collector switch.	Replace brush springs.
	Damaged extension cylinder (internal parts).	Remove extension cylinder and repair as necessary.
	Bent piston rods.	Replace piston rods and seals.
	Excessive material on outrigger beams.	Clean outrigger beams.
	Binding outrigger beam.	Repair or replace outrigger beam.
	Damaged selector valve.	Repair or replace valve.
	Damaged valve coil.	Replace coil.
	Main hydraulic pump cavitation.	Replace or tighten hose and fitting.
	Partially shifted hydraulic selector spool.	Disassemble, clean, and polish spool and valve housing with very fine emery cloth. (Water paper.)
	Insufficient voltage for operation of solenoid valve.	Solenoids require a minimum of 9.5 volts to energize. Check outrigger wiring and electrical swivel coupling collector rings.
	Damaged piston seals.	Replace all cylinder seals.
	Worn or damaged hydraulic pump section.	Repair or replace pump section.
Cylinder extends while machine is roading.	Scored cylinder barrel.	Repair or replace extension.
	Cracked or damaged piston.	Replace piston and all cylinder seals.

(continued)

TABLE 8-4 Outrigger Troubleshooting (*Continued*)

Trouble	Probable cause	Remedy
	Piston loose on piston rod.	Replace all cylinder seals and torque piston locknut.
Outrigger vertical jack cylinder slow or erratic.	Low hydraulic oil.	Replenish oil to proper level.
	Damaged main relief valve.	Repair or replace valve.
	Damaged holding valve seals.	Replace holding valve seals.
	Bent cylinder rod.	Replace cylinder rod and seals.
	Binding outrigger housing.	Repair or replace outrigger housing.
	Damaged "O" rings in swivel.	Replace "O" rings.
	Excessive material on beams.	Clean outrigger beams.
	Sticking solenoid valve spool.	Repair or replace valve spool.
	Damaged wiring to solenoid.	Repair or replace wiring.
	Weak brush springs.	Replace brush springs.
	Collector ring dirty or glazed.	Clean or deglaze collector ring.
	Directional selector switch sticking.	Clean or replace switch.
	Main hydraulic pump cavitation.	Replace or tighten hose and fitting.
	Worn or damaged hydraulic pump section.	Repair or replace pump section.
Jack cylinder retracts under load.	Damaged piston seals.	Replace all cylinder seals.
	Damaged holding valve seals.	Replace seals.
	Damaged holding valve.	Replace valve assembly.
	Scored cylinder barrel.	Repair or replace cylinder.
	Cracked or damaged piston.	Replace piston and all cylinder seals.
	Piston loose on cylinder rod.	Replace all cylinder seals and tighten locknut.
Jack cylinder extends while machine is traveling.	Damaged piston seals.	Replace all cylinder seals.
	Scored cylinder barrel.	Replace jack cylinder.
	Cracked or damaged piston.	Replace piston and seals.

Trouble	Probable cause	Remedy
	Piston loose on cylinder rod.	Replace seal and retorque.
Outrigger system will not activate (from stowed or extended and down position).	Hydraulic oil low.	Replenish system.
	Loose or broken wire on switch.	Repair or replace wiring.
	Clogged, broken or loose lines or fittings.	Clean, tighten, or replace lines or fitting.
	Damaged relief valve.	Repair or replace valve.
	Damaged control valve.	Repair or replace valve.
Outrigger system activates, but selected outrigger will not stow or extend and lower as desired.	Clogged, broken, or loose hydraulic lines or fitting.	Clean, tighten, or replace lines or fittings.
	Loose or broken wire on control switch or solenoid valve.	Repair or replace wiring.
	Damaged solenoid valve.	Repair or replace valve.
	Damaged control switch.	Replace switch.
	Damaged hydraulic cylinder.	Repair or replace cylinder.
Outriggers will not set.	Improper sequence of activation.	Activate individual control switch; then activate system control switch.
Two outriggers activate from single control switch.	Damaged solenoid valves.	Repair or replace valves.
One/two outriggers will not stow.	Hydraulic lock.	Recycle individual outrigger(s).
Individual outrigger will not set or stow.	Damaged piston seals.	Replace seals.
	Damaged check valve.	Repair or replace valve.
	Loose or broken wire on control switch or solenoid valve.	Repair or replace wiring.
	Damaged solenoid valve.	Repair or replace valve.

WARNING: When performing any electrical maintenance, remove all rings, watches and jewelry as serious burns may result from accidental grounding or shorting of live circuits. Be sure batteries are disconnected before starting maintenance. Never use smaller-diameter wire as a replacement.

Batteries The 12-volt, lead-acid storage batteries wired in parallel have the primary function of supplying power for starting the superstructure engine. During inspection, check the batteries and cables for cracks, corrosion, and other visible damage. Check electrolyte level and specific gravity to determine charge state.

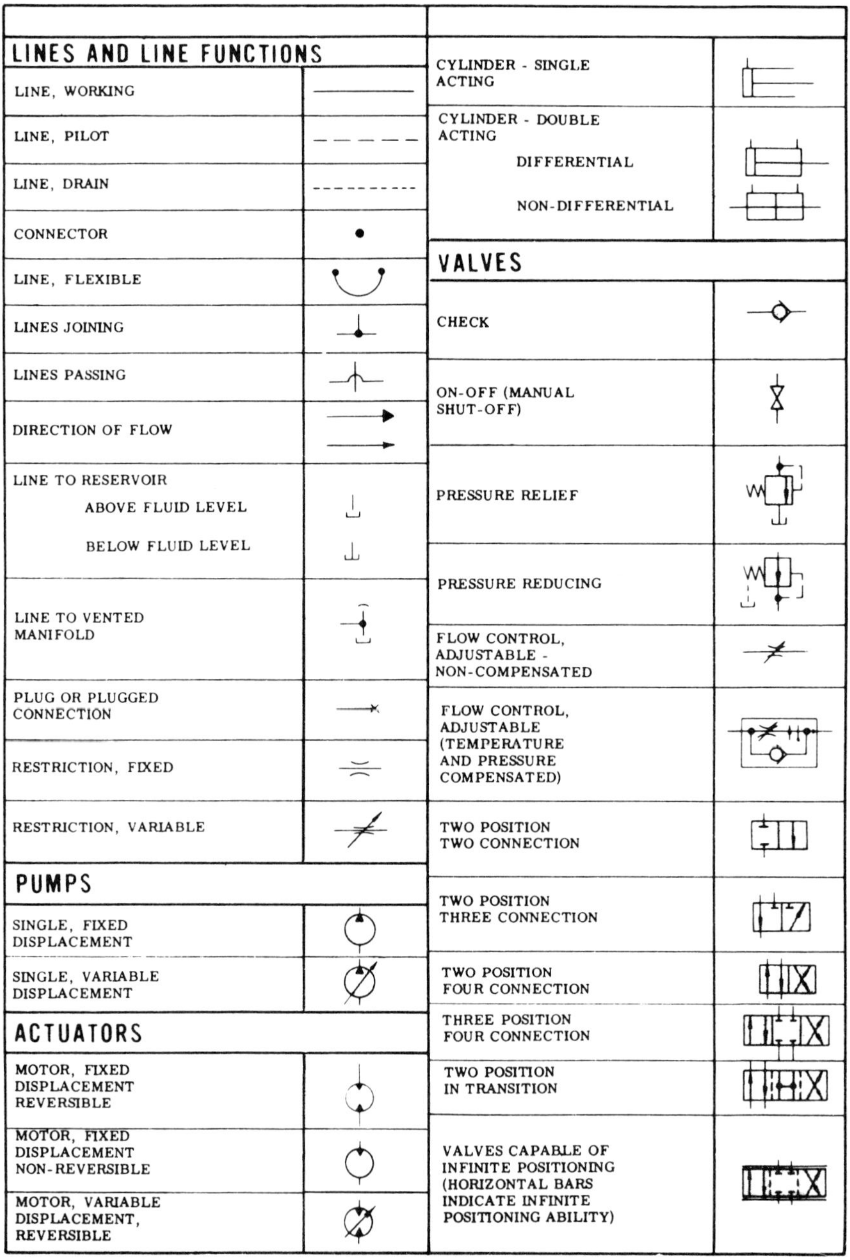

LINES AND LINE FUNCTIONS		CYLINDER - SINGLE ACTING	
LINE, WORKING	————	CYLINDER - DOUBLE ACTING	
LINE, PILOT	— — — —	DIFFERENTIAL	
LINE, DRAIN	- - - - - -	NON-DIFFERENTIAL	
CONNECTOR	•	**VALVES**	
LINE, FLEXIBLE	⌣		
LINES JOINING	⊥	CHECK	—◊—
LINES PASSING	⤴	ON-OFF (MANUAL SHUT-OFF)	⧖
DIRECTION OF FLOW	→		
LINE TO RESERVOIR ABOVE FLUID LEVEL	⌐	PRESSURE RELIEF	
BELOW FLUID LEVEL	⊥⌐		
LINE TO VENTED MANIFOLD		PRESSURE REDUCING	
PLUG OR PLUGGED CONNECTION	—×	FLOW CONTROL, ADJUSTABLE - NON-COMPENSATED	⤳
RESTRICTION, FIXED	≃	FLOW CONTROL, ADJUSTABLE (TEMPERATURE AND PRESSURE COMPENSATED)	
RESTRICTION, VARIABLE	⤳	TWO POSITION TWO CONNECTION	
PUMPS		TWO POSITION THREE CONNECTION	
SINGLE, FIXED DISPLACEMENT	⊙		
SINGLE, VARIABLE DISPLACEMENT	⊘	TWO POSITION FOUR CONNECTION	
ACTUATORS		THREE POSITION FOUR CONNECTION	
MOTOR, FIXED DISPLACEMENT REVERSIBLE		TWO POSITION IN TRANSITION	
MOTOR, FIXED DISPLACEMENT NON-REVERSIBLE	⊙	VALVES CAPABLE OF INFINITE POSITIONING (HORIZONTAL BARS INDICATE INFINITE POSITIONING ABILITY)	
MOTOR, VARIABLE DISPLACEMENT, REVERSIBLE	⊘		

Fig. 8-10 Standard hydraulic system symbols. *(From the "Fluid Power Handbook," Industrial Publishing Co.)*

METHODS OF OPERATION		MISCELLANEOUS	
SPRING		ROTATING SHAFT	
MANUAL		ENCLOSURE	
PUSH BUTTON		RESERVOIR VENTED	
PUSH-PULL LEVER		PRESSURIZED	
PEDAL OR TREADLE		PRESSURE GAUGE	
MECHANICAL		ELECTRIC MOTOR	
DETENT		ACCUMULATOR, SPRING LOADED	
PRESSURE COMPENSATED		ACCUMULATOR, GAS CHARGED	
SOLENOID, SINGLE WINDING		HEATER	
REVERSING MOTOR		COOLER	
PILOT PRESSURE REMOTE SUPPLY		TEMPERATURE CONTROLLER	
INTERNAL SUPPLY		FILTER, STRAINER	

Battery servicing is applicable to both the carrier and the superstructure electrical systems in cranes so equipped.

Specific gravity Proper electrolyte specific gravity is 1.275 for filling new batteries (shipped with plates dry-charged). Although some battery life may be sacrificed in dry humid climates, a more uniform operation will result in using a universal fill from 1.275 to 1.300 specific gravity for a fully charged battery.

Using a siphon hose or syringe, fill all battery cells with the prescribed electrolyte; do not allow electrolyte to be more than three-eighths above the protector on top of the cell separators.

Check Specific Gravity as Follows:
1. Remove all vent caps from battery
2. Using a hydrometer, check the specific gravity reading of each battery cell. A reading in any one cell of 1.250 or less indicates that the battery requires charging or replacement.
3. Be sure to reinstall battery vent caps.

Engine instruments Engine instruments are typically installed on the crane cab's console and provide the operator with a visual display of water temperature, oil pressure, battery condition, fuel supply, and engine rpm. Electrical power for operating the instruments is supplied by the main 20-amp circuit breaker in the cab.

Functional Check. Start crane engine and observe indicators. Observe for proper functioning of selected indicator. Further troubleshoot as necessary any system malfunction not corrected by repair or replacement of the indicator or associated wiring.

Electrical switches Two basic functions are performed: completing an electrical circuit, allowing current to flow (continuity), and opening a circuit, preventing a flow of current (circuit isolation). Maintenance and inspection procedures for all crane electrical switches are similar and therefore are covered under one general procedure.

WARNING: Always ensure that batteries are disconnected before performing any maintenance on the electrical system.

Switch Inspection. Visually check switch for cracks, damaged connections, or other damage. Check wiring for damaged insulation or damaged terminals. Perform the following check to determine switch serviceability:
1. Using an ohmmeter or continuity light, check for continuity between switch terminals with switch in ON or activated position.
2. Position switch to OFF. Ohmmeter should register zero (no continuity).

Functional Check. Operate switch as described for proper functioning of selected system. Further troubleshoot as necessary any system malfunction not corrected by repair or replacement of switch or associated wiring.

Solenoid valves Electrically operated control valves (solenoid valves) control the distribution of hydraulic oil for the outriggers. Normal control is from the operator's cab.

Inspection. Visually inspect valves and hydraulic connections for evidence of leaks or other damage. Check security of electrical connection. Inspect wiring for evidence of cracks or breaks.

Removal. Tag and disconnect hydraulic lines to solenoid valve bank. Cap all lines and openings. Tag and disconnect electrical leads; tape lead ends. Remove hardware securing valve bank assembly to the carrier frame; remove valve assembly.
1. Using an ohmmeter or continuity light, check for continuity between switch terminals with switch in ON or activated position.
2. Position switch to OFF. Ohmmeter should register zero (no continuity).

Functional Check. Operate switch and observe proper functioning of selected system. Further troubleshoot as necessary any system malfunction not corrected by repair or replacement of switch or associated wiring.

WINDSHIELD WIPER ASSEMBLY

A two-speed wiper motor is located inside the cab at the center of the windshield. The motor shaft extends through the cab's sheet metal to the exterior for mounting the blade. Control of the motor is provided by the WIPER switch installed on the console.

OUTRIGGER CONTROL PANELS

Activation of the solenoid valves controlling outrigger operation is governed by the outrigger control panel. The panel, typically located in the superstructure cab, receives power for operation from the carrier electrical system. The electrohydraulic swivel coupling transfers hydraulic fluid and electrical impulse between the superstructure and the carrier.

The control panel incorporates eight button-type switches and one spring-loaded, three-position switch. The buttons select the cylinder(s) which will receive hydraulic oil

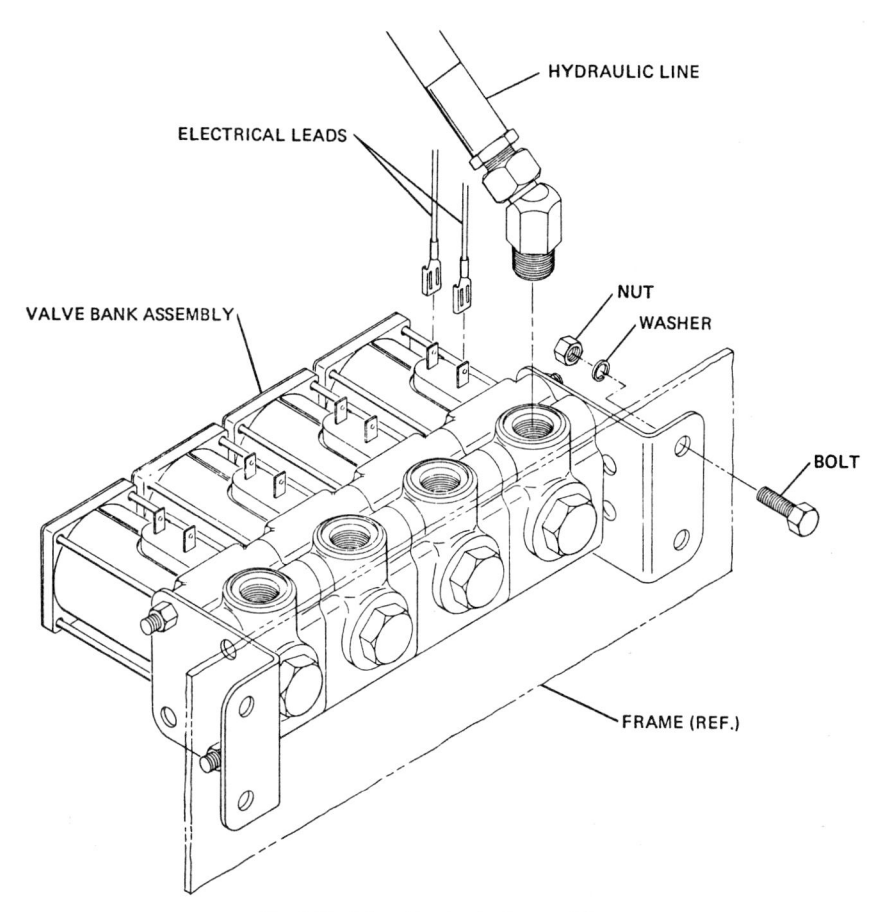

Fig. 8-11 Solenoid valve bank assemblies.

by governing operation of the various outrigger solenoid valves. The three-position EXTEND/RETRACT switch, controlling operation of the outrigger selector valve, activates the system by releasing hydraulic oil for extension and retraction of the selected cylinders. A pressure relief valve is incorporated in the circuit to eliminate the possibility of pressure buildup when activating the outrigger system.

Solenoid valves Electrically operated control valves (solenoid valves) control the distribution of hydraulic oil for the outriggers. They are located in two banks of four valves each, are of the two-way, sliding spool type, and may be actuated from the outrigger control panel. (See Fig. 8-11.)

7-202 Maintenance Programs for Equipment Entities

Inspection. Visually inspect valves and hydraulic connections for evidence of leaks or other damage. Check the security of electrical connections. Inspect wiring for evidence of cracks or breaks.

GENERAL

To assure maximum crane utilization, following prescribed lubrication procedures is of utmost importance. Service intervals specified are for normal operation where moderate temperature, humidity, and atmospheric conditions prevail. In areas of extreme cold

TABLE 8-5 Solenoid Valve Troubleshooting

Trouble	Probable cause	Remedy
Sticking spool	Dirt in system.	Change oil and flush system.
	Distortion caused by tie bolts being overtorqued.	Retorque tie bolts.
	Flow in excess of valve rating.	Limit flow through valve to that recommended; check pump output and cylinder ratio.
	Pressure in excess of valve rating.	Check relief valve setting or pump compensation with that recommended.
	Electrical failure.	Check wiring and solenoids.
External leakage.	Damaged "O" rings or quad rings.	Check for chipped packings and replace.
	Loose tie bolts.	Retorque tie bolts.
	Damaged solenoid.	Replace damaged parts.
Solenoid failure.	No current.	Check power source of at least 85% of coil rating.
	Damaged solenoid assembly.	Replace solenoid.
	Short in solenoid.	Replace coil.
	Loss of solenoid force.	Decrease time of solenoid energization, decrease cycle rate.

temperature, the service periods and lubrication specifications should be altered to meet existing conditions.

Items not equipped with grease fittings, such as linkages, pins, levers, etc., should be lubricated with oil once a week. Motor oil, applied sparingly, will provide the necessary lubrication and help prevent the formation of rust.

REMEMBER: Proper lubrication is a safety factor in any heavy equipment operation.

Suggested Lubrication Standards

Key	Specification
EP(LSB)	A multipurpose-type grease having a minimum dripping point of 350°F (176.6°C), excellent water resistance, and of an extreme pressure type (minimum Timken OK load 40 lb). For above 100°F (37.8°C) temperatures, NLGI #2 or #3 grade.
GL	A straight mineral gear oil of good quality, minimum viscosity index 85, and meeting viscosity requirements of the SAE grade used.
EPGL	A multipurpose extreme-pressure gear oil designed to meet the requirements of military specification MIL-L-2105 (also recommended by commercial vehicle manufacturers).
EPGL (SCL)	An extreme-pressure gear oil compound with sulphur-chlorine-lead additives.

Key	Specification
WGL	Quality oil designed for lubricating worn gears under the ambient temperatures indicated below.
CG	Chassis lubricant. A suitable grease for general chassis lubrication, having good water resistance and adhesiveness qualities, minimum oil viscosity 300 SSU at 100°F (37.7°C). For summer, use NLGI #2 or #3 grade; for winter, use NLGI #1 grade. Higher-quality multipurpose automotive greases may also be used.
EP	Heavy-duty extreme-pressure grease. A grease for heavier service meeting the following requirements: dripping point—minimum 180°F (82.2°C); NLGI grade—as local temperature requirements indicate. Extreme pressure—Timken OK 40 lb; minimum oil viscosity—minimum SSU at 210°F (98.8°C). Must have good water resistance. (*Note:* Same as U.S. Steel No. 350 EP Rolling Mill Grease.)
HO	Hydraulic oil. Use severe-duty-type engine oils meeting API service. MIL-L-2104 or antiwear-type hydraulic oils. Moderate hydraulic service. Use qualified Type A suffix A Automatic Transmission Fluid or General Motors approved Dexron Automatic Transmission Fluid.
OG	A viscous lubricant designed for lubricating open gears and having good water resistance and adhesiveness. Viscosity to be appropriate to operating temperature and to assure adhesion to gear teeth.
ATF	Automatic transmission fluid.

Typical Oil Inspection Criteria

Gravity, °API	29.2
Flash point	405° (201.6°C)
Pour point	−40°F (−40°C)
Viscosity, SSU at 100°F (37.8°C)	207
Viscosity, SSU at 210°F (98.8°C)	50.9
Viscosity index (ASTM D567)	134
Viscosity index (ASTM D2270)	145
Color	Red
Qualification no.	AQ-ATF-2924A

Lubrication requirements of the boom pivot and anchor shafts should be based upon crane usage. It is recommended that they be lubricated every 10 hr of crane operation. *Boom should be positioned on boom rest for pivot shaft lubrication.*

Boom Sections. To ensure that all sections are thoroughly lubricated, the boom should be fully extended (in horizontal position) and brushed (sides and bottom of bottom plate) with extreme-pressure (EP) lubricant.

LUBRICATION CHART

The lubrication chart (Fig. 8-12) in this section reflects a typical machine. Lubrication point locations are general, as changes in manufacturer and vendor component design may result in relocation and/or number of lubrication fittings, fill plugs, drain plugs, etc.

Specific brand name lubricants are not referenced since most all nationally known oil suppliers' products are suitable for use, providing they meet the requirements of the Mil specs and standards appearing in this section.

Should any conflict of information arise between lubricant recommendations for vendor components and those recommended in the applicable vendor's publication, contact the nearest vendor representative, as changes in their recommendations may occur after issuance of this information.

Maintenance Programs for Equipment Entities

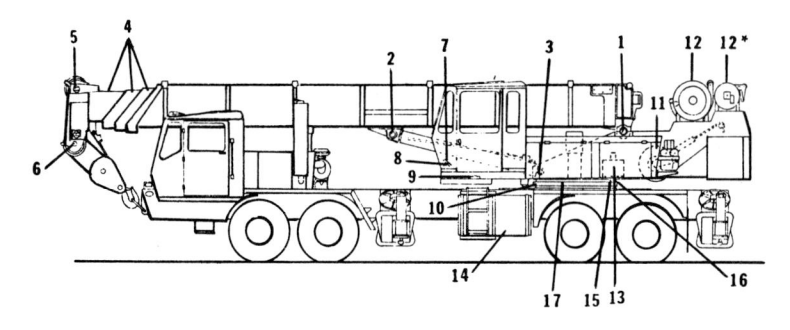

No.	Lubrication periods	Quantity and fitting	Location and instructions	Lubrication type*
	Every 10 operating hours			
1	Boom pivot	2 SAE std. grease	1 on top of each pivot/anchor shaft trunnion block	EP
2	Lift cylinders (top)	4 SAE std. grease	2 on each rod end shaft housing	EP
3	Lift cylinders (base)	4 SAE std. grease	2 on each base end shaft housing	EP
	Every 25 operating hours			
4	Boom sections	None	NOTE: All telescoping boom sections should be extended and coated with grease—sides and bottom of bottom plates.	EP
	Every 50 operating hours			
5	Boom nose idler sheave	1 SAE std. grease	1 on sheave hub	EP
6	Boom nose sheaves	3 SAE std. grease	1 on each end of shaft 1 center of shaft (through access hole)	EP
7	Swing brake pedal	1 SAE std. grease	1 underneath front of cab floor—reached from outside cab	CG
8	Throttle pedal	1 SAE std. grease	*Underneath front of cab floor—reached from outside cab	CG
9	Brake master cylinder	Fill (cap)	Underneath front of cab floor—reached from outside cab	HBF
10	Turntable bearing	2 SAE std. grease	1 behind each lift cylinder on turntable deck top	EP
11	Hose reel	1 access hole	Left side of hose reel, saturate spring with motor oil sprayed under pressure	MO
12	Hoist final drive assembly	1 Fill (pipe plug)	Check/fill oil level to bottom of fill plug	EPGL
13	Swing box gear case	1 level/fill (pipe plug)	Check/fill oil level to bottom of oil level plug	EPGL
	Change when dirty or cloudy			
14	Hydraulic oil reservoir	1 fill (cap) 1 drain (plug) hydraulic oil	Left-hand side of superstructure	HO

No.	Lubrication periods	Quantity and fitting	Location and instructions	Lubrication type*
	Every 100 operating hours			
15	Pinion	1 SAE std. grease	1 under turntable, just above pinion gear	EP
16	Swing box bearing	1 SAE std. grease	1 top of swing box case	EP
17	Turntable pinion gear and bull ring gear	None	Swing boom over the side and coat gear with lubricant	OG
	After first 250 operating hours: thereafter, every 500 operating hours or 12 months, whichever occurs first		Change oil	
12	Hoist final drive assembly	1 fill (pipe plug)	Left side of hoist	EPGL
13	*Every 12 months* Swing box gear case	1 level/fill (pipe plug) 1 drain plug	Change oil	EPGL

*CG: chassis grease; EP: heavy-duty extreme-pressure mill type; HBF: hydraulic brake fluid 70-R1; HO: hydraulic oil, SAE 10 MS; EPGL: extreme-pressure gear lube SAE 90; OG: open gear lubricant; MO: motor oil.

Fig. 8-12 Hydraulic crane lubrication chart.

Chapter **9**

Crushing Equipment Maintenance

DAVID J. SAMEK
Iowa Manufacturing Company
Cedar Rapids, Iowa

ROLL CRUSHER MAINTENANCE

The maintenance data under this heading will deal with the conventional two-roll crusher shown in Fig. 9-1 and the type of three-roll crusher shown in Fig. 9-2. (Jaw Crusher Maintenance and Impact Breaker Maintenance are discussed later in the chapter.)

Lubrication Correct lubrication of a roll crusher is a vital maintenance requirement which must never be neglected. Figure 9-3 shows a typical lubrication instruction sheet for two- and three-roll crushers. A copy of the appropriate lubrication sheet should be kept in a protective transparent cover and posted in the lubricant supply house for ready reference.

Installation and Function of Oil Seals. Most roll crusher shaft bearing assemblies will include one or more oil seals (Fig. 9-4). It is important that the function of these seals be clearly understood by maintenance personnel and that each one is correctly installed at overhaul. If a seal is reversed from its intended attitude, bearing failure can result due to improper lubrication and a lack of protection against dirt.

Seals must be carefully installed so that their sealing lip is not damaged or turned under as it is worked over a step or raised area of the shaft. To pass the lip of the seal rubber over a shaft step, use a piece of shim stock, plastic sheet material, or heavy paper (0.007 in.) wrapped over the step. Work the seal lip over the step, then withdraw the thin material toward the roll. *Do not* simply work the seal lip over with a screwdriver.

If the lip is on the trailing side of the seal at installation, it will work over any step area without special effort.

Check crusher mounting Keep crusher shafts horizontal so that bearing loading is uniform and lubricant distribution is proper. The best method is to periodically place a spirit level vertically against the machined surface of the flywheel, as shown in Fig. 9-5. Correct unlevel condition immediately by shimming.

Lifting crusher Whenever it is necessary to lift the crusher, attach hoist chains or cables *at factory specified lifting points only!* (See Fig. 9-6 for example.) Serious problems can be caused by exerting lifting force on other parts of the crusher.

Rubber tire drive for movable roll

Tire Pressure. Check tire pressures as often as necessary to maintain the proper inflation. Whenever a pressure loss is repeated and a leak is suspected, find the cause and correct it as soon as possible.

Tires must be kept properly inflated at all times during operation. Owing to the variation in distance between roll shafts, which is necessary for producing different particle sizes, tire pressure must be adjusted each time a sizable change of product is made.

7-208 Maintenance Programs for Equipment Entities

If the tires are too soft, slippage can occur under load particularly when hard material is being crushed. When this happens, serious damage to the tires can occur very rapidly due to the heat of friction.

If tires are too hard, bits of rubber will flake off as premature wearing occurs. No crushing advantage is realized unless material being crushed is very hard and slippage would otherwise result.

Typical Tire Inflation Procedure
1. Set crusher discharge opening and adjust hopper liners to rolls.
2. Measure the distance between centers of roll shafts (Fig. 9-7).
3. Use factory recommended tire pressure (Table 9-1).
4. Inflate each tire 4 to 5 lb *less* than operating pressure.
5. Operate crusher under load. Check for signs of slippage or flaking off of rubber particles. At the next shut-down, recheck tire pressure. The heat of friction during operation will normally raise the pressure in each tire 4 to 5 psi. If no allowance is made for this increase, the tires will be overinflated after the warm-up period.

IMPORTANT: Under no circumstances should the pressure in any tire be allowed to rise above 30 psi. This applies to all roll sizes and discharge settings.

Three-roll Crushers. Maintain equal pressure in all tires at all times. Even though there is a difference in the distances between the upper and the lower roll shafts and the fixed roll shaft, the tire pressures must be kept equal to prevent excessive deflection of one tire. Always use the lowest pressure that will drive the rolls without excessive slippage.

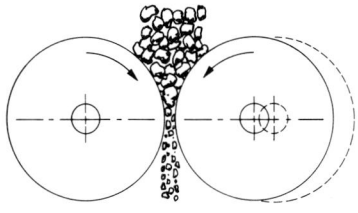

STATIONARY ROLL   MOVABLE ROLL
Fig. 9-1 Conventional two-roll crusher.

Tire Specifications. The tires specified for a typical movable roll drive have a nylon ply construction, traction sipes (slitted treads), and tubes and flaps for low pressure, high deflection. The highest degree of deflection occurs when the rolls are nearest each other for production of the smallest possible particle size.

IMPORTANT: Use only the size and type of tire recommended by the crusher manufacturer. Random substitution will usually result in unnecessary downtime.

Under normal operating conditions, tires will wear evenly and *both should be replaced at the same time.*

TABLE 9-1 Recommended Tire Pressure

Crusher size	Center-to-center working range, in.		
	Minimum 15 psi	Normal 20–25 psi	Maximum 30 psi
2416		$22^{7}/_{8}$–$25^{1}/_{4}$	$25^{7}/_{8}$
3018 3025	$28^{3}/_{4}$–30	30–$32^{1}/_{2}$	$32^{5}/_{8}$
3025	$30^{3}/_{4}$–32	32–34	
3030	$28^{3}/_{4}$–30	30–$32^{1}/_{2}$	$32^{5}/_{8}$
3136	$28^{3}/_{4}$–30	30–$33^{1}/_{4}$	$33^{1}/_{2}$
4026	$37^{3}/_{4}$–$39^{1}/_{2}$	$39^{1}/_{2}$–42	$42^{1}/_{2}$
4130	$38^{1}/_{2}$–40	40–44	$44^{1}/_{4}$
4132	$38^{1}/_{2}$–40	40–44	$44^{1}/_{4}$
4136	$38^{1}/_{2}$–40	40–44	$44^{1}/_{4}$
5530	53–54	54–58	58–$59^{1}/_{4}$

Crushing Equipment Maintenance   **7-209**

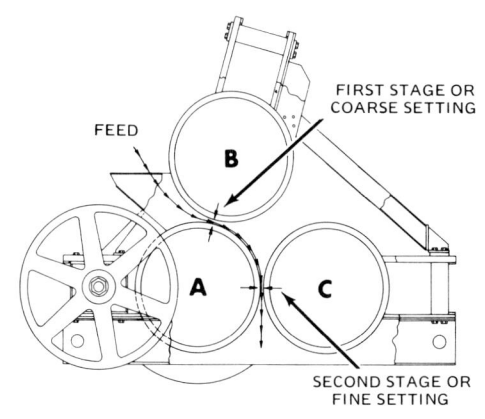

Fig. 9-2 Typical three-roll crusher.

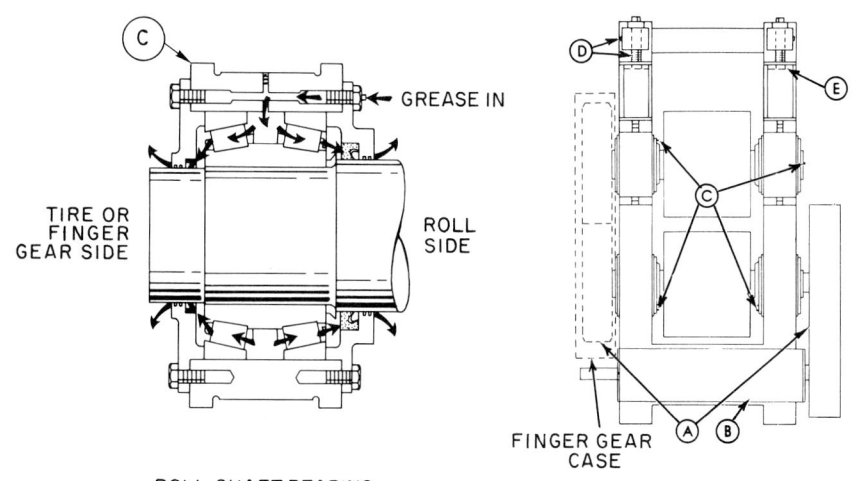

ROLL SHAFT BEARING BLOCK GREASING DETAIL

TWO-ROLL CRUSHER TOP VIEW

Item	Frequency and method	Lubricant recommended
A: Main drive gear case. (NOTE: Also applies to finger gear case when crusher is so equipped.)	Maintain oil supply to height of level plug. Keep filler breather cap cleared for free air passage. Every 1000 hr (or seasonally) remove drain plug when oil is hot. Flush and refill gear case to proper level.	Above 32°: Crater 0 Below 32°: Crater 00 (Flush case with light fuel oil.)
B: Countershaft housing (roller bearings)	Every 48 hr of operation remove level plug and fill plug when lubricant is hot. Add fresh heated lubricant until overflow occurs at level hole. Replace both plugs.	Marfak 00

Fig. 9-3 Lubrication instructions for roll crusher.   (continued)

7-210 Maintenance Programs for Equipment Entities

Item	Frequency and method	Lubricant recommended
	Every 1000 hr (or seasonally) remove drain plug when lubricant is hot and drain housing. Flush and refill to proper level.	(Flush housing with SAE 10 oil heated to 150°F. Run crusher at least 20 min.)
C: Roll shaft bearings and countershaft dust seal	Every 8 hr, or as needed, check the seals for visible grease slick. Grease must extrude from seals at all times to make an effective dust and moisture seal. For average conditions add 1 oz (20 to 25 pumps of average gun) to each bearing. The actual time interval and amount will vary with operating conditions and size of units.	Above 90°F: Marfak 1 32–90°F: Marfak 0 0–32°F: Marfak 00 Below 0°: Regal AFB2
D: Snub bolts and lock screws	As often as required, grease the threads of the snub bolts and lock screws to prevent rusting and binding.	Marfak 0
L: Gas hydraulic cylinders (optional)	Once each month pump grease into the fitting at the piston rod end of each gas hydraulic cylinder until grease extrudes from the hole near the air breather.	Marfak 0

Fig. 9-3 Lubrication instructions for roll crusher (*Continued*).

Tire Guards. Keep tire guards in place at all times during operation to prevent large pieces of rock from jamming between the tires.

Feed rate to crusher affects wear The most efficient operation of roll crusher occurs when the particle size is correct and uniform and the rate of feed keeps the crushing area filled up to the angle of nip, or just slightly higher, across the full width of the rolls. (See Fig. 9-8.)

Too much feed merely increases the surface contact between roll shells and material, resulting in unnecessary wearing of the shells. The wear area extends across the full widths of both roll shells. The hopper may suddenly tend to fill up and resembles a container of water starting to boil over. Jamming and tire slippage also tend to occur.

Size of feed particles affects wear A buildup of material in the feed hopper may be blamed on an excessive feed rate, when actually it may be due to oversized feed.

Fig. 9-4 Oil seals for roll crusher shaft bearings.

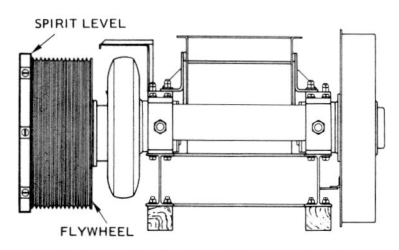

Fig. 9-5 Method of leveling crusher frame.

If particle size is too large, material stays above the angle of nip and is not subjected to any crushing force until the weight of material accumulated above it eventually drives it between the rolls. When an accumulation of material builds up in the hopper, *the revolving roll shells are subjected to a great deal of unnecessary wear as they rub the bottom of the material mass.*

When feed is stopped, material in the hopper should empty out. If a large quantity remains, reduce the feed particle size (see Fig. 9-9) or change roll shells to beaded or other type which will handle the larger size.

Spread of material across rolls If the feed to a roll crusher concentrates material at one portion of the roll width, the shells will wear faster in that area. The result is two concave roll shells which are difficult to build up and difficult to adjust for accurate product sizing.

A spreader bar in the center area of most crusher hoppers diverts material toward the ends of the rolls to prevent concentration of crushing at the center (see Fig. 9-10). An operator should check the feed distribution periodically to detect any concentration of wear. If the spreader bar is not wide enough to prevent concentration, its width can be increased by fastening a strip of rubber belting over the top of the bar.

Some crusher hoppers are equipped with adjustable deflector plates which can be set to produce varying degrees of material deflection to keep the feed distribution and roll shell wear uniform (see Fig. 9-11). Close attention to these deflectors can save hours of roll shell buildup.

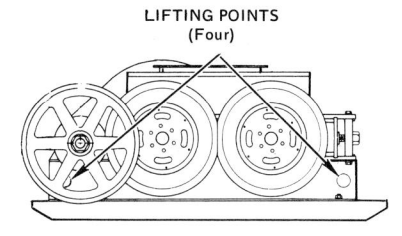

Fig. 9-6 Proper lifting points for raising crusher.

Check for contact between rolls Before starting the roll crusher drive after a change of roll setting or a buildup of roll shells has been completed, jog the drive and carefully check for any direct contact between roll surfaces. If beaded roll shells are used, be sure there is at least $1/16$-in. clearance between directly opposed beads. (See Fig. 9-12.)

Direct contact between rolls can cause severe vibrational stresses.

Keep compression springs tight The springs or spring package which maintains pressure on the movable roll to accomplish the crushing should be kept tight to prevent excessive movement of the roll. Most crushers have a snub bolt behind each compression assembly to fix its position. (Some have hydraulic compression assemblies which may or may not require attention.)

Excessive roll movement or so-called chatter results in rapid wearing of bearing blocks and slide bars (see Fig. 9-13). This tendency is greatest when crushing material to a size $3/4$

Fig. 9-7 Determining center distance of roll shafts.

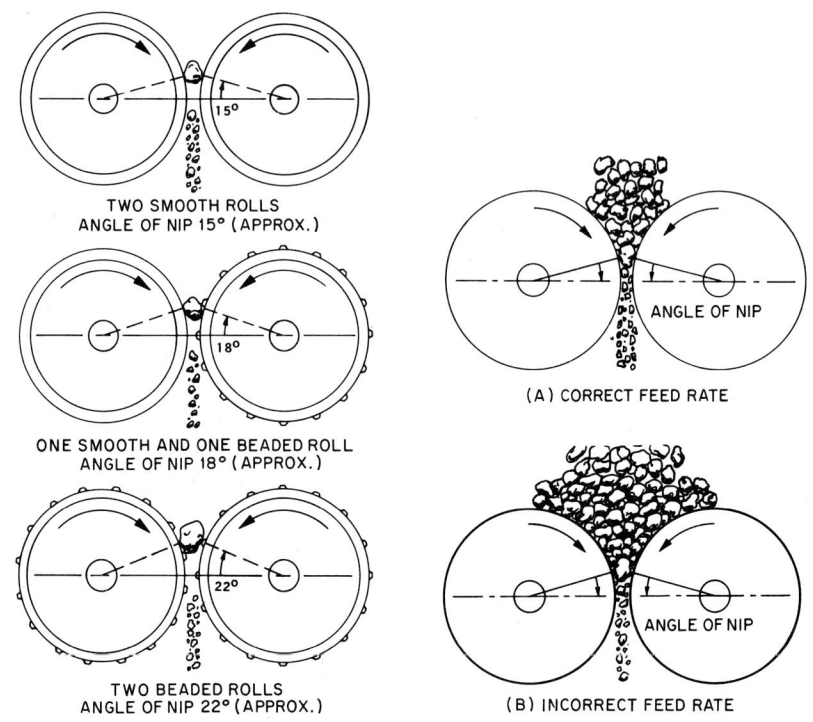

Fig. 9-8 Angle of nip for various roller combinations.

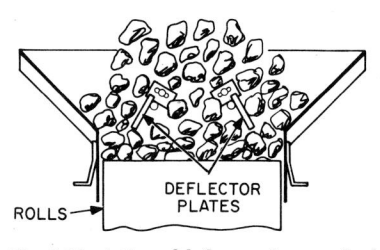

Fig. 9-9 With correct feed rate (A), material builds to level slightly above angle of nip. With incorrect feed rate (B), choke feeding wears roll shells excessively and causes jamming.

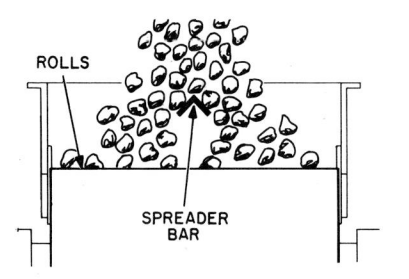

Fig. 9-10 Action of spreader bar on feed.

Fig. 9-11 Action of deflector plates on feed.

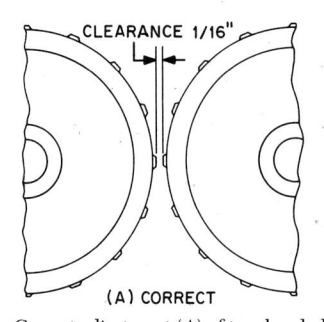

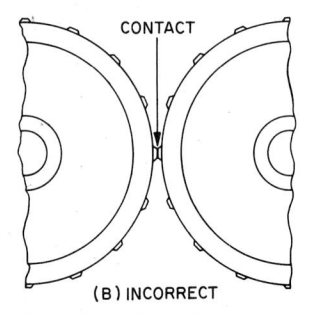

Fig. 9-12 Correct adjustment (A) of two beaded rolls allows at least 1/16-in. clearance when beads are directly opposite each other. Direct contact between roll and shells (B) should *not* be possible.

in. and smaller. Roll chatter also indicates low crushing efficiency, with excessive oversize material being produced.

Keep springs tight!

Compression assemblies Check compression assemblies frequently to be sure they are properly and equally adjusted.

Replace coil-type springs which have lost their life and no longer have an adequate activity range between coils when compressed (see Fig. 9-14).

(A) CORRECT   (B) INCORRECT

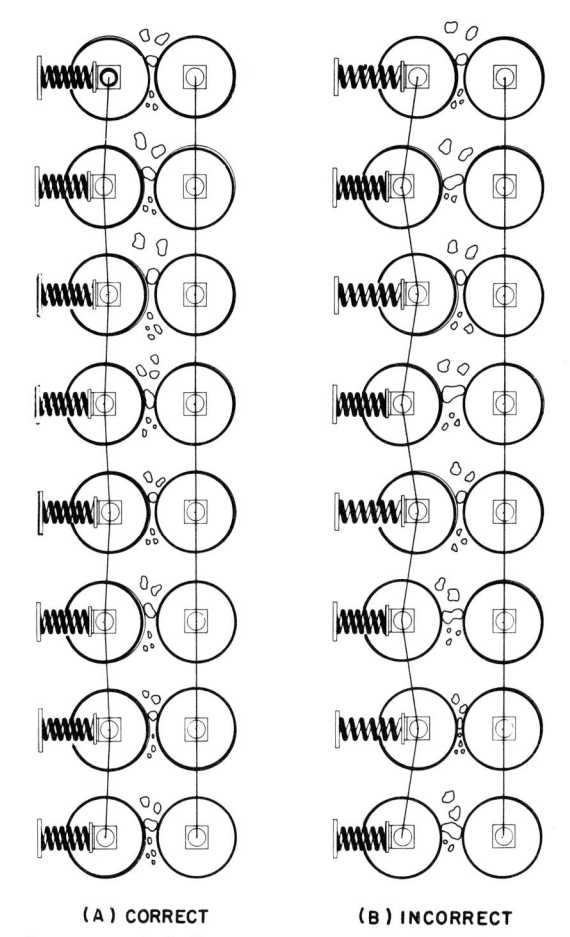

Fig. 9-13 Effect of spring tension of roll movement. Correct tension (A) allows normal roll movement. Low spring tension (B) allows excessive roll movement.

Keep gas-charged springs at the factory recommended pressure. If gas or fluid leakage occurs, overhaul the cylinders *according to the factory instructions.*

Protective shear assemblies Crushers are protected against damage due to passage of uncrushables by various shear members. Figures 9-15 to 9-17 show the details of a typical shear-washer-type assembly.

It is important that the shear members (washer, pin, etc.) be kept under uniform stress during operation. Vibration due to looseness can cause unnecessary premature failures and lost operation time.

Never substitute non-factory replacement parts for the shear member!

Troubleshooting guide—possible cause and solution

Low Capacity

1. Incorrect roll speed: May be due to insufficient drive power or belt slippage. Determine cause and correct.
2. Roll shells worn excessively: In closed circuit, too much oversize recirculates. Build up shells or replace them.
3. Discharge opening is not correct: Set rolls closer together unless production of fines will be too great.
4. Feed to crusher is irregular: Determine the cause and correct the problem.
5. Feed material is oversize: Feed hopper overfills when rolls cannot accept the larger particles. Change screen to reduce opening so that maximum size passing is smaller.

Overheated Bearing

1. Lubrication is incorrect due to one or more of the following:
    a. Too much lubricant
    b. Not enough lubricant
    c. Dirty lubricant. Bearing should be flushed out completely and relubricated.
    d. Incorrect lubricant (see lubrication chart).
2. Operating above recommended roll speed.
3. Bearing beginning to fail: Make a close investigation by listening to the sound of the bearing as the crusher drive is slowly turned by hand when the area is quiet.
4. Out-of-level (countershaft bearings only): Check the crosswise level of the countershaft housing. Out-of-level operation will starve the high-side bearing.

Excessive Roll Shell Wear

1. Operating above recommended roll speed
2. Operating with material level in hopper above the angle of nip: This wears roll

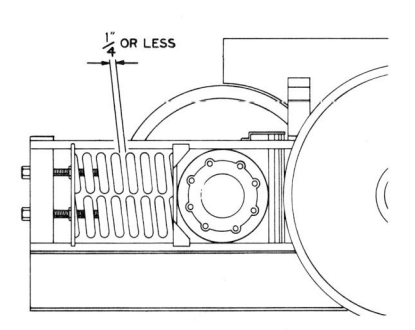

Fig. 9-14 Recommended start-up setting for new spring.

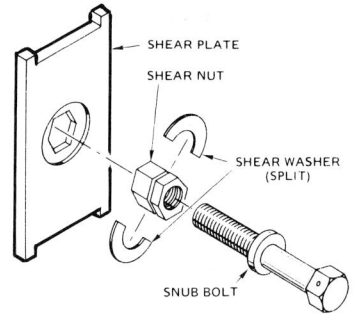

Fig. 9-15 Shear washer components, exploded view.

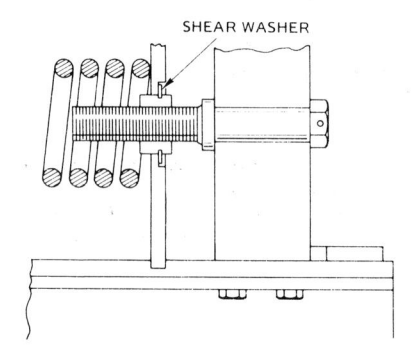

Fig. 9-16 Shear washer holding plate against spring.

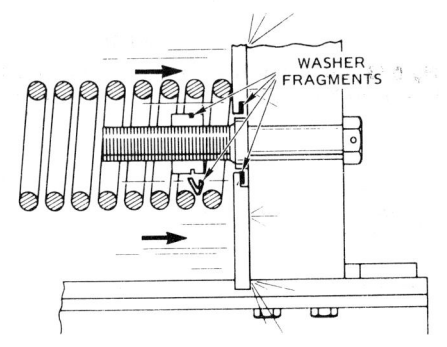

Fig. 9-17 Washer sheared by uncrushable object.

shells by rubbing over broad areas rather than by crushing. Reduce feed rate or size of material fed if larger than recommended.

3. Incorrect weld rod or procedure used for shell buildup.
4. Material is very abrasive.
5. Shells are not genuine factory made: Substitute shells can appear the same but do not have the requirements. Use factory made shells.

Condition of roll shells Roll shells which have worn unevenly will produce an abnormal gradation of product. The usual tendency is a widening of the discharge opening at the midpoint of the rolls where the feed tends to concentrate if corrective measures to obtain a uniform spread were not taken. Figure 9-18 shows the gradation changes when shells worn concave are kept in use without any corrective buildup work.

IMPORTANT: Never close the roll setting to the extent that the roll shells touch.

Roll shell surface buildup The material contact surfaces of roll shells will wear at a rate proportional to the operating capacity and to the abrasive and hardness characteristics of the material being crushed. The most common method of compensating for roll shell wear is applying weld rod to the shell surface by hand or by automatic welding equipment which can be attached to the crusher (see Fig. 9-19). The details of automatic welding are discussed in the following paragraphs.

Roll shell welding information Most roll crusher manufacturers, as the original suppliers of the cast manganese steel roll shells, offer roll shell welding information for the customer to use *at his own risk,* but do *not* honor the original parts guarantee after the shells have been subjected to the welding operation.

Roll shell temperature limitation Care must be exercised in building up roll shells to avoid embrittlement, which will cause the shell to crack, and warping, which can seriously affect seating of roll shells on spider. Never exceed 600°F temperature in welding roll shells. When welding shells in cold temperature, either weld immediately after a day's crushing or preheat shells before starting to weld. This will reduce the temperature differential and the possibility of shell cracking from the welding. The use of a special crayon for visual temperature indication is highly recommended. (See later discussion under Roll Shell Welding Recommendations.)

Do not cool the welded roll shell by artificial methods, such as pouring cold water over shells.

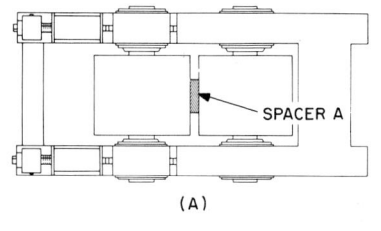

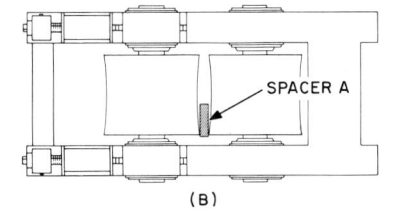

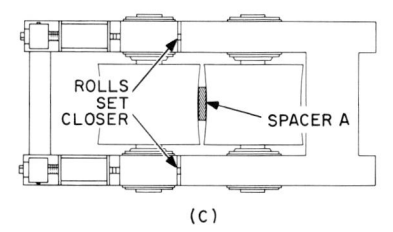

Fig. 9-18 Comparison of gradation differences when roll shells are badly worn. (*A*) Parallel roll shell surfaces produce normal product gradation. (*B*) Roll shell surfaces worn in center area produce excessive oversize. (*C*) Worn roll shells set closer together produce excessive fines.

Roll shell welding details* The majority of roll crusher shells are made of austenitic manganese steel, which in its cast form is brittle and rather useless. It must be heat-treated before it can be used. The standard heat treatment involves heating to 1800 to 1900°F, holding for 1 hr per inch of thickness, and immediately quenching with water.

When austenitic manganese steel is reheated (by welding or other means), the heat-treating process is reversed, and the casting will become brittle again. Both time and temperature factors are involved. Embrittlement due to overheating is accumulative. Once the part has been overheated and embrittlement has taken place, it will never return to its original tough, ductile heat-treated structure unless the original heat-treating process

---

*Acknowledgment to Teledyne-McKay Co., York, Pa., for all data relating to roll shell repair and buildup by the electric welding process.

is repeated. *For these reasons, it is very important to keep the temperature of the roll shell casting below 600°F.*

The high heat intensity of an electric arc (2300°F+) will create high local heat that will cool faster than the slow heat input generated by an oxyacetylene torch. Semiautomatic open arc wire at 400 amps run at relatively high travel speeds will cool faster per unit section than a manual electrode at 250 amps. Heavy shells will cool a weld more rapidly than thin shells. By utilizing the correct electric welding procedure, the rate of cooling in the area can be controlled so that practically no embrittlement occurs and welding can be repeated many times.

Preparation for welding shells When rebuilding roll shells, the following must be considered:

Condition of the Shells. When checking the condition of the rolls, look for cracks at the edges and for loose or spalled metal on the roll surface and thickness of the roll. Repair small cracks with McKay Hardalloy 118 or Chrome-Mang electrodes. Cut out cracks large enough to manipulate the electrode and hand weld from center to edge (Figs. 9-20 and 9-21). Peening of the weld bead will help ensure a successful repair. Do not overheat. If extensive buildup is needed, the wedges which clamp the shell on the spider and shaft should be loosened. Remove loose or spalled metal before rebuilding.

What Alloy or Alloys Should Be Used. The choice of alloys for rebuilding roll shells are many. Nonmagnetic, fully austenitic, work-hardening alloys are preferred for buildup. McKay 218-0 open arc wire or Hardalloy 118 electrode is a good choice. A better choice, at slightly higher initial cost, is McKay AP-0 wire or Chrome-Mang electrode. A good overlay is 240-0 open arc wire or 40 TiC electrode.

The Type of Welding Process. Three welding processes are available: automatic, semiautomatic, and manual arc. The equipment needed for the automatic process is the Roll-O-Matic, PA-1 wire feeder, and a 400-amp power supply. With the semiautomatic

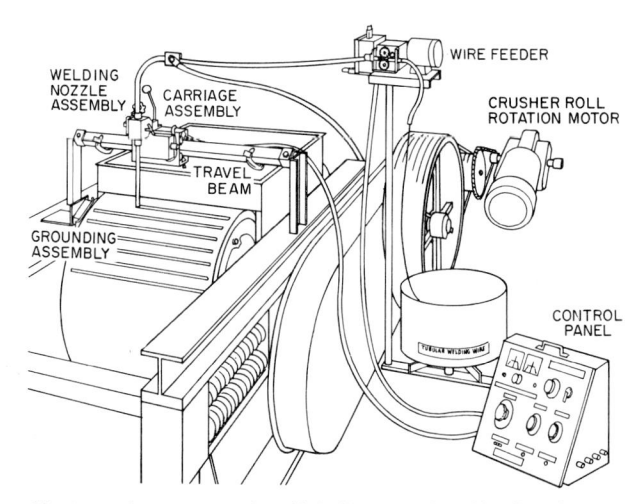

Fig. 9-19 Compensating for roll shell wear with weld rod application.

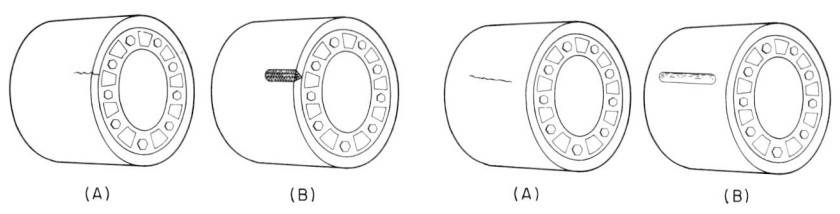

Fig. 9-20 Crack at edge of roll (A) and method of repairing it (B).

Fig. 9-21 Crack in face of roll (A) and method of repairing it (B).

process, a wire drive and power supply is required. When welding with electrodes, all that is needed is the power supply.

When the semiautomatic or manual arc processes are used, care must be taken to prevent localized overheating by concentrating too long in one area. Weld three or four beads, rotate the shell 90°, and repeat the process. Keep the weld area under 600°F.

The automatic process is by far the fastest and least expensive. It is common to deposit 15 to 20 lb/hr without overheating the rolls. Welding is started in the area of the greatest

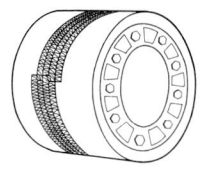

Fig. 9-22 Method of welding bead rows on roll.

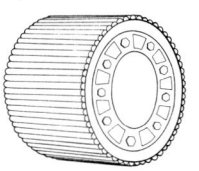

Fig. 9-23 Welding parallel beads all the way around the roll.

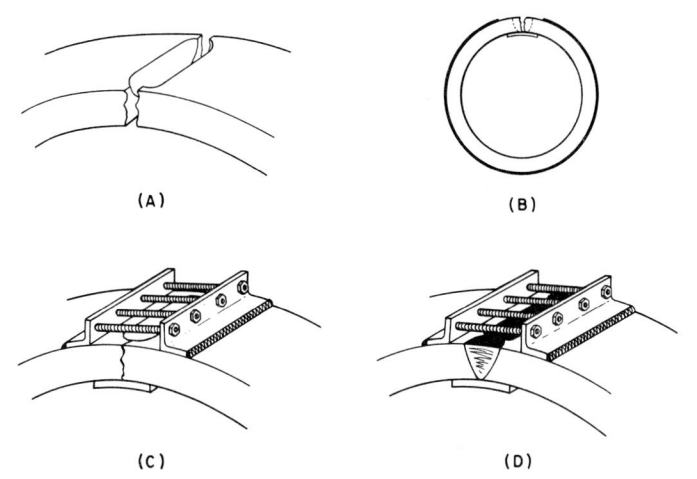

Fig. 9-24 Repairing a cracked roll shell. (A) "V" out length of crack leaving an inch at each end. (B) Use steel (¼ × 4 in.) for weld backup. (C) Weld angle iron on each side, burn holes through each angle, insert bolts (1 in. or larger), pull crack together, snug wedges, and weld from center toward edges. (D) Cut out rest of crack and weld.

wear, usually at the center of the roll. The roll is rotated under a welding gun at a surface speed of 25 to 35 in./min. After the roll has made one complete revolution, the gun moves automatically over the width of the weld bead and deposits a second bead. This process is continued until the roll is brought back to its size (Fig. 9-22). If corrugations are desired, these also can be made automatically. Controls are provided to lock the roll in place and move the gun across the roll and return on the same weld bead. After a double pass has made a complete cycle, the roll automatically starts and rotates a desired amount and the process is repeated to give a series of parallel weld beads all the way around the roll (Fig. 9-23).

New austenitic manganese roll shells should be worn down somewhat before hard surfacing. The working of the rolls helps to stress-relieve the newly heat-treated shells. A day's crushing will usually accomplish this.

Repairing cracked roll shells From time to time, a roll shell will crack and open up. The major causes are defective casting, improper heat treatment, overheating, and improperly aligned wedges. In most cases, the cracked roll shell can be salvaged at much less cost than installing a new shell. (See Fig. 9-24.) The procedure is as follows:

1. Loosen all wedge bolts and wedges. Sometimes the wedges will have to be knocked loose.

2. Using an Arc-Air touch or cutting electrode, remove the hard-facing (if any) over an area 5 to 6 in. on either side of the crack.

3. "V" out along the length of the crack but leave an inch at each end. Clean all slag from cracked area.

4. Use a piece of ¼ × 4 in. steel for a weld backup. Be sure that it does not interfere with the wedges. This backup will help ensure a 100 percent weld.

5. Weld heavy angle iron on each side of the crack. Use Chrome-Mang for welding. Burn a series of holes through both angles. Insert heavy bolts 1 in. or larger, and pull crack together.

6. Once crack is pulled together, snug up wedges, but do not tighten.

7. Start welding using McKay Chrome-Mang or 118. Weld from the center toward the edges. Peening of the weld can be beneficial. Do not overheat. Keep below 600°F.

8. Cut out rest of crack and weld. After the so-called root passes are in, snug wedges again. If desired, the balance of the welding can be done with the semiautomatic using AP-0 or 218-0. Use a short wire extension (not over 1½ in.) and low arc voltage (24 to 28 volts).

After the repair job is completed, finish tightening the wedges and put the rolls back in service as soon as possible. Working the rolls will help to stress-relieve the weld. Tighten the wedges again at the end of the working day.

Typical welding patterns A variety of welding patterns can be achieved, confining buildup to the worn areas, providing more uniform heat distribution, and avoiding excessive buildup in areas not subjected to wear.

Circumferential Pattern. With automatic, fully adjustable step-over, this pattern provides uniform deposit across center areas of greatest wear (Fig. 9-25).

Beaded, or Feeder, Pattern. This pattern, applied over built-up deposits, provides gripping action. The weld automatically stops at the end of each bead before indexing, then starts again at the point of the next bead (Fig. 9-26).

Single-pass Beaded Pattern. Without buildup, the single-pass beaded pattern helps increase gripping action. Automatic indexing is available at one or both sides (Fig. 9-27).

Double-pass Beaded Pattern. This pattern assures extra gripping life. Spacing is adjustable and repeatable for all patterns (Fig. 9-28).

Roll shell welding recommendations

CAUTION: Do not arc weld on any part of the crusher unless the actual part being welded is grounded directly to the welding machine. (Arcing through roller bearings will ruin them.)

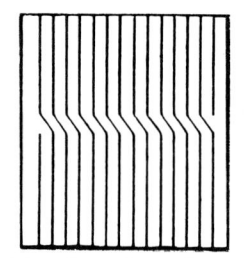

Fig. 9-25 Circumferential pattern.

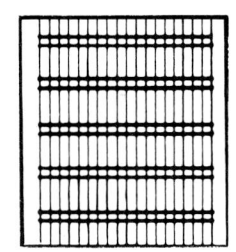

Fig. 9-26 Beaded, or feeder, pattern.

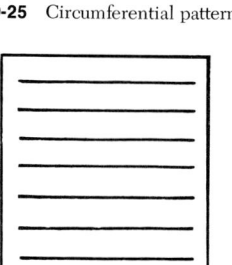

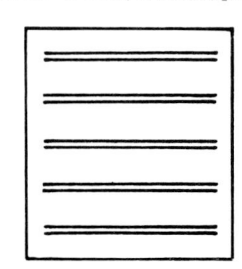

Fig. 9-27 Single-pass beaded pattern.

Fig. 9-28 Double-pass beaded pattern.

1. Maintain original diameter of roll shell and avoid eccentric action of improperly built-up shells by checking distance from surface of shaft to outside diameter of roll shell (see Fig. 9-29). This is important to avoid a large circulating load and excessive movement of the movable roll assembly.

2. Use an accurate method, such as a temperature crayon, to make certain shells are not being overheated during the buildup procedure. Whenever the shell temperature does reach the 600°F limit despite the rotation technique, welding should be stopped so that the shells will air-cool.

The temperature crayon has a melting action at predetermined temperatures. Some of the rated melting temperatures (in °F) available are as follows:

313	363	450	650
325	375	500	700
338	388	550	750
350	400	600	800

Instructions for Use of Temperature Crayon. Stroke workpiece with proper crayon from time to time during heating process. Below its temperature rating, the crayon will leave a dry mark (chalky or charred appearance). When the rated temperature is reached or exceeded, the mark will make a liquid smear. This method can be used for all temperatures listed.

For crusher roll shells which are limited to 600°F, it is satisfactory to make a mark on the shell with the proper crayon before heating begins. When the shell is heated to the rated temperature (600°F), the crayon mark will liquefy.

An important caution is to disregard all color changes. Temperature signal is the appearance of liquid owing to melting of the crayon mark.

Temperature crayons can be secured from most distributors of welding equipment. They are packed in individual tubes. When ordering, give temperature (°F) required and quantity desired.

Fig. 9-29 Measuring roll shells when welding to maintain diametric accuracy.

Roll shell replacement To minimize the time required for roll shell replacement, consider the following requirements and recommendations:

1. Maintain adequate space on each side of the crusher for working and for hoisting the heavy rolls from the frame.

2. Arrange a suitable support for the roll assembly when it is in a vertical position. The full weight of the assembly will be resting on the bearing block at one end of the roll shaft (see Fig. 9-30).

3. Improvise a reliable hoisting apparatus for the shell so that there will be no chance of its being dropped during the removal or installation.

4. Have the required tools for the job on hand. This includes an electric arc or acetyline burning and welding torch, long accurate metal straightedge, adjustable carpenter's square, sledge hammer, box end or open end wrench for tie stud hex nuts, and impact wrench, if available.

Shell Removal Method

NOTE: Each crusher manufacturer will have an exclusive roll assembly design which differs from others in certain details. Most of the commonly used components will have similarities which will make the following instructions applicable. *It is important that each crusher owner follow the roll manufacturer's instructions for roll shell replacement.*

A typical procedure is as follows (see Fig. 9-31):

1. Burn through any tack welds which fasten roll shell to roll spider. If anchor wedges were used to key the shell to the spider, burn through tack welds and pull the anchor wedges.

2. Stand roll in vertical attitude on its bearing block, *with wedge segments at top* (see Fig. 9-30).

3. Remove hex nuts from lower end of each tie stud. Pull all tie studs upward and out of the roll.
4. Exert a slight lifting force on the worn roll shell while bumping it upward with a sledge to loosen the wedge segments.
5. Lift off the wedge segments.
6. Lift off the worn roll shell.

Shell Installation Method
1. Check shell and spider for alignment of anchor wedge grooves (or other keying features which must be correctly arranged at assembly). Mark the alignment on each.
2. Check the tapered mating surfaces of each part to be sure each is free of dirt or metal high spots which could interfere with a uniform wedge fit.
3. Lower the new shell onto the spider, with anchor wedge grooves aligned.
4. Fit the wedge segments into place. Tap with sledge hammer to obtain uniformity all the way around.
5. Lay an accurate metal straightedge across the upper edge of the shell and with an adjustable carpenter's square check from this reference line to the top of each wedge segment. When each measurement is equal, the shell is properly centered.
6. Install the tie studs and hex nuts. Tighten the nuts snug, at first, all the way around. Then finish tightening with impact wrench (if available). Install jam nuts on each end and tighten. Always use new self-locking jam nuts.
7. Lay roll on its side and install the anchor wedges. Tack weld each anchor wedge to prevent loosening.

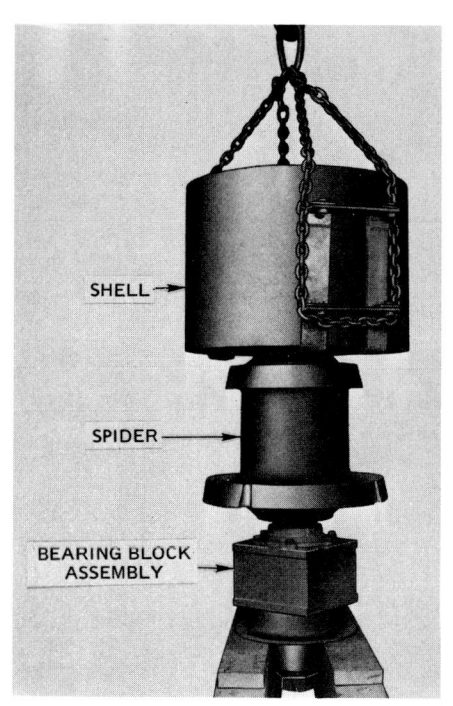

Roll shaft bearing care To provide maximum durability most stationary and movable shaft assemblies are equipped with tapered roller bearings mounted back to back in a housing at each end of the shaft (see Fig. 9-32). Each bearing assembly is protected against entrance of dust, dirt, and moisture by labyrinth-type grease seals in combination with mechanical seals. Sufficient lubricant is permitted to work out through the labyrinth seals to form a dust and moisture barrier.

Bearing temperature check
1. Bearings normally operate at a temperature of 100 to 150°F, periodically running as high as 180°F. Under these conditions and within these limits there is no cause for concern. However, if a bearing maintains a temperature of 212°F or higher, it is considered to be overheated and the cause should be found and corrected to prevent damage.
2. A thermometer provides the best method of checking bearing temperatures. Readings may be taken by placing the thermometer on a convenient spot on the housing.
3. The bearing temperatures can be checked by placing the hand on the bearing cap or block; if the hand can be held there for a few seconds, the bearing is not too warm. WARNING: *Never* attempt to feel bearing temperatures when the crusher is operating.
4. If there is a question whether or not the bearing is too hot, always use a thermometer as the final check. CAUTION: *Never* reduce bearing temperatures by artificial means, such as by pouring water on them. The sudden change in temperature is apt to warp, crack, or crystallize some part of the bearing assemblies.

Fig. 9-30 Roll assembly in vertical position for maintenance.

Causes of bearing overheating

1. The most common cause of overheating is too much lubricant and most generally occurs when the crusher is new or immediately after lubricant has been added. The bearing will heat up and force out excessive lubricant and gradually cool down to the normal operating temperature. Damage can result.
2. The bearing will overheat if the crusher is operating faster than recommended

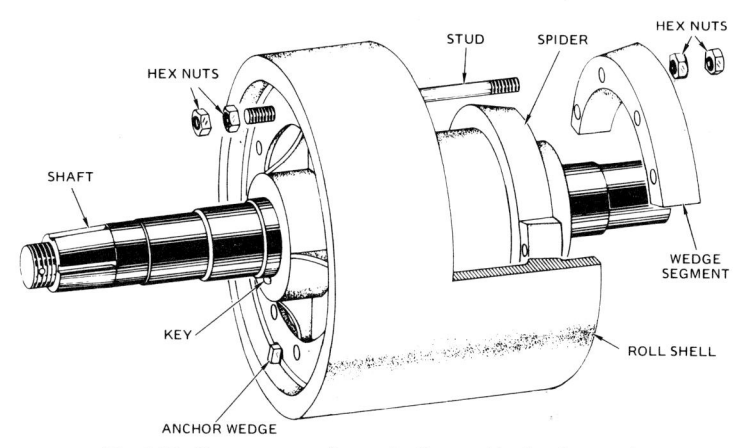

Fig. 9-31 Cutaway view of typical roll assembly (less bearings).

under full load. Check the speed with a speed indicator on the countershaft where the proper speed is indicated. (See crusher instruction manual for speed data.)
3. One of the bearing assemblies may overheat if the unit is operated decidedly out of level. An extremely unlevel position may result in one of the bearing assemblies not receiving its share of lubricant, causing it to get hot. Level unit by checking crusher shaft with a spirit level and compensate for any settlement by the use of cribbing or jacks.
4. If the stud bolts which hold the stationary roll shaft assembly in place on the frame are allowed to work loose, the bearing housing will get out of line and cause premature bearing wear.
5. The tension springs should have the same amount of tension on both springs, with enough tension on both at all times to keep the movable bearing blocks from moving excessively. The same thickness of shims should be installed on each side of the movable roll shaft and bearing assembly, to keep the bearing housings parallel with the slide bars.
6. Arcing through the bearings, when welding up roll shells or any other part of the crusher, will ruin the bearings. *Always ground directly to the part being welded.*

Typical roll shaft bearing removal procedure

1. Remove the six cap screws which attach the inner bearing cap.
2. Move bearing cap and oil seal toward roll spider.
3. Remove outer bearing cap.
4. Install puller screws in the tapped holes of bearing block. Pull block off shaft.

Fig. 9-32 Back-to-back arrangement of tapered roller bearings.

7-222   Maintenance Programs for Equipment Entities

5. Install bearing puller. Pull bearings and thrust collar from shaft. NOTE: Use hydraulic ram capable of high-pressure development.

6. Remove inner bearing cap and oil seal.

Typical roll shaft bearing installation procedures

1. With seal lip toward cap, install oil seal in inner bearing cap.
2. Bolt bearing cap to block.
3. Cool inner bearing cup on dry ice. Oil bore of bearing block. Install by dropping cup into block, *beveled side upward.* Allow to warm up and tighten in block.
4. Oil the roll shaft.
5. Install tapered spreader ring tool to help oil seal pass shoulder of shaft. (Improvise spreader if tool is not available.)
6. Carefully work block and oil seal over shaft shoulder. *Seal lip must not curl under at any point!*
7. Heat thrust collar in 275°F oil and note correct attitude shown in Fig. 9-35 *before* making installation.
8. Install heated thrust collar and hold tight against shaft shoulder until tight on shaft.
9. Heat inner bearing cone in 275°F oil. Install and hold against collar until tight on shaft.

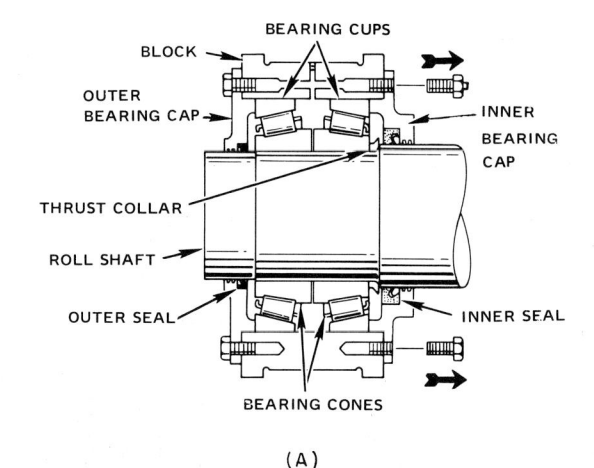

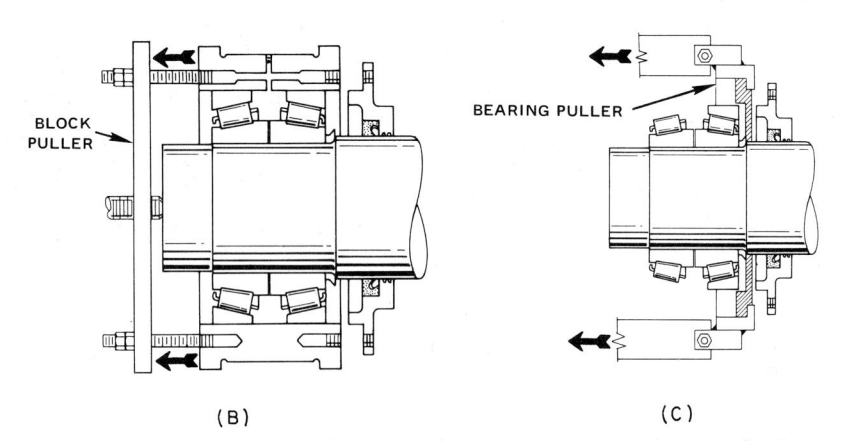

Fig. 9-33 Steps in removing roller shaft bearings. (A) Remove cap screws attaching inner bearing cap; move bearing cap and oil seal toward roll spider; remove outer bearing cap. (B) Install puller screws in tapered holes of bearing block; pull off block shaft. (C) Install bearing puller.

10. Heat outer bearing cone in 275°F oil. Install and hold against inner cone until tight on shaft.
11. Drive wooden wedges between block and roll to remove all bearing clearance.
12. Oil bore of bearing block.
13. Cool outer bearing cup on dry ice. Install with beveled side inward.
14. Hold cup in place until it warms up and is tight in block.
15. Remove wedges. Drive block toward roll with fiber-head mallet to loosen bearings until the block can be turned by hand.

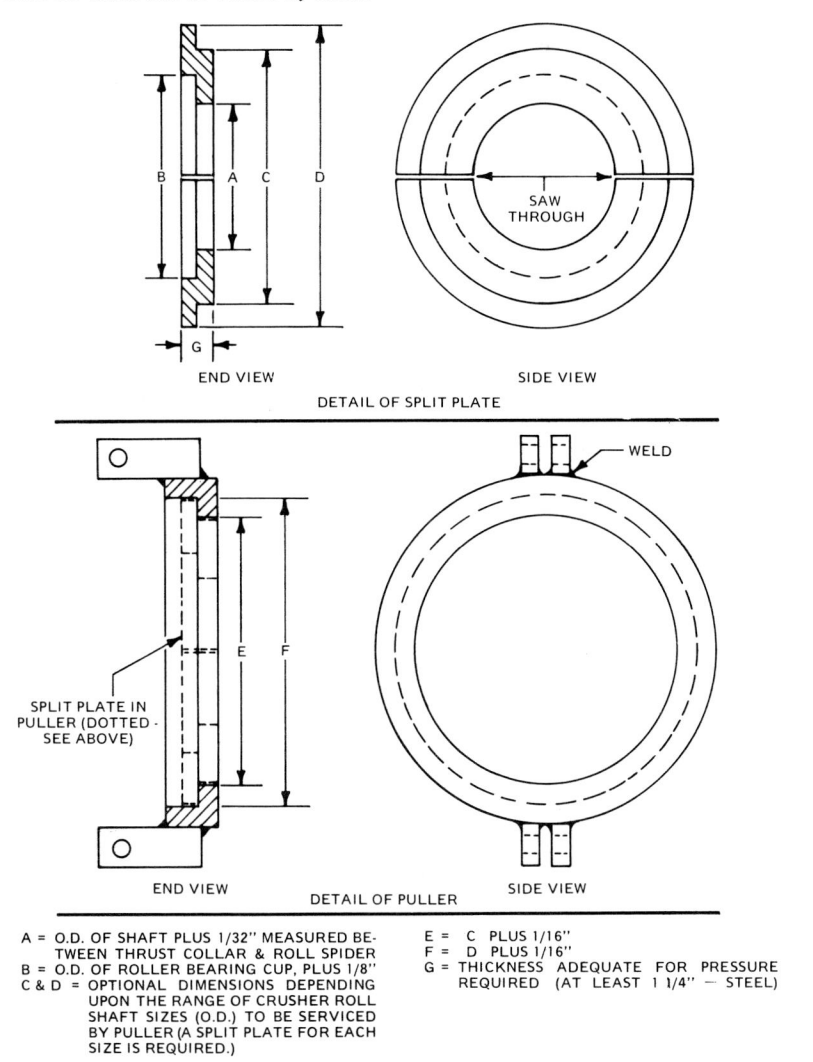

A = O.D. OF SHAFT PLUS 1/32" MEASURED BETWEEN THRUST COLLAR & ROLL SPIDER
B = O.D. OF ROLLER BEARING CUP, PLUS 1/8"
C & D = OPTIONAL DIMENSIONS DEPENDING UPON THE RANGE OF CRUSHER ROLL SHAFT SIZES (O.D.) TO BE SERVICED BY PULLER (A SPLIT PLATE FOR EACH SIZE IS REQUIRED.)
E = C PLUS 1/16"
F = D PLUS 1/16"
G = THICKNESS ADEQUATE FOR PRESSURE REQUIRED (AT LEAST 1 1/4" — STEEL)

Fig. 9-34 Typical puller with variable split plate inserts for removal of shrunk-on bearings and thrust collar from roll shafts and countershaft.

16. Install oil seal in outer bearing cap with seal lip extending downward toward workbench.
17. Install outer bearing cap using only four bolts. Draw bolts snug, but not tight.
18. By trial and error, using various shim thickness, determine clearance between block and outer bearing cap (see Fig. 9-36).

7-224 **Maintenance Programs for Equipment Entities**

19. Remove outer cap and install shim set which is 0.005 to 0.007 in. thicker than the clearance found in step 18.

20. Bolt the outer cover (with shim set) to the block. Tighten all bolts.

21. With 0.005- to 0.007-in. clearance in the bearings, raise heaviest side of block to top. Release block and see if gravity will slowly spin block to move heavy side downward. If it does not, the bearing is too tight and some shim thickness should be added. If block swings very rapidly, the bearing is too loose and some shim thickness should be removed.

22. After bearing clearances have been tested and found to be correct, pump each block full of grease so that excess extrudes past seals.

Heating roller bearings for installation It is important that the heating be done properly, otherwise damage to these vital parts can result. The following instructions should be used when arranging the heating job:

1. Use a nondetergent oil.

2. Provide a clean container which will permit suspension of the bearings on hangers so that they do not rest on the bottom (see Fig. 9-37). *Direct contact with the metal*

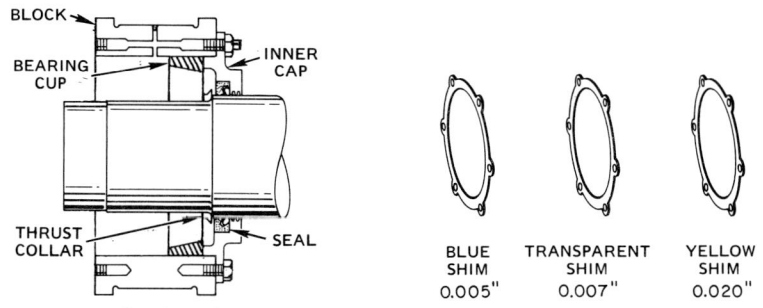

Fig. 9-35 Installing heated thrust collar.

Fig. 9-36 Typical plastic shims.

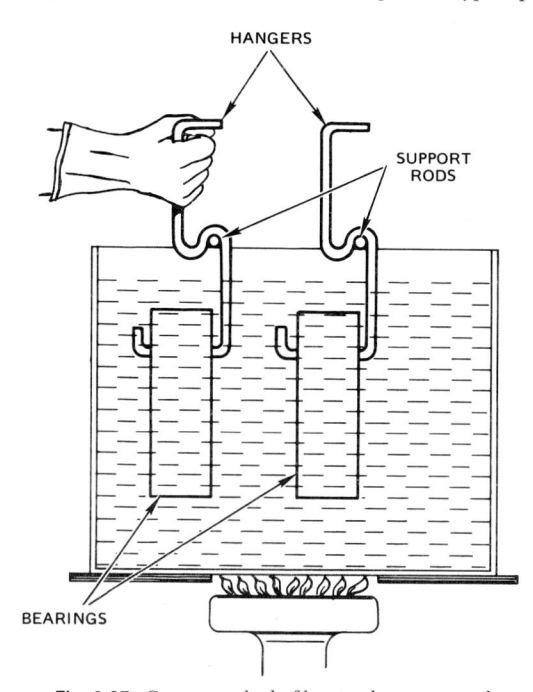

Fig. 9-37 Correct method of heating bearings in oil.

container can raise the bearing temperature to the extent that certain areas are softened and service life is greatly reduced.

3. Heat the oil to 275°F in a safe manner. Use an accurate thermometer to check the temperature. Exercise the proper precautions when working with a container of hot oil.

4. When the oil is at 275°F, immerse the bearings for at least 30 min to be sure they reach that temperature internally. Be sure oil temperature remains at 275°F during this period.

Roll shafts and spiders Most roll shafts are pressed into the spider by the force of a large hydraulic press. Field repair is complex and when attempted should be done only after consultation with the manufacturer's engineering department. Both the shaft and the spider are salvageable items and when one is still in good condition while the other is no longer serviceable, it is often economic to make the repair, either in the field or in the factory.

If field repair is to be attempted, a hydraulic press with the necessary pressure capabilities must be available. The following list of pressure requirements for typical models covers *installation* of the shaft *only!* The pressure required to remove an old shaft will vary, and has been known to reach 400 tons.

Hydraulic Pressure Required for Installation of Typical Roll Shaft
Contact surfaces lubricated

Crusher size	Pressure, tons
1616	25–35
2416	30–40
3018	45–60
3022–3025 (6-in. shaft)	60–70
3025 (7-in. shaft)	80–90
3030–3136	90–110
4022, 4024, 4026	95–130
4130, 4132, 4136, 5530	160–170

The direction of shaft movement in the spider for installation and removal is different for various crusher models. It is vital that the repairman know this detail before placing the assembly in the hydraulic press. Figure 9-38 shows a typical instruction drawing supplied by the factory for a specific crusher roll.

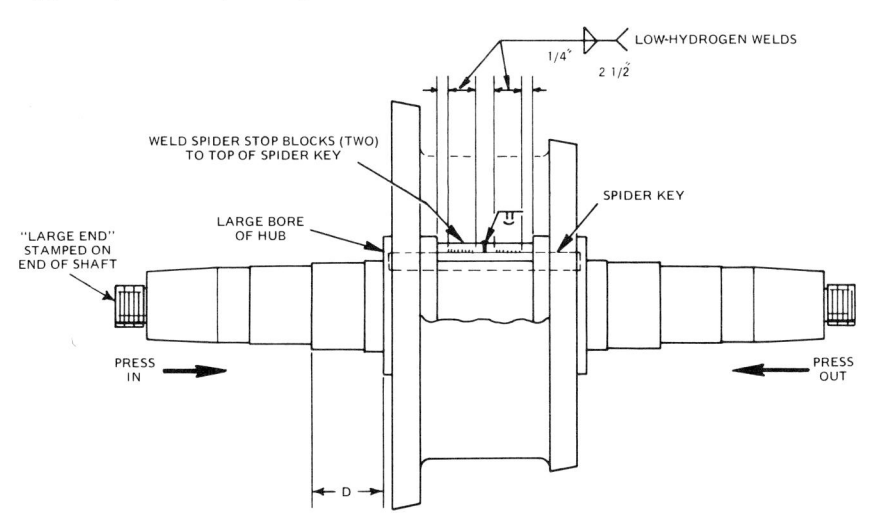

Fig. 9-38 Typical roll shaft and spider assembly.

7-226 Maintenance Programs for Equipment Entities

Each spider is keyed to the shaft to prevent rotational slippage. A bar is sometimes welded on top of the shaft key to prevent lateral shifting of the spider. On some roll types the shaft key extends across the spider and is welded to the hub at the ends. Before any field repair attempt is started, check for welded key or welded blocks on top of the key.

After the shaft has been pressed out, it is necessary to remachine the spider for an oversize shaft or to build it up and remachine it to the original bore. For proper bore size and shaft diameter information contact the manufacturer. The location of the spider on the shaft is also an important detail which will be supplied.

Bearing block slide bars The upper and lower slide bars (four) which contact the movable roll bearing blocks serve as guides and wear strips for the bearing blocks. After a long period of crusher operation, particularly with coil-type compression springs, the slide

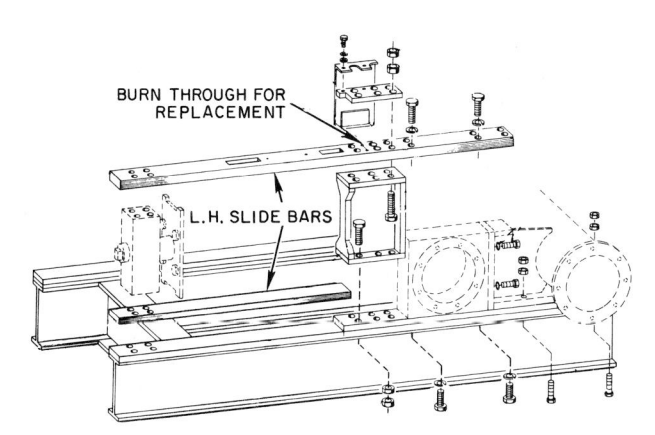

Fig. 9-39 Crusher slide bar arrangement.

bars wear away to the extent that the bearing blocks are no longer tightly held. During operation a noticeable amount of vertical movement and tipping of the blocks occurs. It is recommended that worn slide bars be replaced.

The upper slide bars on most crushers (Fig. 9-39) can be unbolted and have a relieved area where the separation should be made. When the new sections are in place and accurately aligned, they should be welded to the opposite end portion of the bar to again make a one-piece unit.

The lower slide bars are accessible when the spring assemblies, backup blocks, and movable roll are removed from the frame. Again it is necessary to burn off all skip welds which attach the worn bar to the frame so that the end section can be replaced. Align the new bars carefully before skip-welding them to the frame. If any shims were used with the original bars, be sure to use them again as required.

Spur gear and pinion gear removal procedure
1. Remove retaining nut and washer from spur gear shaft.
2. Use a puller to remove the key.
3. IMPORTANT: Screw retaining nut onto shaft to stop travel of gear when it breaks loose. Install puller. Strike puller screw as it is tightened to help jar gear loose (Fig. 9-40). NOTE: Use a torch to heat the rim of the gear if it will not loosen from puller pressure alone. The entire gear expands as the heat migrates toward the hub. Do *not* apply the torch at the hub area!

A hydraulic jack can also be used with the puller instead of the screw arrangement. *Always have the retaining nut on the shaft to keep the loosened gear from coming off the shaft suddenly.*

4. Use an adequate hoist to handle the loose gear.
5. Remove retaining nut and washer from pinion gear shaft.
6. Use a puller to remove the key.
7. IMPORTANT: Screw retaining nut back onto shaft to stop the travel of the loosened

pinion. Install puller, as shown. Strike puller screw to help loosen pinion as screw is tightened (Fig. 9-41).

Spur gear and pinion gear installation procedure
1. Clean shafts and gears. Remove any burrs or nicks. Install gears but omit gear case back plate.
2. Install washers and retaining nuts.
3. Tighten retaining nuts so that gears are fully seated on tapered shafts.
4. Use feeler gauges to check for clearance and backlash. IMPORTANT: Turn spur gear through 360° to be sure the point of minimum clearance is checked.

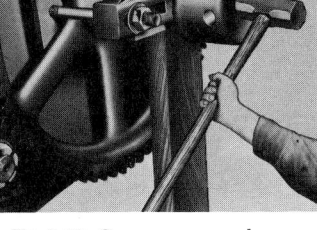

Fig. 9-40 Spur gear removal.

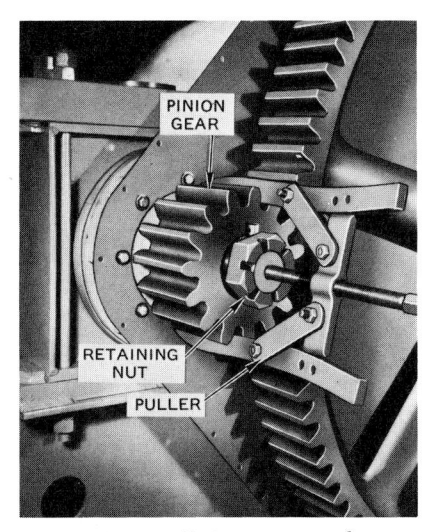

Fig. 9-41 Pinion gear removal.

5. Clearance is the distance between the tip of pinion tooth and the base of spur gear teeth. Backlash is the distance remaining between pinion and spur gear teeth when there is tight contact on one side (Fig. 9-42).

Each manufacturer will specify the clearance and backlash dimensions for his roll crusher. (Contact the factory for this information.) If the proper clearance is not within 0.005 in. of recommended clearance, it will be necessary to turn the bearing adapters. Adapters should be set at matching attitudes.

At the O stamping on the bearing adapter and countershaft housing it is possible to increase or decrease the distance X between spur and pinion gear centers. The first hole on either side of position O on the bearing adapter will increase or decrease the distance between spur and pinion gear from 0.0077 to 0.0115 in., depending on the size of the crusher. The second hole from position O will increase or decrease the distance between 0.014 and 0.021 in., depending on crusher size. The third and fourth position from O will further increase or decrease distance to obtain desired gear clearances.

If the factory found it necessary to move the adapters from the O–O midpoint adjustment to obtain the correct clearance and backlash, an X mark will be stamped on the countershaft housing in line with

Fig. 9-42 Backlash between spur and pinion gears.

the O on the adapters. NOTE: If an adjustment is made in the field to compensate for gear wear after a long period of crusher operation, the housing should be stamped with another code letter such as A opposite the O on the adapter.

If for some reason, such as worn gears, both dimensions require adjustment, set the clearance dimension correctly and let the backlash be the incorrect value. When the backlash exceeds the recommended dimension, the crusher may be noisy, but no harm is

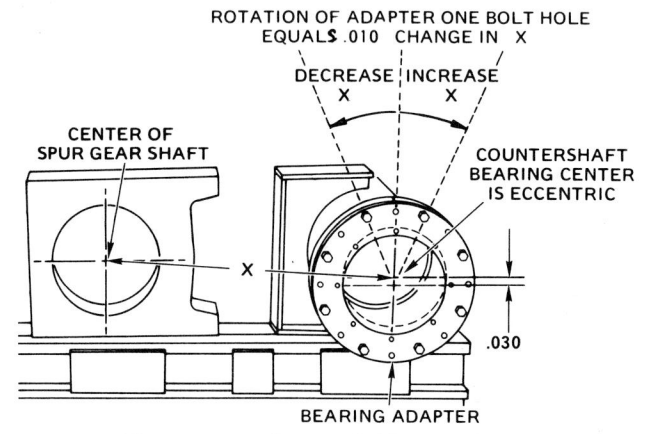

Fig. 9-43 Typical countershaft bearing adapter.

done to the gears and production is not affected. NOTE: At no point should the gears bottom in each other. Check by turning through 360° of rotation.

6. After any type of adapter adjustment and after a bearing, gear, or shaft replacement, recheck the alignment of the two outer faces. They must be parallel. If faces are not parallel a poor pattern of gear wear will be visible on the teeth owing to partial contact instead of full contact all the way across the spur gear. (The pinion gear is wider than the spur gear; therefore, the face surfaces cannot be in line. Use the straightedge measure method given in step 7.)

7. Hold an accurate metal straightedge against the pinion face. Distances from the straightedge to two points diametrically opposite each other on the spur gear face should be equal. When clearance, backlash, and alignment are all correct, tighten all bolts to anchor countershaft housing. Weld shear blocks to frame if change of location was made.

8. Remove retaining nuts.
9. Remove both gears.
10. Install gear guard back cover.
11. Reinstall gears.
12. Install keys. (See Figs. 9-49 and 9-50 and text discussion for fitting keys toward end of this section.)
13. The spur gear is seated on the roll shaft by a special hydraulic ram at the factory. For field installation, heating the spur gear to 200 to 250°F before installation will produce a similar degree of final tightness. Apply heat at the base of the teeth, all the way around.
14. The pinion gear is seated on the countershaft by a special hydraulic ram at the factory. For field installation, heating the pinion gear to 200 to 250°F before installation will produce a similar degree of final tightness.
15. Install washers and retaining nuts. Tighten retaining nut with wrench and sledge. Be sure lock bolt hole is clear when tightening limit is reached. *Never back off nut* so that lock pin can be inserted.
16. Install gear case cover. Fill case to correct level with correct lubricant.
17. After the first day's operation, remove the covers which permit inspection of the spur gear and pinion retaining nuts. With a small hammer, strike the washers on the shafts to see if there is any sign of looseness. If so, remove lock bolt and tighten hex nut.

Spur gear case The spur gear case serves as an oil reservoir for lubrication of the gears. The oil breather cap at the top has two notch-type passages for free movement of air

(Fig. 9-44). These passages must be kept open at all times so that internal case pressure does not force lubricant past the seals and out of the case.

Two special access ports are provided so that roll shell wedge segments can be tightened after a new shell has been in operation for a short time. A bar can be passed through the openings to contact the wedge segments when they are at the top of the roll. By alternately driving on the bar and tightening the stud bolt hex nuts the segments are tightened against the shell and spider. (See Fig. 9-45.)

Inspection ports for the pinion and spur gear hubs are provided so that any looseness of these parts can be detected. A check of the tightness should be made after the first few hours of new crusher operation and after overhaul which requires gear removal.

All gaskets used for oil sealing should be in good condition at the time of installation to prevent serious oil loss.

Removal of countershaft bearings which are shrunk onto shaft Bearings and thrust collars which have been heated and shrunk onto a countershaft can best be removed by using a mechanical puller assembly with variable split plate inserts. (This type of puller is illustrated in Fig. 9-34 and 9-46.) Split plates can be fabricated to fit each of the various countershaft diameters.

Figure 9-46 illustrates the use of a puller in the removal of a countershaft thrust collar and bearing.

Assembly and installation of countershaft with shrunk-on bearings

1. Check bearing and seal contact areas at both ends of shaft for smooth OD. File smooth if required.
2. Heat both thrust collars in 275°F oil. (Note correct attitude on shaft as shown in Fig. 9-47.)
3. Install heated thrust collars tight against shaft shoulders and hold until fixed in place.
4. Heat both roller bearings in 275°F oil. Install each one tight against the thrust collar and hold until fixed in place. (See Fig. 9-37 showing correct method of heating bearings in oil to prevent damage to bearings.)
5. Oil both bearing adapters.
6. Clean entire housing interior thoroughly. Install each adapter with index mark O aligned with O on housing. Use only two cap screws in each.
7. Carefully work one end of shaft and bearing assembly through one of the adapters.
8. Shift support to end of shaft and push countershaft through housing.
9. Support both ends of shaft. Carefully work both bearings through the adapters. Position bearings identically in housings.

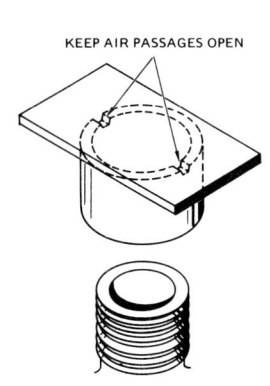

Fig. 9-44 Air breather for gear case.

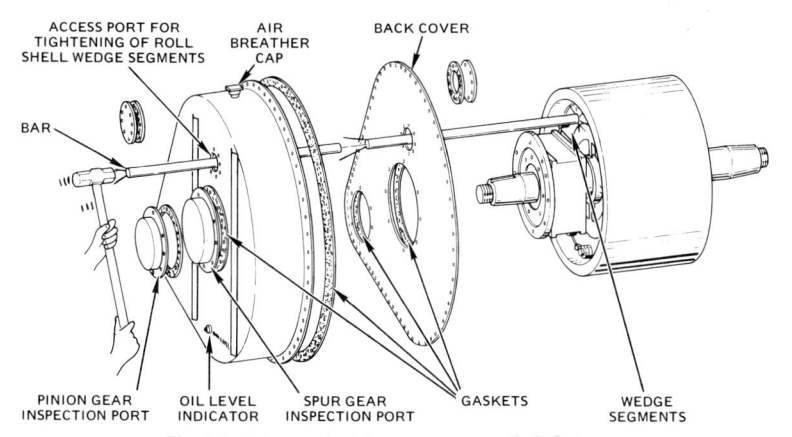

Fig. 9-45 Spur and pinion gear case, exploded view.

10. Install wear sleeves on each end of shaft at the points where the oil seals make contact. These sleeves prevent damage to the shaft from the friction and abrasion of the seal.

11. Install bearing cap at each end of housing using only two cap screws in each. After pinion and spur gears have been adjusted for clearance and backlash by turning the adapters (see step 5 in earlier Spur Gear and Pinion Gear Installation Procedure section), install all cap screws in adapters and bearing caps.

12. Check to be sure all magnetic plugs are in tight, then fill housing to correct level with the proper lubricant. Grease the dust barrier passages if the crusher is so equipped.

Removal of countershaft bearings (four-piece clamped-on type)

1. Drain lubricant from countershaft housing and spur gear case.
2. Remove spur gear case cover.
3. Remove flywheel, spur gear, and pinion gear.
4. Remove gear case back plate.
5. IMPORTANT: Accurately measure and record the amount of shaft projection at each end (Fig. 9-48).
6. Remove cover plates. The rubber slingers will come off with them.
7. Bend prong of lock washer on *pinion gear side* so that locknut can be unscrewed. Unscrew locknut and remove lock washer.
8. Screw locknut back onto bearing sleeve but stop when nut is $1/8$ in. away from bearing. Using a brass drift pin which is wide enough to contact the full thickness of the locknut, drive inward at various points on the locknut face until the tapered sleeve breaks loose from the shaft and slides inward.
9. Again remove the locknut.
10. Support the gear end of the shaft to keep it centered in the housing. Carefully push or lightly bump the shaft to force it and the attached bearing assembly out the opposite end of the housing. Be sure to support the flywheel end as the bearing emerges from the adapter.
11. With the countershaft on a bench, repeat steps 7 through 9 to remove the bearing assembly from the flywheel end.
12. Remove adapter from the housing. Remove spacer and bearing from adapter.

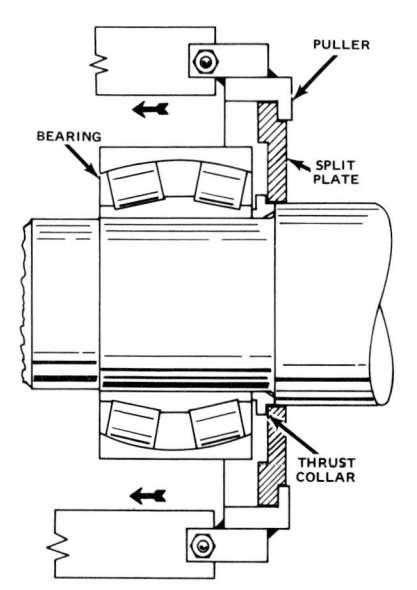

Fig. 9-46 Cross-sectional view of bearing puller arrangement.

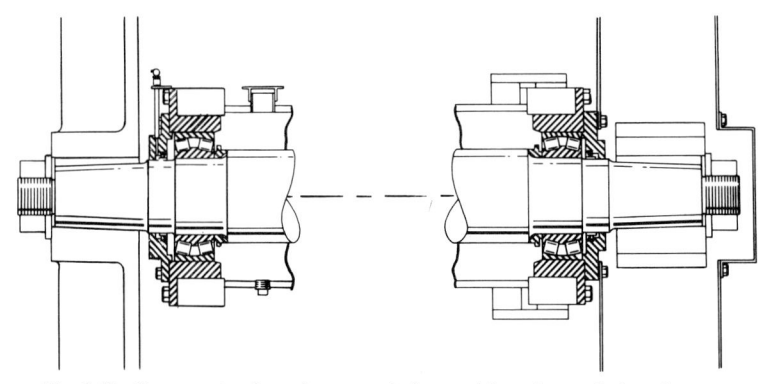

Fig. 9-47 Cross section through countershaft assembly with straight-bore bearings.

Installation of countershaft with (four-piece clamped-on bearings)

1. Clean countershaft housing very thoroughly so that no dirt particles remain inside.
2. Check the bores of bearing adapters. They must not be more than 0.003 in. oversize at the ID (inside diameter), otherwise new bearings will be damaged by the hammering force set up in the assembly during operation. Bearings must fit tight in the adapters. *Replace adapters with oversize bore!*
3. Install adapter with flanged bore at pinion gear end of housing. Install adapter with plain bore at flywheel end. IMPORTANT: Set the stamped match mark on each adapter so that it aligns with the same mark stamped on the housing. Use only two cap screws in each.
4. Insert countershaft. Be sure to have flywheel end at flywheel side.
5. Assemble the four-piece bearings, but do not tighten locknuts. Install bearing on countershaft *at pinion gear end only.* Block up the opposite end to keep the shaft centered in the housing. Using a brass drift pin, carefully drive the outer race of the bearing all the way inward until it bottoms on the flange of the adapter.
6. Install the second bearing. Carefully measure from the shoulder of the shaft to the face of the adapter to obtain the previous amount of shaft extension (dimension A, Fig. 9-48, as recorded in step 5 of the shaft removal procedure). If this dimension is not available,

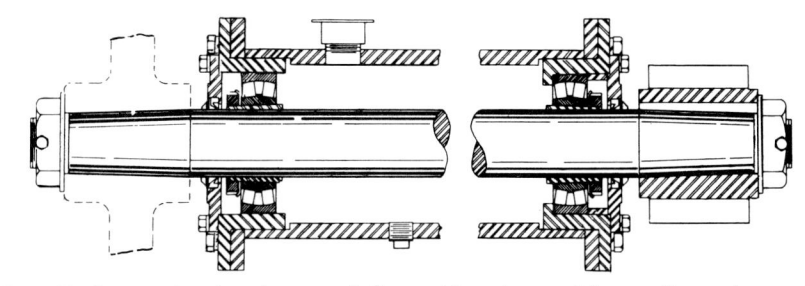

Fig. 9-48 Cross section through countershaft assembly with tapered bore, pull-type sleeve, roller bearings.

it will be necessary to temporarily install the spur gear and pinion so that the countershaft can be moved in the bearings to place the spur gear teeth in the center of the pinion gear. (Pinion is wider.)

7. When the shaft is located laterally, check to be sure the flywheel end bearing is located as shown in Fig. 9-48, with the locknut $1/4$ in. away from a straightedge laid across the face of the adapter. Tighten bearing locknut on pinion side first.

With *tightening wrench* (available as accessory tool) or with brass driving bar, tighten locknut in a clockwise direction and check roller clearance with feeler gauge.

The mounted, operating radial clearance between the outer race of the bearing and the rollers, as checked at the top of the housing, should comply with specific factory recommendations for the crusher. NOTE: The minimums recommended may not be obtainable. Tighten until no further takeup is noted. *Too much tightening* will stretch threads on sleeve or nut. The bearing assembly must be tight enough to keep sleeve from rotating on shaft. The clearances listed are based on *half the unmounted bearing clearance.*

8. When nut is tight, bend one prong of the lock washer to engage nut slot.
9. Tap spacer inward to seat flywheel end bearing against flange of adapter. Tighten locknut and check roller clearance as in step 7. Bend lock washer prong to engage nut slot.
10. Install new oil seals in cover plates. Install cover plates.
11. Install rubber slingers.
12. When gears are installed adjust eccentric housings to obtain correct clearance and backlash. (See step 5 in earlier Spur Gear and Pinion Gear Installation Procedure section for details.) Install all cap screws in adapters and tighten.
13. Fill housing to proper level with the correct lubricant. Be sure magnetic drain plugs are screwed in tight.

Fitting and installing straight keys Straight keys (Fig. 9-49) are intended to fit tightly at the sides of the two keyways with clearance allowed at either top or bottom. They prevent any independent rotary movement of shaft or gear but do not affect any tightness between the two parts. The tightness between parts may be obtained through the use of a tapered shaft OD and tapered bore of the gear.

To fit a straight key to both keyways, lay it side down on the flat cutting surface. Remove material slowly until the key can be driven separately into each keyway. The fit must be tight in both, not tight in one while only snug in the other. (Check keyways if this condition exists.)

Fitting and installing tapered keys Tapered keys (Fig. 9-50) are intended to affect a tightness between a shaft and gear and to prevent any independent rotary movement of the parts. They are intended to fit tight at the sides of both keyways and to fit snug over as much of the top and bottom surfaces of the key when it is driven in to a set limit A. The set limit is equal to the width C of the key.

To fit a tapered key, lay it side down on the flat cutting surface. Remove material slowly until the key fits snugly into each keyway. The fit must be snug in both, not snug in one keyway while somewhat loose in the other (check keyway if this condition exists).

When a correct fit is obtained, install the gear, drive the key in tight and measure dimension A. Remove key and lay it bottom side down on the cutting surface. Remove enough material to reduce dimension A to match dimension C. Remove material slowly and drive the key in tight each time the A dimension is checked. When key is tight at B there should be little or no clearance at the opposite end, indicating that the wedge is tight across a large portion of the keyway.

Finger gear drive The finger-gear-type drive (Fig. 9-51) for roll crushers provides a reliable, positive synchronization of roll movement. With this drive it is possible to use two beaded or two corrugated roll shells in a close setting for fine crushing, with no possibility of contact between rolls due to slippage.

The gear teeth are designed with sufficient length to maintain complete engagement at

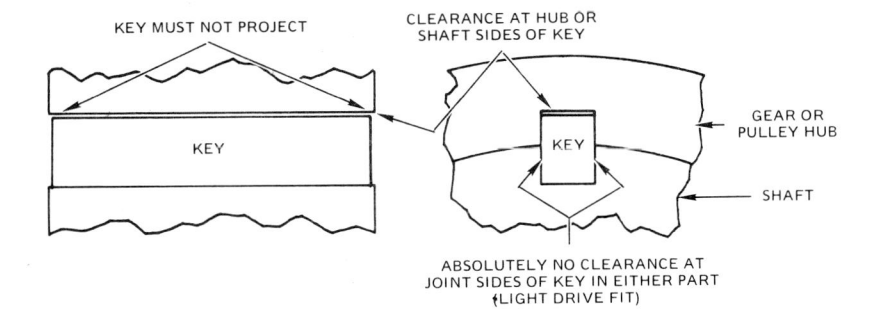

Fig. 9-49 Method of installing straight keys.

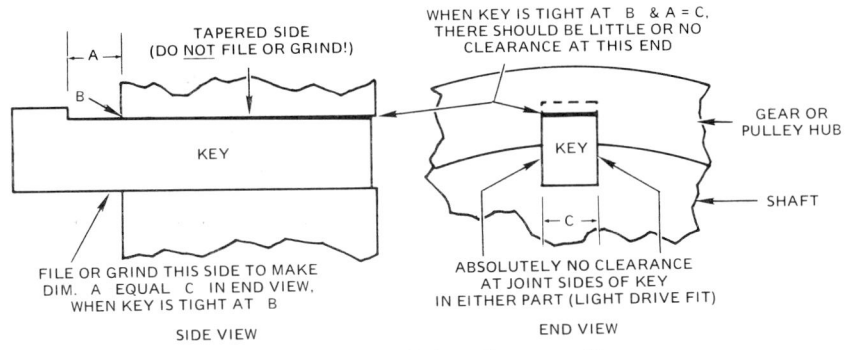

Fig. 9-50 Method of installing tapered key.

the widest roll shaft separation for coarse crushing. The gear teeth are continuously bathed in oil and are maintenance free.

Removal of finger gears

1. Drain lubricant from finger gear case, preferably when warm.
2. Remove cap screws, lock washers, and locknuts which fasten the outer section of the gear case to the back plate of the gear case. NOTE: Use care when removing outer part of gear case so as not to destroy the gaskets. It is advisable to have available a new set of gaskets in case gaskets are destroyed during disassembly.
3. Remove the lock bolt from the retaining nut.
4. Loosen and remove retaining nut and washer.
5. Pull the keys.
6. IMPORTANT: Replace shaft nut to hold the finger gear from slipping off the tapered shaft when puller breaks it loose suddenly.

Fig. 9-51 Finger gear drive.

7. Pull finger gear with a standard roll crusher gear puller, or similar puller.
8. Assemble puller on finger gear by engaging the stud bolts into the two holes located in the finger gears. Two square nuts are welded into the back of the finger gear for puller stud bolts.

Install puller onto the stud bolts, and line up threaded center bolt in puller so the point of the bolt pilots into the center of the shaft.

9. Place socket wrench on puller hex nut. As one man drives on the end of the socket with a sledge, another man takes up the slack in the nut by extending a bar through the socket wrench and exerting downward force. The gears are usually seated very tightly on the shaft, and considerable force is required to break them loose. In some cases it may be necessary to heat the gears with acetylene torch. If this becomes necessary, apply heat evenly around the gear *just below the base of the teeth on gears.* The heat will expand this area of metal and, in turn, expand the bore. Never apply heat to the hub.

Finger gear installation procedure Install gear. Align keyways and drive keys in flush with gear surface. Install washer. Install retaining nut and tighten. Apply heat by directing torch flame to area *at base of gear teeth all the way around the gear.* At the same time, tighten retaining nut and sledge gear hub, or use the special socket wrench to help seat gear.

When nut is at maximum tightness, install lock bolt and self-locking hex nut. IMPORTANT! Never back off nut to expose lock bolt hole. Continue tightening to next position.

When both gears are installed and gear case cover is in place, be sure to fill case to proper level with the correct lubricant before operation.

After the first day's operation at full load, remove inspection covers and check for looseness of the washer. A loose washer will produce a distinctive sound when struck with a hammer. Tighten retaining nuts if necessary.

Finger gear case The typical finger gear case (Fig. 9-52) is equipped with a special slide plate in the back cover which permits the movable roll to travel through its full range without any oil leakage.

Both the case and the back cover are made in two detachable sections so that either of the rolls can be removed for repair without disturbing the case section for the opposite roll. Separation requires the unbolting of the sections from each other.

Inspection ports for both gear hubs are provided so that any looseness of gears can be detected. *A check of gear tightness should be made after the first few hours of new crusher operation and after overhaul which requires gear removal.*

All gaskets used for oil sealing should be in good condition at the time of installation to prevent serious oil loss.

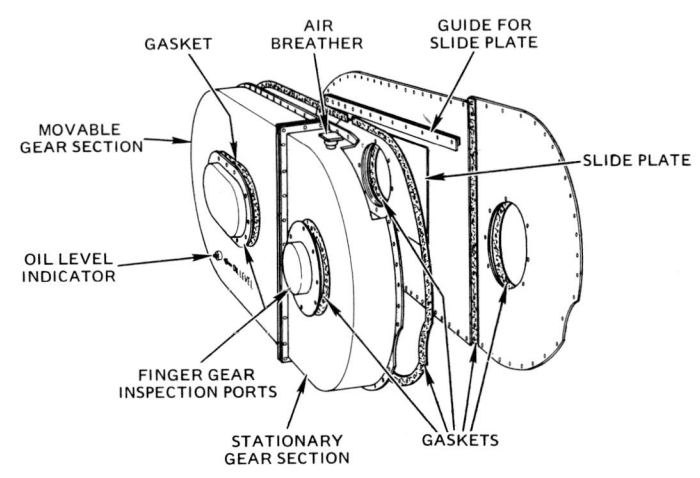

Fig. 9-52 Finger gear case, exploded view.

Records of crusher maintenance Many contractors find it advantageous to do the following kind of record keeping to bid jobs more competitively.

1. Keep a record of output tonnages for each crusher. This is useful in computing the service life of the wearing parts such as rolls, jaws, liners, impeller bars, hammer tips. If different types of material are run, record the hardness characteristics along with the output data.

2. Keep a record of time and expenses involved in hard-facing the wearing parts of a crusher. When these data are compared to output information, production costs per ton are more predictable and maintenance planning is more efficient.

3. Keep a record of part replacements in the noncontact areas such as bearings, shafts, seals, toggle plates. When these data are compared to output information, costs per ton are more accurate and preventative maintenance is more realistic.

JAW CRUSHER MAINTENANCE

The following maintenance data will deal with the conventional single-jaw crusher (Fig. 9-53) and with the conventional twin-jaw crusher (Fig. 9-54).

Lubrication Correct lubrication of a roll crusher is a vital maintenance requirement which must never be neglected. Figure 9-55 shows a typical lubrication instruction sheet for single-jaw crushers. A copy of the appropriate lubrication sheet should be kept posted in the lubricant supply shed for ready reference at all times.

Check crusher mounting Keep crusher shafts horizontal so that bearing loading is uniform and lubricant distribution is proper. Periodically place a spirit level vertically against the machined surface of a flywheel (Fig. 9-56). Correct an unlevel condition immediately by shimming the base.

Lifting the crusher Whenever it is necessary to lift the crusher base, attach hoist chains and cables at factory specified lifting points only (see Fig. 9-57)! Never wrap chain or cable around pitman shafts to lift base.

Lifting the pitman assembly The pitman assembly with bearings, and with or without the movable jaw and the flywheels, can be lifted from the crusher by attaching hoist chains to the pitman shaft.

Keep stationary jaw tight If a crusher is left to operate with a loose stationary jaw, the jaw will wear excessively and the capacity of the crusher is reduced.

On most crushers the stationary jaw is wedged against the base by tapered key plates. After each 100 hr of operation, the key plate retaining bolts should be loosened and the

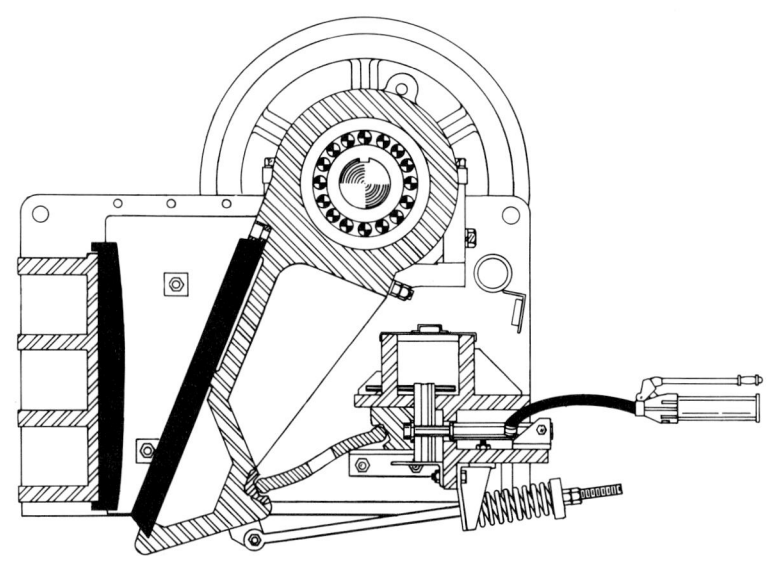

Fig. 9-53 Conventional single-jaw crusher.

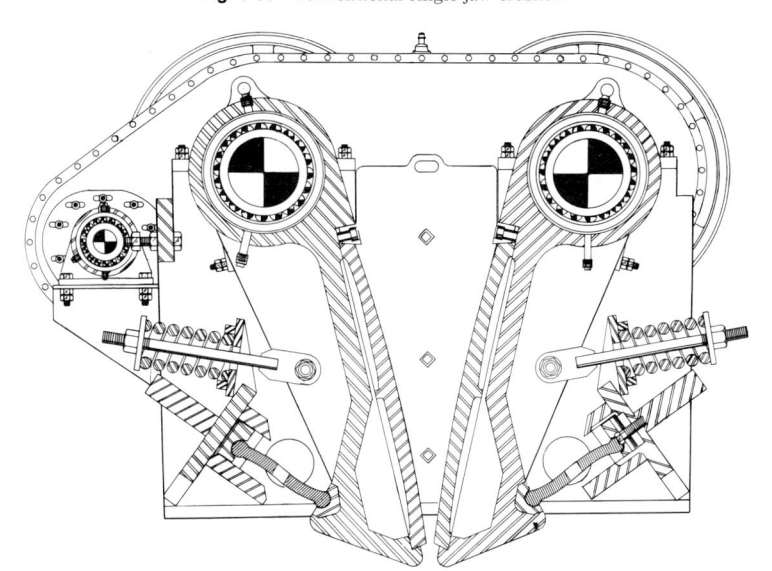

Fig. 9-54 Conventional twin-jaw crusher.

CROSS SECTION   BACK VIEW

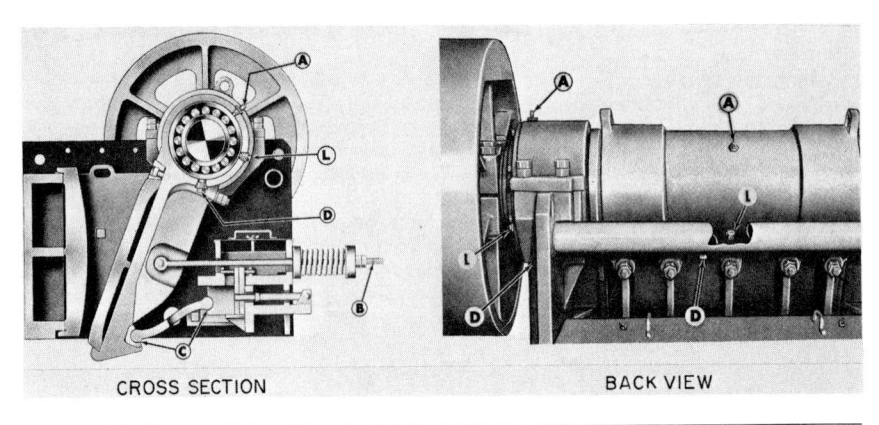

Symbol	Part	Instructions	Lubricant recommended
A	Pitman bearings	*Every 48 hr:* Check seals. Whenever grease gun fails to show at seals lubricate with pressure gun injecting 9 oz of grease in each side bearing and 18 oz in pitman. Grease must show extrusion from labyrinth seal to make effective dust seal.	Above 90°F: Marfak 1 32–90°F: Marfak 0 Below 32°F: Marfak 00
B	*Toggle plate* *Tension rod threads* *Tension rod pin* *Adj. rod threads, if used* *Adj. sprockets, if used* *Adj. chain, if used*	*Every 48 hr:* Lubricate with oilcan.	Regal Oil E (R&O)
C	*Adjusting mechanism*	Lubricate at assembly	Marfak 0
D	*Drain plug*	*Every 1000 hr or seasonally:* Drain, flush, and refill with heated lubricant.	Flush with SAE 10 oil; refill with proper grade of lubricant
L	*Level plugs (if used)*	Level plugs are provided in pitman and side bearings on most of the present sizes to visually check lubricant level.	

Note: After lubrication replace protective caps on side and pitman bearing grease fittings.

Crusher size	Bearing grease capacities, pt or lb	
	Each side bearing	Pitman bearing
1016, 1020, 1024	3	8
1036, 1236, 1524, 1824	5	10
1536, 1836, 2236	8	15
2225	8	12
1242	8	20
2540	10	20
2436	12	20
1648	10	25
3040, 3242, 2248	12	25
4248	16	16 (each bearing)

20% additional after complete overhaul.

Fig. 9-55 Lubrication instructions for jaw crusher.

key plates driven downward by sledging so that tightness of the stationary jaw is assured. The key plate retaining bolts should be kept tight at all times.

Turn stationary jaw for uniform wear The stationary jaw should be turned end for end periodically to obtain the maximum service life from this vital part. The time interval between turnings will vary according to the hardness and abrasive quality of the material being crushed.

SPIRIT LEVEL

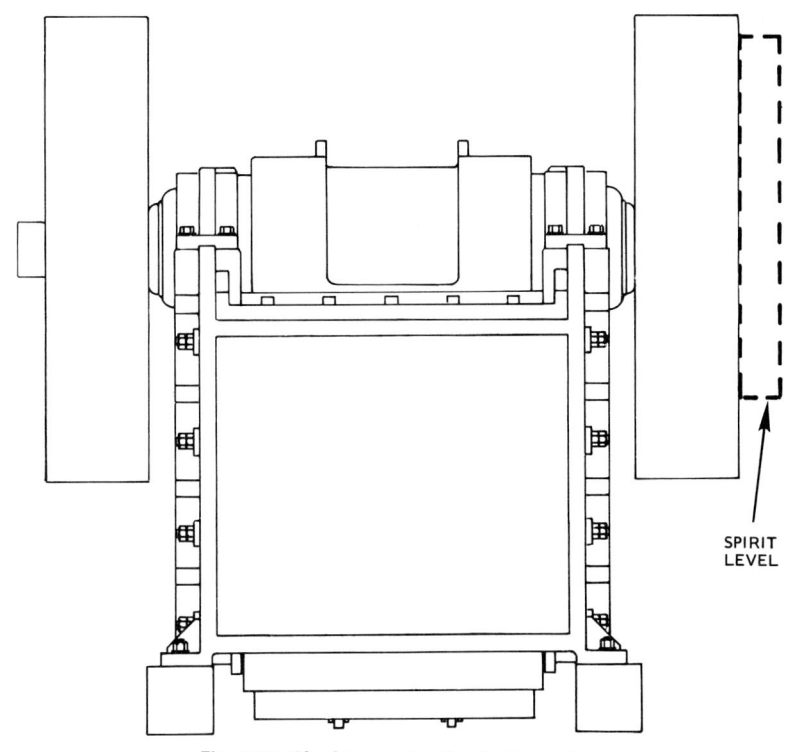

Fig. 9-56 Checking crusher flywheel mountings.

The stationary jaw will usually wear faster and require more frequent turning to balance the jaw than the movable jaw requires.

Each time the jaw is turned, the mating surfaces of jaw and base should be absolutely free of material particles which could hold the jaw off its true seating surface and produce a false tightness. After a brief period of operation, the jaw becomes loose when the material particles are forced out.

Replace worn-out stationary and movable jaws The crushing efficiency of jaws is usually related to the depth of their tooth, groove, or corrugation. As wear occurs, the high points diminish and material fracturing efficiency is reduced (Fig. 9-58). Material between the jaws is lifted by the compression force instead of being crushed. Eventually material is compacted instead of lifted, and extreme force is exerted upon bearings, shaft, toggle plate, and base. Failure of these parts can be prevented by replacement of the worn jaws, before their crushing efficiency reaches the 50-percent-of-new stage. The stationary jaw absorbs more wear than the movable jaw and will usually require more frequent turning and replacement.

Replace worn key plates The key plates which form the sides of the crushing chamber and hold the stationary jaw tightly in place are subjected to the abrasive action of material and will be worn away. Replace key plates as soon as there is an indication that a hole is developing. Never allow material contact with the side plates of the base through a hole in the key plates!

The lower key plates will wear faster than the upper key plates.

Maintenance Programs for Equipment Entities

Keep movable jaw wedge tight The wedge which clamps the movable jaw against the pitman must be kept tight at all times to prevent any loosening of the movable jaw. A loose jaw will wear more rapidly than normal and reduce the capacity of the crusher.

On most crushers the movable jaw wedge is retained by bolts which pass through the pitman. To tighten the wedge, tighten the bolt hex nut as the wedge is being driven in tighter with a sledge. Strike the wedge close to the upper end of each bolt as its nut is tightened. Do the tightening by increments in several passes across the length of the wedge to obtain a uniformity of seating.

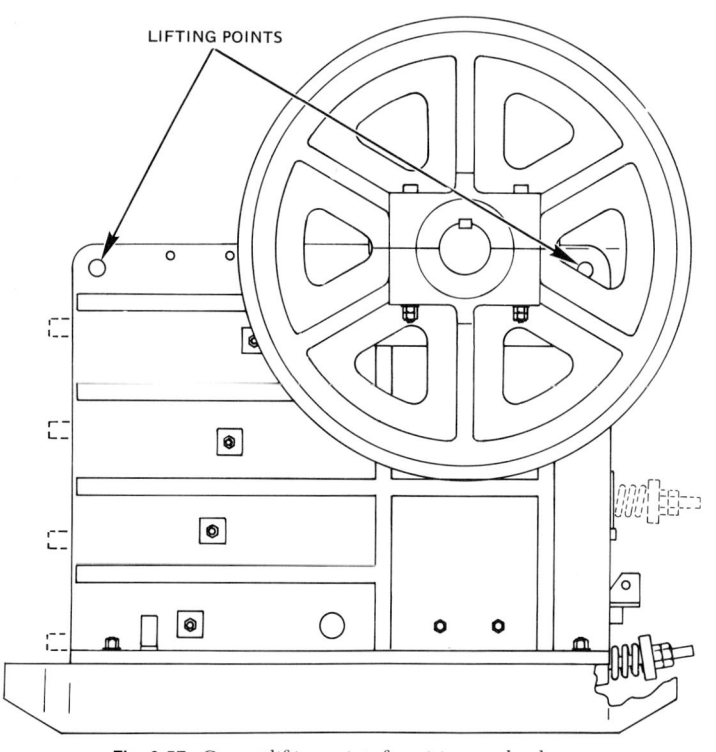

Fig. 9-57 Correct lifting points for raising crusher base.

Fig. 9-58 Example of progressive jaw wear.

If any gap remains between the wedge and the pitman after tightening, the wedge must be shimmed at the top to eliminate the gap completely.

Replace worn toggle plates The toggle plate which holds the lower end of the movable jaw, has two rounded or ball ends which pivot in the toggle seat insert. The toggle plate should be replaced whenever half the original ball diameter has worn away.

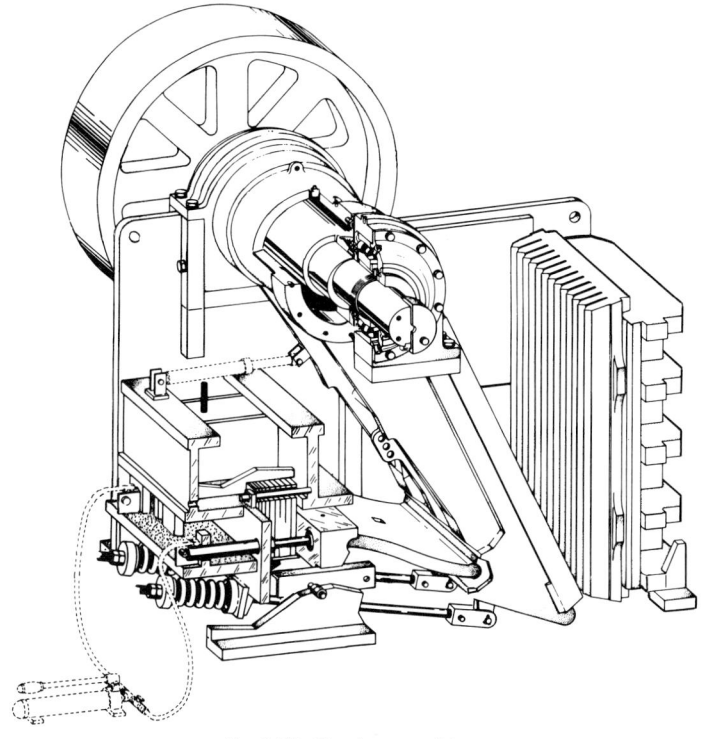

Fig. 9-59 Tension assembly.

By so doing, the wear on the toggle set insert is kept normal. A toggle plate with sharper ends will quickly wear away the inserts.

When replacing a toggle plate be sure to center it in the insert with relation to the pitman width.

Lubrication of each end of the toggle plate as directed by the crusher lubrication instruction sheet will greatly extend the life of a toggle plate by reducing wear at the ball ends.

Keep toggle plate tension spring(s) tight The tension assembly (Fig. 9-59) which holds the pitman tight against the toggle plate should be kept adjusted to a strong spring tension at all times. The coil springs in the assembly(ies) should be kept from one-third to one-half compressed from free length when new. Older springs may require tightening to the three-fourths or nearly completely compressed length.

Loose tension assemblies can be detected by listening for a clattering or pounding noise when the crusher is running empty. The hammering effect on parts will result in premature wear and breakage of toggle plate inserts and seats.

Keep crusher discharge clear The material discharged from a jaw crusher must be removed fast enough to prevent any pileup under the crushing chamber to the height of the pitman. The bottom end of the pitman can be rapidly worn away if it rubs against discharged material.

Make periodic inspections of this area to be sure no such condition has developed.

IMPORTANT: Do not use dynamite to clear large rocks from the crushing chamber. Breakage or premature failure of vital parts can result.

Keep hydraulic pumps and rams in good condition   Periodically inspect hydraulic hoses for cracks and abrasions which warn of impending failure. If pumps or rams are leaking fluid or if their efficiency is low, replace packing and seals to restore good operation.

Keep toggle plate seats and inserts in good condition   Each end of the toggle plate contacts a toggle seat insert which absorbs the wear and stress of toggle plate movement and compression. The inserts are housed in toggle seats which are the removable backup pieces for the two points of highest compression. Because of this stress, it is possible for the inserts and seats to crack.

Make a careful inspection of these parts whenever the toggle plate is removed. Look for cracks and wear. Replace inserts which are in questionable condition.

Keep toggle seat wedges tight   Some crushers are equipped with wedges located at the ends of each toggle seat to keep the seat tight against the base and adjusting shims. The lower wedge at each end is fitted with bolts which regulate the tightness of the wedge. These bolts must be loosened each time the toggle seat is moved for a change of discharge opening (product size).

Keep wedge bolts tight at all times during operation to prevent any loosening of the toggle seat.

Operational troubleshooting
Low Capacity
1. Direction of rotation incorrect
2. Stationary jaw loose
3. Speed incorrect
4. Jaws worn excessively
5. Attempting to feed raw material larger than crushing chamber
6. Incorrect toggle plate for size of crusher discharge opening
7. Lack of raw material or erratic feeding of crusher
8. Insufficient power or V-belt slippage

Bearings Heating
1. Too much lubricant
2. Insufficient lubricant
3. Dirty lubricant
4. Wrong lubricant
5. Operating faster than recommended
6. Out of level
7. Side bearing outer seal and flinger not turning with shaft
8. Insufficient radial clearance between side bearing outer seal and end cap
9. Toggle plate improperly seated
10. Bearings failing
11. Movable jaw rubbing base or key plates
12. Arced through bearings with electric welding machine
13. Dynamited boulders in crusher

Excessive Jaw Wear
1. Stationary jaw loose
2. Closing crusher below minimum recommended discharge opening
3. Operating faster than recommended
4. Not using original manufacturer's jaws
5. Very abrasive material

Difficult Adjustment of Crusher Discharge Opening
1. Tension spring not released
2. Support angles on base not loose

Removal of side bearings
1. Remove hopper and drives from crusher and drain lubricant from side and pitman bearings.
2. Disassemble tension rods, toggle plate, and movable jaw from crusher and allow pitman to hang freely.
3. Remove bolts holding side bearing housing to crusher base. With a crane or power

shovel lift pitman and shaft free from crusher by means of cable attached to lifting lugs on top of pitman.

If parts are to be used again, matchmark the parts as they are removed to be sure of getting them in the same relationship and on the same side of the crusher when reassembled.

4. Remove outer labyrinth seal.

5. Remove cap screws holding end cap and take off end cap.

6. Remove cap screws which hold lock plate and remove lock plate. Unscrew locknut from tapered sleeve.

7. Screw special factory supplied removal nut onto the sleeve until tight against bearing.

8. Screw the hose from a hydraulic pump into the threaded hole in bearing sleeve. Pump and tighten removal nut until sleeve breaks loose from shaft and bearing. NOTE: Some bearing sleeves will be equipped with two threaded holes and will require two separate hydraulic pumps for removal of bearings.

9. When bearing breaks loose from shaft, thread removal nut farther onto bearing sleeve until it is possible to remove both nut and sleeve from the shaft.

10. Thread an eye bolt in top of side bearing housing, and by means of a chain hoist or similar equipment slide the housing with bearing off the shaft. NOTE: If the housing fails to slide easily off the shaft, it may be necessary to bar and wedge the housing off the shaft. The inner side bearing flinger is a fairly tight fit to the shaft, and the housing may have to be walked off the shaft.

11. Remove cap screws holding inner labyrinth seal to housing.

12. Slide or push side bearing out of housing. Remove inner side bearing flinger.

13. Clean all seals, flingers, housing bore, and bearing with cleaning fluid, and inspect for imperfections, scoring, or galled areas. Replace all damaged parts.

Installation of side bearings NOTE: The following procedure is used to replace either right- or left-hand side bearings, or both bearings. The only difference between the right- and left-hand side bearing assemblies is in the inner labyrinth seals and the side bearing end caps. The inside seal and the outer bearing end cap on the right-hand side bearing assembly allow approximately ¼-in. lateral bearing movement for shaft expansion and contraction, while the left-hand side is fixed. The left- and right-hand sides are determined when facing the rear of the crusher, that is, when looking at the adjusting mechanism and tension springs.

1. Make sure all parts are clean and remove any imperfections in grease seals, flingers, and housing.

2. Set bearing housing on its base and bolt inner labyrinth seal to housing on either right- or left-hand, depending on which side bearing is being assembled.

3. Slide inner side bearing labyrinth seal onto shaft.

4. Prefit inner side bearing flinger onto shaft, filing or scraping so it slides easily over shaft. Then, place flinger inside housing so beveled edge is toward pitman.

5. With a lifting device, lift bearing up and slide into housing with large bearing bore outward. Make sure bearing is inserted squarely into the housing bore by checking with a small square.

6. Clean shaft and remove any imperfections. Check keys so they fit tightly in the keyways but slide freely along the length of the keyway.

7. Fit a small clip angle (supplied with crusher) into housing and against bearing to prevent bearing from falling out of the housing during the assembly sequence. IMPORTANT: Before mounting new bearing on shaft, check clearance between rollers and outer race. After bearing is placed in housing and on the shaft, it will be tightened until original unmounted clearance is reduced by 0.007 in.

8. Coat the shaft and inside and outside of the sleeve with powdered graphite. With a hoist, lift the bearing housing with bearing inside and slide onto shaft, being careful to start the flinger over end of shaft. Check to make sure labyrinth seals engage the mating grooves in seal.

9. Place the tapered sleeve on the shaft with the smallest diameter toward the bearing. Expand the sleeve by inserting a thin wedge or large screwdriver into the open slot, and slide the sleeve on the shaft and into the bearing with the threaded end toward the end of the shaft. Push the sleeve and bearing assembly in until the inner race of the bearing strikes the grease flinger. Remove wedge or screwdrive. Start locknut onto

tapered sleeve with beveled edge toward bearing. Leave clearance between locknut and bearing so sleeve can be driven in. Install bearing mounting tool (supplied by factory).

10. Place end plate on shaft and against bearing mounting tool. Tighten cap screws which apply pressure to the end plate. Attach the hose from the hydraulic pump to threaded hole in sleeve. NOTE: Some bearing sleeves will be equipped with two threaded holes and will require two separate hydraulic pumps for assistance in the assembly of bearings.

11. Expand the inner race of the bearing with the hydraulic pump pressure, then tighten cap screws which apply pressure to the end plate.

12. Keep expanding bearing with pump and tightening the cap screws on end plate until the proper bearing clearance is obtained. It may be necessary to drive the end plate on with a sledge to obtain the clearances required.

13. Check clearance at bottom of the bearing with a feeler gauge. When the bearing is sufficiently tightened, there should be 0.007 in. less clearance than when bearing was unmounted. NOTE: To check clearance, slide feeler gauge between outer race of the bearing and the rollers taken at the bottom of the bearing. To obtain a recommended clearance of 0.008 in., the 0.008-in. feeler gauge must be a tight slide fit.

14. Screw locknut onto sleeve with tapered side toward bearing. Screw it tight against bearing race using a long-handled wrench for sufficient torque development.

15. When locknut is tight against the bearing, install lock plate and tighten the two cap screws which retain it. Large tab of lock plate fits the slot of the tapered sleeve and small tabs are bent against cap screws to prevent loosening.

16. Slide the side bearing end cap into place (either right- or left-hand, depending on which side bearing is being assembled). Line up the grease filler hole and the drain hole in the cap. Fasten end cap to bearing housing with cap srews.

17. Slide outer labyrinth seal onto shaft with the grooves mating with those on the end cap. There must be a clearance for a least a 0.010-in. feeler gauge completely around the circumference of the seals. Check clearance four times, rotating shaft 90° before each check.

18. When assembly of both side bearings is completed, place shaft, bearing, and pitman into crusher base.

19. Check to make sure the magnetic drain plug is in each housing. Then fill with specified type and amount of Texaco Marfak grease. Replace filler plug on top of each housing.

20. Clean the exposed ends of eccentric shaft and flywheel bore. Align keyways in end cap with the keyways in shaft. Raise the flywheels with a suitable hoist and slide into place on the crusher shaft. Align and replace flywheel keys. Install the shaft end caps with self-locking cap screws. IMPORTANT: Drive the keys into the flywheel so that the key extends into the slot on the outer labyrinth seal and the shaft end cap has full bearing on the face of the flywheel. NOTE: It is very important that cap screws in the end cap be kept tight, as the lateral tightness of the complete bearing assemblies is maintained by these cap screws.

21. IMPORTANT: When starting a crusher for the first time after an overhaul, oil freely the areas between the pitman and the side bearings where the grease seals revolve in the side bearing housing. *Apply oil until the grease inside the bearing begins to extrude from the grease seals.* This precautionary measure ensures lubrication until the internal lubricating system begins to function. NOTE: A crusher which has been idle for a period of time and exposed to the weather can well afford to receive the same lubrication.

Removal of pitman bearings

1. Remove shaft and pitman assembly.
2. Remove side bearings.
3. Remove pitman labyrinth seal.
4. Remove the cap screws holding the pitman bearing end cap. NOTE: Some of the seals are designed for right- and left side of the pitman shaft. *Be sure to matchmark and designate proper side so parts may be reassembled in the same relationship.*
5. Remove pitman flinger.
6. Attach hydraulic hose to bearing removal hole in shaft, and operate pump until bearing breaks loose from shaft. NOTE: In the event the bearings are very tight, strike the pitman sharply with a sledge at a point in the middle of the pitman between the lifting

lugs. CAUTION: It is possible for the bearing to break loose with such force that it will slide off the end of the shaft. Devise some method of preventing this occurrence.

7. With suitable hoist, lift the loose bearing from shaft.

Installation of pitman bearings NOTE: When assembling both pitman bearings, *the left pitman bearing must be assembled first.* This bearing is fixed while the right pitman bearing is free to float to allow for lateral expansion. By assembling the fixed bearing first, the shaft is positioned to provide the correct location for the free bearing. The right and left sides of the crusher are determined by facing the tension springs at the rear of the crusher.

The following assembly procedure is correct for both pitman bearings.

1. Thoroughly inspect all parts and clean with a suitable cleaning solvent. Replace all damaged or worn parts. Clean inside bore of pitman. IMPORTANT: Before mounting new bearing on shaft, check clearance between rollers and outer race. After bearing is

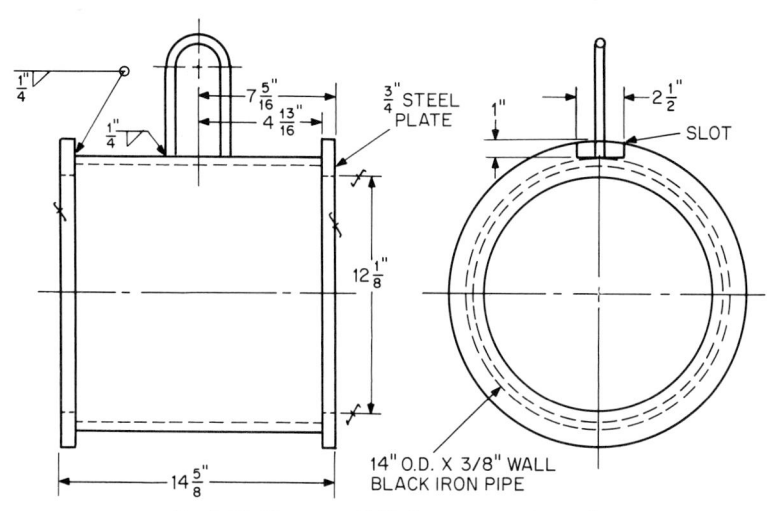

Fig. 9-60 Factory available bearing mounting tool.

placed on shaft and in pitman, it will be tightened until original unmounted clearance is reduced by 0.007 in.

2. Using a sling lift, hoist bearing and slide into place in the pitman bore. Care must be exercised in guiding the pitman bearing into place *so that the outer bearing race is not forced out of line with the inner race.*

3. Slide factory supplied bearing mounting tool over shaft (Fig. 9-60). Place end plate on shaft and against bearing mounting tool. Tighten cap screws on end plate. Attach the hose from hydraulic pump to the threaded hole in shaft.

4. Expand the inner race of the bearing with the hydraulic pump, while tightening cap screws on end plate with a wrench.

5. Keep expanding bearing with pump and tightening the cap screws on end plate until the proper clearance is obtained. It may be necessary to drive the end plate on with sledge to obtain the clearances required.

6. Check clearance at top of the bearing with a feeler gauge. When the bearing is sufficiently tightened, there should be 0.007 in. less clearance than when bearing was unmounted.

7. Use a gun to inject grease into the space between the rollers in the bearing, making certain that it is forced into the center of the bearing and that the grease does not, in effect, short circuit to the outer face of the bearing.

8. Clean all grease from the outer face of the bearing and slide the pitman bearing flinger into position with the beveled edge outward.

9. Prefit the pitman labyrinth seal on the shaft to be sure it slides freely into position

against the grease flinger. Remove seal and assemble pitman end cap on each end of the pitman. Install self-locking cap screws in each end cap. NOTE: Pitman end caps are right and left hand and must be assembled on the correct side.

10. Check the clearance between the outside rim of the seals and the mating grooves in the pitman end caps. There must be clearance for at least a 0.010-in. feeler gauge completely around the circumference of the seals. Check clearance four times, rotating the seals about 90° after each check. If the clearance cannot be reached, loosen the cap screws and jar the pitman end cap toward the side where the clearance is insufficient. Tighten the cap screws and check for the required clearance again.

11. Reassemble the side bearings.

Removal of shaft from pitman It will rarely be necessary to replace the pitman shaft unless complete bearing failure has occurred and scored or marred the shaft. However, it is recommended that the shaft be removed and that both the shaft and the pitman be cleaned and inspected prior to new bearing installation.

1. To lift the shaft, wrap cables around each end being sure to use an adequate protective material between cables and shaft to prevent marring the machine surfaces.

2. With a suitable lifting device, slide shaft out of pitman as far as possible.

3. Place another cable around middle of shaft and continue sliding shaft out of the pitman.

Installation of pitman shaft IMPORTANT: Thoroughly clean shaft and bores in pitman to remove any metal particles or contaminated lubricant.

1. Check pitman bores for oversize and elongation. If the bore is 0.0030 to 0.0035 in. oversize or elongated, it will be necessary to weld up and remachine housing bore. When this extreme condition exists, it is wise to consult factory for proper housing bore and additional information necessary to remachine pitman.

2. NOTE: There is no difference between the right and left ends of the shaft so the shaft may be inserted from either side. However, if the flywheel keyways are suitable for reuse, and have been matchmarked, reinstall shaft in pitman so that the respective shaft ends are on the same side as when disassembled.

3. When shaft has been inserted as far as possible into the pitman, one sling may be removed; then continue sliding shaft through the pitman until it is *centered*.

Installation of pitman assembly

1. After all bearing work has been completed, move the pitman to the crusher and lower pitman into the crusher. Care must be taken not to bump the crusher base.

2. Guide side bearing housings into place in base using long rods through the bolt holes in housing and through corresponding holes in the crusher base (see Fig. 9-61).

3. Install the large cap screws which fasten the bearing house to the base. Before doing final tightening of the cap screws, check the housing to make sure it is square with the shaft.

4. After all the cap screws have been installed, begin to draw them up tighter. Tighten all the cap screws and check the clearance between the outer rim of the outside bearing seals and the main grooves in the side bearing end caps. There must be a clearance for at least a 0.010-in. feeler gauge completely around the circumference of the seals. Check four times, rotating the shaft assembly about 90° after each check. NOTE: It may be necessary to strike the solid part of the bearing housing to get the proper radial clearance. Check the cap screws for tightness and keep them tight. After the crusher has been in operation for several days, it is a good practice to check cap screws and clearances again to detect any rubbing.

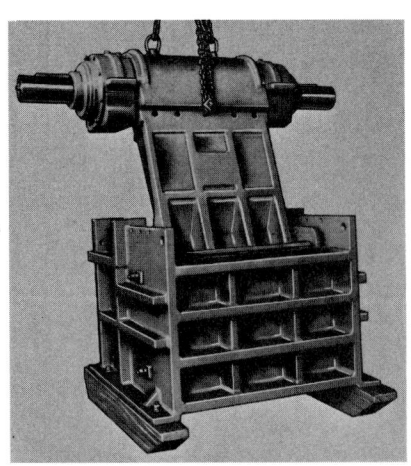

Fig. 9-61 Placing bearing housings.

5. Replace magnetic drain plugs in the housings. Fill with specified type and amount of Texaco Marfak grease. Replace fill plugs on top of the housings.

6. Clean exposed end of shaft and flywheel bore. Coat mating surfaces with light grease and line up keyway in outer labyrinth seal with shaft keyway. With a crane or similar equipment, raise flywheel and slide into place on shaft.

7. Align shaft and flywheel keys. Install keys and drive properly into place. Replace end plate and tighten new self-locking cap screws securely. IMPORTANT: Make sure the

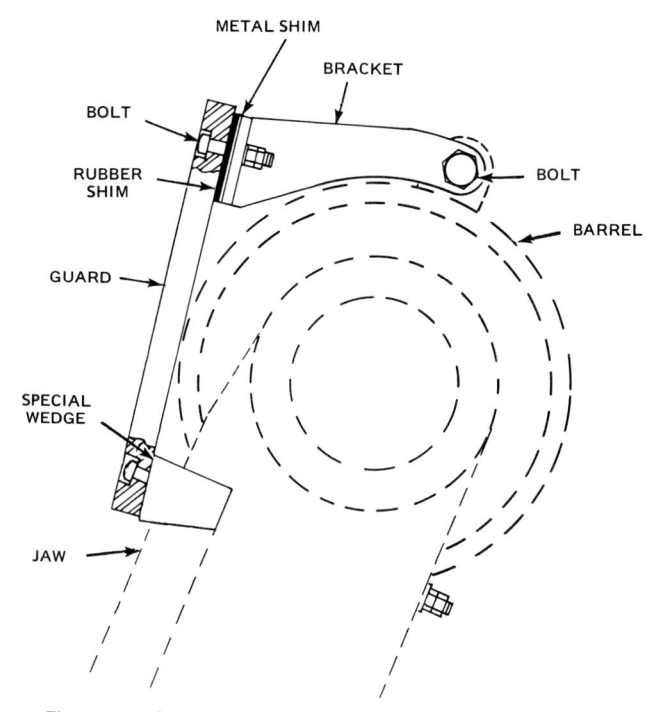

Fig. 9-62 Side view of pitman barrel with guard assembly in place.

keys are driven into the flywheel so that end cap has full bearing on the face of the flywheel.

8. Tighten the bolts in the split hub of the flywheel.
9. Replace movable jaws.
10. Replace toggle seat and inserts, toggle plate, and tension rod and springs.
11. When starting a new crusher for the first time, or after an overhaul, oil freely the areas between the pitman and side bearings where the grease seals revolve in the side bearing housing. Apply oil until the grease inside the bearing begins to extrude from the grease seals. This precautionary measure ensures lubrication until the actual lubricating system begins to function. A crusher which has been idle for a time and has been exposed to the weather can well afford to receive the same lubrication.

Pitman barrel guard Some feed arrangements will result in extreme wear on the pitman barrel even though it is above the crushing chamber. If it is impossible to reduce the wear by rearranging or adjusting the feed, a barrel guard of heavy steel plate can be bolted onto the pitman to absorb the wear and protect the barrel. Figure 9-62 shows a side view of the pitman barrel with guard assembly in place.

IMPACT BREAKER MAINTENANCE

The maintenance data under this heading will deal with the conventional single-impeller impact breaker shown in Fig. 9-63 and the double-impeller breaker shown in Fig. 9-64.

Lubrication Correct lubrication of the impeller shaft bearings is a vital maintenance requirement which must never be neglected. Figure 9-65 shows a typical lubrication instruction sheet for a double-impeller impact breaker. A copy of the appropriate lubrication instructions should be kept posted in the lubricant supply shed for ready reference at all times.

Check breaker mounting Keep breaker shafts horizontal so that bearing loading is uniform and lubricant distribution is proper. Periodically place a spirit level vertically

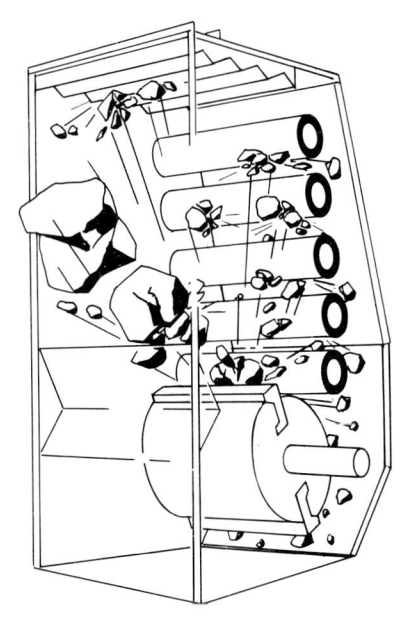

Fig. 9-63 Single-impeller impact breaker.

Fig. 9-64 Double-impeller impact breaker.

against the machined surface of the driven pulley on the impeller shaft. Correct an unlevel condition immediately by shimming the base. Extreme nonsupport at one or more corners of the breaker base can cause a twisting and distortion of the entire assembly so that bearing loads are abnormal and failure occurs rapidly.

Lifting the breaker properly Whenever it is necessary to lift the breaker base, attach hoist chains and cables to the factory specified lifting points only (see Fig. 9-66)! Never wrap chain or cable around impeller shafts to lift base. The impellers themselves can be lifted by passing the chain or cable around the shaft close to the body of the impeller.

Keep discharge from accumulating Check the discharge chute or conveyor beneath the breaker regularly to be sure there is never an accumulation of crushed rock backing up into the impeller bars. This can easily cause a jammed condition of the breaker.

Feed large pieces separately Do not add large pieces of rock or boulders to the crushing chamber when it is still partially filled with smaller pieces. All material fed into the breaker expands greatly (due to space between fragments). A large piece requires adequate space for expansion. When a large piece or boulder approaches the breaker chamber opening, *stop the feed* until the previous material clears, then feed the boulder, and again stop the feed. When the chamber begins to clear, resume feeding. Failure to operate in this manner can result in many unnecessary stoppages due to plugging. Stalling a breaker under full power causes extreme stresses and leads to shaft breakage.

Adjust penetration of feed The tips of the impeller bars should absorb most of the wear. If wear occurs on the face of the bars or if the impeller casting itself begins to show wear due to material abrasion, reduce the velocity of the feed into the breaker chamber (not the volume). The height or angle of feed chutes affects the feed velocity. The velocity should be the minimum at which satisfactory entry into the crushing chamber can be achieved.

Crushing Equipment Maintenance 7-247

Keep feed evenly distributed Arrange the feeding method so that material spreads across the full width of the breaker chamber. The impeller bars will then wear uniformly. A concentration of feed in one area will wear the bars unevenly, and this is a definite indication that the full output capability of the breaker is not being realized.

Keep feed to double-impeller unit centered The feed to a double-impeller-type impact breaker (Fig. 9-64) should be so arranged that rock falls between the two revolving

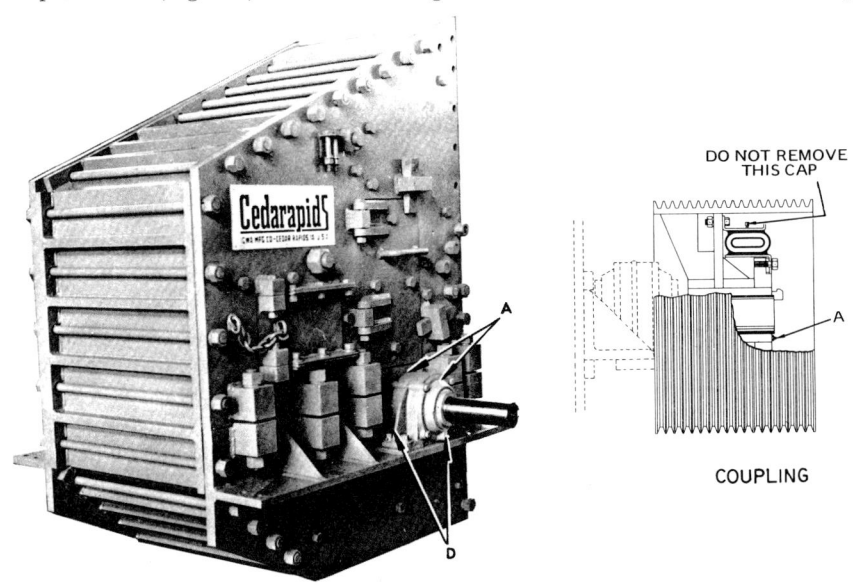

Symbol	Part	Instructions	Lubricant recommended
A	Bearings	*Every 48 hr:* Inject grease in each bearing. Usually five or six pumps of lubricant from a hand-operated grease gun into each bearing is sufficient to maintain proper lubrication.	Regal AFB 2
D	Drain plug	*Every 1000 hr or seasonally, whichever of the two occurs first:* Drain and flush with Texaco Rando AA oil and refill with proper grade of lubricant.	

Fig. 9-65 Lubrication instructions for impact breaker.

impellers and is spread across the full width of the crushing chamber. All impeller bars will then wear uniformly. If rock is directed primarily to one impeller or is concentrated at one side of the chamber, the obvious uneven wear rate is a definite indication that the full output capability of the breaker is not being realized.

Impact breaker buildup by welding There are many sizes and types of impact crushers, but although they vary in size and features, they all have one thing in common. As they produce a product their internal parts wear. The following information outlines how these parts may be protected and their life extended by application of electric welding materials.[1] See Fig. 9-67 for vital wear areas.

Liner Buildup. Crusher *liners* are very easy to maintain but are often neglected. The

---

[1] Acknowledgment to Teledyne-McKay Co., York, Pa., for all data relating to the buildup of impact breaker wear surfaces by the electric welding process.

7-248 Maintenance Programs for Equipment Entities

wear on these parts is gradual, and therefore, some operators feel that replacement is the easy way out. This is false economy. The liners can be hard-surfaced with a crisscross weld (Fig. 9-68) and maintained in about the same length of time required in changing the liners. A periodic inspection and touch-up of wear areas will increase the wear life of the liners indefinitely.

Feed Plate Buildup. The *feed plate* should be hard-surfaced with a series of vertical stripes $\frac{1}{2} \times \frac{3}{4}$ in. apart (Fig. 9-69). When these stripes wear, weld new stripes between the old ones, not on top of them.

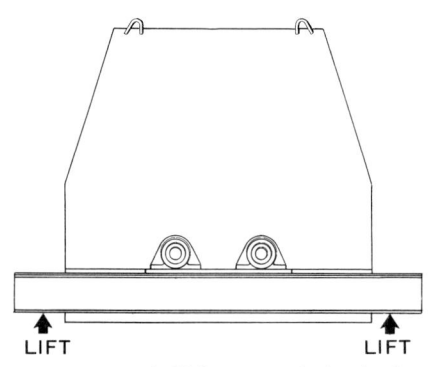

Breaker Bar Buildup. The breaker bars are another area of high neglect. These breaker bars would last much longer with only a little hard-surfacing. All that is required is a series of weld beads $\frac{1}{2}$ to $\frac{3}{4}$ in. apart along these bars (Fig. 9-70). As the weld shows signs of wearing, new beads should be deposited between the old ones. A few weld beads will break up the wear pattern, extending the useful life of the breaker bars.

The Rotor or Impeller Buildup. The rotors or impellers do not require constant attention and with proper maintenance should last indefinitely. Many rotors are hard-surfaced by the manufacturers; if they are not, they should be protected by hard-surfacing as soon as possible. A series of straight beads or a cross-check pattern is effective (Fig. 9-71).

Fig. 9-66 Specified lifting points for breaker base.

If the rotor has a rotor ring, this part should be hard-surfaced. The top of the ring is hard-surfaced in a solid pattern; two layers will give the best results. The vertical leg of the ring should be hard-surfaced with a series of stripes $\frac{1}{2}$ to $\frac{3}{4}$ in. apart (Fig. 9-72).

Hammer or Impeller Bar Buildup. The hammers or impeller bars are the heart of any impact-type crusher. If these are not maintained, the crusher will not produce. When the crusher is new or when a new set of hammers are installed, a template should be made of

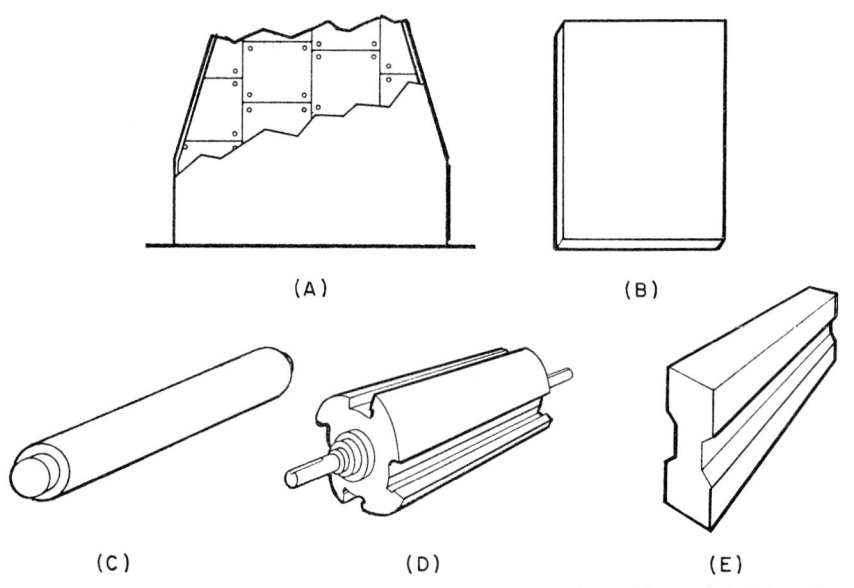

Fig. 9-67 Basic wear areas on impact breakers that can be built up by welding include (A) crusher liners, (B) feed plates, (C) breaker bars, (D) rotors or impellers, and (E) hammers or impeller bars.

the hammer cross section in relation to the rotor. The hammers should always be built back to this configuration.

When the hammers' leading edge is worn to a radius, the hammers become less effective. Rock slides across this radius instead of being broken or picked up and thrown against the breaker bars. The greater this radius becomes, the faster the rate of wear. It is safe to state that if a hammer needs rebuilding each day, the greatest amount of wear occurs the last 2 or 3 hr of crushing. Hammers can be rebuilt with the proper alloys and configuration to help extend their effectiveness.

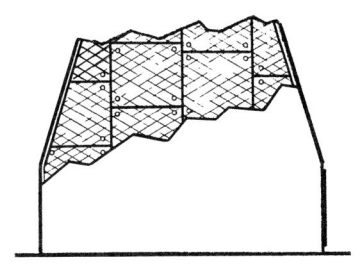

Fig. 9-68 Crisscross weld on liners.

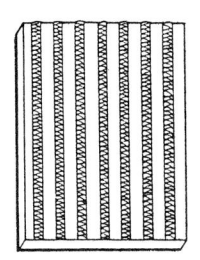

Fig. 9-69 Vertical stripe welding on feed plate.

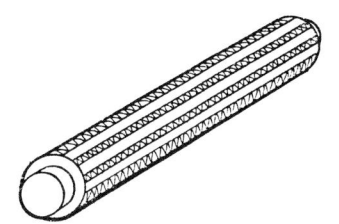

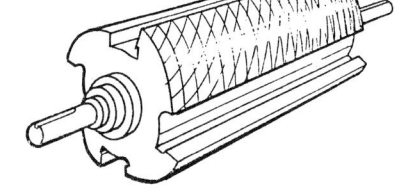

Fig. 9-71 Cross-check weld pattern for rotor or impeller.

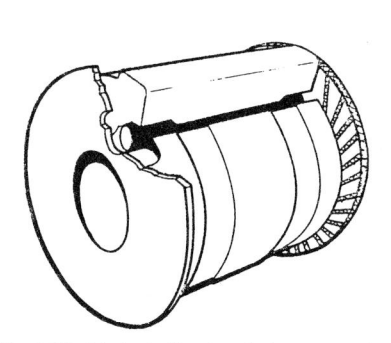

Fig. 9-72 Method of hard-surfacing a rotor ring.

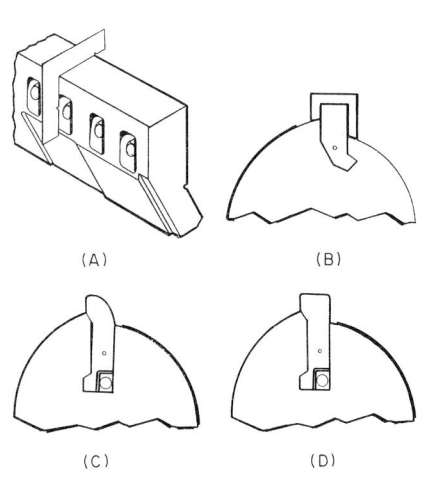

Fig. 9-73 Hammer bars need attention. Use a template to check the hammer's cross section in relation to the rotor (*A* and *B*). When the leading edge is worn to a radius, the hammers are less effective (*C*). For maximum effectiveness, keep a square leading edge (*D*).

New hammers have a square leading edge; for the most effective operation this edge must be maintained. The crusher operator has a choice of two methods of doing this:
1. Change hammers as they wear.
2. Rebuild with electrodes.

Changing hammers is out of the question for most operations and in most cases should not even be considered. The cost of the hammers plus the time required for changing is much too expensive.

Manual Welding. Welding manually with electrodes has its place in certain operations on some of the smaller mills. If a material is being crushed that is not extremely abrasive and welding is only required once or twice a week, manual welding is probably best. An exception is if you have a large mill. This could be too time-consuming, and the semiautomatic or the automatic system should be considered.

Semiautomatic Welding. The semiautomatic system can be used on any size impact crusher. The one recommendation in this type of welding is not to cut corners in purchasing the semiautomatic system. The system should include a good 400-amp or larger power supply (a 300-amp supply can be used if in excellent condition), a good set of welding cables at least 3/0, with 4/0 being preferred, and a heavy-duty wire drive unit. The selection of welding wires for the system will be discussed later.

Automatic Welding. The automatic system for rebuilding impact hammers is the fastest maintenance system for impact crushers on the market today. This unit is designed for use in crushers of 36-in. and over hammer widths. The automatic unit is temporarily mounted inside the crusher and does the welding while the operator controls the operation from outside the crusher. The many advantages in this method are as follows:

1. *Speed.* Less time is required for the total welding job. Up to 25 lb/hr of metal is deposited.

2. *Heat input.* Due to the extremely fast travel speed, 25 to 35 in./min, heat input into the manganese hammers is kept to a minimum. The open-arc system (no shielding gases or external flux) tends to dissipate the arc heat very rapidly; this is partially due to the very light or no slag coverage of the weld deposit and higher deposition at the same current than other welding processes.

3. *Hammer balance.* By welding a given length of time on each hammer, at a given amperage, a similar amount of metal will be deposited on each hammer.

4. *Cost.* Overall cost is made up of many things. Time is a major cost for any maintenance. Due to the extremely high deposition rates with the automatic unit, time is kept to a minimum.

Procedure for Welding Hammers. The hammers almost always round off on the ends. An easy method for repairing these ends is as follows: Determine the amount of wear at the ends; cut a piece of bar stock $1/8$ or $1/4$ in. thick times the depth of wear times the width of the hammer and weld it to the hammer (Fig. 9-74). This will act as a dam to weld against. If welding is done by the manual or semiautomatic method, proceed to fill up this void flush with the rest of the hammer. If the automatic welder is used, deposit one or two layers of weld at the end of the hammer, joining the bar stock securely to the hammer end. The welder will effectively fill the rest of the void. There is no danger of this piece of bar stock being knocked off as the wear factor at this area of the hammer is chiefly abrasion, with little impact. This method of welding hammer ends will save many hours of work and will help keep the rotor from sticking to the chamber wall.

In rebuilding impact hammers, serious thought should be given to the choice of an alloy and the method of doing this job. Price of the welding product is important but should not be the major consideration. For example, an electrode for doing this can be purchased for 50 cents/lb. This electrode is 65 percent efficient—out of every 10 lb purchased 6.5 lb is actually deposited on the hammer. The welder uses 5 lb/hr and deposits 3.25 lb. A hammer rebuilt with this product will last x hr. Another product is bought for 85 cents/lb. The efficiency of this product is the same as the first product, but the wear life is $1.5x$ hr.

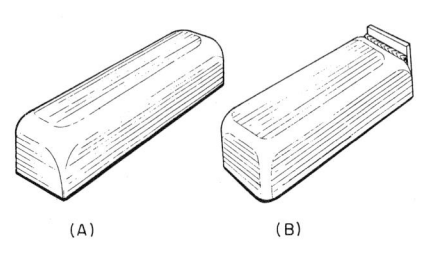

Fig. 9-74 Hammer with rounded edge (*A*) can be repaired with a small piece of bar stock (*B*).

If the semiautomatic system were used and the wire cost was 85 cents/lb, the overall cost would still be less. Semiautomatic wires are 90 percent efficient, and a welder will use about 8 lb/hr. This would give a deposit of 7.2 lb, or twice the amount that could be deposited with electrodes. If the wear were only x hours, the cost would still be less than the 50 cents/lb product.

With the correct automatic welder the operator would burn at least 15 lb/hr for a deposit of 13.5 lb/hr. This would lower still further the overall hammer maintenance cost.

The least expensive electrode or wire to use in a rebuilding program is the product that will last the longest. Hammers are almost always made of manganese. In welding these, a product that is compatible with manganese should be used.

Attach welding ground wire properly *It is important that the ground wire of the welding circuit be attached directly to the item being welded.* There must not be a passage of electric current through any roller bearing due to incorrect grounding. Any arcing through a bearing will cause pitting and damage which will ruin the unit.

Check impeller bar weight balance after welding It is very important that the weight balance between impeller bars be maintained at all times to prevent serious vibration. When weld is applied to the bars during the buildup process, large weight differences can be created. *A careful balance check must be made before breaker operation.* This should be done before the automatic welder (if used) is removed from the chamber, so that additional weld can be applied as needed.

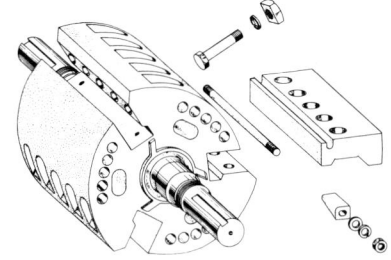

Fig. 9-75 Removal of impeller bars.

Method of Checking Balance

1. Remove drive belts.

2. Turn impeller by hand through one complete revolution. Do this in stages, releasing the impeller after each partial turn. If one bar outweighs the others, it will swing downward when the impeller is released.

3. Add weld in appropriate amounts to the bar(s) at the top of the impeller. When properly balanced, the impeller will remain stationary when released, as the impeller is turned through one revolution by stages.

Check for contact between double impellers after welding When double impellers have been built up and are being weight-balance checked with V belts removed, *be sure to see that no two impeller bars can come into direct contact with each other!* Direct contact can break bearings or shafts instantly.

Removal of impeller bars

1. Open the large inspection doors on each side of breaker.
2. Remove the impeller bar cap screws.
3. Remove the impeller wedge retaining bolt by driving bolt out of the assembly in either direction depending on which side is clear.
4. Drive the impeller wedges loose with a suitable punch and hammer.
5. Drive the impeller bars out of either side of the impeller with a sledge or suitable battering ram.

Installation of impeller bars

1. Before installing impeller bars make sure that they are within 3 *lb* of each other to maintain balance. NOTE: Very large impact breaker bars should be held within 4 *lb* of each other. Lightweight bars can be built up with weld to equalize total weights between bars.
2. Reassemble the impeller right and left wedges.
3. Install the impeller bar cap screws in the impeller and impeller bars.
4. With an impact wrench, or large socket wrench start to tighten the impeller cap screws. After the cap screws are partially tightened, start to draw the wedge bolts into the impeller assembly. Keep alternating; tightening the cap screws and wedge bolts. With the

bolts tight, sledge top of impeller bar to help seat it in the impeller casting. Ater sledging, tighten the cap screws and wedges again.

5. Close and lock the inspection doors.

6. After the first hour of breaker operation, shut down and repeat the tightening procedure. Check bar tightness weekly thereafter.

Impeller and impeller shaft The shaft is a *press fit* in the impeller casting and rarely, if ever, will it be necessary to replace shaft or impeller, if properly maintained and operated. Should it become necessary to do so, refer to the following instructions.

Removal of impeller assembly

1. Remove the upper sections of the breaker so that the complete impeller assembly can be hoisted from the lower section.
2. Remove all impeller bars.
3. Dismantle and remove both bearings.
4. Lift impeller assembly off breaker base.

Removal of impeller shaft If the shaft or the impeller casting is to be salvaged, proceed as follows:

1. Check the ends of the shaft for the stamped-on pressure (in tons) required to press the shaft into the impeller casting. NOTE: The amount of pressure required for shaft *removal* is usually much greater than the pressure required for installation.

2. Locate a facility equipped with a hydraulic press large enough to do the job.

3. Measure and mark the shaft and impeller very accurately so that the new shaft and impeller assembly will be correctly assembled.

4. Remove the flat point setscrews in the impeller. Then remove the pins which are set into the shaft by rotating the shaft so the pins are upside down and can drop out. If pins do not drop out, thread a long bolt into the tapped holes in pin and pull pin out of the assembly.

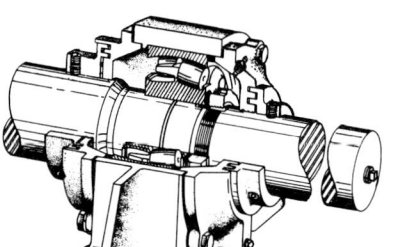

Fig. 9-76 Bearing arrangement for impeller shaft.

5. Place the impeller shaft and impeller in the press. Apply force on the *end of the shaft opposite the key* in the impeller. (Contact factory for information.)

6. Remove the impeller shaft from the impeller.

7. IMPORTANT: Check the dimensions of the part you wish to salvage, using a machinist's gauge for accuracy within thousandths of an inch. Contact the factory to be sure the salvaged part is actually serviceable.

Reassembly of shaft and impeller CAUTION: Before using old impeller check bore dimensions and consult your dealer or factory for proper bore dimensions (see step 7 of disassembly procedure above).

1. Press the new shaft into the impeller in the proper attitude so that the key in the shaft and keyway in the impeller is aligned.

2. With the impeller centrally located on the shaft, spot drill the shaft (in a new area) where the pins and the setscrews are to be installed. Clean the steel shavings out of the drilled areas and install pins and setscrews. Tighten the setscrews.

Impeller shaft bearing maintenance Impact breakers are usually equipped with double-row spherical roller self-aligning bearings. The bearings are protected against the loss of lubricant and the entrance of sand and grit by labyrinth-type grease seals. Sufficient grease is permitted to work through seals to prevent dust and grit from entering the bearing housings. (See Fig. 9-76.)

Check Bearing Temperature. The bearings normally operate at a temperature of 100 to 150°F but in warm climates may operate at 180°F. If one periodically runs as high as 212°F or higher, it is considered as overheating and the cause should be found and corrected to prevent damage.

1. A thermometer provides the best method of checking bearing temperatures. Remove fill plug and insert pencil-type thermometer in housing lubricant and get actual

reading. Readings for the bearings can be taken on the bearing cap by removing the fill plug on the housing and inserting thermometer or by draining out a small amount of lubricant from the drain hole on underside of housing.

2. The bearing temperatures can be checked by placing the hand on the bearing housing. If the hand can be held there for a few seconds, the bearing is not too warm.

3. If there is a question whether or not the bearing is too hot, always use a thermometer as the final check. CAUTION: Never reduce bearing temperatures by artificial means, such as by pouring water on them, as the sudden change in temperature is apt to warp, crack, or crystallize some part of the bearing assemblies.

Bearing Heating Causes

1. The most common cause of overheating is too much lubricant and most generally occurs when the breaker is new or immediately after lubricant has been changed. The bearing will heat up and force out excessive lubricant until the proper amount is left, then will gradually cool down to the normal operating temperature.

2. The bearing will overheat if the breaker is operating faster than recommended under full load. Check the speed with a speed indicator.

3. One of the bearings may overheat if the unit is operated decidedly out of level. An extremely unlevel attitude may result in a bearing carrying more than its share of load. It will begin to get hot. Level the breaker by checking shaft pulley face with a spirit level. Compensate for any settlement by use of cribbing, jacks, or shims.

4. Overheating is also a sign that bearing failure is developing. Check the overheating bearings for condition and wear by jacking up the end of shaft at the affected bearing. Observe the shaft assembly during these operations. *The bearing should not permit perceptible up and down movement.* If movement is visible, bearing is excessively worn and should be replaced.

5. Heat will result if there is insufficient radial clearance between inner labyrinth flinger and inner labyrinth cover and between outer labyrinth flinger and outer labyrinth cover. There must be clearance for at least a 0.015-in. feeler gauge completely around the circumference of the seals. Check four times, revolving the shaft assembly about 90° after each check. If these seals are touching at any point, there apparently is a bearing failure. NOTE: If a seal is allowed to run loose and dirt and grit enters the bearing, rapid bearing failure can result.

Disassembly of Bearing

1. Drain lubricant when it is hot. Check for foreign particles. If steel or brass particles are in evidence, the bearing has failed and must be replaced.

2. Remove sheave if bearing failure is on drive side.

3. Remove the outer labyrinth flinger by removing lock screws and loosening setscrews and sliding flinger off the shaft.

4. Remove bearing housing cover bolt and outer labyrinth cover.

5. Loosen and remove the bearing housing bolts which fasten the bearing assembly to breaker.

6. Straighten the prong on the lock washer that is bent into one of the grooves on the outside of the bearing locknut. Place a short square bar or drift in one of the grooves on the outside of the nut and pound the bar with a hammer, turning the nut counterclockwise about *three or four turns.* Do not remove locknut. This is a precautionary safety measure as the locknut will prevent the bearing assembly from shooting off the shaft when the hydraulic pump is used.

7. Install jack under end of shaft and lift loose end of shaft enough to relieve pressure on the bearing. *Caution* must be taken not to lift the shaft too high to prevent the labyrinth seals on the opposite side of the shaft from breaking. Hold shaft in this position while disassembling bearing and housing.

8. Remove the pipe plug from end of shaft.

9. Screw the hydraulic pump hose fitting into end of shaft and tighten.

10. Work the hydraulic pump until sufficient pressure is reached to release the bearing assembly from the shaft.

11. Remove hydraulic pump hose and replace the pipe plug into end of shaft.

12. Remove locknut and lock washer from shaft and slide the bearing assembly off the end of the shaft. NOTE: The above procedure is used where normal conditions exist, but if the bearing assembly has completely failed in motion and has scored or distorted

some of the parts it will be necessary to pull the housing over the bearing and disassemble each piece separately.

Reassembly of Bearing. NOTE: The following procedure is used to replace either the held or free bearing assembly, or both bearing assemblies. The only difference between held and free bearing assemblies is that on the held bearing assembly a stabilizing ring is installed on each *side* of the bearing. The free bearing assembly has no stabilizing ring, allowing for shaft expansion and contraction while the held bearing is fixed.

1. Thoroughly clean, with suitable safe solvent, all parts of the bearing assembly to be replaced. Prefit all parts, except the bearings. Remove any imperfection that might impair proper seating or mating of grease seal grooves.
2. Slide the inner labyrinth flinger on the shaft as far as possible.
3. Start to partially preassemble the bearing on a table or bench.
4. Carefully place gasket on labyrinth cover. Place labyrinth cover with gasket into housing, being sure the drain plug hole is located at the bottom of housing. Lay the housing with cover on the table or bench.
5. At this point in the reassembly, it is necessary to determine whether it is a *free or held assembly.* For held bearing assembly, slide the stabilizing ring into the housing and against the inner labyrinth cover. The free bearing assembly does *not* require the stabilizing rings.
6. The bearing assembly is now ready to be placed in the housing. Place the bearing assembly on the bearing housing.
7. With a carpenter's combination square, check the bearing to see if it is perpendicular to the back of the bearing block, and with a hard wood block or rubber hammer, tap the bearing into the housing.
8. Put the outside labyrinth cover into the bearing housing and install the long housing bolts.
9. Slide the assembled bearing housing onto the shaft. Place jack under the shaft and raise just enough to allow the bearing housing to slide in position.
10. When both bearings are replaced, it is necessary to center the impeller between the liners. Open the inspection door and check the distance between liner and impeller on both sides and drive tapered wedges between liners and impeller, *being careful to keep same clearance on both sides.*
11. Align the bearing housing bolt holes with holes in lower side plate support.
12. Remove the outer labyrinth cover.
13. Assemble the locknut for tightening the bearing.
14. Proceed to tighten bearing by holding the bearing against the inner bearing stabilizing ring while tightening the locknut by hand (if held side bearing is being installed).
15. The *free side bearing* must be *centrally located* between the labyrinth covers. When the dimension from machined edge of housing to the bearing is $3/4$ in., the bearing is properly positioned. After the bearing is tightened sufficiently to hold it in place, use the factory supplied wrench and tighten to the factory specified clearance.
16. To check the clearance, insert the feeler gauge between the outer race of the bearing and the rollers taken at the *top* of bearing, checking several different rollers by turning the shaft. For example, to obtain a recommended clearance of 0.004 in., the feeler gauge must be a tight slide fit between roller and race. *Do not roll* feeler gauge between the roller and race. IMPORTANT NOTE: In some cases it may *not* be possible to obtain the minimum clearance. Tighten until it is impossible to tighten further with recommended wrench but do not stretch the threads on the sleeve or nut. After the bearing is tight, turn the nut in a clockwise direction to loosen nut and then remove it from shaft.
17. Put on the lock washer, making certain the prongs are pointing away from the bearings and the locking prong is started in the slot of the sleeve or shaft. Place tightening wrench on nut, turning the nut clockwise until it is very tight. Line up one of the outer prongs on the lock washer with a notch on the locknut. Bend the prong over, setting it firmly with a drift pin.
18. Place the outer labyrinth cover with gasket into the bearing housing so the hole for the grease fitting is turned upward. Replace nuts on the long bearing housing bolts and draw up tight.
19. Tighten the bolts which attach the bearing housing to the base.
20. Slide the outer labyrinth flinger onto the shaft and into the labyrinth cover until it

is tight against the cover. Then with a pencil or similar marking tool, *mark the shaft ¹⁄₁₆ in. back* from the labyrinth cover. Slide the labyrinth flinger back on the shaft to the mark and hold while tightening setscrews in the flinger.

Flingers should be inspected periodically to make sure they both remain locked in place properly. If the flingers move, the bearing will extrude considerably more grease than normal. Dirt may also be able to enter the bearing.

After the setscrews are tightened, the lock screw must be threaded down firmly on top of the cup point setscrew. IMPORTANT NOTE: Both inner and outer labyrinth flingers must be located and locked in position as described.

21. After bearing assembly is completely put together and tightened, check the radial clearance between the outer rim of the inside and outside labyrinth cover and the mating grooves in the labyrinth flinger. There must be clearance for at *least a 0.015-in. feeler gauge* completely around the circumference of the seals. Check four times, revolving the shaft assembly 90° after each check. NOTE: If there is not a 0.015-in. clearance between labyrinth cover and labyrinth flinger, loosen the bearing housing bolts and shim under the bearing housing until clearance is obtained. It also may be necessary to parallel the housing bore with a shaft to obtain correct clearance.

22. Replace sheave. Align and replace key, setting it firmly.

23. Replace drain plug in the housing and fill with specified type and amount of grease.

24. General. When starting a new single-impeller impact breaker for the first time or after an overhaul, freely oil the areas between the cover and flinger. Apply oil until the grease inside the bearings begins to extrude from the bearing housing. This precautionary measure ensures lubrication until the actual lubrication system begins to function. An impact breaker which has been idle for a period of time and exposed to the weather can well afford to receive the same lubrication.

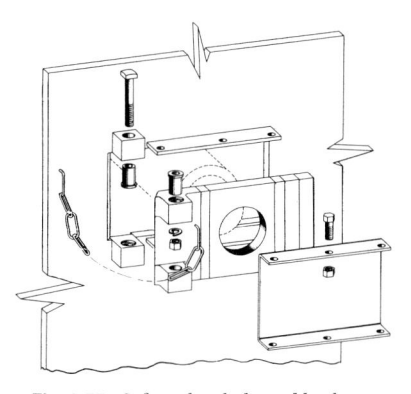

Fig. 9-77 Safety plate bolts and bushings.

Safety plate bolts and bushings The safety plates are located on the sides of the chamber and are attached to the side plates with a strand of chain. The safety plate is designed to shear the special bolts, releasing the plate when tramp iron or uncrushable material enters the breaking chamber.

After a considerable amount of tramp iron and uncrushable material has passed through the breaker and caused the safety plate bolts to shear repeatedly, it will be necessary to drive out bushings and replace them. When the bushings become elongated and considerably larger than the shear bolts, they will allow the bolts to shear from only hard rock or heavy feed in the breaking chamber.

Operational troubleshooting instructions
Excessive Wear on Impeller Bars
1. Improper feed of material
2. Material very abrasive
3. Incorrect speed of impeller
4. Improper welding or welding rod

Low Capacity
1. Impeller bars worn excessively
2. Incorrect speed of impeller
3. Breaker bar too close to impeller bars
4. Improper feed of material

Frame Breakage
1. Breaker not level
2. Frame not supported correctly
3. Impeller bars worn excessively

Vibration
1. Impeller bars not within specified weights of each other
2. Bearing assembly loose
3. Sheave out of balance
4. Bearing failing

Shearing Horizontal Safety Release
1. Installing incorrect diameter shear bolts
2. Shear block adapters worn or broken
3. Tramp iron in breaker

HAMMERMILLS AND IMPACT MILLS

The maintenance data under this heading will deal with the conventional hammermill shown in Fig. 9-78. Many lime mills and impact mills are similar in construction and operation, so the same maintenance recommendations apply.

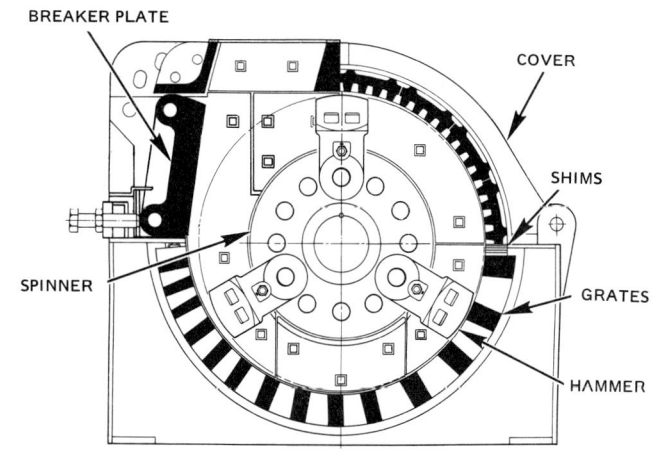

Fig. 9-78 Conventional hammermill.

Lubrication Correct lubrication of a hammermill is a vital maintenance requirement which must never be neglected. Figure 9-79 shows a typical lubrication instruction sheet for these machines. A copy of the appropriate lubrication sheet should be kept posted in the lubricant supply shed for ready reference at all times.

Check hammermill mounting Keep the shaft of the spinner level so that lubricant distribution is proper and bearing loading is uniform. Periodically place a spirit level vertically against the machined surface of the driven pulley. Correct an unlevel condition immediately by shimming the base.

Lifting the mill Whenever it is necessary to lift the mill, attach hoist chains to the lifting lugs provided on the housing. Do *not* wrap hoist chain or cable around the shaft.

Care of the hammermill To ensure efficient operation of the hammermill it is recommended that the operator inspect, lubricate, and schedule necessary adjustments and repairs at regular intervals. The intervals will vary considerably according to the different types of rock and stone, uniformity of feed, and moisture content.

Daily Checks. Throughout the daily operating period the operator should check for unusual sounds or other signs of abnormal operation that warn of future trouble if not promptly corrected.

    1. *Bearings.* Check hammermill bearings for overheating. Do not operate hammermill if bearings overheat.

    2. *Vibration.* Any unusual amount of vibration should be checked and corrected before damaging the frame or bearings.

    3. *Grate pins.* Every 8 or 10 hr grate pins should be driven into body to tighten grates

and then locked in place with setscrews. However, if the hammermill is equipped with small opening grates to make a high percentage of fines, the grate pins should not be used, because they will provide an area between the first grate and body where oversize material can escape before being milled to size.

4. *Bolts.* Check and tighten all fastening and mounting bolts, cap screws, and set screws.

5. *Hammer tips.* Inspect the hammer tips for wear. To maintain efficient operation and gradation, the hammer positions should be changed when worn.

6. *Breaker plate.* Inspect the breaker plate for wear, especially the lower section. Reverse when worn or replace after both ends are worn. If the lower section of the breaker plate is allowed to wear too much, making a concave surface, it will impede the flow of material and decrease the efficiency of the hammermill.

7. *Liners.* Check the liners for wear and tighten liner bolts. Replace liners before they wear completely through and cause damage to the hammermill body.

8. *Grates.* Check the grates for wear and replace when worn. Grates are one of the factors governing the gradation of the product. When worn excessively, away from the tips of the hammers, poor grates allow rock to recirculate within the hammermill, reducing capacity due to loss of the sharp nip action needed for good crushing.

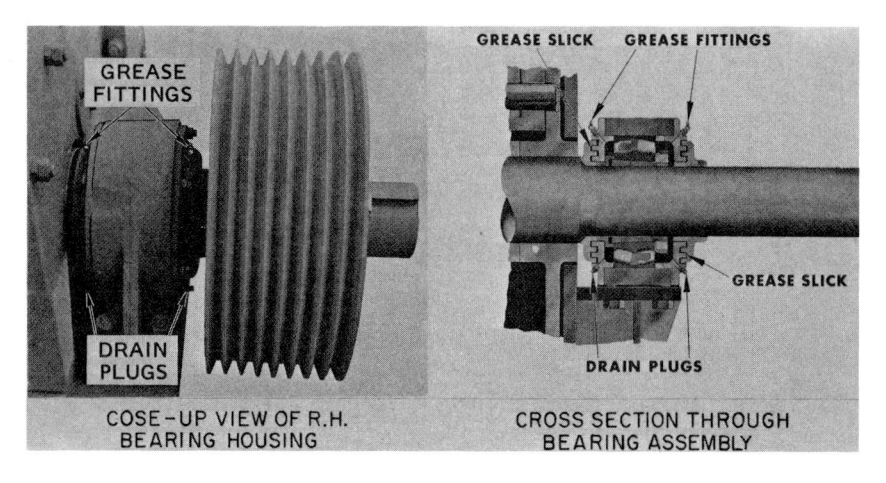

Frequency	Location	Instructions		Lubricant recommended
Check daily but grease only as needed (see instructions column)	Bearing housing grease fittings (four)	Grease only as often as necessary to maintain slick of grease extruding from the seals. Generally 10 pumps of lubricant from a hand-operated grease gun on each grease fitting of the bearing housings is sufficient for lubrication and dust seal (approximately once a week)	Above 90°F 32–90°F 0–32°F Below 0°F	Marfak 1 Marfak 0 Marfak 00 Regal AFB 2
Every 1000 hr or seasonally (whichever occurs first)	Bearing housing grease fittings and drain plugs (four of each)	Drain and flush with Texaco Rando AA oil. Refill with the proper grade of lubricant.		

Fig. 9-79 Lubrication chart for hammermill.

Increasing hammer tip life The Cedarapids Fasturn hammer assembly offers an efficient method that permits the turning of hammer tips to expose the opposite wear surface, without the need for arm or arm pin removal. It is only necessary to drive each tip sideways to detach it from the arm. Tips are interchanged between adjacent arms in a back-to-back arrangement which also helps to lock the tips in place. A U-shaped lockwire is used as a temporary tip retainer until material dust is packed between tips and arms to create a bonding effect.

Installation of Fasturn Hammers
1. Assemble all hammers in the following way (Fig. 9-80):
   a. Slide tip onto arm.
   b. Insert lockwire (No. 11 gauge-steel) as shown in Fig. 9-80.
   c. Tap the lockwire into place with a hammer. The spread of the wire tips holds the wire in place.
2. Weigh hammers for spinner balancing (optional). The weight of hammer assemblies is approximately equal. However there are normally some small variations, and if these happen to occur so that there is an accumulation of weight on one side of the spinner, the mill may vibrate. In view of the small amount of extra work involved, it is strongly recommended that the balancing procedure be carried out as described in the following section.
3. Install the hammers in a back-to-back arrangement as shown in Fig. 9-80. Insert arm pin to retain.
4. Install the retaining rings at the ends of each arm pin.
5. Make a test rotation of the spinner to determine the intensity of any vibration that may be generated. If there is severe vibration and the weight-balancing procedure was *not* used, it is strongly recommended that the hammers be removed and reinstalled using the balancing procedure.

Procedure for turning tips Most mill users turn the tips at intervals to balance the wear (Fig. 9-81). When area no. 1 of the tips is worn away, they are turned and area no. 2 is worn away. When area no. 2 is gone, they are turned to wear area no. 3, and so on. The procedure for turning tips is as follows:
1. Pry out the lockwire using a screwdriver or pliers. Where a section of the wire has been worn away, check to make sure that no portion of the wire remains in the lock grooves.
2. Move arm out of line with the row of hammers, and using a 12-lb (approximate) shop hammer with an 18-in. long handle, strike the tip a square blow on the lockwire side. Repeat this until the tip is jarred approximately $1/4$ to $3/8$ in. and is then loose enough to be removed by hand. Usually three to four blows will loosen a tip.

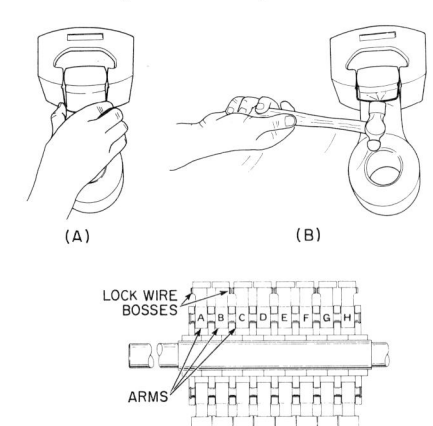

Fig. 9-80 Method of hammer assembly includes inserting lockwire (A), spreading the wire tips to hold the wire in place (B), and installing the hammers in a back-to-back arrangement (C).

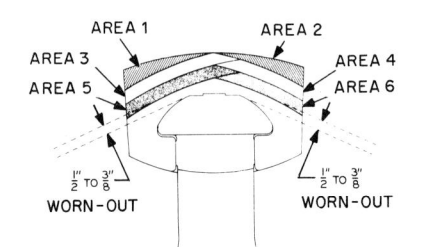

Fig. 9-81 Tip wear and wear limitation for hammer tip.

3. Remove the tip from arm B, driving it in the opposite direction to loosen.

4. Interchange tips between pairs of arms. Be sure to *use new* lockwires each time. At the end of each wear period, change them back to their previous location.

Limit of Fasturn Tip Wear. Fasturn tips should be replaced when the working surface is from $1/2$ to $3/8$ in. from the key slot in the side. This limitation is both for operating safety and ease of tip removal. *Never operate with tips worn beyond the $3/8$-in. limit!*

When tips are worn out and require replacement, change the entire set in the hammermill. Normally new tips can be installed at random. To assure the maintenance of the best dynamic balance, new tips can be weighed and distributed as outlined in following paragraphs.

Arm Service Life Limitation. No arm should remain in use after 25 percent of its new (original) width has been worn away, even though its lockwire grooves are still complete

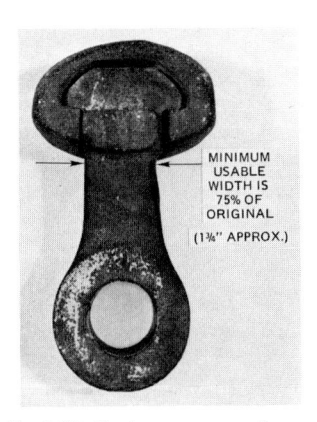

Fig. 9-82 Replacement point for tips.

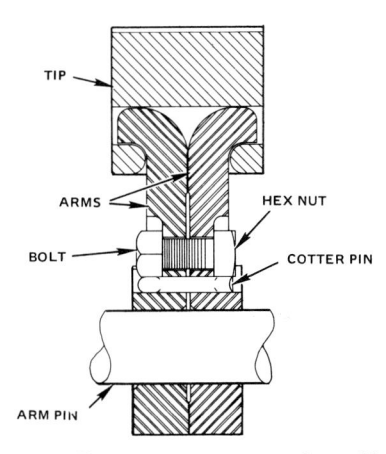

Fig. 9-83 Front view cross section of assembly and installed finger clamp hammer.

(Fig. 9-82). When arms show rapid wear in the shank area, material penetration may be far too great due to the velocity of the feed. Immediate steps should be taken to adjust penetration so that arm shank wear is reduced!

When arms are replaced, the balance of the rotor assembly must be maintained. Never attempt to replace only one arm!

Finger clamp hammers (Clark) The finger clamp hammer assembly, shown in Fig. 9-83, is made up of six separate pieces.

Assembly Method

1. Lay the tip on one of its impact faces, with the four finger holes exposed on the sides and the bottom hole toward you.

2. Hook one arm section into the right-hand finger holes, and one into the left-hand finger holes.

3. Move arm sections together and install bolt and nut. Tighten nut.

4. Insert cotter pin through arm holes under bolt and spread split end to keep it in place. NOTE: Always use a cotter pin which is in new or good condition.

Disassembly Method

1. Close ends of cotter pin and pull or drive pin out.
2. Remove bolt.
3. Spread arm sections and unhook each from tip.

Installation of Assembled Hammers

1. Check to be sure all hammers are properly assembled.
2. Weigh hammers for spinner balancing (optional). The weight of hammer assemblies is approximately equal. However, there are normally some small variations, and if

these happen to occur so that there is an accumulation of weight on one side of the spinner, the mill may vibrate. In view of the small amount of extra work involved, it is strongly recommended that the balancing procedure be carried out as described in the following paragraphs.

Limit of Hammer Tip Wear (Clark). Hammer tips are completely worn out and should no longer be used when only ¼ in. of material remains above the outer corners of the finger holes (see Fig. 9-85).

Any use of worn-out tips is dangerous to personnel, and a potential cause of severe damage to the mill. Continued use is poor economy.

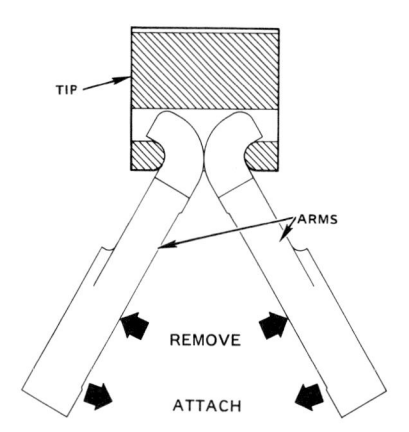

Fig. 9-84 Attaching or removing hammer tip.

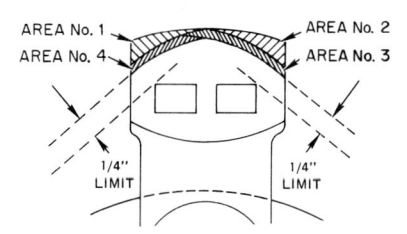

Fig. 9-85 Tip wear pattern and wear limit for finger clamp hammers.

Hammer wear   Hammers should be carefully inspected at frequent intervals to determine the rate and nature of tip wear for each specific material being crushed. As the tips wear, they should be turned 180° so that their trailing edge becomes their leading edge and wear occurs over a new area. By so doing, tips maintain a uniformity of shape and do not reach the worn-out condition on one side only.

In addition to the turning feature to increase tip service life, the spinner is also equipped with radially graduated sets of holes for the arm pins. This permits adjustment of worn tips ¼ in. closer to the fixed grates so that material gradation can be maintained despite tip wear. There are three possible adjustments of ¼ in. for mills having two rows of hammers on the spinner (see Fig. 9-86).

Spinner   The spinner assembly is made up of the main shaft, roller bearings, disks, arm pins, and hammers (Fig. 9-87). The double-row roller bearings at each end of the shaft support the assembly. Bearing clearance is adjustable to permit compensation for wear, so that shaft movement is only rotary.

The nine steel disks are shrunk onto the shaft for positive positioning. Each disk has 12 holes for arm pin insertion. The holes are arranged so that the arm pins (and hammer tips) can be set farther from the shaft axis to increase the service life of the tips while maintaining product particle size.

The arm pins which hold the hammer assemblies are locked in place by split retaining rings (snap rings) which fit into a machined groove in each of the two end disks.

Hammer weight-balancing procedure (two-row spinner only)   The following procedure will produce a minimum vibration of the impact mill, as the spinner will be both statically and dynamically balanced within normal standards and requirements.

1. Weigh each of the completely assembled hammers and write the weight on the hammer tip.

2. Lay the weighted hammers in a straight-line sequence of *diminishing weight*. The heaviest hammer will be at one end, and the lightest hammer at the other.

3. Mark the heaviest hammer no. 1. Continue down the line, numbering each hammer in sequence until no. 16, the lightest hammer, is marked.

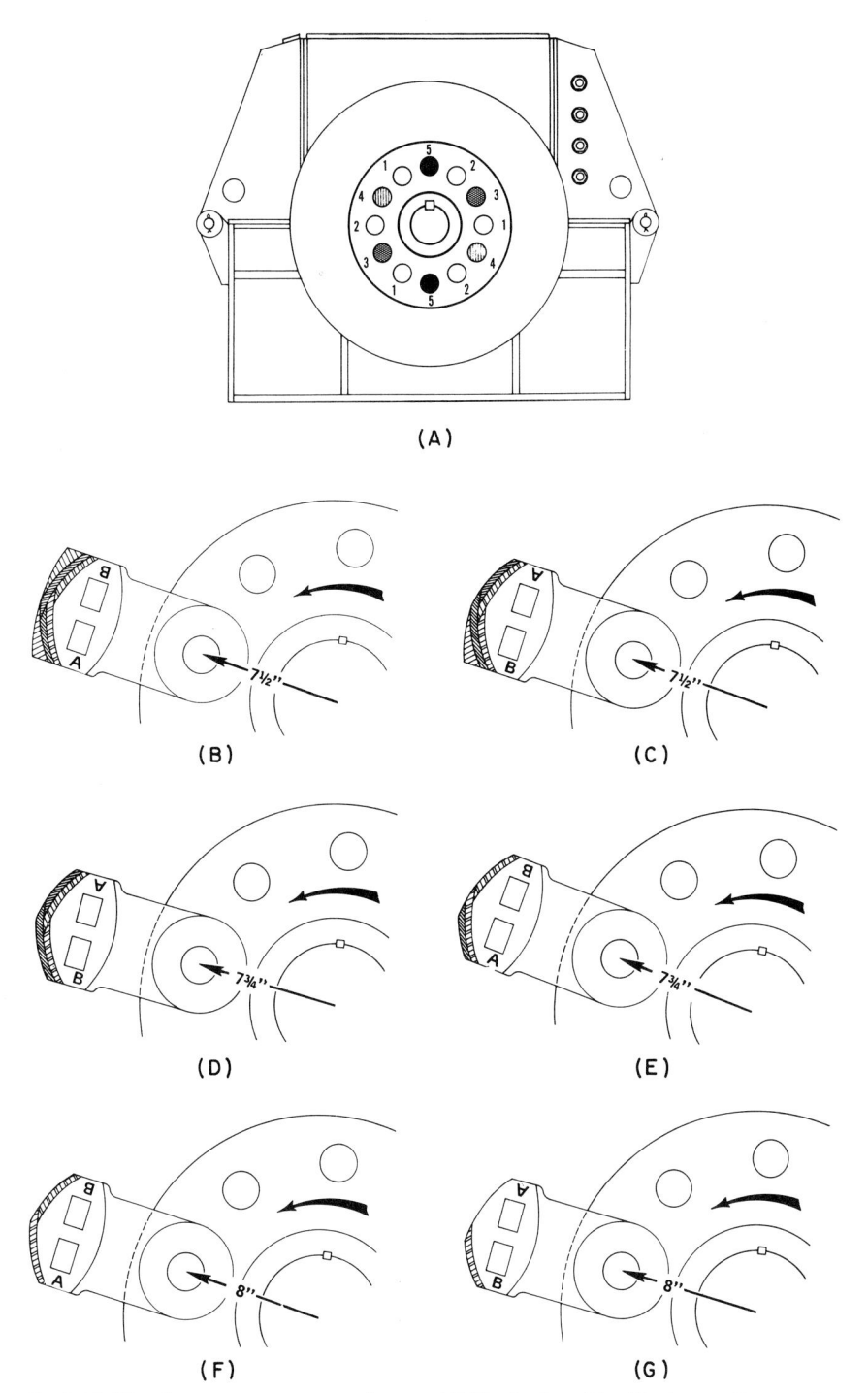

Fig. 9-86 Tip adjustment to compensate for wear. (A) Flywheel on right side. Arm positions no. 4 = 7½-in. radius (new tips), no. 3 = 7¾-in. radius, no. 5 = 8-in. radius (final wear). (B) First wear area, (C) second wear area (tip turned), (D) third wear area (arm pin moved), (E) fourth wear area (tip turned), (F) fifth wear area (arm pin moved), and (G) sixth wear area (tip turned).

4. Select the two holes in the spinner disk where arm pins will be inserted. Mark one Y and one Z.

5. Insert arm pin Y and install hammers numbered 13, 10, 7, 1, 4, 6, 11, and 16, in that order.

6. Insert arm pin Z from the same end of the spinner and install hammers numbered 14, 9, 8, 2, 3, 5, 12, and 15, in that order.

7. Total the weights of each row. IMPORTANT: There must not be more than 2 lb difference in total weights. If there is, interchange two or more weights between the rows so that the difference is less than 2 lb. Always keep the heavier hammers in the middle of the spinner.

NOTE: In the event that the mill is to be operated at a spinner speed greater than 1000 rpm, a more accurate weight-balancing effort is necessary. At higher speeds, even small amounts of dynamic unbalance create intolerable frame vibrations which can rapidly shake the spinner assembly loose. In the event of spinner weight balance problems, contact the factory at once.

Hammer weight-balancing procedure (three-row spinner only) The following procedure will produce minimum vibration of the impact mill, as the spinner will be both statically and dynamically balanced within normal standards and requirements.

1. Weigh each of the completely assembled hammers and write the weight on the hammer tip.

2. Lay the weighed hammers in a straight-line sequence of diminishing weight.

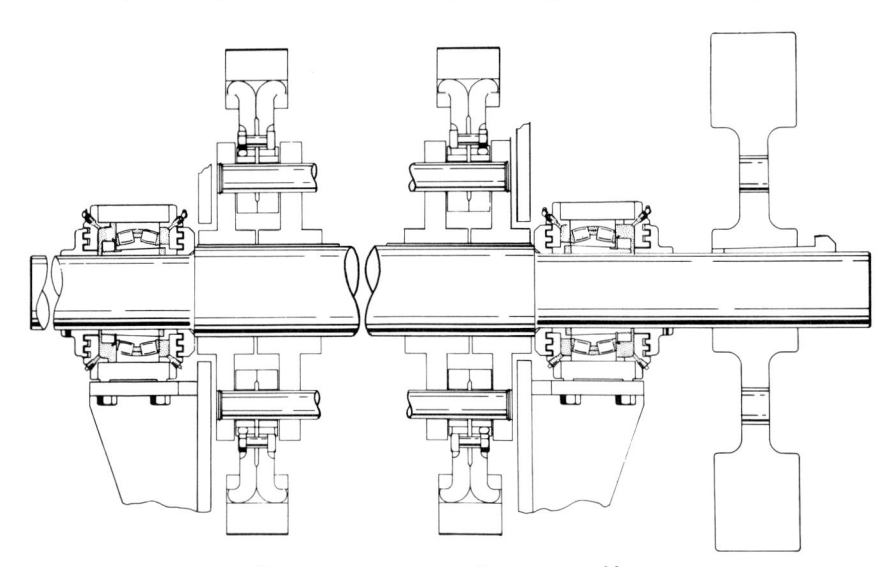

Fig. 9-87 Cross section of spinner assembly.

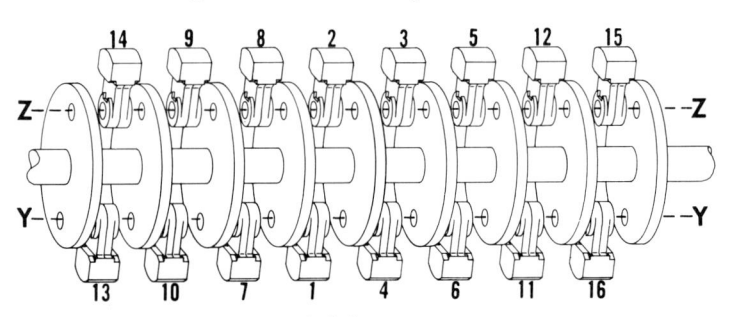

Fig. 9-88 Weight-balancing a two-row spinner.

3. Mark the heaviest hammer no. 1. Continue down the line, numbering each hammer in sequence until no. 24, the lightest hammer, is marked.

4. Select the three holes in the spinner disk where arm pins will be inserted. Mark one V, one W, and one X as shown in Fig. 9-89.

5. Insert arm pin V and install hammers numbered 24, 15, 7, 2, 5, 11, 18, and 19, in that order.

6. Insert arm pin W from the same end of the spinner and install hammers numbered 22, 13, 8, 3, 4, 12, 17, and 20, in that order.

7. Insert arm pin X from the same end of the spinner and install hammers numbered 23, 14, 9, 1, 6, 10, 16, and 21, in that order.

8. Total the weights of each row. IMPORTANT: There must not be more than 2 lb difference between any two of the rows. If there is, interchange two or more hammers

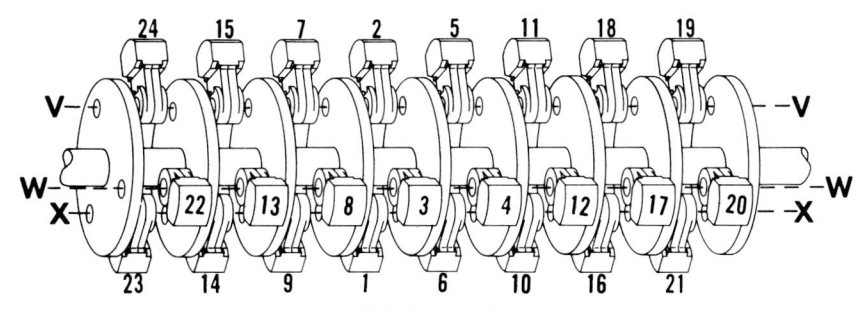

Fig. 9-89 Weight-balancing a three-row spinner.

between rows so that the difference is less than 2 lb. Recheck all three totals against each other after making changes. Always keep the heavier hammers in the middle of the spinner.

NOTE: In the event that the mill is to be operated at a spinner speed greater than 1000 rpm, a more accurate weight-balancing effort is necessary. At higher speeds, even small amounts of dynamic unbalance create intolerable frame vibrations which can rapidly shake the spinner assembly loose. In the event of spinner weight balance problems, contact the factory at once.

Spinner repair The disks of the spinner assembly are very durable parts which normally will be serviceable for several years. In time the abrasive action of the material will reduce their diameter and thickness. The hammer tips will deform their rim when the hammers are forced back against it by large or hard pieces of material or by an excessive feed rate. The holes for the hammer arm pins will become elongated and cracks toward the rim may develop. Any time the spinner cannot be kept in balance despite careful hammer weighing and distribution, the spinner disk should be replaced.

NOTE: Spinner disk replacement is a complex job which requires that the individual disks be shrunk onto the shaft with proper aligning bars used for accurate positioning. It is strongly recommended that this work be done at the factory. Contact the factory for complete details.

Spinner shaft bearings The mill is equipped with double-row spherical and self-aligning bearings. The bearings are protected against the loss of lubricant and the entrance of sand and grit by labyrinth-type grease seals. Sufficient grease is permitted to work through seals to prevent dust and grit from entering the bearing housings.

Check Bearings. The bearings normally operate at a temperature of 100 to 150°F but in warm climates may operate at 180°F, periodically running as high as 200°F. Under these conditions and within these limits there is no cause for alarm. However, if a bearing maintains a temperature of 212°F or higher, it is considered as overheating and the cause should be found and corrected to prevent damage.

1. A reliable thermometer provides the best method of checking bearing temperatures. Remove fill plug and insert pencil-type thermometer in housing lubricant and get actual reading. Readings for the bearings can be taken on the bearing cap by removing the fill plug on the housing and inserting thermometer or by draining out a small amount of lubricant from the drain hole on the underside of the housing.

2. The bearing temperatures can be checked by placing the hand on the bearing pillow block. If the hand can be held there for a few seconds, the bearing is not too warm. CAUTION: Mill must not be in operation when this is done.

3. If there is a question whether or not the bearing is too hot, always use a reliable thermometer as the final check.

Never reduce bearing temperatures by artificial means, such as by pouring water on them, as the sudden change in temperature is apt to warp, crack, or crystallize some part of the bearing assemblies.

Bearing Heating Causes

1. The most common cause of overheating is *too much* lubricant and most generally occurs when the mill is new or immediately after lubricant has been changed. The bearing will heat up and force out the excessive lubricant until the proper amount is left and then gradually cool down to the normal operating temperature.

2. The bearing will overheat if the mill is operating faster than recommended under full load. Check the speed with a speed indicator.

3. One of the bearings may overheat if the unit is operated decidedly out of level. An extremely unlevel position may result in the bearing not receiving its share of load and one bearing will begin to get hot. Level unit by checking shaft with a spirit level and compensate for any settlement by use of cribbing, jacks, or shims.

4. Overheating is also a sign that bearing failure is developing. Check the overheating bearings for condition and wear by jacking up the end of shaft at the affected bearing. Observe the assembly during these operations. *The bearing should not permit perceptible up and down movement.* If movement is visible, bearing is excessively worn and should be replaced.

5. Heat will result if there is insufficient radial clearance between the bearing cap and bearing outer grease seal.

There must be clearance for at least a 0.020-in. feeler gauge completely around the circumference of the seals. Check four times, revolving the shaft assembly about 90° after each check. If these seals are touching at any point, there apparently is a bearing failure. Also, if the seal is allowed to run loose and dirt and grit enter the bearing, this will have a serious effect on the bearing.

Operational troubleshooting guide

Excessive Wear on Hammers
1. Improper feed of material
2. Incorrect speed
3. Incorrect grate combination
4. Incorrect hammer position
5. Material very abrasive
6. Not using genuine manufacturer's parts

Low Capacity
1. Hammers or grates worn excessively
2. Incorrect speed
3. Grates too close to hammers
4. Insufficient power or V-belt slippage
5. Improper feed of material

Bearings Heating
1. Too much lubricant
2. Insufficient lubricant
3. Dirty lubricant
4. Wrong lubricant
5. Operating at higher rpm than recommended
6. Out of level
7. Outer bearing grease seal not turning with shaft
8. Insufficient lateral clearance between bearing cap and outer grease seal
9. Bearing failing
10. Arced through bearing with welding machine

Vibration
1. Hammer locked between disks
2. Using hammer assemblies of different weights
3. Out of level
4. Bearing failing

Spinner installation procedure
1. Set spinner assembly on base and install hold-down bolts. Do *not* tighten.
2. Set the rim of the two outer disks an equal distance from inner surface of base (four measurements).
3. Lay a straightedge along the front and back surfaces of base, and center the shaft accurately between these surfaces at both ends.
4. Use a feeler gauge to check for clearance between both grease seals and the bearing housing. There must be no contact (see Fig. 9-90)!
5. If necessary, shift housing enough to produce side clearance for seals.
6. If necessary, add shims at corners of housing to produce top or bottom clearance for seals.
7. When housing is correctly located, tighten hold-down bolts, then recheck seal clearances.
8. Install jam nuts and tighten.
9. With left-hand bearing assembly still loose on shaft, pry housing outward as much as bolts will allow. Draw all four bolts snug.
10. Remove bearing cap and locknut. Install lock washer and force outer race of bearing tight against inner bearing cap.

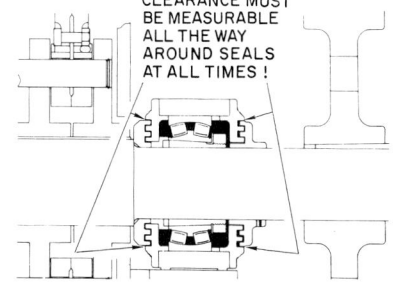

Fig. 9-90 Checking clearance between grease seal and bearing housing.

Repeat steps 4 through 8 and 10 through 14 of the Bearing Installation procedure sequence. Omit step 9 as floating side has no spacer.
11. Install upper section of mill body.
12. Remove bearing drain plugs. Pump the proper lubricant into housing through both fittings until it extrudes from both drain holes. Replace drain plugs.

Bearing removal
1. Remove flywheel key and flywheel (if flywheel side).
2. Remove outer grease seal.
3. Remove outer bearing cap.
4. Remove spacer ring (if flywheel side).
5. Straighten prong of lock washer so that locknut can be loosened.
6. Unscrew locknut and remove lock washer.
7. Screw locknut back onto sleeve until its outer face is just flush with the outer end of the sleeve.
8. Using a soft drift bar, drive the sleeve inward enough to loose it from the bearing.
9. Remove the locknut, and pull the bearing out of the housing. The end of the shaft must be supported to free the bearing.
10. Use a screwdriver to spread the split sleeve so that it can be pulled off the shaft.

Bearing installation
1. If required, install new grease seals on spinner shaft. Heat to 300°F in oil, install, and hold tight against shoulder until fixed (see Fig. 9-91).
2. Install bearing assemblies on shaft with hex nuts *outward*. Force inner bearing cap tight against inner grease seal.
3. Remove outer bearing cap and locknut. Install lock washer and force outer race of bearing tight against inner cap.
4. Reinstall locknut and tighten against lock washer.
5. Use a spanner wrench for final tightening. (Be sure outer bearing race is all the way in against cap.)
6. Tighten locknut to obtain clearance shown in step 7. If you cannot obtain 0.005 in. even when nut is very tight, *do not* continue tightening and stretch sleeve threads!

7. Check for proper clearance with feeler gauge (0.004 to 0.005 in.). (See Fig. 9-92.)
8. Bend lock washer prong to engage locknut notch.
9. Install spacer. (Flywheel side only is fixed side.)
10. Install outer bearing cap (grease fitting upward).
11. Install tie bolt hex nuts. Draw nuts just snug all the way around, then tighten.
12. Install outer grease seal. Force it tight against bearing housing.
13. Scribe a pencil line around the shaft using outer seal to guide the point.
14. Pull grease seal outward just enough to hide the pencil line. Tighten setscrew while in that position. IMPORTANT: Whenever excessive grease is extruding from seals, reset the outer seals in this manner.

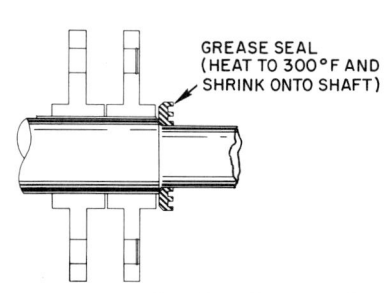

Fig. 9-91 Applying heated grease seal.

Fig. 9-92 Checking clearance with feeler gauge.

Bearing assembly (on bench)
1. Install bearing cap on housing with grease fitting up. Insert all tie bolts.
2. Turn housing so that bearing cap is on bottom side.
3. Install spacer with small end fitted in bore of bearing cap.
4. Fit tapered sleeve into bearing.
5. Screw locknut onto tapered sleeve. Do *not* tighten!
6. Lower bearing assembly into housing with locknut upward. Start bearing squarely and tap with plastic mallet if necessary.
7. Check to be sure bearing contacts spacer at bottom.
8. Install outer bearing cap with grease fitting at top of housing.
9. Install hex nuts on tie bolts.

CRUSHER LUBRICATION RECOMMENDATIONS

The environmental conditions under which crushers operate are among the most severe. Fine mineral particles are a threat to ball and roller bearings, and special precautions are necessary to obtain satisfactory service life from these very vital parts. The following suggestions and recommendations will help to establish sound lubrication practices:

1. Provide a good storage and distribution center for lubricants. It should be an area that can be kept as free of dust accumulation as possible. It should have adequate shelving for segregation of lubricants and efficient means of labeling containers or compartments. Labels can include identification of machines requiring the lubricant and even details of the exact application, taken from machine manufacturer's instructions.

If the manufacturer's instructions are in the form of special illustrated charts or manual pages, it is also highly recommended that each of these be encased in a plastic envelope for protection and permanently posted in the storage room for ready reference.

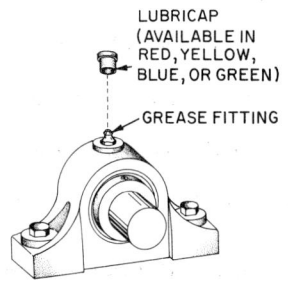

Fig. 9-93 Using plastic Lubri-caps.

2. Maintain an inventory and use chart in the storage center which will show at a glance the approximate

rate at which each lubricant is used. This not only helps to prevent shortages but serves as a check by supervisors on the efficiency of the lubrication job being performed.

3. Maintain a good supply of industrial wipers so that the oiler is encouraged to keep grease guns, fill can spouts, funnels, and grease fittings wiped clean before each use. Any introduction of dust particles along with lubricant will nullify the lubrication efforts by actually causing bearing failures.

The use of plastic Lubricaps (Fig. 9-93) is highly recommended. These snap-on coverings make it unnecessary to wipe the fitting before the grease gun is applied. If the cap is held during the greasing and then reinstalled, there is no chance for contamination. The gun tip must be wiped each time before application. The caps come in several colors and the color-code method of identifying lubrication requirements is fast and efficient.

4. Use cartridge-type grease guns rather than those filled from bulk supply. There is much less chance for contamination with the cartridge method.

5. Use lubricants specified by machinery manufacturers for each lubrication point. Many times the lubricant specified has special qualities which enable it to perform under conditions of extreme heat, pressure, or chemical stress without damaging effects on any of the machine components. Substitution of lubricants thought to be equal is risky unless a complex comparative study is made by lubrication experts.

6. Establish a foolproof system of communicating any unusual lubrication problems or information, from operators or mechanics who see equipment in operation and in the dismantled state to oilers who perform their work when the plant is shut down and deserted. Failure to do so can result in frequent unnecessary breakdowns. A simple note board in the lubricant storage shed can accomplish this exchange of information.

Chapter **10**

Asphalt Plant Maintenance

G. F. RITTER
Service Manager, Asphalt and Stabilization Products
Barber-Greene Company
Aurora, Ill.

INTRODUCTION

Modern asphalt plants can produce highest-quality mixes at rates which were considered astronomical just a few years ago. Hourly rates can vary from 20 to 1000 tons/hr or more. Obviously, high production coupled with heat and dust which are inherent in asphaltic concrete manufacture make proper maintenance procedures even more important than they were in the past.

Put another way, plant production breakdowns are much more costly than just the loss of the mix. Clearly, if no mix is available, hauling units, finishers, and crews are idled not only on your job but on outside jobs to which you are supplying mix.

Obviously, then, there is no substitute for sound maintenance. Many operators who do give careful attention to proper maintenance report they have gone through entire seasons without a single significant plant breakdown.

Asphalt plants range from portable, small (Fig. 10-1) and large (Fig. 10-2) to large, permanent drum mixers (Fig. 10-3) and to pugmill batch plants (Fig. 10-4). The smallest consists of little more than a mixer whereas the largest is a collection of different types of equipment: conveying, screening, drying, mixing, storage, and controls. Some are fully automated.

Because plant types are so different, each requires different maintenance procedures. For simplicity, this chapter is divided into a discussion of the various components that make up most asphalt plant installations.

A schedule for the periodic inspection, cleaning, lubrication, and adjustment of plant components should be established by the operator on the basis of past experience and severity of service. If such a procedure is not provided for in your operator's manual, some kind of maintenance checklist should be developed. Perhaps, too, a daily logbook should be established to record what maintenance was done and what adjustments or repairs were made.

Always remember that asphalt plants operate under the most adverse conditions. They handle highly abrasive materials that generate a lot of dust and grit which are hard on machinery. This places special emphasis on the importance of good housekeeping—of keeping components clean.

INSPECTION AND MAINTENANCE

Daily inspection of plant components is just as important as lubrication. This procedure will enable you to catch problems before they become serious.

Inspection should be made before the start of the day's work and at the close of the day as well. Never assume that because the plant worked well all day that something cannot go wrong. Also, don't assume that because everything was in order the evening before that something hasn't happened to the plant overnight. There is always the danger of vandalism, for example. If not discovered in time, acts of vandalism, such as sand in gearboxes, can cause major damage to the plant. Each morning walk around the plant, checking for signs of tampering.

Fig. 10-1 Portable continuous asphalt mixer.

Fig. 10-2 Portable batch plant.

Fig. 10-3 Drum mixer.

Fig. 10-4 Fully automated batch plant.

Vibrating Screen

Aggregate is carried over the vibrating screen (Fig. 10-5) with a minimum of screen action. A minimum throw is recommended so that the aggregate will be carried over to the screen with very little lifting action, giving each particle the maximum number of chances to pass through a screen cloth opening.

Counterweights installed at the factory normally give the best results for screening most types of aggregate commonly used in paving mixes.

Yet, due to the variation in fracture of aggregate particles in the many types of paving aggregate encountered, the screen may not give maximum screening efficiency at any given counterweight setting. Therefore, adjustments to vary the amount of screen action may be necessary to gain maximum efficiency.

Elongated and tapering aggregate particles may be encountered that will not screen

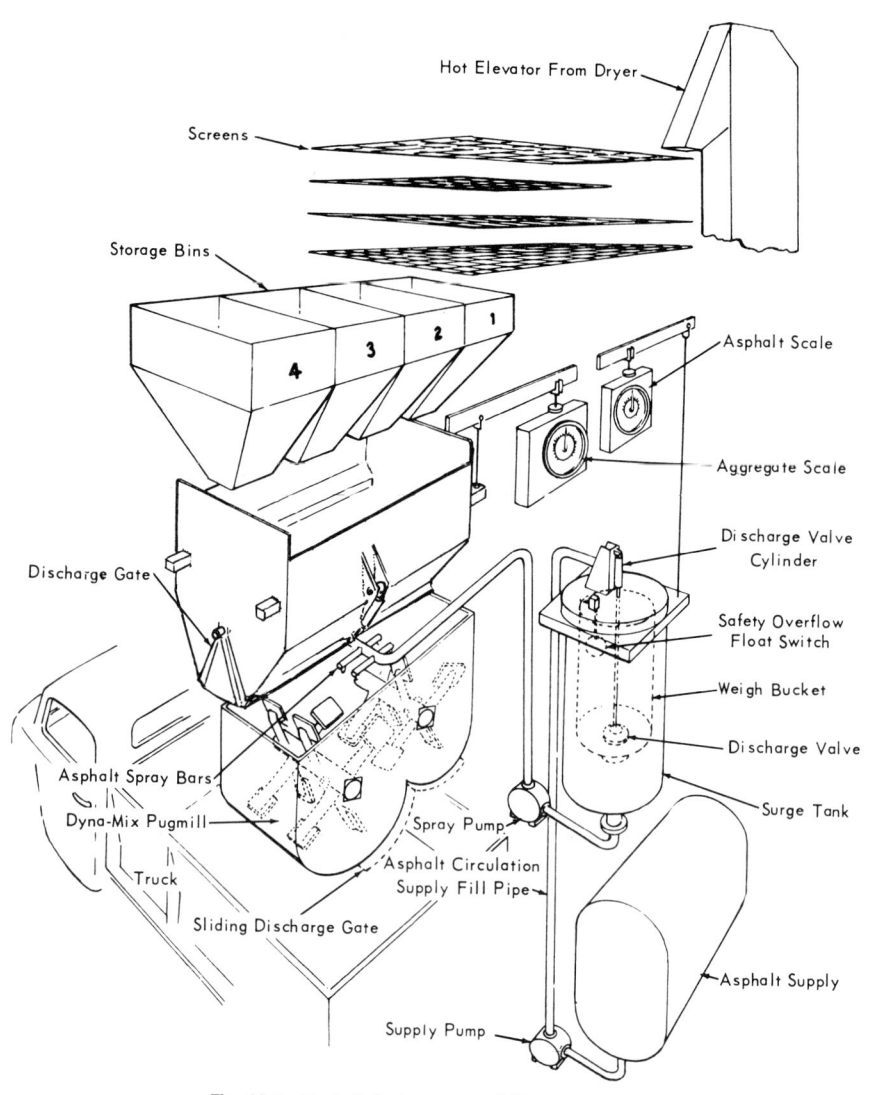

Fig. 10-5 Exploded view, material flow components.

properly. These slivers of stone may become lodged in the screen openings and blind the screen. This blinding will cause a carry-over of material from a given cloth opening, and a high percentage of overrun will show up in the gradation tests. To correct this, weights will have to be adjusted to create greater screen action and bounce the aggregate so it will not become lodged in the screen cloth openings and blind the screen.

Adjusting the stroke or amplitude The stroke or amplitude of the horizontal screen can be adjusted on the job site by altering the amount of counterweights (Fig. 10-6) to increase or decrease the throw of the unit.

To increase the stroke, add the counterweights to the outer offset portion of both shafts. To decrease the stroke, remove counterweights from the outer offset portion of both

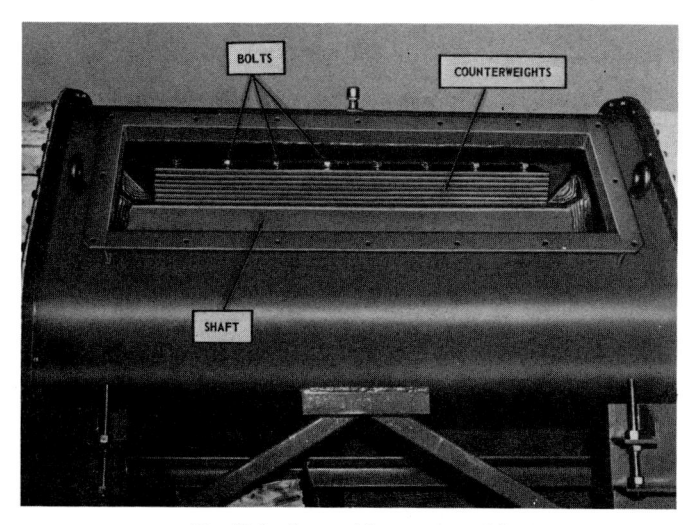

Fig. 10-6 Screen drive counterweights.

shafts. If the shafts on your particular unit are equipped with drilled holes or counterweights on the inner portion of the shafts, it is possible to reduce the stroke even more by removing all the counterweights from the offset portion and adding counterweights at the inner point. However, we must caution individuals who do want to alter the stroke that it is of vital importance to ensure that the proper balance of the unit is maintained. Each shaft should normally contain the same amount of counterweights.

A certain thickness or blanket of aggregate must be carried over the screen at all times for proper screening action. This is usually twice the thickness of the aggregate being screened. If the blanket is too thick, the screen's efficiency will be cut, causing overrun. To correct this condition the amount of aggregate being discharged onto the screen will have to be reduced.

Replacing screen cloths Screen cloths are replaceable to provide various aggregate gradations. It is always desirable to select screen sizes that will separate aggregate equal proportionately to the number of bins.

For best operation, screen cloths should be kept drumhead-tight. Check for tightness on a weekly basis. Also check the screen cloth clamps for loose bolts. Periodically inspect the screen for worn or damaged screen cloths.

If the plant is to be shut down for any length of time, as during the winter season, run the screen a few minutes every month to maintain a film of oil on the bearings.

Springs Periodically check for weak or broken support and snubber springs. Replace immediately.

Screen drive Periodically check screen drive sheaves (Fig. 10-7) to see that they are secure on the shaft. Should they become loose, it may be necessary to replace the key. Also, check sheave alignment with a straightedge across the flat of the drive and driven sheaves.

7-274 Maintenance Programs for Equipment Entities

Charging end wear plate Some types of aggregate will cause more rapid wear of the charging end wear plate than others. A periodic inspection should be made to establish the rate of wear and interval when a new wear plate should be installed. It should be replaced when worn thin before the support plate is exposed to wear. A new plate should be kept on hand, available when needed.

The flow selector gate is also replaceable and should be replaced when wear is indicated.

Vibrating unit

Disassembly. The wear rate of vibrating unit components (Fig. 10-8) is far more rapid than that of other components of an asphalt plant because of the extreme shock and

Fig. 10-7 Screen drive.

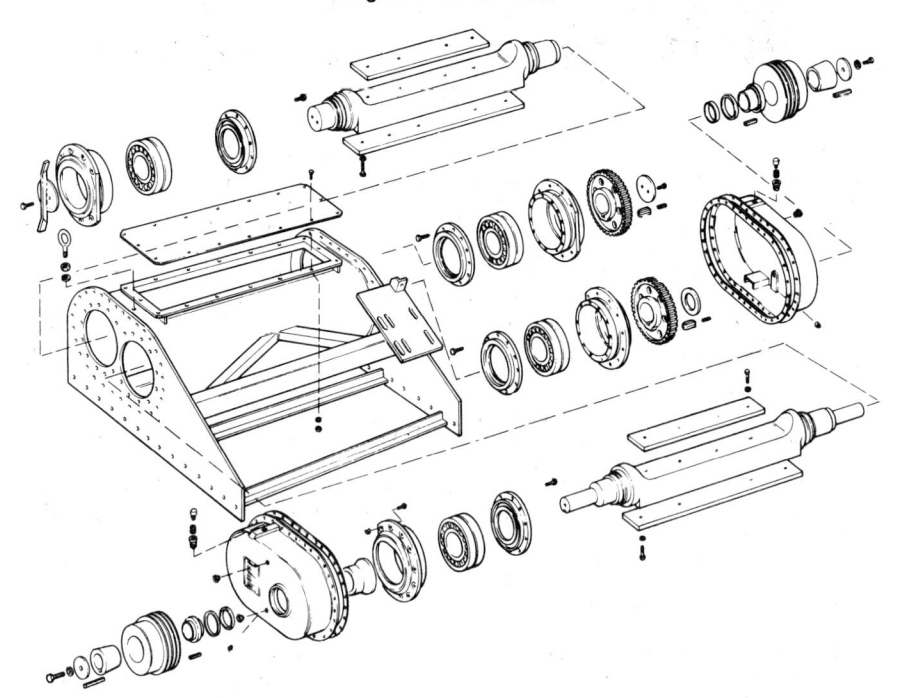

Fig. 10-8 Vibrating unit assembly.

agitation to which they are submitted. Repairs, however, should not be done in the open but rather in a shop, garage, or other dust-free and weather-proof shelter. Nor should the work be done by anyone who is not thoroughly familiar with the construction of the vibrating unit.

Extreme care should be taken to prevent the entry of dirt into bearings or other components of the unit. Before starting disassembly, clean the exterior of the unit thoroughly.

1. Use extreme care, also, in handling gears and shafts. Avoid nicking or marring the surfaces.
2. Handle bearings and oil seals cautiously. Do not cut or damage seals when reassembling.
3. If shafts, bearings, or other parts with machined surfaces are to be kept out of the unit for any length of time, apply a protective coating of rust preventative.
4. Keep all parts of the gear end of the unit separate as they should be installed in the same side from which they were removed.
5. Inspect gears thoroughly under a strong light. Replace if gear teeth are broken, pitted, galled, scored, cracked, or heavily discolored. When any doubt exists, it is better to replace gears rather than risk early failure after repair of the unit.
6. Thoroughly inspect all other parts for signs of wear. When one part of the vibrating unit has failed, there is always the possibility that other parts were subjected to excessive loads. Check to prevent reassembly of partially damaged parts. Look for scratches, nicks, or burrs. If any are found, dress them neatly with a file and emery cloth. Before assembly, fit all parts to their mating surfaces to make sure they match properly.
7. Replace mechanical friction-type oil seals with new seals each time the unit is assembled.
8. Replace any gaskets if they show any signs of leakage or damage.

Reassembly. Strict standards of cleanliness should be maintained when reassembling the vibrating unit. With the exception of new bearings, all other parts should be thoroughly washed, cleaned, and then protected to keep out contaminants.

Bearings. The most common cause of bearing failure is the introduction of dirt or other foreign materials into the bearing before or during assembly or in operation. Here are a few rules to follow when handling bearings.

1. New bearings should never be removed from their protective wrappers until you are ready to use them. Install just as they come from the package without washing.
2. Work with clean tools in clean surroundings.
3. Use clean solvents and flushing oils.
4. Remove all exterior dirt from unit before exposing bearings.
5. Handle bearings with clean, dry hands.
6. Use clean, lint-free rags.
7. Wash used bearings in clean solvent.
8. Blow the bearings dry with filtered, moisture-free compressed air.
9. Immediately after drying, dip the bearings in a light spindle oil to prevent rusting.

Fines Feeding and Measuring System

Elevator The chordal action of the elevator bucket line chain (Fig. 10-9) causes a so-called hopping action of the elevator foot shaft. Coil springs have been provided to dampen this hopping action, but from time to time they may have to be adjusted.

Never completely collapse springs. Instead, adjust lower springs until the foot shaft is level and springs are just partially collapsed, putting tension on the foot shaft. Measure from take-up bolt support angles to center line of foot shaft to assure correct sprocket alignment. Then, tighten upper springs to provide slight tension on the lower springs.

Screw conveyor To adjust the drive chain (Fig. 10-10), loosen idler mounting bolts and move the idler until the chain has the proper tension.

Weigh Hopper

Storage bins Periodically check the strike-off gates (Fig. 10-11) for wear. If a gate is sticking, it may be warped or worn and need replacement.

Storage bin indicator Check indicator counterweight arms periodically to see that they operate freely and do not bind.

Weigh Hopper Scale

If you have reason to believe that the scale is giving inaccurate readings, test with a given amount of known weights. Frequently, though, inaccurate readings are an indication that some components of the weighing system are not level. Follow leveling procedure as outlined in the owner's manual for your plant.

Should the scale still give inaccurate readings, adjustment may be required. This procedure calls for extreme care and should be attempted only by an experienced

MINERAL FILLER (FINES) SECTION AND SERVICE INFORMATION

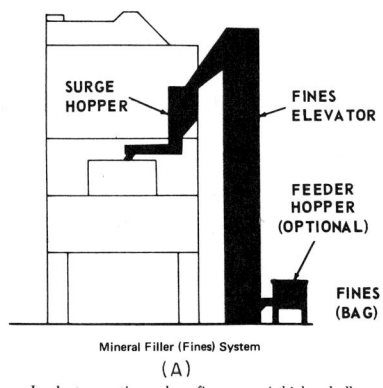

(A) Mineral Filler (Fines) System

In plant operations where fines usage is high, a bulk fines system using a storage silo for maintaining several days supply of fines is used,

MINERAL FILLER (FINES) SYSTEM & SERVICE INFORMATION (OPTIONAL)

The mineral Filler (fines) feeding and measuring system is optional equipment. The fines are fed with a ground mounted feeder and silo, Figure A The fines feeder discharges into a bucket elevator which elevates the fines to a surge hopper, mounted inside of the supply bin section of the tower, Figure B

The surge hopper is kept full by the continuous operation of the fines feeder and elevator. The surge hopper overflow returns to the fines elevator.

A screw conveyor delivers the fines from the surge hopper into the weigh-hopper.

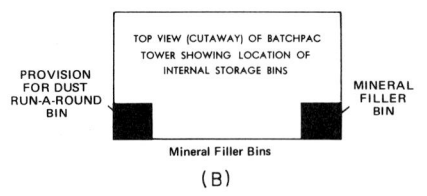

(B) Mineral Filler Bins

The screw conveyor is driven by a motor with speed reduction. The motor is equipped with an automatic brake. When power is applied to the motor, the brake is automatically disengaged. As the motor power is broken or cut off, the brake is applied by spring tension, causing the screw to stop without coasting.

A flop gate is provided from emptying the surge hopper for screw maintenance or hopper clean out.

A roto-bindicator (optional), Figure C indicates amount of fines in the surge hopper.

ROTO-BINDICATOR (OPTIONAL)

The roto-bindicator consists of a motor, micro switch and a driven paddle wheel. The motor is powered constantly but is stalled by fines Material holding the paddle wheel from turning. When the paddle wheel is held immobile, one micro switch is actuated indicating the fines bin is full (Pilot light out). If there is insufficient material in the bin the wheel turns free, the pilot light indicates a low bin.

(C) Fines System

Fig. 10-9 Mineral filler (fines) system.

mechanic. After the scale has been adjusted, it is likely that it will have to be sealed by the local department of weights and measures.

Asphalt Weigh Bucket

Always allow the asphalt heating system sufficient time to warm up before actuating the asphalt weight bucket dump valve. Sticking of the valve owing to cold asphalt can cause damage.

Spraying of asphalt while surge tank is being filled may indicate a worn or damaged dump valve.

Asphalt Spray

Spray pump (optional) Periodic inspection should be made of the packing gland to ensure that the packing is adjusted properly. When all take-up of the packing gland has been used, new packing should be installed. Also, occasionally check tension of V-belts which drive the pump.

Fig. 10-10 Screw conveyor drive.

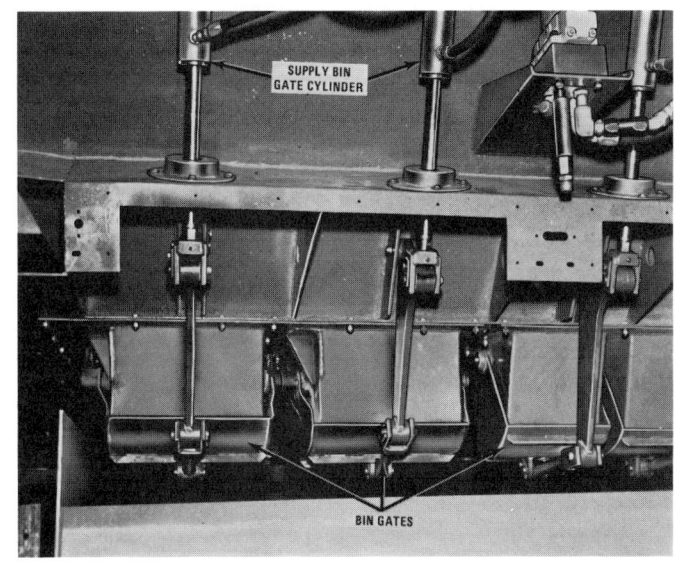

Fig. 10-11 Supply bin strike-off gates.

Gravity spray (optional) In normal operation a gravity spray system (Fig. 10-13) requires little attention. From time to time check the slots in the spray bar for plugging.

Asphalt Sump and Strainer

In normal operation the screen should be removed and cleaned after every 40 hr of operation. However, with some types of asphalt containing an excess amount of foreign materials the screen should be checked and cleaned more often.

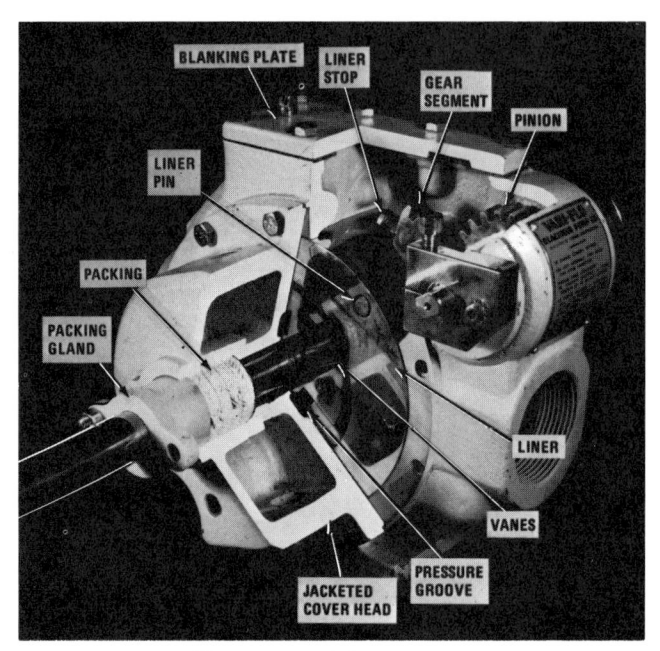

Fig. 10-12 Asphalt metering pump.

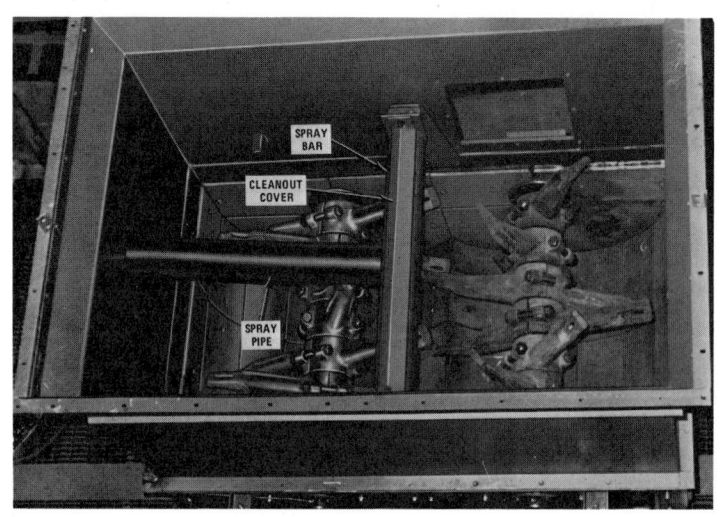

Fig. 10-13 Gravity spray bar.

Asphalt Transfer Pump

A periodic inspection of the transfer pump (Fig. 10-14) packing gland should be made to see that the packing is adjusted properly. When all take-up of the packing gland has been used, new packing should be installed.

CAUTION: Do not overtighten gland nut as overheating and scoring will occur. A slight leakage is preferred.

It is always wise when replacing the packing to check the shaft in the area of the stuffing box for scoring. Replacement with another set of packing will be ineffective in controlling the leakage if the previous packing has critically damaged the shaft. In that case the motor and shaft assembly must be replaced.

Asphalt Transfer Valve

Little maintenance of the valve is required. When worn, the head seal will need replacement.

NOTE: Some valve designs require grease for sealing.

Asphalt Metering Pump

On drum-type mixers periodically inspect the packing gland for leaks. When take-up on the packing glands is used up, new packing should be installed.

Fig. 10-14 Asphalt transfer pump.

If leakage appears from between the casing and the head, the head should be removed, sealing compound cleaned off, and the inner face inspected for burrs, dirt, or other surface imperfections. Reassemble, using a fresh coat of sealing compound on the faces. It is advisable to allow the compound to dry for 2 hr before operating the unit.

If there is any leakage of steam or liquid between the cover and the head even after tightening the cap screw firmly, the cover should be removed and its gasket replaced. Leakage around the shaft should be controlled by proper adjustment of the packing glands. If leakage is excessive even with the proper adjustment or the glands overheat when the leakage is controlled, packing should be removed and completely replaced.

Overheating of bearings Abnormally hot bearings may result from:

1. The packing being worn out or adjusted too tightly.
2. Misalignment of the unit. Remove the inboard bearing from the stuffing box. If the adjustment is proper, it should reenter the stuffing box without force.
3. Pumping a hot liquid. Bearing temperatures will actually run somewhat hotter than the liquid being pumped. When handling a high-temperature liquid, care should be taken that a lubricant is used which will not melt and run out of the bearing.
4. Insufficient lubrication. Bearings with standard grease cups should be serviced regularly, and lubricant levels in the automatic type of lubricators should be inspected and kept adequately full. New bearings may be damaged permanently by poor attention to their conditioning.
5. Noise. When a pump has been completely drained and is full of air, some noise may occur when the pump is restarted and the air is purged out of the system. The noise, although seldom present, is usually of short duration and causes no damage.

Supply pump Periodically inspect the packing gland to see that the packing is adjusted properly. When take-up on the packing gland is used up, new packing should be installed.

Packing gland nuts should be tightened only as much as needed to stop seepage. Even a small amount of seepage is not objectionable. It is much better than to have the packing too tight, causing heating and damage to the pump shaft.

Should the pump fail to operate satisfactorily, check for the following possible causes:

1. Air leak in suction line. Make sure all fittings in the suction line are tight and not pulling air into the line.
2. Cold spot in line. Too great a length of unjacketed suction line allows bitumen, in

effect, to freeze in the line. A cold spot can be detected by feeling the temperature of the line beginning at the storage tank and working toward the pump. To cure, heat the cold spot.

3. Pump needs priming. Screw the male end of a street elbow into the bitumen drain cock and screw a piece of pipe into the female end of the street elbow. With pump running, immerse the pipe in a bucket of bitumen. The suction created will pull the bitumen out of the pail and cause a sufficient prime. Close the cock before bitumen in the pail lowers sufficiently to allow air to be pulled into the priming pipe.

Continuous Mixer Pump Drive

Pump drives are protected by a breaker bolt located in the hub of the drive sprocket (Fig. 10-15). Periodically check the breaker bolts for worn shanks. Replace if badly worn.

Should either breaker bolt break while the pumps are in operation, the mixer should be stopped immediately. Find and remove the cause of breakage and install a new bolt.

Fig. 10-15 Metering and supply pump drive, continuous mixer.

Meter counter Do not attempt to adjust counter until all other possibilities for error in asphalt amount has been checked. Air in the system, from a leak on the suction side of the transfer pump, is a major cause of error.

Do not attempt to repair measuring unit. Return it to the distributor or factory for repair.

Pugmill

Since the function of the pugmill-type mixer (Fig. 10-16) is to mix all materials into a thoroughly coated mixture, it is subjected to a great deal of wear by abrasion.

The pugmill should be cleaned out thoroughly each time the plant is shut down. This is especially important when mixing the heavier asphalts. Any mixed material left in the pugmill may harden to such an extent that damage to the pugmill drive may result when the plant is again started.

Pugmill paddle tips The paddle tips especially are subject to high abrasive wear and should be checked weekly to determine the extent of wear. Should the outer edge of the paddle tip show considerable wear while the face of the tip has worn only slightly, the tip can be adjusted or removed and turned end for end to obtain maximum service.

Paddle tips should always be replaced before the end of the paddle arm itself is allowed to wear.

NOTE: Check for loose belts periodically.

Paddle arms Paddle arm bolts should be checked periodically to be sure they are tight and paddle arms are secure on the shaft. To give the best mix and longest life of paddle tips and liner plates, the setting of paddle arms should be exactly as specified in your owner's manual.

NOTE: Bolts used in the paddle shaft assembly are a special heat-treated type. Do not substitute. Check periodically for loose bolts.

Pugmill liner plates The bottom, sides, and charging end of the pugmill are protected from wear by replaceable abrasion-resisting liner plates (Fig. 10-17). Check the liner plates at weekly intervals to determine the extent of wear. Some types of aggregate can cause more extensive wear than others. The liners at the discharge end will wear faster because more mixing pressure is built up there than at the charging end.

Moderately worn discharge end liners should be interchanged with charging end liners. When any liner wears through, it must be replaced immediately to prevent damage to the pugmill. Keep new liner plates on hand for availability when needed.

All liners are bolted to the pugmill or gate frame. When installing new liners, be sure they fit the curvature of the frame. It may be necessary to shim with washers between liner plates and frame. Apply caulking around bolt holes and liner edges.

Pugmill gate The pugmill gate (Fig. 10-18) is mounted on rollers that roll on guide rails located on each side of the gate at the bottom of the pugmill. The only time the gate should require adjustment is when new gate liners or seal bars are installed.

Pugmill gate liners are made of a different type of steel than that used in the ends and side of the pugmill. To replace gate liners and seal bars:

1. Apply a small strip of silicone sealant around the liner plate mounting holes on the gate.
2. Place a liner plate on the side of the pugmill gate and bolt it loosely in position.
3. Apply a small strip of sealant in the V-slot.
4. Place the second plate in position and bolt loosely. Repeat this procedure until all liners are in position.

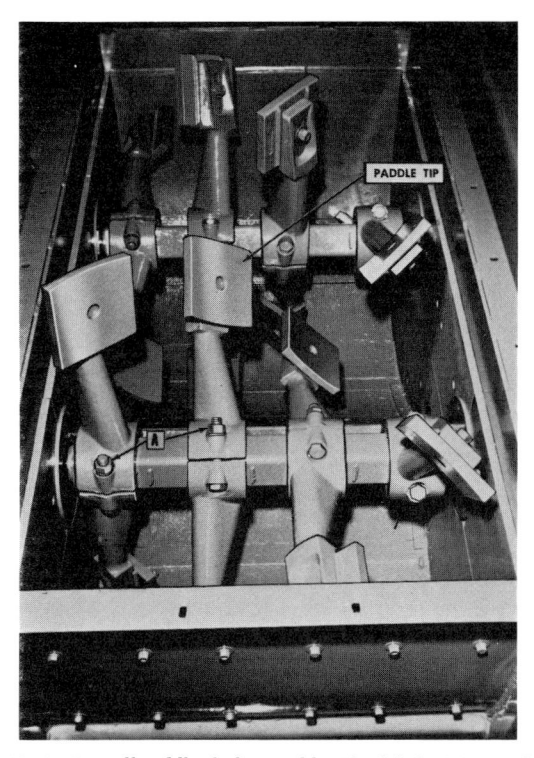

Fig. 10-16 Pugmill paddle shaft assembly. Check bolts (A) periodically.

7-282   Maintenance Programs for Equipment Entities

5. After all liners are positioned, tighten bolts.
6. Apply sealant to seal bar and bolt bar into position at the bottom of the pugmill.

Pugmill gear box   Inspect the pugmill gear box at least once a week to make sure lubricant is at the proper level.

Pugmill shaft seals   Seals are located on the pugmill drive shafts (Fig. 10-19) between the charging end of the pugmill and the pugmill gearbox. These seals prevent seepage of material through the end of the pugmill. From time to time it will become necessary to replace them.

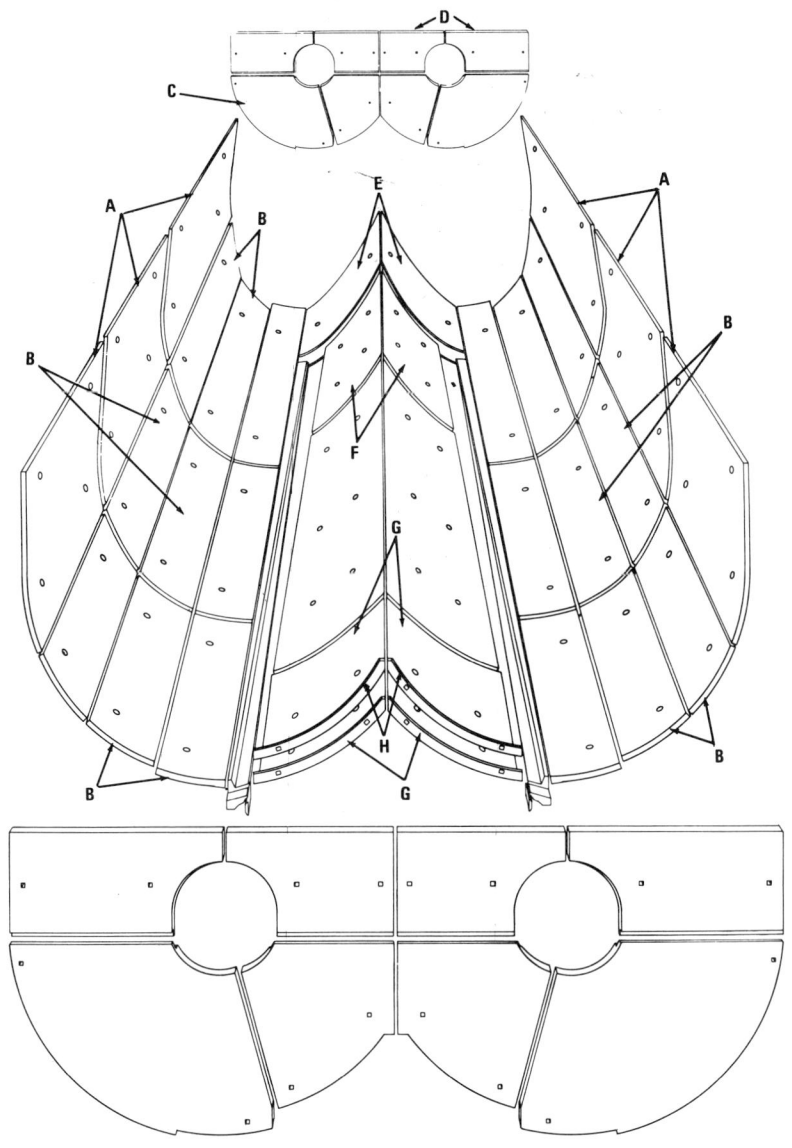

Fig. 10-17   Pugmill and pugmill gate liner plates. (A & B) side liner plates; (C & D) end liner plates; (E & H) end liner shoes; (F & G) pugmill gate liners.

NOTE: When replacing seals it may be necessary to add shims to keep the large seals from rubbing against the small seals.

Loose-fitting keys After the plant has run awhile, there may be a tendency for the keys which anchor the paddle arms to the rotating shafts to become loose in the keyways. A few thousandths of an inch looseness in the key may be reflected as several thousandths

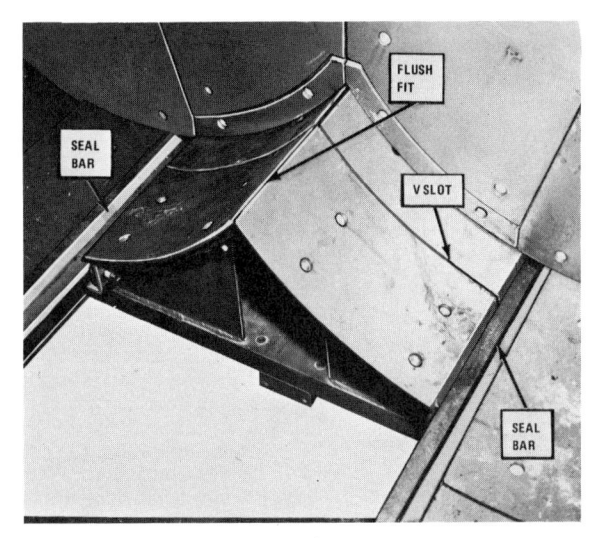

Fig. 10-18 Gate liner installation.

of an inch of free travel at the paddle tips. If there is any looseness in the keys whatsoever, new ones should be installed before the keyways become damaged. New keys should require a light drive fit.

Drum Mixer

Trunnion rollers With most drum mixers the drum is supported by two steel tires bolted to the outside. They ride on and are supported by the trunnion rollers.

The natural tendency of the rotating drum is to work downhill. One of the purposes of the trunnion rollers is to neutralize that downhill thrust.

When the trunnions are properly adjusted, the drum tire should contact either thrust roller only slightly, causing them to turn intermittently. If the thrust roller bearings should overheat, excessive thrust against that roller is indicated. Adjust the trunnions immediately to relieve this condition.

The forces between the riding rings and trunnions reach such magnitudes that no degree of design precaution is sufficient to compensate for wear and breakdown resulting from misalignment. Therefore, proper adjustment of the trunnions on rotating equipment is mandatory for best equipment life span.

The following conditions must be met before starting trunnion roll training:

 1. Check main frame for level. Adjust jacklegs to remove any sag or twist.
 2. Check the trunnion roll and trunnion tire peripheral surfaces to be sure they make line contact. Remove or install shims under bearings to achieve this.
 3. Check for parallel by placing a straightedge along the edge of the trunnion and checking points to be sure they are equal. Shift roll with adjustment screws to make correction.

NOTE: As a final check to ensure that trunnions are realigned properly and belts are tensioned evenly, check the amperage draw of each motor. This applies to friction drive where each trunnion is powered individually.

Normally, the alignment procedure should be sufficient for proper operation of the drum. If after material has been introduced into the drum and the drum has run 2 to 3 hr it

Maintenance Programs for Equipment Entities

is evident that the drum is riding with too much force on the lower thrust roll, it will be necessary to skew the trunnion rolls. See your operator's manual for procedure.

Improperly trained trunnion rolls may run for weeks without evidence of distress. However, once the distress shows up, the effect is cumulative and the trouble usually advances rapidly toward a critical stage.

A cylinder with trunnion roll shafts out of parallel can appear to float satisfactorily between the thrust rolls but chew up one set of trunnion rolls after another.

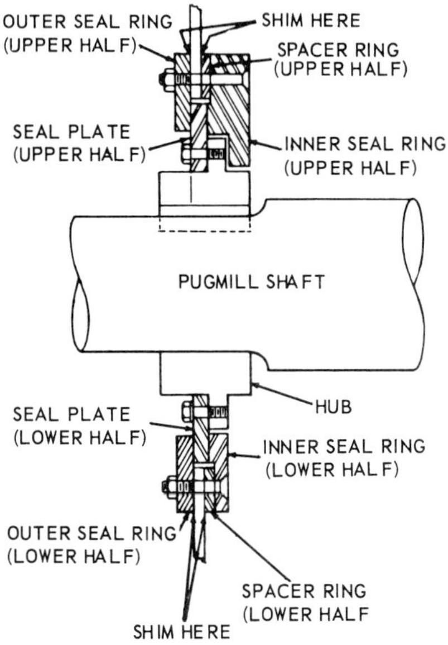

When a new dryer is put into service, watch the surface of the riding rings and trunnion rolls. If they become smooth and bright *without* ripple marks or little, seemingly slight pattern figures repeated over and over as the ring revolves, change daily inspection to weekly inspection.

If small pattern figures or score marks repeat themselves around the periphery of a riding ring or trunnion roll, look for trouble without delay. Odds are that the rolls have shafts out of parallel. It does not take the small score or pattern marks long to get rough around the riding ring or trunnion roll. Soon deep scores develop and flaking begins, followed by the urgent need for new trunnion rolls and riding rings.

Drum Check drum for excessive warping which will cause the charging end to whip around and rub against the stack. Examine flights for general good condition. Make sure hold-down bolts are tight when flights are installed.

Fig. 10-19 Pugmill shaft seals.

Discharge chute Inspect the discharge chute liner plates for excessive wear. Replace when worn thin before the outer chute is exposed to wear.

Hot Elevator

Bucket line A chordal action of the hot elevator bucket line chain on the elevator foot shaft causes a so-called hopping action of the foot shaft. Coil springs (Fig. 10-20) have been provided to dampen this hopping action.

When springs are properly adjusted, they minimize the chordal action. Do not adjust too tightly and never completely collapse springs.

Check the slack in the bucket line through the inspection doors. When bucket line slack reaches a point where buckets are apt to strike the elevator housing, it should be adjusted.

Do not tighten bucket line too tight. Allow as much slack as possible to prevent excess strain on head shaft bearings, sprockets, and bucket chain.

Periodically inspect for missing cotter pins and loose buckets. Tighten chain only enough to avoid chain striking elevator housing. Keep discharge plate adjusted close to dumping buckets to prevent spillage into the chain and lower traction wheel.

Elevator drive chain Proper adjustment is maintained by a spring-loaded idler sprocket. To check for proper adjustment, the chain should be observed while the elevator is in operation. When in operation there should be no slack at any point in the chain.

Discharge chute liner plate The elevator discharge chute has replaceable liner plates that should be inspected periodically for wear. Some types of aggregate will have a more rapid wear rate on their liners than other aggregates. An inspection should be made to establish rate of wear and determine a time interval when new lining should be installed. It should be replaced when worn thin, before the outer housing is exposed to wear. New liner plates should be kept on hand.

Drive belt Proper drive belt tension (Fig. 10-21) is obtained by moving the torque arm assembly with the tumbuckle provided.

WARNING: Do not loosen turnbuckle completely with material in buckets. Gearbox will whip around if buckets are full.

Asphalt Plant Maintenance 7-285

Fig. 10-20 Bucket line adjustment, showing coil springs (A & B).

Fig. 10-21 Drive belt adjustment.

V-belt Drives

The life of a V-belt drive can be increased considerably by proper care and maintenance. Improper alignment is one of the most common causes of excessive belt and sheave wear. Damaged sheaves are another common cause of belt wear. If belts show a tendency to roll or climb out of sheave grooves, the cause is usually misalignment, worn sheave grooves, or insufficient tension.

Alignment Indications of misalignment are wear on one side of the belt only or sheave grooves that are more highly polished on one side than the other. Continuous, regular squeaking caused by belts rubbing the sides of the sheave grooves is generally due to misalignment. If bearings become overheated, bearing and shaft may be subjected to unusually hard wear by a misaligned drive. To check adjustment place a straightedge across the face of the sheaves. If the sheaves are in line, the straightedge will squarely contact the face of both driving and driven sheaves.

Sheave grooves Burrs and rough spots in sheave grooves or along the rim are also disastrous to V-belts. Sheaves with worn grooves should be replaced, since old sheaves give uneven traction, putting excessive strain on some parts of the belt. V-belts can easily bottom in badly worn sheaves.

See that sheave grooves are kept clean and free from dirt. Inaccurate and dirty grooves in a sheave can cause variation in the amount of V-belt sag, making them appear to be unmatched for the drive.

Cleaning To clean belts, wipe with a dry cloth. The safest way to remove stubborn dirt and grime is to wash with soap and water and rinse well. If the belts accidentally become grease- or oil-splattered, remove the spots with cleaning material.

A simple test which can be made to check V-belt drive tension is illustrated above. Press down firmly on each individual belt. When the top can be depressed so that it is in line with the bottom of the other belts on the drive, the correct amount of tension has been applied. Each belt should be given this test individually.

Another good method for checking proper tension of V-belts is by "striking" the belt with your fist. Slack V-belts feel dead under this test, while properly adjusted V-belts vibrate and feel alive.

Fig. 10-22 Checking V-belt tension.

Replacement When new belts are necessary, install a complete set. If some new belts are put on a drive with worn or stretched belts, the new belts, being shorter, will carry more of the load.

Always order replacement belts by part number. When new belts are installed, make sufficient slack in the drive so that the new belts can be placed easily in the proper grooves by hand. Under no circumstances should belts be forced onto the sheaves with crowbars, wedges, screwdrivers, or bars of any kind.

When assembling new belts, tighten them to about two times normal tension. There will be a rapid drop in tension during the run-in period (first 24 to 48 hr) while the belts seat themselves in sheave grooves. After the first day or two, check for the correct amount of tension in each belt. If belt deflection force is over 1.5 times normal, the belts are too tight. If the force is below normal belt tension, they are too loose.

Tension It is not necessary to pull V-belts excessively tight. Tighten (Fig. 10-22) only enough to take out slack and undue sag. When belts are too tight, bearings are apt to burn out or wear out faster even if well-lubricated. Excessive tension also stretches and weakens belts.

On the other hand, if belts are too loose, they slip and slide easily under load increases. The result is excessive wear to both belt and sheave.

Belt squeal on starting or stopping indicates slipping belts. Excessive vibration on stopping may also be caused by loose belts although it is usually caused by incorrect timer adjustment.

The original tension put on belts when they are installed should be maintained by using the take-up on the driver unit whenever necessary.

A simple test to check V-belt tension is to press down firmly on each individual belt.

When the top can be depressed so that it is in line with the bottom of the other belts on the drive, the correct amount of tension has been applied. Each belt should be given this test individually.

Another good method for checking proper tension of V-belts is by striking the belt with your fist. Slack V-belts feel dead under this test, whereas properly adjusted belts vibrate and feel alive.

Adjustment Procedure
1. Measure the span of your drive.
2. At the center of the span apply a force perpendicular to the span, large enough to deflect one belt on the drive 1/64 in. per inch of span length from its normal position.

TABLE 10-1   Recommended Belt Deflection Forces, Pounds per Belt

V-belt cross section	Avg small sheave diam. range, in.	Drive ranges small sheave, rpm	Speed ratio range	For normal tension, lb	For 1.5 times normal tension, lb	For 2 times normal tension, lb
3V	2.65–3.35	1200–3600	2.00–4	3	4½	6
3V	4.75–6.0	900–1800	2.00–4	4	6	8
5V	7.1 –9.0	600–1500	2.00–4	8	12	16
5V	12.5 –16.0	400–800	2.00–4	10	15	20
8V	18.0 –22.4	200–700	2.00–4	20	30	40

CONTROLS

Actions of the various components of an asphalt plant are either pneumatically, hydraulically, or electrically controlled. Thus the control systems are vital elements of the plant.

Main causes of problems with control systems are dust, dirt, and moisture. Constant good housekeeping is essential.

Hydraulic System

Generally, a hydraulic system (Fig. 10-23) will operate smoothly with little maintenance providing the proper hydraulic oil is used, all connections are kept tight, and the oil strainer and filter are kept clean. Dirt and grit are the principal causes of failure of hydraulic systems to operate properly. A very small piece of grit may cause a valve to become inoperative, for example. Dismantling and thoroughly cleaning a component may enable the system to function without replacement of the component. If, however, valve spools or the housing are found to be scored or the valve is defective in some other way, it will have to be replaced.

A daily inspection of the hydraulic system which takes only a few minutes will warn you of potential trouble. Here are the points to check:
1. Fluid level in reservoir
2. External leakage
3. Unusual noise
4. Proper cycle operation
5. Fluid temperature.

CAUTION: Always relieve hydraulic pressure before attempting to work on any hydraulically actuated unit, and be sure the pump cannot be started.

Reservoir   Conditions inside the tank are particularly important. Fluid level should not be permitted to drop below the inlet line opening. The end of the return line should also be below fluid level to avoid splashing and the resulting aeration. Occasionally check the surface of the fluid while the system is operating to see if there is excessive foaming.

Change the fluid whenever it shows signs of deterioration. Flush tank, filter, and strainers thoroughly before replacing fluid. It is most important to keep the fluid perfectly clean and free of dust and sludge. This will prolong the life of pump, valves, and cylinders and maintain an efficient operating system.

Under certain atmospheric conditions, condensation may occur in the reservoir. The moisture can then collect in valves and filter. If this occurs, drain and clean the system thoroughly.

When changing fluid, don't merely drain and clean the tank. Break the lines and drain them. Drain and clean the cylinders as well. Unless these components are thoroughly cleaned, contamination will remain in the system.

If the plant operates during cold weather (below 32°F), allow the hydraulic system to circulate for 15 to 30 min or until the oil manifold feels warm to the touch. It may be desirable to install electric head heater bolts in the reservoir to take the initial chill off the

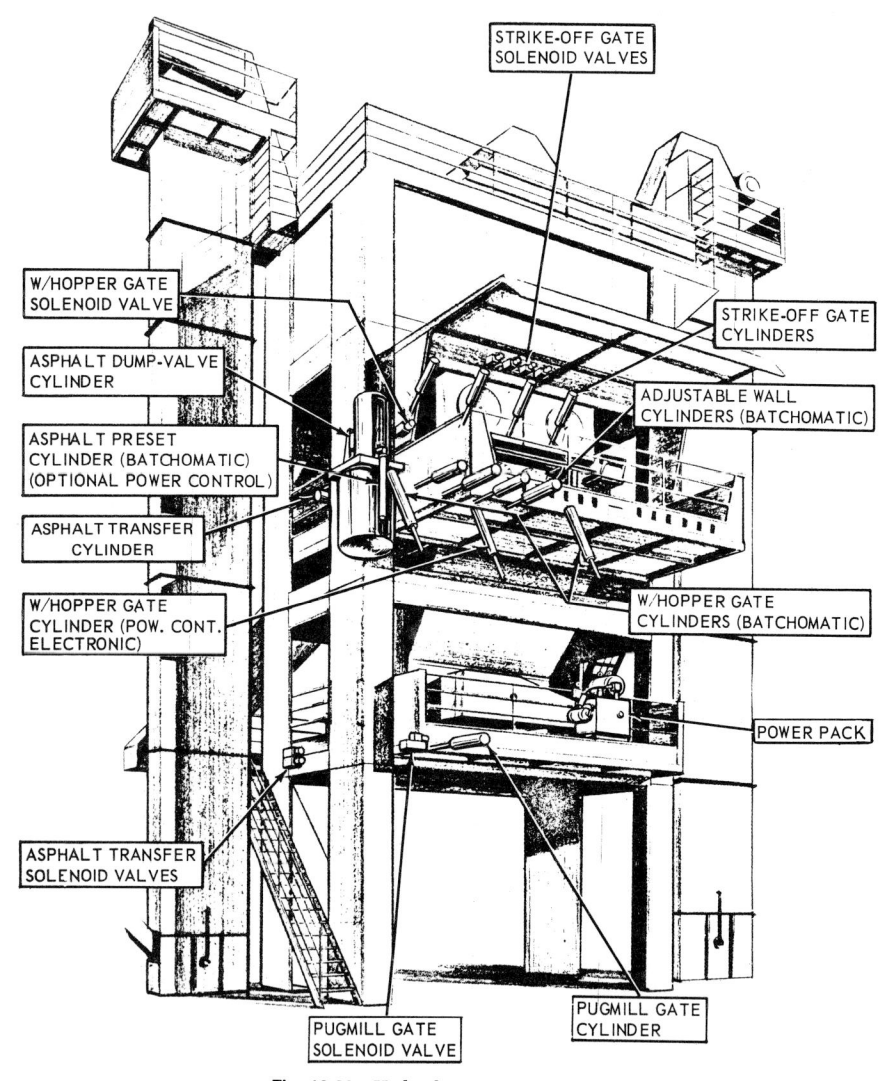

Fig. 10-23 Hydraulic component locations.

oil. If continuous winter operation is anticipated, it is advisable to use a lighter viscosity oil.

Hydraulic oil strainer The strainer is located inside the hydraulic reservoir in the suction line to the pump. Periodically remove the strainer and clean it thoroughly.

Suction filter The filter element should be replaced after the first and second week's operation of the plant. Thereafter, inspect the filter weekly and replace at least every 500 hr.

Oil pressure filter After each of the first two week's operation of the plant, install a new filter element. Thereafter, inspect the element weekly and replace when necessary. A 10-micron cleanout filter is available for cleaning the system. Run the system 2 to 3 hr with a cleanout filter. Then, replace with a new 40-micron filter.

Pump Under average conditions the modern pump operates efficiently for a long time. An early failure usually indicates trouble somewhere else in the system. Possible causes include:

    Rapid wear caused by aeration or cavitation.

    Scoring or seizing caused by contaminants or inadequate lubrication.

    Running the pump backward.

    Neglect of filters.

Relief valve Do not operate the hydraulic system beyond recommended limits. Ordinarily, if this precaution is observed no maintenance of the relief valve is required. However, if erratic or improper cycling is experienced, first inspect the hydraulic oil filter element. A dirty or clogged filter will cause malfunction of the unloading valve.

Next, check the entire system for leaks. Then, if the erratic cycling continues, the relief or unloading valve probably is the cause. If necessary, dismantle and clean it.

Solenoid-operated hydraulic control valves These valves are a precision-fit. With the exception of installing new seals and "O" rings, field repairs are not recommended.

Flow control valves If necessary, adjustable flow control valves may be dismantled and cleaned. If after cleaning and inspection the valve still does not operate properly, it must be replaced.

Directional control valve

Manual (Optional). The manual directional control valve may be dismantled for cleaning and replacing of "O" ring seals if necessary.

Automatic. With the exception of installing new seals and "O" rings, field repairs are not recommended.

Pressure gauge Check operating pressure several times daily. The unloading valve is preset to open at a certain operating pressure. It will close when pressure drops to allowable limits.

Don't operate the hydraulic system beyond recommended limits. Therefore, you shouldn't adjust the unloading valve setting without proper operating instructions. If the valve has been set correctly and adequate pressure is not developing, the trouble usually is at some other point.

Accumulator The accumulator (Fig. 10-24) performs two functions in the hydraulic system: counteracts internal leakage and acts as a shock absorber for the system.

Check the precharge pressure of the accumulator once a month. If pressure is high, oil has leaked into the gas side of the cylinder. If pressure is low, gas has leaked into the gas end of the accumulator.

Cylinders Most asphalt plant cylinders are the double-acting type with an adjustable packing gland nut. Periodically inspect the packing gland nuts for leakage. When tightening, tighten packing nuts only as much as is required to stop leakage. When adjustment room has been used up, install new packing.

A sluggish, slow-acting cylinder with high internal leakage, indicated by rapid cycling of the unloading valve when the hydraulic system is operating, is a good indication of worn or leaking piston packing.

Breather vent The air breather vent on the tank contains a filter. This filter requires weekly cleaning to maintain atmospheric pressure in the tank and to assure clean air. You must take particular care in maintaining the air vent and filter.

Recommended Grades of MS Oils

Hydraulic system operating range, min to max	SAE viscosity	API service classification
0–180°F	10W	MS
15–210°F	20-20W	MS
32–230°F	20	MS
0–210°F	102-30W	MS

7-290 Maintenance Programs for Equipment Entities

Two other filtering tips:
1. Filter new fluid before placing it in the system to guard against contamination which may have taken place during handling.
2. Periodically clean the inlet line strainer in the reservoir.

Batch Controls

Limit switches Check actuating arms to make sure they are free of accumulations of dirt which might hinder operation. To adjust limit switches (Figs. 10-25 to 10-27) be sure the switch is located on the mounting bracket so that it is in the proper relation to the

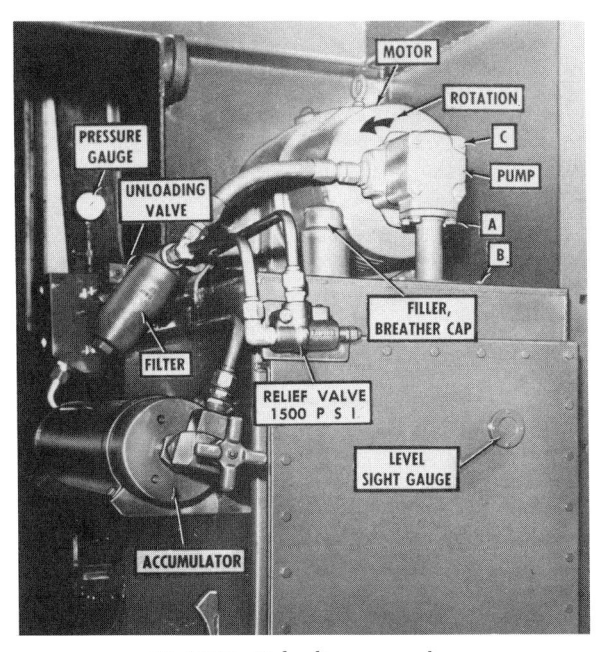

Fig. 10-24 Hydraulic power pack.

Fig. 10-25 Asphalt fill valve, closed limit.

Fig. 10-26 Asphalt dump valve, closed limit, and asphalt overflow switch.

actuator. The arm may be removed by taking out the Allen head setscrew and taking the arm off the shaft. Replace the arm on the shaft so that as the switch is actuated there is an audible click. A further check is an ohmmeter across the normally open contact. As the switch is actuated the contact should close and have a continuity reading. Replace setscrew and be sure switch will not be damaged by too much overtravel of the actuator.

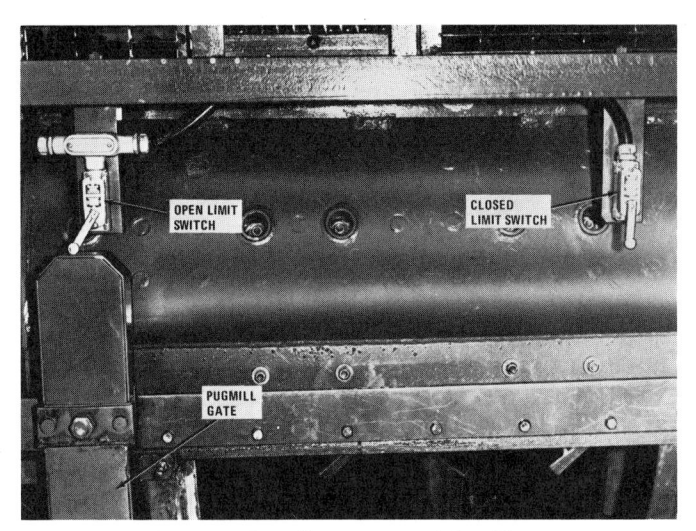

Fig. 10-27 Pugmill gate, open and closed limits.

Supply bin level indicators Check indicator wires periodically for fraying or hanging where excessive wear might occur. Also inspect switch holder clips and lead wire. Test action of telltale arms to make sure they operate freely and do not bind.

Solenoid valves Check solenoid valves frequently to make sure they work freely with no binding. To test operation, depress the solenoid armature manually while the hydraulics are running.

Wiring From time to time check tightness of all terminal connections in the main tower junction box.

Main control console Keep the controls dry and clean. If the relay contacts, dry and wet timers, or batch counter should need cleaning, use a good spray-on liquid burnisher. Check cable harnesses for wire insulation chafing.

Panel buttons may experience sticking if exposed to dust conditions. Buttons may be removed for cleaning. However, when replacing buttons, apply even pressure so buttons will seat properly in the operator.

Bayonet-type bulbs may be replaced from the front after removing the lens. A bayonet bulb puller is available for removal along with spare bulbs.

Automatic weighing console Keep the control dry and clean. If the plug-in relay contacts should need cleaning, use a good spray-on liquid burnisher. Check cable harnesses for wire insulation chafing.

Electrical System

Regardless of the type of problem involved, the following approach should be used in troubleshooting electrical and electronic problems.

1. *Observe.* Make a visual observation of the problem.
2. *Analyze.* Use the knowledge of operation procedures and theory of the circuitry involved to formulate possible causes.
3. *Isolate.* Make an analysis of the problem and known facts, that is, use of test equipment to determine continuity or proper voltage.
4. *Remedy.* Perform the necessary repair or replacement.
5. *Report.* Write a record of the problem for reference.

Proper inspection and observation lead to the correct analysis of the trouble and minimize the amount of isolation that must be accomplished. Isolation separates the good from the bad and is a fundamental function of effective, successful troubleshooting.

For proper troubleshooting it is recommended to have:
1. *Test equipment:* Volt-ohmmeter with proper voltage and resistance ranges.
2. *Tools:* One screwdriver, one needlenose pliers, one wire cutter, one wire stripper, a set of Allen wrenches, and a set of Bristol wrenches.

Burner

Nozzle assembly Inspect and clean oil tube and nozzle assembly at least twice each operating season and more often if conditions are considered very dusty. Remove oil tube and nozzle assembly to check for foreign material that may be plugging nozzle holes. Clean any accumulated dirt from primary and secondary air passages. Dusty operating conditions may cause oil deposits to form on nozzle and burner housing. Soak nozzle in a good solvent to loosen deposits, and scrape nozzle body and holes with wooden tools only.

Linkage rods Lubricate bearing ends of swivel linkage rods periodically and check all setscrews to make sure they are tight.

Spark plug Remove the spark plug from the pilot piping to inspect for carbon deposits. Replace if necessary. Check gap setting to make sure it conforms to the proper setting prescribed in the operator's manual for your plant.

Combustion chamber The combustion chambers of some rotary mixers consist of cone-shaped cylinders mounted at the charging end of the mixers. The cone's function is to obtain complete combustion of the fuel.

Other rotary mixers may be equipped with a combustion chamber line with firebrick. All missing or damaged brick must be replaced as soon as possible to prevent burning of the combustion chamber steel drum. The combustion chamber can be rebuilt or a complete new assembly can be ordered.

Refractory Pugmill-type plants will have ignition- or combustion-type chambers. If large cracks or holes develop in the chambers, they must be repaired immediately. Temporary repairs may be made with a plastic-type refractory material. High-bond plastic refractory, rated at 3000°F, may be obtained from a refractory supplier.

Oil basket strainer Either basket not covered by a handle may be lifted out for cleaning while in operation. Remove the basket from the well and clean with a brush or soak it in a solvent. Do not strike the basket to clean it because that may cause damage. Frequency of cleaning will depend on the degree of contamination of the fuel oil which often varies from grade to grade.

Control motor A minimum of maintenance is required since the control motor and gear train are submerged in oil for both continuous lubrication and cooling. If it is necessary to refill the actuator with oil, always use immersion oil which is available in 1-qt cans. For best performance, the oil level (with the actuator upright) should be up to the edge of the oil fill hole which is located in the front case of the actuator. At frequent intervals check radius arms and linkage rod connections of control motor (Fig. 10-28) for tightness.

Preheater Since mechanical cleaning is difficult and ineffective, it is not recommended in this area. To clean, drain the shell side of the heater unit. Mix an emulsion of one part Oakite No. 9 or equivalent and four parts kerosene. Fill the heater with the fluid, allowing it to remain for 24 to 48 hr.

Drain the emulsion and prepare a mixture of 8 oz of Okemco, or equivalent, per gallon of water. Circulate this liquid through the heater shell for 24 hr at a temperature of approximately 200°F by admitting hot oil to the tube side of the heater.

Automatic gas valve Check the valve visually from time to time to make sure it is opening and closing properly. If it is not working right, the system should be shut down and the valve returned to the factory for repairs (Fig. 10-29).

Burner Control Automatic Recording

Limit switches

Draft Limit. Keep the sensor clean, using a brush to wipe away accumulated dust and dirt. Keep copper tubing clear of dirt and large dust particles.

Thermocouple. Keep the sensor clean, using a brush to remove dust and dirt.

Asphalt Plant Maintenance 7-293

Low Fire Limit. Check actuating arm to make sure it is free from the accumulation of dirt.

Flame safeguard system

Control. Keep the control clean and dry. If the relay contacts require servicing, use a good spray-on burnisher.

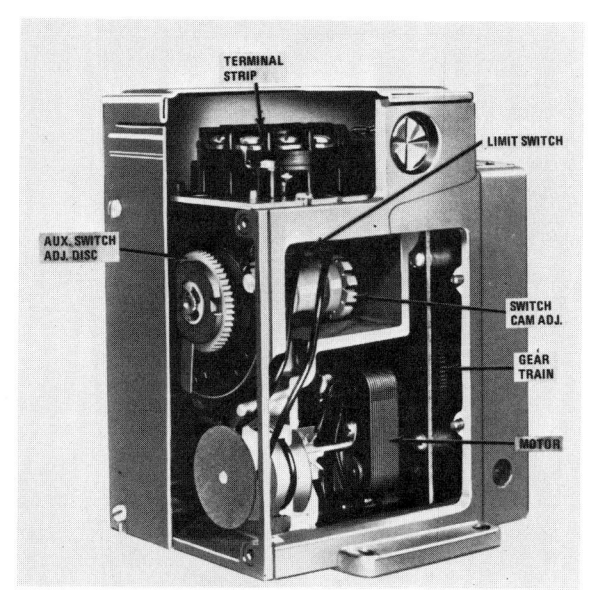

Fig. 10-28 Control motor.

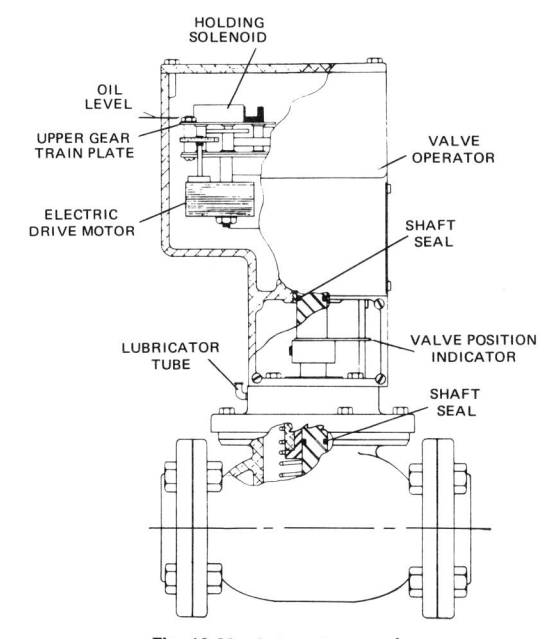

Fig. 10-29 Automatic gas valve.

7-294 Maintenance Programs for Equipment Entities

Scanner. Keep the scanner ultraviolet tube clean. Use a clean cloth with a detergent as often as operating conditions require.

Safety Check. Test the safeguard system at least once a month. This test should verify flame failure safety shutdown and positive fuel cutoff when the fuel valve is deenergized.

Flame Meter. If the meter should need a zero adjust, use the screw on the meter face for adjusting meter to zero with no flame established (Fig. 10-30).

Thermocouple Rotate the protector well at least once a month to reduce excessive wear on any one side. To rotate, simply loosen the setscrew in the support coupling and turn the well. Tighten down the setscrew afterward (Fig. 10-31).

Fig. 10-30 Flame guard control.

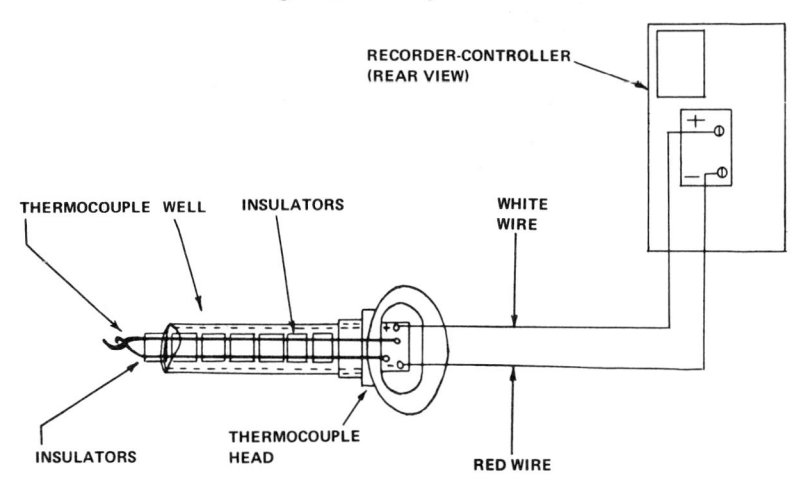

Fig. 10-31 Thermocouple.

Recording and indicating components

Scale. Clean the scale as required. Use a soft cloth with a soap and water solution to remove dirt, grease, and ink. Do *not* use any alcohol-base solvents as they may attack the finish of the scale.

Ink Reservoir. Check ink level periodically. Clean the capillary system and ink reservoir every 3 months unless a blockage occurs.

Pen. Clean the tip once a month. Clean the passage with pen cleaner or a fine wire. Wipe dirt and other foreign matter from the outside of the top.

Gears. Use a stiff bristle brush and trichlorethylene or a similar solvent to clean the gears in the chart drive system.

LUBRICATION

Nothing can add to the life of an asphalt plant more than thorough lubrication of moving parts in the correct manner at proper intervals. There is no excuse for plant breakdowns due to improper lubrication when they can be so easily avoided.

Detailed instructions for lubricating your plant should be included in the operator's manual which came with your plant. Lubrication instructions for electric motors are generally included in metal plates attached to the motors.

If your plant is powered by an internal combustion engine, the manual provided by the engine manufacturer will give lubrication instructions.

Don't guess. Always consult the lubrication charts for your plant to make sure you are hitting all lubrication points and using the right lubricant for each one.

When draining gear cases, do so when the lubricant is hot so that the draining fluid will carry off the sludge.

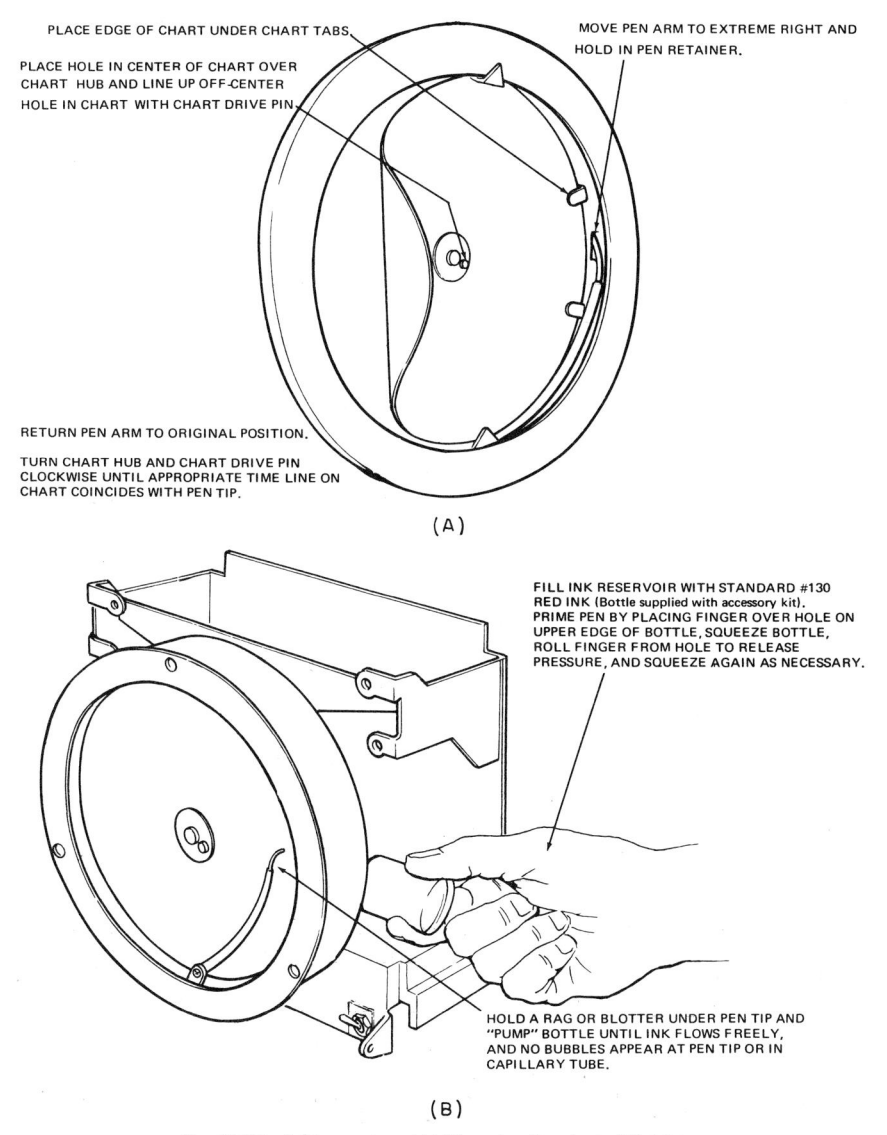

Fig. 10-32 Inking system. (A) Changing the chart. (B) Inking.

Use Specified Lubricants

Generally, asphalt plant manufacturers specify a particular lubricant for each lubrication point, because certain lubricant qualities are required for each lubrication point. For example, the qualities of the lubricants used at lubrication points exposed to high heat are quite different than the lubricants required for lubrication points exposed to cool outdoor temperatures.

Never substitute a lubricant of lesser grade or value for the one specified by the plant manufacturer. It is all right, of course, to substitute one brand of lubricant for another providing it has the same specified qualities and characteristics.

It is false economy to use a general-purpose grease or anything other than the specified lubricant. Money or time saved by using the wrong lubricant can be extremely costly.

Keep each lubricant container clearly labeled to eliminate the danger of using the wrong lubricant in the wrong place.

Use the Correct Amount

Don't overdo. Give each fitting only the exact number of shots of grease specified—no more and no less.

Lubricant in all gear cases should also be kept to the exact level specified. Too low a level can result in bearing or gear failure. Too high a level may lead to a high-temperature buildup at high operating temperatures. Excess lubricant can leak into other areas, causing damage.

CAUTION: Never use a power-operating grease gun on antifriction bearings. As a rule, antifriction bearings have been lubricated at the factory. The grease cannot be seen since it is concealed within the bearing by grease retainer seals. Overgreasing distorts and damages these seals, allowing dirt to enter and greatly shorten the life of the bearing.

REMEMBER: The number of shots of grease from a manual grease gun is based on greasing with a standard 13-oz hand gun delivering 1 oz of grease to 54 shots of the gun, using the recommended grease. In other words, one shot of the gun equals $\frac{1}{54}$ oz of grease.

Follow Proper Intervals

Frequency of lubrication should never be extended beyond the manufacturer's recommendations. On the other hand, under unusual or severe operating conditions it may be necessary to lubricate some points more often than specified for normal conditions.

Another factor is the effectiveness of bearing seals. This will vary from one make of bearing to another. Also, as the bearings get older the effectiveness of the seals will be reduced and lubrication will have to be done more frequently.

Keep It Clean

Keep the plant as clean as possible, especially around lubrication points. Always remove oil and grease which may accumulate around lubrication points before applying lubricant.

With some types of bearing applications it is possible to flush out the old grease while replacing it with new. A container should be positioned below the bearing to catch the old grease so that it does not build up on machinery and become a dirt catcher.

Keep the grease gun clean and wipe each grease fitting with a clean cloth to prevent pumping grit into the bearing along with the grease.

Keep grease and oil containers clean and their covers tightly in place when not in use to keep dust and dirt out of the lubricant.

Running Gear

Tires If the portable plant is to be set up on a job over a period of 3 to 4 months, it is advisable to remove the tire and wheel assemblies and store them inside.

Brakes A schedule for the periodic adjustment, cleaning, inspection, and lubrication of brake equipment should be established. It should be based on job experience and the severity of operations.

To compensate for this wear, brakes should be adjusted as frequently as required to maintain satisfactory operation and maximum safety.

Tag axle Periodically drain the moisture from the air reservoirs. Also, inspect the suspension including the tightness of all bolts and U-bolts.

SATELLITE EQUIPMENT
Gradation Unit

Screen cloths   Periodically check the screen for worn or damaged screen cloths. Also check the screen cloth clamps for loose bolts. Cloths should be kept drumhead-tight for best operation.

If the plant is to be shut down for any length of time, as during the winter season, fill the screen bearing housings full of new oil to completely immerse the bearings. Before starting the machine, drain the oil down to the normal operating level.

Screen drive sheave   Periodically check the screen drive sheave and taper lock hub to see that they are secure on the shaft. Should they become loose, they can be tightened by simply tightening the setscrews in the hub.

Screen operating speeds are set at the factory according to the application for which the screen is to be used. With some makes the maximum speed for good bearing life and

Fig. 10-33   Apron feeder skirt plates.

normal stresses in the screen frame is stamped on the nameplate. The correct amount of counterweight is installed on the eccentric shafts to produce the correct vibrating amplitude. Increasing the speed may shorten the lives of the bearings and screen frame. Contact the manufacturer if you desire to change screen throw or speed.

Apron Feeder

Check periodically for worn or bent flights.

Apron feeder chain   The apron feeder chain should be kept adjusted so that it has the proper amount of slack recommended by the manufacturer. Check the apron feeder periodically for worn or bent flights.

Apron feeder skirt and divider plates   Plates (Fig. 10-33) are adjustable. Clearance between skirt and feeder plates must be $1/16$ in.

Belt flashing   Flashing should be adjusted so that it just contacts the belt. Too much contact will cause premature belt wear.

Feeder control gate liners   The feeder control gates have replaceable wearing plates. Replace worn plates before the gates are exposed to wear.

Bin level telltales   Check telltales daily to make sure they operate freely. Lubricate according to lubrication charts provided with your machine.

Cold Feeder

Conveyor drive chain   From time to time check slackness in the drive chain. When the chain becomes too slack, it can be tightened by moving the motor and speed reducer farther from the conveyor head shaft.

Feeder drive belts   See the V-belt Drives section in this chapter for instructions on the maintenance and adjustment of V-belt drives.

Conveyor belt Check the belt periodically for holes and cuts and also for signs of excessive wear. Cuts and tears should be repaired immediately by lacing, vulcanizing, or filling with rubber cement. Which repair to make will be determined by the severity of the damage (Fig. 10-34).

Excessive wear may be due to improper alignment, too much tension, exposure to oil or grease, damage to idlers, buildup of material on idlers, or friction from external causes. To

Fig. 10-34 Conveyor and feeder belts.

solve such problems see Section 4, Chapter 13, on Maintenance of Belt Conveyors and Conveying Equipment.

Conveyor rollers Check rollers periodically to make sure they turn freely. Failure to turn without resistance may be due to buildup of material on the rollers or to inadequate lubrication.

Belt wiper The wiper must be adjusted so that the rubber blade contacts the gathering conveyor belt firmly and evenly. Too much pressure on the belt, however, will result in excessive belt wear and premature breakdown of the wiper itself.

Feeder belt flashing Belt flashing should be adjusted so that it contacts the belt lightly (Fig. 10-35).

Dust Collectors

Strict environmental protection laws have required the development of efficient dust collection equipment. Collector types include fabric filter, vertical and horizontal cyclone, and wet.

Fabric filter bag wear When a bag becomes worn in a small area it may be patched. The patch may be part of an old bag or a piece of material of the same type as the bag. For durability sew with Nomex thread or equivalent.

Bag life may range from 1 to 4 years, depending upon severity of service. For longest bag life, frequency of pulsing should be adjusted and readjusted as often as necessary to obtain the desired pressure drop (1 to 4 in. water gauge). Allow the pressure drop to become stable before making the final adjustment in the cleaning cycle. The frequency should be as long as possible, still being consistent with good ventilation required at the dryer.

Static pressure drop on the dust collector will generally range from 1 to 4 in. water gauge. Abnormal wear can be caused by a number of factors. The following are a few of the common failures and some causes:

1. Very abrasive material, such as sand, will sometimes cause wear at the bottom of the bag.
2. Bag or bags are too large, allowing them to abraid on each other when pulsing.
3. If the temperature exceeds the maximum temperature allowed (425°F), the bags will deteriorate rapidly.

Visible dust in collection of draft fan discharge Generally, visible dust in the discharge indicates one or a number of damaged bags in the collector. An inspection of the

Fig. 10-35 Feeder belt flashing.

bags should be made to pinpoint failure. The bag or bags should be replaced and then repaired (if possible) for reuse.

It is not uncommon to notice a slight haze in the filtered air at start-up or after all the bags have been replaced with new ones or just after a module has been cleaned. This haze is not an indication of a broken bag. It is known as *bleed through*. The fabric will allow fine dust to filter through momentarily after it has been cleaned or when it is brand new.

Moisture infiltration into the collector Since the collector is operating under negative static pressure (suction), moisture will leak into the collector at areas that are not sealed properly at erection. It is not practical to disassemble areas that leak. The best solution is to caulk with epoxy or silicone rubber sealant after the area is properly cleaned.

Ducts Inspect ducts and clean when necessary. Excessive moisture in the aggregate will cause the scavenger to clog with damp dust, requiring more frequent cleaning.

Oil in collector Improper combustion in a dryer oil burner, especially one using heavy oil, creates carbon that will collect on the bags and cause plugging. Flame safety equipment is mandatory for safe operation of a fabric filter. The burner should be checked frequently and cleaned—especially nozzles—to prevent improper air-oil mixture.

Oil in the fines material in the dust collector is an indication of improper oil combustion. A good check for oil in the fine is:
1. Take a sample of fines.
2. Dilute with water until soupy.
3. Let settle. If oil is present, it will float to the surface.

Fan Inspect the fan monthly for wear of parts. Should excessive fan vibration develop, check for the following possibilities:
1. Buildup of dirt or other foreign matter on the wheel.
2. Loose bolts on bearings, housings, or driver.
3. Improper V-belt alignment, belt tension, or unbalanced sheaves.
4. Inadequate bearing clearance and alignment.

5. Loose wheel.
6. Foreign matter causing damage to the wheel, shaft, or bearings.
7. Vibration from a source other than the fan.
8. Improper clearance between the wheel and inlets.

To check whether vibration is coming from another source, stop the fan and see if the vibration still exists. If that test is unsatisfactory, disconnect the drive from the fan and operate the fan by itself to determine if it produces vibration.

If the fan is to remain idle for an extended period, it is recommended that the exposed surfaces be protected with a protective coating. The shaft should be rotated periodically in the bearings to prevent corrosion.

NOTE: The hanger bearings inside the house are dry-running bearings and have a tendency to squeak. Do not lubricate.

Dryers

Generally speaking, all aggregate dryers look and work pretty much alike. They are essentially a long rotating drum with inside lifting flights which continuously drop a veil of aggregate through the hot gases. The drum is inclined, the amount of slope determining the time required for the aggregate particles to pass through the drum.

A feeding hopper for the aggregate is located at the high end, and the oil or gas burner is located at the low or discharge end. This arrangement gives a counterflow of the gases to the aggregate so that the highest temperature is located at the point of aggregate discharge.

Trunnion and thrust rollers These subjects are covered earlier in this chapter in connection with Drum Mixers.

Rotary charging chute The only adjustment necessary on the rotary charging chute is to the trunnions which keep the chute centered in the inlet. To adjust, add or remove shims.

A wiper blade, located in the top of the chute, may be adjusted to the drum as wear indicates.

Rotary charging chute hopper liner plates Some types of aggregate will produce a more rapid rate of wear on liner plates than others. Make a periodic inspection to establish a rate of wear and the interval when liners should be replaced. Liners should be replaced when worn thin but before the outer housing is exposed to wear.

Discharge chute Inspect the discharge chute liner plates for excessive wear and replace when worn thin before the outer chute is exposed to wear. Inspect periodically and keep a supply of new liner plates on hand to be able to make immediate replacements when necessary.

Dryer drum Check the drum for excessive warping which will cause the charging end to whip around and rub against the stack.

Examine the sprocket teeth for wear. Check hold-down bolts for tightness.

Examine flights to see that they are not bent or broken. Make sure hold-down bolts are tight.

Combustion chamber Inspect the lining of the combustion chamber to make sure all firebrick is secure and unbroken. All missing or damaged firebrick must be replaced as soon as possible to prevent burning of the combustion chamber steel drum (Fig. 10-36).

Drum drive chain Proper adjustment of the drum drive chain is maintained by a spring-loaded idler sprocket. To check for proper adjustment the chain should be observed while the dryer is in operation. When in operation there should be no slack at any point in the chain.

At the end of each season's operation the drum drive chain should be removed from the machine and thoroughly washed and cleaned with a solvent. Reoil the chain after cleaning with a light meter oil, making sure the oil penetrates the chain bushings.

Drum drive chain oiler A drip-type oil lubricator is provided to keep the drive chain rollers lubricated. The lubricator should be kept filled with an SAE 30 weight engine oil and the drippers adjusted to drip approximately two drops per minute to provide adequate lubrication to the rollers of the drive chain (Fig. 10-37).

NOTE: Flush drive chain once a week with 1 gal of SAE 30 weight engine oil. Rotate the dryer for 10 min after flushing.

Drum seals The dryer drum is sealed at both ends with spring-loaded cast iron seals. The seals are mounted with slotted brackets for aligning. The only maintenance necessary

Asphalt Plant Maintenance   7-301

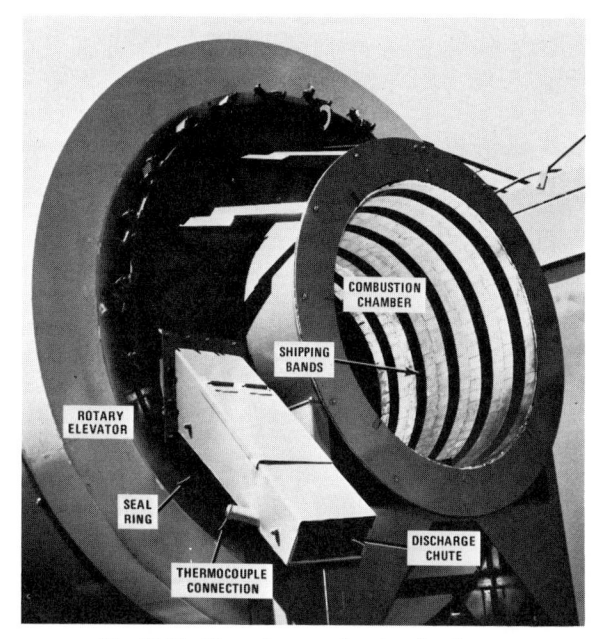

Fig. 10-36 Dryer drum combustion chamber.

Fig. 10-37 Drum drive chain oiler.

is a periodic check for wear and spring tension. The seals should be replaced when they are worn to approximately 3/16 in. The seal rings against which the seals ride should also be watched for wear.

Surge Storage System

Slat conveyor Inspect the chain, slat attachments, and sprockets weekly. The segmented sprockets can be replaced without breaking the chain. However, loosening the

Fig. 10-38 Slat conveyor.

chain take-ups will make sprocket replacement easier. Check the chain rollers for excessive wear and replace as required (Fig. 10-38).

Slat conveyor return rollers Inspect the chain return rollers monthly for surface grooving. Replace or reface rollers as required.

Conveyor bed plates Inspect conveyor bed plates for holes at periodic intervals. Holes will allow material to leak into and plug the heat chamber.

To check wear, remove bed plate hold-down bars and check with a straightedge across the bed plate. Measure at several locations. Plates should be replaced when wear exceeds 3/8 in.

Air cylinders Depending upon the application, your cylinder may eventually require some maintenance and possible replacement of parts. Whenever a cylinder is disassembled, carefully check and replace static and moving seals. Look for worn or scored parts which may need replacing.

Air oiler Inspect batcher and discharge gate air oilers daily. Air cylinders that are not lubricated properly have relatively short life and require extra maintenance.

Main air filter Daily inspection and draining of the air filter will prevent problems in valves and cylinders caused by rust. For cold weather operation, give extra attention to prevent valve freezing.

Chapter **11**

Trenching and Ditching Equipment Maintenance

THE STAFF
Cleveland Trencher Company,
A Division of the American Hoist & Derrick Company,
St. Paul, Minn.

Manufacturers over the years have developed trenching equipment which gives high production with a high reliability factor through the use of carefully selected materials, properly sized components, and sound engineering principles. Today's trenching machine embodies the latest thinking in drive components, metallurgy, and hydraulics. Trenching equipment can be kept available for productive work through the regular use of good maintenance procedures.

The need for good maintenance is well-understood: it has a direct effect on profitability and availability of the equipment on the job. The well-maintained machine is a safe machine. It is a machine which helps reduce operator fatigue and promotes high job morale. No one enjoys running a machine that needs to be nursed along.

Regular maintenance keeps your machine fit for duty. An athlete does not compete without first having spent many hours preparing for an event. A machine should not be placed on the job unless it is fit for duty. An unfit machine causes delays, loss of profits, and endangers the safety of workers.

DEVELOPING AN EFFECTIVE MAINTENANCE PROGRAM

Trenching machines are carefully inspected prior to delivery. Your dealer has performed certain predelivery inspections which ensure that the machine is ready for work. He has checked all fluid-containing compartments for proper level with correct fluids. He has checked all systems for proper operation, has operated the machine, and has made whatever adjustments were indicated.

Today's profit-making contractor insists on good maintenance programs. He knows that it must be performed at regular intervals by properly trained personnel. Checklists remind maintenance personnel of all items which must be serviced, checked, or inspected. Machine records should have information on past inspections, giving date and service meter readings. They should have notations on what was done, particularly on items that were not on the checklist. Notations by maintenance personnel on items such as the progression of wear, wetting of hoses and fittings, or the fraying of drive belts will be particularly helpful since they will draw attention at the next inspection. Therefore it is a good idea to read the record of the last inspection before performing a machine inspection.

Schedules for regular inspections are best developed by the contractor according to his operations. However, many items must be checked on a daily basis. These relate mostly to lubrication of the machine and a vigilance of all fluid compartments. An operator should not begin the day's work unless he has gone through his daily checklist and made a visual inspection of the machine for loose fasteners and possible hazardous conditions developing on the machine. The service meter on the trenching machine is an invaluable aid for providing data relative to maintenance intervals. Be sure that it is always working.

A weekly maintenance interval may be well-suited for routine cleaning of the machine and replacement of parts on which progression of wear has been observed. Engine air cleaner systems, hydraulic filter elements, and adjustments to belts and drive chains requiring removal of guards can be conveniently made at this time interval.

A program should be developed for off-season maintenance. This is when major overhauls can be scheduled, and it is a good time to make conversions and possible updating of components on your machine. It also properly prepares your machine for off-season storage and permits you to get it ready with a minimum amount of time for the next season's work.

TRENCHER ENGINE MAINTENANCE PROCEDURES

Performing a daily inspection of the engine will pay dividends by spotting problems in the making. For this reason, engine cleanliness is important. Clean away accumulations of dirt. This not only simplifies spotting difficulties, it helps improve air circulation and heat transfer from the engine to the atmosphere.

Trenchers are required to work under varied conditions. Lubrication programs should meet the requirements of these operating conditions. Under hot and dusty conditions, more frequent oil changing will be required. Consult your engine maintenance manual for the proper grade of engine oil for the particular climatic conditions. Consult with your oil supplier for the oils which meet your engine manufacturer's requirements. Always use clean containers for the addition of lubricants to your engine.

Check engine oil level daily; do not allow oil level to get below the ADD mark. This is particularly important when operating on steep slopes. Close observation of the oil pressure gauge during operation will also help ward off trouble due to low oil level. Under no condition should an engine be operated where oil pressure has been allowed to fall to a dangerous level.

Engines are equipped with one or more lubricant oil filters. Consult the engine maintenance manual for proper replacement elements. Many lubricant oil filters have a decal which give time intervals and element part numbers. Keep a supply of clean filter elements on hand in their original containers until they are ready for use. Do not allow them to bounce around in the back of a pickup truck or otherwise be carelessly handled.

Poor maintenance of the electrical system can give great difficulty, causing poor starting, weak spark, and inability to crank the engine properly. Probably the one single item in the electrical system which can give the most difficulty and requires closest observation is the battery. Trenchers are equipped with heavy-duty industrial-type batteries. For gasoline engine installations, single 12-volt batteries rated at 72 amp·hr are used. For smaller displacement diesel applications, two 12-volt 72 amp-hr batteries are connected in parallel. For larger displacement diesel engines, two 12-volt 200 amp-hr batteries are connected in series. In most cases the negative side of the battery is grounded. Some grounds are made directly to the trencher frame while others are made at the starter. These have an auxiliary ground to the frame to provide a ground for other electrical accessories.

Check the electrolyte level in the batteries weekly. If necessary to add water, be sure to add clean distilled water in sufficient quantities to just cover the plates. Overfilling can prevent proper elimination of gases formed during the charging process. Keep battery terminals clean and free of corrosion. A good cleaning solution can be made by mixing 2 tablespoons of baking soda and approximately 1 pint of water. Once the terminals and battery case are cleaned, rinse with clean water. It is a good idea to apply some grease or petroleum jelly to the terminals and battery clamps to help prevent further corrosion. Also pay careful attention to battery hold-downs. A corroded or loose hold-down will allow a battery to vibrate excessively and can cause shorting of plates.

Closely monitor the battery charge, particularly if freezing temperatures are anticipated. If a battery is nearly completely discharged, the electrolyte will freeze near 32°F. A cracked battery case can be the result. The best way to check the battery charge is to take specific gravity readings of the electrolyte. For temperature climates and a fully charged battery, electrolytes should have a specific gravity reading of approximately 1.28. A battery at 50 percent charge will read 1.21, and a discharged battery will read 1.11. If you find that your battery is repeatedly at a low state of charge, troubleshoot the electrical system and have the condition corrected immediately. Pay particular attention to generator or alternator drive belts. A loose belt may be the only problem.

Keep a close eye on the wiring in the engine electrical system. Check for loose connections, frayed wires, and abrading of wires against engine components. Be sure wires are dressed away from exhaust manifolds and other frame members. Wires should be clamped adequately to prevent their movement.

The engine cooling system also warrants close attention. Check the cooling level daily before the start of the day's work. It is recommended that a premixed solution of 50% ethylene glycol type antifreeze and water be added to the cooling system at all times of the year. This will give adequate protection against freezing down to the level of 34°F below zero and also provide good rust-inhibiting and water pump lubrication. For temperatures lower than −34°F antifreeze up to 60% solutions may be used.

On a weekly basis, check all radiator and hose connections. Pay particular attention to the condition of the hoses; a hose that feels soft and mushy is on the road to failure. Inspect the radiator for wet spots which may indicate cracks developing. Also, be sure to keep trash and bugs from accumulating on the radiator core. Check the mounting of fan shrouds and be sure that the air has a clear and unobstructed flow through the radiator core. Do not install auxiliary equipment, trash guards, or other paraphernalia in front of radiators to obstruct the flow of air. On trenchers, blower-type fans keep hot air from blowing on the operators. This type of system is more sensitive to obstructions in front of the radiator.

Overheating of an engine can be caused by loose fan drive belt, low water level, improperly operating thermostat, air recirculation, and obstructions to air flow through the radiator core. Do not operate engines with the thermostats removed. The thermostats provided by the engine manufacturer gives the best overall performance and economy. Drain and flush the engine cooling system at least once a year. Early fall may be the best time for this. Discard old coolant. Fresh coolant ensures good rust-inhibiting and water pump lubrication.

Without a proper fuel supply, your engine cannot operate. The importance of clean fuel cannot be overstressed, particularly for diesel engines. Fuel injection pumps are equipped with closely fitting parts, and minute particles of dirt can cause serious problems. It is therefore important that proper handling and storage of fuel be maintained. Although engines are equipped with fuel filters which can handle occasional foreign matter, it should not be overtaxed by feeding it dirt-laden fuel.

The presence of water in the fuel is also detrimental. Be sure that you have provisions for draining off condensation in your storage tanks. Also, it is a good idea to fill your tank on the machine at the end of each work day to prevent the entrance of moisture-laden air into the fuel tank and to help keep condensation to a minimum. Once a week drain off some fuel from the bottom of the fuel tank, or if your engine is equipped with a sediment bowl, water may be removed at this point. Be sure all connections in your fuel system are tight. Dripping fuel is not only expensive but hazardous.

Engines are air-breathing mechanisms. Without an adequate supply of oxygen they cannot properly and efficiently burn the energy-providing fuel. Air induction systems merit close surveillance and good maintenance practice.

Probably the most important item in the air induction system is the air cleaner. Inspect the trays or cups of oil bath air cleaners on a weekly basis. Under extremely dusty and dirty operation conditions, do this daily. Remove the cup, clean, and refill with engine crankcase oil to the proper level. SAE 30 grade oil is usually recommended. If you should be operating on extreme grades, be sure to watch the oil level in your air cleaner as it may be sucked into the engine.

Engines equipped with dry-type air cleaners may also be equipped with air restriction indicators. Check these daily. Many types of dry-type filter elements may be cleaned. Follow the filter manufacturer's recommendations very carefully should you elect this course. It is a good idea, however, to always have a spare filter element on hand. Whenever reinstalling filter elements, be sure that all gaskets, "O" rings, and other closure devices are in good serviceable condition. Remember that the smallest leak in an air induction system can allow the entrance of enough dirty air to doom an engine to failure.

Inspect all connections in the air induction system daily, paying particular attention to flexible hoses. Wear or deterioration in these hoses indicates that immediate replacement is required. Be sure that hoses do not contact frame members or other components which can abrade them. Carefully inspect hoses at the air horn of carburetors. Sometimes repeated flooding of a carburetor may cause gasoline to accumulate in a hose, causing deterioration.

Check exhaust systems on a weekly basis. Be sure all connections are tight and pipes and clamps are in sound condition. Check condition of rain caps to make sure that they are operating properly. A stuck rain cap can provide excessive back pressure which may eventually damage the valves in the engine.

Do not overlook routine maintenance of engine controls. Be sure that governor levers, governor control linkages, emergency shutdown systems, choke controls, and all instruments are in good working order. Do not allow dirt to accumulate on linkages. For engines equipped with safety shutdown devices, be sure that these are all operative and will perform properly should they be required.

TRANSMISSIONS

Engine transmission Your trencher is equipped with either a four- or five-speed and reverse truck-type transmission. Some of these may have the fifth speed blocked out and others may have reverse blocked out, depending on the model. Some models have all speeds plus reverse available for use.

There is very little required in the way of maintenance for this transmission other than regular checking of the fluid levels. Inspect the oil level on a monthly basis. Clean around the filler plug before inspection and add sufficient oil to maintain the correct level.

Several types of lubricants may be successfully used in your engine transmission. Your selection will depend on the climatic condition in which you are operating. For temperatures 20°F and above, the following may be used: (1) MIL L-2105-B or of an API classification GL-5 Grade 90, (2) heavy-duty engine oil SAE 50 service class MS, SD, or SE, and (3) straight mineral gear lubricant Grade 90. Do not use extreme-pressure lubricants other than MIL L-2105-B or of an API classification GL-5. Some EP lubricants contain chemical compounds which may cause severe corrosion, residual deposits, and poor lubrication. Use of these EP lubricants may result in failure or impair operation of your transmission. For temperatures from −20 to 60°F the following may be used: (1) MIL L-2105-B or of an API classification GL-5 Grade 80, (2) heavy-duty engine oil SAE 30 service class MS, SD, or SE, and (3) straight mineral gear lubricant Grade 80. From temperatures from −60 to 0°F, MIL L-10324A lubricants are recommended.

Drain the oil in your engine transmission after 500 hr of use, more frequently in extremely dusty and hot operations. Draining the transmission will help remove microscopic particles of metal which are caused by normal wear and service. Also, chemical changes occur in your lubricant during use. Drain oil from the transmission when the transmission is thoroughly warm.

After draining, flushing the transmission with a light flushing oil is a good idea. This may be done by filling the transmission to the correct level and driving the transmission at fast idle in a manner in which the gears are not under load. Drain all the flushing oil before adding new oil to the transmission.

Refill the transmission with one of the lubricants mentioned above, according to your climatic requirements.

Crawler transmission The trencher crawler transmission is basically two transmissions in one housing. It contains a three-speed section and a four-speed section plus a forward and reverse compartment. Combinations of these gear selections give up to 12 speeds for each speed of the engine transmission in both directions. The gears in the transmission are simple spur gears with pointed teeth to facilitate shifting. Gears are selected by three levers while the vehicle is standing still. Service the crawler transmission according to the same schedule as the engine transmission. Similar procedures may be used and the same lubricants may be used in the crawler transmission as specified for the engine transmission.

Boom hoisting and conveyor elevating transmissions The boom hoisting and conveyor elevating transmissions are basically worm and gear drive units with self-centering double clutches on the input to provide a directional section in either forward or reverse. The boom hoisting transmission is equipped with two independent output shafts, one for the front end of the boom and one for the rear end of the boom; thus, complete and independent control of the boom may be had.

The conveyor elevating transmission is basically just one-half of the boom hoisting transmission and contains only one output shaft. Input to both transmissions is through a chain and sprocket drive. Little maintenance for the input clutches is required except occasional adjustment for wear. This may be accomplished as follows (see Fig. 11-1). The clutch must be adjusted whenever the lock-in pressure is lost, or whenever the hoist will not raise the boom.

1. Remove cover of case and drain or pump out oil until clutches are exposed.
2. With shifter sleeve C in neutral, lift spring A with tool E resting on sleeve C. Lift spring just high enough for lip D to clear teeth on collar B.
3. Turn collar B one notch at a time, by hand. Turn clockwise to tighten, counterclockwise to loosen, as viewed from each clutch shifter sleeve C.

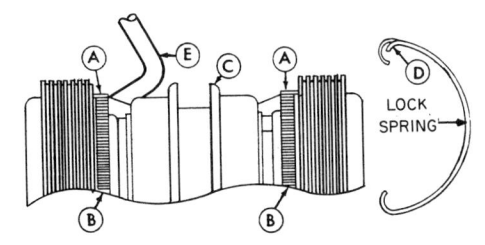

Fig. 11-1 Duplex hoisting clutch. (A) Spring, (B) collar, (C) shifter sleeve, (D) lip, (E) tool.

4. Adjust each clutch so that it will just lock in when lever is applied in either direction.
5. A hoist transmission clutch can be restored to original condition (after all adjustment is used) by the addition of two plates, one inner and one outer.
6. After clutches are adjusted, refill case and replace cover. Test operate.

Lubricate daily the control linkage for the boom hoisting transmission and the elevating transmission. Be sure that the spring centering device returns the control levers to neutral.

Check on a monthly basis the fluid levels in both the boom hoisting and conveyor elevating transmission. The level plug in these transmissions is high enough to be sure that input clutches are completely submerged in oil.

Some later machines are equipped with boom hoist transmissions which have hydraulic motor inputs instead of chain-driven inputs, and clutches are not required. The same procedure for servicing the chain driven transmissions applies to these transmissions. The oil level in the hydraulically driven transmissions is somewhat lower than for the chain-driven transmissions.

Conveyor drive transmission The conveyor drive transmission is basically a miter gearbox equipped with a sliding gear for forward and reverse selections. Input to the transmission as well as output is by means of a chain and sprocket. This transmission requires little in the way of routine maintenance. Check oil on a monthly basis and make oil changes at approximately 1000 hr of operation. Lubricant similar to those used for the engine transmission may also be used in the conveyor drive transmission.

Conveyor pulley transmission With the advent of hydraulically driven conveyors on trencher machines, a unique conveyor drive pulley (Figs. 11-2 to 11-4) was developed by The Cleveland Trencher Company. It embodies an internal gear reduction with a hydraulic motor completely contained within the shell of the pulley, thus eliminating hung-on gearboxes. Later designed conveyors are equipped with hydraulic pulleys on which the motor is mounted outboard. This provides for easier servicing plus the ability to interchange motors of various sizes.

The pulley shell is divided in two compartments. One is the transmission side, the other is a shell with a hollow shaft passing through. This shaft provides a bearing support as well as a vent for the transmission side of the pulley. The transmission side of the pulley contains the drive gears and the hydraulic pulley motor, which is bolted to the conveyor frame and provides the torque reaction for the drive pulley. Oil passage is into and out of the end of the hydraulic pulley motor.

Conveyors are generally equipped with two hydraulic pulleys having their motors connected in series, thus providing a drive for the conveyor belt in either direction.

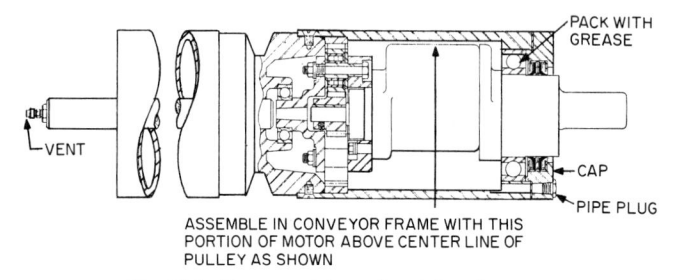

Fig. 11-2 Conveyor drive pulley, inboard motor.

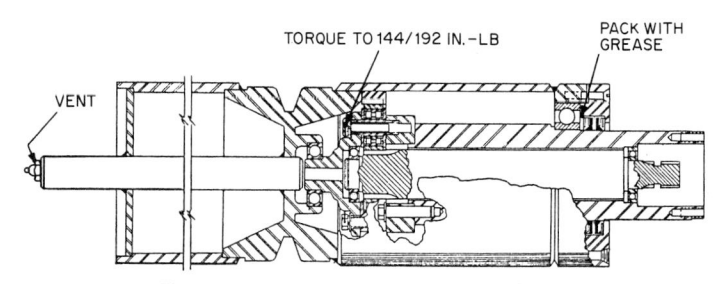

Fig. 11-3 Conveyor drive pulley, outboard motor.

Inboard-mounted conveyor pulley motors are internally drained and do not require external drain lines. Outboard-mounted motors are provided with an additional drain line. These lines are connected in series and returned to the reservoir.

The transmission side of the conveyor pulley is filled to the level plug with SAE 80 mineral gear oil. The level plug should be coincident with the horizontal centerline of the pulley assembly whenever checking the oil level. Check the oil level on a monthly basis. Clean and flush the transmission on a yearly basis.

If continuous weeping is noted out of the vent, this indicates the possibility of a defective motor shaft seal. This will then require disassembly of the pulley and replacement of the defective seal.

Crawler and digging differentials Your trenching machine may be equipped with two truck-type differentials. These may be of the spiral bevel (Fig. 11-5) or hypoid gear design. In each case, the differential provides a means of equal distribution of power to either the digging wheel or the crawler tracks of the trencher. The crawler differential further provides a means for steering the machine. Braking of one output shaft provides a drive to a single track, thus causing the machine to steer in a direction opposite of the moving track. This is a simple and effective means for making the trencher a highly maneuverable machine.

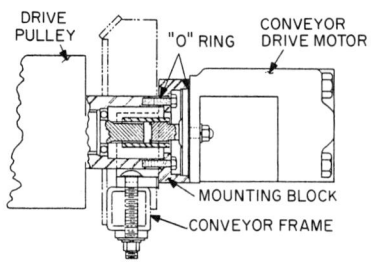

Fig. 11-4 Interconnection of drive pulley and motor.

Little routine maintenance is required for

both the crawler and digging differential. Check oil levels on a monthly basis. Use extreme-pressure lubricants according to the schedule provided for the engine transmission. Drain and replace oil on a yearly basis.

CLUTCHES

Engine clutch Your trencher is equipped with either a single- or a two-plate clutch. This clutch disconnects the engine from the entire drive line. It is held in engagement by spring pressure on the pressure plate, and it is held out of engagement by a toggle linkage. The reaction of the springs is taken by the main thrust bearings in the engine during

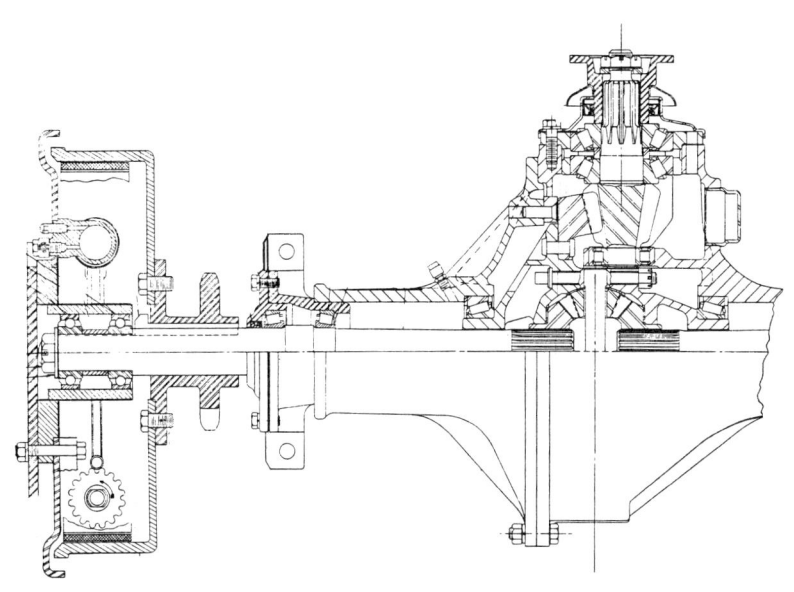

Fig. 11-5 Crawler differential, spiral bevel unit.

disengagement. Therefore it becomes very important that if the engine is to be allowed to idle for a prolonged period of time, the engine transmission should be shifted to neutral and the clutch reengaged, thus relieving the reaction on the crankshaft main bearings. This is necessary since under idling conditions, it is possible that your engine main bearings may not receive adequate lubrication to support the added thrust from the clutch.

Grease the clutch release bearing daily with a small amount of high-temperature grease. Too much lubricant will foul the clutch plates and cause slipping.

Earlier designs of engine clutches required periodic adjustment for wear. Later designs automatically compensate for wear. For those clutches requiring adjustment, the procedure is as follows.

Adjustment of engine clutch It is recommended that the clutch be kept in proper adjustment by a periodic check of dimension A (as shown on Fig. 11-6) between clutch release bearing and ground surface of clutch cover. Check this dimension while the engine is stopped and the clutch engaged.

CAUTION: Never wait for the clutch to slip before adjusting it. Once the facing is burned through slippage, it quickly disintegrates. Do not adjust clutch linkage to restore free travel until after clutch is adjusted by pulling shims as outlined in the following procedure.

1. With engine stopped and clutch engaged, place gauge between clutch release sleeve and clutch release bearing as shown. Obtain dimension A by removing shims, following the procedure below.

2. Disengage clutch with hand lever. This prevents bending of the adjusting straps and facilitates loosening of the adjusting nuts.

3. Loosen adjusting nuts five full turns, then engage clutch. This will allow adjusting plate to move back for easy removal of shims.

4. Remove shims using sharp nosed pliers. Removing one shim from each of the studs reduces dimension A by 3/32 in. Make sure that an equal number of shims remain on each of the studs before proceeding.

5. Disengage clutch lever and tighten adjusting nuts. Recheck dimension A.

6. Check clearance between clutch release sleeve and clutch release bearing. If clearance is not 1/8 in., adjust linkage to obtain correct clearance. This can easily be done

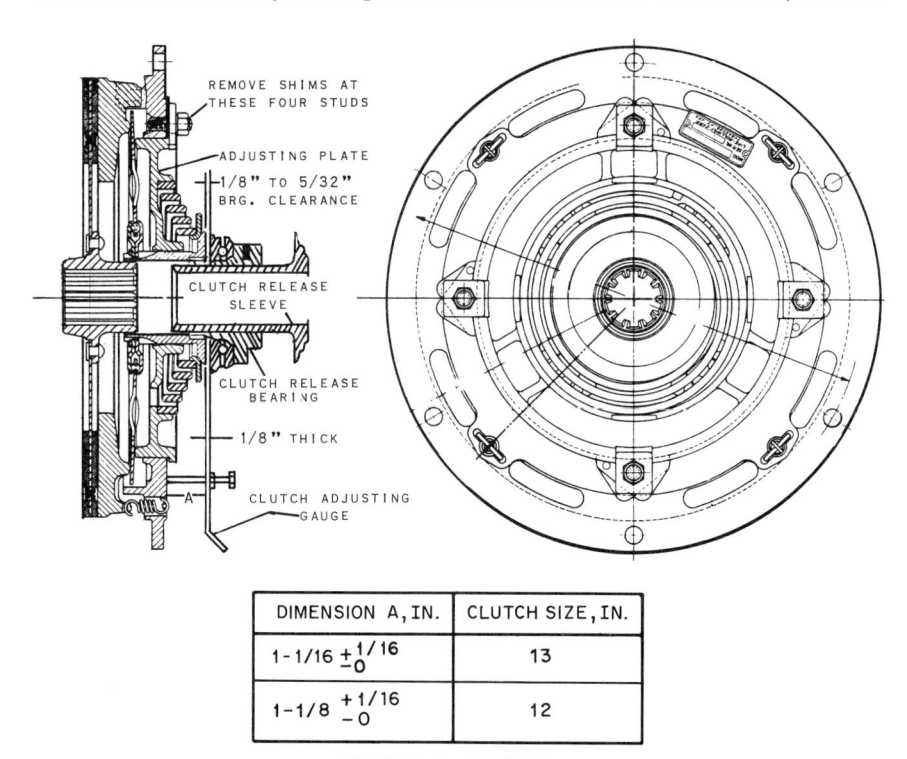

DIMENSION A, IN.	CLUTCH SIZE, IN.
1-1/16 +1/16 / -0	13
1-1/8 +1/16 / -0	12

Fig. 11-6 Engine clutch.

by disconnecting clutch rod at clutch lever and screwing in or out on yoke end. Reassemble and check for 1/8-in. clearance.

Digging and crawler clutches Trenchers are equipped with a digging clutch, and some models are also equipped with a crawler clutch. The digging clutch disconnects power to the digging wheel, while the crawler clutch disconnects power to the crawler transmission. When the operator wants to interrupt forward progress of the machine, he merely disengages the crawler clutch and in the meantime all other functions of the machine continue. This saves time in selecting the crawler gears and relieves the operator of the need for picking up the load on the digging wheel and conveyor when resuming forward progress.

The digging clutches are of a multiple disk design and may either be 8-in. two-plate clutches, 8-in. three-plate clutches, or 10-in. three-plate clutches (Fig. 11-7). Friction members of these clutches are divided into three segments so that as wear progresses they may be replaced without disassembling the clutch from the main shaft. These clutches are held in engagement by a toggle mechanism activated by a lever and a linkage system. Since these clutches are exposed to weather, it is very important that the linkage be kept well-lubricated and the formation of rust be prevented. For prolonged storage of the machine it would be wise to make a weatherproof covering for the clutch.

Clutch adjustment If the clutch slips, heats up, or disengages during use, it requires adjusting, and possibly servicing.

Adjustment is made by holding the locking pin out of engagement and rotating the adjusting yoke assembly either in or out (clockwise or counterclockwise) as required, until clutch snaps into positive engagement from a distinct pressure applied on the hand lever. After correct adjustment has been obtained, release the locking pin and rotate the adjusting yoke assembly until the locking pin seats in the nearest hole in the floating plate. A new or rebuilt clutch may require several adjustments during the wearing-in period.

Adjustment to give positive lock-in or drive is recommended for all normal digging. However, under abnormal conditions where underground obstructions may be encountered, the clutch may be adjusted to allow slippage in case the digging teeth hook onto something. This, of course, may cause the clutch to heat up and wear more rapidly, but may prevent other serious damage.

Replacing friction disks Loosen locking collar on the main shaft and slide it toward engine. The cone collar assembly may then be shifted away from the clutch. By holding out on the locking pin the adjusting yoke assembly may be backed out by unscrewing. This will then permit access to the friction disks which are made up of three segments for ease of servicing. If space does not permit removal of the friction disks per the above method, the driving ring may be removed. It is not necessary to remove the clutch from the main shaft to replace the friction disks. However it may become necessary for replacement of other parts.

Digging clutch line up and lubrication The clutch must be aligned between the main shaft and digging differential for proper operation. To check alignment, with engine off, engage clutch. Without rotating clutch, check distance from clutch friction disk teeth to front edge of the driving ring. If properly set up, the distance will be the same, top, bottom, and sides. If misaligned vertically, shimming under the differential is required. If misaligned laterally, squaring of the main shaft is required.

Grease clutch throw-out bearing daily with all-purpose grease. Grease pilot bearing after every 200 hr of operation with a very small amount of grease. If disassembled, be sure to pack pilot bearing with grease before reassembling.

MAIN SHAFT ASSEMBLY

Your trencher is equipped with a main shaft assembly (Fig. 11-8) which provides a means for distributing power to the crawler transmission, boom hoist transmission, conveyor elevating transmissions, and digging differential. It consists of an alloy steel shaft, mounted in pillow blocks, and carries sprockets, digging clutch, and crawler clutch in some cases. The main shaft requires very little routine maintenance.

Pillow blocks The main shaft is supported by self-aligning pillow blocks equipped with locking collars. Lubricate pillow blocks daily with multipurpose grease. Make a daily

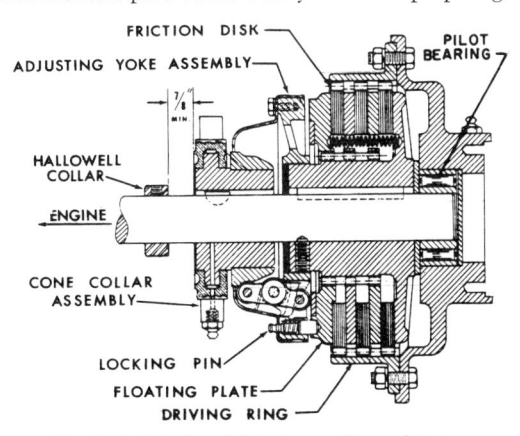

Fig. 11-7 Digging clutch, two- or three-plate type.

Maintenance Programs for Equipment Entities

check of the locking collars. To tighten the collars turn them in the direction of shaft rotation and tighten the setscrew. If for any reason the main shaft must be removed from the machine, it is imperative that upon reassembly the same number of shims be reinstalled under the pillow blocks as were removed. This will guarantee alignment and squareness with the digging differential and relieve strain on the shaft and bearings.

Sprockets The drive to the crawler transmission and boom hoisting transmission is by roller chain and sprockets. Daily lubrication of these chains will prolong life. Use an

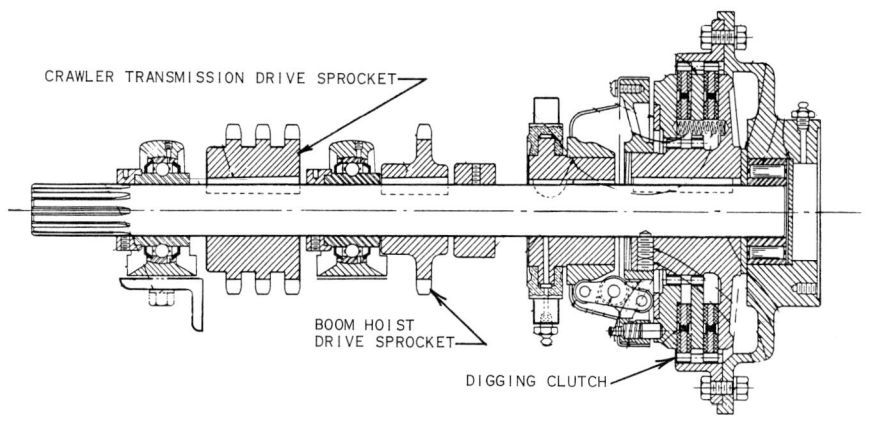

Fig. 11-8 Main shaft assembly.

oil can and light engine oil. If chains become rusted, use a mixture of diesel fuel and engine oil. Lubricate these chains under all operating conditions.

CAUTION: Never lubricate chains while they are moving.

If guards are removed to facilitate lubrication, replace them immediately. Under no condition should a machine be operated with chain guards not installed.

Check for proper alignment of sprockets to be sure they have not moved off their keys. Misalignment of the sprockets causes undue loading on the chains, accelerated wear, and increased loading on bearings and shafts. Make a weekly visual inspection of chain alignment and chain tension.

DIGGING DRIVE SHAFT ASSEMBLY

The digging drive shaft assembly (Fig. 11-9), also known as the number 1 shaft, is mounted on the boom and provides the turning force for the digging wheel through a system of split sprockets and segmented teeth mounted on the digging rims.

The digging wheel drive shaft is supported by two specially designed pillow blocks containing self-aligning bearings. Bearings may be of the ball, spherical roller, or hour glass roller type. The ball bearings are retained on the shaft with eccentric collars. Check these daily for tightness. To tighten, rotate the collar in the direction of rotation and tighten. Ball-type bearings are also provided with their own seals. Spherical-type bearings are retained on the shaft by a bearing cap and are sealed with lip-type seals. Daily lubrication with a multipurpose grease is required.

Cleaning accumulations of trash and dirt around the pillow blocks will help prevent damage to seals. Occasionally barbed wire picked up during digging operations will wind itself tightly around the digging wheel drive shaft.

Power input to the digging drive shaft assembly is by means of roller chains and sprockets. The sprockets may be directly mounted on the digging wheel drive shaft and held in place with tapered keys, or they may be bolted to split hubs which are mounted on a digging drive shaft and held in place with tapered keys.

In either case, the alignment of these sprockets is important. They should be in line with sprockets on the digging differential and located as close to the bearings as possible without interfering with frame members. This will minimize loading on the bearings on both the digging differential and the digging wheel drive assembly.

For machines equipped with a conveyor drive transmission, the digging wheel drive shaft assembly provides the input with a second sprocket mounted directly to one of the digging drive shaft drive sprockets. This sprocket is held in place by a system of spacers, cap screws, and nuts. Check these fasteners daily to be sure that they do not come loose. Once again, alignment of this sprocket with the input socket on the conveyor transmission is important. Chain tension is regulated by an idler take-up sprocket assembly, force being provided by a spring. Lubricate this idler take-up daily with light engine oil where the take-up is mounted on its shaft. Also, the sprocket in the take-up mechanism is supported by two tapered roller bearings. Lubricate these daily with multipurpose grease.

The actual drive to the digging wheel is accomplished through a pair of drop-forged split sprockets mounted on hubs which are keyed to the digging drive shaft. Alignment of these sprockets is carefully controlled by the manufacturer to guarantee that they mesh equally with the segments on both rims. It is imperative therefore that the entire digging wheel drive shaft assembly be kept square with the boom frame. On some machines, blocks are welded on both ends of the pillow blocks to stabilize the location of the digging wheel drive shaft assembly and relieve shearing forces on the bolts which hold the digging drive shaft assembly in place. Make a daily visual inspection of these details. Proper root clearance between the segment and the split sprockets must be maintained. This is usually 3/16 in. (Fig. 11-10). This clearance is necessary to guarantee free meshing of the split sprocket with the segments and reduce scrubbing action of the split sprockets with the segments. The wear patterns of the split sprockets and segments should be carefully observed. Any misalignment, loose mounting, or poor engagement should be corrected immediately.

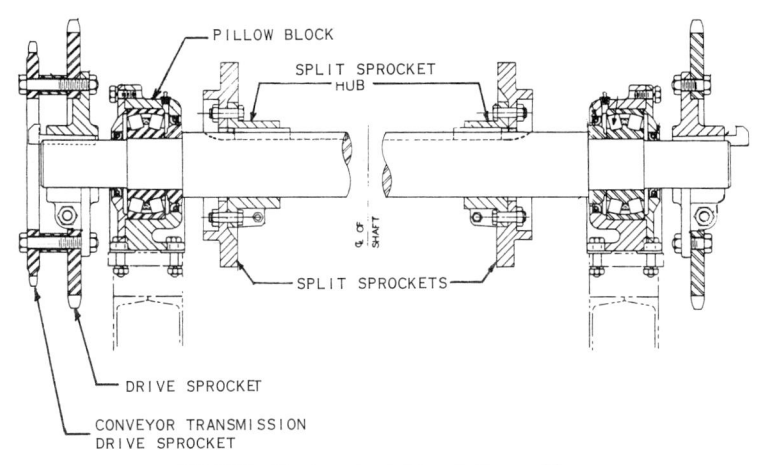

Fig. 11-9 Digging wheel drive shaft assembly.

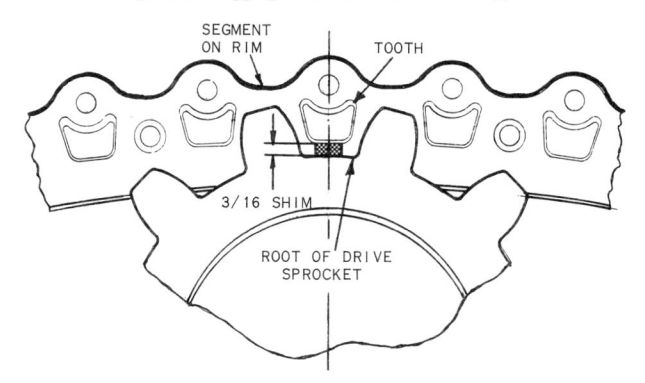

Fig. 11-10 Split sprocket clearance.

Many trencher owners rebuild their split sprockets by welding. The economics of this is questionable. Improper welding can destroy the accuracy of the sprockets and ruin the heat treating that they have received. This can go a long way toward accelerating wear and premature requirements for replacements of more costly segments on the digging rims. Split sprockets are designed to be easily renewable and may be regarded as expendable items. Allow them to wear until the teeth are nearly gone and then replace. Always replace in pairs. Never replace split sprockets on one side only as this will not provide even driving of the digging rims.

Side clearance between the faces of the split sprockets and the segments is also important. Enough side clearance should always be maintained to prevent scrubbing of the split sprockets on the segment faces. Do not allow this side clearance to exceed approximately $3/32$ in.

DIGGING WHEEL ASSEMBLY

The digging wheel assembly is where the action takes place. It is the whole reason for the trencher's existence. It consists basically of two high-tensile steel, single-piece rims having a plurality of buckets mounted on the outside diameter in specific arrangements and the entire assembly being supported on a system of wheels attached to a boom or frame member which may be vertically regulated to control depth. Cutting width is controlled by the width of the buckets mounted on the rims. Each trencher machine has a range of buckets which may be used.

Buckets Buckets are made from the same material as the digging rims and are usually one-piece hot-formed. A variety of buckets may be used on each trencher machine. There are round buckets, square buckets, stripping buckets, and occasionally specially shaped buckets for special applications. All are available in various widths.

Buckets have two mounting lugs for securing to the digging rims by means of high-strength flathead cap screws and special locknuts. The lugs are welded on the outside of the buckets so that flush mounting of the inside of the bucket with the inside of the rim is guaranteed, allowing free dumping of the material in the buckets. These lugs are carefully located to guarantee tight contact with the corners of the buckets on the edge of the rim. This is of the utmost importance in all digging conditions.

Check mounting hardware on buckets daily for proper tightness. Should any hardware need replacement, be sure to use at least Grade 5 or better. Use locking nuts as specified by the manufacturer and do not use lock washers. Tighten the hardware according to the following table.

Bucket Fastening Torques

Bolt size, in	Torque, lb·ft
$1/2$	90
$3/4$	300

During the manufacturing procedure, each bucket is individually fitted to the rims to guarantee as tight a fit as possible. The buckets and rims are stamped to indicate the position of the bucket on the rim. Do not interchange these positions. Each bucket also has a specific arrangement of rooters. A series of buckets is required to completely cover the cut of the trench. On some machines two buckets in series will cover the cut and on others three, and as many as six buckets may be required depending on the digging conditions. Rooter arrangements have been scientifically devised to provide high penetrating forces and equal loading for all buckets. Again, it is very important to keep the buckets in proper sequence as mixing of the buckets will seriously hamper the cutting efficiency of the machine.

Close attention must be paid to the rooters or the digging teeth mounted on the buckets. Trenchers can be equipped with a wide variety of digging teeth. All modern trenchers have renewable digging teeth. In some cases the shanks or the tooth holders are bolted to the buckets, and in other cases they are welded to the buckets. Generally, for easy or normal digging conditions bolted shanks are quite satisfactory, but for very heavy-duty or rugged digging conditions, as in coral, frost, caliche, or hardpan, welded shanks will give superior performance. For buckets equipped with bolted shanks, check these daily or even more frequently for proper tightness. Should a shank become broken off by a rock or other underground obstruction, replace it immediately as a lack of a tooth on a bucket will impair the cutting efficiency of that particular bucket and overload the subsequent bucket.

Bolted side shanks give the ability to vary the cutting width of the machine within certain limits. Never reduce the cutting width of a bucket to the point to where the side shanks provide less than 1 in. of cutting clearance for the segments mounted on the rims. Also, observe the wear of the side shanks to make sure that they are not worn to a point where the ends of the segments are beginning to rub the side of the trench. In particularly abrasive digging, rapid segment wear will be the result, requiring costly replacement of the segments. In situations where you may be digging against a pavement, outfit the rims with wear bars which extend beyond the ends of the segments to guarantee that the segments do not rub against the pavement. Wear bars are bolted onto the rims and are easily renewable, while segments are welded to the rims and are costly to replace.

Maintain constant surveillance of the rooters. Rooters are marked with a wear line. Operation of the trencher with teeth worn beyond this line greatly reduces the cutting efficiency of the machine, and proper cutting clearance for the bucket is not being maintained. Digging under these conditions for prolonged periods of time will accelerate lip wear and heel wear of the buckets, causing a reduction in strength at the lip and eventual replacement. If a tooth is knocked off or broken during digging operations, replace it immediately. Failure to replace this tooth will not only impair digging efficiency of the machine but will cause wear on the shank, leading to destruction of the shape of the shank and requiring replacement.

In some cases it may be beneficial to remove rooter tips and interchange their position on the buckets to provide for additional wear. This is at the discretion of the user, however, as the people cost may not warrant such a procedure. It may be more economical to completely wear the teeth and then replace entirely.

Buckets may also be equipped with narrower, picklike teeth known as rock teeth. These teeth are used in frost, caliche, hardpan, coral, rocky digging, and other extreme digging conditions. The secret of success in these situations is sharp teeth, particularly where the material is tight and homogeneous. Dull teeth, even with great digging forces applied, will not penetrate material such as permafrost. Under these conditions, specially hard-faced digging teeth may be used.

Digging in coral is a different situation again. In some cases wear is less marked. However, maintenance of a good sharp edge on the cutting teeth is essential. When the teeth become rounded at the ends, it will pay dividends to renew the edge on the tooth. Specially sharpened teeth may also be purchased from the manufacturer.

Bucket backs Bucket backs are generally bolted between the rims at the rear ends of the buckets, providing a closure of the bucket so that material may be conveyed up and dumped onto the conveyor for deposition to the side of trench. Bucket backs take the most battering of any component in the entire digging wheel assembly. They are subjected to severe bending forces caused by rocks and heavy roots. Mounting bolts for bucket backs are sized so that they will shear before serious damage is done to the bucket back. If a bucket back becomes badly bent, it should be removed, straightened, and reinstalled. A loose or badly bent bucket back does not provide for the proper support of the digging rim. This can contribute to poor mesh of the segments, and support of the rims will be reduced, increasing bending stresses in the rims. If repeated bending and damage to bucket backs exists, give consideration to installing a rear rock guard. These guards prevent entrance of boulders into the bucket and minimize battering of the back.

On some machines in special digging situations, such as coral and permafrost, bucket backs having great weight are used. These backs increase the inertia of the wheel and provide a higher penetrating force for the digging teeth. It is very unlikely that you will bend one of these backs, but pay close attention to mounting since if bolts are allowed to become loose, the backs can provide enough force to deform holes in the digging rims, causing premature failures. Also, since the backs are strong enough, engaging obstacles such as boulders and logs may cause damage to other members of the wheel end assembly such as conveyor frames and truck shafts.

Wheel suspension The digging wheel assembly rotates on a system of replaceable wheels known as *trucks*. The modern wheel-type trencher is equipped with a front upper truck assembly, a rear upper truck assembly, and a lower truck assembly. The front and rear truck assemblies are adjustable. The lower truck is initially set and requires no further adjustment. The front and rear upper trucks are carried on shafts, mounted on self-aligning bearings. Provision for spacing the trucks to accommodate various widths of buckets is incorporated in the truck assembly. Also, replacement of the truck wheels can usually be done without removing the truck assembly from the machine or disassembly of the complete truck assembly, thus saving wear and tear on the bearings. The wearing surfaces on the truck wheels are the treads and the flanges. Replacement is indicated when the treads are worn to the point where the OD of the flange begins contacting the bottom of the segment teeth.

The order of the adjustment of the truck wheels is quite important. The first step that should be followed is to adjust the front upper truck to provide the proper root clearance between the split sprocket and the bottom of the segment teeth. As noted earlier, this is generally 3/16 in. Next, the rear upper truck should be adjusted so that the clearance between the tread of the lower truck and the tread of the segment is from 1/8 to 3/16 in. If this cannot be obtained, it may be necessary to shim or adjust the rear stay rods to bring the lower truck to closer contact with the rim. Establishment of these clearances permits a free-running wheel and helps accommodate any out of roundness or irregularities found in this type of structure without putting undue bending loads on truck shafts and overloading bearings.

Lubricate upper truck assembly bearings daily with a multipurpose-type grease. Inspection for presence and tightness of all fasteners may be done during lubrication. Lower trucks are equipped with tapered roller bearings and double-lip flexible seals. Also lubricate these daily with a multipurpose grease. Some later model machines are equipped with lower trucks having tapered roller bearings and metallic face seals similar to seals used in tractor rollers. These are filled with SAE 30 engine oil and do not require further attention until major overhauling. Do not use any lubricants containing EP additives as the chemicals present will attack the elastomer members of the seal assembly.

Be sure that all fasteners for the lower truck and stay rods system are kept tight. Loose trucks and stay rods do not properly support the digging wheel assembly during trenching operations. Also bent stay rods should be either straightened or replaced as they do not accurately support the lower truck.

Segments and split sprockets   The power to the digging wheel is transmitted directly by the split sprockets to segments having teeth extending from their faces. Segments are either riveted or welded to the digging rims. The split sprockets segments are drop-forged alloy-steel heat-treated for maximum wear and toughness. These parts must have the ability to transmit power and absorb digging shocks and still provide good resistance to wear. The rate of wear on split sprockets and segments will vary widely depending on the abrasiveness of the material being dug.

Maintain surveillance of the progression of wear on both the split sprockets and the segments. Wear patterns should be uniform on both sides of the digging wheel assembly. Should this not be the case, examine the entire wheel suspension system for loose or damaged parts. Correct this condition immediately to guarantee equal wear on segments on both rims and equal driving of the digging rims.

Allow split sprockets to wear until they are no longer serviceable. It is a mistake to begin building up with weld on the surfaces of the teeth as soon as some wear is apparent. This destroys the metallurgy and heat treatment of the part, and the economics of this procedure is questionable. During the digging season, always keep several sets of split sprockets on hand. Always replace these sprockets in sets. Keep individual split sprockets in pairs as received from the factory. This is done to allow even wear and driving of both digging rims. The split sprockets are held in place on hubs with heat-treated bolts in closely fitting holes. Check tightness of these fasteners daily.

Segments on the digging wheel require no maintenance other than observation of wear patterns. If excess wear is indicated on the end of the teeth, it means that side rooters are not providing adequate clearance for the segments. Maintain close surveillance of rooter teeth mounted on the side shanks.

Allow segments to wear until they are no longer serviceable. Building up of teeth and tread surfaces is not recommended. For digging rims having welded segments, the procedure outlined below may be used for replacement. Generally this job should be scheduled for off-season maintenance as it will take the unit out of service for several days.

Welding segments to rims in field

1. Remove a section of four to five segments from each rim and grind welds, rivets, and high spots from ID and outside face of rim.
2. Bolt new segments in place with ³⁄₈ × 1½-in. bolts. NOTE: When replacing segments on riveted-type rims, make sure to force all segments toward leading edge of buckets (both sides) to minimize line-up error.
3. Preheat rim adjacent to segment approximately 200°F, and preheat one segment at a time, 500 to 600°F. (Check temperature with 500°F Tempilstick.) While segment is still hot, weld both sides with ⅛-in. diameter *low-alloy* low-hydrogen electrode (AWA-Class E8016-B2 or E8018-B2) ³⁄₁₆-in. continuous weld on outside and ⅛-in. continuous weld on inside.
4. Remove bolts, grind welds flush with inside face of rims, and grind high spots at joints on ID tread of segments to blend contour before proceeding with next section. Continue as outlined in steps 1 to 4.

For rims without locating holes, the following procedure may be used.

1. Wire-brush the gap between two segments thoroughly. Check gap with feeler gauge and note thickness.
2. Remove four segments 90° apart. Grind welds flush on ID and face of rims. Clamp new segments in place. Keep gap equal between ends of new segments and old segments. Tack four segments two places on both sides and remove clamps.
3. Remove 90° section of old segments between tacked segments on both rims. Grind welds flush on ID and face. Use feeler gauge to obtain proper gap between segments. Clamp and tack each segment two places on both sides.
4. Preheat rim adjacent to segment approximately 200°F, and preheat one segment at a time, 500 to 600°F. (Check temperature with 500°F Tempilstick.) While segment is still hot, weld both sides with ⅛-in. diameter *low-alloy* low-hydrogen electrode (AWS-Class E8016-B2 or E8018-B2) ³⁄₁₆-in. continuous weld on outside and ⅛-in. continuous weld on inside.
5. Grind welds flush with inside face of rims and grind high spots at joints on ID of tread of segments to blend contour before proceeding with next 90° section. Continue as outlined in steps 3 to 5.

REFERENCE: It is advisable to drill 25 to 64 holes through rims after welding, locating from hole in segment to facilitate future segment installation.

Rim maintenance Digging rims on the trenching machine are flame cut in single pieces from low-alloy, high-strength, constructional-type steel. They require very little in the way of routine maintenance other than close inspection for the beginning of cracks and damage to bucket mounting holes. If a crack appears, do not permit it to progress to the point where the rim fails. V it out and repair with weld using low-hydrogen rods in the 60,000 to 70,000 psi tensile range.

Digging operations where bucket backs are not used or where extremely shallow digging is being done can put very severe bending strains into the digging rims. When digging under these conditions, be especially vigilant about rim inspection. If repeated problems occur, it is quite possible that the trencher should be equipped with a heavier set of rims. Most models are available with rims in several thicknesses.

CONVEYOR ASSEMBLY

Conveyors on trenching machines are asked to do a tremendous job. They generally have belts that are narrow and travel at high speeds. They are subjected to extremely severe dumping conditions. Loading is usually done 90° to belt travel and from above. Sharp rocks and cyclical loading as each bucket dumps its load are also present. Tumbling of rocks on a conveyor can be a serious problem.

Several styles of conveyors are available on trenching machines. Arc-type conveyors have frame work which follows an exact radius. V-type conveyors generally have an arc for only a short distance in the center from which the conveyor frame extends tangentially to the ends of the frame. Larger trenchers are generally equipped with elevating-type conveyors with extensions of varying lengths, depending on the cutting width of the buckets.

Belt widths on conveyors vary generally from 18 to 36 in. Some belts are driven with roller chains, and others are driven hydraulically with speed variation in steps or continuous variation from zero to maximum.

Conveyor belts should be given very close attention and inspections of their condition should be made daily. Particular attention should be devoted to the belt lacing. Generally if trouble is going to be had, this is where it will be. Examine the lacing for snags and damage by rocks; pay particular attention to the ends of the lacing and the portion which passes beneath the side belts or side skirts. Bent frames and bent side belt supports can cause serious damage to belt lacings.

Training of conveyor belts is of the utmost importance. The belt that wanders off to the side is sure to give problems and eventually destroy itself from wear on the edges. Some belts are equipped with metal clips in the form of a V uniformly spaced and fastened on the underside of the belt on the center line. These clips engage V-grooves on the drive pulleys, thereby helping to keep the belt from shifting sideways. More recent conveyor belts have a vulcanized V-guide on the underside of the belt, engaging a V-groove in the drive pulleys. There are several advantages in the vulcanized V-guide system. Guiding of the belt is continuous, and wear to the drive pulley is minimal. The greatest advantage, however, is to eliminate extra hardware which may become loose and become trapped between the underside of the belt and the drive pulleys. Also, the necessity for holes in the belt to accommodate the hardware is removed. Many belts have been punctured by metal V-guides which have come loose at one end and forced their way through the belt.

Some conveyor belts are equipped with metal cleats fastened to the top surface of the belt, spaced approximately 1½ ft apart. These cleats are intended to aid movement of the material in the direction of spoiling. Cleats are generally fastened to the conveyor belt with elevator bolts, in some cases with two bolts, one at each end. In other cases where the alignment happened to be coincident with the V-guide clip, there will be an additional two bolts at the center of the cleat. It is difficult to keep bolts holding these cleats from becoming loose during hard digging operations. The operator must be on the alert constantly for this condition. If one end of the cleat is allowed to come loose, it can strike the boom as it passes and can rip holes in the conveyor belt. Should digging conditions dictate the need for cleats, use a belt equipped with vulcanized cleats. Conveyor belts with metal guides and metal cleats require daily inspection and daily maintenance. Keep a supply of clips and cleats and fastening bolts on hand.

It is of the utmost importance that the end drive pulleys be mounted square and parallel. Squareness with the conveyor frame may be checked by measurement from some convenient reference point on the frame. This also helps guarantee parallelism of the drive pulleys. A good check on this, however, is to measure from center to center of the drive pulleys at both ends and compare the readings. All conveyor drive pulleys are adjustable at all four points of attachment to the conveyor frame. Once proper alignment of the end drive pulleys is established, they should be adjusted to give proper tension for the belt. The belt should be tightened so that it just drives without slipping. On arc-type and V-type conveyors, excessive belt tension will cause the belt to hump up at the center and pull out from beneath the side skirts or side belts. This can be troublesome when the belt is not loaded with dirt.

In some cases, particularly on elevating conveyors, edge guides are provided for the conveyor belt. Adjust these so there is approximately $3/8$-in. clearance between the belt and the guide. They are provided only to help relieve side forces on the belt when setting in and the boom is not level.

Trencher conveyors are equipped with side skirts or side belts which help keep material from spilling over the edge of the belt. Side belt supports are particularly vulnerable to damage in the dumping area of the conveyor, for it is here that rocks are thrown off, landing on the side belts. Inspect the condition of side belts and side belt supports daily. The side belt support that is bent downward may be forcing backing strips onto the conveyor belt which will very shortly cut through the belt and reduce the width of the belt to the point where it is no longer serviceable. If a conveyor belt shows any signs of grooving, shut down the machine immediately and inspect for side belt supports rubbing. The material used in the side belt is of a soft, abrasion-resisting-type rubber. It contains no duck or fabric of any kind. Use of homemade side belts which contain fabric or duck will be detrimental and cause wear to the top cover of the conveyor belt. Side belts made from anything but the recommended material can be very costly.

Make a weekly inspection of other conveyor components. These include scrapers, pulley scrapers, dump plates, and idler pulleys. Belt scrapers are installed to keep the underside of the belt from accumulating mud and dirt. Examine these for free tracking on the surface of the belt. Also, belt scraper belts should be in good condition and not worn.

Pulley scrapers are provided to prevent accumulation of dirt and mud on drive pulley surfaces. Adjust these to maintain the proper clearance between the scraper and the pulley surface. Generally 1/8 in. should be maintained. Dump plates are installed to provide additional support for the belt, particularly beneath the digging rims in the dumping area. Inspect these for worn corners, for bending, and for the development of razor-edges due to abrasive material being dragged over by the belts. Idler pulleys are installed to support the load as it is being carried off by the belt. These idler pulleys are driven tangentially by the belt surfaces; therefore dirt should not be allowed to accumulate on the pulleys as they may become stalled. Should an idler pulley become stalled, the belt will eventually wear through it. Inspect idler pulleys for bent shafts, loose pulleys. Lubricate idler pulley bearings with a multipurpose grease.

Conveyor shift systems Many models of trenching machines are equipped with shiftable conveyors so that spoil may be deposited on both sides of the trench and placement of the spoil may be controlled. On older model machines, conveyors were shifted by hand cranking and usually required an additional person or the operator to get down from his position. Modern trenchers have power conveyor shifting mechanisms that are controlled directly from the operator's position.

Very little is required in the way of maintenance for hand cranking systems. Usually it is only necessary to lubricate the cranking mechanisms on a monthly basis. Condition of the pin wheels which engage holes in the lower runners in the conveyor frame and accomplish the shifting should be checked. A set of pin wheels will usually give many years of satisfactory service on a trencher with very little attention.

Several systems of power shifting conveyors are in common use today. An extension of the hand crank system has been devised by merely removing the hand crank and installing a hydraulic motor. These are usually high-torque, low-speed-type motors which require little attention other than routine inspection of the hydraulic fittings for leaks and tightness. Alignment of the hydraulic motor with the pin wheel shaft and mounting of the motor should be observed on an occasional basis.

Trenchers employing hydraulic cylinder and systems of wire ropes and sheaves for shifting the conveyor require closer observation and inspection. Wire rope clips and fastening hardware must be properly installed and kept tight to prevent the development of slack. Excess slack will cause lost motion and restrict the shifting travel of the conveyor. Also, wire ropes can jump the sheaves and become entangled in the framework of the machine. The cylinders themselves require little in the way of routine maintenance other than inspection for leaking fittings, alignment of hoses so that they do not abrade and wear against supporting members, and condition of cylinder rods and packing. Keep cylinder travel cages tightly anchored. If they are damaged by rocks and other entrapped material, they should be removed and straightened to allow free travel of the cylinder.

For conveyors equipped with cylinder systems for shifting, specific procedures must be used to ensure equal shifting travel in both directions. The procedure for accomplishing this on Cleveland V and J series machines is as follows:

1. Operate shift cylinder control valve to extend the cylinder. Do not permit sheave cage on cylinder rod to reach the end of the track. It may be necessary to lengthen turnbuckle at cylinder end of conveyor to accomplish this.

2. Loosen turnbuckle at end of conveyor on operator's side. Check spring for initial setting of 1/2-in. preload.

3. On operator's side, measure from outside of boom to end of conveyor frame. This dimension should be approximately 10 in. on the J26, JS36, J36, and JS30 (see Fig. 11-11 for other models).

4. Tighten turnbuckle on cylinder end of conveyor to decrease above dimension. Loosen turnbuckle at cylinder end and tighten turnbuckle at operator's end to increase dimension.

5. Tighten turnbuckle at cylinder end until conveyor just begins to move. Tighten turnbuckle at other end until spring is compressed and nut portion of eyebolt is off stop by 1/8 in.

6. Shift conveyor to operator's side and check dimension from boom to end of conveyor on other side. Check for clearance of sheave cage and boom. Readjust turnbuckles to obtain equal travel if required.

Crawlers Crawler-mounted trenchers generally use two types of track systems. The first type employs cast track pads with integrally cast roller surfaces using unbushed pins. The rollers employed in this system are generally cast or forged and are equipped with plain bearings. Lubrication is accomplished by greasing through a drilled hole in the roller supporting shaft. This type of track is usually used on lower-weight trenchers used on short-run jobs.

Some larger trenchers may use components from crawler tractor vehicles. Other larger models employ variations of this, as is the case with Cleveland machines. The track belts on these vehicles are composed of links forged from alloy steel having machined rail surfaces. The link is thorough-hardened for long wear and improved toughness. The bushings are low-carbon steel, carburized and hardened for wear and good core qualities.

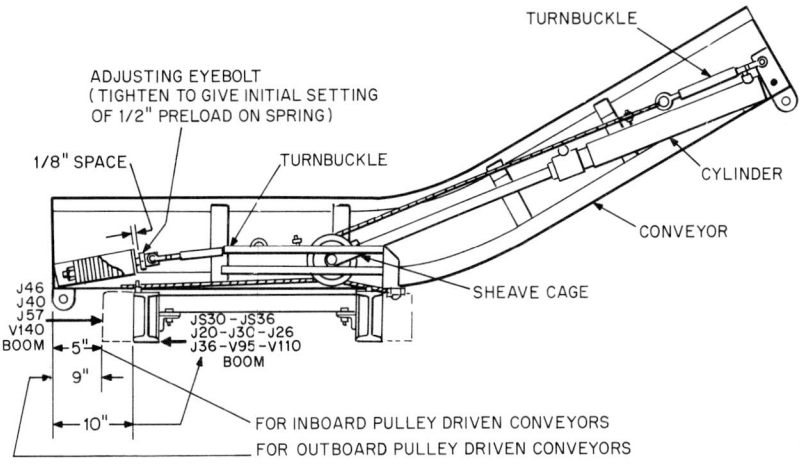

Fig. 11-11 Conveyor shift arrangement.

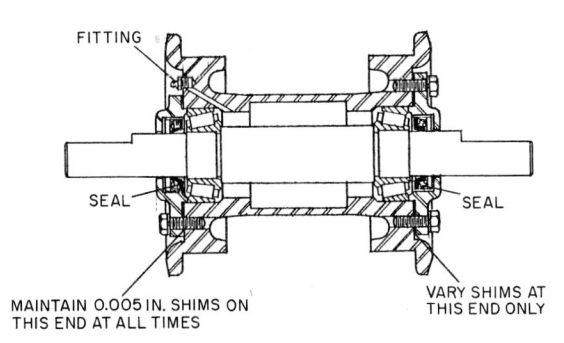

Fig. 11-12 Track roller assembly.

Track pins are of high-carbon steel, induction-hardened to resist wear and retain desirable toughness in the core. Track roller assemblies are forged having alloy steel shafts and mounted on antifriction-type bearings. Seals are of the flexible-lip-type design using synthetic materials. Maintenance for track rollers having antifriction bearings is as follows:

   1. The track roller bearings are lubricated through a hydraulic fitting installed in the outside end of the roller. Every 1000 hr fill reservoir in roller with SAE 90 gear oil if disassembled. In the field use multipurpose grease for lubrication.
   2. Inspect for leaking out past the seals, which would indicate excessive seal wear. It is more economical to replace a seal than a complete track roller.
   3. The track roller bearings are adjusted by varying the thickness of the shims under the bearing cap. Remove or add shims until a torque of 16 to 30 lb·in. is required to turn the shaft when the cap is securely bolted.

4. Periodic inspection and adjustment of track roller bearings will greatly increase bearing and seal life by eliminating end play and wobble of the shaft in the seal.

Crawler sprockets and crawler idler wheels are cast of medium-carbon material, heat-treated for toughness and long wear. They are supported on alloy steel shafts, using double-row ball bearings. Lubrication is by greasing through a fitting mounted in the hub of the sprocket or wheel. Use a multipurpose grease once a week.

The crawler components are mounted on roller frames which are usually welded to the subframe of the trencher. The roller frames also supply the support system for various means of adjusting track tension and compensating for wear in drive chains and in the track. Trenchers using simple unbushed track pads generally feature a spring-loaded take-up system. Track tension should be maintained so that a minimum of $3/4$ in. of sag is maintained in the top side of the crawler between the supporting rollers. When in digging conditions conducive to dirt buildup and packing of material on track pads, roller surfaces, idler wheel surfaces, and crawler sprocket surfaces, tension of the track should be very closely observed as undue loading of the bearings and components will occur if the correct slack is not maintained in the crawler system. For systems using spring take-ups, be sure that the idler wheel is squarely aligned with the roller frame.

Some machines are equipped with automatic track adjusters which are loaded hydraulically. One variation of this system employs a hydraulic relief valve which will permit buildup of material to a point below a dangerous level and then relieve oil from behind the track-adjusting cylinders to reinstate the proper tension. Another variation of this system employs hydraulic accumulators which act as a hydraulic spring in maintaining proper tension on the track. These systems all automatically compensate for track wear and allow relief of track tension should debris become entrapped in the track system. Systems using the accumulators for spring action require regular bimonthly inspection of the accumulator to be sure that the charge has not been lost. Gauge kits for accomplishing this are available. Accumulators may be charged with dry nitrogen at your local cylinder gas distributor.

Do not permit track systems to become packed and encrusted with dry mud and dirt. Clean off accumulated dirt at the earliest opportunity. Track shoes should be inspected daily for proper tightness of mounting hardware. Tracks having fabricated track shoes are welded to the track links and are free of loose hardware problems. When a crack develops in the welds between the shoe and track links, special procedures must be used for repairing these cracks. Since track links are of high-carbon material, the use of preheating, postheating, and low-hydrogen rods is of the utmost importance. Gouge out cracks, and preheat the shoe and the link to 500°F. Pack the weld with asbestos or other insulating material and allow to cool.

Probably the highest wearing item in the crawler system is the crawler drive sprocket. Do not build up sprocket teeth with weld as soon as wear is observed. Allow wear to progress to the point where the sprocket is no longer serviceable and then replace the sprocket. Building up the teeth very often will cause unequal driving forces on the ends of the pins as the integrity and spacing of the tooth has not been preserved. If track pins become worn through the outer hardened surfaces, press out and rotate the pins for extended life. It should be many years before such a condition becomes apparent on a trencher. Progress of wear will, of course, be relative to the abrasiveness of the material being worked in.

Antifriction mounted track rollers are lubricated at the factory with gear oil, and the assemblies are provided with a grease-type fitting. Track rollers generally do not require relubrication for at least 1000-hr intervals. Should it become necessary to relubricate a track roller, it is preferable to use gear oil but grease may also be used. Use a No. 2 multipurpose grease. Track rollers are bolted to the roller frame with high-strength hardware using locking-type nuts. Also, on roller frames having beveled surfaces, be sure that beveled washers are in place so that bending strains are not transmitted to the heads of the cap screws.

Steering brakes Steering of crawler-mounted trenching machines is accomplished by braking one track and allowing the other track to rotate and steer the machine in the direction of the stalled track. Braking of the track may be accomplished by band-, drum-, or disk-type brakes.

Adjustment of band brakes is important. If wear of band linings is allowed to progress too far, excess travel in the linkage will be experienced. For band brakes, pressure is applied to a drum by a band actuated through a toggle mechanism. This multiplies the operator's foot effort many times. Proper adjustment of the toggle and linkage must be maintained.

When installing new brakes, there are five points of adjustment to be taken care of:

1. Adjust the toggle so the main toggle lever just clears other parts of the toggle mechanism when the brake is fully applied (see A on Fig. 11-13). This permits obtaining the maximum force from the toggle action. Make adjustment by removing pin (see B on Fig. 11-13) from yoke and (see C on Fig. 11-13) from short toggle arm and unscrewing yoke end to tighten or screwing in on same to loosen. After adjusting, reinstall pin and hold in with a cotter pin.

2. Then adjust rod length to obtain the proper pedal movement. Make this adjustment in a similar manner to that just described for the toggle adjustment. If toggle and rod

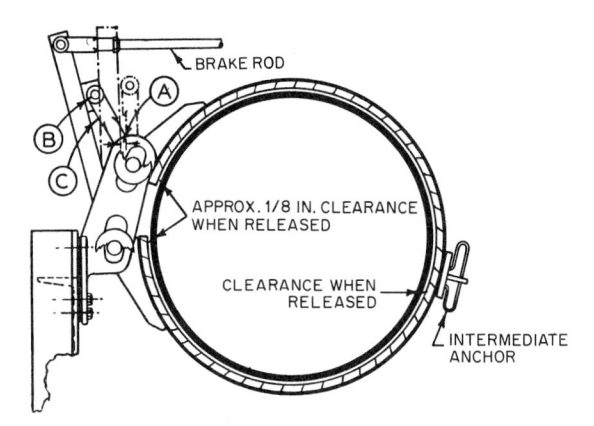

Fig. 11-13 Steering brake and toggle (band type).

adjustments have been properly made, when pedal is released both ends of the band should pull away from the drum approximately $1/8$ in.

3. Then check the intermediate anchor to see that the spring pulls the band free of the drum at this location. The entire function of this anchor is to prevent the band from riding on the drum and thereby causing unnecessary wear. Proper adjustment of this anchor can ordinarily be made by loosening the cap screws holding it to the machine frame, then driving a screwdriver between anchor and band to move the anchor approximately $1/8$ in. away from band, thereby compressing the intermediate anchor spring a like amount. Retighten cap screws. In extreme cases it may be necessary to bend the anchor slightly to achieve the desired result. A further check is advisable to ensure that the band ends pull away from the drum at least $1/8$ in.

4. If the proper clearance is not obtained, it may then be necessary to add or take out shims between the main anchor and the supporting frame.

5. In making the previous adjustments, if it is found that both ends of the bands do not pull away from the drum an equal amount, it may be necessary to bend the main anchor casting as it is essential that the same freedom of movement be obtained at both ends of the band. Make the adjustment by bending the main anchor casting, which is made of malleable iron, at a point just at the top of the machine frame crossmember. Usually this bending can be done without removing the casting from the machine by using a crowbar. NOTE: On machines built since April 1951, main anchor casting must be correctly shimmed at top and bottom instead of bending the casting.

For trenchers equipped with drum-type hydraulic brakes, make the adjustment when it is apparent that foot travel for engagement is excessive. Adjust according to the following procedure:

Master Cylinder and Pedal Linkage. The pedal (or lever) and master cylinder assembly (Fig. 11-14) require particular care in the initial adjustment. This is a factory setting, but it is repeated here so that proper adjustment can be made by the operator.

1. Position pedal against stop.

2. Release locking nut and turn adjusting nut in or out until a clearance of $1/8$ to $3/16$ in. is obtained between push rod and piston. This can be determined by feeling motion of pedal. Pedal will have $1/2$ to 1 in. free movement before pressure stroke starts when push rod is properly adjusted.

3. Tighten locking nut.

Bleeding hydraulic lines Whenever a tubing line has been disconnected it is necessary to bleed the hydraulic system to expel all air.

Fill the master cylinder supply tank with heavy-duty brake fluid before beginning this operation. Keep tank at least half full at all times. Perform the bleeding operations as follows:

1. Bleed system by opening bleed screw on wheel cylinder.
2. Slowly depress and release brake pedal until all air is expelled through the bleed screw and the fluid stream is solid.
3. Tighten bleed screw.
4. Refill master cylinder reservoir.

Brake shoe linings may be of either organic or sintered metal material. Inspect brake shoes on an annual basis for lining wear. Do not permit linings to wear below rivet head as drums will become severely scored. For brake shoes having sintered metal linings, replacement shoes with the linings already welded on are available.

Minor adjustments (to compensate for lining wear only) Brake drums should be at approximately room temperature when adjustments are being made. If brakes are adjusted when drums are hot and expanded, the shoes may drag when the drums cool and contract.

1. See that brake lever is in the fully released position.

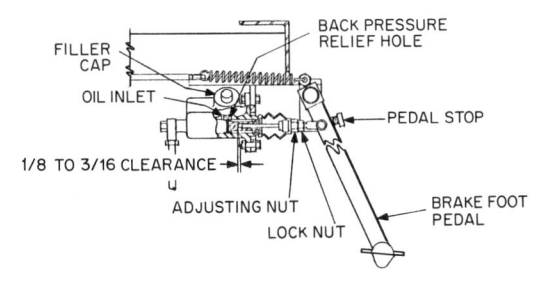

Fig. 11-14 Master cylinder and pedal linkage.

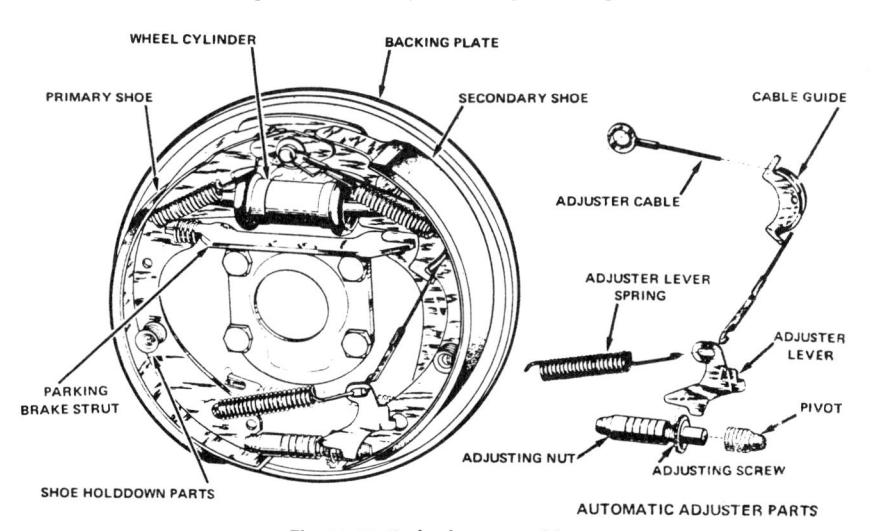

Fig. 11-15 Brake drum assembly.

2. Check brake lever adjustment to make sure lever travels approximately ½ to 1 in. before master cylinder piston starts to move.

REPEAT THE FOLLOWING OPERATIONS AT EACH DRUM:

3. Check anchor bolt nuts with a 16-in. wrench to make sure that they are tight. If an anchor bolt is found loose, reset anchor pin according to instructions under following Major Adjustments section.
4. Remove adjusting hole cover and feeler gauge covers from backing plate. Turn adjusting screw (moving handle of tool or screwdriver) toward the axle until brake shoes are expanded tightly against the drum.
5. Back off adjusting screw (moving handle of tool away from axle) until a clearance of 0.010 to 0.015 in. is found at feeler gage holes.

Major adjustments The following adjustments are necessary when minor adjustments fail to give satisfactory results or when replacing shoe and lining assemblies.

1. Disconnect copper tube at wheel cylinder fitting.
2. Unbolt brake anchor arm at frame and remove cap, cotter pin, and nut from end of differential shaft. The entire brake assembly can now be slipped off the shaft for reconditioning. Any brake fluid leaks at the wheel cylinder must be corrected by replacing or reconditioning cylinder. If shoes are to be removed it would be well to use a clamp or wire around the ends of the wheel cylinder to prevent the internal parts from accidentally falling out of the cylinder.
3. Inspect each drum braking surface and rebore or replace drum if necessary; if drums are rebored remove only sufficient metal to provide a smooth surface. If excess material is removed, the drum may be weakened to the extent that erratic braking and rapid lining wear will result.
4. Thoroughly clean shoes and brake backing plates with a wire brush. All bearing surfaces should be thoroughly cleaned, and a thin coat of Bendix Brake Lubricant should be applied.
5. Inspect bolts that hold backing plates to the hub: tighten if found loose. (This, of course, must wait until reassembly.)
6. When connecting the centralizer mechanism to a pair of relined shoes, be sure that the notched wheel adjusting screw is at the right-hand shoe when the shoes are assembled to each brake backing plate. This applies to all brakes having centralizer mechanism. Release the adjusting screw several notches to provide clearance for drum during installation. NOTE: When installing new shoes, it may be necessary to turn the eccentric anchor bolt to allow clearance between the brake shoes and the drum for ease of installation.
7. Reinstall hub and drum and install brake assembly.
    a. *Necessity for brake anchor arm alignment.* To prevent strains and eccentric loading on the bearings and shafts in the crawler differential, it is essential that the brake anchor arms be properly located with respect to the bolting pad on the machine frame. When this is properly done, tightening of the bolts which secure the anchor arm to the frame will not distort the arm. This alignment is done by means of shims and should be checked each time the brake anchor arm is disassembled for any reason.
    b. *Procedure for brake anchor arm alignment.* First rotate the brake anchor arm into position and insert a bolt to hold it in place. Next push the anchor arm tight against its mounting pad. This may take a heavy push on the brake at the hub to slide the entire assembly. Observe the fit of the brake anchor arm to the frame bolting pad to determine if there is a gap between the two. If a gap exists it should be fitted with shims at each bolt location and all the bolts tightened.
8. Install and tighten nut: replace cotter pin, cap, and hydraulic lines. After installing hub, drum, and brake assemblies, repeat the following operations at each brake:
9. Remove brake adjusting hole and feeler gauge hole covers in backing plate rim.
10. Be sure that the shoe centralizer mounting bolt nuts are just free of lock washer tension so that the centralizer can float freely.
11. Turn adjusting screw, moving handle of tool or screwdriver toward the axle until brake shoes are expanded tightly against the drum.

12. Tap backing plate near centralizer with a light hammer to ensure centralizer takes correct position between shoe ends.
13. Tighten centralizer mounting bolt nuts.
14. Back off adjusting screw (moving handle of tool away from the axle) and check the secondary shoe lining to drum clearance at both ends of the lining. A clearance of 0.010 to 0.015 in. at both ends of the secondary shoe indicates correct shoe position and is obtained by adjusting the anchor pin.
15. To adjust the anchor pin, loosen each anchor bolt nut about one turn, turn the eccentric bolt in the direction required to secure the correct secondary shoe position. It may be necessary to turn the adjusting screw when resetting the anchor pin to obtain 0.010- to 0.015-in. clearance at both ends of the secondary shoe.
16. Tighten anchor bolt nuts securely with a 16-in. wrench while carefully holding the eccentric anchor bolt with a wrench to prevent it from turning during the tightening operation. After tightening the anchor bolt nuts, again check the secondary shoe clearance to make certain that the anchor position has not changed in the nut tightening operation.
17. Reset shoe centralizer by repeating operations described in steps 11 to 14; then release the adjusting screws until feeler gauge reading of 0.010 to 0.015 in. is obtained.
18. Replace adjusting screw hole cover feeler gauge hole covers. Bleed lines.
19. Test adjustments by roading machine both for response to steering and for possible drag on one or both brakes when brakes are released.

PRECAUTIONS: Use only alcohol to clean rubber parts or inside of cylinders. Kerosene or gasoline will cause trouble. Do not allow grease, oil, paint, or brake fluid to come in contact with brake lining. Do not allow the supply tank to become less than half-full of brake fluid. Do not attempt to reline brakes having metal-type lining that is welded to the shoes. When reassembling master or wheel cylinder, apply brake fluid to all parts.

Required maintenance for hydraulically actuated disk-type brakes is somewhat different because of the nature of the components used. On slow-moving trencher machines simple flat-type discs may be used as conversion of energy into heat is not a requirement. All that is required is an application of sufficient force to stall a track. Disks are pilot-mounted on the output sprocket of the crawler differential. Attention should be paid to fasteners to be sure that they are tight. A weekly inspection of these fasteners would be in order.

Alignment of the disk and caliper mechanism must be maintained. Good alignment will guarantee even puck wear and prevention of leaks in the actuating piston of the caliper mechanism.

Inspect condition of brake fluid lines and fittings on a weekly basis. Check for wetness at the fittings and abrasion on tubing. Keep tubes clamped and routed along frame members so that vibration will not cause wear. Check the fluid level of the brake fluid once a month. Should it be necessary to add brake fluid, always use a heavy-duty brake fluid.

Crumbing shoe Many trenching machines are equipped with crumbing shoes mounted at the rear end of the boom behind the digging wheel. Their function is to provide a support for the wheel so that accurate grades may be held and also to sweep loose material or crumbs into the buckets to maintain a clean trench bottom. Trenchers used in congested areas such as municipalities are generally equipped with crumbing shoes which may be swung out of the way to permit close set-ins and reduce the distance the trencher must travel before full depth is achieved.

There are two varieties of crumbing shoes in common use: a so-called mechanical crumbing shoe which must be swung out of the way manually or a hudraulic crumbing shoe which can be swung out of the way hydraulically and controlled from the operator's position. The hydraulic crumbing shoe offers the maximum in ease of operation and does not require extra personnel. It is an asset in tree lawn trenching where many driveways may be encountered.

Proper adjustment of the crumbing shoe is important. An improperly adjusted shoe will prevent the wheel from reaching the correct grade. Expansion plates not properly adjusted for width will allow loose material to accumulate behind the crumbing shoe or the plates will scrape the side of the trench, placing unnecessary loads on the shoe body and the boom structure. Height adjustments are usually made by selecting the proper holes in a series of holes provided in the shoe mounting plates on the boom. Expansion plates are adjustable through the use of horizontal slots provided in the plates. Expansion plates should be adjusted so that there is approximately 1-in. clearance between the edge of the plate and the side wall of the trench. This clearance will allow easy turning of the machine, minimum damage to the expansion plates because of rock outcroppings in the trench wall, and a minimum of crumbs bypassing the crumbing shoe.

Proper clearance must be maintained between the face of the crumbing shoe and the rooter teeth. If this clearance is too close, there is a possibility of snagging the crumbing shoe with one of the rooters. If the clearance is too great, large amounts of dirt will accumulate on the body of the crumbing shoe, adding strain to boom members and increasing the weight of the wheel when the boom and wheel are lifted out of the trench. The proper clearance is generally from 3 to 4 in.

Many crumbing shoes are equipped with adjustable shoe bottoms. The shoe bottom may be of the flat type or a round type conforming to the shape of the trench bottom. In both cases, the toe of the shoe is pivot-mounted at the bottom of the crumbing shoe body and supported with adjustable braces at the heel, allowing vertical adjustment of the heel with respect to the toe.

The heel of the crumbing shoe should be set so that the vertical distance from grade is approximately $7/8$ to 1 in. less than the cutting depth of the wheel. The toe of the crumbing shoe should be approximately 2 in. above this point. This then provides the proper slope for the crumbing shoe to allow a sledding action of the bottom over the crumbs. Also it minimizes the possibility of snagging rocks at the bottom of the trench with the toe of the crumbing shoe. Should it occur, there is a good possibility of ripping the crumbing shoe off and possibly doing damage to the boom frame supporting members. Keeping the heel of the crumbing shoe somewhat above digging depth removes a great portion of the supporting strain for the boom and wheel structure from the crumbing shoe and places it on the wheel. This also helps minimize transfer of torsional strains to the boom frame members through the lever action of the crumbing shoe body.

The wear pattern on the crumbing shoe bottom should be closely observed. A properly adjusted shoe bottom will show contact with the bottom of the trench at the heel of the shoe bottom and not at the toe. Crumbing shoe bottoms are equipped with abrasive-resisting wear plates which are renewable. Inspect these plates at least weekly during the trenching season and do not permit them to wear to the point where the actual shoe bottom is beginning to wear. The rate of wear is dependent on the abrasiveness of the material being worked.

Hydraulic crumbing shoes are equipped with latching mechanisms which hold the shoe body in place and permit release of the shoe during raising. Adjustment of the latch is quite important; if it is allowed to come free during the operations, undue strains will be transmitted to the hydraulic actuator and the linkage system. Hydraulic crumbing shoes are equipped with pressure relief valves to guard against this contingency. Relief valves are generally set 200 to 300 psi higher than system operating pressures. The hydraulic circuit is also equipped with pilot operated double-check valves to guard against leakage of hydraulic fluid past the spool from the directional control valve. The hydraulic hose leading to the cap end of the actuating cylinder is fitted with a restrictor to prevent rapid dropping of the crumbing shoe body when it is being lowered from the raised position.

Crumbing shoes may also be fitted with large flat plates called *tile shields* which help support the side wall of the trench behind the crumbing shoe body. This allows a man to sit between these plates and accurately place tile at the bottom of the trench. Tile shoe bottoms are also equipped with a groover on the center line of the bottom which leave a small V-groove in the bottom of the trench, thus greatly simplifying the alignment of tile.

The same adjustments which apply to standard and hydraulic crumbing shoes also apply to tiling crumbing shoes. The only other adjustment that is required is the setting of the proper width for the tile shields. If the tile shields are set too widely, they will drag along the side of the trench and will certainly cause failure of the boom supporting members. If they are set too narrowly, they will permit passage of crumbs and also restrict the seating space of the tile layer unnecessarily.

Trenchers may also be equipped with stripping shoes when it is necessary to excavate and salvage an old pipeline. These crumbing shoes are equipped with concave bottoms conforming generally to the shape of the pipe being excavated. These shoes are used in conjunction with stripping buckets which also conform to the shape of the pipe but actually dig down alongside the diameter of the pipe so that it may be lifted from the trench bottom. Obviously it is important that depth be very carefully set on these shoe bottoms.

Crumbing shoes normally require only routine maintenance except for observation of damage and wear and replacement of damaged expansion plates and other components of the crumbing shoe system.

Frames Basically trenchers are equipped with three frames which support all the working components of the machine. These include the main frame, the boom frame, and the conveyor frame.

The main frame provides the support for the engine, main shaft, digging differential, crawler differential, crawler transmission, boom hoisting, transmission conveyor hoisting transmission, operator's platform, boom mast, and crawler system. Frames are of welded construction and have a high percentage of structural-type members. Construction of frames is deliberately kept open to allow for easier service and replacement of the various machinery components. Main frames usually require very little maintenance. A routine inspection on a monthly basis for beginnings of cracks and welds in frame members should be made. In very severe digging operations, inspection may be increased to once a week. Pay particular attention to mast members as this is where digging strains will be transmitted to the main frame. Should repairs be indicated, normal welding and repairing techniques are all that are required. Heat may be applied to straighten bent members and welding may be done with low-hydrogen rods in the 60,000 psi tensile range.

The boom frame provides the support for the wheel, conveyor, and crumbing shoe. Boom frames can be subjected to considerable damage during digging, particularly in rocky digging. Areas which must be closely observed are the four corners and hoisting sheave supporting members. The only maintenance that is required other than routine inspection is repair to damaged components. Normal repairing techniques may be applied quite successfully.

Of all the trencher frame members, the conveyor frame is probably subject to the greatest structural damage for it is here that rocks land, logs become entangled, dirt piles up, and trash accumulates. Maintenance of the conveyor should include regular cleaning. The need for repairs to structural members of the conveyor will be indicated by bent members, misalignment of members, and broken welds. Do not allow these conditions to prevail as the result will certainly be accelerated belt wear and eventual failure.

Machines equipped with side-shifting wheels have a fourth major frame member, the movable mast. It provides a support for the boom and wheel as well as the boom hoisting mechanisms and digging differential. Daily lubrication of the mast slider surfaces is required. Greases having high molybdenum disulfide content are recommended. The only other maintenance required to the movable mast member is the routine inspection for structural damage and regular lubrication of the rollers on a weekly basis with multipurpose grease in the fittings provided.

Hydraulic system The use of hydraulic systems on trenchers has greatly increased over the past 20 years. Today there is no trencher which does not have at least some hydraulics on it. Indeed, many trenchers have hydrostatic drives for digging, traveling, conveyor spoiling, and boom and wheel hoisting. A wide variety of components are used, ranging from gear pumps to variable-displacement piston pumps and working in pressures from 1500 up to 5500 psi. Hydraulic systems for trenchers are subjected to all the problems common to hydraulic systems on other construction equipment. They may not have the duty cycle and the high flows required on machines such as hydraulic backhoes, but they do have requirements for variable and continuous motor drive. It is not unusual for hydraulic conveyors on cross-country trenchers to run 4 and 5 hr continuously under load. These conditions require the same sound maintenance practices that would be used on machines having highly sophisticated hydraulic systems.

Whenever anyone talks about a hydraulic system, the need for cleanliness is always mentioned. In no way can this need be overemphasized. Contamination of hydraulic systems is beyond a doubt the greatest contributor to failures. Thus, it becomes extremely important that good filtration, good hydraulic fluid management, and regular maintenance procedures be followed.

Hydraulic reservoirs on trencher machines are usually not of large capacity. Most machines use systems vented to the atmosphere and fitted with filling caps having air filters. The level of the fluid in the reservoir is usually kept so that it is just visible in the bottom of the filling strainer. Reservoirs are also equipped with removable covers so that cleaning can be easily managed. Most reservoirs have fill caps as part of the cover.

Do not permit dust, dirt, and debris to accumulate on top of the reservoir or around the vent cap. Should it become necessary to add hydraulic fluid to the reservoir, be sure to always clean the cap and surrounding area before removing the cap. Remove accumulations of dust from the fill cap by rinsing in flushing oil. Inspect covers of reservoirs for tightness and to ensure they are not permitting the entrance of air and water.

Trenching machines are usually equipped with return-line filters in the 10- to 25-micron range, depending on the design of the hydraulic system. Variable-displacement piston pumps operating at higher pressures should be provided with at least 10-micron filtration. When a trencher is new, filter elements should be replaced after the first 40 hr of operation. Subsequent replacement should be made after each 1000 hr of operation. When replacing filter elements, clean the filter housing carefully to avoid contaminating remaining fluid inside the filter housing. Be sure that replacement filter elements are in good condition and do not permit them to bounce around in a pickup truck or to be carelessly stored. Do not use a filter element which has had its protective plastic covering damaged. Renew gaskets and "O" rings when filter changes are being made.

The pump may be regarded as the heart of any hydraulic system. It has the job of converting mechanical energy into hydraulic energy. Regular inspection of pump mounting, suction lines, and fittings will pay great dividends in extended pump life. Pumps directly coupled in line with their driving mechanisms have the most favorable mounting conditions. Torsional loading of the drive shaft in a well-designed system is the only loading present. Bending loads which cause shaft deflection, added strain on bearings, and misalignment of sealing surfaces are held at a minimum. For pumps having side-mounted gear drives, maintain proper backlash between the pump drive gears. For pumps having universal joint drives, check the condition of the universal joint weekly. Universal joints driving hydraulic pumps should be balanced and have fittings for lubrication of the bearings. Pumps equipped with companion flanges so that universal joint attachment may be made have proven most satisfactory. This permits removal of universal joints without pounding and pulling on the pump shaft. Never pound on the shaft of a pump to remove a universal joint or a companion flange.

Close monitoring of the fittings in the suction line is essential. Leaks should not be permitted to exist in this line at any time. Under vacuum conditions which sometimes prevail in pump suction lines, dirt and air will be sucked into the hydraulic system. Aeration of hydraulic fluid is a major contributor to pump failure. Oil containing air will contribute to pump cavitation. Cavitation can usually be detected by an increase in noise level and reduced output from the pump. Continued operation of a pump under these conditions will lead to complete failure.

Relief valves are included in the hydraulic circuit to protect the pump and other circuit components. A malfunctioning relief valve can cause a variety of problems. A relief valve that is stuck open or partially open will cause sluggish operation of the machine, reduction in speed of travel in actuators, and contribute to heat buildup in the system. A relief valve that is stuck closed or late in opening can have devastating results in the hydraulic system. Closed relief valves can cause rupture of lines or failure of any hydraulic component in the circuit. A late-opening relief valve is extremely dangerous as it may go undetected for some time. This condition permits a buildup of hydraulic shocks which are directly felt by the pump and if allowed to persist will eventually cause failure of the pump. Relief valves should never be reset to improve performance of the machine. The manufacturer has carefully sized and selected all components in the hydraulic system to operate within certain pressure capabilities.

If continuous and repeated flow of oil over the relief valves is experienced, take steps to seek out the problem and eliminate it. Conveyor drive circuits may be overloaded when easy digging conditions are experienced. Digging in sandy, free-flowing material will allow great forward progress of the machine, thus depositing heavy loads on the conveyor, which will eventually become flooded and stall. Under such conditions it is wiser to reduce the forward speed than to risk the loss of expensive hydraulic components. A system with oil continuously passing over the relief valves causes a loss in production and eventual failure of hydraulic components.

Relief valve settings should be checked when trouble is indicated on a machine. They should be checked when hydraulic fluid has reached operating temperature and generally with pump flows at about one-half their maximum. In the case of a fixed-displacement pump, this may be controlled by varying the engine speed. For variable-displacement pumps, speed control levers should be set at the halfway mark. Always use gauges which have a range of approximately twice the hydraulic pressure experienced in the system; that is, for a 2000 psi system, use gauges that have a range of 4000 psi. Gauge readings are always most accurate at midscale, and this also helps guard against the possibility of destroying a gauge because of relief valves which may be set too high. Further, be sure that your gauges are accurate. Most hydraulic circuits are equipped with gauge ports, and pressure readings may be easily taken.

Directional control valves found on trenchers are usually of the stack type. Many of them have multiple inlet sections with separate relef valves. Systems of this type are used to guarantee flow to various components of the hydraulic systems without reducing flow to others. Valves may be equipped with pipe ports or SAE straight thread ports. Spools are chrome-plated to prevent rusting. In some cases control levers of unequal heights are used to permit rapid identification.

Routine inspections of valves should include an examination of all fittings for leaks and proper tightness. For valves equipped with pipe ports, great damage can be done by overtightening pipe thread fittings. Spools may become sticky or cracking of the housing may occur. If a leak develops around a pipe fitting and the fitting is tight, find the cause of the leak and correct it. Use Teflon sealing tape and other pipe sealants judiciously. Use of these materials can lead to overtightening of pipe thread fittings. For fittings having straight thread bosses, further tightening will not always cure a leak. This type of fitting depends on sealing with an "O" ring. If repeated leaking is experienced, remove the fitting, examine the "O" ring for nicks and possible hardening, and also examine the "O" ring seat for nicks and burrs.

Leaking between mating sections of the hydraulic control valve may be caused by loose bolts holding the sections together or damaged gaskets or "O" rings between the sections. The only solution to this problem is to disassemble the valve and replace the damaged "O" ring or gasket seals. Be sure to follow manufacturer's recommendations for retorquing tie bolts. Uneven torquing or excessive torquing may contribute to valve warping and spools sticking.

A wide variety of hydraulic cylinders may be found on trenching machines. Their bores may vary from 1½ in. up to 7 or 8 in., and their strokes may vary from just a few inches to up to 5 or 6 ft. Most are double-acting cylinders. They are used for boom hoisting, conveyor shifting, wheel and boom traverse, wheel and boom tilting, and hydraulic crumbing shoe actuating. All cylinders have chrome-plated rods and some have high-alloy heat-treated rods. Cylinders are equipped with a variety of packings and rod wipers. One popular design is that of a V-packing having controlled stack height. This greatly simplifies maintenance, and adjustment of packings is not required.

Hydraulic cylinders require little maintenance unless some difficulty is indicated. This could be leaking from the rod packing or excessive drifting. Drifting may be caused by damaged packings or scored cylinder walls. If this should be the case, the reason for the packing damage or the scoring of the cylinder wall must be determined and remedied. Leaking from the cylinder rod indicates a worn or damaged rod packing. Inspect cylinder rods regularly to detect nicks and rust. If small nicks appear, stone them off immediately so that their passage through the rod packing does not damage it. Whenever a machine is parked overnight or for a lengthy storage, retract cylinder rods. This keeps rods from unnecessary exposure to weather and dirt. Observe the condition of the cylinder mounting. Poor mountings can cause eccentric loading on cylinder rods and contribute to rod bending.

As with pumps, a variety of motors may be found on trenching machines. These range from gear- and vane- to piston-type motors. Displacements and pressure requirements also vary widely. Motors may be used singly or connected in parallel or series. Whatever the case, similar procedures for pump inspection and pump maintenance may be followed on motors. For motors equipped with external drains, inspect drain lines for kinking and unencumbered routing to the reservoir. Obstructed drain lines will cause excessive back pressure and damage low-pressure shaft seals. This may be noted by wetness around the output shaft of the motor. In cases where the motor is directly coupled into a housing such as in a conveyor drive pulley, this condition will be evidenced by a continual leaking from the vent of the pulley since hydraulic fluid is building up in the enclosure. Be sure that drains from other parts of the circuit which contain relief valves are not interconnected directly with motor drain lines. Motor drain lines are usually not sized to accommodate high flows as may be experienced over a relief valve. This high flow will put undue back pressure on the motor low-pressure seal and cause damage. Check fittings for proper tightness and hoses for good routing and do not permit hoses to become loose where they may be jammed in moving parts or abraded against a structural member of the machine.

Good management of hydraulic fluids is of the utmost importance for the successful operation of any hydraulic system. Trencher hydraulic systems are no exception. A good grade of industrial-type hydraulic fluid should be used. The oils should have additives for foam depressing, rust and corrosion inhibiting, and antiwear. Detergent-type oils should not be used as they tend to keep contaminants in suspension.

The condition of hydraulic fluid should be constantly monitored. Oil change requirements will vary greatly depending on the use of the machine. It is not difficult to detect oil breakdown. Overheated oil has a dark appearance and a burnt odor. If oil is contaminated with water, it will generally have a milky appearance. All these conditions detract seriously from the oil's ability to perform satisfactorily. Viscosity and lubricity are greatly reduced. Acids may be formed which can attack and corrode metal parts of pumps, valves, and cylinders. Sludges and varnishes can accumulate, causing sticky valve operation and plugging of control orifices. For a trenching machine, 1000 hr of operation should be the maximum without changing hydraulic fluid. This type of maintenance can be scheduled to be performed between jobs or possibly during a bad weather period.

Good management of hydraulic fluid includes storage and handling of the fluid. Generally contractors will use one or two hydraulic fluids which are suitable for their entire fleet of machines. If oil is stored in drums, the drums should be kept on their sides so that water may not be drawn in through the bungs. Hydraulic fluid should be transported in clean containers. In no case should hydraulic fluid be added to a system unless it has gone through at least a 25-micron filter. Oil, as received from the refinery, is not necessarily clean enough to be used in a hydraulic system.

No other maintenance program for any part of the machine will pay greater dividends than a good working maintenance program of a hydraulic system. Once again, the best advice that can be given is to keep it clean.

Drive chains American Standard roller chains and engineering class chains may be found on trenching machines. Roller chains vary in pitch from ½ to 2 in. They may be single- or multiple-strand chains, depending on the application. Some chains are totally enclosed and run in oil, while others are fairly open and merely guarded and are subjected to dirt, dust, and other airborne contaminants. Engineering class chains may be found in pitches of 3 in. and over. These are generally used for propelling the track system of the machine. They are generally fairly open and subjected to mud, dirt, and water contamination.

In any discussion of chain maintenance, the question of lubrication under varying conditions always comes up. It is best to keep chains lubricated as much as possible under all conditions. Lubrication of chains should be part of the daily maintenance task. It may be necessary to lubricate even more frequently under particularly dusty and dirty digging conditions. Chains may be lubricated with a light engine oil. Always use an oil can and never lubricate any chain while it is moving. For chains equipped with oilers, be sure that these oilers are filled at the start of each day's work.

Chains which have become rusted should be lubricated with a mixture of diesel fuel and engine oil until they once again run freely and do not kink.

As with chains, a wide range of pitches may be found in sprockets on trenching machines. Sprockets may be machine-cut, flame-cut, forged, or cast, depending on the pitch and their particular application. Generally finer pitch sprockets are machine-cut while greater pitches are cast, forged, or flame-cut. Sprockets may be heat-treated to improve wear and extend the life of the sprocket. Maintenance of the sprocket should include observation of wear patterns, tightness of mounting on shafts and hubs, and alignment. Misaligned sprockets will put severe strains on all components of a drive system, including bearings, shafts, and the chain. Sprocket alignment should be such that chains run free with proper clearance with frame members and guards. Chains should not be permitted to saw against any member of the machine.

Many chains on a trencher machine, such as the digging wheel drive chains and conveyor drive chains, must move through widely varying center distances. These requirements can be generally met with the assistance of chain take-ups. Sprockets and rollers and take-up mechanisms may be equipped with ball bearings or tapered roller bearings. Lubricate these daily with multipurpose grease. Alignment of take-ups is equally as important as alignment of sprockets. Maintain spring tension on take-up mechanisms at its proper level. If spring tension is too low, the slack side of the chain will not be properly kept under control. If spring tension is too high, excessive strain will be placed on the spring when the sprockets move to their greatest center distance. This puts unnecessary strain on the chain and may cause failure of the take-up mechanism. Other than routine lubrication of take-ups, only observation on a regular basis that they are working properly is required.

Wire ropes and sheaves Wire ropes are used on trenching machines to control the attitude and position of the digging wheel. These ropes may be used in conjunction with boom hoisting transmissions, which act as winches, or hydraulic cylinders. Ropes are usually improved plow steel and are of independent wire rope construction. Diameters of ropes may vary from $3/8$ to $1\frac{1}{4}$ in.

During digging operations, particularly on machines equipped with crumbing shoes, ropes controlling the rear end of the boom are generally held slack. Should an operator become careless and allow too much slack to accumulate, problems may be expected in spooling, ropes jumping off sheaves, and possibly kinking and entrapment of ropes between sheaves and frame members. Therefore the operator should constantly monitor the condition of the ropes. Ropes which are allowed to spool unevenly on boom hoisting transmissions will cause out-of-level booms, place unnecessary strain on boom sliders and masts, and kinking and flattening of the ropes on their spools.

Hardware used with ropes should be checked on a weekly basis. These should be part of the weekly inspection. Ropes should be lubricated with a light oil as required, probably on a monthly basis. Ropes should be swabbed with oil, using a brush. At this time, the rope can also be inspected for broken wires, crushing, doglegs and kinks.

Guards Guards are used fairly extensively on trenching machines; usually three types may be found: dirt guards, chain guards, and rock guards. Dirt guards keep dirt and rocks and other trash away from working machinery on the trencher and help prevent entrapment of material in moving parts. They also help in preventing chains being knocked off by falling rocks, limbs, and other debris. Chain guards offer protection to the operator and others working around the machine as well as keep rocks and branches away from moving chains. Rock guards are installed in the lower rear portion of the wheel between the rims and prevent the entrance of rocks that may damage and bend bucket backs.

When using rock guards, particularly where the population of the rocks is extremely heavy, rocks will accumulate on the rock guard. The wheel then must be raised periodically and cleared of these rocks. Should rocks be allowed to accumulate to an excessive level, difficulty in hoisting the boom and wheel out of the trench may be experienced. In hoisting systems, equipped with hydraulic cylinders for hydraulic motor drives, relief valve settings may be exceeded and the boom and wheel will not move. The operator will soon learn to what level he can permit rocks to build up before clearance is required.

Maintenance of guards should include a regular inspection for damage and the integrity of mounting hardware. Guards should not be permitted to rub against moving machinery or chains. Also, safety regulations require that chain guards be in good condition and installed whenever operating the machine. Safety regulations aside, operators should never operate a machine without guards properly installed as a matter of their own safety and the safety of others.

Chapter **12**

Drills and Drilling Equipment Maintenance

THE STAFF
The Compressed Air and Gas Institute
Cleveland, Ohio

Good operating and maintenance practices on rock drills, paving breakers, and air-operated percussion tools return dividends in lowered costs, elimination of time-consuming breakdowns, and continuing peak performance. The most important recommendations to get best results are:

1. Lubricate adequately with proper quantity and quality of lubricants.
2. Inspect front end and chuck parts frequently, and replace worn components promptly.
3. Operate properly to avoid steel and drill rod misalignment.
4. Use steels, rods, and striker bars with proper dimensions and tolerances.
5. Keep equipment clean.
6. Keep bolts tight.

Following these six basic recommendations ensures long, trouble-free life for air tools and eliminates, to a large extent, costly on-the-job breakdowns. And since failures tend to be progressive, the prompt replacement of worn parts results in a major economy.

LUBRICATION

Use of the proper grade and quantity of lubricant cannot be overemphasized. No other phase of maintenance and operation has as great an effect on performance and tool life. And, as vital as good lubrication is to the small tools, paving breakers, and hand-held rock drills, it becomes increasingly critical with the larger-bore, mounted drifter drills which generally are operated much more continuously.

A great deal of the wear and subsequent parts difficulties are traceable to poor or too little oil. Inadequate lubrication causes excess heat, which creates heat cracks, metal fatigue, and premature breakage.

Historically, almost all paving breakers, hand-held rock drills, and similar tools have been designed with so-called built-in lubricators. Having necessarily small capacity, they require frequent refilling and depend on the operator—on the job—for servicing. The chances for introducing dirt or for forgetting to lubricate are very real.

To lubricate properly, an air line oiler should be placed in the air line or attached to the compressor at the air supply. As a further safeguard, oilers that shut off the air supply to the tool when oil is exhausted are available.

WORN CHUCKS

The need to maintain alignment of the moil, drill steel, or striker bar in the front end of the tool is extremely important to good, economical operation. This requires frequent checking of tolerances on all parts which guide the shank—front head or chuck, front head or chuck bushing, and on the mounted drills, chuck housing cap, or washer.

Misalignment of steel is a prime cause of air tube and water tube breakage in the large-bore mounted rock drill. It leads to chipping and breakage of steel shanks, the striking end

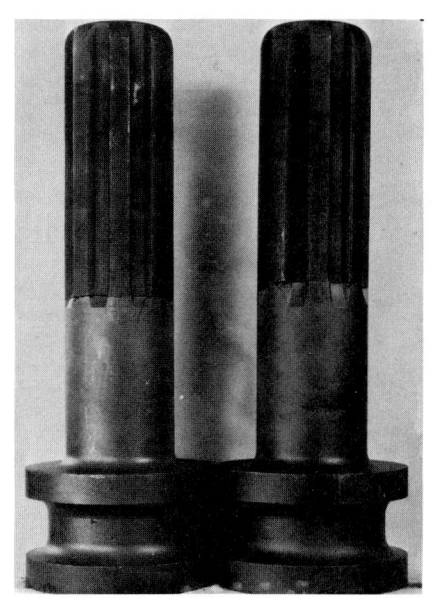

Fig. 12-1 Failures in the radar stem are characteristic of friction scoring and checking due to inadequate lubrication. Upper halves in photo were acid-etched to more clearly show the friction cracks. *(Ingersoll-Rand Co).*

Fig. 12-2 Piston chipping and spalling around the hole caused by a combination of crowned and excessively hard shanks. *(Ingersoll-Rand Co.)*

of pistons and tappets, and anvil blocks. Further, misalignment imposes an excessive load on rock drill rotation parts which slows or even stops rotation, generates excessive heat, breaks down lubricants, and accelerates parts wear and breakage.

MISALIGNMENT OF DRILL WITH DRILL STEEL

Although misalignment of the drill with the drill steel is primarily an operating problem, it is included here since it is a dangerous, damaging practice that has a direct effect on machine maintenance.

Such practices as doglegging, riding the drill with a steel through the handle, and the loosening of saddle clamps on mounted machines create a serious hardship on the drill by cramping the working parts all the way from the backhead to the chuck. The uncalled-for side strains produced by misalignment on all rotating parts is a common cause of excessive rifle bar and rifle nut wear and quite often cracks or breaks either of these parts. Misalignment not only slows down the number of blows the piston delivers, it reduces the foot-pound strength of each blow as well and is frequently the cause of scuffing or seizing of the piston in the cylinder. In the front head of the drill, failure to keep the drill and steel lined up produces needless chuck wear and allows the piston to strike the end of the steel shank on an angle. This is conducive to spalling of the piston and premature breakage of the drill steel. If, by chance, one side rod is looser than the other, misalignment frequently causes the tight side rod to break just behind the nut. In many instances, misalignment is directly related to the condition of the chuck; when the chuck is worn, it permits some degree of misalignment and one condition aggravates the other until a breakdown occurs.

On power-feed drills, especially those on crawler mountings, it is extremely important to watch alignment carefully. The combination of power feed and, on some models, independent rotation on a nonrigid feed mount make it rather easy to force the drill and steel out of alignment.

Maintaining machine alignment with the drill steel while drilling is a simple preventive maintenance measure which should be practiced by every driller. A good operator will not fail to line up his drill as soon as the hole is collared, and from then on he will maintain uniform pressure on each handle until the hole is bottomed. This will assure straight, true holes which will go down faster with less effort and, most important, will not cause unnecessary wear and damage to the drill.

OUT-OF-STANDARD DRILL STEEL SHANKS

Because the drill steel functions as part of the rock drill itself, it is an instrumental factor in the overall efficiency of the machine. Frequent inspection and prompt repair of steel shanks is as important to good rock drill performance as replacement of worn parts. This becomes more apparent when it is realized that out-of-standard shanks actually cause damage to bits and working parts of the drill. *A drill steel shank is out of standard when it does not meet all the following requirements.*

Fig. 12-3 New shank used in worn-out bushing or chuck results in damaged pistons from limited contact. *(Ingersoll-Rand Co.)*

Striking end flat and square A square, flat striking face on the shank end is essential to prevent piston damage. A crowned shank reduces the area of contact between the shank end and the striking face of the piston so that the full force of the piston blow is highly concentrated. This results in chipping and rapid cupping of the striking face of the piston. Off-square shank ends also have a small area of contact with the striking face of the piston and frequently cause chipping or spalling of the piston.

Fig. 12-4 *Trouble.* A western ore mine was having difficulty with drifter pistons spalling on the striking face. *Cause.* Shanks shown in this photo are too heavily beveled on the outer edge and around the water tube hole. As a result, the force of the piston's blow is concentrated on too small an area, a condition which naturally causes spalled pistons. This excessive beveling was found to be an aftermath of an epidemic of soft shanks. *Remedy.* Bevel the outer edge of the shank only $\frac{1}{16}$ in. The bevel on the water tube hole was made by driving the punch pin into the steel until its shoulder formed the bevel. The correct procedure is to use a shank end former to shape the bevel, or to use a small pointed grinding wheel to remove burrs. *(Ingersoll-Rand Co.)*

7-336 Maintenance Programs for Equipment Entities

All shanks become more and more crowned in service. Shanks should be checked periodically and brought back to standard with a machined or ground-flat-and-square striking end, properly chamfered to prevent sharp edges from chipping.

Properly formed collar or lugs Since the end of the chuck or chuck bushing receives the impact of the blow from the rebounding drill steel, improperly formed lugs or collars—or lugs or collars which have become battered through misuse—damage the chuck. Misalignment while drilling adds to the problem and tends to cause further damage to shanked steels and chuck bushings. An incorrectly formed collar also changes the striking point of the shank end with reference to the piston, and both piston and shank end then suffer.

Shank lugs and collars must be within standardized tolerances for shanked steels; if not, they should be discarded or reconditioned. Shanking dies should be checked frequently to make sure they do not exceed permissible limits.

Correct length, collar to striking end Shank ends that are overlength extend too far into the chuck and cause short stroking of the piston, with resultant loss of drilling power. Conversely, if the shank is too short, long stroking occurs and the blow of the piston is partially absorbed by the air cushion. This results in poor drilling speed and possible overheating of front end parts. If the front head parts of the drill are so worn that no cushion is present, a short shank allows the piston to strike the buffer ring with its full force, and the piston will break.

Fig. 12-5 Spalled pistons caused by the use of the worn-out chuck bushing shown in Fig. 12-4. The use of wear limit or discard gauges would prevent this condition. *(Ingersoll-Rand Co.)*

Correct shank length is particularly important for the reasons given above. Never use a steel with a short or overlength shank. Shank gauges are available for checking length from under the collar or lug to the striking face—use them! Remember too, that worn and battered collars or lugs tend to increase shank length and cause short stroking.

Center hole clear, punched, and chamfered A center hole plugged with rock cuttings or dirt restricts blowing action and interferes with hole cleaning. In a wet drill a plugged center hole may cause water to flood the drill and wash away the lubricant. When the center hole is not enlarged by punching, and punched deep enough, lack of clearance for the water tube may cause the shank to batter the end of the tube and close it up. The sharp edges of an unchamfered center hole may likewise damage the water tube or even shear it off.

Before using each drill steel, examine it to see that the center hole is unobstructed, enlarged, or punched on the shank end and chamfered to remove sharp edges. If possible, avoid laying steels down in the mud and dirt—stand them up with the bit end down.

Fig. 12-6 Improperly made shanks, excessively beveled and highly crowned and spalled. This condition causes early damage to piston faces. *(Ingersoll-Rand Co.)*

Correct temper Hard shanks cause chipping from the steel, and small pieces then become embedded in the piston striking face. Often, small particles of steel find their way into the drill and score the cylinder wall. A soft shank caused by incorrect tempering will be battered and upset from the constant pounding of the piston and may actually be riveted in the chuck. The shank must be uniformly hard enough to absorb the force of the piston blow and still be softer than the striking end of the piston.

Correct size An undersize steel shank has a sloppy fit in the chuck and contributes to premature chuck wear. The use of an undersize shank also damages the piston striking face because of poor alignment between these parts. Blowing power and overall perfor-

Fig. 12-7 New shank in worn-out check bushing, showing limited contact that would occur on piston face. (*Ingersoll-Rand Co.*)

Fig. 12-8 Example of overheated and burnt shanks. The large grains indicate an excessive forging temperature was used, plus possible soaking too long at the high temperature. The steel is ruined and cannot be restored by any subsequent heat treatments. (*Ingersoll-Rand Co.*)

mance of the drill is reduced when a small shank is used since it permits air to escape between the chuck liner and the shank. There is danger in using an oversize shank too—it may wedge itself in the chuck or it may even split the chuck.

Check steel shanks frequently to see that they fall within standard tolerances. Discard or recondition steel with undersize or oversize shanks.

On larger drifter drills, 4½-in. bore and larger, the use of shank pieces or striking bars is standard. The manufacturer's recommendations and specifications should be followed exactly in obtaining replacement striking bars, with particular care exercised in specifying the air tube size to establish the center hole diameter. You will find that these strikers are generally manufactured to correct dimensions of the proper steel and hardness.

KEEP THE ROCK DRILL CLEAN

Because of the very nature of its work and the adverse conditions under which a rock drill operates, it is difficult to keep dust and dirt from entering the machine. Rock drills can be kept clean, however, and preventing impurities from entering the drill pays off in improved operation and less lost time for repairs.

Dirt and abrasive foreign particles enter a rock drill chiefly from two sources: dirt and dust in the intake air to the compressor, and dirt, pipe scale, or flakes of rubber from deteriorated hose in the transmission. Throwing the drill down in the muck, leaving the drill in the blast area, or not plugging off openings of an idle drill are all invitations to trouble. Servicing a drill with oil that has been left standing in open containers in dusty surroundings is another bad practice, equally as bad as not blowing out the air hose before coupling up. Impurities which enter a drill speed abrasive wear, shorten the life expectancy of the drill, and contribute to operating difficulties.

Impurities which enter the compressor through the air intake can be reduced by installing intake air filters and, where possible, by locating the compressor in clean surroundings. The transmission line to the drill should always be blown out before connecting the drill to the air line. This precautionary measure is perhaps the most important in reducing the amount of dirt that enters the drill. Another effective way to minimize the danger of dirt getting into the drill is to install an air strainer in the air hose before the line oiler. This will catch coarse solids and pipe scale which would otherwise find their way into the drill. Some rock drills are equipped with a fine wire screen strainer in the gooseneck, which serves the same purpose as an air line strainer. Some operators remove air strainers, complaining that strainers impair drilling efficiency. This is only because the strainers are not cleaned often enough and become clogged with solid matter, thus restricting the flow of air. Air strainers and filters are good insurance against excessive rock drill wear, and they should be cleaned regularly at intervals determined by operating conditions.

In many instances, drills are returned to the shop for repair only because they are too full of dirt to run. This can readily happen when the drill is left too close to the blast area without protection from flying debris. Abuse of this type is particularly hard on drills and can be avoided easily by moving the drill a safe distance away before blasting. When this is not practical, plug the chuck opening, exhaust ports, and air inlet to keep dirt out, or at least cover the drill with a canvas.

Even when drills receive reasonable care and protection against dirt, they should be rotated to the shop at regular intervals for cleaning. If for any reason a drill cannot be completely dismantled and cleaned at definite intervals, a fairly good job of cleaning can be accomplished by flushing it with kerosene or a suitable solvent. Immediately after flushing, pour a few ounces of rock drill oil in the air inlet and run the drill briefly at low throttle to distribute oil to all parts. This is important to prevent rusting.

KEEP SIDE RODS TIGHT

Keeping the side rods tight is a simple precaution, easily overlooked, and yet it prevents sluggish operation, and more important, prevents damage to working parts. When side rods are loose, the excessive play between parts causes premature wear, and because parts get out of line with each other, bad strains are set up which often lead to broken parts. Beyond this, loose side rods allow air leakage between the back head and cylinder and between the cylinder and chuck housing, and frequently strip the threads on the rods.

Uneven tension on side rods frequently cause the tight rod to break just behind the nut. This should serve as a warning to keep nuts tight and at equal tension to prevent the following more serious trouble. When one side rod is tighter than the other, the normal alignment of parts throughout the length of the drill is disturbed and the working parts are pulled together and cramped along one side of the machine. This handicaps rotation of parts because the tight side rod causes the piston to bind in the chuck nut and buffer ring. If not corrected, the piston progressively wears the buffer ring until the protective air cushion between the piston head and buffer ring is destroyed. The loss of compression between these parts lets the piston strike the front head with its full force, and breakage of the piston results.

To avoid the damage which may be caused by looseness between parts, side rod nuts must be kept tight and at equal tension. Side rod nuts should be tightened once a shift with a wrench of the proper size and length to prevent setting the nuts too tight and possibly stripping the threaded rod ends. On drills equipped with side rod springs, weak or broken springs cramp working parts owing to the uneven tension on the side rods; replace weak or broken springs with a complete new set. When the nuts are tightened, tension on each spring should be the same and care should be taken to see that the coils are not completely closed.

NOTE: A bent side rod will have the same effect on parts alignment as uneven tension on side rods. Straighten or replace a bent side rod promptly.

FIELD MAINTENANCE

Extensive maintenance and repair of rock drills should never be attempted in the field, simply because facilities are not available to permit doing the work properly. Field maintenance should be limited to applying preventive maintenance measures which can be performed readily by the drill runner or field service crew. These are (1) attention to proper and regular lubrication, (2) protecting the drill against dirt, and (3) care in keeping all connections and nuts and bolts secure, with special emphasis on keeping side rods tight. Proper operation is still another important phase of preventive field maintenance which should not be overlooked. The manner in which a drill is operated has a direct bearing on machine life and the amount of maintenance required to keep the drill productive. Maintenance costs can be sharply reduced by ensuring that drills are not needlessly abused by operating personnel.

SHOP MAINTENANCE

Never leave a rock drill on the job just because it still runs. Drills should be rotated to the shop for cleaning, inspection, and repair at regular intervals short of the time when breakdowns are apt to occur. Because of varied operating conditions, inspection periods cannot be determined arbitrarily—they can only be arrived at through the accumulated experience of the user.

Disassembly

Observe these few precautions when dismantling rock drills for inspection or repairs:
 1. Clean off exterior of machine before starting disassembly.
 2. Provide a clean work area; cleanliness is important.
 3. Use a lead or babbitt hammer to drive off front head and back head. Use a soft drift for removing interior parts.
 4. Handle parts carefully; hardened parts may chip or break if dropped on a hard surface.
 5. Clean all disassembled parts in a suitable solvent. Probe portings in back head, valve parts, cylinder, etc., to loosen and clean out foreign matter. Place small parts in a clean box to prevent loss.

Assembly

Observe the following suggestions when assembling a reconditioned drill:
 1. Keep hands and tools free from dirt.
 2. Wipe a film of clean oil over all working parts as they are assembled.
 3. Allow no dirt or chips from soft hammers to get into the drill.
 4. Do not use force. Except for press fit bushings, parts should fit together easily. If force is required, something is out of alignment and must be corrected to prevent binding and premature damage.
 5. Service the integral oil reservoir with rock drill oil.
 6. Insert a steel in the chuck after assembling. The steel should turn freely by hand in the direction of rotation.

USE GENUINE REPLACEMENT PARTS

Through the accumulated experience of many years of research and development, together with efficient production methods, drill manufacturers are able to supply repair parts with field-proved dependability at prices consistent with normal trade practices. Don't handicap equipment by using nongenuine replacement parts. Many times, nonstandard parts have a decided effect on machine efficiency, and if breakage or damage does occur as a result of using nonstandard parts, drill manufacturers' warranties are generally voided and damaged parts will not be replaced. Original equipment manufacturers' warehouses also maintain a complete stock of repair parts ready for prompt delivery when repairs are needed, thus avoiding unnecessary production delays. The use of genuine spare parts made by the original drill manufacturer will maintain the original performance built into your drill.

MAXIMUM ALLOWABLE WEAR TOLERANCES

In general, the automatic valve, piston, and buffer ring are the only parts that wear out in the sense that they are no longer usable because of improper cycling. All other parts actually wear out; that is, metal is removed until the part breaks or no longer performs its function (for example, rifle bar, rifle nut, chuck nut).

Although maximum allowable wear tolerances for rock drill parts vary somewhat according to the make and size of drill, certain average wear limits can be given for the automatic valve, piston, cylinder, and buffer ring. The information given here should be helpful in determining the general wear condition of most drills up to $3\frac{1}{2}$-in. bore; however, you are urged to consult the drill manufacturer for specific information pertaining to wear limits for your particular drill.

Notes for measuring clearances If feeler gauges or plug gauges are employed, it is important to realize that the clearance indicated by a given size of feeler or plug gauge is more than the actual clearance. For example, if the 0.003-in. feeler can be inserted between a valve and valve chest, the clearance is more than 0.003 in. To be certain the clearance is not more than 0.003 in., the largest permissible feeler used should be 0.0025 in. or, in general, the gauge should be 0.0005 in. less in size than the allowable clearance.

Measuring the IDs and ODs with micrometers and subtracting is the best means of determining clearances. This method requires properly calibrated micrometers. It is true, however, that even though micrometers are available, feeler gauges will probably be more advantageous since they are less trouble to use.

Check the wear condition of the following parts whenever the rock drill is returned to the shop for overhaul, inspection, or repair.

Automatic valve The total allowable diametral clearance between the valve and valve chest or between the valve and valve guide is 0.003 in. When more than 0.003 in. is measured between the valve and valve chest or between the valve and valve guide, one or both parts are worn and should be replaced. To determine which part is excessively worn, various combinations of new valve parts should be assembled. The worn parts can then be determined by a feeler gauge.

Piston, cylinder, and buffer ring The total allowable diametral clearance between the piston head and cylinder and between the piston stem and buffer ring is 0.004 to 0.006 in., depending on the size of the drill.

When more than 0.004 to 0.006 in. is measured between the piston head and cylinder, replacement is in order. To determine whether the piston or cylinder is worn, insert a new piston in the cylinder and check the clearance with a feeler gauge.

The buffer ring should be discarded, or a new lining installed, when a clearance of more than 0.004 to 0.005 in. (depending on the size of the drill) is measured between the piston stem and buffer ring bore.

Some manufacturers have oversize pistons available. If it is considered advisable to use an oversize piston, the cylinder must be reground. In all cases the manufacturer's representative should be consulted.

PISTON AND CYLINDER

Piston There are many reasons for piston damage but the greatest of these are lack of lubrication, bad shanks, and the continued use of worn chucks.

Adequate lubrication is the dominant factor in realizing increased life from rock drill pistons. The continuous presence of *clean* oil between piston head and cylinder, between buffer ring lining and piston extension, and between piston splines and chuck nut offers protection against excessive wear, overheating and scoring, possible piston seizure, and the loss of close clearances so essential to good rock drill performance. When oil is not present, the increased friction between piston splines and the chuck nut overheats the piston stem and causes small cracks to develop on the splines which soon leads to fractures and a broken piston and chuck nut.

Bad drill steel shanks contribute to piston damage, shank ends which are crowned,

Drills and Drilling Equipment Maintenance 7-341

Fig. 12-9 Section of cupped piston ruined because it was not face-ground soon enough. *(Ingersoll-Rand Co.)*

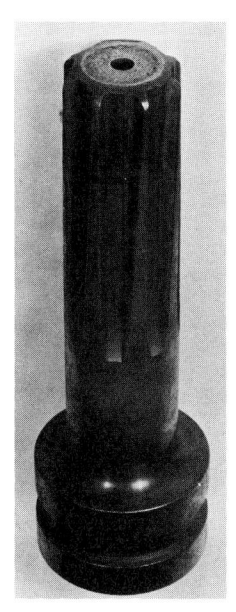

Fig. 12-10 Excessive piston flute wear caused by dirt getting into front end. *(Ingersoll-Rand Co.)*

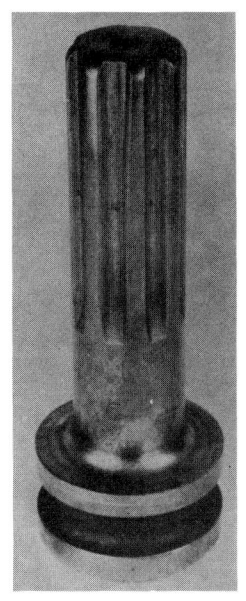

Fig. 12-11 Piston showing the effects of corrosive mine water. *(Ingersoll-Rand Co.)*

Fig. 12-12 Piston damage caused by excessively hard and crowned shanks. *(Ingersoll-Rand Co.)*

chipped, too hard, or not flat and square on the striking end will ruin a piston in a comparatively short time. If the striking face of a piston is cupped from the use of crowned shanks, substituting a perfectly flat and square shank makes matters worse because the total area of contact between shank and piston is then very small, and usually the piston striking face and shank both suffer.

Replace worn chucks promptly to avoid premature piston damage. The use of an excessively worn chuck allows the drill steel to get out of alignment with the piston, and this misalignment immediately reduces the contact area between the shank end and the piston striking face.

This highly localized contact will soon chip and spall the piston. Drill operators can increase piston life materially by maintaining alignment between the drill and drill steel at all times while operating. This will also help to reduce chuck wear.

Dull bits also contribute to piston damage. When dull bits are used, most of the force of the piston blow is absorbed by the piston and drill steel rather than by the rock being drilled. This puts abnormally high impact stresses on the piston which will cause premature piston failure.

Cylinder Above-normal frictional wear caused by abrasive dirt, and/or lack of lubrication, is the most potent enemy of rock drill cylinders. Dirt which enters the drill with contaminated oil or with impure air from the compressor causes abrasive wear that enlarges the ID of the cylinder and leads to the destruction of other parts. The combination of insufficient lubrication and dirty air will ruin a rock drill cylinder in a comparatively short time.

Fig. 12-13 Piston spalling caused by badly worn chuck bushings. *(Ingersoll-Rand Co.)*

To prevent needless wear and damage of rock drill cylinders, clean dry air and due care in servicing the drill with the proper type of rock drill oil at regular intervals are mandatory. Clean air must originate at the compressor, and the drill should be further protected from dirt by installing an air strainer in the supply line. The usual precaution of blowing out the air hose before coupling up should also be taken. The importance of

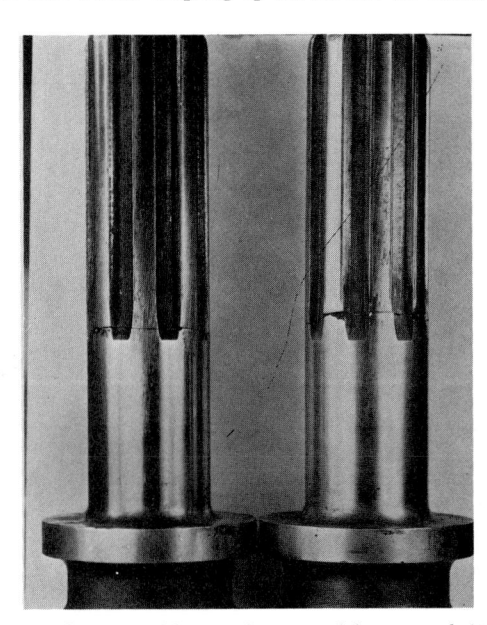

Fig. 12-14 Piston breakage caused by use of improper lubricating oil. *(Ingersoll-Rand Co.)*

proper lubrication—not only for protection against excessive cylinder and piston wear, but against wear of all working parts—has been stressed. Use oil only from containers which have been kept covered or otherwise protected from contamination. The oil should be of the correct viscosity for operating conditions and must have properties which allow it to emulsify with water to provide rust protection when the drill is idle.

Presence of moisture in a drill is a constant danger. A moist air supply will leave the cylinder wall wet with water and cause rusting when the drill is idle unless the oil being used mixes well with water. When the drill is put into operation, the rust will polish off, but in the process the cylinder wears oversize. In wet drills the problem of rusting and corrosion from mine water is even greater, and sufficient lubrication with the proper type of oil assumes more importance than in dry drills.

Loss of front head cushion Extensive piston damage due to loss of front head cushion can be corrected by replacing the buffer ring, provided that tolerances between piston head and cylinder are up to standard. If excessive clearance has developed between piston and cylinder, air will escape past the piston as it moves forward and consequently no cushion will be formed. It should be noted here that the piston extension must also have a good fit in the buffer ring, otherwise air will escape out the front end unrestricted. The cushion test can be applied to determine if it is safe to use a new piston with a buffer ring that has been in service for some time.

Cushion test To test for front end cushion, drop the free piston (with oil wiped off) in the cylinder. The piston should bounce or cushion on the air trapped between the piston head and buffer ring. If, instead, the piston strikes the buffer ring hard, either the piston or cylinder (or both) is worn beyond safe limits, and the part(s) must be replaced or reconditioned. This test also gives an indication of buffer ring wear. With a worn buffer ring, excess clearance between the piston stem and buffer ring lining will prevent an air cushion from being formed.

Fig. 12-15 Piston pits caused by corrosive mine water. *(Ingersoll-Rand Co.)*

Reconditioning Cylinder and Pistons

Inspect the cylinder and piston head for scuffs, scratches, and signs of scoring from overheating and lack of lubrication. If the piston splines are not cracked from overheating

Fig. 12-16 Stripped threads in drill piston caused by loose rifle nut. Looseness caused by worn rotation parts. *(Ingersoll-Rand Co.)*

Fig. 12-17 Piston breakage due to no cushion caused by worn piston stem bearing. *(Ingersoll-Rand Co.)*

and the face is in good condition, and depending on the amount of wear or damage to the piston head and cylinder, it is sometimes possible to recondition these parts. Scuffs and burrs on the piston head can be removed with a fine emery stone and the damaged surface can be smoothed with a whetstone. Similarly, the cylinder bore can be brought back to a smooth finish by grinding and polishing. Many times, however, refinishing may prove impractical since the clearance between piston head and cylinder will then be too large. This leaves several alternatives. The most obvious would be to replace the cylinder and piston with new parts. Second, some manufacturers can supply an oversize piston to fit a

Fig. 12-18 Piston breakage caused by insufficient lubrication. *(Ingersoll-Rand Co.)*

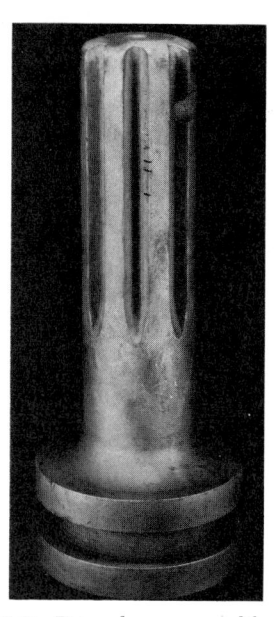

Fig. 12-19 Piston damage caused by insufficient lubrication. *(Ingersoll-Rand Co.)*

cylinder enlarged by wear. This sometimes presents a problem because it interferes with interchangeability between like machines. And before an oversize piston can be installed, the cylinder bore must be ground and lapped at both ends where the amount of wear has been very small. Finally, the bore of a worn cylinder can be built up by chrome plating and then refinished to its original ID.

Repairing piston striking face Regular inspection of the piston striking face is important. A piston which is chipped or cupped on the striking end can be repaired, providing damage has not progressed too far. If discovered in time, a slightly cupped or chipped piston can be trued up by careful grinding performed slowly to avoid overheating and annealing of the hardened case. No more than $\frac{1}{16}$ in. of the metal should be removed. There are two reasons for this. If the core under the hardened case is exposed, the soft surface will peen and flow from impact with the steel shank. In addition, removing more than $\frac{1}{16}$ in. will change the original striking point of the piston and long stroking will occur, reducing the efficiency of the drill.

BUFFER RING*

Neglect in renewing a worn buffer ring can lead to extreme damage. A close fit between buffer ring and piston is necessary to trap air on the forward stroke of the piston so that an air cushion is formed between these parts. The trapped air cushions the blow and prevents the piston from traveling all the way forward and striking the buffer ring with its

*Or cylinder liner, piston stem bearing, spacer bushing, cylinder front washer, etc.

full force. (The cushion also produces a certain amount of bounce which assists the piston on its rearward travel.) A worn buffer ring allows air to escape between the piston stem and buffer ring lining, and the heavy impact of the unretarded piston causes considerable damage to the front head parts and frequently breaks the piston. In addition to this, a worn buffer ring simply wastes air and greatly reduces drilling speed.

A good cushion is particularly important when pulling steels. In the normal process of drilling, the drill steel shank and the rock absorb the heavy blow of the piston, but when withdrawing the steel, the steel rests on the retainer and the piston does not come in contact with the shank of the steel. Consequently, in a drill with little or no front end cushion, the resultant impact of the piston against the buffer ring causes extensive damage.

Besides the damage that can be caused by loss of front end cushion, an excessively worn buffer ring used with a worn chuck does not furnish sufficient guide for the piston and the piston strikes the drill steel shank on an angle. This chips and spalls the striking face of the piston and damages the drill steel shank.

Inadequate lubrication is the principal reason for abnormal buffer ring wear. When sufficient oil is used, excessive wear can also be caused by abrasive dust that enters the drill through the air line or finds its way into the drill with dirty, gritty oil.

Still another serious cause of excessive wear and parts failure is the dieseling of the lubricating oil in the cylinder between the buffer ring and the piston hammer. This occurs when the hammer is run at full throttle while removing drill steel from the blast hole, generally when the drill operator is fighting stuck steel. Since the drill string is hanging in the chuck or in the retainer, the energy of the hammer is absorbed in the front air cushion. This causes a high compression which fires and burns the lubricating oil.

The first steps, then, in preventive maintenance are to ensure that the drill is run with clean air and an abundant supply of clean oil. Obviously, the buffer ring should be replaced before it has worn to such an extent that the drill begins to break pistons or front head parts. Whenever the drill is dismantled for cleaning or repair, it should be cushion-tested by dropping the free piston in the cylinder. If the piston does not catch on compression, the buffer ring or buffer ring lining may require replacement. Before assuming that the buffer ring lining is worn, insert the stem end of the piston in the buffer ring through the front end of the machine. If the lining is good there will be little clearance between these parts and the loss of cushion may be due to excess clearance between piston and cylinder.

Because of variations in design and construction it is not feasible to cover the many methods used for removing and installing buffer ring linings. In general, arbors used for removing bushings should be hardened to avoid upsetting and should have flat, square faces and sharp corners for maximum bearing surface. Some manufacturers supply special tools and devices to aid in extracting the old lining and installing the new lining. In any case, manufacturers can furnish informative literature describing the correct procedures for renewing buffer ring linings, as well as other press fit bushings.

ROTATION PARTS

Rifle bar Lack of lubrication and abrasive wear caused by dirt are the principal reasons for rifle bar failure. Without lubrication the rotating and sliding action of the rifle bar in the rifle nut wears the rifle bar flutings rapidly and cuts the rifle nut. Bronze cuttings from the rifle nut then harm other parts of the drill, and when drilling stresses become severe the rifle bar may break. Overheating due to lack of oil (or the wrong type of oil) also discolors and blues the steel in the areas of greatest stress, and the flutings will be heat-checked and broken out. Abrasive dirt and gritty particles which enter the drill also promote wear of rifle bar flutings and the rifle nut much the same as inadequate lubrication. When the rifle bar and rifle nut are worn, the piston travels rearward farther than normal before turning the drill steel; on the forward stroke the piston travels a greater than normal distance before rotating the pawls. This adds up to poor rotation and reduced drilling efficiency.

Heavy rotational loads caused by stuck steel, misalignment, or drilling in bad ground overheat the rifle bar and damage the flutings. Overheating in the presence of sufficient oil is indicated by black carbon deposits on the flutes, and more than likely there will also be small transverse cracks on the sides of the flutes.

Good lubrication practices conscientiously carried out will make rifle bar and rotation parts last longer. Along with an abundant supply of clean oil, attention must be given to furnishing the drill with clean air to halt needless abrasive wear.

Inspection and Repair of Rotation Parts

Rifle bar and rifle nut Regular inspections to determine the wear condition of these parts are advisable. Small cracks on the rifle bar flutes, discoloration caused by overheating, or flutes worn around on the edges indicate damage from lack of oil or the use of oil not suited for rock drill lubrication. Either condition warrants installing a new part. If preventive maintenance records show an unusual number of rifle bars and nuts being replaced, look to your present oil supplier for the solution or consider a different oil supplier. Because flutes on rifle bars are much smaller than those on the piston stem, rotational pressures are much greater in psi on the rifle bar flutes than on any other area in the drill. This area *must* have an extreme-pressure lubricant. On hand-held drills and light drifters, discard the rifle bar or rifle nut when the flutes are worn away approximately $1/16$ in. On larger drifters and other large-bore drills, wear up to $1/8$ in. can be tolerated on the flutes of the rifle bar and rifle nut. The amount of wear can be determined by placing the rifle nut on the rifle bar. If a $1/8$-in. shim (small machines) or $3/16$-in. shim (large machines) can slide between the flutes, then either the rifle bar or rifle nut (or both) is worn excessively. Always replace the rifle nut when a new rifle bar is installed; rotation power will not improve unless this is done.

The use of bits larger than recommended can result in premature failure of rotation parts.

Ratchet pawls and ratchet ring Weak or retarded rotation is evident when the ratchet pawls are worn on the bearing edges, causing them to slip in the ratchet ring. (Worn teeth in the ratchet ring will also cause the pawls to slip.) Pawls should be reversed in the slots of the rifle bar when worn to a $1/16$-in. radius on one edge and discarded when worn on both edges. Worn ratchet pawls wear the notches in the ratchet ring, and either condition allows the pawls to slip and impair the rotation power of the drill. The ratchet ring should be replaced when the teeth are worn to a $1/16$-in. radius. Check pawl springs for elasticity by comparing with a new spring, and replace weak or broken springs. Weak or broken pawl springs cause incomplete engagement with the ratchet ring teeth, resulting in chipping and rounding off on both the pawl bearing edges and the teeth. Once again, pawls can be reversed in single-rotation hammer drills, and some manufacturers use reversible ratchet rings. New pawls should not be used with worn ratchet rings, and, obviously, new ratchet rings should only be installed with new (or nearly new) pawls.

It may be noted here that rotation troubles frequently develop when side rods are not tightened evenly. The misalignment of parts thus produced causes the rifle bar to bind in the valve chest on one end and in the rifle nut on the opposite end. This seriously impairs ratcheting power, may stall the drill, and accelerates wear of all rotation parts. The simple precaution of keeping side rods tight and at equal tension adds to the life of rotation parts. Check side rod nuts daily to see that they are snug.

Chuck nut (drills not equipped with reverse rotation) Since the piston bears on the splines in the chuck nut in the direction of rotation only, considerable wearing away of the splines can be tolerated. However, in the interest of good maintenance practices, at inspection times replace the chuck nut when the splines are worn halfway through. The amount of allowable chuck nut wear depends also on the wear condition of other rotation parts. The foregoing information is generally true concerning chuck nuts in larger hammer drills having reverse rotation for uncoupling sectional drill steel; however, wear must be checked on both sides of the chuck nut splines, and experience will have to guide the drill repairer on allowable wear due to wide variations of necessary rotational torque in the many rock formations throughout the world.

A handy tool for removing the rifle nut and chuck nut can be made from an old rifle bar or piston with the stem end cut off to approximately the same length as the rifle nut or chuck nut. An extension handle can then be welded to the part for applying leverage. To remove the rifle nut, clamp the splined end of the piston in a vise equipped with soft jaws and unscrew the nut. Similarly, the chuck nut can be removed by holding the rotating chuck in a vise, then using the piston tool to screw out the chuck nut. Both the rifle nut and chuck nut have left-hand threads. On those machines with a press fit chuck bushing, the bushing can be removed by using a hammer and drift, arbor press, or gear pulling device.

A new rifle nut should have a tight fit in the piston, or vibration of the machine will tend to unscrew the nut. Moreover, a rifle nut or chuck nut with a loose fit does not receive sufficient support from its mating part for the work it is doing and will wear excessively and may crack or break. Generally, a new rifle nut should not screw into the piston more than two or three threads by hand.

Independently rotated rock drills In independent rotation rock drills, common maintenance practices concerning gears and their backlash, bearings and their wear, and loose-fitting component parts should be observed. The efficiency or the air consumption of the independent rotation motor is related to the wear which may be found on the geared or vane motor parts. When the air consumption becomes high, the parts are worn beyond further use and an economic decision must be made to replace them.

Wear problems on the independent air motor and the gear train are confined to the simple, common type of wear which occurs on gears, bearings, and oil seals. There are solid bronze bearings, tapered roller bearings, and ball bearings included in this type of rotation mechanism in addition to various types of seals. Excessive backlash in the gears, looseness in any of the bearings, or leakage past seals is cause for inspection and replacement, if necessary. Gear motor and gear train bearings, etc., are usually lubricated by air mist lubrication through the line oiler.

CHUCK PARTS

Chucks for hexagon and quarter octagon steels Worn chucks cause trouble. When the chuck or chuck bushing is worn, the steel can no longer be held in line with the piston, and the loss of proper support for the drill steel shank allows the piston to strike the shank on an angle, damaging the piston, the shank, and in wet machines the water tube. Chucks worn oversize or bell-mouthed should never be returned to service. The use of wear gauges determine the wear condition of the chuck or chuck bushing. In lieu of a suitable wear gauge, a good steel shank inserted in the chuck should have a close fit to hold the shank in proper alignment with the piston.

Chucks for round lug steels Chucks for round lug steels generally show evidence of wear on the driving lugs of the mating parts of the two-piece chuck and on the driving surfaces provided for the lugged shank. Frequently, wear notches are formed on the end of the chuck bushing from continual impact with the lugs on the steel shank. Examine these parts to see that they are not excessively chipped, battered, or worn. The limits of chuck bushing wear can be determined by using a wear gauge. It should be noted that wear on the driving lugs increases rapidly after the hardened case is destroyed (usually about 0.09 in. deep). The chuck bushing should not be so deeply notched that it causes short stroking of the machine. If not unduly worn the chuck bushing can be extracted, then rotated and replaced in the chuck to present a new wearing surface. Usually, chucks for round lug steels do not wear bell-mouthed and cause misalignment problems because of the longer bearing surfaces offering better support for the steel as compared to hexagon and quarter octagon machines.

AUTOMATIC VALVE ASSEMBLY

Valve parts are manufactured to extremely close tolerances to provide the precise metering of air needed for efficient rock drill performance. Abrasive dust and dirt in the oil or air supply damage the highly finished surfaces and frequently plug air portings or cause the valve to stick in either of its extreme positions. Anything that hampers the free action of the valve makes a drill lose power, run erratically, or cease operation. Neglect of lubrication causes abnormal wear and rapidly destroys the close clearances needed for perfect operation of the valve.

Clean air, clean oil—these are the two requirements for trouble-free valve operation. If the drill is kept clean and lubricated correctly the automatic valve assembly will give long service. Good insurance against valve malfunctioning includes blowing out the air hose before connecting it to the drill, using an air strainer in the line to the drill, and capping or plugging the air inlet swivel when the drill is idle.

A rock drill oil of the correct viscosity oils used in low temperatures, for example, will retard valve action and slow up drilling speed. Only use oils intended for rock drill service.

Valve Reconditioning

Axial end-seating (or kicker port) valves The seating surfaces on the valve seat and valve cover become concave or dished after long service, resulting in something less than a perfect seal with the main valve. This condition disturbs the precise metering of air and positive control over the piston stroke. Although these surfaces can be refinished by grinding, it is usually more economical to replace the worn parts. Care must be used in grinding to avoid removing too much of the metal since this will increase the travel of the valve and, in turn, retard the reciprocating speed of the piston.

Butterfly valves The flat faces of the valve and valve seats of a butterfly valve can be scratched and pitted by dirt particles which enter the valve assembly with dirty oil or impure air. The faces can be rehoned to restore the smooth seating surfaces by using a fine grade grinding compound on a perfectly flat cast iron surface plate, then patiently moving the valve part over the plate with a circular motion. The angle across the diameter of a butterfly valve must not be altered when resurfacing the valve or valve cover. If the angle is changed, the flat surfaces will not be tangent across the entire seating surface and the valve mechanism will operate erratically, waste air, and impair operation of the drill.

IMPORTANT: Whether or not it is advisable to recondition valve parts depends on the extent of wear or damage, the repair facilities available, and the skill of the person making the repairs. The cost of making repairs should be weighed carefully against the cost of parts replacement.

AIR AND WATER TUBES

In wet machines the water tube extends down the centerline of the drill beyond the end of the piston and into the drill steel shank far enough to ensure that water cannot escape into the chuck parts. A worn chuck, however, disturbs the normal alignment of piston, water tube, and steel shank, and the shank will meet the piston on an angle. This misalignment exerts a side thrust on the end of the water tube. When this happens, the tube chafes on the wall of the shank center hole, and in a short time the tube will be battered and misshapen to such an extent that water no longer flows through the drill steel but splashes out the end of the piston into the chuck parts. Frequently, the shank or piston shears off the end of the water tube entirely, allowing water to flood the machine and wash away the lubricant. In stoppers, air leg drills, and drifters equipped with tappets, a worn tappet chuck produces these same results, but the danger of extensive damage from the washing effect of water is increased because the water can drain back into the cylinder and other working parts.

Breakage of water tubes also occurs when the shank center hole is not enlarged sufficiently to accept the water tube, when it is not punched deep enough, or when the hole is off center. Moreover, a shank hole which is not chamfered to remove the sharp edge damages the water tube.

Ice or frost formation on the exhaust port indicates that water is getting into the drill, and the drill may freeze up because of the refrigerating effect of the expanding air. Freezing may be caused by a moist air supply or by water being forced into the cylinder through a split or cracked water tube.

Inspection and Replacement

Air and water tubes must be straight and true, not pinched, split, or damaged in any way. Damaged water tubes and worn or deteriorated rubber packing should be replaced promptly to avoid water leakage and accelerated wear and rusting. At the same time, worn chucks and tappet sleeves must be renewed or the damage will repeat itself. Drill steels with shanks having improperly formed center holes should be discarded or reconditioned.

In large-bore blast hole drills (drifters) with striking bar[1] front end construction, a cup seal is used in the striking bar to prevent air from escaping around the air tube and creating a back pressure on the piston. This cup seal (older striking bars used an "O" ring) also centers the tube and holds it free from the walls of the shank. Drills operated with cup seals break air tubes repeatedly. Replace shank cup seals if they are torn, worn badly, or simply lost; otherwise tube breakage and loss of penetration will be experienced. When the tip of the air tube is inserted in the striking bar shank, the cup seal should provide a

[1] Or driver rod, shank adapter, bonus bar.

perfect seal around the air tube. Air leakage here reduces the effective penetration rate of the drill and hampers hole-cleaning ability of the machine.

STEEL RETAINER

The steel retainer should be replaced when it becomes worn to the extent that it no longer holds the collar of the drill steel shank. Yoke springs or rubber buffers should be replaced when they have weakened or deteriorated and no longer hold the yoke firmly in place.

THROTTLE VALVE

The valve plug should have a close fit in the back head and should turn easily with very little axial play. If it sticks, chances are the bore or plug has been scored by abrasive dirt. Scratches or abrasions on the valve or valve bore can be removed by using an oil hone, but if further damage is apparent, a new valve plug should be installed. It is not advisable to attempt to lap the valve plug in the bore since this will reduce the diameter of the plug, enlarge the bore, and cause further difficulty. After long service the detents or positioning notches on the valve plug may become so worn that vibration causes the valve to change position during operation. If this is the case, a new valve plug is needed. The detent plunger should also be checked for wear and the plunger spring replaced if weak or broken.

TESTING A RECONDITIONED DRILL

To check that a reconditioned drill is actually in perfect running order, it should be tested. Before connecting the air line, service the drill reservoir and line oiler with rock drill oil and add a small amount of oil in the air inlet swivel to be sure of immediate lubrication. The drill should start with very little air pressure and the piston should reciprocate smoothly. Allow the drill to run in slowly at reduced pressure long enough to determine that it is in good working order. If the drill should stall, turn off the air immediately—this indicates binding of parts due to snug fits or perhaps because side rods are not tightened evenly. Check side rod tension first, then start the drill again. After a short period of low-pressure operation, the drill should develop a definite rhythm, evidenced by an even sound at the exhaust. The machine may become warm in the vicinity of the buffer ring, but it should not overheat. If erratic operation persists or frequent stalling occurs, the drill should be dismantled and checked for indication of binding between parts.

After the initial running in at reduced pressure, check the performance of a reconditioned drill by comparing its drilling speed with the drilling speed of a new drill under similar operating conditions, using normal air pressure.

After testing, insert rags in the chuck opening, exhaust port, and air inlet to keep dirt out of the machine during temporary storage and until it reaches its working location.

MAINTENANCE AND PERFORMANCE RECORDS

Adequate records should be kept on each rock drill which will show reasons for repair or part renewals, cost of parts and labor, and footage drilled since the machine was last

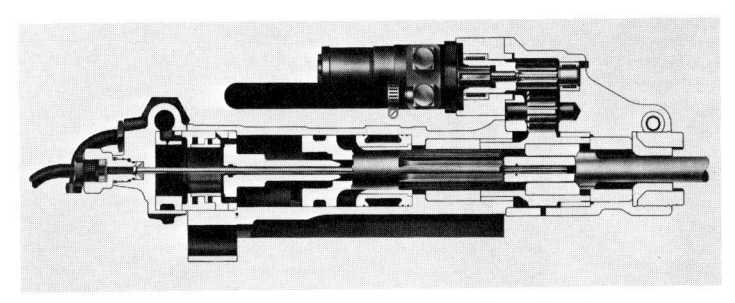

Fig. 12-20 URD 350 section. *(Ingersoll-Rand Co.)*

repaired. Over a period of time these records give an indication of those parts being repaired more often than normally expected. Periodic examination by the drill repairer or maintenance supervisor will suggest practices which should be corrected and point out incorrect practice to the shift boss or operating supervisor, which should correct the situation and help to reduce maintenance costs.

Although a rock drill can be kept running indefinitely by repair or replacement of parts, accurate records will indicate a point at which it is no longer economical to keep the drill in service. Consideration should then be given to replacement with a new machine.

Section **8**

Pump and Compressor Maintenance in Construction

Chapter **1**

General Pump Maintenance

ROBERT J. PORTER
General Service Manager, The Gorman-Rupp Company,
Mansfield, Ohio

A cardinal rule in performing maintenance on pumps is to follow procedures established by manufacturers in manuals written specifically for their equipment.

In a business where money is made when equipment is off the premises, downtime can be shortened by performing a few simple tasks each time a pump is returned. By doing this, equipment will be ready when an emergency arises.

Inspection, parts replacement, and tolerances First, clean the pump. Then, inspect the exterior for loose bolts, nuts, and other attached hardware. Next, replace any damaged parts or parts that are worn to such an extent that performance (head and capacity) is affected. Critical parts are the impeller, wearplate or wear rings, and volute casing and seal. It is best to use parts designed and produced by the manufacturer of the pump.

Check tolerances between the impeller face and wearplate or wear ring. Most pump manufacturers provide a means whereby adjustment can be made easily. Head and capacity can be severely reduced if tolerances are not correct.

Lubrication Proper lubrication can add life and extend the use of the pump. The two most important areas are bearings and seals.

Most pumps use bearings lubricated with grease and normally need no further lubrication until overhaul or after 5000 hr. When adding grease, use the grade and technique recommended by the manufacturer.

Under normal operating conditions, oil-lubricated bearings should be drained and refilled annually; however, the oil level should be checked frequently.

Do not lubricate bearings sooner than necessary. Overlubricating will cause excessive preloading and heating which will shorten bearing life.

Seals A pump may have any one of four types of seals: oil-lubricated mechanical, grease-lubricated mechanical, self-lubricated mechanical, or stuffing box. Each requires different attention.

Oil-Lubricated Mechanical Seals. The oil level in an oil-lubricated mechanical seal should be checked frequently. The cavity may become diluted with the liquid being pumped after 2 to 3 months of operation. When the oil becomes approximately 50 percent diluted or becomes milky in color, drain the seal cavity and refill with SAE No. 30 motor

8-2 Pump and Compressor Maintenance in Construction

oil (see Fig. 1-1). Fill the seal cavity only until the oil level can be seen in the fill hole. Do not overfill the cavity. Be sure that the vented fill plug on top of the housing is not clogged with dirt. Use only nondetergent oil in the seal cavities.

Grease-Lubricated Mechanical Seals. The grease-lubricated mechanical seal applies grease to the seal from a spring-loaded grease cup. When the plunger bottoms on the grease-cup cap, the grease cup must be refilled (Fig. 1-2). Use only a good soft grade of No. 2 pressure gun grease. Do not use a hard grease since it will not flow into the seal and

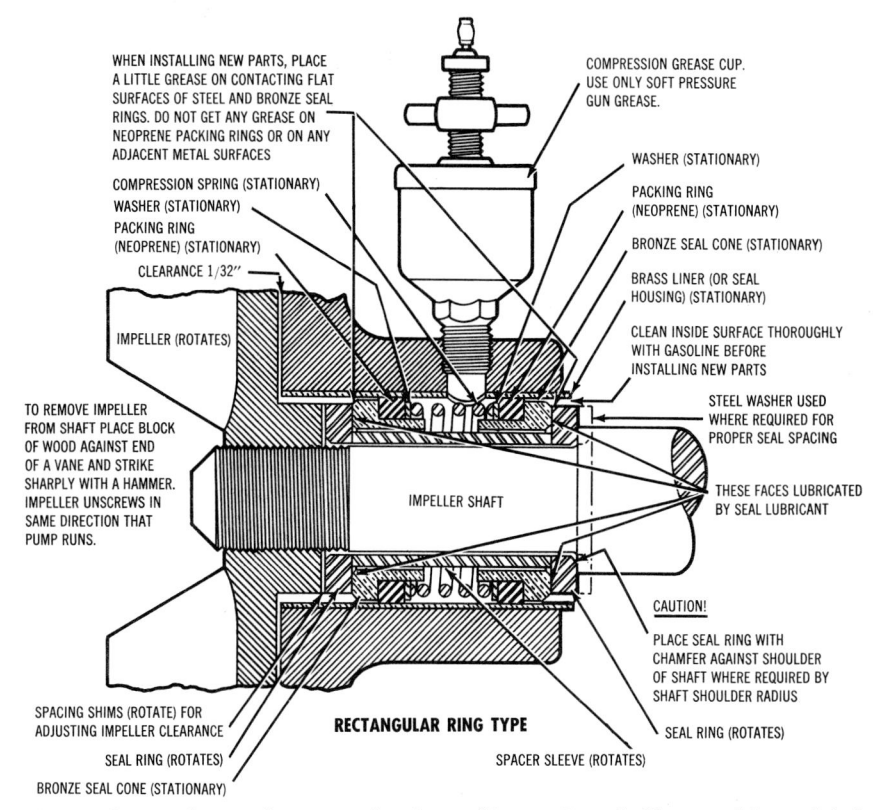

Fig. 1-1 Cutaway drawing shows internal workings of Gorman-Rupp double grease-lubricated shaft seal. Maintenance of most pumps at these points must be heavily emphasized.

Fig. 1-2 Actual seal components point up the complexity of such devices and the need for close attention.

seal failure may result. To fill the grease cup, turn the cross arm clockwise to raise the plunger and compress the spring. While in this position, the cup can be filled using a grease gun fitting on the top of the plunger. Fill the cup until the grease comes out of the relief hole. The grease cup can also be filled by hand by removing the cap. During pump operation, turn the cross arm counterclockwise until it is at the top of the threaded plunger to apply spring pressure to the grease to automatically lubricate the seal.

General Pump Maintenance 8-3

Self-Lubricated Mechanical Seals. Check by removing the seal from the pump. Clean it by washing metallic parts in cleaning solvent, and dry thoroughly. Inspect the mating surfaces of the seal for wear, scoring, grooves, or any other damage which could cause leakage. If any of the seal parts are worn, replace the entire seal. Never mix new and old parts of a seal.

Stuffing-Box Seals
 1. Lubrication. Some stuffing-box seals are lubricated by a spring-loaded grease cup in the same manner as the grease-lubricated seal.
 2. Leakage adjustment. Excessive leakage may occur from a stuffing-box seal if the packing gland is not properly secured. To adjust the gland, slightly loosen the nuts that secure the packing gland and then retighten them finger tight only. Start the pump and tighten evenly while the pump is pumping liquid. Do not tighten the gland so much that all leakage from the seal stops. A slight leakage from the seal is necessary for proper lubrication. After the gland is adjusted, the pump shaft should rotate freely by hand. If the shaft does not rotate easily, the gland is too tight. If the seal leakage cannot be controlled by adjusting the packing gland, the packings must be replaced.

 Vacuum test, speed, and accessories After checking the pump, replacing parts, and making adjustments, it is recommended that a maximum vacuum test be made. Fill the volute casing with liquid and block the suction. Install a vacuum gauge on the suction side and operate the pump. A good vacuum reading indicates the pump is okay. Readings will vary according to the altitude of the test location.
 It is desirable at this point to also tack the pump speed to be sure the pump is operating per the manufacturer's curve.
 Accessories and fittings used with pumps can also affect overall performance. Damaged suction hose with air leaks or collapsible suction hose will cause priming loss. Suction strainers usually supplied as standard equipment with pumps should be used at all times to prevent foreign materials from entering the pump and damaging parts.

 Maintenance timetable Generally, the following timetable can be used for pump maintenance if it is impractical to check the pump each time it is returned. (Remember that pumps used for tough applications may require a more frequent schedule; see Fig. 1-3.)

 Check the Pump Each Month. Priming, speed, capacity, noise, seal leakage, and gasket leaks should be checked. If engine driven, check the crankcase for clean oil, see if the spark plug is intact, and check noise, speed, and carburetor adjustment.

 Remove the Impeller Every 6 Months. Replace if the vanes are worn seriously. Check tolerances. Replace the shaft seal or packing and replace the shaft sleeve if worn. Clean the casing and recirculation device. Overhaul the engine and replace worn parts. Reassemble with new gaskets. Clean out seal grease and renew.

 Open up the Pump Once a Year. Check and clean the interior parts. Clean engine carbon in the head and flush and refill the engine crankcase.

 Troubleshooting checklist To assist in locating reported trouble, refer to the following checklist.

Pump Does Not Prime Properly
 1. Pump casing is not filled with water.
 2. Leak in suction line or connections—check to be sure that all fittings are tight in the suction line, and make sure there are no leaks in the hose.
 3. Pump is clogged.
 4. Pump seal is worn and leaks air.
 5. Inlet valve rubber is frozen to seat.
 6. Pump is running too slowly.
 7. Clearance between impeller and pump body is greatly worn. Refer to manufacturer's istruction manual for proper adjustment.
 8. Suction lift is too high.
 9. Suction line or suction strainer is clogged.
 10. Water is too warm for the suction lift being used. As the temperature of water increases above 60°F, the practical maximum suction lift of the pump will decrease.
 11. When water in the pump becomes too warm and the pump is unable to create a sufficient vacuum, it may be necessary to replace the water in the pump case with fresh cold water.

8-4 Pump and Compressor Maintenance in Construction

(A)

(B)

Fig. 1-3 A and B Tough application units like this self-priming centrifugal trash-type pump always demand thorough and continuous preventive maintenance.

Not Enough Water Delivered
1. Engine is not running at rated speed.
2. Strainer or inlet valve or line is clogged.
3. Suction line or fittings leak air.
4. Pump seal is worn and leaks air.
5. Suction check valve is not seating.
6. Too much clearance between impeller and pump body due to wear. For good performance, refer to manufacturer's instruction manual for proper adjustment.
7. Lining in suction hose is collapsing—the rubber lining inside the fabric layers has caused a stoppage by pulling away from the fabric and pulling together under the vacuum created by the pump. It is possible that a new suction hose will do this at times.
8. Suction lift is too high—at a 25-ft lift, the pump delivers only about 50 percent of the water it delivers at a 10-ft lift. The suction hose is too long. A long suction hose will cause excessive friction loss and correspondingly reduce the capacity of the pump.
9. Discharge head is too high—check hose or pipe friction losses. A larger hose or pipe may correct this condition.

Not Enough Pressure
1. Engine is not running at rated speed.
2. Leaking seal.
3. Too much clearance between impeller and pump body due to wear. For good performance, refer to manufacturer's instruction manual for proper adjustment.

Chapter **2**

Centrifugal Pump Maintenance

H. W. LINNEMAN
Gardner-Denver Company
Quincy, Ill.

Successful and efficient operation of centrifugal pumps depends greatly upon proper selection and installation. This chapter treats primarily maintenance; however, the selection of the correct pump for each application or service is essential to efficient and trouble-free performance. To ensure the most efficient operation and the least maintenance, submit complete data on the prospective application to the pump manufacturer so that he can properly select a pump to fulfill requirements.

Most pump manufacturers supply instruction books covering installation, operation, and maintenance of their pumps; so the following information is general in character to apply to all makes.

Installation Locate the pump in a place that is easily accessible for regular inspection during operation. The pump should be placed as near the liquid supply as possible to permit use of short and direct suction pipe (see Suction Piping below). Ample headroom should be provided for a crane, hoist, or tackle. Pits in which pumps are placed should be safeguarded against floods.

It is of paramount importance that the pump be placed on a good foundation, preferably of concrete. Foundation bolts should be placed according to the method shown in Fig. 2-1 and according to dimensions furnished on a certified drawing. Pipe sleeves used should be about two and one-half diameters larger than the size of the anchor bolts being used.

Alignment Pumps are properly aligned at the factory by leveling the base and bringing the pump and driving unit into exact alignment with shims. Experience has proved, however, that all bases, no matter how rugged, will spring and twist during shipment. Therefore, there is no guarantee that the original alignment will be maintained. Consequently, it is necessary that the factory alignment be reproduced when the unit is erected on its foundation.

    1. Place the pump unit on its foundation, allowing approximately 1 in. between top of foundation and bottom of base, using wedges to obtain proper spacing (see Fig. 2-1).

    2. Remove coupling pins (if used) and check top of base for degree of level, using the wedges for adjustment.

    3. Tighten foundation bolts evenly and firmly so that the base rests solidly on wedges.

    4. Check the alignment at the coupling by placing a straightedge across the coupling flanges. This should be done at four points on the coupling, the points being 90° apart. The distance between the faces of the coupling halves should also be checked at four points with a thickness gauge. See coupling sketches, Fig. 2-2. The coupling halves are to be brought into perfect alignment by adjusting the wedges under the base, the base at the same time being level.

5. If the pump is to be connected to the prime mover by gears or chains, the alignment should be checked by a straightedge across the faces of the gears or sprockets. This should be done in two directions at an angle to each other as large as permitted by the relative size of the gears or sprockets. When pumps or prime movers are to be heated in operation, that is, steam-driven prime movers or pumps handling hot water, the unit should be aligned under the thermal conditions in order that contraction and expansion due to these changes in the temperature may be taken into account.

6. Build a form or dam around the foundation and fill it with grout to a point about 1 in. above the bottom of the base. Allow grout 48 hr to set.

Connecting piping Pipes must line up naturally. Do not force them into place with flange bolts, for this may draw the pump out of alignment. Pipes should be supported independently of the pump so as not to put any strain on the pump casing. After the piping has been installed, alignment should be checked again and, if necessary, correction made. For unusually long discharge lines, a packed slip joint should be provided to compensate for elongation of pipe due to pressure. Also, when piping is subject to temperature changes, it should be arranged so expansion and contraction do not place a strain on the pump casing. Air-conditioning and service pumps installed in buildings where any noise is objectionable should be insulated from the steelwork and walls in such a way that vibration cannot be transmitted to the building, and the discharge pipe should be insulated from the pump so that no noise or vibration can be transmitted to it.

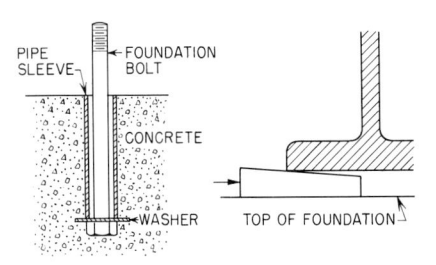

Fig. 2-1 Method of placing foundation bolts.

Discharge piping To protect the pump, a gate valve and a check valve should be installed in the discharge pipe close to the pump. The check valve should be placed between the pump and the gate valve.

If increasers are used on the discharge side to increase the size of discharge piping, they should be placed between the check valve and the pump. The selection of the discharge pipe should be made with due reference to friction losses. The discharge pipe should never be smaller than the pump discharge and preferably should be one or two sizes larger.

Suction piping The suction piping should be as direct and short as possible. It should be at least one or two sizes larger than the pump nozzle. Length and size are determined by the maximum allowable suction lift, which should never exceed 15 ft (friction included). If changes from one pipe to another are necessary, standard ASME suction reducers should be used. Hot liquids must flow to the point of pump suction by gravity. Pipe should be laid out so that air pockets are eliminated. Refer to Figs. 2-3 to 2-6 showing correct and incorrect methods of installing piping at the pump. Pipe should be tested with pressure for leaks.

A foot valve in the suction line will keep the pump primed. The net area of the foot valve should be at least equal to the area of pump suction but preferably larger.

Care should be taken with new suction lines to see that no foreign material, such as chips and rocks, is in the piping or near the entrance to it, as this debris will be drawn into the pump and cause damage and trouble. To protect the pump from being clogged with foreign material, a strainer should be installed with a net area of at least three or four times the area of the suction pipe.

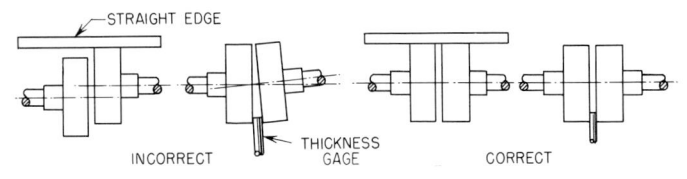

Fig. 2-2 Checking coupling alignment.

Centrifugal Pump Maintenance 8-9

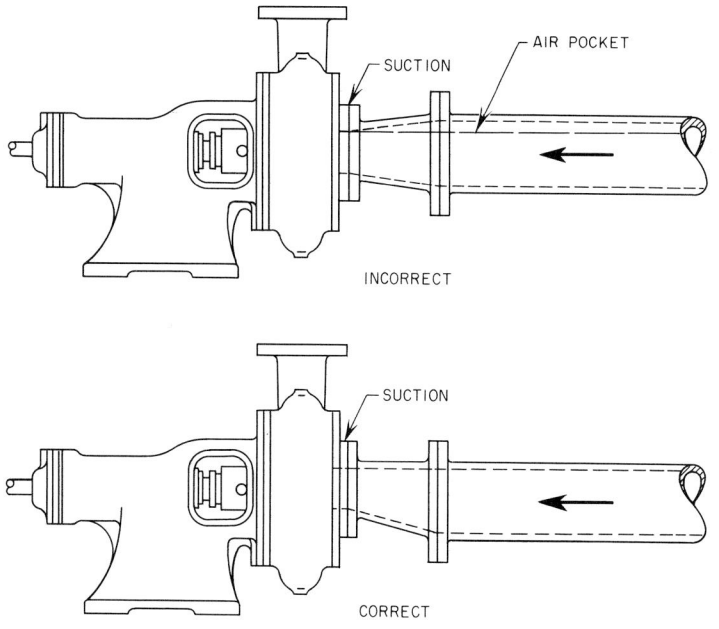

Fig. 2-3 Correct and incorrect methods of installing piping at pump.

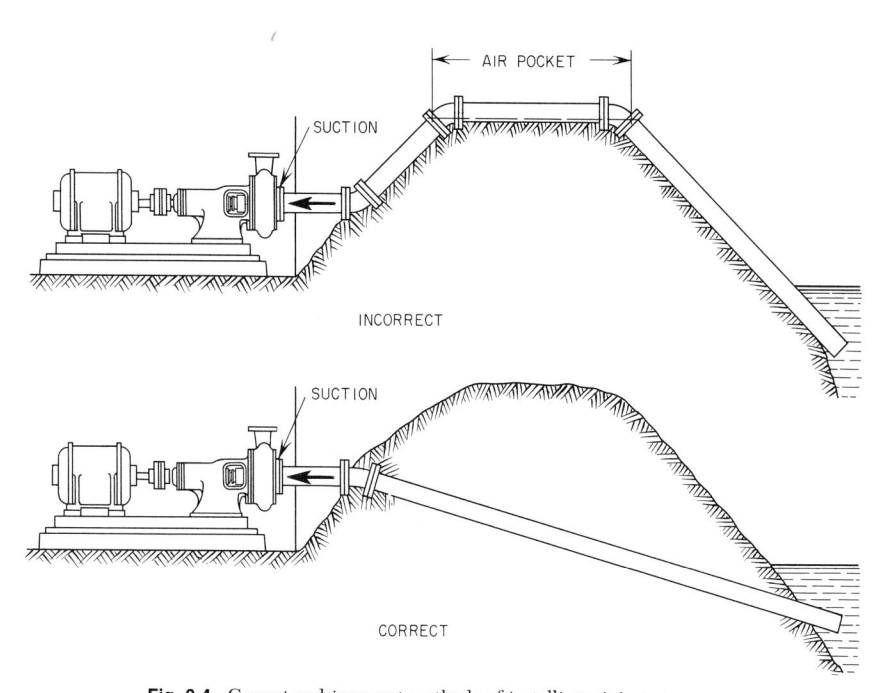

Fig. 2-4 Correct and incorrect methods of installing piping at pump.

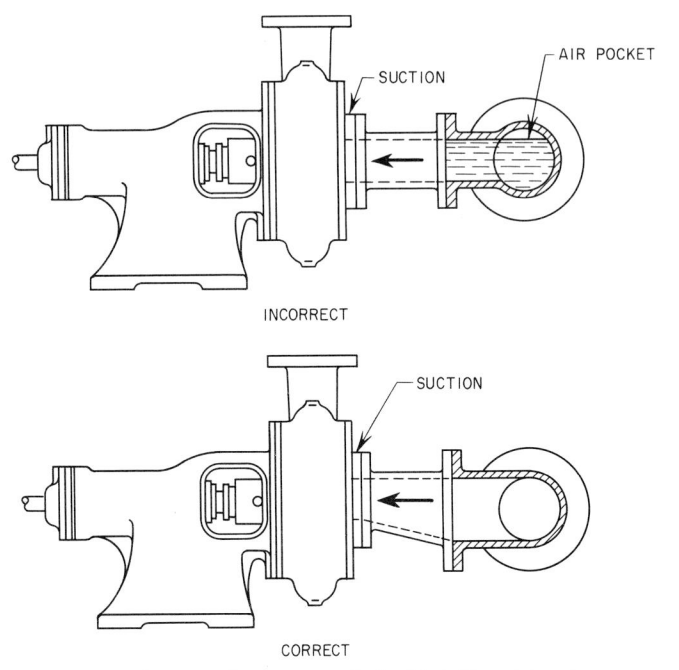

Fig. 2-5 Correct and incorrect methods of installing piping at pump.

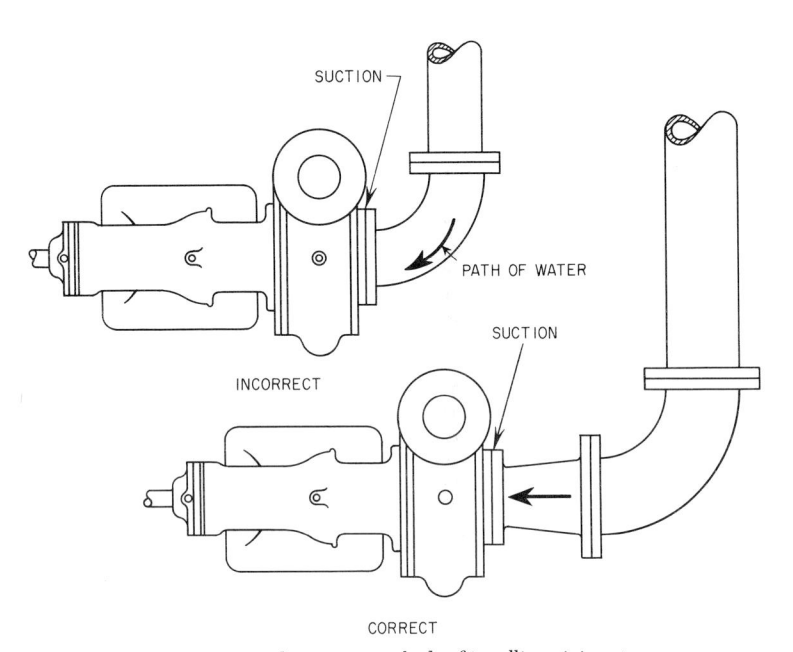

Fig. 2-6 Correct and incorrect methods of installing piping at pump.

Final check on alignment Check the alignment after the piping has been completed, using the straightedge and thickness-gauge method. As the unit has been aligned before completing the piping, the chances are that piping strains are the cause of any misalignment found, and changes should be made accordingly. If the stuffing box is properly adjusted and the pump drives are properly aligned, the unit can be turned over easily by hand.

Rotation The pump must be run in the direction indicated by an arrow on the casing, which is always toward the discharge nozzle. Rotation, right-hand or left-hand, is determined by facing the pump from the drive end. See Fig. 2-7, showing rotation toward the discharge nozzle and the rotation of the impeller vanes indicated by dotted lines. You will note that the impeller rotates in the direction away from the vane curvature.

Starting Fill the pump with water (this is called *priming*). Before starting a centrifugal pump, the casing and the suction pipe must be completely filled; unless this is done, the pump will not operate, as air will be pumped instead of water. Centrifugal pumps can be primed in three ways:

1. By filling the casing and suction pipe with water and holding the water with a foot valve
2. With a vacuum pump
3. With a steam, air, or water ejector

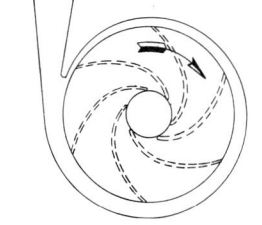

Fig. 2-7 Direction of rotation of inpeller vane.

When starting the unit, the discharge valve should be set so that the least load is thrown on the driver when the pump is started. For radial or Francis-type impellers this occurs at shutoff or when the discharge valve is closed, and for mixed-flow or propeller-type pumps the valve should be fully open. Opening a closed valve should be gradual to avoid throwing a large sudden load on the driver and to prevent a sudden surge in the discharge line.

Stopping Before stopping the prime mover, the discharge valve should be in the same position as when starting, so that less horsepower is dropped from the line and any sudden surges in the pipe system are avoided.

Locating causes for faulty operation In operating a centrifugal pump, apparently serious troubles may arise, but close and careful inspection usually will reveal the fault to be some minor oversight, and investigation for irregular conditions should be made. See Table 2-1.

Belt drive When installing a pump for V-belt drive, belts must be in perfect alignment; any slight misalignment will cause excessive belt wear. This will shorten materially the life and use of the belts. V-belts should be only tight enough to prevent slippage.

Bearings It would be impossible to overemphasize the importance of proper lubrication of bearings. For grease-lubricated ball bearings, it is recommended that regular ball-bearing grease be used. A good general-purpose grease for ball-bearing lubrication has the following characteristics: It is clean and neutral, with a mineral soap base free from acids, alkalies, fillers, or impurities. It has good film strength and a consistency of about No. 1 to No. 2, which is a little stiffer than petroleum jelly. It should be chemically stable, and there should be no separation of the oil from the soap base in the container or during use. It does not oxidize from standing or in service. It has low internal friction, good adhesive qualities, is water-resistant, and has an operating temperature range from -40 to $200°F$.

Ball bearings require only a small amount of lubricant, and the lubrication intervals generally are long. How long a bearing can run without grease being added or replaced depends upon the grease properties, the size and design of the bearing and housing, the speed, and other operating conditions. It is not possible to establish any general rule as to when new grease must be added. The reason for this is that grease in the bearing does not suddenly lose its lubricating ability or life; rather, it is reduced gradually. For pumps operating under severe service, perhaps greasing is required every 3 months; and for normal service, 1 year. To be on the safe side, the addition of grease should be determined from experience.

Heating of bearings invariably means too much grease instead of an insufficiency, and careful inspection to determine the trouble should be made before more grease is added.

Great care should be exercised to keep the bearing housing immaculately clean, and only clean grease should be used. Under no circumstances should grease which has been

TABLE 2-1 Locating Causes for Faulty Operations

Trouble	Cause	Correction
No water being delivered	Pump may not be primed.	Refer to text paragraph on starting.
	Speed may be too low.	Check whether motor is directly across the line and receiving full voltage. In case of steam turbine check governor and determine if receiving full steam pressure.
	Discharge head too high.	Check operation conditions. See that pipe friction and suction and discharge heads are as specified.
	Suction lift too high.	Check with gauges. Normal suction should not exceed 15 ft.
	Impeller and/or piping may be plugged.	Inspect piping, suction strainer, and impeller.
	Impeller may be rotating in wrong direction.	Refer to text paragraph on rotation.
Not enough water being delivered	Air leaks may exist in suction line or stuffing box.	Plug inlet and put line under pressure. A gauge in line will indicate leakage with a drop in pressure. A 1% air leak may cause the capacity to decrease 10%.
	Speed may be too low.	Refer to text paragraph on starting.
	Discharge head may be higher than anticipated.	Check operating conditions. See that pipe friction and suction and discharge heads are as specified.
	Suction lift may be too high.	Check with gauges. Normal suction should not exceed 15 ft.
	Impeller or suction line may be partially plugged.	Inspect piping, suction strainer, and impeller.
	May not be sufficient suction head for hot liquid.	Hot liquids in almost all cases must flow by gravity and have sufficient head or submergence to eye of impeller. Refer to pump manufacturer for complete information on suction piping, size and type of liquid, and amount of submergence available.
	Wearing rings may be worn.	Refer to text paragraph on wearing rings.
	Impeller may be damaged.	Repair or replace.
	Foot valve may be too small.	Inspect. Net area should be at least equal to area of pump suction but preferably larger. Suction-strainer area should be at least three or four times the area of suction pipe.

Centrifugal Pump Maintenance 8-13

Trouble	Cause	Correction
Not enough pressure.	Casing packing may be defective.	Replace all worn packing.
	Foot valve or suction opening may not be submerged enough.	Submerge entrance of suction pipe at least 3 ft below surface of the liquid.
	Speed may be too low.	Check whether motor is directly across the line and receiving full voltage. In case of steam turbine, check governor and determine if receiving full steam pressure.
	May be air in the water.	Plug inlet and put line under pressure. A gauge in line will indicate leakage with a drop in pressure. A 1% air leak may cause the capacity to decrease 10%.
	Wearing rings worn.	Refer to text paragraph on wearing rings.
	Impeller damaged.	Repair or replace.
	Casing packing defective.	Replace all worn packing.
Pump works for a while and then loses suction	May be a leak in the suction line.	Plug inlet and put line under pressure. A gauge in line will indicate leakage with a drop in pressure. A 1% air leak may cause the capacity to decrease 10%. (An 8 to 10% air leak will cause pump to lose its prime.)
	Water seal may be plugged.	Inspect line and position of seal cage in stuffing box.
	Suction lift may exceed 15 ft.	Check for obstruction in suction line and for low water level.
	Air or gas may be found in the liquid.	Vent suction back to source of supply.
Pump takes too much power	Speed too high.	Check speed of driver or, in case of belt drive, sheave or pulley diameters.
	Head is lower than rating, and pump capacity increases.	Have pump manufacturer calculate impeller diameter required, and then turn impeller outside diameter.
	Liquid may be heavier than water.	Check the specific gravity and also viscosity of the liquid.
	Mechanical defects such as a bent shaft may be present.	Check runout of shaft. Total runout allowed depends upon pump design and speed. Approximately 0.003 in. for high-speed and 0.006 in. for slow-speed units.
	Rotating elements may be binding.	Check for too tight stuffing boxes, wearing-ring fit, and defective packing.

used before be applied. Foreign solids or liquids invading the housing can completely ruin the bearings in a short time. It is important to use clean instruments and cloths when cleaning housing. The housing should be flushed clean, using gasoline or a high grade of water-free kerosene.

For oil-lubricated ball bearings, mineral oil of the best quality should be used, such as automobile- and aircraft-engine oil of the better grades. For general use, an SAE 30 oil usually will be satisfactory.

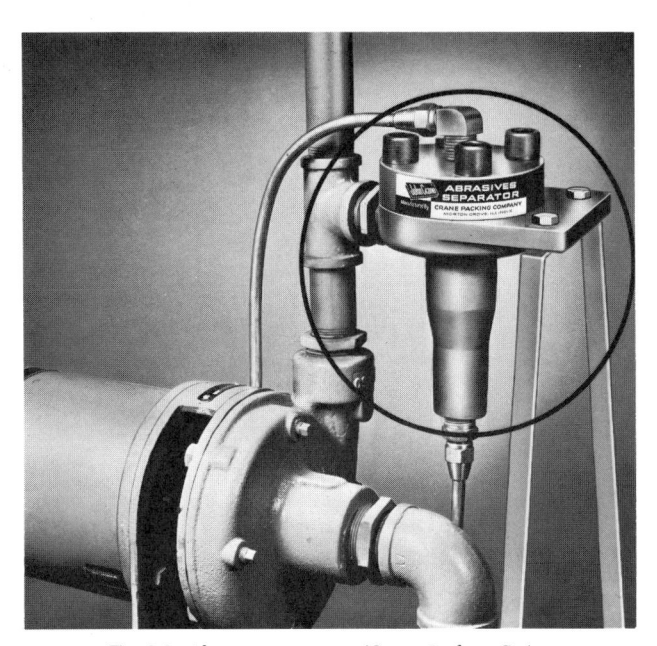

Fig. 2-8 Abrasives separator. *(Crane Packing Co.)*

Bearing housings with oil-bath lubrication or with an oil sump which is to be filled to a given level ordinarily are equipped with oil gauges. Oil is added when the oil level, owing to loss, has dropped below the established low limit. In general, the oil level should never reach higher than the center of the lowest rolling element when the bearing is not rotating.

How soon the oil must be drained and new oil added depends on the operating conditions. For temperatures below 120 to 140°F, if there is no contamination, a lubrication interval of 1 year can be considered normal. For higher temperatures, the oil must be replaced after a shorter time, ordinarily 2 to 3 months.

For oil-bath lubrication, drain off the old oil and flush the bearings out with an oil of low viscosity.

Wearing rings Wearing-ring clearances should be checked from time to time, depending upon the liquid handled. Liquids containing gritty or corrosive materials may make monthly inspection necessary, whereas when handling clear cold water, annual inspection may be sufficient.

When the wearing-ring clearances increase, a loss in capacity and head is caused. If the clearance is approximately twice the original, or if the loss in capacity and head does not meet requirements, it is time to replace the rings.

Seal and packing protection When it is not possible to use an outside source of clear cold water to protect seal faces, packing, and shaft sleeve, an abrasive separator can be employed. It provides a simple yet extremely effective method for keeping dirt and other abrasives from the sealing faces—a method that can increase seal or packing life many times over to greatly reduce replacement and maintenance costs.

Connected into the discharge side of the pump (see Fig. 2-8), the separator takes the laden fluid and removes these foreign bodies completely. The clear fluid is then injected into the gland housing over the seal faces or in the packing-seal ring area to keep them free of abrasives.

This is particularly important in plant start-ups where foreign particles are invariably present in the liquid. It also permits the taking of service water from streams, rivers, lakes, and other sources containing abrasives without the usual wear to seals or packings.

Protection from dry running To protect mechanical seals, impellers, rings, and other sensitive pump components against damage as the result of dry running, an automatic shutdown device is employed. A typical pressure switch with wiring diagram is illustrated in Fig. 2-9.

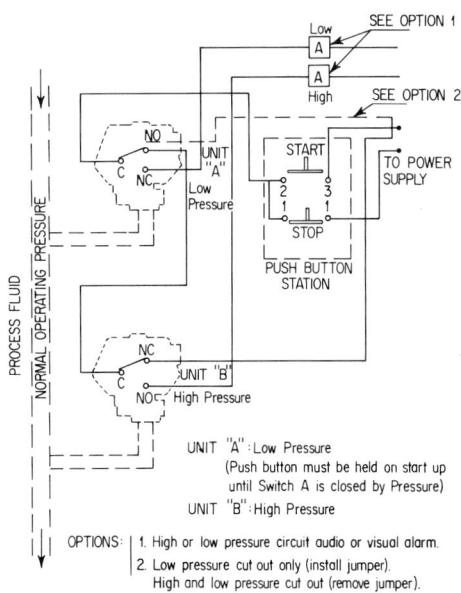

Fig. 2-9 Pressure-sensitive switching device for protecting mechanical seals. (*Crane Packing Co.*)

It can be readily installed in a system so as to shut off operation instantly and/or actuate an audio or visual alarm when a pressure drop occurs.

The pressure-sensitive switch is especially valuable in systems engaged in batch system transfer cycles, loading or unloading tanks and vessels, or similar operations. It also protects against maintenance failure in systems dependent on flow through a filter or where an independent flushing source is used.

Shaft sleeves Shaft sleeves should be replaced when it becomes difficult to control leakage without tightening the glands excessively. When pumping water containing gritty particles, such as sand, silt, or slurries, the life of sleeves can be increased by the following procedure:

1. Plug the water-seal passage to the stuffing box, as illustrated in Fig. 2-10 for single suction pumps.

2. Provide either an outside source of clear cold water or grease lubrication. The pressure for sealing and flushing must be slightly greater than stuffing-box pressure. Care should be exercised in determining this, as some types of pumps have stuffing boxes operating at pressures approaching the discharge pressures.

3. Select the sleeve material for the best wearing qualities. Special materials with hardness of approximately 400 to 450 Brinell are available.

Figure 2-11 illustrates a typical water-seal arrangement for the stuffing box of horizontal split-case pumps. The stuffing boxes on single-stage horizontal split-case pumps operate under suction pressure. The above procedure is applicable for horizontal split-case pumps except for the substitution of lubrication. If the pump is operating under a suction lift, air leakage into the pump through the stuffing box may be expected unless seal pressure is provided. Therefore, an outside source of water with pressure slightly greater than suction pressure is required. In cases where an outside source of clear water is not available, an alternate method would be to install a filter in the water-seal tubing lines.

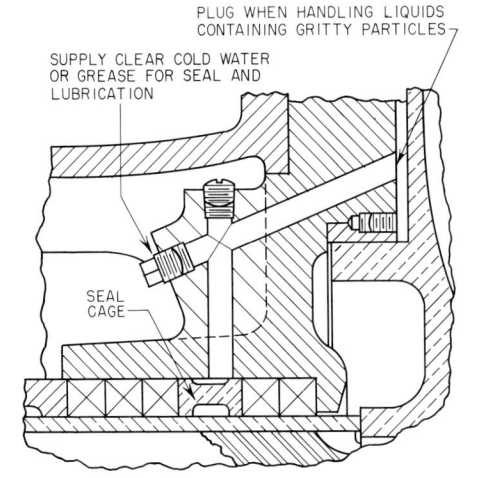

Fig. 2-10 Plug water-seal passage as shown when pumping water containing grit.

Packing The packing normally recommended for clear cold water is long-fiber asbestos, square-braided and well impregnated with oils and graphite. When handling liquids other than water, special packing is required. Consult the pump manufacturer or a manufacturer of packing for recommendations.

The following procedure should be followed in repacking a pump: After the glands have been removed, the packing, cut to the proper length and compressed just enough to slide readily without being smashed while placing, is inserted into the stuffing box. Pressure with the hand and fingers should be sufficient for pushing the rows of packing into place. If it is not, either the packing is too large or some obstruction exists. The rings are spaced so that the splices are staggered. After all rings are placed, glands are put into position and inserted into the stuffing box tight enough to permit just drops of water to drip out per minute. This slight amount of water helps to lubricate the pump shaft at the packing joints.

The packing should be replaced after it becomes hard and tends to score the shaft. Also, it is good practice always to fully repack the pump; never add only one or two rings. Be sure the water-seal cage is withdrawn to remove packing at the bottom of the stuffing box. Make note of the number of rings so the water-seal cage will be installed in the correct position. This must be in line with the water-seal passage as shown in Fig. 2-10.

Mechanical seals Mechanical seals are steadily gaining in popularity and acceptance and, in many cases, replacing packing as standard equipment. They are particularly recommended when absolute control of leakage is required. When properly installed, mechanical seals provide a sealing method which eliminates the human element. The

Centrifugal Pump Maintenance   8-17

only maintenance required is lubrication for seal faces, which may be either oil or grease, depending upon the design and service of the seal. Method of lubrication and type of lubricant depend upon service and, therefore, the recommendation of the pump manufacturer should be followed.

There are many applications that require no lubrication other than the liquid being pumped.

Because of the numerous types of seals, no general rule for replacing seals can be established. However, extreme care should be exercised and instructions as provided by the pump manufacturer followed carefully. Some seals are installed in seal cavities requiring no adjustment (see Fig. 2-12), while others employing locking collars require

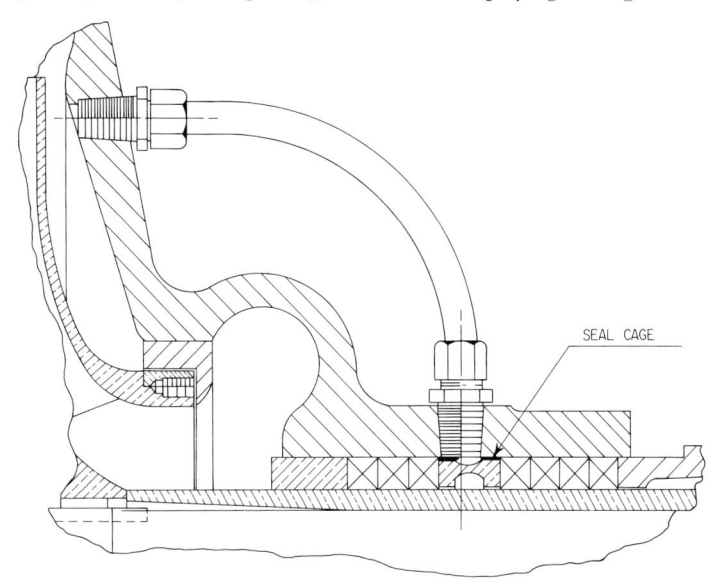

Fig. 2-11 Typical water-seal arrangement of stuffing box of horizontal split-case pump.

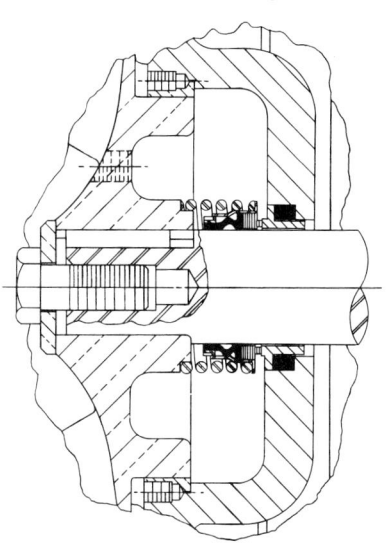

Fig. 2-12 Seal cavity requiring no adjustment.

positioning. Before disassembly, note the position of the seal on the shaft or make sure you know what dimensional setting is required. It is essential that seals be installed in the correct position; otherwise there may be too much load on seal faces, causing rapid wear or too little load and causing seals to leak. Assuming that correct design and materials are used for the application, the seal should give carefree service. For severe service, an average seal life of 3 to 4 months may be expected, and for clear liquids approximately 24 months.

Chapter **3**

Concrete Pump Maintenance

ROBERT P. WEATHERTON
General Manager, Concrete Pump Division,
Challenge-Cook Bros., Inc.
Industry, Calif.

RECOMMENDED CONCRETE PUMP MAINTENANCE

Operating conditions vary so widely that to recommend one schedule of preventive maintenance for all concrete pumps is not possible. Yet some sort of inspection and lubrication on a regular basis along with accurate record keeping must be planned and followed. Both piston-type pumps (Figs. 3-1 and 3-2) and squeeze-type pumps (Fig. 3-3 A and B) are practical for concrete delivery.

Observing the pump manufacturer's recommended maintenance procedures—with regional adjustments to accommodate geographic location, weather conditions, and materials available for pumping—is one of the best ways to keep a pump in good working order and thereby prevent costly breakdowns.

As for daily maintenance, a pump must be thoroughly cleaned after each pumping operation and it should be checked for any structural damage or failure, hydraulic leaks, or evidence of excessive wear. Minor routine adjustments can be made at this time.

A record of operating hours and cubic yards of concrete pumped should be kept in a log book. Maintenance of valves, cylinders, and hydraulic components should be programmed according to a fixed number of pump operating hours. As specific periods of operation elapse, the pump should be brought in for maintenance of these various items. Proper lubrication is essential to good maintenance. (Fig. 3-4.)

It is a good idea to establish a pump utilization factor based on the time a machine is actually pumping. For example, some pumpers consider operating 5 hr per day, 3 days per week to be 100 percent utilization. Suppose they normally conduct their major maintenance work once a month. This maintenance routine is governed by the utilization factor. If the pump is utilized 200 percent instead of 100 percent, then major maintenance is conducted twice per month. Likewise, 50 percent utilization would cut back major maintenance to once every 2 months.

Coordinate the maintenance program to meet varying degrees of pumping severity. Must the pump handle long distances at high pressures? Is the aggregate harsh? Is the gradation poor? All these items contribute to a high degree of severity. On the other hand, pumping a good mix design right off the end of the boom develops a low degree of severity. Don't expect a bimonthly maintenance program that is geared to handle end-of-the-boom pumping to preserve the life of a pump when it is being used to pump 80 yd/hr through 700 ft of pipe 8 hr a day. Make continual adjustments for the basic scheduled maintenance program, pump utilization factor, and degree of pumping severity.

A permanent record should be kept of all periodic inspections and maintenance. Documentation starts with written operator reports that record daily maintenance per-

8-20 Pump and Compressor Maintenance in Construction

formed by the operator, as well as any pump malfunctions and other observations. The operator can provide a wealth of information with very little paperwork if the report form includes such questions as

1. How many hours did the pump operate today?
2. How many cubic yards were pumped?
3. Does this machine require any maintenance, such as structural or fabrication repairs?
4. Was there anything abnormal—noise, leaks, or incorrect pumping action?
5. Was this pour at a fast or slow rate?

The last item (5) is one very few people are even aware of. But it is as important as the other questions. With the capacity of concrete pumps currently available, rate of placement plays an important part in establishing the frequency of regularly scheduled maintenance. Pumping 5000 yd at 80 yd/hr might very well create system (pipeline and hose) wear equal to 10,000 yd pumped at 40 to 50 yd/hr.

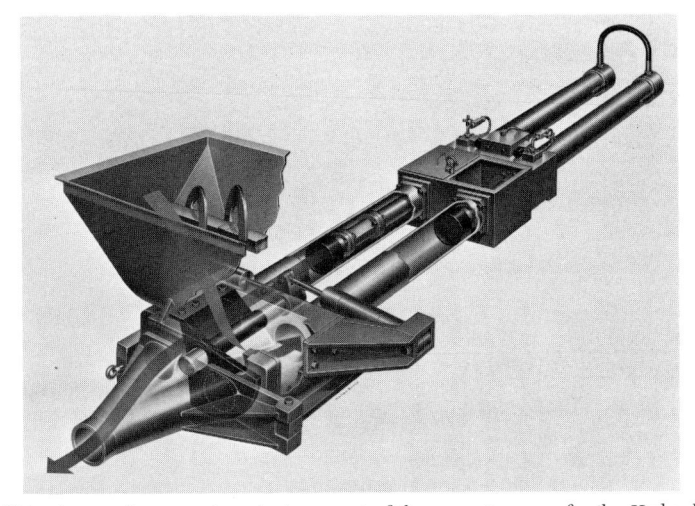

Fig. 3-1 Piston-type units are an important segment of the concrete pump family. Hydraulic rams drive rubber pistons into material cylinders. The material is drawn in from remixing hopper through a valve arrangement that alternates position in sequence with the strokes of the pistons. Note the water box which provides clean water to lubricate both pistons and cylinders.

Fig. 3-2 In service, the trailer-mounted piston-type concrete pump affords a convenient and compact tool.

Concrete Pump Maintenance 8-21

Operator reports become part of a pump's history record, which is maintained by pump serial number and should also contain shop repair work orders, part replacement records, and pump usage records. The parts book for the specific serial number pump and engine should be kept in this file.

A review of all pump history records will help in the prudent stocking of spare parts and

25–27 inches of vacuum inside pump chamber immediately restores pumping tube to normal shape permitting a continuous flow of concrete

ROTATING ROLLERS   MATERIAL HOSE
PRESSURE
RE-MIXER HOPPER
PUMPING CHAMBER
SUCTION
PUMPING TUBE

Rollers squeeze concrete through pumping tube into material hose

Rotating Blades assist concrete into pumping tube

(A)

(B)

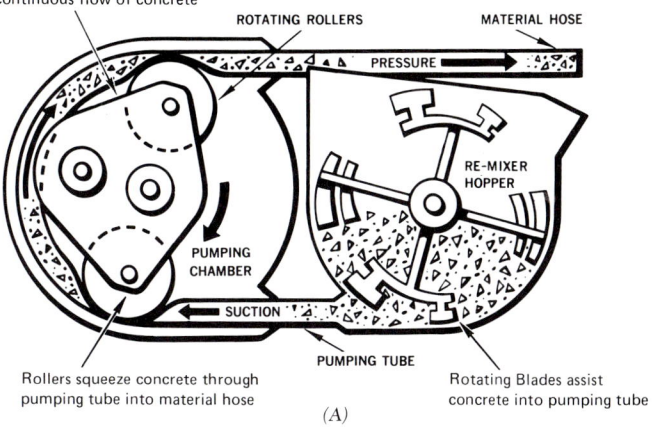

Fig. 3-3 A squeeze-type concrete pump, shown here (A) in cross section and (B) in sk... view, affords a second practical approach to the problem of concrete delivery.

spare assemblies. Quick replacement of a complete major assembly by a st... by spa... may prove much more economical (despite the initial cost of acquiring the as... bly) th... keeping the machine out of service while the assembly is being torn dow... a routi... overhaul or maintenance check. Each machine should be equipped with emerge... parts kit at all times.

RECOMMENDED MAINTENANCE PROCEDURES

The following procedures can help prevent unnecessary downtime and extend the life of concrete pumps, placement booms, steel pipeline, rubber placement hoses, and related pumping accessories.

Agitator shaft bearings and seals These parts can be costly to replace if they are not properly lubricated. Normally, seals on ends of the agitator shaft should be lubricated twice a day. Pack grease heavily behind the seals at all times and check the condition of

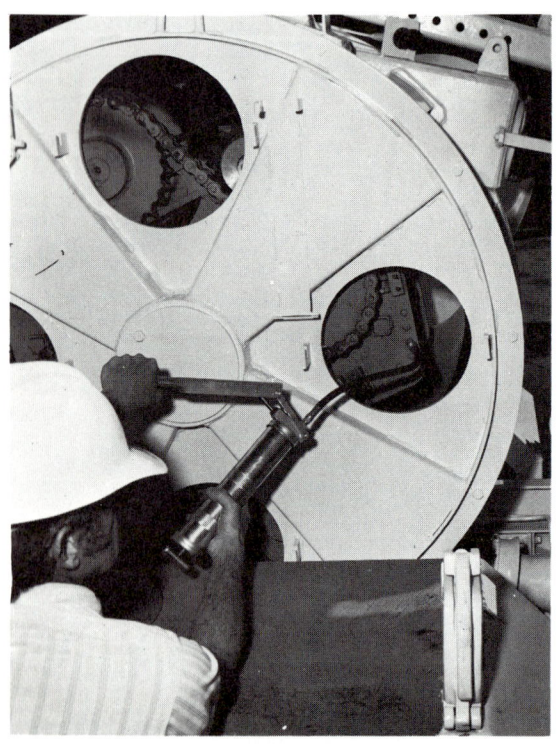

Fig. 3-4 Lubrication is vital to concrete pump life and performance. The manufacturer of this unit has grouped twelve of the pump's fourteen grease fittings in one area. The lubricant is distributed through copper feed lines to the bearings.

the seals frequently. This procedure is important to keep grout from getting into shaft bearings and shortening their lives.

It is not unusual for an excessively worn agitator blade to break off and travel through a pump system, only to become lodged in an elbow. Some have been wedged so tightly as to require a cutting torch for removal. This problem is easily eliminated by periodic inspections and by maintenance of 1 1/2 in. clearance between the hopper barrel and the agitator. Paddle supports are also subject to wear and should be built up to ensure good operation. (Fig. 3-5.)

Piston heads The life of concrete pistons in a piston-type concrete pump can be greatly extended by adhering to good pumping practice. The pump should never be operated without an adequate supply of water in the water box. Running at full stroking speed without water will destroy a set of pistons. The water in the water box should be changed often and must be kept clean. The concrete cylinders should be hosed down before initially stroking the pistons. (Fig. 3-6.)

Inspect piston heads often. Concrete that accumulates on the piston face will cause excessive wear to sealing edges. A worn piston that allows sand and cement to pass will rapidly wear the cylinder wall. This sand and cement will also plug the lubricating water system, causing the pistons to cycle dry and burn. When installing new pistons, always lubricate the pistons and the cylinder walls.

Pump tubes The pump tube in a squeeze-type concrete pump should be removed if the pump is to remain idle, thereby preventing the formation of a permanent set under the drive rollers. Tube life will be increased and replacement normally takes only a few minutes. (Fig. 3-7.)

Procedure for cleaning a squeeze-type pump is simple (Fig. 3-8). Following clean-up in freezing weather, the pump tube should be removed to prevent the possibility of water freezing in the bottom of the tube. The tube should then be stored in a warm, dry location to keep it pliable and ready for reinstallation.

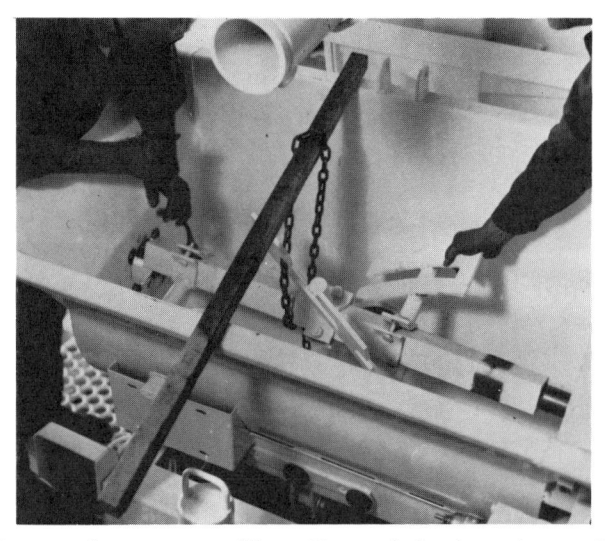

Fig. 3-5 Replacement of worn remixer paddles and hopper shaft seals must be part of any scheduled maintenance job.

Fig. 3-6 Pistons for the concrete cylinder.

Fig. 3-7 To install a pump tube into a squeeze-type pump, simply insert the tube in the discharge opening and let the running drive rollers work it into position. Water is used to lubricate the tube.

Fig. 3-8 Cleaning a squeeze-type concrete pump is simple and easy. Once concrete is exhausted from the remixing hopper, wash the residue into the pump. Then shut it down, insert a rubber sponge into the intake, and partially fill the hopper with clear water. Restart the pump and let the sponge drive the remaining concrete in the tube and the discharge line out of the system. The clear water following the sponge acts as a rinse.

Extreme pumping pressures Any high-pressure pump that is run at pressures above those specified by the manufacturer is capable of bursting hose, pipe, and reducers. To be safe, use pipe and hose that have a higher pressure rating than the maximum pressure rating of your pump. Also, regarding excessive pump pressures, keep in mind that nobody ever pumped out a plugged pipeline! Once the line becomes plugged, it is necessary to go in and clean it out; it can't be pumped out. Revving the engine and kicking the pump into gear will only damage the pump and place undue stress on all the fittings. Likewise, running the pump at pressures below those specified by the manufacturer may result in sluggish performance.

Figures 3-9 through 3-13 illustrate some special maintenance procedures for piston-type pumps.

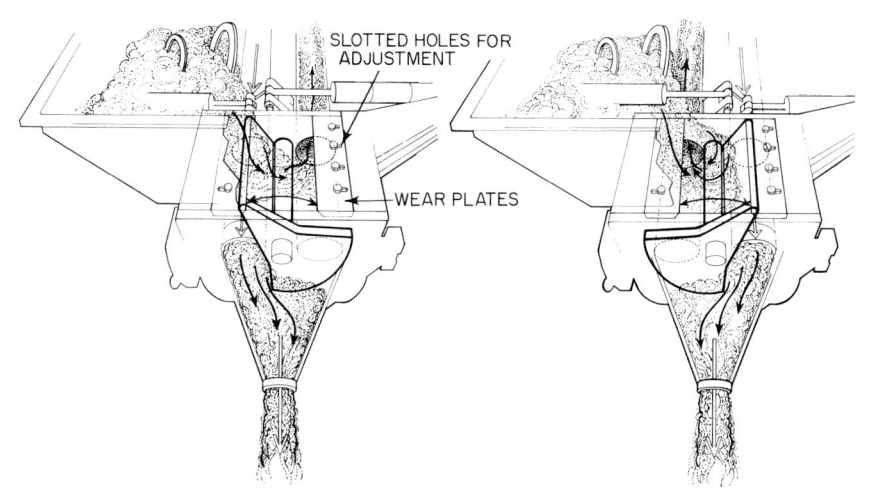

Fig. 3-9 In a conventional piston-type pump, the inherent engineering difficulty in pumping a plastic concrete mixture of cement, sand, gravel, and coarse aggregate is a function of valve arrangement. Most modern units use a simple single-piece valve that simultaneously controls intake and exhaust on the same stroke. Adjustable wear plates help to take up the inevitable wear of the harsh mix on the valve. Note the slotted holes and set of four bolts holding each wear plate in position and making adjustment easy.

Hydraulic lines and systems Check hydraulic lines and fittings for leaks both during and after every pumping operation. Keep all connections tight to avoid leaks, but *never* attempt to loosen *or* tighten a hydraulic line when there is pressure on the hydraulic system. Tightening a line that is leaking because of a split fitting will only make it worse. Protect hydraulic lines from contact with the pump to prevent chafing that will result in wear.

Any repair or replacement of hydraulic parts should be handled by experienced personnel only. If performed by unqualified people, attempts to repair even the smallest problem may lead to needless complications and more expense.

Hydraulic oil filtration Many pump owners regularly remove all the hydraulic oil from their pumps once a month or every 200 pumping hours and replace it with oil that has been completely filtered through a 10-micron filter system. Usually located in the maintenance shop, this separate oil-filtering system incorporates three holding tanks. Old oil is pumped into the first tank, filtered into the second tank, and then transferred to the third tank for storage until it is recirculated into the concrete pump's hydraulic system. (Fig. 3-14.)

Oil that only looks clean may leave contaminants when allowed to remain in these filtering tanks for 4 days. Because water cannot be filtered from oil economically, it is a good idea to allow the oil to remain in the first tank until all water has settled to the bottom of the tank before starting the filtering process.

Pump system versus delivery system Today, concrete pump technology surpasses related delivery system technology. Looking at pump maintenance and delivery system

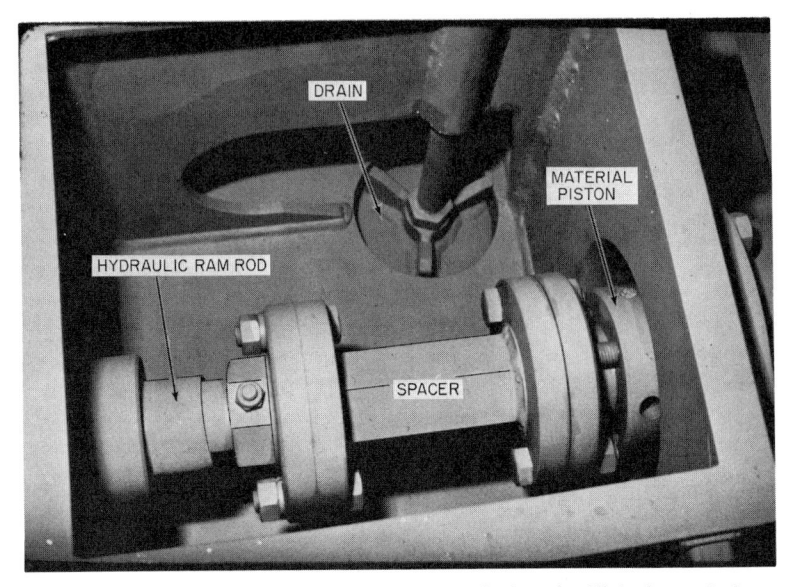

Fig. 3-10 Interior of a water box on a piston pump. Drain, flush, and refill the box with clean water every shift. Keeping the percentage of suspended solids in the water low helps to prolong the life of the pistons, cylinders, and polished ram shafts.

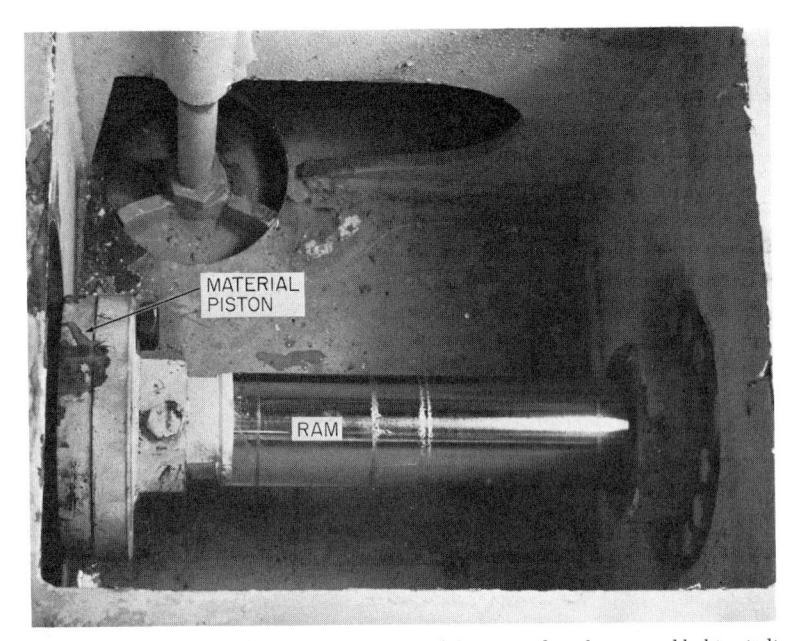

Fig. 3-11 By unbolting the spacer and removing it, and then extending the ram and bolting it directly to the piston (as shown), the ram can be used to pull the piston out of the material cylinder and into the water box where it can be inspected for signs of wear or where it can be quickly removed and replaced if badly worn.

Fig. 3-12 Ram retracted to pull piston into easy access area of water box where it can be inspected or replaced if badly worn.

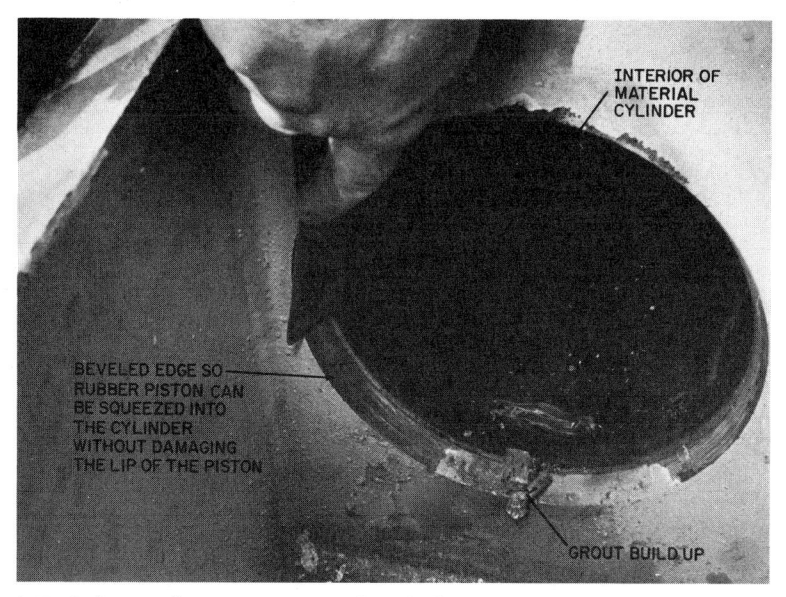

Fig. 3-13 Before installing new pistons in the cylinders, use a putty knife and scrape away any accumulated buildup of grout in the water box so as not to damage the lips of the pistons upon installation.

maintenance as two separate procedures, it can be seen that delivery systems now account for a larger percentage of maintenance costs than in the past.

Steel pipeline versus rubber hose One way to increase the efficiency of a delivery system is to use steel pipeline wherever possible. Instead of using four 25-ft sections of rubber hose, use 75 ft of steel pipeline with 25 ft of rubber hose on the end for ease of placement. Rubber hose develops 2.2 times as much line friction as steel pipeline. This greater line resistance increases pumping pressures. More power is needed to move the same amount of material and this means more wear and tear on the pump. The contractor

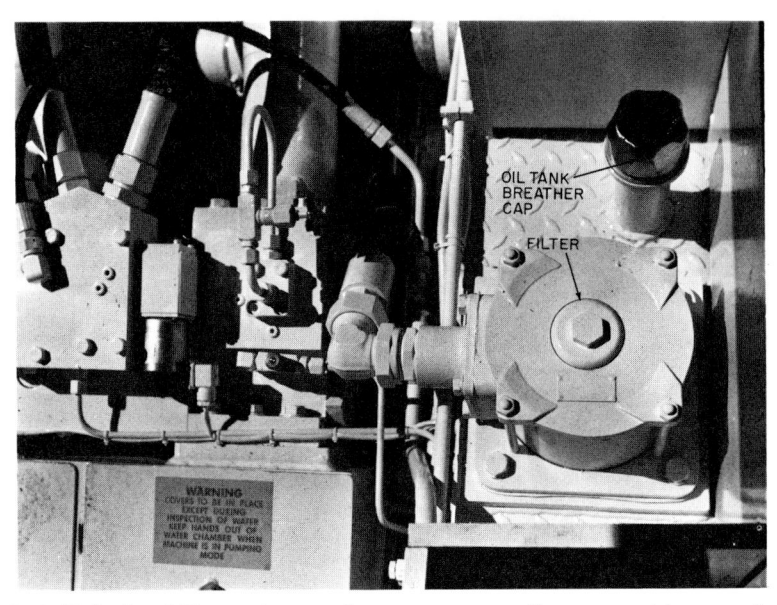

Fig 3-14 Hydraulic oil filters on better-grade concrete pumps will remove particles as small as 10 microns. Frequent cleaning or replacement of the filter element is an important ingredient of sound pump maintenance.

who chooses to pump through steel pipeline instead of rubber hose can pump 2.2 times farther with the same amount of line resistance.

Large pipeline versus small pipeline The minimum-size pipeline must be large enough to accommodate 3 times the maximum-size aggregate to be pumped. (Fig. 3-15.) For example, a mix containing 1 $\frac{1}{4}$-in. aggregate requires at least a 4-in. diameter pipeline (1$\frac{1}{4}$ in. × 3 = 3$\frac{3}{4}$ in.). However, a larger pipeline would be preferable since it causes less wear on the pump. Going from 4-in. line to 5-in. line while pumping material at the same velocity reduces pumping pressure by 50 percent. Smaller-size line might save some hand labor, but the additional pressure it places on the pump and the resultant downtime from plugged lines are more costly in the long run.

Benefits of larger pipe include

1. More volume at the same hydraulic pressure, since that pressure is used to achieve volume rather than to overcome friction.

2. Reduced pump capability requirement for a given job. More jobs can be handled with a given pump.

3. Ability to handle marginal mixes which will not transfer at higher pressures, for example, a lightweight mix where increased pressure merely forces water and cement into porous aggregate and creates a dry packing to plug the line.

4. Ability to pump longer distances.

5. Less maintenance on pump's valve and hydraulic system and much longer life of the pipe itself.

The simple graph in Fig. 3-16 shows the origin of many of these benefits. The relative friction of pipes from 3 to 6 in. is plotted against pump volume, taking a 4-in. pipe as the median. Note that at a given pumped volume the friction of the 5-in. line is *half* that of the 4-in. line.

Care of pipe, hose, and couplings Steel pipeline should always be clean and free of concrete, thin-wall areas, and dents. Rubber hose should be free of ragged interiors, kinks,

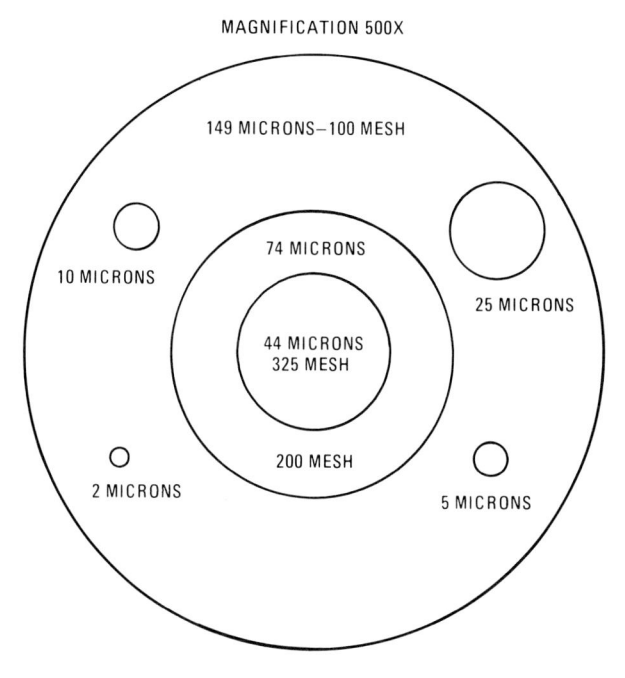

MAGNIFICATION 500X

149 MICRONS—100 MESH
10 MICRONS
74 MICRONS
25 MICRONS
44 MICRONS 325 MESH
200 MESH
2 MICRONS
5 MICRONS

SIZES OF FAMILIAR OBJECTS

SUBSTANCE	MICRON	INCH
GRAIN OF TABLE SALT	100	0.004
HUMAN HAIR	70	0.0027
LOWER LIMIT OF VISIBILITY	40	0.00158
WHITE BLOOD CELLS	25	0.001
TALCUM POWDER	10	0.0004
RED BLOOD CELLS	8	0.0003
BACTERIA (AVERAGE)	2	0.00008

SCREEN SIZES

U.S. SIEVE NO.	OPENING IN INCHES	OPENING IN MICRONS
50	.0117	297
60	.009	238
70	.0083	210
100	.0059	149
140	.0041	105
200	.0029	74
270	.0021	53
325	.0017	44
PAPER	.00039	10
PAPER	.00019	5

Fig. 3-15 Relative size of particles and comparison of dimensional units.

8-30 Pump and Compressor Maintenance in Construction

and deep weather checking. Badly worn hose, with loose fabric or cord hanging inside, increases line resistance, requiring greater pump power to push material through the line. To protect the hose, keep it away from contact with oil and grease. Store hose on full-length horizontal racks to prevent permanent kinks or sets.

Coupling gaskets are normally made of rubber or some pliable synthetic material. If these gaskets are worn, leaks will occur. Therefore, they should be examined periodically and replaced as required. To preserve the gaskets, clean and grease them regularly. Couplings should also be cleaned and lubricated to prevent formation of rust and concrete buildup.

Maintenance of concrete pump booms The concrete pump boom is a working structure. As such, it requires a *mandatory* maintenance program. Failure to safely maintain the boom and its related systems may result in liability losses, should the structural stability fail and cause property damage and possible injuries or loss of life.

Daily inspection (similar to customary crane in-service) should be a thorough procedure in accordance with the manufacturer's recommendations. Once every 2 weeks, the pump operator or maintenance man should physically inspect the entire

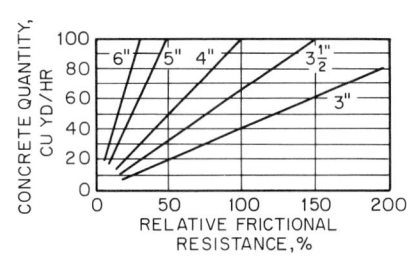

Fig. 3-16 Frictional resistance and concrete quantity effects for various pipe sizes.

Fig. 3-17 Stresses on the boom of a concrete pumping system emphasize the importance of maintenance in this area. The boom here has 110 ft of vertical reach, using a 4-in. slick line.

boom, including the distribution pipe, outriggers, and boom controls. He should note any defects, such as stretch marks in the paint, cracks, or leaking cylinders, enter these items in the pump's maintenance log book, and sign his name. (See Fig. 3-17.)

Once a month, the boom should be load-tested, using a weight equivalent to concrete in the boom plus the tail end hose and using the manufacturer's recommended safety factor. The operation of holding valves must be checked at the time of the load test.

Once a year, the entire boom should be magnafluxed to reveal any cracks beneath the paint, just as crane companies do with their equipment. A pumper who operates cranes as well as concrete pumps with booms had been magnafluxing the cranes but not the booms. But, when one boom developed a visible crack, he had it magnafluxed along with the cranes. Underneath the paint, they discovered that 29 more surface cracks had developed in that one boom during 9 years of operation. Now he magnafluxes all his booms regularly.

Any program to keep boom pumps working must include the operator. No amount of maintenance can keep the booms operational if operators, either through ignorance or lack of concern, abuse equipment.

Boom delivery system To avoid costly on-the-job service calls, a boom delivery system should be inspected periodically. Worn elbows, pipe, and reducers should be repaired or replaced immediately. Elbows may be patched if they are properly bonded. However, if a patched elbow is used very long, pumping pressure may cause a blowout around the patched area. System pipe in a saddle or U-bolt clamp should be reinforced to avoid wear to pipe from constant rubbing against the saddle.

Boom outriggers Hydraulically actuated outriggers have check valves or safety devices to lock them into place. Maintenance should include inspection for bent or twisted cross beams, worn hinge pins, cracked or broken pads, leaks in hydraulic check valves, damaged hydraulic oil suppy lines, and damaged safety lock devices.

Manual pullout-type outriggers must be kept in good working order and cleaned of concrete and dirt. When pullouts are not in good working order, operators often neglect to install safety devices to lock the outriggers in position. Neglecting to use safety locks may cause truck slippage, resulting in outriggers buckling or losing their footing and causing the boom to collapse.

Boom hinge pivots If not properly inspected and maintained, boom hinge pivots may cause serious problems. Pins should be carefully inspected and kept in good condition. As a result of thorough inspection, pins thought to be in excellent condition have been found to be badly worn when pulled.

Extendable jibs Safety pins and clips must be inspected regularly. Most jibs will slide out of the boom if these pins are not in place. The last section of pipe on the jib or boom may be dangerous because it is not supported by the boom. As the section of pipe wears and becomes thin, breakage may occur. This section should be equipped with safety chains and inspected daily to avoid needless accidents.

Remote control units Boom remote controls are often a neglected item. They get yanked, dropped, run over, washed with water, or splashed by spilled concrete. Dropping the control box causes loose connections and bent switches. Is it any wonder that booms sometimes take off on their own?

Control box covers should be removed and inspected for loose wire connections and worn switches. Most boxes are equipped with waterproof gaskets and should be kept tightly fastened to prevent moisture from seeping into the controls.

THE RIGHT COMBINATION

Concrete pumps have become a common tool of the construction industry. They allow the contractor to move concrete efficiently and economically from the truck mixer to the point of placement.

The modern concrete pump is a sophisticated machine. Some people worry that it is beyond the skill of people available to use it. However, well-trained operators have made pumping the best of all high-volume handling systems. Naturally, the best system requires the best people to run it—but it requires far fewer people.

The first requirement for any successful pumping project is a well-trained operator. The second requirement is proper application of equipment—selecting the right pump for the job at hand. The third requirement is a good maintenance program that will ensure top performance throughout a job. And the fourth requirement is a pumpable-mix design. The successful combination of these requirements has made pumping the preferred method of concrete placement.

Chapter **4**

Rotary Helical-Screw Air Compressor Maintenance

HARRY W. ALLEN
Manager, Rotary Compressor Product Maintenance,
Gardner-Denver Company,
Quincy, Ill.

Introduced to the United States Construction market in the early 1960s, the oil-injected rotary helical-screw air compressor has supplanted the reciprocating and other rotary-type compressors in many areas. The advantages of high capacity in a small package, high reliability, low vibration, fewer parts, and ease of installation and maintenance have contributed to the great success of the rotary screw compressor.

This compressor is sold as a unit, a system package with supporting equipment, requiring only air and electrical connection to begin operation (see Fig. 4-1). It is a precision-built machine and should be treated as such. There are specific maintenance requirements, which should be met at regular intervals to ensure satisfactory and continuing operation.

All major compressor manufacturers supply an installation, operation, and maintenance manual with each unit. This instruction book should be read carefully by the personnel charged with operating and maintaining the compressor unit, so that routine maintenance, adjustments, and emergency repairs can be made without calling upon the manufacturer's field service.

Installation   The compressor unit should be installed in a clean, well-lighted, well-ventilated area with ample working space all around and overhead for maintenance. Too often, compressor units are installed so that it is nearly impossible to service air and oil filters and oil separators. Major compressor manufacturers supply outline drawings which show minimum clearances around the unit. These dimensions should be followed closely to reduce maintenance time and cost.

Ventilation.   It is essential to provide adequate cooling air for the motor and air-cooled oil cooler as well as air to the compressor inlet. Figure 4-2 shows a typical compressor unit ventilating system. To estimate the amount of ventilation air required, sum the air flow capacities in cubic feet per minute of the unit components: compressor, fan-cooled oil cooler, fan-cooled aftercooler, and driving motor. It will be noted that the compressor capacity (the air that is required for compression) will be a small part of the total; any ventilation system based on compressor capacity alone would be badly undersized. The manufacturer's instruction manual will list the air flow required either by component parts or a total for the entire package unit.

Foundation.   Requirements for the rotary helical-screw compressor-unit foundation are minimal. Usually a smooth solid surface of sufficient strength to support the weight of the unit is all that is necessary. The unit should be installed as nearly level as possible. Unit base flanges should be installed as nearly level as possible. These flanges should

8-34 Pump and Compressor Maintenance in Construction

be supported as continuously as possible; voids or depressed areas should be filled. Use shims or wedges where necessary to ensure positive contact with the supporting surface. Base-to-floor mounting bolts are not normally required; however, piping rigidity, danger of shifting due to outside vibrations, or accidental movement by passing vehicles may require at least one mounting bolt at each corner of the unit base.

Oil-Reservoir Drains. Drains should be considered when installing a compressor unit since a fairly large amount of lubricant will require periodic removal and replacement. Consult the manufacturer's outline drawing for oil drain locations and plan a draining method. Drain possibilities include portable containers sized to fit under the drain, a sump and gravity line to move the oil to a disposal location, a sump and pump to lift the oil into disposal containers, or a process for pumping the oil directly from the reservoir to a disposal location.

Inlet Lines. Occasionally, inlet lines are required to reach a source of clean, cool, dry air away from a compressor unit. Depending on ease of servicing, the air filter may remain on the unit with the inlet line running from the filter to the air source, or the air filter may be remotely located with the line leading to the unit. The best inlet line is the shortest and most direct possible. Long inlet lines or lines with many bends restrict air flow and increase the horsepower required for compression. A good general rule for adequate inlet

Fig. 4-1 Air-cooled rotary screw compressor.

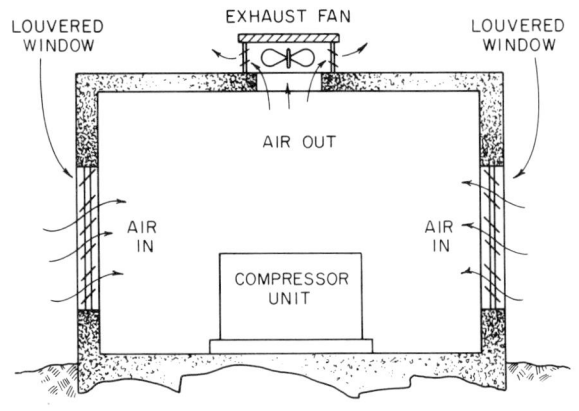

Fig. 4-2 Typical compressor-unit ventilating system.

size is shown in Table 4-1. For inlet lines longer than 38 ft or with many bends, pipe should be sized so that air velocity is within 4000 to 6000 fpm. Metal inlet lines must be clean, free of scale, and coated internally with a moisture- and oil-proof sealer. There is little point in filtering air which will be contaminated by pipe scale or burdening air filters with unnecessary debris as a pipe ages. An inlet line must also be airtight to prevent ingesting fine particles through poorly made connections in the piping or at the compressor and air filter. Heavy-gauge plastic pipes can be used as inlet lines, but care must be taken that the connectors are properly sealed and that the plastic will withstand extremes of temperature and atmosphere where it is used.

Discharge Lines. It is important that the discharge-line connection to the air system be correctly sized. Usually, continuing the unit discharge-pipe size for short distances is adequate. Long discharge pipelines are subject to considerable pressure drop and require special consideration for correct sizing. Tables and equations for correctly sizing discharge air lines are published in handbooks covering compressed air.

Manifolding of several rotary screw compressors into a common header requires a check valve between the manifold and the compressor unit oil reservoir to prevent pressurizing of inactive units. Most manufacturers supply the check valve as part of the unit discharge lines; but, if it is not present, good practice dictates installation of one. If

TABLE 4-1 Determining Pipe Size by Inlet Line Length

Inlet line length, ft	Pipe size
0–10	Same as compressor inlet
10–17	One size larger than inlet
17–38	Two sizes larger than inlet

rotary screw compressor units and reciprocating compressors are to be manifolded into a common system, an adequately sized air receiver should be placed so that each type of compressor is individually piped directly to the air receiver. Pulsations of the reciprocating compressor not damped by an air receiver can cause abnormal vibration and noise in the rotary compressor units. Be sure to install a safety valve in each section of a manifolded system that may be individually shut off from discharging compressed air to the system. All safety valves should be correctly sized for the system operating pressure.

Air Receivers. Normally, a rotary screw compressor unit does not require an air receiver since the discharged air is practically pulse-free and does not require damping. There are some control systems, such as automatic start-stop, which require an air receiver to prevent rapid cycling. Air receivers can also be useful in precipitating small amounts of moisture which would otherwise pass to the air system. If a large amount of moisture is present, an aftercooler with a separator and trap should be used. Figure 4-3 is an ideal system, listing all the components required.

Rotary helical-screw compressor The heart of the rotary screw compressor package unit is the compressor itself. Descriptively, the compressor is a single-stage positive-displacement rotary machine, using meshing helical rotors to effect compression. The actual design of the rotor-supporting structure varies with the manufacturer and occasionally with the application. All screw compressors have the rotors and the rotor chamber, or cylinder, in common. Typically, as shown in Fig. 4-4, rotors are supported and located by antifriction bearings and are enclosed in a cylinder. The cylinder provides fixed porting and is closed at each end by plates which are ported to match the cylinder. While a single-stage machine is by far the most common, two-stage machines that use two rotor sets within two compression chambers in a single cylinder structure are to be found.

Compression Principle. Compression is accomplished by synchronously meshing rotors enclosed in a cylinder. The main rotor has a number of equally spaced helical lobes which mesh with matching helical cavities on the second rotor. The main rotor has fewer lobes than the number of secondary rotor cavities; a common ratio is 4:6—four main lobes and six secondary cavities.

Figure 4-5 illustrates the compression cycle; the view is inverted from a normal compressor position to show both inlet and discharge ports. The air inlet opening is located on top of the compressor near one end. The discharge port is located near the bottom of the cylinder at the opposite end from the inlet opening. The compression cycle

Fig. 4-3 Aftercooler and air receiver system.

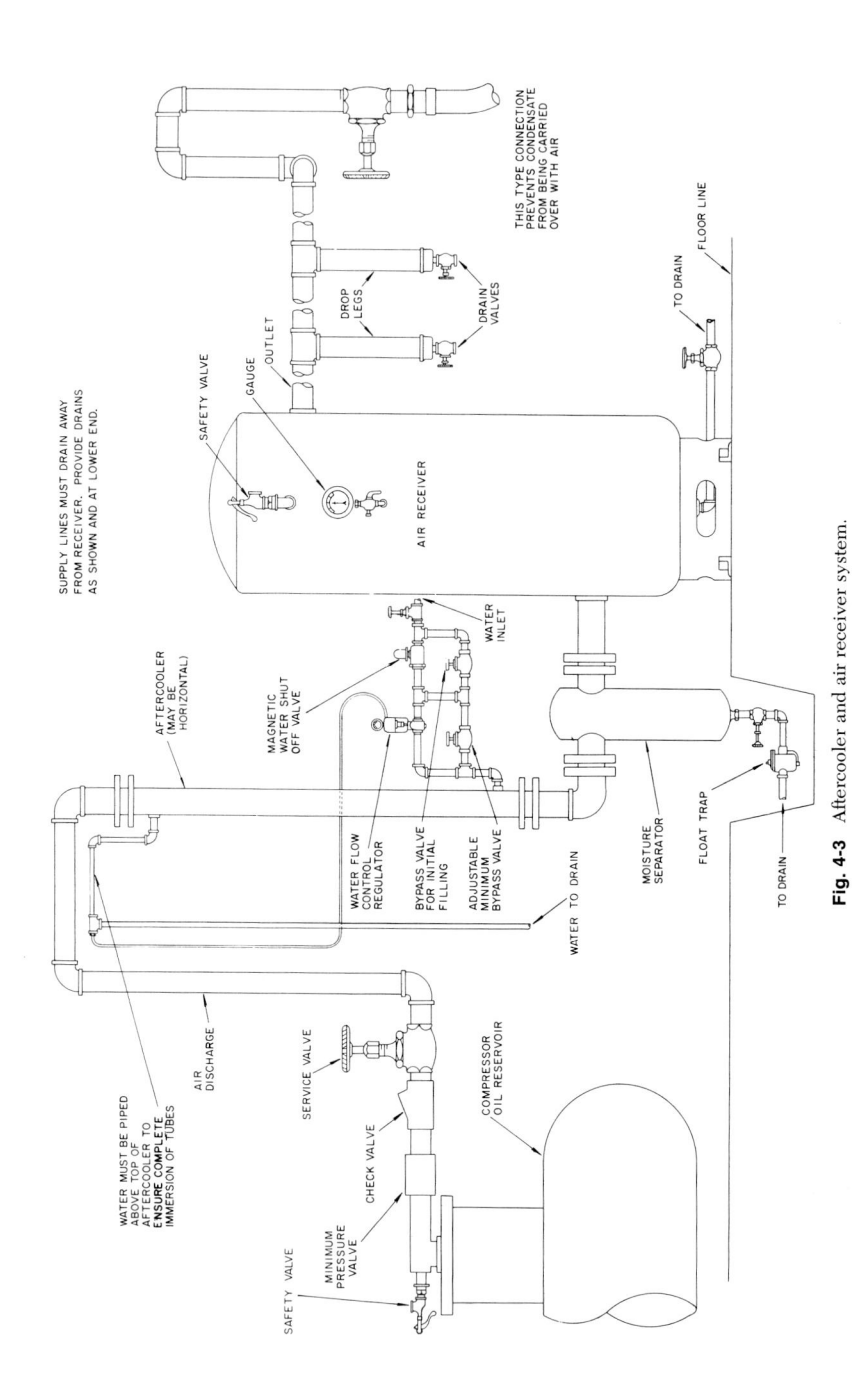

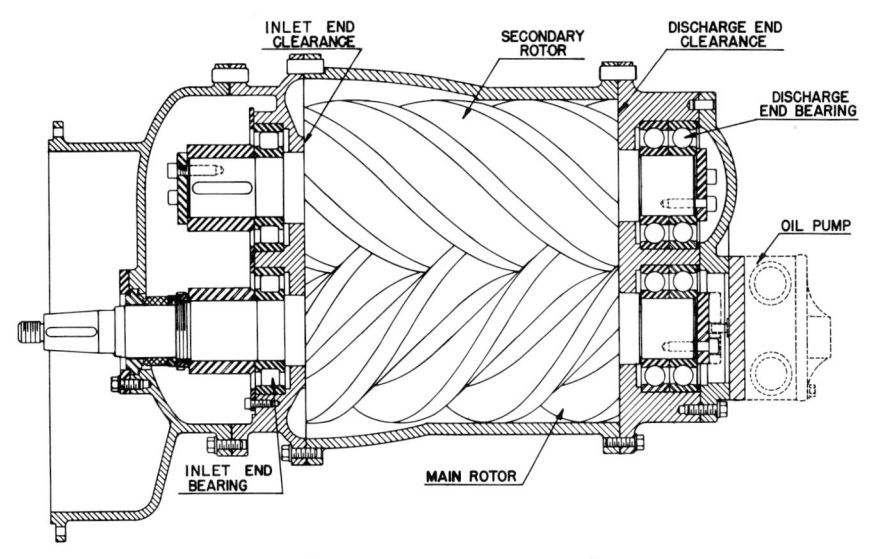

Rotary Helical-Screw Air Compressor Maintenance   8-37

Fig. 4-4 Rotary screw compressor sectional view.

(A)

(B)

(C)

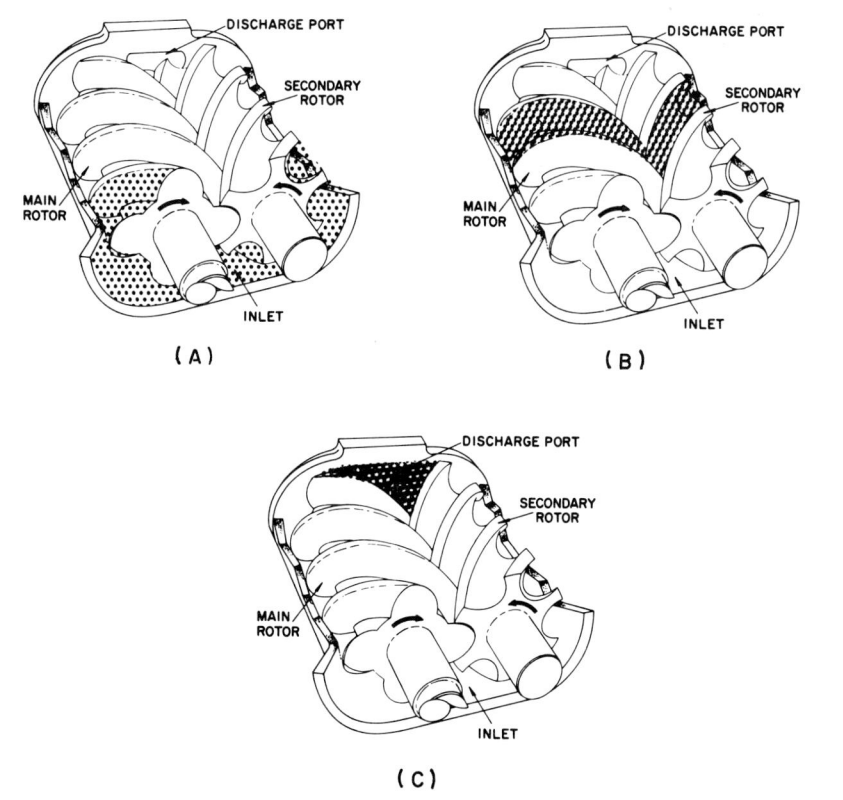

Fig. 4-5 Rotary screw compression cycle.

8-38   Pump and Compressor Maintenance in Construction

begins as rotors unmesh at the inlet port and air is drawn into the spaces between the main rotor lobes and the secondary rotor cavities (see view A, Fig. 4-5). When the rotors pass the inlet-port cut-off, the air is trapped in the interlobe cavity and flows axially with the meshing rotors (see view B, Fig. 4-5). As rotation continues, more of the main rotor lobe enters the secondary rotor cavity, normal volume is reduced, and pressure increases. Oil is injected into the cylinder to remove the heat of compression and to seal the internal

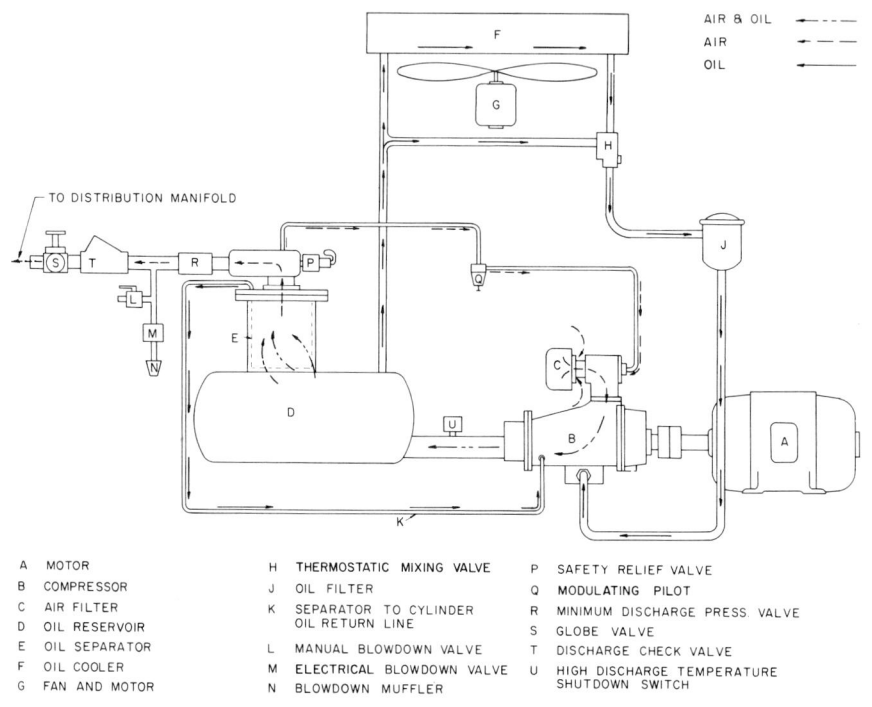

A	MOTOR	H	THERMOSTATIC MIXING VALVE	P SAFETY RELIEF VALVE
B	COMPRESSOR	J	OIL FILTER	Q MODULATING PILOT
C	AIR FILTER	K	SEPARATOR TO CYLINDER OIL RETURN LINE	R MINIMUM DISCHARGE PRESS. VALVE
D	OIL RESERVOIR			S GLOBE VALVE
E	OIL SEPARATOR	L	MANUAL BLOWDOWN VALVE	T DISCHARGE CHECK VALVE
F	OIL COOLER	M	ELECTRICAL BLOWDOWN VALVE	U HIGH DISCHARGE TEMPERATURE SHUTDOWN SWITCH
G	FAN AND MOTOR	N	BLOWDOWN MUFFLER	

Fig. 4-6   Air-oil system for a rotary screw compressor.

clearances. Volume reduction and pressure increase continue until the rotors pass the discharge port and the air-oil mixture trapped in the interlobe cavity is released to the oil reservoir (see view C, Fig. 4-5). Each rotor cavity follows the same fill-compress-discharge cycle in rapid succession to produce a discharge air flow that is continuous, smooth, and shock-free.

Figure 4-6 illustrates a typical, complete air-oil system for a rotary screw compressor. Air enters the air filter and passes through the inlet valve to the compressor. After compression, the air-oil mixture passes into the oil reservoir where most of the entrained oil is removed by velocity change and impingement and drops back into the reservoir. The air and remaining oil then pass through the oil separator; the separated oil is returned to the oil system through tubing connecting the separator and the compressor. The air finally passes through the reservoir discharge manifold and into the air-distribution system.

For lubricating, cooling, and sealing of the compressor, a lubricant is forced by air or pump pressure from an oil reservoir through a temperature-controlled oil-cooling and filtering system into the compressor. A portion of the oil is directed through internal passages to the bearings, gears, and shaft oil seal. The balance of the oil is injected directly into the compression chamber to remove the heat of compression, to seal internal clearances, and to lubricate the rotors.

Compressor Design.   The design of the compressor takes many forms, for example, direct drive through either of the rotors or a variety of drives through a separate shaft connected by gearing to one of the rotors. A typical direct-driven design is shown in Fig. 4-4. Note the main rotor which absorbs about 97 percent of the input power. The secondary rotor, sometimes called *idler,* or *gate,* rotor uses little power, mainly following the main rotor and providing a containing cavity for the air. The main rotor can be visualized as a piston traveling in a cylinder, the secondary rotor, to effect compression.

Large antifriction bearings support and locate each rotor since working clearances must be accurately established and maintained. Clearances must be small within the compression chamber of the compressor. Only a few inches separate zones of large pressure differentials, and high leakage rates past working surfaces would destroy a compressor's efficiency. Discharge end clearances are critical and are usually held well under 0.010 in. between the rotor and endplate faces. Interlobe clearances between meshing rotor surfaces are also critical and are held by close machining tolerances to about the same range as the discharge end clearance in a given compressor size. Cylinder-to-rotor clearances, enjoying the benefit of a copious supply of lubricant from centrifugal force, are less critical but are usually held below 0.025 in. The inlet end clearance, least critical of all because of negligible pressure differentials, may range beyond 0.030 in.

Bearings are lubricated by the same lubricant as the rotors. Because they are load-carring members, they are subject to wear from abrasives and from the effects of low-viscosity oil. For this reason, air- and oil-filter and cooling-system maintenance is most important. Rotors themselves suffer little from abrasives in oil but can contact cylinders or endplates if bearing wear is excessive.

The shaft oil seal is also a wearing part, lubricated and cooled by the lubricant stream. Abrasives or low-viscosity or high-temperature lubricants will shorten the life of the seal. All mechanical seals must seep to be effective, but the seepage is generally not noticeable. When a noticeable leak develops, it usually signals other problems and not only just oil-seal wear.

Note that the oil pump is shown in dotted lines in Fig. 4-4 to indicate it is not used in every design. Some compressor manufacturers use an oil pump to force the lubricant through the system. Other manufacturers rely on air pressure in the oil reservoir to force the lubricant through the system.

Starting the compressor When a new or repaired rotary screw compressor is to be started for the first time, some careful checks and servicing will ensure a smoother start-up experience. The compressor unit has been assembled, serviced, and tested at the factory, but it has traveled some distance and been subjected to changing environments and less-than-considerate handling. Prestart inspection and checks always pay dividends.

1. Read the instruction manual supplied by the manufacturer.
2. Be sure the unit is reasonably level and that connecting piping is properly sized and secured.
3. Check the lubricant level. If necessary, add some. Be sure the lubricant used is that recommended by the manufacturer.
4. Be sure the oil-reservoir filler plugs, oil-filter covers, and similar pressure-vessel covers are tight.
5. Remove any shipping covers or tapes from the air filter and be certain that it is tightly assembled.
6. Check all fasteners, especially compressor inlet and discharge connections.
7. Check coupling alignment according to the manufacturer's specifications. Be sure coupling bolts and setscrews are tight.
8. Rotate the compressor by hand several times, noting any tight spots or unusual noise.
9. Check electrical connections for proper size and voltage and secure attachment. Be sure all fuses and overloads are in place and of the correct size. Reset all manual circuit breakers and resets.
10. Jog the motor once to check that the direction of compressor rotation is correct.
11. Open any water shut-off valves on water-cooled units.
12. Start the compressor unit and close the air-line valve to build air pressure to the operating level.
13. Set any operating controls according to the manufacturer's instructions.

Safety devices All rotary screw compressor units are protected by several, if not all, of the listed safety devices. An understanding of these devices is essential to the proper maintenance of the compressor unit. The manufacturer's instruction manual should be studied and the devices should be treated not as mystical black boxes, but as integral working parts that perform a necessary, if only an occasional service.

In addition to the driving-motor electrical protective devices, such as starter overload heaters, fuses, and motor thermistors, several devices are used to protect the compressor and its lubrication system.

High-Air-Temperature Shutdown. A thermally actuated switch protects the compressor from lubrication failure or abnormally high temperature operation. The thermal element is usually inserted in the air-discharge line near the compressor. The switch may be mounted directly on the element or attached to it by a capillary tube. The switch is wired into the motor starter circuit so that, if a temperature exceeding the switch setting occurs, the switch will open the starter circuit and shut the compressor unit down.

In the event of a high-temperature shutdown, the switch must be manually reset before operation can resume. Be careful—open the main circuit breaker, reset the switch, then close the main circuit breaker, and restart the compressor. Just resetting the switch will cause an immediate compressor start while the operator is dangerously close to the rotating equipment. If a high-temperature shutdown occurs several times over a short period, do not continue to reset the high-temperature shutdown switch; find and correct the problem before resuming operation. Do not attempt to adjust the temperature setting of the switch. A small amount of movement can cause a large change in set temperature and allow the compressor to operate at dangerously high temperatures or shut the compressor down before a normal operating temperature is reached.

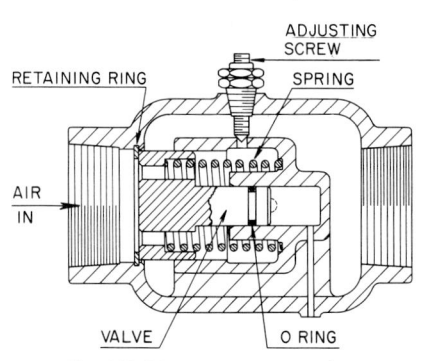

Fig. 4-7 Minimum pressure valve.

Low-Oil-Pressure Switch. A pressure-sensing switch is connected to the compressor lubrication supply line to sense the loss of, or abnormally low, lubricant pressure. The switch mechanism is usually remotely mounted and piped to the sensing point. The switch is wired into the motor starter circuit and, if lubricant pressure falls below the switch setting, the switch will open the starter circuit and shut the compressor down. The switch does not require manual resetting; it will automatically reset on the next pressure rise. Some switch systems will have a timing relay in the electrical circuit to temporarily bypass the switch so that lubricant pressure may be built up during the first seconds of start-up, and keep the switch from sensing the low pressure and immediately stopping the compressor unit. Other switch systems require holding the start button down for a few seconds to override the switch and allow lubricant pressure to build up. Never hold the start button down for a longer time than recommended by the manufacturer; severe electrical circuit and motor damage can occur. Do not continue to restart a compressor unit that has shut down several times in a short period due to low lubricant pressure; find and correct the trouble. Do not exceed the manufacturer's recommended pressure setting for the lubrication system. Too high a pressure setting will cause difficult and longer-than-necessary start-up periods. Too low a pressure setting can cause operation with too little lubricant and high compressor temperatures.

Minimum Pressure Valve. A valve is supplied by some major compressor manufacturers to maintain a positive pressure in the oil reservoir and on the compressor oil system. This valve is used in place of the oil pump found in some compressor unit designs. A typical valve (Fig. 4-7) contains a spring-loaded valve and a means of limiting air flow by fixed valve orifices and an adjustable body orifice. When the orifices will no longer allow sufficient air to escape, the upstream pressure rises, overcomes the spring force and the downstream air pressure, and allows the valve to open. When the valve opening and orifices allow enough air to pass, the system is in balance and holds the set minimum upstream pressure. Always observe the manufacturer's recommended valve-pressure setting to ensure proper lubricant pressure for the compressor.

Automatic Blowdown Valve. This valve is installed in the final air-discharge line and wired into the motor starter circuit to release pressure from the oil reservoir and oil system each time the compressor unit is shut down. Automatic release of reservoir pressure ensures loadless starting at next start-up for long motor life and operator safety in that no pressure is retained in the system after shutdown. The valve is a simple electrical solenoid-operated two-way valve. It closes when the compressor unit is started and current is applied to the solenoid and opens when current is removed on compressor shutdown. Manual blowdown valves are usually also supplied as a safety backup to the automatic valve.

Safety Valve. This mandatory device in any pressurized line or vessel protects both men and machines against accidental overpressuring of the air system. A safety valve is usually installed in the final air-discharge line. A general rule is to set it at 110 percent of system operating pressure but never higher than the working pressure of the lowest-rated pressure vessel in the system. Check periodically to ensure proper relieving action. Never operate any compressor unit or compressed air system without a proper safety valve.

Two general rules apply to all safety devices: (1) never disconnect, jumper, or render inoperative any safety device that protects the compressor unit or operating personnel; (2) never continue to restart the unit if the same malfunction occurs within a short period of time—find and correct the trouble.

Instruments As important as safety devices, instruments which visually monitor and display the operation of the compressor unit should be checked. Be sure they are in good working order and are accurate. Much needless troubleshooting and repair work are caused by a faulty gauge.

Air-Pressure Gauge. This kind of gauge indicates the final discharge pressure of a unit. The compressor is usually sized to use nearly all of the available horsepower of the driving motor at the rated pressure of the unit. Consequently, the air-pressure gauge is an important guide for setting the correct operating pressure to use the full output of the motor yet to prevent overloading. The air-pressure gauge should be checked periodically with a gauge known to be accurate to avoid air-system problems due to low pressure or motor-overload problems due to high pressure. Replace any defective air-pressure gauges before unit damage can occur. Never exceed the manufacturerer's recommended operating-pressure limits for any compressor unit.

Temperature Gauges. Such gauges usually indicate temperature levels at which the compressor operates most efficiently or safely. Commonly monitored temperatures include oil inlet to the compressor, air-oil discharge from the compressor, and cooling-water outlet from a water-cooled unit. All these temperatures indicate the general health of the compressor. Temperatures outside the manufacturer's recommended operating range should be investigated at once. Temperature gauges should be checked periodically against a gauge known to be accurate and, if defective, reset or replaced. Temperature gauges which read too low can cause settings to be made for operation at higher-than-normal temperatures and nuisance shutdowns by the high-discharge-temperature switch. Temperature gauges which read too high can cause settings to be made for operation at lower-than-normal temperatures with resultant moisture accumulation in the lubricant system.

Differential Pressure Gauges. These gauges are usually supplied to indicate the condition of the oil filter and oil separator and the approaching time for change. The gauge is attached to the upstream and downstream sides of the monitored device. By an increasing pressure differential, the guage shows resultant restriction. Always change the oil-filter element or the oil separator at the pressure differential recommended by the manufacturer. Failure to change the oil-filter element will cause it to bypass contaminants to the compressor. Failure to change the oil separator may cause it to rupture with resultant excessive oil carryover into the air system.

Air-filter service indicators are a special type of differential pressure gauge. The indicator is occasionally included on the instrument panel. It can be attached directly on the air-filter or the air-inlet line. See the air-filter section for a discussion of service indicators.

Hourmeter. Seemingly not a key part of an instrument panel, an hourmeter is essential for scheduling routine maintenance such as lubricant changes or motor servicing as well as major overhaul times. Additionally, in multiple compressor installations, the hourmeter aids in accumulating equal operating hours on each unit. The hourmeter should be protected against accidental damage during operation or servicing of adjacent components. Hourmeters, with their encapsulated construction, cannot be repaired. An inoperative hourmeter should be replaced at the earliest opportunity.

Air filter An air filter is the important first link in ensuring long, satisfactory service from any compressor. The rotary screw compressor mixes every particle of air intimately with the fluid used for lubrication and circulates it to all parts of the machine. If airborne dirt is allowed to enter the compressor, bearing wear or worse is a certain result. Of several available types of air filters, the dry type with replaceable element is the most commonly used because of ease of servicing and high efficiency in removal of small particles. The compressor manufacturer's operating manual usually contains a large amount of information on the type used on the compressor unit and strict observance of these recommendations will pay dividends. In general, dry-type air filters (Fig. 4-8) will

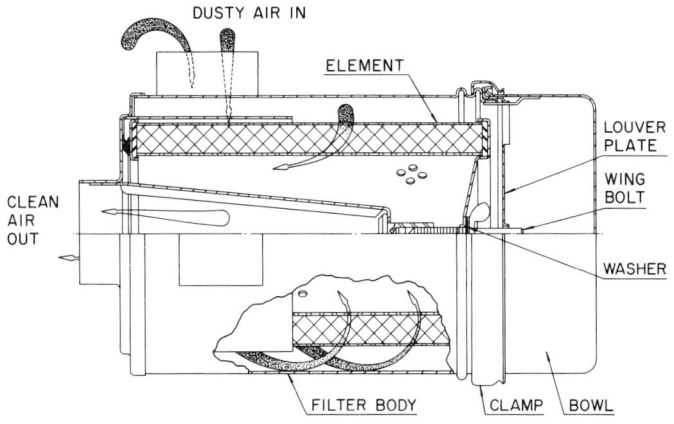

Fig. 4-8 Dry-type air filter.

remove particles above 10 microns in size. If smaller particles are found—flour dust and cement fines average just below 10 microns, fly ash about 1 micron, and tobacco smoke about 0.1 micron—a high-efficiency filter recommendation should be secured from the manufacturer.

Servicing Interval. The frequency of servicing for any air filter depends upon the cleanliness of the surrounding atmosphere; it may range from as short a time as 4 hr to as long as once a week. Service intervals can be established only by observation of dust conditions and experience. Higher-than-normal motor current or loss of compressor capacity is a direct indication of the need for air-filter service. When servicing any air filter use care not to damage the canister, cup, or gaskets on removal or reinstallation. Be sure the new or cleaned element does not have breaks and that it is properly seated and tightened in the canister. Check the outlet tube for signs of dirt passing the element; if found, check the element for breaks, improper seating, or damaged gaskets. If the amount of dust collected by the air filter requires unusually frequent servicing, centrifugal precleaners can be installed or the intake point for the air filter can be moved to a cleaner location. The cost of proper air-filter maintenance is minor compared with that for major repairs and downtime caused by neglect.

Service Indicators. For air filters, these indicators are inexpensive and efficient devices to signal time for service of the air-filter element. Most compressor manufacturers supply them as standard equipment on large-capacity units. They are available as optional equipment for all sizes of compressor units. A visual reminder that air-filter service is required is well worth the small cost. Rotary screw-unit air filters should be serviced or changed when the pressure differential across the element reaches 20 in. of water. A typical indicator (Fig. 4-9) has a stem which gradually rises as the pressure differential across the air-filter element increases. When the differential reaches 20 in. of water, the stem locks in the fullup position, indicating a need for service. After service, the indicator stem is released manually to return to the lower position. Once the indicator stem has risen and locked, it cannot be lowered until the air filter has been serviced and the pressure differential is normal for a clean filter. Remember that the service indicator cannot function if the air-filter element has breaks or leaking gaskets since the pressure differential which causes the stem to rise will not be present.

Inlet valve All compressors require some means of stopping the air flow to the system once the system operating pressure is reached. In the rotary screw compressor, this is accomplished by shutting off the inlet and is commonly called *inlet throttling*. Some compressor units use a butterfly valve operated by an external air cylinder, but by far, the most widely used is the self-contained piston-actuated inlet valve; a typical design is shown in Fig. 4-10. The valve operates on an air-pressure signal relayed from the air system by a pilot or a pressure switch through a pressure regulator. Full air-system pressure is not sent directly to the valve piston since the normally high pressure would cause the valve to slam and severely damage the resilient valve seat.

Control systems Several control systems are applicable to individual rotary screw compressor units; they vary with the application. The most widely used is the constant-speed pilot system. In second place is the dual-control combination system. Not widely used other than in special applications are the timed system, the automatic start-stop system, and the on-off pilot or pressure-switch system.

Constant speed. This control is used where air requirements are high and are maintained for long periods of time. Since the rotary screw compressor operates best at sustained speed, full load, and full pressure, the constant-speed control offers the most simple, economical, and easily maintained system. In this system, the driving motor runs continuously, and the inlet valve is throttled according to the demand for air. At normal air usage from the system within the compressor capacity, the inlet valve will modulate slightly and maintain a nearly constant system air pressure. Usually an air receiver is not needed in a constant-speed system. This ability to maintain constant pressure and continuous running of the driving motor is the source of the name *constant speed*.

The system uses an inlet valve and a modulating pilot for control of the air flow through

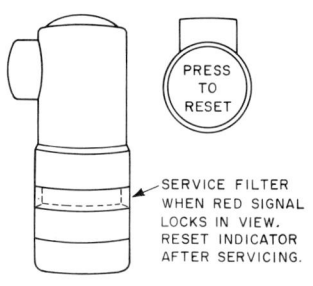

Fig. 4-9 Air-filter service indicator.

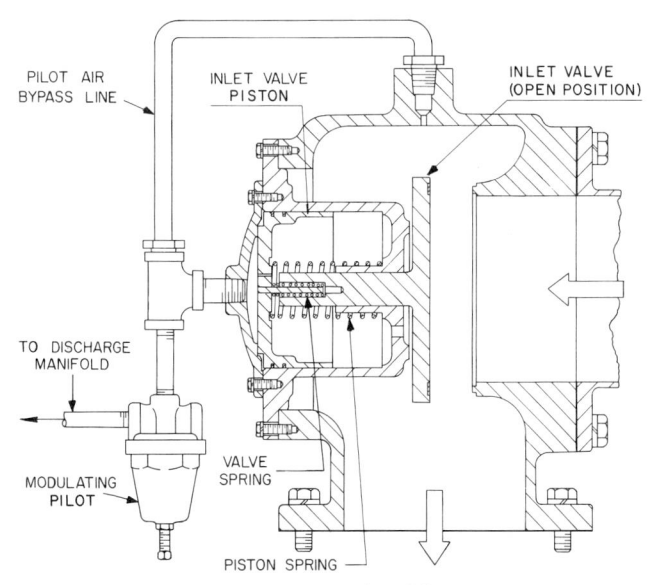

Fig. 4-10 Inlet valve and modulating pilot.

the compressor based on air pressure in the discharge line of the compressor. On start-up, the inlet valve is wide open since there is no air pressure in the system. The pilot remains closed—no pressure is passed to the inlet valve piston—until air system pressure exceeds the low point of the pilot range, usually about 15 psi below full system pressure. As air system pressure rises above the pilot range low point, pressure is passed to the inlet valve piston at a 1:1 ratio; that is, a 1 psi increase in air system pressure increases pressure on the inlet valve piston 1 psi, and the piston gradually moves the inlet valve toward the inlet housing seat. At full system pressure, the inlet valve is fully closed and no air is being compressed. At this point the compressor is said to be *unloaded*. As air system pressure falls, the pilot pressure is gradually removed from the piston and the piston spring moves the piston away from the valve stem. Atmospheric pressure on the valve face can now overcome the pressure on the piston and air will flow into the compressor. When air flows through the inlet valve to the compressor, the compressor is said to be *loaded*. The degree of loading varies according to the amount of opening of the inlet valve. Partially closed positions of the inlet valve reduce air flow and power required proportionately. A fully closed inlet valve and the compressor unloaded against full system pressure require about 70 percent of full load power.

Usually, inlet throttling becomes effective at about one-half the modulating pilot range, or 8 psi on the piston. With inlet throttling, air flow and horsepower are reduced. Therefore, to obtain full compressor capacity and full horsepower use, set the modulating pilot about 8 psi above the normal operating pressure. As the system pressure rises above the operating pressure set point, the inlet valve will throttle the air flow and reduce the horsepower required. Do not exceed the manufacturer's recommended operating pressure by more than one-half the pilot range. The pilot range is fixed (about 15 psi) and cannot be changed; only the point at which the pilot begins to pass pressure can be adjusted.

A unique characteristic of the modulating-pilot inlet-valve system is that it seeks the optimum operating-pressure level for the system within the compressor capacity. As air-system pressure rises or falls, the inlet valve closes or opens to maintain steady pressure. Of course, if air usage exceeds compressor capacity, system pressure will continue to fall. Or, if all air usage is stopped, the compressor will build up system pressure to the pilot set point and unload. The optimum pressure level characteristic of the rotary screw compressor is in direct contrast to the reciprocating compressor system which will load, build up to full air receiver pressure, then unload until the pressure falls sufficiently to load again. Usually, the load-unload range of a reciprocating compressor is 10 to 15 psi. It is not unusual to observe a rotary screw compressor hold a constant system pressure, say 95 psi, for an entire shift, if air usage stays somewhat constant.

An air bypass line from the valve piston connection to the inlet valve housing completes the control system. When the compressor unloads, air from the modulating pilot is bypassed behind the closed inlet valve to aerate injected oil and prevent hydraulic noise. Experience has shown that no mechanical damage is caused by hydraulic noise; however, it is somewhat disquieting to the uninitiated and is best eliminated. The small amount (1 to 3 percent of compressor capacity) of bypass air does not raise system pressure enough to actuate the safety valve, even during extended periods of unloaded operation.

The constant-speed control should not be applied to every system. Applications where air demand is low, causing the compressor to remain unloaded for long periods, should be avoided. Low air use in a constant-speed system can cause moisture accumulation in the lubricant system due to cooling of the system during long unloaded periods. Moisture will foul control lines, causing erratic operation or complete malfunction of the inlet valve and gauge components. The best practice is to stop the compressor and restart when necessary if long unloaded periods are anticipated. Constant-speed systems work best when the compressor is loaded 75 percent or more of the working period.

Automatic Start-Stop. These systems are used where the air demand is intermittent and the air system is large enough to support normal requirements—once the system is charged—for long periods of time. Automatic start-stop systems always require an air receiver to prevent rapid cycling. The automatic start-stop system uses a conventional inlet valve and a pressure switch connected to the air-discharge line, or air receiver, and to the driving-motor control.

At start-up, the inlet valve is wide open since there is no air pressure in the system.

When system pressure reaches the operating cutoff point as set on the pressure switch, the switch opens, interrupting power to the motor control and stopping the motor and compressor. Without inlet air flow to overcome the inlet valve spring, the valve closes and system air pressure is contained. When the system air pressure falls, the pressure switch closes, restarting the motor and compressor. With rotation, a low pressure is established in the compressor; the higher atmospheric pressure overcomes the inlet valve spring and air flows through the compressor.

Automatic start-stop systems should be used with care. Application should be made only on small compressor units, usually below 100 hp, where the driving motor can allow repeated starts during a normal work period. The compressor manufacturer should be consulted on the maximum allowable motor starts per day on the specific-size compressor unit being considered for automatic start-stop. Low usage inherent with this system promotes moisture accumulation due to minimum heating of the system when loaded. As always, controls, gauges, and lubricant suffer damage from continued moisture accumulation. Automatic start-stop units should be operated at higher system temperatures than constant-speed units. If possible, the unit should be operated at a normal system pressure and air flow for about $\frac{1}{2}$ hr a day, wasting air if necessary, to remove the entrained moisture. Moisture can be handled and removed beyond the compressor lubricant system; it is a nuisance and a hazard when retained.

A variation of the automatic start-stop system uses a timer in the motor control circuit to allow the motor to continue to run for a specified period after full system pressure has been reached. This eliminates the repeated motor starting problem, if the timer interval is set sufficiently long. Usually, the motor run interval is overlapped by a period of air demand which resets the timer. In effect, this variation is a full-open–full-closed inlet valve with the motor running constantly.

Any automatic start-stop system uses considerably more power for the amount of compressed air delivered than the modulating constant-speed system. Either system requires the same unloaded horsepower for a fully closed inlet valve and an equal capacity. However, the automatic start-stop has a fully open inlet valve, admitting full compressor capacity and using full horsepower, rather than throttling the inlet to match the air-system demand with reduced air flow and horsepower.

Dual Control. These systems combine the best features of the constant-speed and automatic start-stop systems. A single multiple-contact switch is used to select the mode of operation. The constant speed is used during normal air-use operation time. When air demand falls at lunch time, or during an off-peak work shift, the automatic start-stop is used to maintain full air pressure at the reduced demand. Alternating use of the constant-speed mode overcomes the moisture-accumulation problem usually associated with automatic start-stop, because it allows periodic long-time normal system temperatures. Dual-control systems should not be used in compressor units over 100 hp unless precautions on repeated motor starting and moisture accumulation are taken, as noted earlier in the automatic start-stop section.

Timed Systems with Oil-Reservoir Blowdown. These are the so-called low-unloaded horsepower systems; they provide two basic modes of operation: constant speed and automatic start–timed stop. The timed systems have the lowest horsepower use of any conventional method which allows the motor to run while the compressor is unloaded. A conventional inlet valve is used with a pressure switch connected to the discharge line, or air receiver, and driving-motor control. An air receiver is usually required in the system to prevent rapid cycling.

The constant-speed mode uses a null position in the timer to allow the motor to run continuously. On start-up, the inlet valve will be wide open with no air pressure in the system. When system pressure reaches the pressure switch setting, the switch electrically signals a three-way control valve to admit reduced system air pressure to the inlet valve piston and close the inlet valve. At the same time, the electrical signal opens a small solenoid-operated valve in the oil-reservoir discharge line and the air pressure in the oil-reservoir and compressor unit system is evacuated. External air-system pressure is held by a check valve between the oil reservoir and the external system. When the reservoir is blown down to atmospheric pressure with the inlet valve closed, the compressor operates at minimum horsepower—about 20 percent of that required at full system pressure. When air-system pressure falls to the pressure switch low setting, an electrical signal to the three-way valve bleeds the pressure away from the inlet valve piston, allowing the inlet valve to open and load the compressor. The electrical signal causes the oil-reservoir blowdown valve to close and the unit air system builds pressure again.

The automatic start–timed stop mode uses an operator-set time on the timer to stop the motor if system air pressure stays above the pressure switch low setting for the chosen period—3 to 30 minutes. Operation of this mode duplicates the constant-speed mode as long as the air demand causes the control to function through repeating load-unload-blowdown cycles. If air system pressure remains above the pressure switch low setting for a time longer than that set on the timer, motor control power is interrupted and the motor stops.

The timed system application should be chosen with care. Air demand should be low, intermittent, and in regular cycle. Time between periods requiring the compressor to load should be, at the minimum, long enough to allow complete reservoir blowdown. Periods of loading before complete reservoir blowdown cause oil foaming, with oil carryover into the air system and inefficient use of the low-unloaded horsepower feature. The system is best applied on compressor units below 100 hp, where driving motors can allow repeated starts. The compressor manufacturer should be consulted for the maximum number of motor starts allowed per day for the size of the compressor unit being used. Timed control systems should be operated at full load for as long as possible to minimize moisture accumulation in the lubricant system. Prevention of moisture accumulation is most critical in a timed system because of the larger number of control lines and valves which can be fouled and then cause system malfunction.

Capacity Control Systems. The most recent addition to the rotary screw compressor-control systems, capacity controls cannot be applied to conventional fixed-port compressor cylinders. Capacity control requires a specially designed cylinder with a means of changing the position of the inlet porting within the cylinder. The length of cylinder/rotor, other things being equal, determines compressor capacity. Changing the inlet-port position, in effect, changes the length of the cylinder/rotor and the capacity of the compressor. An automatic control system, responding to system air-pressure changes, positions the inlet porting to match the amount of air being used from the system. A capacity control system will usually be effective from 40 to 100 percent of the given compressor capacity. Since the horsepower required is directly proportional to air flow through the compressor, capacity matched to system requirements is more economical than a fixed-capacity-operated start-stop, throttled inlet, or any conventional means. Capacity control compressor units can operate with power savings up to 25 percent of the overall power usage of equal-size fixed-capacity compressor units.

Lubrication An oil-injected rotary screw compressor depends on its lubrication system for cooling, sealing internal clearances, and lubricating rotating parts. Since the lubricant touches every part of the compressor system and is literally its lifeblood, more than usual attention should be paid to the selection of a lubricant and to maintenance of the system. Figure 4-6 shows a typical lubrication system for a rotary screw compressor. For the balance of this discussion *lubricant* and *oil* are used interchangeably—either means that fluid which is used in the system for compressor lubrication. Oil-system components vary with the manufacturer and the compressor application. Not all components shown will be on every compressor unit. Some manufacturers use components which are not shown, such as an oil pump to assist or to meter the oil-flow rate.

In general, the oil suction line is connected to the oil reservoir. Oil is forced by air pressure in the reservoir through an oil cooler, a mixing valve, and the oil filter to the compressor. In the compressor, the oil seals the internal clearances, removes the heat of compression, and lubricates the rotating parts. After compression, the oil, now mixed intimately with the air, is discharged back to the oil reservoir where most of the oil is removed by velocity change and impingement. The air and a small amount of remaining oil pass through a final oil separator. The separated oil is returned to the oil system and the air passes to the external air system.

Lubricants. All major compressor manufacturers specify one or more types of lubricant as being satisfactory for their rotary screw compressor designs. Lengthy testing and evaluation have preceded the selection of that particular type of lubricant as the one which the manufacturer can recommend. Always follow those recommendations and enjoy long compressor unit service.

Although types of lubricant recommended vary among manufacturers, some characteristics are common. All lubricants should have high levels of corrosion inhibitor, oxidation inhibitor, and foam inhibitor. Corrosion inhibitors control the attack on metal parts exposed to the lubricant. Oxidation inhibitors control the breakdown of the lubricant at elevated temperatures with resultant varnish formation. Foam inhibitors control the foaming of the lubricant to provide an adequate fluid film for rotating parts and to prevent foam blanketing of the oil separator with eventual oil carryover to the air system. Since every particle of oil is intimately mixed with every particle of air at elevated temperatures, the circumstances are excellent for oil degradation in a short time. Only high-quality lubricants are suitable for rotary screw compressor service. Many brands of lubricants are marketed as suitable for rotary screw compressor service—the proof of the lubricant is how well it performs and how long it maintains that performance. A compressor manufacturer can only specify operating limits; how well a lubricant performs in an application is the responsibility of the oil supplier.

Synthetic Fluids. Recent additions to the lubricants available for the rotary screw compressor, the main claim for any synthetic fluid is longer use before changing. Length of service must still be evaluated by the user and supplier. The compressor manufacturer's minimum requirements for his recommended lubricant—viscosity range, corrosion, oxidation, and form characteristics—must be met by any chosen synthetic fluid. Most synthetic fluids do not have any effect on the materials used in the ordinary screw compressor unit and its oil system. Some synthetics attack gasket materials and certain interior coatings as well as exterior air line fittings. Always check the suitability and compatibility of any synthetic fluid with the compressor manufacturer before use. Most major compressor manufacturers have evaluated the common synthetics and have approval lists readily available. As with any lubricant, the quality of the synthetic and performance in the application rest with the fluid supplier.

Oil Levels. These levels are usually indicated by a gauge on the oil reservoir. It is important that the oil level specified by the manufacturer be maintained. Too low an oil level may cause loss of oil circulation, compressor overheating, and shutdown. Too high an oil level may cause oil carryover into the air system. The oil level should always be observed when the compressor unit is operating at full load. Some oil systems may indicate an overfull condition when not operating, owing to system drain back. Oil systems may indicate a low oil level if observed immediately after shutdown and before reservoir blowdown and system drain back is completed. Never open any plugs or connections in the oil reservoir or the oil system until all pressure has been relieved.

Oil Change Intervals. The time between oil changes is specified by the compressor manufacturer and should be followed closely. The interval between changes is dependent on intake air quality, regularity of air-filter and oil-filter maintenance, ability of the cooling system to hold oil temperature within specified levels, operating cycle of the compressor unit, and last but never least in importance, oil quality. Most compressor manufacturers specify at least one type of lubricant which performs well under most conditions as well as an average period of time for which the lubricant will perform satisfactorily. Lengthy tests and evaluation of many lubricants have preceded the recommendation—it is to the users' best advantage to adhere to the specification.

The manufacturer's change interval may be altered by special conditions, including (but not limited to) extreme conditions of humidity, dust, or high operating temperatures. Operating conditions should be critically surveyed and oil changes planned accordingly. A good method to establish an optimum drain interval is for the oil supplier to analyze oil drained at specific intervals—say 250, 500, 1000, 1500 operating hours, and up—and to advice on the continued use of the lubricant. After establishing an oil-change interval, be sure that compressor operating conditions remain relatively the same. Changing operating conditions may require a different change interval or even a different lubricant.

Draining and Cleaning the Oil System Properly. This procedure is important. Stop the compressor unit and be sure no air pressure is in the oil reservoir or connecting oil lines. Always drain the complete system—reservoir, oil filter(s), oil cooler(s), and low points such as the compressor cylinder. Draining oil when it is hot tends to carry away contaminants in the oil and prevent varnish deposits. The oil drain operation presents a good opportunity to observe general compressor health. Look at the drained oil, oil-filter element(s), and oil system pipe plugs or covers for signs of wear metal, an unusual amount of sediment, or a varnish deposit. Wear metal may indicate internal metal part wear due to a low-viscosity lubricant. Sediment may indicate a need for more frequent air-filter servicing or change of air-filter type or location. Varnish deposits may indicate overlong use of a lubricant, high temperature in an oil system, or an incorrect lubricant.

Filling the Oil System. Carefully wipe away all dirt from filler plug(s) or cover(s). Check that the proper grade, specification, and amount of lubricant are being used. Use only clean containers, funnels, or pump equipment to fill the system. Provide for clean storage of the oil and filling equipment. Do not leave partially used lubricant in containers open to the atmosphere.

Oil filter The compressor oil filter is the final element in an oil system designed to provide a proper lubricant to the compressor unit and is vital to maintaining a trouble-free compressor. The oil filter removes dirt and abrasives from the circulating lubricant. Most oil filters have replaceable elements and will remove particles down to 10 microns in size. Because of relatively small particle retention, oil-filter elements tend to clog rapidly. Bypass valves are built into most oil filters used on rotary compressor units and will bypass the oil at about 15 psi pressure differential rather than shut off oil flow. This bypass feature, while a safety device to ensure constant oil flow, presents a hazard in bypassing contaminated oil when the oil-filter element is not regularly serviced. Most compressor manufacturers specify either the maximum recommended time of oil-filter element use or provide a gauge which indicates the time for change based on pressure differential. In either case, always follow the manufacturer's recommendation.

Oil cooler system Cooling of the compressor system lubricant is important to maintaining proper compressor operating temperatures. Both air-cooled and water-cooled oil coolers are used, the choice depending on quality, temperature, and amount of cooling media available.

Air-Cooled Oil Coolers. These coolers are generally finned tube radiators which use ambient air forced across the fin/tube section to remove the heat from oil circulating through the tubes. Some oil coolers are mounted directly on or just adjacent to the compressor unit; other coolers may be mounted some distance from the unit. The oil cooler fans are driven by direct connected electric motors mounted on the cooler structure. Some very compact units use a shaft extension on the compressor or on the compressor driving motor to mount and drive the fan; occasionally, belt-driven fans are used to obtain optimum fan rotative speeds. Whatever the style of mounting, it is important to provide an unobstructed air flow to and from the cooler. Allow at least 2 ft around all sides of the cooler. If possible, duct the exhaust air away from the cooler to prevent recirculation of heated air and artificially high oil temperatures. Provide the coolest possible air to the fan. Always follow the manufacturer's air flow recommendations for adequate cooling. In operation, keep the radiator section clean and free of dirt, lint, and debris which will reduce cooling efficiency.

Thermostatic Mixing Valves. An integral part of most air-cooled oil cooler systems, these valves function somewhat like the thermostat in an automobile radiator, but rather than only controlling volume of flow, they provide mixing of high- and low-temperature oil streams (see Fig. 4-11). On start-up, with the compressor unit cold, the thermostatic element is closed to the oil cooler and allows oil to circulate directly from the oil reservoir to the compressor. As the oil warms up, the element gradually opens to the oil cooler and closes to the bypass to allow the mixing of the two streams in proportion to the temperature of the oil passing over the element. When the unit attains operating temperature, the valve is fully open to the oil cooler. In normal ambient air temperatures below the design temperature for air to the fan, the mixing valve will maintain the system design oil temperature. As ambient air rises above the design air temperature, the mixing valve is fully open and the oil temperature will rise degree for degree with the ambient temperature. In a properly maintained oil cooler system, the discharge air-oil temperature will be about 100°F above the ambient temperature. Thermostatic elements are built to individual operating temperatures and cannot be adjusted. If adjustments are necessary, a different range element is installed. Elements are usually supplied at some temperature range judged by the manufacturer to cover the majority of operating conditions, with special ranges available on request. Standard range elements are usually satisfactory except in high-humidity applications where the oil must be

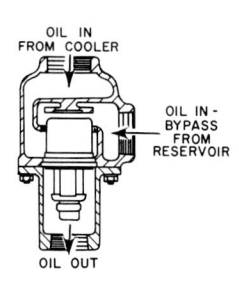

Fig. 4-11 Thermostatic mixing valve.

maintained at higher-than-usual temperatures to avoid moisture condensation in the oil system. Once the element is fully open, the oil temperature depends on the cooling ability of the oil cooler, fan, and air temperature to the fan. The thermostatic valve cannot prevent the compressor from operating at an abnormally low temperature.

Water-Cooled Oil Coolers. Such coolers are of the shell-and-tube-type heat exchanger. Water passing through the tubes cools the oil passing through the shell. Generally, to conserve space, these oil coolers use multipass arrangements on the water side with one pass on the oil side. A self-operated water-control valve (Fig. 4-12) is used to maintain oil temperature at a specified level. The valve is mounted on the water outlet pipe and the temperature-sensing

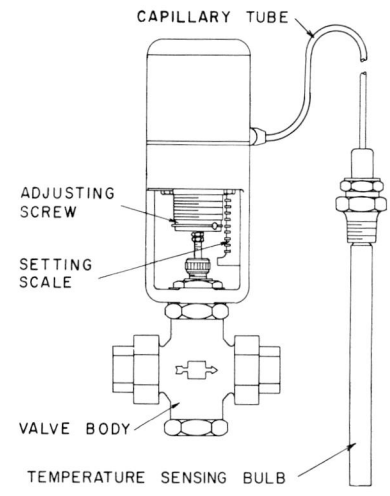

Fig. 4-12 Water-control valve.

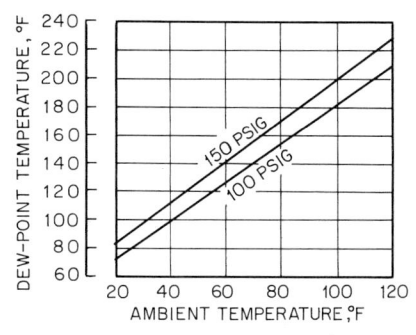

Fig. 4-13 Dew-point versus ambient air temperatures.

bulb is installed in the heat exchanger oil outlet piping. As the oil temperature rises or falls at the sensing bulb, the valve opens or closes to change the water flow and to regulate the oil temperature within the set range specified by the compressor manufacturer.

Water-cooled oil coolers are vulnerable to water quality. Fouling of water tubes and loss of cooling efficiency and erosion and corrosion with eventual tube failure are results of neglecting to survey and control the water supplied to the oil cooler. A local water treatment concern is best equipped to test the water and recommend treatment if a problem exists. The need for water treatment may involve only filtration to remove sand, silt, or debris. Chemical treatments may be necessary to inhibit corrosion or unusual scale deposits, control acidity, or prevent growth of microorganisms. A prudent normal maintenance program will include periodic inspection and cleaning of the oil cooler tubes. Remember that water quality can change; any changes noted in discharge water or cooler tube appearance should be investigated before oil-cooling problems develop.

The water flow control valve is also vulnerable to water quality. Check the valve stem periodically for scale or corrosion. Careful removal of foreign material, a drop or two of oil on the stem, and finger tightening or very light wrench tightening of the packing nut constitute a good preventive maintenance procedure. Do not tighten the packing nut so tightly that the stem is unable to move. Use care when working on the valve. Do not bend or scratch the valve stem. Do not bend or kink the capillary tubing from the thermal element to the valve head. Valve failures can generally be traced to a packing nut tightened too tightly, a valve stem corroded or bent, foreign material clogging the valve internally, or a damaged thermal (capillary) system. If foreign material is likely, use a strainer in the inlet water line. Any protection provided for the oil cooler likewise benefits the water control valve.

Moisture in the Oil System. Moisture should not be allowed to accumulate since it degrades the lubricant and fouls controls, causing erratic operation. Moisture can be prevented by keeping the oil system temperature, particularly the oil reservoir, above the pressure dew point of the air in the system. Figure 4-13 shows the dew-point temperature versus the ambient (intake) air temperature at 100 percent relative humidity. If the discharge air temperature is maintained above the dew-point temperature for a given intake air and discharge pressure, no moisture will condense within the oil system.

Care should be taken not to allow oil-reservoir walls to be cooled below the dew-point temperature since moisture will condense on the cool vessel walls, even though the discharge air is above the dew-point temperature. To preclude this problem, oil reservoirs located in direct cool air flow paths should be protected by baffling. Oil reservoirs located in air-conditioned rooms should be insulated.

Ordinary amounts of moisture contained in intake air cannot be prevented from entering a compressor system but can be handled by proper removal equipment. Excessive moisture should be prevented from entering the compressor system since a large amount

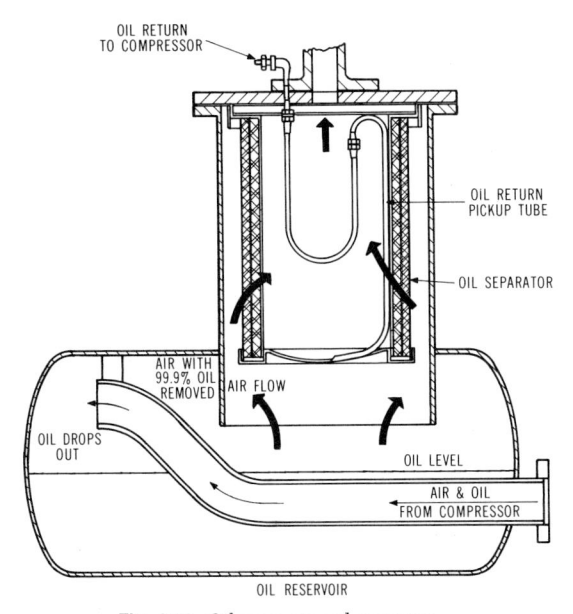

Fig. 4-14 Oil reservoir and separator.

will compound removal problems. Unwanted moisture will enter the system if the air filter is located near a steam- or other moisture-carrying vent line. Avoid locating an air filter near such a vent, but if not practical, raise the air filter above and upwind of the vent.

Care in air-filter placement, operation at a discharge temperature above the dew point, and protection of the oil reservoir from unusual cooling effects will prevent moisture in the oil system and allow the condensate to be removed easily downstream.

Oil separation and oil separators Separation or removal of entrained oil from the discharge air stream is the final action which occurs in the oil system. While some small amount of lubricant is beneficial to downstream devices, the amount of oil injected into a rotary screw compressor, if allowed to continue downstream, would be objectionable. Oil injected into the compressor ranges around 5 to 10 gal per 100 cu ft of compressor capacity. Even if such a large amount could be tolerated downstream, the makeup oil logistics would be overwhelming in a large air system. A reasonable removal and return system, called an *oil separator,* is necessary.

Oil separator design and material vary with the compressor manufacturer as does the design of the vessel(s) in which the separation occurs. Figure 4-14 shows a typical construction to illustrate the oil-removal sequence.

The air-oil stream from the compressor enters the oil reservoir through a pipe of about the same diameter as the compressor discharge. As the stream emerges from the pipe into the large volume of the oil reservoir above the oil, velocity is slowed. With reduced velocity, the heavier oil particles fall out to the liquid oil. The lighter oil particles continue until they impinge on the oil-reservoir wall and are carried by gravity to the liquid oil. At these two points, 99 percent of the oil removal occurs. The remaining air-oil aerosol flows to the periphery of the oil separator; here, impingement of oil particles occurs again with return of the larger droplets to the liquid oil by gravity. The air-oil aerosol continues through the oil separator. Small oil particles collect on the fibers of the oil separator and drain by gravity to a collecting point. A tube connects the separator interior with a lower pressure point in the oil system, usually the compressor cylinder, and returns all scavenged oil to the system. Single separator systems of this type will pass about 20 parts of oil per million parts of air. Dual, or two-stage, separators will pass only about 2 parts of oil per million parts of air.

Oil separator systems are efficient as long as the components are intact. Oil foaming, defective gaskets, too high an oil level, a damaged separator element, or a defective return tube will all cause unusual and readily apparent oil carryover.

Aftercooling After the clean air has been discharged into the external system lines, some applications require moisture removal. In dry climates and where the normal discharge air temperature of about 70 to 100°F over the intake air temperature is not objectionable, drip legs drained periodically are sufficient. More-humid and less-temperature-tolerant applications require some type of aftercooling. Depending on local conditions, either air- or water-cooled aftercoolers can be used. These devices will reduce the air stream temperature to within 10 to 20°F of the cooling media. Moisture content of the air stream leaving the aftercooler depends on the temperature of the leaving air. Moisture separators with automatic dump traps are almost always necessary, owing to large amounts of moisture removed. For applications where moisture control and temperature control are critical, a refrigerated air dryer is needed.

Heat recovery A very real source of recoverable heat exists in the air-cooled rotary screw compressor unit. Recovery rates from an air-cooled oil cooler alone range about 800 Btu/min per 100 cu ft of compressor capacity. Air-cooled aftercoolers can add 150 Btu/min per 100 cu ft of compressor capacity. Figure 4-15 illustrates a typical simple system. When using any recovery system, be sure the duct work does not add enough static resistance so that the fan air flow and cooling capacity is reduced. Good heating and air-conditioning practice should be used when designing ducts. If static resistance exceeds fan capacity, a booster fan must be used. Always consult the compressor manufacturer before adding a heat recovery system to any compressor cooling system.

Maintenance logs Such logs are a valuable tool in maintaining an efficiently operating compressor unit. All lubricant additions or changes, air-filter service, oil-filter service, lubrication and service to accessories, major overhaul, and even periods of operation or downtime can be recorded in a well-designed log, such as in Fig. 4-16. Many compressor manufacturers on request will furnish a maintenance log for a specific compressor unit.

Troubleshooting If operating troubles develop, the following general guide may allow adjustments or repairs to be made by operating personnel. One rule should be the

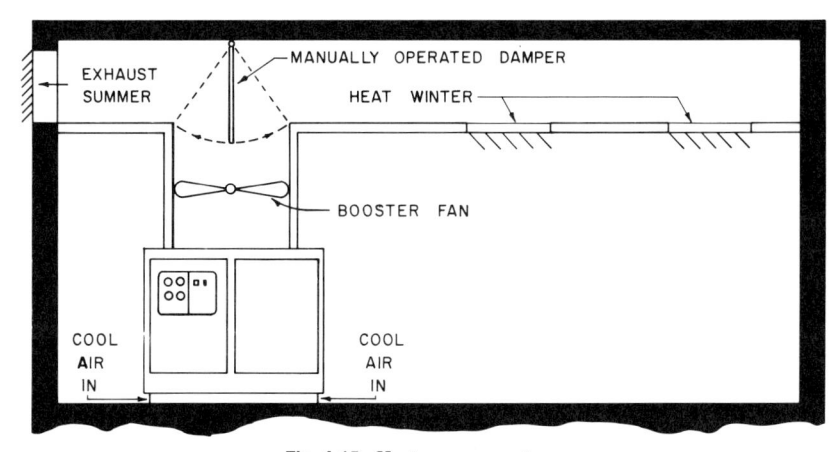

Fig. 4-15 Heat recovery system.

guiding principle in troubleshooting: If the same malfunction occurs repeatedly or if the same safety device stops the unit repeatedly within a short period, do not continue to adjust or reset the device; find and correct the trouble. In short, don't treat the symptom and neglect the disease.

Unit Fails to Start. Check the following:
1. Wiring for wrong lead connections
2. Temperature switch manual reset button
3. Fuses in control enclosure or starter enclosure
4. Compressor motor starter overload heaters and adjusting knob setting
5. Oil cooler fan motor overload heaters and adjusting knob setting
6. Low oil pressure

GARDNER-DENVER COMPANY

ELECTRA-SCREW® & ELECTRA-SAVER®

COMPRESSOR MAINTENANCE LOG

Model_____ Serial Number_____

REFER TO THE COMPRESSOR INSTRUCTION MANUAL FOR COMPLETE MAINTENANCE INFORMATION AND SCHEDULE. THE COMPRESSOR INSTRUMENT PANEL LISTS A BRIEF MAINTENANCE OUTLINE.

DO NOT REMOVE LOG FROM THIS UNIT

Record date in Date Column and hourmeter reading in the column under each item on which maintenance is performed. Record pressure differential gauge readings when any other maintenance is performed to provide running record of approaching oil filter or separator change.

Date	Air Filter		Pressure Drop		Oil Filter Changed	Oil Changed	Oil Separator Changed	Oil Cooler Cleaned	
	Cleaned	Changed	Oil Filter	Oil Separator				Radiator Core	Heat Exchanger Tubes

Fig. 4-16 Typical maintenance log.

7. Defective low-oil-pressure switch or improper setting
8. Contacts on timing relay for low-oil-pressure shutdown stuck open
9. Timing relay for proper time delay setting
10. Defective timing relay

Unit Starts but Stops after a Short Run. Check the following:
1. High-air-discharge temperature caused by
   a. Low compressor oil level
   b. Clogged oil cooler or oil filters
   c. Thermostatic mixing valve inoperative
   d. Dirt on oil cooler core faces
   e. Poor ventilation of unit and/or oil cooler
   f. Water control valve inoperative
   g. Water inlet temperature too high
   h. Water-flow restricted
   i. Magnetic water shut-off valve inoperative
   j. On remote oil cooler unit, oil stop valve or check valve in piping to oil cooler inoperative
2. Temperature switch manual reset button
3. Fuses in control panel enclosure or starter enclosure
4. Compressor motor starter overloads and adjusting knob setting
5. Oil cooler fan motor overload heaters and adjusting knob setting
6. Low oil pressure
7. Defective low-oil-pressure shutdown switch or improper setting
8. Time delay on timing relay for low-oil-pressure shutdown set for too short a time
9. Defective timing relay

Compressor Does Not Unload. Check the following:
1. Control lines for restriction or moisture
2. Air leaks in control system
3. Inlet valve stuck or valve spring broken
4. Pilot or pressure switch adjustment
5. Pilot or pressure switch for dirt or leaking diaphragm

Blowdown Valve Continues to Pass Air. Check for the following:
1. Loose wiring to the blowdown valve
2. Coil failure on the blowdown valve

Excessive Oil Consumption. Check for the following:
1. Oil carryover through discharge caused by
   a. Overfilling the reservoir
   b. Clogged, broken, or loose oil return lines
   c. Ruptured oil separator element(s)
   d. Loose assembly
   e. Incorrect oil causing foam
   f. Inoperative minimum pressure valve
2. Oil leaks at all fittings and gaskets

Compressor Low on Delivery and Pressure. Check for the following:
1. Clogged air filter
2. Restricted inlet valve
3. Broken inlet valve spring
4. Binding inlet valve piston
5. Incorrect motor speed
6. Pilot adjustment and/or malfunction
7. Automatic blowdown valve leaking
8. Partially closed minimum pressure valve
9. External system piping leaks or open valves

Moisture in Oil. Check for the following:
1. Defective thermostatic mixing valve
2. Wrong temperature element in mixing valve
3. High-humidity exhaust near air inlet
4. Oil temperature too low for humidity conditions
5. Oil-reservoir walls cooled below air dew point

6. Leaking heat exchanger tubes

Oil Temperature too Low. Check for the following:
1. Mixing valve element stuck open
2. Wrong temperature element in mixing valve
3. Water flow not controlled through oil cooler

If the oil temperature is too high, see high-air-discharge temperature.

Chapter 5

Reciprocating Air Compressor Maintenance

M. F. BAECKER[1]
Engineering Department
Gardner-Denver Company
Quincy, Ill.

An adequate and dependable supply of air is always necessary for continuous and economical operation of air tools. A specific compressor-maintenance program will go a long way toward obtaining the maximum efficiency from a compressor and eliminating unnecessary shutdown periods. The modern compressor is a precision-built machine, and it should be operated and maintained as such. Too many compressors are installed in out-of-the-way locations and are practically forgotten until trouble develops.

Each major air-compressor manufacturer furnishes an installation, operation, and service instruction book with each unit. Many hours of preparation and years of experience are represented in these books. They are included with the compressors so that owners and operators will have sufficient information to install, operate, and maintain the equipment for maximum efficiency. Read the instruction book carefully and become familiar with the compressor construction so that minor adjustments and emergency repairs can be made. Also know whom to contact should serious difficulty develop.

Location For good maintenance a clean, well-lighted location should be selected with enough space allowed to dismantle any parts that may need to be removed for servicing. Too often compressors are located so that it is impossible to remove the pistons, rods, or cylinders without breaking through a wall or moving the compressor. Outline and foundation drawings show the necessary service clearance. Maintenance and costs are materially reduced where these recommendations are followed.

Foundation An adequate compressor foundation (Fig. 5-1) is a necessity for satisfactory operation and maintenance of a compressor. A foundation that is designed without sufficient mass and bearing surface will cause vibration of the compressor, resulting in discharge-, suction-, and water-line breakage and excessive wear of compressor parts.

For compressors requiring concrete foundations, the compressor vendor furnishes prints showing the foundation above the floor line plus the weights of the parts to be mounted on the foundation and the out-of-balance forces that must be absorbed by the foundation. The amount of foundation will depend on the type of soil upon which it is being set. To determine the depth and size of the foundation below the floor line. a competent foundation engineer should be consulted who will take test cores and from these calculate the soil-carrying capacity. With this information, along with the weights and out-of-balance forces, a foundation can be designed for satisfactory compressor operation.

[1]The author is now retired.

Many small vertical compressors are installed on existing concrete floors and usually operate very well this way, as the large area of the floor forms a more than sufficient mass to offset any out-of-balance forces of the compressor.

At some locations it is impossible to set the compressor on a foundation or concrete floor that is poured on the ground. It must be located on a floor that does not have a solid base under it. For this type of installation, isolation dampers are used under the base supporting the compressor and its driver. Suction, discharge, and water lines should be attached

Fig. 5-1 Compressor on proper foundation.

with good flexible connections to prevent vibration and noises from being carried through the building. There are many manufacturers of isolation dampers, and their engineers should be consulted for recommendations for this type of application.

Air cleaners and suction lines Every compressor must be equipped with an air cleaner, which should be the most efficient type made for the service it is applied to. The air cleaner must be located so that an adequate supply of cool, clean, and acid-free air will be had at all times, with explicit instructions for servicing the air cleaner posted where maintenance personnel will always be reminded of the regular servicing required for good maintenance (Fig. 5-2).

At some locations it is necessary to place the air cleaners away from the compressor because of unfavorable surrounding conditions. Care must be used in providing a suction line to a compressor. It must be tight, free of dirt, chips, and scale, and of correct size for the length necessary to reach compressor suction. Normally the shortest possible suction line is preferable.

The time interval for cleaning an air filter depends on the type of cleaner and its location, and must be determined by checking the cleaner for dirt accumulation. No set time can be made to cover all installations and types of air cleaners.

Air-receiver location and capacity Air receivers often are considered accessories to air compressors and, for many applications, are not correctly installed or properly sized.

Proper installation and proper sizing are very important for both compressor and air-line systems. An air receiver absorbs pulsations in the discharge line from the compressor and smooths the flow of air to the service lines. It serves as a reservoir for the storage of compressed air to take care of sudden and unusual momentary demands in excess of the capacity of the compressor. Another of its functions is to precipitate moisture that may be condensed in the receiver and prevent it from being carried into the air-distribution system.

The preferable location for an air receiver is as near the compressor as possible so that the discharge line can be of minimum length, eliminating pressure drop between the receiver and compressor. Many receivers are located outside the compressor room and are exposed to the weather, offering difficulties when the temperature drops low enough to cause freezing. An ordinary top-outlet safety valve can be frozen shut, creating a hazard; the valve should be placed with opening down, thereby keeping water out and allowing the valve to function if necessary. Should the compressor be shut down, allowing no air to pass through the receiver, the drain valve or mechanism can freeze, perhaps breaking the parts making up the drain.

The size of the receiver usually is recommended by the compressor vendor, who has charts listing the necessary receiver sizes for various compressor sizes. Start-and-stop compressors require larger receivers than do continuously operated compressors, to keep them from starting too often. Each start requires electrical inrush to the motor, which can cause expense by increasing electrical requirements beyond normal electrical demand.

Air from the compressor should flow into the receiver at the bottom and out at the top. Condensate will stay near the bottom, giving drier air from the top opening. If excess condensate is a troublesome factor in the system, use an efficient water-cooled aftercooler and separator between the compressor and receiver. The aftercooler will condense the moisture and collect most of it in the separator, which can be drained by hand or automatically. The aftercooler dries and cools air, which promotes efficiency and safety. Most aftercoolers will cool the air to within 15°F of the incoming cooling water. Where water supply is short or expensive, air-cooled aftercoolers are available. They are not as efficient as the water-cooled but, if properly sized and of good quality, usually will cool to within 20 to 30°F of the ambient temperature.

Always consult the compressor vendor about receiver problems. Many states are exacting about pressure-vessel requirements; pressure vessels must meet the codes for safety and pass inspection by the insurance companies.

Starting a new compressor Before a new or repaired compressor is started, careful check must be made of the lubricating system, making certain all places needing lubrication have been oiled per manufacturer's requirements. On compressors having a forced mechanical lubricator, crank or pump by hand until it is certain the oil is getting to the parts requiring lubrication, as some initial lubrication is required before the unit is

Fig. 5-2 Air cleaners mounted on compressor.

started. Tighten all bolts, nuts, and cap screws. Turn the compressor over by hand wherever possible to determine that there is no interference or binding of working parts.

In the case of compressors requiring cooling water from a water main, turn on the water and check for leaks and for circulation through all parts requiring cooling water. For compressors having a self-contained water-cooling system, fill it and check to see that all air is out of the cooling system.

Check the discharge line from the compressor to the receiver, and if there are any globe, gate, or check valves anywhere between the compressor and receiver, be sure the valves are open and that there is a safety valve between the compressor and valves. The safety valve is a necessity, as it is possible that a valve could be left closed and the compressor started, resulting in an explosion should there be sufficient power in the driver, or should the overload protection fail to act.

If all points have been checked, apply driving power momentarily and let the machine coast to rest. Close observation during the coasting period will reveal any excessive tightness in the moving parts. The time that the unloaded machine continues to roll after driving power has been removed gives a fair indication of no-load friction; if no trouble is evident, the unit can be run without load.

After running from 1 to 2 hr unloaded, with periodic stops to check for any heating of bearings or other working parts, apply partial load and build up to maximum load and pressure gradually. The entire breaking-in period should consume a minimum of 4 hr.

The importance of break-in run cannot be stressed too strongly. The time and care spent in giving the running surfaces a polished finish pay dividends by increasing compressor life. After the initial run, compressor operation resolves itself into maintaining a clean air supply, feeding sufficient cooling water, and supplying adequate lubrication.

Operating a water-cooled compressor with too much cooling water through the system will cause excess condensate and cylinder wear because a cold cylinder will not lubricate properly; and because lubrication is affected, excess horsepower is required, adding to both maintenance and operating costs. A good rule is to hold the outlet temperature of the water between 120 and 130°F. This range will allow for good cooling and lubrication and also will keep condensation in the cylinder to a minimum.

All the foregoing requirements are necessary to get a compressor ready for efficient operation and to hold maintenance costs to the minimum. Routine maintenance must now be set up and definite pattern followed.

Lubrication The most important check for any compressor is the lubrication system (Fig. 5-3). Keep the compressor well lubricated; check the oil level at least once every 24 hr of operation. Use only oil and greases as recommended by the compressor manufacturer. The oil used should have a low carbon-forming tendency and sulfur content and contain an oxidation inhibitor. It is important to use the correct weight of oil, consistent with existing temperatures. The instruction book lists these conditions.

Because dust, dirt, and atmospheric conditions are different at various locations, it is not practical to state definitely how often the oil should be changed in the crankcase or power end of an air compressor. Oil will become contaminated with foreign materials held in suspension and will also oxidize. The time for oil changes is regulated by local conditions and must be determined by the discoloration and physical condition of the oil.

When oil changes are made, it will always pay to remove a handhole or coverplate and wipe the inside of the crankcase or power end clean with lint-free rags. If impossible to wipe out, use a good grade of flushing oil to remove any particles that may have settled on the crankcase floor. When refilling the compressor oil sump, be certain the filling container is free of all dirt, grit, or dust. This simple point often is overlooked.

Valves In a reciprocating compressor the valves must be kept in first-class operating condition, as leaking or inoperative valves cause loss of net air delivered. Heating often overloads the driver. It is therefore important to check the valves periodically and be certain they are always in good operating condition (Fig. 5-4).

The checking time for valves depends on several conditions, such as efficiency of the air cleaner, carbon-forming tendency of oil used, and the overall condition of the compressor. If the air cleaner is efficient and regularly serviced, excess dirt will be kept out of the airstream and dirt will not lodge in the valves. By using low-carbon-forming oil, the carbon buildup on the valves is held to a minimum. For single-acting vertical compressors, the pistons, rings, and cylinder walls should be kept in good condition so that excess oil will not pass these parts. Low oil consumption adds to valve life by eliminating unnecessary carbon deposit. No set checking time can be recommended; it will need to

Fig. 5-3 Lubrication points.

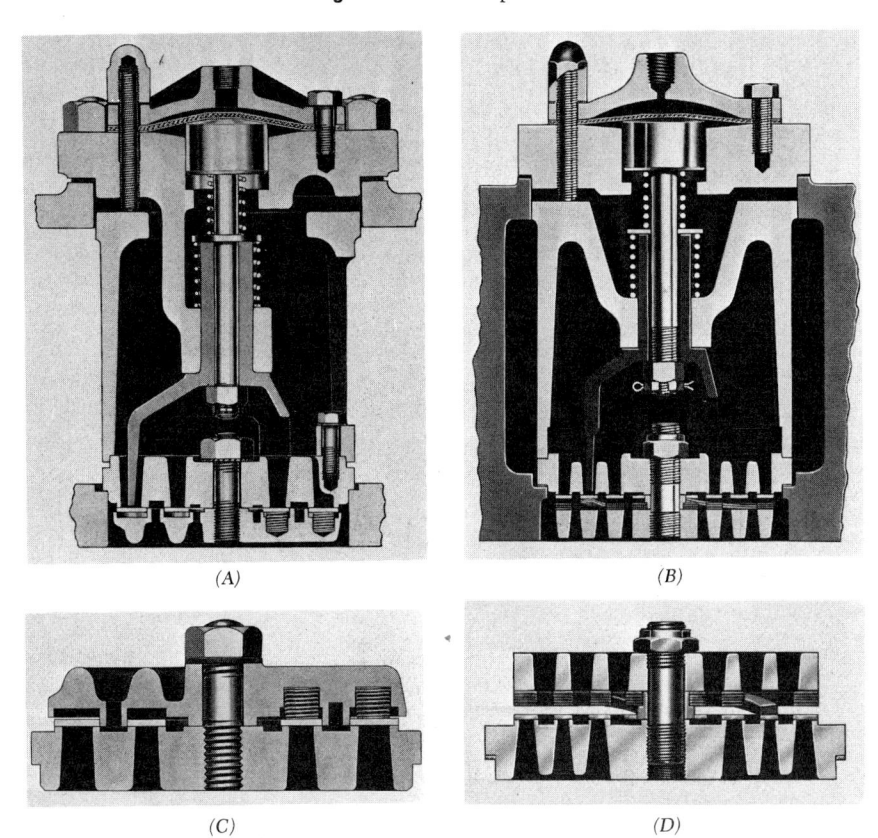

(A) *(B)*

(C) *(D)*

Fig. 5-4 Compressor valves: (*A* and *B*) Different designs of suction-unloading valve assemblies; (*C*) compressor valve with individual disks and coil springs; (*D*) compressor valve with plate and disk and finger springs.

8-59

be determined by actual investigation by the maintenance personnel. On a new unit the valves should be checked after 200 hr of operation.

Many compressor owners have found it helpful to have a spare set of valves so that a change of valves can be made immediately, and the replaced set reconditioned when time allows.

When valve troubles occur, there are several means of locating the valve or valves causing the difficulty. The first symptoms usually are low net air delivery and heating around the valve compartments. On a single-stage compressor, the usual method used is to feel the valve cover plates and examine the valve under the cover plate that is the hottest. If suction valves are leaking, a definite blow-back noise can be heard in the air cleaner when the compressor is operating under load.

On two-stage compressors, the intercooler pressure gauge is used as a guide to locate defective valves. When low intercooler pressure occurs, examine the valves on the low-pressure cylinders, and when high intercooler pressure is found, examine those on the high-pressure cylinders. By feeling the valve cover plates, the defective valve can be located under the cover plate that is the hottest. If high-pressure suction valves are leaking, the intercooler-gauge hand will fluctuate above normal intercooler pressure and the intercooler safety valve will pop. If high-pressure discharge valves are leaking, the intercooler-gauge hand will rise steadily and pressure will build up in the intercooler until the intercooler safety valve will release it.

When low-pressure suction valves leak, the air will blow back through the suction line and air cleaner if the compressor is operating under load. Leaking low-pressure discharge valves will cause the intercooler pressure gauge to fluctuate below normal intercooler pressure.

Since the valves are such an important part of the compressor, the information given in the instruction book must be followed when removing and installing them.

Wear between the valve disks or plate and the valve seat appears as indentations in the valve disks or plate, leaving a shoulder. The valve disks or plate are normally replaced if they show any amount of wear.

Most worn valve seats can be resurfaced. On some types of valves it is necessary to check the lift of the valve after resurfacing the seat, and if found to be more than recommended by the vendor, the bumper will need to be cut down to get the correct lift. Too much lift causes rapid wear and breakage.

Most valves usually have raised valve seats, and when the seat is refinished it is not necessary to do anything to the bumper, as the lift will still be to manufacturer's specifications.

Whenever a valve has been overheated, replace all the valve disks or plates and springs, because excessive temperature resulting from this heat will reduce the life of these parts and may result in breakage, causing damage to the compressor.

Most compressor valves have a gasket under the seat. This gasket must be in first-class condition; should it show any imperfection, replace it, as a leaking valve-seat gasket will eventually blow out.

The cover-plate gaskets also are important, and when installing valve cover plates, be sure the gaskets are in good condition. It is imperative that the valve cover-plate nuts or cap screws be pulled down evenly. Do not completely tighten one side and then the opposite side, as this will cause uneven gasket pressure, resulting in leaks or sprung cover plates.

Several types or designs of valves are used by different compressor manufacturers, and in order to get the proper installation in the compressor, refer to the instruction book that was furnished with the compressor. Too much care cannot be used when installing the valves and the component parts.

Piston rings Valves often are the cause for lost compressor efficiency, but should the valves be known to be satisfactory, the lost efficiency could well be in the piston rings. Piston-ring wear usually is very slow when the rings are properly lubricated, but operating time will eventually wear them so that the gap increases and the piston-ring lands wear to the point where some of the ring valving action is lost, allowing for blow-by through the gap and around in back of the ring.

One way to check rings is to put air pressure on top of the piston and listen or feel for blow-by past the piston and rings. For checking a double-acting cylinder, one valve can be removed on one end and air applied at the other end, and then check for blow-by on the end with the valve removed.

CAUTION: Do not put a hand in the cylinder, as air pressure on the opposite end may cause the piston to move, resulting in a crushed hand and arm.

When excess blow-by is found, remove the pistons and check the piston-to-cylinder-wall clearance and the piston rings for the amount of wear, determining the parts that should be rebored to a standard oversize, then new pistons and rings fitted. Scored cylinders always will allow excess blow-by, adding to operating costs due to lost horsepower and fast wear.

Standard oversize pistons and rings usually are available in 0.005-, 0.010-, 0.020-, and 0.020-in. oversizes. When reboring over 0.030-in. oversize, it is well to check with the compressor manufacturer, even though most cylinders can be rebored considerably over 0.062-in. oversize. This is particularly true for horizontal cylinders, which have a relief in

Fig. 5-5 Connecting rod with automotive-insert-type crankpin bearings.

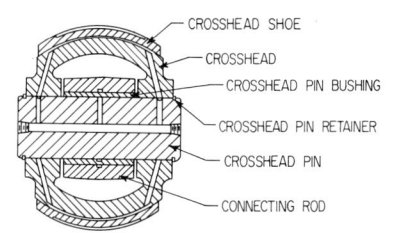

Fig. 5-6 Crosshead, crosshead pin, and crosshead pin bushing.

the valve section to clear the portion of the piston that extends beyond the piston-ring travel. Too much oversize can cause the piston to strike the valves.

Compressors having automotive-type pistons should have the wrist-pin fits checked when new piston rings are installed, and if found loose, the pin bushings should be replaced. Often the added drag on the cylinder walls caused by new piston rings will result in a pin knock when too much clearance is allowed.

Bearings Crankpin bearings on vertical and horizontal compressors are usually the automotive-insert type (Fig. 5-5). To correct problems with the insert type, the installation of new inserts will serve. Should the crankpin be damaged, it can be reground undersize, and undersize inserts can be used to get correct fit.

Horizontal double-acting compressors have crosshead pins which operate in crosshead-pin bushings which have no adjustment. If a failure occurs, the crosshead-pin bushings must be replaced. Because of different fit requirements of different compressor manufacturers, the running fit must be obtained from the compressor instruction book (Fig. 5-6).

Many different constructions are used for the main bearings in both vertical and horizontal compressors, no matter whether they are sleeve or antifriction bearings. Antifriction single-row tapered-roller-bearing adjustment is made by removing or adding shims. Double-row tapered roller bearings have an adjusting nut locked on the shaft. Unlock the nut and turn it to move the cone in on the cup. For trial purposes use a feeler gauge and get about 0.002 in. over the free rolls. Check bearings for heat and noise after starting, as it may be necessary either to tighten or to loosen them slightly (Fig. 5-7).

Intercoolers and aftercoolers These are important compressor parts that often are neglected to the extent that they become inefficient. The most important maintenance is simple, and that is the proper draining of the moisture traps or compartments. Any type of cooler is a condenser, and the condensate, if not drained regularly, will build up until water is carried over to the high-pressure cylinders in the case of an intercooler, and on into the air receiver and air lines in the case of an aftercooler. Coolers should be drained regularly, according to existing humidity condition. The surest way to ensure draining is the use of automatic drain traps on the intercooler, aftercooler, and air receiver.

Tube-type intercoolers and aftercoolers are subject to buildup from the mineral content in water, which, if not removed, will eventually affect cooling; therefore, these coolers need inspection for deposit removal.

Air-cooled intercoolers and radiators must have the core sections cleaned on the outside, because dirt will lodge in the core, reducing heat dissipation. For removal of dust, air blown through in a direction opposite the usual flow will do; but in case the dirt is contaminated with oil, a solvent should be applied, allowed to soak for a while, and then blown clean.

Cleaning An important item for proper compressor maintenance is keeping the compressor clean on the outside surfaces. Dirt and oil will make an insulation which hinders heat dissipation to atmosphere; this is especially true for an air-cooled compressor, which must depend on all heat dissipation through the cylinder and cylinder-head surfaces. When dirt is allowed to accumulate on the surfaces of a compressor, it is certain

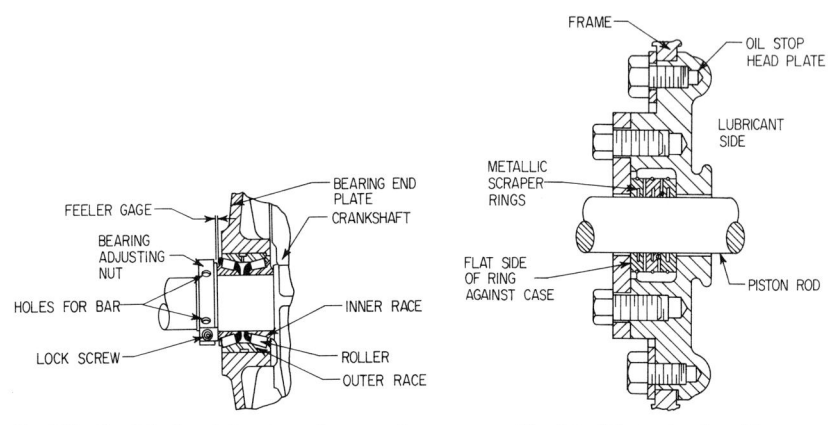

Fig. 5-7 Crankshaft main-bearing adjustment for double-row tapered roller bearings.

Fig. 5-8 Oil-stop-head packing.

some will find its way into the working parts. A well-kept clean compressor will pay dividends with a good appearance plus reduced operating and maintenance costs.

Unloading Practically every compressor manufacturer has his own type of air-unloading and control system; to cover all types would require complete data for each system. Some compressor vendors use several types; so for servicing the unloading system and its control, the instruction book should be referred to.

Some common unloading systems are suction unloading valves, suction throttling device, centrifugal unloaders, and bypass systems. Most of these controls are operated by means of a pressure switch and a three-way valve actuated by the solenoid. Another means of actuating the unloading device is a pneumatic pilot, of which there are several types on the market.

Packing Double-acting compressors using piston rods have oil-stop-head packing and cylinder-head packing which require periodic checking. The oil-stop-head packing usu-

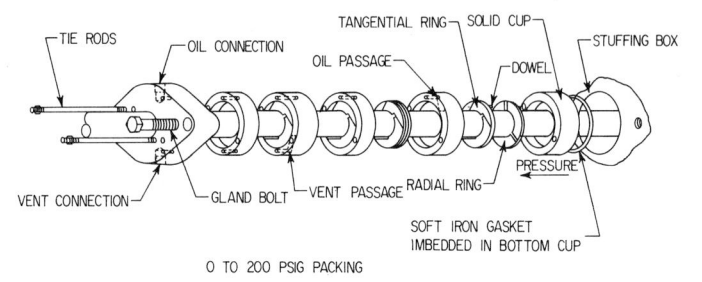

Fig. 5-9 Cylinder-head packing.

ally is a set of metallic scraper rings. They require very little attention because they are designed to scrape oil off the rod, yet get excellent lubrication. Should the piston rod become damaged, the packing will be ruined and new packing required. Never put new scraper rings on a piston rod that is nicked, scratched, or worn (Fig. 5-8).

The cylinder-head packing usually is the full floating design with self-adjusting packing rings. The material, style, and quanity of the packing rings and numbers of lubrication lines depend on the type of gas being compressed and the discharge pressure. Some applications require special packing, such as vented and/or the elimination of nonferrous packing-ring and gasket materials (Fig. 5-9).

Metallic packing, after it is installed and worn in, requires little attention. However, the piston rod should be checked where it passes through the packing and, if any scratches are present, the packing must be removed and inspected for embedded material causing the scratches. As long as the packing does not leak or show any signs of marking the rod, it should not be disturbed, as the metal rings are self-adjusting for the slight wear that occurs under normal operation.

The service check chart in Table 5-1 lists the common causes of malfunctions of mechanical parts of compressors.

TABLE 5-1 Service Check Chart, Mechanical Parts

1. Low oil pressure
    a. Low oil level
    b. Plugged oil-pump strainer
    c. Leaks in suction or pressure lines
    d. Worn-out bearings
    e. Defective oil pump
    f. Dirt in oil-filter check valve
    g. Broken oil-filter-check-valve spring
    h. Oil-pressure-bypass leaks
2. High oil pressure
    a. Plugged oil-pressure lines
    b. Defective oil-filter mechanism
    c. Excessive spring tension on filter check valves
    d. Excessive spring tension on oil-pressure adjusting mechanism
3. Incorrect delivery of mechanical lubricator
    a. Dirty or gummed valves
    b. Broken spring in check valve at cylinder
    c. Leak in lines or sight feed
    d. Low oil level
    e. Plugged vent in lubricator reservoir
4. Overheated low-pressure cylinder
    a. Insufficient cooling water
    b. Scored piston or cylinder
    c. Broken valves or valve springs
    d. Excessive carbon deposits
    e. Packing too tight
    f. Insufficient lubrication
    g. Corroded or clogged cylinder water passages
5. Overheated high-pressure cylinder
    a. Insufficient cooling water
    b. Scored piston or cylinder
    c. Broken valves or valve springs
    d. Excessive carbon deposits
    e. Insufficient lubrication
    f. Packing too tight
    g. Corroded or clogged cylinder water passages
6. Water in cylinders
    a. Leaking head gaskets
    b. Cracked cylinder or head
    c. Condensate caused by too much cooling water
7. High intercooler pressure
    a. Broken or leaking high-pressure valves
    b. Defective gauge
    c. Defective or leaking valve-seat gaskets

TABLE 5-1 Service Check Chart, Mechanical Parts *(Continued)*

8. Low intercooler pressure
   a. Broken or leaking low-pressure valves
   b. Leak in intercooler
   c. Piston-rod-packing leaking
9. Knocks
   a. Excessive carbon deposits
   b. Scored piston or cylinder
   c. Defective lubricator
   d. Foreign material in cylinder
   e. Piston hitting cylinder head
   f. Loose piston or piston pin
   g. Burned-out or worn rod bearings
   h. Loose main bearings
   i. Scored crosshead or crosshead guides
10. Scored cylinder, liner, or piston
    a. Foreign material
    b. Dirty or inefficient air cleaners
    c. Lack of lubrication
    d. Too much and too cold cooling water causing excess condensate and washing out lubrication
    e. Excessive heat
    f. Plugged water jackets
11. Broken valves and springs
    a. Too much condensation, causing rust
    b. Carbon deposits
    c. Foreign materials not removed by air cleaners
    d. Incorrect assembly
    e. Acid condition prevailing at location of suction air inlet
12. Control trouble
    a. Suction-valve unloader stuck open or closed
    b. Pressure switch defective
    c. Solenoid burned out
    d. Foreign material in three-way valves
    e. Vibration of control
    f. Voltage drop or loss of power
    g. Plugged air line or strainer
    h. Incorrect voltage or cycle
13. Incorrect operation of suction-valve unloaders
    a. Leaks in unloader line
    b. Foreign material in guides or seats
    c. Worn plungers
    d. Leaking or ruptured diaphragms
    e. Broken springs
    f. Manual shutoff partly closed
    g. Wrong pressure-switch setting

Note: Remember to read the instruction book carefully and to keep it and the parts list in an accessible place so that when information to make adjustments and repair is needed, shutdown time can be held to a minimum.

Index

Index

Accounting practices, 3-3 to 3-7
　equipment costing data sheet, typical, figure, 3-6
　fixed costs, 3-3
　　depreciation hourly charge, 3-3
　　depreciation hourly rate, formula, 3-4
　　depreciation method, 3-3
　　depreciation vs. repair costs, figure, 3-4
　　purchase cost percentage, annual, formula, 3-4
　　salvage value, 3-4
　　salvage value determination, 3-4
　operating costs (*see* variable costs, *below*)
　ownership costs (*see* fixed costs, *above*)
　variable costs, 3-1
　　filters and supplies, 3-5
　　fuel, 3-4
　　grease and oil, 3-5
　　labor, 3-4
　　repair parts, 3-4
　　schedule of accounts, example, 3-5
　　wages of operators, 3-5
　(*See also* Budgets)
Acids and alkalis for cleaning, 4-5
Adjustable wrenches, 2-42
Air-acetylene soldering (*see* Welding, gas, air-acetylene soldering, heating, and brazing)
Air compressors:
　reciprocating, 8-55 to 8-64
　　aftercoolers, 8-61
　　air cleaners, 8-56
　　air receiver, location and capacity, 8-56
　　air unloading, 8-62
　　bearings, 8-61
　　cleaning, 8-62
　　foundations, 8-55
　　intercoolers, 8-61
　　location, 8-55
　　lubrication, 8-58
　　packing, 8-62
　　piston rings, 8-60
　　starting, 8-57
　　troubleshooting, table, 8-63
　　valves, 8-58

Air compressors (*Cont.*):
　rotary helical-screw, 8-33 to 8-54
　　aftercooling, 8-51
　　air filters, 8-42
　　　frequency of servicing, 8-42
　　　service indicators, 8-41, 8-42
　　control systems, 8-43 to 8-46
　　　automatic start-stop, 8-44
　　　capacity control, 8-46
　　　constant-speed, 8-43
　　　dual-control, 8-45
　　　timed, with oil-reservoir blowdown, 8-45
　　heat recovery, 8-51
　　inlet valves, 8-43
　　installation, 8-33 to 8-35
　　　air receivers, 8-35
　　　discharge lines, 8-35
　　　foundations, 8-33
　　　inlet lines, 8-34
　　　oil-reservoir drains, 8-34
　　　ventilation, 8-33
　　instrumentation, 8-41
　　　air-filter service indicators, 8-41, 8-42
　　　air-pressure gauge, 8-41
　　　differential pressure gauge, 8-41
　　　hourmeter, 8-41
　　　temperature gauge, 8-41
　　lubrication, 8-46 to 8-48
　　　drainage and cleaning, 8-47
　　　filling the system, 8-48
　　　lubricants, 8-46
　　　oil-change frequency, 8-47
　　　oil levels, 8-47
　　　synthetic fluids, 8-47
　　oil cooling systems, 8-48
　　oil filter, 8-48
　　oil separation, 8-50
　　pipe size determination, table, 8-35
　　principles of, 8-35 to 8-39
　　　aftercooler system, figure, 8-36
　　　air receiver system, figure, 8-36
　　　compression principle, 8-35
　　　design of, 8-38
　　record keeping on, 8-51
　　safety devices, 8-29 to 8-41
　　　blowdown valve, automatic, 8-40

3

4 Index

Air compressors, rotary helical-screw, safety devices (*Cont.*):
    high-air-temperature shutdown, 8-40
    minimum-pressure valve, 8-40
    safety valve, 8-41
    starting of, 8-39
    troubleshooting, 8-51 to 8-54
Air tools (*see* Portable power tools, air tools)
Apprentice training (*see* Training maintenance-trades personnel, apprentice training)
Apprenticeship schedules (*see* Training maintenance-trades personnel, apprenticeship schedules)
Arc welding (*see* Welding, arc)
Asphalt paving equipment (*see* Paving equipment, asphalt)
Asphalt plants:
    controls for, 7-287 to 7-294
        batch controls, 7-290
            automatic weighing console, 7-291
            limit switches, 7-290
            main console, 7-290
            solenoid valve, 7-281
            supply bin level indicator, 7-291
            wiring, 7-291
        burner, 7-292
            automatic gas valve, 7-292
            combustion chamber, 7-292
            control motor, 7-292
            linkage rods, 7-292
            nozzle assembly, 7-292
            oil basket strainer, 7-292
            preheater, 7-292
            refractory, 7-292
            spark plug, 7-292
        burner control automatic recording, 7-292 to 7-294
            flame safeguard system, 7-293
            limit switches, 7-292
            recording and indicating components, 7-294
            thermocouple, 7-294
        electrical system, 7-291 to 7-292
        hydraulic system, 7-287
            accumulator, 7-289
            breather vent, 7-289
            cylinders, 7-289
            directional valve, 7-289
            flow control valve, 7-289
            oil pressure filter, 7-289
            oil strainer, hydraulic, 7-288
            pressure gauge, 7-289
            pump, 7-289
            relief valve, 7-289
            solenoid-operated control valves, 7-289
            suction filter, 7-288

Asphalt plants (*Cont.*):
    inspection (*see* maintenance and inspection, *below*)
    lubrication (*see* Lubrication, asphalt plants)
    maintenance and inspection, 7-269 to 7-287
        continuous mixer and pump drive, 7-280
        drum mixer, 7-283
            discharge chute, 7-284
            drum, 7-284
            trunnion rollers, 7-283
        fines feeding and measuring system, 7-275
            elevator, 7-275
            screw conveyor, 7-275
        general considerations, 7-269 to 7-271
        hot elevator, 7-284
            bucket line, 7-284
            figure, 7-285
            chute liner plate, 7-284
            drive belt, 7-284
            figure, 7-285
            drive chain, 7-284
        metering pump, 7-279
        pugmill, 7-280 to 7-283
            arms, 7-280
            gate, 7-281
            figure, 7-291
            gear box, 7-282
            keys, loose-fitting, 7-283
            liner plates, 7-281
            paddle tips, 7-280
            shaft seals, 7-282
        spray, asphalt, 7-277
            gravity spray, 7-278
            spray pump, 7-277
        sump and strainer, 7-278
        transfer pump, 7-279
        V-belt drives, 7-286
            adjustment procedure, 7-287
            alignment, 7-286
            belt deflection forces, recommended, table, 7-287
            cleaning, 7-286
            replacement, 7-286
            sheave grooves, 7-286
            tension, 7-286
        vibrating screen, 7-272 to 7-275
            amplitude adjustment, 7-273
            charging end wear plate, 7-274
            screen cloth replacement, 7-273
            screen drive sheaves, 7-273, 7-297
            figure, 7-274
            springs, 7-273
            vibrating unit, 7-274
            weigh bucket, 7-276

Asphalt plants, maintenance and
  inspection (*Cont.*):
   weigh hopper, 7-275
    storage bin indicator
     counterweight arms, 7-275
    storage bin strike-off gates, 7-275
   weigh hopper scale, 7-276
  satellite equipment for, 7-297 to 7-302
   apron feeder, 7-297
    belt flashing, 7-297
    bin level telltales, 7-297
    chain, 7-297
    control gate liners, 7-297
    divider plates, 7-297
   cold feeder, 7-297
    belt flashing, 7-298
    belt wiper, 7-298
    conveyor belt, 7-298
    conveyor drive chain, 7-297
    conveyor rollers, 7-298
    drive belts, 7-297
   dryers, 7-300 to 7-302
    combustion chamber, 7-300
     figure, 7-301
    discharge chute, 7-300
    drum drive chain, 7-300
    drum drive chain oilers, 7-300
     figure, 7-301
    drum seals, 7-300
    dryer drum, 7-300
    rotary charging chute, 7-300
    trunnion and thrust rollers (*see*
     drum mixer, *above*)
   dust collectors, 7-298 to 7-300
    ducts, 7-299
    fabric filter bag wear, 7-298
    fan, 7-299
    moisture infiltration, 7-299
    oil present, 7-299
    visible dust, 7-299
   gradation unit, 7-297
    screen cloths, 7-297
    screen drive sheave, 7-273, 7-297
     figure, 7-274
   surge storage system, 7-302
    air cylinders, 7-302
    air filter, main, 7-302
    air oiler, 7-302
    conveyor bed plates, 7-302
    return roller, 7-302
    slat conveyor, figure, 7-302
Atomic-hydrogen welding, 4-41
Augers, hard-facing, 4-123
Automobile-body soldering, 4-82

Backhoes (*see* Excavators, hydraulic, and
  backhoes)
Bearings:
 general (*see* Mechanical power
  transmission equipment, bearings)

Bearings (*Cont.*):
 plain, 4-125 to 4-134
  babbitt thickness as a measure of
   bearing life, figure, 4-127
  bearing temperature as a determinant
   of life, figure, 4-130
  case tolerances, table, 4-126
  cast-babbitt liners, renewal of, 4-133
  connecting-rod tolerances, table, 4-126
  design of, 4-125 to 4-127
   grooving, 4-127
   loads, 4-125
   lubrication, 4-127
   materials, 4-125
   tolerances, 4-125
  housing reclamation, 4-133
  inspection and reconditioning of, 4-129 to 4-130
   bearing replacement, 4-130
   connecting rods, 4-129
   journals, 4-129
  load-carrying capacities for various
   materials, figure, 4-128
  maintenance and care of, 4-128
   cleanliness, 4-128
   lubricant selection, 4-127
   pressure, 4-128
   temperature, 4-128
  oil-clearance values, recommended,
   table, 4-127
  reassembly of, 4-131 to 4-133
   bolt torque, 4-131
   crush, 4-131
   end clearance, 4-132
   final checking, 4-133
   free rotation, 4-133
   measure of clearance, 4-132
   oil clearance, 4-132
   preliminary lubrication, 4-133
  shaft tolerances, table, 4-126
 rolling, 4-135 to 4-154
  bearing grease dropping-point
   temperatures for various materials,
   4-154
  boundary dimensions, 4-135
  design and nomenclature for, 4-135
  general considerations, 4-135
  load ratings for, 4-138
  lubrication of, 4-150 to 4-154
   grease lubrication, 4-153
   oil lubrication, 4-151
  mounting and dismounting methods,
   4-144 to 4-150
   cold mounting, 4-144
   dismounting procedures, 4-149
   general considerations, 4-144
   hydraulic removal, 4-149
   tapered-bore bearing mounting, 4-146

Bearings, rolling (*Cont.*):
    mountings of, **4**-139 to **4**-143
    series breakdown of, **4**-136
    shaft and housing fits, **4**-139
    temperature mountings, **4**-144
        hot-oil bath, **4**-145
        hot plate, **4**-145
        introduction heaters, **4**-145
        temperature-controlled ovens, **4**-145
        temperature/viscosity relationships, figure, **4**-153
    specific (*see under type of unit*)
Beaufort scale of wind force, xiv to xv
Belt conveyors (*see* Conveyors, conveyor belts)
Blades:
    dozer (*see* Moldboard, dozer)
    scraper (*see* Scrapers and scraper blades)
Boilermaker, apprenticeship schedules for, **1**-22
Brazing (*see* Welding, gas)
Bucket teeth, **6**-38
    (*See also* Buckets and bucket teeth)
Buckets and bucket teeth, **6**-31 to **6**-42
    components: bucket, alloy steel welding of, **6**-38 to **6**-40
        bucket teeth, **6**-38
        dot hard-facing, **6**-39
        electrodes, **6**-39
        ESCO alloy-12 series, **6**-38
        manganese steel, **6**-39
        moisture control, **6**-39
        preparation for, **6**-39
        structural attachments, **6**-39
        techniques for, **6**-39
        welding rod choices, **6**-39
    dippers, shovel and hoe, **6**-34 to **6**-36
        adapter installation, **6**-36
        clamshell bucket adapter installation, **6**-36
        integral nose tooth base adapter installation, **6**-37
        tooth horn adapter installation, **6**-34
        Whistler adapter maintenance, **6**-35
    dragline buckets (*see* dippers, shovel and hoe, *above*)
    front end loaders, **6**-31 to **6**-34
        adapters: one-leg, welding of, **6**-33
        two-leg, welding of, **6**-32
        types of, **6**-32
        locking devices, **6**-32
        nose types, **6**-32
        point selection, **6**-32
        tooth spacing, **6**-31
        weld sizes, **6**-34
        welding points, **6**-32
    rebuilding operations, **6**-40 to **6**-42
        conical lip tooth bases, **6**-42

Buckets and bucket teeth, rebuilding operations (*Cont.*):
    integral noses, **6**-41
    lip tooth bases, **6**-41
    nonconical lip tooth bases, **6**-41
    Whistler lip adapter bearing pads, **6**-40
Budgets, **3**-1 to **3**-2
    cost recording for, **3**-2
    monitoring, **3**-2
    policy establishment through, **3**-1
    systems for controls and, **3**-2
    (*See also* Accounting practices)
Bureau of Apprenticeship and Training, U. S. Department of Labor, **1**-19

C-clamps, **2**-59
Carbon-arc welding, **4**-42
Carpenter, apprenticeship schedules for, **1**-23
Celsius and Fahrenheit scales, table of temperature equivalents, xiv
Centralized lubrication systems (*see* Lubrication, centralized systems)
Centrifugal pumps (*see* Pumps, centrifugal)
Chain hoists (*see* Hoists and slings, chain)
Chain pipe tongs, **2**-32
Chain pipe vise, **2**-33
Chain wrenches, **2**-32
Chemical cleaning (*see* Cleaning, chemical)
Chisels, **2**-53
Circular saws, **2**-23
Cleaning:
    chemical, **4**-5 to **4**-9
        vs. alternative cleaning methods, **4**-8
        application methods, **4**-6
            fill and empty, **4**-6
            flow-through vessels, **4**-6
            large hollow vessels, **4**-7
            new ideas, **4**-7
            pigs, plugs, balls, and jets, **4**-7
        data and decisions for planning a job, **4**-8
        disposal problems, **4**-7
        materials for cleaning, **4**-5
            acids and alkalis, **4**-5
            organic solvents, **4**-6
            sequestrants, **4**-5
            specialty cleaning products, **4**-6
            synthetic detergents and acid inhibitors, **4**-6
        planning a job, data and decisions for, **4**-8
        references, **4**-9
    steam and hot water, **4**-245 to **4**-250
        auxiliary equipment for, **4**-249
        ChemJet, **4**-249

Cleaning, steam and hot water, auxiliary equipment for (*Cont.*):
  degasser, 4-249
  parts washer, 4-249
  SandJet, 4-249
  tank cleaner, 4-249
  cleaning unit operation for, 4-247
  hot water high-pressure washer, 4-248
  steam cleaner, 4-248
  portability vs. permanence, 4-246
  pressure washer, 4-245
  process of, 4-245
  program establishment for, 4-246
    daily cleaning, 4-246
    on-site machinery inspection, 4-246
    premaintenance cleanup, 4-247
  size of unit for, 4-246
  steam cleaner, 4-245
  washing techniques for, 4-247
Combination wrenches, 2-42
Compactors:
  static weight, 7-91 to 7-106
  general considerations, 7-91
  preventive maintenance procedures for, 7-94 to 7-97
    daily checkpoints, 7-94
    figure, 7-95
    100-hr checkpoints, 7-96
    2-week checkpoints, 7-96
    figure, 7-97
  preventive maintenance programs for, 7-91
  preventive maintenance requirements for, 7-93
  storage, winter, 7-105
  troubleshooting major systems on, 7-97 to 7-105
    clutch, 7-103
    cooling, 7-103
    diesel fuel, 7-104
    engine, 7-98
    final drive, 7-103
    fuel, 7-104
    hydraulic, 7-100
    hydrostatic transmission, 7-98
    figure, 7-99
    steering, 7-104 to 7-105
    torque converter, 7-103
    transmission, 7-103
  winter storage, 7-105
  vibratory drum, 6-43 to 6-50
    cold weather start-up procedures, 6-48
    erratic rpm, causes of, 6-49 to 6-50
      air in fluid, 6-49
      belts loose, 6-50
      relief valve setting off, 6-50

Compactors, vibratory drum (*Cont.*):
  field troubleshooting (*see* troubleshooting, field *below*)
  general maintenance considerations for, 6-43
  major repair of, 6-50
  pressure levels, maintaining, 6-45
  record keeping for maintenance of, 6-43
    figures, 6-44, 6-45
  repair, major, 6-50
  schematic guide to, figure, 6-46
  start-up procedures: cold weather, 6-48
    steering system, 6-45 to 6-48
    vibration system, 6-45 to 6-48
  steering system start-up, 6-45 to 6-48
  troubleshooting, field, 6-44 to 6-45
    air in system, 6-45
    broken control cable, 6-45
    filter, plugged, 6-44
    parking brake engaged, 6-44
    plugged filter, 6-44
    pump drive disconnected, 6-45
    suction line clogged, 6-44
  vibration circuit, troubleshooting, 6-49
  vibration system start-up, 6-45 to 6-48
Compressors (*see* Pumps and compressors)
Computer, role of, in maintenance cost, 3-9 to 3-13
  equipment repair cost sheet, figure, 3-12
  foreman's daily report, figure, 3-10
  mechanic's time sheet, figure, 3-11
  schedule of machinery and equipment, figure, 3-13
  voucher authorization, figure, 3-12
Concrete pumps (*see* Pumps, concrete)
Conversion factors, table of, xiii to xv
Conveying equipment (*see* Conveyors)
Conveyors, 4-203 to 4-221
  conveyor belts, 4-205 to 4-212
    cutting, procedure for, 4-208
    fasteners, applying, figure, 4-210
    inspection of, 4-207
    installation of, 4-211
    training of, 4-211
    troubleshooting for, table, 4-209
    wear, causes of, 4-205
  gas-driven unit adjustment, 4-216
    countershaft drive belt, 4-216
    drive shaft V-belts, 4-216
    head shaft drive chain, 4-216
  introduction to, 4-203
  lubrication, 4-216
    bearings: antifriction, 4-221
      plain, 4-221
    chain guards, oil-tight, 4-221

8   Index

Conveyors, lubrication (*Cont.*):
   criteria for, 4-204
   drive, 4-221
   engines, gasoline, 4-221
   idlers, 4-219
   motors, electric, 4-221
   oil-tight chain guards, 4-221
   safety, 4-219
   operating precautions (*see* precautions, operating, *below*)
   permanent, description of, 4-203
   portable, description of, 4-203
   precautions, operating, 4-212 to 4-216
     belt scraper, 4-216
     belt scraper adjustments, 4-216
     capacity measurement, belt, 4-213
     conveyor hopper flashing, 4-214
     feeding, 4-213
     general, 4-212
     hand hydraulic pump, 4-214
     hydraulic hoist assembly, 4-214
     hydraulic ram, 4-215
     overloading, conveyor, 4-213
     tires and wheels, 4-216
     V-belt tension adjustment, 4-214
        figure, 4-215
   safety considerations, 4-204, 4-219
Corrosion control, 4-1 to 4-4
   inhibitors, 4-3
   methods of stopping corrosion, 4-2 to 4-3
   nonmetallic materials, 4-3
     plastic, 4-3
     rubber and elastomers, 4-3
   protective coatings, 4-3
   references, 4-4
   types of corrosion, 4-1
     crevice corrosion, 4-2
     erosion, 4-1
     exfoliation and selective leaching, 4-2
     galvanic corrosion, 4-1
     intergranular corrosion, 4-2
     pitting, 4-2
     selective leaching and exfoliation, 4-2
     stress-corrosion cracking, 4-2
Costs, maintenance:
   budgets, 3-1 to 3-2
   computer role in, 3-9 to 3-13
   fixed and variable, 3-1 to 3-5
Cranes, hydraulic, 7-177 to 7-205
   circulatory system (*see* hydraulic circulatory system, *below*)
   components of, 7-178
     boom elevating system, 7-178
     boom swing system, 7-178
     boom telescope system, 7-178
     hydraulic hoist system, 7-178
     hydraulic outrigger system, 7-178
     superstructure, 7-178

Cranes, hydraulic (*Cont.*):
   control valves, hydraulic, 7-181 to 7-184
     causes of faulty, 7-182
     contamination scoring of, 7-181
     faulty, causes of, 7-182
     inspection of, 7-181
     sticking valve spools, 7-181
     troubleshooting on, table, 7-183 to 7-184
     valve leakage checks for, 7-181
   electrical systems, 7-193 to 7-200
     batteries, 7-197
     instruments, engine, 7-200
     solenoid valves, 7-200
     specific gravity, electrolyte, 7-200
     switches, 7-200
   filter, hydraulic system, 7-182
   functions of, 7-177
   general description of, 7-177
   hoist, hydraulic (winch), 7-190 to 7-192
     alignment, 7-190
     functional check, 7-192
     inspection, 7-190
     installation, 7-191
     removal, 7-190
   hydraulic circulatory system, 7-178
     arctic conditions, 7-179
     draining, 7-179
     fluid selection for performance, 7-178
     flushing, 7-179
     flushing procedures, 7-179
     hydraulic fluid specification, 7-178
   hydraulic oil breakdown, 7-181 to 7-182
   hydraulic symbols, table, 7-198 to 7-199
   lubrication plan, 7-203 to 7-205
   lubrication standards, 7-202 to 7-203
   oil breakdown, hydraulic, 7-181 to 7-182
   oil inspection criteria, 7-203
   outrigger control panels, 7-201
   outrigger troubleshooting, table, 7-195 to 7-197
   outriggers, 7-193
   pumps, hydraulic, 7-179 to 7-181
     functional check of, 7-180
     inspection of, 7-179
     installation of, 7-180
     removal of, 7-180
     troubleshooting for, table, 7-180
   reels, hydraulic hose, 7-192
   solenoid valve troubleshooting, table, 7-202
   swing bearings, hydraulic, 7-184 to 7-187

Cranes, hydraulic (*Cont.*):
   swing system, hydraulic, 7-184
   swing system troubleshooting, table, 7-185
   swivel couplings, hydraulic, 7-187
     functional check, 7-188
     inspection, 7-187
     installation, 7-188
     reassembly, 7-187
     removal, 7-187
   symbols, hydraulic, table, 7-198 to 7-199
   telescoping-boom assembly, 7-188
     alignment, 7-188
     drift checks of boom elevation cylinders, 7-189
     inspection, 7-188
     removal, 7-188
     servicing, 7-188
   winch (*see* hoist, hydraulic, *above*)
   windshield wiper assembly, 7-200
Crawler clutch, trencher, 7-310 to 7-311
Crawler tractor master pin removal set, 2-11
Crawler tractor sprocket removal set, 2-11
Crushers:
   hammermills and impact mills, 7-256 to 7-266
     bearing: bench assembly of, 7-266
       installation of, 7-265
       removal of, 7-265
     hammer tip, increasing life of, 7-258
     hammermill, care of, 7-256
     hammers: fasturn, installation of, 7-258
       finger clamp type, 7-259
       wear of, 7-260
     lifting of, for maintenance, 7-256
     lubrication, 7-256
     lubrication plan for, figure, 7-257
     mounting check, 7-256
     spinner, 7-260
       installation of, 7-265
       repair of, 7-263
     spinner shaft bearings, 7-263
     tips, turning of, procedure, 7-258
     troubleshooting on, 7-264
     weight balancing, hammer: two-row spinner, 7-260
       three-row spinner, 7-262
   impact breaker type, 7-245 to 7-256
     bearing, impeller shaft, maintenance of, 7-252
     bolts and bushings, safety plate, 7-255
     discharge accumulation prevention, 7-246
     double-impeller contact check, 7-251
     feed, even distribution of, 7-247

Crushers, impact breaker type (*Cont.*):
     feed penetration adjustment, 7-246
     feed to double-impeller, need to center, 7-247
     ground wire attachment, welding, 7-251
     impact breaker buildup, by welding, 7-247
     impeller and impeller shaft, 7-252
       reassembly of, 7-252
     impeller bar: installation of, 7-251
       removal of, 7-251
     impeller bar weight balance check, 7-251
     impeller shaft removal, 7-252
     large piece feed, restrictions on, 7-246
     lifting of, for maintenance, 7-246
     lubrication, 7-246
     mounting check, 7-246
     safety plate bolts and bushings, 7-255
     troubleshooting on, 7-255
     welding ground wire attachment, 7-251
   impact type (*see* Crushers, hammermills and impact mills)
   jaw type, 7-234 to 7-245
     discharge clearance, need for, 7-239
     jaw replacement, stationary and movable, 7-237
     key plate replacement, 7-237
     lifting of, for maintenance, 7-235
     lubrication, 7-234
     lubrication plan for, figure, 7-236
     mounting check, 7-234
     pitman assembly: installation of, 7-244
       lifting of, 7-235
     pitman barrel guard, 7-245
     pitman bearing: installation of, 7-243
       removal of, 7-242
     pitman shaft: installation of, 7-244
       removal of, 7-244
     pumps, hydraulic, 7-240
     rams, hydraulic, 7-240
     shaft removal, from pitman, 7-244
     side bearing: installation of, 7-241
       removal of, 7-242
     stationary jaw rotation, 7-237
     stationary jaw tightness, 7-235
     toggle plate inserts, 7-240
     toggle plate replacement, 7-239
     toggle plate seats, 7-240
     toggle plate tension spring, tightness of, 7-239
     troubleshooting on, 7-240
       bearing heating, 7-240
       capacity low, 7-240

Crushers, jaw type, troubleshooting on (*Cont.*):
    discharge adjustment difficult, 7-240
    jaws worn, 7-240
    wedge, movable jaw, tightness of, 7-238
    toggle plate, tightness of, 7-240
  roll type, 7-207 to 7-234
    bearing(s): care of, 7-220
      installation of, 7-222 to 7-224
      overheating of, causes of, 7-221
      removal of, 7-221
      roller, heating of, 7-224
      shrunken-on, removal of, 7-229
    bearing block slide bars, 7-226
    bearing temperature check, 7-220
    compression assemblies, 7-213
    compression spring tightness, 7-211
    countershafts: assembly and installation of, 7-229
      removal of, 7-230
    cracked shell repair, 7-217
    feed particle size as wear cause, 7-210
    feed rate as wear cause, 7-210
    finger gear: installation of, 7-233
      removal of, 7-233
    finger gear case, 7-234
    finger gear drive, 7-232
    gears: spur, cases for, 7-228
      spur and pinion: installation of, 7-227
        removal of, 7-226
    keys: straight, fitting and installing, 7-232
      tapered, fitting and installing, 7-232
    lifting of, for maintenance, 7-207
    lubrication, 7-207
    lubrication plan for, 7-209 to 7-210
    material spread to curb wear, 7-211
    mounting checks, 7-207
    protective shear assemblies, 7-213
    record keeping on, 7-234
    roll contact check, 7-211
    roll shaft pressure requirements, figure, 7-225
    roll shell, 7-215
      condition of, 7-215
      surface buildup, 7-215
      temperature limitations, 7-215
      welding details for, 7-215
      welding information on, 7-215
    rubber tire drive for, 7-207
    shafts and spiders, 7-225
    shell replacement, 7-219
    shell welding: preparation for, 7-216
      recommendations for, 7-218
    spiders and shafts, 7-225

Crushers, roll type (*Cont.*):
  tire pressures, recommended, table, 7-208
  troubleshooting on, 7-214
    bearing overheating, 7-214
    capacity low, 7-214
    roll shell wear excessive, 7-214
  wear, material spread to curb, 7-211
  welding patterns, typical, 7-218

Depreciation, machinery or equipment (*see* Accounting practices, fixed costs)
Diesel engines, 5-9 to 5-31
  air intake and exhaust, 5-23 to 5-25
    exhaust system, 5-24
    intake system, 5-23
    turbocharger, 5-24
  cold-weather starting, 5-11
  cooling system, 5-17 to 5-20
    aeration, 5-19
    clogging, 5-17
    coolant, 5-17
    exhaust gas leakage, 5-20
    external leaks, 5-18
    fan and belts, 5-19
    filters, 5-19
    flushing cooling system, 5-20
    hoses, 5-19
    internal leaks, 5-18
    radiator, 5-17
    radiator cap, 5-17
    thermostats, 5-18
    water pump, 5-19
  electrical system, 5-25 to 5-28
    alternator, 5-27
    batteries, 5-25
    gauges, 5-28
    generator, 5-27
    starting motor, 5-27
  engine stopping precautions, 5-12
  fuel system, 5-12 to 5-17
    bleeding, 5-16
    filters, 5-13
    fuel-injection nozzles, 5-14
      troubleshooting, table, 5-15
    fuel lines, 5-14
    fuel storage, 5-16
    fuel-transfer pump, 5-13
    governors, 5-16
    injection pumps, 5-14
    troubleshooting, 5-12
  lubrication of, 5-20 to 5-23
    additives, 5-21
    oil consumption, 5-22 to 5-23
    oil contamination, 5-20 to 5-21
    oil coolers, 5-22
    oil filters, 5-21 to 5-22
    pressure-regulating valve, 5-22
    ventilation, 5-22

Diesel engines (*Cont.*):
  start-up considerations for, **5**-11 to **5**-12
  troubleshooting, by malfunction, **5**-28 to **5**-31
  visual inspection of, **5**-9 to **5**-11
    air supply, **5**-10
    coolant leaks, **5**-10
    electric systems, **5**-10
    fuel systems, **5**-10
    oil leaks, **5**-10
    turbocharged engine checks, **5**-11
Diesel tractors (*see* Tractors, diesel)
Digging clutch, trencher, 7-310 to 7-311
Ditchers (*see* Trenchers and ditchers)
Dozers (*see* Moldboard, dozer; Tractors, diesel)
Drill (portable electric tool), **2**-20
Drilling equipment (*see* Drills and drilling equipment)
Drills and drilling equipment, 7-333 to 7-350
  air and water tubes, 7-348
  automatic valve assembly, 7-347
    general considerations, 7-347
    reconditioning of, 7-348
      axial end-seating valves, 7-348
      butterfly valves, 7-348
  buffer ring, 7-344
  chuck parts, 7-347
    prismatic steels, 7-347
    rock drills, 7-347
  chuck wear, 7-334
  cleanliness of, 7-337
  cushion tests, 7-343
  cylinders, maintenance of, 7-342
  cylinders and pistons, reconditioning of, 7-343
  drill and drill steel misalignment, 7-334
  field maintenance of, 7-339
  front end cushion, loss of, 7-343
  general considerations for maintenance of, 7-333
  lubrication of, 7-333
  misalignment, drill and drill steel, 7-334
  piston, maintenance of, 7-340
  piston striking face, repair of, 7-344
  reconditioned drill, testing of, 7-349
  record keeping for maintenance on, 7-349
  replacement parts for, 7-339
  rifle bar, care of, 7-345
  rotating parts, inspection of, 7-346
    chuck nut, 7-346
    ratchet pawls, 7-346
    ratchet ring, 7-346
    rifle bar and rifle nut, 7-346
    rock drills, 7-347
  shanks, steel, out-of-standard, 7-335 to 7-337
    center hole cleanness, 7-336

Drills and drilling equipment, shanks, steel, out-of-standard (*Cont.*):
    collar formation, 7-336
    collar length, 7-336
    size, 7-337
    striking end configuration, 7-335
    temper, 7-337
  shop maintenance of, 7-339
    assembly, 7-339
    disassembly, 7-339
  side rods, tightness of, 7-338
  steel retainer, 7-349
  throttle valve, 7-349
  tubes, air and water, 7-348
  wear tolerance, allowable maximum, 7-340
    automatic valve, 7-340
    clearances: measuring of, 7-340
      piston, cylinder, and buffer ring, 7-340

Earthmoving buckets and bucket teeth (*see* Buckets and bucket teeth)
Economics of maintenance, **3**-1 to **3**-26
  accounting, **3**-3 to **3**-7
  budgets, **3**-1 to **3**-2
  costs, computer in, **3**-9 to **3**-13
  preventive maintenance in, **3**-15 to **3**-26
Electric motors (*see* Motors, electric)
Electric tools (*see* Portable power tools, electric tools)
Electrical power systems (*see* Power systems, electrical)
Electron-beam welding, **4**-43
Electroslag welding, **4**-41
Engine clutch, trencher, 7-309 to 7-310
Engines (*see* Diesel engines; Gasoline engines)
Excavators, hydraulic, and backhoes, 7-107 to 7-120
  air system, 7-115 to 7-117
    air filter, 7-116
    air line lubricator, 7-116
    alcohol evaporator, 7-116
    moisture ejector, 7-116
    reservoir, 7-115
  bearings, 7-117
  chains, drive, for tracks, 7-118
  cleanliness and housekeeping of, 7-108 to 7-110
    cab, 7-110
    hydraulic system, 7-108
    undercarriage, 7-110
  compounds, special, use of, 7-119
  daily inspection of, 7-107
  drive chains for tracks, 7-118
  hydraulic fluids for, 7-112
  hydraulic pressure relief valves, 7-120

Excavators, hydraulic, and backhoes (*Cont.*):
    hydraulic systems, 7-113 to 7-115
        components, 7-114
        filters, 7-114
        lines, 7-114
        priming of, 7-114
        tanks for, 7-113
        venting of, 7-114
    logbook for, 7-104
    lubrication of, 7-110 to 7-112
        amount, 7-111
        application, 7-112
        automatic units for, 7-112
        types, 7-110
    parts replacement, 7-119
    pressure relief valves, hydraulic, 7-120
    pump drive case, 7-116
    ring gear, 7-117
    sprockets, drive, for tracks, 7-118
    torque ratings, 7-118
    track drive chains, 7-118
    track drive sprockets, 7-118
Explosive welding, 4-43

Fahrenheit and Celsius scales, table of temperature equivalents, xiv
Feeler gauge, 2-35
Fixed costs (*see* Accouting practices, fixed costs)
Flange-Jacks, 2-35
Flaring tool, 2-34
Flash welding, 4-42
Flow welding, 4-43
Friction disk replacement, trencher, 7-311
Friction welding, 4-43
Fuel, cost of, 3-4

Gas welding (*see* Welding, gas)
Gasoline engines, 5-33 to 5-37
    air cleaner: dual-element, servicing, 5-35
        function and design, 5-34
        oil-foam, servicing, 5-34
    combustion chamber cleanout, 5-36
    cooling system cleaning, 5-37
    fuel for, 5-35
    general considerations for maintenance of, 5-33
    lubrication of, 5-35
    spark plug blast cleaning, 5-35
    storage of, 5-37
Generators, electric, 5-1 to 5-7
Graders (*see* Motor graders)
Graphite (*see* Lubricants, solids, characteristics of, 4-17)
Grease (*see* Lubricants, grease)

Hammer welding, 4-43
Hammers, 2-24, 2-52 to 2-53
Hand tools (*see* Tools, hand)
Heat recovery, rotary helical-screw air compressor, 8-51
Heavy tools (*see* Tools, heavy)
Hoists and slings, chain, 4-187 to 4-201
    electric, 4-189
    general considerations, 4-187
    instruction check diagram, 4-191
    instructions for operators, 4-193
    lever-operated, 4-189
    lubrication of, 4-198, 4-200
    preventive maintenance of, 4-190 to 4-193
        frequency, formula for, 4-191
        inspection, 4-192
            frequent, 4-192
            idle-hoist, 4-193
            initial, 4-192
            periodic, 4-192
            procedure for, 4-192
            records, 4-192
    preventive maintenance vs. breakdown maintenance, 4-200
    rigger ratchet, 4-187
    selection of, 4-190
    service, 4-192
    spur-geared (*see* spur-geared, *under* types of *below*)
    testing, 4-193
        load, 4-193
        operating, 4-193
    troubleshooting on: electric, all, table, 4-196 to 4-198
        electric, two-speed, table, 4-199
        spur-geared, table, 4-194 to 4-195
    types of, 4-187 to 4-190
        electric, 4-189
        lever-operated, 4-189
        rigger ratchet, 4-187
        spur-geared, 4-187, 4-189
            cyclone and satellite, 4-187
            low-headroom trolley, 4-187
            modern, 4-189
    welded link load and hand chain, 4-200
Hose, hydraulic, 4-223 to 4-244
    characteristics of, by stock type, table, 4-224
    components of, 4-223
        cover, 4-226
        reinforcement, 4-224
        tube, 4-223
    couplings, 4-228 to 4-231
        general, 4-228
        permanent, 4-229
    fittings and adapters, 4-231 to 4-244
        adapters, 4-243

Hose, hydraulic, fittings and adapters (*Cont.*):
   assemblies: assembly length, overall, 4-237
      coupling installation equipment, 4-239
      determining, 4-237
      installing, 4-243
   flanged fittings, 4-232
      definitions for, 4-232
      iron pipe, 4-232
      shapes, 4-232
      standard screw, 4-232
      thread identification, 4-232
   flow capacity of, table, 4-227
   selection of, 4-236
   threaded fittings, 4-231
Hose fittings (*see* Hose, hydraulic)
Hot water cleaning (*see* Cleaning, steam and hot water)
Human factors in maintenance management, 1-35 to 1-42
   confidence in others, 1-40
   confidence meter, figure, 1-41
   group behavior, 1-38
   individual motivation, 1-38
   individuality of the maintenance worker, 1-37
   interdepartmental relations, 1-40
   motivation from work, 1-39
      ability utilization, 1-39
      achievement recognition, 1-39
      action freedom, 1-40
      challenge, 1-39
      extended responsibility, 1-39
      goal setting, involvement in, 1-40
      information access, 1-39
      planning, involvement in, 1-40
      problem solving, involvement in, 1-40
   motivational needs, 1-37
      to be alive, 1-37
      to be social, 1-37
      to do work we like, 1-37
      to feel safe, 1-37
      to feel worthy and respected, 1-37
   nonmotivators, 1-39
   power of participation, 1-38
   references, 1-41
   uniqueness of maintenance work, 1-35
   uniqueness of the work force, 1-36
      at the bottom, 1-36
      at the craft level, 1-36
      at the semiskilled level, 1-36
      at the top, 1-36
Hydraulic cranes (*see* Cranes, hydraulic)
Hydraulic hose (*see* Hose, hydraulic)
Hydraulic systems, 5-39 to 5-51
   actuators for, 5-46

Hydraulic systems (*Cont.*):
   ball check valve, figure, 5-45
   circuitry for, figure, 5-40
   cylinder, nomenclature and parts of, figure, 5-46
   external gear pump, figure, 5-44
   filters, micronic-particle, figure, 5-49
   fluid, viscosity of, table, 5-47
   fundamentals of, 5-39
   globe valve, figure, 5-45
   internal gear pump, figure, 5-45
   maintenance considerations, 5-46 to 5-51
      filtration, 5-48
      fluid factors, 5-47
      reservoir, 5-49
      start-up, 5-51
      troubleshooting, 5-50
   micronic-particle filters, figure, 5-49
   needle valve, figure, 5-45
   nomenclature and parts of cylinder, figure, 5-46
   pumps for, 5-40
      classification of, 5-41
      external gear pump, figure, 5-44
      gear types, 5-41
      internal gear pump, figure, 5-45
      piston types, 5-41
      radial pump, figure, 5-44
      vane pump, figures, 5-44
   radial pump, figure, 5-44
   reversing directional (four-way) valve, figure, 5-46
   symbols for, figure, 5-42 to 5-43
   valves for, 5-45
      ball check valve, figure, 5-45
      globe valve, figure, 5-45
      needle valve, figure, 5-45
   vane pump: balanced, figure, 5-44
      unbalanced, figure, 5-44
   viscosity of fluid, table, 5-47

Impact wrenches, 2-21
Induction welding, 4-43
Inhibitors, corrosion, 4-3
Inspection, role in preventive maintenance, 3-18 to 3-24

Jacks, flange, 2-35
Jaw type crushers (*see* Crushers, jaw type)
Jets in cleaning:
   ChemJet, 4-249
   SandJet, 4-249
   water, 4-7
Journals, bearing, 4-129

Keys, crusher (*see* Crusher, roll type, keys)

Labor, cost of, 3-4, 3-5
Ladders (*see* Scaffolds and ladders)
Lubricants, 4-11 to 4-29
  additives for, 4-15
    anticorrosion additives and rust preventives, 4-18
    antioxidants (inhibitors), 4-15
    engine-cleanliness, 4-18
    extreme-pressure, 4-18
    foam depressants, 4-18
    pour-point depressants, 4-18
    viscosity-index improvers, 4-15
  endurance value factors for, 4-10
    adhesions, 4-19
    emulsification, 4-19
    interfacial tension, 4-19
    saponification, 4-19
    surface tension, 4-19
    wetting ability, 4-19
  grease, 4-12, 4-15
    characteristics of, 4-12
      base, 4-12
      dropping point, 4-12
      melting point, 4-12
      penetration, 4-12
    definition and discussion of, 4-15
    protection of, 4-18
      against fire, 4-19
      need for proper, 4-18
      storage in, 4-19
  solids, characteristics of, table, 4-17
    bentones, table, 4-17
    boron nitride, table, 4-17
    fullers earth, table, 4-17
    graphite, table, 4-17
    mica, table, 4-17
    molybdenum disulfide, table, 4-17
    talc, table, 4-17
    zinc oxide, table, 4-17
  synthetics, characteristics of, table, 4-16
    esters, table, 4-16
    fluorocarbons, table, 4-16
    glycols (polyethers), table, 4-16
    hydrocarbons, table, 4-16
    polyalkylene oxides, table, 4-16
    polyethers (glycols), table, 4-16
    silicones, table, 4-16
  tests of, 4-11
    carbon-residue content, 4-11
    demulsibility, 4-12
    emulsification, 4-12
    flash point, 4-11
    neutralization number, 4-12
    pour-point, 4-11
    saponification number, 4-12
    viscosity, 4-11

Lubricants (*Cont.*):
  types of, 4-12
    circulating oils, 4-13
    gear oils, 4-13
    greases (*see* grease, *above*)
    machine (engine) oils, 4-14
    steam-cylinder oils, 4-14
    synthetics and solids, 4-15
Lubrication (*see* centralized systems, *below;* general considerations for, *below*)
  air compressors, reciprocating, 8-58
  air compressors, rotary (*see* Air compressors, rotary helical-screw, lubrication)
  asphalt plants, 7-294 to 7-296
    amount of lubricant, 7-296
    cleanliness during, 7-296
    frequency of, 7-296
    running gear, 7-296
    type specified, 7-296
  centralized systems, 4-21 to 4-25
    advantages of, 4-21
    basic system of, 4-21
    front end loader example, 4-22
    large drill or shovel example, 4-23
    savings through use of, 4-24
  conveyors (*see* Conveyors, conveyor belts, lubrication)
  cranes, hydraulic, 7-202 to 7-205
    plan for, 7-203 to 7-205
    standards for, 7-202 to 7-203
  crushers, all types, 7-266 to 7-267
    (*See also under specific type*)
  diesel engines (*see* Diesel engines, lubrication of)
  drills and drilling equipment, 7-333
  excavators (*see* Excavators, hydraulic, and backhoes, lubrication of)
  gasoline engines, 5-35
  general considerations for, 4-11 to 4-20
    approach to, 4-18
    failures attributed to, 4-20
    lubricant types (*see* Lubricants)
    personnel selection for, 4-20
    procedures for, 4-20
    timing of, 4-20
    tools of, 4-20
      hand-pressure grease guns, 4-20
      power guns, 4-20
  mechanical power transmission equipment: babbitted and bronze sleeve bearings, 4-158
    chain drive, 4-164
    dry fluid drives and couplings, 4-169
    mounted bearings, 4-157
    shaft-mounted speed reducers, 4-163
  motors, electric (*see* Motors, electric, bearings and lubrication)

Lubrication (*Cont.*):
  portable power tools, air tools, 2-28
  pumps, general, 8-1
    (*See also* Pumps and compressors, seals)
  scrapers and scraper blades: engine lubrication system, 7-33
    scraper bowl lubrication, 7-44
  shovels, power: shovel crowd clutch lubrication schedule, 7-155
    travel base lubrication plan, figure, 7-153
    typical lubricants for, table, 7-124
    upper machinery and rotary base, 7-125
    upper machinery lubrication plan, figure, 7-147
    table, 7-145 to 7-146
  tractors, diesel (dozers), 7-2 to 7-4
  trenchers and ditchers, digging clutch, 7-311

Machinist, apprenticeship schedules for, 1-24
Maintenance:
  construction equipment, theory and practice of, 1-1 to 1-2
  evolution of, 1-1
  handbook as a guide for, 1-2
  interaction as a key to, 1-2
  preventive (*see* Preventive maintenance)
Maintenance budgets (*see* Budgets)
Maintenance cost recording (*see* Accounting practices; Computer, role of, in maintenance cost)
Maintenance facility (*see* Shop layout for maintenance facility)
Maintenance function, establishment of, 1-3 to 1-10
  action plan for, 1-3
    assessing goals, 1-3
    checking tools, 1-4
    enlisting help, 1-5
    inventorying personnel, 1-4
    measuring the target, 1-4
    scheduling work, 1-5
  analyzing for profit in, 1-8
    corrective maintenance, 1-9
    maintainability of equipment, 1-9
    preventive maintenance to curb downtime, 1-8
      (*See also* Preventive maintenance)
    spare parts, 1-9
  record keeping as an adjunct to, 1-6
    charging work, 1-6
    recording repair history, 1-8
    requesting work, 1-6

Maintenance information, system for organizing, 1-11 to 1-14
  by categories, 1-12
  reasons for, 1-11
  use of system, 1-11
Maintenance management, human factors in (*see* Human factors in maintenance management)
Maintenance-trades training (*see* Training maintenance-trades personnel)
Mechanical power transmission equipment, 4-155 to 4-169
  bearings, babbitted and bronze-sleeve type, 4-158
    adverse operations of, 4-159
    alignment of, 4-158
    correct load direction of, 4-158
    load rating limits of, 4-158
    lubrication of, 4-158
    shaft journal surface finish of, 4-158
    temperature limits on, 4-159
    wear inspection of, 4-159
  bearings, mounted, 4-158
    alignment of, 4-157
    flingers for, 4-157
    load direction for, 4-157
    lubrication of, 4-157
    mounting of, 4-156
      adapter type, 4-156
      eccentric collar type, 4-156
      setscrew type, 4-156
    seals for, 4-157
    shaft tolerances for, table, 4-157
    shafting for, 4-157
    troubleshooting on, 4-158
  bushings, tapered, 4-159
    cleanliness for, 4-159
    size recommendations of, 4-159
    tightening of mounting screws, 4-159
    wall thickness considerations for, 4-159
  chain drives, 4-163
    installation of, 4-163
      alignment of, 4-163
      chain installation, 4-163
      chain tensioning, 4-163
      component cleaning, 4-163
      sprocket mounting, 4-163
    maintenance of, 4-164
      cleaning, 4-164
      inspecting, 4-164
      lubricating, 4-164
      replacing, 4-165
      troubleshooting, 4-165
        table, 4-166 to 4-167
  dry fluid drives and couplings, 4-165
    adverse operating conditions of, 4-168
    changing characteristics of, 4-168
    erratic acceleration of, 4-169

Mechanical power transmission equipment, dry fluid drives and couplings (*Cont.*):
  frequent starting, steps to take, 4-169
  high-speed operation, 4-168
  lubrication, 4-169
  overload protection, 4-168
  slippage, 4-169
 flexible couplings, 4-165
  alignment of, 4-165
  clamping bolt tightening for, 4-165
  visual inspection of, 4-165
 speed reducers, shaft-mounted, 4-161
  installation of, 4-161
   belt alignment in, 4-163
   driven shaft check for, 4-161
   input sheave mounting for, 4-163
   lubrication, initial, for, 4-163
   positioning and tightening of, 161
   tape removal for, 4-161
  operational maintenance of, 4-163
   regular oil changes, 4-163
   routine inspection, 4-163
   seasonal oil changes, 4-163
   troubleshooting, 4-163
 V-belt drives, 4-160
  installation of, 4-160
   alignment checking for, 4-160
   belt mounting during, 4-160
   belt selection for, 4-160
   belt tensioning for, 4-160
   sheave inspection for, 4-160
   sheave mounting during, 4-160
  maintenance of, 4-161
   belt dressing not proper in, 4-161
   dirt accumulation prevention, 4-161
   oil and grease buildup prevention, 4-161
   troubleshooting, 4-161
   table, 4-162
Metal resurfacing, 4-107 to 4-114
 alloy selection criteria, 4-110
  forms of alloy, 4-113
  types of alloy, 4-113
   build-up, Group 1, 4-111
   high-alloy ferrous, Group 3, 4-112
   low-alloy ferrous, Group 2, 4-112
   nonferrous, Group 5, 4-113
   tungsten-carbide, Group 4, 4-112
  types of wear, 4-110
   abrasion, 4-110
   corrosion, 4-110
   heat, 4-111
   impact, 4-110
   stress-related, 4-111
 applications of hard-facing, 4-120
  augers, 4-123
  baffle plates, 4-121
  cable sheaves, 4-121

Metal resurfacing, applications of hard-facing (*Cont.*):
  chutes, 4-121
  conveyor screws, 4-123
  crusher roll shells, 4-123
  engine valves, 4-122
  shafts, 4-121
  swing hammers, 4-121
  teeth, 4-122
 functions of rebuilding and hard-facing, 4-108
 methods of rebuilding and hard-facing, 4-114
  base metal use, 4-114
  preheating, 4-115
  welding preparations, 4-114
  welding procedures, 4-115
   automatic rebuilding, 4-117
   manual arc hard-facing, 4-116
   oxyacetylene hard-facing, 4-115
   semiautomatic hard-facing, 4-117
   thermal spraying, 4-118
 original equipment surface hardening, 4-107
  diffusion alloying, 4-107
  flame hardening, 4-107
  hard chrome plating, 4-108
 patterns of hard-facing, 4-120
 processes of rebuilding and hard-facing, 4-109
  automatic welding, 4-110
  manual arc welding, 4-110
  oxyacetylene welding, 4-109
  semiautomatic welding, 4-110
 surface checks, 4-119
 welded overlays, 4-108
Millwright, apprenticeship schedules for, **1-22**
Moldboard, dozer, **6-23** to **6-27**
 cutting edges of, **6-25** to **6-27**
 end bits, **6-25** to **6-27**
 installation of, **6-25**
 types of, **6-23** to **6-24**
  angle, **6-23**
  full-U, **6-24**
  semi-U, **6-23**
  straight, **6-23**
Motor graders, **7-53** to **7-90**
 axles, **7-84** to **7-90**
  bevel gear adjustment, **7-88**
  general considerations for, **7-84** to **7-88**
  pinion gear adjustment, **7-88**
 blade upkeep, **7-72** to **7-78**
  circle, **7-72**
  cutting edge, **7-78**
  drawbar, **7-72**
  manual blade shift and tilt, **7-74**
  moldboards, **7-78**
  rams, hydraulic, **7-77**

Motor graders, blade upkeep (*Cont.*):
  worm gear circle drive, **7-74**
  worm gear circle motor, **7-72**
  troubleshooting, table, **7-75**
engines (*see* power trains, *below*)
hydraulic systems, **7-58** to **7-68**
  aeration, **7-60**
  cavitation, **7-63**
  rams, **7-67**
  service brakes, **7-68**
  troubleshooting: excess heat, table, **7-61**
    excessive noise, table, **7-60**
    faulty operation, table, **7-63**
    general considerations, **7-58** to **7-63**
    incorrect flow, table, **7-62**
    incorrect pressure, table, **7-62**
  types of, **7-58**
  valve banks, **7-66**
initial considerations for, **7-53** to **7-55**
  cleanliness, **7-54**
  depth desired by owner, **7-54**
  design, **7-53**
  function, **7-53**
  preventive maintenance programming, **7-54**
power trains, **7-78**, **7-84**
  air cleaner, **7-79**
  contaminants, **7-82**
  cooling system, **7-80**
  general factors for, **7-78**
  torque converter operation, **7-80**
  torque converter troubleshooting, **7-81**
  transfer cases, **7-83**
  transmission operation, **7-80**
  transmission troubleshooting, **7-81**
preventive maintenance (PM), **7-55** to **7-58**
  blade, **7-57**
  circle field, **7-57**
  engine, **7-55**
  steering system, **7-55**
  tire pressure, **7-56**
steering circuit system, **7-68** to **7-72**
  general considerations, **7-68**
  power steering, **7-69**
  ram adjustment, **7-71**
wheels (*see* axles, *above*)
Motors, electric, **5-53** to **5-88**
  ac induction type, **5-54** to **5-59**
    rotor windings, **5-77**
    slip, **5-54**
    stator windings, **5-57**
    torque, **5-54**
    voltage and frequency, **5-56**
  bearings and lubrication, **5-73** to **5-79**
    adding lubricant, **5-75**
    bearing and journal clearance setting, procedure for, **5-78**

Motors, electric, bearings and lubrication (*Cont.*):
    bearing and journal reassembly, procedures for, **5-78**
    bearing reassembly, procedures for, **5-75**
    motor-oil properties, table, **5-77**
    relubrication periods, guide for, table, **5-74**
    viscosities of motor oils, table, **5-76**
  check charts, **5-79** to **5-87**
    ac and dc motors, **5-79**
    ac motors, **5-85** to **5-87**
    dc motors, **5-80** to **5-85**
  dc type, **5-59** to **5-66**
    armature handling, **5-59**
    brush and brush holder, **5-62**
    brush pressure, **5-62**
    commutation, variables affecting, **5-62** to **5-66**
      brush chatter, **5-62**
      brush life, **5-66**
      brush wear, **5-66**
      commutator films, **5-65**
      copper pickup, **5-66**
      flashing, **5-65**
      neutral brush setting, effect of, **5-65**
      sparking, **5-63**
      streaking of commutator surface, **5-63**
      threading of commutator surface, **5-63**
    commutator, **5-59**
      conditions for sound operation, **5-59** to **5-62**
      importance of, **5-59**
  general approach to maintenance of, **5-53**
  good maintenance practices, table, **5-78**
  insulation and its care, **5-66** to **5-73**
    cleaning of, **5-69**
    cleanliness requirements, **5-67**
    coolness requirements, **5-67**
    drying of, **5-69**
    dryness requirements, **5-67**
    power supply for, determining, **5-68**
    testing: dielectric, **5-71**
      general, **5-72**
  plan for maintaining, **5-53**
  references, **5-88**
  suggested schedule, table, **5-56**

Open-end wrenches, **2-39**
Organic solvents for cleaning, **4-6**

18 Index

Oxyacetylene welding (*see* Welding, gas, oxyacetylene welding, cutting, gouging, and hard-facing)

Pavers (*see* Paving equipment, asphalt)
Paving equipment, asphalt, 7-161 to 7-175
    auger and conveyor drive chain adjustment, 7-168
    checkout, preoperational, 7-162
    conveyor and auger drive chain adjustment, 7-168
    conveyor adjustment on, 7-166 to 7-168
    general maintenance considerations for, 7-161
    maintenance practices for, 7-164 to 7-166
    off-season storage of, 7-166
    preoperational checkout, 7-162
    road widener adjustment, 7-168 to 7-169
        general, 7-168
        new belt, 7-169
        retensioning, 7-169
    screed adjustment, 7-171 to 7-175
        angle of attack, 7-172
        crown adjustment, 7-175
        general factors in, 7-171
        head of material, 7-173
        paving speed, 7-172
    storage of, off-season, 7-166
    traction drive chain adjustment, 7-169
    washdown, daily procedure for, 7-163
Percussion welding, 4-42
Pipefitter, apprenticeship schedules for, 1-23
Plasma-arc welding, 4-42
Pliers, 2-36
PM (*see* Preventive maintenance)
Portable power tools, 2-17 to 2-29
    air tools, 2-24 to 2-29
        characteristics of, 2-26
            percussion types, 2-26
            rotating types, 2-26
        components for, 2-24
            blades, 2-25
            cylinder liner, 2-24
            end plates, 2-24
            governors, 2-26
            rotors, 2-25
        failure diagnosis, 2-26
        operating arrangement, 2-27
            air pressure, 2-27
            flow through the orifice, 2-28
            lubrication, 2-28
            piping, 2-27
        safety, 2-29
        test specifications for, 2-29
        troubleshooting, 2-26, 2-29

Portable power tools (*Cont.*):
    electric tools, 2-17 to 2-24
        inspection of, 2-20
            brushes and commutators, 2-20
            cable, 2-20
            conductive dust, 2-20
            failure causes, 2-20
            ground test at 500 volts, 2-20
            ventilating openings, 2-20
        parts maintenance for, 2-17
            bearings, 2-17
            brushes, 2-18
            cordless, 2-19
            double insulated, 2-20
            electric cord, 2-18
            gearing, 2-20
            motor, 2-18
            switches, 2-18
            variable speed, 2-20
        specific tools, maintenance of, 2-20
            circular saws, 2-23
            drills, 2-20
            hammers, 2-24
            impact wrenches, 2-21
            reciprocating saws, 2-23 to 2-24
            routers, 2-24
            sanders, 2-23
            screwdrivers, 2-20
                figure, 2-21
            shears and nibblers, 2-23
Power shovels (*see* Shovels, power)
Power systems, electrical, 5-1 to 5-7
    air cleaner servicing, 5-5
    cylinder head servicing, 5-5
    engine troubleshooting, figure, 5-4
    exhaust smoke indicators, 5-7
    fuel system servicing, 5-7
    generator switching, 5-7
    generator troubleshooting, table, 5-6
    low compression troubleshooting, table, 5-5
    preventive maintenance on, 5-3
    routine servicing of, 5-1
    scheduling routine service, table, 5-2
    spark plug analysis, 5-4
    tests on, 5-4
        compression, 5-5
        crankcase vacuum, 5-4
    tuneup practices, 5-3
Power tools (*see* Portable power tools)
Power transmission equipment (*see* Mechanical power transmission equipment)
Pressure welding, 4-43
Preventive maintenance (PM), 3-15 to 3-26
    description and definition of, 3-15
    downtime, PM as tool for curbing, 1-8
    frequency of inspection determination, 3-20
        engineering analysis for, 3-20

Preventive maintenance (PM), frequency of inspection determination (*Cont.*):
  gradual refinement of, 3-20
  statistical checks of, 3-21
  initial considerations for implementation of, 3-16
    allowing sufficient time, 3-17
    inventorying conditions, 3-17
    selling the concept, 3-16
  inspection areas for a PM program, 3-19
  inspection exclusions in a PM program, 3-18
  inspection inclusions in a PM program, 3-18
  inspectors (*see* personnel for carrying out, *below*)
  installing a PM program, 3-17
    basic problems, 3-18
    mastering principles, 3-18
    point of application, 3-18
  need for, 3-15
  paperwork for, 3-24
    five basic forms of, 3-26
    general considerations for, 3-26
  personnel for carrying out, 3-22
    inspection checking, 3-23
    inspection manuals, 3-24
    inspection methods, 3-24
    inspection reports, 3-23
    qualifications of inspectors, 3-22
    report routing, 3-23
  schedules for, 3-21
    applying to jobs, 3-22
    types of, 3-21
  (*See also under specific unit*)
Projection welding, 4-42
Protective coatings, corrosion, 4-3
Pullers, 2-55
Pumps:
  centrifugal, 8-7 to 8-18
    alignment, 8-7
      final check, 8-11
    bearings, 8-11
    belt drives for, 8-11
    connecting piping, 8-8
    discharge piping, 8-8
    installation, 8-7
    packing, 8-16
    packing protection, 8-14
    piping: connecting, 8-8
      discharge, 8-8
      suction, 8-8
    rotating, 8-11
    seals: mechanical, 8-16
      protection of, 8-14
        from dry running, 8-15
    sleeves, shaft, 8-16
    starting, 8-11
    stopping, 8-11

Pumps, centrifugal (*Cont.*):
  suction piping, 8-8
  troubleshooting, table, 8-12 to 8-13
  wearing-ring clearances, 8-14
concrete, 8-19 to 8-31
  bearings, agitator shaft, 8-22
  boom delivery system, 8-30
  boom hinge pivots, 8-31
  boom outriggers, 8-31
  booms for, maintenance of, 8-30
  couplings, care of, 8-29
  vs. delivery systems, 8-25
  extendable jibs, 8-31
  filtration, hydraulic oil, 8-25
  hinge pivots, boom, 8-31
  hose, care of, 8-29
  hydraulic fluid (oil) filtration, 8-25
  hydraulic lines, 8-25
  hydraulic systems, 8-25
  jibs, extendable, 8-31
  maintenance approach to, 8-19 to 8-21
  operation of, as an aid to use of, 8-31
  outriggers, boom, 8-31
  particle size characteristics, figure, 8-29
  pipe, care of, 8-29
  pipeline sizes compared, 8-28
  pipeline types compared, 8-28
  piston heads, 8-22
  pumping pressures, extreme, 8-25
  remote control of, 8-31
  seals, agitator shaft, 8-22
Pumps and compressors, 8-1 to 8-5
  accessories as a factor in, 8-3
  inspection, 8-1
  lubrication, 8-1
  parts replacement, 8-1
  seals, 8-1
    components of, figure, 8-2
    grease-lubricated mechanical, 8-2
    oil-lubricated mechanical, 8-1
    self-lubricating mechanical, 8-3
    stuffing-box type, 8-3
  speed tests, 8-3
  tolerances for, 8-1
  troubleshooting, 8-3
  vacuum tests, 8-3
Punches, 2-55

Reciprocating air compressors (*see* Air compressors, reciprocating)
Reference tables, xiii to xv
  Beaufort scale of wind force, xiv to xv
  conversion factors, xiii to xiv
  temperature equivalents, Fahrenheit and Celsius scales, xiv
Repair parts, cost of, 3-5

Resistance welding, 4-41
Resurfacing (see Metal resurfacing)
Rigger, apprenticeship schedules for, 1-24
Ripper tooth and dozer moldboard, 6-24 to 6-29
    dozer moldboard (see Moldboard, dozer)
    hard-facing, 6-29
    installation of, 6-25
    ripper points, 6-27 to 6-28
        failure of, 6-28
        point angle, 6-28
        selection of, 6-28
    ripper shanks, 6-27
        curved, 6-27
        double-offset, 6-27
        position of, 6-27
        repair kits for, 6-27
        straight, 6-27
    types of, 6-24 to 6-25
        parallelogram, 6-24
        radial, 6-25
        straight bar, 6-24
Rotary helical-screw air compressors (see Air compressors, rotary helical-screw)

Salvage value, machinery or equipment, 3-4
Sanders, 2-23
Saws:
    circular, 2-23
    reciprocating, 2-23 to 2-24
Scaffolds and ladders, 4-171 to 4-186
    aluminum tube and coupler scaffolds, 4-181
    general considerations, 4-171
    OSHA scaffolding checklist, 4-183 to 4-185
    safety roles for ladders, 4-175 to 4-178
    scaffolding applications, 4-185
    scaffolding maintenance, 4-186
    scaffolds, 4-178 to 4-183
        safe use and safety rules for, 4-183
        safety requirements for, 4-181
        safety swinging, 4-181
        special-design, 4-181
        tube and coupler (steel and aluminum), 4-181
        welded aluminum, 4-178
        welded sectional steel, 4-179
    steel tube and coupler scaffolds, 4-181
    stepladders and extension ladders, 4-172 to 4-175
        basic ladder groups, 4-173
        choice of materials for, 4-172
        extension, 4-173
        metal, 4-175
        precautionary measures for, 4-175

Scaffolds and ladders, stepladders and extension ladders (*Cont.*):
    single, 4-173
    special-purpose, 4-174
    stepladders, 4-173
    tube and coupler scaffolds (steel and aluminum), 4-181
Scraper blades (see Scrapers and scraper blades)
Scrapers and scraper blades, 7-29 to 7-52
    air system, 7-40 to 7-42
    axle, 7-36
    brakes, 7-42
    differential, 7-36
    drivelines, 7-35
    electrical system, 7-43
    elevator, 7-47 to 7-50
        chain centering, 7-48
        chain slack adjustment, 7-47
        ejector stopper, 7-50
        end gate, 7-49
        gearbox, 7-48
        rolling floor, 7-48
        structural upkeep, 7-50
    engine, 7-29 to 7-34
        air intake system, 7-29
        cooling system, 7-31
        fuel system, 7-30
        lubrication system, 7-33
    engine upkeep, 7-34 to 7-35
        transmission, general, 7-34
        transmission air shift system, 7-35
    final drive, 7-36
    hydraulic system, 7-39
    painting, 7-52
    patching structure, 7-50
    power take-off, 7-37
    reinforcing structure, 7-51
    scraper bowl, 7-44 to 7-47
        blade installation, 7-47
        bowl side wear blades, 7-47
        lubrication, 7-44
    steering system, 7-37 to 7-39
    straightening the structure, 7-51
    tires, 7-43
    welding, 7-52
    wheels, 7-43
Screwdrivers, 2-20, 2-35
    figure, 2-21
Seam welding, 4-42
Shears and nibblers, 2-23
Sheet-metal welding, 4-54, 4-56
Shielded-metal-arc welding (see Welding, arc, shielded-metal-arc)
Shop layout for maintenance facility, 2-1 to 2-8
    basic tools needed for shop, 2-4
    field shop, 2-3
    rebuild shop, 2-5
    trailer shop layout, 2-3

Shovels, power, 7-121 to 7-160
　attachments for, 7-154 to 7-160
　　A-frame, 7-157
　　adjustments of, 7-154
　　boom assembly, 7-157
　　dipper and stick, 7-159
　　front drum, 7-156
　　shovel crowd clutch, 7-156 to 7-158
　　shovel crowd clutch lubrication schedule, table, 7-155
　　wire rope inspection and lubrication, 7-160
　lubricants for, typical, table, 7-124
　maintenance considerations, 7-121 to 7-123
　　depth of, 7-122
　　design, 7-122
　　function, 7-122
　　general, 7-121
　preventive maintenance for, 7-123 to 7-125
　rotating base (*see* upper machinery and rotary base, *below*)
　travel base, 7-148 to 7-151
　　adjustments, 7-148
　　base component servicing, 7-150
　　crawler assembly, 7-148
　　crawler track adjustment, 7-149
　travel base lubrication plan, figure, 7-153
　travel base troubleshooting, table, 7-152
　upper machinery and rotary base, 7-125 to 7-140
　　adjustments on, 7-125
　　air supply system, 7-136
　　batteries, 7-139
　　boom hoist (*see* hoist, boom, *below*)
　　centralized oil system, 7-137
　　daily checks of, 7-126
　　engine power train, 7-125
　　equalizer rollers, 7-138
　　hoist, boom, 7-129 to 7-132
　　　independent, 7-132
　　　nonindependent, 7-132
　　lubrication of, 7-125
　　oil system, centralized, 7-137
　　power train, engine, 7-125
　　rollers, rotating and equalizer, 7-138
　　shaft care, 7-127 to 7-136
　　　drum, 7-128
　　　horizontal reverse, 7-127
　　　rotating, 7-134
　　　vertical reverse, 7-128
　upper machinery lubrication plan, table, 7-145 to 7-146
　　figure, 7-147

Shovels, power (*Cont.*):
　upper machinery troubleshooting, table, 7-140 to 7-145
Slings (*see* Hoists and slings, chain)
Socket wrenches, 2-43
Soldering (*see* Welding, gas, air-acetylene soldering, heating, and brazing)
Solid lubricants (*see* Lubricants, solids, characteristics of)
Solvents, organic, for cleaning, 4-6
Special wrenches, 2-50
Static weight compactors (*see* Compactors, static weight)
Steam cleaning (*see* Cleaning, steam and hot water)
Stud welding, 4-42
Submerged-arc welding, 4-53 to 4-54
Supervisory training (*see* Training supervisory personnel)
Synthetic lubricants (*see* Lubricants, synthetics, characteristics of)

Teeth, bucket (*see* Buckets and bucket teeth)
Temperature equivalents, table of, Fahrenheit and Celsius scales, xiv
Thermit welding, 4-43
Tires, maintenance and upkeep of, 6-1 to 6-14
　adjusted inflation pressure, table, 6-13
　flotation, 6-4
　ground-contact area, 6-4
　handling rules, 6-5
　inflation pressure, adjusted, table, 6-13
　inflation pressure determination, 6-9 to 6-12
　　dozers, 6-12
　　dumpers, 6-9
　　front axle, 6-10
　　front tires, 6-10
　　graders, 6-12
　　lift and carry, loaders, 6-12
　　loaders, 6-10
　　rear axle, 6-12
　　rear tire, 6-10
　　scrapers, 6-9
　load material weights, table, 6-11
　mounting, 6-5 to 6-9
　premature deterioration, causes of, 6-14
　pressure corrections, high temperature, 6-12
　selection, 6-1 to 6-3
　　miscellaneous considerations, 6-2
　　site factors, 6-2
　　size and load index factors, 6-2

22 Index

Tires, maintenance and upkeep of, selection (*Cont.*):
    surface type, **6-2**
    vehicular conditions, **6-1**
    selection aids, **6-3**
    speed restrictions, unbanked curves, table, **6-14**
    storage considerations, **6-5**
Tool bits, **2-58**
Tool holders and blades, **2-55**
Tools:
    air (*see* Portable power tools, air tools)
    electric (*see* Portable power tools, electric tools)
    general, **2-9**
    hand, **2-31** to **2-62**
        C-clamps, **2-59**
        care of, **2-32**
        chain pipe tongs, **2-32**
        chain pipe vise, **2-33**
        chain wrenches, **2-32**
        chisels, **2-53**
        feeler gauge, **2-35**
        Flange-Jacks, **2-35**
        flaring tool, **2-34**
        hammers, **2-52** to **2-53**
        pliers, **2-36**
        pullers, **2-55**
        punches, **2-55**
        range of, **2-31**
        screwdrivers, **2-35**
        selection of, **2-31**
        special wrenches, **2-50**
        tool bits, **2-58**
        tool holders and blades, **2-55**
        torque multipliers, **2-48**
        torque wrench testers, **2-49**
        torque wrenches, **2-47**
        tubing cutter, **2-34**
        wrenches, **2-39**
            adjustable, **2-42**
            combination, **2-42**
            open-end, **2-39**
            socket, **2-43**
            special, **2-50**
            torque, **2-47**
    heavy, **2-9** to **2-15**
    power (*see* Portable power tools)
    special, **2-11**
        crawler tractor master pin removal set, **2-11**
        crawler tractor sprocket removal set, **2-11**
        track pin press, **2-14**
Tooth, ripper (*see* Ripper tooth and dozer moldboard)
Torque multipliers, **2-48**
Torque wrench testers, **2-49**
Torque wrenches, **2-47**

Track maintenance (*see* Undercarriage, wear and maintenance)
Track pin press, **2-14**
Tractors, diesel (dozers), **7-1** to **7-27**
    attention areas, **7-1** to **7-6**
        drive train maintenance, **7-4**
        engines, inspections and adjustments, **7-4**
        fuel system maintenance, **7-1**
        lubrication, **7-2** to **7-4**
        routine coverage, **7-1**
    organizing for maintenance, **7-8** to **7-15**
        frequency levels, **7-12**
        record keeping, **7-14** to **7-15**
        skills, **7-9** to **7-12**
            levels, **7-9**
            requirements, **7-11**
    planning for maintenance, **7-15** to **7-27**
        approach to, **7-15**
        component rebuilding, **7-24**
        component replacement, **7-21**
        cumulative costs, figure, **7-27**
        design trends affecting, **7-15**
        future forecasting, **7-17**
        in-shop inspection, **7-25**
        inspection, general, **7-17**
        parts exchange, **7-23**
        parts inventory, **7-22**
        prefailure replacement decisions, **7-19**
        repair alternatives, **7-22**
        repair facilities, **7-22**
        repair vs. replacement decisions, **7-25**
        shop cost analysis, figure, **7-23**
        tools for repair, **7-22**
    programming for maintenance, **7-4** to **7-8**
        information flow, **7-4** to **7-6**
            figure, **7-5**
        job-tailored repair, **7-7** to **7-8**
        packaged repair, **7-6** to **7-7**
        problem reporting, **7-4**
Training maintenance-trades personnel, **1-15** to **1-28**
    apprentice training, **1-15**
        changes in, **1-17**
        definition of, **1-16**
        location for, **1-16**
    apprenticeship schedules, **1-22** to **1-25**
        boilermaker, **1-22**
        carpenter, **1-23**
        machinist, **1-24**
        millwright, **1-22**
        pipefitter, **1-23**
        rigger, **1-24**
    attitudes toward, **1-16**
        by apprentices, **1-16**
        by craft tradesmen, **1-16**

Training maintenance-trades personnel (*Cont.*):
  basic provisions for, 1-18
  basic standards for apprenticeship, 1-19
  Bureau of Apprenticeship and Training, U.S. Department of Labor, 1-19
  conducting programs in, 1-21
    audiovisual aids, 1-28
    classrooms and laboratories for, 1-21
    evaluating classroom training, 1-25
    evaluating shop and/or field training, 1-26
    instructors, 1-21
    job skills, 1-25
    shop and/or field training, 1-25
    trainee recognition, 1-28
    training-material sources, 1-27
  joint apprenticeship committees, function of, 1-19
  labor agreements, 1-18
    contract, 1-18
    writing the agreement, 1-18
  need for, 1-18
  references, 1-17
  research into, 1-15
  selecting qualified trainees, 1-20
    need for, 1-20
    retesting procedure, 1-20
    test scores, 1-20
    test types, 1-20
      achievement or equivalency, 1-20
      aptitude, 1-20
      general ability, 1-20
      interest, 1-20
      personality, 1-20
Training supervisory personnel, 1-28 to 1-33
  analyzing the need for, 1-30
  determining training needs of supervisors, 1-30
  evaluation of training, 1-32
  interpersonal skills for, 1-29
  managerial skills for, 1-29
  methodologies for, 1-31
    action maze, 1-31
    case studies, 1-31
    games and simulation techniques, 1-31
    group discussion, 1-31
    job rotation, 1-31
    laboratory training, 1-31
    programmed learning, 1-31
    role playing, 1-31
  objectives and goals of, 1-30
  program sources for, 1-29
  resources for, 1-33
    books, 1-33
    journals, 1-33

Training supervisory personnel (*Cont.*):
  starting a program, 1-29
  technology's role in, 1-29
Trenchers and ditchers, 7-303 to 7-331
  clutch, 7-309 to 7-311
    adjustment of, 7-311
    digging: line up of, 7-311
      lubrication of, 7-311
    digging and crawler, 7-310
    disk, friction, replacement of, 7-311
    engine, 7-309
    maintenance of, 7-309
    friction disk, replacement of, 7-311
    lubrication, digging clutch, 7-311
  conveyor assembly, 7-317 to 7-331
    adjustments: major, 7-324
      minor, 7-323
    bleeding hydraulic systems, 7-323
    brakes, steering, 7-321
    chains, drive, 7-330
    crawlers, 7-319
    crumbling shoe, 7-325
    drive chains, 7-330
    frames, 7-327
    general considerations, 7-317 to 7-319
    guards, 7-331
    hydraulic system, 7-327
    sheaves, 7-331
    shifting system, 7-319
    steering brakes, 7-321
    wire ropes, 7-331
  digging drive shaft assembly, 7-312 to 7-314
  digging wheel assembly, 7-314 to 7-317
    bucket backs, 7-315
    buckets, 7-314
    rim, maintenance of, 7-317
    segments, welding, 7-317
    segments and split sprockets, 7-316
    suspension for, 7-315
  engine, maintenance of, 7-304 to 7-306
  main shaft assembly, 7-311
  maintenance program for, 7-303
  shaft assembly: digging drive, 7-312 to 7-314
    main, 7-311
    pillow blocks, 7-311
  sprockets, 7-312
  transmission, maintenance of, 7-306 to 7-309
    boom hoisting, 7-307
    conveyor drive, 7-307
    conveyor elevating, 7-307
    conveyor pulley, 7-307
    crawler, 7-307
    crawler and digging differentials, 7-308
    engine, 7-306

Troubleshooting (*see under specific unit*)
Tubing cutter, 2-34

Undercarriage, wear and maintenance, 6-15 to 6-22
  maintenance, 6-20 to 6-22
    general factors for, 6-20
    optional equipment for, 6-20
    rebuilding vs. replacement, 6-21
  operator actions to reduce wear, 6-19
  soil conditions as a wear factor, figure, 6-17
  wear, 6-15 to 6-19
    forward drive side, 6-15
    front idler, 6-16
    link, 6-16
    pin, 6-15, 6-16
    pitch, 6-15
    reverse drive side, 6-15
    roller, 6-16
    sprocket, tooth tip, 6-16
    track frame misalignment as a cause of, 6-16
    track stretch, 6-15
    wear-causing factors, 6-17
    wear patterns related to use, table, 6-18
Upset welding, 1-12

Variable costs (*see* Accounting practices, variable costs)
Vibratory drum compactors (*see* Compactors, vibratory drum)
Viscosity:
  hydraulic fluids, table, 5-47
  lubricants, tests of, 4-11
  motor oils, table, 5-76
  viscosity-index improvers, 4-15

Welding:
  arc, 4-27 to 4-78
    abrasive wear resistance, 4-56
    alloy steels, 4-45 to 4-47
      chromium, 4-47
      high-manganese, 4-47
      high-tensile low-alloy, 4-45
      stainless, 4-46
      stainless clad, 4-47
    carbon arc: average welding conditions for, table, 4-64
      hand, maximum currents for, table, 4-64
    carbon-arc welding, 4-63
      torch for, 4-64

Welding, arc (*Cont.*):
  carbon steels, 4-44 to 4-45
    cast iron, 4-45
    high-carbon, 4-45
    low-carbon, 4-44
    medium-carbon, 4-45
    preferred analysis range of, table, 4-44
  condensers, 4-75
  delay relays, 4-75
  distortion control, 4-48 to 4-53
    buildup and manganese-steel electrodes, 4-52
    causes and remedies of, 4-48
    cellulose-coated electrodes (EXX10 and EXXII), 4-49
    lime-covered low-hydrogen electrodes, 4-52
    shielded-metal-arc welding, 4-48
    titania-coated electrodes (EXX12 and EXX13), 4-50
    type(s) E6010 and E6011, 4-50
    type(s) E6012 and E6013, 4-51
    type E6027, 4-52
    type E7014, 4-52
    type E7015, 4-53
    type E7016, 4-53
    type E7018, 4-53
    type E7024, 4-52
    type E7028, 4-53
  equipment handling, operational practices for, 4-71 to 4-75
    abuse avoidance, 4-72
    cleanliness of, 4-71
    coolness of, 4-71
    maintenance of, 4-73
  equipment installation, 4-69 to 4-71
  equipment selection and maintenance, 4-67 to 4-69
    accessories, 4-68
    machines, 4-67
  general considerations, 4-27
  hard-facing: guide to, figure, 4-60 to 4-61
    selecting material for, 4-57
  hard-surfacing, 4-56
    with submerged-arc process, 4-62
  impact wear resistance, 4-57
  input cable wire sizes: for ac/dc unit, table, 4-71
    for motor-generator units, table, 4-71
  metal cutting with welding equipment, 4-64
  nonferrous metals, 4-47
    aluminum, 4-47
    copper and copper alloys, 4-47
  partial wear surface checks, 4-59
  processes of, 4-41 to 4-43
    atomic-hydrogen, 4-41

Index

Welding, arc, processes of (*Cont.*):
  carbon-arc, 4-42
  electron-beam, 4-43
  electroslag, 4-41
  explosive, 4-43
  flash, 4-42
  flow, 4-43
  friction, 4-43
  hammer, 4-43
  induction, 4-43
  percussion, 4-42
  plasma-arc, 4-42
  pressure, 4-43
  projection, 4-42
  seam, 4-42
  stud, 4-42
  thermit, 4-43
  upset, 4-42
 references, 4-78
 resistance welding, 4-41
 sheet-metal electrode sizes, table, 4-57
 sheet-metal welding, 4-54
  currents for, table, 4-56
 shielded-metal-arc, 4-30 to 4-41
  distortion control, 4-48
  fully automatic flux-cored, 4-32
  general, gas-shielded, 4-33
  manual shielded metal, 4-32
  metal, gas-shielded, 4-39
   CO_2, 4-40
  self-shielded metal, 4-31
  semiautomatic flux-cored, 4-32
  spot, gas-shielded, 4-40
  submerged, 4-32
  tungsten: equipment for, 4-37
   gas-shielded, 4-33
 spot welding, 4-41
 steel-electrode, classification, characteristics, and uses of, table, 4-49
 submerged-arc welding, 4-53
  electrodes, 4-54
  equipment for, 4-53
  fluxes, 4-54
 surfacing electrodes, types of, 4-57
 torch for carbon-arc welding, 4-64
 troubleshooting for, table, 4-76 to 4-78
 weldability of metals, 4-43
 welding cable sizes: for ac/dc units, table, 4-72
  for motor-generator units, table, 4-72

Welding, arc (*Cont.*):
 welding procedures, checks on, 4-59
 welding processes: electric-arc, 4-28
  general, 4-28
  manual and automatic, 4-29
 gas, 4-79 to 4-106
  air-acetylene soldering, heating, and brazing, 4-79 to 4-86
   automobile-body soldering, 4-82
   electrical connections, 4-82
   miscellaneous applications, 4-85
   paint burning, 4-84
   precautions and safe practices, 4-86
   sheet-metal working, 4-81
   soft solders, common, table, 4-81
   soldering, 4-79
   soldering fluxes, 4-81
   sweat-type fitting installation, 4-83
  oxyacetylene welding, cutting, gouging, and hard-facing, 4-87 to 4-106
   blowpipe motion, 4-90
   braze welding, 4-88
   codes, specifications, and standards, 4-98 to 4-102
   cutting, 4-87, 4-95
    preparation for, 4-86
   flame adjustment, 4-88
   fusion welding: cast iron, 4-82
    general, 4-89
   gouging, 4-94
   hard-facing, 4-97
   hard-facing cast iron, 4-104
   hard-facing deposit finishing, figure, 4-105
   hard-facing rods, characteristics of, tables, 4-103
   hard-facing steel, 4-97
   heavy braze welding, 4-88
   outfit setup, 4-87
   oxygen cutting, 4-92
   oxygen cutting equipment, 4-94
   references, 4-106
   weld making, 4-90
   welding and brazing, 4-86
   welding methods: ferrous, table, 4-93
    nonferrous, table, 4-94
Wind force, Beaufort scale of, xiv to xv
Work request form, design of, figure, 1-7
Wrenches (*see* Tools, hand, wrenches)